Lecture Notes in Computational Science and Engineering **103**

Editors:

More information about this series at
http://www.springer.com/series/3527

Assyr Abdulle • Simone Deparis • Daniel Kressner •
Fabio Nobile • Marco Picasso
Editors

Numerical Mathematics and Advanced Applications ENUMATH 2013

Proceedings of ENUMATH 2013, the 10th European Conference on Numerical Mathematics and Advanced Applications, Lausanne, August 2013

Springer

Editors
Assyr Abdulle
Simone Deparis
Daniel Kressner
Fabio Nobile
Marco Picasso
Ecole Polytechnique Fédérale de Lausanne
Lausanne
Switzerland

ISSN 1439-7358 ISSN 2197-7100 (electronic)
ISBN 978-3-319-38383-5 ISBN 978-3-319-10705-9 (eBook)
DOI 10.1007/978-3-319-10705-9
Springer Cham Heidelberg New York Dordrecht London

Mathematics Subject Classification (2010): 65XX, 74XX, 76XX, 78XX

Cover Photo: Campus of École polytechnique fédérale de Lausanne. By courtesy of Alain Herzog/EPFL.

Printed on acid-free paper

Springer is part of Springer Science+Business Media (www.springer.com)

Preface

The European Conference on Numerical Mathematics and Advanced Applications (ENUMATH) is a series of conferences held every 2 years to provide a forum for discussion on recent aspects of numerical mathematics and scientific and industrial applications. The previous ENUMATH meetings took place in Paris (1995), Heidelberg (1997), Jyvaskyla (1999), Ischia (2001), Prague (2003), Santiago de Compostela (2005), Graz (2007), Uppsala (2009), and Leicester (2011).

This book contains a selection of invited and contributed lectures of the ENUMATH 2013 conference organised by the Mathematical Institute of Computational Science and Engineering (MATHICSE), EPFL, and held in Lausanne, Switzerland, August 26–30, 2013. It gives an overview of recent developments in numerical analysis, computational mathematics, and applications by leading experts in the field. The conference attracted around 400 participants from around the world including 11 invited talks by:

- Ruth Elizabeth Baker (University of Oxford, UK), on "Developing multiscale models for exploring biological phenomena"
- Eric Cancès (CERMICS, Ecole des Ponts ParisTech, France), on "Electronic structure calculation"
- Omar Ghattas (ICES, University of Texas at Austin, USA), on "Stochastic Newton MCMC methods for Bayesian inverse problems with application to ice sheet dynamics"
- Ernst Hairer (Université de Genève, Switzerland), on "Long-term analysis of numerical and analytical oscillations"
- Jan Hesthaven (Brown University, USA), on "High-order reduced basis multiscale finite element methods"
- Petr Knobloch (Univerzita Karlova, Czech Republic), on "Finite element methods for convection dominated problems"
- Dmitri Kuzmin (Friedrich Alexander Universität Erlangen-Nürnberg, Germany), on "Vertex-based limiters for continuous and discontinuous Galerkin methods"
- Ilaria Perugia (Università degli studi di Pavia, Italy), on "Trefftz-discontinuous Galerkin methods for time-harmonic wave problems"

- Rolf Stenberg (Aalto University, Finland), on "Mixed finite element methods for elasticity"
- Martin Vetterli (EPFL, Switzerland), public lecture on "Inverse problems regularized by sparsity"
- Barbara Wohlmuth (TU München, Germany), on "Interfaces, corner and point sources"

There were 24 minisymposia and numerous contributed talks covering a broad spectrum of numerical mathematics. This ENUMATH 2013 proceeding will be useful for a wide range of readers giving them a state-of-the-art overview of advanced techniques, algorithms, and results in numerical mathematics and scientific computing. Advances on finite element methods, time integrators, multiscale methods, numerical linear algebra, and discretisation techniques for fluid mechanics and optics are presented. This book contains a selection of 79 papers by the invited speakers and from the minisymposia as well as the contributed sessions. It is organised in 11 parts as follows:

Part I Space Discretisation Methods for PDEs
Part II Time Integration Schemes
Part III A Posteriori Error Estimation and Adaptive Methods
Part IV Numerical Linear Algebra
Part V Multiscale Modeling and Simulation
Part VI Reduced Order Modeling
Part VIII Uncertainty, Stochastic Modeling, and Applications
Part IX Solvers, High Performance Computing, and Software Libraries
Part X Computational Fluid and Structural Mechanics
Part XI Computational Electromagnetics

We would like to thank all the participants for their valuable contributions and scientific discussions during the conference and to the minisymposium organisers for helping to shape the core structure of the meeting. The members of the Scientific Committee have helped us tremendously in reviewing the contributions to this proceedings. This conference would not have been possible without all the work and guidance provided by the Programme Committee: Franco Brezzi, Miloslav Feistauer, Roland Glowinski, Gunilla Kreiss, Yuri Kuznetsov, Jacques Periaux, Alfio Quarteroni, Rolf Rannacher, Endre Süli. We also thank our sponsors for their generous support: the School of Basic Sciences and the Centre Interfacultaire Bernoulli from EPFL, the Center for Advanced Modeling Science (CADMOS), MathWorks and Springer. Last but not least we would like to acknowledge the tireless effort of Virginie Ledouble leading the administration tasks, Corinne Craman who coordinated the edition of this proceedings, all the secretaries of MATHICSE for their tremendous help in organising this conference, and our PhDs and Post-Docs that have helped us in many ways.

We hope that this volume reflects the inspiring talks and lively scientific exchanges that took place at EPFL during the ENUMATH 2013 meeting.

Lausanne, Switzerland
August 2014

Assyr Abdulle
Simone Deparis
Daniel Kressner
Fabio Nobile
Marco Picasso
Alfio Quarteroni

Contents

Part XI Computational Electromagnetics

Part I
Space Discretisation Methods for PDEs

Weakly Symmetric Mixed Finite Elements for Linear Elasticity

Rolf Stenberg

Abstract The approximation of the equations of linear elasticity by so-called weakly symmetric mixed methods is considered. It is shown that the technique of mesh dependent norms yields a natural and elementary error analysis of the methods. The technique is applied to several families of methods.

1 Introduction

During the last decade the theory of mixed finite element methods have been recast with the aid of differential geometry, cf. [6]. This was first done for methods for scalar second order elliptic equation, i.e., the Raviart-Thomas-Nédélec [29, 30] and the Brezzi-Douglas-Marini-Duran-Fortin [14, 15] families. Lately, the theory has been extended to methods for linear elasticity [17]. Both methods with a symmetric approximation for the stress tensor [5, 19, 24, 33] and methods where the symmetry is imposed weakly [1, 4, 6, 18, 20, 21, 31], have been analyzed.

The purpose of this paper is to highlight an alternative and more elementary way of analysis, which, nevertheless, gives optimal error estimates. The approach is that of using mesh dependent norms, first used by Babuška, Osborn and Pitkäranta [11]. In this, the norm used for the "stress" variable is the L_2-norm, which has the physical meaning of energy. For the "displacement" variable the broken H^1-norm (now well-known from Discontinuous Galerkin Methods) is used. The stability of the methods follows directly from local scaling arguments. The second ingredient is the so-called "equilibrium condition," which the methods fulfill. Using these, the quasi-optimal error estimate for the stress follows by the classical saddle point theory.

Our exposition follows the talk at the conference, and we consider mixed methods for weakly symmetric elasticity elements. We write down the continuous form of the problem, for which we first prove the uniqueness. Then we prove the stability in the natural energy norms. Then we turn to the mixed method, and we follow exactly the same approach to show the uniqueness and stability of the

---

R. Stenberg (✉)
Department of Mathematics and Systems Analysis, Aalto University, Esbo, Finland
e-mail: rolf.stenberg@aalto.fi

A. Abdulle et al. (eds.), *Numerical Mathematics and Advanced Applications ENUMATH 2013*, Lecture Notes in Computational Science and Engineering 103,
DOI 10.1007/978-3-319-10705-9_1

method. In this we use only elementary finite element techniques, but nevertheless, we obtain the optimal error estimates.

For the displacement the analysis yields a superconvergence result for the distance between the L_2 projection onto the discrete space and the finite element solution. This is utilized to postprocess the displacement yielding an approximation of two polynomial degrees higher, with an optimal convergence rate.

2 The Equations of Elasticity

We consider the equations of linear elasticity for which we, for simplicity, assume a unit Young's modulus $E = 1$, and a vanishing Poisson ratio $\nu = 0$. The unknown are the symmetric stress tensor $\sigma \in \mathbb{R}^{d \times d}$, and the displacement $u \in \mathbb{R}^d$.

The displacement gradient is the sum of its symmetric and skew-symmetric parts, the strain and the rotation tensors:

$$\nabla u = \varepsilon(u) + \omega(u)$$

with

$$\varepsilon(u) = \frac{1}{2}(\nabla u + \nabla u^T)$$

and

$$\omega(u) = \frac{1}{2}(\nabla u - \nabla u^T).$$

Note that symmetric and skew-symmetric tensors are orthogonal. The linear elasticity problem in mixed form is then [23]: Find σ and u such that

$$\begin{aligned} \sigma - \varepsilon(u) &= 0, \\ \operatorname{div} \sigma + f &= 0 \ \text{ in } \Omega, \\ u &= 0 \ \text{ on } \partial\Omega. \end{aligned} \tag{1}$$

Mixed finite elements based on this formulation are rather complicated to construct, cf. [5, 19, 24, 33]. The simplest elements are composite, and the requirement of pure polynomials lead to elements of high degree [2, 7].

It is, however, possible to design simpler elements if one treats the symmetry of the stress tensor as an independent equation [4, 18, 31]. Mechanically, this is natural, since the symmetry of the stress tensor is the condition of moment equilibrium [23].

With one more equation, an additional unknown is needed, and this is the skew-symmetric rotation, which we denote by ρ. The system of equations is then

$$\begin{aligned} \sigma + \rho - \nabla u &= 0, \\ \sigma - \sigma^T &= 0, \\ \operatorname{div} \sigma + f &= 0 \text{ in } \Omega, \\ u &= 0 \text{ on } \partial\Omega. \end{aligned} \tag{2}$$

Comparing (1) and (2) we see that they are equivalent with $\rho = \omega(u)$.

The weak formulation is: Find $(\sigma, \rho, u) \in [L^2(\Omega)]^{d\times d} \times [L^2(\Omega)]^{d\times d}_{\text{skw}} \times [H_0^1(\Omega)]^d$ such that

$$\begin{aligned} (\sigma, \tau) + (\rho, \tau) - (\nabla u, \tau) &= 0 \quad \forall \tau \in [L^2(\Omega)]^{d\times d} \\ (\sigma, \eta) &= 0 \quad \forall \eta \in [L^2(\Omega)]^{d\times d}_{\text{skw}} \\ (\sigma, \nabla v) &= (f, v) \quad \forall v \in [H_0^1(\Omega)]^d. \end{aligned} \tag{3}$$

The first thing to check, is the uniqueness of solution to these equations.

Theorem 1 *The solution to* (3) *is unique.*

Proof We have to prove that $f = 0$ implies $u = 0,\ \rho = 0$ and $\sigma = 0$. To this end, we first choose $\eta = \sigma - \sigma^T$ in the second equation. The orthogonality of symmetric and skew-symmetric tensors then yields

$$0 = (\sigma, \eta) = (\sigma, \sigma - \sigma^T) = \frac{1}{2}\|\sigma - \sigma^T\|_0^2$$

implying

$$\sigma = \sigma^T.$$

Next, choosing $\tau = \sigma$ in the first equation and $v = u$ in the third, we get (as $f = 0$), again using the orthogonality,

$$\|\sigma\|_0^2 - (\varepsilon(u), \sigma) = 0 \text{ and } (\sigma, \nabla u) = (\sigma, \varepsilon(u)) = 0.$$

Hence

$$\sigma = 0.$$

The first equation now reduces to

$$(\rho, \tau) - (\nabla u, \tau) = 0,$$

the symmetric and skew-symmetric parts of which are

$$(\tau_{\mathrm{sym}}, \varepsilon(u)) = 0 \quad \text{and} \quad (\rho - \omega(u), \tau_{\mathrm{skw}}) = 0.$$

Choosing $\tau_{\mathrm{sym}} = \varepsilon(u)$ and $\tau_{\mathrm{skw}} = \rho - \omega(u)$ then yields

$$\varepsilon(u) = 0 \text{ and } \rho - \omega(u) = 0.$$

Hence, the deflection u is a rigid displacement, and since it vanishes on the boundary, it vanishes in the whole of the domain. From above, it then follows that also the rotation ρ vanishes.

For the analysis we need stronger results, i.e., stability in proper norms, and these we will choose as energy type norms.

Theorem 2 *There exists a positive constant C such that*

$$\|\sigma\|_0 + \|\varepsilon(u)\|_0 + \|\rho - \omega(u)\|_0 \le C\|f\|_* \tag{4}$$

and

$$\|u\|_1 + \|\rho\|_0 \le C\|f\|_{-1}, \tag{5}$$

with

$$\|f\|_* = \sup_{v \in [H_0^1(\Omega)]^d} \frac{(f, v)}{\|\varepsilon(v)\|_0}.$$

Proof Define the bilinear form

$$B(\sigma, \rho, u; \tau, \eta, v) = (\sigma, \tau) - (\rho - \nabla u, \tau) - (\eta - \nabla v, \sigma)$$

so that the variational form is

$$B(\sigma, \rho, u; \tau, \eta, v) + (f, v) = 0 \quad \forall (\tau, \eta, v) \in [L^2(\Omega)]^{d\times d} \times [L^2(\Omega)]^{d\times d}_{\mathrm{skw}} \times [H_0^1(\Omega)]^d.$$

It now holds

$$B(\sigma, \rho, u; \sigma, -\rho, -u) = \|\sigma\|_0^2.$$

Define

$$\tau = \rho - \nabla u.$$

By the orthogonality $(\rho - \omega(u), \varepsilon(u)) = 0$ we get

$$\|\tau\|_0^2 = \|\rho - \omega(u)\|_0^2 + \|\varepsilon(u)\|_0^2$$

and

$$B(\sigma, \rho, u, \tau, 0, 0) = (\sigma, \tau) + \|\rho - \omega(u)\|_0^2 + \|\varepsilon(u)\|_0^2.$$

Since $|ab| \leq (a^2 + b^2)/2,\ a, b \in \mathbb{R}$, Schwarz inequality gives

$$\begin{aligned}(\sigma, \tau) + \|\rho - \omega(u)\|_0^2 + \|\varepsilon(u)\|_0^2 &\geq -\|\sigma\|_0\|\tau\|_0 + \|\rho - \omega(u)\|_0^2 + \|\varepsilon(u)\|_0^2 \\ &\geq -\frac{1}{2}\|\sigma\|_0^2 - \frac{1}{2}\|\tau\|_0^2 + \|\rho - \omega(u)\|_0^2 + \|\varepsilon(u)\|_0^2 \\ &= -\frac{1}{2}\|\sigma\|_0^2 + \frac{1}{2}(\|\rho - \omega(u)\|_0^2 + \|\varepsilon(u)\|_0^2).\end{aligned}$$

With $(\varphi, \eta, v) = (\sigma + \tau, -\rho, -u)$ we now get

$$B(\sigma, \rho, u; \varphi, \eta, v) \geq \frac{1}{2}\big(\|\sigma\|_0^2 + \|\rho - \omega(u)\|_0^2 + \|\varepsilon(u)\|_0^2\big).$$

Hence, it holds

$$\|\sigma\|_0^2 + \|\rho - \omega(u)\|_0^2 + \|\varepsilon(u)\|_0^2 \leq 2|(f, v)| \leq 2\|f\|_*\|\varepsilon(v)\|_0 = 2\|f\|_*\|\varepsilon(u)\|_0.$$

The arithmetic-geometric mean inequality gives

$$2\|f\|_*\|\varepsilon(u)\|_0 \leq 2\|f\|_*^2 + \frac{1}{2}\|\varepsilon(u)\|_0^2.$$

Combining the two inequalities above gives

$$\|\sigma\|_0^2 + \|\rho - \omega(u)\|_0^2 + \frac{1}{2}\|\varepsilon(u)\|_0^2 \leq 2\|f\|_*^2,$$

and the stability estimate (4) follows. Korn's inequality and the standard definition of $\|\cdot\|_{-1}$ then gives (5).

Remark 1 In the theorem we proved that the stability of the bilinear form B follows from the L_2-ellipticity of the bilinear form $(\cdot,\cdot)$ and the "inf-sup" condition (which we also verified)

$$\sup_{\tau \in [L^2(\Omega)]^{d\times d}} \frac{(\tau, \eta - \nabla v)}{\|\tau\|_0} \geq C(\|\eta - \omega(v)\|_0 + \|\varepsilon(v)\|_0) \quad \forall (\eta, v) \in [L^2(\Omega)]^{d\times d}_{\mathrm{skw}} \times [H_0^1(\Omega)]^d,$$

i.e., the main result of the Babuška-Brezzi [8–10, 13] theory applied for this particular problem.

3 Mixed Finite Element Methods

The mixed method is not directly based on the variational formulation of the previous section. Instead, a formulation in which the integration by parts

$$(\nabla v, \tau) = -(\operatorname{div} \tau, u) + \langle \tau n, v \rangle_{\partial \Omega}$$

is used. The stress is sought in

$$H(\operatorname{div}:\Omega) = \{ \tau \mid \tau \in [L^2(\Omega)]^{d\times d}, \ \operatorname{div} \tau \in [L^2(\Omega)]^d \}$$

and the displacement in $[L^2(\Omega)]^d$. The Dirichlet boundary condition for the displacement now becomes a natural boundary condition, and the mixed formulation is then: Find $(\sigma_h, \rho_h, u_h) \in S_h \times K_h \times V_h \subset H(\operatorname{div}:\Omega) \times [L^2(\Omega)]^{d\times d}_{\text{skw}} \times [L^2(\Omega)]^d$ such that

$$\begin{aligned} (\sigma_h, \tau) + (\rho_h, \tau) + (\operatorname{div} \tau, u_h) &= 0 \quad \forall \tau \in S_h, \\ (\sigma_h, \eta) &= 0 \quad \forall \eta \in K_h, \\ (\operatorname{div} \sigma_h, v) + (f, v) &= 0 \quad \forall v \in V_h. \end{aligned} \tag{6}$$

By defining

$$b(\tau; v, \eta) = (\operatorname{div} \tau, v) + (\tau, \eta).$$

the problem in saddle point form is: Find $(\sigma_h, \rho_h, u_h) \in S_h \times K_h \times V_h \subset H(\operatorname{div}:\Omega) \times [L^2(\Omega)]^{d\times d}_{\text{skw}} \times [L^2(\Omega)]^d$ such that

$$\begin{aligned} (\sigma_h, \tau) + b(\tau; u_h, \rho_h) &= 0 \quad \forall \tau \in S_h, \\ b(\sigma_h; v, \eta) + (f, v) &= 0 \quad \forall (v, \eta) \in V_h \times K_h. \end{aligned}$$

By the Babuška-Brezzi theory the stability is a consequence of an "inf-sup" condition

$$\sup_{\tau \in S_h} \frac{b(\tau; v, \eta)}{\|\tau\|} \geq C \|v, \eta\| \quad \forall (v, \eta) \in V_h \times K_h$$

for some norms. The traditional approach is to use the $H(\operatorname{div}:\Omega)$ norm for the stress and the L_2 norm for both the displacement and rotation. With this choice, a direct application of the Babuška-Brezzi theory does not give optimal error estimate. Furthermore, when posing the stability in these norms, Korn's inequality has been used, and hence the discrete stability cannot be proved by local scaling arguments.

First in the next section we will consider specific finite element spaces. Now we will only assume that they are piecewise polynomials on the finite element mesh $\mathscr{C}_h$. The collection of edges/faces on the mesh is denoted by Γ_h.

In our approach we proceed in analogy to (4). The norm chosen for the stress is a mesh dependent L_2 norm

$$\|\tau\|_{0,h}^2 = \|\tau\|_0^2 + \sum_{E\in\Gamma_h} h_E \|\tau n\|_{0,E}^2.$$

This is paired with the broken norm:

$$\|v,\eta\|_h^2 = \sum_{K\in\mathscr{C}_h} \|\varepsilon(v)\|_{0,K}^2 + \sum_{E\in\Gamma_h} h_E^{-1} \|[\![v]\!]\|_{0,E}^2 + \sum_{K\in\mathscr{C}_h} \|\eta - \omega(v)\|_{0,K}^2.$$

Here $[\![v]\!]$ denotes the jump of v when E is in the interior of the domain, and the value of v when $E \subset \partial\Omega$.

Lemma 1 *It holds*

$$|b(\tau; v,\eta)| \le \|\tau\|_{0,h} \|v,\eta\|_h \quad \forall (\tau,\eta,v) \in S_h \times K_h \times V_h.$$

Proof Integrating by parts on each element and using Schwarz inequality yields

$$\begin{aligned}
b(\tau; v,\eta) &= (\operatorname{div}\tau, v) + (\tau,\eta)\\
&= \sum_{K\in\mathscr{C}_h} (\operatorname{div}\tau, v)_K + (\tau,\eta)\\
&= -\sum_{K\in\mathscr{C}_h} (\tau,\nabla v)_K + \sum_{E\in\Gamma_h} \langle \tau n, [\![v]\!]\rangle_E + (\tau,\eta)\\
&= -\sum_{K\in\mathscr{C}_h} (\tau,\varepsilon(v))_K + \sum_{E\in\Gamma_h} \langle \tau n, [\![v]\!]\rangle_E + \sum_{K\in\mathscr{C}_h} (\tau,\eta-\omega(v))_K\\
&\le \sum_{K\in\mathscr{C}_h} \|\tau\|_{0,K}\|\varepsilon(v)\|_{0,K} + \sum_{E\in\Gamma_h} \|\tau n\|_{0,E}\|[\![v]\!]\|_{0,E} + \sum_{K\in\mathscr{C}_h} \|\tau\|_{0,K}\|\eta-\omega(v)\|_{0,K}\\
&\le \|\tau\|_{0,h}\|v,\eta\|_h.
\end{aligned}$$

The stability condition we require for the method is hence

$$\sup_{\tau\in S_h} \frac{b(\tau; v,\eta)}{\|\tau\|_{0,h}} \ge C\|v,\eta\|_h \quad \forall (v,\eta) \in V_h \times K_h. \tag{7}$$

With the broken H^1 norm

$$\|v\|_{1,h}^2 = \sum_{K\in\mathscr{C}_h} \|\nabla v\|_{0,K}^2 + \sum_{E\in\Gamma_h} h_E^{-1}\|[\![v]\!]\|_{0,E}^2, \tag{8}$$

the following discrete Korn's inequality holds in the subspace V_h [12, 28]:

$$\sum_{K\in\mathscr{C}_h} \|\varepsilon(v)\|_{0,K}^2 + \sum_{E\in\Gamma_h} h_E^{-1}\|[\![v]\!]\|_{0,E}^2 \geq C\|v\|_{1,h}^2. \tag{9}$$

The triangle inequality then gives

$$\|v,\eta\|_h \geq C(\|v\|_{1,h} + \|\eta\|_0). \tag{10}$$

The stability condition can thus be written

$$\sup_{\tau\in S_h} \frac{b(\tau;v,\eta)}{\|\tau\|_{0,h}} \geq C(\|v\|_{1,h} + \|\eta\|_0) \quad \forall (v,\eta)\in V_h\times K_h. \tag{11}$$

In addition to the stability condition the discrete spaces have to satisfy the equilibrium condition.

$$\operatorname{div} S_h \subset V_h. \tag{12}$$

For the L_2 projection $P_h : [L_2(\Omega)]^d \to V_h$, it then holds

$$(\operatorname{div}\tau, v - P_h v) = 0, \quad \forall \tau\in S_h,\ \forall v\in [L_2(\Omega)]^d. \tag{13}$$

We then have the following error estimate.

Theorem 3 *Suppose that the stability condition* (11) *and the equilibrium condition* (12) *are valid. Then there exists a positive constant C such that*

$$\|\sigma-\sigma_h\|_{0,h} + \|\rho-\rho_h\|_0 + \|P_h u - u_h\|_{1,h} \leq C\big(\inf_{\tau\in S_h}\|\sigma-\tau_h\|_{0,h} + \inf_{\eta\in K_h}\|\rho-\eta\|_0\big).$$

Proof Define the bilinear form

$$B(\sigma,\rho,u;\tau,\eta,v) = (\sigma,\tau) + b(\tau;\rho,u) + b(\sigma;\eta,v).$$

Let $(\tau,\eta)\in S_h\times K_h$. The stability implies that exits $(\varphi,\gamma,z)\in S_h\times K_h\times V_h$, with

$$\|\varphi\|_{0,h} + \|\gamma\|_0 + \|z\|_{1,h} \leq C,$$

such that

$$\|\sigma_h-\tau\|_{0,h} + \|\rho_h-\eta\|_0 + \|u_h-P_h u\|_{1,h} \leq B(\sigma_h-\tau,\rho_h-\eta,u_h-P_h u;\varphi,\gamma,z).$$

The discrete and exact solutions satisfy

$$\begin{aligned}
B(\sigma_h - \tau, \rho_h - \eta, u_h - P_h u; \varphi, \gamma, z) &= -(f, z) - B(\tau, \eta, P_h u; \varphi, \gamma, z) \\
&= B(\sigma, \rho, u; \varphi, \gamma, z) - B(\tau, \eta, P_h u; \varphi, \gamma, z) \\
&= B(\sigma - \tau, \rho - \eta, u - P_h u; \varphi, \gamma, z) \\
&= (\sigma - \tau, \varphi) + b(\varphi; \rho - \eta, u - P_h u) + b(\sigma - \tau; \gamma, z) \\
&= (\sigma - \tau, \varphi) + (\varphi; \rho - \eta) + (\operatorname{div} \varphi, u - P_h u) + b(\sigma - \tau; \gamma, z) \\
&= (\sigma - \tau, \varphi) + (\varphi; \rho - \eta) + b(\sigma - \tau; \gamma, z),
\end{aligned}$$

where we in the last step used (13). By the Schwarz inequality we have

$$\begin{aligned}
(\sigma - \tau, \varphi) + (\varphi, \rho - \eta) + b(\sigma - \tau; \gamma, z) \ &\leq (\|\sigma - \tau\|_{0,h} + \|\rho - \eta\|_0)\|\varphi\|_0 \\
&\quad + \|\sigma - \tau\|_{0,h}(\|\gamma\|_0 + \|z\|_{1,h}),
\end{aligned}$$

and by combining the above inequalities, the assertion is proved.

4 Finite Element Families

In this section we discuss concrete families of elements. In all of them the elements are triangles or tetrahedra. We start with the one introduced by us in 1988.

The Stenberg family [31].

We will partly use different notation as in [31]. For $K \in \mathscr{C}_h$ we define the bubble function $b_K \in P_{d+1}(K)$ by

$$b_K = \prod_{i=0}^{d} \lambda_i,$$

where $\lambda_0, \ldots, \lambda_d$, are the barycentric coordinates on K. For a vector valued function z in $\mathbb{R}^3$ we define $\operatorname{curl} z = \nabla \times z$ and for a scalar function z we let

$$\operatorname{curl} z = (\frac{\partial z}{\partial x_2}, -\frac{\partial z}{\partial x_1}).$$

Define

$$\begin{aligned}
S_{k+d-1}(K) = \{\, \tau = \{\tau_{ij}\},\ i, j = 1, \ldots, d \mid (\tau_{i1}, \ldots, \tau_{id}) = \operatorname{curl}(w^i b_K) \\
w^i \in [P_{k-1}(K)]^3 \text{ for } d = 3, \text{ and } w^i \in P_{k-1}(K) \text{ for } d = 2,\ i = 1, \ldots, d \,\}.
\end{aligned}$$

Here the index $k + d - 1$ is equal to the polynomial degree of the space (the degree of the bubble b_K is $d + 1$, w^i is of degree $k - 1$, and taking the curl lowers the degree by one). Note that these "stabilizing" degrees of freedom satisfy

$$\operatorname{div} \tau = 0 \text{ on } K, \text{ and } \tau n = 0 \text{ on } \partial K, \ \forall \tau \in S_{k+d-1}(K). \tag{14}$$

The family of [31] is then defined for the degree $k \geq 2$ by

$$\begin{aligned} S_h &= \{ \tau \in H(\operatorname{div}:\Omega) \mid \tau|_K \in [P_k(K)]^{d\times d} + S_{k+d-1}(K) \ \forall K \in \mathscr{C}_h \}, \\ K_h &= \{ v \in [L^2(\Omega)]^{d\times d}_{\text{skw}} \mid v|_K \in [P_k(K)]^d_{\text{skw}} \ \forall K \in \mathscr{C}_h \}, \\ V_h &= \{ v \in [L^2(\Omega)]^d \mid v|_K \in [P_{k-1}(K)]^d \ \forall K \in \mathscr{C}_h \}. \end{aligned} \tag{15}$$

Note that the stress space consists of d copies of the BDM/BDDF space augmented by the space $S_{k+d-1}(K)$ on each element.

We now in analogy with the uniqueness proof of Theorem 1, prove the uniqueness of the finite element solution.

Theorem 4 *The solution of* (6) *with the finite element spaces* (15) *is unique.*

Proof The uniqueness of the stress σ_h is immediate as for the continuous case. Hence, as in the proof of Theorem 1 we have to verify that the condition

$$b(\tau; u_h, \rho_h) = 0 \quad \forall \tau \in S_h$$

first implies that u_h is a rigid body motion and that $\rho_h = \omega(u_h)$, and then by the boundary conditions that both of them vanish. To this end, let K be arbitrary and choose τ such that $\tau = 0$ in $\Omega \setminus K$ and $\tau|_K \in S_{k+d-1}(K)$. Due to (14) it then holds

$$b(\tau; u_h, \rho_h) = (\rho_h, \tau) + (\operatorname{div} \tau, u_h) = (\rho_h, \tau)_K.$$

We proceed slightly differently in two and three dimensions. For $d = 2$ let

$$\rho_h = \begin{pmatrix} 0 & z \\ -z & 0 \end{pmatrix}.$$

We then choose

$$(\tau_{i1}, \tau_{i2}) = \operatorname{curl}\,(\frac{\partial z}{\partial x_i} b_K).$$

Integrating by parts, this gives

$$(\rho_h, \tau)_K = \int_K [-\frac{\partial}{\partial x_1}(b_K \frac{\partial z}{\partial x_1}) - \frac{\partial}{\partial x_2}(b_K \frac{\partial z}{\partial x_2})] z = \int_K b_K |\nabla z|^2.$$

The condition $b(\tau; u_h, \rho_h) = 0$ then implies that z equals a constant. Hence, on each K the rotation is constant ρ_K. Let $R(K)$ be the space of rigid body motions on K. For each K there is a $r_K \in R(K)$ such that

$$\rho_K = \omega(r_K) = \nabla r_K. \tag{16}$$

For $d = 3$ we choose

$$(\tau_{i1}, \tau_{i2}, \tau_{i3}) = \text{curl}(\text{curl}\, \rho_h^i\, b_K),$$

where ρ_h^i is the i-th row of ρ_h. Integrating by parts gives

$$0 = (\tau, \rho_h)_K = \sum_{i=1}^{3} \int_K b_K |\text{curl}\, \rho_h^i|^2,$$

showing that $\text{curl}\, \rho_h^i = 0$ on K. Since ρ_h is skew-symmetric, this implies that it is constant. Hence, we conclude that also for $d = 3$ there is a $r_K \in R(K)$ such that (16) holds.

In the sequel we use the subspace of S_h consisting of the Raviart-Thomas-Nédélec spaces

$$S_h^{RTN} = \{ \tau \in H(\text{div}:\Omega) \mid \tau|_K \in [P_{k-1}(K)^d \oplus z\tilde{P}_{k-1}(K)]^d \ \forall K \in \mathscr{C}_h \}$$

where $\tilde{P}_{k-1}(K)$ denotes homogeneous polynomials of degree $k-1$.

The degrees of freedom for this spaces are

$$\langle \tau n, z \rangle_E \qquad \forall z \in P_{k-1}(E)^d,\ E \subset \partial K, \tag{17}$$

$$(\tau, z)_K \qquad \forall z \in [P_{k-2}(K)]^{d\times d}, \tag{18}$$

for each $K \in \mathscr{C}_h$.

Now choose $\tau \in S_h^{RTN} \subset S_h$ such that $\tau = 0$ in $\Omega \setminus K$. Then it holds

$$\begin{aligned} b(\tau; u_h, \rho_h) &= (\rho_h, \tau) + (\text{div}\, \tau, u_h) = (\rho_h, \tau)_K + (\text{div}\, \tau, u_h)_K \\ &= (\rho_h, \tau)_K - (\tau, \nabla u_h)_K \\ &= (\nabla r_K, \tau)_K - (\tau, \nabla u_h)_K \\ &= (\nabla(r_K - u_h), \tau)_K. \end{aligned}$$

Now $\nabla(r_K - u_h)|_K \in [P_{k-2}(K)]^{d\times d}$, and hence the degrees of freedom (18) show that the condition $b(\tau; u_h, \rho_h) = 0$ implies that

$$\nabla(r_K - u_h) = 0. \tag{19}$$

The symmetric and skew-symmetric parts of this are

$$\varepsilon(u_h) = 0 \text{ and } \omega(u_h) = \omega(r_K) = \rho_h|_K.$$

Hence $u_h|_K$ is a rigid body motion and there is a constant vector c_K such that

$$u_h|_K = r_K + c_K.$$

For u_h and ρ_h it thus holds

$$b(\tau; u_h, \rho_h) = \sum_{E \in \Gamma_h} \langle \tau n, [\![u_h]\!] \rangle_E.$$

In the condition $b(\tau; u_h, \rho_h) = 0 \ \forall \tau \in S_h$, the degrees of freedom (17) for τ show that the jump $[\![u_h]\!]$ vanishes along interior edges, i.e. u_h is continuous and a global rigid body motion, and $\rho_h = \omega(u_h)$.

Finally, since, $[\![u_h]\!] = u_h$ on the boundary $\partial\Omega$, the degrees of freedom (17) show that $u_h = 0$, and hence also $\rho_h = 0$.

We note that in the uniqueness proof above, we have used local arguments element by element, and edge by edge. The stability can thus be built in the same manner and the norms and the bilinear form scales in the same way. Hence, we have essentially already proved the stability estimate.

Theorem 5 *There is a positive constant C such that*

$$\sup_{\tau \in S_h} \frac{b(\tau; v, \eta)}{\|\tau\|_{0,h}} \geq C \|v, \eta\|_h \quad \forall (v, \eta) \in V_h \times K_h. \tag{20}$$

Theorem 3 then gives the following error estimate.

Theorem 6 *There is a positive constant C such that*

$$\|\sigma - \sigma_h\|_{0,h} + \|\rho - \rho_h\|_0 + \|P_h u - u_h\|_{1,h} \leq C h^{k+1}\big(\|\sigma\|_{k+1} + \|\rho\|_{k+1}\big).$$

Let us continue by discussing this family. First, for an implementation in mind, the sum of local spaces in the definition of the stress should be replaced by a direct sum. To this end, we define

$$\hat{S}_{k+d-1}(K) = \{\, \tau = \{\tau_{ij}\},\ i, j = 1, \ldots, d \mid (\tau_{i1}, \ldots, \tau_{id}) = \operatorname{curl}(w^i b_K),$$
$$w^i \in [\hat{P}_{k-1}(K)]^3 \text{ for } d{=}3, \text{ and } w^i \in \hat{P}_{k-1}(K) \text{ for } d{=}2,\ i{=}1, \ldots, d \,\},$$

with

$$\hat{P}_{k-1}(K) = \{\, v \in P_{k-1}(K) \mid (v, w)_K = 0 \ \forall w \in P_{k-n}(K) \,\}.$$

Since all degrees of freedom in $S_{k+d-1}(K) \setminus \hat{S}_{k+d-1}(K)$ are contained in $[P_k(K)]^{d\times d}$, the stress space is

$$S_h = \{ \tau \in H(\mathrm{div}:\Omega) \mid \tau|_K \in [P_k(K)]^{d\times d} \oplus \hat{S}_{k+d-1}(K) \ \forall K \in \mathcal{C}_h \}.$$

Next, we note from the proof that in the space $\hat{S}_{k+d-1}(K)$ there are more degrees of freedom that what is used to get the stability. We also note that this space contains polynomials of degree $k+1$ for $d = 2$, and $k+2$ for $d = 3$, and these do not contribute to the accuracy.

These drawbacks are fixed in the
The Gopalakrishnan-Guzmán family [22].

In two dimensions the family is obtained from our family by restricting the stabilizing degrees of freedom to those which are actually needed. From the proof above, we see that in $\mathbb{R}^2$ the additional degrees of freedom are

$$\hat{S}_{k+1}(K)=\{ \tau=\{\tau_{ij}\},\ i,j = 1,2 \mid (\tau_{i1},\tau_{i2}) = \mathrm{curl}\,(\frac{\partial z}{\partial x_i} b_K)\ i = 1,2,\ z \in \tilde{P}_k(K) \}.$$

In three dimension they were able to reduce the degree of the additional degrees of freedom with one. They define the matrix bubble by

$$B_K = \sum_{l=0}^{3} \lambda_{l-3}\lambda_{l-2}\lambda_{l-1}(\nabla\lambda_l)^t \nabla\lambda_l,$$

where the index is modulo 4, and $\nabla\lambda_l$ is considered as a row vector. In [16] it is shown that this is symmetric and positively definite, and it can then be used as a weight for an inner product on tensors. Defining the curl of a tensor as the tensor in which each row is the curl of the corresponding row in the original tensor, the space is defined as

$$\hat{S}_{k+1}(K) = \{ \tau \in [L^2(K)]^{d\times d} \mid \tau = \mathrm{curl}\,(\mathrm{curl}\,(\xi) B_K)\ \xi \in [\tilde{P}_k(K)]^d_{\mathrm{skw}} \}.$$

Since

$$(\mathrm{curl}\,(\mathrm{curl}\,(\xi) B_K), \xi)_K = (\mathrm{curl}\,(\xi) B_K, \mathrm{curl}\,(\xi))_K$$

these degrees of freedom can be used in the proof of the stability.

The family is then defined by

$$\begin{aligned}
S_h &= \{ \tau \in H(\mathrm{div}:\Omega) \mid \tau|_K \in [P_k(K)]^{d\times d} + \hat{S}_{k+1}(K) \ \forall K \in \mathcal{C}_h \},\\
K_h &= \{ v \in [L^2(\Omega)]^{d\times d}_{\mathrm{skw}} \mid v|_K \in [P_k(K)]^d_{\mathrm{skw}} \ \forall K \in \mathcal{C}_h \},\\
V_h &= \{ v \in [L^2(\Omega)]^d \mid v|_K \in [P_{k-1}(K)]^d \ \forall K \in \mathcal{C}_h \}.
\end{aligned} \tag{21}$$

Theorem 7 *For $k \geq 2$ the family* (21) *is stable and the following error estimate holds*

$$\|\sigma - \sigma_h\|_{0,h} + \|\rho - \rho_h\|_0 + \|P_h u - u_h\|_{1,h} \leq Ch^{k+1}\big(\|\sigma\|_{k+1} + \|\rho\|_{k+1}\big).$$

Remark 2 The family of [16] is based on the Raviart-Thomas-Nédélec elements and not the Brezzi-Douglas-Fortin-Duran-Marini elements as above.

The third family to be considered is
The Arnold-Falk-Winther family [22].

In this the polynomial degree for the rotation is decreased by one. We then note that the there is no need to include additional degrees of freedom in order to obtain stability for the rotation, the degrees of freedom are already included in the stress space. The family is then the following.

$$\begin{aligned} K_h &= \{ v \in [L^2(\Omega)]^{d\times d}_{\text{skw}} \mid v|_K \in [P_{k-1}(K)]^{d\times d}\ \forall K \in \mathscr{C}_h \} \\ V_h &= \{ v \in L^2(\Omega)^d \mid v|_K \in [P_{k-1}(K)]^d\ \forall K \in \mathscr{C}_h \}, \\ S_h &= \{ \tau \in H(\text{div}:\Omega) \mid \tau|_K \in [P_k(K)]^{d\times d}\ \forall K \in \mathscr{C}_h \}. \end{aligned} \tag{22}$$

The advantage of this space is that pure polynomial degrees of freedom are used. The disadvantage is, however, that the convergence rate is decreased by one and the full approximation power of the stress space is not achieved.

Theorem 8 *For $k \geq 2$ the family* (22) *is stable and the following error estimate holds*

$$\|\sigma - \sigma_h\|_{0,h} + \|\rho - \rho_h\|_0 + \|P_h u - u_h\|_{1,h} \leq Ch^{k}\big(\|\sigma\|_{k} + \|\rho\|_{k}\big).$$

In our analysis we have assumed that $k \geq 2$ so that the rigid body motions are included in the local displacement spaces. The method are, however, stable also for $k = 1$, but the analysis has to be modified. This case can be found in the recent paper [25].

Finally, let us remark that the estimates for $\|P_h u - u_h\|_{1,h}$ are superconvergence results that enables a local postprocessing of the displacement, cf. [3, 31, 32]. The postprocessed displacement is crucial for a posteriori estimates [26, 27].

References

1. M. Amara, J.M. Thomas, Equilibrium finite elements for the linear elastic problem. Numer. Math. **33**(4), 367–383 (1979)
2. D.N. Arnold, G. Awanou, R. Winther, Finite elements for symmetric tensors in three dimensions. Math. Comput. **77**(263), 1229–1251 (2008)
3. D.N. Arnold, F. Brezzi, Mixed and nonconforming finite element methods: implementation, postprocessing and error estimates, RAIRO Modél. Math. Anal. Numér. **19**(1), 7–32 (1985)

4. D.N. Arnold, F. Brezzi, J. Douglas Jr., PEERS: a new mixed finite element for plane elasticity. Jpn. J. Appl. Math. **1**(2), 347–367 (1984)
5. D.N. Arnold, J. Douglas Jr., C.P. Gupta, A family of higher order mixed finite element methods for plane elasticity. Numer. Math. **45**(1), 1–22 (1984)
6. D.N. Arnold, R.S. Falk, R. Winther, Finite element exterior calculus, homological techniques, and applications. Acta Numer. **15**, 1–155 (2006)
7. D.N. Arnold, R. Winther, Mixed finite elements for elasticity. Numer. Math. **92**(3), 401–419 (2002)
8. I. Babuška, Error-bounds for finite element method. Numer. Math. **16**, 322–333 (1970/1971)
9. ———, The finite element method with Lagrangian multipliers. Numer. Math. **20**, 179–192 (1972/1973)
10. I. Babuška, A.K. Aziz, Survey lectures on the mathematical foundations of the finite element method, in *The Mathematical Foundations of the Finite Element Method with Applications to Partial Differential Equations: Proceedings of the Symposium*, University of Maryland, Baltimore (Academic, New York, 1972, With the collaboration of G. Fix and R. B. Kellogg), pp. 1–359
11. I. Babuška, J. Osborn, J. Pitkäranta, Analysis of mixed methods using mesh dependent norms. Math. Comput. **35**(152), 1039–1062 (1980)
12. S.C. Brenner, Korn's inequalities for piecewise H^1 vector fields. Math. Comput. **73**(247), 1067–1087 (2004)
13. F. Brezzi, On the existence, uniqueness and approximation of saddle-point problems arising from Lagrangian multipliers. Rev. Française Automat. Informat. Recherche Opérationnelle Sér. Rouge **8**(R-2), 129–151 (1974)
14. F. Brezzi, J. Douglas Jr., R. Durán, M. Fortin, Mixed finite elements for second order elliptic problems in three variables. Numer. Math. **51**(2), 237–250 (1987)
15. F. Brezzi, J. Douglas Jr., L.D. Marini, Two families of mixed finite elements for second order elliptic problems. Numer. Math. **47**(2), 217–235 (1985)
16. B. Cockburn, J. Gopalakrishnan, J. Guzmán, A new elasticity element made for enforcing weak stress symmetry. Math. Comput. **79**(271), 1331–1349 (2010)
17. R. Falk, Finite elements for linear elasticity, in *Mixed Finite Elements, Compatibility Conditions, and Applications* ed. by D. Boffi, F. Brezzi, L.F. Demkowicz, R.G. Durán, R.S. Falk, M. Fortin (Springer, Berlin Heidelberg, 2008), pp. 159–194
18. B. Fraejis de Veubeke, Stress function approach, in *Proceedings of the World Congress on Finite Element Methods in Structural Mechanics*, Dorset, 12–17 Oct 1975, vol. 1
19. ———, Displacement and equilibrium models in the finite element method, in *Stress Analysis*, ed. by O. Zienkiewics, G. Holister (Wiley, London/New York, 1965), pp. 145–197
20. B.M. Fraeijs de Veubeke, Discretization of rotational equilibrium in the finite element method, in *Mathematical Aspects of Finite Element Methods: Proceedings of the Conference, Consiglio Naz. delle Ricerche (C.N.R.)*, Rome. Lecture Notes in Mathematics, vol. 606 (Springer, Berlin, 1977), pp. 87–112
21. B.M. Fraeijs de Veubeke, A. Millard, Discretization of stress fields in the finite element method. J. Frankl. Inst. **302**(5–6), 389–412 (1976). Basis of the finite element method
22. J. Gopalakrishnan, J. Guzmán, A second elasticity element using the matrix bubble. IMA J. Numer. Anal. **32**(1), 352–372 (2012)
23. M.E. Gurtin, in *An Introduction to Continuum Mechanics*. Mathematics in Science and Engineering, vol. 158 (Academic, New York, 1981)
24. C. Johnson, B. Mercier, Some equilibrium finite element methods for two-dimensional elasticity problems. Numer. Math. **30**(1), 103–116 (1978)
25. M. Juntunen, J. Lee, A mesh dependent norm analysis of low order mixed finite element for elasticity with weakly symmetric stress. Math. Models Methods Appl. Sci. **24**(11), 2155–2169 (2014)
26. K.-Y. Kim, Guaranteed a posteriori error estimator for mixed finite element methods of linear elasticity with weak stress symmetry. SIAM J. Numer. Anal. **49**(6), 2364–2385 (2011)

27. C. Lovadina, R. Stenberg, Energy norm a posteriori error estimates for mixed finite element methods. Math. Comput. **75**(256), 1659–1674 (2006). (Electronic)
28. K.-A. Mardal, R. Winther, An observation on Korn's inequality for nonconforming finite element methods. Math. Comput. **75**(253), 1–6 (2006)
29. J.-C. Nédélec, Mixed finite elements in $\mathbf{R}^3$. Numer. Math. **35**(3), 315–341 (1980)
30. P.-A. Raviart, J.M. Thomas, A mixed finite element method for 2nd order elliptic problems, in *Mathematical Aspects of Finite Element Methods: Proceedings of the Conference, Consiglio Naz. delle Ricerche (C.N.R.)*, Rome. Lecture Notes in Mathematics, vol. 606 (Springer, Berlin, 1977), pp. 292–315
31. R. Stenberg, A family of mixed finite elements for the elasticity problem. Numer. Math. **53**(5), 513–538 (1988)
32. ———, Postprocessing schemes for some mixed finite elements. RAIRO Modél. Math. Anal. Numér. **25**(1), 151–167 (1991)
33. V. Watwood, B. Hartz, An equilibrium stress field model for finite element solution of two-dimensional elastostatic problems. Int. J. Solids Struct. **4**, 857–873 (1968)

Energy-Corrected Finite Element Methods for Scalar Elliptic Problems

Thomas Horger, Markus Huber, Ulrich Rüde, Christian Waluga, and Barbara Wohlmuth

Abstract In this work, we consider the finite element solution of several scalar elliptic problems with singularities in two dimensions. We outline recent theoretical developments in energy corrected approaches and demonstrate numerically that by local and easy to implement modifications of the discrete operators, optimal convergence orders in weighted Sobolev norms can be recovered.

1 Introduction

We consider boundary value problems in an open and bounded domain $\Omega \subset \mathbb{R}^2$, involving the second order linear elliptic operator

$$L := -\mathrm{div}(K \nabla u),$$

where $0 < K_0 \leq K \in L^\infty(\Omega)$ is a known coefficient, e.g., a diffusivity or the permeability of a porous medium. If not mentioned otherwise, we will set $K = 1$ for simplicity. However, we will also consider the case of a heterogeneous coefficient having jumps. It is known that numerical methods applied to such problems often suffer from suboptimal convergence due to singularities in the solution which dictate the regularity. In this paper, we consider modified finite element methods that allow to deal with singular solution components in an efficient and effective way.

T. Horger • C. Waluga (✉) • B. Wohlmuth
Institute for Numerical Mathematics, Technische Universität München, Boltzmannstraße 3, 85748 Garching b. München, Germany
e-mail: horger@ma.tum.de; waluga@ma.tum.de;wohlmuth@ma.tum.de

M. Huber • U. Rüde
Universität Erlangen-Nürnberg, Lehrstuhl für Informatik 10, Cauerstraße 11, 91058 Erlangen, Germany
e-mail: markus.huber@fau.de; ulrich.ruede@fau.de

A. Abdulle et al. (eds.), *Numerical Mathematics and Advanced Applications ENUMATH 2013*, Lecture Notes in Computational Science and Engineering 103,
DOI 10.1007/978-3-319-10705-9_2

As a first example for such singularities, we consider the case of a non-convex Lipschitz domain Ω having a re-entrant corner with interior angle $\pi < \theta < 2\pi$. In general, the solution to the problem

$$Lu = f \quad \text{in } \Omega, \qquad u = 0 \quad \text{on } \partial\Omega \tag{1}$$

will then be composed of smooth components as well as singular components of type $s_k = r^{k\pi/\theta} \sin(k\pi/\theta\phi)$, $k \in \mathbb{N}$, where ϕ denotes the angle in polar coordinates, and r stands for the distance to the re-entrant corner of Ω. This can be observed even when the data are smooth [18,23,27]. While the convergence order in the H^1-norm is the same as the order of the best approximation, this does not hold for the L^2-norm, where a gap of $1 - \pi/\theta$ can be observed due to the non-smoothness of the singular component $s_1 \notin H^\alpha(\Omega)$, $\alpha \geq 1 + \pi/\theta$. This effect is commonly referred to as *pollution* [9, 10, 32]. A similar, but even worse situation occurs in interface problems, see, e.g., [21, 22], when the interfaces between subdomains $\Omega_i \subset \Omega$ with different coefficient K intersect in one point. Here we typically see singular components of type r^ϵ, $0 < \epsilon \ll 1$.

In the literature, many different approaches have been proposed to deal with singular solution components, such as graded meshes [2], the enrichment of the finite element space with singular functions [6, 9, 13, 14, 16, 26, 32], or first order system least squares approaches, which add discrete versions of the singular basis functions to standard finite element spaces in a least-squares framework [7, 8, 11, 17, 24].

Most of the aforementioned approaches have in common the aim to improve the finite element approximation nearby the singularity. However, in some applications, the quantity of interest can be computed by excluding or relaxing the influence of the neighborhood of the singularities, e.g., stress intensity factors, eigenvalues or the flux at some given interface not including the singular points. Here, an accurate representation of the solution is not required near the singularity. This motivates the use of energy correction schemes that do not enrich the finite element spaces [20,30, 31,33]. The basic idea was originally introduced in the context of finite difference methods in [31, 33] and applied to finite elements in [29]. It was then analyzed in [20] for more general meshes, and it was proved that a careful modification of the energy in the original method can drastically improve the convergence. In the following, we briefly sketch the main ideas.

2 Re-entrant Corners

We consider a weak form of the boundary value problem (1). The corresponding bilinear form is given by $a(v, w) := \int_\Omega \nabla v \cdot \nabla w \, dx$, $v, w \in H^1(\Omega)$. To define the energy correction, we introduce $a_{i,h}(v, w) := \int_{\omega_{i,h}} \nabla v \cdot \nabla w \, dx$, where $\omega_{i,h}$ denotes the union of the ith layer of elements in $\mathcal{T}_h$ around the re-entrant corner x_c, i.e.,

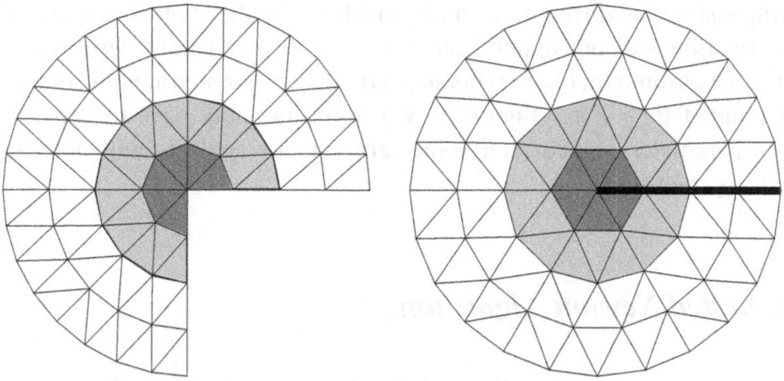

Fig. 1 *left*: Triangulation of the circular L-shape domain with $\theta = \frac{3\pi}{2}$; *right*: circular slit domain with $\theta = 2\pi$. The elements marked in *dark gray* belong to $\omega_{1,h}$ and the elements in *light gray* are in $\omega_{2,h}$

$\omega_{1,h} := \bigcup_{T \in \mathcal{T}_h, x_c \in \partial T} \overline{T}$ and $\omega_{i,h} := \bigcup_{T \in \mathcal{T}_h, \partial T \cap \overline{\omega}_{i-1,h} \neq \emptyset} \overline{T}$ for $i > 1$; cf. Fig. 1 for some illustration. Here $\mathcal{T}_h$ stands for a family of quasi-uniform simplicial meshes with mesh-size h. For given $\gamma \in \mathbb{R}^n$, $n \in \mathbb{N}$ fixed, we then define the bilinear form

$$a_{ec}(v, w) := a(v, w) - \sum\nolimits_{i=1}^{n} \gamma_i \, a_{i,h}(v, w). \tag{2}$$

The energy-corrected finite element form of (1) then reads: Find $u_h(\gamma) \in V_h^p$ s.t.

$$a_{ec}(u_h(\gamma), v) = (f, v), \quad v \in V_h^p, \tag{3}$$

where $(\cdot, \cdot)$ denotes the standard L^2-scalar product, and $V_h^p \subset H_0^1(\Omega)$ is the conforming piecewise polynomial finite element space of degree $p > 0$ associated with $\mathcal{T}_h$. For $\gamma = 0$ the standard finite element solution is recovered. As we will see in Sect. 3, our approach is not restricted to Dirichlet boundary conditions but also applies for Neumann boundary conditions. Note that the Laplace operator can be used to model a membrane, i.e., the effect of the modification with the parameters γ can be regarded as a softening ($\gamma_i \in (0, 1)$) or stiffening ($\gamma_i < 0$) of the material in the vicinity of the re-entrant corner.

The modification (2) does not change the structure of the stiffness matrix. Hence, it is cheap and easy to implement into existing codes, provided that γ_i and $\omega_{i,h}$ are known. Theoretical results for linear finite elements [20] require that we apply the correction in a union of elements $\omega_h \subset B_{k_0 h}$, where $B_{k_0 h}$ is a ball with radius $k_0 h$ and k_0 sufficiently large, with center at the re-entrant corner. Numerical results show that fixing $n = 1$ and setting $\omega_h = \omega_{i,h}$ is sufficient for the case $p = 1$, see also [30]. Our choice of $\omega_{i,h}$ is motivated by this observation indicating that for one parameter $\gamma \in [0, 1)$, it is sufficient to impose the correction only in those elements directly attached to the singularity.

We emphasize that, different from other techniques, the number of nodes affected by the correction will not depend on h as long as n does not depend on h. For simplicity, we assume that the layers $\omega_{i,h}$ are mirror-symmetric with respect to the singular point if $\theta \geq \frac{3}{2}\pi$. Otherwise, we would possibly have to consider more correction parameters, since then some terms in the analysis will not cancel by symmetry arguments.

2.1 A Nested Newton Algorithm

Let us first consider the case of linear finite elements and set $n = 1$: Assuming that the union of elements defined by $\omega_{1,h}$ is large enough, the quality of $u_h(\gamma) \in V_h^1$ is only determined by the choice of γ. In [20] it has been shown that for each $h > 0$ a proper subinterval of $[0, 1)$ exists such that no pollution occurs, and second order convergence in a suitably-weighted Sobolev norm can be recovered. The length of the subinterval tends to zero as the mesh-size does. Thus asymptotically exactly one correction parameter exists such that optimality can be observed. Here, we define the correction parameter as the unique root of the non-linear scalar-valued *energy defect function*

$$g_h(\gamma) := a(s_1, s_1) - a_{ec}(s_{1,h}(\gamma), s_{1,h}(\gamma)), \qquad \text{for } \gamma \in \mathbb{R}. \tag{4}$$

where we recall that s_1 denotes the first singular function and $s_{1,h}(\gamma)$ its modified finite element approximation. In [30] we developed and analyzed Newton algorithms for the calculation of an accurate enough correction parameter γ in a multi-level context. Such methods will be used frequently in the numerical results in this article to determine suitable correction parameters.

Next, we consider the extension to quadratic finite elements (or, analogously, linear elements on non-symmetric meshes) and $n = 2$. It turns out that a good choice of $\gamma \in \mathbb{R}^2$ is the root of the vector-valued energy defect function

$$g_h(\gamma) := \begin{pmatrix} a(s_1, s_1) - a_{ec}(s_{1,h}(\gamma), s_{1,h}(\gamma)) \\ a(s_2, s_2) - a_{ec}(s_{2,h}(\gamma), s_{2,h}(\gamma)) \end{pmatrix}, \qquad \text{for } \gamma \in \mathbb{R}^2. \tag{5}$$

By similar considerations as in the case of one parameter, we can derive a nested one-step Newton algorithm on a family of uniformly refined meshes $\mathcal{T}_l$. The mesh $\mathcal{T}_{l+1}$ is obtained by decomposing each element of $\mathcal{T}_l$ into four sub-elements. Given the initial guess $\gamma_0 = (0,0) \in (-1,1)^2$ on the coarse mesh $\mathcal{T}_0$, we set for $l = 0, 1, \ldots$

$$\gamma_{l+1} = [\nabla g_h(\gamma_l)]^{-1} \begin{pmatrix} a(s_{1,h}(\gamma_l), s_{1,h}(\gamma_l)) - a(s_1, s_1) \\ a(s_{2,h}(\gamma_l), s_{2,h}(\gamma_l)) - a(s_2, s_2) \end{pmatrix}, \tag{6}$$

with

$$\nabla g_h(\gamma) = \begin{pmatrix} a_{1,h}(s_{1,h}(\gamma), s_{1,h}(\gamma)) & a_{2,h}(s_{1,h}(\gamma), s_{1,h}(\gamma)) \\ a_{1,h}(s_{2,h}(\gamma), s_{2,h}(\gamma)) & a_{2,h}(s_{2,h}(\gamma), s_{2,h}(\gamma)) \end{pmatrix}. \tag{7}$$

Note that the index l denotes the refinement level, and on each level typically only one Newton step is carried out. The start value for the Newton on level $l+1$ is the value computed on level l. For the nested one-step Newton method we see that

$$[g_h]_i = \mathcal{O}\left(h^{\frac{2i\pi}{\omega}}\right) \qquad \text{and} \qquad [\nabla g_h]_{ij}^{-1} = \mathcal{O}\left(h^{\frac{-2j\pi}{\omega}}\right),$$

hence $\gamma_{l+1} \in \mathcal{O}(1)$, i.e., the values of γ stay bounded independent of the mesh-level. However, a detailed analysis of this algorithm is beyond the scope of this paper.

In the following section, we demonstrate the efficiency of the nested Newton method in numerical experiments.

2.2 Numerical Examples with Second Order Elements

In the following examples, we consider the Poisson problem in a circular domain with re-entrant corners of angle $\theta = \frac{3}{2}\pi$ and $\theta = 2\pi$ and assume that $\omega_{1,h}$ is mirror-symmetric on the coarsest mesh. We set a homogeneous right hand side and Dirichlet boundary conditions are chosen such that the exact solution is given by $u = s_1 + s_2 + s_3$. Note that the third singular component does not reduce the convergence rates of the uncorrected approach due to sufficient regularity. We conduct our numerical experiments on a series of uniformly refined meshes (cf. Fig. 1 for the initial triangulations and correction domains) and compare the errors of corrected vs. uncorrected finite elements in weighted L^2 norms

$$\|u - u_h\|_{0,\alpha} := \|r^\alpha (u - u_h)\|_{L^2(\Omega)}, \tag{8}$$

where $r = |x - x_c|$ denotes the distance to the re-entrant corner x_c, and α is the weighting parameter. Note that for $\alpha = 0$ we recover the standard L^2 norm. In each case, the roots of the energy correction function (5) are determined by the nested Newton-method discussed above where the initial guess on level 0 is always set to zero. The contour lines of the energy correction function for the initial meshes are plotted in Fig. 2. The bullet in the pictures denotes the unique root of the energy correction function.

The correction parameters for the L-shape and slit domain are listed in Table 1. As it can be seen easily both parameters converge with respect to the level and for both domains $\gamma_{1,opt}$ is positive while $\gamma_{2,opt}$ is negative.

In our convergence study, we consider $p = 2$ and different choices of the correction parameter for illustration, namely, the energy correction with the level-

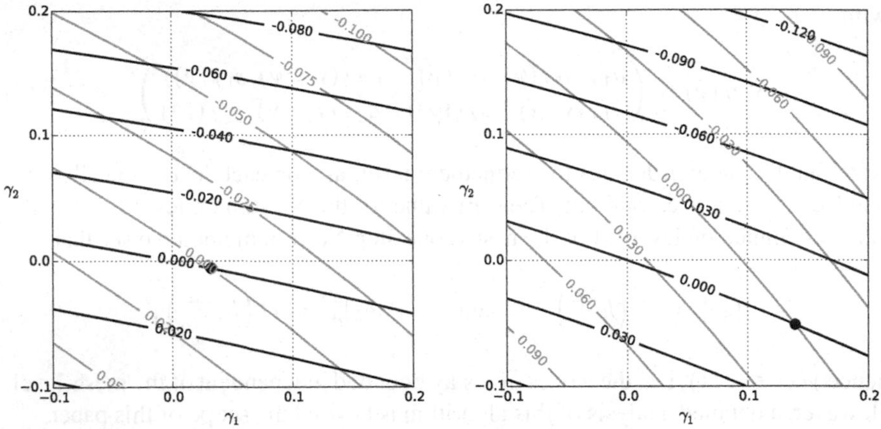

Fig. 2 Contour lines and root (*bullet*) of the energy functional (5) in the L-shape (*left*) and slit domain (*right*); *gray lines* are associated with the first and *black* ones with the second component of the energy functional

Table 1 Approximate roots of $g_h(\gamma)$ obtained by a nested Newton method applied to subsequent refinements of the respective initial meshes of Fig. 1

Level	$\gamma_{1,opt}$	$\gamma_{2,opt}$
1	0.0320706	−0.0055412
2	0.0315346	−0.0055392
3	0.0315229	−0.0055351
4	0.0315208	−0.0055342
L-shape		

Level	$\gamma_{1,opt}$	$\gamma_{2,opt}$
1	0.1493019	−0.0497673
2	0.1395878	−0.0510151
3	0.1395358	−0.0510148
4	0.1395348	−0.0510143
Slit		

dependent parameters of Table 1, and the choice $(\gamma_1, 0.0)$ with γ_1 determined by the one-step Newton method of [30]. Moreover, we compare the standard quadratic finite element method resulting from the choice $(\gamma_1, \gamma_2) = (0.0, 0.0)$ as well as two manually chosen values, where one is closer to the actual optimal parameter than the other. The latter experiments are given to demonstrate that the energy correction can yield solutions of much better quality by manual fine-tuning of parameters, even when the actual asymptotically correct values are unknown.

In Tables 2 and 3, we list the results of our convergence study for the L-shape and slit domains, respectively. The errors are measured in the weighted L^2 norm with weighting $\alpha \approx 1.38$ for the L-shape and $\alpha \approx 1.55$ for the slit.

In both cases, we observe that the asymptotically optimal convergence order of $\mathcal{O}(h^3)$ is only recovered for the correction approach with both parameters chosen according to the root of the energy defect function. Restricting the correction to only one parameter, however, still improves the rate to $\mathcal{O}(h^2)$, since the effects of the stronger singularity s_1 can still be compensated by this simpler correction. From the results for the manually tuned parameters it can be concluded that our approach can significantly improve the quality of the solution, even when the exact parameters are

Table 2 L-shape: Errors $10^4 \times \|u - u_h(\gamma_1, \gamma_2)\|_{0,\alpha\approx 1.38}$ for different parameters

(γ_1, γ_2)	$(\gamma_{1,opt}, \gamma_{2,opt})$		$(\gamma_2, 0)$		$(0, 0)$		$(0.031, -0.005)$		$(0.02, -0.006)$	
Level	Error	Rate	Error	Rate	Error	Rate	Error	Rate	Error	Rate
1	1.8697	–	1.9402	–	4.9912	–	1.7893	–	2.8886	–
2	0.2031	3.20	0.2843	2.77	1.9475	1.36	0.2017	3.10	1.0433	1.47
3	0.0240	3.08	0.0425	2.74	0.7716	1.34	0.0257	2.94	0.4100	1.35
4	0.0030	3.00	0.0064	2.72	0.3061	1.33	0.0047	2.46	0.1625	1.34

Table 3 Slit domain: Errors $10^3 \times \|u - u_h(\gamma_1, \gamma_2)\|_{0,\alpha\approx 1.55}$ for different parameters

(γ_1, γ_2)	$(\gamma_{1,opt}, \gamma_{2,opt})$		$(\gamma_2, 0)$		$(0, 0)$		$(0.139, -0.051)$		$(0.15, -0.06)$	
Level	Error	Rate	Error	Rate	Error	Rate	Error	Rate	Error	Rate
1	1.1790	–	1.3055	–	2.4165	–	1.0874	–	1.2833	–
2	0.1372	3.10	0.2994	2.12	1.1968	1.01	0.1372	2.99	0.1749	2.88
3	0.0168	3.03	0.0738	2.02	0.5966	1.00	0.1720	3.00	0.0312	2.49
4	0.0021	3.02	0.0184	2.01	0.2978	1.00	0.0275	2.65	0.0112	1.48

unknown. However, to observe optimal asymptotic convergence rates, we require a high accuracy of the parameter on the finer meshes. Let us remark here that for multiple re-entrant corners, the correction parameters can be determined by solving local problems for each corner in a preprocessing step. The optimal correction parameters depends on the interior angle, as already seen from the numerical results, but also on $\omega_{1,h}$. More precisely, the number and the local shape of elements in $\omega_{1,h}$ influence the optimal correction parameters.

3 Eigenvalue Problems

Next, we consider the eigenvalue problem with homogeneous Neumann boundary conditions

$$Lu_m = \lambda_m u_m \qquad \text{in } \Omega.$$

As before, we choose an L-shaped domain $\Omega := (-1, 1)^2 \backslash ([0, 1] \times [-1, 0])$ and a slit-domain $\Omega := (-1, 1)^2 \backslash ([0, 1] \times \{0\})$ for our numerical results. For comparison with reference values given in the literature [12, 19], we use different meshes in this example. Moreover, we also present results for a domain with multiple re-entrant corners, in which case we compute a reference solution on a finer mesh. The meshes are always constructed such that we obtain perfectly symmetric isosceles triangles around the singular points; cf. Fig. 3.

It is well-known from the literature [3–5, 12, 25], that the above mentioned pollution effect can be observe, but that it occurs only for all eigenfunctions and eigenvalues, for which a non-smooth singular component is present. For sufficiently

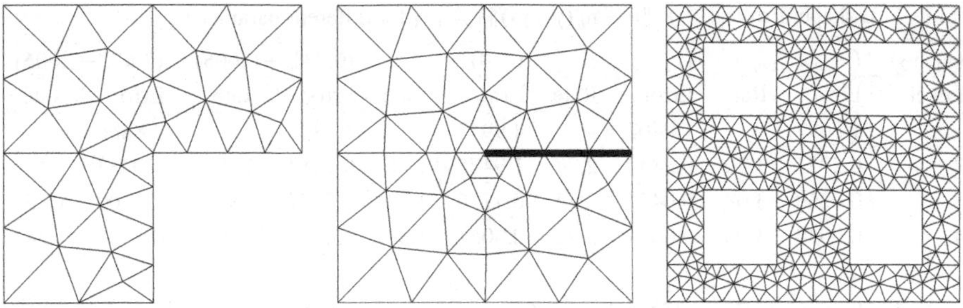

Fig. 3 L-shape, slit, and domain with multiple re-entrant corners

smooth eigenfunctions, a quadratic convergence rate for the eigenvalues can be observed in case of linear finite elements, i.e., $p = 1$.

Our modified finite element formulation reads: find the discrete eigenvalues $\lambda_{m,h} \in \mathbb{R}$ and the eigenfunctions $u_{h,m}(\gamma) \in V_h^1$ such that

$$a_{ec}(u_{h,m}(\gamma), v) = \lambda_{h,m}(u_{h,m}(\gamma), v), \quad v \in V_h^1, \tag{9}$$

where $0 \leq \lambda_{h,1} \leq \lambda_{h,2} \leq \ldots$. For simplicity of notation, we use the same symbol as before for the finite element space although no homogeneous boundary conditions are imposed on the space. Let us next briefly outline the convergence analysis for this modified scheme.

3.1 Convergence Analysis

In this subsection, we focus on the convergence analysis of the discrete eigenvalues $\lambda_{h,m}$ and follow the lines of [28] in the conforming setting. To do so, we introduce the eigenvalue problem: find $\lambda \in \mathbb{R}$ and $w \in H^1(\Omega)$ such that $a(w, z) = \lambda(w, z)$ for all $z \in H^1(\Omega)$. The non-negative eigenvalues are ordered such that $0 \leq \lambda_1 \leq \lambda_2 \leq \ldots$, and the associated eigenfunctions are denoted by w_i with the normalization $(w_i, w_j) = \delta_{i,j}$. Now define the m-dimensional space V_m by $V_m := \text{span}\{w_i, i \leq m\}$. Further for each $v \in V_m$ let the modified Galerkin projection R_h onto V_h^1 be defined by $a_{ec}(R_h v, v_h) = a(v, v_h)$ for all $v_h \in V_h^1$. We recall that R_h depends on the specific choice of γ. In terms of R_h, we define $E_{m,h} := R_h V_m$ and note that $\dim E_{m,h} = m$ for $h \leq h_0$ small enough.

For the sake of presentation, let us first state the main result and subsequently develop the ingredients needed for its proof.

Theorem 1 *Let* $1 - \frac{\pi}{\omega} < \alpha < 1$. *If the above mentioned modification is used with* $n = 1$ *and* γ_{opt}, *the following upper and lower bound for* $\lambda_{h,m}$ *hold,*

$$\lambda_m(1 - Ch^2\lambda_m^{1+\alpha}) \leq \lambda_{h,m} \leq \lambda_m(1 + Ch^2\lambda_m^{\alpha+1}).$$

Our proof is based on the following two technical results which are provided without a detailed proof.

Lemma 1 *Let* $1 - \frac{\pi}{\omega} < \tilde{\alpha} < \alpha < 1$ *then it holds*

$$\|r^{-\tilde{\alpha}} v\|_0 \leq C \, \|v\|_0^{1-\alpha} \, \|v\|_1^{\alpha}, \quad v \in H^1(\Omega).$$

The upper bound in Lemma 1 can be obtained by using the Hölder inequality in combination with interpolation arguments and standard Sobolev embedding results. Combining Lemma 1 with [28, Lemma 6.4-2], [20, Theorem 2.4] and some straightforward computations yield the following bounds.

Lemma 2 *Let* $v \in V_m$ *with* $(v, v) = 1$. *Then* $v = \sum_{i=1}^m \beta_i w_i$ *with* $\sum_{i=1}^m \beta_i^2 = 1$, *and it satisfies* $a(v, z) = (f_v, z)$ *for all* $z \in H^1(\Omega)$ *with* $f_v := \sum_{i=1}^m \beta_i \lambda_i w_i$. *Moreover, we have* $r^{-\alpha} f_v \in L^2(\Omega)$ *for* $1 - \frac{\pi}{\omega} < \alpha < 1$, *and the following bounds hold with constants independent of the mesh-size*

$$\begin{aligned} |a(v, v) - a_{ec}(R_h v, R_h v)| &\leq C h^2 \lambda_m^{2+\alpha}, \\ (R_h v, R_h v) &\geq 1 - C h^2 \lambda_m^{\alpha+1}. \end{aligned}$$

Now we are prepared to provide the proof of the main result.

Proof (Theorem 1) The proof is based on the characterization of the eigenvalues by the Rayleigh quotient. We start with the upper bound. Using $E_{h,m}$ as defined above together with Lemma 2, we get the following upper bound:

$$\begin{aligned} \lambda_{h,m} &\leq \max_{v \in E_{m,h}} \frac{a_{ec}(v, v)}{(v, v)} = \max_{v \in V_m} \frac{a_{ec}(R_h v, R_h v)}{(R_h v, R_h v)} \\ &= \max_{v \in V_m} \frac{a(v, v) + a_{ec}(R_h v, R_h v) - a(v, v)}{(R_h v, R_h v)} \\ &= \max_{v \in V_m} \frac{a(v, v)}{(v, v)} \max_{v \in V_m} \frac{(v, v)}{(R_h v, R_h v)} + \max_{v \in V_m} \frac{a_{ec}(R_h v, R_h v) - a(v, v)}{(R_h v, R_h v)} \\ &\leq \lambda_m \frac{1 + C h^2 \lambda_m^{2+\alpha}}{1 - C h^2 \lambda_m^{\alpha+1}} \lesssim \lambda_m (1 + C h^2 \lambda_m^{\alpha+1}) + C h^2 \lambda_m^{2+\alpha} (1 + C h^2 \lambda_m^{\alpha+1}). \end{aligned}$$

As next step it remains to show the lower bound. In contrast to the uncorrected scheme, we do not have the trivial bound $\lambda_m \leq \lambda_{m,h}$. The proof of the lower bound follows basically the lines of the upper bound but requires the use of a different m-dimensional space. Firstly we define a new space given by $E_m := \operatorname{span}\{\widetilde{w_i}, i \leq m\}$ where $\tilde{w}_i \in H^1(\Omega)$ is defined by $a(\tilde{w}_i, z) = (w_{i,h}, z)$ for all $z \in H^1(\Omega)$. Secondly, we note that $R_h E_m = E_{m,h}$ and thus for h small enough we have $\dim E_m = m$.

Now, similar arguments as before yields

$$\begin{aligned}\lambda_m &\le \max_{v\in E_m} \frac{a(v,v)}{(v,v)} = \max_{v\in E_m} \frac{a_{ec}(R_h v, R_h v) + a(v,v) - a_{ec}(R_h v, R_h v)}{(v,v)} \\ &= \max_{v\in E_m} \frac{a_{ec}(R_h v, R_h v)}{(R_h v, R_h v)} \max_{v\in E_m} \frac{(R_h v, R_h v)}{(v,v)} + \max_{v\in E_m} \frac{a(v,v) - a_{ec}(R_h v, R_h v)}{(v,v)} \\ &\le \lambda_{h,m}(1 + Ch^2 \lambda_{h,m}^{1+\alpha}).\end{aligned}$$

Combining the upper bounds for λ_m and $\lambda_{h,m}$ yields the lower bound for $\lambda_{h,m}$. □

The steps outlined in this section show the flexibility and potential of the ideas of [20, 30] to eigenvalue problems. A more detailed analysis provides also optimal bounds for the eigenfunction convergence. Let us next support our theoretical ideas by numerical results.

3.2 Numerical Computation of Eigenvalues

We conduct convergence studies for the eigenvalue problem defined on the geometries depicted in Fig. 3. We first compare the numerical results obtained without correction to those obtained with a suitable modification parameter.

In this example, we make use of the Neumann fit tabulated in [30, Table 5.3], which provides a simple heuristic approach to determine modification parameters in case of meshes consisting of isosceles triangles around the singularity. This purely geometric assumption is satisfied for our meshes by construction (see Fig. 3).

The correction parameter γ for the L-shape ($\theta = \frac{3}{2}\pi$) and the multiple re-entrant corners domain is given by $\gamma \approx 0.1478$. We note that in each case four isosceles triangles are attached to the singularity, and thus the correction parameter is the same for all re-entrant corners. For the slit domain ($\theta = 2\pi$), we count six adjacent elements at the singular vertex, and hence we determine our correction parameter to $\gamma \approx 0.2716$ (more precise values are given in the respective tables).

In Tables 4–6 we list the results for our convergence study for the L-shape, the slit domain and the domain with multiple re-entrant corners, respectively. Without correction, we observe suboptimal rates for some eigenvalues in each of the three cases. However, using the modified method, the asymptotically optimal convergence of $\mathcal{O}(h^2)$ for all given eigenvalues is obtained in the three cases. Note that we excluded the results for some eigenvalues for the slit domain in Table 5. This is because the corresponding eigenfunctions do not include singular components strong enough to affect the optimal rate for linear elements. Hence, for these eigenvalues, a convergence rate of $\mathcal{O}(h^2)$ can be reached already by using the non-corrected method.

Table 4 Convergence rates for eigenvalues in the L-shaped domain with and without energy correction

No correction $\gamma = 0$:

	1.EV		2.EV		3.EV		4.EV		5.EV	
	Exact: 1.47562		Exact: 3.53403		Exact: 9.86960		Exact: 9.86960		Exact: 11.38948	
Level	Value	Rate	Value	Rate	Value	Rate	Value	Rate	Value	Rate
1	1.55008	–	3.63939	–	10.79641	–	10.90447	–	12.67323	–
2	1.50014	**1.60**	3.56116	1.96	10.10623	1.97	10.12814	2.00	11.71975	1.96
3	1.48402	**1.55**	3.54091	1.98	9.92994	1.97	9.93496	1.98	11.47442	1.96
4	1.47861	**1.49**	3.53576	1.99	9.88483	1.99	9.88604	1.99	11.41101	1.98
5	1.47672	**1.44**	3.53447	2.00	9.87342	2.00	9.87372	2.00	11.39489	1.99

Correction $\gamma = 0.147850426060652$:

	1.EV		2.EV		3.EV		4.EV		5.EV	
	Exact: 1.47562		Exact: 3.53403		Exact: 9.86960		Exact: 9.86960		Exact: 11.38948	
Level	Value	Rate	Value	Rate	Value	Rate	Value	Rate	Value	Rate
1	1.51075	–	3.62603	–	10.78432	–	10.89390	–	12.64990	–
2	1.48457	**1.97**	3.55913	1.87	10.10560	1.95	10.12758	1.99	11.71674	1.95
3	1.47785	**2.00**	3.54060	1.93	9.92990	1.97	9.93493	1.98	11.47396	1.95
4	1.47617	**2.03**	3.53571	1.96	9.88482	1.99	9.88604	1.99	11.41094	1.98
5	1.47575	**2.06**	3.53446	1.98	9.87342	2.00	9.87372	2.00	11.39488	1.99

Table 5 Convergence rates for eigenvalues in the slit domain with and without energy correction

No correction $\gamma = 0$:

	1.EV		2.EV		5.EV		7.EV		8.EV	
	Exact: 1.03407		Exact: 2.46740		Exact: 9.86960		Exact: 12.26490		Exact: 12.33701	
Level	Value	Rate	Value	Rate	Value	Rate	Value	Rate	Value	Rate
1	1.14032	–	2.52918	–	10.78301	–	13.97705	–	14.17015	–
2	1.07951	**1.23**	2.48307	1.98	10.10730	1.94	12.75729	**1.80**	12.81216	1.95
3	1.05478	**1.13**	2.47135	1.99	9.93045	1.97	12.43849	**1.50**	12.44434	2.15
4	1.04392	**1.07**	2.46839	2.00	9.88497	1.99	12.32633	**1.50**	12.36411	1.99
5	1.03887	**1.04**	2.46765	2.00	9.87346	1.99	12.28924	**1.34**	12.34381	1.99

Correction $\gamma = 0.271607294328175$:

	1.EV		2.EV		5.EV		7.EV		8.EV	
	Exact: 1.03407		Exact: 2.46740		Exact: 9.86960		Exact: 12.26490		Exact: 12.33701	
Level	Value	Rate	Value	Rate	Value	Rate	Value	Rate	Value	Rate
1	1.06167	–	2.49235	–	10.76626	–	13.85168	–	13.88581	–
2	1.04116	**1.96**	2.47377	1.97	10.10651	1.92	12.66340	**1.99**	12.73749	1.95
3	1.03583	**2.01**	2.46902	1.97	9.93041	1.96	12.36581	**1.98**	12.43961	1.96
4	1.03449	**2.06**	2.46781	1.99	9.88497	1.98	12.29021	**1.99**	12.36294	1.98
5	1.03417	**2.14**	2.46750	1.99	9.87346	1.99	12.27120	**2.01**	12.34352	1.99

Table 6 Convergence rates for eigenvalues in the domain with multiple re-entrant corners with and without correction

No correction $\gamma = 0$:

	1.EV		2.EV		3.EV		4.EV		5.EV	
	Exact: 0.11422		Exact: 0.11422		Exact: 0.23460		Exact: 0.31626		Exact: 0.31626	
Level	Value	Rate	Value	Rate	Value	Rate	Value	Rate	Value	Rate
1	0.11609	–	0.11609	–	0.23841	–	0.32303	–	0.32323	–
2	0.11489	1.48	0.11489	1.48	0.23595	1.51	0.31861	1.53	0.31867	1.53
3	0.11446	1.46	0.11446	1.46	0.23509	1.47	0.31709	1.49	0.31711	1.50
4	0.11431	1.42	0.11431	1.42	0.23478	1.44	0.31656	1.45	0.31657	1.45
5	0.11425	1.39	0.11425	1.39	0.23467	1.40	0.31637	1.41	0.31637	1.42

Correction $\gamma = 0.147850426060652$:

	1.EV		2.EV		3.EV		4.EV		5.EV	
	Exact: 0.11422		Exact: 0.11422		Exact: 0.23460		Exact: 0.31626		Exact: 0.31626	
Level	Value	Rate	Value	Rate	Value	Rate	Value	Rate	Value	Rate
1	0.11472	–	0.11473	–	0.23574	–	0.31869	–	0.31887	–
2	0.11435	1.85	0.11436	1.85	0.23492	1.85	0.31690	1.92	0.31695	1.92
3	0.11425	1.97	0.11425	1.96	0.23469	1.96	0.31642	1.99	0.31643	1.99
4	0.11422	2.04	0.11422	2.03	0.23462	2.02	0.31630	2.04	0.31630	2.03
5	0.11422	2.11	0.11422	2.10	0.23461	2.09	0.31627	2.09	0.31627	2.08

4 Jumping Coefficients

Next, let us study the influence of heterogeneous coefficients with jumps. We consider again the scalar elliptic problem (1) but in contrast to the previous sections, we now assume that K is piecewise constant on disjoint subsets $\Omega_i \subset \Omega := (-1,1)^2$, i.e., we define $\Omega_1 := (0,1)^2$, $\Omega_2 := (-1,0) \times (0,1)$, $\Omega_3 := (-1,0)^2$ and $\Omega_4 := (0,1) \times (-1,0)$, and set $K = 1$ in Ω_2, Ω_4 and $K = a$ in Ω_1, Ω_3 for finite $a > 0$. Whenever $a \neq 1$ one obtains a discontinuity at the origin, which causes solutions in $H^{1+\epsilon}(\Omega)$ for $0 < \epsilon < 1$ and possibly $\epsilon \ll 1$; cf. [15, 21]. As before, this severely limited regularity bound results in poor L^2-accuracy of the uncorrected discrete solutions. The solution of (1) again admits a singular function representation, with singular components of the form

$$s_k(r,\phi) = r^{\lambda_k}\theta_k(\phi), \quad \lambda_k \leq \lambda_{k+1}, k \in \mathbb{N}. \tag{10}$$

Here the exponent λ_k can be determined with help of an auxiliary Sturm–Liouville eigenvalue problem [15]. We note that (r,ϕ) are the polar coordinates with respect to the origin $(0,0)$. Typically, the exponent λ_1 is very small, i.e., $\lambda_1 \ll 1$, which renders s_1 as the dominating singular component in the problem. Since by [21, Lemma 3.3] there holds $\lambda_1 + \lambda_2 = 2$ for the case outlined above, the energy correction with only one correction parameter per singular vertex is feasible. Hence, we introduce a modification of the weak form at the singularity by

$$a_{ec}(v,w) := \int_{\Omega} K\nabla v \cdot \nabla w \, \mathrm{d}x - \gamma \int_{\omega_h} K\nabla v \cdot \nabla w \, \mathrm{d}x. \tag{11}$$

where $\gamma \in [0, 1)$, and ω_h is a union of elements in $\mathscr{T}_h$. As before, we choose ω_h as the union of those elements adjacent to the vertex at which the singularity is located. The modified finite element method for (1) is then again given in the form of (3), and the corrections are again determined as the root of the energy functional (4). We remark that a similar correction has been proposed to recover optimal multigrid convergence rates for elliptic problems with intersecting interfaces in [1].

4.1 Numerical Results

As a first benchmark, we consider non-homogeneous Dirichlet boundary conditions such the exact solution is given by $u = s_1$. We conduct a convergence study on a series of uniformly refined meshes and, as before, we compare the energy corrected approach to the standard finite element method. The first two mesh levels and correction domains are depicted in Fig. 4 for illustration.

We consider two values of a, i.e., $a = 10$ and $a = 1{,}000$, and in a preprocessing step we solve the Sturm–Liouville problem to obtain the first eigenvalue as $\lambda_1 \approx 0.38996$ and $\lambda_1 \approx 0.04025$, respectively. The other singular components are of higher regularity (i.e., $s_i \in H^2(\Omega), i > 1$) and are therefore not considered in the following. For both cases we then approximate the roots of the energy defect function (4) by a classical Newton method until the relative error between two consecutive iterates is at most $1.48 \cdot 10^{-8}$. The results are listed in Fig. 5 alongside with nonlinear fits against the function $\gamma_\infty + ch^{2(1-\lambda_1)}$. For $a = 1{,}000$, we observe

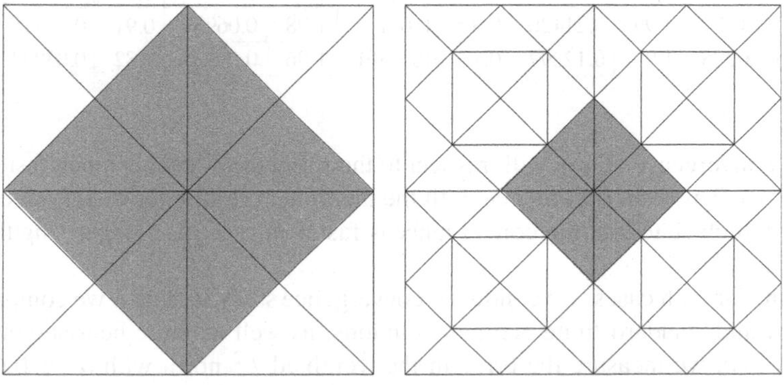

Fig. 4 Mesh levels 1 and 2 of the unit square with correction domains ω_h in *gray*

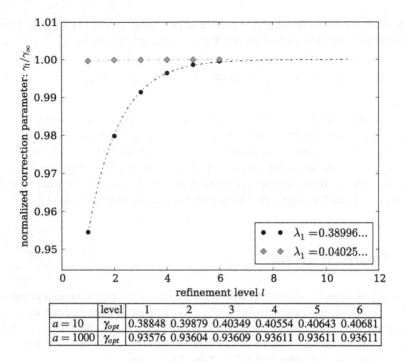

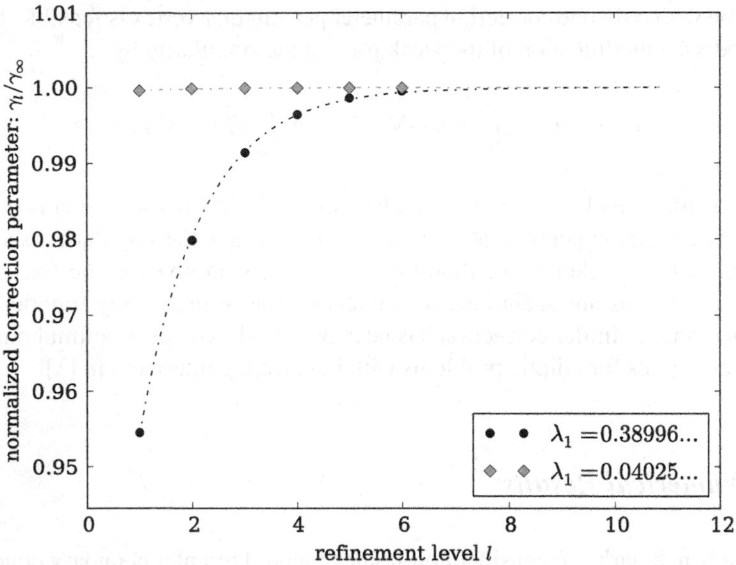

	level	1	2	3	4	5	6
$a = 10$	γ_{opt}	0.38848	0.39879	0.40349	0.40554	0.40643	0.40681
$a = 1000$	γ_{opt}	0.93576	0.93604	0.93609	0.93611	0.93611	0.93611

Fig. 5 Approximate roots of the energy defect function (4) obtained on subsequent refinements of the initial meshes of Fig. 4 for jumping material coefficients. *Dashed lines* indicate the nonlinear fits against the function $\gamma_\infty + ch^{2(1-\lambda_1)}$

Table 7 Errors $10^2 \times \|u - u_h(\gamma)\|_{0,\alpha\approx 0.66004}$ for jumping coefficients with $a = 10$

γ	γ_{opt}		0		0.406		0.3		0.5	
Level	Error	Rate	Error	Rate	Error	Rate	Error	Rate	Error	Rate
1	2.72860	–	3.06090	–	2.71390	–	2.80460	–	2.63900	–
2	0.71951	1.92	1.38890	1.14	0.71293	1.93	0.84889	1.72	0.68297	1.95
3	0.18359	1.97	0.69770	0.99	0.18253	1.97	0.29198	1.54	0.23705	1.53
4	0.04636	1.99	0.37922	0.88	0.04627	1.98	0.13016	1.17	0.11776	1.01
5	0.01170	1.99	0.21420	0.82	0.01175	1.98	0.06939	0.91	0.06718	0.81
6	0.00295	1.99	0.12297	0.80	0.00301	1.96	0.03938	0.82	0.03924	0.78

that the convergence of γ_{opt} with respect to the refinement level is much faster than for $a = 10$. This is in accordance with the theoretical results for re-entrant corners of [30], which state that the convergence is faster in case of stronger singularities (Table 7).

Again, for both cases, we conduct a convergence study in which we compare the modified and standard finite element solutions, as well as three heuristic choices. For $a = 10$, we measure the error in the weighted L^2-norm with $\alpha \approx 0.66004$. For the modified approach with correction parameters as listed in Fig. 5, an optimal convergence rate of $\mathcal{O}(h^2)$ is obtained, while the convergence of the standard finite element method is clearly limited due to the singular components. For the manually tuned choices we again observe that the solution quality is much better than without correction. Although on coarse meshes nearly optimal rates can only be seen for the choice $\gamma = 0.406$, which is close enough to the optimal values of Fig. 5, asymptotically the convergence rate will deteriorate. This effect can be seen more dominantly for the choices $\gamma = 0.3$ and $\gamma = 0.5$. A not optimal choice of γ will yield asymptotically the same poor convergence rate as the uncorrected method has.

For comparison, we list the results for the case $a = 1{,}000$ in Table 8, where the error is measured in the weighted L^2-norm with $\alpha \approx 0.96475$. Also here we see that using the previously determined parameters an optimal asymptotic convergence rate is recovered despite the strongly singular component.

Finally, let us consider a more complex numerical example in $\Omega = [-2,2] \times [-2,2]$. We define $i = \lfloor x \rfloor$ and $j = \lfloor y \rfloor$, and set

$$K = \begin{cases} 1{,}000 & \text{if } (-1)^{i+j} > 0, \\ 1 & \text{if } (-1)^{i+j} < 0, \end{cases}$$

thus, we identify 9 singularities which are corrected as depicted in Fig. 6. The boundary conditions are chosen as $u = 1$ on the bottom boundary, $u = 0$ on the top and $\nabla u \cdot \mathbf{n} = 0$ elsewhere.

Table 8 Errors $10^2 \times \|u - u_h(\gamma)\|_{0,\alpha\approx 0.96475}$ for jumping coefficients with $a = 1{,}000$

γ	γ_{opt}		0		0.936		0.9		0.95	
Level	Error	Rate	Error	Rate	Error	Rate	Error	Rate	Error	Rate
1	0.41774	–	0.51892	–	0.43095	–	0.43531	–	0.42932	–
2	0.11332	1.88	0.32494	0.68	0.11944	1.85	0.12422	1.81	0.11825	1.86
3	0.02917	1.96	0.24001	0.44	0.03142	1.93	0.03725	1.74	0.03178	1.90
4	0.00738	1.98	0.19177	0.32	0.00813	1.95	0.01876	0.99	0.01073	1.57
5	0.00186	1.99	0.15902	0.27	0.00209	1.96	0.01587	0.24	0.00704	0.61
6	0.00047	1.99	0.13485	0.24	0.00054	1.96	0.01475	0.11	0.00637	0.14

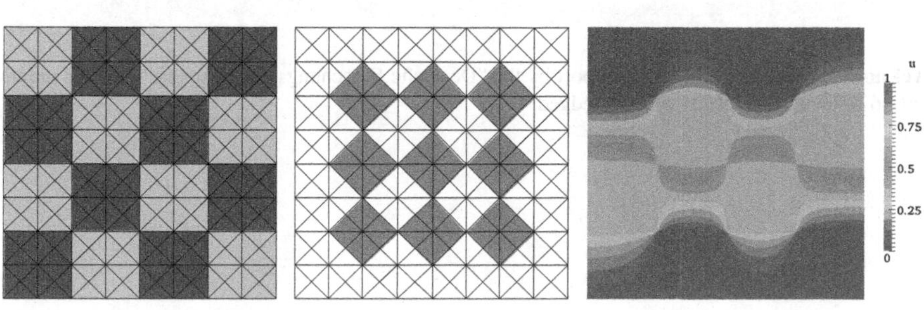

Fig. 6 *Left*: coefficient $K = 1$ (*dark*), $K = 1{,}000$ (*light*); *center*: correction on level 1 in *light gray*; *right*: reference solution (with energy correction on level 8)

Table 9 Errors $10 \times \|u - u_h(\gamma)\|_{0,\alpha\approx 0.96475}$; Convergence analysis of the energy correction method and standard finite element method for the example with 9 singularities

γ	γ_{opt}		0	
Level	Error	Rate	Error	Rate
1	0.50254	–	1.15720	–
2	0.09690	2.07	0.92582	0.32
3	0.02472	1.97	0.77083	0.26
4	0.00636	1.96	0.65581	0.23

Due to the construction of the jumping coefficient K each of the 9 singularities has the same first eigenvalue $\lambda_1 = 0.04025$, thus, also the correction parameter in the neighborhood of each singularity has the same value, namely, $\gamma \approx 0.93611$. Since we have no exact solution at hand for this example, we compute a reference solution with the energy corrected method on a high resolution mesh for comparison. In Table 9, we study the convergence to the reference solution, starting with the mesh of Fig. 6. The error is measure in a weighted L^2 norm with a weight of $\alpha \approx 0.96475$ around each singularity. Again, we observe optimal convergence rates for the energy corrected method. Hence, the correction parameters for each singularity arising in a practical application can be determined independently in a preprocessing step.

Conclusion

In this work, we discussed the extension of energy corrected finite element methods from the special case of linear finite elements on domains with re-entrant corners to second order finite elements, eigenvalue problems, and singularities resulting from jumping material parameters in two space dimensions. We demonstrated that given the correction parameters, the modified methods can dramatically improve the numerical solutions in presence of strong singular solution component. Quasi-optimal convergence rates are recovered. Hence, they provide an appealing alternative to graded or adaptive meshes and function space enrichment for applications in which a high accuracy in vicinity of the singular regions is not required.

Acknowledgements Financial support by the "Deutsche Forschungsgemeinschaft" through grant WO-671/13-1 is gratefully acknowledged.

References

1. R.E. Alcouffe, A. Brandt, J.E. Dendy Jr., J.W. Painter, The multi-grid method for the diffusion equation with strongly discontinuous coefficients. SIAM J. Sci. Stat. Comput. **2**(4), 430–454 (1981)
2. T. Apel, A. Sändig, J. Whiteman, Graded mesh refinement and error estimates for finite element solutions of elliptic boundary value problems in non-smooth domains. Math. Methods Appl. Sci. **19**, 63–85 (1996)
3. I. Babuška, B. Guo, J. Osborn, Regularity and numerical solution of eigenvalue problems with piecewise analytic data. SIAM J. Numer. Anal. **26**(6), 1534–1560 (1989)
4. I. Babuška, J. Osborn, Finite element-Galerkin approximation of the eigenvalues and eigenvectors of selfadjoint problems. Math. Comput. **52**(186), 275–297 (1989)
5. ———, Eigenvalue problems, in *Handbook of Numerical Analysis*, vol. 2 (North Holland, Amsterdam, 1991), pp. 643–787
6. I. Babuška, M. Rosenzweig, A finite element scheme for domains with corners. Numer. Math. **20**, 1–21 (1972)
7. M. Berndt, T. Manteuffel, S. McCormick, Analysis of first-order system least-squares (FOSLS) for elliptic problems with discontinuous coefficients: part II. SIAM J. Numer. Anal. **42**, 409–436 (2005)
8. M. Berndt, T. Manteuffel, S. McCormick, G. Starke, Analysis of first-order system least-squares (FOSLS) for elliptic problems with discontinuous coefficients: part I. SIAM J. Numer. Anal. **42**, 386–408 (2005)
9. H. Blum, M. Dobrowolski, On finite element methods for elliptic equations on domains with corners. Computing **28**, 53–63 (1982)
10. H. Blum, R. Rannacher, Extrapolation techniques for reducing the pollution effect of reentrant corners in the finite element method. Numer. Math. **52**, 539–564 (1988)
11. P. Bochev, M. Gunzburger, Finite element methods of least-squares type. SIAM Rev. **40**(4), 789–837 (1998)
12. D. Boffi, Finite element approximation of eigenvalue problems. Acta Numer. **19**(01), 1–120 (2010)
13. M. Bourlard, M. Dauge, M.-S. Lubuma, S. Nicaise, Coefficients of the singularities for elliptic boundary value problems on domains with conical points iii. Finite element methods on polygonal domains. SIAM J. Numer. Anal. **29**, 136–155 (1992)
14. S. Brenner, Multigrid methods for the computation of singular solutions and stress intensity factors. I: corner singularities. Math. Comput. **68**(226), 559–583 (1999)
15. S. Brenner, L. Sung, Multigrid methods for the computation of singular solutions and stress intensity factors. III: interface singularities. Comput. Methods Appl. Mech. Eng. **192**, 4687–4702 (2003)
16. Z. Cai, S. Kim, A finite element method using singular functions for the Poisson equation: corner singularities. SIAM J. Numer. Anal. **39**, 286–299 (2001)
17. C. Cox, G. Fix, On the accuracy of least squares methods in the presence of corner singularities. Comput. Math. Appl. **10**, 463–475 (1984)
18. M. Dauge, *Elliptic Boundary Value Problems on Corner Domains: Smoothness and Asymptotics of Solutions* (Springer, Berlin, 1988)
19. ———, Benchmark computations for Maxwell equations (2003). http://perso.univ-rennes1.fr/monique.dauge/benchmax.html
20. H. Egger, U. Rüde, B. Wohlmuth, Energy-corrected finite element methods for corner singularities. SIAM J. Numer. Anal. **52**(1), 171–193 (2014)
21. R.B. Kellogg, On the Poisson equation with intersecting interfaces. Appl. Anal. **4**(2), 101–129 (1974)
22. R. Kellogg, Singularities in interface problems, in *Numerical Solution of Partial Differential Equations, II (SYNSPADE 1970): Proceedings of the Symposium*, University of Maryland, College Park, 1970 (1971), pp. 351–400

23. V. Kondratiev, Boundary value problems for elliptic equations in domains with conical or angular points. Trans. Mosc. Math. Soc. **16**, 227–313 (1967)
24. E. Lee, T. Manteuffel, C. Westphal, Weighted norm first-order system least-squares (FOSLS) for problems with corner singularities. SIAM J. Numer. Anal. **44**, 1974–1996 (2006)
25. X. Liu, S. Oishi, Verified eigenvalue evaluation for the Laplacian over polygonal domains of arbitrary shape. SIAM J. Numer. Anal. **51**, 1634–1654 (2013)
26. J.M.-S. Lubuma, K.C. Patidar, Towards the implementation of the singular function method for singular perturbation problems. Appl. Math. Comput. **209**, 68–74 (2009)
27. V.G. Maz′ja, B.A. Plamenevskii, The coefficients in the asymptotic expansion of the solutions of elliptic boundary value problems to near conical points. Dokl. Akad. Nauk SSSR **219**, 286–289 (1974)
28. P. Raviart, J. Thomas, *Introduction à l'analyse numérique des équations aux dérivées partielles* (Masson, Paris, 1983)
29. U. Rüde, Local corrections for eliminating the pollution effect of reentrant corners. Technical report TUM-INFO-02-89-I01, Institut für Informatik, Technische Universtät München, 1989
30. U. Rüde, C. Waluga, B. Wohlmuth, Nested Newton strategies for energy-corrected finite element methods. SIAM J. Sci. Comput. **36**(4), A1359–A1383 (2014). doi: 10.1137/130935392
31. U. Rüde, C. Zenger, On the treatment of singularities in the multigrid method, in *Multigrid Methods II*, ed. by W. Hackbusch, U. Trottenberg. Lecture Notes in Mathematics, vol. 1228 (Springer, Berlin/Heidelberg, 1986), pp. 261–271. 10.1007/BFb0072651
32. G. Strang, G. Fix, *An Analysis of the Finite Element Method*, 2nd edn. (Wellesley-Cambridge Press, Wellesley, 2008)
33. C. Zenger, H. Gietl, Improved difference schemes for the Dirichlet problem of Poisson's equation in the neighbourhood of corners. Numer. Math. **30**, 315–332 (1978)

Stabilized Galerkin for Linear Advection of Vector Fields

Holger Heumann and Ralf Hiptmair

Abstract We present a stabilized Galerkin method for linear advection of vector fields and prove, for sufficiently smooth solutions, optimal a priori error estimates for $\boldsymbol{H}(\mathbf{curl}, \Omega)$ and $\boldsymbol{H}(\mathrm{Div}, \Omega)$-conforming approximation spaces.

1 Introduction

The focus of this article is the following linear advection problem for a vector field $\mathbf{A}$:

$$\begin{aligned} \alpha\mathbf{A} + \mathbf{curl}\,\mathbf{A} \times \boldsymbol{\beta} + \mathbf{grad}\,(\mathbf{A}\cdot\boldsymbol{\beta}) = \mathbf{f} \quad \text{in } \Omega\,, \\ \mathbf{A}_{|\Gamma_{\mathrm{in}}} = \mathbf{A}_0 \quad \text{on } \Gamma_{\mathrm{in}}\,. \end{aligned} \tag{1}$$

Here $\Omega \subset \mathbb{R}^3$ is a bounded domain with inflow boundary $\Gamma_{\mathrm{in}} \subset \partial\Omega$, and $\boldsymbol{\beta} = \boldsymbol{\beta}(\boldsymbol{x})$ and $\alpha = \alpha(\boldsymbol{x})$ are given parameters of which we assume $\boldsymbol{\beta} \in \boldsymbol{W}^{1,\infty}(\Omega)$ and $\alpha \in L^\infty(\Omega)$.

This advection problem is an important model problem for devising reliable numerical methods for problems in electromagnetics and fluid dynamics, when vector fields such as electromagnetic fields or vorticity are advected in some flow. Since the natural space of such quantities is either $\boldsymbol{H}(\mathbf{curl}, \Omega)$ or $\boldsymbol{H}(\mathrm{Div}, \Omega)$, it is desirable to have stabilized methods for appropriate conforming finite element spaces.

H. Heumann (✉)
Zentrum Mathematik – M16, Technische Universität München, Boltzmannstraße 3, 85747 Garching, Germany
e-mail: holger.heumann@tum.de

R. Hiptmair
Seminar for Applied Mathematics, ETH Zürich, Rämistrasse 101, CH-8092 Zürich, Switzerland
e-mail: Ralf.Hiptmair@sam.ethz.ch

A. Abdulle et al. (eds.), *Numerical Mathematics and Advanced Applications ENUMATH 2013*, Lecture Notes in Computational Science and Engineering 103,
DOI 10.1007/978-3-319-10705-9_3

Problem (1) owes its name to the following consideration: Let $\mathbf{X}_t(\mathbf{x})$ denote the flow associated with the given vector field $\boldsymbol{\beta}(\mathbf{x})$, C a one-dimensional manifold and $\mathbf{X}_t(C)$ the image of C under the flow. The transformation rule for line integrals yields

$$\frac{d}{dt}\int_{\mathbf{X}_t(C)} \mathbf{A}(\mathbf{y})\cdot d\mathbf{S}(\mathbf{y}) = \frac{d}{dt}\int_C D\mathbf{X}_t^T(\mathbf{x})\mathbf{A}(\mathbf{X}_t(\mathbf{x}))\cdot d\mathbf{S}(\mathbf{x}),$$

and a lengthy calculation verifies:

$$\frac{d}{dt}\left(D\mathbf{X}_t^T(\mathbf{x})\mathbf{A}(\mathbf{X}_t(\mathbf{x}))\right)_{|t=0} = \mathbf{curl}\,\mathbf{A}(\mathbf{x})\times\boldsymbol{\beta}(\mathbf{x}) + \mathbf{grad}\,(\mathbf{A}(\mathbf{x})\cdot\boldsymbol{\beta}(\mathbf{x})).$$

Hence, the first order differential operator in (1) is a generalization of the material derivative of scalar functions u that are integrated over volumes M, i.e.

$$\frac{d}{dt}\int_{\mathbf{X}_t(M)} u(\mathbf{y})d\mathbf{y} = \frac{d}{dt}\int_M \det(D\mathbf{X}_t(\mathbf{x}))\,u(\mathbf{X}_t(\mathbf{x}))d\mathbf{x} \tag{2}$$

and

$$\frac{d}{dt}(\det(D\mathbf{X}_t(\mathbf{x}))\,u(\mathbf{X}_t(\mathbf{x})))_{|t=0} = \mathrm{Div}(\boldsymbol{\beta}(\mathbf{x})u(\mathbf{x})).$$

It is the framework of differential forms [4] that embeds this advection idea in a general setting, *the Lie derivative formalism*, and we would like to refer to [2, 9, 17] for recent applications of this formalism in devising new numerical methods.

The advection problem (1) can be regarded as the hyperbolic limit case of an advection-diffusion type problem, where a **curl curl**-operator doubles for the diffusion. Such models appear for electromagnetic problems within a quasi-magnetostatic setting. This was the main motivation in [11] to define and analyse stabilized Galerkin methods for the linear advection problem (1) that rely on $\boldsymbol{H}(\mathbf{curl},\Omega)$-conforming finite element spaces. The theoretical convergence theory in [11] yields the same approximation results as a more classical stabilized Discontinuous Galerkin method for Friedrichs' operators [5, 6, 16], which employs approximation functions that are discontinuous across element interfaces. In contrast, the functions of $\boldsymbol{H}(\mathbf{curl},\Omega)$-conforming finite element spaces, sometimes called *edge elements* or *Whitney forms* [1, 12, 18, 19], have continuous tangential components and discontinuous normal components at the element interfaces. The classical stabilized methods that work with globally continuous finite element functions, add certain stabilization terms to the standard Galerkin variational formulations, that enhance the stability but do not destroy the consistency of the methods [14, 15]. The stabilization effect of the method in [11] does not rely on such additional stabilization terms, but uses the upwinding idea of the Discontinuous Galerkin method. We consider this to be a remarkable advantage of $\boldsymbol{H}(\mathbf{curl}, \Omega)$-conforming approximation spaces for linear advection of vector fields: *a similar simple stabilization as in Discontinuous Galerkin methods, but fewer degrees of freedom.*

In light of this point of view, it appears reasonable to ask for stabilized Galerkin methods for (1) that use $\boldsymbol{H}(\mathrm{Div}, \Omega)$-conforming finite element spaces since the functions of these spaces have discontinuous tangential and continuous normal components at the element interfaces. Besides this conceptual motivation we also emphasize that the advection operator $(\boldsymbol{\beta} \cdot \mathbf{grad})\mathbf{u}$ in linearized Navier-Stokes problems can be rephrased in terms of the advection operator in (1). And, $\boldsymbol{H}(\mathrm{Div}, \Omega)$-conforming finite element spaces are frequently used for such kind of problems [3, 7].

In the next section, Sect. 2 we present the method and state stability and consistency. This is followed, in Sect. 3, by a short summary on previous convergence result. In Sect. 4 we state and prove the main result.

2 Stabilized Galerkin

Standard well-posedness results for (1) (see e.g. [10, Section 3]) require the following assumption.

Assumption 1 *We assume that $\alpha \in L^\infty(\Omega)$ and $\boldsymbol{\beta} \in \boldsymbol{W}^{1,\infty}(\Omega)$ are such that $\lambda_{\min}\left\{(2\alpha - \mathrm{Div}\,\boldsymbol{\beta})\mathbf{I}_3 + D\boldsymbol{\beta} + (D\boldsymbol{\beta})^T\right\} \geq \alpha_0$, almost everywhere in Ω for some $\alpha_0 > 0$. $D\boldsymbol{\beta}$ is the Jacobi matrix and λ_{min} the smallest eigenvalue.*

Let us first introduce some notation that is similar the notation used in Discontinuous Galerkin methods.

Let $\mathscr{T}$ be a regular partition of Ω into tetrahedral elements T; h_T is the diameter of T, and $h = \max_{T\in\mathscr{T}} h_T$. The boundary of each element is decomposed into four triangles, called *facets*. We assume that each facet f has a distinguished normal $\mathbf{n}_f$. If a facet f is contained in the boundary of some element T then either $\mathbf{n}_f = \mathbf{n}_{\partial T|_f}$ or $\mathbf{n}_f = -\mathbf{n}_{\partial T|_f}$. Then, if $\mathbf{u}$ is a piecewise smooth vector field on $\mathscr{T}$, $\mathbf{u}^+$ and $\mathbf{u}^-$ denote the two different restrictions of $\mathbf{u}$ to f, e.g. $\mathbf{u}^+ := \mathbf{u}_{|_{T^+}}$ where element T^+ has outward normal $\mathbf{n}_f$. With these restrictions we define also the jump $[\mathbf{u}]_f = \mathbf{u}^+ - \mathbf{u}^-$ and the average $\{\mathbf{u}\}_f = \frac{1}{2}(\mathbf{u}^+ + \mathbf{u}^-)$. For $f \subset \partial\Omega$ we assume f to be oriented such that $\mathbf{n}_f$ points outwards. Let $\mathscr{F}^\circ$ and $\mathscr{F}^\partial$ be the set of interior and boundary facets. $\mathscr{F}^\partial_-, \mathscr{F}^\partial_+ \subset \mathscr{F}^\partial$ are the sets of facets on the inflow and outflow boundary, respectively.

We define the bilinear mapping, $(\mathbf{u}, \mathbf{v})_{f,\boldsymbol{\beta}} := \int_f (\boldsymbol{\beta} \cdot \mathbf{n}_f)(\mathbf{u} \cdot \mathbf{v})\,\mathrm{d}S$, the advection operator $\mathsf{L}_{\boldsymbol{\beta}}\,\mathbf{u} := \mathbf{grad}(\boldsymbol{\beta} \cdot \mathbf{u}) + \mathbf{curl}\,\mathbf{u} \times \boldsymbol{\beta}$ and its formal adjoint $\mathscr{L}_{\boldsymbol{\beta}}\,\mathbf{u} := \mathbf{curl}(\boldsymbol{\beta} \times \mathbf{u}) - \boldsymbol{\beta}\,\mathrm{Div}\,\mathbf{u}$. Hence, for smooth $\mathbf{u}$ and $\mathbf{v}$ we have $\left(\mathsf{L}_{\boldsymbol{\beta}}\,\mathbf{u}, \mathbf{v}\right)_\Omega - \left(\mathbf{u}, \mathscr{L}_{\boldsymbol{\beta}}\,\mathbf{v}\right)_\Omega = (\mathbf{u}, \mathbf{v})_{\partial\Omega,\boldsymbol{\beta}}$. In the following $\mathbf{V}_h$ denotes some space of piecewise polynomial vector fields that are continuous on each element of the mesh $\mathscr{T}$. For the moment we do not specify, which components are continuous and which components are discontinuous at the element interfaces. The *stabilized Galerkin method* reads as follows:

Find $\mathbf{u} \in \mathbf{V}_h$, such that:

$$\mathsf{a}(\mathbf{u},\mathbf{v}) = (\mathbf{f},\mathbf{v})_\Omega - \sum_{f\in\mathscr{F}^\partial_-} (\mathbf{g},\mathbf{v})_{f,\boldsymbol{\beta}}\,, \quad \forall \mathbf{v} \in \mathbf{V}_h\,, \tag{3}$$

with

$$\begin{aligned}
\mathsf{a}(\mathbf{u},\mathbf{v}) = {}& (\alpha\mathbf{u},\mathbf{v})_\Omega + \sum_T (\mathbf{curl}\,\mathbf{u}\times\boldsymbol{\beta},\mathbf{v})_T - (\mathbf{u},\boldsymbol{\beta}\mathrm{Div}\,\mathbf{v})_T \\
&+ \sum_{f\in\mathscr{F}^\circ} \int_f \boldsymbol{\beta}\cdot\{\mathbf{u}\}_f[\mathbf{v}]_f\cdot\mathbf{n}_f\,\mathrm{d}S - \int_f \left([\mathbf{u}]_f\times\mathbf{n}_f\right)\cdot\left(\{\mathbf{v}\}_f\times\boldsymbol{\beta}\right)\mathrm{d}S \\
&+ \sum_{f\in\mathscr{F}^\circ} \int_f c_f\boldsymbol{\beta}\cdot[\mathbf{u}]_f[\mathbf{v}]_f\cdot\mathbf{n}_f\,\mathrm{d}S + \int_f c_f\left([\mathbf{u}]_f\times\mathbf{n}_f\right)\cdot\left([\mathbf{v}]_f\times\boldsymbol{\beta}\right)\mathrm{d}S \\
&+ \sum_{f\in\mathscr{F}^\partial\setminus\mathscr{F}^\partial_-} \int_f (\boldsymbol{\beta}\cdot\mathbf{u})(\mathbf{v}\cdot\mathbf{n}_f)\,\mathrm{d}S - \sum_{f\in\mathscr{F}^\partial_-} \int_f \left(\mathbf{u}\times\mathbf{n}_f\right)\cdot\left(\mathbf{v}\times\boldsymbol{\beta}\right)\mathrm{d}S\,,
\end{aligned} \tag{4}$$

We refer to [11, Section 2] for a detailed derivation of this method. There too, it is shown that the method is consistent and stable in the mesh dependent norm (with $\|\cdot\|^2_{f,\boldsymbol{\beta}} := (\mathbf{u},\mathbf{u})_{f,\boldsymbol{\beta}}$)

$$\|\mathbf{u}\|^2_h := \|\mathbf{u}\|^2_{L^2(\Omega)} + \sum_{f\in\mathscr{F}^\circ} \|[\mathbf{u}]_f\|^2_{f,c_f\boldsymbol{\beta}} + \sum_{f\in\mathscr{F}^\partial\setminus\mathscr{F}^\partial_-} \|\mathbf{u}\|^2_{f,\frac{1}{2}\boldsymbol{\beta}} + \sum_{f\in\mathscr{F}^\partial_-} \|\mathbf{u}\|^2_{f,-\frac{1}{2}\boldsymbol{\beta}}\,,$$

when the parameter c_f fulfills the following positivity condition.

Assumption 2 *Assume the parameters c_f in the definition* (3) *satisfy for all faces f the positivity condition $c_f\boldsymbol{\beta}\cdot\mathbf{n}_f > K|\boldsymbol{\beta}\cdot\mathbf{n}_f|$ for positive $K\in\mathbb{R}$.*

Lemma 1 *Let Assumptions 1 and 2 hold. Then we have for all $\mathbf{u}\in\mathbf{V}_h$:*

$$\mathsf{a}(\mathbf{u},\mathbf{u}) \geq \min(\frac{1}{2}\alpha_0, 1)\|\mathbf{u}\|^2_h\,.$$

3 Previous Results

If we choose $\mathbf{V}_h$ in (3) to be a space of vector fields that have neither continuous tangential nor continuous normal components at the element interfaces our method coincides with the Discontinuous Galerkin method for Friedrichs' operators [11, Section 3]. The choice $c_f = \frac{\boldsymbol{\beta}\cdot\mathbf{n}_f}{|\boldsymbol{\beta}\cdot\mathbf{n}_f|}$ yields the classical upwind methods, and we can cite the following convergence result from [5, Theorem 4.6 & Corollary 4.7], [16, Theorem 50 & Corollary 12] or [11, Theorem 3.1]

Theorem 1 *Let Assumptions 1 and 2 hold. Let* $\mathbf{V}_h$ *be the finite element space of discontinuous piecewise polynomial vector fields:*

$$\mathbf{V}_h = \mathbf{V}^r_{\text{dis}} := \{\mathbf{v} \in \boldsymbol{L}^2(\Omega), \mathbf{v}_{|T} \in (P_r(T))^3, T \in \mathscr{T}\}, \tag{5}$$

where P_r*,* $r \geq 0$ *is the space of polynomials of degree r or less. Let* $\mathbf{u} \in \boldsymbol{H}^{r+1}(\Omega)$ *and* $\mathbf{u}_h \in \mathbf{V}_h$ *be the solutions to the advection problem* (1) *and its variational formulation* (3)*. We get with* $C > 0$ *depending only on* α*,* $\boldsymbol{\beta}$*,* K*, the polynomial degree and the shape regularity*

$$\|\mathbf{u} - \mathbf{u}_h\|_h \leq C h^{r+\frac{1}{2}} \|\mathbf{u}\|_{\boldsymbol{H}^{r+1}(\Omega)}.$$

Surprisingly, the same rate of convergence can be shown, if $\mathbf{V}_h$ in (3) is a space of vector fields that have continuous tangential but discontinuous normal components at the element interfaces [11, Theorem 4.2].

Theorem 2 *Let Assumptions 1 and 2 hold.* P_r*,* $r \geq 0$ *is the space of polynomials of degree r or less. Let then* $\mathbf{V}_h$ *be a finite element space of* $\boldsymbol{H}(\mathbf{curl}, \Omega)$*-conforming piecewise polynomial vector fields of degree r or less:*

$$\mathbf{V}_h = \mathbf{V}^r_{\text{cnf},1} := \{\mathbf{v} \in \boldsymbol{H}(\mathbf{curl}, \Omega), \mathbf{v}_{|T} \in (P_r(T))^3, T \in \mathscr{T}\},$$

such that best approximation estimates

$$\min_{\mathbf{w}_h \in \mathbf{V}_h} \|\mathbf{u} - \mathbf{w}_h\|_{H^s(T)} \leq C h^{r+1-s} \|\mathbf{u}\|_{\boldsymbol{H}^{r+1}(T)}, \; s = 0, 1, \forall \mathbf{u} \in \boldsymbol{H}^{r+1}(\Omega)$$

hold with constants depending only on shape regularity of the mesh, e.g., $\mathbf{V}_h$ *can belong to one of the two families of spaces proposed in [18] and [19]. Let* $\mathbf{u}$ *and* $\mathbf{u}_h \in \mathbf{V}_h$ *be the solutions to the advection problem* (1) *and its discrete variational formulation* (3)*. Then, with* $C > 0$ *depending only on* α*,* $\boldsymbol{\beta}$*,* K *the polynomial degree and shape regularity, we get*

$$\|\mathbf{u} - \mathbf{u}_h\|_h \leq C h^{r+\frac{1}{2}} \|\mathbf{u}\|_{\boldsymbol{H}^{r+1}(\Omega)},$$

provided that h is sufficiently small.

4 $\boldsymbol{H}(\mathbf{Div}, \boldsymbol{\Omega})$-Conforming Approximation

In this section we prove the main result, the optimal convergence of our method (3) when $\mathbf{V}_h$ is a space of $\boldsymbol{H}(\text{Div}, \Omega)$-conforming vector fields. The proof relies on the so-called *averaging interpolation operators* mapping piecewise polynomial non-conforming vector fields to piecewise polynomial $\boldsymbol{H}(\mathbf{curl}, \Omega)$-conforming or $\boldsymbol{H}(\mathrm{Div}, \Omega)$-conforming vector fields. Similar to $\mathbf{V}^r_{\mathrm{cnf},1}$ in Theorem 2 we introduce $\mathbf{V}^r_{\mathrm{cnf},2} := \{\mathbf{v} \in \boldsymbol{H}(\mathrm{Div}, \Omega),\, \mathbf{v}_{|T} \in (P_r(T))^3\,,\, T \in \mathscr{T}\}$, the space of $\boldsymbol{H}(\mathrm{Div}, \Omega)$-conforming finite elements.

Proposition 1 *Let* $\mathbf{u} \in \mathbf{V}^r_{\mathrm{dis}}$. *Then there exist* $\mathbf{u}^{c,1} \in \mathbf{V}^r_{\mathrm{cnf},1}$ *and* $\mathbf{u}^{c,2} \in \mathbf{V}^r_{\mathrm{cnf},2}$ *such that*

$$\left\|\mathbf{u} - \mathbf{u}^{c,1}\right\|^2_{L^2(\Omega)} \leq C_1 \sum_{f \in \mathscr{F}^\circ} h_f \int_f \left|[\mathbf{u}]_f \times \mathbf{n}_f\right|^2 \mathrm{d}S \tag{6}$$

and

$$\left\|\mathbf{u} - \mathbf{u}^{c,2}\right\|^2_{L^2(\Omega)} \leq C_2 \sum_{f \in \mathscr{F}^\circ} h_f \int_f \left|[\mathbf{u}]_f \cdot \mathbf{n}_f\right|^2 \mathrm{d}S\,, \tag{7}$$

where h_f *is the diameter of facet* f *and* C_1 *and* C_2 *depend only on the shape-regularity and the polynomial degree* r*, and, in particular, are independent of the mesh size.*

The proof of (6) can be found in [13, Proposition 4.5] and the second assertion follows by similar arguments (see also [8, Proposition 4.1.2]).

Theorem 3 *Let Assumptions 1 and 2 hold.* P_r, $r \geq 0$ *is the space of polynomials of degree* r *or less. Let then* $\mathbf{V}_h$ *be a finite element space of* $\boldsymbol{H}(\mathrm{Div}, \Omega)$*-conforming piecewise polynomial vector fields of degree* r *or less:*

$$\mathbf{V}_h = \mathbf{V}^r_{\mathrm{cnf},2} := \{\mathbf{v} \in \boldsymbol{H}(\mathrm{Div}, \Omega),\, \mathbf{v}_{|T} \in (P_r(T))^3\,,\, T \in \mathscr{T}\}\,,$$

such that best approximation estimates

$$\min_{\mathbf{w}_h \in \mathbf{V}_h} \|\mathbf{u} - \mathbf{w}_h\|_{\boldsymbol{H}^s(T)} \leq C h^{r+1-s} \|\mathbf{u}\|_{\boldsymbol{H}^{r+1}(T)}\,,\ s = 0, 1\,, \forall \mathbf{u} \in \boldsymbol{H}^{r+1}(\Omega)$$

hold with constants depending only on shape regularity of the mesh. Let $\mathbf{u}$ *and* $\mathbf{u}_h \in \mathbf{V}_h$ *be the solutions to the advection problem* (1) *and its discrete variational formulation* (3)*. Then, with* $C > 0$ *depending only on* α, $\boldsymbol{\beta}$, K *the polynomial degree and shape regularity, we get*

$$\|\mathbf{u} - \mathbf{u}_h\|_h \leq C h^{r+\frac{1}{2}} \|\mathbf{u}\|_{\boldsymbol{H}^{r+1}(\Omega)}\,,$$

provided that h *is sufficiently small.*

Proof Let $\bar{\mathbf{u}}_h$ denote the global L^2-projection of $\mathbf{u}$ onto $\mathbf{V}_h$ and define $\boldsymbol{\eta} := \mathbf{u} - \bar{\mathbf{u}}_h$ and $\boldsymbol{\gamma}_h := \mathbf{u}_h - \bar{\mathbf{u}}_h$. With this, the error $\|\mathbf{u} - \mathbf{u}_h\|_h$ is bounded by two terms:

$$\|\mathbf{u} - \mathbf{u}_h\|_h \leq \|\eta\|_h + \|\boldsymbol{\gamma}_h\|_h\,.$$

For the first term, by the assumptions of the theorem, we have

$$\|\boldsymbol{\eta}\|_{L^2(T)} \leq C h^{r+1} \|\mathbf{u}\|_{H^{r+1}(T)} ,$$

and for the second term, by stability, consistency and $\boldsymbol{\gamma}_h \in \mathbf{V}^r_{\mathrm{cnf},2}$:

$$\min(\frac{1}{2}\alpha_0, 1)\|\boldsymbol{\gamma}_h\|_h^2 \leq \mathsf{a}(\boldsymbol{\eta}, \boldsymbol{\gamma}_h) .$$

Next, we add and subtract the Lie-derivative with respect to a piecewise constant velocity field $\boldsymbol{\beta}_h \in \mathbf{V}^0_{\mathrm{dis}}$ that is the L^2-projection of $\boldsymbol{\beta}$ onto $\mathbf{V}^0_{\mathrm{dis}}$:

$$\begin{aligned}\mathsf{a}(\boldsymbol{\eta}, \boldsymbol{\gamma}_h) = {} & (\alpha\boldsymbol{\eta}, \boldsymbol{\gamma}_h)_\Omega + \sum_T (\boldsymbol{\eta}, (\mathscr{L}_{\boldsymbol{\beta}} - \mathscr{L}_{\boldsymbol{\beta}_h})\boldsymbol{\gamma}_h)_T + (\boldsymbol{\eta}, \mathscr{L}_{\boldsymbol{\beta}_h}\boldsymbol{\gamma}_h)_T \\ & + \sum_{f \in \mathscr{F}^\partial \backslash \mathscr{F}^\partial_-} (\boldsymbol{\eta}, \boldsymbol{\gamma}_h)_{f,\boldsymbol{\beta}} + \sum_{f \in \mathscr{F}^\circ} \left(\{\boldsymbol{\eta}\}_f, [\boldsymbol{\gamma}_h]_f\right)_{f,\boldsymbol{\beta}} + \left(c_f [\boldsymbol{\eta}]_f, [\boldsymbol{\gamma}_h]_f\right)_{f,\boldsymbol{\beta}} .\end{aligned}$$

Yet, as $\sum_T (\boldsymbol{\eta}, \mathscr{L}_{\boldsymbol{\beta}_h}\boldsymbol{\gamma}_h)_T \neq 0$, the difficult part is now to show

$$\left|\sum_T (\boldsymbol{\eta}, \mathbf{curl}(\boldsymbol{\gamma}_h \times \boldsymbol{\beta}_h) + \boldsymbol{\beta}_h \operatorname{Div} \boldsymbol{\gamma}_h)_T\right| \leq C h^{-\frac{1}{2}} \|\boldsymbol{\eta}\|_{L^2(\Omega)} \|\boldsymbol{\gamma}_h\|_h . \tag{8}$$

Let $\mathbf{w}^{c,1} \in \mathbf{V}^r_{\mathrm{cnf},1}$ and $\mathbf{w}^{c,2} \in \mathbf{V}^r_{\mathrm{cnf},2}$ be the conforming approximations of $\boldsymbol{\beta}_h \times \boldsymbol{\gamma}_h \in \mathbf{V}^r_{\mathrm{dis}}$ and $\boldsymbol{\beta}_h \operatorname{Div} \boldsymbol{\gamma}_h \in \mathbf{V}^r_{\mathrm{dis}}$. Since $\boldsymbol{\eta} = \mathbf{u} - \bar{\mathbf{u}}_h$ and both $\mathbf{curl}\,\mathbf{w}^{c,1} \in \mathbf{V}^r_{\mathrm{cnf},2}$ and $\mathbf{w}^{c,2} \in \mathbf{V}^r_{\mathrm{cnf},2}$ we find

$$\left|\sum_T (\boldsymbol{\eta}, \mathbf{curl}(\boldsymbol{\gamma}_h \times \boldsymbol{\beta}_h))_T\right| \leq C_0 h^{-1} \|\boldsymbol{\eta}\|_{L^2(\Omega)} \left\|\boldsymbol{\gamma}_h \times \boldsymbol{\beta}_h - \mathbf{w}^{c,1}\right\|_{L^2(\Omega)}$$

and

$$\left|\sum_T (\boldsymbol{\eta}, \boldsymbol{\beta}_h \operatorname{Div} \boldsymbol{\gamma}_h)_T\right| \leq \|\boldsymbol{\eta}\|_{L^2(\Omega)} \left\|\boldsymbol{\beta}_h \operatorname{Div} \boldsymbol{\gamma}_h - \mathbf{w}^{c,2}\right\|_{L^2(\Omega)} .$$

The approximation results (6) and (7) give

$$\left\|\boldsymbol{\gamma}_h \times \boldsymbol{\beta}_h - \mathbf{w}^{c,1}\right\|^2_{L^2(\Omega)} \leq C_1 h \sum_{f \in \mathscr{F}^\circ} \left\|[\boldsymbol{\gamma}_h \times \boldsymbol{\beta}_h]_f \times \mathbf{n}_f\right\|^2_{L^2(f)}$$

and

$$\left\|\boldsymbol{\beta}_h \operatorname{Div} \boldsymbol{\gamma}_h - \mathbf{w}^{c,2}\right\|^2_{L^2(\Omega)} \leq C_2 h \sum_{f \in \mathscr{F}^\circ} \left\|[\boldsymbol{\beta}_h \operatorname{Div} \boldsymbol{\gamma}_h]_f \cdot \mathbf{n}_f\right\|^2_{L^2(f)}.$$

Inverse inequalities, approximation properties of $\boldsymbol{\beta}_h$, normal continuity of $\boldsymbol{\gamma}_h$ and tangential continuity yield for the right hand sides of the last two equations:

$$\begin{aligned}
\left\|[\boldsymbol{\gamma}_h \times \boldsymbol{\beta}_h]_f \times \mathbf{n}_f\right\|_{L^2(f)} &\leq \left\|[\boldsymbol{\gamma}_h \times (\boldsymbol{\beta}_h - \boldsymbol{\beta})]_f \times \mathbf{n}_f\right\|_{L^2(f)} + \left\|[\boldsymbol{\gamma}_h \times \boldsymbol{\beta}]_f \times \mathbf{n}_f\right\|_{L^2(f)} \\
&\leq C_3 h^{\frac{1}{2}} \|\boldsymbol{\gamma}_h\|_{L^2(T_1 \cup T_2)} + \left\|\boldsymbol{\beta} \cdot \mathbf{n}_f \, [\boldsymbol{\gamma}_h]_f - [\boldsymbol{\gamma}_h]_f \cdot \mathbf{n}_f \boldsymbol{\beta}\right\|_{L^2(f)} \\
&\leq C_3 h^{\frac{1}{2}} \|\boldsymbol{\gamma}_h\|_{L^2(T_1 \cup T_2)} + \left\|\boldsymbol{\beta} \cdot \mathbf{n}_f \, [\boldsymbol{\gamma}_h]_f\right\|_{L^2(f)}
\end{aligned}$$

and

$$\begin{aligned}
\left\|[\boldsymbol{\beta}_h \operatorname{Div} \boldsymbol{\gamma}_h]_f \cdot \mathbf{n}_f\right\|_{L^2(f)} &\leq \left\|[(\boldsymbol{\beta}_h - \boldsymbol{\beta}) \operatorname{Div} \boldsymbol{\gamma}_h]_f \cdot \mathbf{n}_f\right\|_{L^2(f)} + \left\|\boldsymbol{\beta} \cdot \mathbf{n}_f \, [\operatorname{Div} \boldsymbol{\gamma}_h]_f\right\|_{L^2(f)} \\
&\leq C_4 h^{-\frac{1}{2}} \|\boldsymbol{\gamma}_h\|_{L^2(T_1 \cup T_2)} + C_5 h^{-1} \left\|\boldsymbol{\beta} \cdot \mathbf{n}_f \, [\boldsymbol{\gamma}_h]_f\right\|_{L^2(f)},
\end{aligned}$$

with constants C_3, C_4 and C_5 independent of h, and T_1 and T_2 those elements that share f. Hence we deduce (8), and the assertion follows.

We refer to [8, Section 4.1.4] for detailed numerical experiments for test cases with both smooth and non-smooth solutions.

References

1. A. Bossavit, Whitney forms: a class of finite elements for three-dimensional computations in electromagnetism. IEEE Proc. A Phys. Sci. Meas. Instrum. Manag. Educ. Rev. **135**(8), 493–500 (1988)
2. ———, Extrusion, contraction: their discretization via Whitney forms. COMPEL **22**(3), 470–480 (2004)
3. F. Brezzi, M. Fortin, *Mixed and Hybrid Finite Element Methods*. Springer Series in Computational Mathematics, vol. 15 (Springer, New York, 1991)
4. H. Cartan, *Differential Forms* (Hermann, Paris, 1970)
5. A. Ern, J.-L. Guermond, Discontinuous Galerkin methods for Friedrichs' systems. I. General theory. SIAM J. Numer. Anal. **44**(2), 753–778 (2006)
6. R.S. Falk, G.R. Richter, Explicit finite element methods for symmetric hyperbolic equations. SIAM J. Numer. Anal. **36**(3), 935–952 (1999) (electronic)
7. V. Girault, P. Raviart, *Finite Element Methods for Navier-Stokes Equations*. Springer Series in Computational Mathematics (Springer, New York, 1986)

8. H. Heumann, Eulerian and semi-Lagrangian methods for advection-diffusion of differential forms, Ph.D. thesis, ETH Zürich, 2011
9. H. Heumann, R. Hiptmair, Eulerian and semi-Lagrangian methods for convection-diffusion for differential forms. Discret. Cont. Dyn. Syst. **29**(4), 1471–1495 (2011)
10. H. Heumann, R. Hiptmair, Convergence of lowest order semi-Lagrangian schemes. Found. Comput. Math. **13**(2), 187–220 (2013) (English)
11. H. Heumann, R. Hiptmair, Stabilized Galerkin methods for magnetic advection. ESAIM: Math. Model. Numer. Anal. **47**, 1713–1732 (2013)
12. R. Hiptmair, Finite elements in computational electromagnetism. Acta Numer. **11**, 237–339 (2002)
13. P. Houston, I. Perugia, D. Schötzau, Mixed discontinuous Galerkin approximation of the Maxwell operator: non-stabilized formulation. J. Sci. Comput. **22/23**, 315–346 (2005)
14. T.J.R. Hughes, A. Brooks, A multidimensional upwind scheme with no crosswind diffusion, in *Finite Element Methods for Convection Dominated Flows*. AMD, vol. 34 (American Society of Mechanical Engineers, New York, 1979), pp. 19–35
15. T.J.R. Hughes, L.P. Franca, G.M. Hulbert, A new finite element formulation for computational fluid dynamics. VIII. The Galerkin/least-squares method for advective-diffusive equations. Comput. Methods Appl. Mech. Eng. **73**(2), 173–189 (1989)
16. M. Jensen, Discontinuous Galerkin methods for Friedrichs systems with irregular solutions. Ph.D. thesis, University of Oxford, 2005
17. P. Mullen, A. McKenzie, D. Pavlov, L. Durant, Y. Tong, E. Kanso, J. Marsden, M. Desbrun, Discrete Lie advection of differential forms. Found. Comput. Math. **11**(2), 131–149 (2011)
18. J.-C. Nédélec, Mixed finite elements in $\mathbf{R}^3$. Numer. Math. **35**(3), 315–341 (1980)
19. ———, A new family of mixed finite elements in $\mathbf{R}^3$. Numer. Math. **50**(1), 57–81 (1986)

Finite Element-Boundary Element Methods for Dielectric Relaxation Spectroscopy

Stephan C. Kramer and Gert Lube

Abstract We apply the finite element-boundary element method (FEM-BEM) for a smooth approximation of a curvilinear interior interface in a finite domain. This avoids unphysical singularities at the interface due to a piece-wise linear boundary. This type of FEM-BEM coupling arises from simulating the biophysical problem of dielectric relaxation spectroscopy of solvated proteins. Boundary elements convert the linear Poisson problem due to the intramolecular charges of the protein into a boundary condition at the protein-solvent interface. The electro-diffusion of ions in the solvent is modeled as a set of convection-diffusion equations. The spatial distributions of the ion species induce an electrostatic potential which solves a Poisson problem. The gradient of the potential constitutes the convective flow field. The link to experiments is given by computing the stationary ionic current through the system. This requires Robin-type boundary conditions at the electrodes.

1 Introduction

The coupling of finite and boundary element methods, (FEM) and (BEM), is commonly used for interface problems on unbounded domains. Finite elements are applied to bounded "regions of interest" which contain non-linearities, inhomogeneities and other properties which need a well-resolved volume mesh. The BEM part models unboundedness and physical effects described by a homogeneous partial differential equation (PDE) with constant coefficients. In this work we

S.C. Kramer (✉)
Institut für Numerische und Angewandte Mathematik der Universität Göttingen, Lotzestraße 16-18, D-37073 Göttingen, Germany

Max-Planck Institut für biophysikalische Chemie, Am Faßberg 11, D-37077 Göttingen, Germany
e-mail: stkramer@math.uni-goettingen.de; stephan.kramer@mpibpc.mpg.de; sck.goe@googlemail.com

G. Lube
Institut für Numerische und Angewandte Mathematik der Universität Göttingen, Lotzestraße 16-18, D-37073 Göttingen, Germany
e-mail: lube@math.uni-goettingen.de

A. Abdulle et al. (eds.), *Numerical Mathematics and Advanced Applications ENUMATH 2013*, Lecture Notes in Computational Science and Engineering 103,
DOI 10.1007/978-3-319-10705-9_4

discuss how to employ BEM to exclude a subdomain from a finite domain. This accurately models the geometric shape of an interior, curvilinear and smooth interface and unifies the computational domain for the components of a PDE model.

The investigation of this subclass of FEM-BEM coupling arises from the need to simulate the biophysical problem of dielectric relaxation spectroscopy (DRS) of solvated proteins, in particular ubiquitin, which plays a fundamental role in cell biology. The discovery of ubiquitin-mediated protein degradation won the nobel prize in chemistry in 2004. The physical basis of DRS is the polarizability of non-conducting materials in the presence of an external electric field. Polarization is the material-specific part of the dielectric displacement which is proportional to the electric field. The proportionality is given by the complex dielectric permittivity ε^*. In the frequency domain it quantifies the dynamic response of molecular dipoles (contributing at high frequencies) and mobile charge carriers (predominant influence at low frequencies). The DRS technique allows to measure dielectric properties in the range of 10^{-6}–10^{12} Hz. For a detailed review see the monograph [1]. The typical experimental setup is a parallel plate capacitor with the dielectric sample in between the plates [1, Chapter 2]. Application of an alternating voltage yields the dielectric loss spectrum, i.e. the conductivity-corrected imaginary part of ε^* as function of frequency. In case of ubiquitin in aqueous solution this spectrum is dominated by the "γ" peak at about 10 GHz, which represents the reorientation of the dipoles of water molecules in the bulk, and the "β" peak at roughly 10 MHz which accounts for the tumbling motion of the protein molecule while its molecular dipole aligns with the applied electric field [2]. This is sketched in Fig. 1a. The peak positions reveal the time scales on which the relaxation processes take place. Recent DRS studies on ubiquitin [3] suggest that the dynamics of conformational sampling, i.e. a protein's ability to switch between different molecular conformations (indicated by the different positions of the intramolecular charges in Fig. 1b, c), influence the direct current component of the dielectric loss spectrum and can be observed as the "sub-β" peak. This important discovery provides a direct experimental access to the rates of the intramolecular dynamics, which are mostly inaccessible to nuclear magnetic resonance (NMR) spectroscopy, the most frequently used experimental

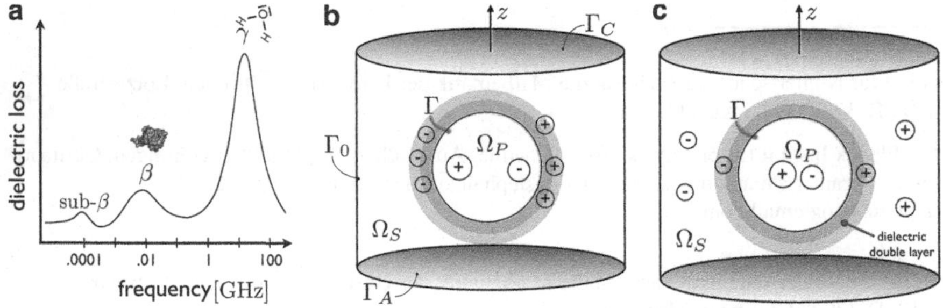

Fig. 1 (**a**) Dielectric loss spectrum of ubiquitin. (**b, c**) Charge configurations in protein (domain Ω_P) in the DRS cell $\Omega = \Omega_S \cup \Omega_P$. Details see text

technique to characterize protein dynamics. For the detailed explanation of many biomolecular processes, e.g. of protein-protein recognition [4], the exact knowledge of the kinetics of conformational sampling is decisive.

In this paper we apply the theory of FEM-BEM methods for infinite domains to the case of using the BEM part to exclude a subdomain from a finite domain in order to develop a deeper understanding of the origin of the "sub-β" peak. We use BEM to retain the smooth shape of the protein-solvent interface. The proper incorporation of a stationary current by means of Robin-type boundary conditions (BCs) provides the link to a comparison with experimental data. The equations are solved by the geometric multigrid (GMG) method [5] from deal.II [6].

2 Poisson-Nernst-Planck Model

Initial theoretical studies explained the "sub-β" peak by a 2-state, ratchet-like stochastic model for the conformational dynamics coupled to a Fokker-Planck model for the mobile ions [3, supplementary material]. Depending on its conformation the ubiquitin molecule may bind a varying number of ions in its dielectric double layer thus influencing the density of mobile ions responsible for the direct current component. Although it explains the essential features of the "sub-β" peak, this stochastic model neither includes spatial inhomogeneities nor BCs.

For the effects at the protein-solvent interface Γ we need at least a generic anion and cation species with densities c_- and c_+, respectively, with charges of equal strength. To incorporate a stationary current through the DRS cell (Fig. 1), we have to take into account the redox reaction $\mathrm{I}^+ + \mathrm{e}^- \leftrightarrow \mathrm{N}$ for converting a cation I^+ into a neutral particle N at the cathode Γ_C or the anode Γ_A. Thus, we have to incorporate the density c_0 of the neutral particles. The stochastic description of the ion dynamics is replaced by the Poisson-Nernst-Planck equations

$$\partial_t c_a = -\nabla \cdot \mathbf{j}_a\,, \tag{1a}$$

$$\mathbf{j}_a = -(\nabla c_a + a c_a \nabla \Phi) \tag{1b}$$

$$-\nabla \cdot (\varepsilon_r(\mathbf{r}) \nabla \Phi) = -(c_+ - c_-)\chi_{\Omega_S} + \rho_f \tag{1c}$$

for non-dimensional ion densities $c_a : \Omega_S \to \mathbb{R}$, $a \in \{+, 0, -\}$, electro-diffusive fluxes $\mathbf{j}_a : \Omega_S \to \mathbb{R}^3$ and electrostatic potential $\Phi : \Omega \to \mathbb{R}$. The charge density on the right-hand side of Eq. (1c) comprises the mobile ions in the subdomain of the solvent Ω_S, indicated by its characteristic function χ_{Ω_S}, and the intramolecular, conformation-specific charge distribution ϱ_f, indicated by the index f. Here, the protein is a dipole with two point charges immersed in a spherical, dielectric domain $\Omega_P = \Omega \backslash \Omega_S$, $\Omega_P \cap \Omega_S = \emptyset$. The function ε_r in Eq. (1c) is piece-wise constant and denotes the relative permittivities, i.e. $\varepsilon_r = \varepsilon_S \approx 80$ on Ω_S, $\varepsilon_r = \varepsilon_P \approx 2$ on Ω_P. Note the different computational domains for ions, Ω_S, and potential, Ω.

A realistic description of DRS requires BCs for the $\mathbf{j}_a$ and Φ capable of modeling an applied current. Usually, the redox reaction rates K_R and K_O are described by Butler-Volmer kinetics, including the Frumkin correction due to the Stern layer [7]. As discussed in [8], for Φ Dirichlet BCs, $\Phi\big|_{\Gamma_C} = \Phi_C$, $\Phi\big|_{\Gamma_A} = \Phi_A$, suffice. The redox reaction implies a balance of in- and outward fluxes at the electrodes

$$\mathbf{n}\cdot\mathbf{j}_+\big|_{\Gamma_C} = K_R\, c_+\big|_{\Gamma_C} = -\mathbf{n}\cdot\mathbf{j}_0\big|_{\Gamma_C}, \tag{2}$$

$$-\mathbf{n}\cdot\mathbf{j}_+\big|_{\Gamma_A} = K_O\, c_0\big|_{\Gamma_A} = \mathbf{n}\cdot\mathbf{j}_0\big|_{\Gamma_A}, \tag{3}$$

where $\mathbf{n}$ is the outer normal of the surface $\partial\Omega_S$ and $\cdot|_B$ is the trace on some part $B \subset \partial\Omega_S = \Gamma_C \cup \Gamma_0 \cup \Gamma_A \cup \Gamma$. The rates are treated as constants, especially their dependence on Φ is neglected. The anions do not contribute to the current transport and fulfill $\mathbf{n}\cdot\mathbf{j}_-|_{\Gamma_A} = \mathbf{n}\cdot\mathbf{j}_-|_{\Gamma_C} = 0$. The hull Γ_0 of the cell and the protein surface Γ are impermeable for all ions, $\mathbf{n}\cdot\mathbf{j}_a|_{\Gamma_0} = \mathbf{n}\cdot\mathbf{j}_a|_{\Gamma} = 0$, $a \in \{+, 0, -\}$. For Φ we have $\mathbf{n}\cdot\nabla\Phi|_{\Gamma_0} = 0$ and Γ is a dielectric interface with continuity and the jump relations

$$\lim_{\delta\to 0}\Phi(\mathbf{x}-\delta\mathbf{n}) = \lim_{\delta\to 0}\Phi(\mathbf{x}+\delta\mathbf{n}), \quad \varepsilon_P(\mathbf{x})\mathbf{n}\cdot\nabla\Phi = \varepsilon_S(\mathbf{x})\mathbf{n}\cdot\nabla\Phi \quad \forall\mathbf{x}\in\Gamma \tag{4}$$

One goal in computational biochemistry is to model molecular surfaces of proteins in a smooth manner [9]. Instead of an accurate sub-cell resolution of the dielectric interface Γ we convert the interior constant-coefficient-Poisson equation into a boundary integral equation (BIE) on Γ. The protein becomes an excluded volume Ω_P of constant dielectric permittivity $\varepsilon_r = \varepsilon_P$ containing point charges $\{q_k\}$ at fixed positions $\{\mathbf{x}_k\}$. To do this, we apply the discussion of the BIE formulation for linear interior Neumann boundary value and interface problems in [10]. The Johnson-Nédélec coupling [11] needs the normal component of the electric field w.r.t. to the outer normal $\mathbf{n}_P$ ($\mathbf{n}_P = -\mathbf{n}$ on Γ) of Ω_P as independent variable $t^P := -\partial_{\mathbf{n}}\Phi$. Potential theory shows that on C^1-smooth, closed surfaces Γ the intramolecular part Φ_P of the potential at $\mathbf{x}\in\Gamma$ fulfills

$$\frac{1}{2}\Phi(\mathbf{x}) + \oint_\Gamma\left[\Phi(\mathbf{x}')\frac{\partial G_{\mathbf{x}}}{\partial\mathbf{n}'_P}(\mathbf{x}') - G_{\mathbf{x}}(\mathbf{x}')\frac{\partial\Phi}{\partial\mathbf{n}'_P}(\mathbf{x}')\right]d\Gamma(\mathbf{x}') = \frac{1}{\varepsilon_P}\int_{\Omega_P} G_{\mathbf{x}}(\mathbf{x}')\rho_f(\mathbf{x}').$$

Here, $G_{\mathbf{x}}(\mathbf{y}) := 1/(4\pi|\mathbf{x}-\mathbf{y}|)$ is the Green's function of the Laplace equation. The right-hand side defines the Newton potential ϕ^C. We define the single layer boundary integral operator (BIO) $V : H^{-1/2}(\Gamma) \to H^{1/2}(\Gamma)$ and the double layer BIO $K : H^{1/2}(\Gamma) \to H^{1/2}(\Gamma)$ as in [12, Secs. 6.2 and 6.4]. Instead of $\mathbf{n}_P$ we use the outward normal $\mathbf{n} = -\mathbf{n}_P$ relative to Ω_S

$$(Vt^P)(\mathbf{x}) := \oint_\Gamma G_{\mathbf{x}}(\mathbf{x}')t^P(\mathbf{x}')d\Gamma(\mathbf{x}'), \qquad (K\Phi_P)(\mathbf{x}) := \oint_\Gamma \frac{\partial G_{\mathbf{x}}}{\partial\mathbf{n}(\mathbf{x}')}(\mathbf{x}')\Phi(\mathbf{x}')d\Gamma(\mathbf{x}').$$

From Eq. (4) follows $\varepsilon_S \partial_{\mathbf{n}} \Phi\big|_\Gamma = -\varepsilon_P t^P$ and we get

$$\left(\frac{1}{2}I - K\right)\Phi + \frac{\varepsilon_S}{\varepsilon_P} V t^P = \phi^C . \tag{5}$$

This is the basis for the FEM-BEM method for the potential, reducing its computational domain to $\Omega_S = \Omega \backslash \Omega_P$. The distribution of the ions is governed by convection-diffusion equations with either Neumann or Robin but no Dirichlet BCs. The link to experiments is the direct current I_{dc} created by a potential difference $\eta := \Phi_C - \Phi_A$. Due to the redox reaction at the electrodes $I_{dc} = \int_{\Gamma_C} \mathbf{n} \cdot \mathbf{j}_+ d\Gamma_C = K_R \int_{\Gamma_C} c_+ d\Gamma_C$. The conformational sampling introduces a time-dependence on I_{dc}. This is modeled by a two-state telegraph process, i.e. a random switching between two stationary states. This makes Eq. (1a) formally time independent. For details cf. [8]. To validate the hypothesis about the origin of the "sub-β" peak we have to compute two different values for I_{dc} from the time independent version of Eq. (1).

3 Weak Formulation and Discretization

We do not solve Eq. (1) in its mixed form, but reduce it to a set of convection-diffusion equations by inserting Eq. (1b) into Eq. (1a), eliminating the currents.

Let $(\cdot,\cdot)_D$ be the L^2 inner product on a domain D and $\|\cdot\|_X$ the norm of a function space X. For $D \equiv \Omega_S$ we drop the index. The weak form of Eq. (1) is derived by multiplying with test functions, integrating by parts and inserting all flux BCs. The Dirichlet BCs for the potential Φ are built into the solution space X for the FEM part. We define X as a direct product of a space $X^c := [H^1(\Omega_S)]^3$ for the densities and $X^\Phi := \{\Phi \in H^1(\Omega_S) : \Phi|_{\Gamma_A} = 0, \Phi|_{\Gamma_C} = \eta\}$ for Φ. For the BCs for Φ on Γ we need the space $Y := H^{-1/2}(\Gamma)$. The final solution space is $V := X \times Y$. The FEM part of the solution is $\mathbf{u} := (c_+, c_0, c_-, \Phi)$ and the test function is $\mathbf{v} := (s, u, v, w)$. Except for the interface term for Φ on Γ the weak form is a semilinear form $a(\cdot;\cdot) : X \times X \to \mathbb{R}$ which is nonlinear in its first argument. The terms in $a(\cdot;\cdot)$ can be grouped to reflect, after linearizing, the block structure of the matrix using scalar test functions as block row and trial functions as block column indexes. Diagonal terms are in $a_D(\cdot;\cdot)$, linear upper off-diagonal terms in $a_U^l(\cdot;\cdot)$, nonlinear drift terms in $a_U^d(\cdot;\cdot)$ and lower off-diagonal terms in $a_L(\cdot;\cdot)$, i.e.

$$a_D(\mathbf{u};\mathbf{v}) := (\nabla s, \nabla c_+) + (\nabla u, \nabla c_0) + k_O(u, c_0)_{\Gamma_A} + (\nabla v, \nabla c_-) + \varepsilon_S(\nabla w, \nabla \Phi),$$

$$a_U^l(\mathbf{u};\mathbf{v}) := -k_O(s, c_0)_{\Gamma_A}, \quad a_U^d(\mathbf{u};\mathbf{v}) := (\nabla s, c_+ \nabla \Phi) - (\nabla v, c_- \nabla \Phi),$$

$$a_L(\mathbf{u},\mathbf{v}) := -k_R(u, c_+)_{\Gamma_C} - (w, c_+ - c_-).$$

After linearizing $a_U^d(\cdot;\cdot)$ w.r.t. c_+ and c_-, the associated matrices are A_D, A_U and A_L, respectively. The left-hand side of the weak form of Eq. (5), with associated matrices B_K and B_V, is a sum of the two bilinear forms

$$
\begin{aligned}
b_K(\psi, \Phi) &:= \left(\psi, \left(\tfrac{1}{2}I - K\right)\Phi\right)_\Gamma &&: H^{1/2}(\Gamma) \times H^{1/2}(\Gamma) \to \mathbb{R}\,, \\
b_V(\psi, t^P) &:= \tfrac{\varepsilon_S}{\varepsilon_P}\left(\psi, V t^P\right)_\Gamma &&: H^{-1/2}(\Gamma) \times H^{1/2}(\Gamma) \to \mathbb{R}\,.
\end{aligned}
$$

We use conformal discretizations $X_h \subset X$ and $Y_h \subset Y$ by globally continuous Lagrange elements for which we use deal.II's `FE_Q<dim>` class. In practice, the trial functions in Y_h are given by the traces of those in X_h because we treat the normal derivative as independent variable. This is due to the way finite elements are implemented in deal.II. The same holds for the test functions ψ in the dual space $Y_h' \subset H^{1/2}(\Gamma)$. Then, the discretized variational problem is: *Find* $(\mathbf{u}_h, t_h^P) \in X_h \times Y_h$ *s. t.*

$$
\forall \mathbf{v}_h \in X_h : \quad a(\mathbf{u}_h; \mathbf{v}_h) + (w, \varepsilon_P t^P)_\Gamma = 0\,, \tag{6a}
$$

$$
\forall \psi_h \in Y_h' : \quad b_K(\psi_h, \Phi_h) + b_V(\psi_h, t_h^P) = \left(\psi_h, \phi^C\right)_\Gamma\,. \tag{6b}
$$

Several numerical problems arise in solving the discretized problem. Only the potential Φ is unambiguous since it is subject to Dirichlet BCs at the electrodes Γ_A and Γ_C. The equation for the density of the neutral particles c_0 effectively is a pure Neumann Laplace problem. Its average merely enters via the boundary terms in Eq. (3) for the cations c_+. Particle numbers, and thus average densities, are conserved $\int_{\Omega_S} c_a \, d\Omega_S = const,\ a \in \{+, 0, -\}$ in the stationary state. This is enforced by adding a pseudo-time dependence, i.e. $-\nabla^2 u = f$ becomes $[\delta I - \nabla^2]u^{n+1} = f + \delta u^n$, where δ is an inverse time step and I is the identity operator.

We solve by interleaving successive mesh refinement, pseudo-time stepping and reassembly of the nonlinear terms. This introduces a sequence of finite-dimensional subspaces $X_h \subset X$, parametrized by the cell diameter h. On a given mesh, i.e. in FE space $V_h^\ell \subset V$, $V_h^\ell \subset V_h^{\ell+1}$ we run a few steps in pseudo-time (while $\|u^{n+1} - u^n\|_{\ell_2} \geq Tol$). In each time step we reassemble the drift terms in A_U after solving the linear algebraic problem by deal.II's GMRES solver with left-preconditioning.

For the numerical solution of Eq. (6) we have considerably extended the GMG example step-16 of deal.II, v7.2.0. When computing the matrices B_K, B_V from the bilinear forms $b_K(.,.)$ and $b_V(.,.)$ the double integration is avoided by using the support points of the test functions for collocation. With $t_h^P = \sum_i t_i^P \phi_i \in Y_h$, e.g. the entries of the matrix representing the single layer BIO are formally given by $B_{V,ij} = (\psi_i, V\phi_j)$. Let $\mathbf{x}_i$ be the support point of DoF i, then collocation at $\mathbf{x}_i$ can be interpreted as $B_{V,ij} = (\delta(\mathbf{x} - \mathbf{x}_i), V\phi_j)$, i.e. $B_{V,ij}$ is computed as

$$
B_{V,ij} = \oint_{\partial\Omega_S} G_{\mathbf{x}_i}(\mathbf{x}')\phi_j(\mathbf{x}')d\Gamma(\mathbf{x}')\,. \tag{7}
$$

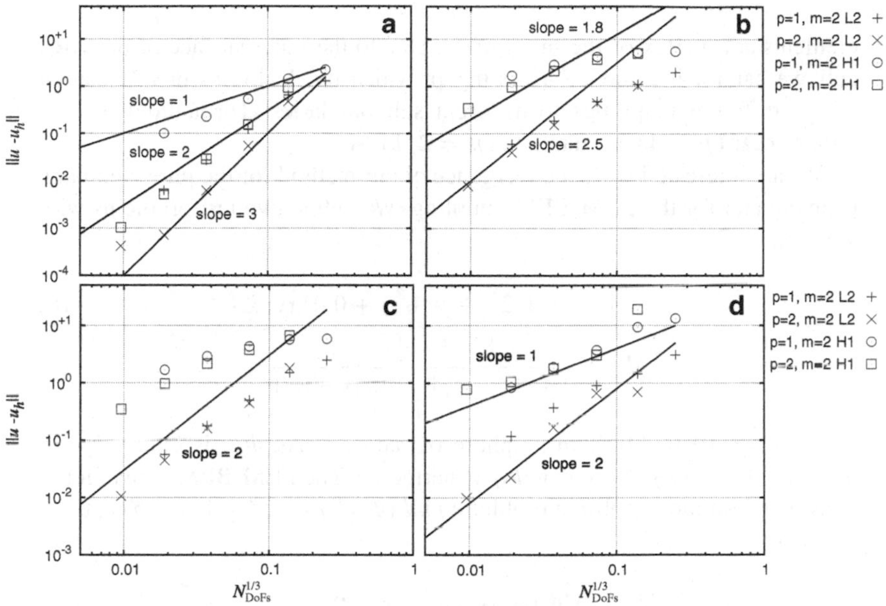

Fig. 2 Convergence of the Neumann problem, Eq. (11), with (**a**) Eq. (8), (**b**) Eq. (9) as solution. (**c**) FEM and (**d**) BEM error for dipole test case, Eq. (9), on the FEM-BEM problem, Eq. (10). Figures share axis labels and legends

To minimize the costs of matrix assembly we compute bulk and boundary mass matrices only once. The definitions of X^c and X^Φ require the assembly of two different Laplacians and hence to setup two GMG preconditioners P^c_{MG} and P^Φ_{MG}. Cell contributions get reused when building global matrices which differ only in the BCs. The costs of assembling the matrices by numerical quadrature are roughly equal to two Poisson equations with variable coefficients as the data for the linear Laplacians can be reused to a great extent for the drift terms. The matrix A for the linearized DRS problem, Eq. (6), is stored as `dealii::BlockMatrixArray` and the preconditioner P_A as `dealii::BlockTrianglePecondition` which acts like a block Gauss-Seidel method. Its diagonal blocks are $(P^c_{MG}, P^c_{MG}, P^c_{MG}, P^\Phi_{MG}, P^V)$, where P^V preconditions B_V and is the identity matrix. The upper off-diagonal blocks of P_A are void. The lower off-diagonal blocks are those of A, i.e. A_L and B_K.

Results and Conclusion

In our tests we model the boundary piece-wise by polynomials of order $m = 2$, cf. legend of Fig. 2. This numerical boundary is not C^1-smooth. According to our tests, it approximates the curved surface of a sphere sufficiently well such that we do not have to consider the solid angle subtended by the surface

(continued)

elements at a vertex of the mesh of Γ. Due to the outer surface of the DRS cell we cannot use the C^1-mapping provided by deal.II as deal.II cannot assign different mappings to different subboundaries. Throughout we use either linear ($p = 1$) or quadratic ($p = 2$) FEM.

We are interested in the convergence of our method for the pure Neumann problem and for the FEM-BEM coupling. We define two test problems with solutions

$$\Phi_{SP} := 0.1(2x + y + z) + 0.01xyz\,, \tag{8}$$

$$\Phi_D := \frac{1}{4\pi|\mathbf{x} - \mathbf{x}_+|} - \frac{1}{4\pi|\mathbf{x} - \mathbf{x}_-|}\,, \tag{9}$$

with $\mathbf{x}_\pm = (0, 0, \pm 0.5)$ in a sphere of radius 1. As Φ_{ref} is either Φ_{SP} or $\Phi_{DSP} := \Phi_D + \Phi_{SP}$. Note that Φ_{SP} is harmonic. The FEM-BEM convergence is assessed on the simplified problem: *find* $(\Phi, t^P) \in X^\Phi \times Y$ *s.t.* $\forall(v, \psi) \in X^\Phi \times Y'$:

$$(\nabla v, \varepsilon_S \nabla \Phi) + (v, \varepsilon_S t^P)_\Gamma = 0\,, \tag{10a}$$

$$b_K(\psi_h, \Phi_h) + b_V(\psi_h, t_h^P) = \left(\psi, \phi^C\right)_\Gamma\,, \tag{10b}$$

with $\Phi|_{\Gamma_A \cup \Gamma_0 \cup \Gamma_C} = \Phi_{ref}$. The test for pure Neumann BCs is: *find* $\Phi \in H^1(\Omega_S)$ *s.t.*

$$(\nabla v, \nabla \Phi) = (v, \partial_{\mathbf{n}} \Phi_{ref}) \quad \forall v \in H^1(\Omega_S)\,. \tag{11}$$

To measure the error we use the standard $L^2(\Omega_S)$- and $H^1(\Omega_S)$-norm for the FEM part. For the BEM part we measure the L^2 error of the trace of Φ on Γ $\|\Phi_{ref} - \Phi_h\|^2_{L^2(\Gamma)}$, and the L^2 error in the trace of $\partial_n \Phi \equiv -t^P$ on Γ. Here, denoted as $H^1(\Gamma)$ semi-norm $|\Phi_{ref} - \Phi_h|^2_{H^1(\Gamma)} := \|\partial_n \Phi_{ref} + t_h^P\|^2_{L^2(\Gamma)}$. In case of Eq. (11) and $\Phi_{ref} = \Phi_{SP}$ convergence is as expected. For FEM of order p we get $\|u - u_h\|^2_{L^2(\Omega_S)} = O(h^{p+1})$ and $\|u - u_h\|^2_{H^1(\Omega_S)} = O(h^p)$ independent of the order of the boundary approximation m, cf. Fig. 2a. Figure 2b shows that for $\Phi_{ref} = \Phi_{DSP}$ we roughly lose half an order which we attribute to the right-hand side of the BEM part containing the δ-distributions for the point charges. Figure 2c shows the convergence of the FEM part of Eq. (10). For Lagrange finite elements of order $p = 2$ the $L^2(\Omega_S)$ and $H^1(\Omega_S)$ error have the same asymptotic behavior. The error in the BEM part, Fig. 2d, is as expected for linear elements ($p = 1$). Due to the collocation the decay of the error in t^P does not improve. The error of $\Phi|_\Gamma$ partly profits from higher order elements. Figure 3 shows that local inhomogeneities of the cation density in

(continued)

the vicinity of the protein surface are resolved and the current-carrying species (cations and neutral particles) are distributed opposite to each other.

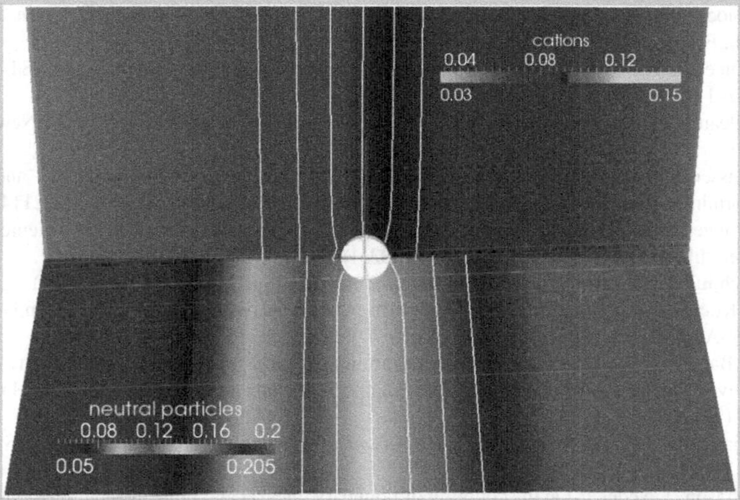

Fig. 3 Distribution of cations and neutral particles in the DRS cell

To conclude, we have derived a mathematical model for the detailed simulation of the electro-diffusive processes in dielectric relaxation spectroscopy of proteins in solution including boundary effects inaccessible in previously derived stochastic models. The key feature is the modeling of the protein-solvent interface as excluded volume with a smooth surface by taking into account its electrostatic properties by means of a boundary integral equation. For the efficient solution of the resulting FEM-BEM model we have extended the geometric multigrid example of the deal.II library (step-16) to vector-valued problems and higher order elements. Unlike the strategy proposed in deal.II's step-34 for boundary elements we have to use the traces of the finite elements as boundary elements. Most of the equations in the DRS model are pure Neumann problems and subject to a conservation of particle numbers. To assure their unique solvability we implemented an interleaved pseudo-time stepping/mesh refinement strategy which avoids the saddle-point problems arising from Lagrange multipliers. The convergence is as expected. The convergence of the FEM-BEM method depends on the particular test case but is consistent with the literature. Applied to the full DRS problem our numerical results indicate the validity of the proposed explanation of the origin of the "sub-β" peak.

References

1. F. Kremer, A. Schönhals, *Broadband Dielectric Spectroscopy* (Springer, Berlin/New York, 2003)
2. A. Knocks, H. Weingärtner, The dielectric spectrum of ubiquitin in aqueous solution. J. Phys. Chem. B **105**(17), 3635–3638 (2001)
3. D. Ban et al., Kinetics of conformational sampling in ubiquitin. Angew. Chem. Int. Ed. **50**(48), 11437–11440 (2011)
4. C. Kleanthous, *Protein-Protein Recognition* (Oxford University Press, Oxford/New York, 2000)
5. B. Janssen, G. Kanschat, Adaptive multilevel methods with local smoothing for H^1- and H^{curl}-conforming high order finite element methods. SIAM J. Sci. Comput. **33**(4), 2095–2114 (2011)
6. W. Bangerth, R. Hartmann, G. Kanschat, deal.II – a general purpose object oriented finite element library. ACM Trans. Math. Softw. **33**(4), 24/1–24/27 (2007)
7. W. Schmickler, E. Santos, *Interfacial Electrochemistry* (Springer, New York, 2010)
8. S.C. Kramer, Cuda-based scientific computing – tools and selected applications, Ph.D. thesis, Georg-August Universität Göttingen, 2012
9. C.L. Bajaj, G. Xu, Q. Zhang, A fast variational method for the construction of resolution adaptive C^2-smooth molecular surfaces. Comput. Methods Appl. Mech. Eng. **198**(21), 1684–1690 (2009)
10. S. Rjasanow, O. Steinbach, *The Fast Solution of Boundary Integral Equations* (Springer, New York/London, 2007)
11. C. Johnson, J. Nédélec, On the coupling of boundary integral and finite element methods. Math. Comput. **35**(152), 1063–1079 (1980)
12. O. Steinbach, *Numerical Approximation Methods for Elliptic Boundary Value Problems*. Texts in Applied Mathematics (Springer, New York/London, 2007)

On the Local Mesh Size of Nitsche's Method for Discontinuous Material Parameters

Mika Juntunen

Abstract We propose Nitsche's method for discontinuous parameters that takes the local mesh sizes of the non-matching meshes carefully into account. The method automatically adapts to the changing material parameters and mesh sizes. With continuous parameters, the method compares to the classical Nitsche's method. With large discontinuity, the method approaches assigning Dirichlet boundary conditions with Nitsche's method.

1 Introduction

Suppose the computational domain is divided along the material edges yielding material parameters that are discontinuous over the subdomain interfaces. If the discontinuity in the material parameters is moderate, the Nitsche's method in [3] applies. Some of the problems with large parameter discontinuities are avoided using the harmonic average of the material parameters to create a weighted average flux over the interface [1, 2, 5, 7–9, 11, 13, 14].

In this article we propose Nitsche's method that takes both the material parameters and the mesh sizes carefully into account in the bilinear form. Both the average flux and the stabilizing term are modified to depend on the material parameters and the mesh sizes, similar to [2]. As a result, the proposed method automatically adapts to the material parameters and mesh sizes. If there is no discontinuity over the interface and the mesh sizes are of the same order, we have the classical Nitsche's method. If the mesh sizes or material parameters have very large contrast over the interface, the method reduces to assigning Dirichlet boundary conditions with Nitsche's method.

M. Juntunen (✉)
Department of Mathematics and Systems Analysis, Aalto University School of Science, P.O. Box 11100, 00076 Aalto, Finland
e-mail: mika.juntunen@aalto.fi

A. Abdulle et al. (eds.), *Numerical Mathematics and Advanced Applications ENUMATH 2013*, Lecture Notes in Computational Science and Engineering 103,
DOI 10.1007/978-3-319-10705-9_5

2 Model Problem

Consider a domain $\Omega \subset \mathbb{R}^d$, $d = 2, 3$, with a piecewise smooth boundary $\partial\Omega$. Assume that the domain is divided into two non-overlapping subdomains Ω_1 and Ω_2. The subdomains cover the whole domain $\bar{\Omega} = \bar{\Omega}_1 \cup \bar{\Omega}_2$ and they share an interface $\Gamma = \partial\Omega_1 \cap \partial\Omega_2$. We solve the Poisson problem such that

$$
\begin{aligned}
-\nabla \cdot k_i \nabla u_i &= f \quad \text{in } \Omega_i, \quad i = 1, 2, && (1)\\
u_i &= 0 \quad \text{on } \partial\Omega, \\
u_1 - u_2 &= 0 \quad \text{on } \Gamma, && (2)\\
k_1 \frac{\partial u_1}{\partial n_1} + k_2 \frac{\partial u_2}{\partial n_2} &= 0 \quad \text{on } \Gamma, && (3)
\end{aligned}
$$

in which $k_i \in \mathbb{R}$, $0 < k_{\min} < k_i < k_{\max}$, $i = 1, 2$, are the material parameters and $f \in L^2(\Omega)$ is the load function. We denote with $k_i \frac{\partial u_i}{\partial n_i} = k_i \nabla u_i \cdot \mathbf{n}_i$ the normal flux and with $\mathbf{n}_1$ and $\mathbf{n}_2$ the outward normals of the subdomains. We also use $\mathbf{n} = \mathbf{n}_1 = -\mathbf{n}_2$ and

$$
\frac{\partial u_1}{\partial n} = \frac{\partial u_1}{\partial n_1} \quad \text{and} \quad \frac{\partial u_2}{\partial n} = \frac{\partial u_2}{\partial n_1} = -\frac{\partial u_2}{\partial n_2}.
$$

Let the subdomains be divided into sets of non-overlapping elements denoted by $\mathcal{T}_1^h$ and $\mathcal{T}_2^h$, in which h denotes the maximum diameter of elements. Let $\mathcal{E}_1^h$ and $\mathcal{E}_2^h$ denote the edges or faces of the meshes $\mathcal{T}_1^h$ and $\mathcal{T}_2^h$, respectively. Let h_K denote the diameter of an element $K \in \mathcal{T}_i^h$ and h_E the diameter of $E \in \mathcal{E}_i^h$.

Suppose the solutions u_i belong to V_i such that

$$
V_i = \left\{ v \in H^1(\Omega_i) : \frac{\partial v_i}{\partial n_i}|_\Gamma \in L^2(\Gamma),\, v|_{\partial\Omega} = 0 \right\}, \quad i = 1, 2.
$$

Let the finite element spaces be

$$
V_i^h = \{ v \in H^1(\Omega_i) : v|_{\partial\Omega} = 0,\ v|_K \in \mathcal{P}^p(K)\ \forall K \in \mathcal{T}_i^h \}, \quad i = 1, 2,
$$

in which $\mathcal{P}^p$ denotes the polynomials of degree $\leq p$. We assume $p \geq 1$. We use the notation $V = V_1 \times V_2$ and $V^h = V_1^h \times V_2^h$. Respectively, we use $v = (v_1, v_2) \in V$ and $v^h = (v_1^h, v_2^h) \in V^h$ to denote the pair of functions defined in the subdomains.

3 The Proposed Method

In this section we propose Nitsche's method that takes the possible discontinuity of the material parameters and mesh sizes into account. The method is similar to [2].

For simplicity of notation we define the functions $h_i : \bar{\Omega}_i \to \mathbb{R}$, $i = 1, 2$ such that

$$h_i(x) = \begin{cases} h_K & \text{if } x \in K,\ K \in \mathcal{T}_i^h, \\ h_E & \text{if } x \in E,\ E \in \mathcal{E}_i^h. \end{cases}$$

At the interface Γ, for $v \in V$ we use $[\![v]\!] = v_1 - v_2$ to denote the jump over the interface.

Let $(\cdot, \cdot)_G$ denote the L^2-inner product over a domain G. Multiplying Eq. (1) with a test function $v \in V_i$ and integrating by parts gives

$$(k_i \nabla u_i, v_i)_{\Omega_i} - \left(k_i \frac{\partial u_i}{\partial n_i}, v_i\right)_\Gamma = (f, v_i)_{\Omega_i}.$$

By Eq. (3) it holds that

$$\left(\frac{k_1 h_2}{k_2 h_1 + k_1 h_2}\left(-k_1 \frac{\partial u_1}{\partial n} + k_2 \frac{\partial u_2}{\partial n}\right), v_1\right)_\Gamma = 0,$$

$$\left(\frac{k_2 h_1}{k_2 h_1 + k_1 h_2}\left(k_1 \frac{\partial u_1}{\partial n} - k_2 \frac{\partial u_2}{\partial n}\right), -v_2\right)_\Gamma = 0,$$

and by Eq. (2) it holds that

$$\left(\frac{k_1 k_2}{k_2 h_1 + k_1 h_2}[\![u]\!], [\![v]\!]\right)_\Gamma = 0.$$

Adding the equations above and introducing a stability parameter $\gamma > 0$ gives the weak form of the proposed Nitsche's method: Find $u^h \in V^h$ such that

$$\mathcal{B}^h(u^h, v^h) = \mathcal{F}(v^h) \quad \forall v^h \in V^h.$$

The bilinear form is

$$\mathcal{B}^h(w, v) = \sum_{i=1}^{2} (k_i \nabla w_i, \nabla v_i)_{\Omega_i} - \left(\left\{\!\!\left\{ k \frac{\partial w}{\partial n} \right\}\!\!\right\}, [\![v]\!]\right)_\Gamma - \left(\left\{\!\!\left\{ k \frac{\partial v}{\partial n} \right\}\!\!\right\}, [\![w]\!]\right)_\Gamma + \gamma \left(\frac{k_1 k_2}{k_2 h_1 + k_1 h_2}[\![w]\!], [\![v]\!]\right)_\Gamma,$$

in which

$$\left\{\!\!\left\{ k\frac{\partial v}{\partial n} \right\}\!\!\right\} = \alpha_1 k_1 \frac{\partial v_1}{\partial n} + \alpha_2 k_2 \frac{\partial v_2}{\partial n},$$

$$\alpha_1 = \frac{k_2 h_1}{k_2 h_1 + k_1 h_2}, \qquad \alpha_2 = \frac{k_1 h_2}{k_2 h_1 + k_1 h_2},$$

denotes the weighted average flux. The linear functional is simply

$$\mathscr{F}(v) = \sum_{i=1}^{2} (f, v_i)_{\Omega_i}.$$

By the derivation above it is clear that the proposed method is consistent with the strong form.

3.1 A Priori Analysis

Following [3, 10, 12, 13] we use the mesh dependent norms in the analysis. Let $\|\cdot\|_G$ denote the L^2 norm over a domain G. The parameters k_1 and k_2 are explicitly shown in the norms on V:

$$\|v\|_{1,h}^2 = \sum_{i=1}^{2} \left\| k_i^{\frac{1}{2}} \nabla v_i \right\|_{\Omega_i}^2 + \left\| \left(\frac{k_1 k_2}{k_2 h_1 + k_1 h_2} \right)^{\frac{1}{2}} [\![v]\!] \right\|_{\Gamma}^2,$$

$$|\!|\!|v|\!|\!|_{1,h}^2 = \|v\|_{1,h}^2 + \sum_{i=1}^{2} \left\| (h_i k_i)^{\frac{1}{2}} \frac{\partial v_i}{\partial n} \right\|_{\Gamma}^2.$$

Clearly $\|v\|_{1,h} \le |\!|\!|v|\!|\!|_{1,h}$ for all $v \in V$. The converse, $|\!|\!|v^h|\!|\!|_{1,h} \le C \|v^h\|_{1,h}$ with a $C > 0$, holds for any $v^h \in V^h$. This follows using the trace inequality [4, 6]

$$\left\| (h_i k_i)^{\frac{1}{2}} \frac{\partial v_i^h}{\partial n} \right\|_{\partial K}^2 \le C_I \left\| k_i^{\frac{1}{2}} \nabla v_i^h \right\|_K^2 \qquad \forall K \in \mathscr{T}_i^h,\ v_i^h \in V_i^h, \tag{4}$$

for a $C_I > 0$. Consequently, the norms are equivalent in V^h independent of the mesh sizes h_i and the parameters k_i: There exists $c, C > 0$ such that

$$c \|v^h\|_{1,h} \le |\!|\!|v^h|\!|\!|_{1,h} \le C \|v^h\|_{1,h} \quad \forall v^h \in V^h.$$

Theorem 1 *The bilinear form $\mathscr{B}^h$ is continuous in V with the norm $|||\cdot|||_{1,h}$ and, assuming the stability parameter satisfies $\gamma > 2C_I$, the bilinear form $\mathscr{B}^h$ is coercive in V^h with the norm $|||\cdot|||_{1,h}$.*

Proof Recall the definition of the weighted average flux and observe that

$$\begin{aligned}
\left(\left\{\!\!\left\{ k\frac{\partial w}{\partial n} \right\}\!\!\right\}, [\![v]\!]\right)_\Gamma &= \left(\alpha_1 k_1 \frac{\partial w_1}{\partial n} + \alpha_2 k_2 \frac{\partial w_2}{\partial n}, [\![v]\!]\right)_\Gamma \\
&= \left((\alpha_1 k_1 h_1)^{\frac{1}{2}} \frac{\partial w_1}{\partial n} + (\alpha_2 k_2 h_2)^{\frac{1}{2}} \frac{\partial w_2}{\partial n}, \left(\frac{k_1 k_2}{k_2 h_1 + k_1 h_2}\right)^{\frac{1}{2}} [\![v]\!]\right)_\Gamma \\
&\le \left(\left\|(\alpha_1 k_1 h_1)^{\frac{1}{2}} \frac{\partial w_1}{\partial n}\right\|_\Gamma + \left\|(\alpha_2 k_2 h_2)^{\frac{1}{2}} \frac{\partial w_2}{\partial n}\right\|_\Gamma\right) \left\|\left(\frac{k_1 k_2}{k_2 h_1 + k_1 h_2}\right)^{\frac{1}{2}} [\![v]\!]\right\|_\Gamma \\
&\le \left(\left\|(k_1 h_1)^{\frac{1}{2}} \frac{\partial w_1}{\partial n}\right\|_\Gamma + \left\|(k_2 h_2)^{\frac{1}{2}} \frac{\partial w_2}{\partial n}\right\|_\Gamma\right) \left\|\left(\frac{k_1 k_2}{k_2 h_1 + k_1 h_2}\right)^{\frac{1}{2}} [\![v]\!]\right\|_\Gamma \qquad (5)
\end{aligned}$$

for all $w, v \in V$. Using (5) it is easy to see that the bilinear form $\mathscr{B}^h$ is continuous.

Applying the trace inequality (4) to (5) and using Young's inequality with a parameter $\epsilon > 0$ gives

$$\begin{aligned}
&\left(\left\{\!\!\left\{ k\frac{\partial w^h}{\partial n} \right\}\!\!\right\}, [\![v^h]\!]\right)_\Gamma \\
&\le \left(C_I^{\frac{1}{2}} \left\|k_1^{\frac{1}{2}} \nabla w_1^h\right\|_{\Omega_1} + C_I^{\frac{1}{2}} \left\|k_2^{\frac{1}{2}} \nabla w_2^h\right\|_{\Omega_2}\right) \left\|\left(\frac{k_1 k_2}{k_2 h_1 + k_1 h_2}\right)^{\frac{1}{2}} [\![v^h]\!]\right\|_\Gamma \\
&\le \frac{C_I}{2\epsilon} \left\|k_1^{\frac{1}{2}} \nabla w_1^h\right\|_{\Omega_1}^2 + \frac{C_I}{2\epsilon} \left\|k_2^{\frac{1}{2}} \nabla w_2^h\right\|_{\Omega_2}^2 + \epsilon \left\|\left(\frac{k_1 k_2}{k_2 h_1 + k_1 h_2}\right)^{\frac{1}{2}} [\![v^h]\!]\right\|_\Gamma^2
\end{aligned}$$

for $v^h, w^h \in V^h$. With this we get that

$$\mathscr{B}^h(v^h, v^h) \ge \left(1 - \frac{C_I}{\epsilon}\right) \sum_{i=1}^{2} \left\|k_i^{\frac{1}{2}} \nabla v_i^h\right\|_{\Omega_i}^2 + (\gamma - 2\epsilon) \left\|\left(\frac{k_1 k_2}{k_2 h_1 + k_1 h_2}\right)^{\frac{1}{2}} [\![v^h]\!]\right\|_\Gamma^2.$$

By the assumption $\gamma > 2C_I$ we can choose $C_I < \epsilon < \gamma/2$, which shows that the bilinear form $\mathscr{B}^h$ is coercive. □

Let $u^h \in V^h$ denote the finite element solution and $u \in V$ the exact solution of the problem. The coercivity, consistency and continuity show that

$$\|u^h - v^h\|_{1,h}^2 \leq C\mathscr{B}^h(u^h - v^h, u^h - v^h) = C\mathscr{B}^h(u - v^h, u^h - v^h)$$
$$\leq C\,|||u - v^h|||_{1,h}\|u^h - v^h\|_{1,h}$$

for any $v^h \in V^h$. Using the triangle inequality and the equivalence of norms we get

$$|||u - u^h|||_{1,h} \leq C \inf_{v^h \in V^h} |||u - v^h|||_{1,h}.$$

Applying the interpolation results to the estimate above, we get the a priori result

$$|||u - u^h|||_{1,h} \leq Ch^{s-1}\left(\sum_{i=1}^{2}\left\|k_i^{\frac{1}{2}}u_i\right\|_{s,\Omega_i}^2\right)^{\frac{1}{2}}$$

for $u_i \in H^s(\Omega_i)$ with $i = 1, 2$ and $2 \leq s \leq p + 1$.

4 Observations on the Method

The method adapts automatically and continuously with respect to the parameters. The relation between k_2h_1 and k_1h_2, or equivalently between k_1/h_1 and k_2/h_2, determines the behavior of the method.

Suppose that $k_2h_1 = k_1h_2$ at the interface Γ. Denoting $k/h = k_1/h_2 = k_2/h_2$, the bilinear form is

$$\mathscr{B}^h(w, v) = \sum_{i=1}^{2}(k_i\nabla w_i, \nabla v_i)_{\Omega_i} - \left(\frac{1}{2}\left(k_1\frac{\partial w_1}{\partial n} + k_2\frac{\partial w_2}{\partial n}\right), [\![v]\!]\right)_\Gamma$$
$$- \left(\frac{1}{2}\left(k_1\frac{\partial v_1}{\partial n} + k_2\frac{\partial v_2}{\partial n}\right), [\![w]\!]\right)_\Gamma + \gamma\left(\frac{k}{2h}[\![w]\!], [\![v]\!]\right)_\Gamma.$$

In other words, the proposed method reduces to the method designed for continuous material parameters [3]. This indicates that the method in [3] should work for discontinuous material parameters too as long as the mesh sizes such that $k_2h_1 \approx k_1h_2$.

Suppose now that $k_2h_1 \gg k_1h_2$ at the interface Γ due to k_2 being very large. At the limit $k_2 \to \infty$, the coefficients of the method become

$$\lim_{k_2\to\infty} \alpha_1 = 1, \qquad \lim_{k_2\to\infty} \frac{k_1k_2}{k_2h_1 + k_1h_2} = \frac{k_1}{h_1},$$
$$\lim_{k_2\to\infty} \alpha_2 = 0,$$

and the bilinear form is

$$\mathscr{B}^h(w,v) = \sum_{i=1}^{2}(k_i \nabla w_i, \nabla v_i)_{\Omega_i} - \left(k_1 \frac{\partial w_1}{\partial n}, [\![v]\!]\right)_{\Gamma} - \left(k_1 \frac{\partial v_1}{\partial n}, [\![w]\!]\right)_{\Gamma} + \gamma \left(\frac{k_1}{h_1}[\![w]\!], [\![v]\!]\right)_{\Gamma}.$$

The interpretation of the bilinear form above is: In the subdomain Ω_1, the method enforces continuity at the interface Γ using Nitsche's method.

References

1. C. Annavarapu, M. Hautefeuille, J.E. Dolbow, A robust Nitsche's formulation for interface problems. Comput. Methods Appl. Mech. Eng. **225–228**, 44–54 (2012)
2. N. Barrau, R. Becker, E. Dubach, R. Luce, A robust variant of NXFEM for the interface problem. Comptes Rendus Math. **350**(15–16), 789–792 (2012)
3. R. Becker, P. Hansbo, R. Stenberg, A finite element method for domain decomposition with non-matching grids. Math. Model. Numer. Anal. **37**(2), 209–225 (2003)
4. S.C. Brenner, L.R. Scott, *The Mathematical Theory of Finite Element Methods*. Texts in Applied Mathematics, vol. 15, 2nd edn. (Springer, New York, 2002)
5. E. Burman, P. Zunino, A domain decomposition method based on weighted interior penalties for advection-diffusion-reaction problems. SIAM J. Numer. Anal. **44**(4), 1612–1638 (2006)
6. D.A. Di Pietro, A. Ern, *Mathematical Aspects of Discontinuous Galerkin Methods*. Mathématiques et Applications, vol. 69 (Springer, Berlin/Heidelberg, 2012)
7. M. Dryja, On discontinuous Galerkin methods for elliptic problems with discontinuous coefficients. Comput. Methods Appl. Mech. Eng. **3**(1), 76–85 (2003)
8. A. Ern, A. Stephansen, A posteriori energy-norm error estimates for advection-diffusion equations approximated by weighted interior penalty methods. J. Comput. Math. **26**(4), 488–510 (2008)
9. A. Ern, A.F. Stephansen, P. Zunino, A discontinuous Galerkin method with weighted averages for advection-diffusion equations with locally small and anisotropic diffusivity. IMA J. Numer. Anal. **29**(2), 235–256 (2008)
10. M. Juntunen, R. Stenberg, Nitsche's method for general boundary conditions. Math. Comput. **78**(267), 1353–1374 (2009)
11. T.A. Laursen, M.A. Puso, J. Sanders, Mortar contact formulations for deformable–deformable contact: past contributions and new extensions for enriched and embedded interface formulations. Comput. Methods Appl. Mech. Eng. **205–208**, 3–15 (2012)
12. R. Stenberg, On some techniques for approximating boundary conditions in the finite element method. J. Comput. Appl. Math. **63**(1–3), 139–148 (1995)
13. ———, Mortaring by a method of J.A. Nitsche, in *Computational Mechanics*, Buenos Aires, 1998 (Centro Internac. Métodos Numér. Ing., Barcelona, 1998)
14. P. Zunino, L. Cattaneo, C.M. Colciago, An unfitted interface penalty method for the numerical approximation of contrast problems. Appl. Numer. Math. **61**(10), 1059–1076 (2011)

Robust Local Flux Reconstruction for Various Finite Element Methods

Roland Becker, Daniela Capatina, and Robert Luce

Abstract We present a uniform approach to local reconstructions of the gradient of primal approximations by conforming, nonconforming and totally discontinuous finite elements of arbitrary order. We start from a hybrid formulation which covers all considered methods and whose Lagrange multipliers approximate the normal fluxes. It turns out that the multipliers can be computed locally and are next used to define local corrections of the flux. We also show that the DG solution and reconstructed flux with stabilisation parameter γ converge uniformly in h with the convergence rate $1/\gamma$ towards the CG or NC ones, depending on the stabilisation.

1 Introduction

This article is concerned with local reconstructions of the gradient of primal finite element approximations. We consider conforming, nonconforming and totally discontinuous (Galerkin) methods (abbreviated next as CG, NC, DG) of arbitrary order.

We consider the Poisson problem on a polygonal bounded domain $\Omega \subset \mathbb{R}^2$

$$-\Delta u = f \quad \text{in } \Omega, \qquad u = 0 \quad \text{on } \Gamma^{\mathrm{D}}, \qquad \frac{\partial u}{\partial n} = g \quad \text{on } \Gamma^{\mathrm{N}} \tag{1}$$

with boundary $\partial\Omega = \Gamma^{\mathrm{D}} \cup \Gamma^{\mathrm{N}}$, $f \in L^2(\Omega)$, $g \in L^2(\Gamma^{\mathrm{N}})$. The generalisation to non-homogeneous Dirichlet condition rises no difficulty.

Let $\sigma_h \in H(\mathrm{div}, \Omega)$ verify $\mathrm{div}\,\sigma_h = -\pi_h f$ in Ω and $\sigma_h \cdot n_\Omega = \pi_h^N g$ on Γ^N, where π_h and π_h^N are $L^2(\Omega)$, respectively $L^2(\Gamma^N)$ projections on piecewise polynomial finite element spaces. In order to be of practical interest, the reconstruction has to be discrete, which requires σ_h to be an element of a $H(\mathrm{div}, \Omega)$-conforming finite element space. We will throughout use the Raviart-Thomas spaces

R. Becker • D. Capatina • R. Luce (✉)
LMAP CNRS UMR 5142, Université de Pau, IPRA BP 1155, 64013 Pau, France
e-mail: roland.becker@univ-pau.fr; daniela.capatina@univ-pau.fr; robert.luce@univ-pau.fr

A. Abdulle et al. (eds.), *Numerical Mathematics and Advanced Applications ENUMATH 2013*, Lecture Notes in Computational Science and Engineering 103,
DOI 10.1007/978-3-319-10705-9_6

$\mathbf{RT}_h^m$ containing complete m-th order polynomials [7]. In addition, the computation of σ_h should be local. This means that a correction $\tau_h = \sigma_h - \nabla u_h$ can be computed on patches of cells.

The importance of local reconstructions for a posteriori error estimators has been noted at many places in the literature, see for example [3–5, 8]. We have $\|\nabla(u - u_h)\| \le \|\sigma_h - \nabla u_h\| + c\mu_h$, which means that the correction τ_h can be used as an error estimator with a sharp upper bound, up to a often higher-order term μ_h related to the data approximation. Such results have been largely developed for various finite element methods by sometimes seemingly different approaches.

The DG case is the simplest one and has been discussed in detail in the literature. In the case of NC and piecewise-constant data the Marini relation [6] between the solutions of the nonconforming and mixed finite element methods yields a cheap reconstruction. However, a more involved procedure is necessary in the general case of arbitrary data and polynomial degrees, see below. Finally, our approach for CG is related to the hypercircle method [2].

Our first aim in this article is to present a uniform approach to flux reconstruction. We start from a hybrid formulation covering all considered finite element methods. The Lagrange multipliers compensating for the different weak continuity conditions yield approximations to the normal fluxes. It turns out that they can be computed locally in all cases on patches defined by the support of the lowest-order basis functions. Then these multipliers are used to define interpolations in broken $\mathbf{RT}_h^m$ spaces which yield the local corrections τ_h. Our second aim is to study relations between the methods, see also [1]. We prove that the DG method converges uniformly in h with convergence rate $1/\gamma$ towards the CG or NC solution, depending on the form of stabilisation. The same convergence result holds true for the reconstructed fluxes and, therefore, also for the error estimators.

A regular triangular mesh h consists of cells $\mathscr{K}_h$, such that $\partial\Omega = \mathscr{S}_h^D \cup \mathscr{S}_h^N$. Let $\mathscr{S}_h^{int}$ the set of interior sides and $\mathscr{S}_h = \mathscr{S}_h^{int} \cup \mathscr{S}_h^D$. The measure of $S \in \mathscr{S}_h$ is denoted by $|S|$, the diameter of $K \in \mathscr{K}_h$ by d_K and $d = \max d_K$. For an interior side S, n_S is a fixed unit vector normal to S. If the side S lies on $\partial\Omega$, we set $n_S = n_\Omega$, the outward unit normal. We denote by π_S^l and π_K^l the $L^2(S)$ and $L^2(K)$ orthogonal projections on P^l. We define the spaces:

$$D_h^k := \left\{v_h \in L^2(\Omega) : v_h|_K \in P^k(K)\ \forall K \in \mathscr{K}_h\right\},$$

$$M_h^l := \left\{\mu_h \in L^2(\mathscr{S}_h) : \mu_h|_S \in P^l(S)\ \forall S \in \mathscr{S}_h\right\}.$$

Let $u \in D_h^k$. We define for $S \in \mathscr{S}_h^{int}$ and $x \in S$: $u_S^{\text{in}}(x) = \lim_{\varepsilon \searrow 0} u(x - \varepsilon n_S)$, $u_S^{\text{ex}}(x) = \lim_{\varepsilon \searrow 0} u(x + \varepsilon n_S)$. Next we define the jump and the mean for $x \in S$ by $[u](x) := u_S^{\text{in}}(x) - u_S^{\text{ex}}(x)$, $\{u\}(x) := \frac{1}{2}\left(u_S^{\text{in}}(x) + u_S^{\text{ex}}(x)\right)$. For a boundary side we set $[u]_S = \{u\}_S = u_S^{\text{in}}$.

2 Unified Framework, Stability and Robustness

Let $\gamma > 0$ and let us introduce a hybrid formulation covering the DG, CG and NC cases. For this purpose, we define the bilinear and linear forms

$$a(u_h, v_h) = \int_{\mathcal{K}_h} \nabla u_h \cdot \nabla v_h - \int_{\mathcal{S}_h} \{\partial_n u_h\}[v_h] - \int_{\mathcal{S}_h} [u_h]\{\partial_n v_h\},$$

$$b(\mu_h, v_h) = \int_{\mathcal{S}_h} \mu_h[v_h],\ c(\theta_h, \mu_h) = \int_{\mathcal{S}_h} |S|\,\theta_h \mu_h,\ l(v_h) = \int_{\Omega} f v_h + \int_{\Gamma^N} g v_h.$$

We consider the following problem as well as its limit version as $\gamma \to \infty$: Find $(u_h^\gamma, \theta_h^\gamma)$ and $(u_h^\infty, \theta_h^\infty)$ in $D_h^k \times M_h^l$ such that $\forall (v_h, \mu_h) \in D_h^k \times M_h^l$,

$$\begin{aligned} a(u_h^\gamma, v_h) + b(\theta_h^\gamma, v_h) &= l(v_h) \\ b(\mu_h, u_h^\gamma) - \frac{1}{\gamma} c(\theta_h^\gamma, \mu_h) &= 0, \end{aligned} \qquad (2)$$

$$\begin{aligned} a(u_h^\infty, v_h) + b(\theta_h^\infty, v_h) &= l(v_h) \\ b(\mu_h, u_h^\infty) &= 0, \end{aligned} \qquad (3)$$

For $l = k-1$, the solution u_h^γ satisfies a DG formulation whereas in the limit case $\gamma \to \infty$, the solution u_h^∞ satisfies the classical nonconforming primal formulation. For $l = k$, it turns out that the space M_h^k has to be reduced in order to obtain uniqueness of the multiplier in the limit case. Since $\theta_h^\gamma = \dfrac{\gamma}{|S|}[u_h^\gamma]$ on $S \in \mathcal{S}_h^{int}$ and since the jump satisfies a node-wise identity, we are lead to introduce

$$M_h^{k,*} := \left\{ \mu_h \in M_h^k : \sum_{S \in \mathcal{S}_N} \alpha_S(N) \mu_h(N) = 0 \quad \forall N \in \mathcal{N}_h \right\}. \qquad (4)$$

Here above, $\mathcal{S}_N$ denotes the set of sides S sharing the node N and $\alpha_S(N) = |S|\,\mathrm{sign}(n_S, N)$, where $\mathrm{sign}(n_S, N)$ equals 1 or -1 depending upon the orientation of n_S with respect to the clockwise rotation sense around N. $\mathcal{N}_h$ represents the set of nodes which are interior to the domain or to Γ^D.

We now consider problem (2) with M_h^l replaced by $M_h^{k,*}$; its solution u_h^γ actually satisfies the well-known DG formulation with interior penalty stabilisation. The limit problem as $\gamma \to \infty$ corresponds now to the CG formulation.

It is useful to introduce the forms $\tilde{b}(\cdot,\cdot)$, $\tilde{c}(\cdot,\cdot)$ obtained from $b(\cdot,\cdot)$, $c(\cdot,\cdot)$ by reduced integration. Let the Newton-Cotes formula on $S \in \mathcal{S}_h$, with $(w_i, N_i^S)_{1 \le i \le k+1}$ the weights and equidistant nodes: $I_h(g) = \sum_{i=1}^{k+1} w_i\, g(N_i^S)$. The weights are uniquely defined, independent of S and strictly positive. Then we define

$$\tilde{b}(\mu_h, v_h) = \sum_{S \in \mathcal{S}_h} |S| I_h(\mu_h[v_h]), \quad \tilde{c}(\theta_h, \mu_h) = \sum_{S \in \mathcal{S}_h} |S|^2 I_h(\theta_h \mu_h).$$

We also consider the corresponding DG and CG variants: Find $(\tilde{u}_h^\gamma, \tilde{\theta}_h^\gamma)$, respectively $(\tilde{u}_h^\infty, \tilde{\theta}_h^\infty)$ in $D_h^k \times M_h^{k,*}$ such that for all $(v_h, \mu_h) \in D_h^k \times M_h^{k,*}$,

$$\begin{aligned} a(\tilde{u}_h^\gamma, v_h) + \tilde{b}(\tilde{\theta}_h^\gamma, v_h) &= l(v_h), \\ \tilde{b}(\mu_h, \tilde{u}_h^\gamma) - \frac{1}{\gamma}\tilde{c}(\tilde{\theta}_h^\gamma, \mu_h) &= 0, \end{aligned} \quad (5) \qquad \begin{aligned} a(\tilde{u}_h^\infty, v_h) + \tilde{b}(\tilde{\theta}_h^\infty, v_h) &= l(v_h), \\ \tilde{b}(\mu_h, \tilde{u}_h^\infty) &= 0. \end{aligned} \quad (6)$$

Let the space M_h^l denote from now on M_h^{k-1} for $l = k-1$, and $M_h^{k,*}$ for $l = k$. We introduce the norms on D_h^k and M_h^l respectively:

$$|||v_h||| := \left(\int_{\mathcal{K}_h} |\nabla v_h|^2 + \int_{\mathcal{S}_h} |S|^{-1} \pi_S^l [v_h]^2 \right)^{1/2}, \quad \|\mu_h\|_{\mathcal{S}_h} := c(\mu_h, \mu_h)^{1/2}.$$

The forms a, b, $\tilde{b}$ and l are clearly uniformly continuous and $\tilde{c}$ is an equivalent scalar product on $M_h^{k,*}$. It is important to note that $\operatorname{Ker} b$ coincides with the P^k nonconforming space for $l = k-1$, respectively with the P^k conforming space for $l = k$ (also $\operatorname{Ker} \tilde{b}$). Therefore, in the conforming case it follows that $u_h^\infty = \tilde{u}_h^\infty$ since they both satisfy the primal CG formulation, although θ_h^∞ and $\tilde{\theta}_h^\infty$ are different.

The uniform coercivity of a on $\operatorname{Ker} b$, with respect to h and γ, is obvious. The technical point concerns the inf-sup conditions, given without proof.

Theorem 1 *There exist $\beta > 0$, $\tilde{\beta} > 0$ independent of h and γ such that*

$$\inf_{\mu_h \in M_h^l} \sup_{v_h \in D_h^k} \frac{b(\mu_h, v_h)}{\|\mu_h\|_{\mathcal{S}_h} |||v_h|||} \geq \beta, \qquad \inf_{\mu_h \in M_h^{k,*}} \sup_{v_h \in D_h^k} \frac{\tilde{b}(\mu_h, v_h)}{\|\mu_h\|_{\mathcal{S}_h} |||v_h|||} \geq \tilde{\beta}.$$

The Babuška-Brezzi theorem yields the following result.

Theorem 2 *The systems (2), (3) and (5), (6) are uniformly well-posed. Their solutions satisfy the following a priori error estimates:*

$$|||u - u_h^\gamma||| + \|\theta_h^\gamma\|_{\mathcal{S}_h} \leq C d^k |u|_{k+1,\Omega}, \quad |||u - u_h^\infty||| + \|\theta_h^\infty\|_{\mathcal{S}_h} \leq C d^k |u|_{k+1,\Omega}.$$

Next we analyse the limit of the hybrid formulations (2), (5) as $\gamma \to \infty$.

Theorem 3 *There exists a constant C independent of h and γ such that*

$$|||u_h^\gamma - u_h^\infty||| + \|\theta_h^\gamma - \theta_h^\infty\|_{\mathcal{S}_h} \leq \frac{C}{\gamma}, \quad \|\tilde{u}_h^\gamma - \tilde{u_h}^\infty\| + \|\tilde{\theta}_h^\gamma - \tilde{\theta}_h^\infty\|_{\mathcal{S}_h} \leq \frac{C}{\gamma}.$$

Proof We consider the orthogonal decomposition $D_h^k = \operatorname{Ker} b \oplus \operatorname{Ker} b^\perp$, which yields $u_h^\gamma - u_h = (u_h^\gamma - u_h)^0 + (u_h^\gamma - u_h)^\perp$. We substract the mixed problems and we first take $v_h = (u_h^\gamma - u_h)^0$. By using the coercivity of $a_h(\cdot,\cdot)$, the fact that $\|\theta_h^\gamma\|_{\mathscr{S}_h}$ is bounded and the inf-sup condition in the second variational equation, we obtain

$$|||(u_h^\gamma - u_h)^0||| \leq C_1|||(u_h^\gamma - u_h)^\perp||| \leq \frac{C_2}{\gamma}\|\theta_h^\gamma\|_{\mathscr{S}_h} \leq \frac{C_3}{\gamma}.$$

The inf-sup condition in the first equation yields the result for the multipliers. □

3 Local Computation of Multipliers and Fluxes

A remarkable feature of the approach is that for the three types of finite element methods, the multiplier θ_h can be computed locally.

The DG case is the simplest: the second equation of any of the hybrid formulations yields, for $l = k, k-1$,

$$\theta_h^\gamma = \frac{\gamma}{|S|}\pi_S^l[u_h^\gamma] \quad \text{on } S \in \mathscr{S}_h. \tag{7}$$

The NC and CG methods are less trivial, since θ_h^∞ is now computed from the first variational equation, that is by solving a global system. For simplicity of notation, we introduce the righthand side term $L(\cdot) := l(\cdot) - a(u_h^\infty, \cdot)$.

We write θ_h as the sum of local contributions θ_ω associated to patches ω representing the support of a finite element basis function. The unknown θ_ω lives only on the interior sides of the patch ω and is null elsewhere.

We first treat the NC formulation. To any side $S \in \mathscr{S}_h$, we associate a patch $\omega_S = \text{supp}\psi_S^j$ where ψ_S^j is the basis function associated to the jth of the k Gauss points on S. For $\{S\} = \partial\bar{K}^{in} \cap \partial\bar{K}^{ex}$, we recall that $\omega_S = K^{in} \cup K^{ex}$. Then θ_S lives only on S, belongs to P^{k-1} and is defined by

$$b(\theta_S, \psi_S^j\chi_K) = L(\psi_S^j\chi_K), \quad 1 \leq j \leq k, \quad \forall K \subset \omega_S \tag{8}$$

where χ_K denotes the characteristic function of K. This allows to uniquely define $\theta_S \in P^{k-1}$ since the nonconforming solution u_h^{NC} satisfies, for all $S \in \mathscr{S}_h^{int}$ and $1 \leq j \leq k$: $L(\psi_S^j) = \sum_{K\subset\omega_S} L(\psi_S^j\chi_K) = 0$.

Theorem 4 *Let* $\theta_h = \sum_{S\in\mathscr{S}_h} \theta_S$. *It satisfies* $b(\theta_h, v_h) = L(v_h)$ *for all* $v_h \in D_h^k$. *Hence,* θ_h *is the unique solution* θ_h^{NC} *of (3) with* $l = k-1$.

Proof The key point is that for another side S' sharing the triangle K with S one has $b(\theta_{S'}, \psi_S^j\chi_K) = 0$, since $\theta_{S'}\psi_S^j \in P^{2k-1}$ is integrated exactly on $S \cup S'$ by the Gauss formula with k points. Therefore, θ_h satisfies for any $K \in \mathscr{K}_h$ and $S \subset \partial K$:

$$b(\theta_h, \psi_S^j\chi_K) = \sum_{S'\in\mathscr{S}_h} b(\theta_{S'}, \psi_S^j\chi_K) = L(\psi_S^j\chi_K), \quad 1 \leq j \leq k. \tag{9}$$

Next, let us note that $P^1(K) = \text{vect}\{\psi_S;\ S \subset \partial K\}$ whereas for $2 \leq k \leq 3$, one has $P^k(K) = \text{vect}\left\{\psi_S^j;\ S \subset \partial K,\ 1 \leq j \leq k\right\} \oplus \text{vect}\{b_k\}$ where b_k is the quadratic, respectively cubic bubble function on K. Taking $v_h = b_k \chi_K$ in (3) one gets $L(b_k \chi_K) = 0$; since $b(\theta_h, b_k \chi_K) = 0$, the conclusion follows. □

Finally, we consider the two CG approximations. To any node N we associate the patch $\omega_N = \text{supp}\,\varphi_N$, where φ_N is the P_1 -basis function associated to N. We introduce the hierarchical basis of P^k on a side S of vertices N and M; besides φ_N and φ_M, it contains for $k \geq 2$ the functions φ_S^j $(1 \leq j \leq k-1)$ associated to the interior Lagrange nodes of S.

For (3) with $l = k$, we define θ_N living on the sides $S \in \mathscr{S}_N$ of vertices $N,\ M \in \mathscr{N}_h$. For all $K \subset \omega_N$, we impose:

$$b(\theta_N, \varphi_N \chi_K) = L(\varphi_N \chi_K), \tag{10}$$

$$b(\theta_N, \varphi_S^j \chi_K) = \frac{1}{2} L(\varphi_S^j \chi_K), \quad S \in \mathscr{S}_h \cap \partial K, \quad 1 \leq j \leq k-1, \tag{11}$$

$$b(\theta_N, \varphi_M \chi_K) = 0, \quad M \in \mathscr{N}_h \cap \partial K, \ M \neq N. \tag{12}$$

We look for $(\theta_N)_{|S}$ in P^k. Note that the conforming solution u_h^{CG} satisfies

$$L(\varphi_N) = L(\varphi_S^j) = 0, \quad \forall N \in \mathscr{N}_h^{int}, \forall S \in \mathscr{S}_h^{int}, \quad 1 \leq j \leq k-1. \tag{13}$$

Hence, the previous system is compatible. However, it has a one-dimensional kernel $\mathscr{K}_N$ characterized by $\text{sign}(n_S, N) \int_S \theta_N \varphi_N = const$, for all $S \in \mathscr{S}_N$.

In conclusion, there exists a solution θ_N, unique up to an element of $\mathscr{K}_N$.

Theorem 5 *Let* $\theta_h = \sum_{N \in \mathscr{N}_h} \theta_N$. *Then* θ_h *is equal to the solution* θ_h^{CG} *of (3) with* $l = k$, *up to an element* θ_h^{Ker} *of the kernel of* b.

Note that the flux σ_h reconstructed thanks to θ_h (see the next section) has the desired divergence and belongs to $H(\text{div}, \Omega)$, independently of the choice of θ_h^{Ker}. However, in order to obtain that it also represents the limit of the DG flux as γ tends towards infinity, we have to determine the correction θ_h^{Ker} such that θ_h belongs to $M_h^{k,*}$, by solving a global problem. In order to overcome this drawback and to preserve both the locality of the computation and the robustness with respect to γ, we turn to the formulation (6). We define $\tilde{\theta}_N$ by the same local system (10)–(12) but using $\tilde{b}(\cdot,\cdot)$ instead of $b(\cdot,\cdot)$. The new kernel is defined by $\tilde{\mathscr{K}}_N = \text{vect}\{\alpha_S(N)\varphi_N\}$. By imposing now that $\tilde{\theta}_N$ is orthogonal to the kernel, we obtain that the global multiplier $\tilde{\theta}_h = \sum_{N \in \mathscr{N}_h} \tilde{\theta}_N$ belongs to $M_h^{k,*}$, therefore $\tilde{\theta}_h = \tilde{\theta}_h^{CG}$.

We are next interested in defining a flux $\sigma_h \in H(\text{div}, \Omega)$ from θ_h. For $l = k-1$ it is natural to use the space $\mathbf{RT}_h^m$ with $m = k-1$, since its normal traces are $(k-1)$-th order polynomials. However, for $l = k$ we can either use $m = k$ or $m = k-1$.

For any method, we impose the degrees of freedom on the edges as below. On the Neumann boundary, we set $\sigma_h \cdot n_S = \pi_S^m g$ while on $S \in \mathscr{S}_h$ we impose:

$$\int_S \sigma_h \cdot n_S \varphi = \int_S \{\partial_n u_h\}\varphi - b_S(\theta_h, \varphi) \quad \text{if } S \in \mathscr{S}_h^{int}, \tag{14}$$

$$\int_S \sigma_h \cdot n_S \varphi = -b_S(\theta_h, \varphi) \qquad \text{if } S \in \mathscr{S}_h^D. \tag{15}$$

The test-functions φ stand for the basis functions of each finite element method, that is: any $\varphi \in P^l(S)$ for DG, ψ_S^j for NC and $\varphi_N, \varphi_S^j, \varphi_M$ for CG, with N, M the vertices of S. The bilinear form $b_S(\cdot,\cdot)$ represents the restriction of the corresponding $b(\cdot,\cdot)$ to the side S and is defined by

$$b_S(\theta_h, \varphi) = \int_S \varphi\theta_h, \quad \tilde{b}_S(\theta_h, \varphi) = |S| I_h(\theta_h\varphi)$$

for the DG and NC problems (2), (3), respectively the DG and CG problems (5), (6). For $m \geq 1$, we also define interior degrees of freedom as follows:

$$\int_K \sigma_h \cdot r = \int_K \nabla u_h \cdot r - \frac{1}{2}\int_{\partial K \cap \mathscr{S}_h^{int}} [u_h]\, r \cdot n_S - \int_{\partial K \cap \mathscr{S}_h^D} u_h r \cdot n_S$$

for the DG formulations, while for the NC and CG formulations we set

$$\int_K \sigma_h \cdot r = \int_K \nabla u_h \cdot r, \quad \forall r \in (P^{m-1})^2.$$

Theorem 6 *For any method, the reconstructed flux satisfies*

$$\mathrm{div}\sigma_h|_K = -\pi_K^m f, \quad \forall K \in \mathscr{K}_h. \tag{16}$$

Moreover, there exists a constant C independent of h and γ such that

$$\|\sigma_h^\gamma - \sigma_h^{NC}\| \leq \frac{C}{\gamma}, \quad \|\tilde{\sigma}_h^\gamma - \tilde{\sigma}_h^{CG}\| \leq \frac{C}{\gamma}.$$

Proof By taking as test-function in any of the hybrid formulations $v\chi_K$ with $v \in P^m$ and by using the integration by parts formula:

$$\int_K \mathrm{div}\sigma_h v = -\int_K \sigma_h \cdot \nabla v + \int_{\partial K} \sigma_h \cdot n_K v,$$

we immediately get the first property. The convergence with respect to γ follows from Theorem 3 and from a standard scaling argument which yields

$$\|\sigma_h^\gamma - \sigma_h^{NC}\|_{0,K} \leq c \sum_{S\subset\partial K} |S|^{1/2} \|(\sigma_h^\gamma - \sigma_h^{NC})\cdot n_S\|_{0,S}.$$

□

Remark 1 In [2], Braess and Schöberl propose a flux correction for the CG method inspired by the hypercircle method and defined as the sum of local contributions on ω_N. Each local correction is obtained as the solution of a mixed method. However, the divergence of the global flux is equal to $\pi_K^{k-1} f$ due to the property $\sum_{N\in\mathcal{N}_h} \varphi_N = 1$.

4 Numerical Experiment

We solve the Poisson equation with Dirichlet condition on the square $]-1,1[^2$. The data are chosen such that $u(x) = e^{-\frac{\|x-x_0\|}{\delta}}$, where $x_0 = (0.5, 0.5)$ and $\delta = 0.03$, is the solution. We take $k = 1$ and we focus on the behavior as $\gamma \to \infty$ on a fixed mesh.

We show in Table 1 the errors for γ between the DG and the CG, respectively NC solutions. In the first table, we consider both variants of DG and CG methods. We numerically obtain the expected $O(1/\gamma)$ convergence rate for the variant with reduced integration. As regards the initial variant without numerical integration (the classical interior penalty method), we find that $u_h^\gamma - u_h^{CG}$ behaves again like $1/\gamma$ but, as predicted, $\theta_h^\gamma - \theta_h^{CG}$ does not converge. Table 2 shows that the DG method (2) converges towards the NC one with $O(1/\gamma)$, and this convergence occurs at rather small values of γ.

Table 1 Convergence of DG towards CG as $\gamma \to \infty$

γ	$\|\|\|\tilde{u}_h^\gamma - \tilde{u}_h^{CG}\|\|\|$	$\|\|\tilde{\theta}_h^\gamma - \tilde{\theta}_h^{CG}\|\|$	$\|\|\|u_h^\gamma - u_h^{CG}\|\|\|$	$\|\|\theta_h^\gamma - \theta_h^{CG}\|\|$
10	1.08439e−01	5.66067e−02	2.25960e−01	3.14170e−01
10^2	1.17825e−02	6.15432e−03	3.00036e−02	4.66233e−01
10^3	1.19242e−03	6.23495e−04	3.10670e−03	4.88739e−01
10^4	1.19388e−04	6.24228e−05	3.11780e−04	4.91095e−01
10^5	1.19403e−05	6.23418e−06	3.11891e−05	4.91332e−01
10^6	1.19385e−06	6.29765e−07	3.11907e−06	4.91356e−01

Table 2 Convergence of DG towards NC as $\gamma \to \infty$

γ	$\|\|\|u_h^\gamma - u_h^{NC}\|\|\|$	$\|\|\theta_h^\gamma - \theta_h^{NC}\|\|$
10	5.66994e−03	1.39880e−03
10^2	4.14435e−04	1.00756e−04
10^3	4.04056e−05	9.79224e−06
10^4	4.03050e−06	9.74755e−07

References

1. R. Becker, D. Capatina, J. Joie, Connections between discontinuous Galerkin and nonconforming finite element methods for the Stokes equations. Numer. Methods Part. Differ. Equ. **28**(3), 1013–1041 (2012)
2. D. Braess, J. Schöberl, Equilibrated residual error estimator for edge elements. Math. Comput. **77**(262), 651–672 (2008)
3. A. Ern, S. Nicaise, M. Vohralík, An accurate $H(\text{div})$ flux reconstruction for discontinuous Galerkin approximations of elliptic problems. C. R. A. S. **345**(12), 709–712 (2007)
4. K.Y. Kim, A posteriori error analysis for locally conservative mixed methods. Math. Comput. **76**(257), 43–66 (2007)
5. R. Luce, B. Wohlmuth, A local a posteriori error estimator based on equilibrated fluxes. SIAM J. Numer. Anal. **42**(4), 1394–1414 (2004)
6. L.D. Marini, An inexpensive method for the evaluation of the solution of the lowest order Raviart-Thomas mixed method. SIAM J. Numer. Anal. **22**(3), 493–496 (1985)
7. P.-A. Raviart, J.-M. Thomas, A mixed finite element method for 2nd order elliptic problems, in *Mathematical Aspects of Finite Element Methods: Proceedings of the Conference*, Consiglio Naz. delle Ricerche, Rome, 1975 (Springer, Berlin, 1977), pp. 292–315
8. M. Vohralík, Residual flux-based a posteriori error estimates for finite volume and related locally conservative methods. Numer. Math. **111**(1), 121–158 (2008)

On the Use of Reconstruction Operators in Discontinuous Galerkin Schemes

Václav Kučera

Abstract This work is concerned with the introduction of reconstruction operators as known from higher order finite volume (FV) schemes into the discontinuous Galerkin (DG) method. This operator constructs higher order piecewise polynomial reconstructions from the lower order DG scheme. The result is the increase in accuracy of the DG scheme which is cheaper than directly using standard DG schemes of very high orders. We discuss the reconstruction operators and their construction, the relation to DG and present numerical experiments which demonstrate the increased accuracy of this approach.

1 Problem Formulation and Notation

We treat a nonlinear nonstationary scalar hyperbolic equation in a bounded domain $\Omega \subset I\!R^d$ with a Lipschitz-continuous boundary. We seek $u : \Omega \times [0, T] \to I\!R$ such that

$$\frac{\partial u}{\partial t} + \operatorname{div} \mathbf{f}(\mathrm{u}) = 0 \quad \text{in } \Omega \times (0, \mathrm{T}) \tag{1}$$

along with an appropriate initial and boundary condition. Here $\mathbf{f} = (f_1, \cdots, f_d)$ and $f_s, s = 1, \ldots, d$ are Lipschitz continuous fluxes in the direction x_s.

Let $\mathscr{T}_h$ be a triangulation of the closure $\overline{\Omega}$ into a finite number of closed simplices $K \in \mathscr{T}_h$. By ∂K we denote the boundary of an element $K \in \mathscr{T}_h$ and set $h = \max_{K\in\mathscr{T}_h} \operatorname{diam}(K)$. By $\mathscr{F}_h$ we denote the system of all faces of all elements $K \in \mathscr{T}_h$. For each $\Gamma \in \mathscr{F}_h$ we define a unit normal vector $\mathbf{n}_\Gamma$, such that for $\Gamma \in \mathscr{F}_h^B$, $\mathbf{n}_\Gamma$ is the outer unit normal to $\partial\Omega$. For each interior face $\Gamma \in \mathscr{F}_h^I$ there exist two

V. Kučera (✉)
Faculty of Mathematics and Physics, Charles University in Prague, Sokolovská 83, Praha 8, 186 75, Czech Republic
e-mail: vaclav.kucera@email.cz

A. Abdulle et al. (eds.), *Numerical Mathematics and Advanced Applications ENUMATH 2013*, Lecture Notes in Computational Science and Engineering 103,
DOI 10.1007/978-3-319-10705-9_7

neighbours $K_\Gamma^{(L)}, K_\Gamma^{(R)} \in \mathscr{T}_h$ such that $\Gamma \subset K_\Gamma^{(L)} \cap K_\Gamma^{(R)}$. For a piecewise H^1-function defined on $\mathscr{T}_h$ and $\Gamma \in \mathscr{F}_h^I$ we introduce the following notation:

$$v|_\Gamma^{(L)} = \text{trace of } v|_{K_\Gamma^{(L)}} \text{ on } \Gamma, \quad v|_\Gamma^{(R)} = \text{trace of } v|_{K_\Gamma^{(R)}} \text{ on } \Gamma$$

and the *jump* $[v]_\Gamma = v|_\Gamma^{(L)} - v|_\Gamma^{(R)}$. For $\Gamma \in \mathscr{F}_h^B$ we define $v|_\Gamma^{(R)} = [v]_\Gamma := v|_\Gamma^{(L)}$.

Let $n \geq 0$ be an integer. We define the space of discontinuous piecewise polynomial functions

$$S_h^n = \{v;\ v|_K \in P^n(K), \forall K \in \mathscr{T}_h\},$$

where $P^n(K)$ is the space of all polynomials on K of degree $\leq n$. Specifically,

- S_h^0: is the space of piecewise constant functions used in the FV method,
- S_h^n, $n > 0$: the DG space of piecewise nth degree polynomials,
- S_h^N, $N > n$: the space of higher order reconstructed DG solutions.

2 Discontinuous Galerkin (DG) Formulation

We multiply (1) by an arbitrary $\varphi_h^n \in S_h^n$, integrate over an element $K \in \mathscr{T}_h$ and apply Green's theorem. By summing over all $K \in \mathscr{T}_h$ and rearranging,

$$\frac{d}{dt}\int_\Omega u(t)\,\varphi_h^n\,dx + \sum_{\Gamma\in\mathscr{F}_h}\int_\Gamma \mathbf{f}(u)\cdot\mathbf{n}\,[\varphi_h^n]\,dS - \sum_{K\in\mathscr{T}_h}\int_K \mathbf{f}(u)\cdot\nabla\varphi_h^n\,dx = 0. \tag{2}$$

The boundary convective terms will be treated similarly as in the finite volume method, i.e. with the aid of a *numerical flux* $H(u, v, \mathbf{n})$, cf. [2, 3, 6]:

$$\int_\Gamma \mathbf{f}(u)\cdot\mathbf{n}\,[\varphi_h^n]\,dS \approx \int_\Gamma H(u^{(L)}, u^{(R)}, \mathbf{n})[\varphi_h^n]\,dS. \tag{3}$$

Finally, we define the *convective form* $b_h(\cdot,\cdot)$ defined for $v, \varphi \in H^1(\Omega, \mathscr{T}_h)$:

$$b_h(v,\varphi) = \int_{\mathscr{F}_h} H(v^{(L)}, v^{(R)}, \mathbf{n})[\varphi]\,dS - \sum_{K\in\mathscr{T}_h}\int_K \mathbf{f}(v)\cdot\nabla\varphi\,dx.$$

Definition 1 (Standard DG scheme) We seek $u : [0, T] \to S_h^n$ such that

$$\frac{d}{dt}\big(u_h(t), \varphi_h^n\big) + b_h\big(u_h(t), \varphi_h^n\big) = 0, \quad \forall \varphi_h^n \in S_h^n,\ \forall t \in (0, T). \tag{4}$$

We note that if we take $n = 0$, i.e. $u_h : (0, T) \to S_h^0$, then from the definition of b_h the DG scheme (4) is equivalent to the standard FV method.

3 Reconstructed Discontinuous Galerkin (RDG) Formulation

For $v \in L^2(\Omega)$, we denote by $\Pi_h^n v$ the $L^2(\Omega)$-projection of v on S_h^n:

$$\Pi_h^n v \in S_h^n, \quad \left(\Pi_h^n v - v, \varphi_h^n\right) = 0, \qquad \forall\, \varphi_h^n \in S_h^n. \tag{5}$$

The basis of the proposed method lies in the observation that (2) can be viewed as an equation for the evolution of $\Pi_h^n u(t)$, where u is the exact solution of (1). In other words, due to (5), $\Pi_h^n u(t) \in S_h^n$ satisfies the following equation for all $\varphi_h^n \in S_h^n$:

$$\frac{d}{dt}\int_\Omega \Pi_h^n u(t)\,\varphi_h^n\,dx + \int_{\mathscr{F}_h} \mathbf{f}(u)\cdot\mathbf{n}\,[\varphi_h^n]\,dS - \sum_{K\in\mathscr{T}_h}\int_K \mathbf{f}(u)\cdot\nabla\varphi_h^n\,dx = 0. \tag{6}$$

Let $N > n$ be an integer. We assume that there exists a piecewise polynomial function $U_h^N(t) \in S_h^N$, which is an approximation of $u(t)$ of order $N+1$, i.e.

$$U_h^N(x,t) = u(x,t) + O(h^{N+1}), \quad \forall x \in \Omega,\ \forall t \in [0,T], \tag{7}$$

which is possible if u is sufficiently regular in space. Now we incorporate the approximation $U_h^N(t)$ into (6): the exact solution u satisfies

$$\frac{d}{dt}\left(\Pi_h^n u(t), \varphi_h^n\right) + b_h\left(U_h^N(t), \varphi_h^n\right) = E(\varphi_h^n, t), \quad \forall \varphi_h^n \in S_h^n,\ \forall t \in (0,T), \tag{8}$$

where $E(\varphi_h^n)$ is an error term defined as

$$E(\varphi_h^n, t) = b_h\left(U_h^N(t), \varphi_h^n\right) - b_h\left(u(t), \varphi_h^n\right). \tag{9}$$

Lemma 1 ([5]) *The following estimate holds for all $t \in [0,T]$:*

$$E(\varphi_h^n, t) = O(h^N)\|\varphi_h^n\|_{L^2(\Omega)}. \tag{10}$$

It remains to construct a sufficiently accurate approximation $U_h^N(t) \in S_h^N$ to $u(t)$, such that (7) is satisfied. This leads to the following problem.

Definition 2 (Reconstruction problem) Let $v : \Omega \to I\!R$ be sufficiently regular. Given $\Pi_h^n v \in S_h^n$, find $v_h^N \in S_h^N$ such that $v - v_h^N = O(h^{N+1})$ in Ω. Define the reconstruction operator $R : S_h^n \to S_h^N$ by $R\,\Pi_h^n v := v_h^N$.

By setting $U_h^N(t) := R\,\Pi_h^n u(t)$ in (8), we obtain the following semidiscrete, formally Nth order scheme for the $L^2(\Omega)$-projections of u onto S_h^n:

$$\frac{d}{dt}\big(\Pi_h^n u(t), \varphi_h^n\big) + b_h\big(R\,\Pi_h^n u(t), \varphi_h^n\big) = O(h^N)\|\varphi_h^n\|_{L^2(\Omega)}, \quad \forall \varphi_h^n \in S_h^n. \tag{11}$$

By neglecting the right-hand side and approximating $u_h^n(t) \approx \Pi_h^n u(t)$, we arrive at the following *reconstructed discontinuous Galerkin* (RDG) scheme.

Definition 3 (Reconstructed DG scheme) Seek $u_h^n : [0, T] \to S_h^n$ s.t.

$$\frac{d}{dt}\big(u_h^n(t), \varphi_h^n\big) + b_h\big(Ru_h^n(t), \varphi_h^n\big) = 0, \quad \forall \varphi_h^n \in S_h^n,\ \forall t \in (0, T). \tag{12}$$

We point out several facts about the RDG scheme:

- The derivation of the RDG scheme follows the methodology of higher order FV schemes. The basis of these schemes is an equation for the evolution of averages of the exact solution on individual elements (i.e. an equation for $\Pi_h^0 u(t)$). Equation (11) is a direct generalization for the case of higher order $L^2(\Omega)$-projections $\Pi_h^n u(t)$, $n \geq 0$.
- Both $u_h^n(t)$ and φ_h^n lie in S_h^n. Only $Ru_h^n(t)$, lies in the higher dimensional space S_h^N. Despite this fact, Eq. (11) indicates that we may expect $u - Ru_h^n = O(h^{N+1})$, although $u - u_h^n = O(h^{n+1})$.
- Numerical quadrature must be employed to evaluate surface and volume integrals in (12). Since test functions are in S_h^n, as compared to S_h^N in the corresponding Nth order standard DG scheme, we may use lower order (i.e. more efficient) quadrature formulae as compared to standard DG.

As in the case of higher order FV, we use an explicit time stepping method. For simplicity, we formulate the forward Euler method, however in Sect. 4, higher order Adams–Bashforth methods are used.

Let us construct a partition $0 = t_0 < t_1 < t_2 \ldots$ of the time interval $[0, T]$ and define the time step $\tau_k = t_{k+1} - t_k$. We use the approximation $u_h^n(t_k) \approx u_h^{n,k} \in S_h^n$. The forward Euler scheme is given by:

Definition 4 (Explicit RDG scheme) Seek $u_h^{n,k} \in S_h^n,\ k = 0, 1, \ldots$ s.t.

$$\left(\frac{u_h^{n,k+1} - u_h^{n,k}}{\tau_k}, \varphi_h^n\right) + b_h\big(Ru_h^{n,k}, \varphi_h^n\big) = 0, \quad \forall \varphi_h^n \in S_h^n,\ k = 0, 1, \ldots, \tag{13}$$

where $u_h^{n,0} = u_{h,0}$ is an S_h^n approximation of the initial condition u^0.

The upper limit on stable time steps, given by a CFL-like condition, is more restrictive with growing N. However, in the RDG scheme, stability properties are inherited from the lower order scheme, therefore a larger time step is possible as compared to the corresponding Nth order DG scheme.

3.1 Construction of the Reconstruction Operator

As in the finite volume method [4, 6], a stencil (a group of neighboring elements and the element under consideration) is used to build an Nth-degree polynomial approximation to u on the element under consideration [4, 6]. In the FV method, the von Neumann neighborhood of an element is used as a stencil to obtain a piecewise linear reconstruction, cf. Fig. 1a. However, for higher order reconstructions, the size of the stencil increases dramatically, cf. Fig. 1b, rendering higher degrees than quadratic very time consuming. In the case of the RDG scheme, we need not increase the stencil size to obtain higher order accuracy, it suffices to take the von Neumann neighborhood and increase the order of the underlying DG scheme.

In analogy to the FV method, the reconstruction operator R is constructed on each stencil independently and satisfies that $R\Pi_h^n$ is in some sense *polynomial preserving*. Specifically, for each element K and its corresponding stencil S_K, we require that for all $p \in P^N(S_K)$

$$\left((R\Pi_h^n)|_{S_K} p\right)\Big|_K = p|_K. \tag{14}$$

This requirement allows us to study approximation properties of R using the Bramble–Hilbert technique as in the standard finite element method. It seems that the simplest construction of R is to take the piecewise P^n function $\Pi_h^n p$ and perform an $L^2(S_K)$-projection onto $P^N(S_K)$. However such a procedure does not preserve p, as we see from the following:

Example Let $K_1 = [0, 1], K = K_2 = [1, 2], K_3 = [2, 3]$, hence $S_K = [0, 3]$. If we take $p(x) = x^2$ and project onto the space of piecewise constants, we get $(\Pi_h^0 p)|_{K_1} = \frac{1}{3}, (\Pi_h^0 p)|_{K_2} = \frac{7}{3}$ and $(\Pi_h^0 p)|_{K_3} = \frac{19}{3}$. If we take this piecewise constant function and perform an $L^2([0, 3])$-projection onto $P^2([0, 3])$, we obtain the function $\frac{40}{81}x^2 + \frac{32}{27}x - \frac{7}{27}$, whose graph on $[0, 3]$ is similar to that of the original function x^2, however we have failed to reconstruct it exactly.

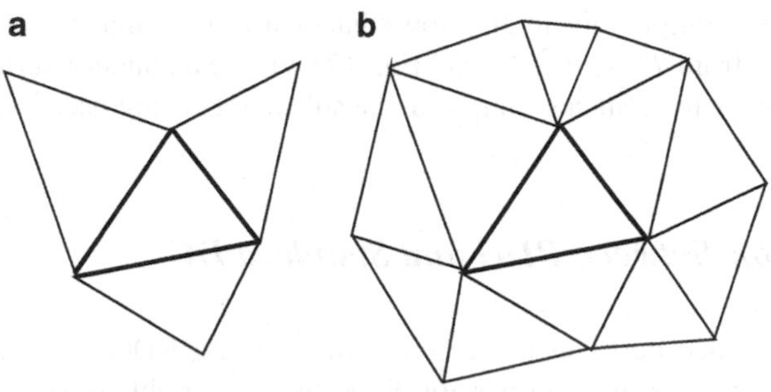

Fig. 1 (**a**) FV stencil for (**a**) linear and (**b**) quadratic reconstruction

Now we shall describe the construction of R satisfying (14). For simplicity, we shall deal with the 1D case. Let $K \in \mathscr{T}_h$, then the reconstruction stencil S_K consisting of the von Neumann neighborhood of K contains three elements K_1, K_2, K_3 numbered from left to right, hence $K_2 = K$. We denote by $\Pi^n_{K_i}$ the $L^2(K_i)$–projection onto the space $P^n(K_i)$, which is a restriction of the operator Π^n_h defined on $\mathscr{T}_h$. The goal is to construct a reconstruction operator $R_K : \bigoplus_{i=1}^3 P^n(K_i) \to P^N(S_K)$ such that (14) is satisfied, i.e.

$$R_K\big(\Pi^n_{K_1} p, \Pi^n_{K_2} p, \Pi^n_{K_3} p\big) = p, \quad \forall p \in P^N(S_K). \tag{15}$$

Let $p(x) = \sum_{j=0}^N \beta_j \Phi_j(x)$, where $\Phi_j, j = 0, \ldots, N$ is some basis of $P^N(S_K)$. Let $\Pi^n_{K_i} p = \sum_{j=0}^n \alpha_j^{K_i} \varphi_j^{K_i}(x), i = 1, 2, 3$, where $\varphi_j^{K_i}, j = 0, \ldots, N$ is some basis of $P^n(K_i)$. By definition, $(\Pi^n_{K_i} p, \varphi) = (p, \varphi)$ for all $\varphi \in P^n(K_i)$, thus

$$\sum_{j=0}^N \alpha_j^{K_i} (\varphi_j^{K_i}, \varphi_k^{K_i}) = \sum_{j=0}^N \beta_j (\Phi_j, \varphi_k^{K_i}) \quad \forall i = 1, 2, 3, \ \forall k = 0, \ldots, n. \tag{16}$$

Given $\alpha_j^{K_i}$ for all $i = 1, 2, 3$ and $j = 0, \ldots, n$, i.e. given the $L^2(K_i)$-projections of p or some other general function f, Eq. (16) represents a system of linear algebraic equations for unknowns $\beta_j, j = 0, \ldots, N$. If $3 * (n + 1) = N + 1$, i.e. $N = 3n + 2$, this is a system with a square matrix which can be solved for β_j and the reconstruction operator is then given by

$$R_K\big(\Pi^n_{K_1} p, \Pi^n_{K_2} p, \Pi^n_{K_3} p\big) = \sum_{j=0}^N \beta_j \Phi_j(x). \tag{17}$$

Clearly, if one constructs R_K as in (17), condition (14) will be satisfied. We note that (16) can be simplified if $\{\varphi_j^{K_i}\}_{j=0}^n$ is an orthonormal basis.

The presented construction can be straightforwardly generalized to higher dimensions, but the resulting system of linear equations will not have a square matrix in general. A direct computation in 2D shows that on a von Neumann neighborhood we can reconstruct from P^n to P^{2n+1} and Eq. (17) will be an underdetermined system with n free parameters. Such systems can be solved e.g. using pseudoinverses.

3.2 Relation Between RDG and Standard DG

The only difference between the DG scheme (4) and RDG scheme (12) is the presence of the reconstruction operator R in the first variable of $b_h(\cdot, \cdot)$. While the error analysis of (4) is well understood (at least for convection-diffusion problems [3]), the analysis of (12) or (13) poses a new challenge. The problem lies in the fact that we cannot test (12) with $\varphi_h^n := Ru_h^{n,k}$ or something similar, since $Ru_h^{n,k} \notin S_h^n$. Therefore, we need to establish a relation between (12) and Nth order DG, instead of only nth order DG.

Definition 5 (Auxiliary problem) We seek $\tilde{u}_h^{N,k} \in S_h^N$ such that

$$\left(\frac{\tilde{u}_h^{N,k+1} - \tilde{u}_h^{N,k}}{\tau_k}, \varphi_h^N\right) + b_h\big(R\Pi_h^n \tilde{u}_h^{N,k}, \varphi_h^N\big) = 0, \quad \forall \varphi_h^N \in S_h^N, \ k = 0, 1, \ldots, \tag{18}$$

where $\tilde{u}_h^{N,0}$ is an S_h^N approximation of the initial condition u^0.

The auxiliary problem has a direct connection to Nth order DG, as can be seen from the following lemma proved in [5].

Lemma 2 *Let $u_h^{n,0} = \Pi_h^n \tilde{u}_h^{N,0}$. Then $u_h^{n,k} \in S_h^n$, the solution of (13) and the solution $\tilde{u}_h^{N,k} \in S_h^N$ of (18) satisfy*

$$u_h^{n,k} = \Pi_h^n \tilde{u}_h^{N,k}, \quad \forall k = 0, 1, \cdots. \tag{19}$$

As a corollary, error estimates for the auxiliary problem imply error estimates for the RDG scheme. Problem (18) is basically the Nth order DG scheme with the operator $R\Pi_h^n$ in the first variable of $b_h(\cdot,\cdot)$. Therefore, sufficient knowledge of the properties of $R\Pi_h^n$ (which is polynomial preserving) and standard DG error estimates would imply estimates for the RDG scheme.

4 Numerical Experiments

We consider the advection of a 1D sine wave with unit speed on uniform meshes with a periodic boundary condition. Experimental orders of accuracy α in various norms on meshes with N elements are given in Tables 1 and 2. Here $e_h = u - Ru_h^n$ at t corresponding to ten periods. The increase in accuracy due to reconstruction

Table 1 1D advection of sine wave, P^1 RDG scheme with P^5 reconstruction

N	$\|\|e_h\|\|_{L^\infty(\Omega)}$	α	$\|\|e_h\|\|_{L^2(\Omega)}$	α	$\|e_h\|_{H^1(\Omega,\mathcal{T}_h)}$	α
4	5.82E−03	–	3.49E−03	–	3.65E−02	–
8	7.53E−05	6.27	4.43E−05	6.30	1.06E−03	5.11
16	9.07E−07	6.38	5.95E−07	6.22	3.58E−05	4.89
32	1.82E−08	5.64	8.70E−09	6.10	1.16E−06	4.95
64	3.41E−10	5.74	1.33E−10	6.03	3.67E−08	4.98

Table 2 1D advection of sine wave, P^2 RDG scheme with P^8 reconstruction

N	$\|\|e_h\|\|_{L^\infty(\Omega)}$	α	$\|\|e_h\|\|_{L^2(\Omega)}$	α	$\|e_h\|_{H^1(\Omega,\mathcal{T}_h)}$	α
4	2.90E−03	–	1.85E−03	–	1.63E−02	–
8	7.75E−06	8.55	3.56E−06	9.02	1.03E−04	7.30
16	2.10E−08	8.53	6.64E−09	9.07	4.34E−07	7.89
32	7.21E−11	8.18	4.02E−11	7.37	1.76E−09	7.94

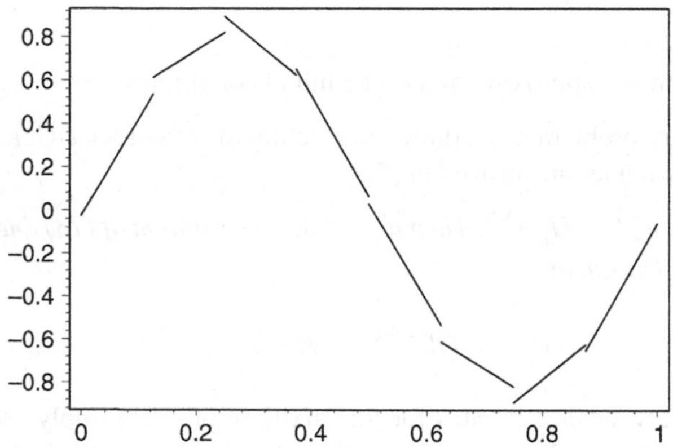

Fig. 2 Standard DG solution, P1 elements

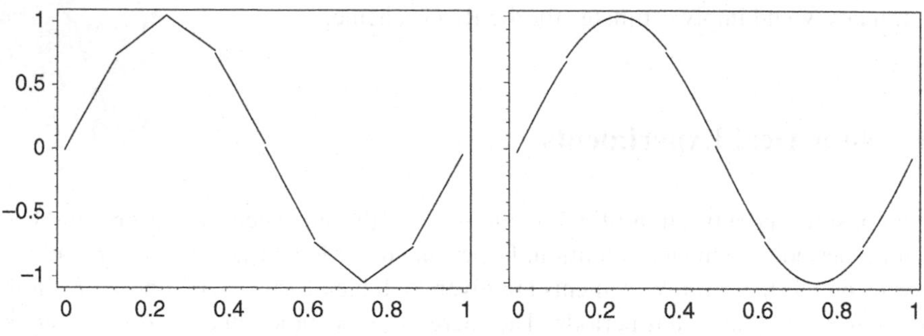

Fig. 3 RDG solution, P1 elements (*left*) with P5 reconstruction (*right*)

is clearly visible. In Fig. 2, we have plotted the P^1 standard DG solution for the considered problem at time $t = 5$. In Fig. 3, the P^1 RDG solution is plotted along with its P^5 reconstructions at $t = 5$. We have chosen a very crude mesh consisting of only eight elements in order to show the increased quality of the RDG scheme.

Conclusions

We have presented a possible generalization of higher-order reconstruction operators as used in the FV method to the DG method. Such possibilities were already considered in [1]. The resulting scheme has many advantages over standard DG and FV schemes:

- To increase the order of the scheme, the reconstruction stencil need not be enlarged, we may simply increase the order of the underlying DG scheme.
- Test functions are from the lower order space, hence more efficient quadratures may be used than in the corresponding higher order DG scheme.
- Since the RDG scheme is basically a lower order DG scheme with higher order reconstruction, the CFL condition is less restrictive than for the corresponding higher order DG scheme.

Acknowledgement This work is a part of the research project P201/11/P414 of the Czech Science Foundation (CSF) and partially supported by project P201/13/00522S of CSF. The author is a researcher in the University Centre for Mathematical Modelling, Applied Analysis and Computational Mathematics (Math MAC).

References

1. M. Dumbser, D. Balsara, E.F. Toro, C.D. Munz, A unified framework for the construction of one-step finite volume and discontinuous Galerkin schemes. J. Comput. Phys. **227**, 8209–8253 (2008)
2. M. Feistauer, J. Felcman, I. Straškraba, *Mathematical and Computational Methods for Compressible Flow* (Oxford University Press, Oxford, 2003)
3. M. Feistauer, V. Kučera, Analysis of the DGFEM for nonlinear convection-diffusion problems. Electron. Trans. Numer. Anal. **32**(1), 33–48 (2008)
4. D. Kröner, *Numerical Schemes for Conservation Laws* (Wiley und Teubner, Chichester/New York, 1996)
5. V. Kučera, Higher-order reconstruction: from finite volumes to discontinuous galerkin, in *Proceedings of FVCA 6, Finite Volumes for Complex Applications VI Problems and Perspectives*, Prague (Springer, Berlin/Heidelberg, 2011), pp. 613–621
6. R.J. LeVeque, *Finite Volume Methods for Hyperbolic Problems* (Cambridge University Press, Cambridge, 2002)

Adaptive Discontinuous Galerkin Methods for Nonlinear Diffusion-Convection-Reaction Equations

Bulent Karasözen, Murat Uzunca, and Murat Manguoğlu

Abstract In this work, we apply the adaptive discontinuous Galerkin method (DGAFEM) to the convection dominated nonlinear, quasi-steady state convection diffusion reaction equations. We propose an efficient algorithm to solve the sparse linear systems iteratively arising from the discretized nonlinear equations. Numerical examples demonstrate the effectiveness of the DGAFEM to damp the spurious oscillations for the convection dominated nonlinear equations.

1 Introduction

Many engineering problems such as chemical reaction processes, heat conduction, nuclear reactors, population dynamics are governed by coupled convection-diffusion-reaction partial differential equations (PDEs) with nonlinear source or sink terms. It is a significant challenge to solve such PDEs numerically when they are convection/reaction-dominated, which is the case in our study. As a model problem, we consider the coupled quasi-stationary model arising from the time discretization of the time dependent nonlinear diffusion-convection-reaction equations

$$\alpha u_i - \epsilon_i \Delta u_i + \mathbf{b}_i \cdot \nabla u_i + r_i(\mathbf{u}) = f_i \quad \text{in } \Omega, \quad u_i = g_i \quad \text{on } \Gamma, \quad i = 1, \ldots, m \tag{1}$$

B. Karasözen (✉)
Department of Mathematics and Institute of Applied Mathematics, Middle East Technical University, 06800 Ankara, Turkey
e-mail: bulent@metu.edu.tr

M. Uzunca
Department of Mathematics, Middle East Technical University, 06800 Ankara, Turkey
e-mail: uzunca@metu.edu.tr

M. Manguoğlu
Department of Computer Engineering, Institute of Applied Mathematics, Middle East Technical University, 06800 Ankara, Turkey
e-mail: manguoglu@ceng.metu.edu.tr

A. Abdulle et al. (eds.), *Numerical Mathematics and Advanced Applications ENUMATH 2013*, Lecture Notes in Computational Science and Engineering 103,
DOI 10.1007/978-3-319-10705-9_8

with Ω is a bounded, open, convex domain in $\mathbb{R}^2$ with boundary $\partial\Omega = \Gamma$, $0 < \epsilon_i \ll 1$ are the diffusivity constants, $f_i \in L^2(\Omega)$ are the source functions, $\mathbf{b}_i \in \left(W^{1,\infty}(\Omega)\right)^2$ are the velocity fields, $g_i \in H^{1/2}(\Gamma_D)$ are the Dirichlet boundary conditions and $\mathbf{u}(x,t) = (u_1, \ldots, u_m)^T$ denotes the vector of unknown solutions. The coefficients of the linear reaction terms, $\alpha > 0$, stand for the temporal discretization, corresponding to $1/\Delta t$, where Δt is the discrete time step. For the uniqueness of the solution of (2), we assume that the nonlinear reaction terms are bounded, locally Lipschitz continuous and monotone, i.e. satisfy the following conditions [3]

$$r_i \in C^1(\mathbb{R}_0^+), \quad r_i(0) = 0, \quad r_i'(s) \geq 0, \quad \forall s \geq 0, \quad s \in \mathbb{R}$$

Such models describe chemical processes and they are strongly coupled as an inaccuracy in one unknown affects all the others. Hence, an efficient numerical approximation of these systems is needed. For the convection/reaction-dominated problems, the standard Galerkin finite element methods are known to produce spurious oscillations, especially in the presence of sharp fronts in the solution, on boundary and interior layers. In contrast to standard Galerkin conforming finite element methods, DG methods produce stable discretizations without the need for stabilization strategies, and overcome the spurious oscillations due to the convection domination. For linear convection dominated problems, the streamline upwind Petrov-Galerkin(SUPG) method is capable of stabilizing the unphysical oscillations [3, 4]. Nevertheless, in nonlinear convection dominated problems, spurious oscillations are still present in crosswind direction. Therefore, SUPG is used with the anisotropic shock capturing technique (SUPG-SC) for reactive transport problems [3,4].

Similar to the stabilized conforming finite elements, discontinuous Galerkin finite element methods (DGFEMs) damp the unphysical oscillations for linear convection dominated problems. In [9], several nonlinear convection dominated problems of type (1) are solved with DGFEM and DG-SC, discontinuous Galerkin method with the shock-capturing technique. The main advantages of DGFEMs are the flexibility in handling non-matching grids and in designing hp-refinement strategies. An important drawback is that the resulting linear systems are more dense than the ones in continuous finite elements and ill-conditioned. The condition number grows rapidly with the number of elements and with the penalty parameter. Therefore, efficient solution strategies such as preconditioning are required to solve the linear systems. In this paper, an adaptive discontinuous Galerkin method (DGAFEM) is developed for the convection dominated nonlinear problems of type (1) using the modification of a posteriori error estimates for linear convection dominated problems in [6]. We show the effectiveness and accuracy of DGAFEM capturing boundary and internal layers very sharply and without significant oscillations.

In the next two sections, we give the DGFEMs discretization and describe the residual based adaptivity for nonlinear diffusion-convection-reaction problems.

Section 4 deals with an efficient solution technique to handle the linear systems arising from the DGAFEM. In Sect. 5, we demonstrate on two examples the efficiency of the adaptivity for handling the sharp layers.

2 DG Discretization

The weak formulation of the scalar equation ($m = 1$) of (1) reads as

$$\int_{\Omega}(\epsilon\nabla u\cdot\nabla v+\mathbf{b}\cdot\nabla uv+\alpha uv)dx+\int_{\Omega}r(u)vdx=\int_{\Omega}fvdx \quad \forall v\in V \tag{2}$$

where the solution space U and the test function space V are given by

$$U=\{u\in H^1(\Omega)\ :\ u=g \text{ on } \Gamma\},\quad V=\{v\in H^1(\Omega)\ :\ v=0 \text{ on } \Gamma\}$$

The variational form of the scalar equation (1) is discretized by the symmetric discontinuous interior penalty Galerkin (SIPG) method with upwinding for the convection part [1,2]

$$a_h(u_h,v_h)+b_h(u_h,v_h)=l_h(v_h)\ ,\qquad \forall v_h\in V_h\subset H^1(\Omega) \tag{3}$$

$$\begin{aligned}
a_h(u_h,v_h)=&\sum_{K\in\xi_h}\int_K\epsilon\nabla u_h\cdot\nabla v_h dx+\sum_{K\in\xi_h}\int_K(\mathbf{b}\cdot\nabla u_h+\alpha u_h)v_h dx\\
&-\sum_{e\in\Gamma_0\cup\Gamma}\int_e\{\epsilon\nabla v_h\}\cdot[u_h]ds-\sum_{e\in\Gamma_0\cup\Gamma}\int_e\{\epsilon\nabla u_h\}\cdot[v_h]ds\\
&+\sum_{K\in\xi_h}\int_{\partial K^-\setminus\Gamma}\mathbf{b}\cdot\mathbf{n}(u_h^{out}-u_h^{in})v_h ds-\sum_{K\in\xi_h}\int_{\partial K^-\cap\Gamma^-}\mathbf{b}\cdot\mathbf{n}u_h^{in}v_h ds\\
&+\sum_{e\in\Gamma_0\cup\Gamma}\frac{\sigma\epsilon}{h_e}\int_e[u_h]\cdot[v_h]ds,\\
b_h(u_h,v_h)=&\sum_{K\in\xi_h}\int_K r(u_h)v_h dx,\\
l_h(v_h)=&\sum_{K\in\xi_h}\int_K fv_h dx+\sum_{e\in\Gamma}\int_e g\left(\frac{\sigma\epsilon}{h_e}v_h-\epsilon\nabla v_h\cdot\mathbf{n}\right)ds\\
&-\sum_{K\in\xi_h}\int_{\partial K^-\cap\Gamma^-}\mathbf{b}\cdot\mathbf{n}gv_h ds,
\end{aligned}$$

where the finite dimensional solution and test function spaces are the same. ∂K^- and Γ^- denote the inflow parts to an element boundary ∂K and the domain boundary Γ, respectively. The jump and average terms for u_h and v_h across the edges are denoted by $[\cdot]$ and $\{\cdot\}$, respectively. The parameter $\sigma \in \mathbb{R}_0^+$ is called the penalty parameter which should be sufficiently large for SIPG [6].

3 Adaptivity

We apply the residual based adaptive strategy in [6] which is robust, i.e. independent of the Péclet number, for linear diffusion-convection equations. We include in the a posteriori error estimates the nonlinear reaction terms as local contributions to the cell residuals and not to the interior/boundary edge residuals [Chp. 5.1.4, [8]]. Let the constant $\kappa \geq 0$ satisfies

$$\alpha(x) - \frac{1}{2}\nabla \cdot \mathbf{b}(x) \geq \kappa \,, \qquad \| - \nabla \cdot \mathbf{b} + \alpha \|_{L^\infty(\Omega)} \leq \kappa^* \kappa$$

for a non-negative κ^*, to easily have the efficiency of the a posteriori error estimator. We define the local error indicator for each element $K \in \xi_h$

$$\eta_K^2 = \eta_{R_K}^2 + \eta_{E_K^0}^2 + \eta_{E_K^D}^2,$$

$$\begin{aligned}
\eta_{R_K}^2 &= \rho_K^2 \| f - \alpha u_h + \epsilon \Delta u_h - \mathbf{b} \cdot \nabla u_h - r(u_h) \|_{L^2(K)}^2, \\
\eta_{E_K^0}^2 &= \sum_{e \in \partial K \cap \Gamma_0} \left(\frac{1}{2} \epsilon^{-\frac{1}{2}} \rho_e \| [\epsilon \nabla u_h] \|_{L^2(e)}^2 + \frac{1}{2} (\frac{\epsilon \sigma}{h_e} + \kappa h_e + \frac{h_e}{\epsilon}) \| [u_h] \|_{L^2(e)}^2 \right), \\
\eta_{E_K^D}^2 &= \sum_{e \in \partial K \cap \Gamma} (\frac{\epsilon \sigma}{h_e} + \kappa h_e + \frac{h_e}{\epsilon}) \| g - u_h \|_{L^2(e)}^2,
\end{aligned}$$

with the weights ρ_K and ρ_e on an element K are defined for $\kappa \neq 0$

$$\rho_K = \min\{h_K \epsilon^{-\frac{1}{2}}, \kappa^{-\frac{1}{2}}\}, \; \rho_e = \min\{h_e \epsilon^{-\frac{1}{2}}, \kappa^{-\frac{1}{2}}\}.$$

When $\kappa = 0$, we set $\rho_K = h_K \epsilon^{-\frac{1}{2}}$ and $\rho_e = h_e \epsilon^{-\frac{1}{2}}$. Our adaptive algorithm is based on the standard adaptive finite element (AFEM) iterative loop: SOLVE → ESTIMATE → MARK → REFINE. The mesh is marked at each iteration using the Dörfler strategy and refined using the longest edge bisection method [5]. For coupled problems, the elements in the set of union of each component are refined.

4 Efficient Solution of Linear Systems

Because the stiffness matrices obtained by DGFEM become ill-conditioned and more dense with increasing polynomial degree [2], several preconditioners are developed for an efficient and accurate solution of linear diffusion-convection equations under DG discretization. Here we apply the matrix reordering and partitioning technique in [7], which uses the largest eigenvalue and corresponding eigenvector of the Laplacian matrix. This reordering reflects very well the block structure of the underlying sparse matrix.

The solution to (3) has the form $u_h = \sum_{i=1}^{dof} U_i \phi_i$ where ϕ_i's are the basis functions spanning the DGFEM space V_h, and U_i's are the unknown coefficients. Then, the discrete residual of (3) can be given as

$$R(U) = SU + h(U) - L \tag{4}$$

where $U = (U_1, U_2, \ldots, U_{dof})^T$ is the vector of unknown coefficients, $S \in \mathbb{R}^{dof \times dof}$ is the stiffness matrix with the entries $S_{ij} = a_h(\phi_j, \phi_i)$, $h \in \mathbb{R}^{dof}$ is the vector function of U with the entries $h_i = b_h(u_h, \phi_i)$ and $L \in \mathbb{R}^{dof}$ is the vector to the linear form with $L_i = l_h(\phi_i)$, $i, j = 1, 2, \ldots, dof$. We start with a non-zero initial vector U^0. The nonlinear system of equations (4) are solved by Newton-Raphson method. The linear system arising from ith-Newton-Raphson iteration step has the form $Jw^i = -R^i$, where J is the Jacobian matrix to $R(U^0)$ (i.e. $J = S + h'(U^0)$ and it remains unchanged among the iteration steps), $w^i = U^{i+1} - U^i$ is the Newton correction, and R^i denotes the residual of the system at U^i ($R^i = R(U^i)$). Next, we construct a permutation matrix P for the Jacobian matrix J as described in [7]. Then, we apply the permutation matrix P to obtain the permuted system $Nw = b$ where $N = PJP^T$, $w = Pw^i$ and $b = -PR^i$. After solving the permuted system, the solution of the unpermuted linear system can be obtained by applying the inverse permutation, $w^i = P^T w$. The permuted and partitioned linear system can be solved via the block LU factorization in which the coefficient matrix has the form

$$N = \begin{bmatrix} A & B \\ C^T & D \end{bmatrix} = \begin{bmatrix} A & 0 \\ C^T & S \end{bmatrix} \begin{bmatrix} I & U \\ 0 & I \end{bmatrix}$$

where $U = A^{-1}B$ and S is the Schur complement matrix: $S = D - C^T U$. For the right hand side vector $b = (b_1, b_2)^T$ and the reordered solution $w = (w_1, w_2)^T$, solution of the block LU factorized system can be obtained in three steps as follows

$$Az = b_1, \quad Sw_2 = b_2 - C^T z, \quad w_1 = z - Uw_2 \tag{5}$$

with both the matrices A and S are well-conditioned compared to the coefficient matrix of the unpermuted system shown in Table 1.

Table 1 Condition numbers of the stiffness matrices corresponding to the systems obtained by the model in Example 5.1 on a uniform mesh

Degree	1	2	3	4
dof	24,576	49,152	81,192	122,880
J	299.1	660.8	1,596.2	3,399.0
S	139.9	279.1	911.7	1,485.7
A	5.3	18.3	37.7	79.7

5 Numerical Results

Example 5.1 We consider the problem in [3] on $\Omega = (0,1)^2$ with $\epsilon = 10^{-6}$, $\mathbf{b} = \frac{1}{\sqrt{5}}(1,2)^T$, $\alpha = 1$ and $r(u) = u^2$. The source function f and the Dirichlet boundary condition are chosen so that $u(x,y) = \frac{1}{2}\left(1 - \tanh \frac{2x_1 - x_2 - 0.25}{\sqrt{5\epsilon}}\right)$ is the exact solution. The problem is characterized by an internal layer of thickness $\mathcal{O}(\sqrt{\epsilon} \mid \ln \epsilon \mid)$ around $2x_1 - x_2 = \frac{1}{4}$. This problem was solved using SUPG-SC in [3] and SIPG-SC in [9]. Similar to those results, the mesh is locally refined by DGAFEM around the interior layer and the spurious solutions are damped out in Fig. 1, right similar to [3] using SUPG-SC, in [9] with SIPG-SC. On adaptively and uniformly refined meshes, from the Fig. 2, left, it is evident that the adaptive meshes save substantial computing time. On uniform meshes, the SIPG is slightly more accurate than the SUPG-SC in [3]. The error reduction by increasing the degree of the polynomials is remarkable on finer adaptive meshes (Fig. 2, right). For solving the sparse linear systems, we present the results for the *BiCGStab* iterative method of MATLAB with the stopping criterion as $\|r_k\|_2/\|r_0\|_2 \leq tol$ for $tol = 10^{-4}$ (r_i is the residual of the corresponding linear system at the ith iteration) applied to the original unpermuted system and Schur complement system with and without preconditioner. As a preconditioner, the incomplete LU factorization of the Schur complement matrix S (ILU(S)) is used. The linear systems with the coefficient matrix A are solved directly. Table 2 shows that solving the problem by block LU factorization where the Schur complement system is solved iteratively using the preconditioner ILU(S) is the fastest and has the least number of iterations. We use an adaptive mesh by quadratic elements with dof 85,488 at the final refinement level of the 16 refinement levels. The time to obtain the reordered matrix N and computing the permutation in sum among the refinement levels takes 45.18 s, whereas, it takes 1.42 s to compute the Schur complement matrix S and ILU(S), on a PC with Intel Core-i7 processor and 8 GB 1066 MHz DDR3 RAM.

Remark *We note that since the Jacobian matrix does not change during the nonlinear iterations, the permutation, the Schur complement matrix and ILU(S) are computed only once for each adaptive refinement level. In Table 2, we give the number of Newton-Raphson iterations, the average number of BiCGStab iterations for each adaptive mesh refinement level and the total time to solve the problem*

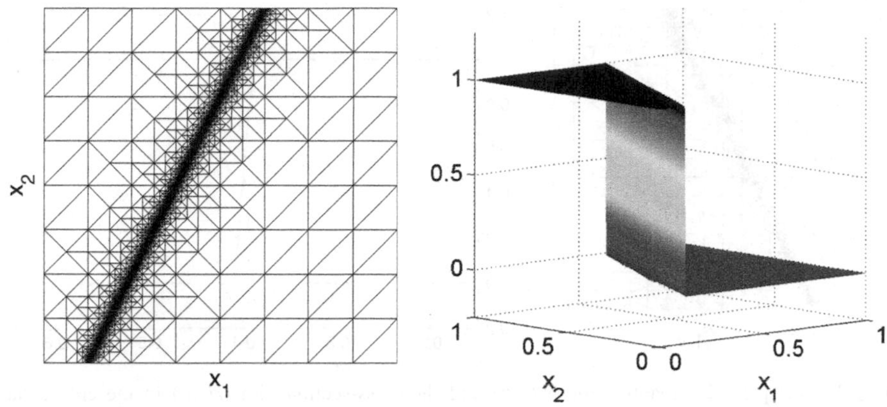

Fig. 1 Example 5.1, Adaptive mesh (*left*) and adaptive solution (*right*), quadratic elements with dof 85,488

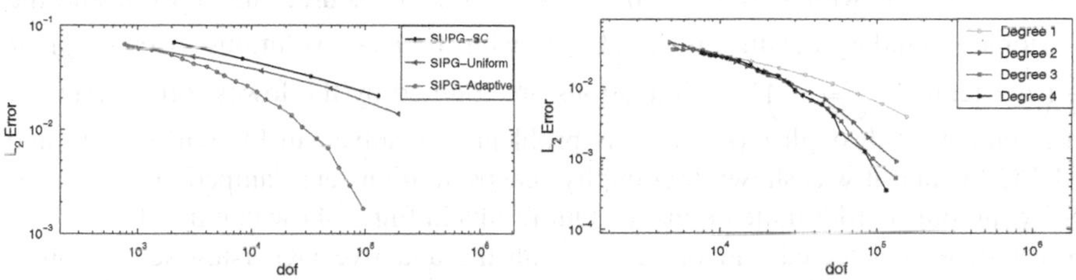

Fig. 2 Example 5.1, Global errors: comparison of the methods by quadratic elements (*left*), adaptive DG for polynomial degrees 1–4 (*right*)

Table 2 Example 5.1, Efficiency results for the sparse linear solver technique for adaptive mesh refinement levels

Linear solver	# Newton its.	# BiCGStab its.	Time (s)
BiCGStab w/o prec. (Unpermuted)	10–11	49–1,143	879.6
Block LU + (BiCGStab w/o prec.)	10–12	33–1,162	454.4
Block LU + (BiCGStab w/ prec. ilu(S))	10–14	4–82	144.9

just including the computation time for the reordered matrix N, the permutation P, Schur complement matrix S, $ILU(S)$ and solving the linear systems among all adaptive mesh refinement levels.

Example 5.2 We consider the two component quasi-steady problem from [4]

$$u_i + \mathbf{b} \cdot \nabla u_i - \nabla \cdot (\epsilon \nabla u_i) + u_1 u_2 = f_i \qquad i = 1, 2$$

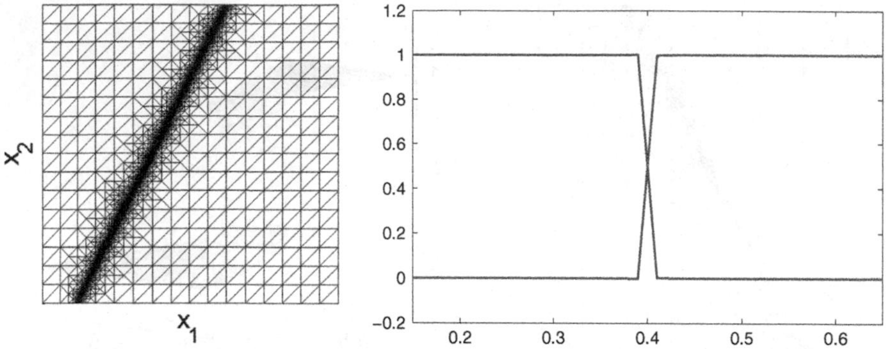

Fig. 3 Example 5.2, Adaptive mesh (*left*) and the cross-section plot (*right*) in the crosswind direction $x_1 + 2x_2 = 1.5$, quadratic elements with dof 144678

on $\Omega = (0,1)^2$ with $\epsilon = 10^{-8}$, $\mathbf{b} = \frac{1}{\sqrt{5}}(1,2)^T$. The source functions f_i and the Dirichlet boundary conditions are chosen with the exact solutions $u_{1,2}(x_1, x_2) = \frac{1}{2}\left(1 \pm \tanh \frac{2x_1 - x_2 - 0.25}{\sqrt{5\epsilon}}\right)$. The equations are coupled by the lowest order terms of the unknowns through $r_i(u, x)$. This problem was solved in [4] with SUPG and SUPG-SC and it was shown that unphysical oscillations are damped using SUPG-SC with fourth order finite elements. Our results in Fig. 3 show that the sharp fronts are very well detected and preserved with the adaptive DG using second order elements. As a results, there is no over or under prediction, and artificial mixing due to discretization will not occur.

We have shown that DGAFEM with the sparse linear solver is an efficient method for solving nonlinear convection dominated problems accurately and avoids the design of the parameters in the shock capturing technique as for the SUPG-SC and DG-SC. The MATLAB programs can be obtained from `http://www.ceng.metu.edu.tr/m̃anguoglu/MatLab.zip`.

Acknowledgement

This work has been partially supported by Turkish Academy of Sciences Distinguished Young Scientist Award TÜBA-GEBIP/2012-19, TÜBITAK Career Award EEAG111E238 and METU BAP-07-05-2013-004.

References

1. D. Arnold, F. Brezzi, B. Cockburn, L. Marini, Unified analysis of discontinuous Galerkin methods for elliptic problems. SIAM J. Numer. Anal. **39**, 1749–1779 (2002)

2. B. Ayuso, L.D. Marini, Discontinuous Galerkin methods for advection-diffusion-reaction problems. SIAM J. Numer. Anal. **47**, 1391–1420 (2009)
3. M. Bause, Stabilized finite element methods with shock-capturing for nonlinear convection-diffusion-reaction models, in *Numerical Mathematics and Advanced Applications*, ed. by G. Kreiss, P. Lötstedt, A. Møalqvist, M. Neytcheva (Springer, Berlin, 2010), pp. 125–134
4. M. Bause, K. Schwegler, Higher order finite element approximation of systems of convection-diffusion-reaction equations with small diffusion. J. Comput. Appl. Math. **246**, 52–64 (2013)
5. L. Chen, *i*FEM: an innovative finite element method package in MATLAB. Technical report, Department of Mathematics, University of California, Irvine, 2008
6. D. Schötzau, L. Zhu, A robust a-posteriori error estimator for discontinuous Galerkin methods for convection-diffusion equations. Appl. Numer. Math. **59**, 2236–2255 (2009)
7. O. Tarı, M. Manguoğlu, A new sparse matrix reordering scheme using the largest eigenvector of the graph Laplacian. Numerical Linear Algebra with Applications, in review. http://people.maths.ox.ac.uk/ekertl/PRECON13/talks/Talk_Tari_Manguoglu.pdf
8. R. Verfürth, *A Posteriori Error Estimation Techniques for Finite Element Methods* (Oxford University Press, Oxford, 2013)
9. H. Yücel, M. Stoll, P. Benner, Discontinuous Galerkin finite element methods with schock-capturing for nonlinear convection dominated models. Comput. Chem. Eng. **58**, 278–287 (2013)

Harmonic Complete Flux Schemes for Conservation Laws with Discontinuous Coefficients

J.H.M. ten Thije Boonkkamp, L. Liu, J. van Dijk, and K.S.C. Peerenboom

Abstract In this paper we discuss several complete flux schemes for advection-diffusion-reaction problems. We consider both scalar equations as well as systems of equations. For the flux approximations in the latter case, we take into account the coupling between the constituent equations. We study conservation laws with discontinuous diffusion matrix/coefficient and show that the (matrix) harmonic average should be employed in the expressions for the numerical fluxes. The vectorial harmonic complete flux schemes are validated for a test problem.

1 Introduction

Conservation laws are ubiquitous in continuum physics. They occur in disciplines like combustion theory, plasma physics, transport in porous media etc. These conservation laws are often of advection-diffusion-reaction type, describing the interplay between different processes such as advection or drift, (multi-species) diffusion and chemical reactions or impact ionization.

Advection-diffusion-reaction problems are usually quite complex and require sophisticated numerical solution methods. In this contribution we discuss numerical flux approximations for two special cases: first, a scalar conservation law with a rapidly varying or even discontinuous diffusion coefficient, and second, a system of conservation laws coupled through a diffusion matrix. We also allow the diffusion matrix to be discontinuous. The second problem is typical for multi-species diffusion in mixtures or plasmas; see for example [3] for a detailed account.

J.H.M. ten Thije Boonkkamp (✉)
Department of Mathematics and Computer Science, Eindhoven University of Technology, P. O. Box 513, 5600 MB Eindhoven, The Netherlands
e-mail: j.h.m.tenthijeboonkkamp@tue.nl

L. Liu • J. van Dijk • K.S.C. Peerenboom
Department of Applied Physics, Eindhoven University of Technology, P.O. Box 513, 5600 MB Eindhoven, The Netherlands
e-mail: leiliucn@hotmail.com; j.v.dijk@tue.nl; k.s.c.peerenboom@tue.nl

A. Abdulle et al. (eds.), *Numerical Mathematics and Advanced Applications ENUMATH 2013*, Lecture Notes in Computational Science and Engineering 103,
DOI 10.1007/978-3-319-10705-9_9

Due to the nonlinear dependency of the diffusion process on pressure, temperature and plasma composition, diffusion matrices can vary rapidly in space.

Therefore, we consider the one-dimensional scalar model problem $\mathrm{d}f/\mathrm{d}x = s$, where f is the (advection-diffusion) flux and s the source term. The flux f is given by

$$f = u\varphi - \varepsilon\frac{\mathrm{d}\varphi}{\mathrm{d}x}, \tag{1}$$

with u the advection velocity and $\varepsilon > 0$ the diffusion coefficient. The system counterpart reads $\mathrm{d}\boldsymbol{f}/\mathrm{d}x = \boldsymbol{s}$ with $\boldsymbol{f}$ the flux vector given by

$$\boldsymbol{f} = \boldsymbol{U}\boldsymbol{\varphi} - \mathscr{E}\frac{\mathrm{d}\boldsymbol{\varphi}}{\mathrm{d}x}, \tag{2}$$

and $\boldsymbol{s}$ the source term. In relation (2) $\boldsymbol{U}$ is the advection matrix, which is usually diagonal, and $\mathscr{E}$ is the diffusion matrix, which we assume symmetric positive definite. We consider equations with ε and $\mathscr{E}$ discontinuous.

The finite volume method is our discretization method of choice. Thus we cover the domain with a finite set of control volumes (cells) I_j of size Δx and choose the grid points x_j, where the unknown has to be approximated, in the cell centres. Consequently, we have $I_j = [x_{j-1/2}, x_{j+1/2}]$ with $x_{j+1/2} = \frac{1}{2}(x_j + x_{j+1})$. Integrating, for example, $\mathrm{d}\boldsymbol{f}/\mathrm{d}x = \boldsymbol{s}$ over I_j and applying the midpoint rule for the integral of $\boldsymbol{s}$, we obtain the discrete conservation law

$$\boldsymbol{F}_{j+1/2} - \boldsymbol{F}_{j-1/2} = \Delta x\, \boldsymbol{s}_j, \tag{3}$$

with $\boldsymbol{F}_{j+1/2}$ the numerical flux approximating $\boldsymbol{f}$ at the cell interface $x_{j+1/2}$ and $\boldsymbol{s}_j = \boldsymbol{s}(x_j)$. For the numerical flux we adopt the complete flux schemes developed in [4,5]. The complete flux schemes for $\boldsymbol{F}_{j+1/2}$ typically read

$$\boldsymbol{F}_{j+1/2} = \boldsymbol{\alpha}_{j+1/2}\boldsymbol{\varphi}_j - \boldsymbol{\beta}_{j+1/2}\boldsymbol{\varphi}_{j+1} + \Delta x(\boldsymbol{\gamma}_{j+1/2}\boldsymbol{s}_j + \boldsymbol{\delta}_{j+1/2}\boldsymbol{s}_{j+1}), \tag{4}$$

where $\boldsymbol{\varphi}_j$ denotes the numerical approximation of $\boldsymbol{\varphi}(x_j)$ and where the coefficient matrices $\boldsymbol{\alpha}_{j+1/2}$ etc. are piecewise constant and depend on $\boldsymbol{U}$ and $\mathscr{E}$. The goal of this paper is to extend the standard complete flux schemes to equations with discontinuous diffusion matrix/coefficient. We will deduce that the (matrix) harmonic average of the diffusion matrix/coefficient is required in the expressions for the numerical fluxes, which we collectively refer to as harmonic complete flux schemes.

We have organised our paper as follows. In Sect. 2 we modify the standard scalar complete flux scheme for piecewise constant diffusion coefficient ε. Next, in Sect. 3, we extend the scalar scheme to systems of conservation laws, taking into account the coupling between the constituent equations. In Sect. 4 we demonstrate the performance of the vectorial harmonic complete flux schemes, and finally we present conclusions in section "Concluding Remarks".

2 Numerical Approximation of the Scalar Flux

In this section we outline the complete flux scheme for the scalar equation, which is based on the integral representation of the flux. The derivation is a modification of the theory in [4].

The integral representation of the flux $f(x_{j+1/2})$ at the cell edge $x_{j+1/2}$ is based on the following model boundary value problem (BVP) for φ:

$$\frac{\mathrm{d}}{\mathrm{d}x}\Big(u\varphi - \varepsilon\frac{\mathrm{d}\varphi}{\mathrm{d}x}\Big) = s, \quad x_j < x < x_{j+1}, \tag{5a}$$

$$\varphi(x_j) = \varphi_j, \quad \varphi(x_{j+1}) = \varphi_{j+1}. \tag{5b}$$

We like to emphasize that $f(x_{j+1/2})$ corresponds to the solution of the inhomogeneous BVP (5), implying that $f(x_{j+1/2})$ not only depends on the advection-diffusion operator, but also on the source term s. It is convenient to introduce the variables $P(x)$, $p(x)$ and $S(x)$ for $x \in (x_j, x_{j+1})$ by

$$P(x) := \frac{u(x)\Delta x}{\varepsilon(x)}, \quad p(x) := \int_{x_{j+1/2}}^{x} \frac{u(\xi)}{\varepsilon(\xi)}\,\mathrm{d}\xi, \quad S(x) := \int_{x_{j+1/2}}^{x} s(\xi)\,\mathrm{d}\xi. \tag{6}$$

Here, $P(x)$ and $p(x)$ are the Peclet function and integral, respectively, generalizing the well-known (numerical) Peclet number. Integrating the differential equation $\mathrm{d}f/\mathrm{d}x = s$ from $x_{j+1/2}$ to $x \in [x_j, x_{j+1}]$ we get the integral balance $f(x) - f(x_{j+1/2}) = S(x)$. Using the definition of p in (6), it is clear that the flux can be rewritten as $f(x) = -\varepsilon(x)\mathrm{e}^{p(x)}\,\mathrm{d}\big(\varphi\,\mathrm{e}^{-p(x)}\big)/\mathrm{d}x$. Substituting this representation into the integral balance and integrating from x_j to x_{j+1} we find the following expressions for the flux:

$$f(x_{j+1/2}) = f^{\mathrm{h}}(x_{j+1/2}) + f^{\mathrm{i}}(x_{j+1/2}), \tag{7a}$$

$$f^{\mathrm{h}}(x_{j+1/2}) = \big(\mathrm{e}^{-p(x_j)}\varphi_j - \mathrm{e}^{-p(x_{j+1})}\varphi_{j+1}\big) \Big/ \int_{x_j}^{x_{j+1}} \varepsilon^{-1}(x)\mathrm{e}^{-p(x)}\,\mathrm{d}x, \tag{7b}$$

$$f^{\mathrm{i}}(x_{j+1/2}) = -\int_{x_j}^{x_{j+1}} \varepsilon^{-1}(x)\mathrm{e}^{-p(x)}S(x)\,\mathrm{d}x \Big/ \int_{x_j}^{x_{j+1}} \varepsilon^{-1}(x)\mathrm{e}^{-p(x)}\,\mathrm{d}x, \tag{7c}$$

where $f^{\mathrm{h}}(x_{j+1/2})$ and $f^{\mathrm{i}}(x_{j+1/2})$ are the homogeneous and inhomogeneous part, corresponding to the homogeneous and particular solution of (5), respectively.

Next, we assume that u is constant and ε is piecewise constant on $(x_j, x_{j+1}]$, i.e., $u(x) = \bar{u}_{j+1/2} := \frac{1}{2}\big(u_j + u_{j+1}\big)$ and

$$\varepsilon(x) = \begin{cases} \varepsilon_j & \text{if} \quad x_j < x \le x_{j+1/2}, \\ \varepsilon_{j+1} & \text{if} \quad x_{j+1/2} < x \le x_{j+1}. \end{cases} \tag{8}$$

Consequently, the function $p(x)$ is piecewise linear. Likewise, in agreement with the finite volume discretization, we take s piecewise constant. Substituting these approximations in the integral representation (7), and evaluating all integrals involved, we obtain the numerical flux:

$$F_{j+1/2} = F^{\mathrm{h}}_{j+1/2} + F^{\mathrm{i}}_{j+1/2}, \tag{9a}$$

$$F^{\mathrm{h}}_{j+1/2} = \frac{\tilde{\varepsilon}_{j+1/2}}{\Delta x}\big(B(-\bar{P}_{j+1/2})\varphi_j - B(\bar{P}_{j+1/2})\varphi_{j+1}\big), \tag{9b}$$

$$F^{\mathrm{i}}_{j+1/2} = \Delta x\big(\gamma_{j+1/2}s_j + \delta_{j+1/2}s_{j+1}\big), \tag{9c}$$

$$\gamma_{j+1/2} = \tfrac{1}{2}\frac{\psi\big(-\tfrac{1}{2}P_j\big)}{\mathrm{e}^{-\bar{P}_{j+1/2}} - 1}, \quad \delta_{j+1/2} = -\tfrac{1}{2}\frac{\psi\big(\tfrac{1}{2}P_{j+1}\big)}{\mathrm{e}^{\bar{P}_{j+1/2}} - 1}, \tag{9d}$$

where $\tilde{\varepsilon}_{j+1/2}$ is the harmonic average of ε and $\bar{P}_{j+1/2}$ the arithmetic average of P, defined by

$$\frac{1}{\tilde{\varepsilon}_{j+1/2}} := \frac{1}{2}\Big(\frac{1}{\varepsilon_j} + \frac{1}{\varepsilon_{j+1}}\Big), \quad \bar{P}_{j+1/2} := \frac{\bar{u}_{j+1/2}\Delta x}{\tilde{\varepsilon}_{j+1/2}}. \tag{10}$$

Furthermore, the functions $B(z)$ and $\psi(z)$ in (9) are defined by $B(z) = z/\big(\mathrm{e}^z - 1\big)$ and $\psi(z) = \big(\mathrm{e}^z - 1 - z\big)/z$. The flux approximation in (9) is referred to as the piecewise constant complete flux scheme (PCCFS).

Alternatively, we propose to replace the (local) Peclet numbers P_j and P_{j+1} in (9d) by the average Peclet number $\bar{P}_{j+1/2}$. This way we obtain

$$\gamma_{j+1/2} = W_1\big(\bar{P}_{j+1/2}\big), \quad \delta_{j+1/2} = -W_1\big(-\bar{P}_{j+1/2}\big), \tag{11}$$

with $W_1(z) = \big(\mathrm{e}^{-z/2} - 1 + z/2\big)/\big(z(1 - \mathrm{e}^{-z})\big)$. The corresponding flux approximation is referred to as the harmonic complete flux scheme (HCF).

3 Extension to Systems of Conservation Laws

In this section we extend the derivation of the complete flux schemes to systems of conservation laws. The derivation is a modification of the theory in [5] and is detailed in [2].

Analogous to the scalar case, we derive the expression for the numerical flux $\boldsymbol{F}_{j+1/2}$ from the following system BVP:

$$\frac{\mathrm{d}}{\mathrm{d}x}\Big(\boldsymbol{U}\boldsymbol{\varphi} - \mathscr{E}\frac{\mathrm{d}\boldsymbol{\varphi}}{\mathrm{d}x}\Big) = \boldsymbol{s}, \quad x_j < x < x_{j+1}, \tag{12a}$$

$$\boldsymbol{\varphi}(x_j) = \boldsymbol{\varphi}_j, \quad \boldsymbol{\varphi}(x_{j+1}) = \boldsymbol{\varphi}_{j+1}, \tag{12b}$$

assuming that $\boldsymbol{U}(x) = \bar{\boldsymbol{U}}_{j+1/2} := \frac{1}{2}(\boldsymbol{U}_j + \boldsymbol{U}_{j+1})$ is constant and $\mathscr{E}(x)$ is piecewise constant on $(x_j, x_{j+1}]$, i.e.,

$$\mathscr{E}(x) = \begin{cases} \mathscr{E}_j & \text{if} \quad x_j < x \le x_{j+1/2}, \\ \mathscr{E}_{j+1} & \text{if} \quad x_{j+1/2} < x \le x_{j+1}. \end{cases} \tag{13}$$

Recall that $\mathscr{E}$ is symmetric positive definite, and thus regular. We assume the source term $\boldsymbol{s}(x)$ to be piecewise constant. Let m denote the size of the system, thus $\boldsymbol{\varphi}$ and $\boldsymbol{s}$ are m-vectors and $\boldsymbol{U}$ and $\mathscr{E}$ are $m \times m$ matrices.

For the derivation which follows it is convenient to introduce the variables

$$\boldsymbol{A}(x) := \mathscr{E}^{-1}(x)\boldsymbol{U}, \quad \boldsymbol{P}(x) := \Delta x \boldsymbol{A}(x), \quad \boldsymbol{S}(x) := \int_{x_{j+1/2}}^{x} \boldsymbol{s}(\xi)\,\mathrm{d}\xi. \tag{14}$$

The matrix $\boldsymbol{P}$ is referred to as the Peclet matrix $\boldsymbol{P}$. Note that the matrices $\boldsymbol{A}$ and $\boldsymbol{P}$ are piecewise constant on $(x_j, x_{j+1}]$. Moreover, we assume that $\boldsymbol{A}$ has m real eigenvalues λ_i and m corresponding, linearly independent eigenvectors $\boldsymbol{v}_i$ $(i = 1, 2, \ldots, m)$. Since $\boldsymbol{A}$ has a complete set of eigenvectors, its spectral decomposition is given by

$$\boldsymbol{A}\boldsymbol{V} = \boldsymbol{V}\boldsymbol{\Lambda}, \quad \boldsymbol{\Lambda} := \mathrm{diag}(\lambda_1, \lambda_2, \ldots, \lambda_m), \quad \boldsymbol{V} := (\boldsymbol{v}_1, \boldsymbol{v}_2, \ldots, \boldsymbol{v}_m), \tag{15}$$

and based on this decomposition we can compute any matrix function of $\boldsymbol{P}$ as follows

$$g(\boldsymbol{P}) := \boldsymbol{V} g(\Delta x \boldsymbol{\Lambda}) \boldsymbol{V}^{-1}, \quad g(\Delta x \boldsymbol{\Lambda}) := \mathrm{diag}(g(\Delta x \lambda_1), g(\Delta x \lambda_2), \ldots, g(\Delta x \lambda_m)), \tag{16}$$

provided g is defined on the spectrum of $\boldsymbol{A}$ [1].

Integrating the conservation law $\mathrm{d}\boldsymbol{f}/\mathrm{d}x = \boldsymbol{s}$ from the interface at $x_{j+1/2}$ to some arbitrary $x \in [x_j, x_{j+1}]$, we obtain

$$\boldsymbol{f}(x) - \boldsymbol{f}(x_{j+1/2}) = \boldsymbol{S}(x). \tag{17}$$

Next, we substitute the integrating factor formulation of the flux, which for $x \neq x_{j+1/2}$ is given by

$$\boldsymbol{f}(x) = -\mathscr{E} \mathrm{e}^{(x - x_{j+1/2})\boldsymbol{A}} \frac{\mathrm{d}}{\mathrm{d}x}\left(\mathrm{e}^{-(x - x_{j+1/2})\boldsymbol{A}} \boldsymbol{\varphi}\right) \tag{18}$$

in (17), isolate the derivative and subsequently integrate over the interval $[x_j, x_{j+1}]$ to obtain the integral formulation of the flux

$$\int_{x_j}^{x_{j+1}} \mathrm{e}^{-(x-x_{j+1/2})\boldsymbol{A}} \mathscr{E}^{-1}(x)\,\mathrm{d}x\, \boldsymbol{f}(x_{j+1/2}) = \\ \mathrm{e}^{\boldsymbol{P}_j/2}\boldsymbol{\varphi}_j - \mathrm{e}^{-\boldsymbol{P}_{j+1}/2}\boldsymbol{\varphi}_{j+1} - \int_{x_j}^{x_{j+1}} \mathrm{e}^{-(x-x_{j+1/2})\boldsymbol{A}} \mathscr{E}^{-1}(x)\boldsymbol{S}(x)\,\mathrm{d}x, \tag{19}$$

where $\boldsymbol{P}_j = \boldsymbol{P}(x_j)$ etc. In the right hand side, the first two terms correspond to the advection-diffusion operator whereas the integral corresponds to the source term. In order to determine the numerical flux we have to evaluate both integrals in (19).

Consider first the integral in the left hand side of (19) and take $\boldsymbol{S}(x) = \boldsymbol{0}$. Since $\mathscr{E}$ and $\boldsymbol{A}$ are piecewise constant, we split the integral in two parts and find the following relation for the homogeneous numerical flux $\boldsymbol{F}^{\mathrm{h}}_{j+1/2}$:

$$\tfrac{1}{2}\Delta x\Big(\big(\mathscr{E}_j B(\tfrac{1}{2}\boldsymbol{P}_j)\big)^{-1} + \big(\mathscr{E}_{j+1} B(-\tfrac{1}{2}\boldsymbol{P}_{j+1})\big)^{-1}\Big)\boldsymbol{F}^{\mathrm{h}}_{j+1/2} = \\ \mathrm{e}^{\boldsymbol{P}_j/2}\boldsymbol{\varphi}_j - \mathrm{e}^{-\boldsymbol{P}_{j+1}/2}\boldsymbol{\varphi}_{j+1}. \tag{20}$$

Note that this expression is properly defined since the matrices $B(\tfrac{1}{2}\boldsymbol{P}_j)$ and $B(-\tfrac{1}{2}\boldsymbol{P}_{j+1})$ are always regular. Next, consider the integral in the right hand side of (19). Since $\boldsymbol{S}(x)$ is piecewise linear, we can also evaluate this integral. Omitting the first two terms in the right hand side of (19) we obtain the following expression for the inhomogeneous flux $\boldsymbol{F}^{\mathrm{i}}_{j+1/2}$:

$$\Big(\big(\mathscr{E}_j B(\tfrac{1}{2}\boldsymbol{P}_j)\big)^{-1} + \big(\mathscr{E}_{j+1} B(-\tfrac{1}{2}\boldsymbol{P}_{j+1})\big)^{-1}\Big)\boldsymbol{F}^{\mathrm{i}}_{j+1/2} = \\ -\tfrac{1}{2}\Delta x\Big(W_2(\tfrac{1}{2}\boldsymbol{P}_j)\mathscr{E}_j^{-1}\boldsymbol{s}_j - W_2(-\tfrac{1}{2}\boldsymbol{P}_{j+1})\mathscr{E}_{j+1}^{-1}\boldsymbol{s}_{j+1}\Big), \tag{21}$$

where $W_2(z) = \big(\mathrm{e}^z(1-z)-1\big)/z^2$. The complete flux approximation is obviously given by $\boldsymbol{F}_{j+1/2} = \boldsymbol{F}^{\mathrm{h}}_{j+1/2} + \boldsymbol{F}^{\mathrm{i}}_{j+1/2}$, referred to as the piecewise constant complete flux scheme (PCCFS).

PCCFS is a rather complicated and expensive scheme, and therefore we propose the following approximation. Assume first that $\boldsymbol{U}$ is regular, then we can rewrite the expression (20) for the homogeneous flux as

$$\Big(\mathrm{e}^{\boldsymbol{P}_j/2} - \mathrm{e}^{-\boldsymbol{P}_{j+1}/2}\Big)\boldsymbol{U}^{-1}\boldsymbol{F}^{\mathrm{h}}_{j+1/2} = \mathrm{e}^{\boldsymbol{P}_j/2}\boldsymbol{\varphi}_j - \mathrm{e}^{-\boldsymbol{P}_{j+1}/2}\boldsymbol{\varphi}_{j+1}. \tag{22}$$

In general the Peclet matrices in (22) do not commute, so that we have to invoke the Baker-Campbell-Hausdorff formula [1], e.g.,

$$\mathrm{e}^{-\boldsymbol{P}_j/2}\mathrm{e}^{-\boldsymbol{P}_{j+1}/2} = \mathrm{e}^{-\bar{\boldsymbol{P}}_{j+1/2}+\frac{1}{8}\Delta x^2[\boldsymbol{A}_j,\boldsymbol{A}_{j+1}]+\mathscr{O}(\Delta x^3)}, \tag{23}$$

where $[\boldsymbol{A}_j, \boldsymbol{A}_{j+1}] := \boldsymbol{A}_j\boldsymbol{A}_{j+1} - \boldsymbol{A}_{j+1}\boldsymbol{A}_j$ is the commutator of both matrices. Neglecting the $\mathscr{O}(\Delta x^2)$-term in the exponent we can derive the vectorial equivalent of (9b), i.e.,

$$\boldsymbol{F}^{\mathrm{h}}_{j+1/2} = \frac{1}{\Delta x}\tilde{\mathscr{E}}_{j+1/2}\Big(B\big(-\bar{\boldsymbol{P}}_{j+1/2}\big)\boldsymbol{\varphi}_j - B\big(\bar{\boldsymbol{P}}_{j+1/2}\big)\boldsymbol{\varphi}_{j+1}\Big), \tag{24}$$

with $\tilde{\mathscr{E}}_{j+1/2}$ and $\bar{\boldsymbol{P}}_{j+1/2}$ the matrix harmonic average of $\mathscr{E}$ and the average Peclet matrix, respectively, defined by

$$\tilde{\mathscr{E}}^{-1}_{j+1/2} := \tfrac{1}{2}\big(\mathscr{E}^{-1}_j + \mathscr{E}^{-1}_{j+1}\big), \quad \bar{\boldsymbol{P}}_{j+1/2} := \Delta x\tilde{\mathscr{E}}^{-1}_{j+1/2}\bar{\boldsymbol{U}}_{j+1/2}, \tag{25}$$

where obviously $\mathscr{E}^{-1}_j = \mathscr{E}^{-1}(x_j)$ etc.

In case $\boldsymbol{U}$ is singular, and consequently also $\boldsymbol{A}$, we apply a regularization technique to derive (24). Therefore, we replace $\boldsymbol{A}$ by a perturbation $\boldsymbol{A}_\delta = \boldsymbol{A} + \delta\boldsymbol{I}$ for some δ such that $\boldsymbol{A}_\delta$ is regular. This is possible, provided $-\delta \notin \sigma(\boldsymbol{A})$. The matrices $\boldsymbol{P}_\delta$ and $\bar{\boldsymbol{P}}_{\delta,j+1/2}$ are the corresponding perturbations of $\boldsymbol{P}$ and $\bar{\boldsymbol{P}}_{j+1/2}$, respectively. Replacing $\boldsymbol{P}$ by $\boldsymbol{P}_\delta$ in (20) we obtain a similar expression as (24) with $\bar{\boldsymbol{P}}_{\delta,j+1/2}$ instead of $\bar{\boldsymbol{P}}_{j+1/2}$. Since $B(z)$ is continuous for $z = 0$ we can take the limit $\delta \to 0$ to arrive at the expression (24).

Next, for the inhomogeneous flux, we take for $\mathscr{E}$ its matrix harmonic average $\tilde{\mathscr{E}}_{j+1/2}$ and evaluate the integral in the right hand side of (19) to obtain

$$\boldsymbol{F}^{\mathrm{i}}_{j+1/2} = \Delta x\big(W_1(\hat{\boldsymbol{P}}_{j+1/2})\boldsymbol{s}_j - W_1\big(-\hat{\boldsymbol{P}}_{j+1/2}\big)\boldsymbol{s}_{j+1}\big), \tag{26}$$

with $\hat{\boldsymbol{P}}_{j+1/2} := \Delta x\bar{\boldsymbol{U}}_{j+1/2}\tilde{\mathscr{E}}^{-1}_{j+1/2}$; for more details see [2]. The resulting complete flux approximation $\boldsymbol{F}_{j+1/2} = \boldsymbol{F}^{\mathrm{h}}_{j+1/2} + \boldsymbol{F}^{\mathrm{i}}_{j+1/2}$ with $\boldsymbol{F}^{\mathrm{h}}_{j+1/2}$ and $\boldsymbol{F}^{\mathrm{i}}_{j+1/2}$ defined in (24) and (26), respectively, is referred to as the harmonic complete flux scheme (HCFS), as opposed to the standard complete flux scheme, which employs the arithmetic average of the diffusion matrix.

4 Numerical Example

As an example, we apply the vectorial complete flux schemes to the following test problem:

$$\frac{\mathrm{d}}{\mathrm{d}x}\Big(\boldsymbol{U}\boldsymbol{\varphi} - \mathscr{E}\frac{\mathrm{d}\boldsymbol{\varphi}}{\mathrm{d}x}\Big) = \boldsymbol{s}, \quad 0 < x < 1, \tag{27a}$$

$$\frac{\mathrm{d}\varphi_1}{\mathrm{d}x}(0) = 0,\ \varphi_1(1) = \varphi_{1,\mathrm{R}},\ \varphi_2(0) = \varphi_{2,\mathrm{L}},\ \frac{\mathrm{d}\varphi_2}{\mathrm{d}x}(1) = 0, \tag{27b}$$

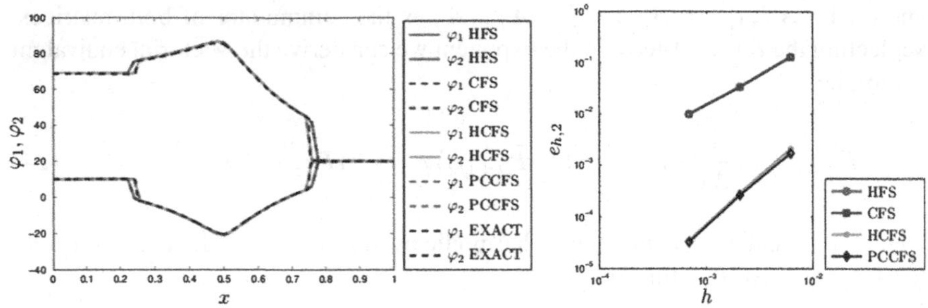

Fig. 1 Numerical solution of (27) (*left*) and discretization errors (*right*). Parameter values are: $u_1 = -1$, $u_2 = 1$, $\alpha = 0.05$ and $s_{\max} = 10^3$. Disretization schemes employed are: standard complete scheme (*CFS*), homogeneous flux scheme (*HFS*), PCCFS and HCFS

with $\boldsymbol{U} = \mathrm{diag}(u_1, u_2)$ and where the diffusion matrix $\mathscr{E}$ and the source term vector $\boldsymbol{s}$ are given by

$$\mathscr{E} = \tfrac{1}{2}\varepsilon \begin{pmatrix} 1+\alpha & 1-\alpha \\ 1-\alpha & 1+\alpha \end{pmatrix}, \quad \varepsilon(x) = \begin{cases} 10^{-2} & \text{if} \quad x \in [0, 0.25) \\ 1 & \text{if} \quad x \in [0.25, 0.75) \\ 10^{-2} & \text{if} \quad x \in [0.75, 1) \end{cases}, \tag{27c}$$

$$\boldsymbol{s}(x) = \frac{s_{\max}}{1 + s_{\max}(2x-1)^2} \begin{pmatrix} 1 \\ 0.2 \end{pmatrix}. \tag{27d}$$

The problem is diffusion dominant in the middle, in $(0.25, 0.75)$, and advection dominant in the remainder of the domain. The parameter α $(0 \le \alpha \le 1)$ determines the coupling between the constituent equations of (27a). The source term has a sharp peak at $x = 0.5$ causing steep interior layers near the discontinuities of ε. A typical solution of (27) is displayed in Fig. 1.

To assess the (order) of convergence of the numerical solutions we first compute a very fine grid solution $\boldsymbol{\varphi}^*$, which to good approximation equals the exact solution. An average discretization error for the second component φ_2, for example, is then given by $e_2(\Delta x) = \Delta x ||\varphi_2^* - \varphi_2||_1$. In Fig. 1 this discretization error as function of the grid size Δx is plotted for various schemes. The standard complete flux and homogeneous flux schemes display only first order convergence, whereas HCFS and PCCFS are second order convergent and have a much smaller discretization error.

Concluding Remarks

In this contribution we have proposed several modifications of the standard complete flux schemes, both scalar and vectorial, for conservation laws with discontinuous diffusion matrix/coefficient. For the numerical flux

(continued)

approximations we employed the (matrix) harmonic average of the diffusion matrix/coefficient, which turned out to be more accurate than the standard schemes. However, more elaborate testing of the modified schemes for realistic applications, such as plasma simulations, is still required.

References

1. N.J. Higham, *Functions of Matrices, Theory and Computation* (SIAM, Philadelphia, 2008)
2. L. Liu, Studies on the discretization of plasma transport equations, PhD thesis, Eindhoven University of Technology, 2013
3. K.S.C. Peerenboom, J. van Dijk, J.H.M. ten Thije Boonkkamp, L. Liu, W.J. Goedheer, J.J.A.M. van der Mullen, Mass conservative finite volume discretization of the continuity equations in multi-component mixtures. J. Comput. Phys. **230**, 3525–3537 (2011)
4. J.H.M. ten Thije Boonkkamp, M.J.H. Anthonissen, The finite volume-complete flux scheme for advection-diffusion-reaction equations. J. Sci. Comput. **46**, 47–70 (2011)
5. J.H.M. ten Thije Boonkkamp, J. van Dijk, L. Liu, K.S.C. Peerenboom, Extension of the complete flux scheme to systems of conservation laws. J. Sci. Comput. **53**, 552–568 (2012)

Analysis of Space-Time DGFEM for the Solution of Nonstationary Nonlinear Convection-Diffusion Problems

Miloslav Feistauer, Monika Balázsová, Martin Hadrava, and Adam Kosík

Abstract The subject of this paper is the analysis of the space-time discontinuous Galerkin method for the solution of nonstationary, nonlinear, convection-diffusion problems. In the formulation of the numerical scheme, the nonsymmetric, symmetric and incomplete versions of the discretization of diffusion terms and interior and boundary penalty are used. Then error estimates derived under a sufficient regularity of the exact solution are briefly characterized. The main attention is paid to the investigation of unconditional stability of the method. An important tool is the concept of the discrete characteristic function. The dominating convection case is not considered. Theoretical results are accompanied by numerical experiments.

1 Continuous Problem

Let $\Omega \subset I\!R^2$ be a bounded polygonal domain and $T > 0$. We consider the initial-boundary value problem to find $u : Q_T = \Omega \times (0,T) \to I\!R$ such that

$$\frac{\partial u}{\partial t} + \sum_{s=1}^{2} \frac{\partial f_s(u)}{\partial x_s} - \operatorname{div}(\beta(u)\nabla u) = g \quad \text{in } Q_T = \Omega \times (0,T), \tag{1}$$

$$u\big|_{\partial\Omega\times(0,T)} = u_D, \tag{2}$$

$$u(x,0) = u^0(x), \quad x \in \Omega. \tag{3}$$

We assume that g, u_D, u^0, f_s are given functions, $f_s \in C^1(I\!R)$, $|f_s'| \le C$, $s = 1,2$, $\beta : I\!R \to [\beta_0, \beta_1]$ with $0 < \beta_0 < \beta_1 < \infty$ and $|\beta(u_1) - \beta(u_2)| \le L|u_1 - u_2|$ for all $u_1, u_2 \in I\!R$.

M. Feistauer (✉) • M. Balázsová • M. Hadrava • A. Kosík
Charles University in Prague, Faculty of Mathematics and Physics, Sokolovská 83, 186 75 Praha 8, Czech Republicat
e-mail: feist@karlin.mff.cuni.cz; b.moncsi@gmail.com; martin@hadrava.eu; adam.kosik@atlas.cz

A. Abdulle et al. (eds.), *Numerical Mathematics and Advanced Applications ENUMATH 2013*, Lecture Notes in Computational Science and Engineering 103,
DOI 10.1007/978-3-319-10705-9_10

2 Space-Time Discretization

We construct a partition in the time interval $[0, T]$ formed by time instants $0 = t_0 < \cdots < t_M = T$ and use the notation $I_m = (t_{m-1}, t_m)$, $\tau_m = t_m - t_{m-1}$, $\tau = \max_{m=1,\ldots,M} \tau_m$. For φ defined in $\bigcup_{m=1}^{M} I_m$ we put $\varphi_m^{\pm} = \varphi(t_m\pm) = \lim_{t\to t_m\pm} \varphi(t)$ (one-sided limits at time t_m), $\{\varphi\}_m = \varphi(t_m+) - \varphi(t_m-)$ (jump).

For each I_m we consider a partition $\mathscr{T}_{h,m}$ of the closure $\overline{\Omega}$ of the domain Ω into a finite number of closed triangles with mutually disjoint interiors. The partitions $\mathscr{T}_{h,m}$ are in general different for different m. The following notation is used: $\mathscr{F}_{h,m}$ – the system of all faces of all elements $K \in \mathscr{T}_{h,m}$, $\mathscr{F}_{h,m}^I$ – the set of all inner faces, $\mathscr{F}_{h,m}^B$ – the set of all boundary faces. Each $\Gamma \in \mathscr{F}_{h,m}$ is associated with a unit normal vector n_Γ. By $K_\Gamma^{(L)}$ and $K_\Gamma^{(R)} \in \mathscr{T}_{h,m}$ we denote the elements adjacent to the face $\Gamma \in \mathscr{F}_{h,m}^I$. Moreover, for $\Gamma \in \mathscr{F}_{h,m}^B$, the element adjacent to this face will be denoted by $K_\Gamma^{(L)}$. We assume that n_Γ is the outer normal to $\partial K_\Gamma^{(L)}$.

Let $C_W > 0$ be a fixed constant. We set

$$h(\Gamma) = \frac{h_{K_\Gamma^{(L)}} + h_{K_\Gamma^{(R)}}}{2C_W} \quad \text{for } \Gamma \in \mathscr{F}_{h,m}^I, \quad h(\Gamma) = \frac{h_{K_\Gamma^{(L)}}}{C_W} \quad \text{for } \Gamma \in \mathscr{F}_{h,m}^B. \tag{4}$$

We introduce the broken Sobolev space

$$H^k(\Omega, \mathscr{T}_{h,m}) = \{v; v|_K \in H^k(K)\ \forall\, K \in \mathscr{T}_{h,m}\}.$$

If $v \in H^1(\Omega, \mathscr{T}_{h,m})$ and $\Gamma \in \mathscr{F}_{h,m}$, then $v_\Gamma^{(L)}$, $v_\Gamma^{(R)}$ will denote the traces of v on Γ from the side of elements $K_\Gamma^{(L)}$, $K_\Gamma^{(R)}$ adjacent to Γ. For $\Gamma \in \mathscr{F}_{h,m}^I$ we set

$$\langle v \rangle_\Gamma = \frac{1}{2}\left(v_\Gamma^{(L)} + v_\Gamma^{(R)}\right), \quad [v]_\Gamma = v_\Gamma^{(L)} - v_\Gamma^{(R)}.$$

Further, let $p, q \geq 1$ be integers. Then for each $m = 1, \ldots, M$ we define the spaces

$$S_{h,m}^p = \left\{\varphi \in L^2(\Omega); \varphi|_K \in P^p(K)\ \forall\, K \in \mathscr{T}_{h,m}\right\}, \tag{5}$$

$$S_{h,\tau}^{p,q} = \left\{\varphi \in L^2(Q_T); \varphi\big|_{I_m} = \sum_{i=0}^{q} t^i \varphi_i \quad \text{with } \varphi_i \in S_{h,m}^p,\ m = 1, \ldots, M\right\}. \tag{6}$$

In the space $S_{h,\tau}^{p,q}$ an approximate solution will be sought.

If $u, \varphi \in H^2(\Omega, \mathscr{T}_{h,m})$, then we define the following forms:

Diffusion form

$$a_{h,m}(u,\varphi) = \sum_{K\in\mathscr{T}_{h,m}} \int_K \beta(u)\nabla u\cdot\nabla\varphi\,\mathrm{d}x \tag{7}$$

$$-\sum_{\Gamma\in\mathscr{F}^I_{h,m}} \int_\Gamma (\langle\beta(u)\nabla u\rangle\cdot n_\Gamma[\varphi] + \theta\langle\beta(u)\nabla\varphi\rangle\cdot n_\Gamma\,[u])\,\mathrm{d}S$$

$$-\sum_{\Gamma\in\mathscr{F}^B_{h,m}} \int_\Gamma (\beta(u)\nabla u\cdot n_\Gamma\,\varphi + \theta\,\beta(u)\nabla\varphi\cdot n_\Gamma\,u - \theta\beta(u)\nabla\varphi\cdot n_\Gamma u_D)\,\mathrm{d}S.$$

Let us note that in integrals over faces we omit the subscript Γ. For $\theta = 1, \theta = 0$ and $\theta = -1$ we get the symmetric (SIPG), incomplete (IIPG) and nonsymmetric (NIPG) variants of the approximation of the diffusion terms, respectively.
Interior and boundary penalty

$$J_{h,m}(u,\varphi) = \sum_{\Gamma\in\mathscr{F}^I_{h,m}} h(\Gamma)^{-1}\int_\Gamma [u]\,[\varphi]\,\mathrm{d}S + \sum_{\Gamma\in\mathscr{F}^B_{h,m}} h(\Gamma)^{-1}\int_\Gamma u\,\varphi\,\mathrm{d}S. \tag{8}$$

Right-hand side form

$$\ell_{h,m}(\varphi) = (g,\varphi) + \beta_0\sum_{\Gamma\in\mathscr{F}^B_{h,m}} h(\Gamma)^{-1}\int_\Gamma u_D\,\varphi\,\mathrm{d}S. \tag{9}$$

Convection form

$$b_{h,m}(u,\varphi) = -\sum_{K\in\mathscr{T}_{h,m}} \int_K \sum_{s=1}^{2} f_s(u)\frac{\partial\varphi}{\partial x_s}\,\mathrm{d}x \tag{10}$$

$$+\sum_{\Gamma\in\mathscr{F}^I_{h,m}} \int_\Gamma H\left(u_\Gamma^{(L)}, u_\Gamma^{(R)}, n_\Gamma\right)[\varphi]\,\mathrm{d}S$$

$$+\sum_{\Gamma\in\mathscr{F}^B_{h,m}} \int_\Gamma H\left(u_\Gamma^{(L)}, u_\Gamma^{(L)}, n_\Gamma\right)\varphi\,\mathrm{d}S.$$

Here $H = H(u, v, n)$ is a numerical flux, which is Lipschitz-continuous in $I\!R^2 \times B_1$, where $B_1 = \{n \in I\!R^2; |n| = 1\}$, consistent:

$$H(u, u, n) = \sum_{s=1}^{2} f_s(u)\, n_s, \quad u \in I\!R, \quad n = (n_1, n_2) \in B_1,$$

and conservative:

$$H(u, v, n) = -H(v, u, -n), \quad u, v \in I\!R, \quad n \in B_1.$$

We set

$$A_{h,m} = a_{h,m} + \beta_0 J_{h,m}, \tag{11}$$

and by $(\cdot, \cdot)$ we denote the scalar product in $L^2(\Omega)$. It induces the norm $\|\cdot\|$ in $L^2(\Omega)$. The space $H^1(\Omega, \mathcal{T}_{h,m})$ will be equipped with the norm

$$\|\varphi\|_{DG,m} = \Big(\sum_{K \in \mathcal{T}_{h,m}} |\varphi|^2_{H^1(K)} + J_{h,m}(\varphi, \varphi) \Big)^{1/2}.$$

In what follows we shall use the notation $U' = \partial U/\partial t, u' = \partial u/\partial t$. Now the *space-time DG approximate solution* is defined as a function $U \in S^{p,q}_{h,\tau}$ such that

$$\begin{aligned} &\int_{I_m} \big((U', \varphi) + A_{h,m}(U, \varphi) + b_{h,m}(U, \varphi) \big)\, \mathrm{d}t + \big(\{U\}_{m-1}, \varphi^+_{m-1} \big) \qquad (12)\\ &= \int_{I_m} \ell_{h,m}(\varphi)\, \mathrm{d}t, \quad \forall\, \varphi \in S^{p,q}_{h,\tau}, \quad m = 1, \ldots, M,\\ &U_0^- = L^2(\Omega)\text{-projection of } u^0 \text{ on } S^p_{h,1}. \end{aligned}$$

3 Error Estimates

The papers [4] and [2] were devoted to the analysis of the STDG method applied to problem in the case of linear diffusion and nonlinear diffusion, respectively. Under the assumptions on the regularity of the exact solution

$$\begin{aligned} &u \in H^{q+1}\big(0, T; H^1(\Omega)\big) \cap C([0, T]; H^{p+1}(\Omega)), \qquad (13)\\ &\|\nabla u(t)\|_{L^\infty(\Omega)} \le C_R \ \text{ for a. e. } t \in (0, T), \end{aligned}$$

using approximation properties of $S^p_{h,m}$- and $S^{p,q}_{h,\tau}$-interpolation operators, assumptions on the properties of the meshes, namely the shape regularity and local quasiuniformity and the condition

$$\tau_m \geq Ch_m^2, \quad m = 1, \dots, M, \tag{14}$$

error estimates in terms of h and τ were proven.

Theorem 1 *There exists a constant $C > 0$ such that*

$$\|e_m^-\|^2 + \frac{\varepsilon}{2}\sum_{j=1}^{m}\int_{I_m} \|e\|^2_{DG,j}\, \mathrm{d}t \tag{15}$$

$$\leq C\left(h^{2p}|u|^2_{C([0,T];H^{p+1}(\Omega))} + \tau^{2q+\alpha}|u|^2_{H^{q+1}(0,T;H^1(\Omega))}\right).$$

Here $\alpha = 2$, if u_D is a polynomial of degree $\leq q$ in t. Otherwise, $\alpha = 0$ under the assumption that the CFL-like condition

$$\tau_m \leq Ch_K, \quad K \in \mathscr{T}_{h,m},\ m = 1, \dots, M, \tag{16}$$

with a constant C independent of h_K, τ_m and M is satisfied for the elements K adjacent to the boundary $\partial\Omega$. (If all meshes $\mathscr{T}_{h,m}$ are identical, then condition (14) can be omitted.)

4 Unconditional Stability of the Space-Time DGM

Our main goal is to show that the space-time DG method is unconditionally stable, which means that the approximate solution U is bounded in suitable norms of data g ($\in L^2(Q_T)$), u^0 ($\in L^2(\Omega)$) and u_D ($\in L^2(\partial\Omega)$) independently of h and τ without condition (16). To this end, we introduce the norm

$$\|v\|_{DGB,m} = \left(\sum_{\Gamma\in\mathscr{F}^B_{h,m}} h^{-1}(\Gamma)\int_\Gamma |v|^2 dS\right)^{1/2}, \quad v \in L^2(\partial\Omega). \tag{17}$$

In what follows, we shall give a sketch of the proof of the unconditional stability, which is rather technical. The details will be contained in [1]. In the stability analysis we start from the relation

$$\int_{I_m}\big((U',U) + a_{h,m}(U,U) + \beta_0\, J_{h,m}(U,U) + b_{h,m}(U,U)\big)\, dt \tag{18}$$

$$+(\{U\}_{m-1}, \varphi^+_{m-1}) = \int_{I_m} \ell_{h,m}(U)\, dt,$$

obtained from (12) putting $\varphi := U$, and estimate individual terms.

We use some auxiliary results:

Coercivity:

$$\int_{I_m} (a_{h,m}(U,U) + \beta_0 J_{h,m}(U,U))\, dt \tag{19}$$

$$\geq \frac{\beta_0}{2} \int_{I_m} \|U\|^2_{DG,m}\, dt - \frac{\beta_0}{2} \int_{I_m} \|u_D\|^2_{DGB,m}\, dt,$$

if C_W is sufficiently large. (Lower bound of C_W is expressed by the constants from the multiplicative trace inequality, inverse inequality and local quasiuniformity of the meshes.)

Bound of the right-hand side form $\ell_{h,m}(U)$: For each $k > 0$ we have

$$\int_{I_m} |\ell_{h,m}(U)|\, dt \tag{20}$$

$$\leq \frac{1}{2} \int_{I_m} (\|g\|^2 + \|U\|^2)\, dt + \beta_0 k \int_{I_m} \|u_D\|^2_{DGB,m}\, dt + \frac{\beta_0}{k} \int_{I_m} \|U\|^2_{DG,m}\, dt.$$

Bound of the form $b_{h,m}$: For each $k > 0$ we have

$$\int_{I_m} |b_{h,m}(U,U)|\, dt \leq \frac{\beta_0}{k} \int_{I_m} \|U\|^2_{DG,m}\, dt + c_b(k) \int_{I_m} \|U\|^2\, dt. \tag{21}$$

On the basis of the above results it is possible to prove the following estimate:

$$\|U_m^-\|^2 - \|U_{m-1}^-\|^2 + \frac{\beta_0}{2} \int_{I_m} \|U\|^2_{DG,m} dt \tag{22}$$

$$\leq c \left(\int_{I_m} \|g\|^2\, dt + \int_{I_m} \|U\|^2\, dt + \int_{I_m} \|u_D\|^2_{DGB,m} dt \right).$$

We see that it is necessary to estimate the expression $\int_{I_m} \|U\|^2\, dt$ in terms of data. The main tool is the concept of the *discrete characteristic function* $\zeta_y \in S^{p,q}_{h,\tau}$ to U for $y \in I_m = (t_{m-1}, t_m)$ defined by

$$\int_{I_m} (\zeta_y, \varphi)\, dt = \int_{t_{m-1}}^{y} (U, \varphi)\, dt \quad \forall \varphi \in S^{p,q-1}_{h,\tau}, \quad \zeta_y(t^+_{m-1}) = U(t^+_{m-1}), \tag{23}$$

introduced, e.g., in [3].

The operator assigning ζ_y to U is continuous, i.e.,

$$\int_{I_m} \|\zeta_y\|^2_{DG,m}\, dt \leq c_q \int_{I_m} \|U\|^2_{DG,m}\, dt, \quad \int_{I_m} \|\zeta_y\|^2\, dt \leq c_q \int_{I_m} \|U\|^2\, dt, \tag{24}$$

where the constant $c_q > 0$ depends only on q.

With the aid of a complicated technical process it is possible to prove the following important inequality: There exists a constant $c > 0$ such that

$$\int_{I_m} \|U\|^2\, dt \le c\,\tau_m \left(\|U_{m-1}^-\|^2 + \int_{I_m} \left(\|g\|^2 + \|u_D\|^2_{DGB,m}\right) dt \right). \tag{25}$$

Now we come to the formulation of the *final main result.*

Theorem 2 *There exists a constant* $c > 0$ *such that*

$$\|U_m^-\|^2 + \frac{\beta_0}{2} \sum_{j=1}^{M} \int_{I_j} \|U\|^2_{DG,j}\, dt$$
$$\le c \left(\|U_0^-\|^2 + \sum_{j=1}^{M} \int_{I_j} \left(\|g\|^2 + \|u_D\|^2_{DGB,j}\right) dt \right),$$
$$m = 1, \dots, M, \quad h \in (0, h_0),$$
$$\|U\|^2_{L^2(Q_T)} \le c \left(\|U_0^-\|^2 + \sum_{j=1}^{M} \int_{I_j} \left(\|g\|^2 + \|u_D\|^2_{DGB,j}\right) dt \right),$$
$$h \in (0, h_0).$$

As we see, this theorem represents the unconditional stability of the space-time DG method in the discrete $L^\infty(L^2)$-norm, energy DG norm and $L^2(L^2)$-norm.

5 Numerical Experiments

In this section, the accuracy and stability of the method is demonstrated by numerical experiments. We consider problem (1) for equation

$$\frac{\partial u}{\partial t} + u\frac{\partial u}{\partial x_1} + u\frac{\partial u}{\partial x_2} = \varepsilon \Delta u + g \quad \text{in} \quad (0,1)^2 \times (0,10) \tag{26}$$

with $\varepsilon = 0.1$ and such initial and Dirichlet boundary conditions that the exact solution has the form $u(x_1, x_2, t) = (1 - e^{-10t}) 2 r^\alpha x_1 x_2 (1 - x_1)(1 - x_2)$, where $r = (x_1 + x_2)^{1/2}$ and $\alpha \in I\!R$ is a constant. It is possible to prove that $u \in H^{q+1}(0, T; H^\beta(\Omega))$ for all $\beta \in (0, \alpha + 3)$. The numerical flux is defined by

$$H(u, v, n) = \begin{cases} \sum_{s=1}^2 f_s(u) n_s, & \text{if } A > 0 \\ \sum_{s=1}^2 f_s(v) n_s, & \text{if } A \le 0, \end{cases}$$

where

$$A = \sum_{s=1}^{2} f_s'\left(\frac{u+v}{2}\right) n_s \quad \text{and} \quad n = (n_1, n_2).$$

Five special triangular meshes having 235, 333, 749, 1,622 and 2,521 elements, space polynomial degrees $p = 1, 2, 3$ and time polynomial degree $q = 2$ were used. We chose the fixed time step $\tau = 0.025$ and the constant $C_W = 100$ in the SIPG version of the diffusion discretization. Figure 1 shows the coarsest mesh, which is refined near the right-hand side of the boundary. This mesh was successively refined.

Tables 1 and 2 show the computational errors in the $L^\infty(L^2)$-norm and the corresponding experimental orders of convergence (EOC). We see that for a sufficiently regular exact solution ($\alpha = 4$) we get the optimal order of convergence $O(h^{p+1})$ for $p = 1, 2, 3$, whereas in the case with irregular solution ($\alpha = -3/2$) the error estimates are of order $O(h^{3/2})$ for $p = 1, 2, 3$. (This result can be proven with the aid of estimates in Sobolev-Slobodetskii spaces.) The presented numerical experiments demonstrate the stability of the numerical process without CFL-like condition (16).

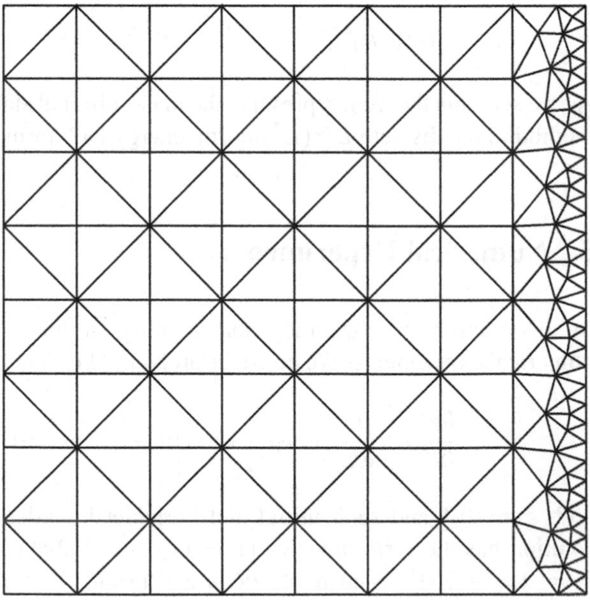

Fig. 1 Coarse mesh with 235 elements

Table 1 Computational errors and the corresponding orders of convergence of the SIPG method for $\alpha = 4$

		$p = 1$		$p = 2$		$p = 3$	
Mesh	h	$\|e_h\|$	*EOC*	$\|e_h\|$	*EOC*	$\|e_h\|$	*EOC*
1	1.768E−01	2.229E−03	–	1.317E−04	–	6.708E−06	–
2	1.414E−01	1.522E−03	1.708	7.272E−05	2.663	2.956E−06	3.672
3	8.839E−02	6.660E−04	1.759	1.994E−05	2.753	5.026E−07	3.770
4	5.657E−02	2.947E−04	1.827	5.634E−06	2.832	9.017E−08	3.850
5	4.419E−02	1.858E−04	1.869	2.770E−06	2.876	3.446E−08	3.897

Table 2 Computational errors and the corresponding orders of convergence of the SIPG method for $\alpha = -3/2$

		$p = 1$		$p = 2$		$p = 3$	
Mesh	h	$\|e_h\|$	*EOC*	$\|e_h\|$	*EOC*	$\|e_h\|$	*EOC*
1	1.768E−01	2.669E−02	–	6.038E−03	–	2.784E−03	–
2	1.414E−01	1.946E−02	1.416	4.330E−03	1.490	2.003E−03	1.475
3	8.839E−02	9.857E−03	1.447	2.149E−03	1.491	9.985E−04	1.481
4	5.657E−02	5.116E−03	1.469	1.103E−03	1.493	5.141E−04	1.488
5	4.419E−02	3.552E−03	1.478	7.630E−04	1.495	3.556E−04	1.493

Acknowledgement This work was supported by the grant No. 13-00522S (M. Feistauer) of the Czech Science Foundation, and by the grant SVV-2014-260106 financed by the Charles University in Prague (M. Balázsová, M. Hadrava and A. Kosík).

References

1. M. Balázsová, M. Feistauer, M. Hadrava, A. Kosík, On the stability of the space-time discontinuous Galerkin method for the numerical solution of nonstationary nonlinear convection-diffusion problems J. Numer. Math. (to appear)
2. J. Česenek, M. Feistauer, Theory of the space-time discontinuous Galerkin method for nonstationary parabolic problems with nonlinear convection and diffusion. SIAM J. Numer. Anal. **30**, 1181–1206 (2012)
3. K. Chrysafinos, N.J. Walkington, Error estimates for the discontinuous Galerkin methods for parabolic equations. SIAM J. Numer. Anal. **44**, 349–366 (2006)
4. M. Feistauer, V. Kučera, K. Najzar, J. Prokopová, Analysis of space-time discontinuous Galerkin method for nonlinear convection-diffusion problems. Numer. Math. **117**, 251–288 (2011)

Space-Time Discontinuous Galerkin Method for the Problem of Linear Elasticity

Martin Hadrava, Miloslav Feistauer, Jaromír Horáček, and Adam Kosík

Abstract The subject of this paper is the numerical solution of the problem of dynamic linear elasticity by several time-discretization techniques based on the application of the discontinuous Galerkin (DG) method in space. In the formulation of the numerical scheme, the nonsymmetric, symmetric and incomplete versions of the discretization of the elasticity term and the interior and boundary penalty are used. The DG space discretization is combined with the backward-Euler, second-order backward-difference formula and DG time discretization. Finally, we present some test problems.

1 Introduction

This paper is concerned with the application of the discontinuous Galerkin (DG) method to the solution of dynamic linear elasticity problem. (For a survey of DG techniques, see, e.g., [2, 4].) The DG space discretization is combined with the time discretization by the backward Euler (BEDG), second-order BDF (BDFDG) or DG scheme in time (STDG).

In [3], the method using the DG technique in time, but conforming finite elements in space is analyzed in the case of a linear wave equation. Here we are not interested in the computation of wave propagation in an elastic body, but our future goal will be to apply the developed method, which is different from the scheme analyzed in [3], to the solution of the interaction of a fluid and an elastic body.

We describe the mentioned methods and apply them to a test problem in order to compare their quality. Numerical experiments show that the STDG method is

M. Hadrava (✉) • M. Feistauer • A. Kosík
Charles University in Prague, Faculty of Mathematics and Physics, Sokolovská 83, 186 75 Praha 8, Czech Republic
e-mail: martin@hadrava.eu; feist@karlin.mff.cuni.cz; adam.kosik@atlas.cz

J. Horáček
Institute of Thermomechanics, The Academy of Sciences of the Czech Republic, v. v. i., Dolejškova 1402/5, 182 00 Praha 8, Czech Republic
e-mail: jaromirh@it.cas.cz

A. Abdulle et al. (eds.), *Numerical Mathematics and Advanced Applications ENUMATH 2013*, Lecture Notes in Computational Science and Engineering 103,
DOI 10.1007/978-3-319-10705-9_11

the most promising. Our further work will be oriented to a deeper analysis of the developed method and its applications to fluid-structure interaction (FSI) problems.

2 Formulation of the Dynamic Elasticity Problem

We consider an elastic body represented by a bounded domain $\Omega \subset \mathbb{R}^2$ with boundary formed by two disjoint parts: $\partial\Omega = \Gamma_D \cup \Gamma_N$. By $\boldsymbol{u} = \boldsymbol{u}(t, \boldsymbol{x}) : [0, T] \times \Omega \to \mathbb{R}^2$ we denote the displacement of the body. The symbol $\nabla \boldsymbol{u} = (\partial u_i / \partial x_j)_{i,j=1}^2$ denotes the gradient of the function $\boldsymbol{u}$. The dynamic elasticity problem is defined as follows: we seek for the displacement function $\boldsymbol{u}$ such that

$$\rho \frac{\partial^2 \boldsymbol{u}}{\partial t^2} + c_M \rho \frac{\partial \boldsymbol{u}}{\partial t} - \operatorname{div} \boldsymbol{\sigma}(\boldsymbol{u}) - c_K \frac{\partial}{\partial t} \operatorname{div} \boldsymbol{\sigma}(\boldsymbol{u}) = \boldsymbol{f} \quad \text{in } (0, T) \times \Omega, \tag{1}$$

$$\boldsymbol{u} = \boldsymbol{u}_D \quad \text{in } (0, T) \times \Gamma_D, \quad \boldsymbol{\sigma}(\boldsymbol{u}) \cdot \boldsymbol{n} = \boldsymbol{g}_N \quad \text{in } (0, T) \times \Gamma_N, \tag{2}$$

$$\boldsymbol{u}(0, x) = \boldsymbol{u}_0(x), \quad \frac{\partial \boldsymbol{u}}{\partial t}(0, x) = z_0(x), \quad \text{in } \Omega. \tag{3}$$

Here $\boldsymbol{f} : (0, T) \times \Omega \to \mathbb{R}^2$ is the outer volume force, $\boldsymbol{u}_D : (0, T) \times \Gamma_D \to \mathbb{R}^2$ is the boundary displacement, $\boldsymbol{g}_N : (0, T) \times \Gamma_N \to \mathbb{R}^2$ is the boundary normal stress, $\boldsymbol{u}_0 : \Omega \to \mathbb{R}^2$ is the initial displacement, $z_0 : \Omega \to \mathbb{R}^2$ is the initial displacement velocity, $T > 0$ is the time interval length and $\rho > 0$ is the constant material density. The expressions $c_M \rho \frac{\partial u}{\partial t}$ and $c_K \frac{\partial}{\partial t} \operatorname{div} \boldsymbol{\sigma}(\boldsymbol{u})$ with $c_M, c_K \geq 0$ represent structural and viscous damping terms. We assume that the material is *isotropic* and *homogeneous* and that the *stress tensor* $\boldsymbol{\sigma}(\boldsymbol{u})$ depends on the *infinitesimal strain tensor* $\boldsymbol{e}(\boldsymbol{u})$ by the relation

$$\boldsymbol{\sigma}(\boldsymbol{u}) := \lambda \operatorname{tr}(\boldsymbol{e}(\boldsymbol{u}))\boldsymbol{I} + 2\mu \boldsymbol{e}(\boldsymbol{u}), \quad \boldsymbol{e}(\boldsymbol{u}) := \frac{1}{2} \left(\nabla \boldsymbol{u} + \nabla \boldsymbol{u}^T \right). \tag{4}$$

We assume that the *Lamè parameters* λ and μ are constant. For most solid materials it holds that $\lambda, \mu > 0$. Finally, $\operatorname{tr}(\boldsymbol{e}(\boldsymbol{u}))$ denotes the trace of the tensor $\boldsymbol{e}(\boldsymbol{u})$.

3 Discretization

In order to introduce the discrete problem we rewrite problem (1)–(3) as a couple of first-order equations in time: find functions $\boldsymbol{u}$ and $z : [0, T] \times \Omega \to \mathbb{R}^2$ such that

$$\varrho \frac{\partial z}{\partial t} + c_M \varrho z - \operatorname{div} \boldsymbol{\sigma}(\boldsymbol{u}) - c_K \operatorname{div} \boldsymbol{\sigma}(z) = \boldsymbol{f}, \tag{5}$$

$$\frac{\partial \boldsymbol{u}}{\partial t} - z = 0 \qquad \text{in } (0, T) \times \Omega, \tag{6}$$

$$u = u_D \quad \text{in } (0,T)\times\Gamma_D, \quad \sigma(u)\cdot n = g_N \quad \text{in } (0,T)\times\Gamma_N, \tag{7}$$

$$u(0,x) = u_0(x), \quad z(0,x) = z_0(x) \quad \text{in } \Omega. \tag{8}$$

3.1 Notation

Let us assume that the computational domain Ω is polygonal. By $\mathscr{T}_h$ we denote a triangulation of the domain Ω with triangular elements $K \in \mathscr{T}_h$ having standard properties from the finite element method, cf. [1].

We say that the elements $K, K' \in \mathscr{T}_h$ are *neighbours*, if the set $\partial K \cap \partial K'$ has positive 1-dimensional measure. We say that $\Gamma \subset \partial K$ is a *face* of K, if it is a maximal connected open subset of either $\partial K \cap \partial K'$, where K' is a neighbour of K or of $\partial K \cap \Gamma_D$ or of $\partial K \cap \Gamma_N$. By $\mathscr{F}_h$ we denote the system of all faces of all elements $K \in \mathscr{T}_h$. Further, we define the set of all boundary, "Dirichlet", "Neumann" and inner faces by

$$\mathscr{F}_h^B = \{\Gamma \in \mathscr{F}_h; \Gamma \subset \partial\Omega\}, \quad \mathscr{F}_h^D = \{\Gamma \in \mathscr{F}_h; \Gamma \subset \Gamma_D\},$$
$$\mathscr{F}_h^N = \{\Gamma \in \mathscr{F}_h; \Gamma \subset \Gamma_N\}, \quad \mathscr{F}_h^I = \mathscr{F}_h \backslash \mathscr{F}_h^B,$$

respectively. We put $\mathscr{F}_h^{ID} = \mathscr{F}_h^I \cup \mathscr{F}_h^D$. For each $\Gamma \in \mathscr{F}_h$ we define a unit normal vector n_Γ. We assume that for $\Gamma \in \mathscr{F}_h^B$ the normal n_Γ has the same orientation as the outer normal to $\partial\Omega$. For each $\Gamma \in \mathscr{F}_h^I$ the orientation of n_Γ is arbitrary, but fixed.

We define the finite dimensional space

$$S_{hp} = \left\{v \in L^2(\Omega); v|_K \in P^p(K), K \in \mathscr{T}_h\right\},$$

where $p \geq 1$ is an integer and $P^p(K)$ denotes the space of all polynomials on K of degree $\leq p$. It is easy to show that $\dim S_{hp} = N_{\mathscr{T}_h}(p+1)(p+2)/2$, where $N_{\mathscr{T}_h}$ is the number of elements in $\mathscr{T}_h$.

Because of the time discretization we introduce a uniform partition $0 = t_0 < \cdots < t_M = T$ of the time interval $[0,T]$ with a constant time step $\tau = t_m - t_{m-1}$, $m = 1,\ldots,M$. Let $p \geq 1$, $q \geq 0$ be integers. By $S_{hp}^{\tau q}$ we denote the space of piecewise polynomial functions

$$S_{hp}^{\tau q} = \left\{ v \in L^2((0,T)\times\Omega); v|_{I_m} = \sum_{i=0}^{q} t^i \varphi_i \text{ with } \varphi_i \in S_{hp}, m = 1,\ldots,M \right\},$$

where $I_m = (t_{m-1}, t_m)$, $m = 1,\ldots,M$. The space $S_{hp}^{\tau q}$ consists of all polynomials of degree less or equal to q in time with coefficients in S_{hp}. As we see, functions from $S_{hp}^{\tau q}$ are, in general, discontinuous on faces $\Gamma \in \mathscr{F}^I$ and at time instants t_m, $m = 1, \dots, M-1$. The dimension of the space $S_{hp}^{\tau q}$ equals $M(q+1) \dim S_{hp}$.

For each $\Gamma \in \mathscr{F}_h^I$ there exist two neighbouring elements $K_\Gamma^{(L)}, K_\Gamma^{(R)} \in \mathscr{T}_h$ such that $\Gamma \subset \partial K_\Gamma^{(L)} \cap \partial K_\Gamma^{(R)}$. We use the convention that $\boldsymbol{n}_\Gamma$ is the outer normal to $\partial K_\Gamma^{(L)}$ and the inner normal to $\partial K_\Gamma^{(R)}$. For $\boldsymbol{v} \in [S_{hp}]^2$ or $[S_{hp}^{\tau q}]^2$ we introduce the following notation:

$$\boldsymbol{v}|_\Gamma^{(L)} = \text{the trace of } \boldsymbol{v}|_{K_\Gamma^{(L)}} \text{ on } \Gamma, \ \boldsymbol{v}|_\Gamma^{(R)} = \text{the trace of } \boldsymbol{v}|_{K_\Gamma^{(R)}} \text{ on } \Gamma,$$
$$\langle \boldsymbol{v} \rangle_\Gamma = \tfrac{1}{2}\left(\boldsymbol{v}|_\Gamma^{(L)} + \boldsymbol{v}|_\Gamma^{(R)}\right), \qquad [\boldsymbol{v}]_\Gamma = \boldsymbol{v}|_\Gamma^{(L)} - \boldsymbol{v}|_\Gamma^{(R)},$$

where $\Gamma \in \mathscr{F}_h^I$. For $\Gamma \in \mathscr{F}_h^B$ there exists an element $K_\Gamma^{(L)} \in \mathscr{T}_h$ such that $\Gamma \subset K_\Gamma^{(L)} \cap \partial\Omega$. Then for $\boldsymbol{v} \in [S_{hp}]^2$ we introduce the following notation:

$$\boldsymbol{v}|_\Gamma^{(L)} = \text{the trace of } \boldsymbol{v}|_{K_\Gamma^{(L)}} \text{ on } \Gamma, \quad \langle \boldsymbol{v} \rangle_\Gamma = [\boldsymbol{v}]_\Gamma = \boldsymbol{v}|_\Gamma^{(L)}.$$

In case that $[\cdot]_\Gamma$, $\langle \cdot \rangle_\Gamma$ and $\boldsymbol{n}_\Gamma$ appear in integrals $\int_\Gamma \dots \, \mathrm{d}S$, $\Gamma \in \mathscr{F}_h$, we omit the subscript Γ and simply write $[\cdot]$, $\langle \cdot \rangle$ and $\boldsymbol{n}$, respectively.

Finally, by $T : S$ we shall denote the tensor inner product, defined by

$$T : S = \sum_{i=1}^{2} \sum_{j=1}^{2} T_{ij} S_{ij} = \operatorname{tr}\left(T^T S\right), \ S, \ T \in \mathbb{R}^{2\times 2}.$$

3.2 Space Discretization

We begin with the space discretization of the dynamic elasticity problem. An approximate solution of problem (5)–(8), i.e., the approximations of the functions $\boldsymbol{u}$, $\boldsymbol{z}$ will be sought in the space $\mathscr{V} := [S_{hp}]^2$ in the finite-difference based schemes or $\mathscr{V} := [S_{hp}^{\tau q}]^2$ in the space-time discontinuous Galerkin method.

In the first step, we multiply Eqs. (5)–(6) by test functions $\boldsymbol{v}$ and $\boldsymbol{w} \in \mathscr{V}$, respectively, integrate the resulting equations over $K \in \mathscr{T}_h$, sum the resulting equations over all $K \in \mathscr{T}_h$ and use the following relations. Using Green's theorem, we obtain the equality

$$-\sum_{K \in \mathscr{T}_h} \int_K \operatorname{div} \boldsymbol{\sigma}(\boldsymbol{u}) \cdot \boldsymbol{v} \, \mathrm{d}x = \sum_{K \in \mathscr{T}_h} \int_K \boldsymbol{\sigma}(\boldsymbol{u}) : \boldsymbol{e}(\boldsymbol{v}) \, \mathrm{d}x$$
$$- \sum_{\Gamma \in \mathscr{F}_h^{ID}} \int_\Gamma (\langle \boldsymbol{\sigma}(\boldsymbol{u}) \rangle \cdot \boldsymbol{n}) \cdot [\boldsymbol{v}] \, \mathrm{d}S - \sum_{\Gamma \in \mathscr{F}_h^N} \int_\Gamma \boldsymbol{g}_N \cdot \boldsymbol{v} \, \mathrm{d}S.$$

The interior and boundary penalty methods incorporate the fact that for the exact solution we have $[\boldsymbol{u}]_\Gamma = 0$ for each $\Gamma \in \mathscr{F}_h^I$ and $\boldsymbol{u}$ satisfies the Dirichlet condition in (7). Hence,

$$\sum_{\Gamma \in \mathscr{F}_h^{ID}} \int_\Gamma \frac{C_W}{h_\Gamma}[\boldsymbol{u}] \cdot [\boldsymbol{v}] \, \mathrm{d}S = \sum_{\Gamma \in \mathscr{F}_h^D} \int_\Gamma \frac{C_W}{h_\Gamma} \boldsymbol{u}_D \cdot \boldsymbol{v} \, \mathrm{d}S$$

for each $\boldsymbol{v} \in \mathscr{V}$, where $C_W > 0$ is a given parameter and h_Γ represents the "magnitude" of Γ as, for example, the length of Γ of $h_\Gamma = \left(h_{K_\Gamma}^{(L)} + h_{K_\Gamma}^{(R)}\right)/2$.

Finally, for the exact solution $\boldsymbol{u}$ and arbitrary $\boldsymbol{v} \in \mathscr{V}$ we have

$$\sum_{\Gamma \in \mathscr{F}_h^{ID}} \int_\Gamma (\langle \boldsymbol{\sigma}(\boldsymbol{v}) \rangle \cdot \boldsymbol{n}) \cdot [\boldsymbol{u}] \, \mathrm{d}S = \sum_{\Gamma \in \mathscr{F}_h^D} \int_\Gamma (\boldsymbol{\sigma}(\boldsymbol{v}) \cdot \boldsymbol{n}) \cdot \boldsymbol{u}_D \, \mathrm{d}S.$$

We now define the forms $a_h(\boldsymbol{u}, \boldsymbol{v}) : \mathscr{V} \times \mathscr{V} \to \mathbb{R}$, $\ell_h^s(\boldsymbol{v}) : \mathscr{V} \to \mathbb{R}$ and $\ell_h^e(\boldsymbol{v}) : \mathscr{V} \to \mathbb{R}$ by

$$\begin{aligned} a_h(\boldsymbol{u}, \boldsymbol{v}) = &\sum_{K \in \mathscr{T}_h} \int_K \boldsymbol{\sigma}(\boldsymbol{u}) : \boldsymbol{e}(\boldsymbol{v}) \, \mathrm{d}x - \sum_{\Gamma \in \mathscr{F}^{ID}} \int_\Gamma (\langle \boldsymbol{\sigma}(\boldsymbol{u}) \rangle \cdot \boldsymbol{n}) \cdot [\boldsymbol{v}] \, \mathrm{d}S \\ &- \theta \sum_{\Gamma \in \mathscr{F}^{ID}} \int_\Gamma (\langle \boldsymbol{\sigma}(\boldsymbol{v}) \rangle \cdot \boldsymbol{n}) \cdot [\boldsymbol{u}] \, \mathrm{d}S + \sum_{\Gamma \in \mathscr{F}^{ID}} \int_\Gamma \frac{C_W}{h_\Gamma} [\boldsymbol{u}] \cdot [\boldsymbol{v}] \, \mathrm{d}S, \end{aligned} \tag{9}$$

$$\ell_h^s(\boldsymbol{v}) = \sum_{K \in \mathscr{T}_h} \int_K \boldsymbol{f} \cdot \boldsymbol{v} \, \mathrm{d}x, \tag{10}$$

$$\begin{aligned} \ell_h^e(\boldsymbol{v}) = &\sum_{\Gamma \in \mathscr{F}^N} \int_\Gamma \boldsymbol{g}_N \cdot \boldsymbol{v} \, \mathrm{d}S - \theta \sum_{\Gamma \in \mathscr{F}^D} \int_\Gamma (\boldsymbol{\sigma}(\boldsymbol{v}) \cdot \boldsymbol{n}) \cdot \boldsymbol{u}_D \, \mathrm{d}S \\ &+ \sum_{\Gamma \in \mathscr{F}^D} \int_\Gamma \frac{C_W}{h_\Gamma} \boldsymbol{u}_D \cdot \boldsymbol{v} \, \mathrm{d}S. \end{aligned} \tag{11}$$

The parameter θ defines the symmetric ($\theta = 1$), incomplete ($\theta = 0$) and nonsymmetric ($\theta = -1$) variant of the interior penalty DG method.

Application of these formulas yields the system

$$\begin{aligned} \rho \left(\frac{\partial \boldsymbol{z}_h}{\partial t}, \boldsymbol{v}_h \right)_\Omega &+ c_M \rho (\boldsymbol{z}_h, \boldsymbol{v}_h)_\Omega + a_h(\boldsymbol{u}_h, \boldsymbol{v}_h) + c_K a_h(\boldsymbol{z}_h, \boldsymbol{v}_h) \\ &= \ell_h^s(\boldsymbol{v}_h) + \ell_h^e(\boldsymbol{v}_h) + c_K \ell_h^{e,dt}(\boldsymbol{v}_h) \quad \forall \boldsymbol{v}_h \in \mathscr{V}, \end{aligned} \tag{12}$$

$$\left(\frac{\partial \boldsymbol{u}_h}{\partial t}, \boldsymbol{w}_h \right)_\Omega - (\boldsymbol{z}_h, \boldsymbol{w}_h)_\Omega = 0 \quad \forall \boldsymbol{w}_h \in \mathscr{V}. \tag{13}$$

The term $\ell_h^{e,dt}(\boldsymbol{v}_h)$ is defined similarly as $\ell_h^e(\boldsymbol{v}_h)$ with the exception that the functions $\boldsymbol{u}_D, \boldsymbol{g}_N$ are replaced with $\partial \boldsymbol{u}_D/\partial t$, $\partial \boldsymbol{g}_N/\partial t$, respectively. By $(\cdot,\cdot)_\Omega$ we denote the $[L^2(\Omega)]^2$-scalar product.

3.3 Time Discretization

We consider two schemes based on finite-difference approximations in time. The process of the derivation of the full discretization is well-known and hence we shall only present the finite-difference approximations here. The backward-Euler (BE) scheme is based on the approximation

$$\frac{\partial \boldsymbol{u}}{\partial t}(t) \approx \frac{\boldsymbol{u}(t) - \boldsymbol{u}(t-\tau)}{\tau}.$$

The second finite-difference scheme is based on the second-order backward-difference formula

$$\frac{\partial \boldsymbol{u}}{\partial t}(t) \approx \frac{3\boldsymbol{u}(t) - 4\boldsymbol{u}(t-\tau) + \boldsymbol{u}(t-2\tau)}{2\tau}.$$

In order to define the space-time discontinuous Galerkin method, let us introduce the one-sided limits and the jump of a function $\boldsymbol{v} \in [S_{h\tau}^{pq}]^2$ at time t_m:

$$\boldsymbol{v}_m^+ = \lim_{s\to 0_+} \boldsymbol{v}(t_m + s), \quad \boldsymbol{v}_m^- = \lim_{s\to 0_+} \boldsymbol{v}(t_m - s), \quad \{\boldsymbol{v}\}_m = \boldsymbol{v}_m^+ - \boldsymbol{v}_m^-. \tag{14}$$

The approximate space-time DG solution of problem (5)–(8) is defined as a couple $\boldsymbol{u}_{h\tau}, \boldsymbol{z}_{h\tau} \in [S_{h\tau}^{pq}]^2$ satisfying

$$\begin{aligned}
&\int_{I_m} \left(\rho \left(\frac{\partial \boldsymbol{z}_{h\tau}}{\partial t}, \boldsymbol{v}_{h\tau} \right)_\Omega + c_M \rho\, (\boldsymbol{z}_{h\tau}, \boldsymbol{v}_{h\tau})_\Omega + a_h(\boldsymbol{u}_{h\tau}, \boldsymbol{v}_{h\tau}) \right. \\
&\qquad \left. + c_K a_h(\boldsymbol{z}_{h\tau}, \boldsymbol{v}_{h\tau}) \right) \mathrm{d}t + (\{\boldsymbol{z}_{h\tau}\}_{m-1}, \boldsymbol{v}_{h\tau}(t_{m-1}+))_\Omega \\
&\qquad = \int_{I_m} \left(\ell_h^s(\boldsymbol{v}_{h\tau}) + \ell_h^e(\boldsymbol{v}_{h\tau}) + c_K \ell_h^{e,dt}(\boldsymbol{v}_{h\tau}) \right) \mathrm{d}t \quad \forall \boldsymbol{v}_{h\tau} \in [S_{h\tau}^{sq}]^2, \\
&\int_{I_m} \left(\left(\frac{\partial \boldsymbol{u}_{h\tau}}{\partial t}, \boldsymbol{w}_{h\tau} \right)_\Omega - (\boldsymbol{z}_{h\tau}, \boldsymbol{w}_{h\tau})_\Omega \right) \mathrm{d}t + (\{\boldsymbol{u}_{h\tau}\}_{m-1}, \boldsymbol{w}_{h\tau}(t_{m-1}+))_\Omega = 0 \\
&\qquad \forall \boldsymbol{w}_{h\tau} \in [S_{h\tau}^{sq}]^2, \quad m = 1, \dots, M.
\end{aligned} \tag{15}$$

The initial states $\boldsymbol{u}_h(0-)$, $\boldsymbol{z}_h(0-) \in [S_{hp}]^2$ are defined by $(\boldsymbol{u}_h(0-), \boldsymbol{v}_h)_\Omega = (\boldsymbol{u}_0, \boldsymbol{v}_h)_\Omega$, $(\boldsymbol{z}_h(0-), \boldsymbol{v}_h)_\Omega = (\boldsymbol{z}_0, \boldsymbol{v}_h)_\Omega$ for all $\boldsymbol{v}_h \in [S_{hp}]^2$.

In all three cases, the resulting linear systems are solved using the direct solver UMFPACK.

4 Numerical Experiments

Here we present numerical results for a simple model problem solved by the STDG method with $\theta = -1$ (nonsymmetric version of the space discretization). We assume that the domain Ω is represented by a rectangular elastic material, which is 2 cm long and 2 mm wide. We consider the following material properties: density $\varrho = 1{,}100\,\mathrm{kg.m}^{-3}$, Young's modulus $E = 10^5\,\mathrm{kg.m}^{-1}.\mathrm{s}^{-2}$, Poisson's ratio $\nu = 0.4$. The Lamè parameters λ and μ can be computed from E and ν by relations $\lambda = \frac{E\nu}{(1+\nu)(1-2\nu)}$, $\mu = \frac{E}{2(1+\nu)}$. These parameters correspond to a very soft, rubber-like material. The material is exposed to a horizontal surface force in the direction of the negative part of the x_1-axis for a short period of time. The lower left corner of the domain is at the point $[-0.001, -0.01]$ and the upper right corner is at the point $[0.001, 0.01]$. On the fixed part of the boundary, where $x_2 = -0.01$ and which is denoted by Γ_D, we prescribe the Dirichlet boundary condition (2) with $\boldsymbol{u}_D = \mathbf{0}$. The Neumann boundary condition (2) is prescribed on the rest of the boundary $\partial\Omega$, denoted by Γ_N, where we put $\boldsymbol{g}_N = (-20, 0)^T$ for $t < 0.02\,\mathrm{s}$, $x_1 = 0.001$ and $x_2 > 0.005$, and $\boldsymbol{g}_N = \mathbf{0}$ otherwise. Finally, we set $T = 0.5\,\mathrm{s}$, $c_M = 0.1\,\mathrm{s}^{-1}$ and $c_K = 0$.

Figure 1 shows the model scheme and the time evolution of the computed displacement at several time instants. Figure 2 shows the evolution of the displacement at the fixed spatial point $[-0.001, 0.01]$ (upper left corner) obtained by the three presented numerical methods. Here STDGM 1 denotes the space-time discontinuous Galerkin method with the time polynomial degree $q = 1$. For all the computations the space polynomial degree p was set to 1, i.e. linear elements in space were used.

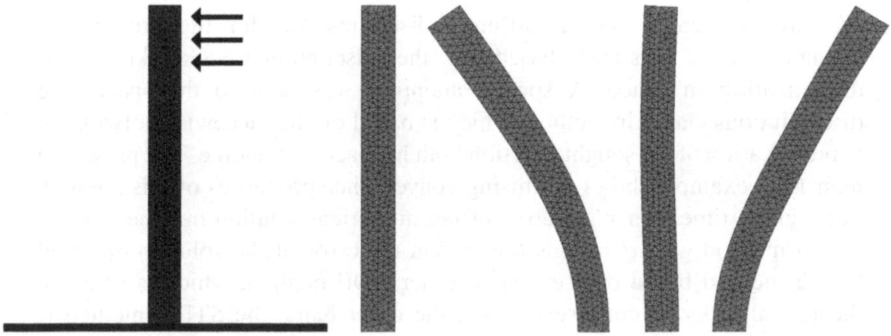

Fig. 1 Schema of the model problem (*left*) and the visualization of the evolution of the displacement function $\boldsymbol{u}$ at the time instants $t = 0, 0.01, 0.02, 0.03$ s

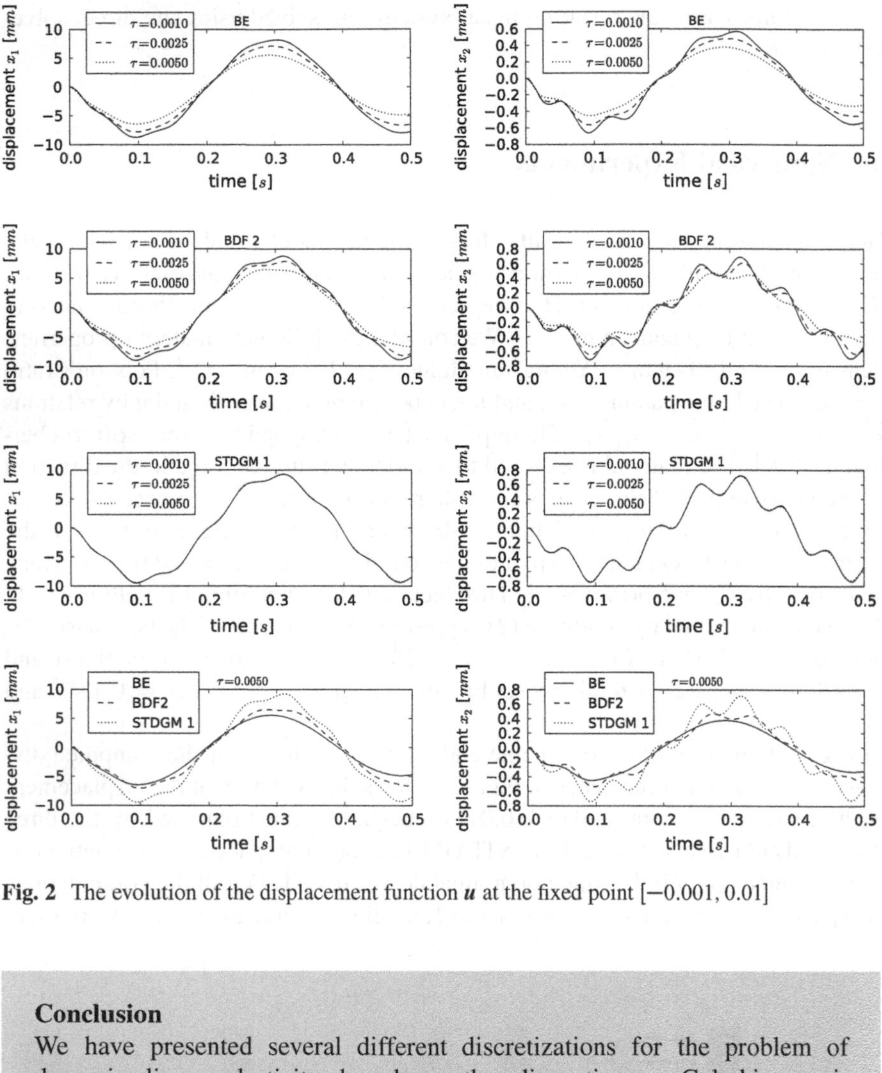

Fig. 2 The evolution of the displacement function $\boldsymbol{u}$ at the fixed point $[-0.001, 0.01]$

Conclusion

We have presented several different discretizations for the problem of dynamic linear elasticity based on the discontinuous Galerkin semi-discretization in space. A special attention was paid to the space-time discontinuous Galerkin method, which is based on the piecewise polynomial approximation of the sought function both in space and in time. The presented numerical example shows promising convergence properties of this method. For a given time step τ the error of the numerical solution obtained by the STDG method with $q = 1$ is lower than the error of the solution obtained by the method based on the second-order BDF method, which is of equal theoretical order of convergence. On the other hand, the STDG method is

(continued)

more expensive in terms of the computational time. This is caused by a larger system of linear algebraic equations, which has to be solved at each time level, and by the quadrature rules, which have to be applied not only in space but also in time.

The future work will be focused towards the analysis of the convergence of the STDG method and its comparison with other methods on more sophisticated test problems.

Acknowledgements This work was supported by the grants P101/11/0207 (J. Horáček) and 13-00522S (M. Feistauer) of the Czech Science Foundation. The work of M. Hadrava was supported by the grant GAChU 549912, the work of A. Kosík was supported by the grant SVV-2014-260106, both financed by the Charles University in Prague.

References

1. P.G. Ciarlet, *The Finite Element Method for Elliptic Problems* (North-Holland, Amsterdam, 1978)
2. B. Cockburn, G.E. Karniadakis, C.W. Shu (Eds.), *Discontinuous Galerkin Methods, Theory, Computation and Applications* (Springer, Berlin, 2000)
3. C. Johnson, Discontinuous Galerkin finite element methods for second order hyperbolic problems. Comput. Methods Appl. Mech. Eng. **107**, 117–129 (1993)
4. B. Rivière, *Discontinuous Galerkin Methods for Solving Elliptic and Parabolic Equations: Theory and Implementation* (SIAM, Philadelphia, 2008)

Mimetic Finite Difference Method for Shape Optimization Problems

P.F. Antonietti, Nadia Bigoni, and Marco Verani

Abstract We test the performance of the Mimetic Finite Difference method applied to a wide class of shape optimization problems. Adaptive strategies based on heuristic error indicators are also considered to validate the effectiveness of the numerical scheme.

1 Introduction

In this paper we are interested in solving shape optimization problems by using the Mimetic Finite Difference (MFD) method. For an introduction of the MFD method applied to elliptic problems we refer to, e.g., [3, 4]. Thanks to a great flexibility allowed in the choice of the grid, the MFD method turns out to be a very promising technology in the context of the approximation of shape optimization problems. Standard numerical methods such as finite elements, finite volumes and spectral elements, usually require a massive use of re-meshing techniques to preserve the geometrical regularity of the computational domain and as a consequence, the computational cost could become prohibitive (see, e.g., [7]). Since the MFD method can deal with grids made of very general polygons/polyhedra, we investigate the possibility of obtain reliable numerical simulations without resorting to any re-meshing strategy.

The paper is organized as follows: in Sect. 2 we study the application of the MFD method to three different shape optimization problems. The first two examples are classical shape optimization problems governed by an elliptic equation and a Stokes equation, respectively, whereas the last example is related to the solution of an elliptic free-boundary problem. Finally, in Sect. 3 we briefly explore the possibility of incorporating an adaptive procedure into the optimization process by showing some numerical results.

P.F. Antonietti • N. Bigoni • M. Verani (✉)
MOX-Dipartimento di Matematica, Politecnico di Milano, P.zza Leonardo da Vinci 32, 20133 Milano, Italy
e-mail: paola.antonietti@polimi.it; nadia.bigoni@polimi.it; marco.verani@polimi.it

A. Abdulle et al. (eds.), *Numerical Mathematics and Advanced Applications ENUMATH 2013*, Lecture Notes in Computational Science and Engineering 103,
DOI 10.1007/978-3-319-10705-9_12

2 MFD Method Applied to Shape Optimization Problems

Let Ω be an open, bounded set of $\mathbb{R}^2$, with a polygonal boundary $\Gamma := \partial\Omega$, and let $\mathcal{J}(\Omega, y(\Omega))$ be a cost functional, which depends on Ω itself and on the solution $y(\Omega)$ of the following boundary value problem

$$Ly(\Omega) = 0 \quad \text{in } \Omega, \tag{1}$$

where L is a differential operator. We are interested in solving the following minimization problem:

$$\text{find } \Omega^* \in \mathcal{A} : \mathcal{J}(\Omega^*, y(\Omega^*)) = \inf_{\Omega \in \mathcal{A}} \mathcal{J}(\Omega, y(\Omega)), \tag{2}$$

where $\mathcal{A}$ is a set of admissible domains in $\mathbb{R}^2$. Assuming that a local minimizer Ω^* of (2) exists, the shape optimization problem (2) can be solved by building a sequence of domains $\{\Omega^{(k)}\}$ for $k > 0$, in such a way that $\Omega^{(k)}$ converges to Ω^* as k approaches infinity (cf. [5]). Roughly speaking, if $\Omega^{(k)}$ is the domain at iteration k, then $\Omega^{(k+1)}$ is updated by

$$\Omega^{(k+1)} = \Omega^{(k)} + \mu^{(k)} V^{(k)},$$

where $\mu^{(k)}$ is a parameter which regulates the "length" of the movement and $V^{(k)}$ is an *admissible* descent direction. In other words, $V^{(k)}$ is such that the *shape derivative* $d\mathcal{J}(\Omega^{(k)}; V^{(k)})$ of the considered functional is negative. For a precise definition of the *shape derivative* and other technical details we refer to [5]. Problem (2) can be solved by means of the steepest-descent like algorithm (cf. [6]), described in Algorithm 2.1.

Algorithm 2.1: Shape optimization algorithm

1: Given the initial domain $\Omega^{(0)}$, set $k = 1$
2: `SOLVE` problem (1) and find $y^{(k)} = y(\Omega^{(k)})$
3: `COMPUTE` the shape derivative $d\mathcal{J}(\Omega; V)$ of the given functional
4: `COMPUTE` a descent direction $V^{(k)}$
5: `FIND` an admissible stepsize $\mu^{(k)}$
6: `UPDATE` $\Omega^{(k+1)} = \Omega^{(k)} + \mu^{(k)} V^{(k)}$, set $k = k + 1$ and GOTO step 2

2.1 Elliptic Problem

In the first example we solve the benchmark problem introduced in [6]. Let us consider the domain $\Omega \subset \mathbb{R}^2$ with $\partial\Omega = \Gamma_f \cup \Sigma_1 \cup \Sigma_2$, where Γ_f, Σ_1 and Σ_2 are disjoint (possibly empty) subset of $\partial\Omega$. Moreover, let D be an open bounded subset of Ω (see Fig. 1). Then, let $y(\Omega)$ be the solution of the following elliptic problem:

$$-\Delta y = 0 \quad \text{in } \Omega, \quad y = 0 \quad \text{on } \Sigma_1, \quad \partial_n y = 0 \quad \text{on } \Sigma_2, \quad \partial_n y = 1 \quad \text{on } \Gamma_f. \tag{3}$$

The cost functional we aim at minimizing is set as

$$\mathscr{J}(\Omega, y(\Omega)) := \frac{1}{2}\int_D \left(y(\Omega) - z_g\right)^2 \, \mathrm{d}x + \frac{\gamma}{2}\left(\int_\Gamma \mathrm{dS} - P\right)^2, \tag{4}$$

where γ and P are positive constants. The set $\mathscr{A}$ of admissible domains is represented by all domains obtained through a deformation of Ω by keeping Σ_1 and Σ_2 fixed and by moving only Γ_f in such a way that $\Gamma_f \cap D = \emptyset$. The mimetic discretization of problem (3) and a detailed discussion of the MFD method for elliptic problems can be found for example in [4].

Let $\mathbf{x} = (x_1, x_2)$, and let $\|\cdot\|$ denote the Euclidean norm. In the numerical test, we choose the region D equal to the half ring $\{2 \le \|\mathbf{x}\| \le 2.5\} \cap \{x_2 > 0\}$ and z_g in (4) as the exact solution of (3) on $\Omega = \{1 < \|\mathbf{x}\| < 3\} \cap \{x_2 > 0\}$. A global minimizer exists and it is exactly given by $\Omega^* = \{1 < \|\mathbf{x}\| < 3\} \cap \{x_2 > 0\}$ (cf. Fig. 1). We point out that in this set of experiments no adaptive nor remeshing technique are used. In Fig. 2 we report the initial computational domain (left) and a zoom of the obtained configuration (right) after four iterations of Algorithm 2.1.

The computed value of the functional (4) starts from 3.2990e − 01 and after four iterations reduces to 9.3176e − 04. A finer grid in the region D is employed in order to accurately approximate the cost functional. We can state that the method seems to be sufficiently robust despite most of the elements around the moving boundary become very stretched.

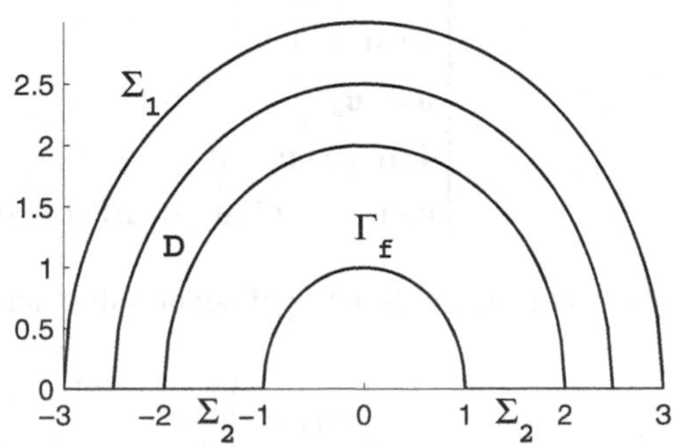

Fig. 1 Sketch of the optimal domain Ω^*

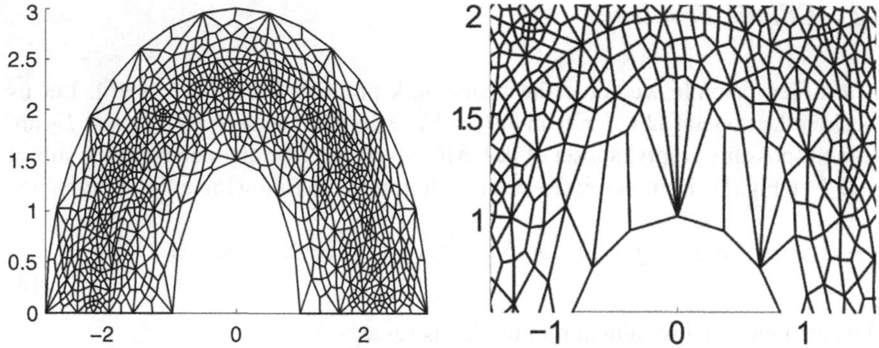

Fig. 2 Shape optimization problem (non-adaptive strategy). Initial computational domain Ω_0 (*left*) and zoom of the final obtained domain (*right*)

2.2 Drag Minimization

In the second example, we are interested in modeling the flow of a fluid around an obstacle. In this numerical test, the initial computational domain is set as follows:

$$\Omega : \{[-1,-1] \times [0,1]\} \cap \{x^2 + y^2 \geq 0.16\},$$

(see Fig. 3 (left)). In this case, the obstacle is represented by the half circle lying on the lower part of the domain and it is denoted by Γ_f. The remaining parts of the boundary are labeled as follows: $\Gamma_{\text{in}} = \{(x,y) : x = -1\}$ and $\Gamma_{\text{out}} = \{(x,y) : x = 1\}$ are the inflow and the outflow layers, respectively while $\Gamma_s = \{(x,y) : y = 0\}$ and $\Gamma_w = \{(x,y) : y = 1\}$ are the lower and upper part of the channel, respectively. The set $\mathscr{A}$ of admissible domains contains all domains obtained through a deformation of Ω by moving only Γ_f and keeping fixed the remaining parts of the boundary. The fluid flow is modeled by the following linear Stokes problem:

$$\begin{cases} -\operatorname{div}(\mathbf{T}(\mathbf{u},p)) = 0 & \text{in } \Omega, \\ \operatorname{div}\mathbf{u} = 0 & \text{in } \Omega, \\ \mathbf{u} = \mathbf{u}_d & \text{on } \Gamma_w \cup \Gamma_f \cup \Gamma_{\text{in}}, \\ \mathbf{T}(\mathbf{u},p)\cdot\mathbf{n} = 0 & \text{on } \Gamma_{\text{out}}, \\ \mathbf{u}\cdot\mathbf{n} = 0 \;\; (\mathbf{T}(\mathbf{u},p)\cdot\mathbf{n})\cdot\mathbf{t} = 0 & \text{on } \Gamma_s, \end{cases} \tag{5}$$

where $\mathbf{T}(\mathbf{u},p) := 2\epsilon(\mathbf{u}) - p\mathbf{I}$ denotes the Cauchy stress tensor. We set

$$\mathbf{u}_d = \begin{cases} [1 - y^2 \; 0]^T & \text{on } \Gamma_{\text{in}} \\ 0 & \text{on } \Gamma_f \cup \Gamma_w\,. \end{cases}$$

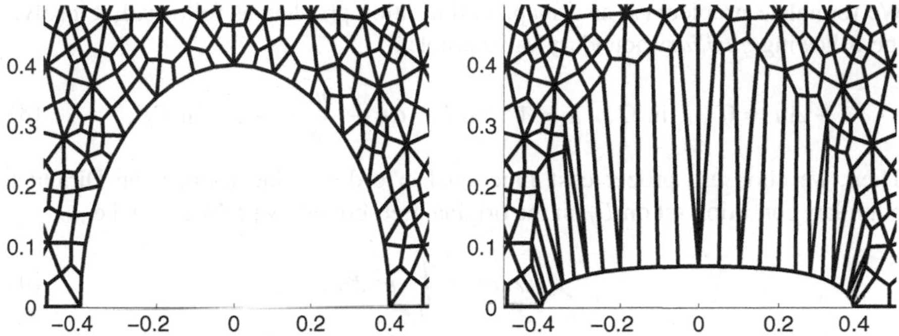

Fig. 3 Drag minimization. Zooms of the initial (*left*) and final (*right*) computed domains

Note that on Γ_s we impose an axial-symmetry boundary condition while on Γ_w we set a non-slip boundary condition. Again, we do not add any details about the mimetic discretization of problem (5) and refer to [2].

In this example, we choose to minimize the following cost functional:

$$\mathscr{J}(\Omega, \mathbf{u}, p) := -\int_{\Gamma_f} (\mathbf{T}(\mathbf{u}, p)\mathbf{n}) \cdot \hat{\mathbf{v}}_\infty \, \mathrm{dS} + \frac{\lambda}{2}\left(|\Omega_0| - \int_\Omega \mathrm{dx}\right)^2, \tag{6}$$

where $(\mathbf{u}, p)$ solves (5), $\hat{\mathbf{v}}_\infty = [1, 0]$ is the direction of the fluid and $|\Omega_0|$ is a given target volume value. The first term of (6) represents the drag of the fluid, while the second one penalizes the volume constraint. In Fig. 3 we plot a zoom of the initial and final computational domains. The computed value of the drag starts from 4.41887e-01 and it reduces progressively to 3.69713e − 02 after five iterations. We note that the obtained final configuration is in agreement with the so-called "rugby-ball" optimal shape known in the literature [8].

2.3 *Free-Boundary Problem*

In the last example, we are interested in solving the free-boundary elliptic problem taken from [9]. We consider an annular domain, where the fixed boundary is $\Gamma = \{\|\mathbf{x}\| = 1\}$ and we want to find the *moving* free-boundary $\Gamma_f := \partial\Omega \setminus \overline{\Gamma}$, so that

$$-\Delta u = 0 \quad \text{in } \Omega, \quad u = 1 \quad \text{on } \Gamma, \quad u = 0 \quad \text{on } \Gamma_f, \quad \frac{\partial u}{\partial n} = -1 \quad \text{on } \Gamma_f. \tag{7}$$

We formulate problem (7) as a shape optimization problem as follows. Let u solve the following *auxiliary* boundary value problem:

$$-\Delta u = 0 \quad \text{in } \Omega, \quad u = 1 \quad \text{on } \Gamma, \quad \alpha u + \frac{\partial u}{\partial n} = -1 \quad \text{on } \Gamma_f\,. \tag{8}$$

Then, we choose a proper cost functional in order to incorporate the Dirichlet boundary condition set on Γ_f in the original free-boundary problem (7), i.e.,

$$\mathscr{J}(\Omega, u) = \int_{\Gamma_f} u^2\,\mathrm{dS}. \tag{9}$$

Since the exact solution of the free-boundary problem (7) is zero on Γ_f, the dumping parameter $\alpha > 0$ appearing in (8) can be chosen freely. However, following [9], it turns out that $\alpha = H$, with H being the mean curvature of Γ_f, is a good choice leading in practice to a robust numerical procedure. We iteratively solve the problem (8) on the half-annulus by imposing proper axial-symmetry boundary conditions on the x-axis (cf. Fig. 4). The final obtained computational domain is reported in Fig. 5 (left). The value of the computed functional reduces progressively, as shown in Fig. 5 (right), where we plot the value of (9) versus the number of total iterations.

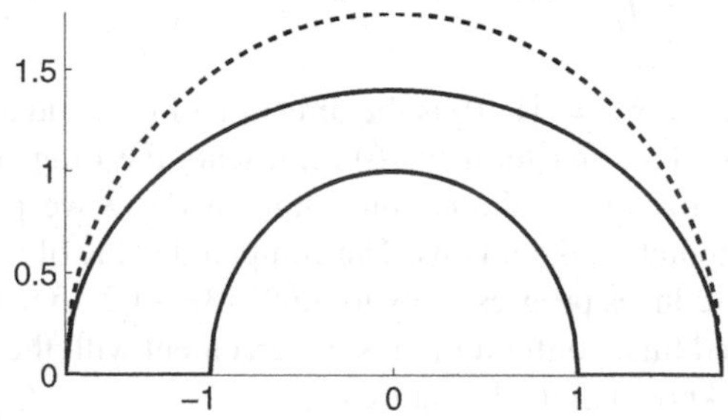

Fig. 4 Free-boundary problem. Sketch of the initial (*solid line*) and final (*dotted line*) domains

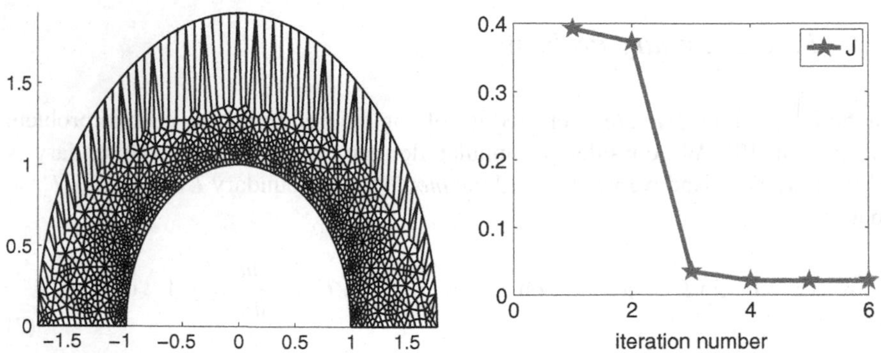

Fig. 5 Free-boundary problem. Final computational domain (*left*) and computed functional (9) versus the number of iteration (*right*)

3 Adaptive Strategy

In this section, we briefly explore the possibility of incorporating mesh adaptivity into the optimization process. An example of a similar approach in the FEM context can be found in [7].

We run the same numerical experiment presented in Sect. 2.1 and we decide a priori to perform an adaptive refinement step every two iterations of the minimization process. To perform the adaptive procedure we employ heuristic indicators, given by the sum of the following two local error indicators:

(η_1) For every polygon $E \subset \Omega_h$ the indicator $\eta_1(E)$ is the local discrete $H^1(E)$ seminorm of the MFD approximate solution to (3);

(η_2) For every polygon $E \subset D$ the indicator $\eta_2(E)$ is the MFD approximation of $\frac{1}{2}\int_E \left(y(\Omega) - z_g\right)^2\, dV$ and is set to zero outside D (cf. Fig. 2 (left)).

The local error indicators $(\eta_1+\eta_2)(E)$ are then used to mark the elements that has to be refined, while the marking procedure relies on the Dörfler strategy with marking parameter $\theta = 0.5$. For a more detailed description of the refinement modulus we refer to [1, Section 4.1].

The optimal discrete configuration is obtained by the algorithm after six iterations (cf. Fig. 1 for the exact optimal domain). In Fig. 6 we plot two zooms of the final configuration. Comparing Fig. 2 (right) with Fig. 6 (right) we can observe that here, thanks to the adaptive procedure, the element close to the moving boundary result less stretched. Moreover, due to the error indicator η_2, the adaptive algorithm correctly refines the elements in the region D (see Fig. 6 (left)).

We analyze the performance of the adaptive and non-adaptive strategies by comparing the obtained value of the cost functional. In Table 1 we report the computed values of $J_1 = \frac{1}{2}\int_D \left(y(\Omega) - z_g\right)^2\, dV$ together with the corresponding number of degrees of freedom. From a closer inspection, it is evident that at comparable number of degrees of freedom the adaptive strategy obtains lower values of $\mathcal{J}$.

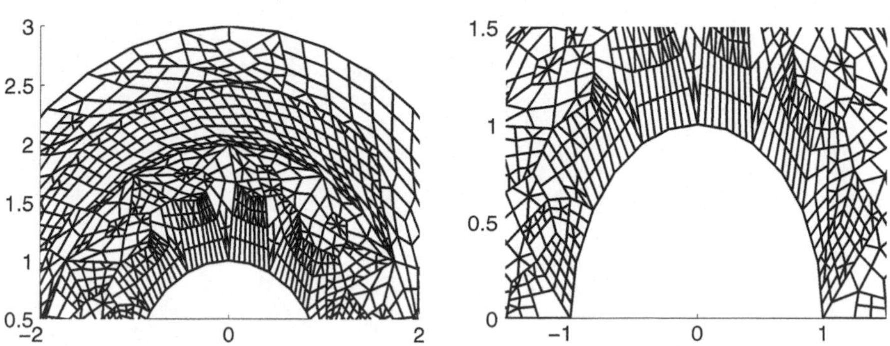

Fig. 6 Shape optimization problem (adaptive strategy). Zooms of the final obtained configuration

Table 1 Adaptive and non-adaptive strategies: cost functional versus dofs

Iteration	Non-adaptive		Adaptive	
	ndofs	J_1	ndofs	J_1
0	1,207	9.990754E−03	157	8.780796E−03
1	1,207	6.501192E−03	157	5.103216E−03
2	1,207	1.493240E−03	283	5.735032E−04
3	1,207	9.618140E−04	283	4.441652E−04
4	1,207	8.911968E−04	564	1.368972E−04
5	–	–	564	5.677177E−05
6	–	–	1,194	7.538973E−05

References

1. P.F. Antonietti, L. Beirão da Veiga, C. Lovadina, M. Verani, Hierarchical a posteriori error estimators for the mimetic discretization of elliptic problems. SIAM J. Numer. Anal. **51**(1), 654–675 (2012)
2. L. Beirão da Veiga, V. Gyrya, K. Lipnikov, G. Manzini, Mimetic finite difference method for the Stokes problem on polygonal meshes. J. Comput. Phys. **228**(19), 7215–7232 (2009)
3. F. Brezzi, A. Buffa, K. Lipnikov, Mimetic finite differences for elliptic problems. M2AN Math. Model. Numer. Anal. **43**(2), 277–295 (2009)
4. F. Brezzi, K. Lipnikov, M. Shashkov, Convergence of the mimetic finite difference method for diffusion problems on polyhedral meshes. SIAM J. Numer. Anal. **43**(5), 1872–1896 (2005) (electronic)
5. M.C. Delfour, J.-P. Zolésio, *Shapes and Geometries*. Advances in Design and Control, vol. 22, 2nd edn. (Society for Industrial and Applied Mathematics (SIAM), Philadelphia, 2011)
6. G. Doğan, P. Morin, R.H. Nochetto, M. Verani, Discrete gradient flows for shape optimization and applications. Comput. Methods Appl. Mech. Eng. **196**(37–40), 3898–3914 (2007)
7. P. Morin, R. Nochetto, M. Pauletti, M. Verani, Adaptive finite element method for shape optimization. ESAIM Control Optim. Calc. Var. **18**(4), 1122–1149 (2012)
8. O. Pironneau, On optimum profiles in Stokes flow. J. Fluid Mech. **59**, 117–128 (1973)
9. T. Tiihonen, Shape optimization and trial methods for free boundary problems. RAIRO Modél. Math. Anal. Numér. **31**(7), 805–825 (1997)

Semi-discrete Time-Dependent Fourth-Order Problems on an Interval: Error Estimate

Dalia Fishelov

Abstract We present high-order compact schemes for fourth-order time-dependent problems, which are related to the "buckling plate" or the "clamping plate" problems. Given a mesh size h, we show that the truncation error is $O(h^4)$ at interior points and $O(h)$ at near-boundary points. In addition, the convergence of these schemes is analyzed. Although the truncation error is only of first-order at near-boundary points, we have proved that the error of these schemes converges to zero as h tends to zero at least as $O(h^{3.5})$. Numerical results are performed and they calibrate the high-order accuracy of the schemes. It is shown that the numerical rate of convergence is actually four, thus the error tends to zero as $O(h^4)$.

1 Introduction

Time-dependent fourth-order differential problems play an important role in various areas of physics. In mechanics they are involved in plate problems, such as the "buckling plate" or the "clamping plate" problem. In fluid dynamics they are used in the Navier-Stokes equations. In this paper we are interested in two time dependent problems, which are related to fourth-order problems.

The first one is

$$u_{xxt} = u_{xxxx} + b\ u_{xx} + c\ u_x + d\ u + f(x,t), \quad 0 < x < 1, \quad t > 0 \tag{1}$$

and the second is

$$u_t = -u_{xxxx} + b\ u_{xx} + c\ u_x + d\ u + f(x,t), \quad 0 < x < 1, \quad t > 0. \tag{2}$$

Both problems are supplemented with boundary conditions

$$u(0,t) = \partial_x u(0,t) = 0, \quad u(1,t) = \partial_x u(1,t) = 0 \tag{3}$$

D. Fishelov (✉)
Afeka-Tel-Aviv Academic College for Engineering, 218 Bnei-Efraim St. Tel-Aviv 69107, Israel
e-mail: daliaf@post.tau.ac.il

A. Abdulle et al. (eds.), *Numerical Mathematics and Advanced Applications ENUMATH 2013*, Lecture Notes in Computational Science and Engineering 103,
DOI 10.1007/978-3-319-10705-9_13

and the initial condition

$$u(x,0) = g(x), \quad 0 \le x \le 1. \tag{4}$$

Consider now the finite interval $I = [0, 1]$ with the grid

$$x_0 = 0 < x_1 < \cdots < x_{N-2} < x_{N-1} < x_N = 1, \tag{5}$$

where $x_j = jh$, $j = 1, \cdots, N$ and $h = 1/N$. In order to approximate the solutions of Problems (1) and (2) one needs to approximate the operators ∂_x^4, ∂_x^2 and ∂_x. We approximate ∂_x^4 by δ_x^4, where δ_x^4 is the three-point compact operator defined by (see [3,4]).

$$\delta_x^4 v_j = \frac{12}{h^2}\left(\frac{v_{x,j+1} - v_{x,j-1}}{2h} - \frac{v_{j+1} + v_{j-1} - 2v_j}{h^2}\right) = \frac{12}{h^2}(\delta_x v_{x,j} - \delta_x^2 v_j), \tag{6}$$

for $1 \le j \le N-1$. The operator ∂_x^2 is a approximated by $\tilde{\delta}_x^2$, where

$$\tilde{\delta}_x^2 v_j = 2\delta_x^2 v_j - \delta_x v_{x,j} = \delta_x^2 v_j - \frac{h^2}{12}\delta_x^4 v_j, \quad 1 \le j \le N-1. \tag{7}$$

Here $v_{x,j}$ is the Hermitian derivative of v at point x_j. It is defined by

$$\frac{1}{6}v_{x,j-1} + \frac{2}{3}v_{x,j} + \frac{1}{6}v_{x,j+1} = \frac{v_{j+1} - v_{j-1}}{2h}, \quad 1 \le j \le N-1. \tag{8}$$

This operator was extensively studied in previous works [2, 4]. Problem (1) is approximated by the semi-discrete finite-difference scheme

$$\frac{d}{dt}\tilde{\delta}_x^2 v_j = \delta_x^4 v_j + b\,\tilde{\delta}_x^2 v_j + c\,v_{x,j} + d\,v_j + f(x_j,t) \tag{9}$$

and Problem (2) by

$$\frac{d}{dt}v_j = -\delta_x^4 v_j + b\,\tilde{\delta}_x^2 v_j + c\,v_{x,j} + d\,v_j + f(x_j,t). \tag{10}$$

Let v, w be two discrete functions, defined on the grid (5) and vanishing at the two endpoints x_0, x_N. We define the discrete inner product $(v,w)_h$ and the discrete norm $|v|_h$ as

$$(v,w)_h = \sum_{j=1}^{N-1} v_j w_j h, \quad |v|_h = \sqrt{(v,v)_h}. \tag{11}$$

2 Consistency for Compact Operators on an Interval

2.1 The Truncation Error

Here we consider the truncation errors related to the operators δ_x^4, $\tilde{\delta}_x^2$ and the Hermitian derivative v_x. Let σ_x be the (Simpson) operator [4]

$$\sigma_x v_j = \frac{1}{6}v_{j-1} + \frac{2}{3}v_j + \frac{1}{6}v_{j+1}. \tag{12}$$

We consider first the truncation error related to δ_x^4. We have the following inequalities (see [4]):

$$|\sigma_x \delta_x^4 u_j^* - \sigma_x (u^{(4)})^*(x_j)| \le Ch^4 \|u^{(8)}\|_{L^\infty}, \quad 2 \le j \le N-2. \tag{13}$$

$$|\sigma_x \delta_x^4 u_j^* - \sigma_x (u^{(4)})^*(x_j)| \le Ch \|u^{(5)}\|_{L^\infty} \quad j = 1, N-1. \tag{14}$$

Consider now the operator $\tilde{\delta}_x^2$

$$-\tilde{\delta}_x^2 u_j = -2\delta_x^2 u_j + \delta_x u_{x,j} = -\delta_x^2 u_j + \frac{h^2}{12}\delta_x^4 u_j. \tag{15}$$

Operating with σ_x on the last equality, we have

$$-\sigma_x \tilde{\delta}_x^2 u_j = -\sigma_x \delta_x^2 u_j + \frac{h^2}{12}\sigma_x \delta_x^4 u_j. \tag{16}$$

Using the truncation error for $-\delta_x^2$, we have $-\delta_x^2 u_j = -\partial_x^2 u(x_j,t) - \frac{h^2}{12}\partial_x^4 u(x_j,t) + O(h^4)$. Thus,

$$-\sigma_x \delta_x^2 u_j = -\sigma_x \partial_x^2 u(x_j,t) - \frac{h^2}{12}\sigma_x \partial_x^4 u(x_j,t) + O(h^4). \tag{17}$$

Inserting the last equality in (16), we have

$$-\sigma_x \tilde{\delta}_x^2 u_j = -\sigma_x \partial_x^2 u(x_j,t) + \frac{h^2}{12}\sigma_x (\delta_x^4 u_j - \partial_x^4 u(x_j,t)) + O(h^4). \tag{18}$$

Combing the above with (13) and (14), we find that

$$-\sigma_x \tilde{\delta}_x^2 u_j = -\sigma_x \partial_x^2 u(x_j,t) + O(h^4), \quad 2 \le j \le N-2. \tag{19}$$

$$-\sigma_x \tilde{\delta}_x^2 u_j = -\sigma_x \partial_x^2 u(x_j,t) + O(h^3), \quad j = 1, N-1. \tag{20}$$

In addition (see [3, 4]), we have

$$u_{x,j} = \partial_x u(x_j, t) + O(h^4), \quad 1 \le j \le N-1. \tag{21}$$

3 The Time-Dependent Case $u_{xxt} = u_{xxxx} + b\, u_{xx} + c\, u_x + d\, u + f(x,t)$

Consider the time-dependent fourth-order problem

$$\begin{cases} u_{xxt} = u_{xxxx} + b\, u_{xx} + c\, u_x + d\, u + f(x,t), \quad 0 < x < 1,\ t > 0 \\ u(0,t) = 0,\ u(1,t) = 0,\ u_x(0,t) = 0,\ u_x(1,t) = 0, \quad t > 0 \\ u(x,0) = g(x), \quad 0 \le x \le 1. \end{cases} \tag{22}$$

The canonical semi-discrete approximation of (22) on the grid (5) is

$$\frac{\partial}{\partial t}\tilde{\delta}_x^2 v_j(t) = \delta_x^4 v_j(t) + b\, \tilde{\delta}_x^2 v_j(t) + c\, v_{x,j} + d\, v_j(t) + f_j(t), \quad j = 1, \dots, N-1,$$

$$v_0(t) = 0,\ v_N(t) = 0,\ v_{x,0}(t) = 0,\ v_{x,N}(t) = 0, \quad t > 0$$

$$v_j(0) = g_j := g(x_j), \quad j = 0, \dots, N. \tag{23}$$

Although the truncation error deteriorate at near-boundary points, we prove in the following Proposition that the convergence of the approximate solution to the exact one is of high accuracy. A similar result was shown in [1, 6, 7] in cases where the accuracy of the scheme deteriorates near the boundary. In [6] and [7] a hyperbolic system of first order and a parabolic problem were analyzed. In [1] it was proved for a parabolic equation that if the scheme is of order $O(h^\alpha)$ at inner points and of order $O(h^{\alpha-s})$ near the boundary, then if $s = 0, 1$ the accuracy of the scheme is $O(h^\alpha)$. However, if $s \ge 2$ then the overall accuracy the scheme is $O(h^{\alpha-s+3/2})$. In our case $\alpha = 4$ and $s = 3$ so this result yields a convergence rate of 2.5, but we actually prove that the convergence rate is at least 3.5.

Theorem 1 *Let $u(x,t)$ be the exact solution of (1) satisfying the boundary conditions (3) and the initial condition (4). Assume that u has continuous derivatives with respect to x up to order eight on* $[0,1]$ *and up to order* 1 *with respect to t. Let $v(t)$ be the approximation to u, given by the (23). Then, the error $e_j(t) = v_j(t) - u(x_j,t)$ satisfies*

$$\max_{0\le t\le T} |e(t)|_h \le \max_{0\le t\le T} |\delta_x^+ e(t)|_h \le C(T)h^{3.5}, \tag{24}$$

where $C(T)$ depends only on f, g and T.

Proof The error $e_j(t)$ satisfies

$$-\tfrac{\partial}{\partial t}\tilde{\delta}_x^2 e_j(t) = -\delta_x^4 e_j(t) - b\,\tilde{\delta}_x^2 e_j(t) - c\,e_{x,j}(t) - d\,e_j(t) + r_j(t), \;\; j = 1,\ldots,N-1,$$

$$e_0(t) = 0,\; e_N(t) = 0,\; e_{x,0}(t) = 0,\; e_{x,N}(t) = 0, \quad t > 0,$$

$$e_j(0) = 0, \quad j = 0,\ldots,N, \tag{25}$$

where $r_j(t)$ is the truncation error at point x_j at time t. Taking the inner product of (25) with $e(t)$ and using

$$\frac{1}{2}\frac{\partial}{\partial t}(-\tilde{\delta}_x^2 e(t), e(t))_h = (-\frac{\partial}{\partial t}\tilde{\delta}_x^2 e(t), e(t))_h, \tag{26}$$

we find that

$$\begin{aligned}\tfrac{1}{2}\tfrac{\partial}{\partial t}(-\tilde{\delta}_x^2 e(t), e(t))_h &= -(\delta_x^4 e(t), e(t))_h - b\,(\tilde{\delta}_x^2 e(t), e(t))_h \\ &- c\,(e_x(t), e(t))_h - d\,(e(t), e(t))_h + (r(t), e(t))_h.\end{aligned} \tag{27}$$

First consider the term $(e_x(t), e(t))_h$. Using the Cauchy-Schwartz inequality, we have

$$|(e_x(t), e(t))_h| \le |e(t)|_h\,|e_x(t)|_h \le \frac{1}{2}|e(t)|_h^2 + \frac{1}{2}|e_x(t)|_h^2. \tag{28}$$

Since $e_x = \sigma_x^{-1}\delta_x e$ and σ_x^{-1} is bounded (see [2] Equation (51)), we have that $|e_x(t)|_h^2 \le C|\delta_x^+ e(t)|_h^2$. By discrete integration by parts $|\delta_x^+ e(t)|_h^2 = -(\delta_x^2 e(t), e(t))_h$. Using the definition (7) of $-\tilde{\delta}_x^2$ and the coercivity of δ_x^4, we have

$$-(\tilde{\delta}_x^2 e(t), e(t))_h \ge -(\delta_x^2 e(t), e(t))_h. \tag{29}$$

Thus, $|e_x(t)|_h^2 \le -C(\tilde{\delta}_x^2 e(t), e(t))_h$. Therefore,

$$|(e_x(t), e(t))_h| \le -C(\tilde{\delta}_x^2 e(t), e(t))_h + \frac{1}{2}(e(t), e(t))_h. \tag{30}$$

Combining (30) with (27) we obtain

$$\begin{aligned}\tfrac{1}{2}\tfrac{\partial}{\partial t}(-\tilde{\delta}_x^2 e(t), e(t))_h &\le -(\delta_x^4 e(t), e(t))_h - C_1(\tilde{\delta}_x^2 e(t), e(t))_h \\ &+ \tilde{C}(e(t), e(t))_h + (r(t), e(t))_h.\end{aligned} \tag{31}$$

Let us now consider the term $\tilde{C}(e(t), e(t))_h$. Using the definition (7) of $-\tilde{\delta}_x^2$, and the coercivity of δ_x^4, we have

$$-(\tilde{\delta}_x^2 e(t), e(t))_h \geq -(\delta_x^2 e(t), e(t))_h = (\delta_x^+ e(t), \delta_x^+ e(t))_h. \tag{32}$$

By the discrete Poincaré inequality, we find

$$(e(t), e(t))_h \leq C\ (\delta_x^+ e(t), \delta_x^+ e(t))_h. \tag{33}$$

Therefore, $(e(t), e(t))_h \leq -C\ (\tilde{\delta}_x^2 e(t), e(t))_h$. Combining (31) with the last inequality, we obtain

$$\frac{1}{2}\frac{\partial}{\partial t}(-\tilde{\delta}_x^2 e(t), e(t))_h \leq -(\delta_x^4 e(t), e(t))_h - C_1\ (\tilde{\delta}_x^2 e(t), e(t))_h + (e(t), r(t))_h. \tag{34}$$

Consider now the term $(r(t), e(t))_h$. Using the Cauchy-Schwartz inequality, we have

$$\begin{aligned} |(r(t), e(t))_h| &= ((\delta_x^{-4})^{1/2} r(t), (\delta_x^4)^{1/2} e(t))_h \\ &\leq |(\delta_x^{-4})^{1/2} r(t)|_h\ |(\delta_x^4)^{1/2} e(t)|_h \leq \tfrac{1}{2}(r(t), \delta_x^{-4} r(t))_h + \tfrac{1}{2}(\delta_x^4 e(t), e(t))_h \\ &= \tfrac{1}{2}(\sigma_x r(t), \sigma_x^{-1}\delta_x^{-4}\sigma_x^{-1}\sigma_x r(t)) + \tfrac{1}{2}(\delta_x^4 e(t), e(t))_h. \end{aligned} \tag{35}$$

Combining the truncation errors (13) and (14) for $\sigma_x \delta_x^4$, (19) and (20) for $\sigma_x \tilde{\delta}_x^2$ and (21) for the Hermitian derivative e_x, we have

$$(PR(t))^T = [O(h), O(h^4), \ldots, O(h^4), O(h)]. \tag{36}$$

Here, P is the matrix representing the operator $6\sigma_x$ and $R(t)$ is the vector corresponding to $r(t)$. As a result of [5] Equations (111) and (116), we have

$$\begin{aligned} &(a) \qquad |(P^{-1}S^{-1}P^{-1}PR)_i| \leq Ch^4, \ \ 2 \leq i \leq N-2, \\ &(b) \qquad |(P^{-1}S^{-1}P^{-1}PR)_i| \leq Ch^5, \ \ i = 1, N-1, \end{aligned} \tag{37}$$

where S the matrix representing δ_x^4. Using (37) (a),(b) and (36), we find that

$$|(PR(t))^T P^{-1} S^{-1} R(t)| \leq Ch^6. \tag{38}$$

Therefore, $|\sigma_x r(t), \sigma_x^{-1}\delta_x^{-4}\sigma_x^{-1}\sigma_x r(t)| \leq Ch^7$. Combining the last inequality with (35), we obtain

$$|(r(t), e(t))_h| \leq \tfrac{1}{2}(\delta_x^4 e(t), e(t))_h + Ch^7. \tag{39}$$

Inserting (39) in (34), we have

$$\begin{aligned}\frac{1}{2}\frac{\partial}{\partial t}(-\tilde{\delta}_x^2 e, e)_h &\le -\frac{1}{2}(\delta_x^4 e(t), e(t))_h - C_1\,(\tilde{\delta}_x^2 e(t), e(t))_h + Ch^7\\ &\le C_1(-\tilde{\delta}_x^2 e, e)_h + Ch^7.\end{aligned} \tag{40}$$

By Gronwall's inequality $-(\tilde{\delta}_x^2 e(t), e(t))_h \le C(t)h^7$. Using the coercivity property

$$-(e(t), \tilde{\delta}_x^2 e(t))_h \ge (\delta_x^+ e(t), \delta_x^+ e(t))_h, \tag{41}$$

and the discrete Poincaré inequality, we obtain the estimate

$$|e(t)|_h \le C|\delta_x^+ e(t)|_h \le C(T)h^{3.5}, \;\; 0 \le t \le T. \tag{42}$$

□

4 The Time-Dependent Case $u_t = -u_{xxxx} + b\,u_{xx} + c\,u_x + d\,u + f(x,t)$

Consider the time-dependent biharmonic problem

$$\begin{cases} u_t = -u_{xxxx} + b\,u_{xx} + c\,u_x + d\,u + f(x,t), \;\; 0 < x < 1,\\ u(0,t) = 0,\; u(1,t) = 0,\; u_x(0,t) = 0,\; u_x(1,t) = 0, \quad t > 0\\ u(x,0) = g(x), \quad 0 \le x \le 1. \end{cases} \tag{43}$$

The canonical semi-discrete approximation of (43) on the grid (5) is

$$\begin{aligned}&\frac{\partial v_j(t)}{\partial t} = -\delta_x^4 v_j(t) + b\,\tilde{\delta}_x^2 v_j + c\,v_{x,j}(t) + d\,v_j(t) + f_j(t), \quad j = 1,\dots,N-1,\\ &v_0(t) = 0,\; v_N(t) = 0,\; v_{x,0}(t) = 0,\; v_{x,N}(t) = 0, \quad t > 0\\ &v_j(0) = g_j := g(x_j), \quad j = 0,\dots,N.\end{aligned} \tag{44}$$

Define the error $e_j(t)$ by $e_j(t) = v_j(t) - u(x_j,t)$.

Theorem 2 *Let $u(x,t)$ be the exact solution of (2) satisfying the boundary conditions (3) and the initial condition (4). Assume that u has continuous derivatives with respect to x up to order eight on $[0, 1]$ and up to order 1 with respect to t. Let $v(t)$ be* *the approximation to u, given by the (44). Then, the error* $e_j(t) = v_j(t) - u(x_j,t)$ *satisfies*

$$\max_{0\le t\le T} |e(t)|_h \le C(T)h^{3.5}, \tag{45}$$

where $C(T)$ *depends only on* f, g *and* T.

Proof The error $e_j(t)$ satisfies

$$\begin{aligned}
&\tfrac{\partial e_j(t)}{\partial t} = -\delta_x^4 e_j(t) + b\,\tilde{\delta}_x^2 e_j(t) + c\ e_{x,j}(t) + d\ e_j(t) - r_j(t), \quad j = 1,\ldots,N-1,\\
&e_0(t) = 0,\ e_N(t) = 0,\ e_{x,0}(t) = 0,\ e_{x,N}(t) = 0, \quad t > 0\\
&e_j(0) = 0, \quad j = 0,\ldots,N,
\end{aligned} \tag{46}$$

where $r_j(t)$ is the truncation error at point x_j at time t. Taking the inner product of (46) with $e(t)$, and using

$$\frac{1}{2}\frac{\partial}{\partial t}(e(t),e(t))_h = (\frac{\partial}{\partial t}e(t),e(t))_h, \tag{47}$$

we find that

$$\begin{aligned}
\tfrac{1}{2}\tfrac{\partial}{\partial t}(e(t),e(t))_h &= -(\delta_x^4 e(t),e(t))_h + b\ (\tilde{\delta}_x^2 e(t),e(t))_h\\
&\quad + c\ (e_x(t),e(t))_h + d\ (e(t),e(t))_h - (r(t),e(t))_h.
\end{aligned} \tag{48}$$

Considering first the term $(e_x(t),e(t))_h$. Combining (30) with (48) we obtain

$$\begin{aligned}
\tfrac{1}{2}\tfrac{\partial}{\partial t}(e(t),e(t))_h &= -(\delta_x^4 e(t),e(t))_h + \tilde{b}\ (\tilde{\delta}_x^2 e(t),e(t))_h\\
&\quad +\tilde{d}\ (e(t),e(t))_h - (r(t),e(t))_h.
\end{aligned} \tag{49}$$

Let us now consider the term $(\tilde{\delta}_x^2 e(t),e(t))_h$. Using the definition of $-\tilde{\delta}_x^2$, we have

$$-(\tilde{\delta}_x^2 e(t),e(t))_h = -(\delta_x^2 e(t),e(t))_h + \frac{h^2}{12}(\delta_x^4 e(t),e(t))_h. \tag{50}$$

Therefore,

$$\begin{aligned}
\tfrac{1}{2}\tfrac{\partial}{\partial t}(e(t),e(t))_h &= -(1+\tilde{b}\tfrac{h^2}{12})(\delta_x^4 e(t),e(t))_h + \tilde{b}\ (\delta_x^2 e(t),e(t))_h\\
&\quad +\tilde{d}\ (e(t),e(t))_h - (r(t),e(t))_h.
\end{aligned} \tag{51}$$

Consider now the term $(\delta_x^2 e(t), e(t))_h$. Using Cauchy-Schwartz inequality, we have

$$-(\delta_x^2 e(t), e(t))_h \le |e(t)|_h \, |\delta_x^2 e(t)|_h \le \frac{1}{2\epsilon}|e(t)|_h^2 + \frac{\epsilon}{2}|\delta_x^2 e(t)|_h^2. \tag{52}$$

Using the coercivity property $|\delta_x^2 e(t)|_h^2 \le C(e(t), \delta_x^4 e(t))_h$, we obtain

$$-(\delta_x^2 e(t), e(t))_h \le \frac{1}{2\epsilon}|e(t)|_h^2 + \frac{C\epsilon}{2}(\delta_x^4 e(t), e(t))_h. \tag{53}$$

Inserting (53) in (51), we obtain that for $h \le h_0$ and $\epsilon \le \epsilon_0$,

$$\tfrac{1}{2}\tfrac{\partial}{\partial t}(e(t), e(t))_h \le -\tfrac{1}{2}(\delta_x^4 e(t), e(t))_h + C_2\,(e(t), e(t))_h - (r(t), e(t))_h. \tag{54}$$

Inserting (39) in (54), we have

$$\frac{1}{2}\frac{\partial}{\partial t}(e(t), e(t))_h \le C_2\ (e(t), e(t))_h + Ch^7. \tag{55}$$

By Gronwall's inequality $|e|_h^2 \le Ch^7$. Thus, $|e(t)|_h \le C(T)h^{3.5}, \ \ 0 \le t \le T.$ □

5 Numerical Results

Consider the exact solution $u(x,t) = e^{-t}e^x$ of the problem

$$\begin{cases} u_{xxt} = u_{xxxx} + u_{xx} + f(x,t), & 0 < x < 1, \quad t > 0, \\ u(0,t) = e^{-t}, \quad u_x(0,t) = e^{-t}, & t > 0, \\ u(1,t) = e^{1-t}, \quad u_x(1,t) = e^{1-t}, & t > 0, \\ u(x,0) = e^x, \quad 0 \le x \le 1. \end{cases} \tag{56}$$

Here u and u_x are given on the boundary points and $f(x,t)$ is chosen as $u(x,t) = e^{-t}e^x$ is the solution of the differential equation above. The results are given in Table 1. They demonstrate the fourth-order accuracy of the scheme.

Table 1 Compact scheme for $u_{xxt} = u_{xxxx} + u_{xx} + f$ with exact solution: $u = e^{-t}e^x$ on $[0,1], t > 0$. We present $|e_h|_h$ the error in u, and $|e_x|_h$ the error in u_x in the l_2 norm at $t = 0.5$

Mesh	$N=8$	Rate	$N=16$	Rate	$N=32$	Rate	$N=64$
$\lvert e\rvert_h$	1.5742(−7)	4.07	9.3544(−9)	4.02	5.7490(−10)	4.00	3.5844(−11)
$\lvert e_x\rvert_h$	1.5500(−6)	4.00	9.7033(−8)	4.00	6.0672(−9)	4.00	3.7893(−10)

Table 2 Compact scheme for $u_{xxt} = u_{xxxx} + u_{xx} + f$ with exact solution: $u = e^{-t}\sin(\pi x)/\pi^2$ on $[0, 1], t > 0$. We present $|e|_h$ the error in u, and $|e_x|_h$ the error in u_x in the l_2 norm at $t = 0.5$

Mesh	$N=8$	Rate	$N=16$	Rate	$N=32$	Rate	$N=64$
$\|e\|_h$	5.2800(−6)	4.01	3.2679(−7)	4.00	2.0381(−8)	4.00	1.2732(−9)
$\|e_x\|_h$	2.0038(−6)	4.07	1.1926(−7)	4.02	7.3451(−9)	4.01	4.5737(−10)

Next we consider the solution $u(x,t) = e^{-t} sin(\pi x)/\pi^2$ of the problem

$$\begin{cases} u_{xxt} = u_{xxxx} + u_{xx} + f(x,t), \quad 0 < x < 1, \quad t > 0, \\ u(0,t) = 0, \quad u_x(0,t) = e^{-t}/\pi, \quad t > 0, \\ u(1,t) = 0, \quad u_x(1,t) = -e^{-t}/\pi, \quad t > 0, \\ u(x,0) = sin(\pi x)/\pi^2, \quad 0 \le x \le 1. \end{cases} \tag{57}$$

Here u and u_x are given at the two boundary points and $f(x,t)$ is chosen such that $u(x,t) = e^{-t}\sin(\pi x)/\pi^2$ if the exact solution of the problem above. The numerical results are shown in Table 2. The calibrate the fourth-order accuracy of the scheme.

References

1. S. Abarbanel, A. Ditkowski, B. Gustafsson, On error bound of finite difference approximations for partial differential equations. J. Sci. Comput. **15**, 79–116 (2000)
2. M. Ben-Artzi, J-P. Croisille, D. Fishelov, Convergence of a compact scheme for the pure streamfunction formulation of the unsteady Navier–Stokes system. SIAM J. Numer. Anal. **44**, 1997–2024 (2006)
3. M. Ben-Artzi, J.-P. Croisille, D. Fishelov, A fast direct solver for the biharmonic problem in a rectangular grid. SIAM J. Sci. Comput. **31**(1), 303–333 (2008)
4. M. Ben-Artzi, J.-P. Croisille, D. Fishelov, *Navier-Stokes Equations in Planar Domains* (Imperial College Press, London, 2013)
5. D. Fishelov, M. Ben-Artzi, J-P. Croisille, Recent advances in the study of a fourth-order compact scheme for the one-dimensional biharmonic equation. J. Sci. Comput. **53**, 55–70 (2012)
6. B. Gustafsson, The convergence rate for difference approximations to mixed initial boundary value problems. Math. Comput. **29**, 386–406 (1975)
7. B. Gustafsson, The convergence rate for difference approximations to general mixed initial boundary value problems. SIAM J. Numer. Anal. **18**, 179–190 (1981)

A Numerical Algorithm for a Fully Nonlinear PDE Involving the Jacobian Determinant

Alexandre Caboussat and Roland Glowinski

Abstract We address the numerical solution of the Dirichlet problem for a partial differential equation involving the Jacobian determinant in two dimensions of space. The problem consists in finding a vector-valued function such that the determinant of its gradient is given point-wise in a bounded domain, together with essential boundary conditions. The proposed numerical algorithm relies on an augmented Lagrangian algorithm with biharmonic regularization, and low order mixed finite element approximations. An iterative method allows to decouple the local nonlinearities and the global variational problem that involves a biharmonic operator. Numerical experiments validate the proposed method.

1 Motivation

Fully nonlinear equations can usually be written $F(\mathbf{u}, \nabla\mathbf{u}, \mathbf{D}^2\mathbf{u}) = 0$, for some function F, in a bounded domain Ω, together with Dirichlet boundary conditions. Several examples and numerical schemes can be found in [1, 2, 5–7, 9], applied mainly to second order equations, such as Monge-Ampère or Pucci's.

Inspired by [3, 4], we consider here a particular equation that involves only the Jacobian of the unknown function. Namely, for a given data f, we want to find $\mathbf{u}$ such that $\det \nabla\mathbf{u} = f$. This example has a geometric partial differential equation flavor, as it corresponds to finding a given deformation. Unlike for the Monge-Ampère equation, it thus does not involve the Hessian $\mathbf{D}^2\mathbf{u}$. The goal in the present work is to provide an alternative, for the computational viewpoint, to the theoretical, explicit, construction of solutions that exists in the literature for simple cases.

A. Caboussat (✉)
Geneva School of Business Administration, University of Applied Sciences Western Switzerland, Route de Drize 7, 1227 Carouge, Switzerland
e-mail: alexandre.caboussat@hesge.ch

R. Glowinski
Department of mathematics, University of Houston, Houston, TX, USA
e-mail: roland@math.uh.edu

A. Abdulle et al. (eds.), *Numerical Mathematics and Advanced Applications ENUMATH 2013*, Lecture Notes in Computational Science and Engineering 103,
DOI 10.1007/978-3-319-10705-9_14

Following previous works on the Monge-Ampère equation [2], a variational approach is advocated. An iterative algorithm, reminiscent of alternating direction implicit methods, allows to alternatively solve linear variational problems and local nonlinear optimization problems. Numerical validation is achieved with simple examples, and convergence results are obtained from a computational perspective.

2 Problem Formulation

Let Ω be a bounded domain of $\mathbb{R}^2$, with Γ its boundary, and $f : \mathbb{R}^2 \to \mathbb{R}$ a given function. The *fully nonlinear partial differential equation involving the Jacobian determinant* we want to solve reads as follows: find $\mathbf{u} : \Omega \to \mathbb{R}^2$ satisfying

$$\begin{cases} \det \nabla \mathbf{u} = f & \text{in } \Omega \\ \mathbf{u} = \mathbf{Id} & \text{on } \Gamma. \end{cases} \tag{1}$$

where **Id** is the identity application. Problem (1) admits a solution, as discussed in [3, 4], under the compatibility condition on the data: $\int_\Omega f d\mathbf{x} = \text{measure}(\Omega)$. Actually, the existence proof has been first made for $f \geq 0$ in [4], and extended to the more general case in [3]. However, the solution to (1) is not necessarily unique; indeed, for instance, if Ω is the unit disc and $f \equiv 1$, $\mathbf{u}_1(\mathbf{x}) = \mathbf{x}$ is an obvious solution, and (denoting the polar coordinates by (ρ, θ)), $\mathbf{u}_2(\rho, \theta) = (\rho\cos(\theta + 2k\pi\rho^2)\,,\, \rho\sin(\theta + 2k\pi\rho^2))^T$ is also a solution.

We assume here that f is non-negative. In order to design a numerical method based on some variational principle, and enforce the uniqueness of the solution to (1), we consider the following problem:

$$\min_{\mathbf{v}\in\mathbf{E}} \frac{1}{2}\int_\Omega |\nabla\mathbf{v} - \mathbf{I}|^2\, d\mathbf{x} \tag{2}$$

where $\mathbf{E} = \{\mathbf{v} \in H^1(\Omega)^2\,,\, \det\nabla\mathbf{v} = f\,,\, \mathbf{v}|_\Gamma = \mathbf{Id}\}$. Here $\mathbf{I}$ denotes the 2×2 identity operator. The Frobenius norm and product are respectively defined by $|\mathbf{T}| = (\mathbf{T} : \mathbf{T})^{1/2}$, $\mathbf{S} : \mathbf{T} = \sum_{i,j=1}^{2} s_{ij}t_{ij}$ for each $\mathbf{S} = (s_{ij})$, $\mathbf{T} = (t_{ij}) \in \mathbb{R}^{2\times 2}$. If $f \in L^1(\Omega)$, then the set $\mathbf{E}$ is not empty.

Let us denote by $\mathbf{u} \in \mathbf{E}$ the solution to (2). The choice of the objective distance function is arbitrary and is made in order to facilitate the decomposition properties of the algorithm discussed below.

3 An Augmented Lagrangian Algorithm

3.1 Regularization and Augmented Lagrangian Functional

We first introduce a *biharmonic regularization* to the variational problem (2). Let $\delta > 0$ be a small parameter. The biharmonic regularization reads:

$$\min_{\mathbf{v}\in\tilde{\mathbf{E}}}\left[\frac{1}{2}\int_{\Omega}|\nabla\mathbf{v}-\mathbf{I}|^2\,d\mathbf{x}+\frac{\delta}{2}\int_{\Omega}\left|\nabla^2\mathbf{v}\right|^2 d\mathbf{x}\right] \tag{3}$$

where $\tilde{\mathbf{E}} = \left\{\mathbf{v}\in H^2(\Omega)^2\,,\ \det\nabla\mathbf{v}=f\,,\ \mathbf{v}|_{\Gamma}=\mathbf{Id}\right\}$. Then we introduce a new variable $\mathbf{p}\in L^2(\Omega)^{2\times 2}$, so that (3) is equivalent to

$$\min_{(\mathbf{v},\mathbf{q})\in\hat{\mathbf{E}}}\left[\frac{1}{2}\int_{\Omega}|\nabla\mathbf{v}-\mathbf{I}|^2\,d\mathbf{x}+\frac{\delta}{2}\int_{\Omega}\left|\nabla^2\mathbf{v}\right|^2 d\mathbf{x}\right] \tag{4}$$

where $\hat{\mathbf{E}} = \left\{\mathbf{v}\in H^2(\Omega)^2\,,\ \det\mathbf{q}=f\,,\ \mathbf{v}|_{\Gamma}=\mathbf{Id}\,,\ \nabla\mathbf{v}=\mathbf{q}\right\}$. With formulation (4), we advocate an *augmented Lagrangian algorithm.* Namely, for $r > 0$ a given parameter, we define the augmented Lagrangian functional

$$\begin{aligned}\mathscr{L}(\mathbf{v},\mathbf{q};\boldsymbol{\mu}) &= \frac{1}{2}\int_{\Omega}|\nabla\mathbf{v}-\mathbf{I}|^2\,d\mathbf{x}+\frac{\delta}{2}\int_{\Omega}\left|\nabla^2\mathbf{v}\right|^2 d\mathbf{x}\\ &+\frac{r}{2}\int_{\Omega}|\nabla\mathbf{v}-\mathbf{q}|^2\,d\mathbf{x}+\int_{\Omega}\boldsymbol{\mu}:(\nabla\mathbf{v}-\mathbf{q})d\mathbf{x}.\end{aligned}$$

and search for a saddle-point of $\mathscr{L}(\mathbf{v},\mathbf{q};\boldsymbol{\mu})$. Thus, after defining the function spaces $\mathbf{V} = \left\{\mathbf{v}\in H^2(\Omega)^2\,,\ \mathbf{v}|_{\Gamma}=\mathbf{Id}\right\}$, $\mathbf{Q} = \left\{\mathbf{q}\in L^2(\Omega)^{2\times 2}\right\}$, and $\mathbf{Q}_f = \{\mathbf{q}\in\mathbf{Q}\,,\ \det\mathbf{q}=f\}$, the saddle-point problem consists in looking for $\{\mathbf{u},\mathbf{p},\boldsymbol{\lambda}\}\in\mathbf{V}\times\mathbf{Q}_f\times\mathbf{Q}$ such that

$$\mathscr{L}(\mathbf{u},\mathbf{p};\boldsymbol{\mu})\le\mathscr{L}(\mathbf{u},\mathbf{p};\boldsymbol{\lambda})\le\mathscr{L}(\mathbf{v},\mathbf{q};\boldsymbol{\lambda}) \tag{5}$$

for all $\{\mathbf{v},\mathbf{q},\boldsymbol{\mu}\}\in\mathbf{V}\times\mathbf{Q}_f\times\mathbf{Q}$.

The addition of the biharmonic regularization is actually not necessary when the problem admits a classical solution and when the data is smooth (which will be the case for the numerical experiments presented in Sect. 5). However, it is incorporated here as it helps, via a smoothing effect, the convergence of the iterative algorithm when the data are less regular or when there is no classical solution. Note that the additional cost of introducing this regularization corresponds to the marginal cost of solving two Poisson problems instead of one at each iteration.

3.2 Iterative Algorithm

In order to solve (5), we advocate an Uzawa/alternating direction iterative algorithm. Let $\mathbf{u}^0 \in \mathbf{V}$ and $\boldsymbol{\lambda}^0 \in \mathbf{Q}$ be given. Then, for $n \geq 1$, we do:

(A) Solve the *constrained nonlinear problem* $\min_{\mathbf{q}\in\mathbf{Q}_f} \mathscr{L}(\mathbf{u}^{n-1}, \mathbf{q}; \boldsymbol{\lambda}^{n-1})$ to obtain $\mathbf{p}^n \in \mathbf{Q}_f$. This is equivalent to the quadratic problem under constraints:

$$\min_{\mathbf{q}\in\mathbf{Q}_f} \left[\frac{r}{2} \int_\Omega |\mathbf{q}|^2 \, d\mathbf{x} - \int_\Omega \mathbf{X}^{n-1} : \mathbf{q} d\mathbf{x} \right], \tag{6}$$

where $\mathbf{X}^n := r\nabla\mathbf{u}^n + \boldsymbol{\lambda}^n \in \mathbf{Q}$. This problem having no derivatives, it can be solved *point-wise* a.e. $\mathbf{x} \in \Omega$ (in practice on each element of a finite element discretization). Namely, for a.e. $\mathbf{x} \in \Omega$, it corresponds to a *constrained quadratic problem*: find $\mathbf{p}^n(\mathbf{x}) \in \mathbb{R}^{2\times 2}$ solution of

$$\min_{\mathbf{q}\in\mathbf{Q}_\mathbf{x}} \left[\frac{r}{2} |\mathbf{q}|^2 - \mathbf{X}^{n-1} : \mathbf{q} \right], \tag{7}$$

where $\mathbf{Q}_\mathbf{x} = \left\{ \mathbf{q} \in \mathbb{R}^{2\times 2}, \det \mathbf{q} = q_{11}q_{22} - q_{12}q_{21} = f(\mathbf{x}) \right\}$.

(B) Solve the *linear variational problem* $\min_{\mathbf{v}\in\mathbf{V}} \mathscr{L}(\mathbf{v}, \mathbf{p}^n; \boldsymbol{\lambda}^{n-1})$ to obtain $\mathbf{u}^n \in \mathbf{V}$. This is equivalent to

$$\min_{\mathbf{v}\in\mathbf{V}} \left[\frac{\delta}{2} \int_\Omega \left|\nabla^2 \mathbf{v}\right|^2 d\mathbf{x} + \frac{1+r}{2} \int_\Omega |\nabla \mathbf{v}|^2 \, d\mathbf{x} - \int_\Omega \nabla \mathbf{v} : \mathbf{Y}^n d\mathbf{x}, \right] \tag{8}$$

where $\mathbf{Y}^n := \mathbf{I} + r\mathbf{p}^n - \boldsymbol{\lambda}^{n-1} \in \mathbf{Q}$. This (linear) problem involves derivatives but does not include any constraints (other than the Dirichlet boundary conditions included in $\mathbf{V}$). The first order optimality conditions corresponding to (8) lead to a *linear variational problem*, of the biharmonic type: find $\mathbf{u} \in \mathbf{V}$ satisfying

$$\delta \int_\Omega \nabla^2 \mathbf{u} \cdot \nabla^2 \mathbf{v} d\mathbf{x} + (1+r) \int_\Omega \nabla \mathbf{u} : \nabla \mathbf{v} d\mathbf{x} = \int_\Omega \mathbf{Y}^n : \nabla \mathbf{v} d\mathbf{x},$$

for all $\mathbf{v} \in (H^2(\Omega) \cap H_0^1(\Omega))^2$.

(C) Update the multipliers $\boldsymbol{\lambda}^n = \boldsymbol{\lambda}^{n-1} + r(\nabla\mathbf{u}^n - \mathbf{p}^n) \in \mathbf{Q}$.

3.3 Numerical Solution of the Constrained Nonlinear Problem

Problem (7) can be rewritten as the following constrained finite dimensional minimization problem:

$$\min_{\mathbf{q}\in\mathbf{E}_c}\left[\frac{1}{2}|\mathbf{q}|^2-\mathbf{b}\cdot\mathbf{q}\right], \tag{9}$$

with $\mathbf{E}_c = \{\mathbf{q}\in\mathbb{R}^4\,,\,q_1q_2-q_3q_4=c(>0)\}$. Actually, here, $c = f(\mathbf{x})$ and $\mathbf{b} = \frac{1}{r}(\mathbf{X}^n_{11},\mathbf{X}^n_{22},\mathbf{X}^n_{12},\mathbf{X}^n_{21})$. Problem (9) is solved with a Lagrangian approach and a Newton algorithm, after a suitable change of variables to take advantage of the structure of the problem (also encountered in incompressible finite elasticity, see, e.g., [8]). Let us denote by $\mathbf{S}$ the 4×4 orthogonal matrix

$$\mathbf{S}=\begin{pmatrix}1/\sqrt{2} & 1/\sqrt{2} & 0 & 0\\ 1/\sqrt{2} & -1/\sqrt{2} & 0 & 0\\ 0 & 0 & 1/\sqrt{2} & 1/\sqrt{2}\\ 0 & 0 & 1/\sqrt{2} & -1/\sqrt{2}\end{pmatrix},$$

and introduce the new variables $\mathbf{z} = \mathbf{S}\mathbf{q}$, together with $\boldsymbol{\beta} = \mathbf{S}\mathbf{b}$. Problem (9) is equivalent to

$$\min_{\mathbf{z}\in\mathbf{F}_c}\left[\frac{1}{2}|\mathbf{z}|^2-\boldsymbol{\beta}\cdot\mathbf{z}\right], \tag{10}$$

with $\mathbf{F}_c = \{\mathbf{z}\in\mathbb{R}^4\,,\,z_1^2-z_2^2-z_3^2+z_4^2=2c(>0)\}$. In order to solve (10), let us introduce the associated Lagrangian functional $\mathscr{L}(\mathbf{z},\mu) = \frac{1}{2}|\mathbf{z}|^2-\boldsymbol{\beta}\cdot\mathbf{z}-\frac{\mu}{2}(z_1^2-z_2^2-z_3^2+z_4^2-2c)$. If $\mathbf{y}$ is a solution of (10), and λ is a related Lagrange multiplier, the first order optimality conditions read:

$$y_1=\frac{\beta_1}{1-\lambda},\quad y_2=\frac{\beta_2}{1+\lambda},\quad y_3=\frac{\beta_3}{1+\lambda},\quad y_4=\frac{\beta_4}{1-\lambda},\quad \frac{\beta_1^2+\beta_4^2}{(1-\lambda)^2}-\frac{\beta_2^2+\beta_3^2}{(1+\lambda)^2}=2c$$

It can be shown (see, e.g., [10]) that the solution of this system of equations corresponds to the unique solution of

$$\frac{\beta_1^2+\beta_4^2}{(1-\lambda)^2}-\frac{\beta_2^2+\beta_3^2}{(1+\lambda)^2}=2c \tag{11}$$

that belongs to $(-1,+1)$. We then solve (11) with a Newton method with initial guess $\lambda^0=0$.

Remark 1 In numerical experiments, the Newton method almost always converges to a root in $(-1,+1)$. When it is not the case, we arbitrarily set $\lambda = 0$ and $y_i = \beta_i, i = 1,\ldots,4$. This procedure does not jeopardize the convergence of the outer iterative algorithm.

3.4 Numerical Solution of the Linear Variational Problem

The first order optimality conditions related to (8) are: find $\mathbf{u}^{n+1} \in \mathbf{V}$ satisfying

$$\delta \int_\Omega \nabla^2 \mathbf{u}^{n+1} \cdot \nabla^2 \mathbf{v} d\mathbf{x} + (1+r) \int_\Omega \nabla \mathbf{u}^{n+1} : \nabla \mathbf{v} d\mathbf{x} = \int_\Omega \mathbf{Y}^n : \nabla \mathbf{v} d\mathbf{x}, \tag{12}$$

for all $\mathbf{v} \in \mathbf{V}_0$, where $\mathbf{V}_0 = \left\{\mathbf{v} \in H^2(\Omega)^2,\ \mathbf{v}|_\Gamma = \mathbf{0}\right\}$. Problem (12) is a classical biharmonic problem, closely related to those encountered when solving the elliptic Monge-Ampère equation in [2]. We observe that this problem is equivalent (if Ω is convex or $\partial\Omega$ smooth enough) to the following second-order variational system: find $\mathbf{w}^{n+1} \in (H_0^1(\Omega))^2$ satisfying

$$\delta \int_\Omega \nabla \mathbf{w}^{n+1} : \nabla \mathbf{v} d\mathbf{x} + (1+r) \int_\Omega \mathbf{w}^{n+1} \cdot \mathbf{v} d\mathbf{x} = \int_\Omega \mathbf{Y}^n : \nabla \mathbf{v} d\mathbf{x}, \quad \forall \mathbf{v} \in (H_0^1(\Omega))^2; \tag{13}$$

followed by: find $\mathbf{u}^{n+1} \in (H^1(\Omega))^2$, $\mathbf{u}^{n+1}|_{\partial\Omega} = \mathbf{g}$, satisfying

$$\int_\Omega \nabla \mathbf{u}^{n+1} : \nabla \mathbf{v} d\mathbf{x} = \int_\Omega \mathbf{w}^{n+1} \cdot \mathbf{v} d\mathbf{x}, \quad \forall \mathbf{v} \in (H_0^1(\Omega))^2. \tag{14}$$

The solution of both second-order elliptic problems can be obtained with many well-known finite element techniques when $\Omega \subset \mathbb{R}^2$.

4 Finite Element Approximation

Finite elements are a natural choice for the discretization of (3) due to the variational flavor of this problem. A piecewise linear and globally continuous approximation of the solution $\mathbf{u}$ and piecewise constant approximations of its gradient $\nabla \mathbf{u}$ and of the additional variable $\mathbf{p}$ over a finite element triangulation of Ω are used here. Let $h > 0$ be a discretization step, and $\mathcal{T}_h$ a conforming triangulation of Ω. We assume for the sake of the discussion that Ω and Γ are exactly approximated by their finite element discretizations. From $\mathscr{T}_h$, we approximate $\mathbf{Q}$ and $\mathbf{Q}_f$ respectively by

$$\mathbf{Q}_h = \left\{\mathbf{q}_h \in L^2(\Omega)^{2\times 2},\ \mathbf{q}_h|_T \in \mathbb{R}^{2\times 2},\ \forall T \in \mathscr{T}_h\right\}$$

$$\mathbf{Q}_{fh} = \left\{\mathbf{q}_h \in \mathbf{Q}_h,\ \det \mathbf{q}_h|_T = \bar{f}_T,\ \forall T \in \mathscr{T}_h\right\},$$

where $\bar{f}_T$ is the value of the piecewise constant approximation of f on $\mathscr{T}_h$ defined as $\bar{f}_T = \frac{1}{3}\sum_{T\ni P_j} f(P_j)$, P_j being the vertices of the triangle T. On the other hand, the space $\mathbf{V}$ is approximated by

$$\mathbf{V}_h = \left\{\mathbf{v} \in C^0\left(\overline{\Omega}\right)^2,\ \mathbf{v}|_T \in (\mathbb{P}_1)^2,\ \forall T \in \mathscr{T}_h,\ \mathbf{v} = \mathbf{Id}_h \text{ on } \Gamma\right\},$$

with $\mathbb{P}_1$ the space of the two-variables polynomials of degree ≤ 1, and $\mathbf{Id}_h$ a piecewise linear interpolant of the identity function on Γ. Similarly, we define $\mathbf{V}_{0h} = \left\{\mathbf{v} \in C^0\left(\overline{\Omega}\right)^2,\ \mathbf{v}|_T \in (\mathbb{P}_1)^2,\ \forall T \in \mathscr{T}_h,\ \mathbf{v} = \mathbf{0} \text{ on } \Gamma\right\}$.

The iterative algorithm in Sect. 3.2 can be re-written at the discrete level. The nonlinear optimization problem (6) is then solved *element-wise* on each triangle T of $\mathscr{T}_h$, with exactly the same method as the one presented in Sect. 3.3 (when replacing $c := f(\mathbf{x})$ by $c := \bar{f}_T$). The discrete version of the variational problem (12) is solved with a sequence of discrete Poisson problems that are the discrete equivalents of (13) and (14), in a similar fashion than in [2].

Remark 2 (On the choice of low order finite element approximations) The solution to (12) is actually a variation of the steady Stokes problem. Indeed, let us denote by $\mathbf{w} = \mathbf{v} - \mathbf{Id}$ and suppose that $\mathbf{w}$ is small and f is close to 1. We have then

$$\nabla\mathbf{w} = \nabla\mathbf{v} - \mathbf{I} \qquad \det\nabla\mathbf{v} = \det(\mathbf{I} + \nabla\mathbf{w}) = 1 + \nabla\cdot\mathbf{w} + \varepsilon(\mathbf{v})$$

where $\varepsilon(\mathbf{w})$ is a (small) residual. This implies that $\nabla\cdot\mathbf{w} = (f-1) - \varepsilon(\mathbf{w})$, meaning that the vector field $\mathbf{w}$ is *nearly* divergence free. The problem we have to solve is thus closely related to the steady Stokes problem. Our approach, where $\mathbf{Q}_h$ is the space of the 2×2 matrix valued functions constant on each triangle T of the triangulation $\mathscr{T}_h$ (used to approximate $(H^1(\Omega))^2$), is therefore close to the $\mathbb{P}_1 - \mathbb{P}_1$ approximation of the Stokes problem, which explains the convergence orders obtained in the next section.

5 Numerical Validation

The purpose of the numerical experiments in this section is to validate the proposed methodology with one well-chosen example, and highlight the convergence orders obtained for a prototypical problem.

Let us consider a validation case, for which (one of) the solution is the identity mapping $\mathbf{u}(\mathbf{x}) = \mathbf{x}$. Thus, for the unit disc $\Omega = \{\mathbf{x} \in \mathbb{R}^2, ||\mathbf{x}||_2 < 1\}$, we consider: find $\mathbf{u} : \Omega \to \mathbb{R}^2$ satisfying

$$\begin{cases} \det \nabla \mathbf{u} = 1 & \text{in } \Omega \\ \mathbf{u}(\mathbf{x}) = \mathbf{x} & \text{on } \Gamma, \end{cases} \tag{15}$$

The set of numerical parameters is given by $r = 10^{-6}$ and $\delta = 10^{-6}$. The tolerance between successive iterates $\mathbf{u}^k$ and $\mathbf{u}^{k+1}$ for the stopping criterion is set to $\varepsilon = 10^{-8}$. The tolerance for the Newton method for local nonlinear problems is set to 10^{-5} on the residual. The mesh is an unstructured Delaunay discretization of Ω. The advocated numerical algorithm converges in less than 20 iterations (actually between 17 and 19 iterations depending on the mesh size). Figure 1 visualizes the solution on one given mesh (with $h \simeq 0.0161$). The most natural solution $\mathbf{u}(\mathbf{x}) = \mathbf{x}$ is correctly approximated, and the radial invariance is appropriately tracked even though the mesh does not guarantee such a symmetry. The determinant of $\mathbf{p}_h$ is exactly equal to one on each element (up to machine precision), while the determinant of $\nabla \mathbf{u}_h$ is nearly everywhere equal to one, implying that the constraint $\nabla \mathbf{u}_h = \mathbf{p}_h$ is weakly satisfied.

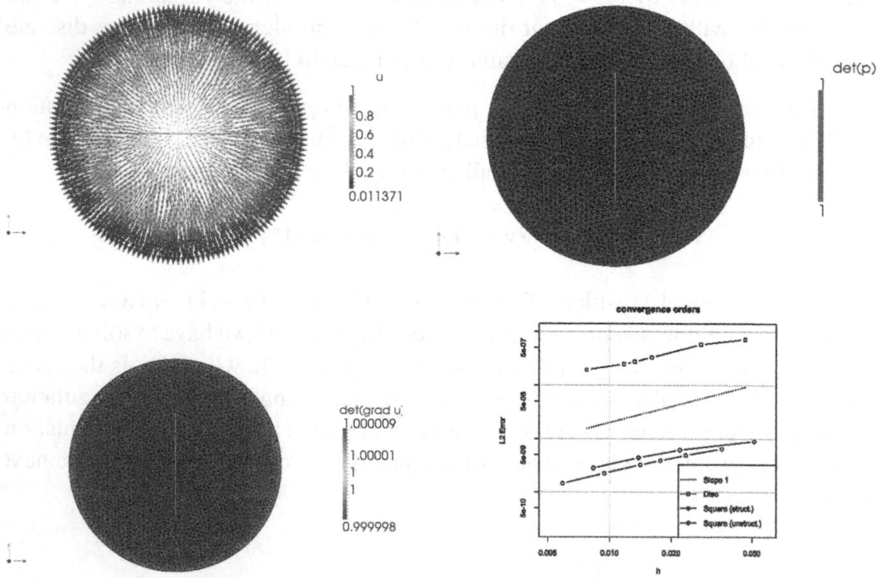

Fig. 1 Validation with the identity mapping. Visualization of the approximated solution obtained with the augmented Lagrangian approach after 19 steps ($h \simeq 0.0161$). Vector field $\mathbf{u}_h$ (*top left*), determinant $\det \mathbf{p}_h$ (*top right*), determinant $\det \nabla \mathbf{u}_h$ (*bottom left*), and (*bottom right*) convergence of the error in L^2 norm, for the unit disc and the unit square

Numerical results are similar when considering the unit square $\Omega = (0,1)^2$, with either a structured or an unstructured mesh. Figure 1 (bottom right) illustrates the convergence of the error between the numerical solution $\mathbf{u}_h$ and the exact solution $\mathbf{u} = \mathbf{Id}$, for the unit disc and the unit square (with both types of meshes). All configurations lead to the convergence with order approximately $\mathcal{O}(h)$ of the numerical approximation towards the exact solution.

Future work will include more complicated test cases, including problems without solutions, non-convex domains, and the generalization to less regular Dirichlet boundary conditions.

Acknowledgements The authors thank Prof. B. Dacorogna (EPFL) for suggesting the investigation of this problem, and for helpful comments and discussions.

References

1. A. Caboussat. *On the numerical solution of the Dirichlet problem for the elliptic (σ_2) equation.* In: Fitzgibbon et al (eds). Modeling, Simulation and Optimization in Science and Technology, ECCOMAS series, (Springer, 2014) pp. 23–40
2. A. Caboussat, R. Glowinski, D.C. Sorensen, A least-squares method for the solution of the Dirichlet problem for the elliptic Monge-Ampère equation in dimension two. ESAIM Control Optim. Calc. Var. **19**(3), 780–810 (2013)
3. G. Cupini, B. Dacorogna, O. Kneuss, On the equation det(Du)=f with no sign hypothesis. Calc. Var. Partial Differ. Equ. **36**, 251–283 (2009)
4. B. Dacorogna, J. Moser, On a partial differential equation involving the Jacobian determinant. Ann. Inst. Henri Poincaré Anal. Non Linéaire **7**, 1–26 (1990)
5. X. Feng, R. Glowinski, M. Neilan, Recent developments in numerical methods for fully nonlinear second order partial differential equations. SIAM Rev. **55**(2), 205–267 (2013)
6. X. Feng, M. Neilan, Vanishing moment method and moment solutions for second order fully nonlinear partial differential equations. J. Sci. Comput. **38**(1), 74–98 (2009)
7. B.D. Froese, A.M. Oberman, Convergent finite difference solvers for viscosity solutions of the elliptic Monge-Ampère equation in dimensions two and higher. SIAM J. Numer. Anal. **49**(4), 1692–1714 (2012)
8. R. Glowinski, P.L. Tallec, *Augmented Lagrangians and Operator-Splitting Methods in Nonlinear Mechanics* (SIAM, Philadelphia, 1989)
9. O. Lakkis, T. Pryer, A finite element method for fully nonlinear elliptic problems. SIAM J. Sci. Comput. **35**(4), A2025–A2045 (2013)
10. D.C. Sorensen, R. Glowinski, A quadratically constrained minimization problem arising from PDE of Monge-Ampère type. Numer. Algorithms **53**(1), 53–66 (2010)

Pattern Formation for a Reaction Diffusion System with Constant and Cross Diffusion

Verónica Anaya, Mostafa Bendahmane, Michel Langlais, and Mauricio Sepúlveda

Abstract In this work, we study a finite volume scheme for a reaction diffusion system with constant and cross diffusion modeling the spread of an epidemic disease within a host population structured with three subclasses of individuals (SIR-model). The mobility in each class is assumed to be influenced by the gradient of other classes. We establish the existence of a solution to the finite volume scheme and show convergence to a weak solution. The convergence proof is based on deriving a series of a priori estimates and using a general L^p compactness criterion.

AMS Subject Classification: 35K57, 35M10, 35A05

1 Introduction

Consider a host population subdivided into three subclasses of individuals, susceptible, infective and recovered with respect to some epidemic disease. The susceptible class consists of individuals who are capable of becoming infected and the infective class consists of individuals who have contracted the disease and are capable of transmitting it. Susceptible individuals can contract the disease from cross contacts with infected ones. Our state variables $(S(t,x), I(t,x), R(t,x))$ represent densities

V. Anaya
Departamento de Matemática and GIMNAP, Universidad del Bío-Bío, Concepción, Chile
e-mail: vanaya@ubiobio.cl

M. Bendahmane • M. Langlais
Institut de Mathématiques de Bordeaux UMR CNRS 5251, Université Victor Segalen Bordeaux 2, Bordeaux, France
e-mail: mostafa.bendahmane@u-bordeaux2.fr; michel.langlais@u-bordeaux2.fr

M. Sepúlveda (✉)
DIM and CI^{2}MA, Universidad de Concepción, Esteban Iturra s/n, Barrio Universitario, Concepción, Chile
e-mail: mauricio@ing-mat.udec.cl

A. Abdulle et al. (eds.), *Numerical Mathematics and Advanced Applications ENUMATH 2013*, Lecture Notes in Computational Science and Engineering 103,
DOI 10.1007/978-3-319-10705-9_15

of susceptible, infective and recovered subclasses for a total host population $H = S + I + R$ at time t and location $x \in \Omega \subset \mathbb{R}^3$. The host population will follow a logistic dynamic with a spatially dependent birth-rate, $b(x)$, identical in each subclass, offspring being susceptible at birth (one assumes the disease to be benign in H). A spatially and density dependent mortality rate, $m(x) + k(x)H$, is considered allowing for a spatially variable carrying capacity. Let $1/\lambda$ be the duration of the infective stage at the end of which a fixed proportion $0 \leq w \leq 1$ of infective individuals become permanently immune, a proportion $0 \leq 1 - w \leq 1$ reentering the susceptible class. Our model system is given by the following set of equations

$$\begin{cases} \partial_t S - \operatorname{div}\Big((d_1 + \alpha_1 S + I + R)\nabla S + S\nabla I + S\nabla R\Big) \\ \qquad = -\sigma SI + (1-w)\lambda I + b(x)H - (m(x) + k(x)H)S, \\ \partial_t I - \operatorname{div}\Big(I\nabla S + (d_2 + S + \alpha_2 I + R)\nabla I + I\nabla R\Big) \\ \qquad = \sigma SI - \lambda I - (m(x) + k(x)H)I, \\ \partial_t R - \operatorname{div}\Big(R\nabla S + R\nabla I + (d_3 + S + I + \alpha_3 R)\nabla R\Big) \\ \qquad = w\lambda I - (m(x) + k(x)H)R, \end{cases} \tag{1}$$

in $Q_T = \Omega \times (0, T)$ the time-space cylinder, where $H = S + I + R$. System (1) is supplemented with no-flux boundary conditions:

$$\nabla S \cdot \eta = 0, \quad \nabla I \cdot \eta = 0, \quad \nabla R \cdot \eta = 0, \tag{2}$$

on $(0, T) \times \partial\Omega$, η being the outer unit normal to Ω along its boundary $\partial\Omega$, and with nonnegative initial data:

$$S(0, x) = S_0(x) \geq 0, \quad I(0, x) = I_0(x) \geq 0, \quad R(0, x) = R_0(x) \geq 0, \tag{3}$$

on Ω. Here, $d_i > 0$ and $\alpha_i > 1$ are supposed constants, for $i = 1, 2, 3$, $m < b$ and $b, m, k \in L^\infty_+(Q_T)$. We assume that individuals move from a higher to a lower density region. Cross-diffusion expresses the dependence of the population flux of one subclass on other subclasses. A positive cross-diffusion term denotes that individuals from a given subclass tends to move in the direction of lower concentration of another subclass. Dynamics of interacting populations with self and cross-diffusion are investigated by several researchers (see [4–8] and the references therein).

1.1 Weak Solution

Before stating our results concerning the weak solution, we collect some preliminary material, including relevant notations and conditions imposed on the data of our problem. Let Ω be a bounded, open subsets of $\mathbb{R}^3$ with a smooth boundary $\partial\Omega$; η is the unit outward normal to Ω on $\partial\Omega$. Next, $|\Omega|$ is the Lebesgue measure of Ω. We denote by $H^1(\Omega)$ the Sobolev space of functions $u : \Omega \to \mathbb{R}$ for which $u \in L^2(\Omega)$ and $\nabla u \in L^2(\Omega;\mathbb{R}^3)$. For $1 \leq p \leq +\infty$, $\|\cdot\|_{L^p(\Omega)}$ is the usual norm in $L^p(\Omega)$. If X is a Banach space, $a < b$ and $1 \leq p \leq +\infty$, $L^p(a,b;X)$ denotes the space of all measurable functions $u : (a,b) \longrightarrow X$ such that $\| u(\cdot) \|_X$ belongs to $L^p(a,b)$. Next T is a positive number and $Q_t = \Omega \times (0,t)$, $\Sigma_t = \partial\Omega \times (0,t)$, for $0 < t \leq T$. Now we define what we mean by weak solutions of the system (1). We also supply our main existence results.

Definition 1 A weak solution of (1)–(3) is a set of nonnegative functions (S, I, R), such that,

$(S,I,R) \in L^2(0,T;H^1(\Omega,\mathbb{R}^3))$, $(\partial_t S, \partial_t I, \partial_t R) \in L^2(0,T;(W^{1,\infty}(\Omega,\mathbb{R}^3))^*)$,
$\xi(0) = \xi_0$ a.e. in Ω, for $\xi = (S,I,R)$ and satisfying

$$\begin{aligned}
&\int_0^T \langle \partial_t S, \phi_1 \rangle\, dt + \iint_{Q_T} ((d_1 + \alpha_1 S + I + R)\nabla S + S\nabla I + S\nabla R) \cdot \nabla\phi_1\, dx\, dt \\
&\quad = \iint_{Q_T} F_1(x,S,I,R)\phi_1\, dx\, dt, \\
&\int_0^T \langle \partial_t I, \phi_2 \rangle\, dt + \iint_{Q_T} (I\nabla S + (d_2 + S + \alpha_2 I + R)\nabla I + I\nabla R) \cdot \nabla\phi_2\, dx\, dt \\
&\quad = \iint_{Q_T} F_2(x,S,I,R)\phi_2\, dx\, dt \\
&\int_0^T \langle \partial_t R, \phi_3 \rangle\, dt + \iint_{Q_T} (R\nabla S + R\nabla I + (d_3 S + I + \alpha_3 R)\nabla R) \cdot \nabla\phi_3\, dx\, dt \\
&\quad = \iint_{Q_T} F_3(x,S,I,R)\phi_3\, dx\, dt,
\end{aligned}$$

for all $\phi_i \in L^2(0,T;W^{1,\infty}(\Omega))$, for $i = 1,2,3$, and where F_1, F_2 and F_3 denote the right hand side terms of (1). Here, $\langle\cdot,\cdot\rangle$ denotes the duality pairing between $W^{1,\infty}(\Omega)$ and $(W^{1,\infty}(\Omega))^*$.

Theorem 1 *If $(S_0, I_0, R_0) \in L^2(\Omega,\mathbb{R}^3)$, then the problem (1)–(3) possesses a weak solution. On the other hand if $(S_0, I_0, R_0) \in C^{2+\theta}(\bar{\Omega},\mathbb{R}^3)$ for some $\theta \in (0,1)$, then the system (1)–(3) has a unique, classical, global nonnegative solution $(S,I,R) \in C^{\frac{2+\theta}{2},2+\theta}([0,+\infty) \times \bar{\Omega},\mathbb{R}^3)$. Furthermore, there is a constant $C > 0$ (dependent upon the initial data and the coefficients) such that, $0 \leq S(t,x), I(t,x), R(t,x) \leq C$ for all $x \in \bar{\Omega}$ and $t > 0$.*

The proof of Theorem 1 is based in a series of a priori estimates of the solutions in Banach spaces, especially the boundness of the solutions in L^∞, and then we apply the Sobolev embedding and standard regularity results of parabolic equations (see e.g. [3]). The proof of Theorem 1 can be seen as a simplification of a more general version in the appendix of [1] (see also [2]).

2 Finite Volume Approximation

To discretize (1), we choose an admissible discretization of Q_T consisting of an admissible mesh $\mathscr{T}_h$ of Ω and of a time step size $\delta t_h > 0$; both δt_h and the size $\max_{K\in\mathscr{T}_h} \mathrm{diam}(K)$ tend to zero as $h \to 0$. We define $N_h > 0$ as the smallest integer such that $(N_h+1)\delta t_h \geq T$, and set $t^n := n\delta t_h$ for $n \in \{0,\ldots,N_h\}$. Whenever δt_h is fixed, we will drop the subscript h in the notation. Furthermore, we denote for $i = 1,2,3$, $F_{i,K}^{n+1} = F_i(x_K^{n+1^+}, S_K^{n+1^+}, I_K^{n+1^+}, R_K^{n+1^+})$, where $K \in \Omega$. To approximate the cross-diffusive terms, we introduce the term $\mathscr{M}_{ij,K,L}^{n+1}$. Herein, we make the choice $\mathscr{M}_{ij,K,L}^{n+1} := \mathscr{M}_{ij}\Big(\min\{S_K^{n+1^+}, S_L^{n+1^+}\}, \min\{I_K^{n+1^+}, I_L^{n+1^+}\}, \min\{R_K^{n+1^+}, R_L^{n+1^+}\}\Big)$, where $\phi^{n+1^+} = \max(0, \phi^{n+1})$ for $\phi = S, I, R$. The discrete initial conditions are given by: $S_K^0 = \frac{1}{|K|}\int_K S_0(x)\,dx$, $I_K^0 = \frac{1}{|K|}\int_K I_0(x)\,dx$, $R_K^0 = \frac{1}{|K|}\int_K R_0(x)\,dx$. We use the following implicit finite volume scheme to advance the numerical solution from t^n to $t^{n+1} = t^n + \delta t$:
Determine $(S_K^{n+1}, I_K^{n+1}, R_K^{n+1})_{K\in\mathscr{T}_h}$, such that

$$|K|\frac{S_K^{n+1}-S_K^n}{\delta t} - \sum_{L\in N(K)} \frac{|\sigma_{K,L}|}{d_{K,L}}\Big[\mathscr{M}_{11,K,L}^{n+1}(S_L^{n+1}-S_K^{n+1}) + \mathscr{M}_{12,K,L}^{n+1}(I_L^{n+1}-I_K^{n+1}) + \mathscr{M}_{13,K,L}^{n+1}(R_L^{n+1}-R_K^{n+1})\Big] = |K|F_{1,K}^{n+1}, \tag{4}$$

$$|K|\frac{I_K^{n+1}-I_K^n}{\delta t} - \sum_{L\in N(K)} \frac{|\sigma_{K,L}|}{d_{K,L}}\Big[\mathscr{M}_{21,K,L}^{n+1}(S_L^{n+1}-S_K^{n+1}) + \mathscr{M}_{22,K,L}^{n+1}(I_L^{n+1}-I_K^{n+1}) + \mathscr{M}_{23,K,L}^{n+1}(R_L^{n+1}-R_K^{n+1})\Big] = |K|F_{2,K}^{n+1}, \tag{5}$$

$$|K|\frac{R_K^{n+1}-R_K^n}{\delta t} - \sum_{L\in N(K)} \frac{|\sigma_{K,L}|}{d_{K,L}}\Big[\mathscr{M}_{31,K,L}^{n+1}(S_L^{n+1}-S_K^{n+1}) + \mathscr{M}_{32,K,L}^{n+1}(I_L^{n+1}-I_K^{n+1}) + \mathscr{M}_{33,K,L}^{n+1}(R_L^{n+1}-R_K^{n+1})\Big] = |K|F_{3,K}^{n+1}, \tag{6}$$

where the matrix coefficients $\mathscr{M}_{ij}$, for $i,j = 1,2,3$ include the cross diffusion terms. The set of values $(S_K^{n+1}, I_K^{n+1}, R_K^{n+1})_{K\in\mathscr{T}_h, n\in[0,N_h]}$ satisfying (4)–(6) is called a discrete solution. The convergence of the discrete solution generated by our scheme is given in the following theorem (see [1, 2]):

Theorem 2 *Let* $(S_0, I_0, R_0) \in (L^2(\Omega,\mathbb{R}^3))^+$ *and* $(S_K^{n+1}, I_K^{n+1}, R_K^{n+1})_{K\in\mathscr{T}_h}^{n\in[0,N_h]}$ *be the solution generated by the finite volume scheme* (4)–(6) *on a family of admissible* *and regular meshes. Then, as $h \to 0$, the discrete solution converges (along a subsequence) a.e. on Ω_T to a limit (S, I, R) which is a weak solution of* (1)–(3).

3 Numerical Results

We take the domain as follow $\Omega = (0,1) \times (0,1)$ We consider here a uniform mesh in the domain, given by a Cartesian grid with $N_x \times N_y$, control volumes. Obviously, it is possible to consider unstructured meshes, but we will take here to an uniform mesh $\Omega_R = \{K_{ij} \in \Omega | K_{ij} = ((i-1)N_x, iN_x) \times ((j-1)N_y, jN_y), i = 1, \ldots, N_x, j = 1, \ldots, N_y\}$, for simplicity of the simulated models.

The discretization in time is given by $N_t = 500$ time steps for $T = 0.5$. That is, $\delta t = T/N_t$ and $m(K) = 1/(N_x N_y)$, with $N_x = N_y = 256$. The parameters of the model are given by $\sigma = 0.8, \omega = 0.1, \lambda = 12, b = 0.03, m = 0.01, k = 0.03$.

3.1 First Example

The initial conditions are given by $S(x, y, 0) = 0.75$, $R(x, y, 0) = 0$ and $I(x, y, 0) = 5 + 5\varrho(x, y)$, with

$$\begin{aligned}\varrho(x,y) &= 1 - (1 + e^{-50(\sqrt{(x-0.75)^2+(y-0.75)^2}-0.18)})^{-1} \\ &\quad -(1 + e^{-50(\sqrt{(x-0.25)^2+(y-0.25)^2}-0.18)})^{-1}\end{aligned}$$

for all $(x, y) \in (0,1) \times (0,1)$, where ϱ represents here two focus of localized initial disease. In the pictures of Figs. 1–3, we observe some patterns obtained by taking different values for the nonlinear diffusion parameters. In Fig. 1, the variation of the nonlinear diffusion coefficient α_1 gives different pattern formation for the susceptible population, but in turn, the behavior of the infected and recovered population is quite insensitive to the variation of this coefficient α_1. Similar behavior occurs with the coefficient α_3, and the recovered population (see Fig. 2): while this parameter α_3 varies, we see different patterns in the recovered population, but not so the other two populations S and I. Finally in Fig. 3, we see that the variation of parameter α_2 affects the pattern of behavior across the three populations S, I and R, getting unusual shapes.

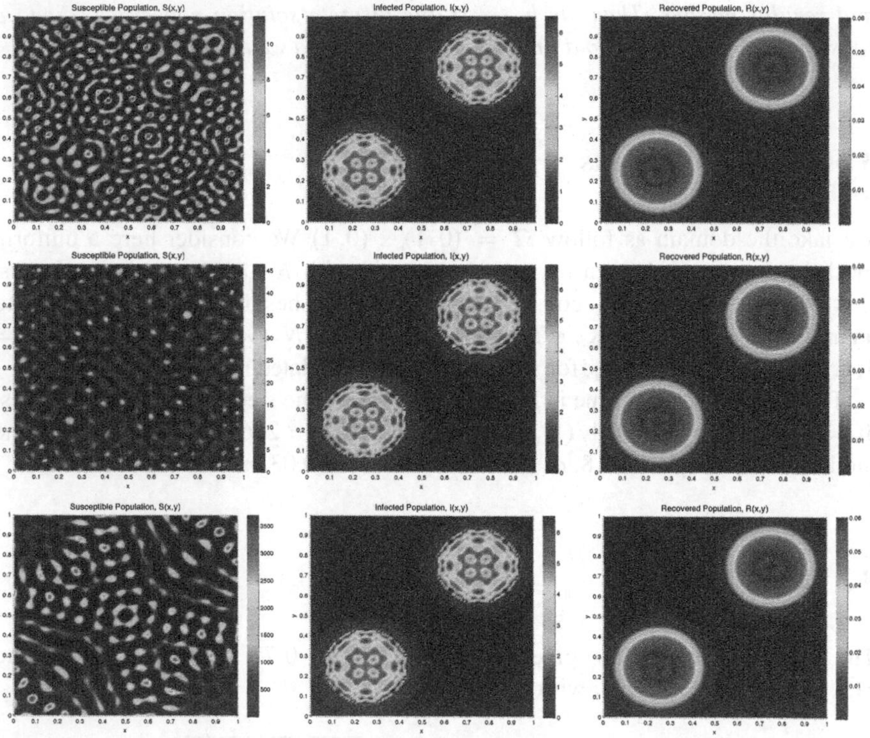

Fig. 1 Variation of the nonlinear diffusion coefficient related with the susceptible population. $\alpha_1 = 500, 700, 900$; $\alpha_2 = \alpha_3 = 100$, $t = 0.01$

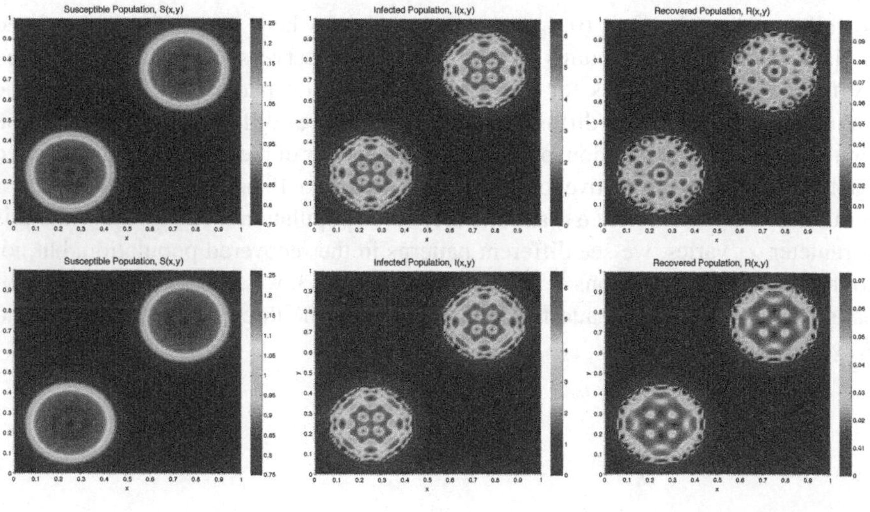

Fig. 2 Variation of the nonlinear diffusion coefficient related with the recovered population. $\alpha_3 = 19{,}000, 21{,}000$; $\alpha_1 = \alpha_2 = 100$, $t = 0.01$

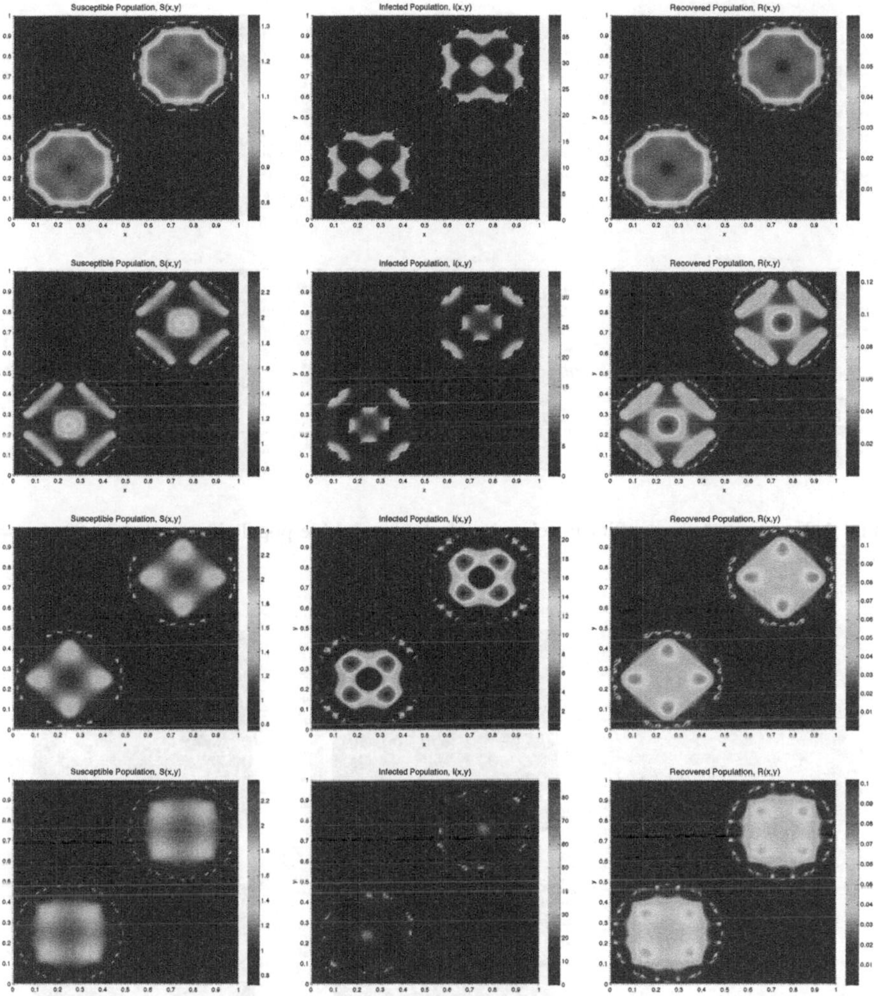

Fig. 3 Variation of the nonlinear diffusion coefficient related with the infective population. $\alpha_2 = 400, 600, 800, 900$; $\alpha_1 = \alpha_3 = 100$, $t = 0.01$

3.2 *Second Example*

We consider now five initial focus of infection given by $S(x, y, 0) = 10$, $R(x, y, 0) = 0$ and $I(x, y, 0) = 40 \sum_{j=1}^{5} sech(15(x - x_j))sech(15(y - y_j))$, with $(x, y) \in (0, 1) \times (0, 1)$, where $(x_1, y_1) = (0.5, 0.5)$, $(x_2, y_2) = (0.3, 0.3)$, $(x_3, y_3) = (0.3, 0.7)$, $(x_4, y_4) = (0.7, 0.3)$, and $(x_5, y_5) = (0.7, 0.7)$. In Figs. 4 and 5, we observe some peculiar patterns taking different values for the nonlinear diffusion parameters α_1 and α_3. In particular in Fig. 4, it shows that if we modify parameter α_1, then different patterns are obtained for the susceptible population.

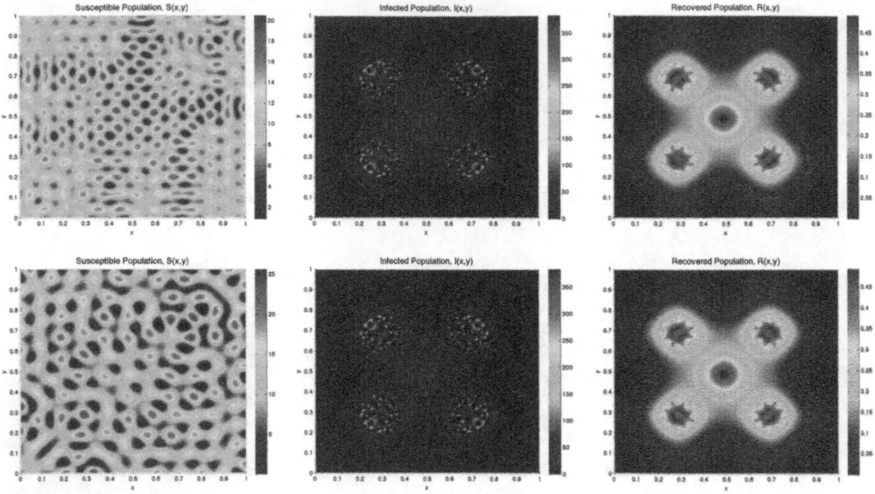

Fig. 4 Variation of the coefficient related with the susceptible population: $\alpha_1 = 70, 90$; $\alpha_2 = \alpha_3 = 10, t = 0.01$

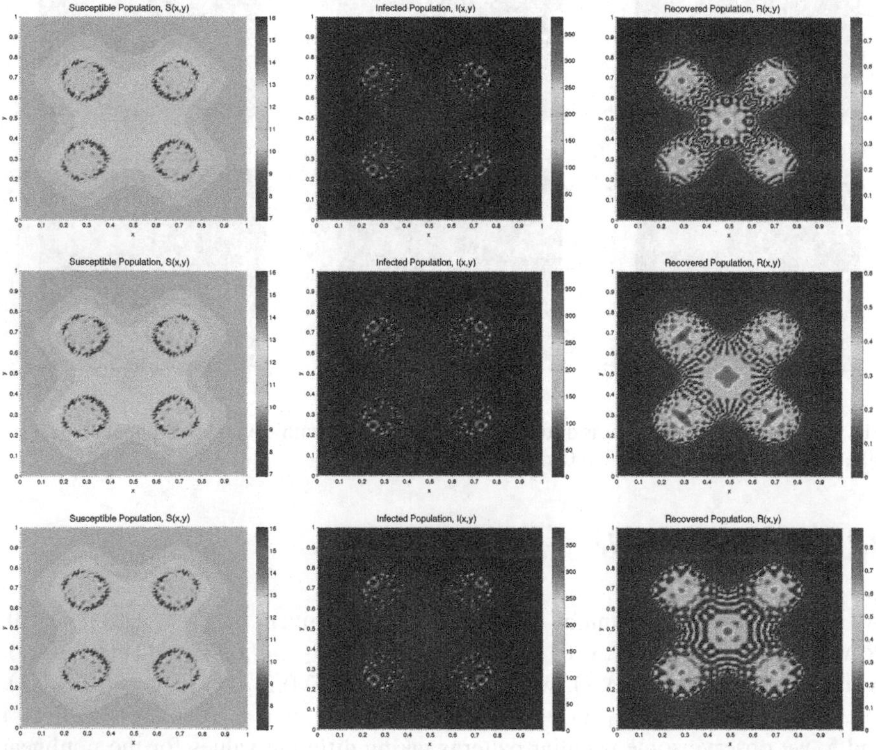

Fig. 5 Variation of the coefficient related with the recovered population: $\alpha_3 = 2{,}000, 2{,}400, 2{,}700$; $\alpha_1 = \alpha_2 = 10, t = 0.01$

Thus, for small values of α_2 and α_3 and higher values of α_1, there is a formation of patterns of "labyrinth" type in the susceptible population. This conformation is very sensitive to the value chosen for α_1, but also is sensitive with respect to the initial condition if we compare with Fig. 1. On the other hand, it is observed in Fig. 5 that for the sensitivity of the recovered population with respect to parameters, it requires very high values for α_3, but with quite amazing images for the recovered population.

Acknowledgements VA thanks support of Fondecyt Postdoc 3120197, CONICYT Proyecto de Inserción 79112012, and DIUBB project 120808 GI/EF. MS thanks support of Fondecyt 1140676, CONICYT project Anillo ACT1118 (ANANUM), REDOC.CTA project UCO1202 at U. de Concepción, Basal, CMM, U. de Chile, and CI^{2}MA, Universidad de Concepción.

References

1. V. Anaya, M. Bendahmane, M. Langlais, M. Sepúlveda, A convergent finite volume method for a model of indirectly transmitted diseases with nonlocal cross-diffusion, Preprint 2013–12, DIM, U. de Concepción
2. V. Anaya, M. Bendahmane, M. Sepúlveda, Numerical analysis for HP food chain system with nonlocal and cross diffusion, Preprint 2011–11, DIM, U. de Concepción
3. E. DiBenedetto, *Degenerate Parabolic Equations*. Universitext (Springer, New York, 1993)
4. G. Galiano, M.L. Garzón, A. Jüngel, Analysis and numerical solution of a nonlinear cross-diffusion system arising in population dynamics. Rev. R. Acad. Cienc. Exactas Fís. Nat. Serie A. Mat. **95**(2), 281–295 (2001)
5. G. Galiano, M.L. Garzón, A. Jüngel, Semi-discretization in time and numerical convergence of solutions of a nonlinear cross-diffusion population model. Numer. Math. **93**(4), 655–673 (2003)
6. Y. Li, C. Zhao, Global existence of solutions to a cross-diffusion system in higher dimensional domains. Discr. Contin. Dyn. Syst. **12**(2), 185–192 (2005)
7. Y. Lou, S. Martinez, Evolution of cross-diffusion and self-diffusion. J. Biol. Dyn. **3**(4), 410–429 (2009)
8. Y. Wang, The global existence of solutions for cross-diffusion system. Acta Math. Appl. Sin. Engl. Ser. **21**(3), 519–528 (2005)

Part II
Time Integration Schemes

Stability of Explicit Runge-Kutta Methods for High Order Finite Element Approximation of Linear Parabolic Equations

Weizhang Huang, Lennard Kamenski, and Jens Lang

Abstract We study the stability of explicit Runge-Kutta methods for high order Lagrangian finite element approximation of linear parabolic equations and establish bounds on the largest eigenvalue of the system matrix which determines the largest permissible time step. A bound expressed in terms of the ratio of the diagonal entries of the stiffness and mass matrices is shown to be tight within a small factor which depends only on the dimension and the choice of the reference element and basis functions but is independent of the mesh or the coefficients of the initial-boundary value problem under consideration. Another bound, which is less tight and expressed in terms of mesh geometry, depends only on the number of mesh elements and the alignment of the mesh with the diffusion matrix. The results provide an insight into how the interplay between the mesh geometry and the diffusion matrix affects the stability of explicit integration schemes when applied to a high order finite element approximation of linear parabolic equations on general nonuniform meshes.

W. Huang
Department of Mathematics, University of Kansas, Lawrence, KS 66045, USA
e-mail: whuang@ku.edu

L. Kamenski (✉)
Weierstrass Institute for Applied Analysis and Stochastics, Berlin, Germany
e-mail: kamenski@wias-berlin.de

J. Lang
Department of Mathematics, Graduate School of Computational Engineering, and Center of Smart Interfaces, Technische Universität Darmstadt, Darmstadt, Germany
e-mail: lang@mathematik.tu-darmstadt.de

A. Abdulle et al. (eds.), *Numerical Mathematics and Advanced Applications ENUMATH 2013*, Lecture Notes in Computational Science and Engineering 103,
DOI 10.1007/978-3-319-10705-9_16

1 Introduction

We consider the initial-boundary value problem (IBVP)

$$\begin{cases} u_t = \nabla \cdot (\mathbb{D}\nabla u), & \boldsymbol{x} \in \Omega, \quad t \in (0,T], \\ u(\boldsymbol{x},t) = 0, & \boldsymbol{x} \in \Gamma_D, \quad t \in (0,T], \\ \mathbb{D}\nabla u(\boldsymbol{x},t) \cdot \boldsymbol{n} = 0, & \boldsymbol{x} \in \Gamma_N, \quad t \in (0,T], \\ u(\boldsymbol{x},0) = u^0(\boldsymbol{x}), & \boldsymbol{x} \in \Omega, \end{cases} \tag{1}$$

where $\Omega \subset \mathbb{R}^d\,(d \geq 1)$ is a bounded polygonal or polyhedral domain, $\Gamma_D \cup \Gamma_N = \partial\Omega$, $\text{meas}_{d-1}\,\Gamma_D > 0$, u^0 is a given function, and $\mathbb{D} = \mathbb{D}(\boldsymbol{x})$ is the diffusion matrix, which is assumed to be time-independent, symmetric and uniformly positive-definite on Ω. If $u^0 \in H_D^1(\Omega) = \{v \in H^1(\Omega) : v = 0 \text{ on } \Gamma_D\}$ and u is sufficiently smooth, then the solution of the IBVP satisfies the stability estimates

$$\begin{cases} \|u(\cdot,t)\|_{L^2(\Omega)} \leq \left\|u^0\right\|_{L^2(\Omega)}, & t \in (0,T], \\ |||u(\cdot,t)||| \leq |||u^0|||, & t \in (0,T], \end{cases}$$

where $|||u||| = \left\|\mathbb{D}^{1/2}\nabla u\right\|_{L^2(\Omega)}$ is the energy norm. We are interested in the stability conditions so that the numerical approximation preserves these stability estimates.

The stability of explicit Runge-Kutta methods depends on the largest eigenvalue of the corresponding system matrix, which, in turn, depends on the mesh and the coefficients of the IBVP. For our model problem this means that we need to estimate the largest eigenvalue of $M^{-1}A$, where M and A are the mass and stiffness matrices for the finite element discretization of the IBVP (1) [4, Theorem 3.1]. For the Laplace operator on a uniform mesh it is well known that $\lambda_{\max}(M^{-1}A) \sim N^{2/d}$, where N is the number of mesh elements. For general meshes and diffusion coefficients, estimates have been derived recently in Huang et al. [4] and Zhu and Du [6, 7] (see also [1–3, 5] for estimates on M and A). All of these works allow anisotropic diffusion coefficients and anisotropic meshes, while the former employs a more accurate measure for the interplay between the mesh geometry and the diffusion matrix and gives a sharper estimate on $\lambda_{\max}(M^{-1}A)$ than the latter. On the other hand, [4] considers only linear finite elements whereas the estimates in [7] are valid for both linear and higher order finite elements.

The purpose of this paper is to extend the result of [4] to high order Lagrangian finite elements as well as provide a mathematical understanding of how the interplay between the mesh geometry and the diffusion matrix affects the stability condition. We show that the main result of [4, Theorem 3.3] holds for high order finite elements as well. The analysis is based on bounds on the mass and stiffness matrices. We follow the approach in [4, 5] and derive simple but accurate bounds for the case of high order Lagrangian finite elements on simplicial meshes (Lemmas 2–5). We also consider the more general case of surrogate mass matrices $\tilde{M}$. The main result (Theorem 1) shows that $\lambda_{\max}(\tilde{M}^{-1}A)$ is proportional to the maximum ratio between the corresponding diagonal entries of the stiffness and surrogate mass matrices. Moreover, $\lambda_{\max}(\tilde{M}^{-1}A)$ is bounded by a term depending only on the number of the mesh elements and the alignment of the shape of the mesh elements with the inverse of the diffusion matrix.

2 Stability Condition for Explicit Time Stepping

Let $\{\mathscr{T}_h\}$ be a family of simplicial meshes for Ω and V^h the Lagrangian $\mathbb{P}_m$ $(m \geq 1)$ finite element space associated with $\mathscr{T}_h$. Let K be an arbitrary element of $\mathscr{T}_h$, $\hat{K}$ the reference element, and ω_i the element patch of the ith vertex (Fig. 1); element and patch volumes are denoted by $|K|$ and $|\omega_i| = \sum_{K\in\omega_i} |K|$. For each $K \in \mathscr{T}_h$ let $F_K \colon \hat{K} \to K$ be an invertible affine mapping and F_K' its Jacobian matrix which is constant and satisfies $\det(F_K') = |K|$ (for simplicity, we assume that $|\hat{K}| = 1$). We further assume that the mesh is fixed for all time steps.

With $V_D^h = V^h \cap H_D^1(\Omega)$, the finite element solution $u^h(t) \in V_D^h$ $(t \in (0,T])$ is defined by

$$\int_\Omega \partial_t u^h v^h \, d\boldsymbol{x} = -\int_\Omega \nabla v^h \cdot \mathbb{D} \nabla u^h \, d\boldsymbol{x}, \qquad \forall v^h \in V_D^h, \quad t \in (0,T], \tag{2}$$

subject to the initial condition

$$\int_\Omega u^h(\boldsymbol{x},0) v^h \, d\boldsymbol{x} = \int_\Omega u^0(\boldsymbol{x}) v^h \, d\boldsymbol{x}, \qquad \forall v^h \in V_D^h. \tag{3}$$

Let N_ϕ be the dimension of the finite element space V_D^h and denote a nodal basis of V_D^h by $\{\phi_1, \ldots, \phi_{N_\phi}\}$, then u^h can be expressed as

$$u^h(\boldsymbol{x},t) = \sum_{j=1}^{N_\phi} u_j^h(t) \phi_j(\boldsymbol{x}).$$

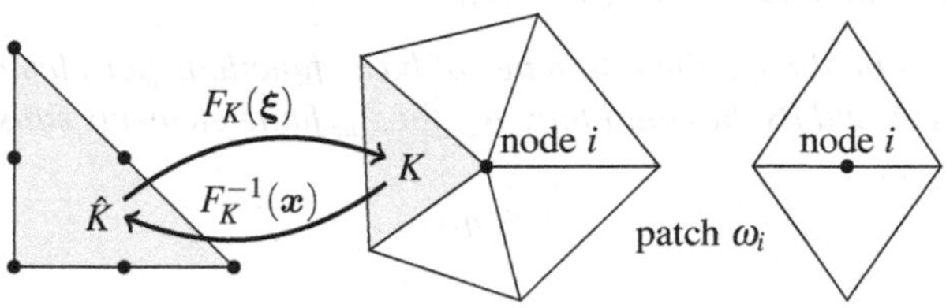

Fig. 1 Example of the standard quadratic FE reference mesh element $\hat{K}$, mapping F_K, the corresponding mesh elements K, nodes and their patches

Using $\boldsymbol{U} = (u_1^h, \ldots, u_{N_\phi}^h)^T$, (2) and (3) can be written into a matrix form

$$M \boldsymbol{U}_t = -A \boldsymbol{U}, \quad \boldsymbol{U}(0) = \boldsymbol{U}_0, \tag{4}$$

where the mass and stiffness matrices M and A are defined by

$$M_{ij} = \int_\Omega \phi_i \phi_j \, d\boldsymbol{x} \quad \text{and} \quad A_{ij} = \int_\Omega \nabla \phi_i \cdot \mathbb{D} \nabla \phi_j \, d\boldsymbol{x}, \qquad i, j = 1, \ldots, N_\phi.$$

We further assume that surrogate mass matrices $\tilde{M}$ considered throughout the paper satisfy

(M1) The reference element matrix $\tilde{M}_{\hat{K}}$ is symmetric positive definite.
(M2) The element matrix $\tilde{M}_K$ satisfies $\tilde{M}_K = |K| \tilde{M}_{\hat{K}}$.

For example, (M1) and (M2) are satisfied for any mass lumping by means of numerical quadrature with positive weights.

Lemma 1 ([4, Theorem 3.1]) *For a given explicit RK method with the polynomial stability function R and a symmetric positive definite surrogate matrix $\tilde{M}$ that satisfies*[1] $c_1 \tilde{M} \le M \le c_2 \tilde{M}$ *for some positive constants c_1 and c_2, the finite element approximation u_n^h at $t_n = n\tau$ satisfies*

$$\|u_n^h\|_{L^2(\Omega)} \le \sqrt{\frac{c_2}{c_1}} \|u_0^h\|_{L^2(\Omega)} \quad \text{and} \quad |||u_n^h||| \le |||u_0^h|||,$$

if the time step τ is chosen such that

$$\max_i |R(-\tau \lambda_i(\tilde{M}^{-1} A))| \le 1.$$

This lemma is proven in [4] for the linear finite element discretization. However, from the proof one can see that it is valid for any system in the form of (4) with symmetric positive definite matrices M and A. Particularly, it can be used for the system (4) resulting from the $\mathbb{P}_m$ finite element discretization. In the following, we establish a series of lemmas for bounds on the stiffness and mass matrices A and $\tilde{M}$ and then develop bounds for $\lambda_{\max}(\tilde{M}^{-1} A)$.

Lemma 2 *Let η be the maximal number of basis functions per element. Then the stiffness matrix A and its diagonal part A_D for $\mathbb{P}_m$ finite elements satisfy*

$$A \le \eta A_D.$$

[1] In the following, the less-than-or-equal-to sign for matrices means that the difference between the right-hand side and left-hand side terms is positive semidefinite.

Proof Notice that for any positive semi-definite matrix S and any vectors $\boldsymbol{u}$ and $\boldsymbol{v}$ we have $\boldsymbol{u}^T S\boldsymbol{v} + \boldsymbol{v}^T S\boldsymbol{u} \le \boldsymbol{u}^T S\boldsymbol{u} + \boldsymbol{v}^T S\boldsymbol{v}$. From this,

$$\begin{aligned}\boldsymbol{u}^T A\boldsymbol{u} &= \sum_{i,j} \int_\Omega (u_i \nabla\phi_i)^T \mathbb{D} \left(u_j \nabla\phi_j\right)\, d\boldsymbol{x} \le \sum_i \eta \int_\Omega (u_i \nabla\phi_i)^T \mathbb{D}\,(u_i \nabla\phi_i)\, d\boldsymbol{x} \\ &= \eta \sum_i u_i^2 \int_\Omega \nabla\phi_i^T \mathbb{D} \nabla\phi_i\, d\boldsymbol{x} = \boldsymbol{u}^T \eta A_D \boldsymbol{u}.\end{aligned}$$

Lemma 3 *Let $\hat{\phi}_i$ be the basis functions on the reference element that correspond to ϕ_i and*

$$C_{H^1} = \max_i |\hat{\phi}_i|^2_{H^1(\hat{K})}.$$

Then the diagonal entries A_{ii} of the stiffness matrix A are bounded by

$$A_{ii} \le C_{H^1} \sum_{K\in\omega_i} |K| \max_{x\in K} \left\| (F_K')^{-1} \mathbb{D} (F_K')^{-T} \right\|_2,$$

Proof From the definition of the stiffness matrix we have

$$A_{ii} = \int_\Omega \nabla\phi_i^T \mathbb{D} \nabla\phi_i\, d\boldsymbol{x} = \sum_{K\in\omega_i} \int_K \nabla\phi_i^T \mathbb{D} \nabla\phi_i\, d\boldsymbol{x}.$$

Let $\hat{\nabla} = \partial/\partial\boldsymbol{\xi}$ be the gradient operator in $\hat{K}$. The chain rule yields $\nabla = (F_K')^{-T}\hat{\nabla}$ and together with $\det(F_K') = |K|$ we obtain

$$\begin{aligned}A_{ii} &= \sum_{K\in\omega_i} |K| \int_{\hat{K}} \hat{\nabla}\hat{\phi}_i^T (F_K')^{-1} \mathbb{D} (F_K')^{-T} \hat{\nabla}\hat{\phi}_i\, d\boldsymbol{\xi} \\ &\le \sum_{K\in\omega_i} |K|\, \|\hat{\nabla}\hat{\phi}_i\|^2_{L^2(\hat{K})} \max_{x\in K} \left\| (F_K')^{-1} \mathbb{D} (F_K')^{-T} \right\|_2 \\ &\le C_{H^1} \sum_{K\in\omega_i} |K| \max_{x\in K} \left\| (F_K')^{-1} \mathbb{D} (F_K')^{-T} \right\|_2.\end{aligned}$$

Lemma 4 *Let $\tilde{M}$ be a surrogate $\mathbb{P}_m$ finite element mass matrix, $\hat{\Lambda}_{\tilde{M}}$ and $\hat{\lambda}_{\tilde{M}}$ be the largest and smallest eigenvalues of the surrogate mass matrix $\tilde{M}_{\hat{K}}$ on the reference element and*

$$\mathscr{W} = \operatorname{diag}\left(|\omega_1|, \ldots, |\omega_{N_\phi}|\right).$$

Then

$$\hat{\lambda}_{\tilde{M}} \mathscr{W} \le \tilde{M} \le \hat{\Lambda}_{\tilde{M}} \mathscr{W}. \tag{5}$$

Proof We have

$$\begin{aligned} \boldsymbol{u}^T \tilde{M} \boldsymbol{u} &= \sum_K \boldsymbol{u}_K^T \tilde{M}_K \boldsymbol{u}_K = \sum_K |K| \boldsymbol{u}_K^T \tilde{M}_{\hat{K}} \boldsymbol{u}_K \le \sum_K |K| \hat{\Lambda}_{\tilde{M}} \|\boldsymbol{u}_K\|_2^2 \\ &= \hat{\Lambda}_{\tilde{M}} \sum_i u_i^2 \sum_{K \in \omega_i} |K| = \hat{\Lambda}_{\tilde{M}} \sum_i u_i^2 |\omega_i| = \hat{\Lambda}_{\tilde{M}} \boldsymbol{u}^T \mathscr{W} \boldsymbol{u}. \end{aligned}$$

The lower bound can be obtained similarly.

Lemma 5 *Let $\tilde{M}_1$ and $\tilde{M}_2$ be two surrogate mass matrices for $\mathbb{P}_m$ finite elements. Then*

$$\frac{\hat{\lambda}_{\tilde{M}_1}}{\hat{\Lambda}_{\tilde{M}_2}} \tilde{M}_2 \le \tilde{M}_1 \le \frac{\hat{\Lambda}_{\tilde{M}_1}}{\hat{\lambda}_{\tilde{M}_2}} \tilde{M}_2.$$

Proof Use Lemma 4 by applying (5) to $\tilde{M}_1$ and $\tilde{M}_2$.

Corollary 1 *Let $\kappa(M_{\hat{K}})$ and $\kappa(\tilde{M}_{\hat{K}})$ be the condition numbers of the full and the surrogate reference element mass matrices. Under the assumptions of Lemma 1 we have*

$$\|u_n^h\|_{L^2(\Omega)} \le \sqrt{\kappa(M_{\hat{K}})\kappa(\tilde{M}_{\hat{K}})} \|u_0^h\|_{L^2(\Omega)} \quad \textit{and} \quad |||u_n^h||| \le |||u_0^h|||.$$

Proof Use $\tilde{M}_1 = M$ and $\tilde{M}_2 = \tilde{M}$ in Lemma 5 and apply Lemma 1.

Corollary 2 *The surrogate mass matrix $\tilde{M}$ for $\mathbb{P}_m$ finite elements and its diagonal part $\tilde{M}_D$ satisfy*

$$\frac{1}{\kappa(\tilde{M}_{\hat{K}})} \tilde{M}_D \le \tilde{M} \le \kappa(\tilde{M}_{\hat{K}}) \tilde{M}_D.$$

Proof Using (5) with the canonical basis vector $\boldsymbol{e}_i$ implies $\hat{\lambda}_{\tilde{M}} \mathscr{W}_{ii} \le \tilde{M}_{ii} \le \hat{\Lambda}_{\tilde{M}} \mathscr{W}_{ii}$, which gives $u_i \hat{\lambda}_{\tilde{M}} \mathscr{W}_{ii} u_i \le u_i \tilde{M}_{ii} u_i \le u_i \hat{\Lambda}_{\tilde{M}} \mathscr{W} u_i$ for any u_i. Since $\tilde{M}_D$ and $\mathscr{W}$ are diagonal matrices, this leads to

$$\hat{\lambda}_{\tilde{M}} \mathscr{W} \le \tilde{M}_D \le \hat{\Lambda}_{\tilde{M}} \mathscr{W}. \tag{6}$$

The statement now follows from Lemma 5 with $\tilde{M}_1 = \tilde{M}$ and $\tilde{M}_2 = \tilde{M}_D$.

Having obtained the preliminary bounds on the stiffness and mass matrices A and $\tilde{M}$, we can now give the estimate for the largest eigenvalue of the system matrix $\tilde{M}^{-1}A$ for $\mathbb{P}_m$ finite elements.

Theorem 1 *The eigenvalues of $\tilde{M}^{-1}A$ are real and positive and the largest eigenvalue is bounded by*

$$\max_i \frac{A_{ii}}{\tilde{M}_{ii}} \leq \lambda_{\max}(\tilde{M}^{-1}A) \leq \eta\,\kappa(\tilde{M}_{\hat{K}}) \max_i \frac{A_{ii}}{\tilde{M}_{ii}}, \tag{7}$$

where η is the maximal number of basis functions per element. Further,

$$\lambda_{\max}(\tilde{M}^{-1}A) \leq \eta \frac{C_{H^1}}{\hat{\lambda}_{\tilde{M}}} \max_i \left\{ \sum_{K\in\omega_i} \frac{|K|}{|\omega_i|} \max_{x\in K} \left\| (F_K')^{-1} \mathbb{D} (F_K')^{-T} \right\|_2 \right\}. \tag{8}$$

Proof Since $\tilde{M}$ and A are symmetric positive definite, the eigenvalues of $\tilde{M}^{-1}A$ are real and positive. The lower bound in (7) is obtained by using the canonical basis vectors $\boldsymbol{e}_i$ and the upper bound follows from Lemmas 2 and Corollary 2,

$$\lambda_{\max}(\tilde{M}^{-1}A) = \max_{v\neq 0} \frac{\boldsymbol{v}^T A \boldsymbol{v}}{\boldsymbol{v}^T \tilde{M} \boldsymbol{v}} \leq \max_{v\neq 0} \frac{\boldsymbol{v}^T \eta A_D \boldsymbol{v}}{\boldsymbol{v}^T \frac{1}{\kappa(\tilde{M}_{\hat{K}})} \tilde{M}_D \boldsymbol{v}} = \eta\,\kappa(\tilde{M}_{\hat{K}}) \max_i \frac{A_{ii}}{\tilde{M}_{ii}}.$$

The geometric bound (8) is a direct consequence of Lemmas 2–4,

$$\begin{aligned}\lambda_{\max}(\tilde{M}^{-1}A) &= \max_{v\neq 0} \frac{\boldsymbol{v}^T A \boldsymbol{v}}{\boldsymbol{v}^T \tilde{M} \boldsymbol{v}} \leq \max_{v\neq 0} \frac{\boldsymbol{v}^T \eta A_D \boldsymbol{v}}{\boldsymbol{v}^T \hat{\lambda}_{\tilde{M}} \mathscr{W} \boldsymbol{v}} \\ &\leq \eta \frac{C_{H^1}}{\hat{\lambda}_M} \max_i \left\{ \sum_{K\in\omega_i} \frac{|K|}{|\omega_i|} \max_{x\in K} \left\| (F_K')^{-1} \mathbb{D} (F_K')^{-T} \right\|_2 \right\}.\end{aligned}$$

Theorem 1 can be used in combination with Lemma 1 or Corollary 1 to derive the stability condition of a given explicit Runge-Kutta scheme, as shown in the next example.

Example 1 (Explicit Euler method) The stability region of the explicit Euler method includes the real interval $[-2, 0]$. Lemma 1 implies that the method is stable if

$$-2 \leq -\tau\lambda_i(\tilde{M}^{-1}A) \leq 0, \qquad i = 1, \ldots, N_\phi.$$

Using Theorem 1, we conclude that the method is stable if the time step τ satisfies

$$\tau \leq \frac{2}{\eta\,\kappa(\tilde{M}_{\hat{K}})} \min_i \frac{\tilde{M}_{ii}}{A_{ii}}$$

or, in terms of mesh geometry,

$$\tau \le \frac{2\hat{\lambda}_{\tilde{M}_{\hat{K}}}}{\eta C_{H^1}} \min_i \left(\sum_{K\in\omega_i} \frac{|K|}{|\omega_i|} \max_{x\in K} \left\| (F_K')^{-1}\mathbb{D}(F_K')^{-T} \right\|_2 \right)^{-1}.$$

Remark 1 Lemmas 2, 3, and Corollary 2 are very general and valid for any mesh, any $\mathbb{D}$ and any surrogate mass matrix $\tilde{M}$ satisfying (M1) and (M2). More accurate bounds can be obtained if more information is available about the mesh or the stiffness and mass matrices.

For example, if A is an M-matrix, then the Gershgorin circle theorem yields $\lambda_{\max}(A) \le 2\max_i A_{ii}$ [4, Remark 2.2] and therefore η in Theorem 1 can be replaced by 2.

If $\tilde{M} = M$ (no mass lumping), then, instead of estimating M_D through (6), a direct calculation for the standard $\mathbb{P}_m$ finite elements yields

$$M_D = C_{L^2}\mathscr{W}, \qquad C_{L^2} = \operatorname{diag}\left(\|\hat{\phi}_1\|_{L^2}^2, \dots, \|\hat{\phi}_{N_\phi}\|_{L^2}^2\right),$$

and

$$\hat{\lambda}_M C_{L^2}^{-1} M_D \le M \le \hat{\Lambda}_M C_{L^2}^{-1} M_D,$$

resulting in a slighly more accurate bound in Corollary 2.

Also, for simplicity, in Lemma 3 we used $C_{H^1} = \max_i |\hat{\phi}_i|_{H^1(\hat{K})}^2$. A slightly more accurate bound can be derived if we use

$$C_{H^1} = \operatorname{diag}\left(|\hat{\phi}_1|_{H^1(\hat{K})}^2, \dots, |\hat{\phi}_{N_\phi}|_{H^1(\hat{K})}^2\right).$$

3 Summary and Conclusion

Theorem 1 states that the largest eigenvalue of the system matrix and, thus, the largest permissible time step can be bounded by a term depending only on the number of mesh elements and the alignment of the mesh with the diffusion matrix.

The bound in terms of matrix entries is tight within a small factor which depends only on the dimension and the choice of the reference element and basis functions but is independent of the mesh or the coefficients of the IBVP. This is valid for any Lagrangian $\mathbb{P}_m$ finite elements with $m \ge 1$.

A similar result is obtained by Zhu and Du [7, Theorem 3.1]. In our notation, it can be written as

$$\lambda_{\max}(M^{-1}A) \lesssim \max_K \left\{ \max_{x\in K} \lambda_{\max}(\mathbb{D}) \left\| (F_K')^{-1}(F_K')^{-T} \right\|_2 \right\}. \tag{9}$$

The significant difference between the new bound (8) and the bound (9) is the factor which represents the interplay between the mesh geometry and the diffusion matrix,

$$\max_{x\in K}\left\|(F_K')^{-1}\mathbb{D}(F_K')^{-T}\right\|_2 \qquad \text{vs.} \qquad \max_{x\in K}\lambda_{\max}(\mathbb{D})\left\|(F_K')^{-1}(F_K')^{-T}\right\|_2.$$

For isotropic $\mathbb{D}$ or isotropic meshes both terms are comparable. However, the former is smaller than the latter in general. In particular, if both $\mathbb{D}$ and K are anisotropic, then the difference between (8) and (9) can be very significant (see [4, Sect. 4.4] for a numerical example in case of $\mathbb{P}_1$ finite elements). In this sense, Theorem 1 can be seen either as a generalization of [4] to $\mathbb{P}_m (m \geq 2)$ finite elements or as a more accurate version of [7] for anisotropic meshes and general diffusion coefficients.

Finally, we would like to point out that a similar result can be established for p-adaptive finite elements without major modifications.

Acknowledgement The work was supported in part by the NSF (U.S.A.) under Grant DMS-1115118, the Excellence Initiative of the German Federal and State Governments, and the Graduate School of Engineering at the Technische Universität Darmstadt.

References

1. Q. Du, D. Wang, L. Zhu, On mesh geometry and stiffness matrix conditioning for general finite element spaces. SIAM J. Numer. Anal. **47**(2), 1421–1444 (2009)
2. I. Fried, Bounds on the spectral and maximum norms of the finite element stiffness, flexibility and mass matrices. Int. J. Solids Struct. **9**, 1013–1034 (1973)
3. I.G. Graham, W. McLean, Anisotropic mesh refinement: the conditioning of Galerkin boundary element matrices and simple preconditioners. SIAM J. Numer. Anal. **44**(4), 1487–1513 (2006)
4. W. Huang, L. Kamenski, J. Lang, Stability of explicit time integration schemes for finite element approximation of linear parabolic equations on anisotropic meshes, WIAS Preprint No. 1869, 2013
5. L. Kamenski, W. Huang, H. Xu, Conditioning of finite element equations with arbitrary anisotropic meshes. Math. Comp. **83**, 2187–2211 (2014)
6. L. Zhu, Q. Du, Mesh-dependent stability for finite element approximations of parabolic equations with mass lumping. J. Comput. Appl. Math. **236**(5), 801–811 (2011)
7. ———, Mesh dependent stability and condition number estimates for finite element approximations of parabolic problems. Math. Comput. **83**(285), 37–64 (2014)

Comparison of Time Discretization Schemes to Simulate the Motion of an Inextensible Beam

Steffen Basting, Annalisa Quaini, Roland Glowinski, and Suncica Canic

Abstract We compare three different time discretization schemes in combination with an augmented Lagrangian method to simulate the motion of an inextensible beam. The resulting saddle-point problem is solved with an Uzawa-Douglas-Rachford algorithm. The three schemes are tested on a benchmark with an analytical solution and on a more challenging application. We found that in order to obtain optimal convergence behavior in time, the stopping tolerance for the Uzawa-type algorithm should be balanced against the time step size.

1 Introduction

The motion of an inextensible beam, while well studied (see, e.g., [4] and references therein), remains to be a challenging problem numerically. The main difficulties stem from the nonlinearity due to the inextensibility condition, and the choice of appropriate time discretization scheme that is stable and accurate (see [7] for a survey on different schemes). In this work, we evaluate the performance of the Houbolt scheme, a generalized Crank-Nicolson scheme, and a Newmark scheme, which are combined with an Uzawa-type algorithm for solving the saddle-point problem associated with an augmented Lagrangian method employed to handle the inextensibility condition.

S. Basting (✉)
Friedrich-Alexander-University Erlangen-Nuremberg, Cauerstr. 11, 91058 Erlangen, Germany
e-mail: basting@math.fau.de

A. Quaini • R. Glowinski • S. Canic
University of Houston, 4800 Calhoun Rd, Houston, TX 77204, USA
e-mail: quaini@math.uh.edu; roland@math.uh.edu; canic@math.uh.edu

A. Abdulle et al. (eds.), *Numerical Mathematics and Advanced Applications ENUMATH 2013*, Lecture Notes in Computational Science and Engineering 103,
DOI 10.1007/978-3-319-10705-9_17

2 Motion of an Inextensible Beam

We consider an inextensible elastic beam in static and dynamic regimes, assuming negligible torsional effects. We will denote by ρ the linear density (i.e. mass per unit length), by L the length, and by EI the flexural stiffness of the beam. We will use the following notation, with s denoting arc length and t time: $\mathbf{y}' = \frac{\partial \mathbf{y}}{\partial s}$, $\dot{\mathbf{y}} = \frac{\partial \mathbf{y}}{\partial t}$, $\mathbf{y}'' = \frac{\partial^2 \mathbf{y}}{\partial s^2}$, $\ddot{\mathbf{y}} = \frac{\partial^2 \mathbf{y}}{\partial t^2}$.

2.1 The Static Problem

We assume that the beam is subject to external forces $\mathbf{f}$ and that the strain-stress relation is linear. The position of the beam at the equilibrium configuration is solution of a non-convex constrained problem:

$$\mathbf{x} = \arg\min_{\mathbf{y}\in K} J(\mathbf{y}), \text{ where } J(\mathbf{y}) = \frac{1}{2}\int_0^L EI\left|\mathbf{y}''\right|^2 ds - \int_0^L \mathbf{f}\cdot\mathbf{y}\, ds, \tag{1}$$

with $K = \left\{\mathbf{y} \in (H^2(0,L))^2, |\mathbf{y}'| = 1, \text{ plus boundary conditions}\right\}$.

To treat the inextensibility condition $|\mathbf{y}'| = 1$, which is a quadratic constraint, we use an augmented Lagrangian method (see, e.g., [1–4]). Let us introduce the following space and set:

$$V = \left\{\mathbf{y} \in (H^2(0,L))^2, \text{ plus boundary conditions}\right\},$$
$$\mathcal{Q} = \left\{\mathbf{q} \in (L^2(0,L))^2, |\mathbf{q}| = 1 \text{ a.e. on } (0,L)\right\}.$$

The static problem (1) is equivalent to

$$\{\mathbf{x}, \mathbf{x}'\} = \arg\min_{\{\mathbf{y},\mathbf{q}\}\in W} J(\mathbf{y}), \quad \text{with} \quad W = \{\mathbf{y} \in V,\ \mathbf{q} \in \mathcal{Q},\ \mathbf{y}' - \mathbf{q} = \mathbf{0}\}.$$

With $r > 0$, we introduce the following augmented Lagrangian functional:

$$\mathcal{L}_r(\mathbf{y}, \mathbf{q}; \boldsymbol{\mu}) = J(\mathbf{y}) + \frac{r}{2}\int_0^L |\mathbf{y}' - \mathbf{q}|^2\, ds + \int_0^L \boldsymbol{\mu}\cdot(\mathbf{y}' - \mathbf{q})\, ds \tag{2}$$

Let $\{\mathbf{x}, \mathbf{p}; \boldsymbol{\lambda}\}$ be a saddle point of $\mathcal{L}_r$ over $(V \times \mathcal{Q}) \times (L^2(0,L))^2$. Then $\mathbf{x}$ is a solution of the static problem (1) and $\mathbf{p} = \mathbf{x}'$. In order to solve the above saddle-point problem, we employ the algorithm called ALG2 in, e.g., [2, 4]. As shown in, e.g., [2], this Uzawa-type algorithm is in fact a 'disguised' Douglas-Rachford operator-splitting scheme. It reads as follow:

Step 0: The initial guess $\{\mathbf{x}_{-1}, \boldsymbol{\lambda}_0\} \in V \times (L^2(0,L))^2$ is given.

Then, for $k \geq 0$, $\{\mathbf{x}_{k-1}, \boldsymbol{\lambda}_k,\}$ being known, proceed with:

Step 1: Find $\mathbf{p}_k \in \mathcal{Q}$ such that:

$$\mathcal{L}_r(\mathbf{x}_{k-1}, \mathbf{p}_k; \boldsymbol{\lambda}_k) \leq \mathcal{L}_r(\mathbf{x}_{k-1}, \mathbf{q}; \boldsymbol{\lambda}_k), \quad \forall \mathbf{q} \in \mathcal{Q}.$$

Step 2: Find $\mathbf{x}_k \in V$ such that:

$$\mathcal{L}_r(\mathbf{x}_k, \mathbf{p}_k; \boldsymbol{\lambda}_k) \leq \mathcal{L}_r(\mathbf{y}, \mathbf{p}_k; \boldsymbol{\lambda}_k), \quad \forall \mathbf{y} \in V_0. \tag{3}$$

Step 3: Update the Lagrange multipliers by:

$$\boldsymbol{\lambda}_{k+1} = \boldsymbol{\lambda}_k + r((\mathbf{x}_k)' - \mathbf{p}_k).$$

If the boundary conditions for problem (1) are $\mathbf{y}(0) = \mathbf{x}_A$ and $\mathbf{y}'(0) = \mathbf{x}_B$, then the test function space at step 2 is defined by:

$$V_0 = \left\{\mathbf{y} \in (H^2(0, L))^2,\ \mathbf{y}(0) = \mathbf{0},\ \mathbf{y}'(0) = \mathbf{0}\right\}.$$

To obtain $\mathbf{p}_k$ at step 1, we have to solve the minimization problem:

$$\min_{|\mathbf{q}|=1} \mathcal{L}_r(\mathbf{x}_{k-1}, \mathbf{q}; \boldsymbol{\lambda}_k), \text{ with the solution } \mathbf{p}_k = \frac{r(\mathbf{x}_{k-1})' + \boldsymbol{\lambda}_k}{|r(\mathbf{x}_{k-1})' + \boldsymbol{\lambda}_k|}. \tag{4}$$

Problem (3) can be stated as the equivalent problem: Find $\mathbf{x}_k \in V$ such that for all $\mathbf{y} \in V_0$:

$$\int_0^L EI\mathbf{x}_k'' \cdot \mathbf{y}'' ds + r \int_0^L \mathbf{x}_k' \cdot \mathbf{y}' ds = \int_0^L \mathbf{f} \cdot \mathbf{y} ds + \int_0^L (r\mathbf{p}_k - \boldsymbol{\lambda}_k) \cdot \mathbf{y}' ds.$$

Steps 1–3 are repeated till the following stopping criterion is satisfied:

$$\frac{||\mathbf{x}_{k+1} - \mathbf{x}_k||}{||\mathbf{x}_k||} < \epsilon. \tag{5}$$

2.2 The Dynamic Problem

Using the virtual work principle, the beam motion for $t \in [0, T]$ is modeled by: Find $\mathbf{x}(t) \in K_t$:

$$\int_0^L \rho\ddot{\mathbf{x}} \cdot \mathbf{y} ds + \int_0^L EI\, \mathbf{x}'' \cdot \mathbf{y}'' ds = \int_0^L \mathbf{f} \cdot \mathbf{y} ds, \quad \forall \mathbf{y} \in dK_t(\mathbf{x}), \tag{6}$$

with

$$K_t = \{\mathbf{y} \in (H^2(0,L))^2, |\mathbf{y}'| = 1, \mathbf{y}(0) = \mathbf{x}_A(t), \mathbf{y}'(0) = \mathbf{x}_B(t)\},$$
$$dK_t(\mathbf{x}) = \{\mathbf{y} \in (H^2(0,L))^2, \mathbf{x}' \cdot \mathbf{y}' = 0, \mathbf{y}(0) = \mathbf{0}, \mathbf{y}'(0) = \mathbf{0}\},$$

and initial conditions $\mathbf{x}(s,0) = \mathbf{x}_0(s)$ and $\dot{\mathbf{x}}(s,0) = \mathbf{x}_1(s)$. Weak formulation (6) assumes that at $s = L$ natural boundary conditions $\mathbf{x}''(L) = \mathbf{0}$ and $\mathbf{x}'''(L) = \mathbf{0}$ are imposed. Note that problem (6) in strong form reads: $\rho\ddot{\mathbf{x}} + EI\mathbf{x}'''' = \mathbf{f}$.

For the time discretization of problem (6), we will consider three schemes: a generalized Crank-Nicolson scheme, the Houbolt scheme [5], and a Newmark scheme (see, e.g., [4, 6]). All these schemes are known to be second order accurate for linear problems. Let Δt be a time discretization step and set $t^n = n\Delta t$, for $n = 1, \ldots, N$, with $N = T/\Delta t$. The time discrete problem reads: Find $\mathbf{x}^{n+1} \in K_{t^{n+1}}$:

$$\int_0^L \rho\ddot{\mathbf{x}}^{n+1} \cdot \mathbf{y} ds + \int_0^L EI\, \tilde{\mathbf{x}}'' \cdot \mathbf{y}'' ds = \int_0^L \tilde{\mathbf{f}} \cdot \mathbf{y} ds, \tag{7}$$

for all $\mathbf{y} \in dK_{t^{n+1}}(\mathbf{x}^{n+1})$. The definition of $\ddot{\mathbf{x}}^{n+1}$, $\tilde{\mathbf{x}}$, and $\tilde{\mathbf{f}}$ in (7) is reported in Table 1 for each scheme under consideration. Time discretization approximates problem (6) by a sequence of quasi-static problems for which ALG2 still applies. For the space discretization of problem (7) we use a third order Hermite finite element method (see, e.g., [1]). For details about the discretization of $\mathbf{p}_k \in \mathcal{Q}$ (4) and $\boldsymbol{\lambda}_k \in (L^2(0,L))^2$ we refer to [4].

Table 1 Definition of $\ddot{\mathbf{x}}^{n+1}$, $\tilde{\mathbf{x}}$, and $\tilde{\mathbf{f}}$ in (7) for the time discretization schemes under consideration: Generalized Crank-Nicolson (GCN), Houbolt, and Newmark with $\beta = 1/4,\ \gamma = 1/2$. For GCN, $0 < \alpha < 1/2$

	GCN	Houbolt	Newmark[a]
$\ddot{\mathbf{x}}^{n+1}$	$\dfrac{\mathbf{x}^{n+1} - 2\mathbf{x}^n + \mathbf{x}^{n-1}}{\Delta t^2}$	$\dfrac{2\mathbf{x}^{n+1} - 5\mathbf{x}^n + 4\mathbf{x}^{n-1} - \mathbf{x}^{n-2}}{\Delta t^2}$	$\dfrac{\mathbf{v}^{n+1} - \mathbf{v}^n}{\Delta t}$
$\tilde{\mathbf{x}}$	$\alpha\mathbf{x}^{n+1} + (1-2\alpha)\mathbf{x}^n + \alpha\mathbf{x}^{n-1}$	$\mathbf{x}^{n+1}$	$\dfrac{\mathbf{x}^{n+1} + \mathbf{x}^n}{2}$
$\tilde{\mathbf{f}}$	$\alpha\mathbf{f}^{n+1} + (1-2\alpha)\mathbf{f}^n + \alpha\mathbf{f}^{n-1}$	$\mathbf{f}^{n+1}$	$\dfrac{\mathbf{f}^{n+1} + \mathbf{f}^n}{2}$

[a] with $\dfrac{\mathbf{v}^{n+1} + \mathbf{v}^n}{2} = \dfrac{\mathbf{x}^{n+1} - \mathbf{x}^n}{\Delta t}$

3 Numerical Results

3.1 Benchmark with Analytical Solution

We consider $s \in [0, \pi/2]$ and $t \in [0, 1]$, and a family of exact solutions which is given by:

$$\mathbf{x}_{ex}(s,t) = (\phi(t))^{-1} \left[\cos(s\phi(t)), \sin(s\phi(t))\right]^T . \tag{8}$$

Notice that solution (8) satisfies the inextensibility condition $|\mathbf{x}|' = 1$ pointwise for every function $\phi(t)$. We chose $\phi(t) = e^t$, for which the solution is a quarter of a circle of initial radius 1 that coils over time as its radius decreases (see Fig. 1). At $s = 0$ and $s = \pi/2$, we impose the values of $\mathbf{x}$ and $\mathbf{x}'$. The forcing term $\mathbf{f}_{ex}$ needed to recover solution (8) is found by plugging $\mathbf{x}_{ex}$ into the governing differential equations (strong form):

$$\rho\ddot{\mathbf{x}}_{ex} + EI\mathbf{x}_{ex}'''' = \mathbf{f}_{ex}. \tag{9}$$

For simplicity, we set $\rho = 1\,\mathrm{Kg/m^3}$ and $EI = 1\,\mathrm{Kg/(m\,s)^2}$. The forcing term $\mathbf{f}_{ex}$ is made up of two contributions: an external body force $\mathbf{f}_b$ and an internal force due to inextensibility $\mathbf{f}_{in}$. To find $\mathbf{f}_{in}$, we notice that problem (6) is equivalent to

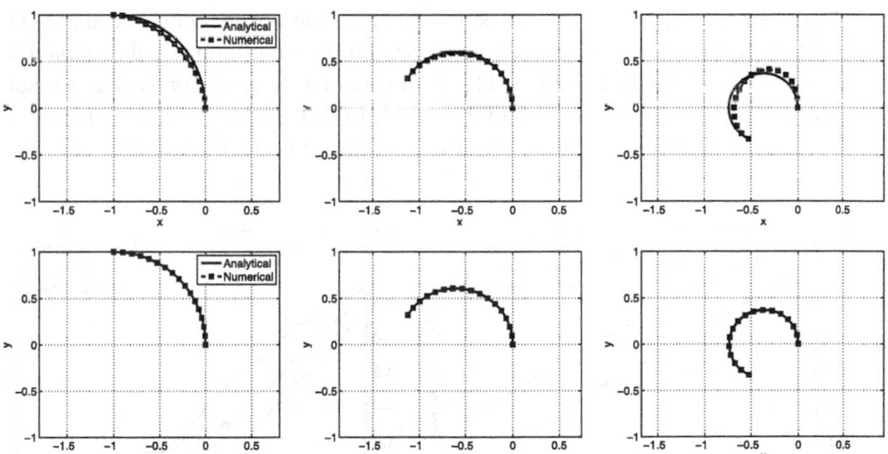

Fig. 1 Comparison between analytical and numerical solution at $t = 0$ s (*left*), $t = 0.5$ s (*center*), $t = 1$ s (*right*) for two values of stopping tolerance: $\epsilon = 10^{-1}$ (*top*) and $\epsilon = 10^{-5}$ (*bottom*). The legend in the subfigures on the *left* is common to all the subfigures

minimization problem $\mathbf{x} = \arg\min_{\mathbf{y}\in K_t} J(\mathbf{y})$, where the total energy of the beam can be written as:

$$J(\mathbf{y}) = \frac{1}{2}\int_0^L \rho|\ddot{\mathbf{y}}|^2 ds + \frac{1}{2}\int_0^L EI\left|\mathbf{y}''\right|^2 ds + \int_0^L \lambda(|\mathbf{y}'|^2 - 1)ds - \int_0^L \mathbf{f}\cdot\mathbf{y}ds,$$

where λ is a scalar function. If the above functional attains its minimum at $\mathbf{x}$, it follows that its Gâteaux derivative must be vanishing at $\mathbf{x}$, leading to

$$\int_0^L \rho\ddot{\mathbf{x}}\cdot\mathbf{y}ds + \int_0^L EI\,\mathbf{x}''\cdot\mathbf{y}''ds = \int_0^L \mathbf{f}\cdot\mathbf{y}ds + \int_0^L (\lambda\mathbf{x}')'\cdot\mathbf{y}ds,$$

for all $\mathbf{y} \in dK_t(\mathbf{x})$. The second integral on the right-hand side (equal to zero if $\mathbf{y} \in dK_t(\mathbf{x})$, which is not the case for the test functions used in the computations) gives the explicit contribution of $\mathbf{f}_{in}$.

We are going to check the convergence rates in time for the three schemes in Table 1 in two cases:

- Linear case: when the forcing term is $\mathbf{f}_{ex}$ the inextensibility condition becomes inactive due to the fact that $\mathbf{f}_{ex}$ is given by (9) and the problem reduces to the linear beam equation;
- Nonlinear case: when then forcing term is $\mathbf{f}_{ex} + (\lambda\mathbf{x}')'$, with, e.g., $\lambda = 1$, the problem becomes nonlinear and the inextensibility is treated via the augmented Lagrangian method described in Sect. 2.

The space resolution Δs is taken to be $\pi/240$. For the generalized Crank-Nicolson scheme, we set $\alpha = 1/4$ since in linear cases this choice leads to an unconditionally stable scheme which possesses a very small numerical dissipation compared, e.g., to Houbolt method [1]. In the nonlinear case, for ALG2 we set stopping tolerance $\epsilon = 10^{-5}$ (5) and $r = 10^2$. In Fig. 2, we plot the L^2 norm of the difference between the exact solution $\mathbf{x}_{ex}$ and the numerical solution $\mathbf{x}_h$ at $t = 1$

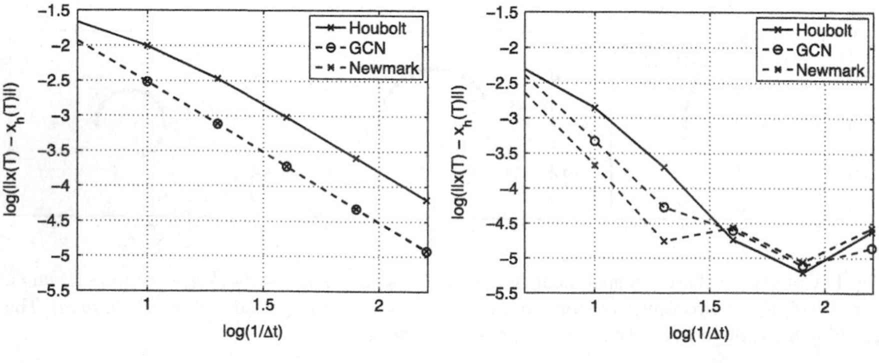

Fig. 2 Convergence rate in time for the generalized Crank-Nicolson (GCN) scheme, the Houbolt scheme, and the Newmark scheme in the linear (*left*) and nonlinear/inextensible (*right*) case

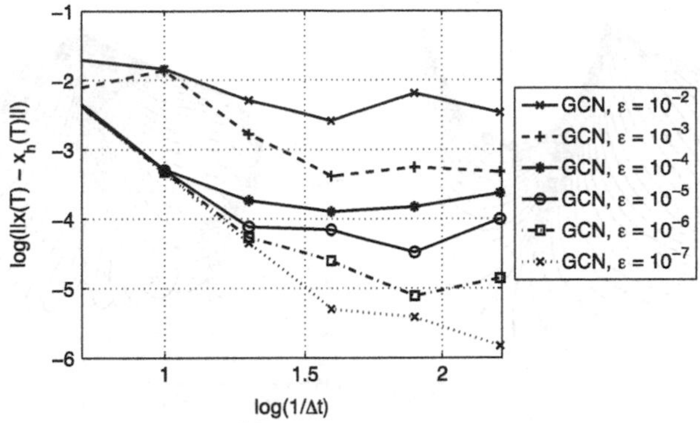

Fig. 3 Convergence rate in time for the generalized Crank-Nicolson (GCN) scheme in the nonlinear case for different values of the stopping tolerance ϵ

against time step ($\Delta t = 0.2, 0.1, 0.05, 0.025, 0.0125, 0.00625$) for the linear and nonlinear cases. The rates predicted by the theory are achieved in the linear case: all the schemes are of second order. We remark that for a given value of Δt the Houbolt scheme is less accurate than the other two. In the nonlinear case, for all the schemes the order of convergence is even larger than 2 provided that Δt is greater than a critical value for which the error reaches the stopping tolerance ϵ. If Δt is less than that critical value, the error remains unchanged or even slightly increases.

As noted earlier, the error depends on the choice of ϵ. To illustrate this, in Fig. 1 we compare analytical solution (8) with the numerical solution at $t = 0, 0.5, 1$ s and for two values of the stopping tolerance: $\epsilon = 10^{-1}$ (top) and $\epsilon = 10^{-5}$ (bottom), every other discretization parameter being the same. For $\epsilon = 10^{-1}$ the difference between analytical and numerical solution is clearly visible, while for $\epsilon = 10^{-5}$ the two solutions are almost superimposed.

Finally, in order to evaluate the dependence of the error on ϵ, we report in Fig. 3 the convergence rates in time for the generalized Crank-Nicolson scheme in the nonlinear case for different values of the stopping tolerance $\epsilon = 10^{-2}, 10^{-3}, 10^{-4}, 10^{-5}, 10^{-6}, 10^{-7}$. The values for Δt and Δs are the same as those used for the results in Fig. 2. We see that at the critical value of Δt the curves reach a plateau for all the values of ϵ, indicating that for a given value of ϵ it does not make sense to choose a time step size that is too small. Our computations seem to indicate that Δt should be larger than $\sqrt{\epsilon}$.

3.2 Swinging Beam

The second test problem we consider involves the two-dimensional motion of a beam subject to gravity, which is a an established test problem [3]. The beam is attached at one extremity (denoted by A here) and free at the other one (B). We aim at comparing our results with those reported in [3]. We have: $L = 32.6$ m,

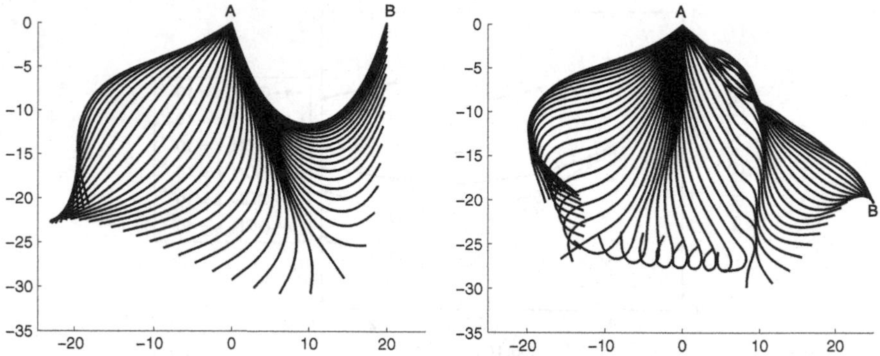

Fig. 4 Position of the beam every 0.1 s for $t \in [0, 5]$ s (*left*) and $t \in [5, 10]$ (*right*)

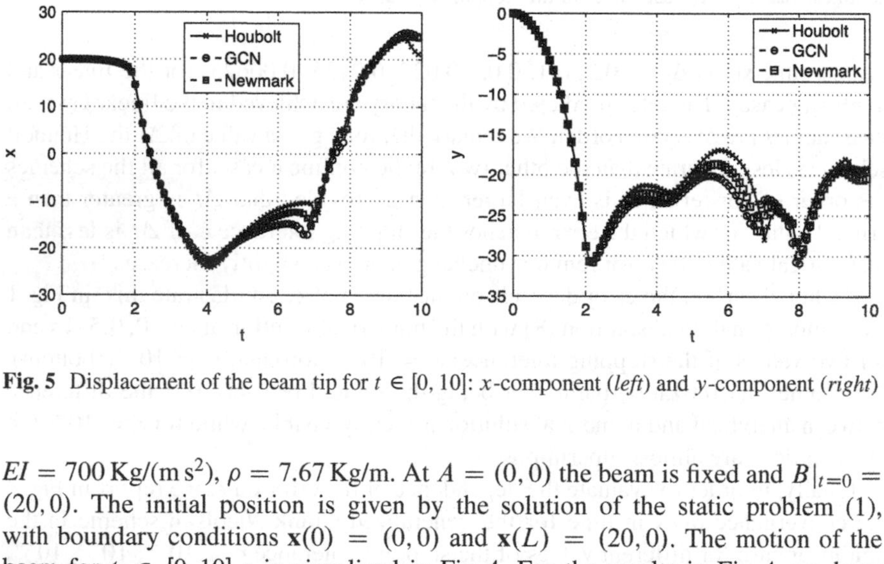

Fig. 5 Displacement of the beam tip for $t \in [0, 10]$: x-component (*left*) and y-component (*right*)

$EI = 700\,\mathrm{Kg/(m\,s^2)}$, $\rho = 7.67\,\mathrm{Kg/m}$. At $A = (0,0)$ the beam is fixed and $B|_{t=0} = (20,0)$. The initial position is given by the solution of the static problem (1), with boundary conditions $\mathbf{x}(0) = (0,0)$ and $\mathbf{x}(L) = (20,0)$. The motion of the beam for $t \in [0,10]$ s is visualized in Fig. 4. For the results in Fig. 4, we have used the generalized Crank-Nicolson scheme ($\alpha = 1/4$) with $\Delta t = 0.01$, and $\Delta s = 32.6/60$. For ALG2, we have set $r = 10^5$ and $\epsilon = 10^{-5}$. Figure 4 is qualitatively very similar to the corresponding pictures in Ref. [3].

Next, we compare the displacement over time of the beam tip given by the generalized Crank-Nicolson scheme, the Houbolt scheme, and the Newmark scheme (see Table 1). The ALG2 and discretization parameters are the same used for the results in Fig. 4. Figure 5 shows the x and y components of the displacement for the three methods. We see that all the schemes are in good agreement, with the Houbolt scheme giving larger oscillations than the other two schemes.

Conclusions
We compared three different time discretization schemes (the Houbolt scheme, a generalized Crank-Nicolson scheme, and a Newmark scheme) in combination with an augmented Lagrangian method to simulate the motion of an inextensible beam. While all these schemes are known to be second order accurate in time for linear problems, for the nonlinear problem considered here, our numerical simulations for a benchmark problem with analytical solution indicate that the accuracy increases when they are combined with an Uzawa-type algorithm to account for inextensibility. Special care has to be taken in selecting the termination criterion. Our computations suggest that the stopping tolerance for the Uzawa-type algorithm should be balanced against the time step size in a rather restrictive manner.

References

1. J.-F. Bourgat, M. Dumay, R. Glowinski, Large displacement calculations of inexstensible pipelines by finite element and nonlinear programming methods. SIAM J. Sci. Stat. Comput. **1**, 34–81 (1980)
2. M. Fortin, R. Glowinski, *Augmented Lagrangian Methods: Application to the Numerical Solution of Boundary Value Problem* (North-Holland, Amsterdam, 1983)
3. R. Glowinski, M. Holmstrom, Constrained motion problems with applications by nonlinear programming methods. Surv. Math. Ind. **5**, 75–108 (1995)
4. R. Glowinski, P.L. Tallec, *Augmented Lagrangian and Operator-Splitting Methods in Nonlinear Mechanics* (SIAM, Philadelphia, 1988)
5. J.C. Houbolt, A recurrence matrix solution for the dynamic response of elastic aircraft. J. Aeronaut. Sci. **17**(9), 540–550 (1950)
6. N.M. Newmark, A method of computation for structural dynamics. J. Eng. Mech. Div. **85**(3), 67–94 (1959)
7. K. Subbaraj, M. Dokainish, A survey of direct time-integration methods in computational structural dynamics II. Implicit methods. Comput. Struct. **32**(6), 1387–1401 (1989)

Multi-value Numerical Methods for Hamiltonian Systems

Raffaele D'Ambrosio

Abstract We discuss the effectiveness of multi-value numerical methods in the numerical treatment of Hamiltonian problems. Multi-value (or general linear) methods extend the well-known families of Runge-Kutta and linear multistep methods and can be considered as a general framework for the numerical solution of ordinary differential equations. There are some features that needs to be achieved by reliable geometric numerical integrators based on multi-value methods: G-symplecticity, symmetry and boundedness of the parasitic components. In particular, we analyze the effects of the mentioned features for the long term conservation of the energy and provide the numerical evidence confirming the theoretical expectations.

1 Hamiltonian Problems

It is the aim of this paper to analyze the effectiveness and the long-term behaviour of multi-value numerical methods for Hamiltonian problems

$$\dot{y}(t) = J^{-1}\nabla H(y), \qquad J = \begin{bmatrix} 0 & I \\ -I & 0 \end{bmatrix}, \tag{1}$$

where the function H, denoted as Hamiltonian or energy of the system, is exactly preserved along the solution of (1). Geometric numerical integrators for (1) (compare [16] and references therein) are able to perform an excellent long-time conservation of the Hamiltonian along the numerical solution: this is classically the case of symplectic (or canonical) Runge-Kutta (RK) methods [1, 16, 20, 21],

R. D'Ambrosio (✉)
Department of Mathematics, University of Salerno, Via Giovanni Paolo II, 132 – 84084 Fisciano, Italy
e-mail: rdambrosio@unisa.it

A. Abdulle et al. (eds.), *Numerical Mathematics and Advanced Applications ENUMATH 2013*, Lecture Notes in Computational Science and Engineering 103, DOI 10.1007/978-3-319-10705-9_18 which are meant to exactly preserve quadratic invariants possessed by (1) along the numerical solution (within round-off error). Moreover, a symplectic numerical method is able to preserve any Hamiltonian function over exponentially long times with an exponentially decreasing error, as proved by Benettin and Giorgilli (see [16], Theorem 8.1, §IX.8).

Symplecticity is a prerogative of certain RK methods, i.e. those satisfying the algebraic constraint [1, 16, 18, 21, 22]

$$b_i a_{ij} + b_j a_{ji} - b_i b_j = 0. \tag{2}$$

Indeed, linear multistep methods cannot be symplectic [23] as well as genuine multi-value methods cannot be symplectic [4, 14, 19].

However, many contributions of the recent literature have been devoted to the analysis and the construction of both multistep and multi-value methods meant to guarantee an excellent near conservation of invariants over long time intervals (compare, for instance, [1–3, 7–9, 15, 16] and references therein). The aim of this paper is that of analyzing the main results achieved so far in the case of multi-value methods and applying them to investigate the long-time behaviour of a method recently developed in [9], both from a theoretical and an experimental point of view.

2 Multi-value Methods

Our attention is focused on the family of multi-value methods, which provides a wide range of methods including multistage methods (e.g. Runge-Kutta and multistep Runge-Kutta methods) and multistep methods (compare [1, 17] and references therein for a complete analysis of known methods regarded as multi-value methods, extended in [10] to the case of second order ODEs).

The numerical scheme given by a multi-value method for the numerical solution of the initial value problem

$$y' = f(t, y), \ t \geq 0 \qquad y(t_0) = y_0, \tag{3}$$

consists in the following three basic steps:

- a *starting* procedure S_h, $y^{[0]} = S_h(y_0)$,
- a *forward* procedure G_h, $y^{[n+1]} = Vy^{[n]} + h\Phi(h, y^{[n]})$,
- a *finishing* procedure F_h, $y_n = F_h(y^{[n]})$.

Thus, the method transfers along the grid a whole vector $y^{[n]}$ containing the approximations of a set of quantities related to the solution of the problem under investigation. At each step, one can always get the numerical approximation of the solution in the current step point by applying the finishing procedure.

Under some basic hypothesis described in details in [16] (compare Theorem 8.1 in Section XV), one can prove that for any given forward and finishing procedures, there exist a unique starting procedure $S_h^*(y)$ and a unique one-step method $y_{n+1} = \Phi_h^*(y_n)$, such that

$$G_h \circ S_h^* = S_h^* \circ \Phi_h^*, \qquad F_h \circ S_h^* = id.$$

Thus, if the starting vector is computed by $Y^{[0]} = S_h^*(y_0)$, then the numerical solution obtained by the multi-value method is (formally) equal to that of the one-step method Φ_h^*. Hence, Φ_h^* is called *underlying one-step method*.

A widely used representation of multi-value methods is usually given by the family of General Linear Methods (GLMs, compare [1, 10, 17] and references therein)

$$\begin{cases} Y_i^{[n]} = h \sum\limits_{j=1}^{s} a_{ij} f(Y_j^{[n]}) + \sum\limits_{j=1}^{r} u_{ij} y_j^{[n]}, \quad i = 1, 2, \ldots, s, \\ y_i^{[n+1]} = h \sum\limits_{j=1}^{s} b_{ij} f(Y_j^{[n]}) + \sum\limits_{j=1}^{r} v_{ij} y_j^{[n]}, \; i = 1, 2, \ldots, r, \end{cases} \tag{4}$$

The formulation (4) is provided in correspondence of the uniform grid $\{t_0 + ih, i = 0, 1, \ldots, N\}$, with $h = (T - t_0)/N$. The vector $y^{[n]} = [y_1^{[n]}, \ldots, y_r^{[n]}]^{\mathsf{T}}$ denotes the vector of external approximations containing all the informations we decide to transfer from step n to step $n+1$, $Y_i^{[n]}$ provides an approximation to the solution of (3) in the internal point $t_n + c_i h \in [t_n, t_{n+1}]$, $i = 1, 2, \ldots, s$, and $F_j = f(Y_j^{[n]})$. A compact representation of GLMs collects their coefficient matrices $A \in \mathbb{R}^{s \times s}$, $U \in \mathbb{R}^{s \times r}$, $B \in \mathbb{R}^{r \times s}$, $V \in \mathbb{R}^{r \times r}$, in the following partitioned $(s+r) \times (s+r)$ matrix

$$\left[\begin{array}{c|c} A & U \\ \hline B & V \end{array}\right].$$

As mentioned in the previous section, even if GLMs cannot be symplectic (unless they reduce to symplectic one step methods, compare [4, 14, 19]), the recent literature has emphasized the possibility to effectively employ GLMs for the numerical treatment of Hamiltonian problem (compare, for instance, [1–3, 7–9] and references therein). In particular, the state-of-art reveals that some specific properties have to be satisfied by multi-value methods in order to accurately approach Hamiltonian problems:

- G-symplecticity (introduced in the first edition of [16], also see [1–3, 5, 7–9]), which ensures conjugate-symplecticity of the underlying one-step method associated to the multivalue method (4);

- symmetry of the numerical scheme [16], which is a suitable property providing the discrete counterpart of the reversibility of the exact flow, in case of reversible dynamical systems;
- boundedness of parasitic components over long times [3, 11, 16], which ensures that the parasitic components generated by the numerical method remain bounded over certain time intervals.

The above mentioned features are considered in the remainder of the treatise.

3 G-Symplecticity

As mentioned, the multivalue nature of GLMs does not allow them to be symplectic, unless they reduce to RK methods. However, a near-conservation property achievable by multivalue methods has been provided and analyzed by the recent literature, defined as follows. If $y^{\mathsf{T}}Ey$ is a quadratic first integral of the differential problem $y' = f(y)$, where E is a symmetric matrix, G-symplecticity assures that

$$y^{[n+1]^{\mathsf{T}}}(G \otimes E)y^{[n+1]} = y^{[n]^{\mathsf{T}}}(G \otimes E)y^{[n]}, \tag{5}$$

(compare [12]), being G a symmetric matrix. Taking into account that any GLM (4) satisfies the following identity (compare [9] and references therein)

$$\begin{aligned} y^{[n+1]^{\mathsf{T}}}(G \otimes E)y^{[n+1]} &= y^{[n]^{\mathsf{T}}}(G \otimes E)y^{[n]} + \sum_{i,j=1}^{r}(G - V^{\mathsf{T}}GV)_{ij}y_i^{[n-1]^{\mathsf{T}}}y_j^{[n-1]} \\ &+ 2h\sum_{i=1}^{s}\sum_{j=1}^{r}(DU - B^{\mathsf{T}}GV)_{ij}y_i^{[n-1]^{\mathsf{T}}}F_j^{[n-1]} \\ &+ h^2\sum_{i,j=1}^{s}(DA + A^{\mathsf{T}}D - B^{\mathsf{T}}GB)_{ij}F_i^{[n-1]^{\mathsf{T}}}F_j^{[n-1]}, \end{aligned}$$

the G-symplecticity property (5) is achieved if the algebraic constraints

$$G = V^{\mathsf{T}}GV, \qquad DU = B^{\mathsf{T}}GV, \qquad DA + A^{\mathsf{T}}D = B^{\mathsf{T}}GB \tag{6}$$

are satisfied [1, 16].

Condition (5) reveals that G-symplectic multivalue method does not preserve quadratic first integrals, but a related quadratic form $y^{[n]^{\mathsf{T}}}(G \otimes E)y^{[n]}$. It was observed in [12] that the first terms of the expansion in powers of h of the quadratic form $y^{[n]^{\mathsf{T}}}(G \otimes E)y^{[n]}$ is $y^{\mathsf{T}}Ey$ (compare [12]): thus, the more h is small, the more the two forms are close each other. observe that conditions (6) of G-symplecticity are equivalent to annihilating the algebraic stability matrix of a GLM (compare [1, 6, 13, 16]).

There is a strong formal relation between G-symplectic and symplectic maps, which is highlighted in [11]. We report here the main result.

Theorem 1 *Consider a G-symplectic multi-value method* (4) *of order p. Then, for every finishing procedure the underlying one-step method is conjugate-symplectic. More precisely, there exists a change of coordinates* $\chi_h(y) = y + O(h^p)$*, such that* $\chi_h \circ \Phi_h^* \circ \chi_h^{-1}$ *is a symplectic transformation.*

In other words, this results asserts that a G-symplectic method has the same behavior of a symplectic one-step method after a global change of coordinates that is $O(h^p)$ close to the identity [12].

4 Control of Parasitism

One-step methods are the only candidates for symplecticity (compare [16, 23] for linear multistep methods and [4, 15, 19] for irreducible multivalue methods). This is due to the fact the multistep and multivalue methods generate parasitic components in the numerical solution which destroy the overall long-time accuracy (see [3, 11, 16]). Hence, if one aims to derive non-symplectic methods which are capable of nearly preserving invariants over the numerical solution, the parasitic behaviour of such methods has to be taken under control over long time intervals [11].

As announced, due to their multivalue nature, GLMs introduce parasitic components in the numerical solution, which have to be controlled in order to achieve a long-term near conservation of the invariants. Rigorous bounds on parasitic solution components have recently been obtained in [11], where the authors have proved that, for carefully constructed methods, the error in the parasitic components typically grows like $h^{p+4} exp(h^2 Lt)$, where p is the order of the method, and L depends on the problem and on the coefficients of the method.

A basic property of boundedness for the parasitic components of multivalue methods is achieved by annihilating the so-called growth parameters [11, 16]

$$\mu_j = \xi_j^{-1} v_j^* B U v_j, \tag{7}$$

where ξ_j are the eigenvalues of the matrix V such that $\xi_j \neq 1$, v_j and v_j^* are the right and left eigenvectors, respectively ($V v_j = \xi_j v_j$ and $v_j^* V = \xi_j v_j^*$) satisfying $v_j^* v_j = 1$. Examples of methods with zero-growth parameters, in the context of multivalue methods, have been provided in [2, 3, 8, 9]. In particular, a G-symplectic, symmetric (i.e. the underlying one-step method is symmetric, compare [16]), order 4 method (4) with zero growth parameter has been introduced in [9]. Denoted by

$$\gamma = 2 + \frac{\sqrt[3]{4}}{2} + \sqrt[3]{2}, \qquad \delta = \left(1 + \sqrt[3]{2}\right)^2, \qquad \varphi = \frac{15}{4} + 2\sqrt[3]{2} + \sqrt[3]{4},$$

such a method depends on the following coefficient matrices

$$\left[\begin{array}{c|c} A & U \\ \hline B & V \end{array}\right] = \left[\begin{array}{ccc|cc} \frac{1}{6}\gamma & 0 & 0 & 1 & \frac{1}{24} \\ \frac{1}{3}\gamma & -\frac{1}{6}\delta & 0 & 1 & \frac{1}{24} \\ \frac{1}{3}\gamma & \frac{1}{6}\delta & \frac{1}{6}\gamma & 1 & \frac{1}{24} \\ \hline \frac{1}{6}\varphi & -\frac{1}{4} - \frac{2\sqrt[3]{2}}{3} - \frac{\sqrt[3]{4}}{3} & \frac{1}{6}\varphi & 1 & \frac{1}{12} \\ 1 & -2 & 1 & 0 & -1 \end{array}\right]. \tag{8}$$

A starting procedure is given in details in [9].

5 Long-Term Behaviour

As explained in the previous section, ideal multi-value methods generate small and bounded parasitic components over long time intervals. In order to derive sharp long-term error estimates for multi-value methods, we have suitably applied in [11] backward error analysis, a powerful tool successfully applied to one-step and linear multistep methods (compare [15, 16] and references therein) which provides a crucial ingredient for the study of the long-time behavior of numerical integrators. In [11], we have derived sharp estimates for the parasitic components and the error in the Hamiltonian numerically computed by a multi-value method: we realized that, for carefully constructed methods (i.e. symmetric and with zero growth parameters) the error in the parasitic components typically grows like $h^{p+4}\exp(h^2Lt)$.

In particular, for the multi-value method (8), the following result holds (compare [11]).

Theorem 2 *If* (8) *is applied to a Hamiltonian system* (1)*, then the energy is nearly preserved according to*

$$H(y_n) - H(y_0) = O(h^4) + O(th^8) + O\big(h^8\exp(h^2Lt)\big)$$

as long as $t = nh = O(h^{-2})$.

Thus, for method (8), parasitic components remain bounded on intervals of steplength $O(h^{-2})$, which is also confirmed by the numerical evidence. We apply

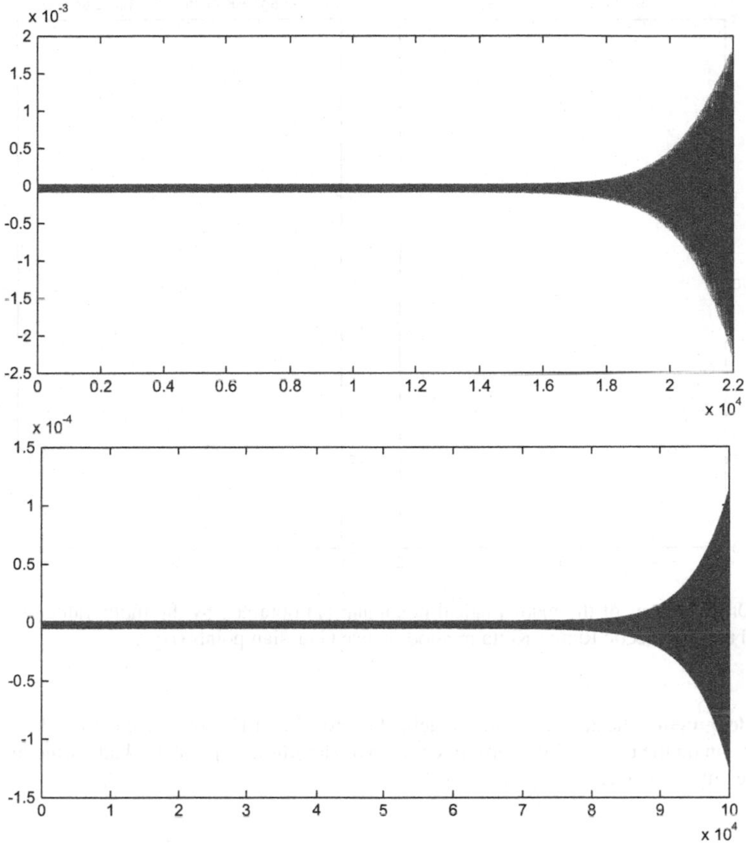

Fig. 1 Error in the Hamiltonian (9) for the method (8) with stepsizes $h = 0.25$ (*top*) and $h = 0.125$ (*bottom*)

method (8) to the simple pendulum problem, depending on the Hamiltonian function

$$H(p,q) = \frac{1}{2}\,p^2 - \cos q, \tag{9}$$

and initial values $q(0) = 3$, $p(0) = 0$.

Figure 1 shows the Hamiltonian error obtained by using the step sizes $h = 0.25$ and $h = 0.125$: confirming the predicted estimate of Theorem 2, the error behaves like $O(h^4)$ on intervals of length $O(h^{-2})$, and then follows an exponential growth. We observe that method (8) is also able to preserve the symplecticity of the phase space, as visible from the orbit pattern in Fig. 2.

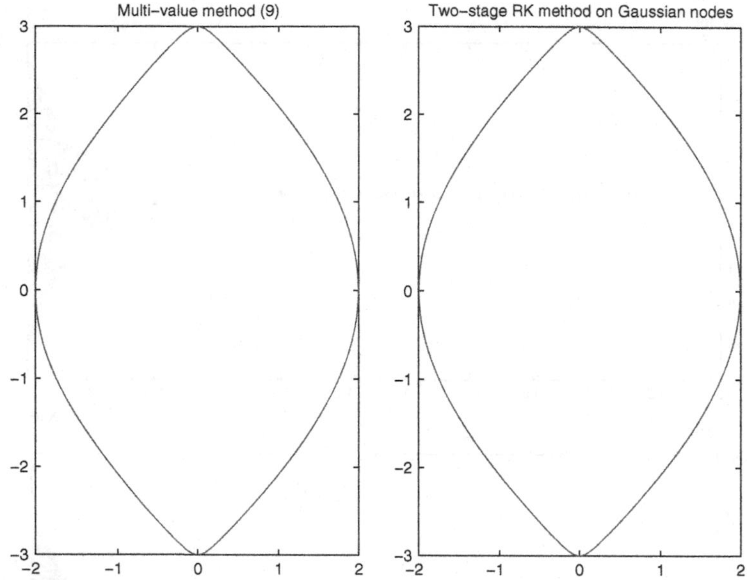

Fig. 2 Orbit patterns of the mathematical pendulum (1) obtained by the multi-value method (8) (*left*) and the symplectic Runge-Kutta method on two Gaussian points (*right*)

Acknowledgment The author is really indebted to prof. Ernst Hairer for the extremely profitable collaboration on the topic of long-term integration of Hamiltonian problems had during his stay at the University of Geneve.

References

1. J.C. Butcher, *Numerical Methods for Ordinary Differential Equations*, 2nd edn. (Wiley, Chichester, 2008)
2. J.C. Butcher, Dealing with parasitic behaviour in G-symplectic integrators, in *Recent Developments in the Numerics of Nonlinear Hyperbolic Conservation Laws*, ed. by R. Ansorge, H. Bijl, A. Meister, T. Sonar. Notes on Numerical Fluid Mechanics and Multidisciplinary Design, vol. 120 (Springer, Heidelberg, 2013) pp. 105–123
3. J.C. Butcher, Y. Habib, A. Hill, T. Norton, The control of parasitism in G-symplectic methods. Submitted for publication
4. J.C. Butcher, L.L. Hewitt, The existence of symplectic general linear methods. Numer. Algorithms **51**, 77–84 (2009)
5. J.C. Butcher, A. Hill, Linear multistep methods as irreducible general linear methods. BIT Numer. Math. **46**(1), 5–19 (2006)
6. D. Conte, R. D'Ambrosio, Z. Jackiewicz, B. Paternoster, Numerical search for algebraically stable two-step continuous Runge-Kutta methods. J. Comput. Appl. Math. **239**, 304–321 (2013)
7. R. D'Ambrosio, On the G-symplecticity of two-step Runge-Kutta methods. Commun. Appl. Ind. Math. (2012). doi: 10.1685/journal.caim.403

8. R. D'Ambrosio, G. De Martino, B. Paternoster, Construction of nearly conservative multivalue numerical methods for Hamiltonian problems. Commun. Appl. Ind. Math. (2012). doi: 10.1685/journal.caim.412
9. R. D'Ambrosio, G. De Martino, B. Paternoster, Numerical integration of Hamiltonian problems by G-symplectic methods. Adv. Comput. Math. **40**(2), 553–575 (2014)
10. R. D'Ambrosio, E. Esposito, B. Paternoster, General linear methods for $y'' = f(y(t))$. Numer. Algorithms **61**(2), 331–349 (2012)
11. R. D'Ambrosio, E. Hairer, Long-term stability of multi-value methods for ordinary differential equations. J. Sci. Comput. (2013). doi: 10.1007/s10915-013-9812-y
12. R. D'Ambrosio, E. Hairer, C.J. Zbinden, G-symplecticity implies conjugate-symplecticity of the underlying one-step method. BIT Numer. Math. **53**(4), 867–872 (2013)
13. R. D'Ambrosio, G. Izzo, Z. Jackiewicz, Search for highly stable two-step Runge-Kutta methods for ODEs. Appl. Numer. Math. **62**(10), 1361–1379 (2012)
14. E. Hairer, P. Leone, Order barriers for symplectic multi-value methods, in *Proceedings of the 17th Dundee Biennial Conference 1997*, ed. by F. Griffiths, D.J. Higham, G.A. Watson. Pitman Research Notes in Mathematics Series, vol. 380 (Longman, Harlow, 1998), pp. 133–149
15. E. Hairer, C. Lubich, Symmetric multistep methods over long times. Numer. Math. **97**, 699–723 (2004)
16. E. Hairer, C. Lubich, G. Wanner, *Geometric Numerical Integration. Structure-Preserving Algorithms for Ordinary Differential Equations*, 2nd edn. (Springer, Berlin, 2006)
17. Z. Jackiewicz, *General Linear Methods for Ordinary Differential Equations* (Wiley, Hoboken, 2009)
18. F.M. Lasagni, Canonical Runge-Kutta methods. Z. Angew. Math. Phys. **39**, 952–953 (1988)
19. P. Leone, Symplecticity and symmetry of general integration methods, Ph.D. thesis, Universite de Geneve, (2000)
20. R.I. McLachlan, G.R.W. Quispel, Geometric integrators for ODEs. J. Phys. A: Math. Gen. **39**, 5251–5285 (2006)
21. J.M. Sanz-Serna, M.P. Calvo, *Numerical Hamiltonian Problems* (Chapman & Hall, London, 1994)
22. Y.B. Suris, Preservation of symplectic structure in the numerical solution of Hamiltonian systems (in Russian). Akad. Nauk SSSR, Inst. Prikl. Mat., 148–160, 232, 238–239 (1988)
23. Y.F. Tang, The simplecticity of multistep methods. Comput. Math. Appl. **25**(3), 83–90 (1993)

Convergence of Parareal for the Navier-Stokes Equations Depending on the Reynolds Number

Johannes Steiner, Daniel Ruprecht, Robert Speck, and Rolf Krause

Abstract The paper presents first a linear stability analysis for the time-parallel Parareal method, using an IMEX Euler as coarse and a Runge-Kutta-3 method as fine propagator, confirming that dominant imaginary eigenvalues negatively affect Parareal's convergence. This suggests that when Parareal is applied to the nonlinear Navier-Stokes equations, problems for small viscosities could arise. Numerical results for a driven cavity benchmark are presented, confirming that Parareal's convergence can indeed deteriorate as viscosity decreases and the flow becomes increasingly dominated by convection. The effect is found to strongly depend on the spatial resolution.

1 Introduction

As core counts in modern supercomputers continue to grow, parallel algorithms are required that can provide concurrency beyond existing approaches parallelizing in space. In particular, algorithms that parallelize in time "along the steps" have attracted noticeable interest. Probably the most widely studied algorithm of this type is Parareal [13], but other important methods exist as well, for example PITA [8] or PFASST [7].

The applicability of Parareal to the Navier-Stokes equations has been studied in [10], where it is shown that Parareal can solve the initial value problem arising from a Finite Element discretization of the Navier-Stokes equations for a Reynolds number of 200 as well as from a Spectral Element discretization for a problem with Reynolds number 7,500. A non-Newtonian problem is studied in [2]. In [17, 18], Parareal is combined with parallelization in space and setups with Reynolds

J. Steiner • D. Ruprecht (✉) • R. Krause
Institute of Computational Science, Università della Svizzera italiana, Via Giuseppe Buffi 13, CH-6900 Lugano, Switzerland
e-mail: johannes.steiner@usi.ch; daniel.ruprecht@usi.ch; rolf.krause@usi.ch

R. Speck
Jülich Supercomputing Centre, Forschungszentrum Jülich, Jülich, Germany
e-mail: r.speck@fz-juelich.de

A. Abdulle et al. (eds.), *Numerical Mathematics and Advanced Applications ENUMATH 2013*, Lecture Notes in Computational Science and Engineering 103,
DOI 10.1007/978-3-319-10705-9_19

numbers up to 1,000 are investigated. While it is confirmed that Parareal can successfully be applied to flow simulations, the attempt to demonstrate its potential to provide speedup beyond the saturation of the spatial parallelization was inconclusive, as either the pure time or pure space parallel approach provided minimum runtimes. A successful demonstration that Parareal can speed up simulations after the spatial parallelization has saturated can be found in [5], where Parareal is used to simulate a driven cavity flow in a cube with a Reynolds number of 1,000. The performance of PFASST for a particle-based discretization of the Navier-Stokes equations on $\mathscr{O}(100{,}000)$ cores is studied in [15].

It has been noted in multiple works that Parareal as well as PITA have stability issues for convection-dominated problems, see [1, 8, 12, 14, 16]. This suggests that Parareal will at some point cease to converge properly for the Navier-Stokes equations if the Reynolds number increases and the problem becomes more and more dominated by advection. This paper discusses results from linear stability analysis and presents a numerical study for two-dimensional driven cavity flow of how the convergence of Parareal is affected as viscosity decreases.

2 Parareal

Parareal is a method to introduce concurrency in the solution of initial value problems

$$u_t = f(u(t), t), \quad u(0) = u_0, \quad 0 \leq t \leq T. \tag{1}$$

It relies on the introduction of two classical one-step time integration methods, one computationally expensive and of high accuracy (denoted by $\mathscr{F}$) and one computationally cheap method of lower accuracy (denoted by $\mathscr{G}$). The former is commonly referred to as the "fine propagator", the latter as the "coarse propagator". Denote by U_n the numerical approximation of the exact solution u of (1) at some point in time t_n. Further, denote as

$$U_{n+1} = \mathscr{F}_{\delta t}(U_n) \tag{2}$$

the result obtained by integrating from an initial value U_n given at a time t_n forward in time to a time t_{n+1} using a time-step δt and the method indicated by $\mathscr{F}$. For a decomposition of $[0, T]$ into N so-called time-slices $[t_n, t_{n+1}]$, $n = 0, \ldots, N-1$, solving (2) time-slice after time-slice corresponds to classical time-marching, running the fine method in serial from $t_0 = 0$ to $t_N = T$. Instead, Parareal approximately computes the values U_n by means of the iteration

$$U_{n+1}^{k+1} = \mathscr{G}_{\Delta t}(U_n^{k+1}) + \mathscr{F}_{\delta t}(U_n^k) - \mathscr{G}_{\Delta t}(U_n^k) \tag{3}$$

where k denotes the iteration counter. For $k \to N$, iteration (3) converges towards the serial fine solution, that is $U_n^k \to U_n$. Once values U_n^k are known, the evaluation of the computationally expensive terms $\mathscr{F}(U_n^k)$ in (3) can be done in parallel on N processors. Then, a correction is propagated serially by evaluating the terms $\mathscr{G}_{\Delta t}(U_n^{k+1})$ and computing U_{n+1}^{k+1}. We refer to e.g. [14] for a more in-depth presentation of the algorithm. The speedup achievable by Parareal concurrently computing the solution on N time-intervals assigned to N processors is bounded by

$$s(N) \leq \min\left\{\frac{N}{N_{\mathrm{it}}}, \frac{C_{\mathscr{F}}}{C_{\mathscr{G}}}\right\} \tag{4}$$

where N_{it} is the number of iterations performed and $C_{\mathscr{F}}$, $C_{\mathscr{G}}$ denote the time required to evaluate $\mathscr{F}_{\delta t}$ and $\mathscr{G}_{\Delta t}$ respectively, see again e.g. [14]. Note that the two bounds are competing in the sense that using a coarser and cheaper method for $\mathscr{G}$ will usually improve the second bound but might cause Parareal to require more iterations to converge, thereby reducing the first bound. In contrast, a more accurate and more expensive $\mathscr{G}$ will likely reduce the iteration number but also reduce the coarse-to-fine runtime ratio $\frac{C_{\mathscr{F}}}{C_{\mathscr{G}}}$.

3 Linear Stability Analysis

In order to illustrate Parareal's stability properties, we apply it to the test equation

$$y'(t) = \lambda_{\mathrm{Re}} y(t) + i\lambda_{\mathrm{Im}} y(t), \quad y(0) = 0, \quad 0 \leq t \leq T. \tag{5}$$

A linear stability analysis of this kind was first done in [16], using RadauIIA methods for both $\mathscr{F}$ and $\mathscr{G}$. Here, in line with the numerical examples presented in Sect. 4, the stability analysis is done for an implicit-explicit Euler method for $\mathscr{G}$ and an explicit Runge-Kutta-3 method for $\mathscr{F}$ with five time steps of $\mathscr{F}$ per two time steps of $\mathscr{G}$. The IMEX scheme treats the real part ("diffusion") implicitly and the imaginary term ("convection") explicitly. Further, $N = 15$ concurrent time slices are used and a time step $\Delta t = 1.0$ for $\mathscr{G}$, so that $T = 15$.

Figure 1 shows the resulting stability domains and isolines of accuracy for the coarse method run serially (a), the fine method run serially (b), and for Parareal with different numbers of iteration (c)–(f). For $N_{\mathrm{it}} = N = 15$, the solution from Parareal is identical to the one provided by $\mathscr{F}$ and thus the stability domains also coincide (not shown). As can be expected because of the stability constraint arising from the explicitly treated imaginary term, the IMEX method used for $\mathscr{G}$ becomes unstable if the imaginary part of λ becomes too dominant. Parareal however ceases to be stable even before reaching the stability limit of the coarse propagator. The analysis confirms again that for problems with imaginary eigenvalues, Parareal can develop instabilities although both $\mathscr{F}$ and $\mathscr{G}$ are stable. Furthermore, the stability domain of Parareal shrinks from $N_{\mathrm{it}} = 1$ to $N_{\mathrm{it}} = 4$ and $N_{\mathrm{it}} = 8$ before expanding again for $N_{\mathrm{it}} = 12$. Note also that for a fixed number of iterations, Parareal becomes less accurate as λ_{Im} increases (in contrast to the serial fine method), corresponding to reduced rates of convergence. This means that achieving the accuracy of the underlying fine method will require more iterations for problems with larger imaginary eigenvalues, therefore reducing the speedup achievable by

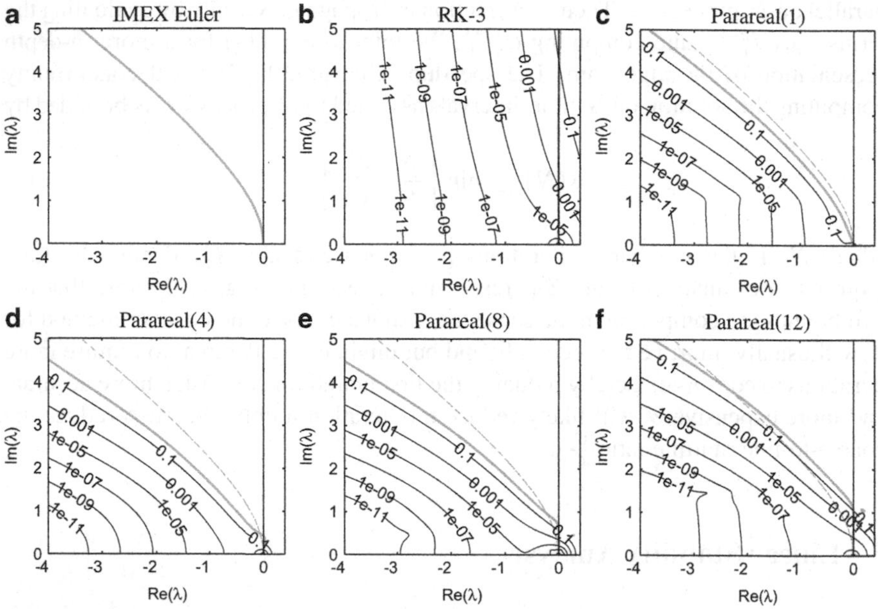

Fig. 1 Stability and accuracy of Parareal using an implicit-explicit Euler for $\mathscr{G}$, a RK3 method for $\mathscr{F}$, $N = 15$ time slices and a ratio of $s = 5/2$ fine to coarse steps in each time-slice. The *thick gray line* indicates where the amplification factor becomes greater than one. The *black lines* indicate error levels. Note that in (**a**) no black lines are visible because the error never drops below 10^{-1}. Note also that $s = 5/2$ means the fine scheme in serial performs five steps per time-slice and the coarse scheme two, so that (**a**) and (**b**) are not identical to the stability function of the respective method with only a single time-step. Figures (**c**)–(**f**) show the stability domain for Parareal with $N_{\text{it}} = 1, 4, 8, 12$ iterations. For comparison, the stability region of $\mathscr{G}$ is also sketched again as a *thin dashed gray line*

Parareal, cf. the estimate (4). Eventually, as convergence becomes too slow, Parareal will no longer be able to achieve speedup at all and will no longer be useful. The mathematical explanation for this behavior is a growing term in the error estimate for Parareal for imaginary eigenvalues that is only compensated for as the iteration number approaches the number of time-slices, see the analysis in [12].

4 Numerical Results for Driven Cavity Flow

In order to investigate if and how the results from the linear stability analysis carry over to the fully nonlinear case, we solve now the non-dimensional, nonlinear, incompressible Navier-Stokes equations in two dimensions

$$\mathbf{u}_t + \mathbf{u} \cdot \nabla \mathbf{u} + \nabla p = \nu \Delta \mathbf{u} \tag{6}$$

$$\nabla \cdot \mathbf{u} = 0 \tag{7}$$

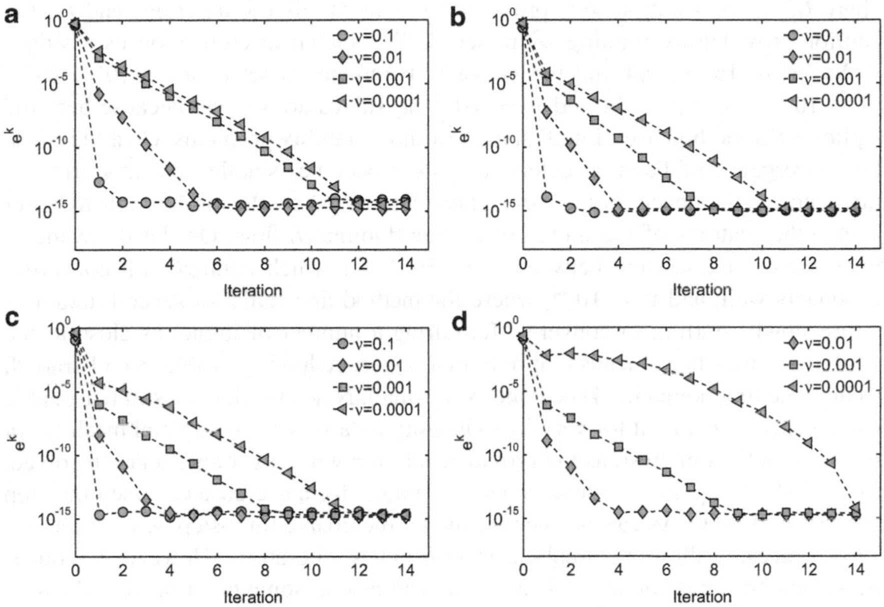

Fig. 2 Convergence of Parareal against the serial fine solution for $\Delta t = 1/200$, different numbers of mesh points N_x and different values for the viscosity ν. (**a**) $N_x = 8$. (**b**) $N_x = 16$. (**c**) $N_x = 32$. (**d**) $N_x = 64$

on a square $[0, 1]^2$. A method-of-lines approach is used to first discretize in space. For the spatial discretization a finite volume method based on a vertex centered scheme is used. On an unstructured or not necessarily structured triangle mesh, control volumes are constructed via a dual mesh. This leads to a non-staggered scheme of velocity and pressure. Therefore, a stabilization based on upwind differences and an incremental version of the Chorin-Temam method for the pressure is used [19]. Parareal is then employed to solve the resulting initial value problem until a final time $T = 15$ with $N = 15$ time-slices. As in the stability analysis above, $\mathscr{G}$ is an implicit-explicit Euler method while $\mathscr{F}$ is an explicit Runge-Kutta-3 method. The time-step for the coarse method is $\Delta t = 1/200$, for the fine method $\delta t = 1/500$, reproducing a rate of $s = 5/2$ fine per coarse steps. Although the driven cavity setup is probably not the most ideal here, since, depending on the viscosity, the solution settles into a steady state rather quickly, its wide use and comparative simplicity still make for a good first test case. Further tests for a more complex vortex shedding setups are currently ongoing.

Figure 2 shows the convergence of Parareal against the solution provided by running $\mathscr{F}$ in serial. Shown is the maximum of the relative error at the end of all time-slices, that is

$$e^k := \max_{n=1,\ldots,N} \left\| U_n^k - U_n \right\|_\infty / \left\| U_n \right\|_\infty \tag{8}$$

where U_n^k is the solution at t_n provided by Parareal after k iterations and U_n the solution provided by running $\mathscr{F}$ in serial. The spatial discretization uses values of $N_x = 8, 16, 32, 64$ and the viscosity parameter is set to $\nu = 10^{-1}, 10^{-2}, 10^{-3}, 10^{-4}$. For $N_x = 64$ and $\nu = 10^{-1}$ no values are shown, because here the explicit RK3 method used for $\mathscr{F}$ started to show stability problems. On all meshes, the convergence of Parareal deteriorates as ν becomes smaller and this effect is much more pronounced for finer spatial resolutions, where the mesh is able to better resolve the features of the more convection dominated flow. On the finest mesh, there is a clear transition between $\nu = 10^{-3}$, for which Parareal still converges reasonably well, and $\nu = 10^{-4}$, where the method first stalls for several iterations before slowly starting to converge. Requiring a number of iterations close to the number of time-slices means that only marginal speedup is possible from Parareal, because the first bound in (4) becomes very small. Note also that the still reasonable convergence of Parareal for very low viscosity on a very coarse spatial mesh is not of great practical interest, as the provided solution will be strongly under-resolved. Figure 3 shows again the convergence of Parareal for a decreased coarse time-step size $\Delta t = 1/400$. As can be seen, reducing the coarse time-step again improves convergence and allows Parareal to converge in fewer iterations. However, it reduces the second speedup bound in (4) and thus will also at some point prevent Parareal from achieving speedup. Therefore, the reduced convergence speed of Parareal for

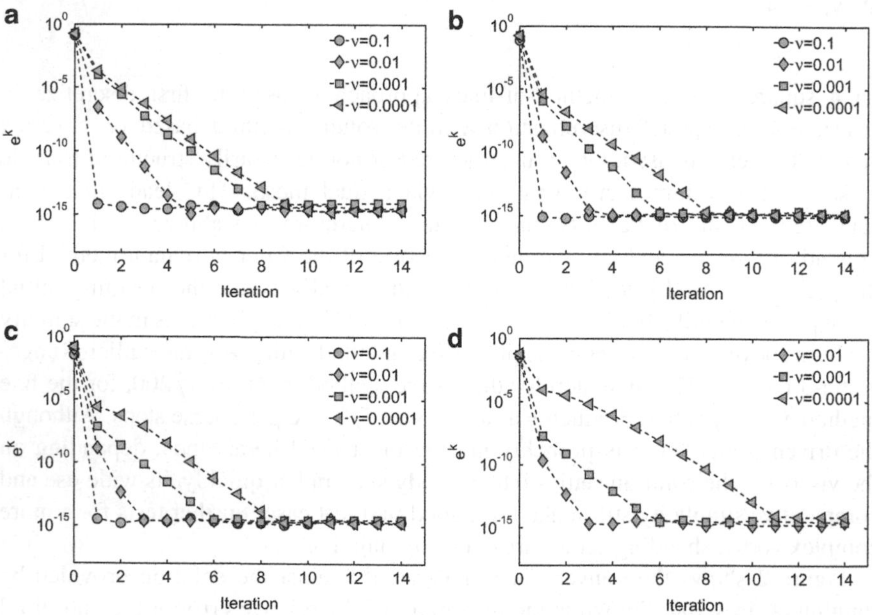

Fig. 3 Convergence of Parareal against the serial fine solution for $\Delta t = 1/400$, different numbers of mesh points N_x and different values for the viscosity ν. (**a**) $N_x = 8$. (**b**) $N_x = 16$. (**c**) $N_x = 32$. (**d**) $N_x = 64$

small viscosities either necessitates a small time-step in the coarse method or a large number of iterations and both choices significantly reduce the achievable speedup. A possible remedy could be the application of stabilization techniques as discussed in [4, 9] for PITA or [3, 6, 11, 14] for Parareal, but so far none of these have been tested for the full Navier-Stokes equations.

Conclusions

The paper presents a numerical study of how the Reynolds number (or, inversely, the viscosity parameter) affects the convergence of the time-parallel Parareal method when used to solve the Navier-Stokes equations. From other works it is known that Parareal can develop a mild instability for problems with dominant imaginary eigenvalues, so it can be expected that as the viscosity is decreased, Parareal will eventually become unstable at some point. A linear stability analysis is performed to motivate this assumption, which is then substantiated by numerical examples, solving a two-dimensional driven cavity problem for different Reynolds numbers and different spatial resolutions. It is confirmed that the convergence of Parareal deteriorates as the viscosity parameter becomes smaller and the flow becomes more and more dominated by convection. This necessitates either the use of a very small time-step in the coarse method or many iterations of Parareal, but both these choices significantly reduce the achievable speedup.

Acknowledgements This work was supported by Swiss National Science Foundation (SNSF) grant 145271 under the lead agency agreement through the project "ExaSolvers" within the Priority Programme 1648 "Software for Exascale Computing" (SPPEXA) of the Deutsche Forschungsgemeinschaft (DFG).

References

1. G. Bal, On the convergence and the stability of the parareal algorithm to solve partial differential equations, in *Domain Decomposition Methods in Science and Engineering*, ed. by R. Kornhuber et al. Lecture Notes in Computational Science and Engineering, vol. 40 (Springer, Berlin, 2005), pp. 426–432
2. E. Celledoni, T. Kvamsdal, Parallelization in time for thermo-viscoplastic problems in extrusion of aluminium. Int. J. Numer. Methods Eng. **79**(5), 576–598 (2009)
3. F. Chen, J. Hesthaven, X. Zhu, On the use of reduced basis methods to accelerate and stabilize the parareal method, in *Reduced Order Methods for Modeling and Computational Reduction*. MS&A – Modeling, Simulation and Applications, vol. 9 (Springer, International Publishing Switzerland, 2014)
4. J. Cortial, C. Farhat, A time-parallel implicit method for accelerating the solution of non-linear structural dynamics problems. Int. J. Numer. Methods Eng. **77**(4), 451–470 (2009)
5. R. Croce, D. Ruprecht, R. Krause, *Parallel-in-Space-and-Time Simulation of the Three-Dimensional, Unsteady Navier-Stokes Equations for Incompressible Flow*. Modeling, Simulation and Optimization of Complex Processes (Springer, Berlin/Heidelberg, 2012, in press)

6. X. Dai, Y. Maday, Stable parareal in time method for first- and second-order hyperbolic systems. SIAM J. Sci. Comput. **35**(1), A52–A78 (2013)
7. M. Emmett, M.L. Minion, Toward an efficient parallel in time method for partial differential equations. Commun. Appl. Math. Comput. Sci. **7**, 105–132 (2012)
8. C. Farhat, M. Chandesris, Time-decomposed parallel time-integrators: theory and feasibility studies for fluid, structure, and fluid-structure applications. Int. J. Numer. Methods Eng. **58**(9), 1397–1434 (2003)
9. C. Farhat, J. Cortial, C. Dastillung, H. Bavestrello, Time-parallel implicit integrators for the near-real-time prediction of linear structural dynamic responses. Int. J. Numer. Methods Eng. **67**, 697–724 (2006)
10. P.F. Fischer, F. Hecht, Y. Maday, A parareal in time semi-implicit approximation of the Navier-Stokes equations, in *Domain Decomposition Methods in Science and Engineering*, ed. by R. Kornhuber et al. Lecture Notes in Computational Science and Engineering, vol. 40 (Springer, Berlin, 2005), pp. 433–440
11. M. Gander, M. Petcu, Analysis of a Krylov subspace enhanced parareal algorithm for linear problems. ESAIM: Proc. **25**, 114–129 (2008)
12. M.J. Gander, S. Vandewalle, Analysis of the parareal time-parallel time-integration method. SIAM J. Sci. Comput. **29**(2), 556–578 (2007)
13. J.-L. Lions, Y. Maday, G. Turinici, A “parareal” in time discretization of PDE’s, Comptes Rendus de l’Académie des Sciences – Series I – Mathematics **332**, 661–668 (2001)
14. D. Ruprecht, R. Krause, Explicit parallel-in-time integration of a linear acoustic-advection system. Comput. Fluids **59**(0), 72–83 (2012)
15. R. Speck, D. Ruprecht, R. Krause, M. Emmett, M. Minion, M. Winkel, P. Gibbon, A massively space-time parallel n-body solver, in *Proceedings of the International Conference on High Performance Computing, Networking, Storage and Analysis*, Salt Lake City (IEEE Computer Society Press, Los Alamitos, 2012), pp. 92:1–92:11
16. G.A. Staff, E.M. Rønquist, Stability of the parareal algorithm, in *Domain Decomposition Methods in Science and Engineering* ed. by R. Kornhuber et al. Lecture Notes in Computational Science and Engineering, vol. 40 (Springer, Berlin, 2005), pp. 449–456
17. J.M.F. Trindade, J.C.F. Pereira, Parallel-in-time simulation of the unsteady Navier–Stokes equations for incompressible flow. Int. J. Numer. Methods Fluids **45**(10), 1123–1136 (2004)
18. ———, Parallel-in-time simulation of two-dimensional, unsteady, incompressible laminar flows. Numer. Heat Transf. Part B: Fundam. **50**(1), 25–40 (2006)
19. H. Versteeg, W. Malalasekera, *An Introduction to Computational Fluid Dynamics: The Finite Volume Method* (Pearson Education, Harlow, England, 2007)

Splitting in Potential Finite-Difference Schemes with Discrete Transparent Boundary Conditions for the Time-Dependent Schrödinger Equation

Alexander Zlotnik, Bernard Ducomet, Ilya Zlotnik, and Alla Romanova

Abstract The time-dependent Schrödinger equation is the key one in many fields. It should be often solved in unbounded space domains. Several approaches are known to deal with such problems using approximate transparent boundary conditions (TBCs) on the artificial boundaries. Among them, there exist the so-called discrete TBCs whose advantages are the complete absence of spurious reflections, reliable computational stability, clear mathematical background and the corresponding rigorous stability theory. In this paper, the Strang-type splitting with respect to the potential is applied to three two-level schemes with different discretizations in space having the approximation order $O(\tau^2 + |h|^k)$, $k = 2$ or 4. Explicit forms of the discrete TBCs are given and results on existence, uniqueness and uniform in time L^2-stability of solutions are stated in a unified manner. Due to splitting, an effective direct algorithm to implement the schemes is presented for general potential.

1 Introduction

The time-dependent Schrödinger equation is important in quantum mechanics, atomic and nuclear physics, wave physics, quantum waveguides, etc. It should be often solved in unbounded space domains. Several approaches were developed to deal with the problems of such kind using approximate transparent boundary

A. Zlotnik (✉) • A. Romanova
National Research University Higher School of Economics, Myasnitskaya 20, 101000 Moscow, Russia
e-mail: alexander.zlotnik@gmail.com; avromm1@gmail.com

B. Ducomet
DPTA/Service de Physique Nucléaire, CEA/DAM/DIF Ile de France, BP 12, F–91297 Arpajon, France
e-mail: bernard.ducomet@cea.fr

I. Zlotnik
Settlement Depository Company, 2-oi Verkhnii Mikhailovskii proezd 9, building 2, 115419 Moscow, Russia
e-mail: ilya.zlotnik@gmail.com

A. Abdulle et al. (eds.), *Numerical Mathematics and Advanced Applications ENUMATH 2013*, Lecture Notes in Computational Science and Engineering 103,
DOI 10.1007/978-3-319-10705-9_20

conditions (TBCs) on the artificial boundaries, see review [1]. Among them, there exist the so-called *discrete TBCs*. Their advantages are the complete absence of spurious reflections, reliable computational stability, clear mathematical background and the corresponding rigorous stability theory, see [2, 3, 5, 7, 9, 12], etc. In this paper, in order to simplify the implementation of schemes with the discrete TBCs in n-dimensional semi-infinite parallelepiped, the Strang-type splitting with respect to the potential is applied to three two-level schemes with different discretizations in space having the approximation order $O(\tau^2 + |h|^k)$, $k = 2$ or 4. Theorems on explicit forms of the discrete TBCs as well as on existence, uniqueness and uniform in time L^2-stability of solutions are stated in a unified manner. An effective direct algorithm to implement the splitting schemes is presented for general potential. The corresponding successful 2D numerical results can be found in [4, 6, 11].

2 The Main Results

We consider the multi-dimensional time-dependent Schrödinger equation

$$i\hbar\frac{\partial\psi}{\partial t} = -\frac{\hbar^2}{2m_0}\Delta\psi + V\psi \text{ for } x = (x_1,\ldots,x_n) \in \Pi_\infty,\ \ t > 0, \tag{1}$$

where Δ is the n-dimensional Laplace operator, $n \geq 2$, and $\Pi_\infty := (0,\infty) \times \Pi_{\hat{1}}$ is the semi-infinite parallelepiped, with $\Pi_{\hat{1}} := (0, X_2) \times \cdots \times (0, X_n)$. Also i is the imaginary unit, $\hbar > 0$ and $m_0 > 0$ are physical constants, $\psi = \psi(x,t)$ is the unknown complex-valued wave function and $V = V(x)$ is the given real potential. Let $c_\hbar := \frac{\hbar^2}{2m_0}$.

We impose the following boundary condition, the condition at infinity and the initial condition

$$\psi(\cdot,t)|_{\partial\Pi_\infty} = 0,\ \ \|\psi(\cdot,t)\|_{L^2(\Pi_\infty)} < \infty,\ \ t > 0, \tag{2}$$

$$\psi|_{t=0} = \psi^0(x) \text{ on } \Pi_\infty, \tag{3}$$

where $\partial\Pi_\infty$ is the boundary of Π_∞. We also suppose that

$$V(x) = V_\infty,\ \ \psi^0(x) = 0 \text{ for } x \in [X_0,\infty) \times \Pi_{\hat{1}} \tag{4}$$

for some (sufficiently large) $X_0 > 0$.

We exploit a uniform mesh $\overline{\omega}_{h,\infty}$ on $\overline{\Pi}_\infty$ with nodes $x_{\mathbf{j}} = (j_1h_1,\ldots,j_nh_n)$, where $j_1 \geq 0, 0 \leq j_2 \leq J_2,\ldots,\ 0 \leq j_n \leq J_n$, and steps $h_1 = \frac{X_1}{J_1},\ldots,h_n = \frac{X_n}{J_n}$, where $X_1 > X_0$ and $h_1 \leq X_1 - X_0$. Let $\omega_{h,\infty} = \{x_{\mathbf{j}},\ j_1 \geq 1,\ 1 \leq j_2 \leq J_2 - 1,\ \ldots, 1 \leq j_n \leq J_n - 1\}$ and $\Gamma_{h,\infty} := \overline{\omega}_{h,\infty} \backslash \omega_{h,\infty}$ be its interior and boundary. Hereafter $h = (h_1,\ldots,h_n)$, $|h|$ is the length of h and $\mathbf{j} = (j_1,\ldots,j_n)$.

We define the backward and forward difference quotients $\bar{\partial}_k$ and ∂_k and the Numerov average $s_{Nk} W_j := \frac{1}{12} W_{j-1} + \frac{5}{6} W_j + \frac{1}{12} W_{j+1}$ acting in x_k.

We also define a non-uniform mesh $\overline{\omega}^\tau$ in time with nodes $0 = t_0 < t_1 < \cdots < t_m < \ldots$, where $t_m \to \infty$ for $m \to \infty$, and steps $\tau_m = t_m - t_{m-1}$. Let $\omega^\tau := \overline{\omega}^\tau \backslash \{0\}$ and $\tau_{\max} = \sup_{m \geq 1} \tau_m$. We define the backward difference quotient $\bar{\partial}_t$ and the average $\overline{s}_t Y = \frac{Y + \check{Y}}{2}$ in t, where $\check{Y}^m = Y^{m-1}$.

We introduce the n-dimensional Numerov-type average operator and the corresponding splitting operator

$$s_N = I + \frac{h_1^2}{12} \partial_1 \bar{\partial}_1 + \cdots + \frac{h_n^2}{12} \partial_n \bar{\partial}_n, \quad \overline{s}_N = s_{N1} \ldots s_{Nn}$$

together with their $n-1$-dimensional versions (excepting the direction x_k)

$$s_{N\hat{k}} = I + \sum_{1 \leq \ell \leq n, \ell \neq k} \frac{h_\ell^2}{12} s_{N\ell}, \quad \overline{s}_{N\hat{k}} = \prod_{1 \leq \ell \leq n, \ell \neq k} s_{N\ell}.$$

We also exploit the simplest $\Delta_h = \partial_1 \bar{\partial}_1 + \cdots + \partial_n \bar{\partial}_n$ and Numerov-type $\Delta_{hN} = s_{N\hat{1}} \partial_1 \bar{\partial}_1 + \cdots + s_{N\hat{n}} \partial_n \bar{\partial}_n$ discretizations of the Laplace operator and the latter one with splitting of the space averages $\bar{\Delta}_{hN} = \overline{s}_{N\hat{1}} \partial_1 \bar{\partial}_1 + \cdots + \overline{s}_{N\hat{n}} \partial_n \bar{\partial}_n$.

We first consider three two-level symmetric in time but different in space discretizations of the Schrödinger equation (1)

$$i\hbar s \bar{\partial}_t \Psi = -c_\hbar \Delta^{(h)} \overline{s}_t \Psi + s\,(V \overline{s}_t \Psi) \text{ on } \omega_{h,\infty} \times \omega^\tau, \tag{5}$$

where $s = I$ (the unit operator), s_N or $\overline{s}_N$ as well as $\Delta^{(h)} = \Delta_h$, Δ_{hN} or $\bar{\Delta}_{hN}$ respectively for the discretizations A (of the Crank-Nicolson type), B (of the Numerov-Crank-Nicolson type) and C (the modified one with splitting of the space averages). The standard discretization A has the second approximation order $O(\tau_{\max}^2 + |h|^2)$ whereas the discretizations B and C have the higher approximation order $O(\tau_{\max}^2 + |h|^4)$ (for the latter one, this is valid owing to formulas $\overline{s}_N = s_N + O(|h|^4)$ and $\bar{\Delta}_{hN} = \Delta_{hN} + O(|h|^4)$). Unfortunately, in the case $n = 3$, the operator s_N has eigenvalues tending to 0 as $h \to 0$ and moreover, in the case $n \geq 4$ and for sufficiently small h, it has the negative ones. Therefore, in the case $n \geq 3$, the discretization B does not possess suitable properties, and below we consider it only in the case $n = 2$. Instead the discretization C is constructed for any $n \geq 2$.

In order to simplify the implementation, we apply the known Strang-type splitting in potential [8] and construct the following three-step discretization

$$i\hbar \frac{\breve{\Psi}^m - \Psi^{m-1}}{\tau_m/2} = \delta V \frac{\breve{\Psi}^m + \Psi^{m-1}}{2} \text{ on } \omega_{h,\infty}, \tag{6}$$

$$i\hbar s \frac{\tilde{\Psi}^m - \breve{\Psi}^m}{\tau_m} = -c_\hbar \Delta^{(h)} \frac{\tilde{\Psi}^m + \breve{\Psi}^m}{2} + s\Big(\tilde{V} \frac{\tilde{\Psi}^m + \breve{\Psi}^m}{2}\Big) + F^m \text{ on } \omega_{h,\infty}, \tag{7}$$

$$i\hbar\frac{\Psi^m-\tilde{\Psi}^m}{\tau_m/2}=\delta V\frac{\Psi^m+\tilde{\Psi}^m}{2}\quad\text{on }\omega_{h,\infty}\tag{8}$$

as well as put the boundary and initial conditions

$$\breve{\Psi}^m|_{\Gamma_{h,\infty}}=0,\ \tilde{\Psi}^m|_{\Gamma_{h,\infty}}=0,\ \Psi^m|_{\Gamma_{h,\infty}}=0,\ \Psi^0=\Psi_h^0\ \text{ on }\overline{\omega}_{h,\infty},\tag{9}$$

for any $m\geq 1$. Hereafter $\delta V:=V-\tilde{V}$, where the auxiliary 1D potential $\tilde{V}=\tilde{V}(x_1)$ is such that $\tilde{V}(x_1)=V_\infty$ for $x_1\geq X_0$ (in particular, $\tilde{V}(x_1)\equiv V_\infty$). We suppose that $\Psi_h^0|_{\Gamma_{h,\infty}}=0$. The free term F^m is added into (7) to study stability in more detail below.

Clearly Eqs. (6) and (8) are reduced to the explicit formulas

$$\breve{\Psi}^m=\mathscr{E}^m\Psi^{m-1},\ \ \Psi^m=\mathscr{E}^m\tilde{\Psi}^m\ \text{ with }\mathscr{E}^m:=\Big(1-i\frac{\tau_m}{4\hbar}\delta V\Big)/\Big(1+i\frac{\tau_m}{4\hbar}\delta V\Big).\tag{10}$$

The main equation (7) is similar to the original one (5) at mth time level, but it is simplified by substituting $\tilde{V}$ for V. $\breve{\Psi}$ and $\tilde{\Psi}$ are the auxiliary unknown functions and Ψ is the main one.

It follows from (10) that $|\breve{\Psi}^m|=|\Psi^{m-1}|$ and $|\Psi^m|=|\tilde{\Psi}^m|$ on $\overline{\omega}_{h,\infty}$; moreover, for $j_1\geq J_1-1$, simply $\breve{\Psi}_{\mathbf{j}}^m=\Psi_{\mathbf{j}}^{m-1}$ and $\Psi_{\mathbf{j}}^m=\tilde{\Psi}_{\mathbf{j}}^m$.

This splitting of Eq. (5) is symmetric in time due to steps (6) and (8) (and the symmetry in time of Eq. (7)), therefore it does not reduce the above approximation orders (that can be checked also more formally).

Let H_h be a Hilbert space of mesh functions $W\colon\overline{\omega}_{h,\infty}\to\mathbb{C}$ such that

$$W|_{\Gamma_{h,\infty}}=0,\ \ \|W\|_{H_h}^2:=\sum_{j_1=1}^{\infty}\sum_{j_2=1}^{J_2-1}\dots\sum_{j_n=1}^{J_n-1}|W_{\mathbf{j}}|^2h_1\dots h_n<\infty.$$

Theorem 1 *Let only the first of conditions (4) be valid. Let $F^m,\Psi_h^0\in H_h$ for any $m\geq 1$. Then there exists a unique solution to the splitting scheme (6)–(9) such that $\Psi^m\in H_h$ for any $m\geq 0$, and the following L^2-stability bound holds*

$$\max_{0\leq m\leq M}\|\Psi^m\|_{H_h}\leq\|\Psi_h^0\|_{H_h}+\frac{2c_0}{\hbar}\sum_{m=1}^{M}\|F^m\|_{H_h}\tau_m\ \ \textit{for any }M\geq 1,$$

where $c_0=1,6$ and $\left(\frac{3}{2}\right)^n$ respectively for the discretizations A, B and C.

For $F=0$, the following mass conservation law holds $\|\Psi^m\|_{H_h}^2=\|\Psi_h^0\|_{H_h}^2$ for any $m\geq 1$.

Scheme (6)–(9) can not be used in practice because of the infinite number of unknowns at each time level. We intend to restrict its solution to a finite space mesh

$\overline{\omega}_h := \{x_{\mathbf{j}} \in \overline{\omega}_{h,\infty};\ 0 \le j_1 \le J_1\}$. Let $\omega_h := \{x_{\mathbf{j}} \in \omega_{h,\infty};\ 1 \le j_1 \le J_1 - 1\}$ and $\partial\omega_h = \overline{\omega}_h \backslash \omega_h$ be its interior and boundary as well as $\Gamma_{1h} := \{x_{\mathbf{j}};\ j_1 = J_1,\ 1 \le j_2 \le J_2 - 1, \ldots, 1 \le j_n \le J_n - 1\}$ and $\Gamma_h = \partial\omega_h \backslash \Gamma_{1h}$ be the boundary parts. Let also $\omega_{h1} := \{j_1 h_1;\ 1 \le j_1 \le J_1 - 1\}$ and $\omega_{h\hat{1}} = \{(j_2 h_2, \ldots, j_n h_n);\ 1 \le j_2 \le J_2 - 1, \ldots, j_n \le J_n - 1\}$. Given a function $W\colon \overline{\omega}_h \to \mathbb{C}$, denote by W_{J_1} its trace on Γ_{1h}.

By definition, *the discrete TBC* is such a (non-local) boundary condition at the artificial boundary Γ_{1h} which allows one to accomplish the above mentioned restriction. To write down it explicitly, we need the following operators

$$s_{\bar{N}1} W_j = \frac{1}{12} W_{j-1} + \frac{5}{12} W_j,\quad s_{\bar{N}} = s_{\bar{N}1} + \frac{h_2^2}{24} \partial_2 \bar{\partial}_2,\quad \overline{s}_{N,\widehat{1k}} := \prod_{2 \le \ell \le n,\, \ell \ne k} s_{N\ell},$$

where $s_{\bar{N}1}$ acts in x_1. Let also $\Delta_{\hat{1}}^{(h)} := \partial_2 \bar{\partial}_2 + \ldots + \partial_n \bar{\partial}_n$ and $\overline{s}_{N\widehat{12}} \partial_2 \bar{\partial}_2 + \ldots + \overline{s}_{N\widehat{1n}} \partial_n \bar{\partial}_n$ respectively for the discretizations A and C. We define the following approximations

$$\mathscr{D}_{1h}(\tilde{\Psi}, \breve{\Psi}) := c_\hbar \bar{\partial}_1 \frac{\tilde{\Psi} + \breve{\Psi}}{2} - \frac{h_1}{2}\left[i\hbar \frac{\tilde{\Psi} - \breve{\Psi}}{\tau} + \left(c_\hbar \Delta_{\hat{1}}^{(h)} - V_\infty I\right) \frac{\tilde{\Psi} + \breve{\Psi}}{2}\right],$$

$$c_\hbar s_{N2} \bar{\partial}_1 \frac{\tilde{\Psi} + \breve{\Psi}}{2} - h_1 \left[i\hbar s_{\bar{N}} \frac{\tilde{\Psi} - \breve{\Psi}}{\tau} + \left(c_\hbar s_{\bar{N}1} \partial_2 \bar{\partial}_2 - V_\infty s_{\bar{N}}\right) \frac{\tilde{\Psi} + \breve{\Psi}}{2}\right],$$

$$c_\hbar \overline{s}_{N\hat{1}} \bar{\partial}_1 \frac{\tilde{\Psi} + \breve{\Psi}}{2} - h_1 s_{\bar{N}1} \left[i\hbar \overline{s}_{N\hat{1}} \frac{\tilde{\Psi} - \breve{\Psi}}{\tau} + \left(c_\hbar \Delta_{\hat{1}}^{(h)} - V_\infty \overline{s}_{N\hat{1}}\right) \frac{\tilde{\Psi} + \breve{\Psi}}{2}\right]$$

for $c_\hbar \frac{\partial}{\partial x_1}$ on Γ_{1h}, respectively for the discretizations A, B and C.

We need the well-known direct and inverse discrete sine Fourier transforms

$$P^{(q)} = (\mathscr{F}_k P)^{(q)} := \frac{2}{J_k} \sum_{j=1}^{J_k - 1} P_j \sin \frac{\pi q j}{J_k},\quad 1 \le q \le J_k - 1,$$

$$P_j = \left(\mathscr{F}_k^{-1} P^{(\cdot)}\right)_j := \sum_{q=1}^{J_k - 1} P^{(q)} \sin \frac{\pi q j}{J_k},\quad 1 \le j \le J_k - 1,$$

in x_k, $2 \le k \le n$. The eigenvalues of the operators $-\partial_k \bar{\partial}_k$ and s_{Nk} (under zero Dirichlet boundary conditions at $x_k = 0, X_k$) equal respectively $\lambda_q^{(k)} = \left(\frac{2}{h_k} \sin \frac{\pi q h_k}{2X_k}\right)^2$ and $\sigma_q^{(k)} = 1 - \frac{1}{3} \sin^2 \frac{\pi q h_k}{2X_k} \in (\frac{2}{3}, 1)$.

Let the time mesh be uniform with the step $\tau > 0$ below. Recall that the discrete convolution of mesh functions $R, Q\colon \overline{\omega}^\tau \to \mathbb{C}$ is given by $(R * Q)^m := \sum_{p=0}^m R^p Q^{m-p}$ for $m \ge 0$.

Theorem 2 *Let $F^m = 0$ and $\Psi_h^0 = 0$ on $\omega_{h,\infty}\backslash\omega_h$ for any $m \geq 1$ and $\Psi_h^0\big|_{j_1=J_1-1} = 0$. The solution of scheme (6)–(9) such that $\Psi^m \in H_h$, for any $m \geq 0$, satisfies the following splitting equations on the finite (space) mesh*

$$i\hbar\,\frac{\breve{\Psi}^m - \Psi^{m-1}}{\tau/2} = \delta V\,\frac{\breve{\Psi}^m + \Psi^{m-1}}{2} \quad \text{on } \omega_h \cup \Gamma_{1h}, \tag{11}$$

$$i\hbar s\frac{\tilde{\Psi}^m - \breve{\Psi}^m}{\tau} = -c_\hbar \Delta^{(h)}\frac{\tilde{\Psi}^m + \breve{\Psi}^m}{2} + s\Big(\tilde{V}\frac{\tilde{\Psi}^m + \breve{\Psi}^m}{2}\Big) + F^m \quad \text{on } \omega_h, \tag{12}$$

$$i\hbar\,\frac{\Psi^m - \tilde{\Psi}^m}{\tau/2} = \delta V\,\frac{\Psi^m + \tilde{\Psi}^m}{2} \quad \text{on } \omega_h \cup \Gamma_{1h}, \tag{13}$$

together with the boundary and initial conditions

$$\breve{\Psi}^m|_{\Gamma_h} = 0,\ \ \tilde{\Psi}^m|_{\Gamma_h} = 0,\ \ \Psi^m|_{\Gamma_h} = 0, \tag{14}$$

$$\mathscr{D}_{1h}(\tilde{\Psi}^m, \breve{\Psi}^m) = c_\hbar \mathscr{S}_{\text{ref}}^m \tilde{\boldsymbol{\Psi}}_{J_1}^m \quad \text{on } \Gamma_{1h}, \tag{15}$$

$$\Psi^0 = \Psi_h^0 \quad \text{on } \overline{\omega}_h, \tag{16}$$

for any $m \geq 1$; here $\tilde{\boldsymbol{\Psi}}_{J_1}^m = \{\tilde{\Psi}^0\big|_{j_1=J_1}, \ldots, \tilde{\Psi}^m\big|_{j_1=J_1}\}$ is a vector-function.

The operator on the right-hand side of the discrete TBC (15) has the form

$$\mathscr{S}_{\text{ref}}^m \boldsymbol{\Phi}^m := \mathscr{F}_2^{-1} \ldots \mathscr{F}_n^{-1}\left[\sigma_{\mathbf{q}} R_{\mathbf{q}} * \Phi^{\mathbf{q}}\right]^m \tag{17}$$

for any $\Phi\colon \omega_{h\hat{1}} \times \overline{\omega}^\tau \to \mathbb{C}$ such that $\Phi^0 = 0$, with $\boldsymbol{\Phi}^m := \{\Phi^0, \ldots, \Phi^m\}$, $\Phi^{\mathbf{q}} := (\mathscr{F}_n \ldots (\mathscr{F}_2\Phi)^{(q_2)} \ldots)^{(q_n)}$ and $\mathbf{q} = (q_2, \ldots, q_n)$. Here $R_{\mathbf{q}}$ can be computed by the recurrent formulas

$$R_{\mathbf{q}}^0 = c_{1\mathbf{q}},\ \ R_{\mathbf{q}}^1 = -c_{1\mathbf{q}}\kappa_{\mathbf{q}}\mu_{\mathbf{q}},$$

$$R_{\mathbf{q}}^m = \frac{2m-3}{m}\kappa_{\mathbf{q}}\mu_{\mathbf{q}} R_{\mathbf{q}}^{m-1} - \frac{m-3}{m}\kappa_{\mathbf{q}}^2 R_{\mathbf{q}}^{m-2} \quad \text{for } m \geq 2,$$

with the coefficients defined by

$$c_{1\mathbf{q}} = -\frac{|\alpha_{\mathbf{q}}|^{1/2}}{2}e^{-i(\arg\alpha_{\mathbf{q}})/2},\ \ \kappa_{\mathbf{q}} = -e^{i\arg\alpha_{\mathbf{q}}},\ \ \mu_{\mathbf{q}} = \frac{\beta_{\mathbf{q}}}{|\alpha_{\mathbf{q}}|} \in (-1, 1),$$

$$\alpha_{\mathbf{q}} = 2a_{\mathbf{q}} + (1 - 4\theta_{\mathbf{q}})h_1^2 a_{\mathbf{q}}^2 \neq 0,\ \ \beta_{\mathbf{q}} = 2\,\mathrm{Re}\,a_{\mathbf{q}} + (1 - 4\theta_{\mathbf{q}})h_1^2|a_{\mathbf{q}}|^2,$$

$$\arg\alpha_{\mathbf{q}} \in (0, 2\pi),\ \ a_{\mathbf{q}} = \frac{V_{\infty,\mathbf{q}}}{2c_{\hbar,\mathbf{q}}} + i\,\frac{\hbar}{\tau c_{\hbar,\mathbf{q}}},\ \ V_{\infty,\mathbf{q}} = V_\infty + c_\hbar \delta V_{\infty,\mathbf{q}}.$$

Moreover, one has

$$\sigma_{\mathbf{q}} = 1,\ \sigma_{q2}^{(2)}\left[1+\left(\frac{h_1h_2\lambda_{q2}^{(2)}}{12\sigma_{q2}^{(2)}}\right)^2\right],\ \sigma_{q2}^{(2)}\ldots\sigma_{q_n}^{(n)};$$

$$\theta_{\mathbf{q}} = 0,\ \frac{1}{12\sigma_{q2}^{(2)}},\ \frac{1}{12};\ \delta V_{\infty,\mathbf{q}} = \lambda_{q2}^{(2)}+\cdots+\lambda_{q_n}^{(n)},\ \frac{\lambda_{q2}^{(2)}}{\sigma_{q2}^{(2)}},\ \frac{\lambda_{q2}^{(2)}}{\sigma_{q2}^{(2)}}+\cdots+\frac{\lambda_{q_n}^{(n)}}{\sigma_{q_n}^{(n)}};$$

$$c_{\hbar,\mathbf{q}} = c_\hbar,\ c_\hbar\left[1+\left(\frac{h_1h_2\lambda_{q2}^{(2)}}{12\sigma_{q2}^{(2)}}\right)^2\right],\ c_\hbar$$

respectively for the discretizations A, B and C.

We introduce the mesh inner product and norms

$$(U,W)_{\omega_{h\hat{1}}} := \sum_{j_2=1}^{J_2-1}\cdots\sum_{j_n=1}^{J_n-1} U_{j_2,\ldots,j_n}W^*_{j_2,\ldots,j_n}h_2\ldots h_n,\ \ \|U\|^2_{\omega_{h\hat{1}}} = (U,U)_{\omega_{h\hat{1}}},$$

$$\|U\|^2_{\omega_h} := \sum_{k=1}^{J_1-1}\|U|_{j_1=k}\|^2_{\omega_{h\hat{1}}}h_1,\ \ \|U\|^2_{\tilde{\omega}_h} := \|U\|^2_{\omega_h} + \|U|_{j_1=J_1}\|^2_{\omega_{h\hat{1}}}h_1.$$

Lemma 1 *The operator $\mathscr{S}^m_{\mathrm{ref}}$ satisfies the important inequality [3]*

$$\mathrm{Im}\sum_{m=1}^{M}\left(\mathscr{S}^m_{\mathrm{ref}}\Phi^m, \bar{s}_t\Phi^m\right)_{\omega_{h\hat{1}}}\tau \ge 0 \ \textit{for any}\ M\ge 1,$$

for any $\Phi\colon \omega_{h\hat{1}}\times\overline{\omega}^\tau \to \mathbb{C}$ such that $\Phi^0 = 0$.

By construction, the splitting scheme on the finite mesh (11)–(17) has a solution. Let us study its uniqueness and stability. For the discretizations B and C, we impose the condition $|\tilde{V}(x_1') - \tilde{V}(x_1)| \le L|x_1' - x_1|^\alpha$, for any $0\le x_1 < x_1' \le X$, with some $\alpha\in[0,1]$ (see [10]). Clearly $L=0$ for $\tilde{V}\equiv$ const.

Theorem 3 *Let $\Psi_h^0\big|_{j_1=J_1-1,\,J_1} = 0$. The solution of the splitting scheme on the finite mesh (11)–(17) is unique and satisfies the following stability bound*

$$\max_{0\le m\le M}\|\Psi^m\|_{\tilde{\omega}_h} \le \|\Psi_h^0\|_{\tilde{\omega}_h} + \frac{2c_0}{\hbar}\sum_{m=1}^{M}\|F^m\|_{\omega_h}\tau \ \textit{for any}\ M\ge 1$$

provided that $L\tau h_1^\alpha < \bar{c}\hbar$, where $\bar{c} = 8$ or 16 for the discretizations B or C.

For the discretization A, the above results are proved by developing the technique from [3] and can be enlarged to the case of a generalized Schrödinger equation with variable coefficients and a non-uniform mesh $\overline{\omega}_h$. For the discretizations B and C, we also apply the approach to stability analysis from [9] as well as the technique for constructing the discrete TBCs and schemes' analysis together with some results from [5, 10, 12]; the uniqueness result is new and relies upon the chosen operator form (16) of the discrete TBC.

The splitting scheme on the finite space mesh (11)–(17) can be effectively implemented. Applying the operator $\mathscr{F}_2 \ldots \mathscr{F}_n$ to Eq. (12) (for brevity, in the case $F = 0$) and to the discrete TBC (15), we get the collection of disjoint 1D finite-difference Schrödinger problems in x_1 for each function $\tilde{\Psi}^{m\mathbf{q}}$

$$i\hbar s_{\theta_{\mathbf{q}}1} \frac{\tilde{\Psi}^{m\mathbf{q}} - \breve{\Psi}^{m\mathbf{q}}}{\tau} = -c_\hbar \partial_1 \bar{\partial}_1 \frac{\tilde{\Psi}^{m\mathbf{q}} + \breve{\Psi}^{m\mathbf{q}}}{2} + s_{\theta_{\mathbf{q}}1}\left(\tilde{V}_{\mathbf{q}} \frac{\tilde{\Psi}^{m\mathbf{q}} + \breve{\Psi}^{m\mathbf{q}}}{2}\right) \text{ on } \omega_{h1}, \tag{18}$$

$$\tilde{\Psi}^{m\mathbf{q}}\Big|_{j_1=0} = 0, \tag{19}$$

$$\left[c_{\hbar,\mathbf{q}} \bar{\partial}_1 \frac{\tilde{\Psi}^{m\mathbf{q}} + \breve{\Psi}^{m\mathbf{q}}}{2} - h_1 s_{\theta_{\mathbf{q}}1}^{-}\left(i\hbar \frac{\tilde{\Psi}^{m\mathbf{q}} - \breve{\Psi}^{m\mathbf{q}}}{\tau} - V_{\infty,\mathbf{q}} \frac{\tilde{\Psi}^{m\mathbf{q}} + \breve{\Psi}^{m\mathbf{q}}}{2}\right)\right]\Big|_{j_1=J_1}$$
$$= c_{\hbar,\mathbf{q}}\left(R_{\mathbf{q}} * \tilde{\boldsymbol{\Psi}}_{J_1}^{\mathbf{q}}\right)^m, \tag{20}$$

where operators $s_{\theta_{\mathbf{q}}1} W_j := \theta_{\mathbf{q}} W_{j-1} + (1 - 2\theta_{\mathbf{q}}) W_j + \theta_{\mathbf{q}} W_{j+1}$ and $s_{\theta_{\mathbf{q}}1}^{-} W_j := \theta_{\mathbf{q}} W_{j-1} + (\frac{1}{2} - \theta_{\mathbf{q}}) W_j$ act in x_1 as well as $\tilde{V}_{\mathbf{q}} := \tilde{V} + c_\hbar \delta V_{\infty,\mathbf{q}}$.

Given Ψ^{m-1}, the direct algorithm for computing Ψ^m comprises five steps.

1. To compute $\breve{\Psi}^m = \mathscr{E}^m \Psi^{m-1}$ on $\omega_h \cup \Gamma_{1h}$ (see (10)).
2. To compute $\breve{\Psi}^{m\mathbf{q}} = \left(\mathscr{F}_n \ldots \left(\mathscr{F}_2 \breve{\Psi}^m\right)^{(q_2)} \ldots\right)^{(q_n)}$ for $1 \le q_2 \le J_2 - 1, \ldots, 1 \le q_n \le J_n - 1$.
3. To compute $\tilde{\Psi}^{m\mathbf{q}}$ by solving the 1D problems (18)–(20) for $1 \le q_2 \le J_2 - 1, \ldots, 1 \le q_n \le J_n - 1$ (this includes the computation of the discrete convolution on the right-hand side of (20) so that the functions $\tilde{\Psi}_{J_1}^{1\mathbf{q}}, \ldots, \tilde{\Psi}_{J_1}^{m-1\,\mathbf{q}}$ have to be stored).
4. To compute $\tilde{\Psi}^m = \mathscr{F}_n^{-1} \ldots \mathscr{F}_2^{-1} \tilde{\Psi}^{m\mathbf{q}}$.
5. To compute $\Psi^m = \mathscr{E}^m \tilde{\Psi}^m$ on $\omega_h \cup \Gamma_{1h}$ (see (10)).

Let $J_2 = 2^{k_2}, \ldots, J_n = 2^{k_n}$, where $k_2, \ldots, k_n$ be natural numbers. Using the fast Fourier transform, it required $O\left(\left(J_1 \log_2 \left(J_2 \ldots J_n\right) + m\right) J_2 \ldots J_n\right)$ or $O((J_1 \log_2 (J_2 \ldots J_n) + M) J_2 \ldots J_n M)$ arithmetic operations for computing Ψ^m respectively at mth time level or at all time levels $m = 1, \ldots, M$. It straightforwardly allows for a parallel implementation.

The results for the discretizations A, B and C are presented in more detail respectively in [6,11,4] including successful results of 2D numerical experiments in the case of both rectangular and smooth potentials. They can be easily extended to the case of a problem like (1)–(4) in the infinite parallelepiped $\mathbb{R} \times \Pi_{\hat{1}}$ (the values $V_{\pm\infty}$ of V as $x_1 \to \pm\infty$ can be different).

The study was supported by The National Research University – Higher School of Economics' Academic Fund Program in 2014–2015, research grant No. 14-01-0014.

References

1. X. Antoine, A. Arnold, C. Besse et al., A review of transparent and artificial boundary conditions techniques for linear and nonlinear Schrödinger equations. Commun. Comput. Phys. **4**(4), 729–796 (2008)
2. A. Arnold, M. Ehrhardt, I. Sofronov, Discrete transparent boundary conditions for the Schrödinger equation: fast calculations, approximation and stability. Commun. Math. Sci. **1**(3), 501–556 (2003)
3. B. Ducomet, A. Zlotnik, On stability of the Crank-Nicolson scheme with approximate transparent boundary conditions for the Schrödinger equation. Part I. Commun. Math. Sci. **4**(4), 741–766 (2006)
4. B. Ducomet, A. Zlotnik, A. Romanova, On a splitting higher order scheme with discrete transparent boundary conditions for the Schrödinger equation in a semi-infinite parallelepiped. Appl. Math. Comput. (2015, in press)
5. B. Ducomet, A. Zlotnik, I. Zlotnik, On a family of finite-difference schemes with discrete transparent boundary conditions for a generalized 1D Schrödinger equation. Kinet. Relat. Models **2**(1), 151–179 (2009)
6. ———, The splitting in potential Crank-Nicolson scheme with discrete transparent boundary conditions for the Schrödinger equation on a semi-infinite strip. ESAIM: Math. Model. Numer. Anal. **48**(6), 1681–1699 (2014)
7. M. Ehrhardt, A. Arnold, Discrete transparent boundary conditions for the Schrödinger equation. Riv. Mat. Univ. Parma **6**, 57–108 (2001)
8. C. Lubich, *From Quantum to Classical Molecular Dynamics. Reduced Models and Numerical Analysis* (EMS, Zürich, 2008)
9. M. Schulte, A. Arnold, Discrete transparent boundary conditions for the Schrödinger equation, a compact higher order scheme. Kinet. Relat. Models **1**(1), 101–125 (2008)
10. A.A. Zlotnik, A.V. Lapukhina, Stability of a Numerov type finite-difference scheme with approximate transparent boundary conditions for the nonstationary Schrödinger equation on the half-axis. J. Math. Sci. **169**(1), 84–97 (2010)
11. A. Zlotnik, A. Romanova, On a Numerov-Crank-Nicolson-Strang scheme with discrete transparent boundary conditions for the Schrödinger equation on a semi-infinite strip. Appl. Numer. Math. (2015)
12. I.A. Zlotnik, Family of finite-difference schemes with approximate transparent boundary conditions for the generalized nonstationary Schrödinger equation in a semi-infinite strip. Comput. Math. Math. Phys. **51**(3), 355–376 (2011)

Part III
A Posteriori Error Estimation and Adaptive Methods

Estimates of Constants in Boundary-Mean Trace Inequalities and Applications to Error Analysis

Sergey Repin

Abstract We discuss Poincaré type inequalities for functions with zero mean values on the whole boundary of a Lipschitz domain or on a measurable part of the boundary. For some basic domains (rectangles, quadrilaterals, and right triangles) exact constants in these inequalities has been found in Nazarov and Repin (ArXiv Ser Math Anal, 2012, arXiv:1211.2224). We shortly discuss two examples, which show that the estimates can be helpful for quantitative analysis of PDEs. In the first example, we deduce estimates of modeling errors generated by simplification (coarsening) of a boundary value problem. The second example presents a new form of the functional type a posteriori estimate that provides fully guaranteed and computable bounds of approximation errors. Constants in Poincaré type inequalities enter these estimates.

1 Introduction

Let $\Omega \in \mathbb{R}^d$ be a bounded connected domain with Lipschitz continuous boundary $\partial\Omega$. By $\tilde{H}^1(\Omega)$ we denote a subspace of $H^1(\Omega)$ that consist of functions satisfying the condition $\{w\}_\Omega = 0$ (here and later on $\{w\}_S$ denotes the mean value of w on the set S). The classical Poincaré inequality [8, 9] reads

$$\|w\|_{2,\Omega} \leq C_P(\Omega)\|\nabla w\|_{2,\Omega} \qquad \forall w \in \tilde{H}^1(\Omega). \tag{1}$$

It is widely used in analysis of PDEs. The constant in (1) is equal to $\lambda^{-\frac{1}{2}}$, where λ is the smallest positive eigenvalue of the problem

$$\begin{aligned} -\Delta u &= \lambda u \quad \text{in} \quad \Omega, \\ \partial_{\mathbf{n}} u &= 0 \quad \text{on} \quad \partial\Omega. \end{aligned} \tag{2}$$

S. Repin (✉)
V. A. Steklov Institute of Mathematics, St. Petersburg, Russia
University of Jyväskylä, Jyväskylä, Finland
e-mail: repin@pdmi.ras.ru; serepin@jyu.fi

A. Abdulle et al. (eds.), *Numerical Mathematics and Advanced Applications ENUMATH 2013*, Lecture Notes in Computational Science and Engineering 103,
DOI 10.1007/978-3-319-10705-9_21

In Payne and Weinberger [6], it was shown that $C_P(\Omega) \leq \frac{\mathrm{diam}\Omega}{\pi}$ for any convex $\Omega \in \mathbb{R}^d$ (for $d > 2$ necessary corrections of the proof are presented in Bebendorf [1]). Numerous publications are concerned with various generalizations of (1) (e.g., in [3] sharp constants in some Sobolev type inequalities were found). These results were used in quantitative analysis of PDEs (in particular, in domain decomposition methods [2] and a posteriori error estimation [10]).

We consider similar estimates for functions in the space

$$\tilde{H}^1(\Omega, \Gamma) = \left\{ w \in H^1(\Omega) \mid \{w\}_\Gamma = 0 \right\},$$

where Γ coincides with $\partial\Omega$ or with a measurable part of it having positive $d-1$ measure. For any $w \in \tilde{H}^1(\Omega, \Gamma)$, we have the estimates

$$\|w\|_{2,\Omega} \leq C_1(\Omega, \Gamma) \|\nabla w\|_{2,\Omega}\,, \tag{3}$$

$$\|w\|_{2,\Gamma} \leq C_2(\Omega, \Gamma) \|\nabla w\|_{2,\Omega}\,. \tag{4}$$

Norm equivalent to the original norm Existence of positive constants $C_1(\Omega, \Gamma)$ and $C_2(\Omega, \Gamma)$ independent of w can be proved by standard compactness arguments. Our first goal is to find exact values of the constants $C_1(\Omega, \Gamma)$ and $C_1(\Omega, \Gamma)$ for rectangular domains and also for some classes of triangles. Having these constants for "basic" domains and using affine mappings it is not difficult to find suitable upper bounds of $C_1(\Omega, \Gamma)$ and $C_1(\Omega, \Gamma)$ for arbitrary simplexes and convex rectangles.

Estimates (3) and (4) can be used in various applications, in particular in analysis of discontinuous Galerkin, finite volume, and mortar type methods (considered or similar inequalities, with the emphasis on the appearing constants, have also been studied in [5, 7, 11] and some other papers cited therein). In this note, we shortly discuss possible applications of (3) and (4) to estimation of modeling and approximation errors in the context of functional type a posteriori estimates.

2 Reduction to Spectral Problems

Finding the constants leads to the problem of finding minimal positive eigenvalues of two spectral problems. It is not difficult to show that the extremal function in (3) is an eigenfunction $u \in \tilde{H}^1(\Omega, \Gamma)$ of the boundary value problem

$$-\Delta u = \lambda u \quad \text{in} \quad \Omega, \tag{5}$$

$$\partial_{\mathbf{n}} u = -\frac{\lambda}{|\Gamma|} \int_\Omega u\, dx \quad \text{on} \quad \Gamma, \tag{6}$$

$$\partial_{\mathbf{n}} u = 0 \quad \text{on} \quad \partial\Omega \setminus \Gamma, \tag{7}$$

which corresponds to the smallest $\lambda > 0$. Analogously, the extremal function in (4) satisfies the system

$$\Delta u = 0 \quad \text{in} \quad \Omega, \tag{8}$$

$$\partial_{\mathbf{n}} u = \lambda u \quad \text{on} \quad \Gamma, \tag{9}$$

$$\partial_{\mathbf{n}} u = 0 \quad \text{on} \quad \partial\Omega \setminus \Gamma. \tag{10}$$

For rectangles, and right triangles the exact values of minimal positive eigenvalues can bc found exactly [4]. Below, we shortly summarize these results. In some cases, the results are obtained fairly easily because the corresponding eigenfunctions are defined explicitly and form a complete system in the energy space of the operator. This is true for rectangles (polyhedrons) and functions with zero mean values on one edge (face) only.

Theorem 1 *If $d = 2$, $\Omega = (0, h_1) \times (0, h_2)$, and*

$$\Gamma = \{x_1 = 0,\ x_2 \in [0,\ h_2]\},$$

then

$$C_1 = \frac{1}{\pi} \max\{2h_1; h_2\} \quad \text{and} \quad C_2 = \left(\frac{\pi}{h_2} \tanh(\frac{\pi h_1}{h_2})\right)^{-\frac{1}{2}}.$$

If $d = 3$, $\Omega = (0, h_1) \times (0, h_2) \times (0, h_3)$, and

$$\Gamma = \{x_1 = 0, x_2 \in [0, h_2], x_3 \in [0, h_3]\},$$

then

$$C_1 = \frac{1}{\pi} \max\{2h_1; h_2; h_3\} \quad \text{and} \quad C_2 = \left(\frac{\pi}{\max\{h_2; h_3\}} \tanh(\frac{\pi h_1}{\max\{h_2; h_3\}})\right)^{-\frac{1}{2}}.$$

If $\Gamma = \partial\Omega$, then due to the biaxial symmetry all the eigenfunctions are either even or odd with respect to the axes x_1 and x_2. Analysis of them leads to the following result.

Theorem 2 *Let $d = 2$, $\Omega = (-\frac{h_1}{2}, \frac{h_1}{2}) \times (-\frac{h_2}{2}, \frac{h_2}{2})$, and $\Gamma = \partial\Omega$. Then, $C_1 = \frac{1}{\pi} \max\{h_1; h_2\}$ and*

$$C_2 = \left(\frac{2z_*(\alpha_0)}{\sqrt{h_1 h_2}} \tanh(\frac{z_*(\alpha_0)}{\alpha_0})\right)^{-\frac{1}{2}},$$

where $\alpha_0 = \sqrt{\frac{\max\{h_1;h_2\}}{\min\{h_1;h_2\}}}$ *and* $z_*(\alpha)$ *is a unique root of the equation*

$$\tanh\left(\frac{z}{\alpha}\right)\tan(z\alpha) = 1,$$

such that $z_*\alpha < \frac{\pi}{2}$.

Finding exact constants for isosceles right triangles requires rather sophisticated analysis (see [4]). The estimates use the constants $\zeta_1 \approx 2.02876$, $\zeta_2 \approx 2.3650$, and $\zeta_3 \approx 0.93755$. Here, ζ_1 is the unique root of the equation $z\cot(z) + 1 = 0$ in the interval $(0, \pi)$ and ζ_2 is the unique root of the equation $\tan(z) + \tanh(z) = 0$ in the interval $(0, \pi)$. The number ζ_3 is the unique root of the equation $\tanh(z)\tan(z) = 1$, in $(0, \frac{\pi}{2})$.

Theorem 3 *Let* Ω *be the isosceles right triangle. 1. If* Γ *coincide with one leg having the length* h*, then*

$$C_1 = \frac{1}{\zeta_1}h \quad \text{and} \quad C_2 = \frac{1}{\sigma_1}h^{1/2}, \quad \sigma_1 = (\zeta_2\tanh(\zeta_2))^{\frac{1}{2}}. \tag{11}$$

2. If Γ *is the union of two legs (each leg has the length* h*), then*

$$C_1 = \frac{h}{\pi} \qquad \text{and} \qquad C_2 = \frac{1}{\sigma_2}h^{1/2}, \quad \sigma_2 = (2\zeta_3\tanh(\zeta_3))^{\frac{1}{2}} \tag{12}$$

3. Let $\Omega = \{0 < |x_2| < x_1 < h\}$ *and* Γ *be the hypotenuse. Then*

$$C_1 = \frac{1}{2\zeta_1}h \qquad \text{and} \qquad C_2 = \frac{1}{\sqrt{2}}h^{1/2}. \tag{13}$$

Now, it is not difficult to find the constants associated with an arbitrary nondegenerate triangle T. First, we note that if $T \subset T'$ and T and T' have the same boundary Γ (where the mean trace vanishes), then $C_2(T,\Gamma) \geq C_2(T',\Gamma)$. Next, let $T = \operatorname{conv}\{(0,0),(h_1,0),(h_2\cos\alpha, h_2\sin\alpha)\}$ and $\Gamma = \{x_1 \in [0,h_1];\ x_2 = 0\}$. For this case, we obtain

$$\|v\|_T \leq C_1 h_1 \|\nabla v\|_T, \quad C_1 = \hat{C}_1\mu^{1/2}(\alpha,\rho), \quad \rho = \frac{h_2}{h_1}, \tag{14}$$

$$\|v\|_\Gamma \leq C_2 h_1^{1/2} \|\nabla v\|_T \quad C_2 = \hat{C}_2\left(\frac{\mu(\alpha,\rho)}{\rho\sin\alpha}\right)^{1/2}, \tag{15}$$

where $\mu(\alpha,\rho) = \frac{1}{2}\left(1 + \rho^2 + (1+\rho^4 + 2\cos(2\alpha)\rho^2)^{1/2}\right)$, and $\hat{C}_1$ and $\hat{C}_2$ are the corresponding constants for the basic right triangle.

3 Errors Generated by Simplification of a Model

Problems with highly oscillating terms and complicated boundary conditions are typical objects of numerical analysis. In this context, it is natural to adjust the initial mesh and accuracy of data representation to the desired tolerance level. For this purpose, we deduce an estimate, which suggests a way to select suitable simplifications of the source term and boundary conditions by means of cheap computations to be performed before solving the boundary value problem.

Consider the problem $\mathscr{P}$: Find u such that

$$\operatorname{div} p + f = 0 \quad \text{in } \Omega, \tag{16}$$

$$p = A\nabla u, \quad \text{in } \Omega, \tag{17}$$

$$u = u_0 \quad \text{on } \Gamma^D, \tag{18}$$

$$A\nabla u \cdot n = F \quad \text{on } \Gamma^N. \tag{19}$$

Let $f \in L^2(\Omega)$, $F \in L^2(\Gamma^N)$, $u_0 \in H^1(\Omega)$, and $A\xi \cdot \xi \geq c|\xi|^2$, where c is a positive constant independent of ξ. The corresponding generalized solution exists and is a unique in the set $V_0 + u_0$, where $V_0 = \overset{\circ}{H}{}^1(\Omega)$.

Assume that Ω is split into a finite set $\mathcal{O}$ of "simple" nonoverlapping subdomains Ω_i. Each Ω_i belongs to one of the following three subsets:

$$\mathcal{O}^D := \{\Omega_i \subset \Omega \mid \partial\Omega_i \cap \Gamma^D =: \Gamma_i^D \neq \emptyset\},$$
$$\mathcal{O}^N := \{\Omega_i \subset \Omega \mid \partial\Omega_i \cap \Gamma^N =: \Gamma_i^N \neq \emptyset\},$$
$$\mathcal{O}^I := \mathcal{O} \setminus (\mathcal{O}^D \cup \mathcal{O}^N).$$

Now, instead of $\mathscr{P}$ we consider a simplified (coarse) problem $\hat{\mathscr{P}}$:

$$\operatorname{div} \hat{p} + \hat{f} = 0 \quad \text{in } \Omega, \tag{20}$$

$$\hat{p} = A\nabla \hat{u}, \quad \text{in } \Omega, \tag{21}$$

$$\hat{u} = \hat{u}_0 \quad \text{on } \Gamma^D, \tag{22}$$

$$A\nabla \hat{u} \cdot n = \hat{F} \quad \text{on } \Gamma^N. \tag{23}$$

In this problem, the functions u_0, f, and F (which may be rather complicated, e.g., oscillating) are replaced by much simpler (e.g., piecewise constant or piecewise affine functions) $\hat{u}_0$, $\hat{f}$, and $\hat{F}$. It is only required that

$$\{u_0 - \hat{u}_0\}_{\Gamma_i^D} = 0 \quad \forall \Gamma_i^D \in \Gamma^D, \tag{24}$$

$$\left\{f - \hat{f}\right\}_{\Omega_i} = 0 \quad \forall\, \Omega_i \in \mathcal{O}^I \cup \mathcal{O}^N, \tag{25}$$

$$\left\{F - \hat{F}\right\}_{\Gamma_i^N} = 0 \quad \forall\, \Gamma_i^N \in \Gamma^N. \tag{26}$$

Our goal is to deduce an easily computable estimate of the corresponding modeling error

$$e_{\rm sim}^2 = \|\!|\, \nabla(u - \hat{u}) \,|\!\|_A^2 := \int_\Omega A\nabla(u - \hat{u}) \cdot \nabla(u - \hat{u})\, dx$$

generated by data simplification. This estimate can be represented in terms of the quantities

$$D_1^2 := \sum_{\Omega_i \in \mathcal{O}} \mathbb{C}_i^2 \|f - \hat{f}\|_{2,\Omega_i}^2,$$

$$D_2^2 := \sum_{\Omega_i \in \mathcal{O}^N} C_2(\Omega_i, \Gamma_i^N)^2 \|F - \hat{F}\|_{2,\Gamma_i^N}^2,$$

where

$$\mathbb{C}_i = \begin{cases} C_P(\Omega_i) & \text{if} \quad \Omega_i \in \mathcal{O}^I \cup \mathcal{O}^N, \\ C_1(\Omega_i, \Gamma_i^D) & \text{if} \quad \Omega_i \in \mathcal{O}^D. \end{cases}$$

Obviously, they are easily computable provided that the constants C_P, C_1, and C_2 associated with the corresponding subdomains are known.

Theorem 4 *Let (24)–(25) hold. Then,*

$$e_{\rm sim}^2 \leq \rho_1 + \sqrt{\rho_2 + \rho_1^2}, \tag{27}$$

where

$$2\rho_1 = \frac{D_1 + D_2}{\sqrt{c}} + \|\!|\, \phi \,|\!\|,$$

$$\rho_2 = \int_\Omega (f - \hat{f})\phi\, dx + \int_{\Gamma^N} (F - \hat{F})\, \phi\, ds,$$

and ϕ is an arbitrary function in $\Phi(\Omega) := \left\{H^1(\Omega), \quad \phi = u_0 - \hat{u}_0 \text{ on} \Gamma^D\right\}$. $H^1(\Omega)$ such that $\phi = u_0 - \hat{u}_0$ on Γ^D.

Remark 1 It is worth outlining that the right hand side of (27) is directly computable and finding ρ_1 and ρ_2 needs a simple integration only. If $\hat{u}_0 = u_0$, then this estimate can be significantly simplified. In this case one can choose $\phi \equiv 0$, and the estimate is reduced to $\|\!|\, u - \hat{u} \,|\!\| \leq \frac{D_1 + D_2}{\sqrt{c}}$.

4 A Posteriori Estimates of Approximation Errors

Let $v \in V_0 + u_0$ (where $V_0 := \{w \in H^1(\Omega) \mid w = 0 \text{ on } \Gamma_N\}$) be an approximation of u (cf. (16)–(19)). Then,

$$\int_\Omega A(\nabla u - \nabla v) \cdot \nabla w \, dx = \int_\Omega (fw - A\nabla v \cdot \nabla w) dx + \int_{\Gamma_N} Fw ds. \tag{28}$$

It is well known that this relation yields a guaranteed upper bound of $|||\, u - v \,|||$ if we transform the right hand side using a suitable vector valued function $y \in H(\Omega, \text{div})$ (see [10] and references therein). Below we show that the estimates (3) and (4) yield computable majorants of the error, which operate with y from a space wider than $H(\Omega, \text{div})$. This additional freedom can be used for a more efficient reconstruction of the dual variable based on local type procedures. As before, Ω is decomposed into a collection of nonoverlapping Lipschitz subdomains Ω_i , $\Gamma_{ij} := \overline{\Omega}_i \cap \overline{\Omega}_j$, and $\Gamma_i^N = \Omega_i \cap \Gamma_N$. The set of all interior edges (faces) is denoted by Γ_{int}. Define the space

$$\hat{H}(\Omega, \mathcal{O}_N, \text{div}) := \Big\{ y \in L^2(\Omega, \mathbb{R}^d) \mid y = y_i \in H(\Omega_i, \text{div}), \text{ in } \Omega_i,$$
$$\int_{\Gamma_{ij}} (y_i - y_j) \cdot n_{ij} \, ds = 0 \text{ if } \Gamma_{ij} \in \Gamma_{int},$$
$$\int_{\Gamma_i^N} (y_i \cdot n_i - F) ds = 0, \quad \{\text{div} y_i + f\}_{\Omega_i} = 0 \quad \forall i = 1, 2, \ldots, N \Big\}$$

In general, functions from $\hat{H}(\Omega, \mathcal{O}_N, \text{div})$ do not belong to $H(\Omega, \text{div})$ because the pointwise continuity of the normal flux is replaced by a weak (integral) continuity.

For all $w \in V_0$ and $y \in \hat{H}(\Omega, \mathcal{O}_N, \text{div})$ the following identity holds:

$$\sum_{i=1}^N \int_{\Omega_i} (y \cdot \nabla w + (\text{div}\, y) w) \, dx =$$
$$\sum_{\Gamma_{ij} \subset \Gamma_{int}} \int_{\Gamma_{ij}} (y_i - y_j) \cdot n_{ij} w ds + \sum_{\Gamma_{Ni} \subset \Gamma_N} \int_{\Gamma_{Ni}} (y_i \cdot n_i - F) w ds.$$

By adding this identity to the right hand side of (28), assembling the terms, and using Hölder estimates, we find that

$$|||\, \nabla(u - v) \,|||_A^2 \le \sum_{i=1}^N \| f + \text{div}\, y_i \|_{\Omega_i} \| w - \{w\}_{\Omega_i} \|_{\Omega_i} + |||\, y - A\nabla v \,|||_{A^{-1}} |||\, \nabla w \,|||_A +$$
$$\sum_{\Gamma_{ij} \subset \Gamma_{int}} \gamma_{ij} \| w - \{w\}_{\Gamma_{ij}} \|_{\Gamma_{ij}} + \sum_{\Gamma_i^N \subset \Gamma^N} \gamma_{i\,0} \| w - \{w\}_{\Gamma_i^N} \|_{\Gamma_i^N},$$

where

$$\gamma_{i\,0} = \|y_i \cdot n_i - F\|_{\Gamma_i^N} \quad \text{and} \quad \gamma_{ij} = \|(y_i - y_j) \cdot n_{ij}\|_{\Gamma_{ij}}.$$

Assume that Ω_i is a polygonal domain, which can be represented as a union of non overlapping simplexes T_{ik}, $k = 1, 2, \ldots, M_i$ and one face of each simplex belongs to Γ_{int} or Γ^N. Then, the last two terms can be estimated with the help of (4). Let C_{2i} be the maximal constant in (4) for all T_{is} forming Ω_i. The quantity $\theta_i^2 := \sum_{k=0}^{M_i} \xi_{ik}^2$, where $\xi_{ik} = \frac{1}{2}\gamma_{ik}$ if $\gamma_{ik} \in \Gamma_{int}$ (the factor $\frac{1}{2}$ is used because the quantity is related to two neighboring subdomains) and $\xi_{ik} = \gamma_{i0}$ if $\gamma_{i0} \in \Gamma^N$ characterizes discontinuity jumps and inconveniences in the Neumann boundary condition associated with Ω_i. Then, the boundary terms related to Ω_i are estimated by the quantity $C_{2i}\theta_i \|\nabla w\|_{\Omega_i}$ and the overall sum of two last terms is estimated by the quantity

$$\Theta(y, \mathcal{O}_N)\|\nabla w\|_\Omega \quad \text{where} \quad \Theta^2(y, \mathcal{O}_N) = \sum_{i=1}^N C_{2i}^2 \theta_i^2$$

Since

$$\sum_{i=1}^N \int_{\Omega_i} (f + \operatorname{div} y_i) w\, dx \le \left(\sum_{i=1}^N C_{Pi}^2 \|f + \operatorname{div} y_i\|_{\Omega_i} \right)^{1/2} \|\nabla w\|_\Omega,$$

we find that

$$\| \nabla(u - v) \|_A \le \| y - A\nabla v \|_{A^{-1}} + \frac{1}{c_1}\left(\Theta(y, \mathcal{O}_N) + \left(\sum_{i=1}^N C_{Pi}^2 \|f + \operatorname{div} y_i\|_{\Omega_i} \right)\right),$$

where the term $\Theta(y, \mathcal{O}_N)$ controls violations of conformity related to $y \cdot n$.

References

1. M. Bebendorf, A note on the Poincaré inequality for convex domains. Z. Anal. Anwend. **22**(4), 751–756 (2003)
2. C.R. Dohrmann, A. Klawonn, O.B. Widlund, Domain decomposition for less regular subdomains: overlapping Schwarz in two dimensions. SIAM J. Numer. Anal. **46**(4), 2153–2168 (2008)
3. V.G. Maz'ja, Classes of domains and imbedding theorems for function spaces. Sov. Math. Dokl. **1**, 882–885 (1960)
4. A. Nazarov, S. Repin, Exact constants in Poincaré type inequalities for functions with zero mean boundary traces. ArXiv Ser. Math. Anal. (2012). arXiv:1211.2224
5. S. Nicaise, A posteriori error estimations of some cell-centered finite volume methods. SIAM J. Numer. Anal. **43**(4), 1481–1503 (2005)

6. L.E. Payne, H.F. Weinberger, An optimal Poincaré inequality for convex domains. Arch. Ration. Mech. Anal. **5**, 286–292 (1960)
7. G.V. Pencheva, M. Vohralik, M.F. Wheeler, T. Wildey, Robust a posteriori error control and adaptivity for multiscale, multinumerics, and mortar coupling. SIAM J. Numer. Anal. **51**(1), 526–554 (2013)
8. H. Poincaré, Sur les équations aux derivées partielles de la physique mathematique. Am. J. Math. **12**, 211–294 (1890)
9. H. Poincaré, Sur les équations de la physique mathematique. Rend. Circ. Mat. Palermo, **8**, 57–156 (1894)
10. S. Repin, *A Posteriori Error Estimates for Partial Differential Equations* (Walter de Gruyter, Berlin, 2008)
11. M.F. Wheeler, I. Yotov, A posteriori error estimates for the mortar mixed finite element method. SIAM J. Numer. Anal. **43**(3), 1021–1042 (2005)

Reliable a Posteriori Error Estimation for Plane Problems in Cosserat Elasticity

Maxim Frolov

Abstract Functional type a posteriori error estimates are proposed for approximate solutions to plane problems arising in the Cosserat theory of elasticity. Estimates are reliable under quite general assumptions and are explicitly applicable not only to approximations possessing the Galerkin orthogonality property. For numerical justification of the approach, the lowest order Arnold-Boffi-Falk approximation is implemented.

1 Introduction

Cosserat continuum [3] is one of interesting generalizations of the classical elasticity theory. Renewed interest to such type of theories appears in the 60s of the twentieth century and is related to names of many famous experts in the field of continuum mechanics: Aero, Eringen, Mindlin, Nowacki, Palmov, Truesdell, and others. At present time, there is a significant amount of publications related to various physical, experimental, mathematical and computational aspects of the theory. More or less complete review of the subject can be found in the literature, for instance, in [1, 4, 6, 7, 9, 16–18] and other papers and books cited therein. On the other hand, there exists an essential lack of publications on the construction of guaranteed a posteriori error estimates for control of accuracy of computed approximations.

The functional approach, which is used here, was formed in the end of 1990s (see [13]) and has been developed for a wide spectrum of boundary-value problems. Theoretical as well as some practical aspects of its implementation are described in monographs [8, 11, 14] and some papers cited therein.

In the present work, the previous result of Repin and Frolov [15] is generalized. Using the duality theory in the Calculus of Variations for one class of plane problems in Cosserat elasticity, functional type a posteriori error estimates have been proposed. In [15], the case of Dirichlet type boundary conditions has been

---

M. Frolov (✉)
Department of Applied Mathematics, St. Petersburg State Polytechnical University, 29 Politekhnicheskaya st., 195251 St. Petersburg, Russia
e-mail: frolov_me@spbstu.ru

A. Abdulle et al. (eds.), *Numerical Mathematics and Advanced Applications ENUMATH 2013*, Lecture Notes in Computational Science and Engineering 103,
DOI 10.1007/978-3-319-10705-9_22

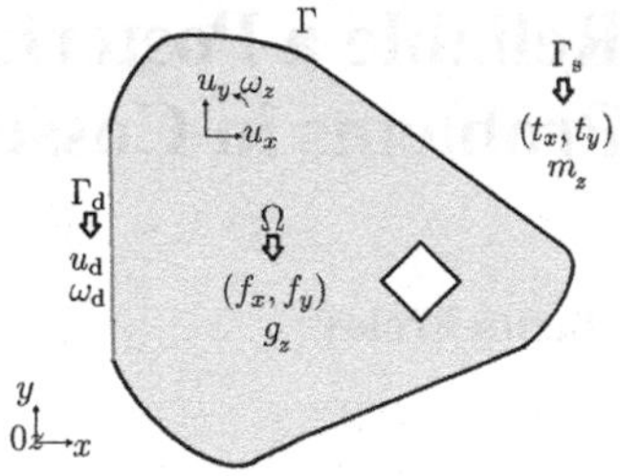

Fig. 1 Main fields, volume and surface loads, and boundary conditions for plane problems in Cosserat elasticity

considered. In the present investigation, the author discusses the case of general boundary conditions including not only given displacements and rotation, but also traction and moment over the out-of-plane axis.

The classical statement in linear elasticity under the plane strain assumption is as follows: find displacements u, stress tensor σ and strain tensor ε that are related by the system of equations

$$\begin{cases} \operatorname{Div} \sigma + f = 0 & \text{in } \Omega, \\ \sigma = L\varepsilon & \text{in } \Omega, \\ \varepsilon(u) = \frac{1}{2}(\nabla u + (\nabla u)^T) & \text{in } \Omega, \end{cases}$$

where domain $\Omega \subset \mathbb{R}^2$ is a bounded connected domain with Lipschitz-continuous boundary, f—volume loads (density), L—the tensor of elastic moduli. This system is completed by boundary conditions in terms of displacements $u = u_{\mathrm{d}}$ on Γ_{d} and in terms of stresses $\sigma n = t$ on Γ_{s}, where Γ_{d} and Γ_{s} are two non-intersecting parts of the boundary Γ, u_{d} and t—given displacements and surface loads, n—the outward normal to the surface of a body.

For Cosserat continuum it is necessary to take into account the microrotation and the couple-stress tensor additionally to the original unknown fields (see Fig. 1). Microrotation ω_z in Ω and ω_{d} on Γ_{d}, volume loads f_x, f_y, g_z, and surface loads t_x, t_y, m_z are depicted in. It is important that, in general case, stresses and strains have no symmetry.

In the rest of the paper, we assume

$$f_x,\ f_y,\ g_z \in \mathbb{L}_2(\Omega);\ \ t_x,\ t_y,\ m_z \in \mathbb{L}_2(\Gamma_{\mathrm{s}});\ \ u_{\mathrm{d}} \in \mathbb{H}^{1/2}(\Gamma_{\mathrm{d}}, \mathbb{R}^2),\ \omega_{\mathrm{d}} \in \mathbb{H}^{1/2}(\Gamma_{\mathrm{d}});$$

$$L_{ijks} = L_{ksij} = L_{jiks}, \quad i, j, k, s = 1, 2;$$

$$\alpha_1 \int\limits_{\Omega} |\tau|^2 \, d\Omega \le \int\limits_{\Omega} L\tau : \tau \, d\Omega \le \alpha_2 \int\limits_{\Omega} |\tau|^2 \, d\Omega;$$

$$\forall \tau \in \mathbb{L}_2(\Omega, \mathbb{M}^{2\times 2}_{sym}), \quad 0 < \alpha_1 \le \alpha_2.$$

Hereafter, standard Lebesgue and Sobolev spaces are denoted by $\mathbb{L}_2$, $\mathbb{H}^{1/2}$, and $\mathbb{H}^1$, respectively.

2 Functional Type a Posteriori Error Estimates

The energy functional for the classical elasticity theory has the following form:

$$J_e(u) = \int_{\Omega} \underbrace{\frac{1}{2} L\varepsilon(u) : \varepsilon(u)}_{\mu\left(u_{x,x}^2 + u_{y,y}^2 + \frac{1}{2}(u_{y,x}+u_{x,y})^2\right) + \frac{\lambda}{2}(u_{x,x}+u_{y,y})^2} d\Omega - \int_{\Omega} f \cdot u \, d\Omega - \int_{\Gamma_s} t \cdot u \, d\Gamma$$

where, for the case of isotropic and homogeneous material (at the macroscopic level), the integrand of the first term depends only on two material constants—Lamé parameters μ and λ. In the case of Cosserat continuum, the energy functional includes additional terms representing both body and surface loads from additional set of variables related to the rotation and moments, namely

$$J(u_x, u_y, \omega_z) = J_e(u) - \int_{\Omega} g_z \omega_z \, d\Omega - \int_{\Gamma_s} m_z \omega_z \, d\Gamma +$$

$$+ \int_{\Omega} \left(\frac{\mu_c}{2} \left(u_{y,x} - u_{x,y} - 2\omega_z\right)^2 + 2B\left(\omega_{z,x}^2 + \omega_{z,y}^2\right)\right) d\Omega$$

where μ_c and B are constants specifying properties of a microstructure. For minimization of the energy functional, it is necessary to choose a proper pair of finite element spaces

$$\inf_{\Upsilon \times \Theta} J(u_x, u_y, \omega_z)$$

$$\Upsilon := \Upsilon^0 + u_{\rm d}, \qquad \Upsilon^0 := \left\{ v^0 \in \mathbb{H}^1(\Omega, \mathbb{R}^2) \mid v^0 = 0 \text{ on } \Gamma_{\rm d} \right\};$$

$$\Theta := \Theta^0 + \omega_{\rm d}, \qquad \Theta^0 := \left\{ \theta^0 \in \mathbb{H}^1(\Omega) \mid \theta^0 = 0 \text{ on } \Gamma_{\rm d} \right\}.$$

Let us assume that some *arbitrary conforming* approximation $(\tilde{u}_x, \tilde{u}_y)$ and $\tilde{\omega}_z$ of the exact solution from $\Upsilon \times \Theta$ is computed. Then, one can introduce the error (deviation from the exact solution) $\xi := (\xi_x, \xi_y, \xi_z)$, where

$$\xi_x := u_x - \tilde{u}_x, \qquad \xi_y := u_y - \tilde{u}_y, \qquad \xi_z := \omega_z - \tilde{\omega}_z.$$

This paper is focused on reliable functional type a posteriori error estimates for the energy norm of the error

$$\|\!|\xi|\!\|^2 := \int\limits_{\Omega} \left(\mu\xi_{x,x}^2 + \frac{\mu}{2}(\xi_{y,x} + \xi_{x,y})^2 + \mu\xi_{y,y}^2 + \frac{\lambda}{2}(\xi_{x,x} + \xi_{y,y})^2 \right) d\Omega +$$

$$+ \int\limits_{\Omega} \left(\frac{\mu_c}{2}(\xi_{y,x} - \xi_{x,y} - 2\xi_z)^2 + 2B|\nabla\xi_z|^2 \right) d\Omega.$$

It is quite simple to show that for any approximate solution, the following relation holds:

$$\begin{aligned} \|\!|\xi|\!\|^2 &= J(\tilde{u}_x, \tilde{u}_y, \tilde{\omega}_z) - J(u_x, u_y, \omega_z) \geq \\ &\geq J(\tilde{u}_x, \tilde{u}_y, \tilde{\omega}_z) - J(u_x^{ref}, u_y^{ref}, \omega_z^{ref}), \end{aligned}$$

where some *reference solution* is taken instead of the exact one.

To obtain error estimates, the relations between displacements, generalized strains γ and nonsymmetric stress tensor σ are considered, namely

$$\gamma_{xx} = u_{x,x}, \quad \gamma_{yy} = u_{y,y}, \quad \gamma_{xy} = u_{y,x} - \omega_z, \quad \gamma_{yx} = u_{x,y} + \omega_z,$$

$$\sigma = (\mu + \mu_c)\gamma + (\mu - \mu_c)\gamma^T + \lambda \operatorname{tr}\gamma\mathbb{I},$$

where $\mathbb{I}$ is the second-rank identity tensor (see, for example, [10,12]). The respective fields, computed from an approximate solution, are also marked by $\tilde{}$

$$\tilde{\sigma}_{xx} = (2\mu + \lambda)\tilde{u}_{x,x} + \lambda\tilde{u}_{y,y}, \quad \tilde{\sigma}_{xy} = \mu(\tilde{u}_{x,y} + \tilde{u}_{y,x}) + \tilde{p},$$

$$\tilde{p} = \mu_c(\tilde{u}_{y,x} - \tilde{u}_{x,y} - 2\tilde{\omega}_z), \ldots \tilde{M} := 4B\nabla\tilde{\omega}_z.$$

Further, as a standard trick of the functional approach, a set of free variables is introduced. It is the set of three variables

$$\tilde{\tau} := (\tilde{\tau}_1, \ \tilde{\tau}_2), \ \ \tilde{s},$$

where

$$\tilde{\tau}_1 = \begin{pmatrix} \tilde{\tau}_{xx} \\ \tilde{\tau}_{yx} \end{pmatrix}, \quad \tilde{\tau}_2 = \begin{pmatrix} \tilde{\tau}_{xy} \\ \tilde{\tau}_{yy} \end{pmatrix}, \quad \tilde{s} = \begin{pmatrix} \tilde{s}_x \\ \tilde{s}_y \end{pmatrix} \quad \in H_{\text{div}}(\Omega, \Gamma_s)$$

with

$$H_{\text{div}}(\Omega, \Gamma_s) := \left\{ \eta \in \mathbb{L}_2(\Omega, \mathbb{R}^2) \mid \operatorname{div}\eta \in \mathbb{L}_2(\Omega), \eta \cdot n \in \mathbb{L}_2(\Gamma_s) \right\}.$$

All of them have clear physical meaning—$\tilde{\tau}$ represents the true nonsymmetric stress tensor and $\tilde{s}$ is an independent approximation of the non-zero components of the couple-stress tensor.

The following estimate is used as an auxiliary one:

$$\frac{1}{C^2} \leq \inf_{(v_x^0, v_y^0, \theta_z^0) \in \Upsilon^0 \times \Theta^0} \frac{\|(v_x^0, v_y^0, \theta_z^0)\|^2}{\Delta^2(v_x^0, v_y^0, \theta_z^0)},$$

where

$$\Delta^2(\xi_x, \xi_y, \xi_z) := \|c_x^{1/2}\xi_x\|_\Omega^2 + \|c_y^{1/2}\xi_y\|_\Omega^2 + \|c_z^{1/2}\xi_z\|_\Omega^2 + \\ + \|d_x^{1/2}\xi_x\|_{\Gamma_s}^2 + \|d_y^{1/2}\xi_y\|_{\Gamma_s}^2 + \|d_z^{1/2}\xi_z\|_{\Gamma_s}^2$$

with six arbitrary constants of properly selected physical dimensions. For instance, the set of constants can be chosen as follows:

$$c_x = c_y = \frac{\mu}{|\Omega|}, \quad c_z = \frac{B}{|\Omega|}, \quad d_x = c_x|\Gamma_s|, \quad d_y = c_y|\Gamma_s|, \quad d_z = c_z|\Gamma_s|.$$

Then, a set of guaranteed functional type a posteriori error estimates for plane problems in Cosserat elasticity has the form

$$\|\xi\|^2 \leq (1+\beta)D^2(\tilde{\tau}_1, \tilde{\tau}_2, \tilde{s}) + (1+\beta^{-1})C^2 R^2(\tilde{\tau}_1, \tilde{\tau}_2, \tilde{s}), \quad \forall \beta > 0; \tag{1}$$

$$D^2 := \int_\Omega \frac{1}{2} L^{-1}(\tilde{\tau} - \tilde{\sigma})^{sym} : (\tilde{\tau} - \tilde{\sigma})^{sym} d\Omega + \\ + \int_\Omega \frac{1}{2\mu_c}\left(\frac{\tilde{\tau}_{xy} - \tilde{\tau}_{yx}}{2} - \tilde{p}\right)^2 d\Omega + \int_\Omega \frac{1}{8B} \mid \tilde{s} - \tilde{M} \mid^2 d\Omega;$$

$$R^2(\tilde{\tau}_1, \tilde{\tau}_2, \tilde{s}) = \frac{1}{4}\Big(\|c_x^{-1/2}(\tilde{\tau}_{xx,x} + \tilde{\tau}_{yx,y} + f_x)\|_\Omega^2 + \\ + \|c_y^{-1/2}(\tilde{\tau}_{xy,x} + \tilde{\tau}_{yy,y} + f_y)\|_\Omega^2 + \|c_z^{-1/2}(\operatorname{div}\tilde{s} + \tilde{\tau}_{xy} - \tilde{\tau}_{yx} + g_z)\|_\Omega^2 + \\ + \|d_x^{-1/2}(t_x - \tilde{\tau}_{xx}n_x - \tilde{\tau}_{yx}n_y)\|_{\Gamma_s}^2 + \|d_y^{-1/2}(t_y - \tilde{\tau}_{xy}n_x - \tilde{\tau}_{yy}n_y)\|_{\Gamma_s}^2 + \\ + \|d_z^{-1/2}(m_z - \tilde{s}\cdot n)\|_{\Gamma_s}^2\Big).$$

This set admits various selections of approximations and parameters, and, for the case $\Gamma = \Gamma_d$, includes as a particular case the result that have been obtained earlier in [15].

3 Numerical Results

To implement the error majorant (1) we use suitable mixed finite element proposed by Arnold, Boffi and Falk [2]. The respective space for it has the form $\mathscr{ABF}_0(\hat{\mathscr{K}}) = \mathscr{P}_{2,0}(\hat{\mathscr{K}})\times\mathscr{P}_{0,2}(\hat{\mathscr{K}})$, where $\mathscr{P}_{i,j}(\hat{\mathscr{K}})$—the space of polynomials over $\hat{\mathscr{K}}$ of power i or less on $\hat{x}_1$ and j—on $\hat{x}_2$ for the reference square $\hat{\mathscr{K}} = (-1,1)\times(-1,1)$, where $\hat{x}_1$ and $\hat{x}_2$ are local coordinates of the reference element.

Example 1 (Square domain with a small hole) Let consider a square domain with a hole (see Fig. 2). For this example, the radius of the hole is comparable with the

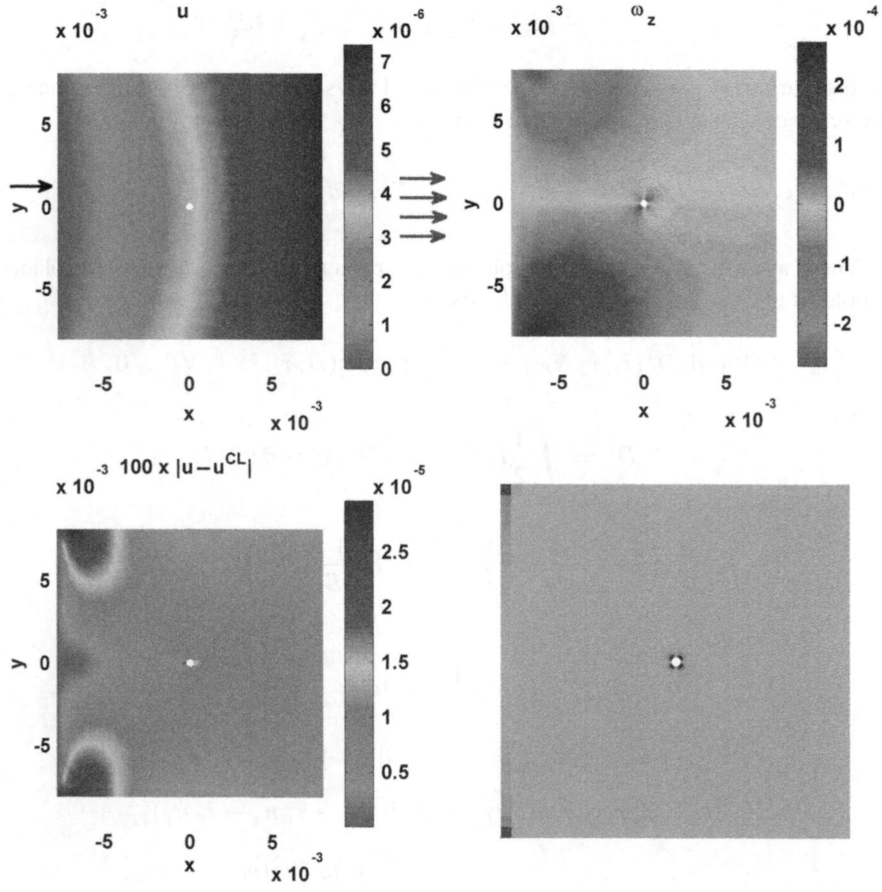

Fig. 2 Geometry and material properties [18]: side 16.2 mm, radius 0.216 mm, $\lambda = 0.11538e10\,\mathrm{N/m^2}$, $\mu = 0.76923e9\,\mathrm{N/m^2}$, $B = 0.31762e2\,\mathrm{N}$, $\mu_c = 0.25638e11\,\mathrm{N/m^2}$, loading 1 MPa, size of particles 0.2 mm; Results (left to right and top to bottom): total displacement and rotation, deviation from the classical solution of ANSYS marked by **CL** superscript, and regions of the stress tensor asymmetry (all results are depicted for Mesh 2)

Table 1 Results for ABF0 approximation in Example 1

MESH	0	1	2	3	REF.[a]
D.O.F.	504	1,872	7,200	28,224	111,744
ERROR (%)	15.8	11.1	7.3	4.0	
I_{eff}	1.23	1.20	1.21	1.34	
β	0.158	0.122	0.087	0.056	

[a]Mesh for computation of the reference solution

Table 2 Results for ABF0 approximation in Example 2

MESH	0	1	2	REF.
D.O.F.	140	747	2,811	198,147
ERROR (%)	28.1	15.2	8.6	Deviation from
I_{eff}	6.9	11.1	13.6	ANSYS
β	1.223	1.396	1.862	0.06 %

size of particles of microstructure. The left edge of the square is fixed and a tensile loading is applied to the right edge. All parameters of the problem are described in the caption of the figure.

Results of error estimation for several steps with uniform mesh refinement are collected in Table 1. The efficiency index as the ratio between the error majorant and the error is used as the main quality measure. From these results we conclude that the efficiency index of estimates remains stable and the approach is reliable—it provides guaranteed upper bounds of the error in the energy norm. Parameter β tends to zero. Practical experience shows that such a behaviour of the parameter is necessary for getting satisfactory results. This example summarizes the group of examples with a significant influence of a microstructure, which produces visible difference between the Cosserat solution and the classical elastic solution (up to 15 %).

Example 2 (About current limitations of the approach) As the second example, we consider some Cosserat continuum in the unit square without holes. Material parameters are selected to get the reference solution with negligible difference from the classical elastic solution of ANSYS (less than 0.1 %). From the results collected in Table 2, one can conclude that, in this case, the error estimate (1) yields worse behaviour of the efficiency index. Reasonable alternative choice can be provided by the functional type a posteriori error estimate for linear elasticity from [14]. The results presented in [5] show that usage of ABF0 approximation for it ensures quite effective approach to compute error estimates.

Finally, we summarize the main conclusions

- Arnold-Boffi-Falk [2] approximation yields good results;
- The proposed method is reliable in any case—estimates are guaranteed (upper) bounds. This statement follows from the theoretical origins of the functional approach. By numerical evidence, it is efficient for a wide spectrum of

problems—efficiency indices are stable, excepting some problems with a negligible microstructure influence. Nevertheless, theoretical justification of this fact is an interesting and important open problem for further investigations.

References

1. J. Altenbach, H. Altenbach, V.A. Eremeyev, On generalized Cosserat-type theories of plates and shells: a short review and bibliography. Arch. Appl. Mech. **80**(1), 73–92 (2010)
2. D.N. Arnold, D. Boffi, R.S. Falk, Quadrilateral H(div) finite elements. SIAM J. Numer. Anal. **42**(6), 2429–2451 (2005)
3. E. Cosserat, F. Cosserat, *Theorie des corps deformables* (A. Hermann et Fils. VI, Paris, 1909)
4. V.A. Eremeyev, L.P. Lebedev, H. Altenbach, *Foundations of Micropolar Mechanics*. Springer-Briefs in Applied Sciences and Technology, Continuum Mechanics (Springer, Berlin, 2013)
5. M.E. Frolov, Application of functional error estimates with mixed approximations to plane problems of linear elasticity. Comput. Math. Math. Phys. **53**(7), 1000–1012 (2013)
6. A.R. Hadjesfandiari, G.F. Dargush, Boundary element formulation for plane problems in couple stress elasticity. Int. J. Numer. Methods Eng. **89**(5), 618–636 (2012)
7. J. Jeong, P. Neff, Existence, uniqueness and stability in linear Cosserat elasticity for weakest curvature conditions. Math. Mech. Solids **15**(1), 78–95 (2010)
8. O. Mali, P. Neittaanmäki, S. Repin, *Accuracy Verification Methods. Theory and Algorithms*. Computational Methods in Applied Sciences, vol. 32 (Springer, Dordrecht, 2014)
9. G.A. Maugin, A historical perspective of generalized continuum mechanics, in *Mechanics of Generalized Continua*. Advanced Structured Materials, vol. 7 (Springer, Berlin/Heidelberg, 2011), pp. 3–19
10. N.F. Morozov, *Mathematical Issues in the Theory of Cracks* [in Russian] (Nauka, Moscow, 1984)
11. P. Neittaanmäki, S.I. Repin, *Reliable Methods for Computer Simulation—Error Control and a Posteriori Estimates* (Elsevier, Amsterdam, 2004)
12. M. Ostoja-Starzewski, I. Jasiuk, Stress invariance in planar Cosserat elasticity. Proc. R. Soc. Lond. Ser. A **451**(1942), 453–470 (1995)
13. S.I. Repin, A posteriori error estimation for variational problems with uniformly convex functionals. Math. Comput. **69**(230), 481–500 (2000)
14. ———, *A Posteriori Estimates for Partial Differential Equations* (Walter de Gruyter, Berlin/New York, 2008)
15. S. Repin, M. Frolov, Estimates for deviations from exact solutions to plane problems in the Cosserat theory of elasticity. Probl. Math. Anal. **62**, 153–161 (2011). Trans. J. Math. Sci. **181**(2), 281–291 (2012)
16. O. Sadovskaya, V. Sadovskii, *Mathematical Modeling in Mechanics of Granular Materials*. Advanced Structured Materials, vol. 21 (Springer, Berlin, 2012)
17. M.A. Wheel, A control volume-based finite element method for plane micropolar elasticity. Int. J. Numer. Methods Eng. **75**(8), 992–1006 (2008)
18. H.W. Zhang, H. Wang, P. Wriggers, B.A. Schrefler, A finite element model for contact analysis of multiple Cosserat bodies. Comput. Mech. **36**(6), 444–458 (2005)

Residual Based Error Estimate for Higher Order Trefftz-Like Trial Functions on Adaptively Refined Polygonal Meshes

Steffen Weißer

Abstract The BEM-based Finite Element Method is one of the promising strategies which are applicable for the approximation of boundary value problems on general polygonal and polyhedral meshes. The flexibility with respect to meshes arises from the implicit definition of trial functions in a Trefftz-like manner. These functions are treated locally by means of Boundary Element Methods (BEM). The following presentation deals with the formulation of higher order trial functions and their application in uniform and adaptive mesh refinement strategies. For the adaptive refinement, a residual based error estimate is used on general polygonal meshes for the higher order, conforming trial functions. The first numerical results, in the context of adaptive refinement, show optimal rates of convergence with respect to the number of degrees of freedom.

1 Introduction

Polygonal and polyhedral meshes attract more and more attention in the discretization of boundary value problems. They appear naturally in geological and biological science and have advantageous properties in several situations.

The BEM-based Finite Element Method is applicable on such general polygonal and polyhedral meshes, and it admits conforming approximation spaces due to the implicit definition of trial functions. These functions are defined in the spirit of Trefftz to fulfil the underlying differential equation locally, and thus they inherit some properties of the unknown solution. In the case of a diffusion problem, the lowest order trial functions coincide with harmonic coordinates studied and applied in computer graphics, see [13, 14]. These functions belong to the class of generalized barycentric coordinates, which are additionally applied in linear elasticity for example, see [20].

S. Weißer (✉)
Universität des Saarlandes, Campus building E1.1, 66041 Saarbrücken, Germany
e-mail: weisser@num.uni-sb.de

A. Abdulle et al. (eds.), *Numerical Mathematics and Advanced Applications ENUMATH 2013*, Lecture Notes in Computational Science and Engineering 103,
DOI 10.1007/978-3-319-10705-9_23

Beside the Finite Element Method there are also other strategies for the numerical approximation of boundary value problems. In this context of polygonal meshes and special trial functions, the Trefftz-discontinuous Galerkin methods [8] have to be mentioned. For conforming approximations on such meshes, there are recent developments in mimetic discretization techniques [3] and within the new methodology of Virtual Element Methods [1,2].

The BEM-based FEM was first introduced in 2009, see [4], and analysed in the following years, see [5, 9, 11, 16]. Since then it has undergone several developments. This includes residual based error estimates for adaptive mesh refinement [21], the application for convection–diffusion problems [12], higher order approximations [18, 23] and mixed formulations with $H(\mathrm{div})$-conforming discretization [7] as well as improved generalizations to three dimensional problems with polyhedral elements [19]. The main results have been gathered in two doctoral theses [10, 22].

The challenge in this presentation is to combine the results from [21] for the adaptive Finite Element Method with the higher order trial functions defined in [18] and extended in [23]. Thus, we aim to give the first results on a higher order adaptive scheme for the BEM-based Finite Element Method on polygonal meshes.

The outline of the paper is as follows. In Sect. 2, the BEM-based FEM with higher order trial functions is formulated for the model problem. The residual based error estimate and the adaptive refinement strategy is given in Sect. 3. And finally, a numerical example is presented in Sect. 4.

2 BEM-Based FEM for the Model Problem

For the discussion in this presentation, we restrict ourselves to a model problem. Let $\Omega \subset \mathbb{R}^2$ be a bounded polygonal domain with boundary $\Gamma = \Gamma_D \cup \Gamma_N$ and $|\Gamma_D| > 0$. For $g_D \in H^{1/2}(\Gamma_D)$, $g_N \in L_2(\Gamma_N)$ and $\alpha \in L_\infty(\Omega)$ with $0 < \alpha_{\min} \leq \alpha \leq \alpha_{\max}$ almost everywhere in Ω, we consider the boundary value problem

$$-\operatorname{div}(\alpha \nabla u) = 0 \quad \text{in } \Omega, \qquad u = g_D \quad \text{on } \Gamma_D, \qquad \frac{\partial u}{\partial n_\Omega} = g_N \quad \text{on } \Gamma_N, \tag{1}$$

where n_Ω denotes the outer unite normal vector to the boundary of Ω. The well known variational formulation reads

$$\text{Seek } u \in g_D + V_D : \quad a(u,v) = (g_N, v)_{\Gamma_N} \quad \forall v \in V_D. \tag{2}$$

Here, $(\cdot,\cdot)_{\Gamma_N}$ denotes the L_2-scalar products over Γ_N and the bilinear form is given by

$$a(u,v) = \int_\Omega \alpha \nabla u \cdot \nabla v.$$

The trial space is given as $V_D = H^1_D(\Omega) = \{v \in H^1(\Omega) : v|_{\Gamma_D} = 0\}$ and $g_D + V_D$ denotes the affine space, where the same symbol g_D is used for the Dirichlet trace and its extension into $V = H^1(\Omega)$. We utilize the standard notation for Sobolev spaces $H^k(\Omega)$, $k \in \mathbb{N}_0$ and their norms $\|\cdot\|_{H^k(\Omega)}$ as well as semi-norms $|\cdot|_{H^k(\Omega)}$.

For the numerical treatment of (2), the domain Ω has to be discretized. Therefore, we allow the decomposition of Ω into open and convex polygonal elements $K \in \mathscr{K}_h$ such that

$$\overline{\Omega} = \bigcup_{K \in \mathscr{K}_h} \overline{K} \quad \text{and} \quad K \cap K' = \varnothing \quad \text{for } K, K' \in \mathscr{K}_h : K \neq K'.$$

Furthermore, we assume that the aspect ratio of the element diameter h_K and the radius ρ_K of the largest inscribed circle is uniformly bounded from above for the whole sequence of meshes in the convergence process, i.e. $h_K/\rho_K < \sigma$. An edge $E \in \mathscr{E}_h$ in the mesh is a straight line which is always situated between two nodes. The set of all nodes is denoted by $\mathscr{N}_h$. We assume that the edge length h_E of each edge E of an element K is uniformly bounded from below by a constant times the element diameter, i.e. $\varsigma h_K < h_E$. Thus, we deal with quite general meshes, where the assumption on the aspect ratio ensures that the elements do not degenerate and the assumption on the edge length ensures that the number of nodes per element is uniformly bounded.

The solution of (2) is approximated by the use of a finite dimensional subspace V_h^k of V, where k gives the order of approximation. To construct this conforming approximation space, we prescribe its basis. For a first order method it is sufficient to utilize nodal basis functions which are defined locally as solutions of boundary value problems. We define for each node $z \in \mathscr{N}_h$ the function ψ_z piecewise for each element $K \in \mathscr{K}_h$ as unique solution of

$$-\Delta\psi_z = 0 \quad \text{in } K, \quad \psi_z(x) = \begin{cases} 1, & x = z, \\ 0, & x \in \mathscr{N}_h \setminus \{z\}, \end{cases} \qquad \psi_z \text{ linear on each } E \in \mathscr{E}_h,$$

and set $V_h^1 = \operatorname{span}\{\psi_z : z \in \mathscr{N}_h\}$. To obtain a higher order approximation space, we follow the hint in [18]. Thus, we have to add additional basis functions which are assigned to edges and basis functions which are assigned to elements. The former ones also fulfil local Laplace problems and the later ones are defined by local Poisson problems with homogeneous Dirichlet data. Since the source term in (1) vanishes, it is sufficient to work with piecewise harmonic trial functions, see [23]. Consequently, the element trial functions are leaved out. We define for each edge $E \in \mathscr{E}_h$ and $i = 2, \dots, k$ the function $\psi_{E,i}$ piecewise as unique solution of

$$-\Delta\psi_{E,i} = 0 \quad \text{in } K, \qquad \psi_{E,i} = \begin{cases} p_{E,i}, & \text{on } E, \\ 0, & \text{on } E' \in \mathscr{E}_h \setminus \{E\}, \end{cases}$$

and set $V_h^k = \operatorname{span}\{\psi_z, \psi_{E,i} : z \in \mathcal{N}_h, E \in \mathcal{E}_h, i = 2,\ldots,k\}$ for $k \geq 2$. Here the Dirichlet data is chosen such that for $E = \overline{z_b z_e}$ with $z_b, z_e \in \mathcal{N}_h$

$$\{\psi_{z_b}|_E, \psi_{z_e}|_E, p_{E,i} : i = 2,\ldots,k\}$$

forms a basis of the space of polynomials of degree smaller or equal to k over E. Due to the continuity of the trial functions across the element boundaries, they are in $H^1(\Omega)$, and thus it holds $V_h^k \subset V$. Furthermore, it is possible to prove that $\mathcal{P}^k_{\mathrm{harm}}(K) \subset V_h^k|_K$, where $\mathcal{P}^k_{\mathrm{harm}}(K)$ is the space of harmonic polynomials of degree smaller or equal to k over K, see [23].

For simplicity, we restrict ourselves to piecewise constant diffusion coefficients in (1), which are resolved by the mesh. Otherwise, the coefficient has to be approximated in a suitable manner for the BEM-based FEM, see [18].

Let us assume that the Dirichlet data in (1) is piecewise polynomial of degree k such that the extension can be chosen as $g_D \in V_h^k$. Consequently, the discrete variational formulation of (2) reads

$$\text{Seek } u_h \in g_D + V_{h,D}^k : \quad a(u_h, v_h) = (g_N, v_h)_{\Gamma_N} \quad \forall v_h \in V_{h,D}^k, \tag{3}$$

where $V_{h,D}^k = V_h^k \cap V_D$. The approximation $u_h \in V_h^k$ fulfils the error estimate

$$\|u - u_h\|_{H^1(\Omega)} \leq ch^k |u|_{H^{k+1}(\Omega)}$$

for sufficiently regular $u \in H^{k+1}(\Omega)$ and mesh size $h = \max\{h_K : K \in \mathcal{K}_h\}$, see [18, 23]. Here, the constant c only depends on k and the mesh parameters σ and ς.

The formulation (3) leads to a system of linear equations with a symmetric positive definite matrix. In the set up of the system matrix the bilinear form $a(\cdot,\cdot)$ has to be applied to nodal and edge basis functions that results in an integration of implicit defined functions. However, by the use of Green's first identity, the bilinear form can be rewritten as

$$a(\psi, \phi) = \sum_{K \in \mathcal{K}_h} \alpha_K \int_{\partial K} \phi \frac{\partial \psi}{\partial n_K} \qquad \text{for } \psi, \phi \in V_h^k,$$

because of the piecewise constant diffusion coefficient with $\alpha(\cdot) \equiv \alpha_K$ on K for $K \in \mathcal{K}_h$. This representation only involves the Dirichlet trace of the trial functions on the element boundaries, which is given by definition, and the Neumann trace. These traces belong to broken Sobolev spaces over element boundaries. The Dirichlet trace $p \in H^{1/2}(\partial K)$ of a harmonic function over K and its corresponding Neumann trace $t \in H^{-1/2}(\partial K)$ are connected by

$$\mathbf{V}_K t = \left(\tfrac{1}{2}\mathbf{I} + \mathbf{K}_K\right) p, \tag{4}$$

where the boundary integral operators $\mathbf{V}_K : H^{-1/2}(\partial K) \to H^{1/2}(\partial K)$ and $\mathbf{K}_K : H^{1/2}(\partial K) \to H^{1/2}(\partial K)$ are the well studied single and double layer potential operators, see [15]. In the realization, these operators have to be approximated. Therefore, a Galerkin scheme is applied to (4) which results in a Boundary Element Method on the given, coarse discretization of the polygonal boundary ∂K, see [17]. Thus, the implicitly defined trial functions can be treated in an efficient way by local Boundary Element Methods.

3 Residual Based Error Estimate and Adaptivity

In real life applications, the problems often do not meet the regularity requirements which are needed to achieve the optimal rates of convergence when refining the mesh uniformly. Thus, it is advantageous to use adaptive mesh refinement strategies which adapt the mesh to the problem. A standard indicator for the local refinement is the residual based error estimator, which measures the jumps of the approximation of the normal flux across the element boundaries, namely

$$[u_h]_E = n_K \cdot \gamma_0^K (a_K \nabla u_h) + n_{K'} \cdot \gamma_0^{K'} (a_{K'} \nabla u_h).$$

Here, $\gamma_0^K : H^1(K) \to H^{1/2}(\partial K)$ denotes the usual trace operator with respect to the element K, and $K, K' \in \mathscr{K}_h$ are the adjacent elements to E. The results of [21] carry over to our situation. Thus, we can prove the reliability of the residual based error estimate for the defined trial space V_h^k on polygonal meshes in the energy norm

$$\|v\|_E = \sqrt{a(v,v)} \quad \text{for } v \in V.$$

Consequently, it holds

$$\|u - u_h\|_E \leq \frac{c}{\sqrt{\alpha_{\min}}} \eta_R$$

with

$$\eta_R^2 = \sum_{K \in \mathscr{K}_h} \eta_K^2 \qquad \text{and} \qquad \eta_K^2 = \sum_{E \in \mathscr{E}_h : E \subset \partial K} h_E \|R_E\|_{L_2(E)}^2,$$

where

$$R_E = \begin{cases} 0 & \text{for } E \subset \Gamma_D, \\ g_N - \gamma_0^K (a_K \nabla u_h) & \text{for } E \subset \Gamma_N \text{ with } E \subset \partial K, \\ -\frac{1}{2}[u_h]_E & \text{else}, \end{cases}$$

for $E \in \mathscr{E}_h$ and the constant c only depends on the mesh parameters σ and ς.

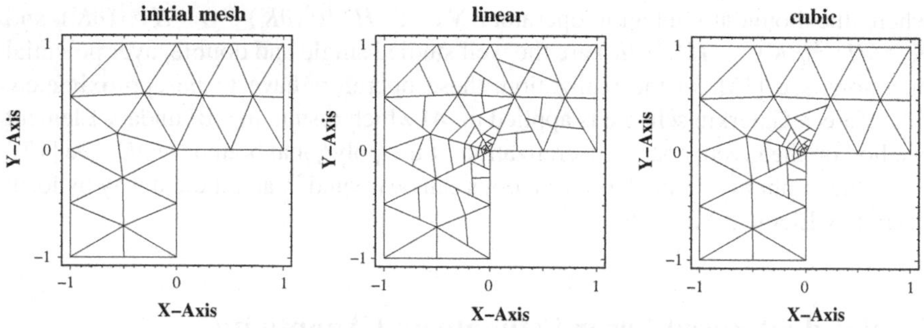

Fig. 1 Initial mesh (*left*) as well as adaptive refined mesh after ten steps for BEM-based FEM with linear (*middle*) and cubic (*right*) approximation order

The local error indicators η_K, which are assigned to elements, can be used to gauge the accuracy of the approximation over each element. Thus, this information is utilized in an adaptive Finite Element Method, which follows the steps

$$\text{SOLVE} \rightarrow \text{ESTIMATE} \rightarrow \text{MARK} \rightarrow \text{REFINE} \rightarrow \text{SOLVE} \rightarrow \cdots .$$

First, we solve the discrete boundary value problem on a given mesh and compute the error estimator η_R and the error indicators η_K for all elements. If the desired accuracy is reached according to η_R, we are done. If not, we mark some elements for refinement using Dörfler's strategy, see [6]. These elements are chosen on the basis of the error indicators η_K. Next, the marked elements are refined, and thus we obtain a new mesh, which is adapted to the problem. So, we can solve the boundary value problem on the refined mesh and continue this procedure until the desired accuracy is achieved.

In the scenario of local refinements, the use of polygonal meshes is advantageous. If one element in divided into two new ones, this might create a new node on a straight line of the boundary of the initial element. Additionally, this means, that a neighbouring element gets one new node on the boundary, cf. Fig. 1. For simplicial meshes, the new node would result in a conditional degree of freedom or in a splitting of the neighbouring element. For polygonal elements, however, this inserted node appears naturally and is treated as every other node on the boundary of the element. Consequently, the refinement is kept very local, as we will see in the numerical experiment.

4 Numerical Experiment

To illustrate the applicability of the residual based error estimate on polygonal meshes for the BEM-based Finite Element Method, we examine a classical test problem for adaptive methods on a domain with reentrant corner.

Thus, let $\Omega = ((-1,1)\times(-1,1)) \setminus ([0,1]\times[0,-1])$ and $\Gamma_D = \partial\Omega$. Using polar coordinates (r,φ), the boundary data g_D is chosen in such a way that

$$u(r\cos\varphi, r\sin\varphi) = r^{2/3}\sin\left(\frac{2\varphi}{3}\right), \quad x = (r\cos\varphi, r\sin\varphi)^\top \in \mathbb{R}^2$$

is the solution of the boundary value problem (1) with diffusion coefficient $\alpha \equiv 1$. The derivatives of u have a singularity in the origin of the coordinate system, and consequently, uniform refinement yields suboptimal rates of convergence. However, we still expect optimal rates of convergence for the adaptive strategy.

In Fig. 1, the initial mesh is visualized as well as the adaptive meshes after ten refinements for the BEM-based Finite Element Method with linear and cubic approximation order. The adaptive strategy obviously detects the singularity and tunes the refinement towards the origin of the coordinate system. The higher order method needs less refinements far from the singularity since it approximates the solution u very well in that region where u is smooth. Furthermore, the true error is smaller than the residual based error estimate, which is proven by the reliability and is observed in Fig. 2. The number of degrees of freedom are proportional to the square of the mesh size h^2 in case of uniform refinement. Thus, we recognize the optimal rates of convergence for the method with first and second order approximation in Fig. 2. The third order method even converges one order faster than expected. This might be due to the simplicity of the problem and is part of further investigations.

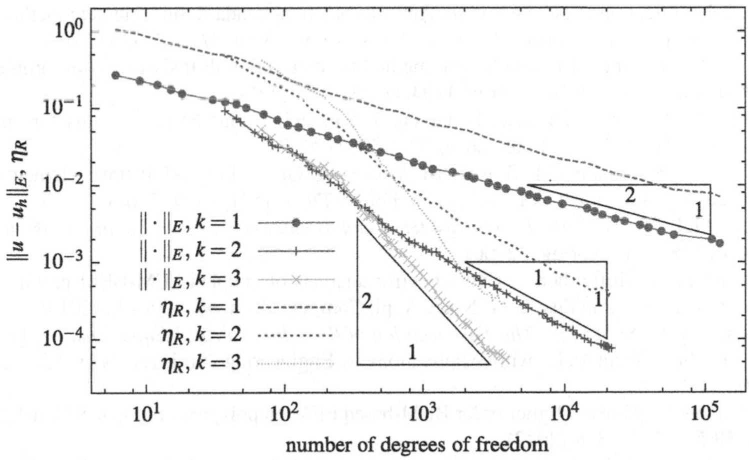

Fig. 2 Convergence graph for the adaptive BEM-based FEM with trial space V_h^k

References

1. L. Beirão da Veiga, F. Brezzi, A. Cangiani, G. Manzini, L. Marini, A. Russo, Basic principles of virtual element methods. Math. Models Methods Appl. Sci. **23**(1), 199–214 (2013)
2. L. Beirão da Veiga, F. Brezzi, L. Marini, Virtual elements for linear elasticity problems. SIAM J. Numer. Anal. **51**(2), 794–812 (2013)
3. L. Beirão da Veiga, K. Lipnikov, G. Manzini, Arbitrary-order nodal mimetic discretizations of elliptic problems on polygonal meshes. SIAM J. Numer. Anal. **49**(5), 1737–1760 (2011)
4. D. Copeland, U. Langer, D. Pusch, From the boundary element domain decomposition methods to local Trefftz finite element methods on polyhedral meshes, in *Domain Decomposition Methods in Science and Engineering XVIII*. ed. by M. Bercovier, M. Gander, R. Kornhuber, O. Widlund. Lecture Notes in Computational Science and Engineering, vol. 70 (Springer, Berlin/Heidelberg, 2009), pp. 315–322
5. D.M. Copeland, Boundary-element-based finite element methods for Helmholtz and Maxwell equations on general polyhedral meshes. Int. J. Appl. Math. Comput. Sci. **5**(1), 60–73 (2009)
6. W. Dörfler, A convergent adaptive algorithm for Poisson's equation. SIAM J. Numer. Anal. **33**(3), 1106–1124 (1996)
7. Y. Efendiev, J. Galvis, R. Lazarov, S. Weißer, Mixed FEM for second order elliptic problems on polygonal meshes with BEM-based spaces, in *Large-Scale Scientific Computing*, ed. by I. Lirkov, S. Margenov, J. Waśniewski. Lecture Notes in Computer Science (Springer, Berlin/Heidelberg, 2014), pp. 331–338
8. R. Hiptmair, A. Moiola, I. Perugia, Error analysis of Trefftz-discontinuous Galerkin methods for the time-harmonic Maxwell equations. Math. Comput. **82**(281), 247–268 (2013)
9. C. Hofreither, L_2 error estimates for a nonstandard finite element method on polyhedral meshes. J. Numer. Math. **19**(1), 27–39 (2011)
10. ———, A non-standard finite element method using boundary integral operators, Ph.D. thesis, Johannes Kepler University, Linz, Dec 2012
11. C. Hofreither, U. Langer, C. Pechstein, Analysis of a non-standard finite element method based on boundary integral operators. Electron. Trans. Numer. Anal. **37**, 413–436 (2010)
12. ———, A non-standard finite element method for convection-diffusion-reaction problems on polyhedral meshes. AIP Conf. Proc. **1404**(1), 397–404 (2011)
13. P. Joshi, M. Meyer, T. DeRose, B. Green, T. Sanocki, Harmonic coordinates for character articulation. ACM Trans. Graph. **26**(3), 71.1–71.9 (2007)
14. S. Martin, P. Kaufmann, M. Botsch, M. Wicke, M. Gross, Polyhedral finite elements using harmonic basis functions. Comput. Graph. Forum **27**(5), 1521–1529 (2008)
15. W.C.H. McLean, *Strongly Elliptic Systems and Boundary Integral Equations* (Cambridge University Press, Cambridge, 2000)
16. C. Pechstein, C. Hofreither, A rigorous error analysis of coupled FEM-BEM problems with arbitrary many subdomains. Lect. Notes Appl. Comput. Mech. **66**, 109–132 (2013)
17. S. Rjasanow, O. Steinbach, *The Fast Solution of Boundary Integral Equations*. Mathematical and Analytical Techniques with Applications to Engineering (Springer, New York/London, 2007)
18. S. Rjasanow, S. Weißer, Higher order BEM-based FEM on polygonal meshes. SIAM J. Numer. Anal. **50**(5), 2357–2378 (2012)
19. ———, FEM with Trefftz trial functions on polyhedral elements. J. Comput. Appl. Math. **263**, 202–217 (2014)
20. A. Tabarraei, N. Sukumar, Application of polygonal finite elements in linear elasticity. Int. J. Comput. Methods **3**(4), 503–520 (2006)
21. S. Weißer, Residual error estimate for BEM-based FEM on polygonal meshes. Numer. Math. **118**(4), 765–788 (2011)

22. ———, Finite Element Methods with local Trefftz trial functions, Ph.D. thesis, Universität des Saarlandes, Saarbrücken, Sept 2012
23. ———, Arbitrary order Trefftz-like basis functions on polygonal meshes and realization in BEM-based FEM. Comput. Math. Appl. **67**(7), 1390–1406 (2014)

Stopping Criteria Based on Locally Reconstructed Fluxes

Roland Becker, Daniela Capatina, and Robert Luce

Abstract We propose stopping criteria for the iterative solution of equations resulting from discretization by conforming, nonconforming, and total discontinuous finite element methods. A simple modification of error estimators based on locally reconstructed fluxes allows to split the estimator into a discretisation-based and an iteration-based part. Comparison of both then leads to stopping criteria which can be used in the framework of an adaptive algorithm.

1 Introduction and General Idea

The idea to couple the accuracy of iterative solvers for partial differential equations to the discretisation error has a long tradition in numerical analysis, and can for example be found in textbooks on multigrid methods, such as [6], where it is often shown that a number of fixed iterations on each grid level is sufficient to achieve optimal convergence of the overall error.

Traditionally, one used in practice an experience-based prescription of iteration parameters, such as the maximal number of iterations or the tolerance of the residual in a certain norm. Through the appearance of a posteriori error estimators, it became possible to replace this by an adaptive stopping criterion, see for example [1].

In recent years, a posteriori error estimators based on locally reconstructed $H(\mathrm{div}, \Omega)$-fluxes have gained considerable interest, since they give (nearly) constant-free upper bounds, see for example [3, 5, 8, 9, 11] and many others. Due to this property, it seems appropriate to use them also to define stopping criteria for iterative solution algorithms, see [4, 7] for a finite volume/element approximation. However, contrarily to residual-based error estimators, these estimators generally require the discrete equations to be solved exactly, since the reconstruction of the fluxes is based on a conservation property intrinsic to the considered method. Therefore the question arises of how to compute them in the case of approximate solution of the discrete system.

R. Becker • D. Capatina (✉) • R. Luce
LMAP CNRS UMR 5142, Université de Pau, IPRA BP 1155, 64013 Pau, France
e-mail: roland.becker@univ-pau.fr; daniela.capatina@univ-pau.fr; robert.luce@univ-pau.fr

A. Abdulle et al. (eds.), *Numerical Mathematics and Advanced Applications ENUMATH 2013*, Lecture Notes in Computational Science and Engineering 103,
DOI 10.1007/978-3-319-10705-9_24

We propose here a simple correction of the right-hand side, which makes it possible to evaluate the estimators and separate them into a discretisation part and an iteration part (the last going to zero as the iterative process converges). This applies to conforming, nonconforming as well as discontinuous Galerkin finite element methods; we only focus here on P^1-continuous elements. In order to do so, we use the uniform framework presented in [2] which yields robust estimators different from those employed in [4, 7]. The proposed algorithm also differs through the prediction of the adaptive tolerance.

We consider the Poisson problem on a polygonal bounded domain $\Omega \subset \mathbb{R}^2$:

$$-\Delta u = f \quad \text{in } \Omega, \qquad u = 0 \quad \text{on } \Gamma^{\mathrm{D}}, \qquad \frac{\partial u}{\partial n} = g \quad \text{on } \Gamma^{\mathrm{N}} \tag{1}$$

with boundary $\partial\Omega = \Gamma^{\mathrm{D}} \cup \Gamma^{\mathrm{N}}$, $f \in L^2(\Omega)$, $g \in L^2(\Gamma^{\mathrm{N}})$. The generalisation to non-homogeneous Dirichlet condition rises no difficulty.

Let $\sigma_h \in H(\mathrm{div}, \Omega)$ belong to the Raviart-Thomas space $\mathbf{RT}_h^m$ [10] and verify $\mathrm{div}\,\sigma_h = -\pi_h f$ in Ω and $\sigma_h \cdot n_\Omega = \pi_h^N g$ on Γ^N where π_h, π_h^N are the $L^2(\Omega)$ and $L^2(\Gamma^N)$ projections on piecewise P^m f.e. spaces. In order to be of practical interest, such a reconstruction has to be local. Then one has

$$\|\nabla(u - u_h)\|_{0,\Omega} \leq \|\sigma_h - \nabla u_h\|_{0,\Omega} + c\mu_h, \tag{2}$$

which means that $\tau_h = \sigma_h - \nabla u_h$ can be used as an error estimator with a sharp upper bound, up to a higher-order term related to data approximation:

$$\mu_h^2 = \sum_{K \in \mathcal{K}_h} |K| \, \|f - \pi_h f\|_K^2 + \sum_{S \subset \Gamma_N} |S| \, \left\|g - \pi_h^N g\right\|_S^2.$$

These reconstructions suppose that the discrete equations are satisfied, which is no longer true when using an iterative solver. We propose here a generalization where the local conservation property of $H(\mathrm{div}, \Omega)$-fluxes is not fulfilled: at each step n, we look for a flux $\sigma_h^n \in H(\mathrm{div}, \Omega)$ satisfying

$$-\mathrm{div}\sigma_h^n = \pi_h f + \rho_h^n \quad \text{in } \Omega, \qquad \sigma_h^n \cdot n_\Omega = \pi_h^N g \quad \text{on } \Gamma_N.$$

The function ρ_h^n is related to the residual of the algebraic linear system to be solved. Then it follows, with the correction $\tau_h^n := \sigma_h^n - \nabla u_h^n$ at step n, that

$$\begin{aligned}
\|\nabla(u - u_h^n)\|_{0,\Omega}^2 &= <\nabla(u - u_h^n), \nabla u - \sigma_h^n>_{0,\Omega} + <\nabla(u - u_h^n), \tau_h^n>_{0,\Omega} \\
&\leq c\mu_h \|\nabla(u - u_h^n)\|_{0,\Omega} + \|u - u_h^n\|_{0,\Omega} \|\rho_h^n\|_{0,\Omega} + \|\nabla(u - u_h^n)\|_{0,\Omega} \|\tau_h^n\|_{0,\Omega},
\end{aligned}$$

which yields the error bound, with C_Ω the constant of Poincaré's inequality,

$$\|\nabla(u - u_h^n)\|_{0,\Omega} \leq \|\tau_h^n\|_{0,\Omega} + C_\Omega \|\rho_h^n\|_{0,\Omega} + c\mu_h.$$

Based on this error estimate, a natural stopping criterion for the iterative solver is to balance the discretization and the iteration errors. The general idea is to stop the iterative process as soon as $\|\rho_h^n\|_{0,\Omega} \leq \alpha(\|\tau_h^n\|_{0,\Omega} + c\mu_h)$ where α is a numerical constant chosen by the user.

2 A Posteriori Estimator for Fully Solved Equations

First let the discrete equations be fully solved. We focus on the construction of an a posteriori error estimator based on locally conservative fluxes.

We consider here a P^1-conforming approximation but the same approach applies to nonconforming and discontinuous Galerkin methods of arbitrary degree. The underlying idea for this uniform approach is to write a hybrid formulation covering all the previous finite element methods. The Lagrange multipliers compensating for the weak continuity conditions yield approximations to the normal fluxes. A priori and a posteriori connections between all these methods have been studied in [2].

A regular triangular mesh consists of cells $\mathscr{K}_h$, such that $\partial\Omega = \mathscr{S}_h^D \cup \mathscr{S}_h^N$. Let $\mathscr{S}_h^{int}$ the set of interior sides and $\mathscr{S}_h = \mathscr{S}_h^{int} \cup \mathscr{S}_h^D$. For an interior side S, n_S is a fixed unit vector normal to S. If the side S lies on $\partial\Omega$, we set $n_S = n_\Omega$, the outward unit normal vector. We denote by π_K^1 the $L^2(K)$ orthogonal projection on P^1. We define the following spaces:

$$D_h = \left\{v_h \in L^2(\Omega);\ v_h|_K \in P^1 \quad \forall K \in \mathscr{K}_h\right\},$$
$$M_h = \left\{v_h \in L^2(\mathscr{S}_h);\ v_h|_S \in P^1 \quad \forall S \in \mathscr{S}_h\right\}.$$

Let $u \in D_h$. We define for $S \in \mathscr{S}_h^{int}$ and $x \in S$: $u_S^{\text{in}}(x) := \lim\limits_{\varepsilon\searrow 0} u(x - \varepsilon n_S)$, $u_S^{\text{ex}}(x) = \lim\limits_{\varepsilon\searrow 0} u(x + \varepsilon n_S)$. Next we define the jump and the mean for $x \in S$ by $[u](x) := u_S^{\text{in}}(x) - u_S^{\text{ex}}(x)$, $\{u\}(x) := \frac{1}{2}\left(u_S^{\text{in}}(x) + u_S^{\text{ex}}(x)\right)$. For a boundary side we set $[u]_S = \{u\}_S = u_S^{\text{in}}$.

2.1 Stable Hybrid Formulation

Let the bilinear and linear forms on $D_h \times D_h$, $D_h \times M_h$, $M_h \times M_h$ and D_h:

$$a(u_h, v_h) = \int_{\mathscr{K}_h} \nabla u_h \cdot \nabla v_h - \int_{\mathscr{S}_h} \{\partial_n u_h\}[v_h] - \int_{\mathscr{S}_h} [u_h]\{\partial_n v_h\}$$

$$b(\mu_h, v_h) = \int_{\mathscr{S}_h} \mu_h[v_h],\ \ c(\theta_h, \mu_h) = \int_{\mathscr{S}_h} |S|\,\theta_h\mu_h,\ \ l(v_h) = \int_\Omega f v_h + \int_{\Gamma^N} g v_h.$$

Noting that the jump satisfies a node-wise identity, we introduce

$$M_h^* := \left\{ \mu_h \in M_h : \sum_{S \in \mathscr{S}_N} \alpha_S(N)\mu_h(N) = 0 \quad \forall N \in \mathscr{N}_h \right\}. \tag{3}$$

Here above, $\mathscr{S}_N$ denotes the set of sides S sharing the node N and $\alpha_S(N) = |S|\,\mathrm{sign}(n_S, N)$, where $\mathrm{sign}(n_S, N)$ equals 1 or -1 depending upon the orientation of n_S with respect to the clockwise rotation sense around N. $\mathscr{N}_h$ represents the set of nodes which are interior to the domain or to Γ^D.

We consider the following hybrid problem: Find (u_h, θ_h) in $D_h \times M_h^*$ such that for all $(v_h, \mu_h) \in D_h \times M_h^*$,

$$\begin{aligned} a(u_h, v_h) + b(\theta_h, v_h) &= l(v_h) \\ b(\mu_h, u_h) &= 0. \end{aligned} \tag{4}$$

The use of the space M_h^* instead of M_h ensures uniqueness of the multiplier.

Note that $\mathrm{Ker}\, b$ coincides with the P^1 conforming space, hence u_h satisfies the primal conforming formulation. We introduce the norms on D_h and M_h^*:

$$|||v_h||| := \left(\int_{\mathscr{K}_h} |\nabla v_h|^2 + \int_{\mathscr{S}_h} |S|^{-1}[v_h]^2 \right)^{1/2}, \quad \|\mu_h\|_{\mathscr{S}_h} := c_h(\mu_h, \mu_h)^{1/2}.$$

The only technical point concerning the stability is the inf-sup condition.

Theorem 1 *There exists a positive constant β independent of h such that*

$$\inf_{\mu_h \in M_h^*} \sup_{v_h \in D_h} \frac{b(\mu_h, v_h)}{\|\mu_h\|_{\mathscr{S}_h} |||v_h|||} \geq \beta.$$

Proof To any $\mu_h \in M_h^*$, we associate $v_h \in D_h$ satisfying $[v_h] = |S|\mu_h$ on $S \in \mathscr{S}_h$ and $|||v_h||| \leq C\|\mu_h\|_{\mathscr{S}_h}$. Its construction is achieved patch-wise: $v_h = \sum_{N \in \mathscr{N}_h} v_h^N$ with v_h^N defined on $\omega_N = \mathrm{supp}\,\varphi_N$ with φ_N the P^1 -basis function associated to the node N.

We first write $\mu_h^S := (\mu_h)_{|S} = \mu_h^S(N)\varphi_N + \mu_h^S(M)\varphi_M$ on each side $S \in \mathscr{S}_h$ of vertices N and M. Then we define the jump of v_h^N on S as follows: $[v_h^N]_{|S} = |S|\mu_h^S(N)\varphi_N$ if both N, M belong to $\mathscr{N}_h$ and $[v_h^N]_{|S} = |S|\mu_h^S$ if $M \in \bar{\Gamma}^N$. Clearly, the global function v_h satisfies $[v_h]_{|S} = |S|\mu_h^S$ such that $b(\mu_h, v_h) = \|\mu_h\|^2_{\mathscr{S}_h}$. We still have to define v_h^N on each $K \subset \omega_N$ from its jumps and to bound it.

The linear system from which we compute it, $[v_h^N]_{|S}(N) = |S|\mu_h^S(N)$ for all $S \in \mathscr{S}_N$, is compatible thanks to (3). Hence we obtain $(v_h^N)_{|K}(N)$ by fixing one of the values, for instance equal to zero. Since the numbers of triangles around each vertex N is bounded, one finally gets that $|||v_h||| \leq c\|\mu_h\|_{\mathscr{S}_h}$ which ends the proof. □

Thus, the hypotheses of the Babuška-Brezzi theorem are uniformly checked, which implies that the system (4) is uniformly well-posed and that its solution satisfies the a priori error estimate: $|||u - u_h||| + \|\theta_h\|_{\mathscr{S}_h} \leq Ch|u|_{2,\Omega}$.

2.2 Local Computation of Fluxes

A remarkable feature is that the multiplier θ_h can be computed locally, up to an element of Ker$b$ which has no influence on the definition of the flux. It is useful to introduce $L(\cdot) = l(\cdot) - a(u_h, \cdot)$.

To any node N we associate the patch $\omega_N = \operatorname{supp} \varphi_N$ and we look for local contributions θ_N associated to each ω_N. The unknown θ_N lives only on the interior sides of ω_N and is null elsewhere. We define $(\theta_N)_{|S}$ in P^1 for any $S \in \mathscr{S}_N$ by imposing, for all $K \subset \omega_N$, $M \in \mathscr{N}_h \cap \partial K$ with $M \neq N$:

$$b(\theta_N, \varphi_N \chi_K) = \int_{\partial K} \theta_N [\varphi_N] = L(\varphi_N \chi_K), \quad b(\theta_N, \varphi_M \chi_K) = \int_{\partial K} \theta_N \varphi_M = 0$$

where χ_K is the characteristic function of K.

Note that the conforming solution u_h satisfies $L(\varphi_N) = 0$ for all $N \in \mathscr{N}_h^{int}$, so the previous system is compatible. However, it has a one-dimensional kernel $\mathscr{K}_N$ characterized by: $\operatorname{sign}(n_S, N) \int_S \theta_N \varphi_N = \text{const}$, for all $S \in \mathscr{S}_N$. In conclusion, there exists a solution θ_N, unique up to an element of $\mathscr{K}_N$.

One can then prove (see also [2]) that $\bar{\theta}_h = \sum_{N \in \mathscr{N}_h} \theta_N$ is equal to the solution θ_h of (4), up to an element $\theta_h^{Ker} \in$ Ker$b$.

We are next interested in defining a flux $\sigma_h \in H(\text{div}, \Omega)$ from $\bar{\theta}_h$. We use the space $\mathbf{RT}_h^1$ and we impose the degrees of freedom as below. On the Neumann boundary, we set $\sigma_h \cdot n_S = \pi_S^1 g$ while on $S \in \mathscr{S}_h$ we impose:

$$\sigma_h \cdot n_S = \{\partial_n u_h\} - \bar{\theta}_h \quad \text{if } S \in \mathscr{S}_h^{int}, \qquad \sigma_h \cdot n_S = -\bar{\theta}_h \quad \text{if } S \in \mathscr{S}_h^D.$$

The interior degrees of freedom are given by $\int_K \sigma_h \cdot r = \int_K \nabla u_h \cdot r$, $\forall r \in (P^0)^2$.

Cf. [2], the reconstructed flux satisfies, independently of θ_h^{Ker},

$$\text{div}\sigma_h|_K = -\pi_K^1 f, \quad \forall K \in \mathscr{K}_h. \tag{5}$$

We are interested in the a posteriori error estimator, that is in the correction $\tau_h = \sigma_h - \nabla u_h$. It clearly satisfies $[\tau_h] \cdot n_S = -[\partial_n u_h]$ and $\{\tau\}_h \cdot n_S = -\bar{\theta}_h$.

One can equivalently compute it as the sum of local contributions on patches, $\tau_h = \sum_{N\in\mathcal{N}_h} \tau_N$ with $\tau_N \cdot n = 0$ on $\partial\omega_N$, $\tau_N|_K \in RT^1$ and

$$\begin{aligned}
&\int_K \tau_N \cdot r = 0, \quad \forall r \in (P^0)^2,\\
&\int_S (\tau_N)|_K \cdot n_S \varphi_N = -\int_S (\frac{1}{2}[\nabla u_h]\cdot n_K + \bar{\theta}_h)\varphi_N, \quad \forall S \in \mathcal{S}_N, \qquad (6)\\
&\int_S (\tau_N)|_K \cdot n_S \varphi_M = 0, \quad \forall S \in \mathcal{S}_N.
\end{aligned}$$

We have used $(\tau_N)|_K \cdot n_S = \frac{1}{2}[\tau_N]\cdot n_K + \{\tau_N\}\cdot n_S$ in order to obtain (6). These relations directly define the d.o.f. of τ_N by means of the multiplier $\bar{\theta}_h$. In order to compute θ_N, we impose its orthogonality to the kernel $\mathcal{K}_N$.

In [3], the authors proposed a flux also inspired by the hypercircle method, but their τ_N is computed as the solution of a local mixed problem.

One can eliminate $\bar{\theta}_h$ from (6) and obtain the following system for τ_N:

$$\begin{aligned}
&\int_S [\tau_N]\cdot n_S \varphi_N = -\int_S [\partial_n u_h]\,\varphi_N, \qquad \int_K (\mathrm{div}\tau_N)\varphi_N = -\int_K f\varphi_N,\\
&\int_S [\tau_N]\cdot n_S \varphi_M = 0, \qquad \int_K (\mathrm{div}\tau_N)\varphi_M = 0.
\end{aligned} \qquad (7)$$

Note that (6) is equivalent to (7), provided that we impose τ_N orthogonal to the one-dimensional kernel of (7).

3 Estimator and Stopping Criterion

Whenever one uses an exact solver, u_h is the Galerkin solution so the system (7) is compatible. This is no longer the case when an iterative solver is employed. In this case, we modify the righthand-side term as follows, such that the system is still comptible at each iteration:

$$\int_S [\tau_N^n]\cdot n_S \varphi_N = -\int_S [\partial_n u_h^n]\,\varphi_N, \qquad \int_K (\mathrm{div}\tau_N^n)\varphi_N = \int_K (-f + \rho_N^n)\varphi_N$$

with

$$\int_{\omega_N} \rho_N^n \varphi_N = \int_{\omega_N} f\varphi_N - \int_{\omega_N} \nabla u_h^n \cdot \nabla\varphi_N =: r_N^n.$$

We built $\rho_N^n \in P^1(\omega_N)$ such that $\rho_N^n = 0$ on $\partial\omega_N$, which amounts to impose $\rho_N^n(N) = \dfrac{6}{|\omega_N|} r_N^n$. Then we define the global correction ρ_h^n on Ω such that for any $K \in \mathscr{K}_h$, $(\rho_h^n)|_K \in P^1$ and

$$\int_K \rho_h^n \varphi_N = \int_K \rho_N^n \varphi_N, \quad \forall N \in \mathscr{N}_h \cup \partial K.$$

This ensures that we finally get $-\mathrm{div}\sigma_h^n = \pi_h^1 f + \rho_h^n$.

An adaptive mesh refinement algorithm consists of four steps: solve the equations, estimate the error, mark the cells and refine the mesh.

Here, we employ as iterative solver the conjugate gradient, denoted by $CG(u^{in}, tol; u^{out}, n)$ but we could use a black box to solve the linear system $Ax = b$, whose stopping criterion is $|r^n| < tol$ with $r^n = Ax^n - b$ the residual. We employ τ_h as error estimator; we do not detail the last two steps.

Let h_k denote the adapted mesh and $u_k^{n_k} \in V_{h_k}$ the converged solution, with n_k the number of iterations. To compute $u_{k+1}^{n_{k+1}}$ from $u_k^{n_k}$, we use the next algorithm with an adaptive tolerance, where $\|\rho\|_h^2 = \sum_{K\in\mathscr{K}_h} h_K^2 \|\rho\|_{0,K}^2 \approx |r|$:

$l := 0, \quad u_{k+1}^l := \mathscr{I}_{k+1}(u_k^{n_k}), \quad$ CONV:= false

while (not CONV) do

  if $\|\rho(u_{k+1}^l)\|_{0,\Omega} \leq \alpha\|\tau(u_{k+1}^l)\|_{0,\Omega}$ then

    $n_{k+1} := l, \qquad u_{k+1}^{n_{k+1}} := u_{k+1}^l, \qquad$ CONV:=true

  else

    $\mathrm{tol} := \dfrac{\alpha\|\tau(u_{k+1}^l)\|_{0,\Omega}}{2\|\rho(u_{k+1}^l)\|_{0,\Omega}} \|\rho(u_{k+1}^l)\|_h, \qquad CG(u_{k+1}^l, \mathrm{tol};\, u_{k+1}^{n_l}, n_l)$

    $l := l + n_l, \qquad u_{k+1}^l := u_{k+1}^{n_l}$

  end if

end while

4 Numerical Experiments

We first solve the Poisson equation with Dirichlet condition on $]-1,1[^2$; the data are such that $u(x) = \mathrm{e}^{-\frac{\|x-x_0\|}{\delta}}$, where $x_0 = (0.5, 0.5)$ and $\delta = 0.03$, is the solution. We compare the previous algorithm with a fixed tolerance algorithm, for

a

N	H1-error	ratio	n_{it}
4	1.378e+00	-	-
16	1.009e+00	1.37	2
64	1.175e+00	0.86	7
256	9.278e-01	1.27	14
1024	6.195e-01	1.50	28
4096	3.298e-01	1.88	57
16384	1.660e-01	1.99	118
65536	8.316e-02	2.00	224
262144	4.160e-02	2.00	435

b

N	$\|\tau_h\|_{0,\Omega}$	ratio	H1-error	ratio	n_{it}
4	2.037e+00	-	1.378e+00	-	-
16	8.036e-01	2.53	1.009e+00	1.37	1
64	7.695e-01	1.04	1.175e+00	0.86	2
256	9.714e-01	0.79	9.278e-01	1.27	3
1024	7.789e-01	1.25	6.196e-01	1.50	2
4096	4.039e-01	1.93	3.299e-01	1.88	5
16384	2.026e-01	1.99	1.661e-01	1.99	10
65536	1.014e-01	2.00	8.317e-02	2.00	21
262144	5.069e-02	2.00	4.161e-02	2.00	23

Fig. 1 Convergence w.r.t. h: uniform mesh refinement (smooth solution). (**a**) Fixed tolerance. (**b**) Adaptive tolerance

a

N	H1-error	ratio	n_{it}
4	1.378e+00	-	-
16	1.009e+00	1.37	2
30	1.175e+00	0.86	6
116	9.278e-01	1.27	11
172	6.260e-01	1.48	15
518	3.493e-01	1.79	26
1262	1.842e-01	1.90	41
3450	1.009e-01	1.83	76
11546	5.487e-02	1.84	135
39504	2.976e-02	1.84	258
137208	1.609e-02	1.85	482

b

N	$\|\tau_h\|_{0,\Omega}$	ratio	H1-error	ratio	n_{it}
4	2.037e+00	-	1.378e+00	-	-
16	8.036e-01	2.53	1.009e+00	1.37	1
30	7.725e-01	1.04	1.175e+00	0.86	1
116	9.712e-01	0.80	9.278e-01	1.27	3
172	7.853e-01	1.24	6.261e-01	1.48	2
505	4.444e-01	1.77	3.494e-01	1.79	7
1286	2.320e-01	1.92	1.832e-01	1.91	7
1286	2.322e-01	1.00	1.832e-01	1.00	3
3486	1.265e-01	1.83	1.002e-01	1.83	19
11898	6.888e-02	1.84	5.417e-02	1.85	36
40766	3.688e-02	1.87	2.919e-02	1.86	68
141560	1.985e-02	1.86	1.581e-02	1.85	127

Fig. 2 Convergence w.r.t. h: adaptive mesh refinement (smooth solution). (**a**) Fixed tolerance. (**b**) Adaptive tolerance

both uniform and adaptive mesh refinement. As expected, one can see in Figs. 1 and 2 that both algorithms yield the expected convergence order w.r.t. h but the adaptive tolerance allows for an important gain in the number of iterations, for a similar error and number of unknowns. Furthermore, when looking at the adaptive tolerance algorithm only, we see that the error for about 262,000 cells and uniform mesh is attained for only 11,800 cells when using mesh adaptivity, the number of iterations being comparable.

Finally, we also test a non-smooth exact solution on a slit domain. We carry on the same comparison and we note in Fig. 3 that the gain is even more important, only about 4,100 cells and 22 iterations are now needed in the case of adaptive mesh refinement and adaptive stopping criterion.

a

N	$\|\tau_h\|_{0,\Omega}$	ratio	H1-error	ratio	n_{it}
16	1.647e+00	-	6.012e-01	-	-
64	1.105e+00	1.49	4.239e-01	1.42	2
256	7.609e-01	1.45	2.992e-01	1.42	3
1024	5.302e-01	1.44	2.116e-01	1.41	5
4096	3.741e-01	1.42	1.495e-01	1.42	15
16384	2.627e-01	1.42	1.058e-01	1.41	15
65536	1.867e-01	1.41	7.479e-02	1.41	43
262144	1.314e-01	1.42	5.285e-02	1.42	28

b

N	$\|\tau_h\|_{0,\Omega}$	ratio	H1-error	ratio	n_{it}
16	1.647e+00	-	6.012e-01	-	-
64	1.105e+00	1.49	4.239e-01	1.42	2
120	7.717e-01	1.43	3.099e-01	1.37	2
176	5.544e-01	1.39	2.363e-01	1.31	3
232	4.086e-01	1.36	1.902e-01	1.24	4
452	2.936e-01	1.39	1.406e-01	1.35	5
986	2.068e-01	1.42	9.942e-02	1.41	7
2134	1.455e-01	1.42	7.013e-02	1.42	15
4168	1.035e-01	1.41	5.066e-02	1.38	22
8232	7.366e-02	1.41	3.646e-02	1.39	32
15418	5.264e-02	1.40	2.648e-02	1.38	40
29570	3.756e-02	1.40	1.914e-02	1.38	61
55472	2.679e-02	1.40	1.382e-02	1.38	81
108312	1.908e-02	1.40	9.938e-03	1.39	116
206142	1.357e-02	1.41	7.126e-03	1.39	153

Fig. 3 Adaptive tolerance algorithm with non-smooth exact solution (slit domain). (**a**) Uniform mesh refinement. (**b**) Adaptive mesh refinement

References

1. R. Becker, C. Johnson, R. Rannacher, Adaptive error control for multigrid finite element methods. Computing **55**, 271–288 (1995)
2. R. Becker, D. Capatina, R. Luce, Robust local flux reconstruction for various finite element methods, in *Numerical Mathematics and Advanced Applications: Proceedings of ENUMATH' 2013, Lausanne*, ed. by A. Abdulle et al. (2014), pp. 65–73. doi:10.1007/978-3-319-10705-9_6
3. D. Braess, J. Schöberl, Equilibrated residual error estimator for edge elements. Math. Comput. **77**(262), 651–672 (2008)
4. A. Ern, M. Vohralík, Adaptive inexact Newton methods with a posteriori stopping criteria for nonlinear diffusion PDE's. SIAM J. Sci. Comput. **35**(4), 1761–1791 (2013)
5. A. Ern, S. Nicaise, M. Vohralík, An accurate $H(\mathrm{div})$ flux reconstruction for discontinuous Galerkin approximations of elliptic problems. C. R. A. S. **345**(12), 709–712 (2007)
6. W. Hackbusch, *Multigrid Methods and Applications*. Springer Series in Computational Mathematics, vol. 4 (Springer, Berlin, 1985)
7. P. Jiráńek, Z. Strakoš, M. Vohralík, A posteriori error estimates including algebraic error and stopping criteria for iterative solvers. SIAM J. Sci. Comput. **32**(3), 1567–1590 (2010)
8. K.Y. Kim, A posteriori error analysis for locally conservative mixed methods. Math. Comput. **76**(257), 43–66 (2007)
9. R. Luce, B. Wohlmuth, A local a posteriori error estimator based on equilibrated fluxes. SIAM J. Numer. Anal. **42**(4), 1394–1414 (2004)
10. P.-A. Raviart, J.-M. Thomas, A mixed finite element method for 2nd order elliptic problems, in *Mathematical Aspects of Finite Element Methods: Proceedings of the Conference Consiglio Naz. delle Ricerche, Rome* (Springer, Berlin, 1977), pp. 292–315
11. M. Vohralík, Residual flux-based a posteriori error estimates for finite volume and related locally conservative methods. Numer. Math. **111**(1), 121–158 (2008)

An a Posteriori Error Estimator for a New Stabilized Formulation of the Brinkman Problem

Tomás Barrios, Rommel Bustinza, Galina C. García, and María González

Abstract We present in this work an a posteriori error estimator for a porous media flow problem that follows the Brinkman model. First, we introduce the pseudostress as an auxiliary unknown, which let us to eliminate the pressure and thus derive a dual-mixed formulation in velocity-pseudostress. Next, in order to circumvent an inf-sup condition for the unique solvability, we stabilize the scheme by adding some appropriate least squares terms. The existence and uniqueness of solution are guaranteed and we derive an a posteriori error estimator based on the Ritz projection of the error, which is reliable and efficient up to high order terms. Finally, we report one numerical example confirming the good properties of the estimator.

1 Introduction

This note deals with the numerical approximation of the velocity and pressure of a porous media flow problem defined on a bounded and simply connected domain Ω in $\mathbb{R}^2$, with polygonal boundary $\Gamma := \partial\Omega$. Indeed, this boundary value problem corresponds to the well-known Brinkman model and reads as follows: Given the

T. Barrios
Universidad Católica de la Santísima Concepción, Casilla 297, Concepción, Chile
e-mail: tomas@ucsc.cl

R. Bustinza (✉)
CI²MA and Departamento de Ingeniería Matemática, Universidad de Concepción, Casilla 160-C, Concepción, Chile
e-mail: rbustinz@ing-mat.udec.cl

G.C. García
Universidad de Santiago de Chile, Casilla 307, Correo 2, Santiago, Chile
e-mail: galina.garcia@usach.cl

M. González
Universidade da Coruña, Campus de Elviña s/n, 15071, A Coruña, Spain
e-mail: mgtaboad@udc.es

A. Abdulle et al. (eds.), *Numerical Mathematics and Advanced Applications ENUMATH 2013*, Lecture Notes in Computational Science and Engineering 103,
DOI 10.1007/978-3-319-10705-9_25

source term $\boldsymbol{f} \in [L^2(\Omega)]^2$ and $\boldsymbol{g} \in [H^{1/2}(\Gamma)]^2$, we look for the velocity $\boldsymbol{u}$ and the pressure p of the fluid occupying the region Ω, such that

$$\alpha \boldsymbol{u} - \nu \Delta \boldsymbol{u} + \nabla p = \boldsymbol{f} \text{ in } \quad \Omega, \quad \mathrm{div}(\boldsymbol{u}) = 0 \text{ in } \quad \Omega, \quad \boldsymbol{u} = \boldsymbol{g} \text{ on } \quad \Gamma, \tag{1}$$

where ν is the kinematic viscosity of the fluid, that we assume constant, α is a positive parameter related to the permeability of the porous media and the datum $\boldsymbol{g}$ satisfies the compatibility condition $\int_\Gamma \boldsymbol{g} \cdot \boldsymbol{n} = 0$, where $\boldsymbol{n}$ stands for the unit outward normal to Γ. In addition, in order to guarantee uniqueness, we look for the pressure $p \in L_0^2(\Omega) := \{q \in L^2(\Omega) : \int_\Omega q = 0\}$.

Probably the most popular finite element discretization of these equations is constructed from a variational formulation where the unknowns are the primitive variables $\boldsymbol{u}$ and p. However, in the last years several studies were made introducing the pseudostress $\boldsymbol{\sigma} := \nu \nabla \boldsymbol{u} - p\boldsymbol{I}$ in Ω as an additional unknown. Up to the authors' knowledge, this was first suggested in the late 1980s in [1], in the context of anisotropic elasticity. In our case, we can eliminate the pressure from (1), since $p = -\frac{1}{2}\mathrm{tr}(\boldsymbol{\sigma})$, and thus deduce a formulation in velocity-pseudostress. On the other hand, it is known also that the unique solvability of a dual formulation involves the proof of a certain inf-sup condition, which limits the choice of discrete approximation spaces. Nevertheless, the use of stabilization techniques allows us to choose a larger class of discretization spaces, and has been the topic of intensive research. For example, a least squares finite element method based on the pseudostress was studied in [5] and different conforming mixed methods applied to the Stokes system in two and three dimensions have been developed in [6, 7, 9]. For related works using the discontinuous Galerkin approach we mention [2] and the references therein. Finally, a quasi-Newtonian Stokes flow is analysed in [8]. Extensions to the Brinkman model can be found in [3] and [10].

In particular, we consider the scheme introduced recently in Remark 3.1 in [3], where the velocity-pseudostress formulation is stabilized by adding suitable least squares terms. Numerical experiments reveal that the scheme is competitive. Then, our interest now is to develop an a posteriori error estimator and use it in an adaptive algorithm to obtain improved numerical approximations. The rest of the paper is organized as follows. In Sect. 2, we recall the velocity-pseudostress formulation described in Remark 3.1 in [3] and establish a new/more general existence and uniqueness result as well as the a priori error estimate. Then, in Sect. 3, we give a brief description of the analysis developed in Section 4 in [4], which let us to obtain a reliable and quasi-efficient a posteriori error estimator. Finally, a numerical example is reported in Sect. 4, validating our theoretical results.

We end this section with some notations to be used throughout the article. Given any Hilbert space H, we denote by H^2 the space of vectors of order 2 with entries in H, and by $H^{2\times 2}$ the space of square tensors of order 2 with entries in H. In particular, given $\boldsymbol{\tau} := (\tau_{ij})$, $\boldsymbol{\zeta} := (\zeta_{ij}) \in \mathbb{R}^{2\times 2}$, we write, as usual, $\boldsymbol{\tau}^{\mathrm{t}} := (\tau_{ji})$, $\mathrm{tr}(\boldsymbol{\tau}) := \tau_{11} + \tau_{22}$, $\boldsymbol{\tau}^{\mathrm{d}} := \boldsymbol{\tau} - \frac{1}{2}\mathrm{tr}(\boldsymbol{\tau})\,\boldsymbol{I}$ and $\boldsymbol{\tau} : \boldsymbol{\zeta} := \sum_{i,j=1}^{2} \tau_{ij}\,\zeta_{ij}$, where $\boldsymbol{I}$ is the identity matrix in $\mathbb{R}^{2\times 2}$. We also use the standard notations for Sobolev spaces and norms. We denote by $\boldsymbol{H}_0 := \{\boldsymbol{\tau} \in H(\mathrm{div}, \Omega) : \int_\Omega \mathrm{tr}(\boldsymbol{\tau}) = 0\}$. We recall that $H(\mathrm{div}, \Omega) = \boldsymbol{H}_0 \oplus \mathbb{R}\boldsymbol{I}$, that is, for any $\boldsymbol{\tau} \in H(\mathrm{div}, \Omega)$ there exists a unique $\boldsymbol{\tau}_0 \in \boldsymbol{H}_0$ and $d := \frac{1}{2|\Omega|}\int_\Omega \mathrm{tr}(\boldsymbol{\tau}) \in \mathbb{R}$ such that $\boldsymbol{\tau} = \boldsymbol{\tau}_0 + d\,\boldsymbol{I}$. Finally, we use C or c, with or without subscripts, to denote generic constants, independent of the discretization parameters, which may take different values at different occurrences.

2 The Discrete Augmented Formulation

In this section we recall the stabilized mixed variational formulation introduced in Remark 3.1 in [3]: Find $(\boldsymbol{\sigma}, \boldsymbol{u}) \in \boldsymbol{H} := \boldsymbol{H}_0 \times [H^1(\Omega)]^2$ such that

$$a((\boldsymbol{\sigma}, \boldsymbol{u}), (\boldsymbol{\tau}, \boldsymbol{v})) = F(\boldsymbol{\tau}, \boldsymbol{v}) \qquad \forall\, (\boldsymbol{\tau}, \boldsymbol{v}) \in \boldsymbol{H}, \tag{2}$$

where the bilinear form $a : \boldsymbol{H} \times \boldsymbol{H} \to \mathbb{R}$, and the linear functional $F : \boldsymbol{H} \to \mathbb{R}$ are defined by

$$a((\boldsymbol{\zeta}, \boldsymbol{w}), (\boldsymbol{\tau}, \boldsymbol{v})) := \int_\Omega \boldsymbol{\zeta}^{\mathrm{d}} : \boldsymbol{\tau}^{\mathrm{d}} + \nu \int_\Omega \boldsymbol{w} \cdot \mathbf{div}(\boldsymbol{\tau}) - \nu \int_\Omega \boldsymbol{v} \cdot \mathbf{div}(\boldsymbol{\zeta})$$

$$+\alpha\nu \int_\Omega \boldsymbol{w} \cdot \boldsymbol{v} + \kappa_1 \int_\Omega (\nu \nabla \boldsymbol{w} - \boldsymbol{\zeta}^{\mathrm{d}}) : \nabla \boldsymbol{v} + \kappa_2 \int_\Omega (\mathbf{div}(\boldsymbol{\zeta}) - \alpha \boldsymbol{w}) \cdot \mathbf{div}(\boldsymbol{\tau}),$$

and $F(\boldsymbol{\tau}, \boldsymbol{v}) := \nu \int_\Omega \boldsymbol{f} \cdot \boldsymbol{v} + \nu \int_\Gamma \boldsymbol{g} \cdot \boldsymbol{\tau n} - \kappa_2 \int_\Omega \boldsymbol{f} \cdot \mathbf{div}(\boldsymbol{\tau})$, for any $(\boldsymbol{\zeta}, \boldsymbol{w}), (\boldsymbol{\tau}, \boldsymbol{v}) \in \boldsymbol{H}$, where κ_1 and κ_2 are positive parameters at our disposal. Now, for the derivation of the corresponding Galerkin scheme, we consider finite element subspaces $H_{0,h}^{\boldsymbol{\sigma}} \subset \boldsymbol{H}_0$ and $H_h^{\boldsymbol{u}} \subset [H^1(\Omega)]^2$. Then, a discrete scheme associated to the variational problem (2) reads: Find $(\boldsymbol{\sigma}_h, \boldsymbol{u}_h) \in \boldsymbol{H}_h := H_{0,h}^{\boldsymbol{\sigma}} \times H_h^{\boldsymbol{u}}$ such that

$$a((\boldsymbol{\sigma}_h, \boldsymbol{u}_h), (\boldsymbol{\tau}_h, \boldsymbol{v}_h)) = F(\boldsymbol{\tau}_h, \boldsymbol{v}_h), \qquad \forall\, (\boldsymbol{\tau}_h, \boldsymbol{v}_h) \in \boldsymbol{H}_h. \tag{3}$$

In order to describe a particular finite element subspace $\boldsymbol{H}_h$, we let $\{\mathcal{T}_h\}_{h>0}$ be a regular family of triangulations of $\bar{\Omega}$. We assume that $\bar{\Omega} = \cup\{T : T \in \mathcal{T}_h\}$ and, given a triangle $T \in \mathcal{T}_h$, we denote by h_T its diameter and define the mesh size $h := \max\{h_T : T \in \mathcal{T}_h\}$. In addition, given an integer $\ell \geq 0$ and a subset S of $\mathbb{R}^2$, we denote by $\mathcal{P}_\ell(S)$ the space of polynomials in two variables defined in S of total degree at most ℓ, and for each $T \in \mathcal{T}_h$, we define the local Raviart-Thomas space of order zero

$$\mathcal{RT}_0(T) := \mathrm{span}\left\{ \begin{pmatrix} 1 \\ 0 \end{pmatrix}, \begin{pmatrix} 0 \\ 1 \end{pmatrix}, \begin{pmatrix} x_1 \\ x_2 \end{pmatrix} \right\} \subseteq [\mathcal{P}_1(T)]^2,$$

where $\begin{pmatrix} x_1 \\ x_2 \end{pmatrix}$ is a generic vector of $\mathbb{R}^2$. Then we define

$$H_h^{\sigma} := \{\tau_h \in H(\text{div}, \Omega) : \tau_h|_T \in [\mathcal{RT}_0(T)^{\mathrm{t}}]^2 \quad \forall\, T \in \mathcal{T}_h\} ,$$

$$H_{0,h}^{\sigma} := \left\{\tau_h \in H_h^{\sigma} : \int_{\Omega} \text{tr}(\tau_h) = 0\right\} ,$$

$$X_h := \{v_h \in \mathcal{C}(\bar{\Omega}) : \quad v_h|_T \in \mathcal{P}_1(T) \quad \forall\, T \in \mathcal{T}_h\} \text{ and } H_h^{u} := X_h \times X_h .$$

2.1 Well-Posedness and a Priori Error Estimate

The proof of the unique solvability of (2) and (3) relies on the application of the well-known Lax-Milgram lemma. To this end, we consider the usual norms on $H(\mathbf{div}; \Omega)$ and $[H^1(\Omega)]^2$, that is

$$||\tau||^2_{H(\mathbf{div};\Omega)} := ||\tau||^2_{[L^2(\Omega)]^{2\times 2}} + ||\mathbf{div}(\tau)||^2_{[L^2(\Omega)]^2} \quad \forall\, \tau \in H(\mathbf{div}; \Omega) ,$$

and

$$||v||^2_{[H^1(\Omega)]^2} := ||\nabla v||^2_{[L^2(\Omega)]^{2\times 2}} + ||v||^2_{[L^2(\Omega)]^2} \quad \forall\, v \in [H^1(\Omega)]^2 .$$

In addition, we introduce the global norm on $H(\mathbf{div}; \Omega) \times [H^1(\Omega)]^2$

$$||(\tau, v)||^2_{H} := ||\tau||^2_{H(\mathbf{div};\Omega)} + ||v||^2_{[H^1(\Omega)]^2} \quad \forall\, (\tau, v) \in H(\mathbf{div}; \Omega) \times [H^1(\Omega)]^2 .$$

Now, proceeding similarly as in the proof of Lemma 3.2 in [3], it is not difficult to check that when $\kappa_1 \in (0, 2\nu)$ and $\kappa_2 \in (0, 2\nu/\alpha)$, the bilinear form a is elliptic on $\boldsymbol{H}$, that is there exists a positive constant $C_{\texttt{ell}}$ such that

$$a((\tau, v), (\tau, v)) \geq C_{\texttt{ell}}\, \|(\tau, v)\|^2_{H} , \quad \forall\, (\tau, v) \in \boldsymbol{H} ,$$

which ensures the well-posedness of problems (2) and (3). The next result establishes the convergence of the method with the optimal rate, provided the exact solution of problem is sufficiently regular.

Theorem 1 *Let* $(\sigma, u) \in \boldsymbol{H}$ *and* $(\sigma_h, u_h) \in \boldsymbol{H}_h$ *be the unique solutions of the continuous and discrete augmented mixed formulations* (2) *and* (3), *respectively. Assuming that* $\sigma \in [H^r(\Omega)]^{2\times 2}$, $\mathbf{div}(\sigma) \in [H^r(\Omega)]^2$ *and* $u \in [H^{r+1}(\Omega)]^2$, *for some* $r \in (0, 1]$, *there exists* $C_{\texttt{opt}} > 0$, *independent of h, such that*

$$\|(\sigma, u) - (\sigma_h, u_h)\|_H$$
$$\leq C_{\texttt{opt}} h^r \left\{\|\sigma\|_{[H^r(\Omega)]^{2\times 2}} + \|\mathbf{div}(\sigma)\|_{[H^r(\Omega)]^2} + \|u\|_{[H^{r+1}(\Omega)]^2}\right\} .$$

Proof Under the assumptions made on the stabilisation parameters κ_1 and κ_2, a Céa-type estimate holds. The rest of the proof is a straightforward application of the approximation theory in Sobolev spaces.

3 A Ritz Projection-Based a Posteriori Error Analysis

The aim of this section is to summarize the analysis developed in Section 4 in [4]. We omit proofs here and, eventually, could give some comments on them. First, we consider the Ritz projection of the error with respect to the inner product of $\boldsymbol{H}$, which is defined as the unique $(\bar{\boldsymbol{\sigma}}, \bar{\boldsymbol{u}}) \in \boldsymbol{H}$ satisfying

$$\langle(\bar{\boldsymbol{\sigma}}, \bar{\boldsymbol{u}}), (\boldsymbol{\tau}, \boldsymbol{v})\rangle_H = a((\boldsymbol{\sigma} - \boldsymbol{\sigma}_h, \boldsymbol{u} - \boldsymbol{u}_h), (\boldsymbol{\tau}, \boldsymbol{v})) \qquad \forall\, (\boldsymbol{\tau}, \boldsymbol{v}) \in \boldsymbol{H}. \tag{4}$$

The existence and uniqueness of $(\bar{\boldsymbol{\sigma}}, \bar{\boldsymbol{u}})$ is guaranteed by the Lax-Milgram lemma. The next result gives us an upper bound for $||(\bar{\boldsymbol{\sigma}}, \bar{\boldsymbol{u}})||_H$, and corresponds to Lemma 4.1 in [4].

Lemma 1 *There exists a positive constant $C = C(\kappa_1, \kappa_2, \nu)$, independent of h, such that*

$$||(\bar{\boldsymbol{\sigma}}, \bar{\boldsymbol{u}})||_H \le C \Big(||\boldsymbol{f} + \mathbf{div}(\boldsymbol{\sigma}_h) - \alpha \boldsymbol{u}_h||_{[L^2(\Omega)]^2} + ||\nu \nabla \boldsymbol{u}_h - \boldsymbol{\sigma}_h^{\mathrm{d}}||_{[L^2(\Omega)]^{2\times 2}} + \nu\, ||\boldsymbol{u}_h - \boldsymbol{g}||_{[H^{1/2}(\Gamma)]^2} \Big).$$

Motivated by the previous result, we define the estimator θ as follows:

$$\theta^2 := \sum_{T \in \mathcal{T}_h} \Big(||\boldsymbol{f} + \mathbf{div}(\boldsymbol{\sigma}_h) - \alpha \boldsymbol{u}_h||^2_{[L^2(T)]^2} + ||\nu \nabla \boldsymbol{u}_h - \boldsymbol{\sigma}_h^{\mathrm{d}}||^2_{[L^2(T)]^{2\times 2}} \Big) + \nu^2\, ||\boldsymbol{u}_h - \boldsymbol{g}||^2_{[H^{1/2}(\Gamma)]^2}.$$

In the next theorem we establish the equivalence between the global error and the (non local) estimator θ (see Theorem 4.1 in [4] for the proof), whose global efficiency is proved, for simplicity, under the assumption $\boldsymbol{f}$ is piecewise polynomial in $\mathcal{T}_h$. Otherwise, (high order) data approximation terms need to be considered for the efficiency of the estimator.

Theorem 2 *Let $(\boldsymbol{\sigma}, \boldsymbol{u}) \in \boldsymbol{H}$ and $(\boldsymbol{\sigma}_h, \boldsymbol{u}_h) \in \boldsymbol{H}_h$ be the unique solutions of problems* (2) *and* (3), *respectively. Then, there exist positive constants $C_{\mathtt{eff}}$, $C_{\mathtt{rel}}$, independent of h, such that*

$$C_{\mathtt{eff}}\theta \le \|(\boldsymbol{\sigma} - \boldsymbol{\sigma}_h, \boldsymbol{u} - \boldsymbol{u}_h)\|_H \le C_{\mathtt{rel}}\theta.$$

Now, in order to obtain a **local** a posteriori error estimator, we first let $\bar{\boldsymbol{u}}_h$ be the unique continuous piecewise-linear function on $\mathscr{T}_h$ such that $\bar{\boldsymbol{u}}_h(\boldsymbol{x}) = \boldsymbol{u}_h(\boldsymbol{x})$ for all node $\boldsymbol{x}$ of $\mathscr{T}_h$ in Ω and $\bar{\boldsymbol{u}}_h(\boldsymbol{x}) = \boldsymbol{g}(\boldsymbol{x})$ for all node $\boldsymbol{x}$ of $\mathscr{T}_h$ on Γ.

Then, we introduce the a posteriori error estimator $\bar{\theta} := \left(\sum_{T\in\mathscr{T}_h} \bar{\theta}_T^2 \right)^{1/2}$, where for each $T \in \mathscr{T}_h$ the **local** a posteriori error estimator $\bar{\theta}_T$ is given by

$$\begin{aligned}\bar{\theta}_T^2 &:= \|\boldsymbol{f} + \mathbf{div}(\boldsymbol{\sigma}_h) - \alpha \boldsymbol{u}_h\|_{[L^2(T)]^2}^2 + \|\nu\nabla\bar{\boldsymbol{u}}_h - \boldsymbol{\sigma}_h^{\mathrm{d}}\|_{[L^2(T)]^{2\times 2}}^2 \\ &+ \|\boldsymbol{u}_h - \bar{\boldsymbol{u}}_h\|_{[L^2(T)]^2}^2 + \sum_{e\in E(T)\cap E_\Gamma} h_e \left\| \frac{\mathrm{d}\boldsymbol{g}}{\mathrm{d}\boldsymbol{t}_T} - \frac{\mathrm{d}\bar{\boldsymbol{u}}_h}{\mathrm{d}\boldsymbol{t}_T} \right\|_{[L^2(e)]^2}^2 .\end{aligned} \tag{5}$$

We conclude this section with our main result in this note. We proved that the a posteriori error estimator $\bar{\theta}$ is reliable and locally efficient in interior triangles. We refer to Lemma 4.2 and Theorem 4.3 in [4] for further details.

Theorem 3 *Assume $\boldsymbol{g} \in [H^1(\Gamma)]^2$. Then, there exists a constant $\bar{C}_{\mathtt{rel}} > 0$, independent of h, such that $\|(\boldsymbol{\sigma} - \boldsymbol{\sigma}_h, \boldsymbol{u} - \boldsymbol{u}_h)\|_{\boldsymbol{H}} \leq \bar{C}_{\mathtt{rel}}\, \bar{\theta}$. If, moreover, $\boldsymbol{f}$ and $\boldsymbol{g}$ are piecewise polynomials in $\mathscr{T}_h$ and E_Γ respectively, then there exists a positive constant, $\bar{C}_{\mathtt{eff}}$, independent of h, such that*

$$\bar{C}_{\mathtt{eff}}\, \bar{\theta}_T \leq \|(\boldsymbol{\sigma} - \boldsymbol{\sigma}_h, \boldsymbol{u} - \boldsymbol{u}_h)\|_{\boldsymbol{H}(T)} + \nu^{1/2} \chi(T)\, \|\boldsymbol{u}_h - \bar{\boldsymbol{u}}_h\|_{[H^1(T)]^2} \qquad \forall\, T \in \mathscr{T}_h ,$$

where $\|(\boldsymbol{\tau}, \boldsymbol{v})\|_{\boldsymbol{H}(T)}^2 := \|\boldsymbol{\tau}\|_{H(\mathbf{div};\, T)}^2 + \|\boldsymbol{v}\|_{[H^1(T)]^2}^2$ and $\chi(T)$ is equal to 1 if $\partial T \cap \Gamma \neq \emptyset$ and is equal to 0 otherwise.

We remark that, for α large enough, $\bar{C}_{\mathtt{rel}} = \mathscr{O}(\alpha)$ and $\bar{C}_{\mathtt{eff}} = \mathscr{O}(\alpha^{-1})$.

4 Numerical Examples

In this section, we show one numerical experiment illustrating the performance of the augmented mixed scheme (3) and the a posteriori error estimator $\bar{\theta}$, defined in (5). To this end we remark that, for implementation purposes, it is very hard to find a suitable basis of $H_{0,h}^{\boldsymbol{\sigma}}$ due to the null media condition required by their elements. We circumvent this difficulty by imposing this requirement through a Lagrange multiplier, which leads to solve the following equivalent discrete scheme (see Theorem 5.1 in [4]): Find $(\boldsymbol{\sigma}_h, \boldsymbol{u}_h, \varphi_h) \in H_h^{\boldsymbol{\sigma}} \times H_h^{\boldsymbol{u}} \times \mathbb{R}$ such that

$$a((\boldsymbol{\sigma}_h, \boldsymbol{u}_h), (\boldsymbol{\tau}, \boldsymbol{v})) + \varphi_h \int_\Omega \operatorname{tr}(\boldsymbol{\tau}) = F(\boldsymbol{\tau}, \boldsymbol{v}) \quad \forall\, (\boldsymbol{\tau}, \boldsymbol{v}) \in H_h^{\boldsymbol{\sigma}} \times H_h^{\boldsymbol{u}},$$
$$\psi \int_\Omega \operatorname{tr}(\boldsymbol{\sigma}_h) = 0 \quad \forall\, \psi \in \mathbb{R}. \tag{6}$$

We now specify the data of the example we present here. We take Ω as the square $]-1,1[^2$. The data $\boldsymbol{f}$ and $\boldsymbol{g}$ are chosen so that the exact velocity and pressure are, respectively, $\boldsymbol{u}(\boldsymbol{x}) = -\sqrt{\alpha}\, \mathrm{e}^{-\sqrt{\alpha}(1-x_1-x_2)^2} \begin{pmatrix} 1 \\ -1 \end{pmatrix}$ and $p(\boldsymbol{x}) = 2\,\mathrm{e}^{x_1} \sin(x_2)$ for all $\boldsymbol{x} = (x_1, x_2) \in \Omega$. We emphasize that in this case $\boldsymbol{u} \in [H^1(\Omega)]^2$ has an inner layer around the line $x_2 = 1 - x_1$.

In what follows, DOF stands for the total number of degrees of freedom (unknowns) of (6), that is, $DOF = 2 \times$ (Numbers of vertices of $\mathscr{T}_h$) + 2 × (Number of edges of $\mathscr{T}_h$) + 1, which leads asymptotically to 4 unknowns per triangle, which reflects the low computational cost, almost the same as the required by considering the $\mathbf{P}_1$–iso$\mathbf{P}_1$ elements for the standard velocity-pressure formulation, whose degrees of freedom are asymptotically 4.5 (unknowns) per triangle. In addition, by setting $p_h := -\frac{1}{2}\operatorname{tr}(\boldsymbol{\sigma}_h)$, we obtain a reasonable piecewise-linear approximation of the pressure $p := -\frac{1}{2}\operatorname{tr}(\boldsymbol{\sigma})$.

Hereafter, we denote the individual errors by $\boldsymbol{e}(\boldsymbol{u}) := \|\boldsymbol{u} - \boldsymbol{u}_h\|_{[H^1(\Omega)]^2}$ and $\boldsymbol{e}(\boldsymbol{\sigma}) := \|\boldsymbol{\sigma} - \boldsymbol{\sigma}_h\|_{H(\mathbf{div},\Omega)}$, the total error by $\boldsymbol{e} := \Big([\boldsymbol{e}(\boldsymbol{u})]^2 + [\boldsymbol{e}(\boldsymbol{\sigma})]^2 \Big)^{1/2}$, and the effectivity index is given by $\boldsymbol{e}/\bar{\theta}$. In addition, if $\boldsymbol{e}$ and $\tilde{\boldsymbol{e}}$ stand for the errors at two consecutive triangulations with N and $\tilde{N}$ degrees of freedom, respectively, then the experimental rate of convergence is given by $r := -2\dfrac{\log(\boldsymbol{e}/\tilde{\boldsymbol{e}})}{\log(N/\tilde{N})}$. The definitions of $r(\boldsymbol{u})$ and $r(\boldsymbol{\sigma})$ are analogous. The numerical results shown below were obtained in a Pentium Xeon computer with a dual processor, using a MATLAB code. Table 1 contains the convergence history for the adaptive refinement algorithm based on $\bar{\theta}$, with $\alpha = 10^6$, $\nu = 0.5$, $\kappa_1 = \nu$ and $\kappa_2 = \nu/\alpha$ (in agreement with the feasible choices described in Sect. 2), when performing the red-blue-green refinement technique. We notice that, looking at the experimental rates of convergence, the order $\mathscr{O}(h)$ is observed for all the unknowns. Moreover, the effectivity index remains bounded, which is in agreement with Theorem 3. From Fig. 1, we observe that the adaptive refinement algorithm converges faster than the uniform one. Finally, we recall that the details of the analysis described here, as well as other numerical experiments, are included in [4].

Table 1 Errors, experimental convergence rates and effectivity index for the adaptive refinement algorithm based on $\bar{\theta}$ with $\alpha = 10^6$ and $\nu = 0.5$

DOF	$e(u)$	$r(u)$	$e(\sigma)$	$r(\sigma)$	e	r	$e/\bar{\theta}$
19	2.117e+03	–	1.957e+06	–	1.957e+06	–	2.815e-03
51	2.183e+03	–	1.369e+06	0.7242	1.369e+06	0.7242	2.529e-03
139	2.446e+03	–	9.448e+05	0.7397	9.448e+05	0.7397	2.317e-03
339	6.778e+03	–	6.856e+05	0.7193	6.856e+05	0.7192	2.394e-03
763	1.121e+04	–	5.212e+05	0.6758	5.213e+05	0.6754	2.869e-03
2,595	1.117e+04	0.0046	4.216e+05	0.3466	4.217e+05	0.3465	5.602e-03
4,451	7.740e+03	1.3610	2.631e+05	1.7479	2.632e+05	1.7476	1.223e-02
12,859	3.761e+03	1.3605	1.521e+05	1.0334	1.521e+05	1.0336	2.599e-02
15,211	3.396e+03	1.2148	1.482e+05	0.3077	1.482e+05	0.3082	3.767e-02
37,923	1.821e+03	1.3648	9.230e+04	1.0365	9.231e+04	1.0367	6.427e-02
53,867	1.653e+03	0.5509	9.336e+04	–	9.337e+04	–	9.700e-02
147,731	9.076e+02	1.1888	5.674e+04	0.9873	5.674e+04	0.9873	1.507e-01

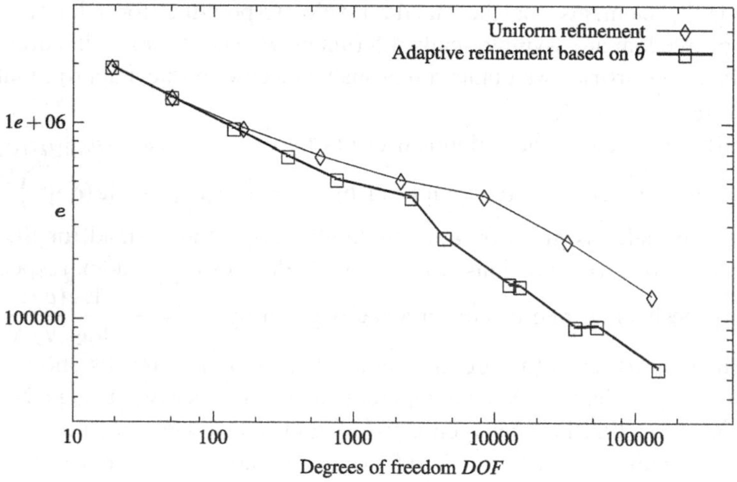

Fig. 1 Total error e vs DOF for uniform and adaptive refinements, with $\alpha = 10^6$ and $\nu = 0.5$

Acknowledgements This work has been partially supported by MICINN project MTM2010-21135-C01-01 and Xunta de Galicia through INCITE09 105339PR, the Dirección de Investigación of the Universidad Católica de la Santísima Concepción, FONDECYT Grants No. 1130158 and 1120560, CONICYT-Chile through BASAL project CMM, Universidad de Chile, by Centro de Investigación en Ingeniería Matemática (CI^{2}MA), Universidad de Concepción, by CONICYT project Anillo ACT1118 (ANANUM).

References

1. D.N. Arnold, R.S. Falk, Well-posedness of the fundamental boundary value problems for constrained anisotropic elastic materials. Arch. Ration. Mech. Anal. **98**(2), 143–165 (1987)
2. T.P. Barrios, R. Bustinza, An augmented discontinuous Galerkin method for stationary Stokes problem, Preprint 2010–20, Departamento de Ingeniería Matemática, Universidad de Concepción, 2010
3. T.P. Barrios, R. Bustinza, G.C. García, E. Hernández, On stabilized mixed methods for generalized Stokes problem based on the velocity-pseudostress formulation: a priori error estimates. Comput. Methods Appl. Mech. Eng. **237–240**, 78–87 (2012)
4. T.P. Barrios, R. Bustinza, G.C. García, M. González, A posteriori error analyses of a velocity-pseudostress formulation of the generalized Stokes problem, Preprint 2013–04, Centro de Investigación en Ingeniería Matemática, Universidad de Concepción, 2013
5. Z. Cai, B. Lee, P. Wang, Least-squares methods for incompressible Newtonian fluid flow: linear stationary problems. SIAM J. Numer. Anal. **42**, 843–859 (2004)
6. Z. Cai, C. Tong, P.S. Vassilevski, C. Wang, Mixed finite element methods for incompressible flow: stationary Stokes equations. Numer. Methods Partial Differ. Equ. **26**, 957–978 (2010)
7. G.N. Gatica, A. Márquez, M.A. Sánchez, Analysis of a velocity-pressure-pseudostress formulation for the stationary Stokes equations. Comput. Methods Appl. Mech. Eng. **199**, 1064–1079 (2010)
8. G.N. Gatica, A. Márquez, M.A. Sánchez, A priori and a posteriori error analyses of a velocity-pseudostress formulation for a class of quasi-Newtonian Stokes flows. Comput. Methods Appl. Mech. Eng. **200**, 1619–1636 (2011)
9. G.N. Gatica, A. Márquez, M.A. Sánchez, Pseudostress-based mixed finite element methods for the Stokes problem in $\mathbb{R}^n$ with Dirichlet boundary conditions. I: a priori error analysis. Commun. Comput. Phys. **12**, 109–134 (2012)
10. G.N. Gatica, L.F. Gatica, A. Márquez, Analysis of a pseudostress-based mixed finite element method for the Brinkman model of porous media flow. Numer. Math. **126**(4), 635–677 (2014)

New a Posteriori Error Estimator for an Augmented Mixed FEM in Linear Elasticity

Tomás P. Barrios, Edwin M. Behrens, and María González

Abstract We consider an augmented mixed finite element method applied to the linear elasticity problem with non-homogeneous Dirichlet boundary conditions and derive an a posteriori error estimator that is simpler and easier to implement than the one available in the literature. The new a posteriori error estimator is reliable and locally efficient in interior triangles; in the remaining elements, it satisfies a *quasi-efficiency* bound. We provide some numerical results that illustrate the performance of the corresponding adaptive algorithm.

1 Introduction

In this work, we consider the augmented dual-mixed method introduced in [5] for the linear elasticity problem with non-homogeneous Dirichlet boundary conditions. The approach in [5] relies on the mixed method of Hellinger and Reissner, that provides simultaneous approximations of the displacement $\mathbf{u}$ and the stress tensor $\boldsymbol{\sigma}$. The symmetry of $\boldsymbol{\sigma}$ is imposed weakly, through the use of a Lagrange multiplier that can be interpreted as the rotation $\boldsymbol{\gamma} := \frac{1}{2}(\nabla\mathbf{u}-(\nabla\mathbf{u})^{\mathtt{t}})$. Then, suitable Galerkin least-squares type terms arising from the constitutive and equilibrium equations, from the relation that defines the rotation in terms of the displacement and from the Dirichlet boundary condition are added. The bilinear form of the resulting augmented variational formulation is coercive in the whole space for appropriate values of the stabilization parameters, with a coercivity constant independent of the Lamé parameter λ. Therefore, the corresponding Galerkin scheme is well-posed and free of locking for *any* choice of finite element subspaces. This fact turns out to be the main advantage of this method.

T.P. Barrios • E.M. Behrens
Universidad Católica de la Santísima Concepción, Casilla 297, Concepción, Chile
e-mail: tomas@ucsc.cl; ebehrens@ucsc.cl

M. González (✉)
Universidade da Coruña, Campus de Elviña s/n, 15071, A Coruña, Spain
BCAM, Alameda Mazarredo 14, 48009, Bilbao, Spain
e-mail: maria.gonzalez.taboada@udc.es

A. Abdulle et al. (eds.), *Numerical Mathematics and Advanced Applications ENUMATH 2013*, Lecture Notes in Computational Science and Engineering 103,
DOI 10.1007/978-3-319-10705-9_26

Concerning the a posteriori error analysis of the augmented scheme presented in [5], an a posteriori error estimator of residual type was introduced and analyzed in [2]. That a posteriori error estimator is both reliable and efficient, and involves the computation of 13 residuals per element, including normal and tangential jumps. In this work, we present a new a posteriori error estimator for the augmented dual-mixed method proposed in [5]. The analysis is based on the use of a projection of the error and allows to derive an a posteriori error estimator that only requires the computation of four residuals per element in the interior triangles, five residuals per element in the triangles with exactly one node on the boundary and six residuals per element in the triangles with a side on the boundary. Moreover, it does not involve the computation of normal nor tangential jumps, which simplifies the numerical implementation. Besides, we prove that the new a posteriori error estimator is reliable and locally efficient in the interior elements.

The rest of the paper is organized as follows. In Sect. 2 we recall the main features of the augmented dual-mixed method introduced in [5] for the linear elasticity problem with non-homogeneous Dirichlet boundary conditions. In Sect. 3 we introduce the new a posteriori error estimator and study its reliability and efficiency. Finally, in Sect. 4 we provide some numerical results that illustrate the performance of the corresponding adaptive algorithm.

2 The Augmented Mixed Finite Element Method

Let $\Omega \subset \mathbb{R}^2$ be a bounded domain with a Lipschitz-continuous boundary Γ, and let $\mathbf{f} \in [L^2(\Omega)]^2$ be a given volume force and $\mathbf{g} \in [H^{1/2}(\Gamma)]^2$ a prescribed displacement on Γ. We denote by $\mathcal{C}$ the elasticity operator determined by Hooke's law, that is,

$$\mathcal{C}\,\boldsymbol{\zeta} \;:=\; \lambda\,\mathrm{tr}(\boldsymbol{\zeta})\,\mathbf{I} \;+\; 2\,\mu\,\boldsymbol{\zeta}\,, \qquad \forall\,\boldsymbol{\zeta} \in [L^2(\Omega)]^{2\times 2}\,, \tag{1}$$

where $\lambda, \mu > 0$ are the Lamé parameters and $\mathbf{I} \in \mathbb{R}^{2\times 2}$ is the identity matrix. We look for the displacement $\mathbf{u}$ and the stress tensor $\boldsymbol{\sigma}$ of a linear elastic material such that

$$\begin{cases} -\mathbf{div}(\boldsymbol{\sigma}) = \mathbf{f} & \text{in } \Omega, \\ \boldsymbol{\sigma} = \mathcal{C}\,\boldsymbol{\varepsilon}(\mathbf{u}) & \text{in } \Omega, \\ \mathbf{u} = \mathbf{g} & \text{on } \Gamma, \end{cases} \tag{2}$$

where $\boldsymbol{\varepsilon}(\mathbf{u}) := \frac{1}{2}\,(\nabla\mathbf{u} + (\nabla\mathbf{u})^{\mathtt{t}})$ is the strain tensor of small deformations.

Let κ_1, κ_2, κ_3 and κ_4 be positive parameters. We denote by $\mathbf{H} := H_0 \times [H^1(\Omega)]^2 \times [L^2(\Omega)]^{2\times 2}_{\mathtt{skew}}$, where $H_0 := \{\boldsymbol{\tau} \in H(\mathbf{div};\Omega) \,:\, \int_\Omega \mathrm{tr}(\boldsymbol{\tau}) = 0\}$ and $[L^2(\Omega)]^{2\times 2}_{\mathtt{skew}} := \{\,\boldsymbol{\eta} \in [L^2(\Omega)]^{2\times 2} \,:\, \boldsymbol{\eta} + \boldsymbol{\eta}^{\mathtt{t}} = \mathbf{0}\,\}$, and define the bilinear form

$A : \mathbf{H} \times \mathbf{H} \to \mathbb{R}$ and the linear functional $F : \mathbf{H} \to \mathbb{R}$ as follows:

$$\begin{aligned} A((\boldsymbol{\sigma},\mathbf{u},\boldsymbol{\gamma}),(\boldsymbol{\tau},\mathbf{v},\boldsymbol{\eta})) &:= \int_\Omega \mathscr{C}^{-1}\boldsymbol{\sigma} : \boldsymbol{\tau} + \int_\Omega \mathbf{u}\cdot\mathbf{div}(\boldsymbol{\tau}) + \int_\Omega \boldsymbol{\tau} : \boldsymbol{\gamma} \\ &- \int_\Omega \mathbf{v}\cdot\mathbf{div}(\boldsymbol{\sigma}) - \int_\Omega \boldsymbol{\sigma} : \boldsymbol{\eta} + \kappa_1 \int_\Omega \left(\boldsymbol{\varepsilon}(\mathbf{u}) - \mathscr{C}^{-1}\boldsymbol{\sigma}\right) : \left(\boldsymbol{\varepsilon}(\mathbf{v}) + \mathscr{C}^{-1}\boldsymbol{\tau}\right) \\ &+ \kappa_3 \int_\Omega \left(\boldsymbol{\gamma} - \frac{1}{2}(\nabla\mathbf{u} - (\nabla\mathbf{u})^{\mathtt{t}})\right) : \left(\boldsymbol{\eta} + \frac{1}{2}(\nabla\mathbf{v} - (\nabla\mathbf{v})^{\mathtt{t}})\right) \\ &+ \kappa_2 \int_\Omega \mathbf{div}(\boldsymbol{\sigma})\cdot\mathbf{div}(\boldsymbol{\tau}) + \kappa_4 \int_\Gamma \mathbf{u}\cdot\mathbf{v}, \end{aligned} \tag{3}$$

$$F(\boldsymbol{\tau},\mathbf{v},\boldsymbol{\eta}) := \int_\Omega \mathbf{f}\cdot(\mathbf{v} - \kappa_2\,\mathbf{div}(\boldsymbol{\tau})) + \langle\boldsymbol{\tau}\,\mathbf{n},\mathbf{g}\rangle_\Gamma + \kappa_4 \int_\Gamma \mathbf{g}\cdot\mathbf{v} + \kappa_1\, c_{\mathbf{g}} \int_\Gamma \mathbf{v}\cdot\mathbf{n}, \tag{4}$$

for any $(\boldsymbol{\sigma},\mathbf{u},\boldsymbol{\gamma})$, $(\boldsymbol{\tau},\mathbf{v},\boldsymbol{\eta}) \in \mathbf{H}$, where $\mathbf{n}$ is the unit outward normal to Γ, $\langle\cdot,\cdot\rangle_\Gamma$ denotes the duality pairing between $[H^{-1/2}(\Gamma)]^2$ and $[H^{1/2}(\Gamma)]^2$ with respect to the $[L^2(\Gamma)]^2$-inner product, and $c_{\mathbf{g}} := \frac{1}{2|\Omega|}\int_\Gamma \mathbf{g}\cdot\mathbf{n}$.

Let us denote by $\|\cdot\|_{\mathbf{H}}$ the product norm of $\mathbf{H}$. The augmented variational formulation introduced in [5] consists in finding $(\boldsymbol{\sigma},\mathbf{u},\boldsymbol{\gamma}) \in \mathbf{H}$ such that

$$A((\boldsymbol{\sigma},\mathbf{u},\boldsymbol{\gamma}),(\boldsymbol{\tau},\mathbf{v},\boldsymbol{\eta})) = F(\boldsymbol{\tau},\mathbf{v},\boldsymbol{\eta}), \qquad \forall\,(\boldsymbol{\tau},\mathbf{v},\boldsymbol{\eta}) \in \mathbf{H}. \tag{5}$$

From now on, we assume that the parameters κ_1, κ_2, κ_3 and κ_4 are chosen independently of λ and such that $\kappa_1 \in (0, 2\,\mu)$, $\kappa_2 > 0$, $0 < \kappa_3 < \left(\frac{\kappa_0}{1-\kappa_0}\right)\kappa_1$ if $\kappa_0 < 1$ or $\kappa_3 > 0$ if $\kappa_0 \geq 1$, and $\kappa_4 \geq \kappa_1 + \kappa_3$, where κ_0 is the constant of a Korn-type inequality (see [5] for more details). Under these assumptions, there exists a positive constant α, independent of λ, such that

$$A((\boldsymbol{\tau},\mathbf{v},\boldsymbol{\eta}),(\boldsymbol{\tau},\mathbf{v},\boldsymbol{\eta})) \geq \alpha\,\|(\boldsymbol{\tau},\mathbf{v},\boldsymbol{\eta})\|_{\mathbf{H}}^2, \qquad \forall\,(\boldsymbol{\tau},\mathbf{v},\boldsymbol{\eta}) \in \mathbf{H}, \tag{6}$$

and the augmented variational formulation (5) is well-posed.

Now, let h be a positive parameter and consider a finite dimensional subspace $\mathbf{H}_h \subset \mathbf{H}$. The Galerkin scheme associated to problem (5) reads: find $(\boldsymbol{\sigma}_h,\mathbf{u}_h,\boldsymbol{\gamma}_h) \in \mathbf{H}_h$ such that

$$A((\boldsymbol{\sigma}_h,\mathbf{u}_h,\boldsymbol{\gamma}_h),(\boldsymbol{\tau}_h,\mathbf{v}_h,\boldsymbol{\eta}_h)) = F(\boldsymbol{\tau}_h,\mathbf{v}_h,\boldsymbol{\eta}_h), \quad \forall\,(\boldsymbol{\tau}_h,\mathbf{v}_h,\boldsymbol{\eta}_h) \in \mathbf{H}_h. \tag{7}$$

The existence and uniqueness of a solution to problem (7) as well as a Céa estimate are established in [5] under the assumptions made above on the stabilization parameters. In order to describe the simplest choice of finite element subspaces for the Galerkin scheme (7), we assume now that Ω is a polygonal region and let $\{\mathscr{T}_h\}_{h>0}$ be a regular family of triangulations of $\bar{\Omega}$ such that $\bar{\Omega} = \cup\{T :$

$T \in \mathscr{T}_h\}$. Given an element $T \in \mathscr{T}_h$, we denote by h_T its diameter and define the mesh size $h := \max\{h_T : T \in \mathscr{T}_h\}$. In addition, given an integer $\ell \geq 0$ and a subset S of $\mathbb{R}^2$, we denote by $\mathscr{P}_\ell(S)$ the space of polynomials in two variables defined in S of total degree at most ℓ, and for each $T \in \mathscr{T}_h$, we define the local Raviart-Thomas space of the lowest order, $\mathscr{RT}_0(T) := \langle \mathbf{e}_1, \mathbf{e}_2, \mathbf{x} \rangle \subseteq [\mathscr{P}_1(T)]^2$, where $\{\mathbf{e}_i\}_{i=1}^2$ is the canonical basis of $\mathbb{R}^2$ and $\mathbf{x} \in \mathbb{R}^2$ is a generic vector. Then, we define the finite element subspaces

$$H_h^{\boldsymbol{\sigma}} := \left\{\boldsymbol{\tau}_h \in H_0 : \boldsymbol{\tau}_h|_T \in [\mathscr{RT}_0(T)^{\mathtt{t}}]^2, \quad \forall\, T \in \mathscr{T}_h\right\} \tag{8}$$

$$H_h^{\mathbf{u}} := \left\{\mathbf{v}_h \in [\mathscr{C}(\bar{\Omega})]^2 : \mathbf{v}_h|_T \in [\mathscr{P}_1(T)]^2, \quad \forall\, T \in \mathscr{T}_h\right\} \tag{9}$$

$$H_h^{\boldsymbol{\gamma}} := \left\{\boldsymbol{\eta}_h \in [L^2(\Omega)]^{2\times 2}_{\mathtt{skew}} : \boldsymbol{\eta}_h|_T \in [\mathscr{P}_0(T)]^{2\times 2}, \quad \forall\, T \in \mathscr{T}_h\right\}. \tag{10}$$

The simplest choice of finite element subspaces for the Galerkin scheme (7) is $\mathbf{H}_h := H_h^{\boldsymbol{\sigma}} \times H_h^{\mathbf{u}} \times H_h^{\boldsymbol{\gamma}}$. In this case, if the solution to problem (5) is sufficiently smooth, we can expect the following rate of convergence:

$$\begin{aligned} \|(\boldsymbol{\sigma}, \mathbf{u}, \boldsymbol{\gamma}) - (\boldsymbol{\sigma}_h, \mathbf{u}_h, \boldsymbol{\gamma}_h)\|_{\mathbf{H}} \leq C_1 h^r \Big(& \|\boldsymbol{\sigma}\|_{[H^r(\Omega)]^{2\times 2}} + \|\mathbf{div}(\boldsymbol{\sigma})\|_{[H^r(\Omega)]^2} \\ & + \|\mathbf{u}\|_{[H^{r+1}(\Omega)]^2} + \|\boldsymbol{\gamma}\|_{[H^r(\Omega)]^{2\times 2}} \Big), \end{aligned} \tag{11}$$

where $r \in (0, 1]$ and $C_1 > 0$ is a constant independent of h and λ.

3 A Posteriori Error Analysis

In this section we present a new a posteriori error estimator for the Galerkin scheme (7). Given an element $T \in \mathscr{T}_h$, we denote by $E(T)$ the set of its edges; E_h denotes the set of all the edges of the mesh $\mathscr{T}_h$ and $E_h(\Gamma) := \{e \in E_h : e \subseteq \Gamma\}$. Finally, h_e stands for the length of edge $e \in E_h$. We also assume that problems (5) and (7) are well-posed.

Let $(\boldsymbol{\sigma}, \mathbf{u}, \boldsymbol{\gamma})$ and $(\boldsymbol{\sigma}_h, \mathbf{u}_h, \boldsymbol{\gamma}_h) \in \mathbf{H}_h$ be the unique solutions to problems (5) and (7), respectively. We define the Ritz projection of the error as the unique element $(\bar{\boldsymbol{\sigma}}, \bar{\mathbf{u}}, \bar{\boldsymbol{\gamma}}) \in \mathbf{H}$ such that

$$\langle (\bar{\boldsymbol{\sigma}}, \bar{\mathbf{u}}, \bar{\boldsymbol{\gamma}}), (\boldsymbol{\tau}, \mathbf{v}, \boldsymbol{\eta}) \rangle_{\mathbf{H}} = A((\boldsymbol{\sigma} - \boldsymbol{\sigma}_h, \mathbf{u} - \mathbf{u}_h, \boldsymbol{\gamma} - \boldsymbol{\gamma}_h), (\boldsymbol{\tau}, \mathbf{v}, \boldsymbol{\eta})), \tag{12}$$

for all $(\boldsymbol{\tau}, \mathbf{v}, \boldsymbol{\eta}) \in \mathbf{H}$. We recall that the existence and uniqueness of $(\bar{\boldsymbol{\sigma}}, \bar{\mathbf{u}}, \bar{\boldsymbol{\gamma}})$ is guaranteed by the Lax-Milgram Lemma. Hence, using the coercivity of $A(\cdot, \cdot)$, we deduce that

$$||(\boldsymbol{\sigma} - \boldsymbol{\sigma}_h, \mathbf{u} - \mathbf{u}_h, \boldsymbol{\gamma} - \boldsymbol{\gamma}_h)||_{\mathbf{H}} \leq \frac{1}{\alpha} ||(\bar{\boldsymbol{\sigma}}, \bar{\mathbf{u}}, \bar{\boldsymbol{\gamma}})||_{\mathbf{H}}. \tag{13}$$

Then, in order to obtain reliable a posteriori error estimates for the discrete scheme (7), it is enough to bound from above the Ritz projection of the error.

Lemma 1 *There exists a constant $C_2 > 0$, independent of h and λ, such that*

$$\begin{aligned} ||(\bar{\boldsymbol{\sigma}}, \bar{\mathbf{u}}, \bar{\boldsymbol{\gamma}})||_{\mathbf{H}} \leq C_2 \Big(& ||\mathbf{f} + \mathbf{div}(\boldsymbol{\sigma}_h)||_{[L^2(\Omega)]^2} + \big|\big|\boldsymbol{\sigma}_h - \boldsymbol{\sigma}_h^{\mathrm{t}}\big|\big|_{[L^2(\Omega)]^{2\times 2}} \\ & + ||\mathbf{g} - \mathbf{u}_h||_{[H^{1/2}(\Gamma)]^2} + \big|\big|\boldsymbol{\varepsilon}(\mathbf{u}_h) - \mathscr{C}^{-1}\boldsymbol{\sigma}_h - c_{\mathbf{g}}\mathbf{I}\big|\big|_{[L^2(\Omega)]^{2\times 2}} \\ & + \left|\left|\boldsymbol{\gamma}_h - \frac{1}{2}(\nabla\mathbf{u}_h - (\nabla\mathbf{u}_h)^{\mathrm{t}})\right|\right|_{[L^2(\Omega)]^{2\times 2}} \Big). \end{aligned} \tag{14}$$

Proof Consider (12) and use that $(\boldsymbol{\sigma}, \mathbf{u}, \boldsymbol{\gamma})$ is the solution to problem (5). Then, use the definitions of the linear functional F and the bilinear form $A(\cdot,\cdot)$, integrate by parts and use the Cauchy-Schwarz inequality and the continuity of $\mathscr{C}^{-1}$. It is not difficult to see that

$$C_2 \leq C_{2,\max} := \max\big(1 + \kappa_1\,(1 + ||\mathscr{C}^{-1}||), 1 + \kappa_2, 1 + 2\,\kappa_3, 1 + \kappa_4\big), \tag{15}$$

where $||\mathscr{C}^{-1}|| := ||\mathscr{C}^{-1}||_{\mathscr{L}([L^2(\Omega)]^{2\times 2},[L^2(\Omega)]^{2\times 2})} \leq \frac{1}{\mu}$. □

Motivated by (13) and Lemma 1, we define

$$\theta := \Big(\sum_{T\in\mathscr{T}_h} \theta_T^2 + ||\mathbf{g} - \mathbf{u}_h||^2_{[H^{1/2}(\Gamma)]^2} \Big)^{1/2}, \tag{16}$$

where

$$\begin{aligned} \theta_T^2 := & ||\mathbf{f} + \mathbf{div}(\boldsymbol{\sigma}_h)||^2_{[L^2(T)]^2} + \big|\big|\boldsymbol{\varepsilon}(\mathbf{u}_h) - \mathscr{C}^{-1}\boldsymbol{\sigma}_h - c_{\mathbf{g}}\mathbf{I}\big|\big|^2_{[L^2(T)]^{2\times 2}} \\ & + \big|\big|\boldsymbol{\sigma}_h - \boldsymbol{\sigma}_h^{\mathrm{t}}\big|\big|^2_{[L^2(T)]^{2\times 2}} + \left|\left|\boldsymbol{\gamma}_h - \frac{\nabla\mathbf{u}_h - (\nabla\mathbf{u}_h)^{\mathrm{t}}}{2}\right|\right|^2_{[L^2(T)]^{2\times 2}}. \end{aligned} \tag{17}$$

The a posteriori error estimator θ is equivalent to the total error. However, the non-local character of the residual $||\mathbf{g} - \mathbf{u}_h||_{[H^{1/2}(\Gamma)]^2}$ makes θ unuseful in an adaptive refinement algorithm.

Let us assume that $\mathbf{g} \in [H^1(\Gamma)]^2$ and that $H_h^{\mathbf{u}}$ is defined by (9). Let $\mathscr{N}_h$ be the set of all the vertices of the triangles in $\mathscr{T}_h$. We denote by $\bar{\mathbf{u}}_h$ the unique continuous piecewise-linear function in $\mathscr{T}_h$ such that $\bar{\mathbf{u}}_h(\mathbf{x}) = \mathbf{u}_h(\mathbf{x})\ \forall\, \mathbf{x} \in \mathscr{N}_h \cap \Omega$ and $\bar{\mathbf{u}}_h(\mathbf{x}) = \mathbf{g}(\mathbf{x})\ \forall\, \mathbf{x} \in \mathscr{N}_h \cap \Gamma$. Let us denote by $\{e_1, e_2, \ldots, e_n\}$ the partition of Γ induced by $\mathscr{T}_h$ and define $\kappa := \max\left\{\frac{h_{e_i}}{h_{e_j}} : e_i \text{ neighbor of } e_j\right\}$. Then, according to Theorem 1 in [4],

$$||\mathbf{g} - \bar{\mathbf{u}}_h||^2_{[H^{1/2}(\Gamma)]^2} \leq C_{1/2}^2 \log(1 + \kappa) \sum_{e\in E_h(\Gamma)} h_e \left|\left|\frac{\mathrm{d}}{\mathrm{d}\mathbf{t}_T}(\mathbf{g} - \bar{\mathbf{u}}_h)\right|\right|^2_{[L^2(e)]^2} \tag{18}$$

where $C_{1/2}$ is a universal constant and $\frac{\mathrm{d}}{\mathrm{d}t_T}$ denotes the tangential derivative. We also have the following Lemma.

Lemma 2 *Assume that* $\mathbf{g}|_{e_i} \in [\mathscr{P}_l(e_i)]^2$, *for* $i = 1, \ldots, n$ *and some* $l \geq 0$. *Then, there exists a constant* $C_3 > 0$, *independent of* h, *such that*

$$h_e \left|\left| \frac{\mathrm{d}}{\mathrm{d}\mathbf{t}_T}(\mathbf{g} - \bar{\mathbf{u}}_h) \right|\right|^2_{[L^2(e)]^2} \leq C_3 \,||\mathbf{u} - \bar{\mathbf{u}}_h||^2_{[H^1(T_e)]^2}, \qquad \forall\, e \in E_h(\Gamma), \tag{19}$$

where $T_e \in \mathscr{T}_h$ *is the triangle that has* e *as an edge.*

Proof Apply the inverse inequality (2.1.30) in [1] with $\tau = -1/2$ and $\sigma = 0$, use the definition of the $H^{-1/2}(e)$-norm and the continuity of the tangential derivative operator from $[H^1(T_e)]^2$ to $[H^{-1/2}(\partial T_e)]^2$. □

At this point, we define $\bar{\theta} := \left(\sum_{T \in \mathscr{T}_h} \bar{\theta}_T^2 \right)^{1/2}$, where

$$\begin{aligned} \bar{\theta}_T^2 &:= \theta_T^2 + ||\mathbf{u}_h - \bar{\mathbf{u}}_h||^2_{[H^1(T)]^2} \\ &+ \log(1+\kappa) \sum_{e \in E(T) \cap E_h(\Gamma)} h_e \left|\left| \frac{\mathrm{d}}{\mathrm{d}\mathbf{t}_T}(\mathbf{g} - \bar{\mathbf{u}}_h) \right|\right|^2_{[L^2(e)]^2} . \end{aligned} \tag{20}$$

Next we establish our main result. Given a triangle $T \in \mathscr{T}_h$, we denote by $\mathbf{H}(T) := H(\mathbf{div}; T) \times [H^1(T)]^2 \times [L^2(T)]^{2\times 2}$.

Theorem 1 *If* $\mathbf{g} \in [H^1(\Gamma)]^2$ *and* H_h^u *is defined by (9), then there exist* $C_{\mathtt{rel}}, C_{\mathtt{eff,int}} > 0$ *such that*

$$||(\boldsymbol{\sigma} - \boldsymbol{\sigma}_h, \mathbf{u} - \mathbf{u}_h, \boldsymbol{\gamma} - \boldsymbol{\gamma}_h)||_{\mathbf{H}} \leq C_{\mathtt{rel}}\, \bar{\theta} \tag{21}$$

$$C_{\mathtt{eff,int}}\, \bar{\theta}_T \leq ||(\boldsymbol{\sigma} - \boldsymbol{\sigma}_h, \mathbf{u} - \mathbf{u}_h, \boldsymbol{\gamma} - \boldsymbol{\gamma}_h)||_{\mathbf{H}(T)}, \quad \forall\, T \in \mathscr{T}_h, \quad T \subset \Omega \tag{22}$$

Moreover, if $T \in \mathscr{T}_h$ *has exactly one vertex on* Γ, *then*

$$C^2_{\mathtt{eff,int}}\, \bar{\theta}_T^2 \leq ||(\boldsymbol{\sigma} - \boldsymbol{\sigma}_h, \mathbf{u} - \mathbf{u}_h, \boldsymbol{\gamma} - \boldsymbol{\gamma}_h)||^2_{\mathbf{H}(T)} + ||\mathbf{u}_h - \bar{\mathbf{u}}_h||^2_{[H^1(T)]^2}. \tag{23}$$

Inequality (23) holds for a different constant $C^2_{\mathtt{eff,b}}$ *in place of* $C^2_{\mathtt{eff,int}}$ *if* $\mathbf{g}|_{e_i} \in [\mathscr{P}_l(e_i)]^2$, *for* $i = 1, \ldots, n$ *and* $l \geq 0$, *and* $T \in \mathscr{T}_h$ *has an edge on* Γ.

Proof The reliability follows from (13), Lemma 1, the triangle inequality, (18) and the trace Theorem. On the other hand, if $T \in \mathscr{T}_h$ is such that $T \subset \Omega$ or has exactly one vertex on Γ, then (22) and (23) follow by using the first equation in (2), the modified constitutive equation satisfied by $\boldsymbol{\sigma} \in H_0$ (see [5]), the symmetry of $\boldsymbol{\sigma}$, the definition of $\boldsymbol{\gamma}$, the triangle inequality and the continuity of $\mathscr{C}^{-1}$. If T has a side on Γ, apply in addition Lemma 2 and the triangle inequality. □

Remark 1 It can be shown that $C_{\mathtt{rel}} \leq \frac{\sqrt{6}}{\alpha} C_{2,\max} \max(1, C_{1/2}, C_\Gamma)$, where C_Γ is the constant of the usual trace inequality between $H^1(\Omega)$ and $H^{1/2}(\Gamma)$. Moreover, it is not difficult to see that $C_{\mathtt{eff,int}}^{-1/2} := 4+2\|\mathscr{C}^{-1}\|^2$ and $C_{\mathtt{eff,b}}^{-2} := 4+2\|\mathscr{C}^{-1}\|^2 + 2\,C_3 \log(1+\kappa)$.

4 Numerical Experiments

In this section we present some numerical results that illustrate the performance of the adaptive algorithm based on $\bar{\theta}$ for the simplest finite element subspace $\mathbf{H}_h$ defined in Sect. 2. We take the Young modulus $E = 1$ and the Poisson ratio $\nu = 0.4900$. The corresponding Lamé parameters are given by $\mu := \frac{E}{2(1+\nu)}$ and $\lambda := \frac{E\,\nu}{(1+\nu)(1-2\nu)}$. We fix $\kappa_1 = \mu$, $\kappa_2 = \frac{1}{2\mu}$, $\kappa_3 = \frac{1}{8}\kappa_1$ and $\kappa_4 = \kappa_1 + \kappa_3$. We let $\Omega =]-0.25, 0.25[^2 \backslash [0, 0.25]^2$ and choose the data $\mathbf{f}$ and $\mathbf{g}$ so that the exact solution is $\mathbf{u}(x_1, x_2) := (u_1(x_1, x_2), u_2(x_1, x_2))^{\mathtt{t}}$, with

$$u_1(x_1, x_2) = u_2(x_1, x_2) = \frac{x_1\,x_2}{(x_1^2 + x_2^2)^{1/3}} + 3\,x_2\,, \quad \forall\,(x_1, x_2) \in \Omega. \tag{24}$$

We remark that $\mathbf{u}$ has a singularity at the boundary point $(0, 0)$. In fact, $\mathbf{div}(\boldsymbol{\sigma}) \in [H^{1/3}(\Omega)]^2$, so that the expected rate of convergence for the uniform refinement is $1/3$.

In Fig. 1 we represent the total error (measured in the $\mathbf{H}$-norm) vs. the total number of degrees of freedom for the usual uniform refinement and for the adaptive

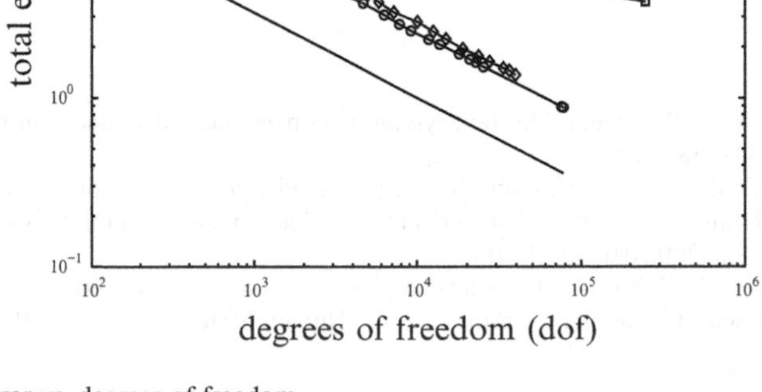

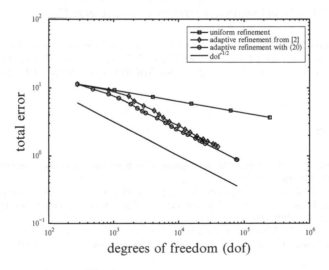

Fig. 1 Total error vs. degrees of freedom

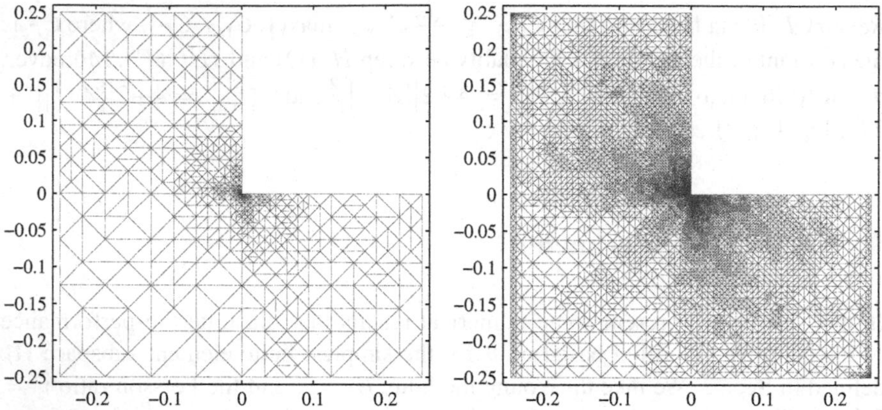

Fig. 2 Adapted meshes obtained after 10 (*left*) and 20 (*right*) iterations, with 1,221 and 15,408 elements, respectively

algorithms based on $\tilde{\theta}$ (the a posteriori error estimator introduced in [2]) and on $\bar{\theta}$. We observe that the errors in the adaptive procedures decrease faster than for the uniform one. Indeed, the experimental convergence rates for the uniform refinement algorithm approach $1/3$ whereas the adaptive refinement algorithm based on $\bar{\theta}$ is able to recover the linear convergence. In addition, in this example the efficiency indices for $\bar{\theta}$ (defined as the ratio between the total error and $\bar{\theta}$) are always in a neighborhood of 0.9.

Finally, we show in Fig. 2 two adapted meshes obtained with the adaptive algorithm based on $\bar{\theta}$. The adaptive algorithm detects the singularity of the solution, since the adapted meshes are highly refined around the origin.

For more details and numerical experiments, we refer to [3].

Acknowledgement This research has been partially supported by Universidade da Coruña, CONICYT-Chile FONDECYT grants 11060014 and 11070085, Dirección de Investigación of Universidad Católica de la Santísima Concepción, and MICINN project MTM2010-21135-C02-01.

References

1. D.N. Arnold, W.L. Wendland, On the asymptotic convergence of collocation methods. Math. Comput. **41**, 349–381 (1983)
2. T.P. Barrios, E.M. Behrens, M. González, A posteriori error analysis of an augmented mixed formulation in linear elasticity with mixed and Dirichlet boundary conditions. Comput. Methods Appl. Mech. Eng. **200**, 101–113 (2011)
3. T.P. Barrios, E.M. Behrens, M. González, Low cost a posteriori error estimators for an augmented mixed FEM in linear elasticity. Appl. Numer. Math. **84**, 46–65 (2014)

4. C. Carstensen, An a posteriori error estimate for a first-kind integral equation. Math. Comput. **66**, 139–155 (1997)
5. G.N. Gatica, An augmented mixed finite element method for linear elasticity with non-homogeneous Dirichlet conditions. ETNA **26**, 421–438 (2007)

Adaptive Numerical Simulation of Dynamic Contact Problems

Mirjam Walloth and Rolf Krause

Abstract We present a new approach for the space-time adaptive solution of dynamic contact problems. By combining ideas from the recently introduced residual-type a posteriori error estimator for static contact problems (Krause et al., An efficient and reliable residual-type a posteriori error estimator for the Signorini problem. Numer. Math. (2014), DOI: 10.1007/s00211-014-0655-8) and the novel discretization scheme with local impact detection (Krause and Walloth, A family of space-time connecting discretization schemes with local impact detection for elastodynamic contact problems. Comput. Methods Appl. Mech. Eng. 200:3425–3438, 2011), a discretization method is constructed which is able to detect and resolve local nonsmooth effects at the contact boundary in space and time. Numerical results in $3D$ illustrate our theoretical findings.

1 Introduction

The simulation of dynamic contact problems provides insight into many complex processes in, e.g., engineering, engineering mechanics, and biomechanics. The more complex the considered process, the greater the need for high accuracy. Besides a good mathematical model and solution method for the arising nonlinear system, the used discretization method is of crucial importance for the accuracy of the numerical simulation. For contact problems, a particular challenge are the non-smooth effects arising at the contact interface, which have to be resolved in space as well as in time.

M. Walloth (✉)
Fachbereich Mathematik, Technische Universität Darmstadt, Dolivostraße 15, D-64293 Darmstadt, Germany
e-mail: walloth@mathematik.tu-darmstadt.de

R. Krause
Institute of Computational Science, University of Lugano, Via Giuseppe Buffi 13, CH-6900 Lugano, Switzerland
e-mail: rolf.krause@usi.ch

A. Abdulle et al. (eds.), *Numerical Mathematics and Advanced Applications ENUMATH 2013*, Lecture Notes in Computational Science and Engineering 103,
DOI 10.1007/978-3-319-10705-9_27

As discussed in preceding works e.g. [2, 3, 5, 9, 11] the discretization in time of dynamic contact problems is challenging. Due to the non-smoothness of the displacements and the discontinuity in the velocities in the moment of impact classical time discretization schemes fail, i.e. energy blow-ups, oscillations in velocity and contact stresses occur.

Although linear finite elements are widely used for the discretization in space of contact problems, there are only a few appropriate a posteriori error estimators, e.g. [4, 10, 13, 14]. However, adaptive mesh refinement is important for the accuracy of the solution, especially for the resolution of the free boundary, i.e., the region between the actual and non-actual contact boundary which is a priori unknown.

In this article we combine the recently presented residual-type a posteriori error estimator for contact problems [10] and a novel discretization scheme with local impact detection [8]. In contrast to the original work [8] the meshes are different in subsequent time-steps, which requires a special treatment of the numerically computed values of the foregoing time step. Further, the a posteriori error estimator [10] originally designed for static contact problems has to be adapted, to meet the conditions of time-discrete contact problems.

Thus, this article presents a discretization method for dynamic contact problems which is able to resolve the local effects at the contact boundary in space and time. This is due to the adaptive discretization in space and a special time-discretization which is able to resolve the local impact times of each node implicitly. At the end of this article numerical results in $3D$ illustrate the performance of the discretization method. In particular, we show the adaptively refined meshes, the stability of contact stresses and velocities and we discuss the course of energy and contact forces.

2 Dynamic Signorini Contact Problem

Our model problem is the dynamic Signorini contact problem. It describes the time-dependent contact of a linear elastic body, represented by the domain $\Omega \subset \mathbb{R}^d$, $d = 2, 3$, with a rigid body. The whole boundary $\Gamma := \partial\Omega$ is divided into three disjoint parts, the potential contact boundary Γ_C, the Neumann boundary Γ_N and the Dirichlet boundary Γ_D. Each material particle in $\bar{\Omega}$ is identified with a point $\boldsymbol{x} = (\boldsymbol{x}_1, \ldots, \boldsymbol{x}_d)^T$. The sought displacements of the linear elastic body are $\boldsymbol{u} : \bar{\Omega} \rightarrow \mathbb{R}^d$. Velocities and accelerations are given by $\dot{\boldsymbol{u}}$ and $\ddot{\boldsymbol{u}}$.

Throughout this work we denote all quantities which refer to tensors including vectors by bold symbols. Their components are printed in normal type and are indicated by subindices, e.g., u_i. The Cartesian basis vectors of $\mathbb{R}^d$ are denoted by $\boldsymbol{e}_i$, $i = 1, \ldots, d$.

As the body is assumed to consist of linear elastic material, the stress tensor $\boldsymbol{\sigma}$ obeys Hooke's law and the strain tensor $\boldsymbol{\epsilon}$ is linearized.

For the ease of presentation the direction of constraints at the potential contact boundary is assumed to be constant and set to $\boldsymbol{e}_1$. In order to avoid penetration of the two bodies we enforce the linearized non-penetration condition $\boldsymbol{u}(\boldsymbol{x},t) \cdot \boldsymbol{e}_1 \leq g(\boldsymbol{x})$ where g is the gap function. The non-penetration condition evokes so-called contact stresses $\hat{\sigma}_1(\boldsymbol{u})$ which is the projection of the boundary stresses $\hat{\boldsymbol{\sigma}}(\boldsymbol{u}) := \boldsymbol{\sigma}(\boldsymbol{u})\boldsymbol{n}$ onto the direction of constraints.

The linear elastic body might be subjected to a volume force density $\boldsymbol{f}$, to surface forces $\boldsymbol{\pi}$, to Dirichlet values $\boldsymbol{u}_D$ and to initial values for displacements $\boldsymbol{u}_0(\boldsymbol{x})$ and velocities $\dot{\boldsymbol{u}}_0(\boldsymbol{x})$ at time $t = 0$.

In the following we state the strong formulation of the dynamic Signorini contact problem where $\rho > 0$ is the density.

$$\rho\ddot{\boldsymbol{u}} - \operatorname{div}\boldsymbol{\sigma}(\boldsymbol{u}) = \boldsymbol{f} \qquad \text{in } \Omega \times [0,T] \tag{1}$$

$$\hat{\boldsymbol{\sigma}}(\boldsymbol{u}) = \boldsymbol{\pi} \qquad \text{on } \Gamma_N \times [0,T] \tag{2}$$

$$\boldsymbol{u} = \boldsymbol{u}_D \qquad \text{on } \Gamma_D \times [0,T] \tag{3}$$

$$u_1 \leq g \qquad \text{on } \Gamma_C \times [0,T] \tag{4}$$

$$\hat{\sigma}_1 \leq 0 \qquad \text{on } \Gamma_C \times [0,T] \tag{5}$$

$$(u_1 - g)\,\hat{\sigma}_1 = 0 \qquad \text{on } \Gamma_C \times [0,T] \tag{6}$$

$$\boldsymbol{u}(\boldsymbol{x},0) = \boldsymbol{u}_0(\boldsymbol{x}), \quad \dot{\boldsymbol{u}}(\boldsymbol{x},0) = \dot{\boldsymbol{u}}_0(\boldsymbol{x}) \qquad \text{in } \Omega \tag{7}$$

Equation (6) is called complementarity condition. It says that the normal stresses are zero if the bodies are not in contact. If the bodies are in contact we expect the velocities $\dot{u}_1$ to be zero. This condition

$$\dot{u}_1\hat{\sigma}_1 = 0 \tag{8}$$

is usually referred to as persistency condition. In order to have a clear presentation we set the Neumann and Dirichlet values to zero, $\boldsymbol{\pi} = 0$, $\boldsymbol{u}_D = 0$ and we set $\rho = 1$. The weak formulation of (1)–(7) is given by: *For every time $t \in (0,T)$ find $\boldsymbol{u}(\cdot,t) \in \mathcal{K}$ with $\ddot{\boldsymbol{u}}(\cdot,t) \in \boldsymbol{L}^2(\Omega)$, such that*

$$\langle \ddot{\boldsymbol{u}}, \boldsymbol{v} - \boldsymbol{u} \rangle + a(\boldsymbol{u}, \boldsymbol{v} - \boldsymbol{u}) \geq \langle \boldsymbol{f}, \boldsymbol{v} - \boldsymbol{u} \rangle \quad \forall \boldsymbol{v} \in \mathcal{K} \tag{9}$$

where $\boldsymbol{f} \in \boldsymbol{L}^2(\Omega)$, $a(\boldsymbol{u},\boldsymbol{v}) := \int_\Omega \boldsymbol{\sigma}(\boldsymbol{u}) : \boldsymbol{\epsilon}(\boldsymbol{v})\,dx$ and $\mathcal{K} := \{\boldsymbol{v} \in \mathcal{H} \,|\, v_1 \leq g$ on $\Gamma_C\}$ is the set of admissible displacements which is a subset of $\mathcal{H} := \{\boldsymbol{v} \in \boldsymbol{H}^1(\Omega) | \boldsymbol{v} = 0 \text{ on } \Gamma_D\}$. The L^2-norm and its scalar product are denoted by $\|\cdot\|$ and $\langle\cdot,\cdot\rangle$.

We define the so-called contact force density $\boldsymbol{F}_{\text{con}}(\boldsymbol{u})$ by means of $\langle \boldsymbol{F}_{\text{con}}(\boldsymbol{u}), \cdot \rangle := \langle \ddot{\boldsymbol{u}}, \cdot \rangle + a(\boldsymbol{u}, \cdot) - \langle \boldsymbol{f}, \cdot \rangle$, which can be obtained as linear residual corresponding to the variational inequality (9). It follows from Green's formula that $\langle \boldsymbol{F}_{\text{con}}(\boldsymbol{u}), \cdot \rangle = \langle \hat{\boldsymbol{\sigma}}(\boldsymbol{u}), \cdot \rangle_{L^2(\Gamma_C)}$. The contact stresses $\hat{\boldsymbol{\sigma}}(\boldsymbol{u})$ and the contact force density $\boldsymbol{F}_{\text{con}}(\boldsymbol{u})$, respectively, are a priori unknown and have to be determined as part of the solution process.

3 Discretization in Space and Time

For the discretization in space we use linear finite elements. The mesh is denoted by $\mathfrak{m}$ and the corresponding space of linear finite elements is denoted by $\mathscr{H}_{\mathfrak{m}}$. We denote the set of nodes $\boldsymbol{p}$ by $\mathscr{N}_{\mathfrak{m}}$ and the subset of nodes at the potential contact boundary by $\mathscr{N}_{\mathfrak{m}}^C$ and the subset of nodes on $\bar{\Gamma}_N$ by $\mathscr{N}_{\mathfrak{m}}^{\bar{N}}$. The nodal basis functions for the finite element spaces are denoted by ϕ_p. The constraints are imposed nodewise, so that the discrete admissible set is given by $\mathscr{K}_{\mathfrak{m}} := \{\boldsymbol{v}_{\mathfrak{m}} \in \mathscr{H}_{\mathfrak{m}} \mid v_{\mathfrak{m},1} \leq g_{\mathfrak{m}} \text{ on } \Gamma_C\}$ where $g_{\mathfrak{m}}$ is the finite element approximation of g.

In the following, the discrete quantities for displacement, velocity, acceleration and force have to be understood as vectors of the dimension $\#\mathscr{N}_{\mathfrak{m}} \cdot d$. The matrices $\boldsymbol{M}_{\mathfrak{m}}$ and $\boldsymbol{A}_{\mathfrak{m}}$ are the mass matrix and the stiffness matrix representing $\langle \cdot, \cdot \rangle$ and $a(\cdot, \cdot)$ in the basis of linear finite elements and the chosen Cartesian coordinate system. We assume the mass matrix to be lumped. The given time step size for the time discretization is denoted by τ and the superscript n stands for the n-th time step, e.g., $\boldsymbol{u}_{\mathfrak{m}}^n$ is the displacement computed in time step n on mesh $\mathfrak{m}$.

In continuum mechanics the classical Newmark scheme is widely-used because it is of second order consistency and conserves the energy. Unfortunately, if contact constraints are imposed the classical Newmark scheme evokes energy blow-ups and oscillations in the contact stresses and velocities, thus spoiling the overall accuracy. Several modifications of the Newmark scheme have been developed in order to overcome this deficiency which can be found in, e.g., [2, 3, 5, 11]. The merits and drawbacks of the different methods are explained and compared in [9]. In [8] we introduced the *improved contact-stabilized Newmark scheme*. The method avoids energy blow-ups. It is even provable dissipative. The contact stresses are stable and the behavior of the velocities in normal direction is motivated by the persistency condition (8).

In [2] an L^2-projection P_{L^2} of $\boldsymbol{u}_{\mathfrak{m}}^n + \tau \dot{\boldsymbol{u}}_{\mathfrak{m}}^n$ onto the admissible set predicts the contact boundary. This L^2-projection is used in the *improved contact-stabilized Newmark scheme* in order to predict the local impact time step sizes $\tau^*(\boldsymbol{p})$ of each node, i.e. $P_{L^2}(\boldsymbol{u}_{\mathfrak{m}}^n + \tau \dot{\boldsymbol{u}}_{\mathfrak{m}}^n) =: \boldsymbol{u}_{\mathfrak{m},\text{pred}}^{n+1} = \boldsymbol{u}_{\mathfrak{m}}^n + \boldsymbol{\tau}^* \dot{\boldsymbol{u}}_{\mathfrak{m}}^n$ where $\boldsymbol{\tau}^*$ is a $(\#\mathscr{N}_{\mathfrak{m}} \cdot d) \times (\#\mathscr{N}_{\mathfrak{m}} \cdot d)$ diagonal matrix. The $d \times d$ diagonal block matrices of $\boldsymbol{\tau}^*$ have the entries

$$\boldsymbol{e}_i(\boldsymbol{p}) \cdot \boldsymbol{\tau}^* \cdot \boldsymbol{e}_i(\boldsymbol{p}) = \begin{cases} \tau^*(\boldsymbol{p}), & \text{if} \quad \boldsymbol{p} \in \mathscr{N}_{\mathfrak{m}}^C \quad \text{and} \quad i = 1 \\ \tau, & \text{otherwise} \end{cases} .$$

For the computation of the local impact time step sizes $\tau^*(\boldsymbol{p})$ we refer to [8].

Under the assumption of a fixed mesh for all time steps the *improved contact-stabilized Newmark scheme* is given by:

$$\boldsymbol{M}_{\mathfrak{m}}\boldsymbol{u}^{n+1}_{\mathfrak{m},\text{pred}} = \boldsymbol{M}_{\mathfrak{m}}\boldsymbol{u}^{n}_{\mathfrak{m}} + \boldsymbol{\tau}^{*}\boldsymbol{M}_{\mathfrak{m}}\dot{\boldsymbol{u}}^{n}_{\mathfrak{m}} \tag{10}$$

$$\boldsymbol{M}_{\mathfrak{m}}\boldsymbol{u}^{n+1}_{\mathfrak{m}} = \boldsymbol{M}_{\mathfrak{m}}\boldsymbol{u}^{n+1}_{\mathfrak{m},\text{pred}} - \frac{\tau^2}{2}\left(\frac{1}{2}\boldsymbol{A}_{\mathfrak{m}}\boldsymbol{u}^{n}_{\mathfrak{m}} + \frac{1}{2}\boldsymbol{A}_{\mathfrak{m}}\boldsymbol{u}^{n+1}_{\mathfrak{m}} - \boldsymbol{M}_{\mathfrak{m}}\boldsymbol{F}_{\text{con}}\left(\boldsymbol{u}^{n+1}_{\mathfrak{m}}\right)\right) \tag{11}$$

$$\boldsymbol{M}_{\mathfrak{m}}\dot{\boldsymbol{u}}^{n+1}_{\mathfrak{m}} = \boldsymbol{M}_{\mathfrak{m}}\dot{\boldsymbol{u}}^{n}_{\mathfrak{m}} - \tau\left(\frac{1}{2}\boldsymbol{A}_{\mathfrak{m}}\boldsymbol{u}^{n}_{\mathfrak{m}} + \frac{1}{2}\boldsymbol{A}_{\mathfrak{m}}\boldsymbol{u}^{n+1}_{\mathfrak{m}} - \boldsymbol{M}_{\mathfrak{m}}\boldsymbol{F}_{\text{con}}\left(\boldsymbol{u}^{n+1}_{\mathfrak{m}}\right) - \boldsymbol{\alpha}\frac{2}{\tau^2}\boldsymbol{M}_{\mathfrak{m}}(\boldsymbol{\tau}^{*} - \boldsymbol{\tau})\dot{\boldsymbol{u}}^{n}_{\mathfrak{m}}\right). \tag{12}$$

We note that $\boldsymbol{F}_{\text{con}}(\boldsymbol{u}^{n+1}_{\mathfrak{m}})$ is a priori unknown. Thus, (11) is a variational inequality in $\boldsymbol{u}^{n+1}_{\mathfrak{m}}$ and $\boldsymbol{M}_{\mathfrak{m}}\boldsymbol{F}_{\text{con}}(\boldsymbol{u}^{n+1}_{\mathfrak{m}})$ its linear residual. In contrast to the classical Newmark scheme $\boldsymbol{u}^{n}_{\mathfrak{m}} + \tau\dot{\boldsymbol{u}}^{n}_{\mathfrak{m}}$ has been replaced by $\boldsymbol{u}^{n+1}_{\mathfrak{m},\text{pred}}$ in the computation of the displacements (11), thus, avoiding artificial oscillations of the contact stresses. This method has been primarily used in [2]. In (12) the local impact time step size is used to correct the velocity. Therefore, a matrix-valued parameter $\boldsymbol{\alpha}$ has to be chosen. In [8] the choice of the diagonal entries $\alpha(\boldsymbol{p})$ has to ensure that the algorithm is dissipative which leads to $0 \leq \alpha(\boldsymbol{p}) \leq 1$ and is further motivated by the persistency condition (8) which says that the velocity at the contact boundary in direction of the constraints has to be zero during a contact phase.

The *improved contact-stabilized Newmark scheme* enables the localization of the impact time of each node. The critical zone where contact occurs is the free boundary zone. Thus, it would be desirable to resolve the free boundary zone by a finer mesh. Further, a finer resolution of the free boundary should lead to less energy loss as follows from the proof of dissipativity between two successive time steps given in [8, Proposition 1]. However, if the mesh is changed between time steps n and $n+1$, dissipativity is no longer guaranteed. For example, let $\mathfrak{m}_n$ be the mesh in time step n and $\mathfrak{m}_{n+1}$ the mesh in time step $n+1$. The solution $\boldsymbol{u}^{n}_{\mathfrak{m}_n}$ is interpolated on mesh $\mathfrak{m}_{n+1}$. However, this interpolated solution $\mathscr{I}^{n+1}_{n}(\boldsymbol{u}^{n}_{\mathfrak{m}_n})$ from the foregoing time step is not necessarily contained in the admissible set. Therefore, we have to apply the L^2-projection $P_{L^2}(\mathscr{I}^{n+1}_{n}(\boldsymbol{u}^{n}_{\mathfrak{m}_n})) =: \boldsymbol{u}^{n}_{\mathfrak{m}_{n+1}}$. Then $\mathscr{E}(\boldsymbol{u}^{n+1}_{\mathfrak{m}_{n+1}}) - \mathscr{E}(\boldsymbol{u}^{n}_{\mathfrak{m}_{n+1}}) \leq 0$. In this sense the algorithm is still dissipative, but $\mathscr{E}(\boldsymbol{u}^{n}_{\mathfrak{m}_{n+1}}) \neq \mathscr{E}(\mathscr{I}^{n+1}_{n}(\boldsymbol{u}^{n}_{\mathfrak{m}_n}))$ in general. By means of the admissible solution of the old time step we compute the old acceleration $\boldsymbol{A}_{\mathfrak{m}_{n+1}}\boldsymbol{u}^{n}_{\mathfrak{m}_{n+1}}$. An important observation is that a simple interpolation of the old acceleration leads to oscillations in the contact stresses.

For the adaptive mesh generation in each time step an a posteriori error estimator is required. A posteriori error estimators for contact problems can be found in [4, 10, 13, 14]. Here, we take the residual-type a posteriori error estimator of [10, 12] which is easy to compute and for which efficiency and reliability is proven in $3D$.

As Eq. (11) is slightly different to the equation of a static contact problem we have to adapt our residual-type a posteriori error estimator to this equation.

To state the error estimator contributions we need some preliminary definitions. The jump terms are either the difference between the stresses $\boldsymbol{J}^I(\boldsymbol{u}_{\mathrm{m}}) := (\boldsymbol{\sigma}|_{\tilde{\mathfrak{e}}}(\boldsymbol{u}_{\mathrm{m}}) - \boldsymbol{\sigma}|_{\mathfrak{e}}(\boldsymbol{u}_{\mathrm{m}}))\boldsymbol{n}$ on two neighboring elements $\mathfrak{e}$ and $\tilde{\mathfrak{e}}$ where $\boldsymbol{n}$ is the unit outward normal to the common side $\mathfrak{s}$ in the interior of Ω or the difference between the given Neumann data and the boundary stress $\boldsymbol{J}^N(\boldsymbol{u}_{\mathrm{m}}) := \boldsymbol{\pi} - \hat{\boldsymbol{\sigma}}|_{\mathfrak{e}}(\boldsymbol{u}_{\mathrm{m}})$ at a Neumann boundary side $\mathfrak{s}$.

As the constraints are solely imposed in direction $\boldsymbol{e}_1$ at the potential contact boundary, we have homogeneous Neumann conditions in the tangential direction. The corresponding jump terms are denoted by $\boldsymbol{J}_T^C(\boldsymbol{u}_{\mathrm{m}}) := -\hat{\boldsymbol{\sigma}}|_{\mathfrak{e}}(\boldsymbol{u}_{\mathrm{m}})$. We denote by ω_p the interior of the union of all surrounding elements and call it patch. The corresponding diameter is abbreviated with $h_p := \operatorname{diam}\omega_p$ and the union of all sides of elements belonging to $\bar{\omega}_p$ is denoted by γ_p. We call the union of all sides in the interior of ω_p, not including the boundary of ω_p, skeleton and denote it by $\gamma_{p,I}$. The intersections of $\partial\omega_p$ for all $\boldsymbol{p} \in \mathcal{N}_{\mathrm{m}}$ with Γ_C and Γ_N are denoted by $\gamma_{p,C} := \Gamma_C \cap \partial\omega_p$ and $\gamma_{p,N} := \Gamma_N \cap \partial\omega_p$.

For Eq. (11) the local estimator contributions $\eta_{2,p}, \eta_{3,p}, \eta_{4,p}$ of [10] change to

$$\eta_{2,p} := \frac{1}{2} h_p^{\frac{1}{2}} \| \boldsymbol{J}^I(\boldsymbol{u}_{\mathrm{m}}^n) + \boldsymbol{J}^I(\boldsymbol{u}_{\mathrm{m}}^{n+1}) \|_{\gamma_{p,I}} \quad \forall \boldsymbol{p} \in \mathcal{N}_{\mathrm{m}}$$

$$\eta_{3,p} := \frac{1}{2} h_p^{\frac{1}{2}} \| \boldsymbol{J}^N(\boldsymbol{u}_{\mathrm{m}}^n) + \boldsymbol{J}^N(\boldsymbol{u}_{\mathrm{m}}^{n+1}) \|_{\gamma_{p,N}} \quad \forall \boldsymbol{p} \in \mathcal{N}_{\mathrm{m}}^{\bar{N}}$$

$$\eta_{4,p} := \frac{1}{2} h_p^{\frac{1}{2}} \| \boldsymbol{J}_T^C(\boldsymbol{u}_{\mathrm{m}}^n) + \boldsymbol{J}_T^C(\boldsymbol{u}_{\mathrm{m}}^{n+1}) \|_{\gamma_{p,C}} \quad \forall \boldsymbol{p} \in \mathcal{N}_{\mathrm{m}}^C .$$

The local error estimator contributions $\eta_{1,p}$ of [10] change to

$$\eta_{1,p} := h^p \left\| \boldsymbol{f} + \frac{1}{2} \operatorname{div} \boldsymbol{\sigma}(\boldsymbol{u}_{\mathrm{m}}^n) + \frac{1}{2} \operatorname{div} \boldsymbol{\sigma}(\boldsymbol{u}_{\mathrm{m}}^{n+1}) + \frac{2}{\tau^2} \boldsymbol{u}_{\mathrm{m,pred}}^{n+1} - \frac{2}{\tau^2} \boldsymbol{u}_{\mathrm{m}}^{n+1} \right\|_{\omega_p} \quad \forall \boldsymbol{p} \in \mathcal{N}_{\mathrm{m}}.$$

Further, we need to adapt the classification of contact nodes in full- and semi-contact nodes which has been given in [10]. The so-called full-contact nodes $\boldsymbol{p} \in \mathcal{N}_{\mathrm{m}}^{\mathrm{fC}}$ for Eq. (11) are those nodes at the potential contact boundary which fulfill the following two criteria: $u_{\mathrm{m},1}^{n+1} = g_{\mathrm{m}}$ and $\frac{1}{2}\hat{\sigma}_1(\boldsymbol{u}_{\mathrm{m}}^n) + \frac{1}{2}\hat{\sigma}_1(\boldsymbol{u}_{\mathrm{m}}^{n+1}) \leq 0$ on $\gamma_{p,C}$. The remaining actual contact nodes which do not fulfill these conditions constitute the set of semi-contact nodes $\mathcal{N}_{\mathrm{m}}^{\mathrm{sC}}$. The local error estimator contributions corresponding to the potential contact boundary are given by

$$\eta_{5,p} := \frac{1}{2} h_p^{\frac{1}{2}} \| \hat{\sigma}_1(\boldsymbol{u}_{\mathrm{m}}^{n+1}) + \hat{\boldsymbol{\sigma}}(\boldsymbol{u}_{\mathrm{m}}^{n+1})_1 \|_{\gamma_{p,C}} \quad \forall \boldsymbol{p} \in \mathcal{N}_{\mathrm{m}}^C \setminus \mathcal{N}_{\mathrm{m}}^{\mathrm{fC}}$$

$$\eta_{6,p} := (s_p d_p)^{\frac{1}{2}} \quad \forall \boldsymbol{p} \in \mathcal{N}_{\mathrm{m}}^{\mathrm{sC}}$$

where s_p is the value of $\boldsymbol{F}_{\text{con}}(\boldsymbol{u}_{\text{m}}^{n+1})$ at the node $\boldsymbol{p}$ and d_p is the mean value of the distance to the obstacle on a strict subset $\tilde{\gamma}_{p,C}$ of $\gamma_{p,C}$:

$$d_p := \int_{\tilde{\gamma}_{p,C}} (g_{\text{m}} - u_{\text{m},1}^{n+1})\phi_p.$$

We note that the restriction to the subset of $\gamma_{p,C}$ is only required for the proof of efficiency. With these definitions of the error estimator contributions the proofs of reliability and efficiency follow as in [10] for Eq. (11) with discrete gap function g_{m}. If the gap function g is not a discrete function and has to be approximated by a linear finite element function g_{m} we have to consider the same additional contributions as in the static case.

4 Numerical Results

We consider a linear elastic unit cube which comes into contact with a rigid unit ball. The elastic modulus is $E = 5 \cdot 10^2$ and the Poisson ratio $\nu = 0.3$. The time step size is $\tau = 0.005$. The arising variational inequalities are solved by the monotone multigrid method explained in [6, 7]. The simulation is implemented in the framework of the finite element toolbox UG [1] and the obstacle toolbox OBSLIB++ [7]. In each third time step we regenerate the mesh by means of our residual-type a posteriori estimator presented in the foregoing section. As the free contact boundary is important for the detection of the local impact times we scale the estimator contributions $\eta_{5,p}$ and $\eta_{6,p}$ so that they are in the some range as $\eta_{1,p}$ and $\eta_{2,p}$ at the beginning. The starting grid consists of hexahedra but due to adaptive remeshing the final grid after seven refinement steps is an unstructured mesh consisting of tetrahedra, pyramids and prisms as well. In Fig. 1 we show the grid of the potential contact boundary. Obviously the free contact boundary is well resolved. The contact stresses and the velocities in time step 10 can be seen in Fig. 2. As already mentioned we cannot expect the energy to be dissipative when we use different meshes for the different time steps. Therefore, we are interested in the course of energy. It can be seen in Fig. 3a. There are slight increases of the energy but we do not get an energy blow-up. In Fig. 3b we show the course of the contact force which is smooth.

Finally, we visualize the error reduction due to adaptive refinement in space. In order to stress the adaptive procedure we take the first time step with a large contact force arising from an initially large velocity. For the computation of the reference solution we add a uniform refinement step to the adaptively refined grid. In Fig. 3c we show the error in the energy norm plotted against the number of nodes in logarithmic scales.

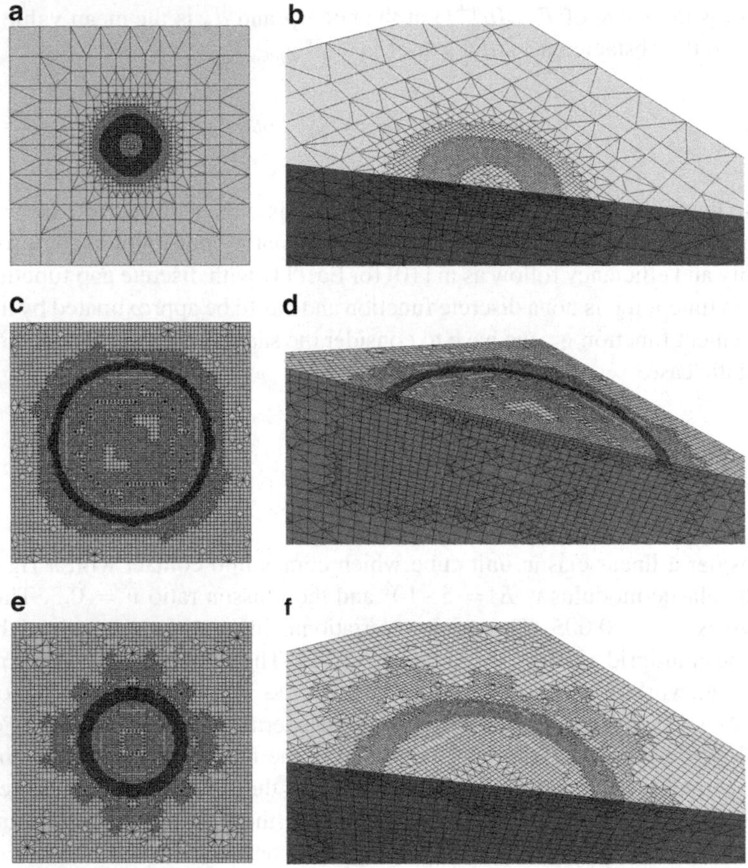

Fig. 1 Adaptively refined grid in different time steps n: potential contact boundary and diagonal cut through the cube. (**a**) $n = 1$. (**b**) $n = 1$; 49,296 dof. (**c**) $n = 19$. (**d**) $n = 19$; 378,315 dof. (**e**) $n = 31$. (**f**) $n = 31$; 255,714 dof

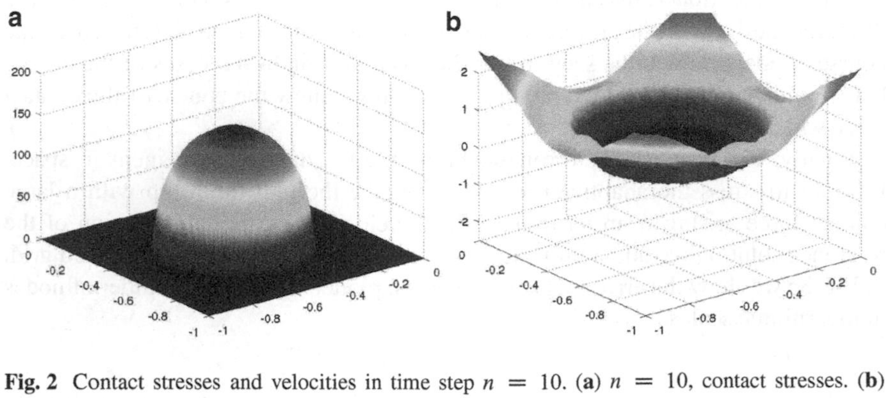

Fig. 2 Contact stresses and velocities in time step $n = 10$. (**a**) $n = 10$, contact stresses. (**b**) $n = 10$, velocities

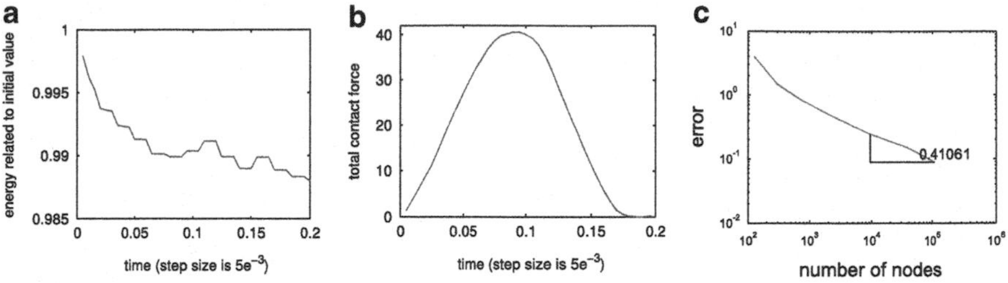

Fig. 3 (**a**) Course of energy; (**b**) course of total contact force; (**c**) error reduction due to adaptive refinement in space

References

1. B. Bastian, K. Birken, S. Lang, N. Neuss, H. Rentz-Reichert, C. Wieners, UG – A flexible software toolbox for solving partial differential equations. Comput. Vis. Sci. **1**, 27–40 (1997)
2. P. Deuflhard, R Krause, S. Ertel, A contact–stabilized Newmark method for dynamical contact problems. Int. J. Numer. Methods Eng. **73**, 1274–1290 (2008)
3. C. Hager, S. Hüeber, B. Wohlmuth, A stable energy conserving approach for frictional contact problems based on quadrature formulas. Int. J. Numer. Methods Eng. **73**, 205–225 (2008)
4. P. Hild, S. Nicaise, Residual a posteriori error estimators for contact problems in elasticity. Math. Model. Numer. Anal. **41**, 897–923 (2007)
5. H. Khenous, P. Laborde, Y. Renard, Mass redistribution method for finite element contact problems in elastodynamics. Eur. J. Mech. A Solids **27**, 918–932 (2008)
6. R. Kornhuber, R. Krause, Adaptive multigrid methods for Signorini's problem in linear elasticity. Comput. Vis. Sci. **4**, 9–20 (2001)
7. R. Krause, Monotone multigrid methods for Signorini's problem with friction, PhD thesis, Freie Universität Berlin, 2000
8. R. Krause, M. Walloth, A family of space-time connecting discretization schemes with local impact detection for elastodynamic contact problems. Comput. Methods Appl. Mech. Eng. **200**, 3425–3438 (2011)
9. R. Krause, M. Walloth, Presentation and comparison of selected algorithms for dynamic contact based on the Newmark scheme. Appl. Numer. Math. **62**, 1393–1410 (2012)
10. R. Krause, A. Veeser, M. Walloth, An efficient and reliable residual-type a posteriori error estimator for the Signorini problem. Numer. Math. (2014), DOI: 10.1007/s00211-014-0655-8
11. T. Laursen, V. Chawla, Design of energy conserving algorithms for frictionless dynamic contact problems. Int. J. Numer. Methods Eng. **40**, 863–886 (1997)
12. M. Walloth, Adaptive numerical simulation of contact problems: Resolving local effects at the contact boundary in space and time, PhD thesis, Rheinische Friedrich-Wilhelms-Universität Bonn, 2012
13. A. Weiss, B. Wohlmuth, A posteriori error estimator and error control for contact problems. Math. Comput. **78**, 1237–1267 (2009)
14. B. Wohlmuth, An a posteriori error estimator for two-body contact problems on non-matching meshes. SIAM J. Sci. Comput. **33**, 25–45 (2007)

An Adaptive Finite Element Method for the Infinity Laplacian

Omar Lakkis and Tristan Pryer

Abstract We construct a finite element method (FEM) for the infinity Laplacian. Solutions of this problem are well known to be singular in nature so we have taken the opportunity to conduct an a posteriori analysis of the method deriving residual based estimators to drive an adaptive algorithm. It is numerically shown that optimal convergence rates are regained using the adaptive procedure.

1 Introduction

Nonlinear partial differential equations (PDEs) arise in many areas. Their numerical simulation is extremely important due to the additional difficulties arising in their classical solution [4]. One such example is that of the *infinity Laplace operator* Δ_∞ defined by

$$\Delta_\infty u := \frac{\sum_{i=1}^d \sum_{j=1}^d \partial_i u \partial_j u \partial_{ij} u}{\sum_{i=1}^d (\partial_i u)^2} = \frac{(\nabla u \otimes \nabla u){:}\mathrm{D}^2 u}{|\nabla u|^2}, \tag{1}$$

for a twice-differentiable function $u : \Omega \to \mathbb{R}$, $\Omega \in \mathbb{R}^d$ open, bounded and connected, where

$$\nabla u := \begin{bmatrix} \partial_1 u \\ \vdots \\ \partial_d u \end{bmatrix}, \quad \boldsymbol{x} \otimes \boldsymbol{y} := \boldsymbol{x}\,\boldsymbol{y}^\intercal, \text{ and } \boldsymbol{X}{:}\boldsymbol{Y} := \operatorname{trace} \boldsymbol{X}^\intercal \boldsymbol{Y} \tag{2}$$

O. Lakkis (✉)
Department of Mathematics, University of Sussex, Brighton, GB-BN1 9QH England, UK
e-mail: lakkis.o.maths@gmail.com

T. Pryer
Department of Mathematics and Statistics, University of Reading, Whiteknights, PO Box 220, Reading, GB-RG6 6AX, England, UK
e-mail: T.Pryer@Reading.ac.uk

A. Abdulle et al. (eds.), *Numerical Mathematics and Advanced Applications ENUMATH 2013*, Lecture Notes in Computational Science and Engineering 103,
DOI 10.1007/978-3-319-10705-9_28

denote, respectively, the gradient, the (algebraic) tensor product of $\boldsymbol{x}, \boldsymbol{y} \in \mathbb{R}^d$, and the Frobenius inner product of two matrices $\boldsymbol{X}, \boldsymbol{Y} \in \mathbb{R}^{d\times d}$. This equation has been popular in classical studies (e.g. [1, 3]) but is difficult to pose numerical schemes due to its nondivergence structure and general lack of classical solvability. The infinity Laplacian, which is in fact a misnomer (*homogeneous infinity Laplacian* is more precise), occurs as the weighted formal limit of a variational problem. A more appropriate terminology would be that of *infinite harmonic* function u being one that solves $\Delta_\infty u = 0$. This is justified, at least heuristically, as being the formal limit of the p-harmonic functions, u_p, $p \geq 1$, $p \to \infty$ where

$$0 = \Delta_p u_p := \operatorname{div}\left(\left|\nabla u_p\right|^{p-2} \nabla u_p\right) = \left|\nabla u_p\right|^{p-2} \Delta u_p + (p-2) \left|\nabla u_p\right|^{p-2} \Delta_\infty u_p. \tag{3}$$

Multiplying by $\left|\nabla u_p\right|^{2-p}/(p-2)$ and taking the limit as $p \to \infty$ it follows that a would be limit $u = \lim_{p\to\infty} u_p$ is infinite harmonic. A rigorous treatment is provided in [6] and is based on the variational observation that the Dirichlet problem for the p–Laplacian is the Euler–Lagrange equation of the following *energy* functional

$$\mathscr{L}_p[u] := \frac{1}{p} \|u\|^p_{\mathrm{L}_p(\Omega)} = \int_\Omega \frac{1}{p} |\nabla u|^p \text{ for } p \in [1, \infty) \tag{4}$$

with appropriate Dirichlet boundary conditions. By analogy, setting

$$\mathscr{L}_\infty[u] := \|\nabla u\|_{\mathrm{L}_\infty(\Omega)} = \operatorname{ess\,sup}_\Omega |\nabla u|, \tag{5}$$

we seek $u \in \operatorname{Lip}(\Omega) = \mathrm{W}^1_\infty(\Omega)$, the space of Lipschitz continuous functions over Ω (Rademacher), with $u = g$ on $\partial\Omega$ such that

$$\mathscr{L}_\infty[u] \leq \mathscr{L}_\infty[v] \quad \forall\, v \in \operatorname{Lip}(\Omega) \text{ and } v = g \text{ on } \partial\Omega. \tag{6}$$

Show that the solution exists and define it to be infinite harmonic. Such a solution is called *absolutely minimising Lipschitz extension of* g, we call it infinite harmonic. The infinity Laplacian is thus considered to be the paradigm of a variational problem in $\mathrm{W}^1_\infty(\Omega)$.

If the solution is smooth, say in C^2 and has no internal extrema, it can be shown to satisfy (3) classically. But an infinite harmonic function is generally not a classical solution (those in $\mathrm{C}^2(\Omega)$ satisfying (1) everywhere). Therefore solutions of (3) must be sought in a weaker sense. The notion of viscosity solution, introduced for second order PDEs in [5] turns out to be the correct setting to seek weaker solutions. Existence and uniqueness of a viscosity solution to the homogeneous infinity Laplacian (1) has been studied [11]. If the domain Ω is bounded, open and connected then (1) has a unique viscosity solution $u \in \mathrm{C}^0(\overline{\Omega})$. In the case $\Omega \subset \mathbb{R}^2$ this can be improved to $u \in \mathrm{C}^{1,\alpha}(\overline{\Omega})$ [9]. A study of existence and uniqueness of viscosity solutions to the inhomogeneous infinity Laplacian can be found in [13].

With Ω defined as before and in addition if $f \in C^0(\Omega)$ and does not change sign, i.e., $\inf_\Omega f > 0$ or $\sup_\Omega f < 0$, one can find a unique viscosity solution.

As to the topic of numerical methods to approximate the infinity Laplacian, to the authors knowledge only two families of methods exist. The first is based on finite differences [14]. The scheme involves constructing monotone sequences of schemes over concurrent lattices by minimising the discrete Lipschitz constant over each node of the lattice. The second is a finite element scheme named the vanishing moment method [10] in which the 2nd order nonlinear PDE is approximated via sequences of biharmonic quasilinear 4th order PDEs.

In this paper we present a finite element method for the infinity Laplacian, without having to deal with the added complications of approximating a 4th order operator. It is based on the nonvariational finite element method introduced in [12]. Roughly, this method involves representing the *finite element Hessian* (see Definition 2) as an auxiliary variable in the formulation, to deal with the nonvariational structure. We also consider the problem as the steady state of an evolution equation making use of a *Laplacian relaxation* technique (see Remark 1) [2, 8] to circumvent the degeneracy of the problem.

The structure of the paper is as follows: In Sect. 2 we examine the linearisation of the PDE and present the necessary framework for the discretisation and state an a posteriori error indicator for the discrete problem. The estimator is of residual type and is used to drive an adaptive algorithm which is studied and used for numerical experimentation is Sect. 3. We choose our simulations in such a way that they can be compared with those given in [10, 14].

2 Notation, Linearisation and Discretisation

We consider the heterogeneous Infinity Laplace problem with Dirichlet boundary conditions on a domain $\Omega \subset \mathbb{R}^d$.

$$\Delta_\infty u = f \quad \text{in } \Omega \quad \text{and} \quad u = g \quad \text{on } \partial\Omega \tag{7}$$

with problem data $f, g \in C^0(\Omega)$ chosen such that f does not change sign throughout Ω. In this case there exists a unique viscosity solution to (7) [13].

2.1 *Linearisation of the Continuous Problem (1)*

The application of a standard fixed point linearisation to (7) results in the following sequence of linear nondivergence PDEs: Given an initial guess u^0, for each $n \in \mathbb{N}$ find u^{n+1} such that

$$\frac{(\nabla u^n \otimes \nabla u^n)}{|\nabla u^n|^2} : D^2 u^{n+1} = f. \tag{8}$$

Due to the degeneracy of the problem we introduce a slightly modified problem which utilises *Laplacian relaxation* [2, 8], the problem is to find u^{n+1} such that

$$\left(\frac{\nabla u^n \otimes \nabla u^n}{|\nabla u^n|^2} + \frac{\mathbf{I}}{\tau}\right):\mathrm{D}^2 u^{n+1} = f + \frac{\Delta u^n}{\tau} \text{ with } \tau \in \mathbb{R}^+. \tag{9}$$

Remark 1 The discretisation proposed in (9) is nothing but an implicit one stage discretisation of the following evolution equation

$$\partial_t(\Delta u) + \Delta_\infty u = f, \tag{10}$$

where Δu is used as shorthand for $\Delta_2 u$, the 2–Laplacian.

With that in mind we must take care with our choice of τ which can be regarded as a timestep. We require a τ that is large enough to guarantee reaching the steady state and small enough such that we do not encounter stability problems.

2.2 Discretisation of the Sequence of Linear PDEs (9)

Let $\mathscr{T}$ be a conforming, shape regular triangulation of Ω, namely, $\mathscr{T}$ is a finite family of sets such that

1. $K \in \mathscr{T}$ implies K is an open simplex (segment for $d = 1$, triangle for $d = 2$, tetrahedron for $d = 3$),
2. For any $K, J \in \mathscr{T}$ we have that $\overline{K} \cap \overline{J}$ is a full sub-simplex (i.e., it is either $\emptyset$, a vertex, an edge, a face, or the whole of $\overline{K}$ and $\overline{J}$) of both $\overline{K}$ and $\overline{J}$ and
3. $\bigcup_{K \in \mathscr{T}} \overline{K} = \overline{\Omega}$.

We also define $\mathscr{E}$ to be the skeleton of the triangulation, that is the set of sub-simplexes of $\mathscr{T}$ contained in Ω but not $\partial\Omega$. For $d = 2$, for example, $\mathscr{E}$ would consist of the set of edges of $\mathscr{T}$ not on the boundary. We also use the convention where $h(\boldsymbol{x}) := \max_{\overline{K} \ni \boldsymbol{x}} h_K$ to be the mesh-size function of $\mathscr{T}$.

Definition 1 (continuous and discontinuous FE spaces) Let $\mathbb{P}^k(\mathscr{T})$ denote the space of piecewise polynomials of degree k over the triangulation $\mathscr{T}$ of Ω. We introduce the *finite element spaces*

$$\mathbb{V}_D(k) = \mathbb{P}^k(\mathscr{T}) \qquad \mathbb{V}_C(k) = \mathbb{P}^k(\mathscr{T}) \cap \mathrm{C}^0(\Omega) \tag{11}$$

to be the usual spaces of discontinuous and continuous piecewise polynomial functions over Ω.

Remark 2 (generalised Hessian) Given a function $v \in \mathrm{H}^1(\Omega)$ and let $\boldsymbol{n} : \partial\Omega \to \mathbb{R}^d$ be the outward pointing normal of Ω then the *generalised Hessian* of v, $\mathrm{D}^2 v$ satisfies the following identity:

$$\langle \mathrm{D}^2 v \,|\, \phi \rangle = -\int_\Omega \nabla v \otimes \nabla \phi + \int_{\partial\Omega} \nabla v \otimes \boldsymbol{n}\, \phi \quad \forall\, \phi \in \mathrm{H}^1(\Omega), \tag{12}$$

where the final term is understood as a duality pairing between $\mathrm{H}^{-1/2}(\partial\Omega) \times \mathrm{H}^{1/2}(\partial\Omega)$.

Remark 3 (nonconforming generalised Hessian) The test functions applied to define the generalised Hessian in Remark 2 need not be $\mathrm{H}^1(\Omega)$. Suppose they are $\mathrm{H}^1(K)$ for each $K \in \mathscr{T}$ then it is clear that

$$\begin{aligned}\langle \mathrm{D}^2 v \,|\, \phi \rangle &= \sum_{K\in\mathscr{T}} \left(-\int_K \nabla v \otimes \nabla \phi + \int_{\partial K} \nabla v \otimes \boldsymbol{n}_K \phi \right) \\ &= \sum_{K\in\mathscr{T}} -\int_K \nabla v \otimes \nabla \phi + \sum_{e\in\mathscr{E}} \int_e \{\!\{\nabla v\}\!\} \otimes [\![\phi]\!] + \sum_{e\in\partial\Omega} \int_e \nabla v \otimes \boldsymbol{n}\, \phi,\end{aligned} \tag{13}$$

where $[\![\cdot]\!]$ and $\{\!\{\cdot\}\!\}$ denote the *jump* and *average*, respectively, over an element edge, that is, suppose e is a $(d-1)$ subsimplex shared by two elements K^+ and K^- with outward pointing normals $\boldsymbol{n}^+$ and $\boldsymbol{n}^-$ respectively, then

$$[\![\eta]\!] = \eta\big|_{K^+} \boldsymbol{n}^+ + \eta\big|_{K^-} \boldsymbol{n}^- \text{ and } \{\!\{\xi\}\!\} = \frac{1}{2}\left(\xi\big|_{K^+} + \xi\big|_{K^-}\right). \tag{14}$$

Definition 2 (finite element Hessian) From Remarks 2 and 3 for $V \in \mathbb{V}_C(k)$ we define the *finite element Hessian*, $\boldsymbol{H}[V] \in [\mathbb{V}_D(k)]^{d\times d}$ such that we have

$$\int_\Omega \boldsymbol{H}[V]\phi = \langle \mathrm{D}^2 V \,|\, \phi \rangle \quad \forall\, \phi \in \mathbb{V}_D(k). \tag{15}$$

We discretise (9) utilising the nonvariational Galerkin procedure proposed in [12]. We construct finite element spaces $\mathbb{V} := \mathbb{V}_C(k)$ and $\mathbb{W}$ which can be taken as $\mathbb{V}_C(k)$, $\mathbb{V}_D(k)$ or $\mathbb{V}_D(k-1)$. Then given $U^0 = \Lambda u^0$, for each $n \in \mathbb{N}_0$ we seek $(U^{n+1}, \boldsymbol{H}[U^{n+1}]) \in \mathbb{V} \times [\mathbb{W}]^{d\times d}$ such that

$$\begin{aligned}&\int_\Omega \left(\frac{\nabla U^n \otimes \nabla U^n}{|\nabla U^n|^2} + \frac{\boldsymbol{I}}{\tau} \right) : \boldsymbol{H}[U^{n+1}]\Psi = \int_\Omega \left(f + \frac{\operatorname{trace} \boldsymbol{H}[U^n]}{\tau} \right) \Psi \\ &\int_\Omega \boldsymbol{H}[U^{n+1}]\Phi = -\int_\Omega \nabla U^{n+1} \otimes \nabla \Phi + \sum_{e\in\mathscr{E}} \int_e \{\!\{\nabla U^{n+1}\}\!\} \otimes [\![\Phi]\!] \\ &\qquad + \sum_{e\in\partial\Omega} \int_e \nabla U^{n+1} \otimes \boldsymbol{n}\, \Phi \quad \forall\, (\Psi, \Phi) \in \mathbb{V} \times \mathbb{W}.\end{aligned} \tag{16}$$

Remark 4 (computational efficiency) Making use of a $\mathbb{V}_D(k)$ or $\mathbb{V}_D(k-1)$ space to represent the finite element Hessian allows us to construct a much faster algorithm in comparison to using a $\mathbb{V}_C(k)$ space for $\mathbb{W}$ due to the local representation of the $\mathrm{L}_2(\Omega)$ projection of discontinuous spaces [7].

Theorem 1 (a posteriori residual upper error bound) *Let u be the solution to the infinity Laplacian* (7) *and U^n be the n-th step in the linearisation defined by* (16). *Let*

$$\boldsymbol{A}[v] := \frac{\nabla v \otimes \nabla v}{|\nabla v|^2} + \frac{\mathbf{I}}{\tau}, \tag{17}$$

then there exists a $C > 0$ such that

$$\begin{aligned}\left\| f + \frac{\Delta U^n}{\tau} - \boldsymbol{A}[U^n]{:}\mathrm{D}^2 U^{n+1} \right\|_{\mathrm{H}^{-1}(\Omega)} \leq C \Bigg(& \sum_{K \in \mathscr{T}} h_K \left\| \mathscr{R}[U^n, U^{n+1}, f] \right\|_{\mathrm{L}_2(K)} \\ & + \sum_{e \in \mathscr{E}} h_K^{1/2} \left\| \mathscr{J}[U^n, U^{n+1}] \right\|_{\mathrm{L}_2(e)} \Bigg)\end{aligned} \tag{18}$$

where the interior residual, $\mathscr{R}[U, \boldsymbol{A}, f]$, over a simplex K and jump residual, $\mathscr{J}[U, \boldsymbol{A}]$, over a common wall $e = \overline{K^+} \cap \overline{K^-}$ of two simplexes, K^+ and K^- are defined as

$$\|\mathscr{R}[U^n, U^{n+1}, f]\|_{\mathrm{L}_2(K)}^2 = \int_K \left(f - \boldsymbol{A}[U^n]{:}\mathrm{D}^2 U^{n+1} + \frac{\Delta U^n}{\tau} \right)^2, \tag{19}$$

$$\|\mathscr{J}[U^n, U^{n+1}]\|_{\mathrm{L}_2(e)}^2 = \int_e \left(\frac{[\![\nabla U^n]\!]}{\tau} - \boldsymbol{A}[U^n]{:}[\![\nabla U^{n+1} \otimes]\!] \right)^2, \tag{20}$$

with

$$[\![\boldsymbol{\xi} \otimes]\!] := \boldsymbol{\xi}|_{K^+} \otimes \boldsymbol{n}^+ + \boldsymbol{\xi}|_{K^-} \otimes \boldsymbol{n}^-, \tag{21}$$

being defined as a tensor jump.

Remark 5 (on the proof of Theorem 1) The proof of Theorem 1 is based on standard residual a posteriori arguments.

3 Numerical Experiments

All of the numerical experiments in this section are implemented using FEniCS and visualised with ParaView . Each of the tests are on the domain $\Omega = [-1, 1]^2$, choosing the finite element spaces $\mathbb{V} = \mathbb{V}_C(1)$ and $\mathbb{W} = \mathbb{V}_D(0)$. This is

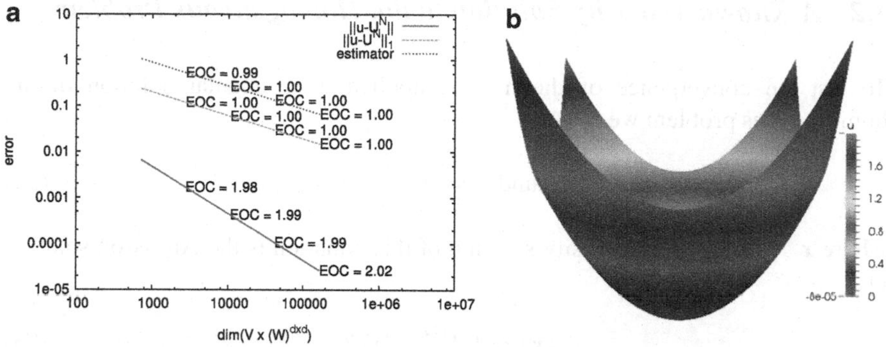

Fig. 1 We benchmark the approximation of a classical solution to the inhomogeneous infinity Laplacian, plotting the log of the error together with its estimated order of convergence. We examine both $\mathrm{L}_2(\Omega)$ and $\mathrm{H}^1(\Omega)$ norms of the error together with the residual estimator given in Theorem 1. The linearisation tolerance is coupled to the mesh-size such that the linearisation is run until $\left\|U^n - U^{n-1}\right\| \leq 10h^2$. The convergence rates are optimal, that is, $\left\|u - U^N\right\| = \mathrm{O}(h^2)$ and $\left|u - U^N\right|_1 = \mathrm{O}(h)$. (**a**) Convergence rates. (**b**) Finite element approximation

computationally the quickest implementation of the nonvariational finite element method and the lowest order stable pair of FE spaces for this class of problem.

3.1 Benchmarking and Convergence: Classical Solution

To benchmark the numerical algorithm we choose the data f and g such that the solution is known and classical. In the first instance we choose $f \equiv 2$ and $g = |\boldsymbol{x}|^2$. It is easily verified that the exact solution is given by $u = |\boldsymbol{x}|^2$. Figure 1 details a numerical experiment on this problem.

Remark 6 (on the value of τ) The optimal values of the *timestep parameter* or tuning parameter τ depend upon the regularity of the solution. For example, for a classical solution, one may choose τ large. In the numerical experiment above we took $\tau = 1{,}000$. Since the linearisation is nothing more than seeking the steady state of the evolution equation (9). The convergence (in n) is extremely quick taking no more than five iterations.

For the examples below one must be careful choosing τ, we will be looking at viscosity solutions that are not $\mathrm{C}^2(\Omega)$, in this case the lack of regularity of the solution will lead to an unstable linearisation for large τ. In each of the cases below $\tau \in [1 : 10]$ was sufficiently small to achieve convergence of the linearisation in at most 20 iterations.

3.2 A Known Viscosity Solution to the Homogeneous Problem

To test the convergence of the method applied to a singular solution of the homogeneous problem we fix

$$f \equiv 0 \text{ and } g = |x|^{4/3} - |y|^{4/3}, \tag{22}$$

where $\boldsymbol{x} = (x, y)^{\mathsf{T}}$. A viscosity solution of this equation is the Aronsson solution [1],

$$u(\boldsymbol{x}) = |x|^{4/3} - |y|^{4/3}. \tag{23}$$

The function has singular derivatives about the coordinate axis, in fact $u \in \mathrm{C}^{1,1/3}(\Omega)$. Figure 2 details a numerical experiment on this problem. In Fig. 3 we conduct an adaptive experiment based on the newest vertex bisection method.

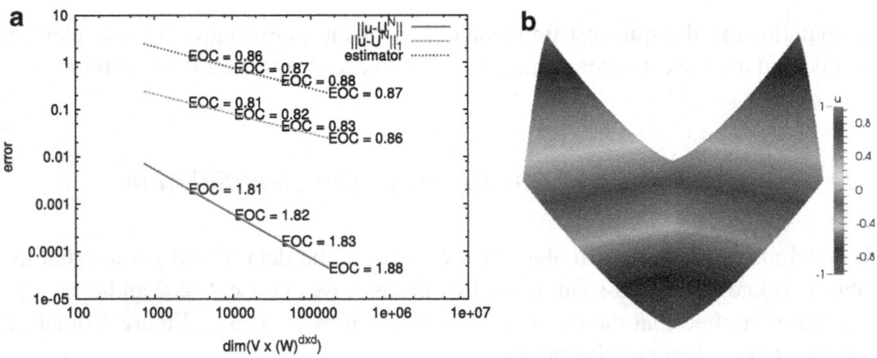

Fig. 2 We benchmark problem (22), plotting the log of the error together with its estimated order of convergence. We examine both $\mathrm{L}_2(\Omega)$ and $\mathrm{H}^1(\Omega)$ norms of the error together with the residual estimator given in Theorem 1. We choose $\tau = 1$ and the linearisation tolerance is coupled to the mesh-size as in Fig. 1. The convergence rates are suboptimal due to the singularity, that is, $\left\| u - U^N \right\| \approx \mathrm{O}(h^{1.8})$ and $\left| u - U^N \right|_1 \approx \mathrm{O}(h^{0.8})$. (**a**) Convergence rates. (**b**) Finite element approximation

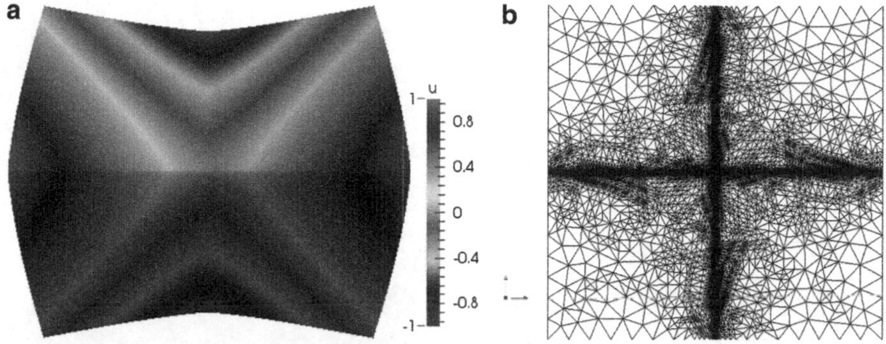

Fig. 3 This is an adaptive approximation of the viscosity solution $u = |x|^{4/3} - |y|^{4/3}$ from (22). The estimator tolerance was set at 0.1 to coincide with the final estimate from the benchmark solution from Fig. 2. The final number of degrees of freedom was 36,325 compared to the uniform scheme which took 165,125 degrees of freedom to reach the same tolerance. We chose $\tau = 0.1$ as the timestep parameter. (**a**) The finite element approximation viewed from the top. (**b**) The underlying mesh

References

1. G. Aronsson, Construction of singular solutions to the p-harmonic equation and its limit equation for $p = \infty$. Manuscr. Math. **56**(2), 135–158 (1986)
2. G. Awanou, Pseudo time continuation and time marching methods for Monge–Ampère type equations (2012). In revision – tech report available on http://www.math.niu.edu/~awanou/
3. E.N. Barron, L.C. Evans, R. Jensen, The infinity Laplacian, Aronsson's equation and their generalizations. Trans. Am. Math. Soc. **360**(1), 77–101 (2008)
4. L.A. Caffarelli, X. Cabré, *Fully Nonlinear Elliptic Equations*. American Mathematical Society Colloquium Publications, vol. 43 (American Mathematical Society, Providence, 1995)
5. M.G. Crandall, H. Ishii, P.-L. Lions, User's guide to viscosity solutions of second order partial differential equations. Bull. Am. Math. Soc. (N.S.) **27**(1), 1–67 (1992)
6. M.G. Crandall, L.C. Evans, R.F. Gariepy, Optimal Lipschitz extensions and the infinity Laplacian. Calc. Var. Partial Differ. Equ. **13**(2), 123–139 (2001)
7. A. Dedner, T. Pryer, Discontinuous Galerkin methods for nonvariational problems (2013). Submitted – tech report available on ArXiV http://arxiv.org/abs/1304.2265
8. S. Esedoglu, A.M. Oberman, Fast semi-implicit solvers for the infinity laplace and p-laplace equations, Arxiv, 2011
9. L.C. Evans, O. Savin, $C^{1,\alpha}$ regularity for infinity harmonic functions in two dimensions, Calc. Var. Partial Differ. Equ. **32**(3), 325–347 (2008)
10. X. Feng, M. Neilan, Vanishing moment method and moment solutions for fully nonlinear second order partial differential equations. J. Sci. Comput. **38**(1), 74–98 (2009)
11. R. Jensen, Uniqueness of Lipschitz extensions: minimizing the sup norm of the gradient. Arch. Ration. Mech. Anal. **123**(1), 51–74 (1993)
12. O. Lakkis, T. Pryer, A finite element method for second order nonvariational elliptic problems. SIAM J. Sci. Comput. **33**(2), 786–801 (2011)
13. G. Lu, P. Wang, Inhomogeneous infinity Laplace equation. Adv. Math. **217**(4), 1838–1868 (2008)
14. A.M. Oberman, A convergent difference scheme for the infinity Laplacian: construction of absolutely minimizing Lipschitz extensions. Math. Comput. **74**(251), 1217–1230 (2005). (Electronic)

Anisotropic Adaptive Meshes for Brittle Fractures: Parameter Sensitivity

Marco Artina, Massimo Fornasier, Stefano Micheletti, and Simona Perotto

Abstract We deal with the Ambrosio-Tortorelli approximation of the well-known Mumford-Shah functional to model quasi-static crack propagation in brittle materials. We employ anisotropic mesh adaptation to efficiently capture the crack path. Aim of this work is to investigate the numerical sensitivity of the crack behavior to the parameters involved in both the physical model and in the adaptive procedure.

1 Introduction to the Problem

The Mumford-Shah functional plays a key role in many applications, from image segmentation to mechanical problems [8]. One such application is the fracture propagation in brittle materials, where no predefined crack path is required. The Mumford-Shah functional is used for the first time by G. Francfort and J.-J. Marigo in [6] to model the quasi-static evolution of such a crack along the critical points of the energy. From a mathematical viewpoint, this leads to minimizing a nonconvex and nonsmooth functional which involves the displacement function u together with a lower dimensional set representing the crack Γ. This is an interesting challenge for both the theoretical analysis and the numerical computation. In the two sections below we address a suitable regularization of the functional proposed by G. Francfort and J.-J. Marigo and a corresponding discrete approximation. This smoothed model will allow us to tackle the numerical approximation via the employment of a suitable anisotropic adapted mesh, able to follow tightly the crack path.

M. Artina (✉) • M. Fornasier
Faculty of Mathematics, Technische Universität München, Boltzmannstrasse 3, 85748, Garching, Germany
e-mail: marco.artina@ma.tum.de; massimo.fornasier@ma.tum.de

S. Micheletti • S. Perotto
MOX – Dipartimento di Matematica "F.Brioschi", Politecnico di Milano, Piazza L. da Vinci 32, I-20133 Milano, Italy
e-mail: stefano.micheletti@polimi.it; simona.perotto@polimi.it

A. Abdulle et al. (eds.), *Numerical Mathematics and Advanced Applications ENUMATH 2013*, Lecture Notes in Computational Science and Engineering 103,
DOI 10.1007/978-3-319-10705-9_29

1.1 The Modified Ambrosio-Tortorelli (MAT) Functional

The functional provided by L. Ambrosio and V.M. Tortorelli is one of the most popular approach to dealing with the intrinsic irregularity of the functional proposed by G. Francfort and J.-J. Marigo [1]. It is given by

$$I(u,v) = \int_{\Omega} (v^2 + \eta)|\nabla u|^2 \, d\mathbf{x} + \kappa \int_{\Omega} \left[\frac{1}{4\varepsilon}(1-v)^2 + \varepsilon |\nabla v|^2 \right] d\mathbf{x}, \tag{1}$$

where $\Omega \subset I\!R^2$ is an open domain, $0 < \eta \ll \varepsilon \ll 1$, $\kappa > 0$ approximates the elasticity constant of the material, while $u : H^1(\Omega) \to I\!R$ and $v : H^1(\Omega) \to [0,1]$ represent the displacement and the crack path, respectively. The first integral takes into account the elastic energy of the material, while the second integral is a fictitious energy spent in propagating the crack inside the material. Furthermore, when $v = 1$, the second integral vanishes, indicating the absence of a crack, whereas, when $v = 0$, only the fictitious energy does contribute and we are in the presence of an actual crack. The Ambrosio-Tortorelli functional enjoys the desirable property of Γ-converging to the Mumford-Shah functional [1].

To drive the crack evolution, an external load $g : \Omega \times [0,T] \to I\!R$ is applied on a subset $\Omega_{D\pm} = \Omega_{D+} \cup \Omega_{D-}$ of Ω, defined as follows

$$g(\mathbf{x},t) = \begin{cases} t & \text{if } \mathbf{x} \in \Omega_{D+}, \\ -t & \text{if } \mathbf{x} \in \Omega_{D-}, \\ 0 & \text{elsewhere}, \end{cases} \tag{2}$$

with $T > 0$ the final time of interest and $\mathbf{x} = (x_1, x_2)^T$. For simplicity, we denote hereafter $g(\mathbf{x},t)$ with $g(t)$. Let us also introduce the space of admissible solutions $\mathscr{A}(g(t)) = \{u \in H^1(\Omega) : u|_{\Omega_{D\pm}} = g(t)|_{\Omega_{D\pm}}\}$. Associated with the time interval $[0,T]$, we define a uniform partition $0 = t_0 < t_1 < \ldots < t_n = T$ of step Δt. According to a quasistatic evolution, at any discrete instant t_k, with $k = 1, \ldots, n$, we solve the following minimization problem

$$(u_\varepsilon(t_k), v_\varepsilon(t_k)) \in \underset{\substack{u \in \mathscr{A}(g(t_k)),\, v \in H^1(\Omega) \\ \text{s.t. } v(\mathbf{x},t_k) = 0,\ \forall \mathbf{x} \in CR_{k-1}}}{\arg\min} I(u,v), \tag{3}$$

with $CR_{k-1} = \{\mathbf{x} \in \overline{\Omega} \,|\, v_\varepsilon(t_{k-1}) < \mathtt{CRTOL}\}$, and where CRTOL is a tolerance controlling somehow the thickness of the crack. At the initial time, we set $CR_{-1} = \emptyset$, i.e., the constraint in (3) is removed since the crack is not yet present. Convergence results relating the actual continuous model with the current time-discrete version can be found in [5].

In [2], we relax the constraints in (3) via suitable penalization terms which allow us to deal with an unconstrained minimization for the Modified Ambrosio-Tortorelli (MAT) functional

$$
\begin{aligned}
I_k^{\mathrm{MAT}}(u,v) &= \int_\Omega \left[(v^2+\eta)|\nabla u|^2\, d\,\mathbf{x} + \frac{1}{4\varepsilon}(1-v)^2 + \varepsilon|\nabla v|^2\right] d\mathbf{x} \\
&+ \frac{1}{\gamma_A}\int_{\Omega_{D\pm}} (g(t_k)-u)^2\, d\mathbf{x} + \frac{1}{\gamma_B}\int_{CR_{k-1}} v^2\, d\mathbf{x},
\end{aligned} \tag{4}
$$

where γ_A and γ_B are (small) penalty constants and $\kappa = 1$ is assumed for convenience. Moreover, as remarked in [2], condition $0 \le v \le 1$ is guaranteed by the minimization process.

1.2 The Discretized MAT Functional

We consider the discrete counterpart of the functional (4) via a finite element approximation. For this purpose, we introduce a family of meshes $\{\mathscr{T}_h\}_h$ of $\overline{\Omega}$, with $h > 0$ the discretization parameter, and denote by $\mathscr{E}_h$ the skeleton of $\mathscr{T}_h$. Moreover, we associate with $\mathscr{T}_h$ the space X_h consisting of continuous piecewise linear functions.

The discretization of the MAT functional based on X_h is

$$
\begin{aligned}
I_{k,h}^{\mathrm{MAT}}(u_h,v_h) &= \int_\Omega \left[\left(P_h(v_h^2)+\eta\right)|\nabla u_h|^2\, d\mathbf{x} + \frac{1}{4\varepsilon}P_h((1-v_h)^2) + \varepsilon|\nabla v_h|^2\right] d\mathbf{x} \\
&+ \frac{1}{\gamma_A}\int_{\Omega_{D\pm}} P_h\left((g_h(t_k)-u_h)^2\right) d\mathbf{x} + \frac{1}{\gamma_B}\int_{CR_{k-1}} P_h\left(v_h^2\right) d\mathbf{x},
\end{aligned}
$$

where $P_h : C^0(\overline{\Omega}) \to X_h$ is the Lagrange interpolant onto X_h, and $g_h(t_k)$ is the $L^2(\Omega_{D\pm})$-projection of $g(t_k)$ onto X_h. In practice, the minimization problem (3) is replaced by the following one

$$
(u_h(t_k), v_h(t_k)) \in \underset{(\hat{u}_h,\hat{v}_h)\in X_h\times X_h}{\arg\min}\ I_{k,h}^{\mathrm{MAT}}(\hat{u}_h,\hat{v}_h).
$$

The critical points of $I_{k,h}^{\mathrm{MAT}}$ satisfy relation $(I_{k,h}^{\mathrm{MAT}})'(u_h, v_h; \varphi_h, \psi_h) = 0$, where $(I_{k,h}^{\mathrm{MAT}})'$ denotes the Gâteaux derivative of the discrete MAT functional given, for any $(\varphi_h, \psi_h) \in X_h \times X_h$, by

$$\begin{aligned}
&(I_{k,h}^{\mathrm{MAT}})'(u_h, v_h; \varphi_h, \psi_h) \\
&= 2\left(\int_\Omega (P_h(v_h^2) + \eta)\nabla u_h \cdot \nabla \varphi_h \, d\mathbf{x} + \frac{1}{\gamma_A}\int_{\Omega_{D^\pm}} P_h\left((u_h - g_h(t_k))\varphi_h\right) d\mathbf{x}\right) \\
&+2\left(\int_\Omega \left[P_h(v_h\psi_h)|\nabla u_h|^2 + \frac{1}{4\varepsilon}P_h((v_h-1)\psi_h) + \varepsilon\nabla v_h \cdot \nabla \psi_h\right] d\mathbf{x} \right.\\
&\left.+\frac{1}{\gamma_B}\int_{CR_{k-1}} P_h\left(v_h\psi_h\right) d\mathbf{x}\right).
\end{aligned}$$

The proof that the condition $0 \le v_h \le 1$ is automatically satisfied also in the discrete case can be found in [2].

2 The Anisotropic a Posteriori Error Estimator

We introduce now the basic setting that we use to enrich the discretization above with mesh adaptation. In particular, we adopt an anisotropic approach due to the highly directional nature of the crack propagation as well as to the expected computational saving led by the employment of anisotropic meshes.
Following, e.g., [7], the geometric information describing a generic stretched element $K \in \mathscr{T}_h$ are based on the spectral properties of the affine map T_K between the reference element $\hat{K}$ and the actual triangle K, whose diameter and area are denoted by h_K and $|K|$, respectively. In practice, we use the unit vectors $\mathbf{r}_{1,K}$ and $\mathbf{r}_{2,K}$ along the directions of the semiaxes of the ellipse, image of the circumscribed circle to $\hat{K}$, and the positive scalars $\lambda_{1,K}, \lambda_{2,K}$, with $\lambda_{1,K} \ge \lambda_{2,K}$, that measure the length of these semiaxes.

Before stating the theoretical tool driving the adaptive procedure in Sect. 3, we anticipate some notation. We introduce the residuals

$$\begin{aligned}
\rho_K^A(v_h, u_h) = \|2v_h\nabla v_h \cdot \nabla u_h\|_{2,K} + \tfrac{1}{2}\|[\![\nabla u_h]\!]\|_{\infty,\partial K}\|v_h^2 + \eta\|_{2,\partial K}\left(\tfrac{h_K}{\lambda_{1,K}\lambda_{2,K}}\right)^{\frac{1}{2}} \\
+\tfrac{\delta_{K,\Omega_D^\pm}}{\gamma_A}\left(\|u_h - g_h(t_k)\|_{2,K} + \|g_h(t_k) - g(t_k)\|_{2,K}\right) \\
+\tfrac{1}{\lambda_{2,K}}\left[\|v_h^2 - P_h(v_h^2)\|_{\infty,K}\|\nabla u_h\|_{2,K} + \tfrac{|K|^{1/2}h_K^2}{\gamma_A}|u_h - g_h(t_k)|_{1,\infty,K}\right],
\end{aligned}$$

$$\begin{aligned}
\rho_K^B(u_h, v_h) = \|(|\nabla u_h|^2 + \tfrac{1}{4\varepsilon})v_h - \tfrac{1}{4\varepsilon}\|_{2,K} + \tfrac{\varepsilon}{2}\|[\![\nabla v_h]\!]\|_{2,\partial K}\left(\tfrac{h_K}{\lambda_{1,K}\lambda_{2,K}}\right)^{\frac{1}{2}} \\
+\tfrac{\delta_{K,\mathrm{cr}_{k-1}}}{\gamma_B}\|v_h\|_{2,K} + \tfrac{h_K^2}{\lambda_{2,K}}\left[\|\,|\nabla u_h|^2 + \tfrac{1}{4\varepsilon}\|_{2,K} + \tfrac{|K|^{\frac{1}{2}}\delta_{K,\mathrm{cr}_{k-1}}}{\gamma_B}\right]|v_h|_{1,\infty,K},
\end{aligned}$$

and the weight

$$\omega_K(w) = \left[\sum_{i=1}^{2} \lambda_{i,K}^2 (\mathbf{r}_{i,K}^T G_{\Delta_K}(w) \mathbf{r}_{i,K}) \right]^{1/2} \qquad \forall w \in H^1(\Omega),$$

where $\delta_{K,\varpi}$ is the Kronecker symbol associated with ϖ, such that $\delta_{K,\varpi} = 1$ if $K \cap \varpi \neq \emptyset$ and $\delta_{K,\varpi} = 0$ otherwise; $G_{\Delta_K} \in I\!R^{2\times 2}$ is the symmetric positive semidefinite matrix with entries $[G_{\Delta_K}(w)]_{i,j} = \int_{\Delta_K} (\partial w/\partial x_i)(\partial w/\partial x_j)\, d\mathbf{x}$, $i = 1, 2$, and with $\Delta_K = \{K' \in \mathcal{T}_h : K' \cap K \neq \emptyset\}$ the patch of elements associated with K; $[\![w_h]\!] = [\partial w_h/\partial n]$ on $\mathcal{E}_h \cap \Omega$ and $[\![w_h]\!] = \partial w_h/\partial n$ on $\mathcal{E}_h \cap \partial\Omega$ is the jump of the normal derivative of $w_h \in X_h$.
The standard notation $\|\cdot\|_{k,p,\varpi}$ is adopted to denote the norm in the Sobolev space $W^{k,p}(\varpi)$, with $\varpi \subset I\!R^d$ for $d = 1, 2$ and where k is omitted when zero.

Proposition 1 *Let $(u_h, v_h) \in X_h \times X_h$ be the critical point of $I_{k,h}^{\mathrm{MAT}}$. Let us assume that $\#\Delta_K \leq \mathcal{N}$ and that* $\mathrm{diam}(T_K^{-1}(\Delta_K)) \leq C_\Delta$. *Then, for all $\varphi, \psi \in H^1(\Omega)$, there exists a constant $C = C(\mathcal{N}, C_\Delta)$ such that*

$$|(I_{k,h}^{\mathrm{MAT}})'(u_h, v_h; \varphi, \psi)| \leq C \sum_{K \in \mathcal{T}_h} \{\rho_K^A(v_h, u_h)\, \omega_K(\varphi) + \rho_K^B(u_h, v_h)\, \omega_K(\psi)\}. \qquad (5)$$

For the proof of this proposition, we refer to [2]. We just observe that an important role is played by the anisotropic estimates for the Clément quasi-interpolant [7]. The actual a posteriori error estimator, say η^{MAT}, involved in the adaptive procedure coincides with the right-hand side of (5) after replacing φ and ψ with u_h and v_h, respectively, and taking $C = 1$.

3 An Optimize-and-Adapt Algorithm

The minimization of the functional $I_{k,h}^{\mathrm{MAT}}$ is not a trivial task since the functional is not convex. However, in [3] a first strategy to deal with it is proposed and further analyzed in [4]. The idea is to resort to a Gauss-Seidel-like algorithm consisting of a two-step procedure, first minimizing with respect to u_h for a fixed v_h, and then to minimize also with respect to v_h using the updated u_h. Moving from this idea, in [2] we couple this optimization step with an anisotropic mesh adaptation procedure in two different ways. In particular, following [7], we employ a metric-based approach relying on estimate (5) with the aim of minimizing the number of mesh elements for a fixed tolerance `REFTOL`$\ll 1$ on η^{MAT}. The first algorithm, optimize-then-adapt, in [2], which is a variant of `ALGORITHM 1` in [4], applies the mesh adaptation after convergence of the minimization algorithm on both u_h and v_h. Since the coupling between optimization and adaptation is not so tight, this algorithm is weak in the presence of a fast evolution of a crack. As a remedy, in the second algorithm, optimize-and-adapt, we introduce a closer alternation of the optimization and mesh adaptation phases, by adapting the mesh just after two steps of the Gauss-Seidel algorithm without waiting for convergence.

In more detail, after fixing a termination tolerance VTOL$\ll$ 1 for the minimization algorithm, a relative tolerance MESHTOL$\ll$ 1 on the change of the mesh cardinality, the optimize-and-adapt algorithm is the following:

Algorithm 3.1: Optimize-and-adapt algorithm

1. Set $k = 0$, $\mathscr{T}_h^{(1)} = \mathscr{T}_h$;
2. If $k = 0$, set $v_h^1 = 1$; else $v_h^1 = v_h(t_{k-1})$;
3. Set $i = 1$; errmesh $= 1$; err$= 1$;

while errmesh $\geq$ & err $\geq$ **do**

4. $u_h^i = \arg\min_{z_h \in X_h^{(i)}} I_{k,h}^{\mathrm{MAT}}(z_h, v_h^i)$;
5. $v_h^{i+1} = \arg\min_{z_h \in X_h^{(i)}} I_{k,h}^{\mathrm{MAT}}(u_h^i, z_h)$;
6. Build the metric-based adapted mesh $\mathscr{T}_h^{(i+1)}$ with tolerance ;
7. err $= \|v_h^{i+1} - v_h^i\|_{\infty,\Omega}$;
8. errmesh $= |\#\mathscr{T}_h^{(i+1)} - \#\mathscr{T}_h^{(i)}|/\#\mathscr{T}_h^{(i)}$;
9. Set $v_h^1 = \Pi_{i\to i+1}(v_h^{i+1})$;
10. $i \leftarrow i + 1$;

end while

11. $u_h(t_k) = \Pi_{i-1\to i}(u_h^{i-1})$; $v_h(t_k) = v_h^1$; $\mathscr{T}_h^k = \mathscr{T}_h^{(i)}$;
12. Set $\mathscr{T}_h^{(1)} = \mathscr{T}_h^k$;
13. $k \leftarrow k + 1$;
14. if $k > n$, stop; else goto 2.

An interpolation step between two successive adapted meshes is employed before restarting any new optimization or time loop. This is carried out by a suitable interpolation operator, $\Pi_{j\to j+1}(w_h)$, which maps a finite element function w_h defined on $\mathscr{T}_h^j$ onto the new mesh $\mathscr{T}_h^{j+1}$. The convergence of the mesh adaptivity is checked by monitoring the change of the number of elements, which is an effective stopping criterion though not rigorously sound.

4 Sensitivity Assessment

In this section we carry out a sensitivity analysis of the optimize-and-adapt algorithm to its main parameters. The test case used for this purpose is the curved crack configuration studied in [2, 4].

We consider a rectangle $\Omega = (0, 2) \times (0, 2.2)$ including the slit $\{1\} \times [1.5, 2.2]$, $2 \cdot 10^{-5}$ wide, and a circular hole of radius 0.2 and center at (0.3, 0.3) (see Fig. 1 (left)). In (2) we choose $\Omega_{D-} = (0, 1) \times (2, 2.2)$ and $\Omega_{D+} = (1, 2) \times (2, 2.2)$. The default values for the parameters are $\varepsilon = 2 \cdot 10^{-2}$, $\eta = 10^{-5}$, $\gamma_A = \gamma_B = 10^{-5}$, $\Delta t = 10^{-2}$, CRTOL $= 3 \cdot 10^{-4}$, VTOL $= 2 \cdot 10^{-3}$, ADAPTOL $= 10^{-2}$, and REFTOL $= 10^{-2}$. In Fig. 1 we show the anisotropic mesh yielded by the algorithm

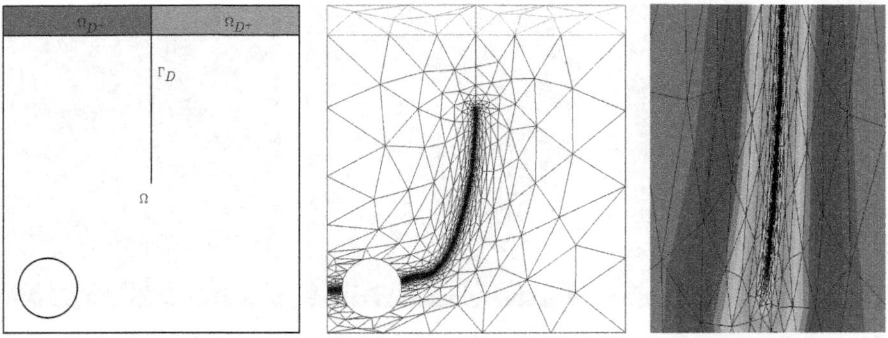

Fig. 1 Computational domain (*left*), final anisotropic adapted mesh (*center*), zoom in (*right*) for the default parameters

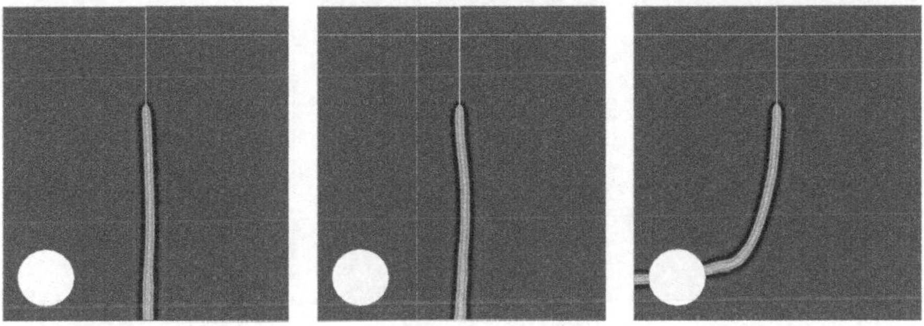

Fig. 2 Sensitivity to the penalty constants: colour plot of the v_h-field for $\gamma_A = \gamma_B = 10^{-4}$ (*left*), $\gamma_A = \gamma_B = 5 \cdot 10^{-5}$ (*center*), $\gamma_A = \gamma_B = 10^{-5}$ (*right*)

at the final time $T = 1.43$ along with a detail around the crack. The number of the elements and the maximum aspect ratio $\max_{K \in \mathcal{T}_h} \lambda_{1,K}/\lambda_{2,K}$ are 15,987 and $3.06 \cdot 10^3$, respectively. The first series of tests check on the sensitivity to the penalty constants $\gamma_A = \gamma_B$, by choosing three pairs of values, i.e., 10^{-4}, $5 \cdot 10^{-5}$, 10^{-5}. From Fig. 2, it is evident that the higher the values of these constants, the larger is the deviation of the crack path with respect to the one assumed as default. In particular, with the first two choices the crack even misses the hole. The two meshes yielding the straight path consist of fewer elements (12,027 and 12,628) than the default mesh in Fig. 1.

The second trial of checks deals with the sensitivity to the tolerance REFTOL involved in the mesh adaptation procedure. We choose both a larger and a smaller value with respect to the default, namely REFTOL$= 10^{-1}$ and REFTOL$= 8 \cdot 10^{-3}$. The associated v_h-field are displayed in Fig. 3. The largest value leads to a wrong path detection with only 8,547 triangles, whereas the choice REFTOL$= 8 \cdot 10^{-3}$ identifies essentially the same path as the default one, but with an excessive number of elements (23,521). Thus, it seems that too small a tolerance just increases the computational effort without improving the crack path tracking.

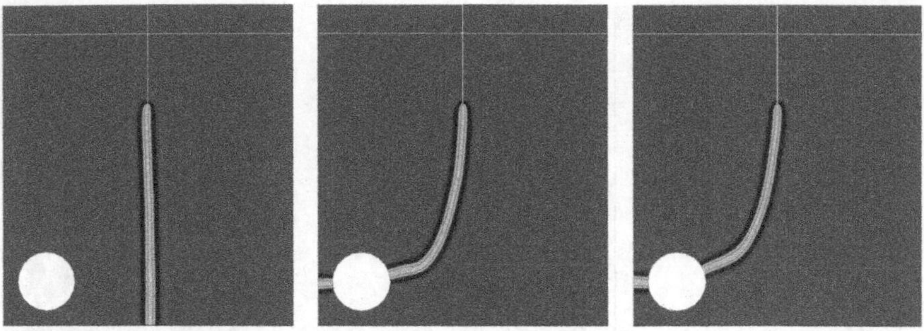

Fig. 3 Sensitivity to REFTOL: plot of the v_h-field for REFTOL= 10^{-1} (*left*), REFTOL= 10^{-2} (*center*), REFTOL= $8 \cdot 10^{-3}$ (*right*)

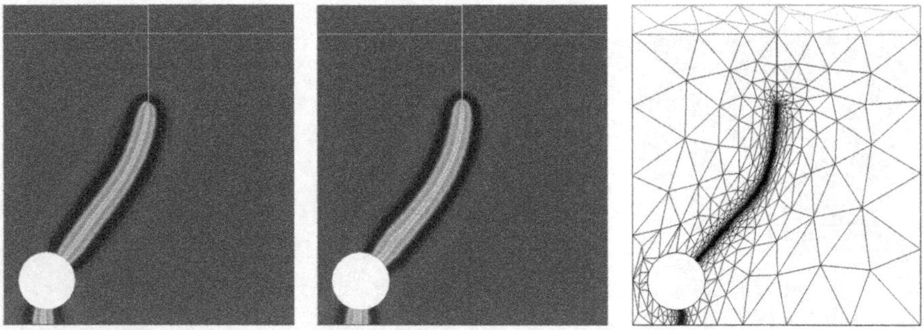

Fig. 4 Plot of the v_h-field for REFTOL= 10^{-1} (*left*), REFTOL= 10^{-2} (*center*), and adapted mesh for $t = 1.43$ and REFTOL= 10^{-2} (*right*)

The last batch of tests assesses the behaviour of the optimize-and-adapt algorithm for a different value of ε, i.e., $\varepsilon = 5 \cdot 10^{-2}$. We observe that ε controls the width of the crack. As expected, the larger value of ε widens the crack boundaries (compare the thickness of the crack in Figs. 3 and 4). Moreover, also the crack trajectory changes considerably. For $\varepsilon = 5 \cdot 10^{-2}$ the crack suddenly turns left entering directly the hole, independently of the two chosen tolerances REFTOL= 10^{-1}, 10^{-2}. Although from a physical viewpoint the behavior seems correct, the bending of the actual path occurs too early and the crack leaves the hole downward instead to the left. A cross-comparison between Figs. 3 and 4 leads to argue that for $\varepsilon = 5 \cdot 10^{-2}$ the value of REFTOL is not so crucial in identifying the actual path of the crack.

The assessment above seems to confirm that there is an actual sensitivity of the crack behaviour to the parameters involved in both the MAT functional and in the optimize-and-adapt algorithm. The employment of an anisotropic mesh adaptation seems strategical to explore the possible scenarios to single out the most reliable one, thanks to the computational saving due to an anisotropic grid.

References

1. L. Ambrosio, V.M. Tortorelli, Approximation of functional depending on jumps by elliptic functional via Γ-convergence. Commun. Pure Appl. Math. **43**(8), 999–1036 (1990)
2. M. Artina, M. Fornasier, S. Micheletti, S. Perotto, *Anisotropic Mesh Adaptation for Crack Detection in Brittle Materials*, Mox-Report 21/2014
3. B. Bourdin, G. Francfort, J.-J. Marigo, Numerical experiments in revisited brittle fracture. J. Mech. Phys. Solids **48**(4), 797–826 (2000)
4. S. Burke, C. Ortner, E. Süli, An adaptive finite element approximation of a variational model of brittle fracture. Soc. Ind. Appl. Math. **48**(3), 980–1012 (2010)
5. G. Dal Maso, R. Toader, A model for the quasi-static growth of brittle fractures based on local minimization. Math. Models Methods Appl. Sci. **12**(12), 1773–1799 (2002)
6. G. Francfort, J.-J. Marigo, Revisiting brittle fracture as an energy minimization problem. J. Mech. Phys. Solids **46**(8), 1319–1342 (1998)
7. S. Micheletti, S. Perotto, Output functional control for nonlinear equations driven by anisotropic mesh adaption: the Navier-Stokes equations. SIAM J. Sci. Comput. **30**(6), 2817–2854 (2008)
8. D. Mumford, J. Shah, Optimal approximations by piecewise smooth functions and associated variational problems. Commun. Pure Appl. Math. **42**(5), 577–685 (1989)

Part IV
Numerical Linear Algebra

Variational Principles for Eigenvalues of Nonlinear Eigenproblems

Heinrich Voss

Abstract Variational principles are very powerful tools when studying self-adjoint linear operators on a Hilbert space $\mathscr{H}$. Bounds for eigenvalues, comparison theorems, interlacing results and monotonicity of eigenvalues can be proved easily with these characterizations, to name just a few. In this paper we consider generalization of these principles to families of linear, self-adjoint operators depending continuously on a scalar in a real interval.

1 Introduction

Let A be a self-adjoint operator on a Hilbert space $\mathscr{H}$ with scalar product $\langle \cdot, \cdot \rangle$, and denote by $\lambda_1 \leq \lambda_2 \leq \ldots$ those eigenvalues of A (if there are any), which are smaller than the minimum of the essential spectrum $\sigma_{\text{ess}}(A)$, each counted according to its multiplicity. Then λ_j can be characterized by three fundamental variational principles [28], namely by Rayleigh's principle [19]

$$\lambda_j = \min\{R(x) : \langle x, x_i \rangle = 0,\ i = 1, \ldots, j-1\} \tag{1}$$

where $R(x) := \langle Ax, x \rangle / \langle x, x \rangle$ is the Rayleigh quotient and $x_1, \ldots, x_{j-1}$ is a set of orthogonal eigenvectors of A (x_i corresponding to λ_i), the minmax characterization by Poincaré [18]

$$\lambda_j = \min_{\dim V = j} \max_{x \in V, x \neq 0} R(x), \tag{2}$$

H. Voss (✉)
Institute of Mathematics, Hamburg University of Technology, D-21071 Hamburg, Germany
e-mail: voss@tuhh.de

A. Abdulle et al. (eds.), *Numerical Mathematics and Advanced Applications ENUMATH 2013*, Lecture Notes in Computational Science and Engineering 103,
DOI 10.1007/978-3-319-10705-9_30

and the maxmin principle due to Courant [5], Fischer [9] and Weyl [30]

$$\lambda_j = \max_{\dim V=j-1} \min_{x \in V^\perp, x \neq 0} R(x) \tag{3}$$

where $V^\perp := \{x \in \mathscr{H} : \langle v, x\rangle = 0 \text{ for every } v \in V\}$.

The purpose of this paper is to survey generalizations of these principles to the nonlinear eigenvalue problem

$$T(\lambda)x = 0 \tag{4}$$

and to trace the history of these generalizations. Here $T(\lambda)$, $\lambda \in J$, is a family of linear self-adjoint and bounded operator on $\mathscr{H}$, and J is a real open interval which may be unbounded. As in the linear case $T(\lambda) := \lambda I - A$ we call $\lambda \in J$ an eigenvalue of $T(\cdot)$ if Eq. (4) has a nontrivial solution $x \neq 0$ and the solution x is called a corresponding eigenelement.

We stress the fact that we are only concerned with real eigenvalues in J although $T(\cdot)$ may be defined on a larger subset of $\mathbb{C}$, and $T(\cdot)$ may have additional eigenvalues in $\mathbb{C} \setminus J$.

2 Overdamped Problems

To receive generalizations of the variational principles to the nonlinear eigenvalue problem (4) the Rayleigh quotient $R(x)$ of a linear problem $Ax = \lambda x$ has to be replaced with some functional. We assume that for every $x \in J$, $x \neq 0$ the real equation $f(\lambda; x) := \langle T(\lambda)x, x\rangle = 0$ has at most one solution in J denoted by $p(x)$. This defines the so called Rayleigh functional p which obviously generalizes the Rayleigh quotient for the linear case.

If the Rayleigh functional p is defined on the entire space $\mathscr{H} \setminus \{0\}$ then the eigenproblem (4) is called overdamped. This term is motivated by the finite dimensional quadratic eigenvalue problem

$$T(\lambda)x = \lambda^2 Mx + \lambda Cx + Kx = 0 \tag{5}$$

governing the damped free vibrations of a system where $M, C, K \in \mathbb{R}^{n\times n}$ are symmetric and positive definite matrices corresponding to the mass, the damping and the stiffness of the system, respectively.

Assume that the damping $C = \alpha\tilde{C}$ depends on a parameter $\alpha \geq 0$. Then for $\alpha = 0$ the system has purely imaginary eigenvalues corresponding to harmonic vibrations of the system. Increasing α the eigenvalues move into the left half plane as conjugate complex pairs corresponding to damped vibrations. Finally they reach the negative real axis as double eigenvalues where they immediately split and move into opposite directions.

When eventually all eigenvalues have become real, and all eigenvalues going to the right are right of all eigenvalues moving to the left the system is called overdamped. In this case the two solutions

$$p_{\pm}(x) = (-\alpha\langle\tilde{C}x, x\rangle \pm \sqrt{\alpha^2\langle\tilde{C}x, x\rangle^2 - 4\langle Mx, x\rangle\langle Kx, x\rangle})/(2\langle Mx, x\rangle).$$

of the quadratic equation

$$\langle T(\lambda)x, x\rangle = \lambda^2\langle Mx, x\rangle + \lambda\alpha\langle\tilde{C}x, x\rangle + \langle Kx, x\rangle = 0 \tag{6}$$

are real, and they satisfy $\sup_{x\neq 0} p_-(x) < \inf_{x\neq 0} p_+(x)$.

Hence, for $J_- := (-\infty, \inf_{x\neq 0} p_+(x))$ Eq. (6) defines the Rayleigh functional p_-, and for $J_+ := (\sup_{x\neq 0} p_-(x), 0)$ it defines the Rayleigh functional p_+.

Duffin [6] proved that all eigenvalues $\lambda_1^- \leq \dots \lambda_n^-$ and $\lambda_1^+ \leq \dots \leq \lambda_n^+$ are maxmin values of the functionals p_- and p_+, respectively, and Rogers [20] generalized it to the finite dimensional overdamped case.

Theorem 1 *Let $T(\lambda) \in \mathbb{R}^{n\times n}$, $\lambda \in J$ be an overdamped family of symmetric matrices depending continuously differentiable on $\lambda \in J$ such that $\langle T'(p(x))x, x\rangle > 0$ for every $x \neq 0$. Then there are exactly n eigenvalues $\lambda_1 \leq \dots \leq \lambda_n$ of $T(\lambda)x = 0$ in J, and it holds*

$$\lambda_j = \min_{\dim V = j} \max_{x\in V, x\neq 0} p(x), \quad j = 1, \dots, n. \tag{7}$$

Infinite dimensional overdamped problems were considered first for quadratic problems $(A - \lambda^2 B - \lambda I)x = 0$ where A and B are bounded, positive definite and compact by Turner [22] and Weinberger [27] who proved all three types of variational characterization by linearization (i.e. taking advantage of the fact that the quadratic problem is equivalent to a linear self-adjoint eigenproblem), and by Langer [15] who proved minmax and maxmin characterizations for the quadratic problem $(\lambda^2 A + \lambda B + C)x = 0$ taking advantage of the theory of J-self-adjoint operators.

The general overdamped problem was considered by Hadeler [11] who proved the following minmax and maxmin theorem:

Theorem 2 *Let $T(\lambda) : \mathcal{H} \to \mathcal{H}$, $\lambda \in J$ be a family of linear self-adjoint and bounded operators such that* (4) *is over-damped, and assume that for $\lambda \in J$ there exists $\nu(\lambda) > 0$ such that $T(\lambda) + \nu(\lambda)I$ is compact.*

Let $T(\cdot)$ be continuously differentiable and suppose that

$$\langle T'(p(x))x, x\rangle > 0 \quad \textit{for ever } x \neq 0. \tag{8}$$

Let the eigenvalues λ_n of $T(\lambda)x = 0$ be numbered in non-decreasing order regarding their multiplicities. Then they can be characterized by the following two variational principles

$$\lambda_n = \min_{\dim V=n} \max_{x\in V,\, x\neq 0} p(x)$$
$$= \max_{\dim V=n-1} \min_{x\in V^{\perp},\, x\neq 0} p(x).$$

Moreover, Hadeler [11] generalized Rayleigh's principle for overdamped problems proving that the eigenvectors are orthogonal with respect to the generalized scalar product

$$[x, y] := \begin{cases} \dfrac{\langle (T(p(x)) - T(p(y)))x, y\rangle}{p(x) - p(y)}, & \text{if } p(x) \neq p(y) \\ \langle T'(p(x))x, y\rangle, & \text{if } p(x) = p(y) \end{cases} \tag{9}$$

which is symmetric, definite and homogeneous, but in general it is not bilinear.

Further generalizations of the minmax and maxmin characterizations were proved for certain overdamped polynomial eigenproblems by Turner [23], and for general overdamped problems by Rogers [21], Werner [29], Abramov [1], and Hadeler [12] who relaxed the compactness conditions on $T(\cdot)$.

Markus [16] and Hasanov [13] (with a completely different proof) considered nonoverdamped problems which depended only continuously on the parameter and they replaced assumption (8) with the condition that $\langle T(\lambda)x, x\rangle$ is increasing at the point $p(x)$ given in condition (A_2) of the next section

3 Nonoverdamped Problems

We consider the nonlinear eigenvalue problem (4), where $T(\lambda) : \mathscr{H} \to \mathscr{H}, \lambda \in J$, is a family of self-adjoint and bounded operators depending continuously on the parameter λ.

We assume that

(A_1) *For every fixed $x \in \mathscr{H}$, $x \neq 0$ the real equation*

$$f(\lambda; x) := \langle T(\lambda)x, x\rangle = 0 \tag{10}$$

has at most one solution $\lambda =: p(x) \in J$.

which defines the Rayleigh functional p of (4) with respect to J, and we denote by $\mathscr{D}(p) \subset \mathscr{H}$ the domain of definition of p.

Generalizing the definiteness requirement for linear pencils $T(\lambda) = \lambda B - A$ we further assume that $\langle T(\lambda)x, x\rangle$ is increasing at the point $p(x)$, i.e.

(A_2) *For every $x \in \mathscr{D}(p)$ and every $\lambda \in J$ with $\lambda \neq p(x)$ it holds that*

$$(\lambda - p(x))f(\lambda; x) > 0. \tag{11}$$

The key to the variational principle in the nonoverdamped case is an appropriate enumeration of the eigenvalues. In general, the natural enumeration i.e. the first eigenvalue is the smallest one, followed by the second smallest one etc. is not reasonable. Instead, the number of an eigenvalue λ of the nonlinear problem (4) is inherited from the location of the eigenvalue 0 in the spectrum of the operator $T(\lambda)$ based on the following consideration (cf. [26]).

For $j \in \mathbb{N}$ and $\lambda \in J$ let

$$\mu_j(\lambda) := \sup_{V \in S_j} \min_{v \in V, v \neq 0} \frac{\langle T(\lambda)v, v\rangle}{\langle v, v\rangle} \tag{12}$$

where S_j is the set of all j dimensional subspaces of $\mathscr{H}$. We assume that

(A_3) *If $\mu_n(\lambda) = 0$ for some $n \in \mathbb{N}$ and some $\lambda \in J$, then for $j = 1, \dots, n$ the supremum in $\mu_j(\lambda)$ is attained, and $\mu_1(\lambda) \geq \mu_2(\lambda) \geq \cdots \geq \mu_n(\lambda)$ are the n largest eigenvalues of the linear operator $T(\lambda)$. Conversely, if $\mu = 0$ is an eigenvalue of the operator $T(\lambda)$, then $\mu_n(\lambda) = 0$ for some $n \in \mathbb{N}$.*

Definition 1 $\lambda \in J$ is an nth eigenvalue of $T(\cdot)$ if $\mu_n(\lambda) = 0$ for $n \in \mathbb{N}$.

Condition (A_3) is satisfied for example if for every $\lambda \in J$ the supremum of the essential spectrum of $T(\lambda)$ is less than 0. The following stronger condition that for every $\lambda \in J$ there exists $\nu(\lambda) > 0$ such that $T(\lambda) + \nu(\lambda)I$ is a compact operator was used in [11].

The following Lemma proved in [25] (and in [26] for $T(\lambda)$ depending differentiable on λ) relates the supremum of p on a subspace V to the sign of the Rayleigh quotient of $T(\lambda)$ on V.

Lemma 1 *Under the conditions (A_1), (A_2) and (A_3) let $\lambda \in J$, and assume that V is a finite dimensional subspace of $\mathscr{H}$ such that $V \cap \mathscr{D}(p) \neq \emptyset$. Then*

$$\lambda \left\{ \begin{matrix} < \\ = \\ > \end{matrix} \right\} \sup_{x \in V \cap \mathscr{D}(p)} p(x) \quad \Leftrightarrow \quad \min_{x \in V} \langle T(\lambda)x, x\rangle \left\{ \begin{matrix} < \\ = \\ > \end{matrix} \right\} 0 \tag{13}$$

Proof [25], Lemma 2.4

Theorem 3 *Assume that the conditions (A_1), (A_2) and (A_3) are satisfied. Then the nonlinear eigenvalue problem $T(\lambda)x = 0$ has at most a countable set of eigenvalues in J, and it holds that:*

(i) For every $n \in \mathbb{N}$ there exists at most one nth eigenvalue, and the following characterization holds:

$$\lambda_n = \min_{\substack{V \in S_n \\ V \cap \mathcal{D}(p) \neq \emptyset}} \sup_{v \in V \cap \mathcal{D}(p)} p(v). \tag{14}$$

(ii) If

$$\lambda_n := \inf_{\substack{V \in S_n \\ V \cap \mathcal{D}(p) \neq \emptyset}} \sup_{v \in V \cap \mathcal{D}(p)} p(v) \in J, \tag{15}$$

then λ_n is an nth eigenvalue of (4), and the infimum is attained, i.e. the characterization (14) holds.

(iii) If there is an m-th and an n-th eigenvalue λ_m and λ_n in J with $m < n$, then J contains a k-th eigenvalue λ_k, $m < k < n$ as well, and

$$\inf J < \lambda_m \leq \lambda_{m+1} \leq \cdots \leq \lambda_n < \sup J.$$

Proof (i) If λ_n is an n-th eigenvalue, then $\mu_n(\lambda_n) = 0$, and

$$\mu_n(\lambda_n) = \max_{\dim V = n} \min_{x \in V,\ \|x\|=1} \langle T(\lambda_n)x, x\rangle = \min_{x \in \tilde{V},\ \|x\|=1} \langle T(\lambda_n)x, x\rangle$$

for the invariant subspace $\tilde{V}$ corresponding to the n largest eigenvalues of $T(\lambda_n)$

Hence, $\min_{x \in V, \|x\|=1} \langle T(\lambda_n)x, x\rangle \leq 0$ for every V with $\dim V = n$, and (13) implies

$$\sup_{x \in V \cap \mathcal{D}} p(x) \geq \lambda_n = \sup_{x \in \tilde{V} \cap \mathcal{D}} p(x).$$

Hence, λ_n is a minmax value of p.

(ii) Was proved in [26] under the condition that $T(\lambda)$ depends differentiable on λ. But the proof uses only the fact that $\mathcal{D}(p)$ is an open set (which follows also from (A_1) and (A_2) considered here; cf. Lemma 2.3 in [25]) and the analogue of Lemma 1. So the proof holds also for the continuous case considered here.

(iii) Follows from the continuity of $\mu_k(\lambda)$ in J (cf. [7]).

Remark 1 We only considered the case that for every $\lambda \in J$ the supremum of the essential spectrum of $T(\lambda)$ is less than 0. In the same way we obtain for the case that for every $\lambda \in J$ the infimum of $T(\lambda)$ exceeds 0 a maxinf characterization of the eigenvalues of $T(\cdot)$ in J if we replace (A_2) with

(A_2') $(\lambda - p(x))f(\lambda; x) < 0$ *for every $x \in \mathcal{D}(p)$ and $\lambda \in J$ such that $\lambda \neq p(x)(p)$*

and (A_3) with

(A_3') *If* $\nu_m(\lambda) := \inf_{V\in S_m} \max_{x\in V, x\neq 0}\langle T(\lambda)x, x\rangle/\langle x, x\rangle = 0$ *for some* $m \in \mathbb{N}$ *and some* $\lambda \in J$*, then for* $j = 1,\dots,m$ *the supremum in* $\nu_j(\lambda)$ *is attained, and* $\nu_1(\lambda) \le \nu_2(\lambda) \le \cdots \le \nu_m(\lambda)$ *are the* m *smallest eigenvalues of the linear operator* $T(\lambda)$*. Conversely, if* $\nu = 0$ *is an eigenvalue of the operator* $T(\lambda)$*, then* $\nu_m(\lambda) = 0$ *for some* $m \in \mathbb{N}$*.*

If the eigenvalues of $T(\cdot)$ are now enumerated in decreasing order, i.e. $\lambda \in J$ is an mth eigenvalue of $T(\cdot)$ if $\nu_m(\lambda) = 0$ for $m \in \mathbb{N}$, then λ_m can be characterized as

$$\lambda_m = \max_{\substack{V\in S_m \\ V\cap\mathcal{D}(p)\neq\emptyset}} \inf_{v\in V\cap\mathcal{D}(p)} p(v).$$

In the following we consider only problem (4) under the conditions (A_1), (A_2) and (A_3), although the analogue results also hold under the conditions (A_1), (A_2') and (A_3') with the modified enumeration given above.

If the extreme eigenvalue λ_1 is contained in J, then the enumeration based on (A_3) is the natural ordering. For this case Barston [3] proved the minmax characterization for some extreme real eigenvalues for the finite dimensional quadratic eigenvalue problem. Abramov [2] and Hasanov [14] derived the minmax and maxmin characterizations for the extreme eigenvalues for pencils of waveguide type, which are certain quadratic eigenvalues problems depending on two parameters.

For the general $T(\cdot)$ it can be shown that the eigenspaces corresponding to eigenvalues in J are contained in $\mathcal{D}(p) \cup \{0\}$. Hence the minmax characterization obtains the following form:

Theorem 4 *Let the conditions* (A_1)*,* (A_2) *and* (A_3) *be satisfied, and assume that* $\lambda_1 = \inf_{x\in\mathcal{D}(p)} p(x) \in J$*, and* $\lambda_n \in J$ *for some* $n \in \mathbb{N}$*.*

If $j \in \{1,\dots,n\}$ *and* $V \in S_j$ *such that* $\lambda_j = \sup_{x\in V\cap\mathcal{D}(p)} p(x)$*, then* $V \subset \mathcal{D}(p) \cup \{0\}$*, and the characterization of* λ_j *can be replaced with*

$$\lambda_n = \min_{\substack{V\in S_j \\ V\subset\mathcal{D}(p)\cup\{0\}}} \sup_{v\in V\cap\mathcal{D}(p)} p(v). \tag{16}$$

The generalization of the maxmin characterization of Courant, Fischer and Weyl is based on the following Lemma which was proved in [24]:

Lemma 2 *Let* $\lambda \in J$*, and let* V *be a finite dimensional subspace of* $\mathcal{H}$ *such that* $V^\perp \cap \mathcal{D} \neq \emptyset$*. Then it holds that*

$$\lambda \begin{Bmatrix} < \\ = \\ > \end{Bmatrix} \inf_{x\in V^\perp\cap\mathcal{D}(p)} p(x) \quad \Leftrightarrow \quad \max_{x\in V^\perp,\ \|x\|=1} \langle T(\lambda)x, x\rangle \begin{Bmatrix} < \\ = \\ > \end{Bmatrix} 0$$

Theorem 5 *Assume that the conditions* (A_1)*,* (A_2) *and* (A_3) *are satisfied. If there exists an n-th eigenvalue* $\lambda_n \in J$ *of* $T(\lambda)x = 0$*, then*

$$\lambda_n = \max_{\substack{V \in S_{n-1} \\ V^\perp \cap \mathscr{D} \neq \emptyset}} \inf_{v \in V^\perp \cap \mathscr{D}} p(v),$$

and the maximum is attained by $W := span\{u_1, \dots, u_{n-1}\}$ *where* u_j *denotes an eigenvector corresponding to the* j*-largest eigenvalue* $\mu_j(\lambda_n)$ *of* $T(\lambda_n)$*.*

Essentially the same variational characterizations of Poincaré and of Courant-Fischer-Weyl type were derived by Mel'nik and Nazarov [17], where $T(\lambda)$ is a set of bounded self-adjoint operators depending continuously differentiable on λ, by Griniv and Mel'nik [10] for $T(\lambda) = A(\lambda) - I$, where $A(\lambda)$ is self-adjoint, and compact, and by Binding, Eschwé and H. Langer [4] for general bounded and self-adjoint $T(\lambda)$ depending continuously on λ. Eschwé and M. Langer [8] obtained these variational characterizations for unbounded operators. In all of these papers the natural enumeration of the eigenvalues is used, but the dimension of the subspace in the characterizations is shifted by the number of the largest eigenvalue of $T(\lambda_1)$.

Hadeler [11] proved Rayleigh's principle for differentiable overdamped problems. For the continuous case the generalized scalar product (9) has to be modified for the case $p(x) = p(y)$ setting $[x, y] := \langle x, y \rangle$. Then the generalized scalar product $[\cdot, \cdot]$ becomes discontinuous for $p(x) = p(y)$, but the continuity is not needed in the proof of Rayleigh's principle which obtains the following form:

Theorem 6 *Under the conditions* (A_1)*,* (A_2)*,* (A_3) *assume that* J *contains* $n \geq 1$ *eigenvalues* $\lambda_1 \leq \dots \leq \lambda_n$ *(where* λ_i *is an* i*th eigenvalue) with corresponding* $[\cdot, \cdot]$ *orthogonal eigenvectors* $x_1, \dots, x_n$*.*

If there exists $x \in \mathscr{D}(p)$ *with* $[x_i, x] = 0$ *for* $i = 1, \dots, n$ *then* J *contains an* $(n + 1)$*th eigenvalue, and*

$$\lambda_{n+1} = \inf\{p(x) : [x_j, x] = 0,\ i = 1, \dots, n\}. \tag{17}$$

References

1. Y.S. Abramov, Variational principles for nonlinear eigenvalue problems. Funct. Anal. Appl. **7**, 317–318 (1973)
2. Y.S. Abramov, Linear operators and spectral theory: pencils of waveguide type and related extremal problems. J. Math. Sci. **64**, 1278–1288 (1993)
3. E.M. Barston, A minimax principle for nonoverdamped systems. Int. J. Eng. Sci. **12**, 413–421 (1974)
4. P. Binding, D. Eschwé, H. Langer, Variational principles for real eigenvalues of self-adjoint operator pencils. Integr. Equ. Oper. Theory **38**, 190–206 (2000)
5. R. Courant, Über die Eigenwerte bei den Differentialgleichungen der mathematischen Physik. Math. Z. **7**, 1–57 (1920)
6. R.J. Duffin, A minimax theory for overdamped networks. J. Rat. Mech. Anal. **4**, 221–233 (1955)

7. J. Eisenfeld, E.H. Rogers, Elementary localization theorems for nonlinear eigenproblems. J. Math. Anal. Appl. **48**, 325–340 (1974)
8. D. Eschwé, M. Langer. Variational principles of eigenvalues of self-adjoint operator functions. Integr. Equ. Oper. Theory **49**, 287–321 (2004)
9. E. Fischer, Über quadratische Formen mit reellen Koeffizienten. Monatshefte für Math. und Phys. **16**, 234–249 (1905)
10. R.O. Griniv, T.A. Mel'nik, On the singular Rayleigh functional. Math. Notes **60**, 97–100 (1996)
11. K.P. Hadeler, Variationsprinzipien bei nichtlinearen Eigenwertaufgaben. Arch. Ration. Mech. Anal. **30**, 297–307 (1968)
12. K.P. Hadeler, Nonlinear eigenvalue problems, in *Numerische Behandlung von Differentialgleichungen* ed. by R. Ansorge, L. Collatz, G. Hämmerlin, W. Törnig, ISNM 27 (Birkhäuser, Stuttgart, 1975), pp. 111–129
13. M. Hasanov, An approximation method in the variational theory of the spectrum of operator pencils. Acta Appl. Math. **71**, 117–126 (2002)
14. M. Hasanov, The spectra of two–parameter quadratic operator pencils. Math. Comput. Model. **54**, 742–755 (2011)
15. H. Langer, Über stark gedämpfte Scharen im Hilbertraum. J. Math. Mech. **17**, 685–705 (1968)
16. A.S. Markus, *Introduction to the Spectral Theory of Polynomial Operator Pencils* (AMS Translations of Mathematical Monographs, Providence, 1988)
17. T.A. Mel'nik, S.A. Nazarov, The asymptotics of the solution to the Neumann spectral problem in a domain of the "dense-comb" type. J. Math. Sci. **85**, 2326–2345 (1997)
18. H. Poincaré, Sur les equations aux dérivées partielles de la physique mathématique. Am. J. Math. **12**, 211–294 (1890)
19. L. Rayleigh, Some general theorems relating to vibrations. Proc. Lond. Math. Soc. **4**, 357–368 (1873)
20. E.H. Rogers, A minimax theory for overdamped systems. Arch. Ration. Mech. Anal. **16**, 89–96 (1964)
21. E.H. Rogers, Variational properties of nonlinear spectra. J. Math. Mech. **18**, 479–490 (1968)
22. R.E.L. Turner, Some variational principles for nonlinear eigenvalue problems. J. Math. Anal. Appl. **17**, 151–160 (1967)
23. R.E.L. Turner, A class of nonlinear eigenvalue problems. J. Func. Anal. **7**, 297–322 (1968)
24. H. Voss, A maxmin principle for nonlinear eigenvalue problems with application to a rational spectral problem in fluid–solid vibration. Appl. Math. **48**, 607–622 (2003)
25. H. Voss, A minmax principle for nonlinear eigenproblems depending continuously on the eigenparameter. Numer. Linear Algebra Appl. **16**, 899–913 (2009)
26. H. Voss, B. Werner, A minimax principle for nonlinear eigenvalue problems with applications to nonoverdamped systems. Math. Meth. Appl. Sci. **4**, 415–424 (1982)
27. H.F. Weinberger, On a nonlinear eigenvalue problem. J. Math. Anal. Appl. **21**, 506–509 (1968)
28. A. Weinstein, W.F. Stenger, *Methods of Intermediate Problems for Eigenvalues* (Academic, New York, 1972)
29. B. Werner, Das Spektrum von Operatorenscharen mit verallgemeinerte Rayleighquotienten. Arch. Ration. Mech. Anal. **42**, 223–238 (1971)
30. H. Weyl, Das asymptotische Verteilungsgesetz der Eigenwerte linearer partieller Differentialgleichungen (mit einer Anwendung auf die Theorie der Hohlraumstrahlung). Math. Ann. **71**, 441–479 (1912)

Tensor Formats Based on Subspaces are Positively Invariant Sets for Laplacian-Like Dynamical Systems

Antonio Falcó

Abstract In this note, we show that the set of tensors with bounded rank are positively invariant sets for linear evolution equations defined by Laplacian-like operators. In consequence, once a trajectory of the system enters to this class of set, it will never leave it again.

1 Introduction

The Proper Generalized Decomposition or, in short, PGD is a technique that reduces calculation and storage cost drastically and presents some similarities with the Proper Orthogonal Decomposition, in short POD. It was initially introduced for the analysis and reduction of statistical and experimental data, the a posteriori decomposition techniques, also known as Karhunen-Loève Expansion, Singular Value decomposition or Principal Component Analysis, are now used in the context of model reduction. It is also related with the so-called n-best term approximation problem. This is one of the main ingredients of PGD framework: the existence of a best approximation by using a tensor decomposition. We would to point out that this result is true only for tensors of order two, because the milestone of the proof is the existence of a best rank-n approximation for this class of tensors. Unfortunately, in [10], it has been proved that tensors of order 3 or higher can fail to have best rank-n approximation, that is, it is an ill-posed problem. In consequence, as shown in [6] only rank-one approximations are available.

On the other hand, it is possible to use a wide class of tensor representations. In particular, Falcó and Hackbusch [3] have proved the existence of a best approximation for tensor representations based in subspaces (see also [8] and [4]). This fact allows to extend the PGD to other type of tensor decompositions.

Mainly, to prove the convergence of a PGD-based algorithm we need to combine the existence of a best approximation and a greedy algorithm. The idea of using

A. Falcó (✉)
Departamento de Ciencias, Físicas, Matemáticas y de la Computación, Universidad CEU Cardenal Herrera, San Bartolomé 55, 46115 Alfara del Patriarca, Valencia, Spain
e-mail: afalco@uch.ceu.es; antonio.falco@gmail.com

A. Abdulle et al. (eds.), *Numerical Mathematics and Advanced Applications ENUMATH 2013*, Lecture Notes in Computational Science and Engineering 103,
DOI 10.1007/978-3-319-10705-9_31

greedy algorithms was introduced by Ammar, Mokdad, Chinesta and Keunings [1], in the context of high-dimensional PDEs. It can be applied to several problems including, among others, the stationary Fokker-Planck equation of the FENE bead-spring chain model [9]. Later on, Ammar, Mokdad, Chinesta and Keunings [2] extend this framework to time-dependent problems. So it seems natural the use of this methodology to solve dynamical problems despite the fact that its convergence, in general, is not guaranteed (a recent result in this approach is given in the paper of Figueroa and Süli [7]).

In order to understand the role of these tensor formats in the dynamical case, the main goal of this paper is to prove, in a tensor Banach space framework, that tensor representations based in subspaces constitute sets which are positively time invariant for a class of dynamical systems related with Laplacian-like operators.

The note is organized as follows. In the next section we introduce some preliminary definitions and results. Finally, Sect. 3 is devoted to the statement and proof of the main result.

2 Preliminary Definitions and Results

In the following, X is a Banach space with norm $\|\cdot\|$. The dual norm $\|\cdot\|_*$ of X^* is

$$\begin{aligned}\|\varphi\|_* &= \sup\{|\varphi(x)| : x \in X \text{ with } \|x\| \leq 1\} \\ &= \sup\{|\varphi(x)| \,/\, \|x\| : 0 \neq x \in X\}.\end{aligned}$$

By $\mathscr{L}(Y,Z)$ we denote the space of continuous linear mappings from Y into Z. The corresponding operator norm is written as $\|\cdot\|_{Z\leftarrow Y}$. Let X be a (complex or) real Banach space and consider one–parameter semigroups of bounded linear operators $T(t)$ on X. By this we understand a subset $\{T(t) : t \in \mathbb{R}_{\geq 0}\}$ of $\mathscr{L}(X,X)$, that we usually write by $\{T(t)\}_{t\geq 0}$ such that

1. $T(0) = id$,
2. $T(s+t) = T(s) \circ T(t)$ for all $s,t \in \mathbb{R}_{\geq 0}$.

Indeed, the map $t \mapsto T(t)$ is a homomorphism from the additive semigroup $(\mathbb{R}_{\geq 0}, +)$ into the multiplicative semigroup $(\mathscr{L}(X,X), \circ)$.

A one–parameter semigroup $\{T(t)\}_{t\geq 0}$ is called *strongly continuous* if the map $t \mapsto T(t)$ is continuous for the strong operator topology on $\mathscr{L}(X,X)$, that is,

$$\lim_{t\to t_0} \|T(t)x - T(t_0)x\| = 0,$$

holds for all $x \in X$ and $t, t_0 \in \mathbb{R}_{\geq}$. A one–parameter semigroup $\{T(t)\}_{t\geq 0}$ is strongly continuous if and only if $\lim_{t\to 0^+} T(t)x = x$.

The most important objects associated to a strongly continuous semigroup $\{T(t)\}_{t\geq 0}$ is its infinitesimal generator, which is obtained as follows. Let $A \in L(X,X)$ a linear (not necessarily bounded) operator defined by

$$Ax := \lim_{h\to 0+} \frac{T(h)x - x}{h},$$

if the limit exists and consider its domain

$$D(A) := \left\{ x \in X : \lim_{h\to 0+} \frac{T(h)x - x}{h} \text{ exists} . \right\}$$

Clearly $D(A)$ is a linear subspace of X and $A : D(A) \to X$ is linear. The following well-known theorem will be useful.

Theorem 1 *Let X be a Banach space. Let A be the generator of a strongly continuous semigroup $\{T(t)\}_{t\geq 0}$ on X. Then the abstract Cauchy problem*

$$\frac{d}{dt}x(t) = Ax(t) \quad x(0) = x_0, \tag{1}$$

has a unique solution $x \in \mathscr{C}^1(\mathbb{R}_{\geq}, X)$ for every $x_0 \in X$. Indeed, this solution is given by

$$x(t) = T(t)x_0. \tag{2}$$

Let X be a Banach space. An operator $A \in L(X,X)$ with domain $D(A)$ is called *closed* if $D(A)$ endowed with the graph norm

$$\|x\|_A := \|x\| + \|Ax\|$$

becomes a Banach space. Thus, $A \in L(X,X)$ is closed if and only if $\{(x, Ax) \in X \times X : x \in D(A)\}$ is closed in $X \times X$, i.e. $\{x_n\}_{n\in\mathbb{N}} \subset D(A)$ where $x_n \to x$ and $Ax_n \to y$ as $n \to \infty$, implies $x \in D(A)$ and $y = Ax$. Let $A \in L(X,X)$, we say that A is *closable* if the closure of its graph in $X \times X$ is the graph of some operator $\overline{A} \in L(X,X)$. The operator $\overline{A}$ is called *the closure of* A. Hence, A is closed if and only if $\overline{A} = A$. It follows easily that A is the restriction of $\overline{A}$ to $D(A)$. A *core* of a closable operator A is a subset C of $D(A)$ such that the closure of the restriction of $\overline{A}$ to C is A.

We will say that a set $\Gamma \subset X$ is said to be *positively invariant* for the dynamical system (1) if $x_0 \in \Gamma$ implies that $x(t) \in \Gamma$ for all time $t \in \mathbb{R}_{\geq}$. This means that once a trajectory of the system enters, it will never leave it again.

Concerning the definition of the algebraic tensor space ${}_a\bigotimes_{j=1}^d V_j$ gene–rated from vector spaces V_j $(1 \leq j \leq d)$, we refer to [8]. As the underlying field we choose $\mathbb{R}$, but the results hold also for $\mathbb{C}$. The suffix 'a' in ${}_a\bigotimes_{j=1}^d V_j$ refers to the 'algebraic' nature. By definition, all elements of

$$\mathbf{V} := {}_a\bigotimes_{j=1}^{d} V_j$$

are *finite* linear combinations of elementary tensors $\mathbf{v} = \bigotimes_{j=1}^d v_j \left(v_j \in V_j\right)$. Next, we introduce the following class of Hilbert spaces.

Definition 1 We say that $\mathbf{V}_{\|\cdot\|}$ is a *Banach tensor space* if there exists an algebraic tensor space $\mathbf{V}$ and a norm $\|\cdot\|$ on $\mathbf{V}$ such that $\mathbf{V}_{\|\cdot\|}$ is the completion of $\mathbf{V}$ with respect a given norm $\|\cdot\|$, i.e.,

$$\mathbf{V}_{\|\cdot\|} := {}_{\|\cdot\|}\bigotimes_{j=1}^{d} V_j = \overline{{}_a\bigotimes_{j=1}^{d} V_j}^{\|\cdot\|}.$$

The following notation and definitions, introduced in [3], will be useful. Let $\mathscr{I} := \{1, \ldots, d\}$ be the index set of the 'spatial directions'. In the sequel, the index sets $\mathscr{I}\backslash\{j\}$ will appear. Here, we use the abbreviations

$$\mathbf{V}_{[j]} := {}_a\bigotimes_{k\neq j} V_k\,, \qquad \text{where} \bigotimes_{k\neq j} \text{means} \bigotimes_{k\in\mathscr{I}\backslash\{j\}}, \tag{3}$$

Similarly, elementary tensors $\bigotimes_{k\neq j} v^{(j)}$ are denoted by $\mathbf{v}_{[j]}$.

The PGD-Galerkin Method [5] is based in the fact that in a tensor space a typical representation format is the tensor subspace or Tucker format

$$\mathbf{u} = \sum_{\mathbf{i}\in\mathbf{I}} \mathbf{a_i} \bigotimes_{j=1}^{d} b_{i_j}^{(j)}, \tag{4}$$

where $\mathbf{I} = I_1 \times \ldots \times I_d$ is a multi-index set with $I_j = \{1, \ldots, r_j\}$, $r_j \leq \dim(V_j)$, $b_{i_j}^{(j)} \in V_j \left(i_j \in I_j\right)$ are basis vectors, and $\mathbf{a_i} \in \mathbb{R}$. Here, i_j are the components of $\mathbf{i} = (i_1, \ldots, i_d)$. The data size is determined by the numbers r_j collected in the tuple $\mathbf{r} := (r_1, \ldots, r_d)$. The set of all tensors representable by (4) with fixed $\mathbf{r}$ is

$$\mathscr{T}_{\mathbf{r}}(\mathbf{V}) := \left\{ \mathbf{v} \in \mathbf{V} : \begin{array}{l} \text{there are subspaces } U_j \subset V_j \text{ such that} \\ \dim(U_j) = r_j \text{ and } \mathbf{v} \in \mathbf{U} := {}_a\bigotimes_{j=1}^d U_j\,. \end{array} \right\} \tag{5}$$

Here, it is important that the description (4) with the vectors $b_i^{(j)}$ can be replaced by the generated subspace $U_j = \operatorname{span}\{b_i^{(j)} : i \in I_j\}$. By using the definition of minimal subspaces in tensor representations [3], it is possible to show for each $\mathbf{v} \in \mathbf{V}$ there exist subspaces $U_j^{\min}(\mathbf{v}) \subset V_j$, where $1 \leq j \leq d$, satisfying:

(a) $\mathbf{v} \in {}_a\bigotimes_{j=1}^d U_j^{\min}(\mathbf{v})$ and
(b) If $\mathbf{v} \in {}_a\bigotimes_{j=1}^d U_j$ for some subspace $U_j \subset V_j$, where $1 \leq j \leq d$, then $U_j^{\min}(\mathbf{v}) \subset U_j$ for $1 \leq j \leq d$.

Define $\operatorname{rank}_j \mathbf{v} := \dim U_j^{\min}(\mathbf{v})$ for $1 \leq j \leq d$. It allows to introduce the tensor rank of $\mathbf{v} \in \mathbf{V}$ by

$$\operatorname{rank} \mathbf{v} := (\operatorname{rank}_1 \mathbf{v}, \ldots, \operatorname{rank}_d \mathbf{v}) \in \mathbb{N}^d.$$

It is possible to extend this definition for every $\mathbf{v} \in \mathbf{V}_{\|\cdot\|}$ (see [3]). However, for $\mathbf{v} \in \mathbf{V}_{\|\cdot\|} \setminus \mathbf{V}$ the property that $\mathbf{v} \in {}_{\|\cdot\|}\bigotimes_{j=1}^d U_j^{\min}(\mathbf{v})$ can be shown only when $U_j^{\min}(\mathbf{v})$ belongs to the Grassmannian of V_j for $1 \leq j \leq d$ and the norm $\|\cdot\|$ is a uniform cross norm (see Theorem 6.29 in [8]).

Now, we recall the definition of uniform cross norm. Any norm $\|\cdot\|$ on ${}_a\bigotimes_{j=1}^d V_j$ satisfying

$$\left\| \bigotimes_{j=1}^d v^{(j)} \right\| = \prod_{j=1}^d \|v^{(j)}\|_j \qquad \text{for all } v^{(j)} \in V_j \ (1 \leq j \leq d) \tag{6}$$

is called a *cross norm*. As usual, the dual norm to $\|\cdot\|$ is denoted by $\|\cdot\|^*$. If $\|\cdot\|$ is a cross norm and also $\|\cdot\|^*$ is a cross norm on ${}_a\bigotimes_{j=1}^d V_j^*$, i.e.,

$$\left\| \bigotimes_{j=1}^d \varphi^{(j)} \right\|^* = \prod_{j=1}^d \|\varphi^{(j)}\|_j^* \qquad \text{for all } \varphi^{(j)} \in V_j^* \ (1 \leq j \leq d), \tag{7}$$

$\|\cdot\|$ is called a *reasonable cross norm*. A norm $\|\cdot\|$ is a *uniform cross norm* if it is a cross norm (cf. (6)) and satisfies

$$\left\| \left(\bigotimes_{j=1}^d A_j \right)(\mathbf{v}) \right\| \leq \left(\prod_{j=1}^d \|A_j\|_{V_j \leftarrow V_j} \right) \|\mathbf{v}\| \tag{8}$$

for all $A_j \in \mathcal{L}(V_j, V_j)$ $(1 \leq j \leq d)$ and all $\mathbf{v} \in {}_a\bigotimes_{j=1}^d V_j$.

3 Statement and Proof of the Main Result

Before state the main result of this note, we introduce the definition of Laplacian-like operator.

Definition 2 Let V_j be a Banach space for $1 \leq j \leq d$. We will say that $\Delta_A \in L(\mathbf{V}, \mathbf{V})$ is a *Laplacian-like operator* if

$$\Delta_A = \sum_{j=1}^{d} id_1 \otimes id_{j-1} \otimes A_j \otimes id_{j+1} \otimes \cdots \otimes id_d$$

where $A_j \in L(V_j, V_j)$ for $1 \leq j \leq d$.

Observe that if Δ_A is a Laplacian-like operator, from Lemma 6.11 in [8], we have

$$D(\Delta_A) = \bigcap_{j=1}^{d} D(A_j) \otimes_a \mathbf{V}_{[j]} = {}_a\bigotimes_{j=1}^{d} D(A_j) .$$

From now one, we will denote $\mathbf{1} := (1, \ldots, 1) \in \mathbb{N}^d$. Moreover, we will say that $\mathbf{r} \leq \mathbf{s}$ for $\mathbf{r}, \mathbf{s} \in \mathbb{N}^d$ if and only if $r_j \leq s_j$ for $1 \leq j \leq d$. The main result of this paper is the following theorem.

Theorem 2 *Assume that $\mathbf{V}_{\|\cdot\|}$ is a Banach tensor space with a uniform cross norm $\|\cdot\|$ and $\Delta_A \in L(\mathbf{V}, \mathbf{V})$ is a Laplacian-like operator such that $A_j \in L(V_j, V_j)$ is the infinitesimal generator of a strongly continuous semigroup $\{T_j(t)\}_{t\geq 0}$ on the Banach space V_j, for $1 \leq j \leq d$. Then Δ_A is the infinitesimal generator of a strongly continuous semigroup on $\mathbf{V}_{\|\cdot\|}$. Moreover, for each $\mathbf{r} \geq \mathbf{1}$ the set $\mathcal{T}_{\mathbf{r}}(\mathbf{V})$ is a positively invariant set for the dynamical system:*

$$\frac{d}{dt}\mathbf{u}(t) = \Delta_A \mathbf{u}(t), \quad \mathbf{u}(0) = \mathbf{u}_0. \tag{9}$$

Proof Is a consequence of two lemmas given below.

Lemma 1 *Assume that $\mathbf{V}_{\|\cdot\|} = {}_{\|\cdot\|}\bigotimes_{j=1}^{d} V_j$ is a Banach tensor space. Let A be the generator of a strongly continuous semigroup $\{T(t)\}_{t\geq 0}$ on $\mathbf{V}_{\|\cdot\|}$, and consider the following abstract Cauchy problem*

$$\frac{d}{dt}\mathbf{u}(t) = A\mathbf{u}(t), \quad \mathbf{u}(0) = \mathbf{u}_0. \tag{10}$$

Assume that $\mathcal{T}_{\mathbf{1}}(\mathbf{V})$ is a positively invariant set for the dynamical system (10). Then

$$\operatorname{rank} \mathbf{u}(t) \leq \operatorname{rank} \mathbf{u}(0), \tag{11}$$

holds for all time $t \in \mathbb{R}_{\geq}$, and hence for each $\mathbf{r} \geq \mathbf{1}$ the set $\mathcal{T}_{\mathbf{r}}(\mathbf{V})$ is also a positively invariant set for the dynamical system (10).

Proof Since $\mathscr{T}_1(\mathbf{V})$ is positively invariant for the dynamical system (10), if $\mathbf{u}_0 = \bigotimes_{j=1}^d u_j^0 \in \mathscr{T}_1(\mathbf{V})$ then, by Theorem 1, $\mathbf{u}(t) = T(t) \otimes_{j=1}^d u_j^0 = \otimes_{j=1}^d u_j(t)$, for all time $t \in \mathbb{R}_{\geq}$. Now, assume that $\operatorname{rank} \mathbf{u}(0) = (r_1, \ldots, r_d)$. Then, we can write

$$\mathbf{u}(0) = \sum_{i_1=1}^{r_1} \cdots \sum_{i_d=1}^{r_d} u_{i_1 \cdots i_d}^{(0)} \bigotimes_{k=1}^{d} v_{i_k}$$

where $u_{i_1 \cdots i_d}^{(0)} \in \mathbb{R}$ for each $1 \leq i_1 \leq r_1, \ldots, 1 \leq i_d \leq r_d$, and $v_{i_k} \in V_k$ for $1 \leq k \leq d$. Thus,

$$\mathbf{u}(t) = T(t)\mathbf{u}(0) = \sum_{i_1=1}^{r_1} \cdots \sum_{i_d=1}^{r_d} u_{i_1 \cdots i_d}^{(0)} T(t) \left(\bigotimes_{k=1}^{d} v_{i_k} \right),$$

and hence $\operatorname{rank}_j \mathbf{u}(t) \leq r_j$, because $\operatorname{rank}_j T(t) \left(\bigotimes_{k=1}^d v_{i_k} \right) \leq 1$ for $1 \leq j \leq d$.

Definition 3 Let X be a Banach space and $\| \cdot \|$ be a norm defined over $\mathbf{V}$. For each $A \in \mathscr{L}(\mathbf{V}, X)$ we will denote by $\overline{A} \in \mathscr{L}\left(\mathbf{V}_{\|\cdot\|}, X\right)$ its unique extension. Recall that $\overline{A}|_{\mathbf{V}} = A$.

Observe that if $\| \cdot \|$ is a uniform cross norm then for all $A_j \in \mathscr{L}(V_j, V_j)$ $(1 \leq j \leq d)$ the map $\overline{\bigotimes_{j=1}^d A_j}$ belongs to $\mathscr{L}(\mathbf{V}_{\|\cdot\|}, \mathbf{V}_{\|\cdot\|})$.

Lemma 2 *Assume that* $\mathbf{V}_{\|\cdot\|} = {}_{\|\cdot\|} \bigotimes_{j=1}^d V_j$ *is a Banach tensor space with a uniform cross norm* $\|\cdot\|$ *and* Δ_A *be a Laplacian-like operator such that* $A_j \in L(V_j, V_j)$ *is the infinitesimal generator of a strongly continuous semigroup* $\{T_j(t)\}_{t\geq 0}$ *on the Banach space* V_j, *for* $1 \leq j \leq d$. *Let us consider*

$$\mathbf{T}_d(t) := \overline{\bigotimes_{j=1}^{d} T_j(t)} \in \mathscr{L}(\mathbf{V}_{\|\cdot\|}, \mathbf{V}_{\|\cdot\|}).$$

Then $\{\mathbf{T}_d(t)\}_{t\geq 0}$ *is a strongly continuous semigroup on* $\mathbf{V}_{\|\cdot\|}$ *and the closure of* Δ_A *defined on* ${}_a \bigotimes_{j=1}^d D(A_j)$ *is its infinitesimal generator.*

Proof The proof is by induction on d. Assume first that $d = 2$. Then it is easy to verify that $\{\overline{T_1(t) \otimes T_2(t)}\}_{t\geq 0}$ is a semigroup of operators on $V_1 \otimes_{\|\cdot\|} V_2$. The strong continuity needs to be only verified at $t = 0$ for an elementary tensor, namely $v_1 \otimes v_2 \in V_1 \otimes_a V_2$. To this end observe

$$\begin{aligned}
&\|T_1(t)v_1 \otimes T_2(t)v_2 - v_1 \otimes v_2\| \\
&\leq \|T_1(t)v_1 \otimes (T_2(t)v_2 - v_2)\| + \|(T_1(t)v_1 - v_1) \otimes v_2\| \\
&= \|T_1(t)v_1\|_1 \|T_2(t)v_2 - v_2\|_2 + \|T_1(t)v_1 - v_1\|_1 \|v_2\|_2.
\end{aligned}$$

It implies that $\lim_{t\to 0^+} \|T_1(t)v_1 \otimes T_2(t)v_2 - v_1 \otimes v_2\| = 0$, and the strong continuity holds. Now it remains to show that the infinitesimal generator of $\{\overline{T_1(t) \otimes T_2(t)}\}_{t\geq 0}$ is obtained as the closure of $A_1 \otimes id_2 + id_1 \otimes A_2$ on $D(A_1) \otimes_a D(A_2)$. To this end, take $v_1 \in D(A_1)$ and $v_2 \in D(A_2)$, then

$$\lim_{h\to 0+} \frac{1}{h} (T_1(t)v_1 \otimes T_2(t)v_2 - v_1 \otimes v_2)$$

$$= \lim_{h\to 0+} \frac{1}{h} (T_1(t)v_1 \otimes (T_2(t)v_2 - v_2) + (T_1(t)v_1 - v_1) \otimes v_2)$$

$$= v_1 \otimes A_2 v_2 + A_1 v_1 \otimes v_2.$$

Since $\mathrm{span}\{v_1 \otimes v_2 : v_1 \in D(A_1) \text{ and } v_2 \in D(A_2)\}$ generates the linear subspace $D(A_1) \otimes_a D(A_2)$ of $V_1 \otimes_{\|\cdot\|} V_2$ and $D(A_i)$ is dense in V_i for $i = 1, 2$ we obtain that $D(A_1) \otimes_a D(A_2)$ is dense in $V_1 \otimes_{\|\cdot\|} V_2$ and invariant under $\{\overline{T_1(t) \otimes T_2(t)}\}_{t\geq 0}$. Hence it is the core of $id_1 \otimes A_2 + A_1 \otimes id_2$.

Next we proceed inductively. Assume that the theorem is true for $d - 1 (\geq 2)$, now we need to show that it is also true for $d (\geq 3)$. The strong continuity for $t = 0$ is not difficult to show. To prove that the infinitesimal generator of $\{\mathbf{T}_d(t)\}_{t\geq 0}$ is obtained as the closure of Δ_A on ${}_a\bigotimes_{j=1}^{d} D(A_j)$ we will use Lemma 3.17 of [3]. To this end write $\mathbf{V}_{[d]} := {}_a\bigotimes_{j=1}^{d-1} V_j$ and consider on $\mathbf{V}_{[d]}$ the uniform cross norm $\|\cdot\|_{[d]}$ given by

$$\|\mathbf{v}_{[d]}\|_{[d]} := \|\mathbf{v}_{[d]} \otimes v_d\| \text{ for some fixed } v_d \in V_d,\ \|v_d\|_d = 1.$$

Thus $\{\mathbf{T}_{d-1}(t)\}_{t\geq 0}$ is a strongly continuous semigroup on ${}_{\|\cdot\|_{[d]}}\bigotimes_{j=1}^{d-1} V_j$ and the closure of $\Delta_A^{(d-1)} = \sum_{j=1}^{d-1} id_1 \otimes id_{j-1} \otimes A_j \otimes id_{j+1} \otimes \cdots \otimes id_{d-1}$ defined on ${}_a\bigotimes_{j=1}^{d-1} D(A_j)$ is its infinitesimal generator. Since $\|\cdot\|$ is a uniform cross norm on $\mathbf{V}_{[d]} \otimes_a V_d$ we obtain that $\{\overline{\mathbf{T}_{d-1}(t) \otimes T_d(t)}\}_{t\geq 0}$ is a strongly continuous semigroup on $\mathbf{V}_{\|\cdot\|}$ and the closure of $\Delta_A^{(d-1)} \otimes id_d + \mathbf{id}_{[d]} \otimes A_d = \Delta_A$, defined on ${}_a\bigotimes_{j=1}^{d} D(A_j)$, is its infinitesimal generator. Since $\overline{\mathbf{T}_{d-1}(t) \otimes T_d(t)} = \mathbf{T}_d(t)$ the theorem follows. □

References

1. A. Ammar, B. Mokdad, F. Chinesta, R. Keunings, A new family of solvers for some classes of multidimensional partial differential equations encountered in kinetic theory modeling of complex fluids. J. Non-Newtonian Fluid Mech. **139**, 153–176 (2006)
2. A. Ammar, B. Mokdad, F. Chinesta, R. Keunings, A new family of solvers for some classes of multidimensional partial differential equations encountered in kinetic theory modeling of complex fluids. Part II: transient simulation using space-time separated representations. J. Non-Newtonian Fluid Mech. **144**, 98–121 (2007)

3. A. Falcó, W. Hackbusch, On minimal subspaces in tensor representations. Found. Comput. Math. **12**(6), 765–803 (2012)
4. A. Falcó, W. Hackbusch, A. Nouy, Geometric structures in tensor representations, Preprint 9/2013 at Max Planck Institute for Mathematics in the Sciences, 2013
5. A. Falcó, L. Hilario, N. Montés, M.C. Mora, Numerical strategies for the Galerkin-Proper generalized decomposition method. Math. Comput. Model. **57**(7–8), 1694–1702 (2013)
6. A. Falcó, A. Nouy, A proper generalized decomposition for the solution of elliptic problems in abstract form by using a functional Eckart-Young approach. J. Math. Anal. Appl. **376**, 469–480 (2011)
7. L. Figueroa, E. Süli, Greedy approximation of high-dimensional Ornstein-Uhlenbeck operators with unbounded drift. Found. Comput. Math. **12**(5), 573–623 (2012)
8. W. Hackbusch, *Tensor Spaces and Numerical Tensor Calculus* (Springer, Berlin/New York, 2012)
9. A. Lozinski, R.G. Owen, T.N. Phillips, The Langevin and the Fokker–Plank Equation in Polymer Rheology, *Handbook of Numerical Analysis*, ed. by P.G. Ciarlet. Numerical Methods for Non-Newtonian Fluids, vol. XVI (Elsevier, Amsterdam, 2011), pp. 211–303
10. V. de Silva, L.-H. Lim, Tensor rank and ill-posedness of the best low-rank approximation problem. SIAM J Matrix Anal. Appl. **30**(3), 1084–1127 (2008)

Accurate Computations for Some Classes of Matrices

Juan M. Peña

Abstract A square matrix is called a P-matrix if all its principal minors are positive. Subclasses of P-matrices with many applications are the nonsingular totally positive matrices and the nonsingular M-matrices. For diagonally dominant M-matrices and some subclasses of nonsingular totally nonnegative matrices, accurate methods for computing their singular values, eigenvalues or inverses have been obtained, assuming that adequate natural parameters are provided. The adequate parameters for diagonally dominant M-matrices are the row sums and the off-diagonal entries, and for nonsingular totally nonnegative matrices are the entries of their bidiagonal factorization. In this paper we survey some recent extensions of these methods to other related classes of matrices.

1 Introduction

Recent research in Numerical Linear Algebra has shown that certain classes of matrices allow us to perform many computations to high relative accuracy, independently of the size of the condition number. For instance, the computation of their singular values, eigenvalues or inverses. These classes of matrices are defined by special sign or other structure and require to know some natural parameters to high relative accuracy, and they are related to some subclasses of P-matrices. Let us recall that a square matrix is called a P-matrix if all its principal minors are positive. Subclasses of P-matrices with many applications are the nonsingular totally nonnegative matrices and the nonsingular M-matrices. Usually, accurate

J.M. Peña (✉)
Dpto. de Matemática Aplicada, Universidad de Zaragoza, Zaragoza, Spain
e-mail: jmpena@unizar.es

A. Abdulle et al. (eds.), *Numerical Mathematics and Advanced Applications ENUMATH 2013*, Lecture Notes in Computational Science and Engineering 103,
DOI 10.1007/978-3-319-10705-9_32

spectral computation (eigenvalues, singular values) or accurate inversion is assured when an accurate matrix factorization with a suitable pivoting is provided. For instance, the bidiagonal decomposition in the case of totally nonnegative matrices (see [17]) or an *LDU* factorization after a symmetric pivoting in the case of diagonally dominant matrices (cf. [9, 20]).

In Sect. 2, we survey accurate computations for diagonally dominant M-matrices and other related classes of matrices. In Sect. 3, we survey accurate computations for nonsingular totally positive matrices and other related classes of matrices parametrized by a bidiagonal decomposition.

2 *M*-Matrices and Diagonal Dominance

Let us start by introducing some classes of matrices used in this section. A real matrix with nonpositive off–diagonal elements is called a *Z–matrix*. We say that a matrix $A = (a_{ij})_{1\le i,j\le n}$ is *row diagonally dominant* if, for each $i = 1,\dots,n$, $|a_{ii}| \ge \sum_{j\ne i} |a_{ij}|$. If A^T is row diagonally dominant, then we say that A is column diagonally dominant. Given a matrix $A = (a_{ij})_{1\le i,j\le n}$, its *comparison matrix* $\mathscr{M}(A) = (m_{ij})_{1\le i,j\le n}$ is the Z–matrix defined by $m_{ii} := |a_{ii}|$ and $m_{ij} := -|a_{ij}|$ if $i \ne j$, $1 \le i,j \le n$. Let us recall that if a Z–matrix A can be expressed as $A = sI - B$, with $B \ge 0$ and $s \ge \rho(B)$ (where $\rho(B)$ is the spectral radius of B), then it is called an *M–matrix*. Let us also recall that a Z–matrix A is a nonsingular M–matrix if and only if A^{-1} is nonnegative. Nonsingular M–matrices present applications to many fields.

Given an algorithm using only additions of numbers of the same sign, multiplications and divisions, and assuming that each initial real datum is known to high relative accuracy, then it is well–known that the output of that algorithm can be computed to high relative accuracy (cf. [8, p. 52]). Moreover, in (well–implemented) floating point arithmetic high relative accuracy is also preserved even when we perform true subtractions when the operands are original (and so, exact) data (cf. p. 53 of [8]).

A crucial tool to derive accurate algorithms for the computation of the singular values of a matrix is provided by the concept of rank revealing decomposition. Let us recall that a *rank revealing decomposition* of a matrix A is defined in [8] as a decomposition $A = XDY^T$, where X, Y are well conditioned and D is a diagonal matrix. In [8] Demmel et al. showed that the singular value decomposition can be computed accurately and efficiently for matrices possessing accurate rank revealing decompositions.

Let us also recall that an idea that has played a crucial role in some recent works on accurate computations has been the need to reparametrize matrices belonging to some special classes. In the class of M–matrices, the natural parameters that permit obtaining accurate and efficient algorithms are the off–diagonal entries and the row sums (or the column sums): see [1,2] and [9], where the class of M–matrices row diagonally dominant was considered. Furthermore, the parameters can have a meaningful interpretation when the matrix arises in a "real" problem. In the field of digital electrical circuits, the column sums are given by the quotient between the conductance and capacitance of each node (see [1]).

An algorithm of [2] computed to high relative accuracy the LDU factorization of an $n \times n$ row diagonally dominant M–matrix A when the off–diagonal entries and the row sums are given. The trick was to modify Gaussian elimination to compute the off–diagonal entries and the row sums of each Schur complement without performing subtractions. On the other hand, let us recall that a symmetric pivoting leading to an LDU-decomposition of A is equivalent to the following factorization of A: $PAP^T = LDU$, where P is the permutation matrix associated to the pivoting strategy. Symmetric complete pivoting was used in [9] in order to obtain well conditioned L and U factors because U is row diagonally dominant and the off-diagonal entries of L have absolute value less than 1. This factorization is a special case of a rank revealing decomposition. To implement symmetric complete pivoting, the algorithm in [9] computes all the diagonal entries and all Schur complements and this increases the cost in $\mathcal{O}(n^3)$ flops with respect to standard Gaussian elimination. In [20] another symmetric pivoting strategy (called diagonally dominant pivoting) was used, also with a subtraction-free implementation and a similar computational cost, but leading to both triangular matrices L and U column and row diagonally dominant, respectively. In [4], an accurate algorithm for the same LDU-decomposition of [20], but requiring $\mathcal{O}(n^2)$ elementary operations beyond the cost of Gaussian elimination, is presented. This method is also valid for diagonally dominant matrices satisfying certain sign patterns: with off–diagonal entries of the same sign or satisfying a chessboard pattern. The problem of computing an accurate LDU decomposition of diagonally dominant matrices has been solved by Ye in [22] (see also [11]). Finally, for a class of $n \times n$ nonsingular almost row diagonally dominant Z-matrices and given adequate parameters, an efficient method to compute its LDU decomposition with high relative accuracy is provided in [6]. It adds an additional cost of $\mathcal{O}(n^2)$ elementary operations over the computational cost of Gaussian elimination.

3 Totally Positive Matrices and Bidiagonal Factorizations

Totally positive matrices have all minors nonnegative. They are also called totally nonnegative matrices and they present many applications in many fields. The bidiagonal factorizations of these matrices have played a crucial role in their study and applications since several decades ago (cf. [12] or [21]). More recently, the bidiagonal factorization has been used to perform accurately many computations with these matrices (see [17]).

Let us define bidiagonal matrices $L^{(k)}$, $U^{(k)}$ by

$$L^{(k)} = \begin{pmatrix} 1 & & & & & & \\ 0 & 1 & & & & & \\ & \ddots & \ddots & & & & \\ & & 0 & 1 & & & \\ & & & l_{n-k}^{(k)} & 1 & & \\ & & & & \ddots & \ddots & \\ & & & & & l_{n-1}^{(k)} & 1 \end{pmatrix}, U^{(k)} = \begin{pmatrix} 1 & 0 & & & & & \\ & \ddots & \ddots & & & & \\ & & 1 & 0 & & & \\ & & & 1 & u_{n-k}^{(k)} & & \\ & & & & \ddots & \ddots & \\ & & & & & 1 & u_{n-1}^{(k)} \\ & & & & & & 1 \end{pmatrix},$$

where $k = 1, \ldots, n-1$.

In this section we shall consider matrices with bidiagonal decompositions of the form presented in the following definition.

Definition 1 Let A be a nonsingular $n \times n$ matrix. Suppose that we can write A as a product of bidiagonal matrices

$$A = L^{(1)} \cdots L^{(n-1)} D U^{(n-1)} \cdots U^{(1)}, \tag{1}$$

where $D = \mathrm{diag}(d_1, \ldots, d_n)$, and, for $k = 1, \ldots, n-1$, $L^{(k)}$ and $U^{(k)}$ are lower and upper bidiagonal matrices with unit diagonal respectively, with off-diagonal entries $l_i^{(k)} := (L^{(k)})_{i+1,i}$ and $u_i^{(k)} := (U^{(k)})_{i,i+1}$, $(i = 1, \ldots, n-1)$ satisfying

1. $d_i \neq 0$ for all i,
2. $l_i^{(k)} = u_i^{(k)} = 0$ for $i < n-k$,
3. $l_i^{(k)} = 0 \Rightarrow l_{i+s}^{(k-s)} = 0$ for $s = 1, \ldots, k-1$ and
   $u_i^{(k)} = 0 \Rightarrow u_{i+s}^{(k-s)} = 0$ for $s = 1, \ldots, k-1$.

Then we denote (1) by $\mathcal{BD}(A)$, a bidiagonal decomposition of A satisfying the conditions of this definition.

A matrix that can be decomposed in terms of bidiagonal matrices can also admit many other bidiagonal factorizations (cf. Chapter 6 of [21]). But the next result of [5] shows that a bidiagonal factorization as in Definition 2.1 is unique.

Theorem 1 *If a $\mathcal{BD}(A)$ exists for some matrix A, then it is unique.*

The following result provides the unique bidiagonal decomposition of a nonsingular totally positive matrix and it is a consequence of Theorem 4.2 of [15].

Theorem 2 *A nonsingular $n \times n$ matrix A is totally positive if and only if there exists a (unique) $\mathcal{BD}(A)$ such that*

1. $d_i > 0$ for all i,
2. $l_i^{(k)} \geq 0$, $u_i^{(k)} \geq 0$ for $1 \leq k \leq n-1$ and $n-k \leq i \leq n-1$.

It is well known that, if we have the $\mathcal{BD}(A)$ of a totally positive matrix with high relative accuracy, then we can perform many computations of A with high relative accuracy, such as computing its inverse or computing its eigenvalues or its singular values (cf. [17]). Therefore, the entries of the bidiagonal factorization (1) are the adequate parameters for nonsingular totally positive matrices. There are several subclasses of nonsingular totally positive matrices for which this factorization can be obtained to high relative accuracy (and so, the computations mentioned previously, too). For instance, Vandermonde positive matrices [10], Bernstein-Vandermond matrices [18], Said-Ball-Vandermonde matrices [19], Pascal matrices [3] or some rational collocation matrices [7]. The factorization is obtained through an elimination procedure called Neville elimination and described below.

Now let us denote by ε the vector $\varepsilon = (\varepsilon_1, \ldots, \varepsilon_m)$ with $\varepsilon_j \in \{\pm 1\}$ for $j = 1, \ldots, m$, which will be called a *signature*.

Definition 2 Given a signature $\varepsilon = (\varepsilon_1, \ldots, \varepsilon_{n-1})$ and a nonsingular $n \times n$ matrix A, we say that A has a signed bidiagonal decomposition with signature ε if there exists a $\mathcal{BD}(A)$ (unique by Theorem 3.2) such that

1. $d_i > 0$ for all i,
2. $l_i^{(k)}\varepsilon_i \geq 0$, $u_i^{(k)}\varepsilon_i \geq 0$ for $1 \leq k \leq n-1$ and $n-k \leq i \leq n-1$.

Bidiagonal decompositions satisfying the properties of Definition 2.1 have been considered in [5] and it was proved that the class of matrices satisfying this definition contains nonsingular totally positive matrices and their inverses. Moreover, in [5] it has been shown that if we have the $\mathcal{BD}(A)$ of a matrix with high relative accuracy, then we can perform many computations of A with high relative accuracy, assuming that A belongs to the class of matrices satisfying the previous definition.

We now present Neville elimination, which provides a constructive way of obtaining bidiagonal factorizations. Neville elimination is an alternative procedure to Gaussian elimination to eliminate nonzeros in a column of a matrix by adding to each row a multiple of the previous one (see [13]). If A is a square matrix of order n, $A = (a_{ij})_{1 \leq i,j \leq n}$ this elimination procedure consists of at most $n-1$ successive major steps, resulting in a sequence of matrices as follows:

$$A = A^{(1)} \to \tilde{A}^{(1)} \to A^{(2)} \to \tilde{A}^{(2)} \to \cdots \to A^{(n)} = \tilde{A}^{(n)} = U, \tag{2}$$

where U is an upper triangular matrix.

On the one hand, $\tilde{A}^{(t)}$ can be obtained by a reordering of the rows of the matrix $A^{(t)}$, moving the rows with a zero entry in column t to the bottom such that $\tilde{a}_{it}^{(t)} = 0$ for $i \geq t$ implies that $\tilde{a}_{ht}^{(t)} = 0$ for $\forall\, h \geq i$. On the other hand, $A^{(t+1)}$ is obtained from $\tilde{A}^{(t)}$ eliminating nonzeros in the column t below the main diagonal by adding an adequate multiple of the ith row to the $(i+1)$th for $i = n-1, n-2, \ldots, t$ according to the following formula

$$a_{ij}^{(t+1)} = \begin{cases} \tilde{a}_{ij}^{(t)}, & \text{if } 1 \leq i \leq j \leq t, \\ \tilde{a}_{ij}^{(t)} - \dfrac{\tilde{a}_{it}^{(t)}}{\tilde{a}_{i-1,t}^{(t)}} \tilde{a}_{i-1,j}^{(t)}, & \text{if } t+1 \leq i, j \leq n \text{ and } \tilde{a}_{i-1,t}^{(t)} \neq 0, \\ \tilde{a}_{ij}^{(t)}, & \text{if } t+1 \leq i \leq n \text{ and } \tilde{a}_{i-1,t}^{(t)} = 0, \end{cases} \tag{3}$$

for all $t \in \{1, \ldots, n-1\}$.

The element

$$p_{ij} = \tilde{a}_{ij}^{(j)}, \quad 1 \leq j \leq i \leq n, \tag{4}$$

is called the (i, j) *pivot* of Neville elimination of A. The Neville elimination can be performed without row exchanges if all the pivots are nonzero. The pivots p_{ii} are called *diagonal pivots*. If all the pivots p_{ij} are nonzero then, by Lemma 2.6 of [13], $p_{i1} = a_{i1}$ for $1 \leq i \leq n$ and

$$p_{ij} = \frac{\det A[i-j+1, \ldots, i | 1, \ldots, j]}{\det A[i-j+1, \ldots, i-1 | 1, \ldots, j-1]} \tag{5}$$

for $1 \leq j \leq i \leq n$. The element

$$m_{ij} = \begin{cases} \dfrac{\tilde{a}_{ij}^{(j)}}{\tilde{a}_{i-1,j}^{(j)}} = \dfrac{p_{ij}}{p_{i-1,j}}, & \text{if } \tilde{a}_{i-1,j}^{(j)} \neq 0, \\ 0, & \text{if } \tilde{a}_{i-1,j}^{(j)} = 0, \end{cases} \tag{6}$$

is called the (i, j) *multiplier* of Neville elimination of A, where $1 \leq j < i \leq n$.

Neville elimination characterizes nonsingular totally positive matrices, as the following result shows. It follows from Theorem 4.2 and p. 116 of [15].

Theorem 3 *A matrix A is nonsingular totally positive if and only if the Neville elimination of A and A^T can be performed without row exchanges, all the mutipliers of the Neville elimination of A and A^T are nonnegative and all the diagonal pivots of the Neville elimination of A are positive.*

Using the previous result as well as results of results of [14] and [15], we can describe bidiagonal decompositions of nonsingular totally positive matrices and their inverses in terms of the diagonal pivots and multipliers of their Neville elimination and the multipliers of the Neville elimination of their transposes.

Theorem 4 *Let A be a nonsingular totally positive matrix. Then A and A^{-1} admit factorizations in the form*

$$A^{-1} = G_1 G_2 \cdots G_{n-1} D^{-1} F_{n-1} \cdots F_1 \quad \textit{and} \quad A = \overline{F}_{n-1} \cdots \overline{F}_1 D \overline{G}_1 \cdots \overline{G}_{n-1}, \tag{7}$$

respectively, where F_i and $\overline{F}_i$, $i \in \{1, \ldots, n-1\}$, are the lower triangular bidiagonal matrices given by

$$F_i = \begin{pmatrix} 1 & & & & & & \\ 0 & 1 & & & & & \\ & \ddots & \ddots & & & & \\ & & 0 & 1 & & & \\ & & & -m_{i+1,i} & 1 & & \\ & & & & -m_{i+2,i} & 1 & \\ & & & & & \ddots & \ddots \\ & & & & & & -m_{n,i} \;\; 1 \end{pmatrix}$$

and

$$\overline{F}_i = \begin{pmatrix} 1 & & & & & & \\ 0 & 1 & & & & & \\ & \ddots & \ddots & & & & \\ & & 0 & 1 & & & \\ & & & m_{i+1,1} & 1 & & \\ & & & & m_{i+2,2} & 1 & \\ & & & & & \ddots & \ddots \\ & & & & & & m_{n,n-i} \;\; 1 \end{pmatrix},$$

G_i and $\overline{G}_i$, $i \in \{1, \ldots, n-1\}$, are the upper triangular bidiagonal matrices whose trasposes are given by

$$G_i^T = \begin{pmatrix} 1 & & & & & & \\ 0 & 1 & & & & & \\ & \ddots & \ddots & & & & \\ & & 0 & 1 & & & \\ & & & -\tilde{m}_{i+1,i} & 1 & & \\ & & & & -\tilde{m}_{i+2,i} & 1 & \\ & & & & & \ddots & \ddots \\ & & & & & & -\tilde{m}_{n,i} \;\; 1 \end{pmatrix}$$

and

$$\overline{G}_i^T = \begin{pmatrix} 1 & & & & & & & & \\ 0 & 1 & & & & & & & \\ & \ddots & \ddots & & & & & & \\ & & 0 & 1 & & & & & \\ & & & \tilde{m}_{i+1,1} & 1 & & & & \\ & & & & \tilde{m}_{i+2,2} & 1 & & & \\ & & & & & \ddots & \ddots & & \\ & & & & & & \tilde{m}_{n,n-i} & 1 \end{pmatrix},$$

and D the diagonal matrix diag$(p_{11} \ldots, p_{nn})$*. The entries* m_{ij}, $\tilde{m}_{ij}$ *are the multipliers of the Neville elimination of A and* A^T*, respectively, and the entries* p_{ii} *are the diagonal pivots of A.*

The results obtained until now assuring accurate computations with some subclasses of totally positive matrices have used the multipliers of Neville elimination as a natural parametrization of the matrices (cf. [3, 7, 10, 16–19]).

References

1. A.S. Alfa, J. Xue, Q. Ye, Entrywise perturbation theory for diagonally dominant M-matrices with application. Numer. Math. **90**, 401–414 (1999)
2. A.S. Alfa, J. Xue, Q. Ye, Accurate computation of the smallest eigenvalue of a diagonally dominant M-matrix. Math. Comput. **71**, 217–236 (2001)
3. P. Alonso, J. Delgado, R. Gallego, J.M. Peña, Conditioning and accurate computations with Pascal matrices. J. Comput. Appl. Math. **252**, 21–26 (2013)
4. A. Barreras, J.M. Peña, Accurate and efficient LDU decompositions of diagonally dominant M-matrices. Electron. J. Linear Algebra **24**, 153–167 (2012)
5. A. Barreras, J.M. Peña, Accurate computations of matrices with bidiagonal decomposition using methods for totally positive matrices. Numer. Linear Algebra Appl. **20**, 413–424 (2013)
6. A. Barreras, J.M. Peña, Accurate and efficient LDU decomposition of almost diagonally dominant Z-matrices. BIT **54**, 343–356 (2014)
7. J. Delgado, J.M. Peña, Accurate computations with collocation matrices of rational bases. Appl. Math. Comput. **219**, 4354–4364 (2013)
8. J. Demmel, M. Gu, S. Eisenstat, I. Slapnicar, K. Veselic, Z. Drmac, Computing the singular value decomposition with high relative accuracy. Linear Algebra Appl. **99**, 21–80 (1999)
9. J. Demmel, P. Koev, Accurate SVDs of weakly diagonally dominant M-matrices. Numer. Math. **98**, 99–104 (2004)
10. J. Demmel, P. Koev, The accurate and efficient solution of a totally positive generalized Vandermonde linear system. SIAM J. Matrix Anal. Appl. **27**, 142–152 (2005)
11. F.M. Dopico, P. Koev, Perturbation theory for the LDU factorization and accurate computations for diagonally dominant matrices. Numer. Math. **119**, 337–371 (2011)
12. S. Fallat, C.R. Johnson, *Totally Nonnegative Matrices*. Princeton Series in Applied Mathematics (Princeton University Press, Princeton, 2011)

13. M. Gasca, J.M. Peña, Total positivity and Neville elimination. Linear Algebra Appl. **165**, 25–44 (1992)
14. M. Gasca, J.M. Peña, A matricial description of Neville elimination with applications to total positivity. Linear Algebra Appl. **202**, 33–53 (1994)
15. M. Gasca, J.M. Peña, On factorizations of totally positive matrices, in *Total Positivity and Its Applications*, ed. by M. Gasca, C.A. Micchelli (Kluver Academic, Dordrecht, 1996), pp. 109–130
16. P. Koev, Accurate eigenvalues and SVDs of totally nonnegative matrices. SIAM J. Matrix Anal. Appl. **27**, 1–23 (2005)
17. P. Koev, Accurate computations with totally nonnegative matrices. SIAM J. Matrix Anal. Appl. **29**, 731–751 (2007)
18. A. Marco, J.J. Martínez, A fast and accurate algorithm for solving Bernstein-Vandermonde linear systems. Linear Algebra Appl. **422**, 616–628 (2007)
19. A. Marco, J.J. Martínez, Accurate computations with Said-Ball-Vandermonde matrices. Linear Algebra Appl. **432**, 2894–2908 (2010)
20. J.M. Peña, LDU decompositions with L and U well conditioned. Electron. Trans. Numer. Anal. **18**, 198–208 (2004)
21. A. Pinkus, *Totally Positive Matrices*. Cambridge Tracts in Mathematics, vol. 181 (Cambridge University Press, Cambridge, 2010)
22. Q. Ye, Computing singular values of diagonally dominant matrices to high relative accuracy. Math. Comput. **77**, 2195–2230 (2008)

Corrected One-Site Density Matrix Renormalization Group and Alternating Minimal Energy Algorithm

Sergey V. Dolgov and Dmitry V. Savostyanov

Abstract Given in the title are two algorithms to compute the extreme eigenstate of a high-dimensional Hermitian matrix using the tensor train (TT)/matrix product states (MPS) representation. Both methods empower the traditional alternating direction scheme with the auxiliary (e.g. gradient) information, which substantially improves the convergence in many difficult cases. Being conceptually close, these methods have different derivation, implementation, theoretical and practical properties. We emphasize the differences, and reproduce the numerical example to compare the performance of two algorithms.

1 Introduction

Actual problems of science, engineering and society can be so complex that their mathematical portrait requires more than three dimensions. Quantum world gives us a perfect example of essentially high–dimensional systems, described by a joint wavefunction (or density matrix) of all particles. A simple system of d spin-$\frac{1}{2}$ particles is an entanglement of $\mathcal{O}(2^d)$ possible states, and should be described by the same amount of numbers, that exceeds the capacity of a typical workstation for $d \gtrsim 30$. Even with a brute force of modern supercomputers, standard numerical methods can not honestly simulate protein–size molecules ($d \sim 10^3$–10^4), since the complexity and storage explode exponentially with d.

To overcome this problem, known as the *curse of dimensionality*, we use data-sparse representations for high-dimensional vectors and matrices, and develop special algorithms to work with them. Proposed in 1992, the *density matrix renormalization group* (DMRG) algorithm [8] and the *matrix product states* (MPS)

S.V. Dolgov
Max Planck Institute for Mathematics in the Sciences, Inselstraße 22, 04103 Leipzig, Germany
e-mail: sergey.v.dolgov@gmail.com

D.V. Savostyanov (✉)
School of Chemistry, University of Southampton, Highfield Campus, SO17 1BJ Southampton, United Kingdom
e-mail: dmitry.savostyanov@gmail.com

A. Abdulle et al. (eds.), *Numerical Mathematics and Advanced Applications ENUMATH 2013*, Lecture Notes in Computational Science and Engineering 103,
DOI 10.1007/978-3-319-10705-9_33

formalism [2] suggest to represent a wavefunction x in the following tensor-product form

$$x = \tau(x^{(1)}, \dots, x^{(d)}) = \sum_{\alpha_1=1}^{r_1} \cdots \sum_{\alpha_{d-1}=1}^{r_{d-1}} x^{(1)}_{\alpha_1} \otimes x^{(2)}_{\alpha_1\alpha_2} \otimes \dots \otimes x^{(d)}_{\alpha_{d-1}}, \quad \text{i.e.}$$
$$x(i_1, \dots, i_d) = \sum_{\alpha_1=1}^{r_1} \cdots \sum_{\alpha_{d-1}=1}^{r_{d-1}} x^{(1)}_{\alpha_1}(i_1) x^{(2)}_{\alpha_1\alpha_2}(i_2) \dots x^{(d)}_{\alpha_{d-1}}(i_d). \tag{1}$$

In numerical linear algebra this format was re-discovered as the *tensor train* (TT) decomposition [5]. A single TT core (or *site*) $x^{(k)} = [x^{(k)}_{\alpha_{k-1}\alpha_k}(i_k)]$ is described by $r_{k-1}n_k r_k$ numbers, where n_k denotes the number of possible states for the k–th particle (the *mode size*), and r_k is the TT rank (or *bond dimension*). The total number of representation parameters scales as $\mathcal{O}(dnr^2)$, $n \sim n_k$, $r \sim r_k$, and is feasible for computations with $d, n, r \lesssim 10^3$.

The DMRG algorithm was originally proposed to find the *ground state*, i.e. the minimal eigenpair of a Hermitian matrix A. This problem is equivalent to the minimization of the Rayleigh quotient $Q_A(x) = (x, Ax)/(x, x)$. Substituting $Q_A(x)$ with $J_{A,b}(x) = (x, Ax) - 2\Re(x, b)$, and applying the same algorithm, we can solve linear systems $Ax = b$ with Hermitian positive definite matrices [3]. This framework can be extended to a broad class of problems.

Since x is a huge high-dimensional vector, the solution is sought in the structured format (1) with some TT ranks r_k, defined a priori or chosen adaptively. The simultaneous optimization over all sites is a highly nonlinear and difficult problem. As it is usual in high-dimensional optimization, we substitute it by a sequence of partial optimizations, each over a particular (small) group of variables. For our problem, it is natural to group the variables according to the tensor format (1), e.g. optimize over the components of a single site $x^{(k)}$ at a time.

The TT format is linear in each site, i.e. $x = \tau(x^{(1)}, \dots, x^{(d)}) = X_{\neq k} x^{(k)}$, where $X_{\neq k}$ is the $(n_1 \dots n_d) \times (r_{k-1}n_k r_k)$ *frame matrix*, which linearly maps the elements of $x^{(k)}$ to the full vector x. This turns every partial optimization into a local problem of the same type, as the original one,

$$x^{(k)}_\star = \arg\min_{x^{(k)}} Q_A(\tau(x^{(1)}, \dots, x^{(k)}, \dots, x^{(d)})) = \arg\min_{x^{(k)}} Q_{A_k}(x^{(k)}), \tag{2}$$

where $A_k = X^*_{\neq k} A X_{\neq k}$ is the $(r_{k-1}n_k r_k)\times(r_{k-1}n_k r_k)$ *reduced matrix*, which inherits the properties of A, i.e. is Hermitian. Since the frame matrix $X_{\neq k}$ has a structured TT representation (which is the same as (1) with $x^{(k)}$ substituted by the identity matrix), the reduced matrix A_k can be assembled avoiding the exponential costs. Finally, introducing simple orthogonality conditions for all sites but $x^{(k)}$, we can make the whole matrix $X_{\neq k}$ orthogonal [7]. As a consequence, A_k becomes better conditioned than A, and the reduced functional writes $Q_{A_k}(x) = (x, A_k x)/(x, x)$ for the ground state problem, and $J_{A_k, X^*_{\neq k} b}(x) = (x, A_k x) - 2\Re(x, X^*_{\neq k} b)$ for the

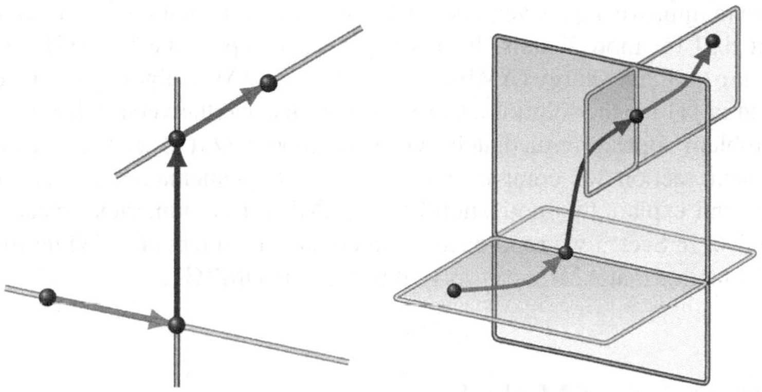

Fig. 1 Sequence of low-dimensional optimizations in subspaces $\mathscr{X}_1$, $\mathscr{X}_2$, $\mathscr{X}_3$, ... (*left*), and $\mathscr{X}_{1,2}$, $\mathscr{X}_{2,3}$, $\mathscr{X}_{3,4}$, ... (*right*)

linear system. Each local optimization (2) is now a simple problem of tractable size, that can be solved by classical algorithms of numerical linear algebra.

Each step (2) finds $\min_{x \in \mathscr{X}_k} Q_A(x)$, where the subspace $\mathscr{X}_k = \operatorname{span} X_{\neq k}$ is of dimension $r_{k-1}n_k r_k$, see Fig. 1 (left). Here and later by span X we denote the subspace of columns of a matrix X. If TT ranks are fixed, the local convergence of such scheme can be analysed using standard methods of multivariate analysis [6]. However, in numerical practice the tensor ranks of the solution are not known in advance, and fixed-rank optimization with wrong ranks would not be efficient. The DMRG scheme with variable TT ranks is more advantageous, but the theoretical analysis is even more difficult.

When we allow TT ranks to grow, the dimensions of subspaces $\mathscr{X}_k$ grow as well, and we can use different strategies to expand the subspaces. Originally, the one-site DMRG scheme (DMRG1) increased the rank r_k by adding (random) orthogonal vectors to $\mathscr{X}_k$, but this algorithm often got stuck far from the ground state. The problem was solved using *two sites* instead of one in the optimization step [8]. The *two-site* DMRG algorithm (DMRG2) merges blocks $x^{(k)}$ and $x^{(k+1)}$, and solves the local optimization problem in $\mathscr{X}_{k,k+1} = \operatorname{span} X_{\neq\{k,k+1\}}$, see Fig. 1 (right). Here $X_{\neq k,k+1}$ is the $(n_1 \dots n_d) \times (r_{k-1}n_k n_{k+1} r_{k+1})$ matrix, which has the same TT representation as (1) with blocks $x^{(k)}$ and $x^{(k+1)}$ replaced by the identities. The DMRG2 converges remarkably well (and is in fact a method of choice) for 1D systems with short–range interactions, but the cost is approximately n times larger than in the DMRG1. For systems with long-range interactions two neighboring sites do not provide sufficient information, and DMRG2 can stagnate as well. To simulate such systems faster and more accurately, better methods to choose search subspaces are required.

The *gradient direction* is central in the theory of optimization methods, and many algorithms use the gradient or its approximate surrogates. In [9], S. White proposed the *corrected one-site* DMRG algorithm (DMRG1c), which adds auxiliary direction to improve the convergence and reduce the computational cost, see [9] and [7, Sect. 6.3] for more details. In this paper we compare the DMRG1c with the *alternating minimal energy* (AMEn) algorithm. The AMEn algorithm was recently proposed in [1] for the solution of linear equations, and the version for the ground state problem appears immediately when we choose $Q_A(x)$ as a target function. In the next section we compare the ideas and implementation aspects of both methods and explain the motivation behind AMEn from numerical linear algebra perspective. In Sect. 3 we reproduce a numerical experiment of S. White from [9], and demonstrate that AMEn can solve it better than DMRG1c.

2 Comparison of Methods

Both DMRG1c and AMEn combine the local optimization (2) with the step that injects the auxiliary information. Both algorithms are *local*, i.e. modify only one block $x^{(k)}$ at a time (cf. the non-local "ALS($t+z$)" algorithm in [1]). Both methods sequentially cycle over TT blocks $(1, 2, \ldots, d, d-1, \ldots)$. In the following we assume that (2) was just solved for $x^{(k)}$, and consider the step that corrects $x^{(k)}$ before the optimization passes to the next block $x^{(k+1)}$. This step does not change the vector $x = \tau(x^{(1)}, \ldots, x^{(d)})$ (for AMEn), or perturbs it slightly (for DMRG1c), and therefore has a minor direct effect on $Q_A(x)$. However, it inserts additional directions to span $X_{\neq k+1}$, that improves the convergence of $Q_A(x)$ to its global minimum.

It is crucial how exactly the block $x^{(k)}$ is modified, and which vectors end up in span $X_{\neq k+1}$ after that. In the following we discuss these details, which constitute the main difference between the DMRG1c and the AMEn.

2.1 Which Vector is Targeted: $p = Ax$ vs. $z = Ax - Q_A(x)x$

Following the *power iteration* method, the DMRG1c algorithm of S. White targets in addition to the solution x the first Krylov vector $p = Ax$. The AMEn algorithm uses the gradient direction $z = Ax - Q_A(x)x$. In exact arithmetics this makes no difference, since $\operatorname{span}\{x, p\} = \operatorname{span}\{x, z\}$. In practical computations both p and z are perturbed by inevitable machine rounding errors, errors of approximation to the tensor format (1), and additional errors that appear when a surrogate formula (like [9, Eq. (14)]) is used to speed up the computations. The DMRG1c algorithm is derived from perturbation arguments, valid in the vicinity of the minimum of $Q_A(x)$. When x approaches the ground state, the angle between x and $p = Ax$ vanishes, and any perturbation in Ax yields a random new direction. This creates a certain gap between the theory supporting the DMRG1c, and the practice.

Following the *steepest descent* algorithm, the AMEn uses orthogonal vectors z and x, and $\operatorname{span}\{x, z\}$ is much more stable to perturbations of z. In general, the

Krylov vectors $\{x, Ax, A^2x, \dots\}$ form an extremely unstable basis, and orthogonalization is crucial. The steepest descent algorithm with z substituted by $\tilde{z}$ converges as long as $(\tilde{z}, z) > 0$. For the linear systems this fact is elegantly proven in [4], and the convergence rate of perturbed method is estimated. An eigenvalue counterpart follows similarly, and the rate of convergence in $\operatorname{span}\{x, \tilde{z}\}$ can be estimated from the spectral range of A. This makes the approach implemented in the AMEn algorithm preferable both theoretically and in practice.

2.2 What Is Approximated: Subspace $\tilde{\mathscr{P}}$ vs. Vector $\tilde{z}$

The computation of full vectors $p = Ax$ and $z = Ax - Q_A(x)x$ is not possible due to their exponentially large size. Since x and A are both in the TT format, we can avoid the curse of dimensionality keeping $p = Ax$ and $z = Ax - Q_A(x)x$ in the TT format. However, the TT ranks of Ax can be as large as the product of TT ranks of A and x, that slows down the calculations.

To reduce these costs, S. White suggests in the **DMRG1c** the following scheme. The TT format (1) is divided in two parts: left blocks (number $1, \dots, k$) are referred to as *system*, and right blocks $(k+1, \dots, d)$ as *environment*. The TT format for the matrix A is written accordingly,

$$A = \sum_{\gamma_1 \cdots \gamma_{d-1}} \underbrace{A^{(1)}_{\gamma_1} \otimes \dots \otimes A^{(k)}_{\gamma_{k-1}\gamma_k}}_{\text{system}} \otimes \underbrace{A^{(k+1)}_{\gamma_k\gamma_{k+1}} \otimes \dots \otimes A^{(d)}_{\gamma_{d-1}}}_{\text{environment}}, \tag{3}$$

or shortly $A = \sum_\gamma A^<_\gamma \otimes A^>_\gamma$. Similarly, Eq. (1) reduces to $x = \sum_\alpha x^<_\alpha \otimes x^>_\alpha$. The targeting of $p = Ax$ is substituted by the targeting of all $p_\gamma = (A^<_\gamma \otimes I)x$.

Although in general $p \notin \cup_\gamma \operatorname{span} p_\gamma$, it can be argued that the set $\{p_\gamma\}$ contains a sufficient subspace information. To show this, we write

$$p = \sum\nolimits_{\alpha,\gamma} \left(A^<_\gamma x^<_\alpha\right) \otimes \left(A^>_\gamma x^>_\alpha\right), \qquad p_\gamma = \sum\nolimits_\alpha \left(A^<_\gamma x^<_\alpha\right) \otimes x^>_\alpha, \tag{4}$$

and consider vectors p and p_γ as system-by-environment matrices P and P_γ of size $(n_1 \dots n_k) \times (n_{k+1} \dots n_d)$. Now $\operatorname{span} P \subset \cup_\gamma \operatorname{span} P_\gamma = \mathscr{P}$, where $A^>_\gamma$ contains the coefficients of the required linear combination—in the exact arithmetics the *system*-related components of p belong to $\mathscr{P}$. Each p_γ is easier to compute than p, because it does not depend on the environment part $A^>_\gamma$.

The total dimension of $\mathscr{P}$ grows in each step, and to keep TT ranks and storage moderate, we have to truncate it. The approximation step in the DMRG1c replaces $\mathscr{P}$ with a subspace $\tilde{\mathscr{P}}$ of a smaller dimension, using the classical *singular value decomposition* (SVD), or Schmidt decomposition technique. The *dominant subspace* $\tilde{\mathscr{P}}$ is spanned by the first singular vectors of the matrix $\left[X \; \sqrt{a_1}P_1 \; \sqrt{a_2}P_2 \dots\right]$, where all target vectors are concatenated with empirically chosen weighting coefficients a_γ. The method assumes that the vector $p = Ax$ is likely to belong to $\tilde{\mathscr{P}}$.

This assumption makes perfect sense if p is a random sample from $\mathscr{P}$—for a random u, Xu is more likely to end up in the dominant subspace of X. However, the target vector $p = Ax$ does not belong to $\tilde{\mathscr{P}}$ in general, for any choice of weights $\sqrt{a_\gamma}$. The reason is that p depends crucially on $A^>_\gamma$, whereas this information is dropped for the sake of faster computations in p_γ and hence $\mathscr{P}$ and $\tilde{\mathscr{P}}$. Selecting $A^>_\gamma$ in (4), we may come across any vector in $\mathscr{P}$, even the smallest singular vector. That is, for each choice of $\sqrt{a_\gamma}$ and x there is a *counterexample* of a Hamiltonian, for which the slightest truncation of $\mathscr{P}$ loses the system-related part of the target vector $p = Ax$.

The **AMEn** approximates $z = Ax - Q_A(x)x$ into its own TT format using any compression tool. Either the SVD–based technique, which computes the approximation $\tilde{z} \approx z$ up to any prescribed tolerance ε, or a faster (but heuristic) *alternating least squares* (ALS) method may be used. In any case, we may generate an approximation $\tilde{z}$ with a desired accuracy, which guarantees the convergence of the steepest descent method with the imperfect direction $\tilde{z}$. This fact provides the theoretical bounds for the global convergence rate of the whole AMEn scheme, similarly to [1].

2.3 How the New Direction Is Used: Averaging vs. Enrichment

The last but not the least detail is how exactly the information about the auxiliary direction is injected in the algorithm. To show this in isolation from the other dissimilarities outlined above, we assume that in both methods we target in addition to x *only one* vector s. To simplify the presentation we also consider the $d = 2$ case, and write $x = \sum_{\alpha=1}^{r_x} x^<_\alpha x^>_\alpha$, and $s = \sum_{\beta=1}^{r_s} s^<_\beta s^>_\beta$, where "<" and ">" denote the first and the second blocks, respectively.

The DMRG1c algorithm *averages* the subspaces $X = \left[x^<_1 \dots x^<_{r_x}\right]$ and $S = \left[s^<_1 \dots s^<_{r_s}\right]$ by computing the dominant subspace span U of the Gram matrix as follows, $G = XX^* + aSS^* \approx UU^*$, where $U = \left[u^<_1 \dots u^<_{r_u}\right]$. As shown in the previous subsection, this procedure does not guarantee that x or s ends up in span$(U \otimes I)$, unless span U = span$\left[X\ S\right]$. The TT core $x^<$ is replaced by the vectors of U, that introduces a $\mathscr{O}(\sqrt{a})$ perturbation to x and probably worsen $Q_A(x)$. It is clear though that a should vanish when we approach the exact solution, but the general recipe is not known.

The AMEn avoids the outlined difficulties by *merging* $U = \left[X\ S\right]$ and zero-padding the second block. Values of x and $Q_A(x)$ are preserved, no rescaling is required, and both $\{x, s\} \in$ span$(U \otimes I)$. The downside is that we choose $r_u = r_x + r_s$ each time we expand the subspaces. However, when we use the approximate gradient direction $s = \tilde{z} \approx z = Ax - Q_A(x)x$ the low-rank $\tilde{z}$ usually suffice, e.g.

with $r_s \approx r_x/2$. We can also truncate the TT-ranks at the end of each iteration and control the perturbation to $Q_A(x)$.

3 Numerical Example

Following S. White [9], we consider the spin-1 periodic Heisenberg chain,

$$\begin{aligned} A &= \mathbf{H}_1 \cdot \mathbf{H}_2 + \mathbf{H}_2 \cdot \mathbf{H}_3 + \ldots + \mathbf{H}_{d-1} \cdot \mathbf{H}_d + \mathbf{H}_d \cdot \mathbf{H}_1, \\ \mathbf{H}_i &= (\mathbf{H}_i^x, \mathbf{H}_i^y, \mathbf{H}_i^z)^\top, \qquad \mathbf{H}_i \cdot \mathbf{H}_j = H_i^x H_j^x + H_i^y H_j^y + H_i^z H_j^z, \\ H_i^{\{x,y,z\}} &= I \otimes \cdots \otimes I \otimes S_{\{x,y,z\}} \otimes I \otimes \cdots \otimes I, \quad S \text{ in position } i, \end{aligned} \tag{5}$$

where $S_{\{x,y,z\}}$, are the 3×3 Pauli matrices for spin-1 particles. The number of spins d is set to 100, i.e. the wavefunction belongs to the 3^{100}-dimensional Hilbert space. This example is particularly illustrating, since the mismatch between the linear TT model (1) and the cycle structure of (5) complicates the problem—the solution has large TT ranks, and both the one– and two–site DMRG converge slowly.

The way how the TT ranks are chosen during the algorithm is also very important. We first adopt the rank selection strategy from [9], and compare the DMRG2, the DMRG1c and the AMEn algorithms. The results are shown in Fig. 2 (top left), which overlays [9, Fig. 3] with the AMEn behavior. In Fig. 2 (top right) the convergence of $\lambda = Q_A(x)$ to the reference value $\lambda_\star = -140.14840390392$ (computed in [9] by the DMRG1c with TT ranks 4,000) is given w.r.t. the cumulative CPU time.

We see that both DMRG methods correctly reproduce the experiment from [9]: the two-site DMRG stagnates at a high error level, while the corrected DMRG converges significantly faster. The AMEn method manifests practically the same efficiency. Since it searches in a larger subspace, it is even more accurate w.r.t. iterations, but becomes slightly slower during the optimization of inner TT blocks. However, letting it to increase the ranks (each fourth iteration) yields sharper error decays.

To release the algorithm from tuning parameter, we prefer to choose the ranks adaptively to the desired accuracy. With this we also avoid artificial rank limitation, which pollutes the convergence. Therefore, in the second experiment we use the same algorithms but perform the truncation of TT blocks via the SVD using the relative Frobenius-norm accuracies $\varepsilon = 10^{-3}$ and $\varepsilon = 10^{-4}$. The results are shown in Fig. 2 (bottom).

We see that when ranks are chosen adaptively, the AMEn rapidly becomes faster than the other algorithms. Even the DMRG1c stagnates relatively early, since the correction p_y (4) contaminates the dominant basis of the ground state. Moreover, since the ε-truncation eliminates the correction if $a \lesssim \varepsilon^2$, it is worthless to decrease the scale a (cf. Fig. 2, top left). Both the adaptivity and speed speak in favour of

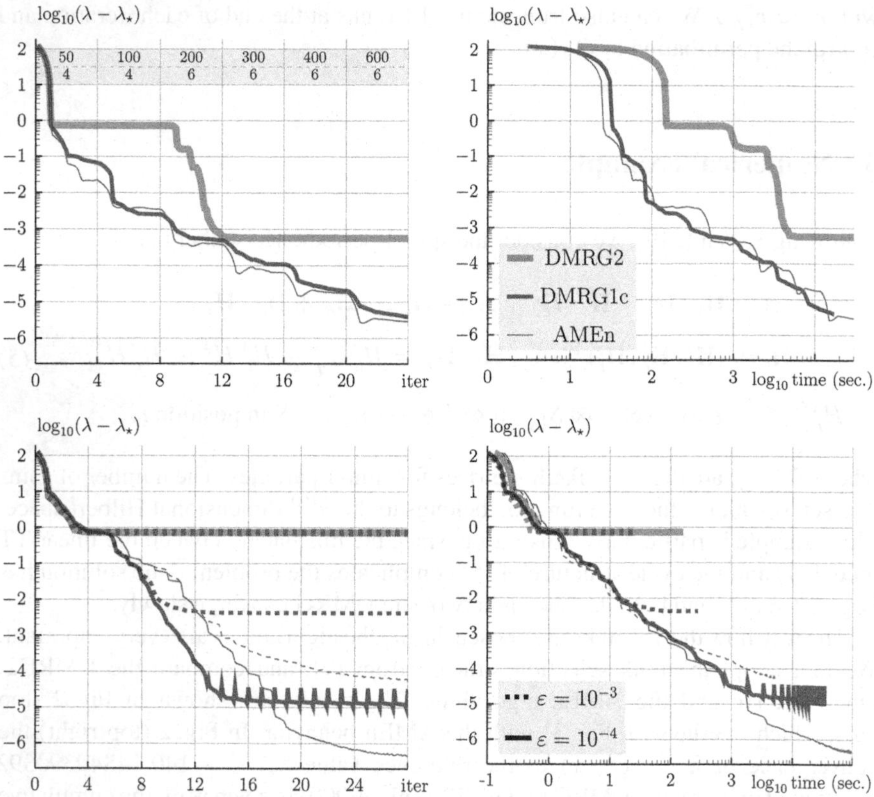

Fig. 2 Error in the eigenvalue vs. iteration (*left*) and CPU time (*right*). Methods: AMEn [1], DMRG1c [9], DMRG2 [8]. *Top*: parameters depend on iteration as shown on top of the left figure (ranks and $\log_{10}(1/a)$, resp.). *Bottom*: $a = 10^{-4}$, ranks depend on accuracies: $\varepsilon = 10^{-3}$ (*dashed lines*), $\varepsilon = 10^{-4}$ (*solid lines*)

such truncation: the same accuracy levels are achieved several times faster than in the fixed-rank experiment (e.g. 10 vs. 100 s. for $\lambda - \lambda_\star \approx 10^{-2}$ and $\varepsilon = 10^{-3}$). Larger time spent by AMEn in the latter iterations is compensated by a significantly better accuracy, which is close to the optimal level $\mathcal{O}(\varepsilon^2)$.

Acknowledgements This work was supported by EPSRC (EP/H003789/2).

References

1. S.V. Dolgov, D.V. Savostyanov, Alternating minimal energy methods for linear systems in higher dimensions. SIAM J. Sci. Comput. **36**(5), A2248–A2271 (2014). doi: 10.1137/140953289
2. M. Fannes, B. Nachtergaele, R. Werner, Finitely correlated states on quantum spin chains. Commun. Math. Phys. **144**(3), 443–490 (1992)
3. E. Jeckelmann, Dynamical density–matrix renormalization–group method. Phys. Rev. B **66**, 045114 (2002)
4. H. Munthe-Kaas, The convergence rate of inexact preconditioned steepest descent algorithm for solving linear systems, Numerical Analysis Report NA-87-04, Stanford University, 1987
5. I.V. Oseledets, Tensor-train decomposition. SIAM J. Sci. Comput. **33**(5), 2295–2317 (2011)
6. T. Rohwedder, A. Uschmajew, Local convergence of alternating schemes for optimization of convex problems in the TT format. SIAM J. Numer. Anal. **51**(2), 1134–1162 (2013)
7. U. Schollwöck, The density-matrix renormalization group in the age of matrix product states. Ann. Phys. **326**(1), 96–192 (2011)
8. S.R. White, Density matrix formulation for quantum renormalization groups. Phys. Rev. Lett. **69**(19), 2863–2866 (1992)
9. ———, Density matrix renormalization group algorithms with a single center site. Phys. Rev. B **72**(18), 180403 (2005)

Approximating the Matrix Exponential of an Advection-Diffusion Operator Using the Incomplete Orthogonalization Method

Antti Koskela

Abstract In this paper we give first results for the approximation of $e^{A}b$, i.e. the matrix exponential times a vector, using the *incomplete orthogonalization method*. The benefits compared to the Arnoldi iteration are clear: shorter orthogonalization lengths make the algorithm faster and a large memory saving is also possible. For the case of three term orthogonalization recursions, simple error bounds are derived using the norm and the field of values of the projected operator. In addition, an a posteriori error estimate is given which in numerical examples is shown to work well for the approximation. In the numerical examples we particularly consider the case where the operator A arises from spatial discretization of an advection-diffusion operator.

1 Introduction

An efficient numerical computation of the product $e^{A}b$ for a matrix $A \subset \mathbb{C}^{n\times n}$ and a vector $b \in \mathbb{C}^n$ is of importance in several fields of applied mathematics. For various applications and numerical methods, see [3].

One large source of problems of this form comes from the implementation of exponential integrators [5]. These integrators have been shown to be particularly efficient for ODEs coming from a spatial semidiscretization of semilinear PDEs. In this case A is usually sparse and has a large norm and dimension. A widely used approach in this case are *Krylov subspace methods*, see e.g. [4] and [9].

Krylov subspace methods are based on the idea of projecting a matrix $A \in \mathbb{C}^{n\times n}$ and a vector $b \in \mathbb{C}^n$ onto a lower dimensional subspace $\mathscr{K}_k(A,b)$ defined by

$$\mathscr{K}_k(A,b) = \text{span}\{b, Ab, A^2b, \ldots, A^{k-1}b\}.$$

A. Koskela (✉)
KTH Royal Institute of Technology, Lindstedtvägen 25, 10044 Stockholm, Sweden
e-mail: akoskela@kth.se

A. Abdulle et al. (eds.), *Numerical Mathematics and Advanced Applications ENUMATH 2013*, Lecture Notes in Computational Science and Engineering 103,
DOI 10.1007/978-3-319-10705-9_34

The Arnoldi iteration performs a Gram–Schmidt orthogonalization for this subspace and gives an orthonormal matrix $Q_k = [q_1, \ldots, q_k] \in \mathbb{C}^{n\times k}$ which provides a basis of $\mathscr{K}_k(A,b)$, and a Hessenberg matrix $H_k = Q_k^* A Q_k \in \mathbb{C}^{k\times k}$, which represents the action of A in the subspace $\mathscr{K}_k(A,b)$. If A is Hermitian or skew-Hermitian, H_k will be tridiagonal and we get the Lanczos iteration. Moreover, the recursion

$$AQ_k = Q_k H_k + h_{k+1,k} q_{k+1} e_k^{\mathsf{T}}$$

holds, where $h_{k+1,k}$ denotes the corresponding entry in H_{k+1} and e_k is the kth standard basis vector in $\mathbb{C}^k$.

Using the basis Q_k and the Hessenberg matrix H_k, the product $\mathrm{e}^A b$ can be then approximated as (see e.g. [9])

$$\mathrm{e}^A b \approx Q_k \mathrm{e}^{H_k} e_1 \|b\|.$$

In case that A is not (skew-)Hermitian the Arnoldi iteration has to be used. The drawback of this approach is that the orthogonalization recursions grow longer which slows down the iteration, and that it needs increasingly memory as k grows. As a remedy for this the *restarted Krylov subspace method* has been proposed [2].

The objective of this paper is to show that the *incomplete orthogonalization method* is a good alternative for approximating the product $\mathrm{e}^A b$ for nonnormal matrices A when long orthogonalization recursions should be avoided. The method has been considered before for eigenvalue problems [7] and for solving linear systems [8]. As the numerical experiments and the short analysis of this paper show, it also provides a good alternative for approximating the matrix exponential.

1.1 Class of Test Problems

A reasonable example of nonnormal large and sparse matrices is obtained from the spatial discretization of the 1-d advection-diffusion equation

$$\partial_t u = \epsilon \partial_{xx} u + \alpha \partial_x u. \tag{1}$$

Choosing Dirichlet boundary conditions on the interval $[0, 1]$ and performing the discretization using central finite differences gives the ordinary differential equation $y' = Ay$, where the operator is the form $A = \epsilon \Delta_n + \alpha \nabla_n \in \mathbb{R}^{n\times n}$ with

$$\Delta_n = \frac{1}{(\Delta x)^2}\begin{bmatrix} -2 & 1 & & & \\ 1 & -2 & 1 & & \\ & \ddots & \ddots & \ddots & \\ & & 1 & -2 & 1 \\ & & & 1 & -2 \end{bmatrix}, \quad \nabla_n = \frac{1}{2\Delta x}\begin{bmatrix} & -1 & & & \\ 1 & & -1 & & \\ & \ddots & \ddots & \ddots & \\ & & 1 & & -1 \\ & & & 1 & \end{bmatrix}, \tag{2}$$

where $\Delta x = 1/(n+1)$. We define the grid Péclet number

$$\mathrm{Pe} = \frac{\alpha \Delta x}{2\epsilon}$$

as in the numerical comparisons of [1]. By the Péclet number the nonnormality of A can be controlled.

Throughout the paper, $\langle \cdot, \cdot \rangle$ denotes the Euclidean inner product and $\|\cdot\|$ denotes the corresponding norm or its induced matrix norm. The Hermitian part of a matrix A is defined as $A^{\mathrm{H}} = (A^* + A)/2$, and the skew-Hermitian part as $A^{\mathrm{S}} = (A^* - A)/2$.

2 The Incomplete Orthogonalization Method

In the *incomplete orthogonalization method* (IOM) (see e.g. [7]), Aq_i is orthogonalized at step i only against m previous vectors $\{q_{i-m+1}, \dots, q_i\}$ instead of all the previous basis vectors. The coefficients h_{ij} are collected as in the Arnoldi iteration. The incomplete orthogonalization method with orthogonalization length m is denoted as IOM(m) for the rest of the paper. As a result of k steps of IOM(m) we get the matrix

$$Q_{k,m} = \begin{bmatrix} q_1 \dots q_k \end{bmatrix}$$

giving the basis of $\mathcal{K}_k(A, b)$, where $q_1 = b/\|b\|$, and the vectors q_i are orthogonal locally, i.e.,

$$\begin{aligned} \langle q_i, q_i \rangle &= 1, \\ \langle q_i, q_j \rangle &= 0, \quad \text{if} \quad |i - j| \leq m, \quad i \neq j. \end{aligned}$$

The iteration also gives a Hessenberg matrix with an upper bandwidth length m,

$$H_{k,m} := \begin{bmatrix} h_{11} \dots h_{1m} & & & 0 \\ h_{21} & \ddots & \ddots & \\ & \ddots & \ddots & h_{k-m+1,k} \\ & & \ddots \quad \ddots & \vdots \\ 0 & & h_{k,k-1} & h_{kk} \end{bmatrix}, \tag{3}$$

where the nonzero elements are given as

$$h_{ij} = \langle Aq_j, q_i \rangle.$$

We can see that by construction the following relation holds

$$AQ_{k,m} = Q_{k,m}H_{k,m} + h_{k+1,k}q_{k+1}e_k^{\mathsf{T}}. \tag{4}$$

It is easy to verify that if $\dim \mathcal{K}_k(A,b) = k$, then $\mathcal{K}_k(A,b) \subset R(Q_{k,m})$, where $R(Q_{k,m})$ denotes the range of $Q_{k,m}$. However, if it happens that $R(Q_{k+1}) = R(Q_k)$, the subdiagonal element $h_{k,k+1}$ will not necessarily be zero like in the case of the Arnoldi iteration.

2.1 Polynomial Approximation Property

Using (4) recursively we see that for $0 \leq j \leq k-1$

$$A^j Q_{k,m} = Q_{k,m}H_{k,m}^j + \sum_{i=0}^{j-1} c_i e_{k-i}^{\mathsf{T}}$$

for some vectors c_i. Multiplying this by e_1 from the right side, we see that for $0 \leq j \leq k-1$ it holds

$$A^j b = Q_{k,m}H_{k,m}^j e_1 \|b\|.$$

This results as the following lemma.

Lemma 1 *Let $A \in \mathbb{C}^{n\times n}$ and let $Q_{k,m}$, $H_{k,m}$ be the results of k steps of IOM(m) applied to A with starting vector b. Then for any polynomial p_{k-1} of degree up to $k-1$ the following equality holds:*

$$p_{k-1}(A)b = Q_{k,m}p_{k-1}(H_{k,m})e_1\|b\|.$$

This leads us to make the approximation

$$\mathrm{e}^A b \approx Q_{k,m}\mathrm{e}^{H_{k,m}}e_1\|b\|. \tag{5}$$

By Lemma 1, the error ϵ_k of this approximation is given as

$$\epsilon_k = \sum_{\ell=k}^{\infty} \frac{A^\ell}{\ell!}b - Q_{k,m}\sum_{\ell=k}^{\infty}\frac{H_{k,m}^\ell}{\ell!}e_1\|b\|. \tag{6}$$

3 Bounds for the Error

To bound the error (6), we consider bounds using the norm and the field of values of the Hessenberg matrix $H_{k,2}$.

Using the representation (see also the analysis of [9])

$$\sum_{\ell=k}^{\infty} \frac{x^\ell}{\ell!} = x^k \int_0^1 \mathrm{e}^{(1-\theta)x} \frac{\theta^{k-1}}{(k-1)!} \, \mathrm{d}\theta$$

and the bound $\|\mathrm{e}^A\| \leq \mathrm{e}^{\mu(A)}$, where $\mu(A)$ is the numerical abscissa of A, i.e., the largest eigenvalue of A^{H} (see [3, Thm. 10.11]), we get the bound

$$\|\epsilon_k\| \leq \frac{\mathrm{e}^{\mu(A)}\|A\|^k + \mathrm{e}^{\mu(H_{k,m})}\|Q_{k,m}\|\|H_{k,m}\|^k}{k!}\|b\|. \tag{7}$$

Note that in the case of the advection-diffusion operator (2), $\mu(A) \leq 0$.

In (7) the norm $\|Q_{k,m}\|$ cannot be bounded, in general. In numerical experiments $\|Q_{k,m}\|$ was found to stay of order 1 for advection-diffusion operators of the form (2) for all values of n, Pe, k and m. A discussion on the effect of the parameter m to the size of $\|Q_{k,m}\|$ can be found in [8].

In the same way as in the analysis of [9], it can be shown that

$$Q_{k,m}\mathrm{e}^{H_{k,m}}e_1\|b\| = p(A)b,$$

where p is the unique polynomial that interpolates the exponential function in the Hermite sense at the eigenvalues of $H_{k,m}$. Then, if the field of values $\mathcal{F}(H_{k,m})$ can be bounded with respect to $\mathcal{F}(A)$, the superlinear convergence of the approximation can be shown in the same way as in the proof of [2, Thm. 4.2] for the case of the restarted Krylov method.

When viewing the incomplete orthogonalization method as an *oblique projection method* [8], also the results [4, Lemma 7 and 8] can be applied.

3.1 Bounds for the Field of Values and the Norm of $H_{k,2}$

In this subsection we show how to bound the norm and the field of values of the Hessenberg matrix $H_{k,2}$, i.e., for the case of IOM(2). The field of values of a matrix $A \in \mathbb{C}^{n\times n}$ is defined as

$$\mathcal{F}(A) = \{x^*Ax \, : \, x \in \mathbb{C}^n, \, \|x\|_2 = 1\}.$$

We first give the following auxiliary lemma.

Lemma 2 *Let $A \in \mathbb{C}^{n \times n}$ be normal, and let the 0-field of values be defined as*

$$\mathscr{F}_0(A) = \{\langle x, Ay \rangle \, : \, \langle x, y \rangle = 0 \, , \, \|x\| = \|y\| = 1\}.$$

Then,

$$\mathscr{F}_0(A) = \{z \in \mathbb{C} \, : \, |z| \leq r\}, \quad \textit{where} \quad r = \frac{1}{2} \max_{\lambda_i, \lambda_j \in \sigma(A)} \left|\lambda_i - \lambda_j\right| .$$

Proof Since A is normal, it is unitary similar to a diagonal matrix with the eigenvalues of A on the diagonal, $A = U \Lambda U^*$. Let $x, y \in \mathbb{C}^{n \times n}$ such that $\langle x, y \rangle = 0 \, , \, \|x\| = \|y\| = 1$. Then $\langle x, Ay \rangle = \langle U^* x, (\Lambda - cI) U^* y \rangle$ for all $c \in \mathbb{C}$. By choosing c to be the center of the smallest disc containing $\sigma(A)$, and by using the Cauchy–Schwartz inequality, we see that

$$|\langle x, Ay \rangle| \leq \|\Lambda - cI\| \leq \frac{1}{2} \max_{\lambda_i, \lambda_j \in \sigma(A)} \left|\lambda_i - \lambda_j\right| .$$

By choosing $x = u_i / \sqrt{2} + u_j / \sqrt{2}$ and $y = u_i / \sqrt{2} - u_j / \sqrt{2}$, where u_i and u_j are eigenvectors of A corresponding to eigenvalues λ_i and λ_j, we see that

$$\langle x, Ay \rangle = \frac{1}{2} (\lambda_i - \lambda_j).$$

Thus, the inequality above is sharp. Since $\mathscr{F}_0(A)$ is a disc centered at the origin [6], the claim follows. □

Using Lemma 2, we may now obtain a bound for the field of values of $H_{k,2}$.

Theorem 1 *Let $A \in \mathbb{C}^{n \times n}$ and let $H_{k,2}$ be the Hessenberg matrix obtained after k steps of IOM(2) applied to A. Then it holds that*

$$\mathscr{F}(H_{k,2}) \subset \{ z \in \mathbb{C} \, : \, d(z, \mathscr{F}(A)) \leq \tfrac{1}{2} (\|A^H\| + \|A^S\|) \}.$$

Proof First, we extend $H_{k,2}$ to a matrix $\tilde{H} \in \mathbb{C}^{(k+2) \times (k+2)}$ by adding zeros such that

$$\tilde{H} = \begin{bmatrix} 0 & \ldots & 0 \\ \vdots & H_{k,2} & \vdots \\ 0 & \ldots & 0 \end{bmatrix}$$

and set $q_0 = q_{k+1} = 0$. It clearly holds that $\mathscr{F}(H_{k,2}) \subset \mathscr{F}(\tilde{H})$.

Let $x = \begin{bmatrix} x_0 \dots x_{k+1} \end{bmatrix}^{\mathrm{T}} \in \mathbb{C}^{k+2}$, $\|x\| = 1$. Then, by inspecting (3), we see that

$$\begin{aligned} x^* \tilde{H} x = &\frac{1}{2} \sum_{i=0}^{k} \begin{bmatrix} x_i \\ x_{i+1} \end{bmatrix}^* \begin{bmatrix} \langle Aq_i, q_i \rangle & \langle Aq_{i+1}, q_i \rangle \\ \langle Aq_i, q_{i+1} \rangle & \langle Aq_{i+1}, q_{i+1} \rangle \end{bmatrix} \begin{bmatrix} x_i \\ x_{i+1} \end{bmatrix} \\ &+ \frac{1}{2} \sum_{i=1}^{k-1} \begin{bmatrix} x_i \\ x_{i+1} \end{bmatrix}^* \begin{bmatrix} 0 & \langle Aq_{i+1}, q_i \rangle \\ \langle Aq_i, q_{i+1} \rangle & 0 \end{bmatrix} \begin{bmatrix} x_i \\ x_{i+1} \end{bmatrix}. \end{aligned} \tag{8}$$

Due to the local orthogonality of the basis vectors $\{q_i\}$ and the convexity of $\mathscr{F}(A)$, we see that the first term of (8) is in $\mathscr{F}(A)$. For the second term, we split $A = A^{\mathrm{H}} + A^{\mathrm{S}}$ and use the Lemma 2 to see that

$$\left| \begin{bmatrix} x_i \\ x_{i+1} \end{bmatrix}^* \begin{bmatrix} 0 & \langle Aq_{i+1}, q_i \rangle \\ \langle Aq_i, q_{i+1} \rangle & 0 \end{bmatrix} \begin{bmatrix} x_i \\ x_{i+1} \end{bmatrix} \right| \leq |x_i| \, |x_{i+1}| \, (\|A^{\mathrm{H}}\| + \|A^{\mathrm{S}}\|).$$

By the inequality $\sum_{i=1}^{k-1} |x_i| \, |x_{i+1}| \leq \sum_{i=1}^{k} |x_i|^2$, the claim follows. □

Using Lemma 2 we now obtain also a bound for $\|H_{k,2}\|$.

Theorem 2 *Let $A \in \mathbb{C}^{n \times n}$ and let $H_{k,2}$ be the Hessenberg matrix obtained after k steps of IOM(2) applied to A. Then it holds that*

$$\|H_{k,2}\| \leq r(A) + \tfrac{1}{2}(\|A^H\| + \|A^S\|),$$

where $r(A) = \max_{z \in \mathscr{F}(A)} |z|$.

Proof Let $x \in \mathbb{C}^n$, $\|x\| = 1$. Then for $1 < i < k$ it holds that

$$(H_{k,2}x)_i = x_i \langle Aq_i, q_i \rangle + \langle Aq_i, (x_{i-1}q_{i-1} + x_{i+1}q_{i+1}) \rangle.$$

By using the triangle inequality, splitting $A = A^{\mathrm{H}} + A^{\mathrm{S}}$, local orthogonality of the vectors $\{q_i\}$ and Lemma 2, the claim follows. □

Although we consider in the analysis of $\mathscr{F}(H_{k,m})$ and $\|H_{k,m}\|$, and also in the numerical comparisons only the case $m = 2$, we note that in numerical experiments the approximation (5) was found to improve for increasing m.

4 A Posteriori Error Estimate

An a posteriori error estimate follows from the relation (4) and can be derived in the same way as the estimate for the Arnoldi iteration, see [9, Thm. 5.1].

Theorem 3 *The error produced by the incomplete orthogonalization method of* $\mathrm{e}^{A}b$ *satisfies the expansion*

$$\mathrm{e}^{A}b - Q_{k,m}\exp(H_{k,m})e_1 = h_{k+1,k}\sum_{\ell=1}^{\infty} e_k^T\varphi_\ell(H_{k,m})e_1 A^{\ell-1}q_{k+1},$$

where $\varphi_\ell(z) = \sum_{k=0}^{\infty}\frac{z^k}{(k+\ell)!}$. In numerical experiments we estimate the error using the norm of the first term, i.e. by using the estimate

$$\|\epsilon_k\| \approx h_{k+1,k}\left|e_k^T\varphi_1(H_{k,m})e_1\right|, \tag{9}$$

which can be obtained with small computational cost by computing the exponential of

$$\tilde{H}_m = \begin{bmatrix} H_{k,m} & e_1 \\ 0 & 0 \end{bmatrix}, \quad \text{since} \quad \mathrm{e}^{\tilde{H}_m} = \begin{bmatrix} \mathrm{e}^{H_{k,m}} & \varphi_1(H_{k,m})e_1 \\ 0 & 1 \end{bmatrix}.$$

5 Numerical Examples

For the first example, we take $A = \epsilon\Delta_n + \alpha\nabla_n \in \mathbb{R}^{n\times n}$, where Δ_n and ∇_n are as in (2). The vector b is taken as a discretization of the function $u_0(x) = 16((1-x)x)^2$, $x \in [0,1]$. We set $n = 400$ and $\epsilon = 1$, and consider the cases of a weak advection and a strong advection. We approximate the product $\mathrm{e}^{hA}b$ using IOM(2) and compare it with the standard Arnoldi iteration and the restarted Krylov method with restarting interval 3. We also compute the estimate (9) for IOM. Figure 1 shows the convergence of the three methods.

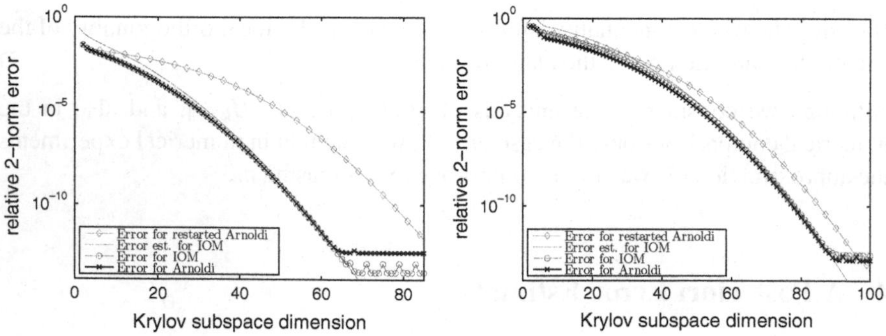

Fig. 1 *Left*: $h = 3\cdot 10^{-4}$, Pe $= 6.2\cdot 10^{-3}$. *Right*: $h = 2\cdot 10^{-4}$, Pe $= 10.0$

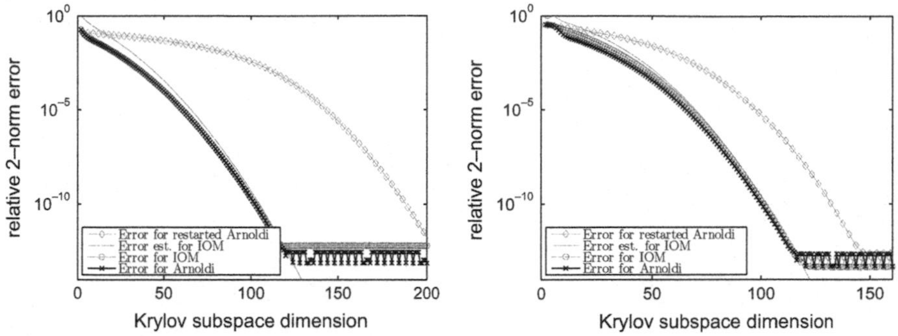

Fig. 2 *Left*: $h = 1 \cdot 10^{-3}$, Pe$= 6.2 \cdot 10^{-3}$. *Right*: $h = 6 \cdot 10^{-4}$, Pe$= 1.3 \cdot 10^{-1}$

In the second example, A, n and ϵ are as above, and b is taken randomly. We compare the methods using larger h for the cases of a weak advection and a mild advection. Figure 2 shows the convergence of the three methods.

The differences in the computational costs come mainly from the differences in the lengths of the orthogonalization recursions, the Arnoldi iteration taking $\mathcal{O}(k^2)$ and the other two methods $\mathcal{O}(k)$ inner products. In these numerical examples, the Arnoldi iteration was for $k = 50$ about 4 times slower and for $k = 100$ about 8 times slower than the other two methods.

Acknowledgements The work was mostly carried out at the University of Innsbruck, Austria, and was supported by the FWF doctoral program 'Computational Interdisciplinary Modelling' W1227.

References

1. M. Caliari, P. Kandolf, A. Ostermann, S. Rainer, Comparison of software for computing the action of the matrix exponential. BIT **54**, 113–118 (2014)
2. M. Eiermann, O.G. Ernst, A restarted Krylov subspace method for the evaluation of matrix functions. SIAM J. Numer. Anal. **44**, 2481–2504 (2005)
3. N.J. Higham, *Functions of Matrices: Theory and Computation* (SIAM, Philadelphia, 2008)
4. M. Hochbruck, C. Lubich, On Krylov subspace approximations to the matrix exponential operator. SIAM J. Numer. Anal. **34**, 1911–1925 (1997)
5. M. Hochbruck, A. Ostermann, Exponential integrators. Acta Numer. **19**, 209–286 (2010)
6. R.A. Horn, C.R. Johnson, *Topics in Matrix Analysis* (Cambridge University Press, Cambridge, 1991)
7. Y. Saad, Variations on Arnoldi's method for computing eigenelements of large unsymmetric matrices. Linear Algebra Appl. **34**, 269–295 (1980)
8. Y. Saad, The Lanczos biorthogonalization algorithm and other oblique projection methods for solving large unsymmetric systems. SIAM J. Numer. Anal. **19**, 485–506 (1982)
9. Y. Saad, Analysis of some Krylov subspace approximations to the matrix exponential operator. SIAM J. Numer. Anal. **29**, 209–228 (1992)

On the Convergence of Inexact Newton Methods

Reijer Idema, Domenico Lahaye, and Cornelis Vuik

Abstract A solid understanding of convergence behaviour is essential to the design and analysis of iterative methods. In this paper we explore the convergence of inexact iterative methods in general, and inexact Newton methods in particular. A direct relationship between the convergence of inexact Newton methods and the forcing terms is presented in both theory and numerical experiments.

1 Introduction

Inexact Newton methods [1] are Newton-Raphson methods in which the Jacobian system $-J(\mathbf{x}_i)\, s_i = \mathbf{F}(\mathbf{x}_i)$ is not solved to full accuracy. Instead, in each Newton iteration the Jacobian system is solved such that

$$\frac{\|\mathbf{r}_i\|}{\|\mathbf{F}(\mathbf{x}_i)\|} \leq \eta_i, \tag{1}$$

where $\mathbf{r}_i$ is the residual vector:

$$\mathbf{r}_i = \mathbf{F}(\mathbf{x}_i) + J(\mathbf{x}_i)\, s_i. \tag{2}$$

The values η_i are called the forcing terms. Over the years a great deal of research has gone into finding good values for η_i, such that convergence is reached with the least amount of computational work. One of the most frequently used methods to calculate η_i is that of Eisenstat and Walker [3].

In this paper, we further study the relationship between the convergence of inexact Newton methods and the choice of forcing terms. We show, both in theory and numerical experiments, that if the iterate $\mathbf{x}_i$ is close enough to the solution, in iteration i the Newton method converges in some norm with a factor $(1 + \alpha)\, \eta_i$, for arbitrarily small $\alpha > 0$.

R. Idema (✉) • D. Lahaye • C. Vuik
Delft Institute of Applied Mathematics, Delft University of Technology, Delft, Netherlands
e-mail: mail@reijeridema.nl; d.j.p.lahaye@tudelft.nl; c.vuik@tudelft.nl

A. Abdulle et al. (eds.), *Numerical Mathematics and Advanced Applications ENUMATH 2013*, Lecture Notes in Computational Science and Engineering 103,
DOI 10.1007/978-3-319-10705-9_35

2 Convergence of Inexact Iterative Methods

Assume an iterative method that, given current iterate $\mathbf{x}_i$, has some way to determine a unique new iterate $\hat{\mathbf{x}}_{i+1}$. If instead an approximation $\mathbf{x}_{i+1}$ of the exact iterate $\hat{\mathbf{x}}_{i+1}$ is used to continue the process, we speak of an inexact iterative method. Inexact Newton methods are examples of inexact iterative methods. Figure 1 illustrates a single step of an inexact iterative method.

Assume that the solution $\mathbf{x}^*$, and the distances ε^c, ε^n, and $\hat{\varepsilon}$ to the solution are unknown, but that the ratio $\frac{\delta^n}{\delta^c}$ can be controlled. In inexact Newton methods this ratio is controlled using the forcing terms. The aim is then to have an improvement of the controllable error impose a similar improvement on the distance to the solution, i.e., that for some reasonably small $\alpha > 0$

$$\frac{\varepsilon^n}{\varepsilon^c} \leq (1+\alpha)\,\frac{\delta^n}{\delta^c}. \tag{3}$$

Define $\gamma = \frac{\hat{\varepsilon}}{\delta^c} > 0$, then we can write

$$\max \frac{\varepsilon^n}{\varepsilon^c} = \frac{\delta^n + \hat{\varepsilon}}{|\delta^c - \hat{\varepsilon}|} = \frac{\delta^n + \gamma\delta^c}{|1-\gamma|\,\delta^c} = \frac{1}{|1-\gamma|}\frac{\delta^n}{\delta^c} + \frac{\gamma}{|1-\gamma|}. \tag{4}$$

Therefore, to guarantee that $\mathbf{x}_{i+1}$ is closer to the solution than $\mathbf{x}_i$, it is required that

$$\frac{1}{|1-\gamma|}\frac{\delta^n}{\delta^c} + \frac{\gamma}{|1-\gamma|} < 1 \Leftrightarrow \frac{\delta^n}{\delta^c} + \gamma < |1-\gamma| \Leftrightarrow \frac{\delta^n}{\delta^c} < |1-\gamma| - \gamma. \tag{5}$$

If $\gamma \geq 1$ this would mean that $\frac{\delta^n}{\delta^c} < -1$, which is impossible. Therefore, to guarantee a reduction of the distance to the solution, we need

$$\frac{\delta^n}{\delta^c} < 1 - 2\gamma \Leftrightarrow 2\gamma < 1 - \frac{\delta^n}{\delta^c} \Leftrightarrow \gamma < \frac{1}{2} - \frac{1}{2}\frac{\delta^n}{\delta^c}. \tag{6}$$

Equation (4) implies that as γ goes to 0, $\max \frac{\varepsilon^n}{\varepsilon^c}$ more and more resembles $\frac{\delta^n}{\delta^c}$. Figure 2 clearly shows that making $\frac{\delta^n}{\delta^c}$ too small leads to oversolving, as there is hardly any return of investment any more. Note that if the iterative method converges to the solution superlinearly, then γ goes to 0 with the same rate of convergence. Thus, for such a method $\frac{\delta^n}{\delta^c}$ can be made smaller and smaller in later iterations

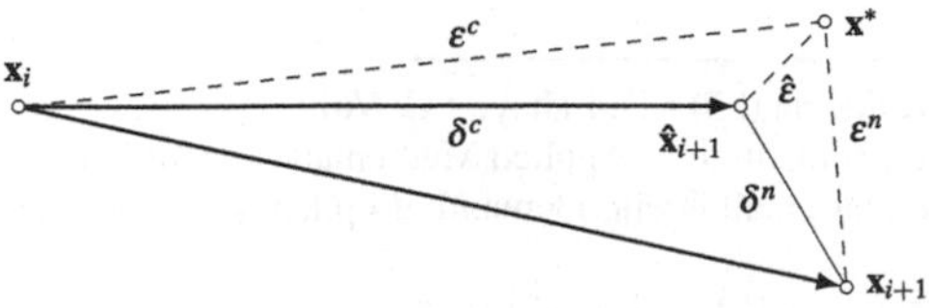

Fig. 1 Inexact iterative step

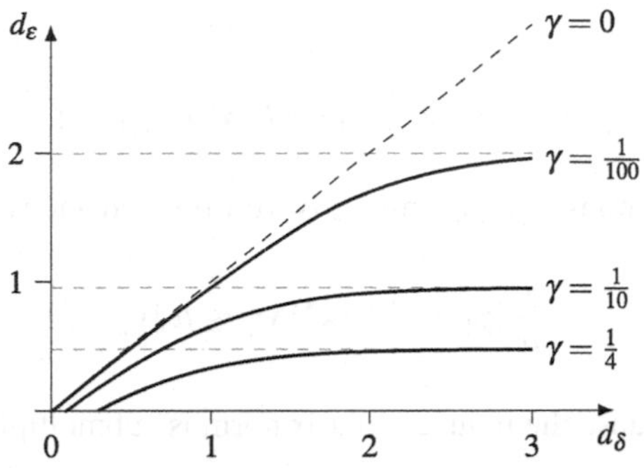

Fig. 2 Plots of equation (4) on a logarithmic scale, for several values of γ. The horizontal axis shows the number of digits improvement in the distance to the exact iterate, and the vertical axis depicts the resulting minimum digits improvement in the distance to the solution, i.e., $d_\delta = -\log \frac{\delta^n}{\delta^c}$ and $d_\varepsilon = -\log\left(\max \frac{\varepsilon^n}{\varepsilon^c}\right)$

without significant oversolving, This is in particular the case for inexact Newton methods, as convergence is quadratic once the iterate is close enough to the solution.

When using an inexact Newton method $\frac{\delta^n}{\delta^c} = \frac{\|\mathbf{x}_{i+1}-\hat{\mathbf{x}}_{i+1}\|}{\|\mathbf{x}_i-\hat{\mathbf{x}}_{i+1}\|}$ is not known, but the relative residual error $\frac{\|\mathbf{r}_i\|}{\|\mathbf{F}(\mathbf{x}_i)\|} = \frac{\|J(\mathbf{x}_i)(\mathbf{x}_{i+1}-\hat{\mathbf{x}}_{i+1})\|}{\|J(\mathbf{x}_i)(\mathbf{x}_i-\hat{\mathbf{x}}_{i+1})\|}$, which is controlled by the forcing terms η_i, can be used as a measure for it. In the next section, this idea is formalized in a theorem that is a variation on Eq. (3).

3 Convergence of Inexact Newton Methods

Consider the nonlinear system of equations $\mathbf{F}(\mathbf{x}) = \mathbf{0}$, where:

- There is a solution $\mathbf{x}^*$ such that $\mathbf{F}(\mathbf{x}^*) = \mathbf{0}$,
- The Jacobian matrix J of $\mathbf{F}$ exists in a neighbourhood of $\mathbf{x}^*$,
- $J(\mathbf{x}^*)$ is continuous and nonsingular.

In this section, theory is presented that relates the convergence of the inexact Newton method for a problem of the above form directly to the chosen forcing terms. The following theorem is a variation on both Eq. (3), and on the inexact Newton convergence theorem presented in [1, Thm. 2.3].

Theorem 1 *Let* $\eta_i \in (0, 1)$ *and choose* $\alpha > 0$ *such that* $(1+\alpha)\eta_i < 1$. *Then there exists an* $\varepsilon > 0$ *such that, if* $\|\mathbf{x}_0 - \mathbf{x}^*\| < \varepsilon$, *the sequence of inexact Newton iterates* $\mathbf{x}_i$ *converges to* $\mathbf{x}^*$, *with*

$$\|J(\mathbf{x}^*)(\mathbf{x}_{i+1} - \mathbf{x}^*)\| < (1+\alpha)\eta_i \|J(\mathbf{x}^*)(\mathbf{x}_i - \mathbf{x}^*)\|. \quad (7)$$

Proof Define

$$\mu = \max[\|J(\mathbf{x}^*)\|, \|J(\mathbf{x}^*)^{-1}\|] \geq 1. \tag{8}$$

Recall that $J(\mathbf{x}^*)$ is nonsingular. Thus μ is well-defined and we can write

$$\frac{1}{\mu}\|\mathbf{y}\| \leq \|J(\mathbf{x}^*)\mathbf{y}\| \leq \mu\|\mathbf{y}\|. \tag{9}$$

Note that $\mu \geq 1$ because the induced matrix norm is submultiplicative.
Let

$$\gamma \in \left(0, \frac{\alpha\eta_i}{5\mu}\right) \tag{10}$$

and choose $\varepsilon > 0$ sufficiently small such that if $\|\mathbf{y} - \mathbf{x}^*\| \leq \mu^2\varepsilon$ then

$$\|J(\mathbf{y}) - J(\mathbf{x}^*)\| \leq \gamma, \tag{11}$$

$$\|J(\mathbf{y})^{-1} - J(\mathbf{x}^*)^{-1}\| \leq \gamma, \tag{12}$$

$$\|\mathbf{F}(\mathbf{y}) - \mathbf{F}(\mathbf{x}^*) - J(\mathbf{x}^*)(\mathbf{y} - \mathbf{x}^*)\| \leq \gamma\|\mathbf{y} - \mathbf{x}^*\|. \tag{13}$$

That such an ε exists follows from [6, Thm. 2.3.3 & 3.1.5].
First we show that if $\|\mathbf{x}_i - \mathbf{x}^*\| < \mu^2\varepsilon$, then Eq. (7) holds. Write

$$\begin{aligned} J(\mathbf{x}^*)(\mathbf{x}_{i+1} - \mathbf{x}^*) = &\left[I + J(\mathbf{x}^*)\left(J(\mathbf{x}_i)^{-1} - J(\mathbf{x}^*)^{-1}\right)\right] \cdot [\mathbf{r}_i + \\ &(J(\mathbf{x}_i) - J(\mathbf{x}^*))(\mathbf{x}_i - \mathbf{x}^*) - (\mathbf{F}(\mathbf{x}_i) - \mathbf{F}(\mathbf{x}^*) - J(\mathbf{x}^*)(\mathbf{x}_i - \mathbf{x}^*))]. \end{aligned} \tag{14}$$

Taking norms gives

$$\begin{aligned} \|J(\mathbf{x}^*)(\mathbf{x}_{i+1} - \mathbf{x}^*)\| \leq &\left[1 + \|J(\mathbf{x}^*)\|\|J(\mathbf{x}_i)^{-1} - J(\mathbf{x}^*)^{-1}\|\right] \cdot [\|\mathbf{r}_i\| + \\ &\|J(\mathbf{x}_i) - J(\mathbf{x}^*)\|\|\mathbf{x}_i - \mathbf{x}^*\| + \|\mathbf{F}(\mathbf{x}_i) - \mathbf{F}(\mathbf{x}^*) - J(\mathbf{x}^*)(\mathbf{x}_i - \mathbf{x}^*)\|], \\ \leq &[1 + \mu\gamma] \cdot [\|\mathbf{r}_i\| + \gamma\|\mathbf{x}_i - \mathbf{x}^*\| + \gamma\|\mathbf{x}_i - \mathbf{x}^*\|], \\ \leq &[1 + \mu\gamma] \cdot [\eta_i\|\mathbf{F}(\mathbf{x}_i)\| + 2\gamma\|\mathbf{x}_i - \mathbf{x}^*\|]. \end{aligned} \tag{15}$$

Here the definitions of η_i and μ were used, together with Eqs. (11)–(13).
Further write, using that by definition $\mathbf{F}(\mathbf{x}^*) = \mathbf{0}$,

$$\mathbf{F}(\mathbf{x}_i) = [J(\mathbf{x}^*)(\mathbf{x}_i - \mathbf{x}^*)] + [\mathbf{F}(\mathbf{x}_i) - \mathbf{F}(\mathbf{x}^*) - J(\mathbf{x}^*)(\mathbf{x}_i - \mathbf{x}^*)]. \tag{16}$$

Again taking norms gives

$$\begin{aligned}\|\mathbf{F}(\mathbf{x}_i)\| &\leq \|J(\mathbf{x}^*)(\mathbf{x}_i-\mathbf{x}^*)\| + \|\mathbf{F}(\mathbf{x}_i)-\mathbf{F}(\mathbf{x}^*)-J(\mathbf{x}^*)(\mathbf{x}_i-\mathbf{x}^*)\| \\ &\leq \|J(\mathbf{x}^*)(\mathbf{x}_i-\mathbf{x}^*)\| + \gamma\|\mathbf{x}_i-\mathbf{x}^*\|. \end{aligned} \tag{17}$$

Substituting Eq. (17) into Eq. (15) then leads to

$$\begin{aligned}&\|J(\mathbf{x}^*)(\mathbf{x}_{i+1}-\mathbf{x}^*)\| \\ &\leq (1+\mu\gamma)\left[\eta_i\left(\|J(\mathbf{x}^*)(\mathbf{x}_i-\mathbf{x}^*)\| + \gamma\|\mathbf{x}_i-\mathbf{x}^*\|\right) + 2\gamma\|\mathbf{x}_i-\mathbf{x}^*\|\right] \\ &\leq (1+\mu\gamma)\left[\eta_i(1+\mu\gamma)+2\mu\gamma\right]\|J(\mathbf{x}^*)(\mathbf{x}_i-\mathbf{x}^*)\|. \end{aligned} \tag{18}$$

Here Eq. (9) was used to write $\|\mathbf{x}_i-\mathbf{x}^*\| \leq \mu\|J(\mathbf{x}^*)(\mathbf{x}_i-\mathbf{x}^*)\|$.

Finally, using that $\gamma \in \left(0, \frac{\alpha\eta_i}{5\mu}\right)$, and that both $\eta_i < 1$ and $\alpha\eta_i < 1$—the latter being a result from the requirement that $(1+\alpha)\eta_i < 1$—gives

$$\begin{aligned}(1+\mu\gamma)\left[\eta_i(1+\mu\gamma)+2\mu\gamma\right] &\leq \left(1+\frac{\alpha\eta_i}{5}\right)\left[\eta_i\left(1+\frac{\alpha\eta_i}{5}\right)+\frac{2\alpha\eta_i}{5}\right] \\ &= \left[1+\frac{2\alpha\eta_i}{5}+\frac{\alpha^2\eta_i^2}{25}+\frac{2\alpha}{5}+\frac{2\alpha^2\eta_i}{25}\right]\eta_i \\ &< \left[1+\frac{2\alpha}{5}+\frac{\alpha}{25}+\frac{2\alpha}{5}+\frac{2\alpha}{25}\right]\eta_i \\ &< (1+\alpha)\eta_i. \end{aligned} \tag{19}$$

Equation (7) follows by substituting Eq. (19) into Eq. (18).

Given that Eq. (7) holds if $\|\mathbf{x}_i-\mathbf{x}^*\| < \mu^2\varepsilon$, we now proceed to prove Theorem 1 by induction.

For the base case

$$\|\mathbf{x}_0-\mathbf{x}^*\| < \varepsilon \leq \mu^2\varepsilon. \tag{20}$$

Thus Eq. (7) holds for $i = 0$.

The induction hypothesis that Eq. (7) holds for $i = 0, \ldots, k-1$ then gives

$$\begin{aligned}\|\mathbf{x}_k-\mathbf{x}^*\| &\leq \mu\|J(\mathbf{x}^*)(\mathbf{x}_k-\mathbf{x}^*)\| \\ &< \mu(1+\alpha)^k\eta_{k-1}\cdots\eta_0\|J(\mathbf{x}^*)(\mathbf{x}_0-\mathbf{x}^*)\| \\ &< \mu\|J(\mathbf{x}^*)(\mathbf{x}_0-\mathbf{x}^*)\| \\ &\leq \mu^2\|\mathbf{x}_0-\mathbf{x}^*\| \\ &< \mu^2\varepsilon. \end{aligned} \tag{21}$$

Thus Eq. (7) also holds for $i = k$, completing the proof. □

In words, Theorem 1 states that for arbitrarily small $\alpha > 0$, and any choice of forcing terms $\eta_i \in (0, 1)$, Eq. (7) holds if the current iterate is close enough to the solution. This does not mean that for a certain iterate $\mathbf{x}_i$, one can choose α and η_i arbitrarily small and expect Eq. (7) to hold, as ε depends on the choice of α and η_i.

If we define oversolving as using forcing terms η_i that are too small for the iterate, in the context of Theorem 1, then the theorem can be characterised by saying that a convergence factor $(1 + \alpha)\,\eta_i$ is attained if η_i is chosen such that there is no oversolving. Using Eq. (10), $\eta_i > \frac{5\mu\gamma}{\alpha}$ can then be seen as a theoretical bound on the forcing terms that guards against oversolving.

A note on preconditioning is in order. Right preconditioning does not change the residual, and thus it does not change the interpretation of the forcing term η_i in Theorem 1. However, left preconditioning changes the residual such that η_i is closer to the ratio $\frac{g^n}{g^c}$. As a result, a theoretical relation closer to Eq. (3) is expected. Indeed, following the proof of Theorem 1 for a left-preconditioned problem, we get

$$\|M^{-1}J\left(\mathbf{x}^*\right)\left(\mathbf{x}_{i+1} - \mathbf{x}^*\right)\| < (1 + \alpha)\,\eta_i\,\|M^{-1}J\left(\mathbf{x}^*\right)\left(\mathbf{x}_i - \mathbf{x}^*\right)\|, \tag{22}$$

where norms of the form $\|M^{-1}J\left(\mathbf{x}^*\right)(\mathbf{y} - \mathbf{x}^*)\|$ are close to $\|\mathbf{y} - \mathbf{x}^*\|$ for a good preconditioner M.

A relation between the nonlinear residual norm $\|\mathbf{F}\left(\mathbf{x}_i\right)\|$ and the error norm $\|J\left(\mathbf{x}^*\right)\left(\mathbf{x}_i - \mathbf{x}^*\right)\|$ can also be derived within the neighbourhood of the solution where Theorem 1 holds. This shows that the nonlinear residual norm is indeed a good measure of convergence of the Newton method.

Theorem 2 *Let* $\eta_i \in (0, 1)$ *and choose* $\alpha > 0$ *such that* $(1 + \alpha)\,\eta_i < 1$. *Then there exists an* $\varepsilon > 0$ *such that, if* $\|\mathbf{x}_0 - \mathbf{x}^*\| < \varepsilon$, *then*

$$\left(1 - \frac{\alpha\eta_i}{5}\right)\|J\left(\mathbf{x}^*\right)\left(\mathbf{x}_i - \mathbf{x}^*\right)\| < \|\mathbf{F}\left(\mathbf{x}_i\right)\| < \left(1 + \frac{\alpha\eta_i}{5}\right)\|J\left(\mathbf{x}^*\right)\left(\mathbf{x}_i - \mathbf{x}^*\right)\|. \tag{23}$$

Proof Using that $\mathbf{F}\left(\mathbf{x}^*\right) = \mathbf{0}$ by definition, again write

$$\mathbf{F}\left(\mathbf{x}_i\right) = \left[J\left(\mathbf{x}^*\right)\left(\mathbf{x}_i - \mathbf{x}^*\right)\right] + \left[\mathbf{F}\left(\mathbf{x}_i\right) - \mathbf{F}\left(\mathbf{x}^*\right) - J\left(\mathbf{x}^*\right)\left(\mathbf{x}_i - \mathbf{x}^*\right)\right]. \tag{24}$$

Taking norms, and using Eqs. (13) and (9), gives

$$\begin{aligned}
\|\mathbf{F}\left(\mathbf{x}_i\right)\| &\leq \|J\left(\mathbf{x}^*\right)\left(\mathbf{x}_i - \mathbf{x}^*\right)\| + \|\mathbf{F}\left(\mathbf{x}_i\right) - \mathbf{F}\left(\mathbf{x}^*\right) - J\left(\mathbf{x}^*\right)\left(\mathbf{x}_i - \mathbf{x}^*\right)\| \\
&\leq \|J\left(\mathbf{x}^*\right)\left(\mathbf{x}_i - \mathbf{x}^*\right)\| + \gamma\|\mathbf{x}_i - \mathbf{x}^*\| \\
&\leq \|J\left(\mathbf{x}^*\right)\left(\mathbf{x}_i - \mathbf{x}^*\right)\| + \mu\gamma\|J\left(\mathbf{x}^*\right)\mathbf{x}_i - \mathbf{x}^*\| \\
&= (1 + \mu\gamma)\,\|J\left(\mathbf{x}^*\right)\left(\mathbf{x}_i - \mathbf{x}^*\right)\|.
\end{aligned} \tag{25}$$

Similarly, it holds that

$$\begin{aligned}\|\mathbf{F}(\mathbf{x}_i)\| &\geq \|J(\mathbf{x}^*)(\mathbf{x}_i - \mathbf{x}^*)\| - \|\mathbf{F}(\mathbf{x}_i) - \mathbf{F}(\mathbf{x}^*) - J(\mathbf{x}^*)(\mathbf{x}_i - \mathbf{x}^*)\| \\ &\geq \|J(\mathbf{x}^*)(\mathbf{x}_i - \mathbf{x}^*)\| - \gamma\|\mathbf{x}_i - \mathbf{x}^*\| \\ &\geq \|J(\mathbf{x}^*)(\mathbf{x}_i - \mathbf{x}^*)\| - \mu\gamma\|J(\mathbf{x}^*)\mathbf{x}_i - \mathbf{x}^*\| \\ &= (1 - \mu\gamma)\|J(\mathbf{x}^*)(\mathbf{x}_i - \mathbf{x}^*)\|. \end{aligned} \tag{26}$$

The theorem now follows from (10). □

4 Numerical Experiments

Both classical Newton-Raphson convergence theory [2, 6], and the inexact Newton convergence theory by Dembo et al. [1], require the current iterate to be close enough to the solution. What exactly is "close enough" depends on the problem, and is in practice generally too difficult to calculate. Decades of practice have shown that the theoretical convergence is reached within a few Newton steps for most problems. Thus the theory is not just of theoretical, but also of practical importance.

In this section, experiments are presented to illustrate the practical merit of Theorem 1. For simplicity, we test an idealised version of relation (7):

$$\|\mathbf{x}_{i+1} - \mathbf{x}^*\| < \eta_i \|\mathbf{x}_i - \mathbf{x}^*\|. \tag{27}$$

The experiments in this section are performed on a power flow problem [4, 5] that results in a nonlinear system of approximately 256k equations, with a Jacobian matrix that has around 2M nonzeros. The linear Jacobian systems are solved using GMRES [7], preconditioned with a high quality ILU factorisation of the Jacobian.

In Figs. 3–5, the results are shown for different amounts of GMRES iterations per Newton step. In all cases two Newton steps with just a single GMRES iteration were performed at the start but omitted from the figure.

Figure 3 has a distribution of GMRES iterations that leads to a fast solution of the problem. Practical convergence nicely follows theory. This suggests that $\mathbf{x}_2$ is close enough to the solution to use the chosen forcing terms without oversolving.

Figure 4 shows the convergence for a more exotic distribution of GMRES iterations, illustrating that practice can also follow theory for such a scenario.

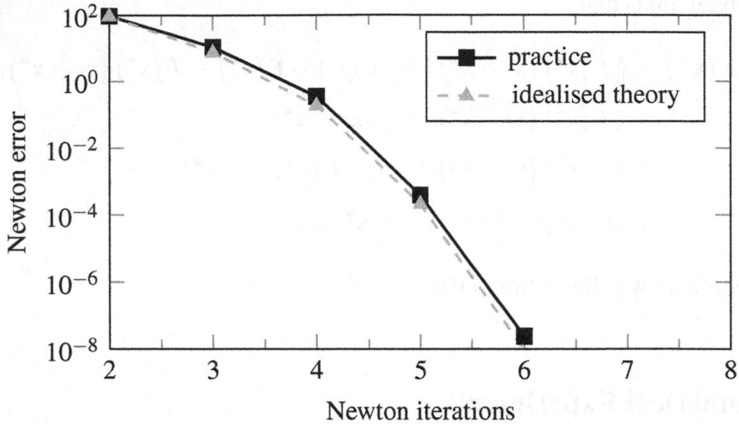

Fig. 3 GMRES iteration distribution 1, 1, 4, 6, 10, 14

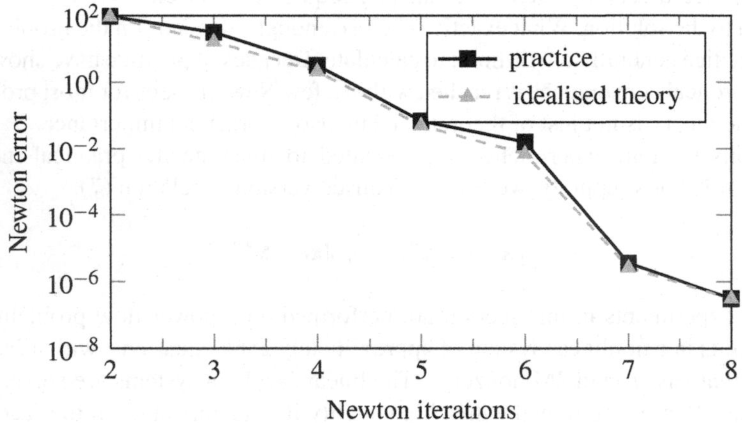

Fig. 4 GMRES iteration distribution 1, 1, 3, 4, 6, 3, 11, 3

Figure 5 illustrates the impact of oversolving. Practical convergence is nowhere near the idealised theory because extra GMRES iterations are performed that do not further improve the Newton error. In terms of Theorem 1 this means that the iterates $\mathbf{x}_i$ are not close enough to the solution to be able to take the forcing terms η_i as small as they were in this example.

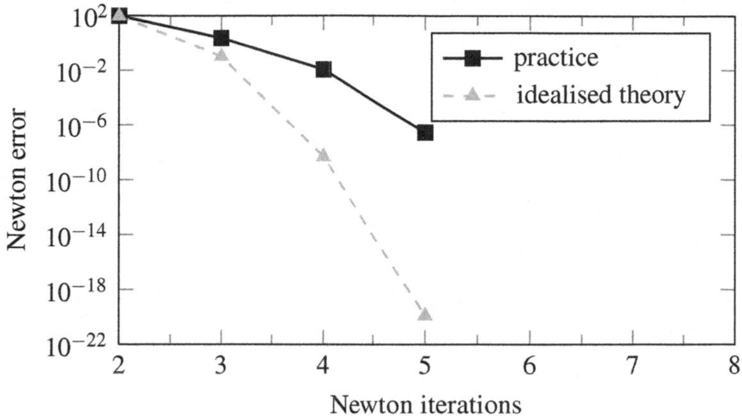

Fig. 5 GMRES iteration distribution 1, 1, 9, 19, 30

Conclusions

A proper choice of tolerances in inexact iterative methods is very important to minimize computational work. In the case of inexact Newton methods these tolerances are called the forcing terms.

In this paper we explored the relation between the choice of tolerances and the convergence of inexact iterative methods, and in particular the relation between the forcing terms and the convergence of inexact Newton methods. We proved that, under certain conditions, in each iteration an inexact Newton method converges with a factor near equal to the forcing term of that iteration, and numerical experiments were used to illustrate the results.

References

1. R.S. Dembo, S.C. Eisenstat, T. Steihaug, Inexact Newton methods. SIAM J. Numer. Anal. **19**(2), 400–408 (1982)
2. J.E. Dennis Jr., R.B. Schnabel, *Numerical Methods for Unconstrained Optimization and Nonlinear Equations* (SIAM, Philadelphia, 1996)
3. S.C. Eisenstat, H.F. Walker, Choosing the forcing terms in an inexact Newton method. SIAM J. Sci. Comput. **17**(1), 16–32 (1996)
4. R. Idema, D.J.P. Lahaye, C. Vuik, L. van der Sluis, Scalable Newton-Krylov solver for very large power flow problems. IEEE Trans. Power Syst. **27**(1), 390–396 (2012)
5. R. Idema, G. Papaefthymiou, D.J.P. Lahaye, C. Vuik, L. van der Sluis, Towards faster solution of large power flow problems. IEEE Trans. Power Syst. **28**(4), 4918–4925 (2013)
6. J.M. Ortega, W.C. Rheinboldt, Iterative solution of nonlinear equations in several variables (SIAM, Philadelphia, 2000)
7. Y. Saad, M.H. Schultz, GMRES: a generalized minimal residual algorithm for solving nonsymmetric linear systems. SIAM J. Sci. Stat. Comput. **7**, 856–869 (1986)

Part V
Multiscale Modeling and Simulation

Multiscale Adaptive Method for Stokes Flow in Heterogenenous Media

Assyr Abdulle and Ondrej Budáč

Abstract We present a multiscale micro-macro method for the Stokes problem in heterogeneous media. The macroscopic method discretizes a Darcy problem on a coarse mesh with permeability data recovered from solutions of Stokes problems around quadrature points. The accuracy of both the macro and the micro solvers is controlled by appropriately coupled a posteriori error indicators, while the total cost of the multiscale method is independent of the pore size. Two and three-dimensional numerical experiments illustrate the capabilities of the adaptive method.

1 Introduction

Fluid flow in porous media is a basic problem in science and engineering. It enters the modeling of geothermal and petroleum reservoirs, subsurface contamination, textile modeling or biomedical materials. Since the pore size is usually much smaller than the considered porous material, global discretization that resolves the pore geometry and standard single-scale techniques such as finite element method (FEM) are extremely expensive.

Averaging techniques such as homogenization of Stokes flow in porous media are thus required in many applications. The homogenization method has been studied by various authors in the past several decades assuming periodic porosity [7, 15, 19, 21]. The effective solution is shown to be given by a Darcy equation where the permeability tensor can be computed from so-called micro problems.

Various multiscale methods have been recently proposed for the numerical approximation of Stokes (or Navier-Stokes) equations in porous media that rely on a Darcy macro problem, recovering the effective permeability from local pore geometries by numerically solving appropriate micro problems. We mention a hierarchical multiscale FEM derived in [10], a two-scale finite element method

A. Abdulle (✉) • O. Budáč
ANMC, Section de Mathématiques, École Polytechnique Fédérale de Lausanne, Lausanne, Switzerland
e-mail: assyr.abdulle@epfl.ch; ondrej.budac@epfl.ch

A. Abdulle et al. (eds.), *Numerical Mathematics and Advanced Applications ENUMATH 2013*, Lecture Notes in Computational Science and Engineering 103,
DOI 10.1007/978-3-319-10705-9_36

proposed in [20], and a control volume heterogeneous multiscale method described in [8].

Most of the aforementioned works discuss a priori convergence rates and assume regularity of the micro problems that might not always hold. Indeed, complicated pore structures in typical applications and non-convexity of microscopic fluid domains result in sub-optimal a priori convergence rates. In this contribution, we give a concise description and illustrate numerically a new adaptive numerical homogenization methods for Stokes flow proposed in [3]. The method is built using the framework of the finite element heterogeneous multiscale method (FE-HMM) [1, 4, 12]. Adaptive FE-HMMs for elliptic problems have been studied in [2, 5, 17]. Our new method relies on adaptive mesh refinement on macro and micro problems and on rigorous residual-based a posteriori error estimates derived in [3]. One challenge is to adequately couple macro and micro error indicators as to achieve optimal accuracy with minimal computational cost.

The paper is organized as follows. We first review the model problem in Sect. 2. We then describe the FE-HMM for Stokes flow in Sect. 3 and the adaptive method in Sect. 4. In Sect. 5 we provide 2D and 3D numerical experiments to test the capabilities of the adaptive method.

2 Model Problem

Let $\Omega \subset \mathbb{R}^d$ be a bounded connected domain, where $d \in \mathbb{N}$ and $d > 1$. Denote by Y the d-dimensional unit cube $(-1/2, 1/2)^d$. For any $x \in \Omega$ let $Y_S^x \subset \overline{Y}$ and denote $Y_F^x = Y \backslash Y_S^x$. The sets Y_F^x and Y_S^x represent the local fluid and solid geometry, respectively. Given a pore size $\varepsilon > 0$, we define the locally periodic porous medium by

$$\Omega_\varepsilon = \Omega \backslash \Big(\varepsilon \bigcup_{m \in \mathbb{Z}^d} (1/2 + m + Y_S^{\varepsilon(1/2+m)}) \Big)$$

and consider the following Stokes problem

$$\begin{aligned} -\Delta \mathbf{u}^\varepsilon + \nabla p^\varepsilon &= \mathbf{f} \quad \text{in } \Omega_\varepsilon, \\ \operatorname{div} \mathbf{u}^\varepsilon &= 0 \quad \text{in } \Omega_\varepsilon, \\ \mathbf{u}^\varepsilon &= 0 \quad \text{on } \partial\Omega_\varepsilon, \end{aligned}$$

for the velocity field $\mathbf{u}^\varepsilon$ and pressure p^ε, where $\mathbf{f}$ is a given force field.

In case of periodic porous media (Y_S^x does not depend on x), the asymptotic behavior of p^ε, $\mathbf{u}^\varepsilon$ as $\varepsilon \to 0$ is studied in [7, 21]. An extension of p^ε and $\mathbf{u}^\varepsilon$ from Ω_ε to Ω is constructed, such that (while keeping the notation for the extensions) $\|p^\varepsilon - p^0\|_{L^2(\Omega)/\mathbb{R}} \to 0$ and $\mathbf{u}^\varepsilon/\varepsilon^2 \to \mathbf{u}^0$ weakly in $L^2(\Omega)$ for $\varepsilon \to 0$, where p^0 and

$\mathbf{u}^0$ are given as follows. Find p^0 such that

$$\begin{aligned} \nabla a^0(\mathbf{f} - \nabla p^0) &= 0 \quad \text{in } \Omega, \\ a^0(\mathbf{f} - \nabla p^0) \cdot \mathbf{n} &= 0 \quad \text{on } \partial\Omega, \end{aligned} \tag{1}$$

where the homogenized tensor $a^0(x)$ is given by the micro problems: Solve

$$\begin{aligned} -\Delta \mathbf{u}^{i,x} + \nabla p^{i,x} &= \mathbf{e}^i \quad \text{in } Y_{\mathrm{F}}^x, & \mathbf{u}^{i,x} &= 0 \quad \text{on } \partial Y_{\mathrm{S}}^x - \partial Y, \\ \operatorname{div} \mathbf{u}^{i,x} &= 0 \quad \text{in } Y_{\mathrm{F}}^x, & \mathbf{u}^{i,x} \text{ and } p^{i,x} & \quad \text{are } Y\text{-periodic}, \end{aligned} \tag{2}$$

for $i \in \{1, \ldots, d\}$, where $\mathbf{e}^i$ is the i-th canonical basis vector in $\mathbb{R}^d$, and define

$$a^0(x) = \int_{Y_{\mathrm{F}}^x} [\mathbf{u}^{1,x}, \mathbf{u}^{2,x}, \ldots, \mathbf{u}^{d,x}] \, \mathrm{d}y. \tag{3}$$

The effective velocity is then defined as $\mathbf{u}^0 = a^0(\mathbf{f} - \nabla p^0)$.

Well-posedness of the model problem (1)–(3) depends on the geometric properties of the micro domains Y_{F}^x and is examined in [3].

3 FE-HMM for Flow in Porous Media

We apply the FE-HMM framework [1, 13] to the problem (1)–(3) following [3]. Let Ω and Ω_ε be open, connected, bounded, and polygonal subsets of $\mathbb{R}^d$ with $\Omega_\varepsilon \subset \Omega$. Let $\mathcal{T}_H$ be a family of conformal, shape-regular triangulations of Ω parametrized by the mesh size $H = \max_{K \in \mathcal{T}_H} H_K$, where $H_K = \operatorname{diam}(K)$. Define the macro FE space

$$S^l(\Omega, \mathcal{T}_H) = \{q^H \in H^1(\Omega) : q^H|_K \in \mathcal{P}^l(K), \ \forall K \in \mathcal{T}_H\},$$

where $\mathcal{P}^l(K)$ is the space of polynomials on K of degree $l \in \mathbb{N}$.

For each element $K \in \mathcal{T}_H$, consider a quadrature formula (QF) with interior quadrature nodes $\{x_{K_j}\}_{j=1}^J$ and positive weights $\{\omega_{K_j}\}_{j=1}^J$, where $J \in \mathbb{N}$. To guarantee the optimal order of accuracy (see [11, Chap. 4.1]), we assume that the QF is exact for polynomials up to order $\max(2l - 2, l)$. Define $Q^K = \{x_{K_j}\}_{j=1}^J$ and $Q^H = \cup_{K \in \mathcal{T}_H} Q^K$.

Let $\delta \geq \varepsilon$. For each $x \in Q^H$ we define the local geometry snapshot by

$$Y_{\mathrm{S}}^{x,\delta} = ((\mathbb{R}^d - \Omega_\varepsilon) \cap (x + \delta\overline{Y}) - x)/\varepsilon, \qquad Y_{\mathrm{F}}^{x,\delta} = (\delta/\varepsilon)Y - Y_{\mathrm{S}}^{x,\delta}.$$

Let $\mathcal{T}_h^x$ be a family of conformal, shape-regular triangulations of $Y_{\mathrm{F}}^{x,\delta}$ parametrized by the mesh size $h = \max_{T \in \mathcal{T}_h^x} h_T$, where $h_T = \operatorname{diam}(T)$. We consider the

Taylor-Hood $\mathbb{P}_{k+1}/\mathbb{P}_k$ FE space with $k \in \mathbb{N}$ and periodic coupling for the micro problems (for other micro FE spaces or couplings see [3]) and define

$$M(Y_{\rm F}^{x,\delta}, \mathcal{T}_h^x) = H^1_{\rm per}(Y_{\rm F}^{x,\delta})^d \cap S^k(Y_{\rm F}^{x,\delta}, \mathcal{T}_h^x),$$
$$X(Y_{\rm F}^{x,\delta}, \mathcal{T}_h^x) = \{\mathbf{v} \in H^1_{\rm per}(Y_{\rm F}^{x,\delta})^d : \mathbf{v} = 0 \text{ on } \partial Y_{\rm S}^{x,\delta}\} \cap S^{k+1}(Y_{\rm F}^{x,\delta}, \mathcal{T}_h^x)^d.$$

The FE-HMM for Stokes flow reads as follows: find $p^H \in S^l(\Omega, \mathcal{T}_H)/\mathbb{R}$ such that

$$B_H(p^H, q^H) = L_H(q^H) \qquad \forall q^H \in S^l(\Omega, \mathcal{T}_H)/\mathbb{R}, \tag{4}$$

where

$$B_H(p^H, q^H) = \sum_{K\in\mathcal{T}_H} \sum_{j=1}^{J} \omega_{K_j} a^h(x_{K_j}) \nabla p^H(x_{K_j}) \cdot \nabla q^H(x_{K_j}),$$

$$L_H(q^H) = \sum_{K\in\mathcal{T}_H} \sum_{j=1}^{J} \omega_{K_j} a^h(x_{K_j}) \mathbf{f}^H(x_{K_j}) \cdot \nabla q^H(x_{K_j}).$$

We observe that the precise knowledge of $\varepsilon > 0$ is not necessary to apply the above method. Here, $\mathbf{f}^H$ is a suitable interpolation of the force field $\mathbf{f} \in L^2(\Omega)^d$ and $a^h(x_{K_j})$ is a numerical approximation of the tensor $a^0(x_{K_j})$ computed by the micro Stokes problems: For any $i \in \{1, \ldots, d\}$ and $x \in Q^H$ find $\mathbf{u}^{i,x,h} \in X(Y_{\rm F}^{x,\delta}, \mathcal{T}_h^x)$ and $p^{i,x,h} \in M(Y_{\rm F}^{x,\delta}, \mathcal{T}_h^x)/\mathbb{R}$ such that[1]

$$\begin{aligned} (\nabla \mathbf{u}^{i,x,h}, \nabla \mathbf{v}) - (\nabla \mathbf{v}, p^{i,x,h}) &= (\mathbf{e}^i, \mathbf{v}) \quad && \forall \mathbf{v} \in X(Y_{\rm F}^{x,\delta}, \mathcal{T}_h^x) \\ (\nabla \mathbf{u}^{i,x,h}, q) &= 0 && \forall q \in M(Y_{\rm F}^{x,\delta}, \mathcal{T}_h^x)/\mathbb{R} \end{aligned} \tag{5}$$

and set

$$a^h(x) = \frac{\varepsilon^d}{\delta^d} \int_{Y_{\rm F}^{x,\delta}} [\mathbf{u}^{1,x,h}, \ldots, \mathbf{u}^{d,x,h}]\, dy.$$

A velocity field approximation can be obtained by interpolation from quadrature points. If the QF has the minimal number of nodes ($J = \binom{l+d-1}{d}$), we know [6] that any tensor $A(x) : Q^H \to \mathbb{R}^d$ uniquely defines an operator $\Pi_A : S_{\rm D}^{l-1}(\Omega, \mathcal{T}_H)^d \to S_{\rm D}^{l-1}(\Omega, \mathcal{T}_H)^d$ such that

$$\Pi_A(\mathbf{v})(x) = A(x)\mathbf{v}(x), \quad \forall x \in Q^H,$$

[1] We use $(\cdot, \cdot)$ for the standard scalar product in $L^2(Y_{\rm F}^{x,\delta})^m$ for any $m \in \mathbb{N}$.

where the space $S_{\mathrm{D}}^{l-1}(\Omega, \mathscr{T}_H)$ is a space of functions $q^H : \Omega \to \mathbb{R}$ such that $q^H \in \mathscr{P}^{l-1}(K)$ for every $K \in \mathscr{T}_H$. We define the reconstructed velocity field by $\mathbf{u}^H = \Pi_{a^h}(\mathbf{f}^H - \nabla p^H)$.

Assuming sufficient regularity, a priori estimates derived in [3] yield

$$|p^0 - p^H|_{H^1(\Omega)} \leq CH^l + r_{\mathrm{mic}} + r_{\mathrm{mod}}, \tag{6}$$

where $|\cdot|_{H^1(\Omega)}$ denotes the standard H^1 seminorm, r_{mod} is a modeling error (vanishing if $Y_{\mathrm{F}}^{x,\delta} = Y_{\mathrm{F}}^x$ is used), and r_{mic} is a micro error. In practice, we expect $r_{\mathrm{mic}} \leq Ch^\theta$ with $\theta < 2$ instead of the ideal $\theta = k + 2$ (see Sect. 5).

4 Adaptive Method

Suboptimal a priori error estimates (for non-convex microscopic fluid domain) suggest to use an adaptive method. In [3], the residual-based FE-HMM error analysis developed in [2, 6] was coupled with an a posteriori error bound for the micro Stokes flow (5) based on [22]. This result is summarized in Theorem 1 and uses the following. Define the *macro residual* η_K by

$$\begin{aligned}\eta_K^2 =& H_K^2 \|\nabla \cdot \Pi_{a^h}(\mathbf{f}^H - \nabla p^H)\|_{L^2(K)}^2 \\ &+ \textstyle\sum_{e \in \partial K} \frac{1}{2} H_e \|[\Pi_{a^h}(\mathbf{f}^H - \nabla p^H) \cdot \mathbf{n}]_e\|_{L^2(e)}^2,\end{aligned}$$

the *data approximation error* $\xi_{\mathrm{data},K}$ by

$$\xi_{\mathrm{data},K}^2 = \|a^0(\mathbf{f} - \nabla p^H) - \Pi_{a}(\mathbf{f}^H - \nabla p^H)\|_{L^2(K)}^2,$$

where $\overline{a}(x) = \lim_{h \to 0} a^h(x)$ and the *micro residual* $\eta_{\mathrm{mic},K}$ by

$$\eta_{\mathrm{mic},K}^2 = \|\mathbf{f}^H - \nabla p^H\|_{L^2(K)}^2 \max_{x \in Q^K} \sum_{i=1}^d \eta_{\mathrm{stokes},x,i}^2,$$

$$\begin{aligned}\eta_{\mathrm{stokes},x,i}^2 = \sum_{T \in \mathscr{T}_h^x} \Bigg(& \sum_{e \in \partial T \setminus \partial Y_{\mathrm{F}}^{x,\delta}} \frac{h_e}{2} \left\| \left[\frac{\partial \mathbf{u}^{i,x,h}}{\partial \mathbf{n}} - p^{i,x,h}\mathbf{n} \right]_e \right\|_{L^2(e)}^2 \\ &+ h_T^2 \|\Delta \mathbf{u}^{i,x,h} - \nabla p^{i,x,h} + \mathbf{e}^i\|_{L^2(T)}^2 + \|\nabla \cdot \mathbf{u}^{i,x,h}\|_{L^2(T)}^2 \Bigg).\end{aligned}$$

Theorem 1 *Assume that $a^0(x)\xi \cdot \xi \geq \lambda|\xi|^2$ and $|a^0(x)\xi| \leq \Lambda|\xi|$ for each $\xi \in \mathbb{R}^d$ and a.e. $x \in \Omega$. Then, there exists a constant C depending only on Ω, λ, Λ, on the shape-regularity of $\mathscr{T}_H$ and $\mathscr{T}_h^x$, and on the Poincaré-Friedrichs and inf-sup* *constants related to* (5), *such that*

$$|p^0 - p^H|^2_{H^1(\Omega)} \leq C \sum_{K\in\mathscr{T}_H} (\eta_K^2 + \eta_{mic,K}^2 + \xi_{data,K}^2).$$

Moreover, if $Y_F^{x,\delta} = Y_F^x$, *then* $\xi_{data,K} = 0$.

Theorem 1 gives a foundation for an adaptive refinement algorithm on both macro and micro problems using the indicators η_K and $\eta_{\mathrm{mic},K}$. The usual refinement cycle solve → estimate → mark → refine is implemented on both scales.

The stopping criterion in the adaptive solution of the micro problems is

$$\eta^2_{\mathrm{stokes},x,i} \leq \mu d^{-1}\eta_K^2 \|\mathbf{f}^H - \nabla p^H\|^{-2}_{L^2(K)} \qquad \forall K \in \mathscr{T}_H, \tag{7}$$

where $\mu > 0$ is a problem dependent constant and can be calibrated as described in [3]. The inequality (7) implies $\eta^2_{\mathrm{mic},K} \leq \mu\eta_K^2$, i.e., the micro error is dominated by the macro error.

4.1 Algorithm

Assume that the user provides Ω, Ω_ε, and δ. Then repeat:

(a) Solve. For each quadrature point $x \in Q^H$ solve the micro problems (5) adaptively using the stopping criteria (7).[2] Then, find p^H by solving (4).
(b) Estimate. Compute η_K and $\eta_{\mathrm{mic},K}$. If (7) is not true, go to (a).
(c) Mark. Using the indicator η_K and the Döfler's bulk-chasing marking strategy E, mark a subset of elements of $\mathscr{T}_H$.
(d) Refine. Refine the marked elements while maintaining conformity [9].

5 Numerical Experiments

In this section we test our adaptive algorithm by presenting two numerical experiments. The implementation is done in Matlab and makes use of gmsh [14]. Sparse saddle point linear systems arising from the micro problems were solved using the Matlab's `mldivide` in two dimensions (2D) and an Uzawa method [18] with algebraic multigrid preconditioning by AGMG [16] in three dimensions (3D).

[2]Since the right hand side of (7) is not known beforehand we use an approximation from the previous solution.

In both experiments we took $\mathbb{P}_2/\mathbb{P}_1$ Taylor-Hood micro FE and $\mathbb{P}_1$ macro FE. We set $\delta = \varepsilon$ (eliminating the modeling error) for the micro domains $Y_{\mathrm{F}}^{x,\delta}$. Variation of Y_{F}^{x} for both examples is depicted in Fig. 1.

5.1 2D Experiment

Let $\Omega = ((0,2) \times (0,3)) \setminus ([1,2] \times [1,2])$ with periodic boundary conditions between the edges $(0,2) \times \{0\}$ and $(0,2) \times \{3\}$ and let $\mathbf{f} = \mathbf{f}^H \equiv (0,-1)$. Setting $\delta = \varepsilon = 10^{-4}$, we performed the adaptive FE-HMM method. Convergence rates and examples of solutions and meshes are displayed in Fig. 2. The global error estimator $\eta_\Omega = (\sum_{K \in \mathscr{T}_H} \eta_K^2)^{-1/2}$ and the error $|p^H - p^0|_{H^1(\Omega)}$ are both following

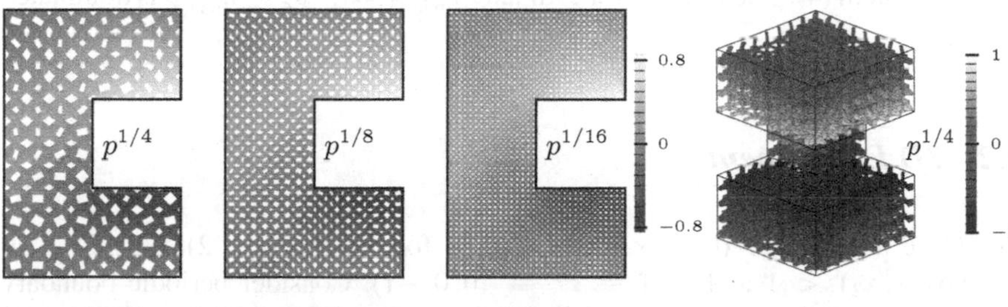

Fig. 1 Plots of p^ε for the 2D (*left*) and 3D (*right*) experiment

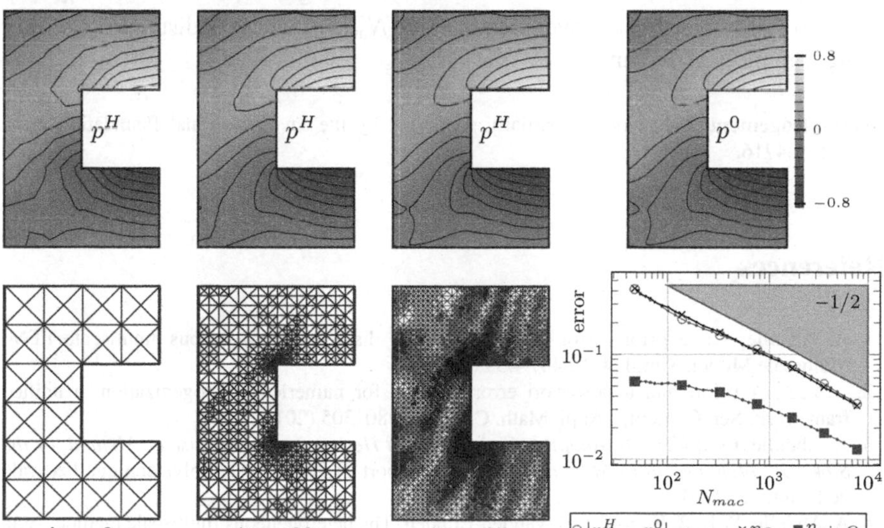

Fig. 2 FE-HMM in 2D: p^H in different stages of refinement (*upper left*); corresponding meshes (*lower left*); p^0 (*upper right*); error and indicators (*lower right*)

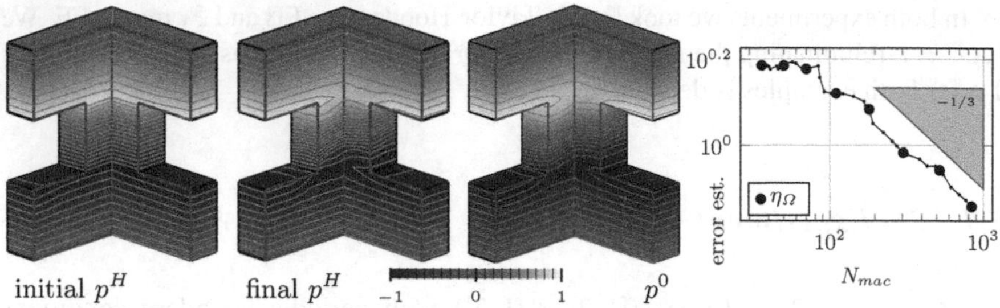

Fig. 3 FE-HMM in 3D: initial and final solution p^H (*left*) and p^0 (*middle*) displayed on a cut domain $\Omega \backslash ((0.5, 2) \times (0.5, 2) \times (0, 3))$; the error indicator (*right*)

the expected rate $O(N_{\mathrm{mac}}^{-1/2})$, where N_{mac} is the number of degrees of freedom of the macro problem (4). The micro error estimator $\eta^2_{\mathrm{mic},\Omega} = \sum_{K \in \mathscr{T}_H} \eta^2_{\mathrm{mic},K}$ is dominated by η^2_Ω.

5.2 3D Experiment

Let Ω be a subset of $(0, 2) \times (0, 2) \times (0, 3)$ for which $(x_3 - 2)(x_3 - 1) > 0$ or $\max(x_1, x_2) < 1$ and let $\mathbf{f} \equiv \mathbf{f}^H \equiv (0, 0, -1)$. Consider periodic boundary conditions on Ω that connect the faces $(0, 2) \times (0, 2) \times \{0\}$ and $(0, 2) \times (0, 2) \times \{3\}$. The adaptive FE-HMM with $\delta = \varepsilon = 10^{-2}$ yields a global error estimate η_Ω that seems to follow the right convergence rate $O(N_{\mathrm{mac}}^{-1/3})$, which is displayed in Fig. 3 along with plots of p^H and p^0.

Acknowledgement This work is partially supported by the Swiss National Foundation grant 200021_134716.

References

1. A. Abdulle, On a priori error analysis of fully discrete heterogeneous multiscale FEM. Multiscale Model. Simul. **4**(2), 447–459 (2005)
2. ———, A priori and a posteriori error analysis for numerical homogenization: a unified framework. Ser. Contemp. Appl. Math. CAM **16**, 280–305 (2011)
3. A. Abdulle, O. Budáč, *An Adaptive Finite Element Heterogeneous Multiscale Method for the Stokes Problem in Porous Media*. MATHICSE Report 41.2013, École Polytechnique Fédérale de Lausanne, 2013
4. A. Abdulle, W.E, B. Engquist, E. Vanden-Eijnden, The heterogeneous multiscale method. Acta Numer. **21**, 1–87 (2012)
5. A. Abdulle, A. Nonnenmacher, Adaptive finite element heterogeneous multiscale method for homogenization problems. Comput. Methods Appl. Mech. Eng. **200**(37–40), 2710–2726 (2011)

6. ———, A posteriori error estimates in quantities of interest for the finite element heterogeneous multiscale method. Numer. Methods Partial Differ. Equ. **29**(5), 1629–1656 (2013)
7. G. Allaire, Homogenization of the Stokes flow in a connected porous medium. Asymptot. Anal. **2**(3), 203–222 (1989)
8. S. Alyaev, E. Keilegavlen, J.M. Nordbotten, *Analysis of Control Volume Heterogeneous Multiscale Methods for Single Phase Flow in Porous Media*. Multiscale Model. Simul. **12**(1), 335–363 (2014)
9. D.N. Arnold, A. Mukherjee, L. Pouly, Locally adapted tetrahedral meshes using bisection. SIAM J. Sci. Comput. **22**(2), 431–448 (2000)
10. D.L. Brown, Y. Efendiev, V.H. Hoang, An efficient hierarchical multiscale finite element method for Stokes equations in slowly varying media. Multiscale Model. Simul. **11**(1), 30–58 (2013)
11. P.G. Ciarlet, *The Finite Element Method for Elliptic Problems*. Studies in Mathematics and Its Applications, vol. 4 (North-Holland, Amsterdam/New York, 1978)
12. W. E, B. Engquist, The heterogeneous multiscale methods. Commun. Math. Sci. **1**(1), 87–132 (2003)
13. W. E, B. Engquist, X. Li, W. Ren, E. Vanden-Eijnden, Heterogeneous multiscale methods: a review. Commun. Comput. Phys. **2**(3), 367–450 (2007)
14. C. Geuzaine, J.-F. Remacle, Gmsh: a three-dimensional finite element mesh generator with built-in pre- and post-processing facilities. Int. J. Numer. Methods Eng. **79**(11), 1309–1331 (2009)
15. E. Marušić-Paloka, A. Mikelić, An error estimate for correctors in the homogenization of the Stokes and Navier-Stokes equations in a porous medium. Boll. Unione Mat. Ital. **10**(3), 661–671 (1996)
16. A. Nonnenmacher, Adaptive finite element methods for multiscale partial differential equations, Ph.D. thesis, École Polytechnique Fédérale de Lausanne, 2011
17. M. Ohlberger, A posteriori error estimates for the heterogeneous multiscale finite element method for elliptic homogenization problems. Multiscale Model. Simul. **4**(1), 88–114 (2005)
18. J. Peters, V. Reichelt, A. Reusken, Fast iterative solvers for discrete Stokes equations. SIAM J. Sci. Comput. **27**(2), 646–666 (2005)
19. E. Sánchez-Palencia, *Non-homogeneous Media and Vibration Theory*. Lecture Notes in Physics, vol. 127 (Springer, Berlin/New York, 1980)
20. C. Sandström, F. Larsson, K. Runesson, H. Johansson, A two-scale finite element formulation of Stokes flow in porous media. Comput. Methods Appl. Mech. Eng. **261**–**262**, 96–104 (2013)
21. L. Tartar, *Incompressible Fluid Flow in a Porous Medium—Convergence of the Homogenization Process*, Appendix of [19], pp. 368–377 (1979)
22. R. Verfürth, A posteriori error estimators for the Stokes equations. Numer. Math. **55**(3), 309–325 (1989)

Homogenization of the One-Dimensional Wave Equation

Thi Trang Nguyen, Michel Lenczner, and Matthieu Brassart

Abstract We present a method for two-scale model derivation of the periodic homogenization of the one-dimensional wave equation in a bounded domain. It allows for analyzing the oscillations occurring on both microscopic and macroscopic scales. The novelty reported here is on the asymptotic behavior of high frequency waves and especially on the boundary conditions of the homogenized equation. Numerical simulations are reported.

1 Introduction

The paper is devoted to the periodic homogenization of the wave equation in a one-dimensional open bounded domain where the time-independent coefficients are $\varepsilon-$periodic with small period $\varepsilon > 0$. Corrector results for the low frequency waves have been published in [2, 7]. These works were not taking into account fast time oscillations, so the models reflect only a part of the physical solution. In [3], an homogenized model has been developed to cover the time and space oscillations occurring both at low and high frequencies. It is comprised with a second order microscopic equation with quasi-periodic boundary conditions but also with a first order macroscopic equation which boundary condition was missing. Therefore, establishing the boundary conditions of the homogenized model is critical and is the goal of the present work. A generalization of the wave equation posed in $\mathbb{R}^n$ has also been considered in [4] but taking into account only ε-periodic oscillations in the space variables resulting in periodic conditions in the microscopic problem. Periodic homogenization of the wave equation have been derived for other asymptotic regime, for instance for long time in [5, 6, 8, 10].

T.T. Nguyen • M. Lenczner (✉)
FEMTO-ST, 26 Chemin de l'Epitaphe, 25000 Besançon, France
e-mail: thitrang.nguyen@femto-st.fr; michel.lenczner@utbm.fr

M. Brassart
Laboratoire de Mathématiques de Besançon, 16 Route de Gray, 25030 Besançon, France
e-mail: matthieu.brassart@univ-fcomte.fr

A. Abdulle et al. (eds.), *Numerical Mathematics and Advanced Applications ENUMATH 2013*, Lecture Notes in Computational Science and Engineering 103,
DOI 10.1007/978-3-319-10705-9_37

To this end, the wave equation is written under the form of a first order formulation and the modulated two-scale transform W_k^ε is applied to the solution U^ε as in [3]. For $n \in \mathbb{N}^*$ and $k \in \mathbb{R}$, the nth eigenvalue λ_n^k of the Bloch wave problem with k-quasi-periodic boundary conditions satisfies $\lambda_n^k = \lambda_n^{-k}$, in addition $\lambda_m^k = \lambda_n^k$ for $k \in \mathbb{Z}/2$, so the corresponding waves are oscillating with the same frequency. The homogenized model is thus derived for pairs of fibers $\{-k, k\}$ if $k \neq 0$ and for fiber $\{0\}$ otherwise which allows to derive the expected boundary conditions. The weak limit of $\sum_{\sigma \in I^k} W_\sigma^\varepsilon U^\varepsilon$ includes low and high frequency waves, the former being solution of the homogenized model derived in [2, 7] and the latter are associated to Bloch wave expansions. Numerical results comparing solutions of the wave equation with solution of the two-scale model for fixed ε and k are reported in the last section.

2 The Physical Problem and Elementary Properties

The physical problem We consider $I = (0, T) \subset \mathbb{R}^+$ a finite time interval and $\Omega = (0, \alpha) \subset \mathbb{R}^+$ a space interval, which boundary is denoted by $\partial\Omega$. Here, as usual $\varepsilon > 0$ denotes a small parameter intended to go to zero. Two functions $(a^\varepsilon, \rho^\varepsilon)$ are assumed to obey a prescribed profile $a^\varepsilon := a\left(\frac{x}{\varepsilon}\right)$ and $\rho^\varepsilon := \rho\left(\frac{x}{\varepsilon}\right)$ where $\rho \in L^\infty(\mathbb{R}), a \in W^{1,\infty}(\mathbb{R})$ are both Y−periodic where $Y = (0, 1)$. Moreover, they are required to satisfy the standard uniform positivity and ellipticity conditions, $0 < \rho^0 \leq \rho \leq \rho^1$ and $0 < a^0 \leq a \leq a^1$, for some given strictly positive numbers ρ^0, ρ^1, a^0 and a^1. We consider $u^\varepsilon(t, x)$ solution to the wave equation with the source term $f^\varepsilon \in L^2(I \times \Omega)$, initial conditions $u_0^\varepsilon \in H^1(\Omega), v_0^\varepsilon \in L^2(\Omega)$ and homogeneous Dirichlet boundary conditions,

$$\begin{aligned} &\rho^\varepsilon \partial_{tt} u^\varepsilon - \partial_x\left(a^\varepsilon \partial_x u^\varepsilon\right) = f^\varepsilon \text{ in } I \times \Omega, \\ &u^\varepsilon(t = 0, .) = u_0^\varepsilon \text{ and } \partial_t u^\varepsilon(t = 0, .) = v_0^\varepsilon \text{ in } \Omega, \\ &u^\varepsilon = 0 \text{ on } I \times \partial\Omega. \end{aligned} \tag{1}$$

By setting: $U^\varepsilon := (\sqrt{a^\varepsilon}\partial_x u^\varepsilon, \sqrt{\rho^\varepsilon}\partial_t u^\varepsilon)$, $A^\varepsilon = \begin{pmatrix} 0 & \sqrt{a^\varepsilon}\partial_x\left(\frac{1}{\sqrt{\rho^\varepsilon}}.\right) \\ \frac{1}{\sqrt{\rho^\varepsilon}}\partial_x\left(\sqrt{a^\varepsilon}.\right) & 0 \end{pmatrix}$, $U_0^\varepsilon := (\sqrt{a^\varepsilon}\partial_x u_0^\varepsilon, \sqrt{\rho^\varepsilon} v_0^\varepsilon)$ and $F^\varepsilon := (0, f^\varepsilon/\sqrt{\rho^\varepsilon})$, we reformulate the wave equation (1) as an equivalent system: $(\partial_t - A^\varepsilon) U^\varepsilon = F^\varepsilon$ in $I \times \Omega, U^\varepsilon(t = 0) = U_0^\varepsilon$ in Ω and $U_2^\varepsilon = 0$ on $I \times \partial\Omega$ where U_2^ε is the second component of U^ε. From now on, this system will be referred to as the physical problem and taken in the distributional sense,

$$\int_{I\times\Omega} F^\varepsilon \cdot \Psi + U^\varepsilon \cdot (\partial_t - A^\varepsilon)\,\Psi dt dx + \int_\Omega U_0^\varepsilon \cdot \Psi(t = 0)\, dx = 0, \tag{2}$$

for all the admissible test functions $\Psi \in H^1(I \times \Omega)^2$ such that $\Psi(t,.) \in D(A^\varepsilon)$ for a.e. $t \in I$ where the domain $D(A^\varepsilon) := \{(\varphi, \phi) \in L^2(\Omega)^2 | \sqrt{a^\varepsilon}\varphi \in H^1(\Omega), \phi/\rho \in H_0^1(\Omega)\}$. As proved in [3], the operator iA^ε with the domain $D(A^\varepsilon)$ is self-adjoint on $L^2(\Omega)^2$. We assume that the data are bounded $\|f^\varepsilon\|_{L^2(I\times\Omega)} + \|\partial_x u_0^\varepsilon\|_{L^2(\Omega)} + \|v_0^\varepsilon\|_{L^2(\Omega)} \le c_0$, then U^ε is uniformly bounded in $L^2(I \times \Omega)$.

Bloch waves We introduce the dual $Y^* = \left(-\frac{1}{2}, \frac{1}{2}\right)$ of Y. For any $k \in Y^*$, we define the space of k−quasi-periodic functions $L_k^2 := \{u \in L_{loc}^2(\mathbb{R}) \mid u(x+\ell) = u(x)e^{2i\pi k\ell}$ a.e. in $\mathbb{R}$ for all $\ell \in \mathbb{Z}\}$ and set $H_k^s := L_k^2 \cap H_{loc}^s(\mathbb{R})$ for $s \ge 0$. The periodic functions correspond to $k = 0$. For a given $k \in Y^*$, we denote by $(\lambda_n^k, \phi_n^k)_{n\in\mathbb{N}^*}$ the Bloch wave eigenelements that are solution to $\mathscr{P}(k) : -\partial_y\left(a\partial_y\phi_n^k\right) = \lambda_n^k\rho\phi_n^k$ in Y with $\phi_n^k \in H_k^2(Y)$ and $\|\phi_n^k\|_{L^2(Y)} = 1$. The asymptotic spectral problem $\mathscr{P}(k)$ is also restated as a first order system by setting $A_k := \begin{pmatrix} 0 & \sqrt{a}\partial_y\left(\frac{1}{\sqrt{\rho}}.\right) \\ \frac{1}{\sqrt{\rho}}\partial_y\left(\sqrt{a}.\right) & 0 \end{pmatrix}$, $n_{A_k} = \frac{1}{\sqrt{\rho}}\begin{pmatrix} 0 & \sqrt{a}n_Y \\ \sqrt{a}n_Y & 0\end{pmatrix}$ and $e_n^k := \frac{1}{\sqrt{2}}\begin{pmatrix} -is_n/\sqrt{\lambda_{|n|}^k}\sqrt{a}\partial_y\left(\phi_{|n|}^k\right) \\ \sqrt{\rho}\phi_{|n|}^k\end{pmatrix}$ where s_n and n_Y denote the sign of $n \in \mathbb{Z}^*$ and the outer unit normal of ∂Y respectively. As proved in [3], iA_k is self-adjoint on the domain $D(A_k) := \{(\varphi, \phi) \in L^2(Y)^2 | \sqrt{a}\varphi \in H_k^1(Y), \phi/\sqrt{\rho} \in H_k^1(Y) \subset L^2(Y)^2\}$. The Bloch wave spectral problem $\mathscr{P}(k)$ is equivalent to finding pairs $\left(\lambda_{|n|}^k, e_n^k\right)$ indexed by $n \in \mathbb{Z}^*$ solution to $\mathscr{Q}(k) : A_k e_n^k = is_n\sqrt{\lambda_{|n|}^k}e_n^k$ in Y with $e_n^k \in H_k^1(Y)^2$. We pose $M_n^k := \{m \in \mathbb{Z}^* | \lambda_{|m|}^k = \lambda_{|n|}^k$ and $s_m = s_n\}$ and introduce the coefficients $b(k,n,m) = \int_Y \rho\phi_{|n|}^k \cdot \phi_{|m|}^k dy$ and $c(k,n,m) = is_n/\left(2\sqrt{\lambda_{|n|}^k}\right)\int_Y \phi_{|n|}^k \cdot a\partial_y\phi_{|m|}^k - a\partial_y\phi_{|n|}^k \cdot \phi_{|m|}^k dy$ for $n, m \in M_n^k$.

The modulated two-scale transform Let us assume from now that the domain Ω is the union of a finite number of entire cells of size ε or equivalently that the sequence ε is exactly $\varepsilon_n = \frac{\alpha}{n}$ for $n \in \mathbb{N}^*$. For any $k \in Y^*$, we define $I^k = \{-k, k\}$ if $k \neq 0$ and $I^0 = \{0\}$. By choosing $\Lambda = (0,1)$ as a time unit cell, we introduce the operator $W_k^\varepsilon : L^2(I \times \Omega)^2 \to L^2(I \times \Lambda \times \Omega \times Y)^2$ acting in all time and space variables, $W_k^\varepsilon := \left(1 - \sum_{n\in\mathbb{Z}^*}\Pi_n^k\right)S_k^\varepsilon + \sum_{n\in\mathbb{Z}^*} T^{\varepsilon\alpha_n^k}\Pi_n^k S_k^\varepsilon$ where the time and space two-scale transforms $T^{\varepsilon\alpha_n^k}$ and S_k^ε, and the orthogonal projector Π_n^k onto e_n^k are defined in [3], see pages 11, 15 and 17, with $\alpha_n^k = 2\pi/\sqrt{\lambda_{|n|}^k}$, and where it is proved that $\|W_k^\varepsilon u\|_{L^2(I\times\Lambda\times\Omega\times Y)}^2 = \|u\|_{L^2(I\times\Omega)}^2$.

We define $(\mathfrak{B}_n^k v)(t,x) = v(t, \frac{t}{\varepsilon\alpha_n^k}, x, \frac{x}{\varepsilon})$ the operator that operates on functions $v(t,\tau,x,y)$ defined in $I \times \mathbb{R} \times \Omega \times \mathbb{R}$. The notation $O(\varepsilon)$ refers to numbers or functions tending to zero when $\varepsilon \to 0$ in a sense made precise in each case. The next lemma shows that $\mathfrak{B}_n^k$ is an approximation of $T^{\varepsilon\alpha_n^k *}S_k^{\varepsilon *}$ for a function which is periodic in τ and k−quasi-periodic in y, where $T^{\varepsilon\alpha_n^k *} : L^2(I \times \Lambda) \to L^2(I)$ and $S_k^{\varepsilon *} : L^2(\Omega \times Y) \to L^2(\Omega)$ are adjoint of $T^{\varepsilon\alpha_n^k}$ and S_k^ε respectively.

Lemma 1 *Let* $v \in C^1(I \times \Lambda \times \Omega \times Y)$ *a periodic function in* τ *and* k*−quasi-periodic in* y*, then* $T^{\varepsilon\alpha_n^k *} S_k^{\varepsilon *} v = \mathfrak{B}_n^k v + O(\varepsilon)$ *in the* $L^2(I \times \Omega)$ *sense. Consequently, for any sequence* u^ε *bounded in* $L^2(I \times \Omega)$ *such that* $T^{\varepsilon\alpha_n^k} S_k^\varepsilon u^\varepsilon$ *converges to* u *in* $L^2(I \times \Lambda \times \Omega \times Y)$ *weakly when* $\varepsilon \to 0$*,*

$$\int_{I\times\Omega} u^\varepsilon \cdot \mathfrak{B}_n^k v \, dtdx \to \int_{I\times\Lambda\times\Omega\times Y} u \cdot v \, dtd\tau dxdy \ \ \text{when } \varepsilon \to 0. \tag{3}$$

Note that for $k = 0$, the convergence (3) regarding each variable corresponds to the definition of two-scale convergence in [1]. The proof is carried out in three steps. First the explicit expression of $T^{\varepsilon\alpha_n^k *} S_k^{\varepsilon *} v$ is derived, second the approximation of $T^{\varepsilon\alpha_n^k *} S_k^{\varepsilon *} v$ is deduced, finally the convergence (3) follows. For a function $v(t, \tau, x, y)$ defined in $I \times \Lambda \times \Omega \times Y$, we observe that

$$A^\varepsilon \mathfrak{B}_n^k v = \mathfrak{B}_n^k \left(\left(\frac{A_k}{\varepsilon} + B \right) v \right) \text{ and } \partial_t \left(\mathfrak{B}_n^k v \right) = \mathfrak{B}_n^k \left(\left(\frac{\partial_\tau}{\varepsilon \alpha_n^k} + \partial_t \right) v \right), \tag{4}$$

where the operator B is defined as the result of the formal substitution of x−derivatives by y−derivatives in A_k.

3 Homogenized Results and Their Proof

For $k \in Y^*$, we decompose

$$\frac{\alpha k}{\varepsilon} = h_\varepsilon^k + l_\varepsilon^k \text{ with } h_\varepsilon^k = \left[\frac{\alpha k}{\varepsilon} \right] \text{ and } l_\varepsilon^k \in [0, 1), \tag{5}$$

and assume that the sequence ε is varying in a set $E_k \subset \mathbb{R}^{+*}$ so that

$$l_\varepsilon^k \to l^k \text{ when } \varepsilon \to 0 \text{ and } \varepsilon \in E_k \text{ with } l^k \in [0, 1). \tag{6}$$

After extraction of a subsequence, we introduce the weak limits of the relevant projections along e_n^k for any $n \in \mathbb{Z}^*$,

$$F_n^k := \lim_{\varepsilon\to 0} \int_{\Lambda\times Y} T^{\varepsilon\alpha_n^k} S_k^\varepsilon F^\varepsilon \cdot e^{2i\pi s_n \tau} e_n^k dyd\tau \text{ and } U_{0,n}^k := \lim_{\varepsilon\to 0} \int_Y S_k^\varepsilon U_0^\varepsilon \cdot e_n^k dy. \tag{7}$$

The next lemmas state the microscopic equation for each mode and the corresponding macroscopic equation.

Lemma 2 *For* $k \in Y^*$ *and* $n \in \mathbb{Z}^*$*, let* U^ε *be a bounded solution of (2), there exists at least a subsequence of* $T^{\varepsilon\alpha_n^k} S_k^\varepsilon U^\varepsilon$ *converging weakly towards a limit* U_n^k

in $L^2(I \times \Lambda \times \Omega \times Y)^2$ when ε tends to zero. Then U_n^k is a solution of the weak formulation of the microscopic equation

$$\left(\frac{\partial_\tau}{\alpha_n^k} - A_k\right) U_n^k = 0 \text{ in } I \times \Lambda \times \Omega \times Y \tag{8}$$

and is periodic in τ and k-quasi-periodic in y. Moreover, it can be decomposed as

$$U_n^k(t, \tau, x, y) = \sum\nolimits_{p \in \mathbb{M}_n^k} u_p^k(t, x)\, e^{2i\pi s_p \tau} e_p^k(y) \text{ with } u_p^k \in L^2(I \times \Omega). \tag{9}$$

Lemma 3 *In the condition of Lemma 2, for each $k \in Y^*$, $n \in \mathbb{Z}^*$, $\varepsilon \in E_k$, for each $\sigma \in I^k$ and $q \in M_n^\sigma$, the macroscopic equation is stated by*

$$\begin{aligned} &\sum\nolimits_{p \in M_n^\sigma} \left(b(\sigma, p, q)\, \partial_t u_p^\sigma - c(\sigma, p, q)\, \partial_x u_p^\sigma\right) = F_q^\sigma \text{ in } I \times \Omega, \\ &\sum\nolimits_{p \in M_n^\sigma} b(\sigma, p, q)\, u_p^\sigma(t = 0) = U_{0,q}^\sigma \text{ in } \Omega, \end{aligned} \tag{10}$$

with the boundary conditions in case where there exists $p \in M_n^k$ such that $c(k, p, q) \neq 0$ and $\phi_{|p|}^k(0) \neq 0$,

$$\sum\nolimits_{\sigma \in I^k, p \in M_n^\sigma} u_p^\sigma \phi_{|p|}^\sigma(0)\, e^{sign(\sigma) 2i\pi \frac{l^k x}{\alpha}} = 0 \text{ on } I \times \partial\Omega. \tag{11}$$

The low frequency part U_H^0 relates to the weak limit in $L^2(I \times \Omega \times Y)^2$ of the kernel part of S_k^ε in the definition of W_k^ε. It has been treated completely, in [2, 3]. Here, we focus on the non-kernel part of S_k^ε, it relates to the high frequency waves and microscopic and macroscopic scales. In order to obtain the solution of the model, we analyze the asymptotic behaviour of each mode through $T^{\varepsilon\alpha_n^k} S_k^\varepsilon$ as in Lemmas 2 and 3. Then the full solution is the sum of all modes. The main Theorem states as follows.

Theorem 1 *For a given $k \in Y^*$, let U^ε be a solution of (2) bounded in $L^2(I \times \Omega)$, for $\varepsilon \in E_k$, as in (5,6), the limit G^k of any weakly converging extracted subsequence of $\sum_{\sigma \in I^k} W_\sigma^\varepsilon U^\varepsilon$ in $L^2(I \times \Lambda \times \Omega \times Y)^2$ can be decomposed as*

$$G^k(t, \tau, x, y) = \chi_0(k)\, U_H^0(t, x, y) + \sum_{\sigma \in I^k, n \in \mathbb{Z}^*} u_n^\sigma(t, x)\, e^{2i\pi s_n \tau} e_n^\sigma(y) \tag{12}$$

where $(u_n^\sigma)_{n,\sigma}$ are solutions of the macroscopic equation, and the characteristic function $\chi_0(k) = 1$ if $k = 0$ and $= 0$ otherwise.

Thus, the physical solution U^ε is approximated by two-scale modes

$$U^\varepsilon(t, x) \simeq \chi_0(k)\, U_H^k\left(t, x, \frac{x}{\varepsilon}\right) + \sum_{\sigma \in I^k, n \in \mathbb{Z}^*} u_n^\sigma(t, x)\, e^{i s_n \sqrt{\lambda_n^\sigma} t/\varepsilon} e_n^\sigma\left(\frac{x}{\varepsilon}\right). \tag{13}$$

The remain of this section provides the proofs of results.

Proof of Lemma 2 The test functions of the weak formulation (2) are chosen as $\Psi^{\varepsilon} = \mathfrak{B}_n^k \Psi(t,x)$ for $k \in Y^*$, $n \in \mathbb{Z}^*$ where $\Psi \in C^{\infty}(I \times \Lambda \times \Omega \times Y)^2$ is periodic in τ and $k-$quasi-periodic in y. From (4) multiplied by ε, since $\left(\frac{\partial_\tau}{\alpha_n^k} - A_k\right)\Psi$ is periodic in τ and $k-$quasi-periodic in y and $T^{\varepsilon\alpha_n^k} S_k^{\varepsilon} U^{\varepsilon} \to U_n^k$ in $L^2(I \times \Lambda \times \Omega \times Y)^2$ weakly, Lemma 1 allows to pass to the limit in the weak formulation, $\int_{I\times\Lambda\times\Omega\times Y} U_n^k \cdot \left(\frac{\partial_\tau}{\alpha_n^k} - A_k\right)\Psi dt d\tau dx dy = 0$. Using the assumption $U_n^k \in D(A_k) \cap L^2(I \times \Omega \times Y; H^1(\Lambda))$ and applying an integration by parts, $\int_{I\times\Lambda\times\Omega\times Y}\left(-\frac{\partial_\tau}{\alpha_n^k} + A_k\right)U_n^k \cdot \Psi dt d\tau dx dy + \int_{I\times\partial\Lambda\times\Omega\times Y} U_n^k \cdot \Psi dt d\tau dx dy - \int_{I\times\Lambda\times\Omega\times\partial Y} U_n^k \cdot n_{A_k} \Psi dt d\tau dx dy = 0$. Choosing $\Psi \in L^2(I \times \Omega; H_0^1 \Lambda \times Y)$ comes the strong form (8). Since the product of a periodic function by a $k-$quasi-periodic function is $k-$quasi-periodic then $n_{A_k}\Psi$ is $k-$quasi-periodic in y. Therefore, U_n^k is periodic in τ and $k-$quasi-periodic in y. Moreover, (9) is obtained, by projection.

Proof of Lemma 3 For $k \in Y^*$, let $\left(\lambda_{|p|}^{\sigma}, e_p^{\sigma}\right)_{p\in M_n^{\sigma}, \sigma\in I^k}$ be the Bloch eigenmodes of the spectral equation $\mathcal{Q}(\sigma)$ corresponding to the eigenvalue $\lambda_{|n|}^k$. We pose $\Psi^{\varepsilon}(t,x) = \sum_{\sigma\in I^k} \mathfrak{B}_n^{\sigma} \Psi_{\varepsilon}^{\sigma} \in H^1(I \times \Omega)^2$ as a test function in the weak formulation (2) with each $\Psi_{\varepsilon}^{\sigma}(t,\tau,x,y) = \sum_{q\in M_n^k} \varphi_{q,\varepsilon}^{\sigma}(t,x) e^{2i\pi s_q \tau} e_q^{\sigma}(y)$ where $\varphi_{q,\varepsilon}^{\sigma} \in H^1(I \times \Omega)$ and satisfies the boundary conditions $\sum_{\sigma\in I^k, q\in M_n^{\sigma}} e^{2i\pi s_q t/(\varepsilon\alpha_q^{\sigma})} \varphi_{q,\varepsilon}^{\sigma}(t,x)$ $\phi_{|q|}^{\sigma}\left(\frac{x}{\varepsilon}\right) = O(\varepsilon)$ on $I \times \partial\Omega$. Note that this condition is related to the second component of Ψ^{ε} only. Since $\alpha_q^{\sigma} = \alpha_n^k$ and $s_q = s_n$ for all $q \in M_n^{\sigma}$ and $\sigma \in I^k$, so $e^{2i\pi s_q t/(\varepsilon\alpha_q^{\sigma})} \neq 0$ can be eliminated. Extracting a subsequence $\varepsilon \in E_k$, using the $\sigma-$quasi-periodicity of $\phi_{|q|}^{\sigma}$ and (5,6), $\varphi_{q,\varepsilon}^{\sigma}$ converges strongly to some φ_q^{σ} in $H^1(I \times \Omega)$, then the boundary conditions are

$$\sum_{\sigma\in I^k, q\in M_n^{\sigma}} \varphi_q^{\sigma}(t,x)\, \phi_{|q|}^{\sigma}(0)\, e^{sign(\sigma)2i\pi\frac{l^k x}{\alpha}} = 0 \text{ on } I \times \partial\Omega. \tag{14}$$

Applying (4) and since $\left(\frac{\partial_\tau}{\alpha_n^{\sigma}} - A_{\sigma}\right)\Psi^{\sigma}=0$ for $\sigma \in I^k$, then in the weak formulation it remains $\sum_{\sigma\in I^k} \int_{I\times\Omega} F^{\varepsilon} \cdot \mathfrak{B}_n^{\sigma}\Psi_{\varepsilon}^{\sigma} + U^{\varepsilon} \cdot \mathfrak{B}_n^{\sigma}(\partial_t - B)\Psi_{\varepsilon}^{\sigma} dt dx - \int_{\Omega} U_0^{\varepsilon} \cdot \mathfrak{B}_n^{\sigma}\Psi_{\varepsilon}^{\sigma}(t=0)\, dx = 0$. Since $(\partial_t - B)\Psi_{\varepsilon}^{\sigma}$ is $\sigma-$quasi-periodic, so passing to the limit thanks to Lemma 1, after using (7) and replacing the decomposition of U_n^{σ}, $\sum_{\sigma\in I^k, \{p,q\}\in M_n^{\sigma}} \left(\int_{I\times\Omega} b(\sigma,p,q)\, u_p^{\sigma} \cdot \partial_t\varphi_q^{\sigma} - c(\sigma,p,q)\, u_p^{\sigma} \cdot \partial_x\varphi_q^{\sigma} - F_q^{\sigma} \cdot \varphi_q^{\sigma}\, dt dx - \int_{\Omega} U_{0,q}^{\sigma} \cdot \varphi_q^{\sigma}(t=0)\, dx\right) = 0$ for all $\varphi_q^{\sigma} \in H^1(I \times \Omega)$ fulfilling (14).

Moreover, if $u_q^{\sigma} \in H^1(I \times \Omega)$ then it satisfies the strong form of the internal equations (10) for each $\sigma \in I^k$, $q \in M_n^{\sigma}$ and the boundary conditions $\sum_{\sigma,p,q} c(\sigma,p,q)\, u_p^{\sigma}\overline{\varphi_q^{\sigma}} = 0$ on $I \times \partial\Omega$ for φ_q^{σ} satisfies (14).

In order to find the boundary conditions of $\left(u_p^{\sigma}\right)_{\sigma,p}$, we distinguish between the two cases $k \neq 0$ and $k = 0$. First, for $k \neq 0$, $\lambda_{|n|}^k$ is simple so $M_n^k = \{n\}$.

Introducing $C = diag\,(c\,(\sigma, n, n))_{\sigma}$, $B = diag\,(b\,(\sigma, n, n))_{\sigma}$, $U = \left(u_n^{\sigma}\right)_{\sigma}$, $F = \left(F_n^{\sigma}\right)_{\sigma}$, $U_0 = \left(U_{0,n}^{\sigma}\right)_{\sigma}$, $\Psi = \left(\varphi_n^{\sigma}\right)_{\sigma}$, $\Phi = \left(\phi_{|n|}^{\sigma}\,(0)\,e^{sign(\sigma)2i\pi l^k x/\alpha}\right)_{\sigma}$, Eq. (10) states under matrix form $B\partial_t U + C\partial_x U = F$ in $I \times \Omega$ and $BU\,(t=0) = U_0$ in Ω which the boundary condition is rewritten as $CU\,(t,x)\,.\overline{\Psi}\,(t,x) = 0$ on $I \times \partial\Omega$ for all Ψ such that $\overline{\Phi}(x).\overline{\Psi}(t,x) = 0$ on $I \times \partial\Omega$. Equivalently, $CU\,(t,x)$ is collinear with $\overline{\Phi}(x)$ yielding the boundary condition $u_n^k \phi_{|n|}^k\,(0)\,e^{2i\pi\frac{l^k x}{\alpha}} + u_n^{-k}\phi_{|n|}^{-k}\,(0)\,e^{-2i\pi\frac{l^k x}{\alpha}} = 0$ on $I \times \partial\Omega$ after remarking that $c\,(k,n,n) \neq 0$ and $c\,(k,n,n) = -c\,(-k,n,n)$. Second, for $k = 0$, $\lambda_{|n|}^0$ is double $\lambda_{|n|}^0 = \lambda_{|m|}^0$ so $M_n^k = \{n, m\}$. With $C = (c\,(0,p,q))_{p,q}$, $B = (b\,(0,p,q))_{p,q}$, $U = \left(u_p^0\right)_p$, $F = \left(F_q^0\right)_q$, $U_0 = \left(U_{0,q}^0\right)_q$, $\Psi = \left(\varphi_q^0\right)_q$, $\Phi = \left(\phi_{|q|}^0\,(0)\right)_q$, the matrix form is still stated as above which the boundary condition is $u_n^0\phi_{|n|}^0\,(0) + u_m^0\phi_{|m|}^0\,(0) = 0$ on $I \times \partial\Omega$ after remarking that $c\,(0,p,p) = 0$ and $c\,(0,n,m) \neq 0$.

Proof of Theorem For a given $k \in Y^*$, let U^{ε} be solution of (2) which is bounded in $L^2(I \times \Omega)$, then $\|W_{\sigma}^{\varepsilon}U^{\varepsilon}\|_{L^2(I\times\Lambda\times\Omega\times Y)}$ is bounded for $\sigma \in I^k$. So there exists $G^k \in L^2\,(I \times \Lambda \times \Omega \times Y)^2$ such that, up to the extraction of a subsequence, $\sum_{\sigma\in I^k} W_{\sigma}^{\varepsilon}U^{\varepsilon}$ tends weakly to $G^k = \chi_0\,(k)\,U_H^0 + \sum_{\sigma\in I^k, n\in\mathbb{Z}^*} U_n^k$ in $L^2\,(I \times \Lambda \times \Omega \times Y)^2$. The high frequency part is based on the decomposition (9) and Lemma 3.

Remark 1 This method allows to complete the homogenized model of the wave equation in [3] for the one-dimensional case. Let $K \in \mathbb{N}^*$, we decompose $\frac{\alpha}{\varepsilon K} = \left[\frac{\alpha}{\varepsilon K}\right] + l_{\varepsilon}^1$ with $l_{\varepsilon}^1 \in [0,1)$ and assume that the sequence ε is varying in a set $E_K \subset \mathbb{R}^{+*}$ so that $l_{\varepsilon}^1 \to l^1$ when $\varepsilon \to 0$ with $l^1 \in [0,1)$. For any $k \in L_K^*$, defined in [3], we denote $p_k = kK \in \mathbb{N}$, so $\frac{\alpha p_k}{\varepsilon K} = p_k\left[\frac{\alpha}{\varepsilon K}\right] + p_k l_{\varepsilon}^1$ and $p_k l_{\varepsilon}^1 \to l^k := p_k l^1$ when $\varepsilon \to 0$ with the same sequence of $\varepsilon \in E_K$.

4 Numerical Examples

We report simulations regarding comparison of physical solution and its approximation for $I = (0,1)$, $\Omega = (0,1)$, $\rho = 1$, $a = \frac{1}{3}(\sin(2\pi y) + 2)$, $f^{\varepsilon} = 0$, $v_0^{\varepsilon} = 0$, $\varepsilon = \frac{1}{10}$ and $k = 0.16$. Since $k \neq 0$, so the approximation (13) comes

$$U^{\varepsilon}\,(t,x) \simeq \sum\nolimits_{\sigma\in I^k, n\in\mathbb{Z}^*} u_n^{\sigma}\,(t,x)\,e^{is_n\sqrt{\lambda_{|n|}^{\sigma}}/\varepsilon}\,e_n^{\sigma}\left(\frac{x}{\varepsilon}\right). \tag{15}$$

The validation of the approximation is based on the modal decomposition of any solution $U^{\varepsilon} = \sum_{l\in\mathbb{Z}^*} R_l^{\varepsilon}\,(t)\,V_l^{\varepsilon}\,(x)$ where the modes V_l^{ε} are built from the solutions v_l^{ε} of the spectral problem $\partial_x\left(a^{\varepsilon}\partial_x v_l^{\varepsilon}\right) = \lambda_l^{\varepsilon}v_l^{\varepsilon}$ in Ω with $v_l^{\varepsilon} = 0$ on $\partial\Omega$. Moreover, in [9], two-scale approximations of modes have been derived on the form of linear

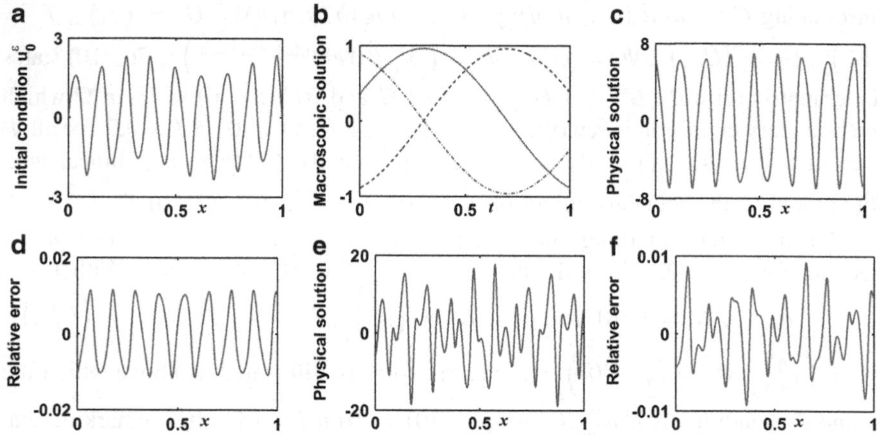

Fig. 1 Numerical results

combinations $\sum_{\sigma\in I^k}\theta_n^\sigma(x)\,\phi_{|n|}^\sigma\left(\frac{x}{\varepsilon}\right)$ of Bloch modes, so the initial conditions of the physical problem are taken on the form $u_0^\varepsilon(x) = \sum_{n\in\mathbb{N}^*,\sigma\in I^k}\theta_n^\sigma(x)\,\phi_n^\sigma\left(\frac{x}{\varepsilon}\right)$. Two simulations are reported, one for an initial condition u_0^ε spanned by the pair of Bloch modes corresponding to $n = 2$ when the other is spanned by three pairs $n \in \{2, 3, 4\}$. In the first case, the first component of U_0^ε approximates the first component of a single eigenvector V_l^ε approximated by (15) where all coefficients $u_n^\sigma = 0$ for $n \neq \pm 2$. Figure 1a shows the initial condition u_0^ε. Figure 1b presents the real part (solid line) and the imaginary part (dashed-dotted line) of the macroscopic solution u_n^k and also the real part (dotted line) and the imaginary part (dashed line) of u_n^{-k} at space step $x = 0.699$ when Fig. 1c, d plot the real part of the first component U_1^ε of physical solution and the relative error vector of U_1^ε with its approximation which $L^2(\Omega)$-norm is equal to 7e−3 at $t = 0.466$. For the second case where $u_n^\sigma = 0$ for $n \notin \{\pm 2, \pm 3, \pm 4\}$, the first component U_1^ε and the relative error vector of U_1^ε with its approximation which $L^2(\Omega)$-norm is 3.8e−3 are plotted in Fig. 1e, f. Finally, for the two cases the $L^2(I)$-relative errors at $x = 0.699$ on the first component are 8e−3 and 3.5e−3 respectively.

References

1. G. Allaire, Homogenization and two-scale convergence. SIAM J. Math. Anal. **23**(6), 1482–1518 (1992)
2. S. Brahim-Otsmane, G. Francfort, F. Murat, Correctors for the homogenization of the wave and heat equations. J. de mathématiques pures et appliquées **71**(3), 197–231 (1992)
3. M. Brassart, M. Lenczner, A two-scale model for the wave equation with oscillating coefficients and data. Comptes Rendus Math. **347**(23), 1439–1442 (2009)

4. J. Casado-Díaz, J. Couce-calvo, F. Maestre, J.D. Martín Gómez, Homogenization and correctors for the wave equation with periodic coefficients. Math. Models Methods Appl. Sci. **24**, 1–46 (2013)
5. W. Chen, J. Fish, A dispersive model for wave propagation in periodic heterogeneous media based on homogenization with multiple spatial and temporal scales. J. Appl. Mech. **68**(2), 153–161 (2001)
6. J. Fish, W. Chen, Space-time multiscale model for wave propagation in heterogeneous media. Comput. Methods Appl. Mech. Eng. **193**(45), 4837–4856 (2004)
7. G.A. Francfort, F. Murat, Oscillations and energy densities in the wave equation. Commun. Partial Differ. Equ. **17**(11–12), 1785–1865 (1992)
8. A. Lamacz, Dispersive effective models for waves in heterogeneous media. Math. Models Methods Appl. Sci. **21**(09), 1871–1899 (2011)
9. T.T. Nguyen, M. Lenczner, M. Brassart, Homogenization of the spectral equation in one-dimension (2013). arXiv preprint arXiv:1310.4064
10. F. Santosa, W.W. Symes, A dispersive effective medium for wave propagation in periodic composites. SIAM J. Appl. Math. **51**(4), 984–1005 (1991)

High-Order Asymptotic-Preserving Projective Integration Schemes for Kinetic Equations

Pauline Lafitte, Annelies Lejon, Ward Melis, Dirk Roose, and Giovanni Samaey

Abstract We study a projective integration scheme for a kinetic equation in both the diffusive and hydrodynamic scaling, on which a limiting diffusion or advection equation exists. The scheme first takes a few small steps with a simple, explicit method, such as a spatial centered flux/forward Euler time integration, and subsequently projects the results forward in time over a large, macroscopic time step. With an appropriate choice of the inner step size, the time-step restriction on the outer time step is similar to the stability condition for the limiting equation, whereas the required number of inner steps does not depend on the small-scale parameter. The presented method is asymptotic-preserving, in the sense that the method converges to a standard finite volume scheme for the limiting equation in the limit of vanishing small parameter. We show how to obtain arbitrary-order, general, explicit schemes for kinetic equations as well as for systems of nonlinear hyperbolic conservation laws, and provide numerical results.

1 Introduction

We study kinetic equations of the form

$$\partial_t f + \frac{v}{\varepsilon^s}\partial_x f = \frac{Q(f)}{\varepsilon^{s+1}}, \tag{1}$$

describing the evolution of the probability density function $f(x, v, t)$ of a particle being at position x, moving with velocity v at time t, and s corresponds to the scaling, which can be either hyperbolic ($s = 0$) or parabolic ($s = 1$). Such

---

P. Lafitte
Laboratoire de Mathématiques Appliquées aux Systémes, Ecole Centrale Paris, Grande Voie des Vignes, 92290 Châtenay-Malabry, France
e-mail: pauline.lafitte@ecp.fr

A. Lejon (✉) • W. Melis • D. Roose • G. Samaey
Department of Computer Science, KU Leuven, Celestijnenlaan 200A, B-3001 Leuven, Belgium
e-mail: firstname.lastname@cs.kuleuven.be

A. Abdulle et al. (eds.), *Numerical Mathematics and Advanced Applications ENUMATH 2013*, Lecture Notes in Computational Science and Engineering 103,
DOI 10.1007/978-3-319-10705-9_38

models frequently arise in the modeling of phenomena in various applications (such as biological systems or traffic flow). The collision kernel $Q(f)$ describes the interactions between particles, which causes a diffusive behavior on longer time scales. The left hand side of the equation represents the advective motion of the particles.

When we are dealing with systems with a large time scale separation (i.e. $0 < \varepsilon \ll 1$), Eq. (1) becomes stiff, with a time step constraint of order ε^{s+1} for classical explicit schemes. Clearly, such a time step restriction is prohibitive when taking the limit of ε going to zero.

In this paper, we propose a numerical scheme that is asymptotic-preserving in the sense that was introduced by Jin [9]. Specifically, we describe a higher-order extension of the projective integration algorithm, developed by Gear and Kevrekidis in [7] and applied to kinetic equations in [12]. The idea of this algorithm is to perform a few small steps with a naive *inner* integrator, which is subject to the time step constraint induced by the stiffness of the problem. In a next step, we then take a large time step with an *outer* integrator.

In the literature, several other possibilities are described to achieve an asymptotic-preserving scheme. For instance, an IMEX scheme [2, 5, 6] relies on a combination of an explicit discretization of the advective term, while the stiff collision kernel is treated implicitly. Recently, the method was adapted by Dimarco et al. [4] to deal with nonlinear collision kernels. To do this, the authors used a penalization with a simpler collision kernel, reducing the complexity of the part that has to be treated implicitly. Furthermore, in [3] Boscarino et al. extended the IMEX scheme to handle hyperbolic systems in a diffusive scaling. Alternative explicit techniques for stiff problems based on state extrapolation have been proposed in [13].

The remainder of the paper is organized as follows. First we introduce the model problems in Sect. 2 that we will consider for the analysis and the numerical experiments. Next, in Sect. 3 we give a description of the general scheme and its stability followed by some numerical experiments in Sect. 4. More details, proofs and a consistency result are in [10, 11].

2 Model Problems

2.1 Simple Linear Kinetic Equations

We study a linear kinetic equation of the form

$$\partial_t f^\varepsilon + \frac{v}{\varepsilon^s}\partial_x f^\varepsilon = \frac{\mathcal{M}(u^\varepsilon) - f^\varepsilon}{\varepsilon^{s+1}}, \tag{2}$$

describing the evolution of the particle distribution function $f^\varepsilon(x, v, t)$, that gives the probability of a particle being at position $x \in [-1, 1)$, moving with velocity

$v \in \mathbb{R}$ at time $t > 0$. The superscript ε denotes the dependence on the small scale parameter ε. The right hand side of Eq. (2) represents a BGK collision operator modeling a linear relaxation of f^ε towards a Maxwellian equilibrium distribution $\mathcal{M}(u^\varepsilon)$, in which $u^\varepsilon(x,t) = \langle f^\varepsilon(x,v,t)\rangle$, where $\langle\cdot\rangle$ denotes the average over velocity space. Throughout the paper, we require the measured velocity space (V,μ) to be discrete:

$$V := \{v_j\}_{j=1}^{J}, \qquad d\mu(v) = \sum_{j=1}^{J} w_j \delta(v - v_j), \tag{3}$$

where J is assumed to be even and V is an odd symmetric velocity space corresponding to : $v_j = -v_{J-j+1}, j = 1, \ldots J/2$. The Maxwellian operator $\mathcal{M}$ is supposed to satisfy the following conditions:

$$\begin{cases} \langle \mathcal{M}(u)\rangle = u, \\ \langle v.\mathcal{M}(u)\rangle = \varepsilon^s A(u). \end{cases} \tag{4}$$

Here, we will consider $\mathcal{M}(u) = u + \varepsilon^s A(u)/v$ as a prototypical example (see [1]). It can be proved via a Chapman-Enskog expansion that the evolution of the system tends to the advection-diffusion equation in the limit of $\varepsilon \to 0$:

$$\partial_t u^\varepsilon + \partial_x(A(u^\varepsilon)) = \varepsilon^{1-s} d\, \partial_{xx} u^\varepsilon + O(\varepsilon^2) \tag{5}$$

in which the constant $d = \langle v^2\rangle$ is the diffusion coefficient.

2.2 A Kinetic Semiconductor Equation

Although the numerical analysis of the presented algorithms is restricted to Eq. (2) with $A(u)$ linear, we also provide numerical results for a problem in which macroscopic advection does not originate from the Maxwellian in the collision operator, but from an external force field F. To this end, we consider a kinetic equation inspired by the semiconductor equation [8],

$$\partial_t f^\varepsilon + \frac{1}{\varepsilon}(v\partial_x f^\varepsilon + F\partial_v f^\varepsilon) = \frac{u^\varepsilon - f^\varepsilon}{\varepsilon^2}, \tag{6}$$

$$F = -\nabla\Phi, \qquad \Delta\Phi = u^\varepsilon. \tag{7}$$

In this equation, an acceleration term appears due to an electric force F resulting from a coupled Poisson equation for the electric potential Φ. The velocity space is given by $V = \mathbb{R}$ endowed with the Gaussian measure $d\mu(v) = (2\pi)^{-1/2}\exp(-v^2/2)dv$.

2.3 *Euler Equations*

In the two previous examples, the interest was in simulating a kinetic equation in a regime close to its macroscopic limit. A second situation in which the proposed methods will prove to be useful is when building a numerical method for nonlinear hyperbolic conservation laws based on the idea of relaxation [9]. In that setting, one starts from a hyperbolic conservation law, and constructs a kinetic equation such that its macroscopic limit corresponds to the given conservation law. Then, one simulates the kinetic equation instead of the original hyperbolic law. The advantage is that the advection term in the kinetic equation has become linear, avoiding the need for specialized (approximate) Riemann solvers. The price to pay is an increase in dimension, as well as the introduction of a stiff source term. It is precisely this stiff term that will be treated by the projective integration method.

As an example, we consider the Euler equations in one space dimension,

$$\begin{cases} \partial_t \rho + \partial_x(\rho \bar{v}) = 0, \\ \partial_t(\rho \bar{v}) + \partial_x(\rho \bar{v}^2 + P) = 0, \\ \partial_t E + \partial_x(E + P\bar{v}) = 0, \end{cases} \tag{8}$$

where ρ denotes the density, P the pressure, E the energy and $\bar{v}$ is the macroscopic velocity of the modeled fluid.

3 Higher Order Projective Integration

The projective integration algorithm we will discuss in this paper is a higher order extension of the projective integration method described in [7, 12]. The resulting algorithms are fully explicit, and will turn out to be asymptotic-preserving, which implies that the system can be stably integrated with a computational cost that is independent of ε. The algorithm relies on a combination of a few small steps with a classical *inner* time-stepping method and a much larger (*projective* or *outer*) time step.

3.1 *Inner Integrator*

The first part of a projective integration algorithm consists of $K + 1$ time steps with an inner integrator, which is described as:

$$f^{n,k+1} = f^{n,k} + \delta t \mathcal{D}_t(f^{n,k}) \quad \forall k = 0, \dots K \tag{9}$$

where $f^{n,k}$ denotes the numerical solution at time $t^{n,k} = n\Delta t + k\delta t$. In case of a large time scale separation (i.e. ε very small) the fast modes will have converged to their equilibrium at this stage of the algorithm.

3.2 Outer Integrator

The approach we pursue to achieve a fast higher order numerical scheme is to implement a Runge-Kutta version with S stages, nodes $c_s (s = 1 \dots S)$ and Runge-Kutta coefficients $a_{sl}(l = 1, \dots s-1, s = 1 \dots S)$ as outer integrator. This is typically represented by a Butcher tableau, as illustrated in Fig. 1 for a few classical choices.

These Runge-Kutta methods need to be modified to take into account the presence of the $K + 1$ inner steps. Specifically, the only modification lies in the calculation of the internal stages of the Runge-Kutta method. For the first Runge-Kutta stage, we have

$$\begin{cases} f^{n,k} = f^{n,k-1} + \delta t \mathcal{D}_t(f^{n,k-1}) \quad \forall 1 \leq k \leq K+1, \\ k_1 = \dfrac{f^{n,K+1} - f^{n,K}}{\delta t}, \end{cases} \tag{10}$$

whereas the remaining stages are obtained via

$$\begin{cases} f^{n+c_s} = f^{n,K+1} + (c_s \Delta t - (K+1)\delta, t) \sum_{l=1}^{s-1} \dfrac{a_{sl}}{c_s} k_l, \\ f^{n+c_s,k} = f^{n+c_s,k-1} + \delta t \mathcal{D}_t \left(f^{n+c_s,k-1}\right), \\ k_s = \dfrac{f^{n+c_s,K+1} - f^{n+c_s,K}}{\delta t}. \end{cases} \tag{11}$$

The stages are combined to obtain

$$f^{n+1} = f^{n,K+1} + (\Delta t - (K+1)\delta t) \sum_{s=1}^{S} b_s k_s, \tag{12}$$

$$\begin{array}{c|c} c & A \\ \hline & b^T \end{array} \qquad \begin{array}{c|cc} 0 & & \\ 1/2 & 1/2 & \\ \hline & 0 & 1 \end{array} \qquad \begin{array}{c|cccc} 0 & & & & \\ 1/2 & 1/2 & & & \\ 1/2 & 0 & 1/2 & & \\ 1 & 0 & 0 & 1 & \\ \hline & 1/6 & 1/3 & 1/3 & 1/6 \end{array}$$

Fig. 1 Butcher tableaux for the Runge-Kutta methods used in Sect. 4

3.3 Stability Analysis

Typically, systems with a large time scale separation (in which case ε is small) have a large gap in their eigenvalue spectrum. Projective integration schemes [7] are designed to capture all the eigenvalues with a time step constraint independent of the stiffness parameter ε. For projective integration methods, stability is expressed in terms of the amplification factors $\tau_{\delta t}$ of the inner integrator. The inner integrator is stable when all amplification factors satisfy $|\tau_{\delta t}| \leq 1$ or $\delta t \sim O(\varepsilon^{s+1})$ for Eq. (2). The projective integration method is stable for a subset of these values, of which the size depends on the ratio $\Delta t/\delta t$ and the number K of inner steps. In [7], it was shown that the stability regions can be described by two discs which are of the form $\mathscr{D}$(center, radius):

$$\mathscr{D}_1^{\mathrm{PFE}} = \mathscr{D}\left(1 - \frac{\delta t}{\Delta t}, \frac{\delta t}{\Delta t}\right) \quad \text{and} \quad \mathscr{D}_2^{\mathrm{PFE}} = \mathscr{D}\left(0, \left(\frac{\delta t}{\Delta t}\right)^{1/K}\right), \tag{13}$$

where the superscript PFE indicates that those are the stability regions for the *Projective Forward Euler method.* We have extended this result to the above-described Runge-Kutta generalization:

Theorem 1 (Stability of projective Runge-Kutta) *Consider a projective Runge-Kutta scheme with S stages with $S \geq 1$, then the stability regions of the scheme contains the stability regions of schemes with L stages with $L \leq S$.*

A proof of this result can be found in [10]. Once the stability regions of the projective Runge-Kutta methods are known, we need to localize the spectrum of the inner integrator (9). We have the following theorem, that is also proved in [10].

Theorem 2 (Spectrum of inner integrator) *The spectrum of the inner integrator* (9) *is located in two clusters:*

$$\mathscr{D}_1\left(1 - \frac{\delta t}{\varepsilon^{s+1}}, \frac{\delta t}{J\varepsilon^s} \max_{j \in \mathscr{J}}(|\alpha_j| + \beta_j|)\right) \cup \{\tau_{\delta t}\}$$

The dominant eigenvalue $\tau_{\delta t}$ is simple and can be expanded as

$$\begin{aligned}\tau_{\delta t} &= \left(1 - \frac{\delta t}{\varepsilon^{s+1}}\right) + \frac{\delta}{\varepsilon^{s+1}}\left(1 + \varepsilon\langle\alpha\rangle + \varepsilon^2\left(\langle(\langle\alpha\rangle - \alpha)^2\rangle - \langle\beta^2\rangle + \delta_s\left\langle\frac{\beta}{v}\right\rangle^2\right)\right) \\ &+ \iota\,\frac{\delta t}{\varepsilon^{s+1}}\left(\varepsilon\delta_s\left\langle\frac{\beta}{v}\right\rangle + \varepsilon^2\left(\delta_s\left\langle\left(\left\langle\frac{\beta}{v}\right\rangle - \beta\right)(\langle\alpha\rangle - \alpha)\right\rangle + \delta_{s-1}\left\langle\frac{\beta}{v}\right\rangle\right)\right),\end{aligned}$$

where the coefficients α and β depend on the chosen spatial discretization.

As a consequence, one can choose the parameters K and Δt independently of ε. Specifically, the PRK-scheme yields a CFL-like time step constraint – i.e.

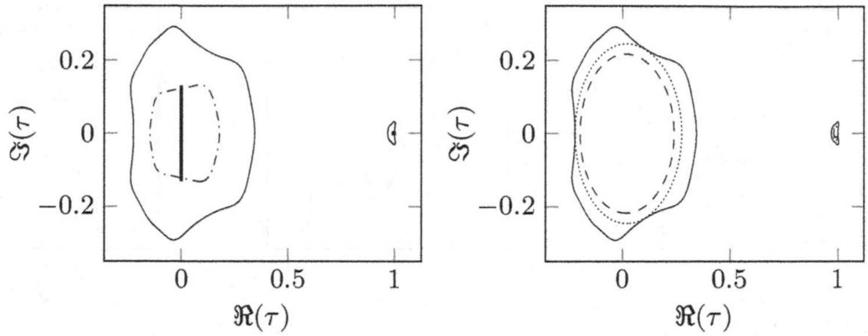

Fig. 2 *Left*: Stability regions corresponding to $\Delta t = 1 \times 10^{-2}$, $\delta t = 1 \times 10^{-4}$, $J = 10$, $\Delta x = 1 \times 10^{-1}$, $K = 3$ (*solid*) and $K = 2$ (*dash dotted*). *Right*: Illustration that the stability regions of schemes with more Runge-Kutta stages contain the stability regions of schemes with less stages

$\Delta t \leq C \Delta x^{s+1}$, where C is a constant dependent on the characteristics of the inner integrator of choice. We refer to [10] for more details (Fig. 2).

4 Numerical Experiments

In the first numerical experiment, we consider a diffusive scaling. We apply the scheme on the semiconductor equation (6), with initial condition

$$f(x, v_j, t) = \exp(-v_j^2/T)\exp(-x^2/0.1) \tag{14}$$

and periodic boundary conditions on the spatial domain. In the discrete velocity space (3) we use Hermitian quadrature points and use no-flux boundary conditions. Figure 3 shows the density evolution. The method parameters are given in the

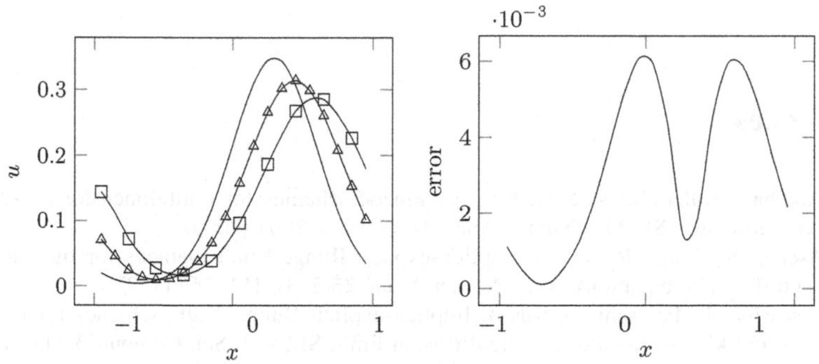

Fig. 3 Long term performance illustrated on semiconductor equation. *Left*: Density calculated with PRK4 scheme, after 400 steps (*solid*), 600 steps (*triangles*) and 800 steps (*squares*). *Right*: Absolute error on the calculated density after 400 steps. Parameters: $\varepsilon = 1 \times 10^{-2}$, $\Delta t = 1 \times 10^{-3}$, $K = 3$, $\phi(-1, t) = 20$ and $\phi(1, t) = 0$

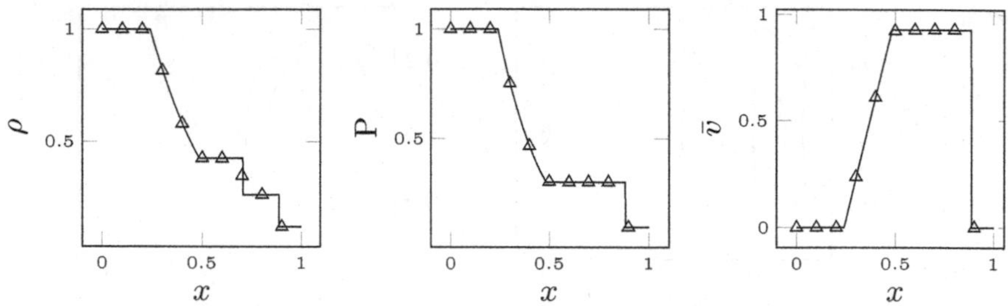

Fig. 4 Evolution of several fluid properties in Sod's shock tube problem obtained with PRK4 with FE as inner integrator and a third order ENO scheme to discretize the spatial derivative. The parameters are chosen as follows: $\Delta x = 5 \times 10^{-3}$, $\varepsilon = 1 \times 10^{-8}$, $\delta t = \varepsilon$, $K = 2$, $J = 4$ and $\Delta t = 0.5$. The *solid line* indicates the analytical solution, while the numerical approximation is denoted by *triangles*

caption. We clearly see stable and accurate results that are obtained with a computational effort that is independent of ε. More information and additional experiments including convergence as a function ε can be found in [10].

In the second numerical experiment, we consider a kinetic equation that results as the relaxation of the Euler equations. We perform Sod's shock tube problem, which is a classical test to check whether the method can deal with shock waves. As initial condition, we have

$$\rho(x,0) = \begin{cases} 1 & x \le L/2 \\ 0.125 & x > L > 2, \end{cases} \qquad P(x,0) = \begin{cases} 1 & x \le L/2 \\ 0.1 & x > L/2, \end{cases} \tag{15}$$

and the initial macroscopic velocity is defined as $\bar{v}(x,0) = 0$. We impose outflow boundary conditions (Fig. 4). Again, the method is shown to capture the expected behavior at a computational cost that is independent of ε. More details can be found in [11].

References

1. D. Aregba-Driollet, R. Natalini, Discrete kinetic schemes for multidimensional systems of conservation laws. SIAM J. Numer. Anal. **37**(6), 1973–2004 (2000)
2. U. Ascher, S. Ruuth, R. Spiteri, Implicit-explicit Runge-Kutta methods for time-dependent partial differential equations. Appl. Numer. Math. **25**(2–3), 151–167 (1997)
3. S. Boscarino, L. Pareschi, G. Russo, Implicit-explicit Runge–Kutta schemes for hyperbolic systems and kinetic equations in the diffusion limit. SIAM J. Sci. Comput. **35**(1), A22–A51 (2013)
4. G. Dimarco, L. Pareschi, Asymptotic preserving implicit-explicit Runge–Kutta methods for nonlinear kinetic equations. SIAM J. Numer. Anal. **51**(2), 1064–1087 (2013)
5. F. Filbet, S. Jin, An asymptotic preserving scheme for the ES-BGK model of the Boltzmann equation. J. Sci. Comput. **46**(2), 204–224 (2010)

6. F. Filbet, S. Jin, A class of asymptotic-preserving schemes for kinetic equations and related problems with stiff sources. J. Comput. Phys. **229**(20), 7625–7648 (2010)
7. C.W. Gear, I. Kevrekidis, Projective methods for stiff differential equations: problems with gaps in their eigenvalue spectrum. SIAM J. Sci. Comput. **24**(4), 1091–1106 (2003)
8. A. Giuseppe, A. Anile, Moment equations for charged particles: global existence results, in *Modeling and Simulation in Science, Engineering and Technology* (Birkhäuser, Boston, 2004), pp. 59–80
9. S. Jin, Efficient asymptotic-preserving (AP) schemes for some multiscale kinetic equations. SIAM J. Sci. Comput. **21**, 1–24 (1999)
10. P. Lafitte, A. Lejon, G. Samaey, A higher order asymptotic-preserving integration scheme for kinetic equations using projective integration, pp. 1–25. http://arxiv.org/abs/1404.6104
11. P. Lafitte, W. Melis, G. Samaey, A relaxation method with projective integration for solving nonlinear systems of hyperbolic conservation laws. (submitted)
12. P. Lafitte, G. Samaey, Asymptotic-preserving projective integration schemes for kinetic equations in the diffusion limit. SIAM J. Sci. Comput. **34**(2), A579–A602 (2012)
13. B. Sommeijer, Increasing the real stability boundary of explicit methods. Comput. Math. Appl. **19**(6), 37–49 (1990)

Reduced Basis Numerical Homogenization Method for the Multiscale Wave Equation

Assyr Abdulle, Yun Bai, and Timothée Pouchon

Abstract A reduced basis numerical homogenization method for the wave equation in heterogeneous media is presented. The method is based on a macroscopic discretization of the physical domain with input data recovered from microscopic problems solved by using reduced basis techniques. A priori error analysis is discussed and the convergence rates are verified by numerical experiments that also illustrate the performance of the method.

1 Introduction

Consider the wave equation in a polygonal domain $\Omega \subset \mathbb{R}^d$

$$\partial_t^2 u^\varepsilon(t,x) - \nabla \cdot (a^\varepsilon(x)\nabla u^\varepsilon(t,x)) = f(t,x) \quad \text{in} \quad (0,T) \times \Omega, \tag{1}$$

where the tensor a^ε has rapid oscillations on a small scale $\varepsilon \ll \mathrm{diam}(\Omega)$. This equation enters the modeling of various applications such as wave propagation in composite material, seismic imaging or medical ultrasound imaging. Together with (1), we set zero Dirichlet boundary condition and initial conditions $u^\varepsilon(0,\cdot) = g_0$, $\partial_t u^\varepsilon(0,\cdot) = g_1$. Furthermore, assume that a^ε is in $\left(L^\infty(\Omega)\right)^{d\times d}$, symmetric, uniformly elliptic and bounded. It follows from [13] that the problem (1) is well-posed, provided sufficient regularity of the functions f, g_0 and g_1.

It is well-known that standard numerical methods based on scale resolution for (1) are prohibitively expensive if ε is small. Fortunately, there is a well-developed mathematical theory, the homogenization theory, that describes an effective equation that captures the macroscopic behavior [7, 12]. In particular for the wave equation, the notions of G or H-convergence [14] have been used in [9] to show that $\{u^\varepsilon\}$

A. Abdulle (✉) • Y. Bai • T. Pouchon
ANMC, Section de Mathématiques, École Polytechnique Fédérale de Lausanne, Station 8, CH-1015 Lausanne, Switzerland
e-mail: assyr.abdulle@epfl.ch; yun.bai@epfl.ch; timothee.pouchon@epfl.ch

A. Abdulle et al. (eds.), *Numerical Mathematics and Advanced Applications ENUMATH 2013*, Lecture Notes in Computational Science and Engineering 103,
DOI 10.1007/978-3-319-10705-9_39

converges in a weak sense as $\varepsilon \to 0$ to the *homogenized solution* u^0 which satisfies the homogenized equation

$$\partial_t^2 u^0(t,x) - \nabla \cdot (a^0(x)\nabla u^0(t,x)) = f(t,x) \quad \text{in} \quad (0,T) \times \Omega. \tag{2}$$

We note that $a^0(x)$ is obtained via micro problems that are usually not explicitly known and various numerical strategies have been proposed in the past several years to compute an approximation of u^ε or u^0 (we refer to [5] for a literature review on such numerical strategies for wave equations). Here we focus on the finite element heterogeneous multiscale method (FE-HMM) introduced in [11] and first proposed for the wave equation in [5].

The basic FE-HMM can be summarized as follows: at the micro scale an effective solver based on numerical quadrature is defined. This solver relies on effective data (located at quadrature points) that needs to be recovered by *micro cell problems*, based on the heterogeneous tensor a^ε. A complete numerical analysis of this method including the contribution of micro errors, modeling error and macro error has been given in [5]. We note that the complexity of this method is independent of ε as the macro partition is independent of this small scale. However the issue for the FE-HMM is the large amount of repeated micro computations that need to be performed. This is a consequence of the fact that these micro problems are located around macroscopic quadrature points. Thus for optimal convergence, as shown in the a priori error analysis in [1, 2, 5], both the number of the micro problems and their complexity increase simultaneously as the macro mesh is refined.

The aforementioned issue triggered the development of a reduced order modeling for the FE-HMM [3, 4]. The main idea for these new methods, called reduced basis finite element heterogeneous multiscale method (RB-FE-HMM), is to combine numerical homogenization method with reduced basis techniques for the micro-problems. The reduced basis method (see [15] and the references therein) gives a methodology to construct low dimensional subspaces of the solutions of parametrized partial differential equations. This can be exploited in numerical homogenization in an offline stage to construct a low dimensional approximation for the micro problems [3, 8]. The approximation of the effective data in the macro RB-FE-HMM (the online stage) is then computed in this low dimensional subspace at a cost comparable to a standard FEM with numerical quadrature for single-scale problems. In this paper, we discuss the use of reduced basis for the numerical homogenization of the wave equation. The paper is organized as follows: the RB-FE-HMM is introduced and discussed in Sect. 2. Numerical experiments confirming the theoretical convergence rates and comparison with the FE-HMM are reported in Sect. 3.

2 RB-FE-HMM for the Wave Equation

The starting point of our numerical method is the FEM with numerical integration for the wave equation. We denote $\mathcal{T}_H$ a macro partition of Ω in simplicial or quadrilateral elements K, where the meshsize $H := \max_{K\in\mathcal{T}_H} \operatorname{diam}(K)$ is independent of ε and in particular $H \gg \varepsilon$ is allowed. For an integer $\ell \geq 1$, we define the macro FE space on $\mathcal{T}_H$,

$$S_0^\ell(\Omega,\mathcal{T}_H) = \{v^H \in H_0^1(\Omega) : v^H|_K \in \mathcal{R}^\ell(K) \ \forall K \in \mathcal{T}_H\}, \tag{3}$$

where $\mathcal{R}^\ell(K)$ is chosen either as $\mathcal{P}^\ell(K)$ the space of polynomials of degree at most ℓ for simplicial elements, or as $\mathcal{Q}^\ell(K)$ the space of polynomials of degree at most ℓ in each variable for quadrilateral elements. For each $K \in \mathcal{T}_H$, we define a quadrature formula (QF) $\{\omega_{Kj}, x_{Kj}\}_{j=1}^J$ and assume that the QF satisfies the standard hypotheses ensuring optimal convergence rates of the FEM in elliptic problems (see [10]). The FEM with numerical integration for the wave equation reads: find $u^H : [0,T] \to S_0^\ell(\Omega,\mathcal{T}_H)$, such that

$$(\partial_t^2 u^H(t), v^H) + B_H(u^H(t), v^H) = f_H(t; v^H) \qquad \forall v^H \in S_0^\ell(\Omega,\mathcal{T}_H), \tag{4}$$

where the bilinear form $B_H(\cdot,\cdot)$ is defined, for $v^H, w^H \in S_0^\ell(\Omega,\mathcal{T}_H)$, as

$$B_H(v^H, w^H) = \sum_{K\in\mathcal{T}_H} \sum_{1\leq j\leq J} \omega_{Kj} a^0(x_{Kj}) \nabla v^H(x_{Kj}) \cdot \nabla w^H(x_{Kj}). \tag{5}$$

Here $f_H(t; v^H)$ is an approximation of $\int_\Omega f(t,x) v^H(x)\,dx$ and the initial conditions are appropriate FE approximations of g_0, g_1.

The method above is however of no practical use in general as $a^0(x_{Kj})$ is unknown. In the FE-HMM presented in [5], $a^0(x_{Kj})$ are computed at *each* quadrature point by solving appropriate micro FEM on a sampling domain K_{δ_j} centered in x_{Kj}. Repeated micro FEM computations can however be expensive as the macro mesh is refined. To address this issue, following [3, 4], we introduce in what follows a low dimensional space spanned by appropriately chosen *reduced basis* to compute an approximation of each $a^0(x_{Kj})$. The RB algorithm is divided in an offline and online stage. While the offline stage is usually only performed once, the output of this process can be used repeatedly in different online procedures, e.g.,

- For adaptive macro mesh refinement [4];
- For different macro solvers (FEMs, finite difference methods, etc.);
- For different source terms, boundary conditions, or initial conditions;
- For different model equations with the same multiscale tensor (stationary or time-dependent problems).

2.1 RB-FE-HMM: Online Stage

We first assume that a set of RB functions $\{\xi_1, \cdots, \xi_N\}$ and the corresponding RB space

$$S_N(Y) := \text{span}\{\xi_1, \cdots, \xi_N\},$$

are already obtained from an offline stage (described in the next subsection) and describe the corresponding macroscopic approximation. We first recall that a fundamental condition for the efficiency of the RB-FE-HMM is that the tensor a^ε has the following affine representation,

$$a_x(y) = \sum_{1\le p\le P} \Theta_p(x) a_p(y), \ \forall y \in Y, \tag{6}$$

where $y := \frac{x}{\varepsilon} \in Y = (-1/2, 1/2)^d$. We note that when (6) is not readily available, one can use the so-called empirical interpolation method (EIM) [6] to provide an affine approximation of the tensor a^ε.

In order to compute an approximation $a_N^0(x)$ of $a^0(x_{Kj})$ at a given quadrature point x_{Kj}, we consider the RB-FE-HMM online cell problems: find $\psi_{N,\tau}(y) = \sum_{j=1}^N \alpha_{\tau,j}\xi_j \in S_N(Y)$ for $\boldsymbol{\alpha}_\tau = (\alpha_{\tau,1}, \cdots, \alpha_{\tau,N}) \in \mathbb{R}^N$, such that

$$\int_Y a_x(y)\nabla\psi_{N,\tau}(y) \cdot \nabla z_N dy = -\int_Y a_x(y)\mathbf{e}_m \cdot \nabla z_N dy, \ \forall z_N \in S_N(Y), \tag{7}$$

where $\{\mathbf{e}_m\}_{m=1}^d$ is the canonical basis of $\mathbb{R}^d$ and τ is the parameter index $\tau = (x, m)$. We see that the cell problems are parametrized by the location $x \in \Omega$ where we want an approximation of $a^0(x)$ and the index m of the canonical basis in $\mathbf{e}_m \in \mathbb{R}^d$.[1] For problems with multiscale tensors of form $a^\varepsilon(x, t)$, the micro problems can be similarly parametrized by (x, t). In this case, the Θ_p functions in (6) are of form $\Theta_p(x, t)$ and the parameter index in (7) is modified as $\tau = (x, t, m)$.

Using (6), we can write (7) as the following linear system

$$\Big(\sum_{1\le p\le P} \Theta_p(x) A_p\Big)\boldsymbol{\alpha}_\tau = -\sum_{1\le p\le P} \Theta_p(x) F_{p,m}, \tag{8}$$

where

$$(A_p)_{mn} = \int_Y a_p(y)\nabla\xi_m(y) \cdot \nabla\xi_n(y) dy, \qquad (F_{p,m})_i = \int_Y a_p(y)\mathbf{e}_m \cdot \nabla\xi_i(y) dy,$$

[1] We recall that for classical homogenization (e.g., for periodic tensors) the homogenized tensor a^0 relies on the solution of d cell problems with a right-hand side of the form $a(y)\mathbf{e}_m$ [7].

are precomputed and stored in the offline stage. The unknown tensor $a^0(x)$ is then estimated from the micro solution $\psi_{N,\tau}$ as

$$\begin{aligned}(a_N^0(x))_{mn} &= \int_Y a_x(y)\,(\nabla\psi_{N,\tau}(y) + \mathbf{e}_m)\cdot\mathbf{e}_n dy,\\ &= \sum_{1\le p\le P} \Theta_q(x)\big(\boldsymbol{\alpha}_\tau \cdot F_{p,n} + (G_p)_{mn}\big), \quad m,n = 1,\cdots,d, \end{aligned} \tag{9}$$

where $(G_p)_{mn} = \int_Y (a_p(y))_{mn} dy$ is also precomputed in the offline stage.

Comparison with the FE-HMM In the FE-HMM, the micro problems (7) are solved in a micro FE space $S^q(K_{\delta_j}, \mathcal{T}_h)$ coupled with periodic or Dirichlet boundary conditions. The FE space $S^q(K_{\delta_j}, \mathcal{T}_h)$ is defined on the cell domain $K_{\delta j} := x_{Kj} + \delta Y, \delta \ge \varepsilon, \forall K \in \mathcal{T}_H$ equipped with a micro mesh size $h/\varepsilon = \hat{h}$ and a FEM with piecewise polynomials of degree q. Hence solving a micro cell problem in the FE-HMM is equivalent to solving a linear system with DOF $\mathcal{O}(\hat{h}^{-d})$, where $\hat{h}$ must be refined simultaneously to the macro meshsize H for optimal accuracy [1, 2]. In contrast, we note that (7) only involves the solution of $N \times N$ linear system (8) with N fixed and usually small [3, 4]. As can be seen in our numerical experiments, the online solver of the RB-FE-HMM for the wave equation is much faster than the FE-HMM.

2.2 *RB-FE-HMM: Offline Stage*

The RB space is constructed in an offline stage, where a small number of representative micro problems which are parametrized by the macro locations and the canonical basis $\mathbf{e}_m$, are selected by a greedy algorithm with accuracy controlled by an a posteriori error estimator. The key steps of the offline procedure are the following.

1. Initial step: For a given offline tolerance *tol* and a natural number N_{train} we consider a randomly chosen training set

$$\Xi_{train} = \{(x_n, \mathbf{e}_\eta);\ x_n \in \Omega, n = 1,\cdots, N_{train}, \eta = 1,\cdots,d\}.$$

2. Iteration step: Assuming that the basis functions $\{\xi_1,\cdots,\xi_{l-1}\}$ are already computed we compute the next RB function ξ_l as described below.

   (a) Select the next target cell problem by an a posteriori estimator $\Delta^\eta_{l-1,n}$:

$$(x_l, \mathbf{e}_l) = \mathrm{argmax}_{(x_n,\mathbf{e}_\eta)\in\Xi_{train}} \Delta^\eta_{l-1,n}.$$

   If $\max_{(x_n,\mathbf{e}_\eta)} \Delta^\eta_{l-1,n} \le tol$ then go to step 3.

(b) Compute the cell problem (7) located at x_l with right-hand side based on $\mathbf{e}_l$ with an *accurate FE solver* in order to have a negligible offline FE discretization error. Add the cell solution to the RB basis $\{\xi_1, \cdots, \xi_{l-1}\}$ after an orthogonalization process.

3. Store the offline output: A_p, $F_{p,m}$, G_p, see (8) and (9).

An appropriate a posteriori estimator is crucial for the offline process and we refer to [3] for details. The accuracy of the RB-FE-HMM for the wave equation is described in the following theorem that can be proved by combining the a priori analyses in [5] and [3].

Theorem 1 *Under appropriate regularity assumptions for the homogenized solution u^0, the following estimate for the RB-FE-HMM solution u^H holds,*

$$\|u^H - u^0\|_{L^\infty(0,T;H^{1-\mu}(\Omega))} \leq C(H^{\ell+\mu} + err_{RB} + err_{mod}),\ \mu = 0, 1,$$

where $H^0(\Omega) := L^2(\Omega)$, err_{mod} is the HMM modeling error analyzed in [5] and $err_{RB} \leq C(\frac{1}{\mathcal{N}^q} + r_{RB})$, where $\mathcal{N}$ is the DOF of the offline FEM used to compute the RB.

As analyzed in [3], provided an exponential Kolmogorov N-width decay for the best N-dimensional approximation subspace of the infinite dimensional space of "cell solutions", the RB a priori error r_{RB} can be bounded as $r_{RB} \leq Ce^{-sN}$, where s is a positive constant and $N = \dim(S_N(Y))$. As can be seen in our numerical tests the offline error err_{RB} is often negligible compared to the macro error $\mathcal{O}(H^{\ell+\mu})$. Finally we note that the RB error r_{RB} can also be controlled by the a posteriori error estimator [3, 8, 15].

3 Numerical Examples

In this section, we consider the model equation (1) in $\Omega \times [0, T] = [0, 1]^2 \times [0, 1]$ and show two numerical examples to verify the performance of the RB-FE-HMM for wave problems in heterogeneous media. In the following, we choose $\varepsilon = 10^{-3}$ and fix the micro sampling domain size $\delta = \varepsilon$. We set $f = 0$ in (1) and choose initial conditions as $g_0(x) = 0.1 \exp\left(\frac{|x-c|^2}{\sigma^2}\right)$, $g_1(x) = 0$, $\sigma = 0.1$, $c = (0.5, 0.5)$. We apply the P1 FEM for the space discretization (for both the offline and the online procedures) and the leap frog scheme for the time discretization. The implementation is done in Matlab 2012a without using a parallel implementation.

2D problem with a continuous tensor We consider the following diagonal tensor applied to (1) (also used for the elliptic test problem in [3])

$$\begin{aligned} a_{11}^\varepsilon(x) &= x_1^2 + 0.2 + (x_2 + 1)(\sin(2\pi x_1/\varepsilon) + 2) \\ a_{22}^\varepsilon(x) &= x_2^2 + 0.05 + (x_1 x_2 + 1)(\sin(2\pi x_2/\varepsilon) + 2). \end{aligned} \tag{10}$$

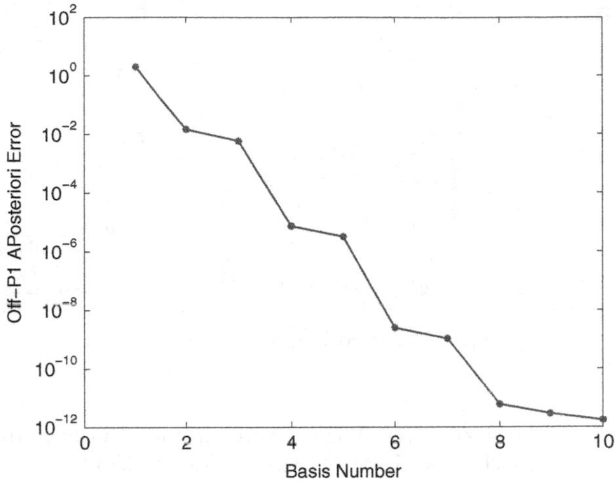

Fig. 1 The decay of the a posteriori error versus reduced basis numbers [3]

Table 1 Comparison between RB-FE-HMM and FE-HMM (the micro DOF N_{MIC} is set to $N_{MIC} = N_{MAC}$)

	RB-FE-HMM			FE-HMM		
N_{MAC}	H^1 error	L^2 error	CPU (s)	H^1 error	L^2 error	CPU (s)
16	0.181	8.8e−3	0.4	0.182	8.8e−3	2.6
32	0.112	4.2e−3	1.1	0.112	4.2e−3	25.1
64	0.041	1.1e−4	8.9	0.041	1.1e−4	412.3
128	0.015	0.3e−4	90.8	0.015	0.3e−4	7,671.5

As we noticed before, the same offline data can be used for different model equations. Therefore we reuse the offline output from the first example in [3, Section 5] for our wave equation. The details of the offline stage can be seen in [3, Section 5] and we just mention here that 10 reduced bases are obtained in the end of the offline stage which are solved on a $1{,}500 \times 1{,}500$ mesh. The offline CPU time is 1,045 s. The fast decay of the a posteriori error in the offline stage is shown in Fig. 1.

In the online stage, we set the time step $\Delta t = 0.1H$ to ensure stability of the time integrator, where $H = 1/N_{MAC}$ and N_{MAC} denotes the DOF in each space direction. We test the error $\|u^H - u^{ref}\|$ in H^1 and L^2 norms as well as the CPU time for both the RB-FE-HMM online stage and the FE-HMM. A reference solution u^{ref} is computed from the homogenized equation with tensor a^0 by P1 FEM on a 256×256 uniform mesh. As shown in Table 1, the RB-FE-HMM and the FE-HMM have almost the same accuracy but the CPU time comparison shows great efficiency advantage using the RB-FE-HMM.

2D problem with a discontinuous tensor In this test, we use a discontinuous multiscale tensor as shown in Fig. 2a for the multiscale wave equation. This tensor is also considered for elliptic equation in [3, Section 5]. As we can see in Fig. 2b,

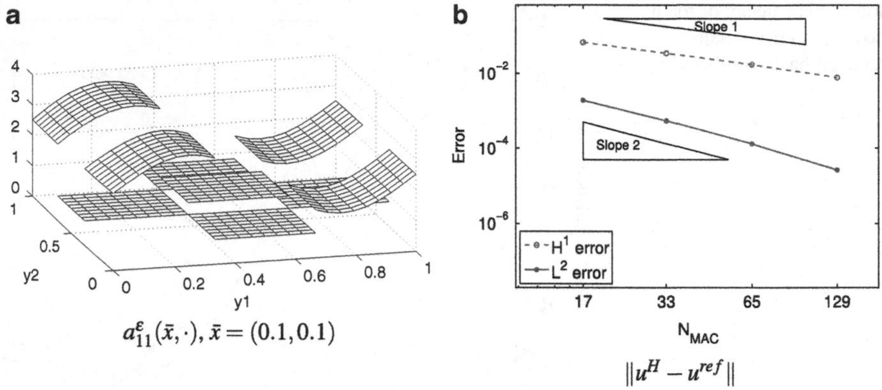

Fig. 2 (**a**) shows $a_{11}^{\varepsilon}(\bar{x},\cdot)$, in reference sampling domain Y. (**b**) shows H^1 and L^2 errors with u^{ref} computed by the FE-HMM ($N_{MAC} = N_{MIC} = 256$)

the H^1 and L^2 errors converge with the rate $\mathcal{O}(H)$ and $\mathcal{O}(H^2)$ respectively. The RB error err_{RB} is negligible here and cannot be observed as the macro discretization error dominates. The convergence rates corroborate the a priori error estimate given in Theorem 1.

Acknowledgement This work is partially supported by the Swiss National Foundation Grant 200021_134716

References

1. A. Abdulle, On a priori error analysis of fully discrete heterogeneous multiscale FEM. SIAM Multiscale Model. Simul. **4**(2), 447–459 (2005)
2. ———, A priori and a posteriori error analysis for numerical homogenization: a unified framework. Ser. Contemp. Appl. Math. CAM **16**, 280–305 (2011)
3. A. Abdulle, Y. Bai, Reduced basis finite element heterogeneous multiscale method for high-order discretizations of elliptic homogenization problems. J. Comput. Phys. **231**(21), 7014–7036 (2012)
4. ———, Adaptive reduced basis finite element heterogeneous multiscale method. Comput. Methods Appl. Mech. Eng. **257**, 201–220 (2013)
5. A. Abdulle, M. Grote, Finite element heterogeneous multiscale method for the wave equation. SIAM Multiscale Model. Simul. **9**(2), 766–792 (2011)
6. M. Barrault, Y. Maday, N. Nguyen, A. Patera, An 'empirical interpolation method': application to efficient reduced-basis discretization of partial differential equations. C. R. Acad. Sci. Paris Ser.I **339**, 667–672 (2004)
7. A. Bensoussan, J.-L. Lions, G. Papanicolaou, *Asymptotic Analysis for Periodic Structures* (North-Holland, Amsterdam, 1978)
8. S. Boyaval, Reduced-basis approach for homogenization beyond the periodic setting. Multiscale Model. Simul. **7**(1), 466–494 (2008)
9. S. Brahim-Otsmane, G.A. Francfort, F. Murat, Correctors for the homogenization of the wave and heat equations. J. Math. Pures Appl. **71**(3), 197–231 (1992)

10. P. Ciarlet, *Basic Error Estimates for Elliptic Problems*. Handbook of Numerical Analysis, vol. 2 (Elsevier Science/North-Holland, Amsterdam/New York, 1991)
11. W. E, B. Engquist, The heterogeneous multiscale methods. Commun. Math. Sci. **1**(1), 87–132 (2003)
12. V. Jikov, S. Kozlov, O. Oleinik, *Homogenization of Differential Operators and Integral Functionals* (Springer, Berlin/Heidelberg, 1994)
13. J.L. Lions, E. Magenes, *Problèmes aux limites non homogènes et applications*. Travaux et recherches mathématiques, vol. 1 (Dunod, Paris, 1968)
14. F. Murat, L. Tartar, H-convergence, topics in the mathematical modeling of composite materials. Progress. Nonlinear Differ. Equ. Appl. **31**, 21–43 (1997)
15. G. Rozza, D. Huynh, A.T. Patera, Reduced basis approximation and a posteriori error estimation for affinely parametrized elliptic coercive partial differential equations. Arch. Comput. Methods Eng. **15**, 229–275 (2008)

Part VI
Reduced Order Modeling

Reduced Order Optimal Control Using Proper Orthogonal Decomposition Sensitivities

Tuğba Akman and Bülent Karasözen

Abstract In general, reduced-order model (ROM) solutions obtained using proper orthogonal decomposition (POD) at a single parameter cannot approximate the solutions at other parameter values accurately. In this paper, parameter sensitivity analysis is performed for POD reduced order optimal control problems (OCPs) governed by linear diffusion-convection-reaction equations. The OCP is discretized in space and time by discontinuous Galerkin (dG) finite elements. We apply two techniques, extrapolating and expanding the POD basis, to assess the accuracy of the reduced solutions for a range of parameters. Numerical results are presented to demonstrate the performance of these techniques to analyze the sensitivity of the OCP with respect to the ratio of the convection to the diffusion terms.

1 Introduction

Optimal control problems for nonlinear and time-dependent partial differential equations (PDEs) depending on a set of parameters are very time consuming. To overcome this, in the last years, POD-ROMs are applied to optimal control of PDEs (see for example [4]). The POD is based on projecting the dynamical system onto subspaces of basis elements using the snapshots computed by finite elements. The finite element solutions are not correlated to the physical properties of the system they approximate, whereas the POD bases express the characteristics of the solutions better. Besides POD, reduced basis methods are also used to obtain efficient ROM solutions for parameterized PDEs (see for example [6]). When ROMs should approximate solutions for a wide range of parameters, the cost of basis selection increases because full data are required. In recent years, sensitivity analysis has been used in the POD basis selection process for fluid dynamics [3]. They rely on the continuous or discrete sensitivities, baseline or reference POD modes and their derivatives with respect to parameters. In this

T. Akman (✉) • B. Karasözen
Department of Mathematics, Institute of Applied Mathematics, Middle East Technical University, 06800 Ankara, Turkey
e-mail: takman@metu.edu.tr; bulent@metu.edu.tr

A. Abdulle et al. (eds.), *Numerical Mathematics and Advanced Applications ENUMATH 2013*, Lecture Notes in Computational Science and Engineering 103,
DOI 10.1007/978-3-319-10705-9_40

work, we extend the parameter sensitivity analysis in [3] to time dependent OCPs constrained by linear diffusion-convection-reaction equations. We compute two new POD bases by extrapolating and expanding the baseline POD basis to assess the accuracy of the reduced solutions for a range of parameters. The optimality system is discretized using space-time dG method. DG time discretization schemes combined with the symmetric interior penalty (SIPG) method in space have the pleasant property that discretization and optimization commute. In addition, dG time-stepping methods require less regularity compared to the finite difference schemes in time [7, Chap. 7].

The paper is organized as follows: In Sect. 2, we give the optimality system for the OCP governed by the unsteady diffusion-convection equation. The fully-discrete optimality system using the space-time dG is given in Sect. 3. The POD-ROM for the OCP and the derivation of POD sensitivities are presented in Sect. 4. Numerical results for an OCP with interior and boundary layers are discussed in Sect. 5.

2 The Optimal Control Problem

We consider the following distributed OCP by the unsteady diffusion-convection-reaction equation without control constraints

$$\begin{aligned}
\underset{u\in L^2(0,T;L^2(\Omega))}{\text{minimize}} \quad & J(y,u) := \frac{1}{2}\int_0^T \left(\|y-y_d\|^2_{L^2(\Omega)} + \alpha\,\|u\|^2_{L^2(\Omega)} \right) dt, \\
\text{subject to } \partial_t y - \epsilon\Delta y + \boldsymbol{\beta}\cdot\nabla y + ry &= f + u \quad (x,t)\in\Omega\times(0,T], \\
y(x,t) &= 0 \qquad (x,t)\in\partial\Omega\times[0,T], \qquad (1)\\
y(x,0) &= y_0(x) \qquad x\in\Omega,
\end{aligned}$$

where Ω is a bounded open, convex domain in $\mathbb{R}^2$ with a Lipschitz boundary $\partial\Omega$ and $I=(0,T]$ is the time interval, $f, y_d \in L^2(0,T;L^2(\Omega))$, $y_0(x)\in H^1_0(\Omega)$, $r \in L^\infty(\Omega)$, $\boldsymbol{\beta}\in (W^{1,\infty}(\Omega))^2$ are given functions and $\epsilon,\alpha>0$ are given scalars. The velocity field $\boldsymbol{\beta}$ does not depend on time and satisfies the incompressibility condition, i.e. $\nabla\cdot\beta = 0$.

In order to write the variational formulation of the problem, we define the bilinear forms $a(y,v)=\int_\Omega(\epsilon\nabla y\cdot\nabla v+\boldsymbol{\beta}\cdot\nabla y v+ryv)\,dx$, $(u,v)=\int_\Omega uv\,dx$, the state and the test space as $Y=V=H^1_0(\Omega)$, $\forall t\in(0,T]$. It is well known that the pair $(y,u)\in H^1(0,T;L^2(\Omega))\cap L^2(0,T;H^1_0(\Omega))\times L^2(0,T;L^2(\Omega))$ is the unique solution of the optimal control problem if and only if there is an adjoint $p\in H^1(0,T;L^2(\Omega))\cap L^2(0,T;H^1_0(\Omega))$ such that (y,u,p) satisfy the following optimality system [8]

$$\begin{aligned}
(\partial_t y, v) + a(y,v) &= (f+u, v) \quad \forall v\in V,\ \ y(x,0)=y_0, \\
-(\partial_t p, \psi) + a(\psi, p) &= -(y-y_d, \psi) \quad \forall \psi\in V,\ \ p(x,T)=0, \qquad (2)\\
\alpha u &= p.
\end{aligned}$$

3 Space-Time Discretization of the Optimal Control Problem

Let $\{\mathscr{T}_h\}_h$ be a family of shape regular meshes such that $\overline{\Omega} = \cup_{K\in\mathscr{T}_h}\overline{K}$, $K_i \cap K_j = \emptyset$ for $K_i, K_j \in \mathscr{T}_h$, $i \neq j$. We use discontinuous piecewise finite element space $V_h = \{y \in L^2(\Omega) : y\mid_K \in \mathbb{P}^1(K) \quad \forall K \in \mathscr{T}_h\}$ for the control, state and adjoint. Here, $\mathbb{P}^1(K)$ denotes the set of all polynomials on $K \in \mathscr{T}_h$ of degree 1. The diffusion term is discretized by the SIPG method and the convection term is discretized by upwinding [2]. Then, the semi-discrete state equation is given as in the study [1]

$$(\partial_t y_h, v_h) + a_h(y_h, v_h) + b_h(u_h, v_h) = (f_h, v_h) \quad \forall v_h \in V_h, \quad t \in (0, T].$$

For time discretization, we also use dG method. Let $0 = t_0 < t_1 < \cdots < t_N = T$ be a subdivision of $I = (0, T)$ with time intervals $I_n = (t_{n-1}, t_n]$ and time steps $k_n = t_n - t_{n-1}$ for $n = 1, \ldots, N$ and $k = \max_{1\leq n\leq N} k_n$. We define the space-time finite element space of piecewise discontinuous functions for test function, state, control and adjoint as

$$V_h^k = \left\{ v \in L^2(0, T; L^2(\Omega)) : v|_{I_m} = \sum_{s=0}^{q} t^s \phi_s, t \in I_m, \phi_s \in V_h, m = 1, \ldots, N \right\}.$$

We use dG(0) method, i.e. $q = 0$, where the approximating polynomials are piecewise constant in time. We define $y_n = y_{hk}|_{I_n}$, $p_n = p_{hk}|_{I_n}$, $u_n = u_{hk}|_{I_n}$ for $n = 1, \cdots, N$, $y_{hk,0}^- = y_0$, $p_{hk,N}^+ = 0$. Then, the fully-discrete state and the adjoint equation are written as

$$(M + kA^s)y_n = My_{n-1} + \frac{k}{2}(f_n + f_{n-1}) + \frac{k}{2}M(u_n + u_{n-1}),$$

$$(M + kA^a)p_{n-1} = Mp_n - \frac{k}{2}M(y_n + y_{n-1}) + \frac{k}{2}(y_n^d + y_{n-1}^d),$$

where M is the mass matrix and A^s, A^a are the stiffness matrices for the state $a_h(y_h, v_h)$ and adjoint equations $a_h(v_h, p_h)$, respectively. We note that the resulting scheme is a variant of the backward Euler method where the temporal terms on the right-hand side of (2) are computed by trapezoidal rule [7, Chap. 7].

4 Reduced-Order Modelling Using POD

In this section, we briefly explain the POD method. Let the matrix W be a real-valued $M \times N$ matrix of rank $d \leq \min(M, N)$ representing the snapshot data. We introduce the correlation matrix $K = \tilde{W}^T\tilde{W}$ with $\tilde{W} = M^{1/2}W$. Then, we compute the coefficients of a POD basis of rank l using the eigenvalue decomposition (EVD) of K as follows

$$\Psi_{:,j} = W\tilde{V}_{:,j}/\sqrt{\lambda_j}, \quad j = 1, \cdots, l,$$

where $\tilde{V}_{:,j}$ is the j-th eigenvector of K and λ_j is the associated eigenvalue. On the other hand, the singular value decomposition (SVD) of the matrix $\tilde{W} = U\Sigma V^T$ can also be used. The POD basis coefficients are computed by solving the linear system $(M^{1/2})^T\Psi_{:,l} = U^l$, with the first l columns of U, for $\Psi_{:,l}$. Then, the l POD basis functions are written as a linear combination of the finite element basis functions, $\psi_j(x) = \sum_{i=1}^{M} \Psi_{ij}\varphi_i(x),\ j = 1, \ldots, l.$

In general, the POD basis generated via the snapshots depending on a parameter μ_0 cannot capture the dynamics of the perturbed problem associated to $\mu = \mu_0 + \Delta\mu$. Motivated by the study of fluid flow equations using POD-ROMs [3], POD sensitivities can be used to enrich the low-dimensional space for a wider range of parameters. In order to derive POD sensitivities, the sensitivity of the snapshot set is required. The sensitivity of a term is defined as the derivative of that term with respect to a quantity of interest. In this study, we are interested in the sensitivities with respect to μ corresponding to the ratio $\mu = \mid \boldsymbol{\beta} \mid /\epsilon$ in the OCP (1). For the computation of the sensitivities, we compare two different approaches: the continuous sensitivity equation (*CSE*) and finite-difference (*FD*) approximation. In *CSE* approach, state, adjoint and control are assumed to be differentiable with respect to μ. The subscript μ denotes the derivative with respect to μ. Then, with the sensitivities $s = y_\mu, q = p_\mu, \nu = u_\mu$, we derive another optimality system depending on s, q and ν,

$$\begin{aligned} (\partial_t s, v) + a(s, v) &= (f_\mu + \nu, v) - (\nabla y, \nabla v), \quad s(x, 0) = (y_0)_\mu, \\ -(\partial_t q, \psi) + a(\psi, q) &= -(s - y^d_\mu, \psi) - (\nabla p, \nabla \psi), \quad q(x, T) = 0, \\ \alpha\nu &= q. \end{aligned} \tag{3}$$

We note that the sensitivity equations are always linear, so *CSE* method would be especially promising for nonlinear problems. The sensitivities s, q and μ can be computed either by inserting the solution of the state and adjoint to the right-hand side of (3) or solving the systems arising from (2) and (3) simultaneously. We use the second approach. In *FD* method, particularly for the centred difference scheme, the solution of the perturbed optimal control problem is required, i.e. depending on $\mu = \mu_0 \pm \Delta\mu$. Then, the sensitivity of the state can be computed via the centred difference as follows

$$y_\mu(\mu_0) \approx \frac{y(\mu_0 + \Delta\mu) - y(\mu_0 - \Delta\mu)}{2\Delta\mu}. \tag{4}$$

We treat each POD mode as a function of both space and the parameter, i.e. $\psi = \psi(x, \mu)$. In order to find POD sensitivities, we differentiate $(M^{1/2})^T \Psi = U^l$ with respect to μ and then solve the resulting equation for Ψ_μ. Then, the sensitivities of the l POD basis functions, namely ψ_μ, are written as a linear combination of the finite element basis functions, $(\psi_j)_\mu = \sum_{i=1}^{M} (\Psi_{ij})_\mu \varphi_i(x),\ j = 1, \ldots, l.$

We have taken the same range of parameters as in [3]. For larger parameter variations, the applicability of this approach might not be useful, because the sensitivities are based on the asymptotic expansion of μ in (4).

The connection between the state and the POD sensitivities is realised through the relation

$$U^l_\mu = (\tilde{W} V^l \Sigma^\dagger)_\mu = \tilde{W}_\mu V^l \Sigma^\dagger + \tilde{W} V^l_\mu \Sigma^\dagger + \tilde{W} V^l \Sigma^\dagger_\mu.$$

For the computation of V^l_μ and $\Sigma^\dagger_\mu$, we consider the equation $B = A^T A$ which leads to the following eigenvalue problem $BV^k = V^k \lambda^k$ with the kth column of V.

After differentiation, one obtains

$$(V^k)^T (B_\mu - \lambda^k_\mu I) V^k = 0. \tag{5}$$

Equation (5) is solved in the least-squares sense and we denote one particular solution by s^k. $\Sigma^\dagger_\mu$ is computed using the relation $\sigma^2 = \lambda$. For details, we refer to [3, Sec. 3.2].

We use the sensitivity information in two ways, i.e. extrapolating POD (ExtPOD) and expanding POD (ExpPOD) basis. In ExtPOD, the POD basis depending on μ is written using the first-order Taylors expansion as follows

$$\psi(x, \mu) = \psi(x, \mu_0) + \Delta\mu \frac{\partial \psi}{\partial \mu}(x, \mu_0) + \mathcal{O}(\Delta\mu^2).$$

In ExpPOD, the POD basis sensitivities are also added to the original POD basis as $[\psi_1, \ldots, \psi_l, (\psi_1)_\mu, \ldots, (\psi_l)_\mu]$ and the reduced order solution is written as

$$y^r_h(x,t) = \sum_{j=1}^{l} y^r_j(t)\psi_j(x) + \sum_{j=l+1}^{2l} y^r_j(t)(\psi_j(x))_\mu,$$

where the dimension of the reduced basis is doubled.

5 Numerical Results

We consider the optimal control problem with

$$Q = (0, 1] \times \Omega,\ \Omega = (0, 1)^2,\ \epsilon = 10^{-2},\ \boldsymbol{\beta} = \frac{1}{\sqrt{2}}(1, 1)^T,\ r = 1,\ \alpha = 1.$$

The source function f, the desired state y_d and the initial condition y_0 are computed from the optimality system (2) using the following exact solutions of the state and control, respectively,

$$y(x,t) = (1 - e^{-t})xye^{-\frac{1-x}{\epsilon}-1}e^{-\frac{1-y}{\epsilon}-1},$$
$$u(x,t) = (1 - t)xy(1 - x)(1 - y)\arctan\left(\frac{x - y}{\epsilon}\right).$$

We observe that the state contains boundary layers along $x = 1$ and $y = 1$, while the control exhibit an interior layer along $x = y$ of the width ϵ. The full problem is solved for $\Delta x = 1/40$, $\Delta t = 1/60$. The conjugate gradient method is used in the optimization step. The error between the full and reduced solution of the control is measured with respect to $L^2(0, T; L^2(\Omega))$.

We choose the parameter range for the ratio $\mu =\mid \boldsymbol{\beta} \mid /\epsilon$ as $1/\epsilon = 80 : 5 : 120$. We compute l POD basis functions associated to the nominal diffusion parameter, i.e. $\epsilon = 1/100$, and compare the resulting error with the ExtPOD and ExpPOD basis. Three different snapshot sets for W are used to generate the POD basis functions, namely state Y, adjoint P and the combination of them $Y \cup P$, as in [5]. The state, adjoint and the control are written in terms of the same POD basis functions associated to W and then the optimality system is projected onto the low-dimensional subspace.

We choose the number of POD basis functions, namely l, according to the relative information content, that is, the ratio of the modelled energy to the total energy contained in the system $\mathscr{E}(l) = \sum_{i=1}^{l} \lambda_i / \sum_{i=1}^{d} \lambda_i$. It is fixed up to $100(1 - \gamma)\%$ by keeping the most energetic POD modes. In this study, we choose 10 POD basis functions setting $\gamma = 10^{-2}$.

Because the velocity field is constant in our example, we proceed with the diffusion term to calculate the sensitivities. In Fig. 1, we present the decrease of the first 15 eigenvalues of the snapshot ensemble $Y, P, Y \cup P$ on the left and their sensitivities $Y_\mu, P_\mu, Y_\mu \cup P_\mu$ on the right. The sensitivities are computed using the centered *FD* quotient and *CSE* method. We observe that *FD* and *CSE* methods yield almost the same eigenvalues. We note that the eigenvalues and their sensitivities are decreasing.

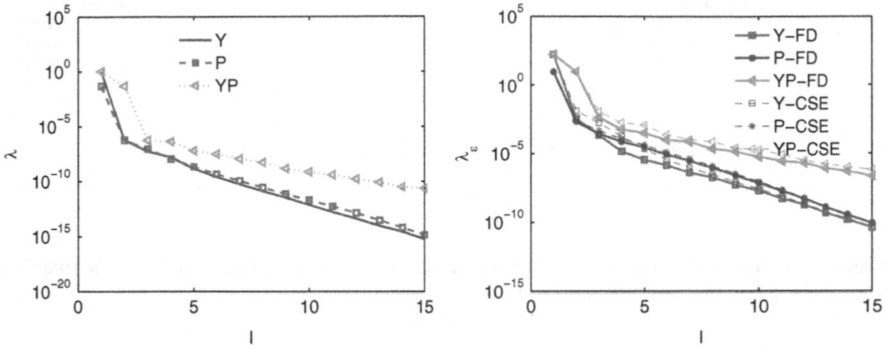

Fig. 1 Eigenvalues(*left*) and their sensitivities(*right*)

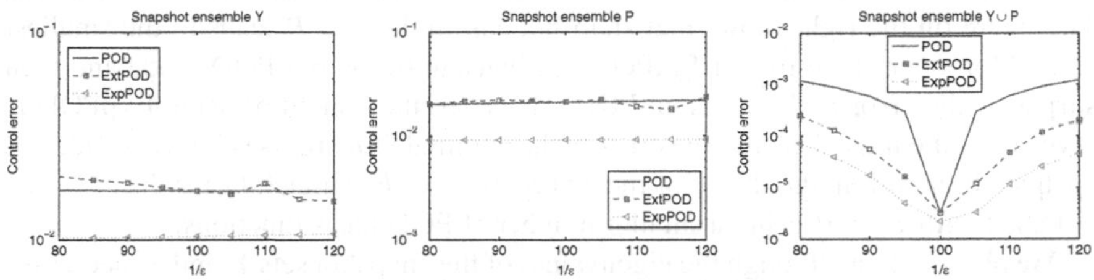

Fig. 2 Error versus parameter for 10 POD basis functions generated with the snapshot set Y (*left*), P (*middle*) and $Y \cup P$ (*right*)

In Fig. 2, we present the error for the control with respect to ϵ with 10 POD bases functions. The control approximated with the POD bases generated from the state solution is poor because the characteristics of the control are totally different from the state solution. The inclusion of the adjoint information in W improves the performance of the method, because the relation between the adjoint and the control is determined through the optimality condition (2). In addition, a good approximation to the control influences the state solution directly due to acting on the right-hand side of the state equation. The figures on the left and in the middle indicate that the snapshot sets Y and P cannot reveal the sensitivity of the control with respect to ϵ. Although ExpPOD gives the smallest error, it is too large for the reduced solution to be accepted. The solution plotted in the right of Fig. 2 is obtained using the snapshot set $Y \cup P$ and it reveals the sensitivity of the problem with respect to ϵ. As we move away the parameter, the error in the reduced solution increases. For the reduced solution of the perturbed problem, ExpPOD basis generated with the snapshot ensemble $Y \cup P$ is the most promising basis among POD and ExtPOD.

In Fig. 3, we present the error for the control with respect to increasing number of POD basis functions by taking $\epsilon = 1/120$. The figure on the left has been obtained with the state snapshots Y and shows that the error decays slowly and is oscillating due to a poor approximation to the control is used. The figure in the middle depicts the error obtained by the POD basis generated with the adjoint snapshot set P.

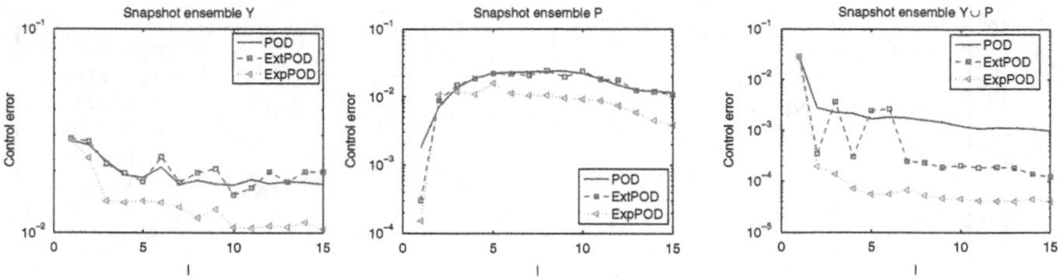

Fig. 3 Error versus the number of POD basis functions for $\epsilon = 1/120$ generated with the snapshot set Y(*left*), P(*middle*) and $Y \cup P$(*right*)

Although the error for the first POD mode is around 10^{-4}, error increases up to 10^{-2}, which is not usual when the number POD basis functions are increased. The figure on the right shows that snapshot ensemble $Y \cup P$ leads to the smallest error. Moreover, the error for ExtPOD oscillates until the 6th POD mode and then surpasses the error in the nominal POD. However, the benefit of using ExpPOD is revealed at the most, because the error in the nominal POD basis is improved almost 2 digits. In addition, the decay of the errors for $Y \cup P$ is much faster than the one obtained with Y or P using a smaller number of POD basis functions.

We observe that although the eigenvalues of the snapshot sets Y and P decreases as shown in Fig. 1, the quality of the reduced-order control obtained using Y or P is not sufficient. The state and adjoint snapshots might be a good choice if associated POD basis is used to approximate the state and adjoint independently. The POD basis generated via the snapshot ensemble $Y \cup P$, containing information about both state and adjoint, give more accurate reduced order solutions and capture the sensitivity of the problem better. For the perturbed problem, expanding the POD basis increases accuracy without solving the nominal problem for each parameter.

Acknowledgements This work has been supported by METU BAP-07-05-2013-003.

References

1. T. Akman, H. Yücel, B. Karasözen, A priori error analysis of the upwind symmetric interior penalty galerkin (SIPG) method for the optimal control problems governed by unsteady convection diffusion equations. Comput. Optim. Appl. **57**(3), 703–729 (2014)
2. D.N. Arnold, F. Brezzi, B. Cockburn, L.D. Marini, Unified analysis of discontinuous Galerkin methods for elliptic problems. SIAM J. Numer. Anal. **39**(5), 1749–1779 (2001/2002)
3. A. Hay, J.T. Borggaard, D. Pelletier, Local improvements to reduced-order models using sensitivity analysis of the proper orthogonal decomposition. J. Fluid Mech. **629**, 41–72 (2009)
4. M. Hinze, S. Volkwein, Proper orthogonal decomposition surrogate models for nonlinear dynamical systems: error estimates and suboptimal control, in *Dimension Reduction of Large-Scale Systems*, ed. by P. Benner, V. Mehrmann, D.C. Sorensen. Lecture Notes in Computational Science and Engineering, vol. 45 (Springer, Berlin, 2005), pp. 261–306

5. M. Hinze, S. Volkwein, Error estimates for abstract linear-quadratic optimal control problems using proper orthogonal decomposition. Comput. Optim. Appl. **39**(3), 319–345 (2008)
6. A. Quarteroni, G. Rozza, A. Manzoni, Certified reduced basis approximation for parametrized partial differential equations and applications. J. Math. Ind. **1**, Art. 3, 44 (2011)
7. V. Thomée, *Galerkin Finite Element Methods for Parabolic Problems*. Springer Series in Computational Mathematics, vol. 25, 2nd edn. (Springer, Berlin, 2006)
8. F. Tröltzsch, *Optimal Control of Partial Differential Equations: Theory, Methods and Applications*. Graduate Studies in Mathematics, vol. 112 (American Mathematical Society, Providence, 2010)

Reduced Basis Approximation of Parametrized Advection-Diffusion PDEs with High Péclet Number

Paolo Pacciarini and Gianluigi Rozza

Abstract In this work we show some results about the reduced basis approximation of advection dominated parametrized problems, i.e. advection-diffusion problems with high Péclet number. These problems are of great importance in several engineering applications and it is well known that their numerical approximation can be affected by instability phenomena. In this work we compare two possible stabilization strategies in the framework of the reduced basis method, by showing numerical results obtained for a steady advection-diffusion problem.

1 Introduction

We show here some recent results about *stabilized reduced basis methods* for the approximation of parametrized advection-diffusion problems with high Péclet number, which expresses the ratio between the advection term and the diffusion one.

Advection-diffusion problems are effectively employed to model a wide range of physical phenomena. Just to give an example, we can recall heat transfer phenomena (with conducion and convection) [11] or diffusion of pollutants in the atmosphere [2, 9]. These equations can depend on several parameters, typically the Péclet number, the advection field direction and the geometry of the domain.

Moreover, parametrized advection-diffusion equations are often used in engineering applications which require very fast evaluations of the solution, given particular values of the parameters. The reduced basis method [8, 12] can effectively

P. Pacciarini
MOX – Modellistica e Calcolo Scientifico, Dipartimento di Matematica F. Brioschi, Politecnico di Milano, P.za Leonardo da Vinci 32, I-20133, Milano, Italy
e-mail: paolo.pacciarini@gmail.com; paolo.pacciarini@polimi.it

G. Rozza (✉)
SISSA mathLab, International School for Advanced Studies, Via Bonomea 265, I-34136 Trieste, Italy
e-mail: grozza@sissa.it
(Acknowledgements: INdAM-GNCS and SISSA NOFYSAS)

A. Abdulle et al. (eds.), *Numerical Mathematics and Advanced Applications ENUMATH 2013*, Lecture Notes in Computational Science and Engineering 103,
DOI 10.1007/978-3-319-10705-9_41

provide a *rapid* approximation of the solution, as well as *rigorous* error bounds, which guarantee the *reliability* of the solution. A very important feature of the reduced basis method is its decomposition in two computational stages. In the first expensive stage, called *Offline* stage, some high-fidelity solutions of the problems are computed, which will become the basis functions for the Galerkin projection performed in the second inexpensive stage, called *Online* stage.

Some applications of the reduced basis method to advection-diffusion problems, such as the Graetz problem or the "thermal fin" problem, can be found in literature, especially the case in which the Péclet number is moderate (i.e. $\sim 10^2$) [3,5,8,11,13].

When the Péclet number takes higher values, the finite elements (FE) approximation of advection-diffusion problems can show significant instability phenomena (see e.g. [10]). To overcome this problem, one can resort to some classical stabilization methods, like the *Streamline/Upwind Petrov Galerkin* (SUPG) method [1]. In this way, it is possible to compute a stable approximated solution suitable to be considered as the *truth* one, i.e, the reference high-fidelity solution for the RB method. A first investigation of the coupling between the stabilized FE formulation and the RB method has been done in [2, 9]. We now base our work on some more recent results given in [6]. Following the latter work, we want to compare two possible strategies of stabilization, by comparing some numerical results in the steady case. The first one, that we will call *Offline-online stabilized* method consists in "stabilize" both the Offline and the Online stages, i.e. using the same stabilized bilinear form in both stages. This method has been actually applied in [2, 9]. The other method, called *Offline-only* consists in "stabilize" only the Offline stage and then perform the Online stage using the standard advection-diffusion operator. To explain the underlying idea, first of all we recall that the RB solution is actually a linear combination of few reduced basis (i.e. the high-fidelity solutions computed during the Offline stage) [12]. It can then be reasonable to expect that if our reduced basis are stable, the reduced solution obtained using the non-stabilized advection-diffusion operator will be stable too. After this brief introduction, in Sect. 2 we recall the stabilized reduced basis method, in Sect. 3 we show some numerical tests and, finally, we draw some conclusions.

2 Stabilized Reduced Basis Method

We take now into account a general parametric advection diffusion problem:

$$-\varepsilon(\boldsymbol{\mu})\Delta u(\boldsymbol{\mu}) + \boldsymbol{\beta}(\boldsymbol{\mu}) \cdot \nabla u(\boldsymbol{\mu}) = 0 \quad \text{on } \Omega. \tag{1}$$

given a parameter value $\boldsymbol{\mu}$ in the parameter domain $\mathscr{D}$ and suitably chosen Dirichlet, Neumann or mixed boundary conditions. We consider a domain Ω which is an open subset of $\mathbb{R}^2$. As regards the coefficients, we consider sufficiently regular functions

$\varepsilon(\boldsymbol{\mu})\colon \Omega \to \mathbb{R}$ and $\boldsymbol{\beta}(\boldsymbol{\mu})\colon \Omega \to \mathbb{R}^2$. The bilinear form associated with the advection-diffusion problem is:

$$a(u, v; \boldsymbol{\mu}) = \int_{\Omega} \varepsilon(\boldsymbol{\mu}) \nabla u \cdot \nabla v + \boldsymbol{\beta}(\boldsymbol{\mu}) \cdot \nabla u \, v \qquad \forall u, v \in H^1(\Omega). \tag{2}$$

Given a triangulation $\mathscr{T}_h$ defined on Ω, with maximum element diameter h, we can set up a FE approximation of the advection-diffusion problem [7]. We denote with $X^{\mathscr{N}}$ the space of piecewise-linear finite elements. It is very well known in literature (see e.g. [7, 10]) that the FE approximation can show instability phenomena when the advective terms dominates the diffusive one. More precisely, we say that a problem is *advection dominated in* $K \subset \Omega$ if the following condition holds:

$$\mathbb{P}\mathrm{e}_K(\boldsymbol{\mu})(\boldsymbol{x}) := \frac{|\boldsymbol{\beta}(\boldsymbol{\mu})(\boldsymbol{x})| h_K}{2\varepsilon(\boldsymbol{\mu})(x)} > 1 \quad \forall \boldsymbol{x} \in K \quad \forall \boldsymbol{\mu} \in \mathscr{D}, \tag{3}$$

where h_K is the diameter of K.

In order to obtain an approximated solution which does not show instabilities, we can resort to some stabilization method. We decided to exploit the classical SUPG method [1]. This consists in substituting, in the FE formulation, the standard advection-diffusion bilinear form (2) with the following one

$$\begin{aligned} a_{stab}(w^{\mathscr{N}}, v^{\mathscr{N}}; \boldsymbol{\mu}) = & \int_{\Omega} \varepsilon(\boldsymbol{\mu}) \nabla w^{\mathscr{N}} \cdot \nabla v^{\mathscr{N}} + (\boldsymbol{\beta}(\boldsymbol{\mu}) \cdot \nabla w^{\mathscr{N}}) v^{\mathscr{N}} \\ & + \sum_{K \in \mathscr{T}_h} \delta_K \int_K L^{\mu} v^{\mathscr{N}} \left(\frac{h_K}{|\boldsymbol{\beta}(\boldsymbol{\mu})|} L^{\mu}_{SS} v^{\mathscr{N}} \right) \end{aligned} \tag{4}$$

with $w^{\mathscr{N}}, v^{\mathscr{N}}$ chosen in $X^{\mathscr{N}}$. In (4) L^{μ} is the advection-diffusion operator $L^{\mu} v^{\mathscr{N}} = -\varepsilon(\boldsymbol{\mu}) \Delta v^{\mathscr{N}} + \boldsymbol{\beta}(\boldsymbol{\mu}) \cdot \nabla v^{\mathscr{N}}$, while L^{μ}_{SS} is its skew-symmetric part. Note that in the case of a divergence free advection field $\boldsymbol{\beta}(\boldsymbol{\mu})$, it holds that $L^{\mu}_{SS} = \boldsymbol{\beta}(\boldsymbol{\mu}) \cdot \nabla v^{\mathscr{N}}$ [10]. The weights δ_K have to be properly chosen in order to ensure the stability and convergence of the SUPG method [7, 10].

We can now consider the RB approximation of the problem (1). As regards the Offline stage, we decided to consider only the stabilized bilinear form (4) and thus we considered as *truth* solution the SUPG stabilized one, that is to find $u^{s\mathscr{N}}(\boldsymbol{\mu}) \in X^{\mathscr{N}}$ such that

$$a_{stab}(u^{s\mathscr{N}}, v^{\mathscr{N}}; \boldsymbol{\mu}) = f_{stab}(v^{\mathscr{N}}; \boldsymbol{\mu}) \quad \forall\, v^{\mathscr{N}} \in X^{\mathscr{N}}. \tag{5}$$

where the right-hand side functional f can be a forcing term or can depend on the imposition of boundary conditions. Considering problem (5), we can set up the Offline stage of the RB method, which produces a reduced space $X^{\mathscr{N}}_N \subset X^{\mathscr{N}}$ with dimension N such that $N \ll \mathscr{N}$.

For the Online stage, we propose two different strategies. The first one, which correspond to the *Offline-Online stabilized* method, consists in using the stabilized bilinear form also during the Online stage. Then the Online problem turns out to be: find $u_N^s(\boldsymbol{\mu}) \in X_N^{\mathcal{N}}$ such that

$$a_{stab}(u_N^s(\boldsymbol{\mu}), v_N; \boldsymbol{\mu}) = f_{stab}(v_N; \boldsymbol{\mu}) \quad \forall\, v_N \in X_N^{\mathcal{N}}. \tag{6}$$

On the contrary in the second method we propose, the *Offline-only stabilized* method, the Online stage is performed using the original advection-diffusion bilinear form (2). The Online problem is then: find $u_N(\boldsymbol{\mu}) \in X_N^{\mathcal{N}}$ such that

$$a(u_N(\boldsymbol{\mu}, v_N; \boldsymbol{\mu}) = f(v_N; \boldsymbol{\mu}) \quad \forall\, v_N \in X_N^{\mathcal{N}}. \tag{7}$$

The right-hand side functional f can in general be different from the one of the stabilized problem, because it does not contain, for example, contributions given by the stabilization term and the lifting of Dirichlet boundary conditions.

3 Numerical Test: Advection-Diffusion Problem with a Boundary Layer

We consider now the following advection-diffusion problem, whose domain $\Omega_o(\boldsymbol{\mu})$ is sketched in Fig. 1,

$$\begin{cases} -\frac{1}{\mu_1}\Delta u(\boldsymbol{\mu}) + \boldsymbol{\beta}\cdot\nabla u(\boldsymbol{\mu}) = 0 \text{ in } \Omega_o(\boldsymbol{\mu}) & \\ u(\boldsymbol{\mu}) = 0 & \text{on } \Gamma_{o,1}(\boldsymbol{\mu}) \cup \Gamma_{o,2}(\boldsymbol{\mu}) \\ \frac{1}{\mu_1}\frac{\partial u}{\partial n}(\boldsymbol{\mu}) = 0 & \text{on } \Gamma_{o,3}(\boldsymbol{\mu}) \\ \frac{1}{\mu_1}\frac{\partial u}{\partial n}(\boldsymbol{\mu}) = 1 & \text{on } \Gamma_{o,4}(\boldsymbol{\mu}) \end{cases} \tag{8}$$

where $\boldsymbol{\mu} = (\mu_1, \mu_2)$ belongs to $\mathcal{D} = [100, 1000]\times[2, 6]$. We choose $\boldsymbol{\beta} = (y, -0.1)$. In order to effectively perform a RB approach, we need to choose a reference domain Ω, as described [8, 12]. We thus set $\Omega = \Omega_o(\mu_2 = 3)$ on which we define the FE triangulation. We also define an affine transformation $T^{\boldsymbol{\mu}}: \Omega \to \Omega_o(\boldsymbol{\mu})$ which maps the reference domain onto the parametrized one, which is $T^{\boldsymbol{\mu}}(x, y) = (\mu_2\, x/3, y)$.

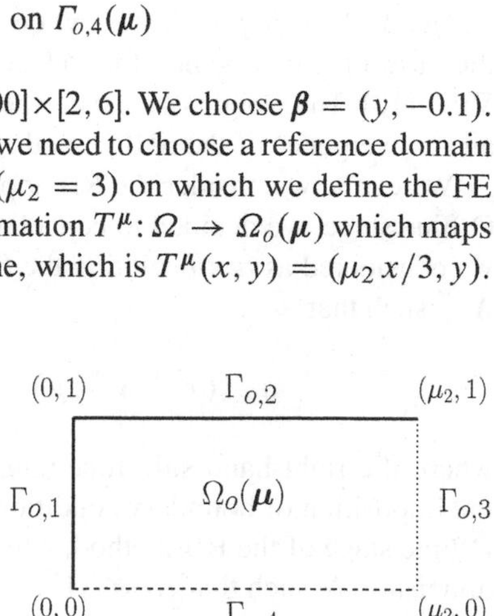

Fig. 1 Domain of problem (8). The boundary conditions are: homogeneous Dirichlet on the *bold sides*, homogeneous Neumann on the *dotted side* and non-homogeneous Neumann on the *dashed side*

Using the transformation T^{μ} we can track back to the reference domain all the bilinear forms defined on the parametrized domain. The transformed advection-diffusion bilinear forms turns out to be:

$$a(w^{\mathcal{N}}, v^{\mathcal{N}}; \mu) = \frac{3}{\mu_1 \mu_2} \int_{\Omega} \partial_x w^{\mathcal{N}} \partial_x v^{\mathcal{N}} + \frac{\mu_2}{3\mu_1} \int_{\Omega} \partial_y w^{\mathcal{N}} \partial_y v^{\mathcal{N}} + \int_{\Omega} y\, \partial_x w^{\mathcal{N}} v^{\mathcal{N}} - \frac{\mu_2}{30} \int_{\Omega} \partial_y w^{\mathcal{N}} v^{\mathcal{N}}, \tag{9}$$

for all $w^{\mathcal{N}}$, $v^{\mathcal{N}}$ in $X^{\mathcal{N}}$. Note that the bilinear form (9), satisfies the *affinity* assumption

$$a(w^{\mathcal{N}}, v^{\mathcal{N}}; \boldsymbol{\mu}) = \sum_{q=1}^{Q_a} \Theta_a^q(\boldsymbol{\mu}) a^q(w^{\mathcal{N}}, v^{\mathcal{N}}) \quad \forall \boldsymbol{\mu} \in \mathscr{D}, \tag{10}$$

where Θ_a^q, $q = 1, \ldots, Q_a$, are functions $\mathscr{D} \to \mathbb{R}$ while a^q, $q = 1, \ldots, Q_a$, are $\boldsymbol{\mu}$-independent bilinear forms on $X^{\mathcal{N}}$. Assumption (10) is crucial for the efficiency of the Offline/Online decomposition of the RB method [8, 12].

As regards the stabilization term, we point out that for piecewise linear approximation we do not have particular restriction on the choice of the weights δ_K [10]. We then set $\delta_K = 1$ for each element K. As piecewise linear functions have null Laplacian inside each element, the stabilization term becomes:

$$s(w^{\mathcal{N}}, v^{\mathcal{N}}; \boldsymbol{\mu}) = \sqrt{\frac{1+\mu_2^2}{10}} \left[\frac{3}{\mu_2} \sum_{K \in \mathscr{T}_h} h_K \int_K y^2 \partial_x w^{\mathcal{N}} \partial_x v^{\mathcal{N}} + \sum_{K \in \mathscr{T}_h} h_K \int_K 2\, y \,(\partial_x w^{\mathcal{N}} \partial_y v^{\mathcal{N}} + \partial_y w^{\mathcal{N}} \partial_x v^{\mathcal{N}}) + \frac{\mu_2}{3} \sum_{K \in \mathscr{T}_h} h_K \int_K \partial_y w^{\mathcal{N}} \partial_y v^{\mathcal{N}} \right]. \tag{11}$$

for all $w^{\mathcal{N}}$, $v^{\mathcal{N}}$ in $X^{\mathcal{N}}$. The term $\sqrt{(1+\mu_2^2)/10}$ has been inserted to keep into account the transformation of the element diameter. In order to ensure the *affinity* assumption (10) also for the stabilization term, with $Q_a \ll \mathcal{N}$, we assumed that each element diameter transforms as the diameter of the whole domain. Considering the exact transformation for each element diameter would have implied a number of affine terms of the order of $\mathcal{N}$ (one affine term per element).

Having defined forms (9) and (11), we can define the stabilized bilinear form $a_{stab} = a + s$. Now we can set up the Offline stage of the RB method, to be performed with respect to the stabilized bilinear form a_{stab}. We applied the Successive Constraint Method (SCM) [4, 12] to build computationally inexpensive lower bounds for the parametric coercivity constants and then we applied the standard RB Greedy algorithm [8, 12].

In our computations, the Offline stage required 311 s (237 s for the SCM) and produced a reduced space with $N = 26$ basis. The tolerance on the Greedy algorithm was $\varepsilon^*_{tol} = 10^{-3}$. This means that we can guarantee that

$$|||u^s_N(\boldsymbol{\mu}) - u^{s\,\mathcal{N}}(\boldsymbol{\mu})|||_{\boldsymbol{\mu},stab} \leq \varepsilon^*_{tol} \quad \forall \boldsymbol{\mu} \in \Xi \tag{12}$$

where Ξ is a sufficiently large subset of $\mathcal{D}$ with finite cardinality (see [12]). In (12), $|||\cdot|||_{\boldsymbol{\mu},stab}$ is the norm induced by the symmetric part of the bilinear form a_{stab}.

We can now compare the *Offline-Online stabilized* method and the *Offline-only stabilized* method. In Fig. 2 we show some *Offline-Online* approximated solutions, while in Fig. 3 we show some *Offline-only* approximated solutions. It is evident that the solutions produced with the *Offline-only stabilized* method can show significant instabilities, as shown in Fig. 3b. We have actually shown that *a Galerkin projection on a subspace spanned by stable functions does not guarantee that the solution does not show instability phenomena.* On the contrary, we observe that the *Offline-Online stabilized* method always produces stable solutions.

In order to understand the bad behaviour of the *Offline-only stabilized* method for our problem, the following upper bound can been proven using the same arguments of [6],

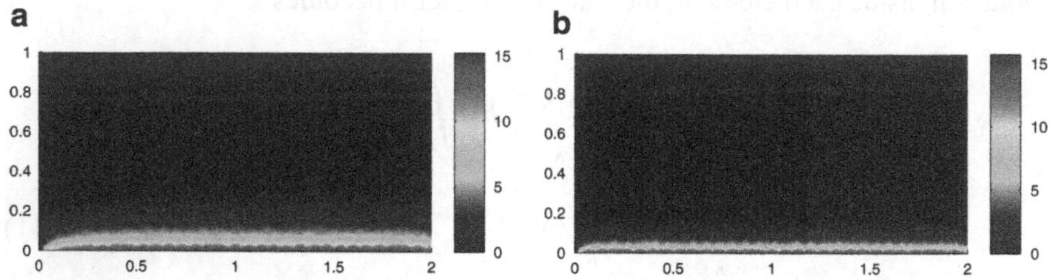

Fig. 2 *Offline-Online stabilized* method. Solutions for some representative values of the parameter. **(a)** $\boldsymbol{\mu} = (200, 3)$. **(b)** $\boldsymbol{\mu} = (900, 3)$

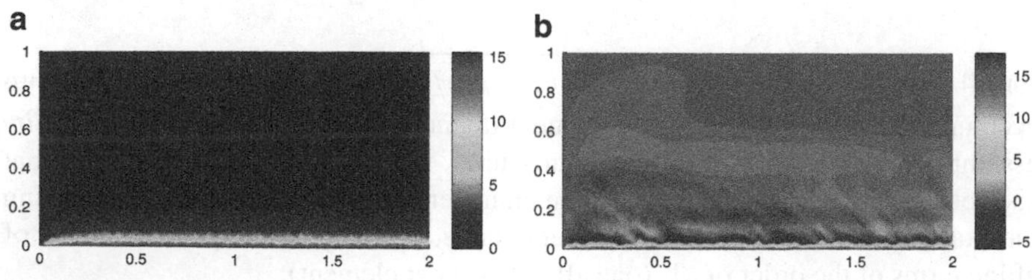

Fig. 3 *Offline-only stabilized* method. Solutions for some representative values of the parameter. **(a)** $\boldsymbol{\mu} = (200, 3)$. **(b)** $\boldsymbol{\mu} = (900, 3)$

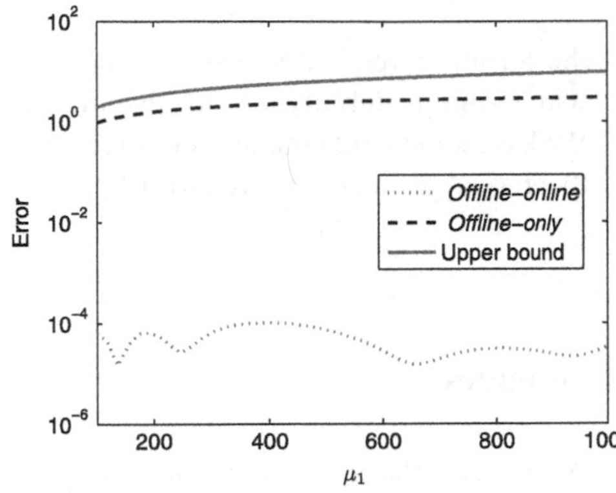

Fig. 4 Approximation errors and upper bound as functions of μ_1, for $\mu_2 = 3$ fixed

Proposition 1 (Upper bound for the *Offline-only* method) *The following estimate of the error between the Offline-only stabilized approximation $u_N(\boldsymbol{\mu})$ and the stabilized FE approximation $u^{s\,\mathcal{N}}(\boldsymbol{\mu})$ holds:*

$$|||u_N(\boldsymbol{\mu}) - u^{s\,\mathcal{N}}(\boldsymbol{\mu})|||_\mu \leq |||u^s_N(\boldsymbol{\mu}) - u^{s\,\mathcal{N}}(\boldsymbol{\mu})||| \\ + h_{max}\sqrt{\mu_1 \tfrac{1+\mu_2^2}{10}}\,\|\boldsymbol{\beta}\cdot\nabla(u^s_N(\boldsymbol{\mu}) + g_h)\|_{L^2(\Omega_o(\boldsymbol{\mu}))} \tag{13}$$

where $u^s_N(\boldsymbol{\mu})$ is the Offline-Online stabilized solution, g_h is the lifting of the Dirichlet boundary condition and $|||\cdot|||_\mu$ is the norm induced by the symmetric part of the bilinear form a. The value h_{max} is the maximum element diameter of the reference mesh $\mathcal{T}_h$.

In Fig. 4 we show a comparison between the *Offline-Online* approximation error, the *Offline-Online* approximation error and the upper bound (13), having fixed $\mu_2 = 3$. The reasonable sharpness shown by the upper bound suggests that in general the *Offline-Only stabilized* method is not a good approximation strategy. We can also highlight that a major component of the *Offline-only* error can be the streamline derivative term in (13). This is also suggested by the fact that, when the streamline derivative term in (13) is "small", e.g. when the advection field and the boundary layer are almost parallel and both the advection field and the gradient of the solution have relatively small modulus, then the *Offline-only stabilized* method can produce satisfactory results too, as shown in [6] for a Graetz problem.

Conclusions

We have investigated the RB approximation of advection dominated RB problems, comparing two possible strategies an *Offline-Online stabilized* method and an *Offline-only stabilized* one. Numerical results have shown that

(continued)

the former gives better results, while the latter produces reduced solutions with strong instability effects, even if the reduced basis functions are stable. We have shown that the numerical results obtained are in accordance with the theoretical estimates proven in [6].

References

1. A. Brooks, T. Hughes, Streamline upwind/Petrov-Galerkin formulations for convection dominated flows with particular emphasis on the incompressible Navier-Stokes equations. Comput. Methods Appl. Mech. Eng. **32**(1–3), 199–259 (1982)
2. L. Dedè, Reduced basis method for parametrized elliptic advection-reaction problems. J. Comput. Math. **28**(1), 122–148 (2010)
3. F. Gelsomino, G. Rozza, Comparison and combination of reduced-order modelling techniques in 3D parametrized heat transfer problems. Math. Comput. Model. Dyn. Syst. **17**(4), 371–394 (2011)
4. D. Huynh, G. Rozza, S. Sen, A. Patera, A successive constraint linear optimization method for lower bounds of parametric coercivity and inf-sup stability constants. C. R. Math. Acad. Sci. Paris **345**(8), 473–478 (2007)
5. F. Negri, G. Rozza, A. Manzoni, A. Quarteroni, Reduced basis method for parametrized elliptic optimal control problems. SIAM J. Sci. Comput. **35**(5), A2316–A2340 (2013)
6. P. Pacciarini, G. Rozza, Stabilized reduced basis method for parametrized advection-diffusion PDEs. Comput. Methods Appl. Mech. Eng. **18**, 1–18 (2014)
7. A. Quarteroni, *Numerical Models for Differential Problems*. MS&A Modeling, Simulation and Applications, vol. 8 (Springer, Milan, 2014)
8. A. Quarteroni, G. Rozza, A. Manzoni, Certified reduced basis approximation for parametrized partial differential equations and applications. J. Math. Ind. **1**, 1(3) (2011)
9. A. Quarteroni, G. Rozza, A. Quaini, Reduced basis methods for optimal control of advection-diffusion problem, in *Advances in Numerical Mathematics*, Moscow/Houston, 2007, ed. by W. Fitzgibbon, R. Hoppe, J. Periaux, O. Pironneau, Y. Vassilevski, pp. 193–216
10. A. Quarteroni, A. Valli, *Numerical Approximation of Partial Differential Equations*. Springer Series in Computational Mathematics, vol. 23 (Springer, Berlin, 1994)
11. G. Rozza, D. Huynh, N. Nguyen, A. Patera, Real-time reliable simulation of heat transfer phenomena, in *ASME – American Society of Mechanical Engineers – Heat Transfer Summer Conference Proceedings*, HT2009 3, San Francisco, pp. 851–860, 2009
12. G. Rozza, D. Huynh, A. Patera, Reduced basis approximation and a posteriori error estimation for affinely parametrized elliptic coercive partial differential equations: application to transport and continuum mechanics. Arch. Comput. Methods Eng. **15**(3), 229–275 (2008)
13. G. Rozza, T. Lassila, A. Manzoni, Reduced basis approximation for shape optimization in thermal flows with a parametrized polynomial geometric map, in *Spectral and High Order Methods for Partial Differential Equations* ed. by J. Hesthaven, E.M. Rønquist. Lecture Notes in Computational Science and Engineering, vol. 76 (Springer, Berlin/Heidelberg, 2011), pp. 307–315

Reduced-Order Modeling and ROM-Based Optimization of Batch Chromatography

Peter Benner, Lihong Feng, Suzhou Li, and Yongjin Zhang

Abstract A reduced basis method is applied to batch chromatography and the underlying optimization problem is solved efficiently based on the resulting reduced model. A technique of adaptive snapshot selection is proposed to reduce the complexity and runtime of generating the reduced basis. With the help of an output-oriented error bound, the construction of the reduced model is managed automatically. Numerical examples demonstrate the performance of the adaptive technique in reducing the offline time. The ROM-based optimization is successful in terms of the accuracy and the runtime for getting the optimal solution.

1 Introduction

Reduced basis methods (RBMs) have been proved to be powerful tools for rapid and reliable evaluation of the output response associated with parameterized partial differential equations (PDEs) [1, 5, 9, 11, 12]. The reduced basis (RB), used to construct the reduced-order model (ROM), is computed from snapshots, that is, the solutions to the PDEs at certain selected samples of parameters and/or chosen time steps. An efficient and rigorous a posteriori error estimation is crucial for RBMs because it enables automatic generation of the RB, and in turn a reliable ROM, with the help of a greedy algorithm.

The efficiency of RBMs is ensured by the strategy of offline-online decomposition. During the offline stage, all full-dimension dependent and parameter-independent terms can be precomputed and a (parametric) ROM is obtained a priori; during the online stage, a reliable output response can be obtained rapidly from the ROM for any given feasible parameter. Although the offline cost is usually not taken into consideration, it is typically high, especially for time-dependent PDEs.

P. Benner • L. Feng • S. Li • Y. Zhang (✉)
Max Planck Institute for Dynamics of Complex Technical Systems, Sandtorstrasse 1, 39106 Magdeburg, Germany
e-mail: benner@mpi-magdeburg.mpg.de; feng@mpi-magdeburg.mpg.de; suzhou@mpi-magdeburg.mpg.de; zhangy@mpi-magdeburg.mpg.de

A. Abdulle et al. (eds.), *Numerical Mathematics and Advanced Applications ENUMATH 2013*, Lecture Notes in Computational Science and Engineering 103,
DOI 10.1007/978-3-319-10705-9_42

To reduce the cost and complexity of the offline stage, we propose a technique of adaptive snapshot selection (ASS) for the generation of the RB. For time-dependent problems, if the dynamics (rather than the solution at the final time) is of interest, the solution at the time instances in the evolution process should be collected as snapshots. However, the trajectory for a given parameter might contain a large number of time steps, e.g. in the simulation of batch chromatography. In such a case, if the solutions at all time steps are taken as snapshots, the subsequent computation will be very expensive because the number of snapshots is too large; if one just trivially selects part of the solutions, i.e. solutions at parts of the time instances (e.g. every two or several time steps), the final RB approximation might be of low accuracy because important information may have been lost due to such a naive snapshot selection. We propose to select the snapshot adaptively according to the variation of the solution in the evolution process. The idea is to make full use of the behavior of the trajectory and discard the redundant (linearly dependent) information adaptively. It enables the generation of the RB with a small number of snapshots but including only "useful" information. In addition, it is easily combined with other algorithms for the generation of RB, e.g. the POD-Greedy algorithm [9].

Batch chromatography is a very important chemical process and widely used in industries. Many efforts have been made for the optimization of batch chromatography over the last decades [4, 6, 7]. Notably, all these studies are based on the finely discretized full-order model (FOM), which must be repeatedly solved in the optimization process, making the runtime of obtaining the optimal solution too long.

In this paper, a RB method is introduced to generate a surrogate ROM for batch chromatography. The nonlinear terms in the FOM are treated by empirical operator interpolation [1, 2]. With the help of the ASS and an output error bound derived in vector space in [13], the ROM is efficiently constructed in a goal-oriented fashion. The resulting ROM is used for the rapid evaluation of the output response during the optimization process.

This paper is organized as follows. A brief review of the RBM is given in Sect. 2. The ASS technique is presented in detail for the construction of the ROM in Sect. 3. Section 4 shows the numerical results. Conclusions are drawn in section "Conclusions and Perspective".

2 Reduced Basis Method and Empirical Interpolation

Consider a parametrized evolution problem defined over the spatial domain $\Omega \subset \mathbb{R}^d$ and the parameter domain $\mathscr{P} \subset \mathbb{R}^p$,

$$\partial_t u(t,x;\mu) + \mathscr{L}[u(t,x;\mu)] = 0, \quad t \in [0,T], \quad x \in \Omega, \quad \mu \in \mathscr{P}, \tag{1}$$

where $\mathscr{L}[\cdot]$ is a spatial differential operator. Let $\mathscr{W}^{\mathscr{N}} \subset L^2(\Omega)$ be an $\mathscr{N}$-dimensional discrete space in which an approximate numerical solution to Eq. (1) is sought. Let $0 = t^0 < t^1 < \ldots < t^K = T$ be $K+1$ time instants in $[0, T]$. Given

$\mu \in \mathscr{P}$ with suitable initial and boundary conditions, the numerical solution at the time $t = t^n$, $u^n(\mu)$, can be obtained by using suitable numerical methods, e.g. the finite volume method. Assume that $u^n(\mu) \in \mathscr{W}^{\mathscr{N}}$ satisfies the following form,

$$L_I(t^n)[u^{n+1}(\mu)] = L_E(t^n)[u^n(\mu)] + g(u^n(\mu), \mu), \tag{2}$$

where $L_I(t^n)[\cdot], L_E(t^n)[\cdot]$ are linear implicit and explicit operators, respectively, and $g(\cdot)$ is a nonlinear μ-dependent operator. To make the RBM feasible, we assume that $L_I(t^n), L_E(t^n)$ are time-independent. By convention, $u^n(\mu)$ is considered as the "true" solution by assuming that the numerical solution is a faithful approximation of the exact (analytical) solution $u(t^n, x; \mu)$.

RBMs aim to find a suitable low dimensional subspace $\mathscr{W}^N \subset \mathscr{W}^{\mathscr{N}}$ and solve the resulting ROM to get the RB approximation $\hat{u}^n(\mu) \in \mathscr{W}^N$. In addition or alternatively to the field variable itself, the approximation of outputs of interest can also be obtained cheaply by $\hat{y}(\mu) = y(\hat{u}(\mu))$. More precisely, given a RB matrix $V := [V_1, \ldots, V_N]$, Galerkin projection is employed to generate the ROM:

$$V^T L_I(t^n)[Va^{n+1}(\mu)] = V^T L_E(t^n)[Va^n(\mu)] + V^T g(Va^n(\mu)), \tag{3}$$

where $a^n(\mu) = (a_1^n(\mu), \ldots, a_N^n(\mu))^T \in \mathbb{R}^N$ is the vector of the weights in the expression $\hat{u}^n(\mu) := Va^n(\mu) = \sum_{i=1}^N a_i^n(\mu) V_i$, and it is the vector of unknowns in the ROM. Thanks to the linearity of the operators L_I and L_E, the ROM (3) can be rewritten as

$$V^T L_I(t^n) V[a^{n+1}(\mu)] = V^T L_E(t^n) V[a^n(\mu)] + V^T g(Va^n(\mu)), \tag{4}$$

where $V^T L_I(t^n) V$ and $V^T L_E(t^n) V$ can be precomputed and stored for the construction of the ROM. However, the computation of the last term in (4), $V^T g(Va^n(\mu))$, cannot be done analogously because of the nonlinearity of g. This can be tackled by using a technique of empirical (operator) interpolation (EI), see e.g. [1, 2] for details.

The POD-Greedy algorithm [9], shown in Algorithm 2.1, is often used for the generation of the RB for time-dependent problems. Note that $\eta_N(\mu_{\max})$ is an indicator of the error of the ROM. As aforementioned, an efficient and rigorous a posteriori error estimator is desired for efficient construction of the ROM. In this paper, we use an output error bound to compute the error indicator $\eta_N(\mu_{\max})$. Due to space limitations, the derivation of the output error bound will be given in a more detailed paper [13]. For some problems, like the batch chromatographic model under consideration, the implementation of Step 4 in Algorithm 2.1 is costly because the number of time steps K is very large. In this work, we propose to use a technique we call ASS to reduce the cost, which is addressed in Sect. 3.

Algorithm 2.1: RB generation using POD-Greedy

Require: $\mathscr{P}_{\text{train}}$, μ_0, $tol_{RB}(< 1)$
Ensure: RB $V = [V_1, \ldots, V_N]$
1: Initialization: $N = 0$, $V = []$, $\mu_{\max} = \mu_0$, $\eta_N(\mu_{\max}) = 1$
2: **while** the error $\eta_N(\mu_{\max}) > tol_{RB}$ **do**
3: Compute the trajectory $S_{\max} := \{u^n(\mu_{\max})\}_{n=0}^K$.
4: Enrich the RB, e.g. $V := [V, V_{N+1}]$, where V_{N+1} is the first POD mode of the matrix $\bar{U} = [\bar{u}^0, \ldots, \bar{u}^K]$ with $\bar{u}^n := u^n(\mu_{\max}) - \Pi_{\mathscr{W}^N}[u^n(\mu_{\max})]$, $n = 0, \ldots, K$. $\Pi_{\mathscr{W}^N}[u]$ is the projection of u onto the current space $\mathscr{W}^N := \text{span}\{V_1, \ldots, V_N\}$.
5: $N = N + 1$
6: Find $\mu_{\max} := \arg\max_{\mu \in \mathscr{P}_{\text{train}}} \eta_N(\mu)$.
7: **end while**

3 Adaptive Snapshot Selection

For the generation of the RB, a training set $\mathscr{P}_{\text{train}}$ of parameters must be determined. On the one hand, the training set is desired to include the information of the parametric system as much as possible. On the other hand, the RB should be efficiently generated.

To construct the RB efficiently, many efforts have been made on adaptively choosing the training set [3,8] in the past years. The authors tried to get an "optimal" training set in the sense that the original manifold $\mathscr{M} = \{u(\mu) | \mu \in \mathscr{P}\}$ can be well represented by the submanifold $\hat{\mathscr{M}} = \{u(\mu) | \mu \in \mathscr{P}_{\text{train}}\}$ induced by the sample set with its size as small as possible.

For time-dependent problems, in spite of an "optimal" training set, the number of snapshots can be huge if the total number of time steps for a single parameter is large. A large number of snapshots means that it is time-consuming to generate the RB because the POD mode in Step 4 in Algorithm 2.1 is hard to compute from the singular value decomposition of $\bar{U}$, due to the large size of $\bar{U}$. As a straightforward way to avoid using the solutions at all time instances as snapshots, one can simply pick out the solutions at certain time instances (e.g. every two or several time steps) as snapshots. However, the results might be of low accuracy because some important information may have been lost during such a trivial snapshot selection.

For an "optimal" or a selected training set, we propose to select the snapshots adaptively according to the variation of the trajectory of the solution, $\{u^n(\mu)\}_{n=0}^K$. The idea is to discard the redundant ((almost) linearly dependent) information from the trajectory. In fact, the linear dependency of two non-zero vectors v_1 and v_2 can be reflected by the angle θ between them. More precisely, they are linearly dependent if and only if $|\cos(\theta)| = 1$ ($\theta = 0$ or π). In other words, the value $1 - |\cos(\theta)|$ is large if the linear relevance between the two vectors is weak. This implies that the quantity $1 - \frac{|\langle v_1, v_2 \rangle|}{\|v_1\| \|v_2\|}$ ($\cos(\theta) = \frac{\langle v_1, v_2 \rangle}{\|v_1\| \|v_2\|}$) is a good indicator for the linear dependency of v_1 and v_2.

Given a parameter μ and the initial vector $u^0(\mu)$, the numerical solution $u^n(\mu)$ ($n = 1, \ldots, K$) can be obtained, e.g. by using the evolution scheme (2). Define an indicator $Ind(u^n(\mu), u^m(\mu)) = 1 - \frac{|\langle u^n(\mu), u^m(\mu) \rangle|}{\|u^n(\mu)\| \|u^m(\mu)\|}$, which is used to

Algorithm 3.1: Adaptive snapshot selection (ASS)

Require: Initial vector $u^0(\mu)$, tol_{ASS}
Ensure: Selected snapshot matrix $S^A = [u^{n_1}(\mu), u^{n_2}(\mu), \ldots, u^{n_\ell}(\mu)]$
Initialization: $j = 1,\ n_j = 0,\ \ S^A = [u^{n_j}(\mu)]$
for $n = 1, \ldots, K$ **do**
  Compute the vector $u^n(\mu)$.
  if $Ind(u^n(\mu), u^{n_j}(\mu)) > tol_{ASS}$ **then**
    $j = j + 1$
    $n_j = n$
    $S^A = [S^A, u^{n_j}(\mu)]$
  end if
end for

Algorithm 3.2: RB generation using ASS-POD-Greedy

Require: $\mathscr{P}_{\text{train}}$, μ_0, $tol_{RB}(< 1)$
Ensure: RB $V = [V_1, \ldots, V_N]$
Initialization: $N = 0$, $V = []$, $\mu_{\max} = \mu_0$, $\eta(\mu_{\max}) = 1$
while the error $\eta_N(\mu_{\max}) > tol_{RB}$ **do**
  Compute the trajectory $S_{\max} := \{u^n(\mu_{\max})\}_{n=0}^{K}$ and adaptively select snapshots using Algorithm 3.1 to get $S_{\max}^A := \{u^{n_1}(\mu_{\max}), \ldots, u^{n_\ell}(\mu_{\max})\}$.
  Enrich the RB, e.g. $V := [V, V_{N+1}]$, where V_{N+1} is the first POD mode of the matrix $\bar{U}^A = [\bar{u}^{n_1}, \ldots, \bar{u}^{n_\ell}]$ with $\bar{u}^{n_s} := u^{n_s}(\mu_{\max}) - \Pi_{\mathscr{W}^N}[u^{n_s}(\mu_{\max})]$, $s = 1, \ldots, \ell$, $\ell \ll K$. $\Pi_{\mathscr{W}^N}[u]$ is the projection of u onto the space $\mathscr{W}^N := \text{span}\{V_1, \ldots, V_N\}$.
  $N = N + 1$
  Find $\mu_{\max} := \arg\max_{\mu \in \mathscr{P}_{\text{train}}} \eta_N(\mu)$.
end while

measure the linear dependency of the two vectors. When $Ind(u^n(\mu), u^m(\mu))$ is large, the linear relevance between $u^n(\mu)$ and $u^m(\mu)$ is weak. Algorithm 3.1 shows the realization of the ASS, $u^n(\mu)$ is taken as a new snapshot only when $u^n(\mu)$ and $u^{n_j}(\mu)$ are "sufficiently" linearly independent, by checking whether $Ind(u^n(\mu), u^{n_j}(\mu))$ is large enough or not. Here, $u^{n_j}(\mu)$ is the last selected snapshot. Note that the inner product $\langle\cdot,\cdot\rangle : \mathscr{W}^{\mathscr{N}} \times \mathscr{W}^{\mathscr{N}} \to \mathbb{R}$ used above is properly defined according to the solution space, and the norm $\|\cdot\|$ is induced by the inner product.

Remark 1 For the linear dependency, it is also possible to check the angle between the tested vector $u^n(\mu)$ and the subspace spanned by the selected snapshots S^A. More redundant information can be discarded but at more cost. However, the data will be compressed further, e.g. by using the POD-Greedy algorithm, we simply choose the economical case shown in Algorithm 3.1. Note that the tolerance tol_{ASS} is prespecified and problem-dependent, and the value at $O(10^{-4})$ gives good results for the numerical examples studied in Sect. 4 based on our observation.

The ASS technique can be easily combined with other algorithms for the generation of the RB and/or the collateral reduced basis (CRB) for EI. For example, Algorithm 3.2 shows the combination with the POD-Greedy algorithm (Algorithm 2.1).

4 Numerical Experiments

In this paper, we use the RBM and ASS presented in the previous sections to generate a surrogate ROM for batch chromatography. The governing equations for batch chromatography can be described as follows,

$$\begin{cases} \frac{\partial c_z}{\partial t} + \frac{1-\epsilon}{\epsilon}\frac{\partial q_z}{\partial t} = -\frac{\partial c_z}{\partial x} + \frac{1}{Pe}\frac{\partial^2 c_z}{\partial x^2}, & 0 < x < 1, \\ \frac{\partial q_z}{\partial t} = \frac{L}{Q/(\epsilon A_c)}\kappa_z(q_z^{Eq} - q_z), & 0 \le x \le 1. \end{cases} \tag{5}$$

Here c_z, q_z ($z = \mathsf{a}, \mathsf{b}$) are the unknowns in the system, and q_z^{Eq} is a nonlinear function of c_a and c_b. A detailed description of model parameters, and the initial and boundary conditions can be found in [13]. The feed flow rate Q and the injection period t_{in} (in the boundary conditions) are considered as the operating variables, denoted as $\mu := (Q, t_{in})$.

In this section, we first illustrate the performance of the ASS for the construction of the ROM, and then show the results of ROM-based optimization. The parameter domain of μ is $\mathcal{P} = [0.0667, 0.1667] \times [0.5, 2.0]$, and $\mathcal{N} = 1{,}000$ for the FOM. We employ the tolerance $tol_{RB} = 1.0 \times 10^{-6}$ and $tol_{ASS} = 5.0 \times 10^{-4}$ unless stated otherwise. All the computations were done on a PC with Intel Core(TM)2 Quad CPU 2.83 GHz and RAM 4.00 GB.

4.1 Performance of the Adaptive Snapshot Selection

To investigate the performance of the ASS, we compare the runtime of the generation of the RB with different threshold values tol_{ASS}. As is shown in Algorithm 3.2, the ASS can be combined with the POD-Greedy algorithm for the generation of the RB. For the computation of the error indicator $\eta_N(\mu_{\max})$ in Algorithm 3.2, EI is involved for an efficient offline-online decomposition. To efficiently generate a CRB, the ASS is also employed. The training set for the generation of the CRB is $\mathcal{P}_{\text{train}}^{CRB} \subset \mathcal{P}$, with 25 uniform sample points. For each $\mu \in \mathcal{P}_{\text{train}}^{CRB}$, Algorithm 3.1 is used to choose the snapshots adaptively. Table 1 shows the results. It is seen that, the larger tolerance is used, the more runtime is saved. Particularly, when $tol_{ASS} = 5.0 \times 10^{-4}$, the runtime is reduced by 93.1 % compared to that without ASS.

Table 1 Runtime for the generation of the CRB with different tol_{ASS}

	No ASS	ASS	ASS	ASS
tol_{ASS}	–	1.0×10^{-4}	5.0×10^{-4}	1.0×10^{-3}
Runtime (h)	25.30 (–)	2.56 (−89.9 %)	1.74 (−93.1 %)	1.04 (−95.9 %)

Table 2 Comparison of the detailed and reduced simulations over a validation set P_{val} with 400 random sample points. FOM: $\mathcal{N} = 1{,}000$; ROM: $N = 43$

Simulations	FOM	ROM (POD-Greedy)	ROM (ASS-POD-Greedy)
Max. error	–	4.0×10^{-7}	7.7×10^{-7}
Aver-runtime (s)/SpF	91.65/-	3.45/27	3.45/27

With the precomputed CRB ($tol_{ASS} = 5.0 \times 10^{-4}$), we perform Algorithms 2.1 and 3.2 to generate the RB with the same tolerance tol_{RB}, respectively. The runtime for the former (using Algorithm 2.1) is 5.24 h, while it is only 3.10 h for the latter (using Algorithm 3.2). The runtime of the RB construction with the ASS is reduced by 40.9 %. Notice that the CRB is obtained a priori, the runtime of it is not included here. The training set is $\mathcal{P}_{\text{train}} \subset \mathcal{P}$, with 64 uniform sample points. Moreover, the resulting ROM with ASS is almost as accurate as that without ASS, as is shown in Table 2.

4.2 *ROM-Based Optimization*

The optimization of batch chromatography aims to find $\mu^{opt} \in \mathcal{P}$ such that

$$\mu^{opt} := \arg\min_{\mu \in \mathcal{P}} \{-Pr(c_z(\mu), q_z(\mu))\},$$

$$\text{s.t. } Rec_{min} - Rec(c_z(\mu), q_z(\mu)) \leq 0, \quad \mu \in \mathcal{P}$$

$$c_z(\mu), q_z(\mu) \text{ are the solutions to the system (5)}, z = \mathrm{a}, \mathrm{b}.$$

More details about the optimization problem, e.g. the definition of the production rate Pr and the recovery yield Rec, can be found in [13].

Before solving the ROM-based optimization, we first assess the reliability of the resulting ROM. We performed the detailed and reduced simulations over a validation set $\mathcal{P}_{val} \subset \mathcal{P}$ with 400 random sample points. From Table 2, it is seen that the average runtime of the detailed simulation is sped up by a factor of 27, and the maximal true error is below the prespecified tolerance.

The global optimizer NLOPT_GN_DIRECT_L [10] is employed to solve the optimization problems. Let μ^k be the vector of parameters determined by the optimizer at the k-th iteration. When $\|\mu^{k+1} - \mu^k\| < \epsilon_{opt}$, the iteration is stopped and the optimal solution is obtained. Table 3 shows the results. It is seen that the optimal solution to the ROM-based optimization converges to the FOM-based one. Moreover, the runtime of getting the optimal solution is largely reduced. The speed-up factor (SpF) is 29.

Table 3 Optimization results based on the ROM and FOM, $\epsilon_{opt} = 1.0 \times 10^{-4}$

Simulations	Obj. (Pr)	Opt. solution (μ)	#Iterations	Runtime (h)/SpF
FOM-based Opt.	0.020271	(0.07969, 1.05514)	211	10.60/-
ROM-based Opt.	0.020276	(0.07969, 1.05514)	211	0.36/29

Conclusions and Perspective
We present a reduced basis method for batch chromatography and solve the underlying optimization efficiently based on a surrogate ROM. The technique ASS is presented for efficient construction of the ROM. Numerical examples demonstrate that it significantly reduces the offline time while not sacrificing the accuracy of the ROM. In addition, the ASS might be applied to other snapshot-based model reduction methods.

References

1. M. Barrault, Y. Maday, N.C. Nguyen, A.T. Patera, An 'empirical interpolation' method: application to efficient reduced-basis discretization of partial differential equations. Comptes Rendus Math. Acad. Sci. Paris Ser. I **339**(9), 667–672 (2004)
2. M. Drohmann, B. Haasdonk, M. Ohlberger, Reduced basis approximation for nonlinear parametrized evolution equations based on empirical operator interpolation. SIAM J. Sci. Comput. **34**(2), 937–969 (2012)
3. J.L. Eftang, D.J. Knezevic, A.T. Patera, An hp certified reduced basis method for parametrized parabolic partial differential equations. Math. Comput. Model. Dyn. Syst. **17**(4), 395–422 (2011)
4. W. Gao, S. Engell, Iterative set-point optimization of batch chromatography. Comput. Chem. Eng. **29**(6), 1401–1409 (2005)
5. M.A. Grepl, Reduced-basis approximation a posteriori error estimation for parabolic partial differential equations, Ph.D. thesis, Massachusetts Institute of Technology, 2005
6. D. Gromov, S. Li, J. Raisch, A hierarchical approach to optimal control of a hybrid chromatographic batch process. Adv. Control Chem. Process. **7**, 339–344 (2009)
7. G. Guiochon, A. Felinger, D.G. Shirazi, A.M. Katti, *Fundamentals of Preparative and Nonlinear Chromatography* (Academic, Boston, 2006)
8. B. Haasdonk, M. Dihlmann, M. Ohlberger, A training set and multiple bases generation approach for parameterized model reduction based on adaptive grids in parameter space. Math. Comput. Model. Dyn. Syst. **17**(4), 423–442 (2011)
9. B. Haasdonk, M. Ohlberger, Reduced basis method for finite volume approximations of parametrized linear evolution equations. ESAIM Math. Model. Numer. Anal. **42**(2), 277–302 (2008)
10. S.G. Johnson, The NLopt nonlinear-optimization package, http://ab-initio.mit.edu/nlopt
11. A.K. Noor, J.M. Peters, Reduced basis technique for nonlinear analysis of structures. AIAA J. **18**(4), 145–161 (1980)

12. A.T. Patera, G. Rozza, *Reduced Basis Approximation and A Posteriori Error Estimation for Parametrized Partial Differential Equations*. MIT Pappalardo Graduate Monographs in Mechanical Engineering. Cambridge, MA (2007). Available from http://augustine.mit.edu/methodology/methodology_book.htm
13. Y. Zhang, L. Feng, S. Li, P. Benner, *Accelerating PDE Constrained Optimization by the Reduced Basis Method: Application to Batch Chromatography*, preprint MPIMD/14-09, Max Planck Institute Magdeburg Preprints, 2014

Output Error Bounds for the Dirichlet-Neumann Reduced Basis Method

Immanuel Martini and Bernard Haasdonk

Abstract The Dirichlet-Neumann reduced basis method is a model order reduction method for homogeneous domain decomposition of elliptic PDEs on a-priori known geometries. It is based on an iterative scheme with full offline-online decomposition and rigorous a-posteriori error estimates. We show that the primal-dual framework for non-compliant output quantities can be transferred to this method. The results are validated by numerical experiments with a thermal block model.

1 Introduction

Recently, several approaches combining the reduced basis (RB) method—a model reduction method for efficient treatment of parametrized partial differential equations (PDEs)—and domain decomposition—a technique for coupling PDEs on adjacent computational domains—have been developed [1–4]. A standard RB approach consists in approximating the solution manifold of a parametrized PDE by a low-dimensional linear space spanned by so-called snapshots—highly accurate solutions computed with Finite Elements (FE) for example—and a Galerkin-projection on this space. In a domain decomposition framework it is no longer necessary to compute detailed solutions on the whole domain. Furthermore, the dimensions of RB approximation spaces on subdomains may be lower than in the monolithic approach.

The Dirichlet-Neumann RB method [4] is based on the Dirichlet-Neumann FE procedure. It represents a well-known iterative domain decomposition method for linear elliptic problems with an offline/online decomposition, which allows solving the PDE in a very fast online-stage. All high-dimensional FE computations are done in the offline-stage. It also includes effective a-posteriori error estimation for RB approximations, that possibly are discontinuous over the internal boundary.

I. Martini (✉) • B. Haasdonk
Institut of Applied Analysis and Numerical Simulation, University of Stuttgart, Pfaffenwaldring 57, 70569 Stuttgart, Germany
e-mail: immanuel.martini@mathematik.uni-stuttgart.de; haasdonk@mathematik.uni-stuttgart.de

A. Abdulle et al. (eds.), *Numerical Mathematics and Advanced Applications ENUMATH 2013*, Lecture Notes in Computational Science and Engineering 103,
DOI 10.1007/978-3-319-10705-9_43

In this contribution we provide an extension of the method regarding computation and error estimation of output quantities. We make use of the primal-dual framework, which is commonly known to produce good output error bounds for non-compliant problems. We refer to [5] and [6] for an introduction into output error estimation for standard RB methods.

2 Problem Definition

Let $\Omega \subset \mathbb{R}^2$ be a domain with Lipschitz–boundary $\partial\Omega$ and $x \in \overline{\Omega}$ the space variable. We introduce a Hilbert space $X \subset H_0^1(\Omega)$ with the norm $\|v\|_X := \|v\|_{H^1(\Omega)}$ which can be either finite or infinite dimensional. We now consider a decomposition of Ω into 2 subdomains, i.e. $\overline{\Omega} = \overline{\Omega}_1 \cup \overline{\Omega}_2$ and $\Omega_1 \cap \Omega_2 = \emptyset$. The interface Γ is defined as $\Gamma := \partial\Omega_1 \cap \partial\Omega_2$. We assume that Ω_1 and Ω_2 have Lipschitz–boundaries and that Γ, $\partial\Omega_1 \setminus \Gamma$ and $\partial\Omega_2 \setminus \Gamma$ have a nonvanishing $(n-1)$-dimensional measure. Several function spaces are defined according to the domain decomposition,

$$X_k := \left\{v_{|\Omega_k} | v \in X\right\},$$
$$X_k^0 := \{v \in X_k | \gamma v = 0\},$$
$$X_\Gamma := \gamma(X_1) = \gamma(X_2),$$

where $k = 1, 2$. The operator γ denotes the trace operator on Γ, where we do not notationally discriminate between the spaces X_1 or X_2, as it will always be clear from the context. It holds $X_1 \subset H^1(\Omega_1)$, $X_2 \subset H^1(\Omega_2)$ and $X_\Gamma \subset H_{00}^{1/2}(\Gamma)$. We equip the Hilbert spaces X_k, $k = 1, 2$ with the norms $\|v\|_{X_k} := \|v\|_{H^1(\Omega_k)}$ and X_Γ with $\|g\|_{X_\Gamma} := \|g\|_{L_2(\Gamma)}$.

Now let $\mathscr{P} \subset \mathbb{R}^P$, $P \in \mathbb{N}$ be the domain of the parameter $\mu \in \mathscr{P}$. We introduce the parametric elliptic variational problem for defining the parameter-dependent primal solution $u(\mu) \in X$ and the output $s(\mu) \in \mathbb{R}$:

$$a(u(\mu), v; \mu) = f(v; \mu), \quad \forall v \in X, \tag{1}$$
$$s(\mu) = l(u(\mu); \mu), \tag{2}$$

with a parametric bilinear form $a : X \times X \times \mathscr{P} \to \mathbb{R}$ and parametric linear forms $f, l : X \times \mathscr{P} \to \mathbb{R}$. We do not assume symmetry in a. Furthermore, the so-called dual problem for defining the dual solution $\psi(\mu)$ reads

$$a(v, \psi(\mu); \mu) = -l(v; \mu), \quad \forall v \in X. \tag{3}$$

The approximation of $\psi(\mu)$ in the RB scheme helps to get good output approximations, although the dual problem is not strictly necessary for the computation of $s(\mu)$.

2.1 Assumptions

We assume that, for all $\mu \in \mathscr{P}$, a is continuous on X and coercive on X with coercivity constant

$$\alpha_X(\mu) := \inf_{v \in X \setminus \{0\}} \frac{a(v, v; \mu)}{\|v\|_X^2} > 0.$$

We also assume that f and l are continuous and that a, f and l are parameter separable, i.e. for all of them exist decompositions of the following type:

$$a(v, w; \mu) = \sum_{q=1}^{Q_a} \Theta_a^q(\mu) a^q(v, w), \quad \forall v, w \in X, \mu \in \mathscr{P},$$

with preferably small integer Q_a and μ–independent continuous bilinear forms a^q.

We assume that the solution $u(\mu)$ of (1) is approximated with an iterative domain decomposition procedure. To this end, symmetric bilinear forms $a_k(v, w; \mu) : X_k \times X_k \times \mathscr{P} \to \mathbb{R}$ ("$a|_{\Omega_k}$") and linear forms $f_k(v; \mu) : X_k \times \mathscr{P} \to \mathbb{R}$ ("$f|_{\Omega_k}$") are given on the subdomains. This enables us also to define a and f on

$$W := X_1 \oplus X_2,$$

which can be identified with a superset of X. For details we refer the reader to [4]. To complete the notational framework we introduce the continuity constant

$$M_W(\mu) := \sup_{v \in W \setminus \{0\}} \sup_{w \in W \setminus \{0\}} \frac{a(v, w; \mu)}{\|v\|_W \|w\|_W} < \infty.$$

3 Reduced Basis Scheme

The approximation of the output $s(\mu)$ defined in (2) for a parameter $\mu \in \mathscr{P}$ consists in an offline-stage, which is done once, and an online-stage, which is performed for every output evaluation. In the offline-stage bases for the RB approximation spaces on the subdomains are generated. This is done in a Greedy-algorithm, using a fastly evaluable a-posteriori error estimate to get the "worst-error" parameter. The bases are extended stepwise by a specific routine, yielding partly orthonormalized bases. For more details we refer the reader to [4]. The primal and dual problem are treated equally in this step, yielding separate primal and dual RB spaces. We concentrate now on the explanation of the online-stage, where the approximations to $s(\mu)$ are actually computed.

We introduce RB spaces $X_{N,k} \subset X_k$, $X^0_{N,k} \subset X^0_k$ and $X^\Gamma_{N,k} \subset X_k$ for $k = 1,2$ with dimensions $N_k := \dim(X_{N,k}) < \infty$, $N^0_k := \dim(X^0_{N,k}) < \infty$, $N^\Gamma_k := \dim(X^\Gamma_{N,k}) < \infty$ for $k = 1,2$ and the following relations:

$$\begin{aligned}
X_{N,k} &\cong X^0_{N,k} \oplus X^\Gamma_{N,k}, \quad k = 1,2 \\
X^0_{N,k} &= X_{N,k} \cap H^1_0(\Omega_k), \quad k = 1,2, \\
\gamma(X^\Gamma_{N,1}) &= \gamma(X^\Gamma_{N,2}).
\end{aligned}$$

Consequently, it holds $N_k = N^0_k + N^\Gamma_k$ for $k = 1,2$ and $N^\Gamma := N^\Gamma_1 = N^\Gamma_2$. Further we define $X_{N,\Gamma} := \gamma(X^\Gamma_{N,1}) = \gamma(X^\Gamma_{N,2})$. This one-to-one correspondence on the interface allows us to transmit values without evaluating traces in $X_{N,k}$ online. It also enables us to define a lifting operator in the following way:

$$R^X_{N,1} : X_{N,\Gamma} \to X_{N,1} : g \mapsto (\gamma|_{X^\Gamma_{N,1}})^{-1} g.$$

We assume that those spaces were built for the approximation of the primal solution. For the approximation of the dual solution we introduce RB spaces $Y_{N,k} \subset X_k$, $Y^0_{N,k} \subset X^0_k$, $Y^\Gamma_{N,k} \subset X_k$ for $k = 1,2$ and $Y_{N,\Gamma} \subset X_\Gamma$ with exactly the same properties. The corresponding dimensions are denoted M_k, M^0_k and M^Γ and the lifting operator $R^Y_{N,1} := (\gamma|_{Y^\Gamma_{N,1}})^{-1}$.

Definition 1 (Primal and dual iteration) Given $\mu \in \mathscr{P}$, $g^0_N(\mu) = 0 \in X_{N,\Gamma}$, $\lambda^0_N(\mu) = 0 \in Y_{N,\Gamma}$ and $\theta^n_N(\mu), \eta^n_N(\mu) \in [0,1]$ for $n \geq 1$. We construct sequences $u^n_{N,1}(\mu) \in X_{N,1}$, $u^n_{N,2}(\mu) \in X_{N,2}$ and $g^n_N(\mu) \in X_{N,\Gamma}$ for $n \geq 1$ satisfying

$$\begin{aligned}
a_1(u^n_{N,1}(\mu), v; \mu) &= f_1(v; \mu), \quad \forall v \in X^0_{N,1}, \\
\gamma u^n_{N,1}(\mu) &= g^{n-1}_N(\mu), \\
a_2(u^n_{N,2}(\mu), v; \mu) &= f_2(v; \mu) + f_1(R^X_{N,1} \gamma v; \mu) \\
&\quad - a_1(u^n_{N,1}(\mu), R^X_{N,1} \gamma v; \mu), \quad \forall v \in X_{N,2}, \\
g^n_N(\mu) &= \left(1 - \theta^n_N(\mu)\right) g^{n-1}_N(\mu) + \theta^n_N(\mu) \gamma u^n_{N,2}(\mu)
\end{aligned}$$

and sequences $\psi^n_{N,1}(\mu) \in Y_{N,1}$, $\psi^n_{N,2}(\mu) \in Y_{N,2}$ and $\lambda^n_N(\mu) \in Y_{N,\Gamma}$ for $n \geq 1$ satisfying

$$\begin{aligned}
a_1(v, \psi^n_{N,1}(\mu); \mu) &= -l_1(v; \mu), \quad \forall v \in Y^0_{N,1}, \\
\gamma \psi^n_{N,1}(\mu) &= \lambda^{n-1}_N(\mu), \\
a_2(v, \psi^n_{N,2}(\mu); \mu) &= -l_2(v; \mu) - l_1(R^Y_{N,1} \gamma v; \mu) \\
&\quad - a_1(R^Y_{N,1} \gamma v, \psi^n_{N,1}(\mu); \mu), \quad \forall v \in Y_{N,2}, \\
\lambda^n_N(\mu) &= \left(1 - \eta^n_N(\mu)\right) \lambda^{n-1}_N(\mu) + \eta^n_N(\mu) \gamma \psi^n_{N,2}(\mu).
\end{aligned}$$

Remark 1 Those infinite sequences are terminated as soon as

$$\|u_{N,1}^n(\mu) - u_{N,1}^{n-1}(\mu)\|_{1,\mu}^2 + \|u_{N,2}^n(\mu) - u_{N,2}^{n-1}(\mu)\|_{2,\mu}^2 \leq \epsilon_{\text{tol}},$$
$$\|\psi_{N,1}^n(\mu) - \psi_{N,1}^{n-1}(\mu)\|_{1,\mu}^2 + \|\psi_{N,2}^n(\mu) - \psi_{N,2}^{n-1}(\mu)\|_{2,\mu}^2 \leq \epsilon_{\text{tol}},$$

for some $\epsilon_{\text{tol}} > 0$ where $\|v\|_{k,\mu} := \sqrt{a_k(v, v; \mu)}$ for all $v \in X_k$. The numbers of actually accomplished iterations are denoted by $n_{u,\text{acc}}(\mu)$ and $n_{\psi,\text{acc}}(\mu)$, respectively.

3.1 Smoothed Solutions

For $n \geq 1$ we define $u_N^n(\mu) := (u_{N,1}^n(\mu), u_{N,2}^n(\mu)) \in W$. In general $u_N^n(\mu) \notin X$ and so we define $\hat{u}_N^n(\mu) \in W$ via

$$\hat{u}_N^n(\mu) = \begin{cases} \hat{u}_{N,1}^n(\mu) := R_1(\gamma u_{N,1}^n(\mu) - \gamma u_{N,2}^n(\mu)) & \text{in } \Omega_1, \\ \hat{u}_{N,2}^n(\mu) := 0 & \text{in } \Omega_2, \end{cases}$$

where $R_1 : X_\Gamma \to X_1$ is an arbitrary but linear lifting operator, that is $\gamma R_1 g = g$ for all $g \in X_\Gamma$. We get the following representation:

$$u_N^n(\mu) = \hat{u}_N^n(\mu) + \bar{u}_N^n(\mu),$$

with a smoothed solution $\bar{u}_N^n(\mu) := u_N^n(\mu) - \hat{u}_N^n(\mu) \in X$ and a part $\hat{u}_N^n(\mu)$ compensating for the jump on the interface. For $n \to \infty$ the solution $u_N^n(\mu)$ converges to a smooth function [4], so $\hat{u}_N^n(\mu)$ tends to zero.

Analogously, we define $\psi_N^n(\mu) := (\psi_{N,1}^n(\mu), \psi_{N,2}^n(\mu)) \in W$, $\hat{\psi}_{N,1}^n(\mu) := 0$, $\hat{\psi}_{N,2}^n(\mu) := R_2(\gamma\psi_{N,2}^n(\mu) - \gamma\psi_{N,1}^n(\mu))$, where $R_2 : X_\Gamma \to X_2$ is an arbitrary but linear lifting operator, and $\bar{\psi}_N^n(\mu) = \psi_N^n(\mu) - \hat{\psi}_N^n(\mu) \in X$. As a result, it holds $\bar{\psi}_N^n(\mu)|_{\Omega_1} = \psi_{N,1}^n(\mu)$ in contrast to $\bar{u}_N^n(\mu)|_{\Omega_2} = u_{N,2}^n(\mu)$. This observation will simplify the offline/online-decomposition of our output approximation.

Definition 2 (Output approximation) Given $\mu \in \mathscr{P}$ and corresponding primal and dual solutions $u_N^{n_u}(\mu)$, $n_u \geq 1$ and $\psi_N^{n_\psi}(\mu)$, $n_\psi \geq 1$ we define the corresponding output approximation

$$s_N^{(n_u,n_\psi)}(\mu) := l(\bar{u}_n^{n_u}(\mu); \mu) - f(\bar{\psi}_N^{n_\psi}(\mu); \mu) + a(\bar{u}_N^{n_u}(\mu), \bar{\psi}_N^{n_\psi}(\mu); \mu). \tag{4}$$

4 Error Estimation

The a-posteriori error estimate of the linear output relies on a-posteriori estimates for the primal and dual solutions. To be more precise, we use estimates for the above defined smoothed solutions $\bar{u}_N^n(\mu)$ and $\bar{\psi}_N^n(\mu)$. To that, we define residuals $r_u^n(\cdot;\mu) \in X'$ and $r_\psi^n(\cdot;\mu) \in X'$ for $n \geq 1$ and $\mu \in \mathscr{P}$ through:

$$r_u^n(v;\mu) := f(v;\mu) - a(u_N^n(\mu), v;\mu), \qquad \forall v \in X,$$
$$r_\psi^n(v;\mu) := -l(v;\mu) - a(v, \psi_N^n(\mu);\mu), \qquad \forall v \in X.$$

Proposition 1 *Given $n \geq 1$ and $\mu \in \mathscr{P}$, the errors $u(\mu) - \bar{u}_N^n(\mu)$ and $\psi(\mu) - \bar{\psi}_N^n(\mu)$ can be estimated in the energy-norm $|||\cdot|||_\mu = \sqrt{a(\cdot,\cdot;\mu)}$ via*

$$|||u(\mu) - \bar{u}_N^n(\mu)|||_\mu \leq \Delta_u^n(\mu), \qquad |||\psi(\mu) - \bar{\psi}_N^n(\mu)|||_\mu \leq \Delta_\psi^n(\mu),$$

where

$$\Delta_u^n(\mu) := \frac{1}{\sqrt{\alpha_X^{\mathrm{LB}}(\mu)}} \|r_u^n(\cdot;\mu)\|_{X'} + \frac{M_W^{\mathrm{UB}}(\mu)}{\sqrt{\alpha_X^{\mathrm{LB}}(\mu)}} \left\|\hat{u}_{N,1}^n(\mu)\right\|_{X_1}, \tag{5}$$

$$\Delta_\psi^n(\mu) := \frac{1}{\sqrt{\alpha_X^{\mathrm{LB}}(\mu)}} \|r_\psi^n(\cdot;\mu)\|_{X'} + \frac{M_W^{\mathrm{UB}}(\mu)}{\sqrt{\alpha_X^{\mathrm{LB}}(\mu)}} \left\|\hat{\psi}_{N,2}^n(\mu)\right\|_{X_2}. \tag{6}$$

Here $\alpha_X^{\mathrm{LB}}(\mu)$ denotes a computable lower bound for the constant $\alpha_X(\mu)$ and $M_W^{\mathrm{UB}}(\mu)$ a computable upper bound for $M_W(\mu)$.

The Proposition 1 for the primal variable was proven in [4]. The proof for the dual variable follows the same lines.

Corollary 1 *Given n_u, $n_\psi \geq 1$ and $\mu \in \mathscr{P}$, the error $|s(\mu) - s_N^{(n_u,n_\psi)}(\mu)|$ can be estimated via*

$$|s(\mu) - s_N^{(n_u,n_\psi)}(\mu)| \leq \Delta_s^{(n_u,n_\psi)}(\mu),$$

where

$$\begin{aligned}\Delta_s^{(n_u,n_\psi)}(\mu) &= \Delta_u^{n_u}(\mu)\Delta_\psi^{n_\psi}(\mu) \\ &= \frac{1}{\alpha_X^{\mathrm{LB}}(\mu)} \left(\|r_u^{n_u}(\cdot;\mu)\|_{X'} + M_W^{\mathrm{UB}}(\mu)\|\hat{u}_{N,1}^{n_u}(\mu)\|_{X_1}\right) \\ &\quad \left(\|r_\psi^{n_\psi}(\cdot;\mu)\|_{X'} + M_W^{\mathrm{UB}}(\mu)\|\hat{\psi}_{N,2}^{n_\psi}(\mu)\|_{X_2}\right).\end{aligned}$$

Thanks to the smoothness of the solutions in the output approximation, the proof of Corollary 1 is analogue to the proof for the standard RB method [5, 6].

4.1 Offline/Online Decomposition

As already mentioned, an efficient offline/online decomposition is essential for our method. The parameter separability (4) is the main ingredient for obtaining such a decomposition. Again we refer to [4] for a detailed explanation of the routine for the primal iteration. Offline/online decomposition of the dual iteration is achieved in the same way. For a decomposition of the output approximation (4) into parameter-dependent coefficients and parameter-independent components we exploit

$$\begin{aligned}
s_N^{(n_u,n_\psi)}(\mu) &= l(\bar{u}_n^{n_u}(\mu)) - f(\bar{\psi}_N^{n_\psi}(\mu);\mu) + a(\bar{u}_N^{n_u}(\mu), \bar{\psi}_N^{n_\psi}(\mu);\mu) \\
&= l(u_N^{n_u}(\mu);\mu) - l_1(R_1\gamma u_{N,1}^{n_u}(\mu);\mu) + l_1(R_1\gamma u_{N,2}^{n_u}(\mu);\mu) \\
&\quad - f(\psi_N^{n_\psi}(\mu);\mu) + f_2(R_2\gamma\psi_{N,2}^{n_\psi}(\mu);\mu) - f_2(R_2\gamma\psi_{N,1}^{n_\psi}(\mu);\mu) \\
&\quad + a(u_N^{n_u}(\mu), \psi_N^{n_\psi}(\mu);\mu) \\
&\quad - a_1(R_1\gamma u_{N,1}^{n_u}(\mu), \psi_{N,1}^{n_\psi}(\mu);\mu) + a_1(R_1\gamma u_{N,2}^{n_u}(\mu), \psi_{N,1}^{n_\psi}(\mu);\mu) \\
&\quad - a_2(u_{N,2}^{n_u}(\mu), R_2\gamma\psi_{N,2}^{n_\psi}(\mu);\mu) + a_2(u_{N,2}^{n_u}(\mu), R_2\gamma\psi_{N,1}^{n_\psi}(\mu);\mu).
\end{aligned}$$

Details on the offline/online decomposition of the error estimate (5), respectively (6) can also be found in [4].

5 Numerical Results

We consider the static heat equation on the unit square in $\mathbb{R}^2$ with a decomposition of the domain into two parts. The heat coefficient $\kappa(x;\mu)$ is piecewise constant and depends on three parameters: $k(\cdot;\mu)|_{B_i} = \mu_i$ for $i = 1,\dots,3$ and $k(\cdot;\mu)|_{B_4} = 1$. Figure 1 shows the blocks $B_1,\dots,B_4$ and the domain decomposition. This model leads to a weak form with

$$a(v,w;\mu) = \int_\Omega \kappa(\mu)\nabla v \cdot \nabla w \, dx, \quad v,w \in X, \mu \in \mathscr{P}.$$

The source consists of two exponential bubbles, with peaks in Ω_1 and Ω_2 and a fourth parameter as a weight between them:

$$\begin{aligned}
f(v;\mu) &= \int_\Omega h(\mu) v \, dx, \quad v \in X, \mu \in \mathscr{P}, \\
h(x;\mu) &= 80\mu_4 \exp\left(-20|x-z_1|^2\right) + 80(1-\mu_4)\exp\left(-20|x-z_2|^2\right),
\end{aligned}$$

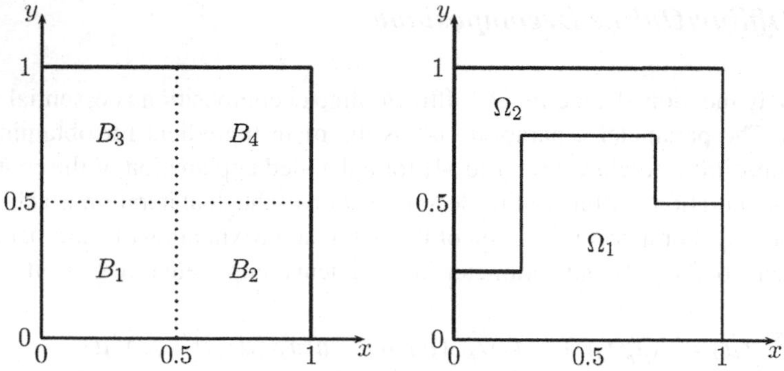

Fig. 1 *Left*: blocks, where $\kappa(\mu)$ is constant in space, *right*: domain decomposition of $\Omega = (0, 1)^2$

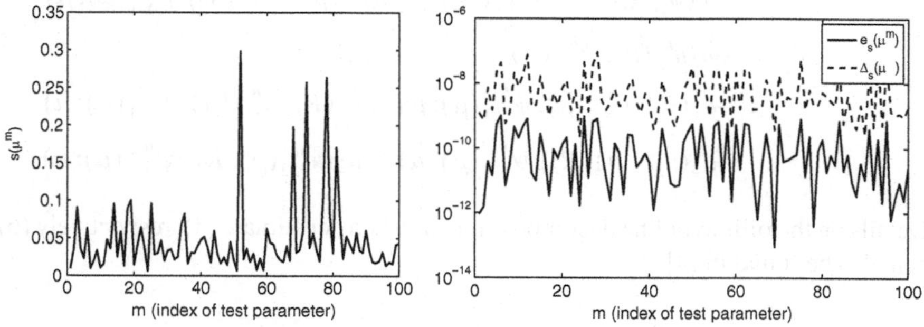

Fig. 2 *Left*: output values $s(\mu)$ on a parameter set of 100 randomly generated parameters. *Right*: investigation of the output error and corresponding estimate on the same parameter set with RB spaces of dimensions $N = 80$, $M = 28$

for $x \in \Omega$, $\mu \in \mathscr{P}$ with $z_1 = (0.5, 0.5)^T$ and $z_2 = (0.875, 0.875)^T$. So the parameter vector is 4-dimensional; $\mathscr{P} \subset \mathbb{R}^4$. The linear output is defined as the mean value of $u(\mu)$ on $\Omega_s = [0, 0.25] \times [0.75, 1]$:

$$s(\mu) = l(u(\mu)) = \frac{1}{|\Omega_s|} \int_{\Omega_s} u(\mu)\, dx, \quad \mu \in \mathscr{P}.$$

The left-hand side of Fig. 2 shows values of $s(\mu)$ for 100 randomly generated parameters. Our basis generation procedure yields bases of different sizes $N = N_1 + N_2$ and $M = M_1 + M_2$ for the primal and the dual approximation space. We define the error $e_s^{n_u,n_\psi}(\mu) = |s(\mu) - s_N^{n_u,n_\psi}(\mu)|$ and the effectivity $\eta_s^{n_u,n_\psi}(\mu) = \Delta_s^{n_u,n_\psi}(\mu)/e_s^{n_u,n_\psi}(\mu)$, where in the following $n_u = n_{u,\mathrm{acc}}(\mu)$ and $n_\psi = n_{\psi,\mathrm{acc}}(\mu)$ for the respective parameter. The right-hand side of Fig. 2 shows that we obtain fairly good approximations and that the estimate is clearly related to the error. The effectivity is at the range of 10^2. Exemplary values of the effectivity are shown in Table 1.

Table 1 Output error $e_s^{n_u,n_\psi}(\mu)$, estimate $\Delta_s^{n_u,n_\psi}(\mu)$ and effectivity $\eta_s^{n_u,n_\psi}(\mu)$ for one randomly generated parameter and different bases sizes

Bases sizes (N, M)	Output error	Estimate	Effectivity
$(61, 13)$	$7.65 \cdot 10^{-7}$	$7.23 \cdot 10^{-5}$	94.55
$(80, 28)$	$4.31 \cdot 10^{-11}$	$3.84 \cdot 10^{-9}$	89.18
$(85, 40)$	$6.94 \cdot 10^{-15}$	$2.78 \cdot 10^{-13}$	40.06

To conclude, the primal-dual framework has been successfully transferred to the Dirichlet-Neumann RB method. The introduction of smoothed solutions in the output approximation allows a-posteriori error estimation in a straight-forward manner. The results meet the expectations to the method.

References

1. D.B.P. Huynh, D. Knezevic, A.T. Patera, A static condensation reduced basis element method: approximation and a posteriori error estimation. Math. Model. Numer. Anal. **47**(1), 213–251 (2013)
2. L. Iapichino, A. Quarteroni, G. Rozza, A reduced basis hybrid method for the coupling of parametrized domains represented by fluidic networks. Comput. Methods Appl. Mech. Eng. **221–222**, 63–82 (2012). doi:10.1016/j.cma.2012.02.005
3. Y. Maday, E. Rønquist, A reduced-basis element method. J. Sci. Comput. **17**, 447–459 (2002)
4. I. Maier, B. Haasdonk, A Dirichlet-Neumann reduced basis method for homogeneous domain decomposition problems. Appl. Numer. Math. **78**, 31–48 (2014). doi:10.1016/j.apnum.2013.12.001
5. C. Prud'homme, D.V. Rovas, K. Veroy, L. Machiels, Y. Maday, A.T. Patera, G. Turinici, Reliable real-time solution of parametrized partial differential equations: reduced-basis output bound methods. J. Fluids Eng. **124**, 70–80 (2002)
6. D. Rovas, Reduced basis output bound methods for parametrized PDE's. PhD thesis, MIT, 2003

One-Dimensional Surrogate Models for Advection-Diffusion Problems

Matteo Aletti, Andrea Bortolossi, Simona Perotto, and Alessandro Veneziani

Abstract Numerical solution of partial differential equations can be made more tractable by model reduction techniques. For instance, when the problem at hand presents a main direction of the dynamics (such as blood flow in arteries), it may be conveniently reduced to a 1D model. Here we compare two strategies to obtain this model reduction, applied to classical advection-diffusion equations in domains where one dimension dominates the others.

1 Introduction

Many applications in scientific computing demand for surrogate models, i.e., simplified models which are expected to be computationally affordable and reliable from a modeling viewpoint. Problems presenting an evident main direction, such as blood flow in arteries, gas dynamics in internal combustion engines, etc., are naturally reduced to 1D equations along the coordinate of the main (or "axial" as opposed to "transverse") direction. Here we consider and compare two different strategies to get surrogate models for this kind of problems. The first procedure stems from an appropriate average of the equation along the transverse direction, combined with (plausible) problem-dependent simplifying assumptions. The second approach comes from a different representation of the axial and of the transverse dynamics, according to what has been called a Hierarchical Model (Hi-Mod) reduction [4].

M. Aletti • A. Bortolossi
Politecnico di Milano, Piazza Leonardo da Vinci 32, I-20133 Milano, Italy
e-mail: matteo.aletti@mail.polimi.it; andrea.bortolossi@mail.polimi.it

S. Perotto (✉)
MOX-Dipartimento di Matematica, Politecnico di Milano, Piazza Leonardo da Vinci 32, I-20133 Milano, Italy
e-mail: simona.perotto@polimi.it

A. Veneziani
Department of Mathematics and Computer Science, Emory University, 400 Dowman Dr., 30322, Atlanta, GA, USA
e-mail: ale@mathcs.emory.edu

A. Abdulle et al. (eds.), *Numerical Mathematics and Advanced Applications ENUMATH 2013*, Lecture Notes in Computational Science and Engineering 103,
DOI 10.1007/978-3-319-10705-9_44

In particular, transverse dynamics are represented by a modal expansion, that is supposed to require just a few modes for the nature of the problem. This leads to solve a system of 1D coupled equations. At the bottom line, when using just one transverse mode, this leads to a genuinely 1D model. For the sake of comparison, these two reduction procedures are applied to the following two-dimensional advection-diffusion problem

$$\begin{cases} -\mu\Delta u + \boldsymbol{b}\cdot\nabla u = f & in\ \Omega \equiv (0,L)\times(-R_0,R_0) \\ u = g & \text{on } \Gamma_{in} \equiv \{0\}\times(-R_0,R_0) \\ \mu\dfrac{\partial u}{\partial n} = 0 & \text{on } \Gamma_{out} \equiv \{L\}\times(-R_0,R_0) \\ \mu\dfrac{\partial u}{\partial n} + \chi u = u_{ext} & \text{on } \Gamma_{lat} \equiv \partial\Omega\backslash(\Gamma_{in}\cup\Gamma_{out}), \end{cases} \tag{1}$$

where x is the main direction, y is the transverse one, and with $\mu \in L^\infty(\Omega)$, $\boldsymbol{b} = (b_1, b_2)^T \in [W^{1,\infty}(\Omega)]^2$, $f \in L^2(\Omega)$, $g \in H^{1/2}(\Gamma_{in})$, $\chi \in L^\infty(\Omega)$, $u_{ext} \in L^2(\Gamma_{lat})$, $\mu\partial u/\partial n$ the conormal derivative of u. Standard notation are adopted for the Sobolev spaces. We distinguish in the domain Ω a supporting fiber Ω_{1D} aligned with the main stream and a set of transverse fibers γ_x, with $x \in \Omega_{1D}$, parallel to the secondary transverse dynamics. Since Ω coincides with a rectangle, $\gamma_x = \gamma$, for each x.
We assume suitable assumptions on the data to guarantee the well-posedness of the weak form of (1), i.e.,

$$\text{find} \quad u \in V \equiv H^1_{\Gamma_{in}}(\Omega) \quad \text{s.t.} \quad a(u,v) = F(v) \quad \forall v \in V, \tag{2}$$

with $a(u,v) = \int_\Omega [\mu\nabla u\cdot\nabla v + \boldsymbol{b}\cdot\nabla u v]\, d\Omega + \int_{\Gamma_{lat}} \chi u v\, d\Gamma$ and $F(v) = \int_\Omega f v\, d\Omega + \int_{\Gamma_{lat}} u_{ext} v\, d\Gamma - a(\rho_g, v)$, ρ_g denoting a lifting of g on Γ_{in}. Problem (1) models, for instance, the oxygen transport inside an artery. In this case, u represents the oxygen partial pressure, μ denotes the diffusivity of oxygen in blood, field $\mathbf{b}$ takes into account the blood dynamics, f is a generic sink or source term, g usually coincides with a concentration profile, the Robin boundary conditions model the absorption of the oxygen through the vessel walls, with χ depending on the absorption properties of the wall and u_{ext} measuring the oxygen partial pressure outside the vessel. For simplicity, we assume μ and χ constant.

When comparing the two approaches mentioned above, we address in particular the combination of models with different accuracy. For the first approach, this leads to what has been called a geometrical multiscale formulation [2]. For Hi-Mod reduction, this is obtained by selecting a different number of modes in different regions of the domain [4, 5].

2 A Transverse Average Model

We particularize the approach in [3] for modeling the transport of solutes in arteries with bifurcations to an elliptic setting. Let us introduce the transverse profile of the solution, given by

$$p(x,y) = \frac{u(x,y)}{U(x)} \quad \text{with} \quad U(x) = \frac{1}{|\gamma|}\int_\gamma u(x,y)\,dy,$$

$U(x)$ denoting the mean of the solution along the transverse (constant) section γ of Ω. As first modeling hypothesis, we assume that the profile p does not depend on x, i.e., only the mean of the solution may vary along the x-direction. Thus, after separation of variables, the solution u can be regarded as a certain profile varying in y tuned by a function varying along x, i.e., $u(x,y) = U(x)p(y)$. By exploiting this representation of u in the assignment of the boundary conditions on Γ_{lat}, i.e., on the boundary of γ, we get

$$\Big(\pm\mu\frac{\partial p(y)}{\partial y}\Big)\Big|_{y=\pm R_0} = \Big(-\chi p(y) + \frac{u_{ext}(x)}{U(x)}\Big)\Big|_{y=\pm R_0}.$$

Consistently with the previous assumption on p, we postulate that the ratio $u_{ext}(x)/U(x)$ is constant along the whole length of the domain. Finally, we constrain the advective field, by assuming $\nabla\cdot\boldsymbol{b} = 0$ and $\boldsymbol{b}|_{\Gamma_{lat}} = 0$. Since $\boldsymbol{b}$ is divergence-free, we can rewrite the full model (1) in a conservative form, as $-\mu\Delta u + \nabla\cdot(\boldsymbol{b}u) = f$. Now, integrating with respect to y along γ, we obtain

$$\begin{aligned} &-\mu\frac{\partial^2}{\partial x^2}\int_\gamma u(x,y)\,dy - \mu\frac{\partial u(x,y)}{\partial y}\Big|_{y=-R_0}^{y=R_0} + \frac{\partial}{\partial x}\int_\gamma \Big[b_1(x,y)u(x,y)\Big]\,dy \\ &+ \big(b_2(x,y)u(x,y)\big)\Big|_{y=-R_0}^{y=R_0} = \int_\gamma f(x,y)\,dy. \end{aligned}$$

By exploiting the Robin conditions and the hypothesis on $\boldsymbol{b}|_{\Gamma_{lat}}$, we have

$$\begin{aligned} &-\mu\frac{\partial^2}{\partial x^2}\int_\gamma u(x,y)\,dy + \chi\big(u(x,R_0) + u(x,-R_0)\big) + \frac{\partial}{\partial x}\int_\gamma \Big[b_1(x,y)u(x,y)\Big]\,dy \\ &= \int_\gamma f(x,y)\,dy + u_{ext}(x,R_0) + u_{ext}(x,-R_0). \end{aligned}$$

Now, we exploit the factorization $u(x,y) = U(x)p(y)$ assumed for the solution u together with the fact that, by definition, the mean of p along γ is equal to one, to get the desired averaged 1D model (the primes denoting x-differentiation)

$$-\mu U''(x) + (U(x)w_r(x))' + \sigma_r U(x) = f_r(x) \quad \text{for} \quad x\in(0,L), \tag{3}$$

with

$$w_r(x) = \frac{1}{|\gamma|}\int_\gamma b_1(x,y)p(y)\,dy, \quad \sigma_r = \chi\frac{p(R_0)+p(-R_0)}{|\gamma|},$$
$$f_r(x) = \frac{1}{|\gamma|}\int_\gamma f(x,y)\,dy + \frac{u_{ext}(x,R_0)+u_{ext}(x,-R_0)}{|\gamma|}. \tag{4}$$

The reduction procedure leads from a 2D advection-diffusion problem to a 1D advection-diffusion-reaction problem. To close the model we need to select a profile p in (4). For simplicity, it may be assumed constant or, more in general, it is suggested by physical considerations. It could be advantageous an automatic criterion to select p. A strategy in such a direction is proposed in the next section.

Remark 1 From a physical viewpoint, the most restrictive hypothesis for deriving model (3) is the independence of p on x. Nevertheless, the numerical validation shows that this surrogate model provides reliable results even when this hypothesis is not strictly guaranteed. The second assumption is reasonable, at least in haemodynamics, since the ratio $u_{ext}(x)/U(x)$ may be reliably considered constant. The two requirements on $\boldsymbol{b}$ are standard in a haemodynamic context. Hypothesis $\nabla\cdot\boldsymbol{b} = 0$ ensures the incompressibility of the blood, while assumption $\boldsymbol{b}|_{\Gamma_{lat}} = 0$ imposes a no-slip condition on Γ_{lat}.

2.1 A Geometrical Multiscale Approach

A geometrical multiscale formulation consists of coupling dimensionally heterogeneous models. The idea is to alternate a full-dimensional model with suitable downscaled models to be associated with the areas characterized by the most complex and by the simplest dynamics, respectively (see, e.g., [2, Chapter 11]). The identification of appropriate matching conditions and the location of the interface between the two models represent the main issues of this approach. We identify the full model with (1) and the downscaled model with (3). We choose $\Omega = (0,10)\times(0,1)$, $\mu = 1$, $\boldsymbol{b} = (20,0)^T$, $f = 10\big((x-1.5)^2+0.4(y-0.5)^2 < 0.01\big)$, $\chi = 1$ and $u_{ext} = 0.02$. We assign a homogeneous Neumann condition on $\Gamma_{out} = \{10\}\times(0,1)$ and a profile compatible with the conditions along Γ_{lat} on Γ_{in}. In Fig. 1 (top-left), we provide the contour plots of the full solution approximated via linear finite elements on a uniform unstructured grid of 8,918 elements. The solution exhibits more significant transverse dynamics in the leftmost part of the domain, where the source term is localized. Conversely, the solution profile is less fluctuating in the rightmost part of Ω, as assumed in the derivation of model (3). This suggests to split Ω into two subdomains, Ω_1 and Ω_2, such that $\overline{\Omega} = \overline{\Omega}_1 \cup \overline{\Omega}_2$. On Ω_1 we solve problem (1), while we resort to (3) in Ω_2. Both the problems are discretized via linear finite elements on uniform meshes. The coupling between the two models is performed via a relaxed Neumann/Dirichlet scheme. In more

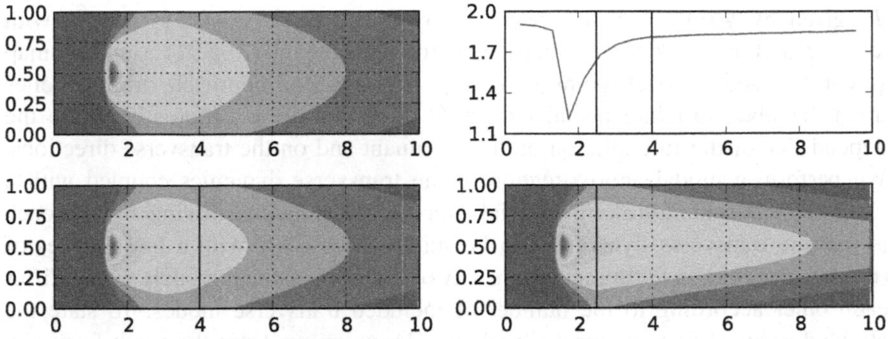

Fig. 1 Geometrical multiscale: full solution (*top-left*); graph of σ_r (*top-right*); coupled solution for Γ_4 (*bottom-left*) and $\Gamma_{2.5}$ (*bottom-right*)

detail, we exploit the derivative of the 1D surrogate solution u_2 to assign a constant Neumann condition on Ω_1 as $\mu \partial u_1/\partial n(x_i, y) = u_2'(x_i)$, where u_1 is the full solution defined on Ω_1 and $\Gamma_{x_i} = \{x_i\} \times (0, 1)$ identifies the interface $\overline{\Omega}_1 \cap \overline{\Omega}_2$. To correctly define the problem on Ω_2, we have to properly select the boundary condition at x_i and the solution profile. As Dirichlet data we assign $u_2(x_i) = |\gamma|^{-1} \int_\gamma u_1(x_i, y)\, dy$, while we follow a new approach to select $p(y)$ at x_i. The idea is to exploit the problem in Ω_1 instead of resorting to an a priori selection. Thus, we pick $p(y) = |\gamma| u_1(x_i, y)/ \int_\gamma u_1(x_i, y)\, dy$. This definition justifies the prescription of a Neumann condition on the left hand side of Γ_{x_i} to allow the solution profile to develop freely. Indeed, the adoption of the surrogate model in Ω_2, implicitly assumes that p is completely developed at Γ_{x_i}. Figure 1 (bottom) compares two couplings associated with different interfaces, i.e., Γ_4 and $\Gamma_{2.5}$, respectively. The second choice introduces the interface where the transverse dynamics are still too significant, thus violating the hypothesis on a fully developed profile. We provide a bidimensional visualization also for the surrogate model simply by using relation $u(x, y) = U(x)p(y)$. In Fig. 1 (top-right) we show the reactive coefficient in (4), computed via the profile of the full solution. Since σ_r strongly depends on p, we argue that when the profile stabilizes, σ_r reaches a constant value. So a possible heuristic way to select Γ_{x_i} is to locate it in a region where σ_r is constant.

3 Hi-Mod Reduction

Hi-Mod reduction is an alternative approach to "compress" high dimensional problems. In this case, a full 2D (or even 3D) model is reduced to a system of 1D coupled differential problems associated with the dominant dynamics [4]. In the geometric setting of a Hi-Mod formulation, for any $x \in \Omega_{1D}$, we introduce a map ψ_x between the generic fiber γ_x and a reference fiber $\hat{\gamma}$, so that the computational domain Ω is mapped into the reference domain $\hat{\Omega} = \Omega_{1D} \times \hat{\gamma}$ via the map

Ψ, given by $\Psi(z) = \hat{z}$, where $z = (x, y) \in \Omega$, $\hat{z} = (\hat{x}, \hat{y}) \in \hat{\Omega}$, with $\hat{x} = x$ and $\hat{y} = \psi_x(y)$. In particular, for the domain Ω in (1) a unique map ψ can be used for each point $x \in \Omega_{1D}$. The Hi-Mod approach strongly relies upon the fiber structure postulated on Ω. The idea is to differently tackle the dependence of the full solution on the dominant and on the transverse directions. We perform a modal approximation of the transverse dynamics coupled with a Galerkin representation along the axial direction. The rationale driving this approach is that the transverse dynamics can be suitably described with a few degrees of (modal) freedom, resulting in a *hierarchy* of one-dimensional models which differ each other according to the number of included transverse modes. To state the Hi-Mod reduced formulation for problem (1), we move from the weak form (2). Now, let V_{1D} be a space spanned by functions defined on Ω_{1D} which properly includes the boundary conditions assigned along Γ_{in} and Γ_{out}, and let $\{\varphi_k\}_{k\in\mathbb{N}^+}$ be a modal basis of functions in $H^1(\hat{\gamma})$, orthonormal with respect to the $L^2(\hat{\gamma})$-scalar product and compatible with the boundary conditions along Γ_{lat}. As a consequence, we look for a reduced solution u_m which belongs to the Hi-Mod reduced space $V_m = \{v_m(x, y) = \sum_{k=1}^{m} \tilde{v}_k(x)\, \varphi_k(\psi(y)), \text{with } \tilde{v}_k \in V_{1D}, x \in \Omega_{1D}, y \in \gamma\}$. A conformity and a spectral approximability hypothesis are introduced on V_m to guarantee the well-posedness and the convergence of u_m to u [4]. We identify the Galerkin representation along Ω_{1D} with a finite element discretization, so that the modal coefficients belong to a finite element space $V_{1D}^h \subset V_{1D}$ associated with a partition $\mathcal{T}_h$ of Ω_{1D}. Thus, the Hi-Mod reduced form for (2) is: for a certain modal index $m \in \mathbb{N}^+$, find $\tilde{u}_k^h \in V_{1D}^h$, with $k = 1, \ldots, m$, such that $\sum_{k=1}^{m} a(\tilde{u}_k^h \varphi_k, \theta_i \varphi_j) = F(\theta_i \varphi_j)$, with $j = 1, \ldots, m$ and $i = 1, \ldots, N_h$, where θ_i denotes the generic finite element basis function in V_{1D}^h and with $N_h = \dim(V_{1D}^h) < +\infty$. From a computational viewpoint, the Hi-Mod formulation leads to solve a system of m coupled 1D advection-diffusion-reaction problems instead of problem (1). As in the derivation of the surrogate model (3), the Hi-Mod reduction procedure yields reactive terms, while no reactive contribution is included in the full model. The system is characterized by an $m \times m$ block matrix, where each block is an $N_h \times N_h$ matrix exhibiting the sparsity pattern typical of the selected finite element space.

The modal index m can be selected a priori moving from some preliminary knowledge of the phenomenon at hand [4] or automatically, driven by an *a posteriori* modeling error analysis [5]. Another important issue is the choice of the modal basis, in particular when Robin boundary conditions are assigned on Γ_{lat} as in (1). We build a specific modal basis able to automatically include these conditions. The idea proposed in [1] is to solve on $\hat{\gamma}$ an auxiliary Sturm-Liouville eigenvalue problem, with conditions on $\partial\hat{\gamma}$ coinciding with the conditions assigned on Γ_{lat}. We call this modal basis *educated basis*.

3.1 Piecewise Hi-Mod Reduction

Now, the idea is to properly exploit the hierarchy of models provided by the Hi-Mod reduced space to couple models with a different accuracy. A different choice for the modal index m identifies a reduced model with a certain level of detail in describing the phenomenon at hand. As a consequence, by properly tuning m over different regions of Ω, we are able to capture the local significant features of the solution with a relatively low number of degrees of freedom. Following [4], we denote this approach by *piecewise* Hi-Mod reduction. This leads to dimensionally homogeneous models (yet with a locally varying level of accuracy), as opposed to the geometrical multiscale approach. For instance, with reference to the test case in Fig. 1, we can preserve the two splittings of the domain identified by Γ_4 and $\Gamma_{2.5}$ and employ a number of modes in Ω_1 higher than in Ω_2, e.g., 5 and 2, respectively. This choice is motivated by the fact that the most complex dynamics are localized in Ω_1 and, consequently, more modes are demanded in this area. To glue the two models we employ a relaxed Neumann/Dirichlet scheme as in the geometrical multiscale formulation. At each iteration of this scheme, we apply a uniform Hi-Mod reduction on Ω_1 and Ω_2, separately, i.e., we solve two systems of coupled 1D problems with a block matrix of order $5N_h^1$ and $2N_h^2$, respectively N_h^i denoting the dimension of the one dimensional finite element space introduced on $\Omega_{1D} \cap \Omega_i$, for $i = 1, 2$. As detailed in [5], to rigorously formalize the piecewise Hi-Mod approach, we introduce a suitable broken Sobolev space, endowed with an integral condition which weakly enforces the continuity of the reduced solution in correspondence with the minimum number of modes common on the whole Ω. This does not necessarily guarantee the conformity of the piecewise reduced solution. This is evident in Fig. 2. The loss of conformity is particularly significant when the interface is located in an area involved by strong transverse dynamics. The reduced solution in Fig. 2 (left) is in good agreement with the full one in Fig. 1 (top-left) and it is very similar to the one in Fig. 1 (bottom-left).

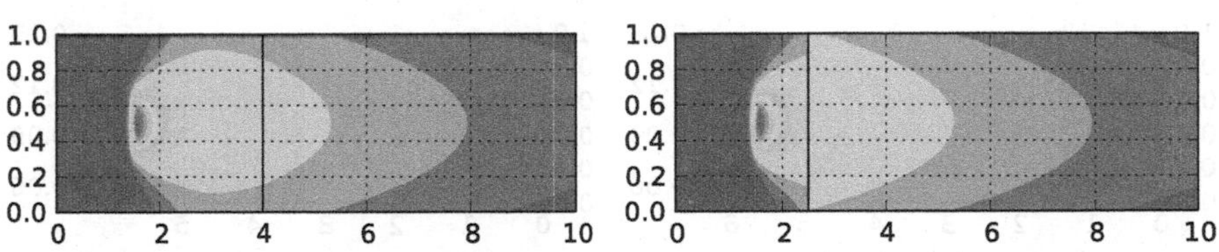

Fig. 2 Piecewise Hi-Mod reduction: reduced solutions associated with $\{5, 2\}$ modes, the interface is Γ_4 (on the *left*) and $\Gamma_{2.5}$ (on the *right*); $h = 0.05$ on Ω_{1D}

4 A Numerical Comparison

For a fair comparison between the two approaches, we consider here a test case where the "low-fidelity" model (the genuine 1D in the geometrical multiscale and a low-mode approximation in the Hi-Mod) are straightforwardly comparable. This means that we employ a single mode in the Hi-Mod approximation. In particular, we consider problem (1) with $\Omega = (0,6) \times (0,1)$, $\mu = 1$, $\mathbf{b} = (20,0)^T$, $f = 10([(x-1.5)^2+0.4(y-0.75)^2 < 0.01]+[(x-1.5)^2+0.4(y-0.25)^2 < 0.01])$, $\chi = 3$ and $u_{ext} = 0.05$. The boundary conditions are as in Sect. 2.1 and the interface is located at $x = 3$. Numerical results are provided in Fig. 3. The top-left panel displays the full solution discretized via standard linear finite elements on a uniform unstructured grid of 5,084 triangles. The top-right panel shows the geometrical multiscale solution. This is fairly accurate even though it suffers from an underestimation of the reactive term induced by the 1D average. This is evident in the contour line associated with the value 0.09. The bottom panels display the Hi-Mod solution, having a "low-fidelity" model with $m = 1$ and two different models for the "high-fidelity" part. In particular, on the left we take $m = 3$ which is clearly not enough to capture reliably the solution in the leftmost domain. In the right-panel, with $m = 5$ we have a pretty accurate solution, where the inaccuracy present in the geometrical multiscale solution as well as the model non-conformity do not pollute significantly the results.

An extensive comparison between the two approaches cannot be clearly completed by these preliminary results. As a matter of fact, the computational advantages of the one approach over the other must be evaluated on 3D more realistic test cases, solved with compiled softwares. However, we may notice that, even though the Hi-Mod approach relies entirely on a "psychologically" 1D representation of the solution within a dimensionally heterogeneous framework, it may provide accurate solution also in presence of significant transverse components. For this reason, we do expect it may lead to easily implemented and manageable solvers, with

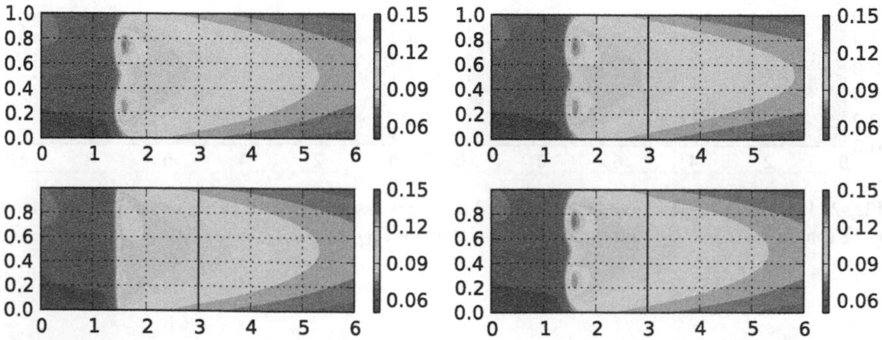

Fig. 3 Full solution (*top-left*); geometrical multiscale solution (*top-right*) and Hi-Mod solution with {3, 1} (*bottom-left*) and {5, 1} (*right*) modes

competitive performances in terms of both accuracy and efficiency. A framework of investigation of practical interest is the blood flow simulation in a network of arteries.

References

1. M. Aletti, S. Perotto, A. Veneziani, (2013 in preparation)
2. L. Formaggia, A. Quarteroni, A. Veneziani, *Cardiovascular Mathematics, Modeling, Simulation and Applications*, vol. 1 (Springer, Berlin/Heidelberg, 2009)
3. A. Krpo, Modélisation par des modèles 1D de l'écoulement du sang et du transport de masse dans des artères avec bifurcation, Master thesis, École Polytechnique Fédérale de Lausanne, 2005
4. S. Perotto, A. Ern, A. Veneziani, Hierarchical local model reduction for elliptic problems: a domain decomposition approach. Multiscale Model. Simul. **8**(4), 1102–1127 (2010)
5. S. Perotto, A. Veneziani, Coupled model and grid adaptivity in hierarchical reduction of elliptic problems. J. Sci. Comput. **60**(3), 505–536 (2014)

Flag Manifolds for the Characterization of Geometric Structure in Large Data Sets

Tim Marrinan, J. Ross Beveridge, Bruce Draper, Michael Kirby, and Chris Peterson

Abstract We propose a flag manifold representation as a framework for exposing geometric structure in a large data set. We illustrate the approach by building pose flags for pose identification in digital images of faces and action flags for action recognition in video sequences. These examples illustrate that the flag manifold has the potential to identify common features in noisy and complex datasets.

1 The Mathematical Challenges of Large Data Sets

Some very intriguing problems faced by scientists today have hints, suggestions, and solutions hidden within large collections of data. Mathematicians, computer scientists and statisticians have a fundamental role to play in developing the theory, tools, and algorithms needed by the general researcher in their quest to extract meaningful information from such data sets. While the type of data can vary drastically, one seeks a range of sufficiently robust tools that can be applied across multiple disciplines.

Finding ways to compactly represent a complicated object (such as a data cloud or a high dimensional array) has allowed for knowledge discovery within massive data sets (e.g. a data set consisting of many data clouds or many high dimensional arrays) [2, 4]. As an example, suppose a data cloud in $\mathbb{R}^n$ clusters along a k-dimensional linear space then, for comparison against other data clouds, one could identify the cloud with this k-dimensional linear space. One could then associate a single point to the data by identifying the k-dimensional linear space with a point on an appropriate Grassmann manifold. Such a map transforms the problem of comparing data clouds in one setting to comparison of points on a Grassmann manifold.

Building on the theme of the previous paragraph, suppose one identifies a portion of the information in a collection of data with a nested sequence of vector spaces.

T. Marrinan • J.R. Beveridge • B. Draper • M. Kirby (✉) • C. Peterson
Colorado State University, Fort Collins, CO, USA
e-mail: marrintp@gmail.com; ross@cs.colostate.edu; draper@cs.colostate.edu; kirby@math.colostate.edu; peterson@math.colostate.edu

A. Abdulle et al. (eds.), *Numerical Mathematics and Advanced Applications ENUMATH 2013*, Lecture Notes in Computational Science and Engineering 103,
DOI 10.1007/978-3-319-10705-9_45

This could be natural in settings involving an ordered sequence of data or as the result of a singular value decomposition. Examples might include the spectral sheets in a hyper-spectral digital image, the frames in a video stream, or the output of a singular value decomposition applied to a data set collected under a variation of state. A nested sequence of vector spaces is known as a *flag*. One could associate a single point to the data by identifying the nested sequence of vector spaces with a point on a flag manifold. Through this representation, comparisons of multiple instances of data can be transformed to comparisons of points on a flag manifold. For both the Grassmann and flag manifolds, there is a rich collection of metrics that can be considered for purposes of comparison.

One goal of the machine learning community is to create algorithms for automated processing and interpretation of the output obtained from a collection of sensors. For example, one may be interested in detecting anomalies in human behavior such as fall detection as an aid for assisted living [8]. Another example, is concerned with automated action recognition in video sequences towards the goal of automated video-to-text algorithms [6]. This paper develops an approach based on the geometry of the flag manifold for exploiting structure and correlations within large data sets.

The paper is structured as follows. Section 2 gives the mathematical background. Section 3 describes an algorithm for producing points on flag manifolds from collections of subspaces. Section 4 provides examples illustrating the flag approach. Section "Conclusions" consists of concluding remarks.

2 Grassmann, Stiefel and Flag Manifolds

A *flag* is a strictly ascending sequence of subspaces of a fixed n-dimensional vector space, V. Given a flag, $V_1 \subset V_2 \subset \cdots \subset V_r \subset V$, the *signature* of the flag is the data $(d_1, d_2, \ldots, d_r, n)$ where d_i denotes the dimension of V_i. The *flag manifold* $FL(d_1, d_2, \ldots, d_r; n)$ is a manifold whose points parameterize all flags with signature $(d_1, d_2, \ldots, d_r, n)$. A flag is called complete if its signature is $(1, 2, 3, \ldots, n-1, n)$. An ordered basis, $v_1, v_2, \ldots, v_n$ of V gives rise to a complete flag by setting V_i equal to the span of $v_1, v_2, \ldots, v_i$. A complete flag in $\mathbb{R}^n$ or $\mathbb{C}^n$ can be used to build an orthonormal basis (unique up to multiplication by a unit length scalar at each step). The general linear group acts transitively on the set of all complete flags in a fixed vector space. A Grassmann manifold $G(k, n)$ is a flag manifold with signature (k, n). The projective space $\mathbb{P}^{n-1}$ is the Grassmann manifold $G(1, n)$. Thus, flag manifolds generalize Grassmann manifolds which generalize projective space.

The *Stiefel Manifold*, $S(k, n)$, parametrizes orthonormal k-frames in a fixed n dimensional inner product space V [7]. It can be viewed as a homogeneous space for the action of a matrix group. There is a natural projection, $F : S(k, n) \to G(k, n)$ where an orthonormal k-frame is sent to its span. The fiber of F over a point $P \in G(k, n)$ is the set of k-frames in the k-dimensional space determined by P. Similarly, there are projection maps from $S(k,n)$ to any flag manifold $FL(d_1, d_2, \ldots, d_r; n)$ with $d_r \leq k$.

If $V = \mathbb{R}^n$, then $S(1,n)$ corresponds to the unit hypersphere S^{n-1} and $S(2,n)$ corresponds to the unit tangent bundle on the unit hypersphere. In general, $S(k,n)$ can be identified with $O(n)/O(n-k)$, $S(n,n)$ corresponds to $O(n)$, and $S(n-1,n)$ corresponds to $SO(n)$. In a similar manner, $Gr(k,n)$ can be identified with $O(n)/O(k) \times O(n-k)$ and $FL(d_1, d_2, \ldots, d_r; n)$ can be identified with $O(n)/O(d_1) \times O(d_2 - d_1) \times \cdots \times O(d_r - d_{r-1}) \times O(n - d_r)$. Through these identifications, concrete descriptions of the tangent and normal bundles to Grassmann, flag and Stiefel manifolds are available [1, 12].

3 Flag Manifolds from Data

There are many approaches for encoding and representing the structure in a data matrix as a point on a Grassmann or flag manifold. At a higher level, what kinds of statistical tools should be developed for the purpose of analyzing or representing a data cloud consisting of points on multiple Grassmann manifolds? An important early step is to find algorithms that produce single points on a flag manifold that represents common structure in such a data cloud. For additional details concerning special manifold statistics, see [9, 11].

3.1 An Algorithm for Computing a Flag from a Collection of Subspaces

Let $[X]$ denote the column space of a matrix X. From a collection of subspaces $\mathscr{D} = \{[X_1], \ldots, [X_N]\}$ of an n-dimensional vector space V, we utilize an optimization algorithm to associate a point on a flag manifold to $\mathscr{D}$. A full description of the optimization algorithm and its properties is found in [5]. The algorithm finds an ordered collection of orthonormal vectors, u^j, by solving

$$\begin{aligned} [u^{(j)}] := \quad & \underset{[u] \in Gr(1,n)}{\arg\min} \sum_{[X_i] \in \mathscr{D}} d_{pF}([u], [X_i])^2 \\ & \text{subject to} \quad [u^{(j)}] \perp [u^{(l)}] \qquad \text{for } l < j, \end{aligned} \tag{1}$$

where $d_{pF}([u], [X_i])$ is the projection Frobenius norm (which can be written in terms of the principal angles between the two subspaces $[u]$ and $[X_i]$). As illustrated by Björck and Golub, the vector of principal angles, Θ, between the subspaces $[X]$ and $[Y]$ can be found as the inverse sines of the singular values of the matrix $Q_X^T Q_Y$, where Q_X and Q_Y are unitary bases for $[X]$ and $[Y]$ respectively [3].

Solving Eq. 1 leads to the set $\{[u^{(1)}], [u^{(2)}], \ldots, [u^{(r)}]\}$ where r is the dimension of the span of the elements in $\mathscr{D}$. Recalling that $d_{pF}([X],[Y]) = \|\sin\Theta\|_2$, the sequence of optimizers can be found analytically as is shown in [5]. From these one-dimensional subspaces, the point $P \in FL(1,2,\ldots,r,n)$ associated with $\mathscr{D}$ is then,

$$P = \left[u^{(1)}\right] \subset \left[u^{(1)}|u^{(2)}\right] \subset \ldots \subset \left[u^{(1)}|\ldots|u^{(r)}\right] \subset V. \tag{2}$$

Principal angles have been widely used for comparing points on Grassmannians. We propose using them to compare points on flag manifolds obtained from the approach above. If individual elements of a flag are of interest, the distance between subspaces can be measured using a variety of metrics, such as the geodesic distance based on arc length, i.e. $d([X],[Y]) = \|\Theta\|_2$. Similarly, we can define metrics between flags with the same signature by taking functions of the principal angles between each of their elements of the appropriate size, such as the sum of the geodesic distances between matching elements.

4 Numerical Experiments

The experiments in this paper are meant to illustrate the ability of the flag representation to organize data with multiple semantically meaningful forms of variation. To this end, we will explore three sets of image and video data. Each experiment will look to isolate one simple form of variation. Success in the task will be measured by percentage of test samples that correctly identified with a flag containing the intended information. A concise description of each data set can be seen in Table 1.

Table 1 Descriptions of the three data sets used for experiments

Data set	Sample size	Number of samples	Classes
Mind's Eye tracklets	$32 \times 32 \times 48$	308 tracklets	'carry', 'gesture', 'leg-motion', 'loiter' 'loiter-group', 'turn', 'walk', 'walk-group'
Images of letters	45×45	60 images	'a', 'f', 'w', 'x' each in 15 fonts
PIE faces	277×299	15,288 images	56 subjects, 21 illuminations, 13 poses

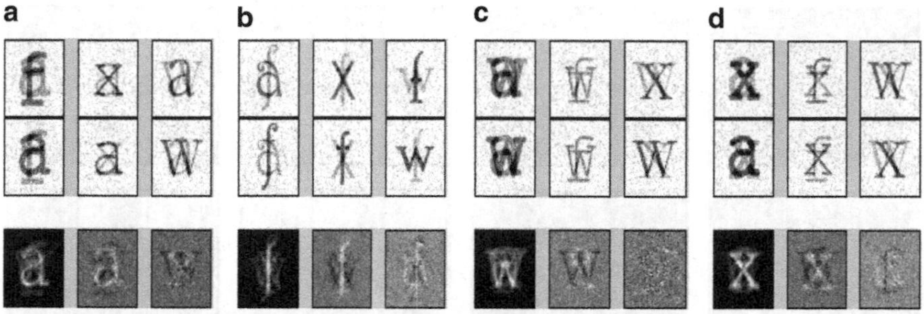

Fig. 1 Four sets of images each with one common feature. Each image is the superposition of two letters and noise, and each column of images is used to create a point on Gr(2, 2025) that spans two letters. Below the sets are the first three images from the associated flag. (**a**) The letter 'a'. (**b**) The letter 'f'. (**c**) The letter 'w'. (**d**) The letter 'x'

4.1 Illustrative Example

The first data set consists of 60 images of the letters 'a', 'f', 'w', and 'x' depicted in 15 fonts. The flag representation will isolate a common letter in a set that contains other distracting letters and noise.

For each letter we randomly choose 6 images from our set of 60; 3 images of one letter and 1 image of each of the other 3 letters. The images are raster-scanned to create vectors in $\mathbb{R}^{2025}$. Three 2-dimensional subspaces of $\mathbb{R}^{2025}$ are produced from the 6 images. Each 2-dimensional subspace is formed as the span of two random linear combinations of an image of the main letter, an image of a different letter, and added Gaussian noise. Thus the 6 images in a set create 3 points on Gr(2, 2025) such that the main letter is a common feature in each Grassmann point. For each set of 3 Grassmann points, we create a flag that helps to expose the similarity between the points. Examples of the four sets and the first three vectors from their associated flags can be seen in Fig. 1.

The flags created from subspaces containing each letter can be used to identify instances of the same letter in fonts that were not used to create the flags. In fact, calculating the closest flag to a test sample did a better job of classifying novel instances of a letter than using a nearest neighbor classifier with the raw images that were used to train the flags. Over 10 trials, classification using the flags had an average success rate of $\approx$83 %, while classification using the nearest image had a success rate of $\approx$65 %.

4.2 Video Sequences

The second example uses portions of video clips, called *tracklets*, that were extracted from larger and longer videos filmed as part of DARPA's Mind's Eye

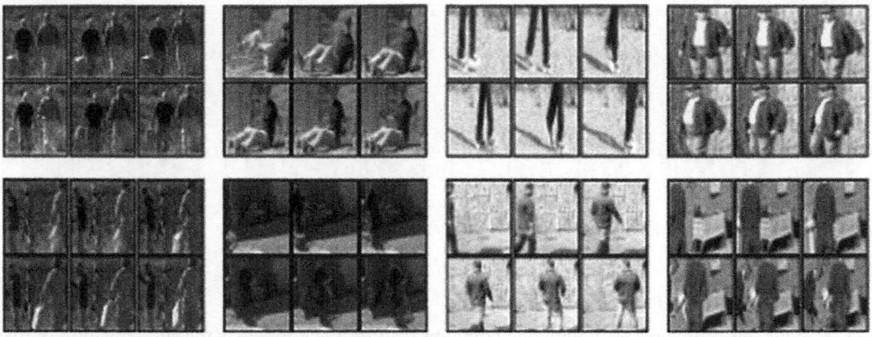

Fig. 2 Mind's Eye video sequences illustrating six frames of eight doubly labeled actions. From left-to-right, top-to-bottom the labels are: 'carry/walk', 'gesture/sit-up', 'leg-motion/walk', 'loiter/walk', 'loiter/group-patdown', 'turn/ride-bike', 'walk/turn', and 'walk-group/bend'

program. The tracklets have been automatically cropped and registered to focus on an action of short duration (48 frames). The goal of this experiment is to automatically recognize a single action contained in a tracklet that depicts two actions being performed simultaneously. The tracklets have been hand labeled with the actions they contain. Examples of frames from some of these tracklets can be seen in Fig. 2. Each video contains at least one of the labels listed in Table 1.

In order to demonstrate the flag's ability to model the dominant form of variation in a set of subspaces, we compare classification accuracy using flags versus a nearest neighbor approach. To begin, each 3-way array is unfolded into a matrix of size $1{,}024 \times 48$ with each column representing a frame of the video. The column space of each matrix is used to represent the tracklet. From the 308 available tracklets, we select K training samples from each action class. It is important to note that some of the training samples for one class may share their second label with another one of the classes being modeled. For example, a video labeled 'walk/turn' could be used as a training sample for either class 'walk' or 'turn', but not both.

Using the K training samples from each of the eight classes, we create a flag for each class. The tracklets that are not used for training make up the test set. Each test video is compared to the 48-dimensional component in each flag. Using the geodesic distance based on arc length, a test sample is given the label of the nearest flag. The classification is considered a success if the label matches one of its given labels. Similarly, each test video is compared to the $8 \times K$ videos that were used to create the flags. In this case a test sample is given the label of the class that contained the nearest video. The results of this experiment can be seen in Fig. 3. The left graph in Fig. 3 shows the accuracy for the classification as the number of training samples, K, increases. Surprisingly, the accuracy is comparable for each method. One would expect superior performance using nearest neighbors given the amount of data required. The graph on the right side of Fig. 3 shows the precision rate for the individual classes. We can see that some classes do much better than

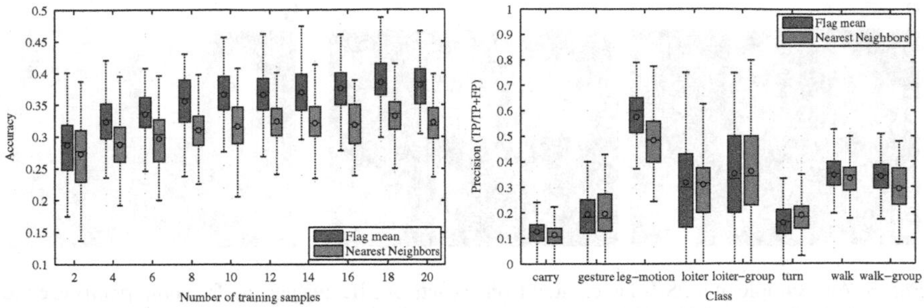

Fig. 3 Mind's Eye data classification: The figure on the left shows overall accuracy vs. number of training samples used per class. The figure on the right shows the precision for each class when four training samples were used

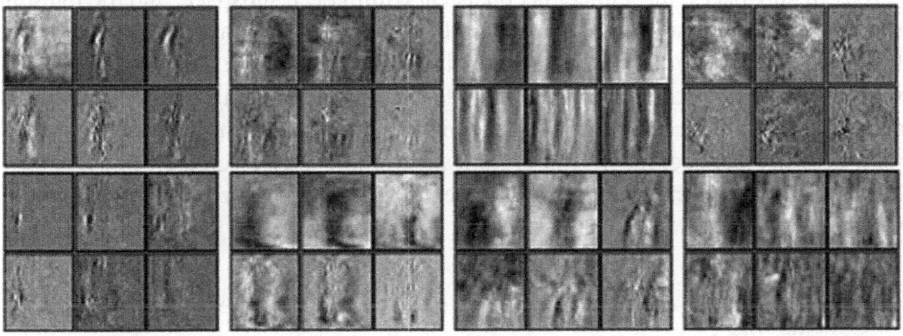

Fig. 4 Mind's Eye flags *left-to-right*, *top-to-bottom*: 'carry', 'gesture', 'leg-motion', 'loiter', 'turn', 'walk', 'walk-group'

others, which is partially a product of the number of samples available from each class. The resulting flags for these actions are shown in Fig. 4.

4.3 Pose Flags

In the final example we create flags to represent poses from a subset of images in the CMU-PIE database [10]. The subset of images used is described in Table 1. The example focuses on the images that have ambient lighting turned on with neutral expressions. By grouping the remaining images in a structured way, we create a flag that represents a single pose.

We find a basis for a collection of vectorized images that share a single pose, a single subject, and whose lighting conditions differ. The span of this basis approximates an illumination subspace. If we then create a flag out of subspaces that share a pose, but contain different subjects, we get a model for that pose that appears independent of subject or illumination. We refer to such a flag as a

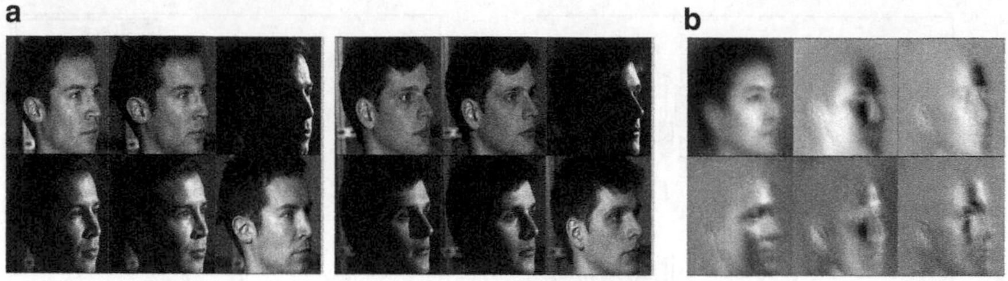

Fig. 5 An example of a PS/I-flag created from 34 sets of PIE images. (**a**) Two *data points* used to create the PS/I flag. (**b**) The PS/I flag

PS/I-flag to indicate the ordering of the variation. That is, the pose is consistent across all subspaces, the subject is consistent within each subspace, and the lighting varies within each subspace. An example of a PS/I-flag can be seen in Fig. 5b. Images from two of the subspaces used to create the flag are shown in Fig. 5a. If we train a flag for each of the 13 poses in the PIE database using half of the subjects and illumination conditions, we can recognize the pose of the remaining images with near perfect accuracy. This organization is one way to create flags from the PIE images. A similar technique can be employed to recognize the other forms of variation as well.

Conclusions

In this paper we investigated the representation of several data sets in terms of flag manifolds. The flag manifolds were used to classify unlabeled patterns and did so with an accuracy comparable to nearest neighbor classification. We infer that the geometric structure characterized by the flag captures information inherent in the data. We note that, in general, nearest neighbor classifiers perform very well, use all the available data, and grow in complexity as more data is collected. In contrast, flags also perform well, serve as prototypes and, as we have seen, have significant representational differences.

Acknowledgements This research was partially supported by the NSF: CDS&E-MSS-1228308, DMS-1322508, DOD-USAF: FA9550-12-1-0408 P00001 and by DARPA: N66001-11-1-4184. Any opinions, findings and conclusions or recommendations expressed in this material are those of the authors and do not necessarily reflect the views of the funding agencies.

References

1. T. Arias, A. Edelman, S.T. Smith, The geometry of algorithms with orthogonality constraints. SIAM J. Matrix Anal. Appl **20**, 303–353 (1998)
2. J.R. Beveridge, B. Draper, J-M. Chang, M. Kirby, H. Kley, C. Peterson, Principal angles separate subject illumination spaces in YDB and CMU-PIE. IEEE TPAMI **29**(2), 351–363 (2008)
3. A. Björck, G.H. Golub, Numerical methods for computing angles between linear subspaces. Math. Comput. **27**, 579–594 (1973)
4. J.-M. Chang, M. Kirby, H. Kley, C. Peterson, J.R. Beveridge, B. Draper, Recognition of digital images of the human face at ultra low resolution via illumination spaces, in *Computer Vision – ACCV 2007*, Tokyo, ed. by Y. Yagi, S.B. Kang, I.S. Kweon, H. Zha. Lecture Notes in Computer Science, vol. 4844 (Springer, 2007), pp. 733–743
5. B. Draper, M. Kirby, J. Marks, T. Marrinan, C. Peterson, A flag representation for finite collections of subspaces of mixed dimensions. Linear Algebra Appl. **451**, 15–32 (2014)
6. T. Geller, Seeing is not enough. Commun. ACM **54**(10), 15–16 (2011)
7. I.M. James, *The Topology of Stiefel Manifolds*. London Math Society Lecture Note Series, vol. 24 (Cambridge University Press, Cambridge/New York, 1977)
8. F. Nater, H. Grabner, L. Van Gool, Exploiting simple hierarchies for unsupervised human behavior analysis, in *IEEE CVPR*, San Francisco, 2010, pp. 2014–2021
9. X. Pennec, Intrinsic statistics on Riemannian manifolds: basic tools for geometric measurements. J. Math. Imaging Vis. **25**, 127–154 (2006)
10. T. Sim, S. Baker, M. Bsat The CMU pose, illumination, and expression (PIE) database, in *Proceedings of 5th International Conference on Automatic Face and Gesture Recognition*, Washington, D.C., 2002
11. P. Turaga, A. Veeraraghavan, A. Srivastava, R. Chellappa, Statistical computations on Grassmann and Stiefel manifolds for image and video-based recognition. IEEE TPAMI **33**(11), 2273–2286 (2011)
12. V.S. Varadarajan, *Lie Groups, Lie Algebras, and their Representations*. Graduate Texts in Mathematics, vol. 102 (Springer, New York, 1984)

Part VII
Optimal Control

Distributed Optimal Control Problems Governed by Coupled Convection Dominated PDEs with Control Constraints

Hamdullah Yücel and Peter Benner

Abstract We study the numerical solution of control constrained optimal control problems governed by a system of convection diffusion equations with nonlinear reaction terms, arising from chemical processes. Control constraints are handled by using the primal-dual active set algorithm as a semi-smooth Newton method or by adding a Moreau-Yosida-type penalty function to the cost functional. An adaptive mesh refinement indicated by a posteriori error estimates is applied for both approaches.

1 Introduction

We investigate a class of distributed optimal control problems governed by a system of convection diffusion partial differential equations (PDEs), arising from chemical processes [1, 3, 6, 13]. In these problems, the constraints are strongly coupled such that inaccuracies in one unknown directly affect all other unknowns. Therefore, prediction of these unknowns is very important for the safe and economical operation of biochemical and chemical engineering processes. Such coupled constraints in large chemical systems have two main issues. One of them is that the reaction terms are assumed to be expressions which are products of some functions of the concentrations of the chemical components and an exponential function of the temperature, called Arrhenius kinetics expression. Therefore, the PDE constraints in our formulation are nonlinear PDEs. The other issue is that the sizes of the diffusion parameters are small compared to the sizes of the velocity fields. Then, such a convection diffusion system exhibits boundary and/or interior layers, localized regions where the derivative of the PDE solution is large. In order to eliminate spurious oscillations emerging from the layers, adaptive mesh refinement is particularly attractive as it usually yields a small number of degrees of freedom. Discontinuous Galerkin methods exhibit better convergence for optimal control

H. Yücel (✉) • P. Benner
Computational Methods in Systems and Control Theory, Max Planck Institute for Dynamics of Complex Technical Systems, Sandtorstr. 1, 39106 Magdeburg, Germany
e-mail: yuecel@mpi-magdeburg.mpg.de; benner@mpi-magdeburg.mpg.de

A. Abdulle et al. (eds.), *Numerical Mathematics and Advanced Applications ENUMATH 2013*, Lecture Notes in Computational Science and Engineering 103,
DOI 10.1007/978-3-319-10705-9_46

problems governed by convection dominated equations. The reason is that the errors in the layers do not propagate into the entire domain [4]. Adaptive discontinuous Galerkin methods are studied in [9–11,13] for convection dominated optimal control problems.

The goal of this paper is to study control constraint optimal control problems governed by coupled convection dominated equations. To handle control constraints, we apply two different approaches. One of them is the primal-dual active set algorithm as a semi-smooth Newton method [2]. The other one is Moreau-Yosida regularization. Although this technique has been used very successfully for state constrained optimization problems, it has also been applied to control constrained problems [6, 8]. For both cases, residual-type error indicators are proposed and applied to a numerical example.

2 The Optimal Control Problem

Let Ω be an open, bounded polygonal domain in $\mathbb{R}^2$ with boundary $\Gamma = \partial\Omega$, let f_i, β_i, α_i, u_d, v_d, g_i be given functions and let $\epsilon_i, \gamma_i, \omega > 0$ be given diffusion, nonlinear reaction, and regularization parameters, respectively, for $i = 1, 2$. We consider a class of distributed optimal control problems governed by a system of convection diffusion PDEs

$$\min\ J(u, v, c) = \frac{1}{2}\|u - u_d\|^2_{L^2(\Omega)} + \frac{1}{2}\|v - v_d\|^2_{L^2(\Omega)} + \frac{\omega}{2}\|c - c_d\|^2_{L^2(\Omega)}, \tag{1}$$

subject to

$$-\epsilon_1 \Delta u + \beta_1 \cdot \nabla u + \alpha_1 u + \gamma_1 uv = f_1 + c \quad \text{in } \Omega, \tag{2a}$$

$$-\epsilon_2 \Delta v + \beta_2 \cdot \nabla v + \alpha_2 v + \gamma_2 uv = f_2 \quad \text{in } \Omega, \tag{2b}$$

$$u = g_1 \quad \text{on } \Gamma, \qquad v = g_2 \quad \text{on } \Gamma, \tag{2c}$$

and the control is constraint to a closed convex set C^{ad}

$$C^{ad} := \{c \in L^2(\Omega) : \ c_a \le c \le c_b \text{ a.e. in } \Omega\} \tag{3}$$

with the constant bounds $c_a, c_b \in \mathbb{R} \cup \{\pm\infty\}$ satisfying $c_a < c_b$. We refer to u and v as the state variables, to c as the control variable and to (2) as the state system. Further, we assume that there exist constants $\kappa_i > 0$ such that $\alpha_i - \frac{1}{2}\nabla \cdot \beta_i \ge \kappa_i > 0$ for $i = 1, 2$.

The optimality conditions of the problem (1)–(3) consist of the coupled state system (2), the coupled adjoint system

$$-\epsilon_1 \Delta p - \beta_1 \cdot \nabla p + (\alpha_1 - \nabla \cdot \beta_1)p + \gamma_1 pv + \gamma_2 qv = -(u - u_d) \quad \text{in } \Omega, \tag{4a}$$

$$-\epsilon_2 \Delta q - \beta_2 \cdot \nabla q + (\alpha_2 - \nabla \cdot \beta_2)q + \gamma_1 pu + \gamma_2 qu = -(v - v_d) \quad \text{in } \Omega, \tag{4b}$$

$$p = 0 \quad \text{on } \Gamma, \qquad q = 0 \quad \text{on } \Gamma, \tag{4c}$$

the gradient equation and complementary conditions with Lagrange multipliers $\lambda_a, \lambda_b \in L^2(\Omega)$

$$\omega(c - c_d) - p - \lambda_a + \lambda_b = 0, \qquad \text{a.e. in } \Omega, \tag{5a}$$

$$\lambda_a \geq 0, \quad c_a - c \leq 0, \quad \lambda_a(c - u_a) = 0 \quad \text{a.e. in } \Omega, \tag{5b}$$

$$\lambda_b \geq 0, \quad c - c_b \leq 0, \quad \lambda_b(c_b - c) = 0 \quad \text{a.e. in } \Omega. \tag{5c}$$

3 Discontinuous Galerkin Discretization

We choose the symmetric interior penalty Galerkin (SIPG) method for the discretization of the problem (1)–(3) due to its symmetric property. It guarantees that "discretize-then-optimize" and "optimize-then-discretize" lead to the same result, see e.g., [12]. We use the notation in [7] and for the spaces of the state, adjoint, control variables and test functions we use piecewise linear functions on the discretized mesh $\mathcal{T}_h$, i.e., $W_h = Y_h = C_h = \{y \in L^2(\Omega) : y\,|_K \in \mathbb{P}^1(K) \ \forall K \in \mathcal{T}_h\}$. Note that the space of state variables Y_h and the space of test functions W_h are identical since DG methods impose boundary conditions weakly.

We now give the upwind SIPG discretization of the state variables u, v in (2) for a fixed distributed control c. This leads to the following formulation

$$a_h^1(u_h, w_h) + \gamma_1 \sum_{K \in \mathcal{T}_h} \int_K u_h v_h w_h \, dx = l_h^1(w_h) + (c_h, w_h), \tag{6}$$

$$a_h^2(v_h, w_h) + \gamma_2 \sum_{K \in \mathcal{T}_h} \int_K u_h v_h w_h \, dx = l_h^2(w_h), \tag{7}$$

where for $i = 1, 2$ and $\forall w \in W_h$, the (bi)-linear forms are defined as

$$\begin{aligned} a_h^i(z,w) &= \sum_{K \in \mathscr{T}_h} \int_K \epsilon_i \nabla z \cdot \nabla w \, dx - \sum_{E \in \mathscr{E}_h} \int_E \{\{\epsilon_i \nabla z\}\} \cdot [\![w]\!] + \{\{\epsilon_i \nabla w\}\} \cdot [\![z]\!] \, ds \\ &+ \sum_{E \in \mathscr{E}_h} \frac{\sigma \epsilon_i}{h_E} \int_E [\![z]\!] \cdot [\![w]\!] \, ds + \sum_{K \in \mathscr{T}_h} \int_K \beta_i \cdot \nabla z w + \alpha_i z w \, dx \\ &+ \sum_{K \in \mathscr{T}_h} \int_{\partial K^- \setminus \Gamma} \beta_i \cdot \mathbf{n}(z^e - z) w \, ds - \sum_{K \in \mathscr{T}_h} \int_{\partial K^- \cap \Gamma^-} \beta_i \cdot \mathbf{n} z w \, ds, \end{aligned} \tag{8a}$$

$$\begin{aligned} l_h^i(w) &= \sum_{K \in \mathscr{T}_h} \int_K f_i w \, dx + \sum_{E \in \mathscr{E}_h^\partial} \frac{\sigma \epsilon_i}{h_E} \int_E g_i \mathbf{n} \cdot [\![w]\!] \, ds - \sum_{E \in \mathscr{E}_h^\partial} \int_E g_i \{\{\epsilon_i \nabla w\}\} \, ds \\ &- \sum_{K \in \mathscr{T}_h} \int_{\partial K^- \cap \Gamma^-} \beta_i \cdot \mathbf{n} \, g_i w \, ds, \end{aligned} \tag{8b}$$

where the jump and average functions across an edge $E \in \mathscr{E}_h$ are denoted by $[[\cdot]]$ and $\{\{\cdot\}\}$, respectively. We refer to [9–11, 13] for the definition of inflow boundaries $\Gamma^-, \partial K^-$. The parameter σ is called the interior penalty parameter which should be sufficiently large independently of the mesh size and the diffusion coefficients ϵ_i to ensure the stability of the SIPG discretization.

4 Primal-Dual Active Set (PDAS) Strategy

We describe the optimality system consisting of a set of nonlinear equations with the notation $\Phi(x) = 0$. Our first approach to solve the system $\Phi'(x)s = -\Phi(x)$ is to use the semi-smooth Newton approach in terms of an active set strategy [2]. For a Newton step, the active and inactive sets are determined by $\mathscr{A}^+ = \{x \in \Omega : p - \omega(c_b - c_d) > 0\}$, $\mathscr{A}^- = \{x \in \Omega : p - \omega(c_a - c_d) < 0\}$ with $\mathscr{I} = \Omega \setminus (\mathscr{A}^+ \cup \mathscr{A}^-)$. Then the discretized Newton system is given by

$$\begin{bmatrix} \mathbf{A} & \mathbf{C} \\ \mathbf{B} & \mathbf{0} \end{bmatrix} \begin{bmatrix} \Delta u \\ \Delta v \\ \Delta c \\ \Delta p \\ \Delta q \end{bmatrix} = - \begin{bmatrix} A_u^T p + \gamma_1 F_{p,v} + \gamma_2 F_{q,v} + Mu - l(u_d) \\ A_v^T q + \gamma_1 F_{p,u} + \gamma_2 F_{q,u} + Mv - l(v_d) \\ \omega\Big(Mc - \chi_{\mathscr{I}} l(c_d) - M\big(\chi_{\mathscr{A}^-} c_a + \chi_{\mathscr{A}^+} c_b\big)\Big) - \chi_{\mathscr{I}} Mp \\ A_u u + \gamma_1 F_{u,v} - Mc - l_u \\ A_v v + \gamma_2 F_{u,v} - l_v \end{bmatrix},$$

where

$$\mathbf{A}=\begin{bmatrix} M & \gamma_1 M_p+\gamma_2 M_q & 0 \\ \gamma_1 M_p+\gamma_2 M_q & M & 0 \\ 0 & 0 & \omega M \end{bmatrix}, \quad \mathbf{C}=\begin{bmatrix} A_u^T+\gamma_1 M_v & \gamma_2 M_v \\ \gamma_1 M_u & A_v^T+\gamma_2 M_u \\ -\chi_{\mathscr{I}} M & 0 \end{bmatrix},$$

$$\mathbf{B}=\begin{bmatrix} A_u+\gamma_1 M_v & \gamma_1 M_u & -M \\ \gamma_2 M_v & A_v+\gamma_2 M_u & 0 \end{bmatrix}.$$

A_u, A_v correspond to the bilinear forms $a_h^1(u,w)$ and $a_h^2(v,w)$, whereas l_u, l_v correspond to the linear forms $l_h^1(w)$ and $l_h^2(w)$. Further, $l_i(z) = \int_\Omega z\varphi_i dx$, $F_{y,z}^i = \int_\Omega yz\varphi_i \, dx$, $M_{ij} = \int_\Omega \varphi_i\varphi_j \, dx$ and $M_z^{i,j} = \int_\Omega z\varphi_i\varphi_j \, dx$.

For each element $K \in \mathscr{T}_h$, our a posteriori error indicators for the states η_K^u, η_K^v and the adjoints η_K^p, η_K^q are

$$(\eta_K^z)^2 = \left[(\eta_{R_K}^z)^2 + (\eta_{E_K}^z)^2\right], \qquad z \in \{u, v, p, q\}, \tag{9}$$

where the interior residual terms are defined by

$$\begin{aligned}
\eta_{R_K}^u &= \rho_{K,1}\| f_h^1 + c_h + \epsilon_1\Delta u_h - \beta_h^1\cdot\nabla u_h - \alpha_h^1 u_h - \gamma_1 u_h v_h\|_{L^2(K)},\\
\eta_{R_K}^v &= \rho_{K,2}\| f_h^2 + \epsilon_2\Delta v_h - \beta_h^2\cdot\nabla v_h - \alpha_h^2 v_h - \gamma_2 u_h v_h\|_{L^2(K)},\\
\eta_{R_K}^p &= \rho_{K,1}\| u_h^d - u_h + \epsilon_1\Delta p_h + \beta_h^1\cdot\nabla p_h\\
&\qquad -(\alpha_h^1 - \nabla\cdot\beta_h^1)p_h - v_h\,(\gamma_1 p_h + \gamma_2 q_h)\,\|_{L^2(K)},\\
\eta_{R_K}^q &= \rho_{K,2}\| v_h^d - v_h + \epsilon_2\Delta q_h + \beta_h^2\cdot\nabla q_h\\
&\qquad -(\alpha_h^2 - \nabla\cdot\beta_h^2)q_h - u_h\,(\gamma_1 p_h + \gamma_2 q_h)\,\|_{L^2(K)},
\end{aligned}$$

and for $z \in \{u, p\}$, $i = 1$ and $z \in \{v, q\}$, $i = 2$, the term measuring the jumps on the edge is defined by

$$\begin{aligned}
(\eta_{E_K}^z)^2 =& \frac{1}{2}\sum_{E\in\partial K\setminus\Gamma} \epsilon_i^{-\frac{1}{2}}\rho_{E,i}\|[\![\epsilon_i\nabla z_h]\!]\|_{L^2(E)}^2 + \Big(\frac{\sigma\epsilon_i}{h_E} + \kappa_i h_E + \frac{h_E}{\epsilon_i}\Big)\|[\![z_h]\!]\|_{L^2(E)}^2\\
&+ \sum_{E\in\partial K\cap\Gamma}\Big(\frac{\sigma\epsilon_i}{h_E} + \kappa_i h_E + \frac{h_E}{\epsilon_i}\Big)\|[\![z_h]\!]\|_{L^2(E)}^2
\end{aligned}$$

with the weights $\rho_{K,i} = \min\{h_K\epsilon_i^{-\frac{1}{2}}, \kappa_i^{-\frac{1}{2}}\}$, $\rho_{E,i} = \min\{h_E\epsilon_i^{-\frac{1}{2}}, \kappa_i^{-\frac{1}{2}}\}$ for $i = 1, 2$. When $\kappa_i = 0$, $\rho_{K,i} = h_K\epsilon_i^{-\frac{1}{2}}$ and $\rho_{E,i} = h_E\epsilon_i^{-\frac{1}{2}}$ are taken.

To obtain a sharp estimator for the control, we divide the domain Ω into the coincidence (contact) set and noncoincidence (non-contact) set, see, e.g., [5] for details. Then, for $K \in \{\mathcal{T}_h\}_h$ the control estimator is defined by

$$\eta_K^c = h_K \|\nabla(\omega(c_h - c_h^d) - p_h)\chi_{\Omega_{c,h}^+}\|_{L^2(\Omega)}, \tag{10}$$

where $\Omega_{c,h}^+ = \Omega \backslash \Omega_{c,h}^-$, $\Omega_{c,h}^- = \{\cup \bar{K} : c|_K = c_a\} \cup \{\cup \bar{K} : c|_K = c_b\}$. The computation of the characteristic function $\chi_{\Omega_{c,h}^+}$ is not straightforward since we usually do not know the position of the free boundary. It can be approximated by the finite element solution as suggested in [5]. For $\mu > 0$, we thus use

$$\chi_{\Omega_{c,h}^+} = \frac{(c_h - c_a)(c_b - c_h)}{h^\mu + (c_h - c_a)(c_b - c_h)}. \tag{11}$$

5 Moreau-Yosida (MY) Regularization Approach

Our second approach to solve the constrained optimization problem (1)–(3) is to penalize the control constraints with a MY-based technique by modifying the objective functional $J(u, v, c)$ (1). Now, we wish to minimize

$$J(u, v, c) + \frac{1}{2\delta}\|\max\{0, c - c_b\}\|_{L^2(\Omega)}^2 + \frac{1}{2\delta}\|\min\{0, c - c_a\}\|_{L^2(\Omega)}^2$$

subject to the state system (2). Here, δ is the regularization parameter. Then, a Newton step reads

$$\begin{bmatrix} \mathbf{A} & \mathbf{B}^T \\ \mathbf{B} & \mathbf{0} \end{bmatrix} s = - \begin{bmatrix} A_u^T p + \gamma_1 F_{p,v} + \gamma_2 F_{q,v} + Mu - l(u_d) \\ A_v^T q + \gamma_1 F_{p,u} + \gamma_2 F_{q,u} + Mv - l(v_d) \\ \omega M c - \omega l(c_d) - Mp - G(c) \\ A_u u + \gamma_1 F_{u,v} - Mc - l_u \\ A_v v + \gamma_2 F_{u,v} - l_v \end{bmatrix},$$

where

$$\mathbf{A} = \begin{bmatrix} M & \gamma_1 M_p + \gamma_2 M_q & 0 \\ \gamma_1 M_p + \gamma_2 M_q & M & 0 \\ 0 & 0 & \omega M + \frac{1}{\delta}\chi_{\mathcal{A}} M \chi_{\mathcal{A}} \end{bmatrix},$$

$$\mathbf{B} = \begin{bmatrix} A_u + \gamma_1 M_v & \gamma_1 M_u & -M \\ \gamma_2 M_v & A_v + \gamma_2 M_u & 0 \end{bmatrix}$$

with $G(c) = \delta^{-1}(\chi_{\mathscr{A}} M \chi_{\mathscr{A}} c - \chi_{\mathscr{A}^-} M \chi_{\mathscr{A}^-} c_a - \chi_{\mathscr{A}^+} M \chi_{\mathscr{A}^+} c_b)$, $\mathscr{A}^+ = \{x \in \Omega : c - c_b > 0\}$, $\mathscr{A}^- = \{x \in \Omega : c - c_a < 0\}$, and $\mathscr{A} = \mathscr{A}^+ \cup \mathscr{A}^-$.

The error estimators of the MY approach are the same as in (9) except for the control indicator. The modified control estimator $\bar{\eta}^c$ is given by

$$\bar{\eta}^c_K = h_K \|\nabla(\omega(c_h - c_h^d) - p_h) + \frac{1}{\delta}\chi_{\mathscr{A}} \nabla c_h\|_{L^2(\Omega)} \tag{12}$$

for any K. The unknown $\chi_{\mathscr{A}}$ is computed similarly as described in (11).

6 Implementation Details

The adaptive procedure consists of successive execution of the steps **SOLVE** $\rightarrow$ **ESTIMATE** $\rightarrow$ **MARK** $\rightarrow$ **REFINE**. The **SOLVE** step is the numerical solution of the optimal control problem with respect to the given triangulation $\mathscr{T}_h$ using the SIPG discretization. For the **ESTIMATE** step, the residual error indicators are defined in (9), (10) and (12). In the **MARK** step, the edges and elements are specified for the refinement by using the a posteriori error indicators and by choosing subsets $\mathscr{M}_K \subset \mathscr{T}_h$ such that the bulk criterion is satisfied for the given marking parameter θ. Finally, in the **REFINE** step, the marked elements are refined by longest edge bisection, whereas the elements of the marked edges are refined by bisection.

We use linear polynomials for the discretization of the state, adjoint and control variables. The penalty parameter in the SIPG method is chosen as $\sigma = 6$ on interior edges and 12 on boundary edges.

Consider the following optimal control problem governed by coupled convection dominated equations. Let $\epsilon_i = 10^{-6}, \beta_i = (2,3)^T, \alpha_i = 1, \gamma_i = 0.1, \delta = 10^{-6}$ and $\omega = 1$ in $\Omega = (0,1)^2$ for $i = 1,2$. The functions f_1, f_2, u_d and v_d and Dirichlet boundary conditions g_1, g_2 are chosen in order to obtain the exact state solutions

$$\begin{aligned} u(x_1,x_2) &= 4e^{\frac{-1}{\sqrt{\epsilon_1}}\left((x-0.5)^2+3(y-0.5)^2\right)} \sin(x\pi)\cos(y\pi), \\ v(x_1,x_2) &= \frac{2}{\pi}\arctan\left(\frac{1}{\sqrt{\epsilon_2}}\left[-\frac{1}{2}x_1 + x_2 - \frac{1}{4}\right]\right) \end{aligned}$$

the exact adjoint solutions

$$\begin{aligned} p(x_1,x_2) &= e^{\frac{-1}{\sqrt{\epsilon_1}}\left((x-0.5)^2+3(y-0.5)^2\right)} \sin(x\pi)\cos(y\pi), \\ q(x_1,x_2) &= 16x_1(1-x_1)x_2(1-x_2) \\ &\quad \times\left(\frac{1}{2} + \frac{1}{\pi}\arctan\left[\frac{2}{\sqrt{\epsilon_2}}\left(\frac{1}{16} - \left(x_1 - \frac{1}{2}\right)^2 - \left(x_2 - \frac{1}{2}\right)^2\right)\right]\right) \end{aligned}$$

and the exact control function

$$c(x_1, x_2) = \max\{0, 2\cos(\pi x_1)cos(\pi x_2) - 1\}.$$

We have tested both approaches using this example, i.e., the PDAS strategy and MY regularization. The adaptively refined meshes are shown in Fig. 1. We observe that the error indicator $\bar{\eta}^u$ (12) picks out the layers of control better than the error indicator η^u (10) using the almost same number of vertices. Therefore, we obtain a better convergence result for the adaptive implementation of the MY approach as shown in Fig. 2. For both approaches, the global errors in L^2 norm on adaptively refined meshes are decreasing faster than the errors on uniformly refined meshes.

Figure 3 shows the computed solutions for the control on adaptively refined meshes by using the PDAS strategy and Moreau-Yosida regularization. We conclude that substantial computing work can be saved by using efficient adaptive meshes for both approaches and the MY technique captures the errors of the control better than the PDAS strategy.

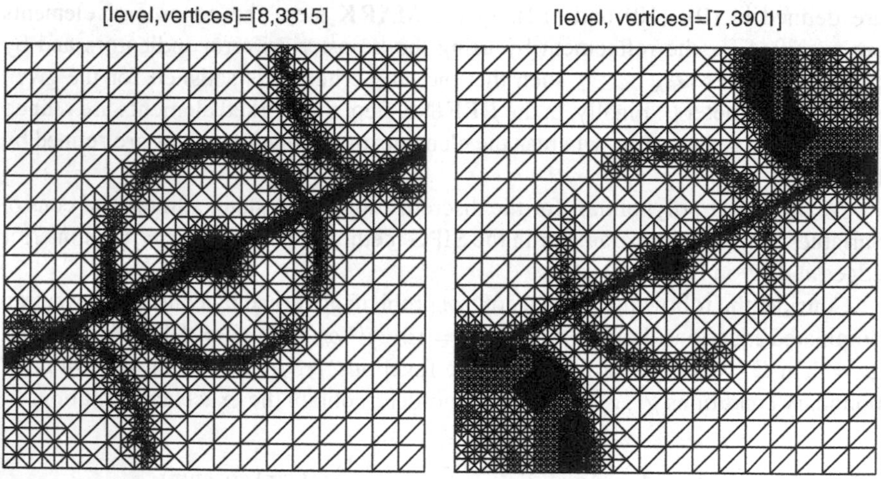

Fig. 1 Adaptively refined meshes for the PDAS and MY approaches, respectively

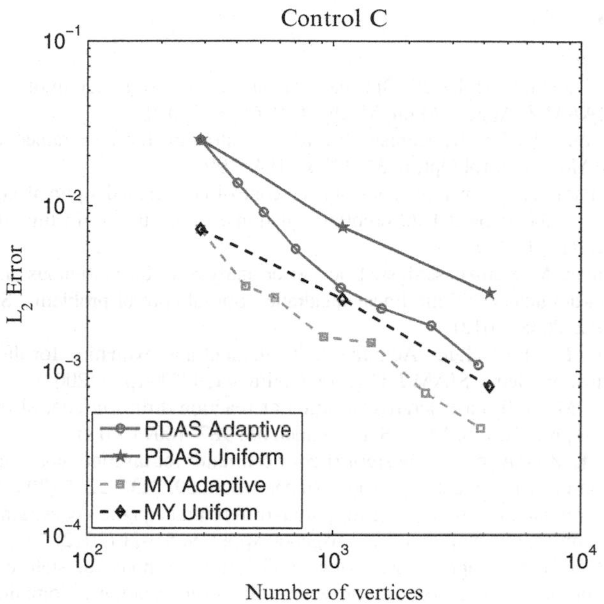

Fig. 2 Global errors with L^2 norm for the control c

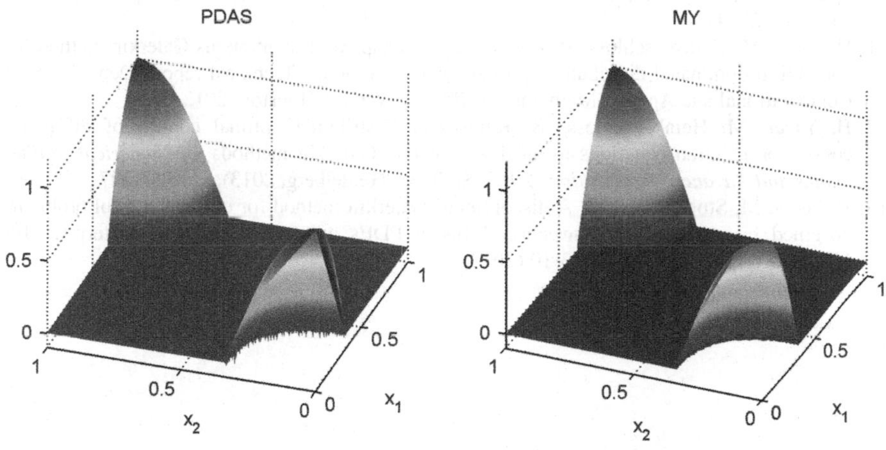

Fig. 3 The computed solutions for the control c obtained by the PDAS on an adaptively refined mesh (3,968 vertices, *left*) and by the MY on an adaptively refined mesh (3,901 vertices, *right*)

Acknowledgement The authors would like to thank Martin Stoll for helpful discussions. This work was supported by the Forschungszentrum Dynamische Systeme (CDS): Biosystemtechnik, Otto-von-Guericke-Universität Magdeburg.

References

1. W. Barthel, C. John, F. Tröltzsch, Optimal boundary control of a system of reaction diffusion equations. ZAAM Z. Angew. Math. Mech. **90**, 966–982 (2010)
2. M. Bergounioux, K. Ito, K. Kunish, Primal-dual strategy for constrained optimal control problems. SIAM J. Control Optim. **37**, 1176–1194 (1999)
3. R. Griesse, Parametric sensivity analysis for control-constrained optimal control problems governed by parabolic partial differential equations, Ph.D. thesis, Institut für Mathematik, Universität Bayreuth, 2003
4. D. Leykekhman, M. Heinkenschloss, Local error analysis of discontinuous Galerkin methods for advection-dominated elliptic linear-quadratic optimal control problems. SIAM J. Numer. Anal. **50**, 2012–2038 (2012)
5. R. Li, W. Liu, H. Ma, T. Tang, Adaptive finite element approximation for distributed elliptic optimal control problems. SIAM J. Control. Optim. **41**, 1321–1349 (2002)
6. J.W. Pearson, M. Stoll, Fast iterative solution of reaction-diffusion control problems arising from chemical processes. SIAM J. Sci. Comput. **35**, 987–1009 (2013)
7. D. Schötzau, L. Zhu, A robust a-posteriori error estimator for discontinuous Galerkin methods for convection-diffusion equations. Appl. Numer. Math. **59**, 2236–2255 (2009)
8. M. Stoll, A. Wathen, Preconditioning for partial differential equation constrained optimization with control constraints. Numer. Linear Algebra Appl. **19**, 53–71 (2012)
9. H. Yücel, P. Benner, Adaptive discontinuous Galerkin methods for state constrained optimal control problems governed by convection diffusion equations. Comput. Optim. Appl., (to appear)
10. H. Yücel, B. Karasözen, Adaptive symmetric interior penalty Galerkin (SIPG) method for optimal control of convection diffusion equations with control constraints. Optimization **63**, 145–166 (2014)
11. H. Yücel, M. Heinkenschloss, B. Karasözen, An adaptive discontinuous Galerkin method for convection dominated distributed optimal control problems, Technical report, Department of Computational and Applied Mathematics, Rice University, Houston, 2012
12. H. Yücel, M. Heinkenschloss, B. Karasözen, Distributed optimal control of diffusion-convection-reaction equations using discontinuous Galerkin methods, in *Numerical Mathematics and Advanced Applications 2011* (Springer, Heidelberg, 2013), pp. 389–397
13. H. Yücel, M. Stoll, P. Benner, A discontinous Galerkin method for optimal control problems governed by a system of convection-diffusion PDEs with nonlinear reaction terms. MPI Magdeburg Preprint MPIMD/13-10 (2013)

Efficient Preconditioning for an Optimal Control Problem with the Time-Periodic Stokes Equations

Wolfgang Krendl, Valeria Simoncini, and Walter Zulehner

Abstract For the optimal control problem with time-periodic Stokes equations a practical robust preconditioner is presented. The discretization of the corresponding optimality system leads to a linear system with a large, sparse and complex 4-by-4 block matrix in saddle point form. We present a decoupling strategy, which reduces the system to two linear systems with a real 4-by-4 block matrix. Based on analytic results on preconditioners for time-harmonic control problems in Krendl et al. (Numer Math 124(1):183–213, 2013), a practical preconditioner is constructed, which is robust with respect to the mesh size h, the frequency ω and the control parameter ν. The result is illustrated by numerical examples with the preconditioned minimal residual method. Finally we discuss alternative stopping criteria.

1 The Model Problem

We consider the following problem: Find the velocity $\mathbf{u}(x,t)$, the pressure $p(x,t)$, and the force $\mathbf{f}(x,t)$ that minimize the cost functional

$$J(\mathbf{u},\mathbf{f}) = \frac{1}{2}\int_0^T \int_\Omega |\mathbf{u}(x,t) - \mathbf{u}_d(x,t)|^2 \, dx \, dt + \frac{\nu}{2}\int_0^T \int_\Omega |\mathbf{f}(x,t)|^2 \, dx \, dt$$

W. Krendl (✉)
Doctoral Program Mathematics, Johannes Kepler University Linz, Altenberger Straße 69, 4040 Linz, Austria
e-mail: wolfgang.krendl@dk-compmath.jku.at

V. Simoncini
Dipartimento di Matematica, Università di Bologna, Piazza di Porta S. Donato 5, 40127 Bologna, Italy
e-mail: valeria.simoncini@unibo.it

W. Zulehner
Institute of Computational Mathematics, Johannes Kepler University Linz, Altenberger Straße 69, 4040 Linz, Austria
e-mail: zulehner@numa.uni-linz.ac.at

A. Abdulle et al. (eds.), *Numerical Mathematics and Advanced Applications ENUMATH 2013*, Lecture Notes in Computational Science and Engineering 103,
DOI 10.1007/978-3-319-10705-9_47

subject to the time-dependent Stokes problem

$$\begin{aligned}
\frac{\partial}{\partial t}\mathbf{u}(x,t) - \Delta\mathbf{u}(x,t) + \nabla p(x,t) &= \mathbf{f}(x,t) \quad \text{in } \Omega \times (0,T), \\
\nabla \cdot \mathbf{u}(x,t) &= 0 \quad \text{in } \Omega \times (0,T), \\
\mathbf{u}(x,t) &= 0 \quad \text{on } \Gamma \times (0,T),
\end{aligned}$$

with time-periodic conditions

$$\mathbf{u}(x,0) = \mathbf{u}(x,T), \; p(x,0) = p(x,T), \; \mathbf{f}(x,0) = \mathbf{f}(x,T) \quad \text{on } \Omega.$$

Here $\Omega \subset \mathbb{R}^d$, $d \in \{2,3\}$ is an open and bounded domain with Lipschitz boundary Γ, $\mathbf{u}_d(x,t)$ is a given target velocity, $\nu > 0$ is a cost or regularization parameter, and $|.|$ denotes the Euclidean norm in $\mathbb{R}^d$. We assume that $u_d(x,t)$ is time-periodic.

For time discretization a truncated Fourier series expansion is used and for space discretization we choose appropriate finite element spaces $\mathbf{V}_h$ of dimension n and Q_h of dimension m for $\mathbf{u}$ and p, respectively, and the same finite element space $\mathbf{V}_h$ for $\mathbf{f}$ as well. The fully discretized problem can then be decoupled in systems, which only depend on one Fourier coefficient. For the Fourier coefficient corresponding to the frequency ω the system reads as follows:

$$J(\underline{\mathbf{u}}, \underline{\mathbf{f}}) = \frac{1}{2}(\underline{\mathbf{u}} - \underline{\mathbf{u}}_d)^*\mathbf{M}(\underline{\mathbf{u}} - \underline{\mathbf{u}}_d) + \frac{\nu}{2}\underline{\mathbf{f}}^*\mathbf{M}\underline{\mathbf{f}} \tag{1}$$

subject to

$$\begin{aligned}
i\omega\,\mathbf{M}\,\underline{\mathbf{u}} + \mathbf{K}\underline{\mathbf{u}} - \mathbf{D}^T\underline{p} &= \mathbf{M}\underline{\mathbf{f}}, \\
\mathbf{D}\underline{\mathbf{u}} &= 0.
\end{aligned}$$

Here the symbol $*$ denotes the conjugate transpose of a vector or a matrix and the real matrices $\mathbf{M}$, $\mathbf{K}$, and $\mathbf{D}$ are the mass matrix, representing the L^2-inner product in $\mathbf{V}_h$, the discretized negative vector Laplacian, and the discretized divergence, respectively. The underlined quantities denote the coefficient vectors of finite element functions relative to a chosen basis.

The Lagrangian functional for this constrained optimization problem is given by

$$\mathscr{L}(\underline{u}, \underline{p}, \underline{f}, \underline{w}, \underline{r}) = J(\underline{\mathbf{u}}, \underline{\mathbf{f}}) + \underline{w}^*\left(i\omega\, M\,\underline{u} + K\underline{u} - \mathbf{D}^T\underline{p} - M\underline{f}\right) + \underline{r}^*\mathbf{D}\underline{u},$$

where $\underline{w}, \underline{r}$ denote the Lagrangian multipliers associated with the constraints. The first-order optimality conditions are $\nabla \mathscr{L}(\underline{u}, \underline{p}, \underline{f}) = 0$, and read in details:

$$\begin{bmatrix} M & 0 & 0 & K - i\omega M & -\mathbf{D}^T \\ 0 & 0 & 0 & -\mathbf{D} & 0 \\ 0 & 0 & \nu M & -M & 0 \\ K + i\omega M & -\mathbf{D}^T & -M & 0 & 0 \\ -\mathbf{D} & 0 & 0 & 0 & 0 \end{bmatrix} \begin{bmatrix} \underline{u} \\ \underline{p} \\ \underline{f} \\ \underline{w} \\ \underline{r} \end{bmatrix} = \begin{bmatrix} M\underline{u}_d \\ 0 \\ 0 \\ 0 \\ 0 \end{bmatrix}. \tag{2}$$

From the third row it follows that $\underline{f} = \nu^{-1}\underline{w}$. So the control $\underline{f}$ can be eliminated. After reordering we obtain the reduced optimality system:

$$\mathscr{M}\underline{x} = \underline{b}, \tag{3}$$

where

$$\mathscr{M} = \begin{bmatrix} A & B^T \\ B & 0 \end{bmatrix}, \quad \underline{x} = \begin{bmatrix} \underline{u} \\ \underline{w} \\ \underline{p} \\ \underline{r} \end{bmatrix} \quad \text{and} \quad \underline{b} = \begin{bmatrix} M\underline{u}_d \\ 0 \\ 0 \\ 0 \end{bmatrix},$$

with

$$A = \begin{bmatrix} \mathbf{M} & \sqrt{\nu}\,(\mathbf{K} - i\omega\mathbf{M}) \\ \sqrt{\nu}\,(\mathbf{K} + i\omega\mathbf{M}) & -\frac{1}{\nu}\mathbf{M} \end{bmatrix} \quad \text{and} \quad B = \begin{bmatrix} 0 & -\mathbf{D} \\ -\mathbf{D} & 0 \end{bmatrix}.$$

2 Transformation to Two Systems with a Real Matrix

Elementary calculations show that:

$$\mathscr{M} = \mathbf{T}^* \mathscr{M}_{\mathrm{T}} \mathbf{T}, \tag{4}$$

where

$$\mathscr{M}_{\mathrm{T}} = \begin{bmatrix} (1+\nu\omega^2)^{1/2}\mathbf{M} & \mathbf{K} & 0 & -\mathbf{D}^T \\ \mathbf{K} & -\nu^{-1}(1+\nu\omega^2)^{1/2}\mathbf{M} & -\mathbf{D}^T & 0 \\ 0 & -\mathbf{D} & 0 & 0 \\ -\mathbf{D} & 0 & 0 & 0 \end{bmatrix},$$

and

$$\mathbf{T} = \begin{bmatrix} T \otimes I_n & 0 \\ 0 & T \otimes I_m \end{bmatrix} \quad \text{with} \quad T = (1+\nu\omega^2)^{-1/4} \begin{bmatrix} (1+\nu\omega^2)^{1/2} & -i \\ 0 & 1 \end{bmatrix}.$$

Here the symbol $\otimes$ denotes the Kronecker product and I_k denotes the identity matrix in $\mathbb{R}^k$. The original system (3) is equivalent to the two systems

$$\mathscr{M}_{\mathbf{T}} \underline{y}_1 = \underline{c}_1 \quad \text{and} \quad \mathscr{M}_{\mathbf{T}} \underline{y}_2 = \underline{c}_2, \tag{5}$$

with $\underline{c} = \underline{c}_1 + i\,\underline{c}_2 = \left(\mathbf{T}^{-1}\right)^* \underline{b}$ and $\underline{y} = \underline{y}_1 + i\,\underline{y}_2 = \mathbf{T}\underline{x}$. So instead of solving one linear system with a complex 4-by-4 block matrix, we have to solve two linear systems with the same 4-by-4 real block matrix $\mathscr{M}_{\mathbf{T}}$, which can be done in parallel.

3 Preconditioning

Our method of choice for solving (5) is the preconditioned MINRES method. As preconditioner $\mathscr{P}$ we consider the block preconditioner constructed in [4]:

$$\mathscr{P} = \begin{bmatrix} P & 0 \\ 0 & R \end{bmatrix}, \quad \text{where} \quad P = \begin{bmatrix} \mathbf{P} & 0 \\ 0 & \frac{1}{\nu}\mathbf{P} \end{bmatrix} \quad \text{and} \quad R = \begin{bmatrix} \nu S & 0 \\ 0 & S \end{bmatrix}, \tag{6}$$

with real and symmetric positive matrices

$$\mathbf{P} = \mathbf{M} + \sqrt{\nu}\,(\mathbf{K} + \omega\,\mathbf{M}) \quad \text{and} \quad S = \mathbf{D}\mathbf{P}^{-1}\mathbf{D}^T. \tag{7}$$

Definition 1 For a matrix N, we denote the eigenvalues of N with minimal and maximal modulus by $\lambda_{\min}(N)$ and $\lambda_{\max}(N)$, respectively.

We have the following estimates.

Theorem 1

$$1/\sqrt{12} \leq |\lambda_{\min}(\mathscr{P}^{-1}\mathscr{M}_T)| \text{ and } |\lambda_{\max}(\mathscr{P}^{-1}\mathscr{M}_T)| \leq (1+\sqrt{5})/2.$$

For the proof, the detailed analysis and further structural spectral results, see [4].

Definition 2 We call a symmetric and positive definite matrix $\mathscr{Q}$ a robust preconditioner for $\mathscr{M}_{\mathbf{T}}$, if $\kappa(\mathscr{Q}^{-1}\mathscr{M}_{\mathbf{T}}) = \lambda_{\max}(\mathscr{Q}^{-1}\mathscr{M}_{\mathbf{T}})/\lambda_{\min}(\mathscr{Q}^{-1}\mathscr{M}_{\mathbf{T}}) \leq C$ with constant C independent of h, ω and ν.

The result of Theorem 1 implies that $\kappa(\mathscr{P}^{-1}\mathscr{M}_{\mathbf{T}}) \leq \sqrt{3}(1+\sqrt{5})$. Hence $\mathscr{P}$ is a robust preconditioner for $\mathscr{M}_{\mathbf{T}}$. Using well known convergence results for the preconditioned MINRES method (see [2]), it follows that the number of iterations, which is needed to decrease the relative error of the k-th residual measured in the $\|\cdot\|_{\mathcal{P}^{-1}}$-norm by a factor $\varepsilon > 0$, is independent of h, ω and ν. Thereby, for a symmetric and positive definite matrix M, the norm $\|\cdot\|_M$ is defined by $\langle M\cdot,\cdot\rangle^{1/2}$, where $\langle\cdot,\cdot\rangle$ denotes the Euclidean inner product.

3.1 The Practical Preconditioner $\tilde{\mathcal{P}}$

The usage of $\mathcal{P}$ as preconditioner for $\mathcal{M}_{\mathrm{T}}$ requires the evaluation of $\mathbf{P}^{-1}\underline{d}$ and $S^{-1}\underline{e}$ for some given vectors $\underline{d}$ and $\underline{e}$ in every step of the MINRES method. These, especially the evaluation of $S^{-1}\underline{e}$, are nontrivial tasks, due to the potentially high number of involved unknowns. To decrease the computational costs we want to replace $\mathbf{P}$ and S by efficient approximations $\tilde{\mathbf{P}}$ and $\tilde{S}$, respectively. This leads to a preconditioner of the form:

$$\tilde{\mathcal{P}} = \begin{bmatrix} \tilde{P} & 0 \\ 0 & \tilde{R} \end{bmatrix} \quad \text{with} \quad \tilde{P} = \begin{bmatrix} \tilde{\mathbf{P}} & 0 \\ 0 & \nu\tilde{\mathbf{P}} \end{bmatrix} \quad \text{and} \quad \tilde{R} = \begin{bmatrix} \tilde{S} & 0 \\ 0 & \frac{1}{\nu}\tilde{S} \end{bmatrix}. \tag{8}$$

Definition 3 For symmetric and positive definite matrices $M, N \in \mathbb{R}^{n\times n}$, we write $M \sim N$, if there exists positive constants γ_1 and γ_2 independent of h, ν and ω, such that $\gamma_1\langle M\underline{v},\underline{v}\rangle \leq \langle N\underline{v},\underline{v}\rangle \leq \gamma_2\langle M\underline{v},\underline{v}\rangle$ for all $\underline{v} \in \mathbb{R}^n$.

Obviously we have that $\tilde{\mathcal{P}}$ is also a robust preconditioner for $\mathcal{M}_{\mathrm{T}}$, if $\tilde{\mathbf{P}} \sim \mathbf{P}$ and $\tilde{S} \sim S$. We will now present a possible choice for $\tilde{\mathbf{P}}$ and $\tilde{S}$ with $\tilde{\mathbf{P}} \sim \mathbf{P}$ and $\tilde{S} \sim S$, which guarantee the robustness of $\tilde{\mathcal{P}}$, and further an efficient evaluation of $\tilde{\mathcal{P}}^{-1}\underline{z}$ for a vector $\underline{z}$:

Choice for $\tilde{\mathbf{P}}$: We replace the evaluation of $\mathbf{P}^{-1}\underline{d}$ by one $V(1,1)$-cycle of a multigrid method with a symmetric Gauß-Seidel smoother as pre- and post-smoother applied to $\mathbf{P}\underline{v} = \underline{d}$, shortly denoted by $\tilde{\mathbf{P}}^{-1}\underline{d}$. In [8] it was shown that $\tilde{\mathbf{P}} \sim \mathbf{P}$.

Choice for $\tilde{S}$: First we replace S by the so called *Cahouet-Chabard preconditioner* $S_{\mathrm{CH}} := (\sqrt{\nu}M_p^{-1} + (1+\sqrt{\nu}\omega)K_p^{-1})^{-1}$, where M_p and K_p denote the mass and stiffness matrices in the finite element space Q_h, respectively. For particular finite elements, e.g. the Taylor-Hood element, we have $S \sim S_{CH}$, see, e.g., [1,5,6] and [7]. In a second step of approximations, we replace the evaluation of $M_p^{-1}\underline{e}$ by one step of a symmetric Gauß-Seidel iteration applied to $M_p\underline{q} = \underline{e}$ and the evaluation of $K_p^{-1}\underline{e}$ by one $V(1,1)$-cycle of the same multigrid method as before applied to $\mathbf{P}\underline{q} = \underline{e}$, shortly denoted by $\tilde{M}_p^{-1}\underline{e}$ and $\tilde{K}_p^{-1}\underline{e}$, respectively. Again from [8] we have $\tilde{M}_p \sim M_p$ and $\tilde{K}_p \sim K_p$. As result of this replacements we obtain

$$\tilde{S}_{\mathrm{CH}} := (\sqrt{\nu}\tilde{M}_p^{-1} + (1+\sqrt{\nu}\omega)\tilde{K}_p^{-1})^{-1}, \tag{9}$$

where $\tilde{S}_{CH} \sim S$ and the inverse of $\tilde{S}_{CH}$ can be applied efficiently. Now we replace the evaluation $S^{-1}\underline{e}$ by applying r-steps (typically $r = 1, 2, 3$) of the preconditioned

Richardson method to the equation $S\underline{q} = \underline{e}$, with scaling parameters $\tau_i > 0$, the preconditioner $\tilde{S}_{CH}$ and the initial vector $q_0 = 0$. The corresponding preconditioner is given by

$$\tilde{S} = S\left(I_{2m} - \prod_{i=1}^{r}(I_{2m} - \tau_i\, \tilde{S}_{CH}^{-1}S)^i\right)^{-1}, \tag{10}$$

In order to guarantee that $\tilde{S}$ is positive definite, it is easy to see that the condition

$$1 - \prod_{i=1}^{r}(1 - \tau_i\, \lambda)^i > 0 \quad \forall \lambda \in (0, 1]. \tag{11}$$

suffices. In particular if we choose $\tau_1 > 0$ fixed and $\tau_i = 1$ for $i \geq 2$, then it follows that $\tilde{S}$ is symmetric, positive definite, and $\tilde{S} \sim S$.

In summary we obtain:

Theorem 2 *$\tilde{\mathscr{P}}$ defined in (8) with the previous presented choices for $\tilde{\mathbf{P}}$ and $\tilde{S}$, is a symmetric and positive definite robust preconditioner for $\mathscr{M}_T$.*

3.2 Numerical Results

We present some numerical examples on the unit square domain $\Omega = (0, 1) \times (0, 1) \subset \mathbb{R}^2$. Following Example 1 in [3] we choose the target velocity $\mathbf{u}_d(x, y) = [(U(x, y), V(x, y)]^T$, given by

$$U(x, y) = 10\,\varphi(x)\varphi'(y) \quad \text{and} \quad V(x, y) = -10\,\varphi'(x)\varphi(y),$$

with $\varphi(z) = \big(1 - \cos(0.8\pi z)\big)(1 - z)^2$. This target velocity $\mathbf{u}_d(x, y)$ is divergence free. The problem was discretized by the Taylor-Hood pair of finite element spaces consisting of continuous piecewise quadratic polynomials for the velocity $\mathbf{u}(x, y)$ and the force $\mathbf{f}(x, y)$, and continuous piecewise linear polynomials for the pressure $p(x, y)$ on a triangulation of Ω. The initial mesh contains four triangles obtained by connecting the two diagonals. The final mesh was constructed by applying ℓ uniform refinement steps to the initial mesh, leading to a mesh size $h = 2^{-\ell}$.

All presented numerical experiments refer to the first of the two systems from (5). The results for the second system are completely identical. Therefore, they are omitted. For each system, the total number of unknowns on the finest level $\ell = 7$ is 1,184,780.

Tables 1 and 2 contain the numerical results produced by the preconditioned MINRES method with the preconditioner $\tilde{\mathscr{P}}$ as described in (8), where we choose

Table 1 $\omega = 10^4$

	ν				
h	10^{-8}	10^{-4}	1	10^4	10^8
2^{-4}	44	46	46	46	46
2^{-5}	48	50	50	50	48
2^{-6}	50	52	52	52	52
2^{-7}	54	56	56	56	56

Table 2 $\nu = 10^{-4}$

	ω				
h	10^{-8}	10^{-4}	1	10^4	10^8
2^{-4}	87	87	87	46	38
2^{-5}	99	99	99	52	34
2^{-6}	101	101	101	51	30
2^{-7}	105	105	105	56	34

Table 3 $h = 2^{-7}$, $\nu = 1$, $\omega = 1$

r	Scaling parameters	$\tilde{k}$	CPU-time (s)
1	$\tau_1 = 1$ (i.e. $\tilde{S} = \tilde{S}_{CH}$)	118	577.65
1	$\tau_1 = 4$ (i.e. $\tilde{S} = 4\,S_{CH}$)	69	341.69
2	$\tau_1 = 4$, $\tau_2 = 1$	46	333.7
3	$\tau_1 = 4$, $\tau_2 = \tau_3 = 1$	41	402.73

$r = 1$ with $\tau_1 = 1$ (i.e. $\tilde{S} = \tilde{S}_{CH}$). The considered values for the mesh size h, the frequency ω, and the regularization parameter ν are specified in the table captions, the first rows and first columns. The other entries of the tables contain the numbers of MINRES iterations that are required for reducing the initial errors in the $\tilde{\mathscr{P}}^{-1}$-norm by a factor of $\varepsilon = 10^{-8}$ with initial vector $x_0 = 0$, respectively.

As expected from the results of Theorem 2, the condition numbers are bounded away from ∞ independent of h, ν and ω, leading to a uniform bound for the number of iterations.

Next, we compare the performance of the practical preconditioner $\tilde{\mathscr{P}}$ with the original (typically better but impractical) preconditioner $\mathscr{P}$ for the particular parameter choice 2^{-7}, $\nu = 1$ and $\omega = 1$. In this case the number of iterations for $\tilde{\mathscr{P}}$ is 118, which is roughly four times higher than the expected number of iterations for $\mathscr{P}$, see Table 1 in [4]. Since the difference is relatively high, it is worthwhile to consider other options for the inner iteration in order to reduce this gap. Table 3 shows the numbers of iterations $\tilde{k}$ and the computational costs, measured in the CPU-time, for $\tilde{S}$ with $r \in \{1, 2, 3\}$, for different values of $\tau_1 \in \{1, 4\}$ and $\tau_i = 1$ for $i \geq 2$. These and similar further numerical experiments show that a significant improvement of the numbers of iterations can be achieved by a proper choice for τ_1 and not so much by a higher number r of inner iterations. It turned out that $\tau_1 = 4$ is a very good choice, also for all other cases.

4 Alternative Stopping Criteria

In our numerical examples the stopping criterion

$$\|r_k\|_{\tilde{\mathscr{P}}^{-1}} \leq \varepsilon \, \|r_0\|_{\tilde{\mathscr{P}}^{-1}} \tag{12}$$

was used. Another natural measure for the error is $\|x - x_k\|_{\tilde{\mathscr{P}}}$. This quantity is not directly computable but can be estimated by using the relation:

$$c \, \|x - x_k\|_{\tilde{\mathscr{P}}} \leq \|r_k\|_{\tilde{\mathscr{P}}^{-1}} \leq C \, \|x - x_k\|_{\tilde{\mathscr{P}}} \tag{13}$$

with $c = |\lambda_{\min}(\tilde{\mathscr{P}}^{-1}\mathscr{M}_{\mathrm{T}})|$ and $C = |\lambda_{\max}(\tilde{\mathscr{P}}^{-1}\mathscr{M}_{\mathrm{T}})|$. Approximations $\tilde{c}$ and $\tilde{C}$ for c and C, respectively, can be computed by using the so called harmonic Ritz values, see [9]. Therefore, the stopping criterion

$$\|x - x_k\|_{\mathscr{P}} \leq \varepsilon \, \|x - x_0\|_{\mathscr{P}} \tag{14}$$

is asymptotically satisfied, if we prescribe (12) with ε replaced by $\varepsilon_* = \tilde{c}/\tilde{C}\, \varepsilon$.

We test the use of the stopping criterion in (14) with a numerical example. For the parameter choice $h = 2^{-7}, \nu = 1$ and $\omega = 1$, we computed the numbers of iterations $\tilde{k}$ produced by the preconditioned MINRES method, for the two different stopping criteria (12) and (14). Thereby we choose for $\tilde{S}$, $r = 1$ with $\tau_1 = 1$ (i.e. $\tilde{S} = \tilde{S}_{\mathrm{CH}}$). As result we obtain $\tilde{k} = 118$ and $\tilde{k} = 130$, using (12) and (14), respectively. The computed approximations are $\tilde{c} = 0.152077$ and $\tilde{C} = 1.60562$.

Standard norm for stopping criterion: Finally we present an analytic convergence result, for the standard norm

$$\|(u, p, w, r)\|_{\mathscr{N}}^2 := \|u\|_{H^1(\Omega)}^2 + \|p\|_{L^2(\Omega)}^2 + \|w\|_{H^1(\Omega)}^2 + \|r\|_{L^2(\Omega)}^2.$$

For $\nu \leq 1$, it is easy to see that:

$$\|x - x_k\|_{\mathscr{N}} / \|x - x_0\|_{\mathscr{N}} \leq 2\,(\max(2,\omega)/\nu)^2 \, \|x - x_k\|_{\mathscr{P}} / \|x - x_0\|_{\mathscr{P}}.$$

This allows to use this standard norm for the stopping criterion in an efficient manner via (14). Using this estimate in combination with the well known convergence results for the preconditioned MINRES method (see [2]), it follows that the number of iterations $k^\star$ which is needed to decrease the initial error by a factor $\varepsilon > 0$, depends only mildly on the parameters ω and ν, namely, logarithmically on $(\max(2,\omega)/\nu)^2$.

Acknowledgement The research of the first and the third author was supported by the Austrian Science Fund (FWF): W1214-N15, project DK12.

References

1. J. Cahouet, J.-P. Chabard, Some fast 3D finite element solvers for the generalized Stokes problem. Int. J. Numer. Methods Fluids **8**(8), 865–895 (1988)
2. A. Greenbaum, *Iterative Methods for Solving Linear Systems*. Frontiers in Applied Mathematics, vol. 17 (SIAM, Philadelphia, 1997)
3. M. Gunzburger, S. Manservisi, Analysis and approximation of the velocity tracking problem for Navier-Stokes flows with distributed control. SIAM J. Numer. Anal. **37**(5), 1481–1512 (2000)
4. W. Krendl, V. Simoncini, W. Zulehner, Stability estimates and structural spectral properties of saddle point problems. Numer. Math. **124**(1), 183–213 (2013)
5. K.-A. Mardal, R. Winther, Uniform preconditioners for the time dependent Stokes problem. Numer. Math. **98**(2), 305–327 (2004)
6. ———, Erratum: uniform preconditioners for the time dependent Stokes problem. Numer. Math. **98**(2), 305–327 (2004); Numer. Math. **103**(1), 171–172 (2006)
7. M.A. Olshanskii, J. Peters, A. Reusken, Uniform preconditioners for a parameter dependent saddle point problem with application to generalized Stokes interface equations. Numer. Math. **105**(1), 159–191 (2006)
8. M.A. Olshanskii, A. Reusken, On the convergence of a multigrid method for linear reaction-diffusion problems. Computing **65**(3), 193–202 (2000)
9. D.J. Silvester, V. Simoncini, An optimal iterative solver for symmetric indefinite systems stemming from mixed approximation. ACM Trans. Math. Softw. **37**(4), 42:1–42:22 (2011)

An Accelerated Value/Policy Iteration Scheme for Optimal Control Problems and Games

Alessandro Alla, Maurizio Falcone, and Dante Kalise

Abstract We present an accelerated algorithm for the solution of static Hamilton-Jacobi-Bellman equations related to optimal control problems and differential games. The new scheme combines the advantages of value iteration and policy iteration methods by means of an efficient coupling. The method starts with a value iteration phase on a coarse mesh and then switches to a policy iteration procedure over a finer mesh when a fixed error threshold is reached. We present numerical tests assessing the performance of the scheme.

1 Introduction

The numerical solution of optimal control problems is a crucial issue for many industrial applications; usually, the final goal is to compute an optimal trajectory for the controlled system and its corresponding optimal control. To solve this problem, we focus our attention on the Dynamic Programming (DP) approach, introduced by Bellman [3] since it produces optimal control in feedback form. However, the synthesis of feedback requires the knowledge of the value function over the whole state space and this is the major bottleneck for the application of DP-based techniques.

It is well-known that the characterization of the value function for optimal control problems by means of DP is obtained in terms of a first order nonlinear Hamilton-Jacobi-Bellman (HJB) partial differential equation. In the last 20 years, this approach has been pursued for all the classical control problems in the

A. Alla
University of Hamburg, Bundesstraße 55, Hamburg, Germany
e-mail: alessandro.alla@uni-hamburg.de

M. Falcone
SAPIENZA – University of Rome, Ple. Aldo Moro 2, Rome, Italy
e-mail: falcone@mat.uniroma1.it

D. Kalise (✉)
RICAM, Austrian Academy of Sciences, Altenberger Straße 69, Linz, Austria
e-mail: dante.kalise@oeaw.ac.at

A. Abdulle et al. (eds.), *Numerical Mathematics and Advanced Applications ENUMATH 2013*, Lecture Notes in Computational Science and Engineering 103,
DOI 10.1007/978-3-319-10705-9_48

framework of viscosity solutions introduced by Crandall and Lions (see [2] for a comprehensive illustration), since the value function of an optimal control problem is known to be only Lipschitz continuous even when the data is regular. Several approximation schemes have been proposed for this class of equations, ranging from finite differences to semi-Lagrangian and finite element methods. Some of these methods converge to the value function but their convergence is slow. Moreover, for grid-based methods, the storage requirements are huge and the complexity of the algorithm increases fast due to the so-called *curse of the dimensionality*; this explains why the problem is very challenging from a computational point of view.

Our main contribution in this article, is a new accelerated algorithm which can produce an accurate approximation of the value function in a reduced amount of time, in comparison to the currently available methods. Furthermore, the proposed scheme can be used over a wide variety of problems connected to static HJB equations, such as infinite horizon optimal control, minimum time control and some cases of pursuit-evasion games. The new method couples two ideas already existing in the literature: the value iteration method (VI), and the policy iteration method (PI) for the solution of Bellman equations. The first is known to be slow but convergent for any initial guess, while the second is known to be fast when it converges (but if not initialized correctly, convergence might be as slow as the value iteration). The approach that we consider relates to multigrid methods (we refer to Santos [10] for a brief introduction to subject in this context), as the coupling that we introduce features an unidirectional, two-level mesh. The work by Chow and Tsitsiklis [5] exploits a similar idea with a value iteration algorithm. However, as far as we know the efficient coupling between the two methods has not been investigated.

To set this paper into perspective, we must recall that algorithms based on the iteration in the space of controls (or policies) for the solution of HJB equations has a rather long history, starting more or less at the same time of dynamic programming. The PI method, also known as Howard's algorithm [6], has been investigated by Kalaba [7], and Pollatschek and Avi-Itzhak [8], who proved that it corresponds to the Newton method applied to the functional equation of dynamic programming. Later, Puterman and Brumelle [9], gave sufficient conditions for the rate of convergence to be either superlinear or quadratic. More recent contributions on the policy iteration method, and some extensions to games can be found in Santos and Rust [11], and Bokanowski et al. [4] (a more complete list of references is given in [1]).

This paper is structured as follows. In Sect. 2, we describe some model problems and the basic building blocks of our approach. In Sect. 3, an accelerated scheme for dynamic programming equations is introduced, and Sect. 4 presents numerical examples assessing its performance.

2 Model Problems and Building Blocks

In this section, we introduce our model problems and summarize the basic results for the two methods which will constitute the building blocks for our new algorithm.

Let the system dynamics be given by

$$\begin{cases} \dot{y}(t) = f(y(t), \alpha(t)) \\ y(0) = x\,, \end{cases} \tag{1}$$

where $y \in \mathbb{R}^n$, $\alpha \in \mathbb{R}^m$ and $\alpha(t) \in \mathscr{A} \equiv L^\infty([0, +\infty[, A)$, and A is a compact subset of $\mathbb{R}^m$. If f is Lipschitz continuous with respect to the state variable and continuous with respect to (x, α), the classical assumptions for the existence and uniqueness result for the Cauchy problem (1) are satisfied.

Let us consider the minimum time problem where we want to compute the time of arrival of the dynamics (1) to a given target $\mathscr{T}$, denoted by $T(x)$. Note that, without additional assumptions, it is not guaranteed that T is finite everywhere, so a crucial role is played by the reachable set $\mathscr{R} = \{x \in \mathbb{R}^n : T(x) < +\infty\}$. By applying the Dynamic Programming Principle, it is possible to obtain the Bellman equation giving the characterization of the solution $T(x)$ on $\mathscr{R}$ and, via the change of variable $v(x) = 1-\exp(-T(x))$ we extend it to $\mathbb{R}^n$. In conclusion, v is the unique viscosity solution of the Dirichlet problem

$$\begin{cases} v(x) + \sup\limits_{a\in A}\{-f(x,a)\cdot Dv(x)\} = 1 & \text{in } \mathbb{R}^n\backslash\mathscr{T} \\ v(x) = 0 & \text{on } \partial\mathscr{T}\,. \end{cases} \tag{2}$$

A similar equation will appear for differential games where two players are acting on the dynamics: player-a wants to hit the target in minimal time, whereas player-b wants to keep the system away from $\mathscr{T}$. Under appropriate assumptions, one can obtain a characterization of the v in terms of the following Dirichlet problem

$$\begin{cases} v(x) + \sup\limits_{a\in A}\inf\limits_{b\in B}\{-f(x,a,b)\cdot Dv(x)\} = 1 & \text{in } \mathbb{R}^n\backslash\mathscr{T} \\ v(x) = 0 & \text{on } \partial\mathscr{T}\,. \end{cases} \tag{3}$$

Value iteration A natural way to solve (2) in the context of semi-Lagrangian schemes, is to consider a pseudotime parameter Δt for the discretization of the system dynamics, to apply a discrete version of the DP, and to consider an approximation in space of the resulting HJB equation. The resulting nonlinear equation is solved by means of a fixed point iteration over the discrete value function, as described in Algorithm 2.1.

Here, we denote by V_i^k the value function at a node x_i of the grid at the k-th iteration and I is an interpolation operator acting on the values of the grid; without loss of generality, throughout this paper we will assume that the numerical grid G is a regular equidistant array of points with mesh spacing denoted by Δx, and that

Algorithm 2.1: Value Iteration for the minimum time problem **(VI)**

Data: Mesh G, Δt, initial guess V^0, tolerance ϵ.
Define $V^0 = 0$ on $\mathscr{T}$, $V^0 = 1$ elsewhere
while $||V^{k+1} - V^k|| \geq \epsilon$ **do**
  forall the $x_i \in G$ **do**

$$V_i^{k+1} = \min_{a \in A}\{e^{-\Delta t} I\left[V^k\right](x_i + \Delta t f(x_i, a))\} + 1 - e^{-\Delta t} \tag{4}$$

  end
  $k = k + 1$
end

Algorithm 2.2: Policy Iteration for the minimum time problem **(PI)**

Data: Mesh G, Δt, initial guess V^0, tolerance ϵ.
Define $V^0 = 0$ on $\mathscr{T}$, $V^0 = 1$ elsewhere
while $||V^{k+1} - V^k|| \geq \epsilon$ **do**
  Policy evaluation step: **forall the** $x_i \in G$ **do**

$$V_i^k = 1 - e^{-\Delta t} + e^{-\Delta t} I\left[V^k\right]\left(x_i + \Delta t f\left(x_i, a_i^k\right)\right) \tag{5}$$

  end
  Policy improvement step:
  forall the $x_i \in G$ **do**

$$a_i^{k+1} = \arg\min_a \left\{e^{-\Delta t} I\left[V^k\right](x_i + \Delta t f(x_i, a))\right\} \tag{6}$$

  end
  $k = k + 1$
end

the interpolant I corresponds to a multilinear interpolation operator. A drawback of this approach resides in the fact that the convergence of the iteration is governed by a contraction constant is given by $e^{-\Delta t}$, thus higher accuracy will dramatically increase the number of iterations. We consider an alternative solution method which circumvents this problem.

Policy iteration In the *approximation over policy space* presented in Algorithm 2.2, we start from an initial guess for the control for every point in the state space. Once the control has been fixed, the Bellman equation becomes linear (no search for the minimum in the control space is performed), and it is solved as an advection equation. Then, an updated policy is computed and a new iteration starts. The resulting sequence of value functions V^k is monotone decreasing at every node of the grid. At a theoretical level, policy iteration can be shown to be equivalent to a Newton method, so under appropriate assumptions, it converges locally with quadratic speed.

3 The Accelerated Scheme for HJB Equations

We present an accelerated iterative algorithm which is constructed upon the building blocks previously introduced. We aim at an efficient formulation exploiting the main computational features of both value and policy iteration algorithms. As it has been stated in [9], there exists a theoretical equivalence between both algorithms, which guarantees a rather wide convergence framework. However, from a computational perspective, there are significant differences between both implementations. A first key factor can be observed in Fig. 1, which shows, for a two-dimensional minimum time problem, the typical situation arising with the evolution of the error measured with respect to the optimal solution, when comparing value and policy iteration algorithms. To achieve a similar error level, policy iteration requires considerably fewer iterations than the value iteration scheme, as quadratic convergent behavior is reached faster for any number of nodes in the state-space grid.

However, a known drawback of policy iteration is its dependence on a good initial guess in order to yield such an efficient behavior. Whereas some guesses will produce quadratic convergence from the beginning of the iterative procedure, others can lead to an underperformant value iteration-like evolution of the error. A final relevant remark can be made from Fig. 1, where it can be observed that for coarse meshes, the value iteration algorithm generates a fast error decay up to a higher global error. This, combined with the fact that value iteration algorithms are rather insensitive to the choice of the initial guess for the value function, are crucial points for the construction of our accelerated algorithm. The accelerated

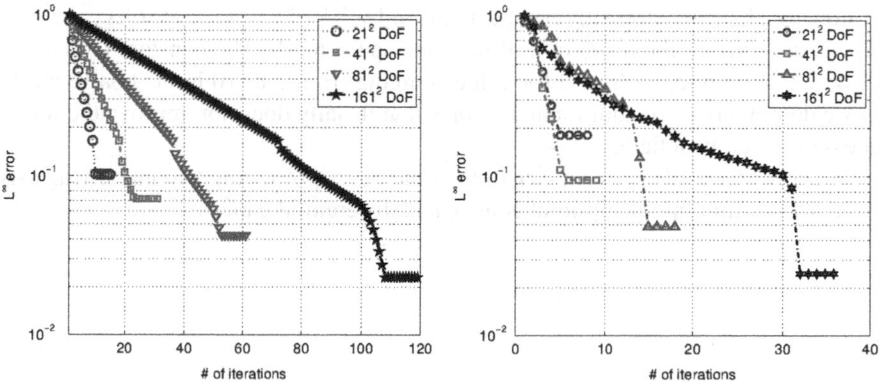

Fig. 1 Error in a 2D problem: value iteration (*left*) and policy iteration (*right*)

Algorithm 3.1: Accelerated Policy Iteration **(API)**

Data: Coase mesh G_c and Δt_c, fine mesh G_f and Δt_f, initial coarse guess V_c^0, coarse-mesh tolerance ϵ_c, fine-mesh tolerance ϵ_f.

begin

  Coarse-mesh value iteration step: perform Algorithm 2.1

  Input: G_c, Δt_c, V_c^0, ϵ_c

  Output: V_c^*

  forall the $x_i \in G_f$ **do**

    $V_f^0(x_i) = I_1[V_c^*](x_i)\ A_f^0(x_i) = \underset{a \in A}{argmin}\ \{e^{-\lambda \Delta t} I_1[V_f^0](x_i + f(x_i, a))\}$

  end

  Fine-mesh policy iteration step: perform Algorithm 2.2

  Input: G_f, Δt_f, V_f^0, A_f^0, ϵ_f

  Output: V_f^*

end

policy iteration (API) Algorithm 3.1, is based on a robust initialization of the policy iteration procedure via a coarse value iteration which will yield to a good guess of the initial control field. The aforementioned accelerated algorithm can lead to a considerably improved performance when compared to value iteration and naively initialized policy iteration algorithms. However, it naturally contains trade-offs that need to be carefully handled in order to obtain a correct behavior. We present some general guidelines for a correct initialization.

Coarse and fine meshes For a good behavior of the PI phase, a good initialization is required, but this should be obtained without deteriorating the overall performance. If we denote by Δx_c and by Δx_f the mesh parameters associated to the coarse and fine grids respectively, numerical findings reported in [1], suggest that for minimum time problems and infinite horizon optimal control, a good balance is achieved with $\Delta x_c = 2\Delta x_f$.

Accuracy Both VI and PI algorithms require a stopping criterion for convergence. Following [11], the stopping criteria is given by $||V^{k+1} - V^k|| \leq C\Delta x^2$, which relates the error to the resolution of the state-space mesh. The constant C is set to $C = \frac{1}{5}$ for the fine mesh, and for values ranging from 1 to 10 in the coarse mesh, as we do not strive for additional accuracy that usually does not improve the initial guess of the control field.

We now assess the performance of the scheme by presenting two numerical tests related to minimum time optimal control and differential games.

4 Numerical Tests

Test 1: Zermelo navigation problem We consider a minimum time problem to the target $\mathcal{T} = \{x \in \mathbb{R}^2 : ||x||_2 \leq 0.2\}$, for states inside the spatial domain $\Omega = [-1, 1]^2$, with system dynamics and parameters given by

$$f(x, y, a) = \begin{pmatrix} 1 + V_b \cos(a) \\ V_b \sin(a) \end{pmatrix}, \quad A = [-\pi, \pi], \quad \Delta t = 0.8\Delta x,$$

where set A is uniformly discretized into 72 values.

We study two cases, setting $V_b = 0.6$ and $V_b = 1.4$, generating solutions with different reachable sets. In the case $V_b = 1.4$, the reachable set corresponds to Ω, whereas for $V_b = 0.6$ there exists a sharp restriction of the region of the state space able to reach the target in finite time, as shown in Fig. 2. For every case, we implement both VI and PI algorithms, and our API iteration procedure. Comparisons of CPU time for different schemes are presented in Table 1; a speedup of 7-8x is observed in the finest meshes for $V_b = 1.4$, and we report that a similar acceleration is achieved in the case $V_b = 0.6$, i.e., in this case the speedup is independent of the regularity of the solution.

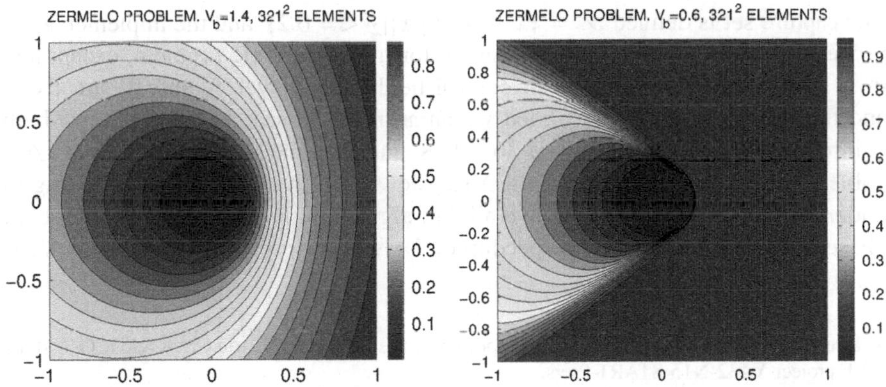

Fig. 2 Zermelo navigation problem: different reachable sets for different choices of the parameter V_b

Table 1 Test 1: CPU time (iterations) for different algorithms with $Vb = 1.4$

Mesh				API		
# nodes	Δx	VI	PI	VI($2\Delta x$)	PI(Δx)	Total
81^2	0.025	32.85 (240)	6.02 (17)	3.1 (36)	0.99 (4)	4.09
161^2	0.0125	184.11 (468)	45.52 (70)	15.47 (115)	4.37 (6)	19.84
321^2	0.00625	1.18e+03 (904)	458.03 (49)	116.9 (339)	33.81 (7)	150.71

Table 2 Test 2: CPU time (iterations) for different algorithms

Mesh				API		
# nodes	Δx	VI	PI	VI($2\Delta x$)	PI(Δx)	Total
41^2	0.05	67 (55)	20.08 (19)	21.06 (21)	7.51 (19)	28.57
81^2	0.025	2.29E02(106)	65.41 (35)	58.19 (48)	15.56 (35)	73.75
# nodes	Δx	VI	PI	PI($2\Delta x$)	PI(Δx)	Total
41^2	0.05	67 (55)	20.08 (19)	10.17 (10)	7.46 (19)	17.63
81^2	0.025	2.29E02(106)	65.41 (35)	20.21 (19)	13.32 (35)	33.53

Test 2: a two-player pursuit-evasion game Although there are cases where the application of policy iteration algorithms for differential games can fail to converge, as shown in [4], there exists a class of pursuit-evasion games where such an approach can lead to convergence to the correct solution in a faster way. We consider a reduced-coordinate system for a two-player pursuit-evasion game in two dimensions. In $\Omega = [-1, 1]^2$, we aim at solving a Hamilton-Jacobi-Bellman-Isaacs equation of the form (3) with

$$f(x, y, a, b) = \begin{pmatrix} \sin(b) - 2\sin(a) \\ \cos(b) - 2\cos(a) \end{pmatrix}, \quad A = \left[-\frac{3}{4}\pi, \frac{3}{4}\pi\right], \quad B = [-\pi, \pi].$$

The capture set is defined $\mathscr{T} = \{x \in \mathbb{R}^2 : ||x||_2 <= 0.2\}$, and the implementation procedure follows the same guidelines as for minimum time problem, except that in a single policy iteration, both control fields (for player A and B) are fixed, and the policy update is performed via an *argmaxmin* search. We discretize both control fields into uniform sets of 36 values, and the time step is set $\Delta t = 0.8\Delta x$. Table 2 shows that in this case, the VI is slow even in a coarse mesh, leading to underperformant results for the overall API algorithm. By replacing the coarse VI phase by a coarse PI phase, a speedup of $7\times$ with respect to the fine mesh VI is recovered.

Acknowledgement This research was supported by the ITN-Marie Curie Grant SADCO, and the FWF project Y432-N15 START-Preis.

References

1. A. Alla, M. Falcone, D. Kalise, An efficient policy iteration algorithm for dynamic programming equations (submitted). Available at `arXiv:1308.2087`
2. M. Bardi, I. Capuzzo-Dolcetta, *Optimal Control and Viscosity Solutions of Hamilton-Jacobi-Bellman Equations* (Birkhäuser, Boston, 1997)
3. R. Bellman, *Dynamic Programming* (Princeton University Press, Princeton, 1957)
4. O. Bokanowski, S. Maroso, H. Zidani, Some convergence results for Howard's algorithm. SIAM J. Numer. Anal. **47**, 3001–3026 (2009)
5. C. Chow, J. Tsitsiklis, An optimal one-way multigrid algorithm for discrete-time stochastic control. IEEE Trans. Autom. Control **36**, 898–914 (1991)
6. R.A. Howard, *Dynamic Programming and Markov Processes.* Technology Press of the Massachusetts Institute of technology. New York; London: J. Wiley, cop. (1960)
7. R. Kalaba, On nonlinear differential equations, the maximum operation and monotone convergence. J. Math. Mech. **8**, 519–574 (1959)
8. M. Pollatschek, B. Avi-Itzhak, Algorithms for stochastic games with geometrical interpretation. Manage. Sci. **15**, 399–415 (1969)
9. M.L. Puterman, S.L. Brumelle, On the convergence of Policy iteration in stationary dynamic programming. Math. Oper. Res. **4**, 60–69 (1979)
10. M.S. Santos, Numerical solution of dynamic economic models, in *Handbook of Macroeconomics*, ed. by J.B. Taylor, M. Woodford (Elsevier Science, Amsterdam/New York, 1999), pp. 311–386
11. M.S. Santos, J. Rust, Convergence properties of policy iteration. SIAM J. Control Optim. **42**, 2094–2115 (2004)

Inverse Problem of a Boundary Function Recovery by Observation Data for the Shallow Water Model

Ekaterina Dementyeva, Evgeniya Karepova, and Vladimir Shaidurov

Abstract In the paper, the shallow water equations are applied to describe the propagation of long waves in the coastal area of an ocean. For a correct formulation of the problem, the equations are closed by boundary conditions involving a function on the open water boundary. In general case this function is unknown. The determination of this function is reduced to the solution of the inverse problem on restoring it with auxiliary data on elevation of the sea surface along some part of the boundary. The solving this (ill-posed) inverse problem is performed by optimal control methods using adjoint operators. To improve the conditioning of the problem, three types of regularization functionals are considered which correspond to higher, deficient, and threshold smoothness of the data involved. The results of their application are illustrated by a numerical example.

1 Introduction

In the monographs [4, 9], for tidal waves in large water areas, various aspects of shallow water equations are outlined taking into account the Earth's sphericity and the Coriolis acceleration. In the papers [2, 8], the correct formulation of the initial-boundary value problem is given for these equations; and the estimates providing its unique solvability are derived. In the paper [5], for the numerical solving this problem, the finite element method with linear elements on triangles is implemented with a priori estimates providing unique solvability of the discrete problem.

However, the right-hand side of the boundary conditions of the problem involves a function that describes the effect of the ocean through the open water boundary. This function is often unknown, whereas such additional (satellite) data are available as the elevation of the water surface along some parts of the boundary. As a result, the inverse problem on restoring this function can be formulated taking into consideration the auxiliary data on elevation of the sea surface along the certain

E. Dementyeva • E. Karepova (✉) • V. Shaidurov
Institute of Computational Modelling of SB RAS, 660036 Akademgorodok, Krasnoyarsk, Russia
e-mail: e.v.dementyeva@icm.krasn.ru; e.d.karepova@icm.krasn.ru

A. Abdulle et al. (eds.), *Numerical Mathematics and Advanced Applications ENUMATH 2013*, Lecture Notes in Computational Science and Engineering 103,
DOI 10.1007/978-3-319-10705-9_49

part of the boundary. To solve it, optimal control methods with adjoint operators are applied [1]. In general, the inverse problem is ill-posed [10]. To improve the conditioning of this problem, in the paper three different types of regularizing functionals [3] are examined which correspond to higher, deficient, and threshold smoothness of data involved in the form of traces of water surface elevation on the part of the boundary. The application and comparison of these regularizers are illustrated by a numerical example.

2 The Differential Formulation of a Problem

Consider the following problem. Let (r, λ, θ) be spherical coordinates with the origin at the terrestrial globe, $0 \le \lambda \le 2\pi$, $-\pi/2 \le \theta < \pi/2$. Here λ means the geographic longitude and instead of the geographic latitude θ we use an angle $\varphi = \theta + \pi/2 \in [0; \pi]$. We put $r = R_E$, where R_E is the radius of the Earth which is assumed to be constant.

We formulate the problem on propagation of the long waves in a water area as follows. Let Ω be some domain on a plane of (λ, φ) variables with the piecewise smooth Lipchitz boundary $\Gamma = \Gamma_1 \cup \Gamma_2$ of the class $C^{(2)}$, where Γ_1 is a part of the boundary passing along a coastline and $\Gamma_2 = \Gamma \setminus \Gamma_1$ is a part of the boundary rounded a water area. Let denote characteristic functions of these parts of the boundary by χ_1 and χ_2, respectively. Without loss of generality we may assume that the points $\varphi = 0$ and $\varphi = \pi$ (poles) are not involved in Ω.

For time discretization we subdivide the segment $[0, T]$ into K equal subintervals: $0 = t_0 < t_1 < \cdots < t_K = T$ with the step $\tau = T/K$.

For the unknown functions $u = u(t_{k+1}, \lambda, \varphi)$, $v = v(t_{k+1}, \lambda, \varphi)$ and $\xi = \xi(t_{k+1}, \lambda, \varphi)$ on the interval (t_k, t_{k+1}) we write the time-discretized equations of the motion and continuity [2, 8] as follows:

$$
\begin{aligned}
&\left(\frac{1}{\tau} + R_f\right) u - lv - mg\,\frac{\partial \xi}{\partial \lambda} = f_1 + \frac{1}{\tau}u^k \quad \text{in } \Omega,\\
&\left(\frac{1}{\tau} + R_f\right) v + lu - ng\,\frac{\partial \xi}{\partial \varphi} = f_2 + \frac{1}{\tau}v^k \quad \text{in } \Omega,\\
&\frac{1}{\tau}\xi - m\left(\frac{\partial}{\partial \lambda}(Hu) + \frac{\partial}{\partial \varphi}\left(\frac{n}{m}Hv\right)\right) = f_3 + \frac{1}{\tau}\xi^k \quad \text{in } \Omega,
\end{aligned}
\tag{1}
$$

where u and v are the components of the velocity vector $\mathbf{U}$ in λ and φ directions respectively; ξ is a deviation of a free surface from the nonperturbed level; $H(\lambda, \varphi) > 0$ is a depth of a water area at a point (λ, φ); the function $R_f = r_* |\mathbf{U}|/H$

takes into account the friction force; r_* is the friction coefficient; $l = -2\omega\cos\varphi$ is the Coriolis parameter; $m = 1/(R_E \sin\varphi)$; $n = 1/R_E$; g is the acceleration of gravity; $f_1 = f_1(t,\lambda,\varphi)$, $f_2 = f_2(t,\lambda,\varphi)$, and $f_3 = f_3(t,\lambda,\varphi)$ are the given functions of the external forces. Here for an arbitrary function $f(t,\lambda,\phi)$ we use $f^k = f(t_k,\lambda,\phi)$, $f = f(t_{k+1},\lambda,\phi) = f^{k+1}$. Further the index $(k+1)$ in the difference expressions is omitted if there is no ambiguity. Base friction $R_f = r_*|\mathbf{U}^k|/H$ is taken from the previous time level.

We consider the boundary conditions in the following form:

$$HU_n + \beta\chi_2\sqrt{gH}\xi = \chi_2\sqrt{gH}d \quad \text{on} \quad \Gamma \tag{2}$$

where $U_n = \mathbf{U}\cdot\mathbf{n}$, $\mathbf{n} = (n_1, \frac{n}{m}n_2)$ is the vector of an outer normal to the boundary in the spherical coordinates; $\beta \in [0,1]$ is a given parameter; $d = d(t,\lambda,\varphi)$ is a function defined on the boundary Γ and equal to zero on the boundary Γ_1.

The system (1)–(2) is the subject of our investigation. In the direct problem (1)–(2) at the time instant t_{k+1}, $k = 0,1,\ldots,K-1$, we should find u, v, ξ when the functions H, f_1, f_2, f_3, d are given. But in general case, the function d is unknown. So we formulate the inverse problem for finding the function d in the problem (1)–(2). In this case, to close the problem (1)–(2), consider the following condition:

$$\xi = \xi_{obs} \qquad \text{on} \quad \Gamma_0 \tag{3}$$

where $\xi_{obs} \in L_2(\Gamma_0)$ is a given function (for example, from an observation data) on the some part of the boundary $\Gamma_0 \subset \Gamma$.

Thus, for the time step t_{k+1}, $k = 0,1,\ldots,K-1$, the differential problem (1)–(3) can be formulated in the following way [2].

Problem 1 (Inverse problem) Assume that at a time instant t_{k+1}, $k = 0,1,\ldots,K-1$, the function ξ_{obs} is defined on Γ_0, the function d is unknown on Γ_2 and vanishes on Γ_1. At the time instant t_{k+1} find u, v, ξ, d, satisfying the system (1), the boundary condition (2), and the closure condition (3).

3 A Problem of Optimal Control

For real vector-functions $\Phi = (u,v,\xi)$, $\hat{\Phi} = (\hat{u},\hat{v},\hat{\xi}) \in (L_2(\Omega))^3$ consider the inner product and the norm [2]

$$(\Phi,\hat{\Phi}) = \int\limits_{\Omega'} R_E^2 \sin\varphi \left(H(u\hat{u} + v\hat{v}) + g\xi\hat{\xi}\right) d\lambda d\varphi, \qquad \|\Phi\| = (\Phi,\Phi)^{1/2} < \infty. \tag{4}$$

For integral posing of the problem (1)–(2) we take the inner product of the system (1) in $(L_2(\Omega))^3$ by an arbitrary vector function $\hat{\Phi} = (\hat{u}, \hat{v}, \hat{\xi}) \in (L_2(\Omega))^2 \times H^1(\Omega) \equiv W$ and perform integration by parts taking into account the boundary condition (2).

Definition 1 A vector-function $\Phi = (u, v, \xi) \in (L_2(\Omega))^2 \times H^1(\Omega) \equiv W$ is called a weak solution of the problem (1)–(2) if the integral equality

$$a(\Phi, \mathbf{W}) = f(\mathbf{W}) + b(d, \mathbf{W}) \tag{5}$$

holds for any vector function $\mathbf{W} = (w^u, w^v, w^\xi) \in W$.

The bilinear forms $a(\Phi, \mathbf{W})$, $b(d, \mathbf{W})$, the linear form $f(\mathbf{W})$ are written in [6, 7].

Note that the boundary condition (2) is natural for the problem (1), hence it imposes no restriction on the spaces of trial and test functions.

In [2] it has been proved for $\beta > 0$ the problem (5) has a unique solution.

Considering the bilinear forms $a(\Phi, \cdot)$ and $b(d, \cdot)$ as the bounded linear functionals defined for any functions $\Phi \in W$ and $d \in L_2(\Gamma_2)$ respectively, the problem (5), (3) can be written in an operator form with the compact operator [6, 7]. Hence the problem is ill-posed. The problem (5), (3) is uniquely and densely solvable if $\Gamma_0 = \Gamma_2$.

The problem (5), (3) is reduced to the family of optimal control problem with a small parameter α and unknown functions d_α and $\Phi_\alpha = (u_\alpha, v_\alpha, \xi_\alpha)$ which satisfy Eq. (5) and minimize one of the cost functional

$$J_\alpha^{(I)}(d_\alpha, \xi_\alpha(d_\alpha)) = \frac{1}{2} g\alpha \int_{\Gamma_2} \sqrt{gH} d_\alpha^2 \, ds + \frac{1}{2} g \int_{\Gamma_0} \sqrt{gH} (\xi_\alpha - \xi_{obs})^2 \, ds, \tag{6}$$

$$J_\alpha^{(II)}(d_\alpha, \xi_\alpha(d_\alpha)) = \frac{1}{2} g\alpha \int_{\Gamma_2} \sqrt{gH} \left(\frac{d}{ds}(d_\alpha) \right)^2 ds + \frac{1}{2} g \int_{\Gamma_0} \sqrt{gH} (\xi_\alpha - \xi_{obs})^2 \, ds, \tag{7}$$

$$J_\alpha^{(III)}(d, \xi_\alpha(d_\alpha)) = \alpha \int_{\Gamma_2} (d_\alpha + \mathscr{A}^{1/2} d_\alpha) d_\alpha \, ds + \frac{1}{2} g \int_{\Gamma_0} \sqrt{gH} (\xi_\alpha - \xi_{obs})^2 \, ds. \tag{8}$$

In (8) the positive and self-adjoint operator $\mathscr{A}^{1/2} : \overset{\circ}{H}{}^{1/2}(\Gamma_2) \to L_2(\Gamma_2)$ is defined as the square root of the operator $\mathscr{A} = -\dfrac{d^2}{ds^2} : H_0^2(\Gamma_2) \to L_2(\Gamma_2)$.

Note that the first term in (6)–(8) is a stabilizing functional. Their form determines smoothness and a search space of the boundary function d. We consider three following cases: deficient smoothness ($d \in L_2(\Gamma_2)$, (6)), higher smoothness ($d \in H^1(\Gamma_2)$, (7)), and threshold smoothness ($d \in \overset{\circ}{H}{}^{1/2}(\Gamma_2)$, (8)).

Applying Euler optimality equation to problems (5) + (6), or (5) + (7), or (5) + (8) we get the following problem.

Problem 2 Let ξ_{obs} be given on Γ_0. For fixed $\alpha > 0$ find the boundary function d_α on Γ_2 and the vector-functions $\Phi_\alpha = (u_\alpha, v_\alpha, \xi_\alpha)$, $\hat{\Phi}_\alpha = (\hat{u}_\alpha, \hat{v}_\alpha, \hat{\xi}_\alpha) \in W$ satisfying the equations

$$a(\Phi_\alpha, \mathbf{W}) = f(\mathbf{W}) + b(d_\alpha, \mathbf{W}) \qquad \forall\, \mathbf{W} = (w^u, w^v, w^\xi) \in W, \tag{9}$$

$$a(\hat{\mathbf{W}}, \hat{\Phi}_\alpha) = g \int_{\Gamma_0} \sqrt{gH}(\xi_\alpha - \xi_{obs}) \hat{w}^\xi \, d\Gamma \quad \forall\, \hat{\mathbf{W}} = (\hat{w}^u, \hat{w}^v, \hat{w}^\xi) \in W, \tag{10}$$

with one of the Conditions:

Condition 2.1

$$\alpha d_\alpha + \hat{\xi}_\alpha = 0 \quad on \quad \Gamma_2; \tag{11}$$

Condition 2.2

$$\alpha \frac{d}{ds}\left(\sqrt{gH}\frac{dd_\alpha}{ds}\right) = \sqrt{gH}\hat{\xi}_\alpha \ \ on \ \Gamma_2, \ d_\alpha(\gamma_0) = d_\alpha(\gamma_1) = 0. \tag{12}$$

Condition 2.3

$$\alpha d_\alpha + \alpha \mathscr{A}^{1/2} d_\alpha = \frac{g}{2}\sqrt{gH}\hat{\xi}_\alpha \quad on \quad \Gamma_2, \quad d_\alpha|_{\partial \Gamma_2} = 0. \tag{13}$$

Here γ_0, γ_1 are the ends of $\Gamma_2 \subset \Gamma$.

The Problem 2 is well-posed for any $\alpha > 0$, its solution converges to a weak solution of the problem (5), (3) as $\alpha \to +0$.

Thus, for computation of a solution u^{k+1}, v^{k+1}, ξ^{k+1}, and d^{k+1} from (9)–(10), and one of (11), (12) or (13) at $(k+1)$ time step, we apply the following iterative process.

Iterative algorithm

1. Take some $d_\alpha^{(0)}$ on Γ_2. From here on, when describing the algorithm, a superscript in parentheses denotes the number of an iteration loop. Put $u_\alpha^{(0)} = u^k$, $v_\alpha^{(0)} = v^k$, $\xi_\alpha^{(0)} = \xi^k$.
2. While given accuracy of the stopping criterion is not achieved an iteration step is performed:

   2.1. Using $d_\alpha^{(l)}$, we solve the direct problem (9) and determine $u_\alpha^{(l)}$, $v_\alpha^{(l)}$, $\xi_\alpha^{(l)}$ (index $(k+1)$ of the time step is omitted as usual).

2.2. Using a solution $\xi_\alpha^{(l)}$ of the direct problem in the boundary condition for the adjoint one, we solve the problem (10) and determine $\hat{u}_\alpha^{(l)}, \hat{v}_\alpha^{(l)}, \hat{\xi}_\alpha^{(l)}$.

2.3. Using a solution $\hat{\xi}_\alpha^{(l)}$ of the adjoint problem, $d_\alpha^{(l)}$ is iteratively refined.

For Condition 2.1. Let $w_\alpha^{(l)}$ be equal to the adjoint solution $\hat{\xi}_\alpha^{(l)}$.

For Condition 2.2. Let $\alpha \equiv 1$ and find a solution $w_\alpha^{(l)}$ of the boundary value problem (12).

For Condition 2.3. Let $\alpha \equiv 1$ and with implementation of discrete approximation of $\mathscr{A}^{1/2}$ in (13), find a solution $w_\alpha^{(l)}$ of the algebraic equation system (13).

After that use the solution $w_\alpha^{(l)}$ of one of the Conditions 2.1–2.3 for the iterative refinement $d_\alpha^{(l)}$ by the scheme:

$$d_\alpha^{(l+1)} = d_\alpha^{(l)} - \gamma_l(\alpha d_\alpha^{(l)} + w_\alpha^{(l)}). \tag{14}$$

Here γ_l, α are parameters of the method.

2.4. Put $d_\alpha^{(l)} = d_\alpha^{(l+1)}, l = l + 1$ and go to point **2**.

Parameter γ_l in the iterative scheme (14) can be chosen by the trial-and-error method as $\gamma_l^1 \equiv const$ for all l. To increase the convergence rate of Iterative algorithm, γ_l can be calculated for all l by the minimal residual method as γ_l^2. In this case in point 2.3 before using iterative scheme it is necessary additionally to solve the direct and adjoint problems with some initial data and then the corresponding refinement equation using obtained adjoint solution. After that γ_l^2 is calculated as the ratio of some norms.

Consider also γ_l chosen with according to the method from the extremum problem theory [1]:

$$\gamma_l^3 = \int\limits_{\Gamma_0} \sqrt{gH}(\xi^{(l)} - \xi_{obs}^{(l)})^2 \, d\Gamma \Bigg/ \int\limits_{\Gamma_0} \sqrt{gH}(w_\alpha^{(l)})^2 \, d\Gamma \,. \tag{15}$$

Thus, on each time interval (t_k, t_{k+1}) for sufficiently large $l = L \gg 0$ and sufficiently small $0 < \alpha \ll 1$, $u^{k+1} \approx u_\alpha^{(L)}$, $v^{k+1} \approx v_\alpha^{(L)}$, $\xi^{k+1} \approx \xi_\alpha^{(L)}$ can be taken as the solution of the differential problem (1)–(3). Note that convergence of the Iterative algorithm is proved.

4 Numerical Tests for Boundary Function Recovery

The numerical solving the direct and adjoint problems is based on the finite element method. Consider a consistent triangulation $\mathscr{T} = \{\omega_i\}\big|_{i=1}^{N_{el}}$ of the domain Ω. The Bubnov-Galerkin method is used for discretization of our problem with respect to space. Linear functions on triangular finite elements are used as trial and test functions. In [5] a priori stable estimation for the discrete analogue is derived and the second order of approximation in internal nodes for an uniform grid is shown.

We consider the water area of the Sea of Okhotsk and a part of the Pacific Ocean near the Kuril Islands as a computational domain. The domain is bounded by a "rectangle": $\Omega = [41°, 62°]$ N. $\times [135°, 162°]$ E.; its liquid boundary Γ_2 passes along $\lambda = 161, 1°$ E. and along $\varphi = 41, 5°$ N. From here on, for convenience, along the λ– and φ–axes instead of the radian measure we use degrees of the eastern longitude and the northern latitude respectively. Test calculations for the computational domain were performed on the grids constructed on the basis of the ETOPO2 open bathymetric data base.

Since in general case for a nonstationary problem the initial data are unknown we use the following procedure for determination of the initial data in Ω. Firstly, we solve a steady-state problem using the function d which is given on the whole "liquid" boundary independently of time. The values ξ from the steady-state solution on some part of the boundary Γ_0 are considered as the "observation" data. Then we "forget" values of d. The aim of the numerical test is recovery of the function d on the whole liquid boundary using our "observation" data. To this end, d is recovered everywhere on the "liquid" boundary with Iterative algorithm starting with $d \equiv 0$.

Using this procedure we obtain a rather smooth function as the "observation" data. However, actual observation data, as a rule, are not so smooth and may contain some gaps. Figure 1 shows the example of recovery of ξ and d by the observation data with gaps including two discontinuous pieces along the boundary. Here we can see at solving Problem 2 with Conditions 2.2 and 2.3 that free surface elevation ξ and boundary function d are recovered on the whole liquid boundary including the segments without the observation data. Note at solving Problem 2 with Condition 2.1 that the recovery of ξ and d is in the points with observations only. Moreover some results of d recovery by the observation data with the superimposed "white noise" and with gaps are considered in [6, 7]. The results about the convergence rate of Iterative algorithm depending on solving problem and the quality of observation data are represented in Table 1. In the numerical tests the smooth, "noisy" observation data and the observation data with gaps are used. Parameter γ_l is chosen by the trial-and-error method (optimal γ_l for the Condition 2.3), the minimal residual method (optimal γ_l for the Condition 2.1), and the method from the extremum problem theory (optimal γ_l for the Condition 2.2).

Table 1 demonstrates: the worse quality of the observation data the more iterations are required by the algorithm for its convergence.

Taking into account the numerical results (Fig. 1 and [6, 7]) and optimal choice of parameter γ_l for Conditions 2.1–2.3 (Table 1) we can conclude the following. Iterative algorithm for solving Problem 2 with Condition 2.1 converges faster than in another cases but it demonstrates the worse quality of the recovered data due to the powerful sensitivity to errors and the gaps in observations. For solving Problem 2 with Condition 2.2 the algorithm converges slower than in another cases but it

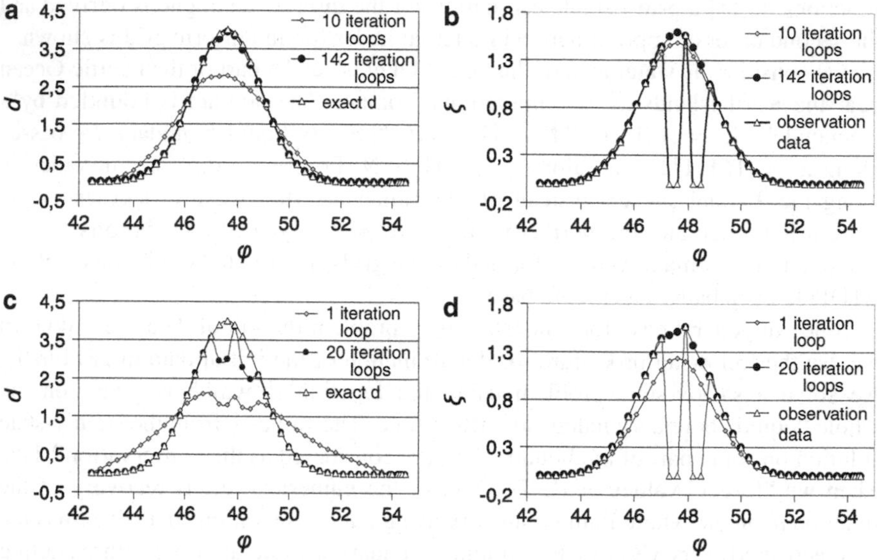

Fig. 1 Dependence of the functions d and ξ upon the number of iteration loops (1, 10, 20, 142) on a liquid boundary of Ω with observation data with gaps: **(a)**–**(b)** with Condition 2.2; **(c)**–**(d)** with Condition 2.3

Table 1 The iterations number for convergence of Iterative algorithm for different choices of parameter γ_l: $\gamma_l^1 = const$, γ_l^2 and γ_l^3 are specified by the minimal residual method and formula (15) respectively

Problem	Observation data	γ_l^1	γ_l^2	γ_l^3
Problem 2	Smooth	9	3	9
with	With gaps	9	3	9
Condition 2.1	With white noise	9	3	19
Problem 2	Smooth	45	57	24
with	With gaps	60	81	29
Condition 2.2	With white noise	50	50	50
Problem 2	Smooth	11	9	60
with	With gaps	13	9	58
Condition 2.3	With white noise	45	32	50

recovers smooth data on the whole boundary by observations with noise or gaps. For solving Problem 2 with Condition 2.3 the algorithm is sensitive to errors in the observations, but it recovers data on the whole boundary including parts without observations; and it converges faster then in the case of Condition 2.2.

The work was supported by Russian Foundation of Fundamental Researches (grant 14-01-00296), and by Project 130 of SB RAS.

References

1. V.I. Agoshkov, *Methods of Optimal Control and Adjoint Equations in Problems of Mathematical Physics* (ICM RAS, Moscow, 2003), 256p
2. V.I. Agoshkov, Inverse problems of the mathematical theory of tides: boundary-function problem. Russ. J. Numer. Anal. Math. Model. **20**(1), 1–18 (2005)
3. H.W. Engl, M. Hanke, A. Neubauer, *Regularization of Inverse Problems* (Kluwer Academic, Dordrecht, 1996), 321p
4. A.E. Gill, *Atmosphere-Ocean Dynamics* (Academic, New York, 1982), 662p
5. L.P. Kamenshchikov, E.D. Karepova, V.V. Shaidurov, Simulation of surface waves in basins by the finite element method. Russ. J. Numer. Anal. Math. Model. **21**(4), 305–320 (2006)
6. E. Karepova, E. Dementyeva, *The Numerical Solution of the Boundary Function Inverse Problem for the Tidal Models*. Lecture Notes in Computer Science (Springer, Berlin/Heidelberg, 2013), pp. 345–354
7. E. Karepova, V. Shaidurov, E. Dementyeva, The numerical solution of data assimilation problem for shallow water equations. Int. J. Numer. Anal. Model. Ser. B **2**(2–3), 167–182 (2011)
8. Z. Kowalik, I. Polyakov, Tides in the Sea of Okhotsk. J. Phys. Oceanogr. **28**(7), 1389–1409 (1998)
9. G.I. Marchuk, B.A. Kagan, *Dynamics of Ocean Tides* (Leningrad, Gidrometizdat, 1983), 471p
10. A.N. Tikhonov, V.Y. Arsenin, *Solution of Ill-Posed Problems* (Winston & Sons, Washington, 1977)

Determination of Extremal Points and Weighted Discrete Minimal Riesz Energy with Interior Point Methods

Manuel Jaraczewski, Marco Rozgiĉ, and Marcus Stiemer

Abstract The asymptotic approximation of continuous minimal s-Riesz energy by the discrete minimal energy of systems of n points on regular sets in $\mathbb{R}^3$ is studied. For this purpose an optimization framework for the numerical solution of the corresponding Gauß variational problem based on an interior point method is developed. Moreover, numerical results for ellipsoids and tori are presented.

1 Introduction

The *Riesz kernel of order* s in $\mathbb{R}^d$ with $d \geq 2$ and $0 \leq s < d$ is defined by

$$R_s(x) := \begin{cases} \|x\|^{-s} & , \ s > 0 \\ -\log \|x\| & , \ s = 0 \, , \end{cases}$$

where $\|\cdot\|$ denotes the Euclidean norm in $\mathbb{R}^d$. For $d \geq 3$ and $0 \leq s < d-2$ the Riesz kernel is superharmonic on $\mathbb{R}^d$, in case $s = d-2$ it is harmonic on $\mathbb{R}^d \setminus \{0\}$, and it is subharmonic for $s > d-2$. In particular for $n = 3$ and $s = 1$ the Riesz kernel coincides with the classical *Newton kernel* [25]. Let $Q : \Omega \to [0, \infty)$ be a continuous non negative function, representing an *external field*. The *weighted energy integral* for a compact set $\Omega \subseteq \mathbb{R}^d$ and a Borel measure μ with total mass $\mu(\Omega) = 1$, is given by

$$I_s^Q(\mu) := \int_\Omega \int_\Omega R_s(x-y)\, d\mu(x) + 2 \int_\Omega Q(x)\, d\mu(x) \, , \tag{1}$$

including the unweighted case with $Q(x) = 0$ for all $x \in \Omega$. The set of all normalized Borel measures is denoted by $\mathcal{M}(\Omega)$. The *weighted s-Riesz energy* $V_{d,s}^Q$ of Ω is defined by

M. Jaraczewski (✉) • M. Rozgiĉ • M. Stiemer
Helmut Schmidt University – University of the Federal Armed Forces Hamburg, Holstenhofweg 85, 22034 Hamburg, Germany
e-mail: manuel.jaraczewski@hsu-hh.de; m.rozgic@hsu-hh.de; m.stiemer@hsu-hh.de

A. Abdulle et al. (eds.), *Numerical Mathematics and Advanced Applications ENUMATH 2013*, Lecture Notes in Computational Science and Engineering 103,
DOI 10.1007/978-3-319-10705-9_50

$$V_{d,s}^Q := V_{d,s}^Q(\Omega) := \inf\left\{I_s^Q(\mu) \,:\, \mu \in \mathcal{M}(\Omega)\right\}. \tag{2}$$

This infimum always exists and is larger than 0 for $s > 0$ and larger than $-\infty$ for $s = 0$, but it may coincide with $+\infty$. The latter is true, e.g., for finite sets. Moreover, we have

$$\inf\left\{I_s^Q(\mu) \,:\, \mu \in \mathcal{M}(\Omega)\right\} = \min\left\{I_s^Q(\mu) \,:\, \mu \in \mathcal{M}(\Omega)\right\}.$$

To avoid some (trivial) particular cases, we assume that the compact sets considered in this work consist of infinitely many points. A measure $\mu_e \in \mathcal{M}(\Omega)$ satisfying $I_s^Q(\mu_e) = V_{d,s}^Q$ is called *equilibrium measure.*

By distributing point charges on Ω, a discretization of the energy integral (1) can be defined. For $P_n := (w_1, \ldots, w_n) \in \Omega^n$, consisting of $n \in \mathbb{N}$ distinct points $w_1, \ldots, w_n$ in Ω, the discrete energy is defined by

$$E_{d,s}^Q(P_n) := \sum_{\substack{j=1 \\ j\neq k}}^{n} \sum_{k=1}^{n} R_s(w_j - w_k) + 2(n-1)\sum_{j=1}^{n} Q(w_j). \tag{3}$$

The discrete counterpart of $V_{d,s}^Q$, i.e., the *discrete weighted n-point s-Riesz energy of* Ω, is consequently defined via

$$\mathcal{E}_{d,s}^Q(n) := \inf_{P_n \in \Omega^n} E_{d,s}^Q(P_n) = \min_{P_n \in \Omega^n} E_{d,s}^Q(P_n). \tag{4}$$

For n fixed, points $w_1, \ldots, w_n$ minimizing (3) are called *extremal points*. The sequence of normalized discrete energies $\left(\frac{1}{n(n-1)}\mathcal{E}_{d,s}^Q(n)\right)_{n\in\mathbb{N}}$ of a compact set $\Omega \subseteq \mathbb{R}^d$ converges to the continuous energy $V_{d,s}^Q$ as n tends to infinity. In the unweighted case $Q = 0$ this is a well known result from potential theory, e.g., [11, 16]. A comprehensive theory of logarithmic potentials (i.e., $d = 2$ and $s = 0$) in the presence of an external field is provided by Saff and Totik [21]. Ohtsuka proved the convergence of the above sequence in the case $Q = 0$ for general lower semicontinuous kernel functions [15]. His argument can be modified to obtain the same result for arbitrary weight functions Q, cf. [9].

Most investigations into minimal discrete energy configurations focus on the sphere or on a torus as *canonical* manifolds, see, e.g., [19, 22] for the sphere and [2, 3, 5, 8] for tori. It should be mentioned, that discrete minimal energy can also be considered in cases, where the continuous energy integral (1) fails to converge, i.e., $s \geq d$, e.g., [7, 10]. In this case local interaction between points dominates over global phenomena, and for $s \to \infty$ minimal energy configurations are given by the midpoints of *best packing spheres*. In this work, however, only the case $0 \leq s < d$ is relevant, and we remain in the realm of potential theory. In Sect. 2 we present an approach to find approximate solutions for the variational problem stated in (2). Using an interior point method the discrete weighted n-point s-Riesz energy and the corresponding point configurations are numerically determined. In the sequel we present some numerical results for several manifolds in Sect. 3.

2 Computing Extremal Points with an Interior Point Method

In this section, we present a flexible method to compute extremal points on a large class of compact sets $\Omega \subseteq \mathbb{R}^d$. Hardin, Saff and Kuijlaars [6, 22] determined numerically extremal points and the corresponding minimal discrete energy on the sphere and on tori using quadratic programming. Minimal energy for more general sets, like a cube and its boundary, has been computed in [17] by Rajon et al. By providing rigorous upper and lower bounds, these methods lead to reliable values for minimal energy and the related capacities. This method has been extended to weighted s-Riesz energy in the presence of external fields in [18]. In contrast, the method presented here is based on an *interior point method*, more precisely on the efficient implementation IPOPT of this method by Wächter and Biegler [23]. The presented approach is easily manageable and can be applied to a huge class of sets. We consider sets $\Omega \subseteq \mathbb{R}^d$ that can be described by a set of finitely many equations or inequalities of the type

$$\varphi_1(x) = 0, \ldots, \varphi_k(x) = 0, \qquad \psi_1(x) \geq 0, \ldots, \psi_\ell(x) \geq 0,$$

where the functions $\varphi_i, \psi_j : \mathbb{R}^d \to \mathbb{R}$, $1 \leq i \leq k$, $1 \leq j \leq \ell$ are assumed to be at least twice continuously differentiable. This general form contains, amongst other sets, smooth compact manifolds of arbitrary (integer) dimensions $\beta \leq d$ and sets which are the union or intersection of a finite number of such manifolds. For d, s and n fixed, we consider $P_n = (w_1, \ldots, w_n) \in \Omega^n$. A set of extremal points of order n on Ω, i.e., points $w_1, \ldots, w_n \in \Omega$ minimizing (3), can be derived by solving the *constrained nonlinear optimization problem*

$$\begin{aligned} &\min_{P_n \in \Omega^n} E_{d,s}^{Q}(P_n) \\ \text{subject to} \quad &\varphi_i(w_\nu) = 0, \qquad i = 1, \ldots, k, \ \nu = 1, \ldots, n, \\ &\psi_j(w_\nu) \geq 0, \qquad j = 1, \ldots, \ell, \ \nu = 1, \ldots, n. \end{aligned}$$

Here, $E_{d,s}^{Q}$ is the *objective function* as given in (3) and the constraints ensure the extremal points to be located in Ω. The, usually non linear, inequalities may be rendered into equalities by subtracting positive *slack variables* $\sigma_j \in \mathbb{R}$, $j = 1,\dots,n\ell$, from each inequality, yielding the following reformulation

$$\min_{P_n \in \Omega^n} E^Q_{d,s}(P_n) \tag{5a}$$

$$\text{subject to} \qquad c(P_n, \sigma) = 0 \tag{5b}$$

$$\sigma_j \geq 0\,, \qquad j = 1,\dots,n\ell\,. \tag{5c}$$

Here $c : \Omega^n \times \mathbb{R}^{n\ell} \to \mathbb{R}^{n(k+\ell)}$ contains the constrained information given by $\varphi_1,\dots,\varphi_k$ and $\psi_j(w_\nu) - \sigma_{j+\ell(\nu-1)}$ for $1 \leq j \leq \ell$, $1 \leq \nu \leq n$ and $\sigma := (\sigma_j)_{1\leq j\leq n\ell}$. We refer to [23, Ch. 3.4] for a more detailed description. In the sequel we use IPOPT, cf. [23], to solve (5). Interior point (or barrier) methods provide a powerful tool for solving nonlinear constrained optimization problems. For an introduction to this field we refer to [14, Ch. 19]. Problem (5) can be transformed to a constrained problem *without* inequality bounds: By converting the bounds into *barrier terms* in the objective function $E^Q_{d,s}$ we obtain

$$\min_{\substack{P_n \in \Omega^n \\ \sigma \in \mathbb{R}^{n\ell}}} \mathscr{B}(P_n, \sigma, \lambda)\,, \qquad \mathscr{B}(P_n, \sigma, \lambda) := E^Q_{d,s}(P_n) - \lambda \sum_{j=1}^{n\ell} \log \sigma_j\,, \tag{6a}$$

$$\text{subject to} \qquad c(P_n, \sigma) = 0 \tag{6b}$$

with the barrier function $\mathscr{B}$ and a barrier parameter $\lambda > 0$. If λ tends to 0, any point fulfilling the *Karush-Kuhn-Tucker conditions* (KKT condition) of problem (6) tends to a KKT point of the original problem (5), see [20] for more details on the relationship of the barrier problem and the original problem. The KKT conditions represent a set of first order necessary conditions for $w_1,\dots,w_n$ to be optimal. If additionally *constraint qualifications* are satisfied [4], the KKT conditions also become sufficient. Let $A_k := \operatorname{grad} c(P_{n,k}, \sigma_k)$ and $W_k := \Delta\mathscr{L}(P_{n,k}, \sigma_k, \omega_k, z_k)$ represent the Hessian with respect to $(P_n, \sigma^\top)^\top$ of the Lagrangian

$$\mathscr{L}(P_n, \sigma, \omega, z) := E^Q_{d,s}(P_n) + c(P_n, \sigma)^\top \omega - z^\top \sigma$$

of the original problem (5) in the kth step with the Lagrange multipliers $\omega \in \mathbb{R}^{n(k+\ell)}$ and $z \in \mathbb{R}^{n\ell}$ for Eqs. (5b) and (5c), respectively. Then, IPOPT solves the optimization problem (5) by applying Newton's method to the barrier problem (6). The system to derive a Newton direction in the kth iteration for a fixed barrier parameter λ reads as

$$\begin{pmatrix} W_k & A_k & -\mathrm{Id} \\ A_k^\top & 0 & 0 \\ Z_k & 0 & X_k \end{pmatrix} \begin{pmatrix} d_k^{(P_n,\sigma)} \\ d_k^\omega \\ d_k^z \end{pmatrix} = - \begin{pmatrix} \operatorname{grad} \mathscr{L}(P_{n,k}, \sigma_k, \omega_k, z_k) \\ c(P_{n,k}, \sigma_k) \\ X_k Z_k \mathbf{1} - \lambda \mathbf{1} \end{pmatrix},$$

yielding the *search directions* $d_k^{(P_n,\sigma)}$, d_k^{ω} and d_k^{z}, which are scaled with an adequate step size and then added to $(P_{n,k}, \sigma_k)$, ω_k, and z_k, respectively, to obtain the corresponding values in the $(k+1)$th iteration step. Here, X_k is a diagonal matrix representing the vectors $P_{n,k}$ and σ_k, i.e., $X_k := \operatorname{diag}\left(P_{n,k}^{\top}, \sigma_k^{\top}\right)^{\top}$, Id represents the identity matrix of adequate size and $\operatorname{grad}\mathscr{L}(P_{n,k}, \sigma_k, \omega_k, z_k)$ the gradient of the Lagrangian with respect to $\left(P_{n,k}^{\top}, \sigma_k^{\top}\right)^{\top}$. Finally, $Z_k := \operatorname{diag}(z_k)$ represents the Lagrange multiplier z_k and $\mathbf{1} := (1, \ldots, 1)^{\top}$. For details about how the step size for the obtained Newton direction is computed within IPOPT we refer to [23]. After each solution of (6) with a current value for the barrier parameter λ, the barrier parameter is decreased (see [23] for the particular algorithm to find a new λ) and IPOPT continues with a further barrier problem based on the approximated solution of the previous one. To solve the KKT system the IPOPT solver requires information about the first and second derivatives of $E_{d,s}^{Q}$ and c to derive search directions proceeding towards the minimal energy.

3 Numerical Results

For a numerical study, we consider here the unweighted case $Q(x) = 0$ for all $x \in \Omega$ with $d = 3$ and $s = 1$ (Newtonian energy). Moreover, we confine ourselves to the case that Ω is a 2-dimensional smooth submanifold of $\mathbb{R}^3$. Then, Ω can be identified by equality constraints $\varphi_i(w_\nu) = 0,\ i = 1, \ldots, k$ [12]. We have implemented the target function $E_{3,1}^{Q=0}$ and the constraints in a MATLAB [13] interface that provides all required information for IPOPT, particularly the corresponding Jacobian and Hessian are conveyed to IPOPT. Thus, the discrete n-point s-energy $\mathscr{E}_{3,1}^{Q=0}(n)$ for different compact sets Ω has been determined. For the unit sphere, the theoretical result is reproduced very well: By direct computation one obtains $\lim\limits_{n\to\infty} \frac{1}{n(n-1)}\mathscr{E}_{3,1}^{Q=0}(n) = \frac{1}{2}$. As it is shown in [10, 24] the error $\left|\frac{1}{n(n-1)}\mathscr{E}_{3,1}^{Q=0}(n) - \frac{1}{2}\right|$ is of the order $\mathbf{O}\left(n^{-\frac{1}{2}}\right)$ if n tends to ∞, which matches the numerical results displayed in Fig. 1. In addition extremal point configurations on further smooth manifolds have been considered, namely on two ellipsoids with different eccentricities (sets Ω_2 and Ω_3 in Table 1) and two different types of tori (Ω_4 and Ω_5). Note, that a torus with major radius $R > 0$ and minor radius $R > r > 0$ is defined by the equation $\left(x_1^2 + x_2^2 + x_3^3 + R^2 - r^2\right)^2 = 4R\left(x_1^2 + x_2^2\right)$. Finally, a more complicated manifold is treated (Ω_6), which is not longer differentiable, being the non-smooth union of two 2-dimensional manifolds. In Fig. 1 the numerical results are presented. Here, the normalized discrete energy $\frac{1}{n(n-1)}\mathscr{E}_{3,1}^{Q=0}(n)$ is derived for different values of $n \in [4{,}500]$. For better visualization of the trend and for an estimation of the continuous minimal energy

$$\frac{1}{2}V_{3,1}^{Q=0} = \lim_{n\to\infty} \frac{1}{n(n-1)}\mathscr{E}_{3,1}^{Q=0}(n)$$

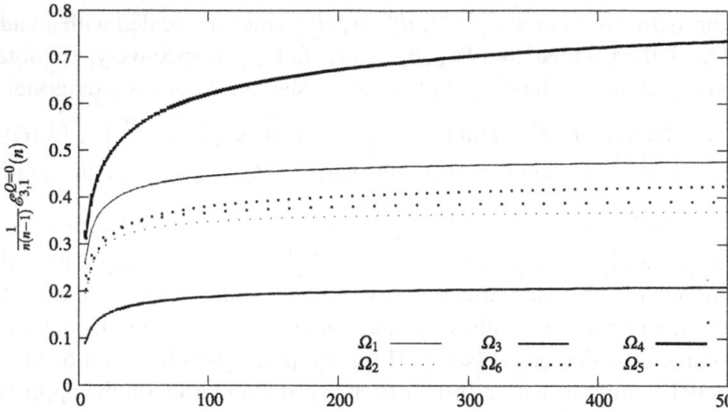

Fig. 1 Discrete Newtonian energy $\frac{1}{n(n-1)}\mathscr{E}_{3,1}^{Q=0}(n)$ for different manifolds in $\mathbb{R}^3$

Table 1 Description of the sets extremal points are numerically computed for

Name	Set
Ω_1	Unit sphere
Ω_2	Ellipsoid with semi-axes $a = 1, b = c = 2^{1/2}$
Ω_3	Ellipsoid with semi-axes $a = 1, b = c = 10^{1/2}$
Ω_4	Torus with major radius $R = 1$, minor radius $r = 1/10$
Ω_5	Torus with major radius $R = 1$, minor radius $r = 1/2$
Ω_6	$\Omega_1 \cup \Omega_2$

a function of the form $x \mapsto a + bx^{-c}$ has been fitted to the data with $a, b, c \in \mathbb{R}$ determined by a least-squares fit. Comparing the asymptotic behavior of $\left(\frac{1}{n(n-1)}\mathscr{E}_{3,1}^{Q=0}(n)\right)$ for the manifolds $\Omega_2, \dots, \Omega_6$ with the results for the unit sphere Ω_1, the same asymptotic behavior of $\left(\frac{1}{n(n-1)}\mathscr{E}_{3,1}^{Q=0}(n)\right)$ for all these manifolds seems to occur. These results give evidence to the hypotheses that the sharp asymptotic behavior observed for the sphere, cf. [1, 24], which is given by

$$\left|\mathscr{E}_{d,s}^{Q=0}(n) - \tfrac{1}{2}V_{d,s}^{Q=0}n^2\right| \leq Cn^{1+\frac{s}{d-1}} \qquad (0 \leq s < d,\ d \geq 2),$$

where C is a constant that may depend on d and s but not on n is also true for general smooth manifolds. In [9] this hypothesis is analyzed in more detail. It is particularly remarkable that the lack of smoothness of Ω_6 does not seem to influence the convergence rate as far as this can be deduced from the computed data. It may be interesting to analyze, whether the observed asymptotic behavior also carries over to lager sets of regular manifolds (e.g., bi-Lipschitz manifolds or even Ahlfors-David

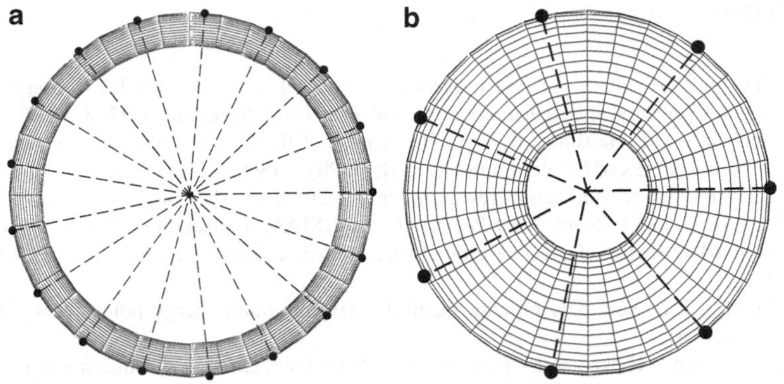

Fig. 2 Top view on the extremal point configuration on the tori Ω_4 for $n = 17$ and on Ω_5 for $n = 7$. The extremal points take the positions of the roots of unity

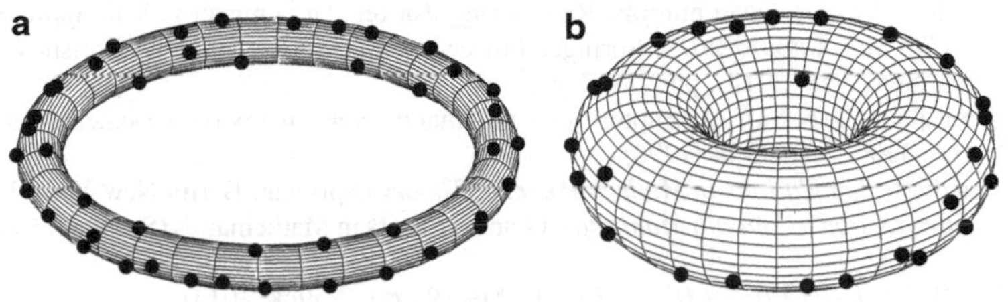

Fig. 3 Extremal point configuration for the tori Ω_4 and Ω_5 for 64 points

regular sets, cf. [9]). A computation of the minimal energy for the different sets yields

$$V_{d,s}^{Q=0}(\Omega_1) = 1\,, \quad V_{d,s}^{Q=0}(\Omega_2) \approx 0.78331\,, \quad V_{d,s}^{Q=0}(\Omega_3) \approx 0.48305\,,$$
$$V_{d,s}^{Q=0}(\Omega_4) \approx 1.40712\,, \quad V_{d,s}^{Q=0}(\Omega_5) \approx 0.82445\,, \quad V_{d,s}^{Q=0}(\Omega_6) \approx 0.89893\,.$$

It is a well known result form potential theory in the complex plane (two dimensional case) that the optimal points on the unit circle coincides with the nth roots of unity [16]. For small numbers of n the extremal points on tori act like the extremal points on the outer boundary of the unit circle. For $n \leq 17$ and $n \leq 7$ for torus Ω_4 and the narrow torus Ω_5, respectively, the extremal points are lying in the plane and their positions coincides (up to rotation) with the nth roots of unity (Fig. 2). As n increases the points *escape* into the third dimension. Thus the patterns in Fig. 3 are obtained.

Acknowledgements The authors like to thank C. Bothner for improving the source code and making the numerical algorithm work efficient. Marco Rozgič would like to express his gratitude towards the German Research Foundation (DFG) for support and funding under the contract PAK 343/2.

References

1. J.S. Brauchart, About the second term of the asymptotics for optimal Riesz energy on the sphere in the potential-theoretical case. Integral Transform. Spec. Funct. **17**, 321–328 (2006)
2. J.S. Brauchart, D.P. Hardin, E.B. Saff, The support of the limit distribution of optimal Riesz energy points on sets of revolution in $\mathbb{R}^3$. J. Math. Phys. **48**(12), 24 (2007)
3. ———, Riesz energy and sets of revolution in $\mathbb{R}^3$. Contemp. Math. **481**, 47–57 (2009)
4. A. Conn, N. Gould, P. Toint, *Trust Region Methods* (SIAM, Philadelphia, 2000)
5. L. Giomi, M.J. Bowick, Defective ground states of toroidal crystals. Phys. Rev. E **78**, 010601 (2008)
6. D.P. Hardin, E.B. Saff, Discretizing manifolds via minimum energy points. Not. AMS **51**, 647–662 (2004)
7. ———, Minimal Riesz energy point configurations for rectifiable d-dimensional manifolds. Adv. Math. **193**, 174–204 (2005)
8. D.P. Hardin, E.B. Saff, H. Stahl, Support of the logarithmic equilibrium measure on sets of revolution in $\mathbb{R}^3$. J. Math. Phys. **48**(2), 022901 (2007)
9. M. Jaraczewski, M. Rozgič, M. Stiemer, Numerical determination of extremal points and asymptotic order of discrete minimal Riesz energy for regular compact sets, in *Approximation Theory XIV: San Antonio 2013*. Springer Proceedings in Mathematics & Statistics, vol. 83 (Springer, New York, 2014), pp. 219–238
10. A.B.J. Kuijlaars, E.B. Saff, Asymptotics for minimal discrete energy on the sphere. Trans. Am. Math. Soc. **350**(2), 523–538 (1998)
11. N.S. Landkof, *Foundations of Modern Potential Theory* (Springer, Berlin/New York, 1972)
12. J. Lee, *Introduction to Smooth Manifolds*. Graduate Texts in Mathematics (Springer, New York, 2006)
13. MATLAB, *Version 8.2.0.701 (R2013b)* (The MathWorks, Natick, 2013)
14. J. Nocedal, S.J. Wright, *Numerical Optimization*. Springer Series in Operations Research (Springer, New York, 1990)
15. M. Ohtsuka, On various definitions of capacity and related notions. Nagoya Math. J. **30**, 121–127 (1967)
16. G. Pólya, G. Szegő, Über den transfiniten Durchmesser (Kapazitätskonstante) von ebenen und räumlichen Punktmengen. J. Reine Angew. Math. **165**, 4–49 (1931)
17. Q. Rajon, T. Ransford, J. Rostand, Computation of capacity via quadratic programming. J. Math. Pure. Appl. **94**, 398–413 (2010)
18. ———, Computation of weighted capacity. J. Approx. Theory **162**(6), 1187–1203 (2010)
19. E.A. Rakhmanov, E.B. Saff, Y.M. Zhou, Minimal discrete energy on the sphere. Math. Res. Lett. **1**, 647–662 (1994)
20. M. Rozgič, M. Jaraczewski, M. Stiemer, Inner point methods: On necessary optimality conditions of various reformulations of a constrained optimization problem, Technical report, Universitätsbibliothek der Helmut-Schmidt-Universität (2014)
21. E.B. Saff, V. Totik, *Logarithmic Potentials with External Fields* (Springer, Berlin/Heidelberg, 1997)
22. E. Saff, A. Kuijlaars, Distributing many points on a sphere. Math. Intell. **19**, 5–11 (1997)
23. A. Wächter, L.T. Biegler, On the implementation of an interior-point filter line-search algorithm for large-scale nonlinear pogramming. Math. Program. **106**, 25–57 (2006)
24. G. Wagner, On means of distances on the surface of a sphere (lower bounds). Pac. J. Math. **144**(2), 389–398 (1990)
25. J. Wermer, *Potential Theory*. Lecture Notes in Mathematics (Springer, Berlin/Heidelberg, 1974)

Part VIII
Uncertainty, Stochastic Modeling and Applications

Sensitivity Estimation and Inverse Problems in Spatial Stochastic Models of Chemical Kinetics

Pavol Bauer and Stefan Engblom

Abstract We consider computational stochastic modeling of diffusion-controlled reactions with applications mainly in molecular cell biology. A complication from the traditional 'well-stirred' case is that our models have a spatial dimension. Our aim here is to put forward a practical algorithm by which perturbations can be propagated through these types of simulations. This is important since the quality of experimental data calls for frequently estimating stability constants. Another use is in inverse formulations which generally relies on being able to effectively and accurately judge the effects of small perturbations. For this purpose we present our implementation of an "all events method" and give two concrete examples of its use. One case studied is the effect of stochastic focusing in the spatial setting, the other case treats the optimization of a small biochemical network.

1 Introduction

In the classical case of non-spatial stochastic modeling of chemical kinetics, the *reaction rates* are understood as transition intensities in a continuous-time Markov chain $X_{t\geq 0}$. When spatial variability is important, space may be discretized in voxels. Between voxels, *diffusion*-, or more generally, *transport rates* become transition intensities in a Markov chain which now takes place in a much larger state space.

This is the point of view taken in the software framework URDME [2, 5] where fairly large-scale spatial stochastic reaction-diffusion models can be simulated. We have developed a solver for sensitivity analysis which allows us to compare single trajectories under arbitrary perturbations of input data and opens up for computing stability estimates as well as optimizing models under various conditions.

P. Bauer • S. Engblom (✉)
Division of Scientific Computing, Department of Information Technology, Uppsala University, Uppsala, Sweden
e-mail: pavol.bauer@it.uu.se; stefane@it.uu.se

A. Abdulle et al. (eds.), *Numerical Mathematics and Advanced Applications ENUMATH 2013*, Lecture Notes in Computational Science and Engineering 103,
DOI 10.1007/978-3-319-10705-9_51

For a given parameter perturbation $c \to c+\delta$ the task is to characterize the mean effect on some function of interest,

$$E[f(X(T,c+\delta)) - f(X(T,c))], \tag{1}$$

for example, by computing a sample average. As a prototypical application, c is a rate constant and f a measure of the molecular population X_t.

The obvious way to carry out this is to conduct two Monte Carlo simulations using independent random numbers and generating N trajectories each of $f(X(T,c+\delta))$ and $f(X(T,c))$, and then taking the average. Two factors can lead to unsatisfactory results with this approach. Firstly, with independent samples, the variance of $f(X(T,c+\delta))$ and $f(X(T,c))$ can be large compared to the difference $f(X(T,c+\delta)) - f(X(T,c))$. This is the *variance reduction problem* which has been discussed in the well-stirred setting by others [12]. Secondly, a slightly more subtle point has been pointed out in [12]; solver algorithms related to Gillespie's Direct Method [8] are not suitable to compute the difference between two processes $X(T,c+\delta)$ and $X(T,c)$ as their coupling is simply not the intended one. This would be a problem for instance, if (1) were to be replaced by

$$E[f(X(T,c+\delta) - X(T,c))], \tag{2}$$

and f some nonlinear function. In fact, a popularly used algorithm for solving spatial stochastic models, the Next Subvolume Method (NSM) [3], belongs to this class of algorithms and can therefore not be used.

In this paper we present the "All Events Method"; a variant of the so-called Common Reaction Path method [12], extended for spatial models in URDME and meeting both the criteria above for an efficient and sound estimation of (1)–(2). In Sect. 2 we give a brief overview of the modeling involved, in the non-spatial as well as in the fully spatial setting, and we also sketch a theory for perturbations, including some implementation aspects of our AEM-solver. In Sect. 3 we discuss two applied examples and show how this solver can be conveniently used in the sense of both forward- and backward formulations.

2 A Viable "All Events Method"-Implementation

After a brief review of stochastic reaction-diffusion modeling we will here summarize the logic behind URDMEs AEM-solver.

2.1 Spatial Stochastic Chemical Kinetics

According to classical *well-stirred* stochastic modeling of chemical kinetics, reactions are transitions between states $x \in \mathbf{Z}_+^D$, counting the number of molecules of each of D distinct species. The transition intensity defines the probability per unit of time for the transition from the state x to $x + S_r$;

$$x \xrightarrow{w_r(x)} x + S_r, \tag{3}$$

where the transition vector $S_r \in \mathbf{Z}^D$ is the rth column in the *stoichiometric matrix* S. Equation (3) defines a continuous-time Markov chain $X_{t \geq 0}$ on $\mathbf{Z}_+^D$.

For spatially extended problems, a stochastic model can be defined by first discretizing space in voxels. Molecular transport can then be handled as a "reaction" which brings a molecule of the lth species from voxel i to j,

$$X_{li} \xrightarrow{a_{ij}\mathbf{x}_{li}} X_{lj}, \tag{4}$$

where $\mathbf{x}_{li}$ is the number of molecules of species l in subvolume i. When space is discretized by general unstructured meshes, suitable rate constants can be obtained by a numerical discretization of the diffusion equation. The consistency in this approach hinges on the fact that the expected value of the concentration converges to the deterministic numerical solution [5].

2.2 Path-Wise Analysis of Perturbations

Without loss of generality, we consider the well-stirred case (3). Let the *state* $X(t) \in \mathbf{Z}_+^D$ count the number of molecules of the D species. The associated Markov chain can be written in the convenient jump SDE form

$$dX_t = S\boldsymbol{\mu}(dt), \tag{5}$$

with counting measure $\boldsymbol{\mu} = [\mu_1, \ldots, \mu_R]^T$. According to this compact notation the time to the arrival of the next reaction of type r is exponentially distributed with intensity $w_r(X_{t-})$. A perhaps more familiar notation is Kurtz's *random time change representation* [6, Chap. 6.2], in which the path is characterized in terms of unit-rate Poisson processes Π_r,

$$X_t = X_0 + \sum_{r=1}^{R} S_r \Pi_r \left(\int_0^t w_r(X_{s-})\, ds \right). \tag{6}$$

This naturally gives rise to the term *operational time* for the argument to each of the R Poissonian processes.

Let a trajectory $Y(t)$ be a *perturbed* version of $X(t)$ in the sense that the former is driven by modified rates $v_r(Y_t)$, but otherwise has an identical reaction topology S. To compare the two trajectories we write

$$dX_t = S\left[\mu^{(0)}(w(X_{t-}), v(Y_{t-});\ dt) + \mu^{(\delta)}(w(X_{t-}), v(Y_{t-});\ dt)\right], \quad (7)$$

$$dY_t = S\left[\mu^{(0)}(w(X_{t-}), v(Y_{t-});\ dt) + \mu^{(\delta)}(v(Y_{t-}), w(X_{t-});\ dt)\right], \quad (8)$$

in terms of the *base* (superscript 0) and *remainder* counting measures (superscript δ), respectively. The intensities for $\mu_r^{(0)}$ and $\mu_r^{(\delta)}$ are given by

$$w_r(x) \wedge v_r(y) \text{ and } w_r(x) - (w_r(x) \wedge v_r(y)). \quad (9)$$

As indicated in the order of the arguments in (7) and (8), there is an asymmetry in the remainder measure.

To analyze $Z_t := \|X_t - Y_t\|^2$ we apply a form of Itô's formula [1, Chap. 4.4.2],

$$dZ_t = 2(X_{t-} - Y_{t-})^T\, S[\mu_{w,v}^{(\delta)} - \mu_{v,w}^{(\delta)}](dt) + S^2[\mu_{w,v}^{(\delta)} + \mu_{v,w}^{(\delta)}](dt). \quad (10)$$

Taking expectation values and ignoring the martingale part we get, after determining the drift parts of the relevant measures,

$$d/dt\, EZ_t = E\left[2(X_t - Y_t)^T\, S[w(X_t) - v(Y_t)] + S^2|w(X_t) - v(Y_t)|\right]. \quad (11)$$

At this point we need some assumption on the dynamics of the process and on the perturbation. Let the rates be *locally Lipschitz* and let the magnitude of the *relative* perturbation be δ. Then for $\|x\| \vee \|y\| \le P$,

$$\|w(x) - v(y)\| \le \|w(x) - w(y)\| + \|w(y) - v(y)\| \quad (12)$$

$$\le L_P\|x - y\| + \delta\|w(y)\| \le C_P(\delta + \|x - y\|). \quad (13)$$

Working similarly, we find from (11) that for some constant C_P,

$$d/dt\, E\|X_t - Y_t\|^2 \le E\, C_P(\delta + \|X_t - Y_t\|^2), \quad (14)$$

where we used the simple observation that for integers n, $\|n\| \le \|n\|^2$. From Grönwall's inequality, assuming $X_0 = Y_0$, we get under a stopping time $t \le \tau_P := \inf_{t\ge 0}\{\|X_t\| \vee \|Y_t\| > P\}$ that

$$E\|X_t - Y_t\|^2 \le \delta(\exp(C_P\, t) - 1). \quad (15)$$

Thus, for *bounded* systems, (15) predicts a RMS perturbation which behaves as $\delta^{1/2}$. For *unbounded* systems, the only immediate generalization is that the limit as $\delta \to 0$ is zero, see [4] and the references therein.

2.3 Simulation Using Consistent Poisson Processes

To motivate our approach to evolving two or more trajectories which can be path-wise compared, consider first the *diffusion approximation* of (5),

$$dX_t = Sw(X_t)\,dt + Sw(X_t)^{1/2}\,d\boldsymbol{W}_t. \tag{16}$$

Two comparable replicas of (16) can clearly be constructed using the *same* Wiener process $\boldsymbol{W}(t)$. In discrete time this boils down to using the same sequence of normal random numbers. This idea can be transferred to the current setting by simply using the same sequence of random numbers when simulating different trajectories, and it leads to the Common Random Numbers method [9].

However, we see from the representation (6) that two trajectories formed by identical Poisson processes are stronger candidates to being similar than any dependency on identical random numbers may generate. This is the motivation behind the Common Reaction Path method [12]. Here all R reaction channels access their own stream of random numbers such that a consistent operational time in the sense of (6) is continuously well-defined. In practise we implement this by storing generator seeds s_i for every channel i and use these for every update of the corresponding Poisson process. For the current case of *spatial* models this implies that all reaction events and all transport events must be associated with a consistent Poisson process. This in contrast to the NSM [3] where only a 'total event' process per voxel is available.

A remark on continuity is made in [12, Appendix B]. When a zero rate is encountered a discontinuity typically forms which is due to the fact that in most implementations, a zero rate will lead to discarding the previous operational time. A new, *uncorrelated* waiting time is drawn whenever the rate becomes non-zero again. In our implementation we circumvent this problem by storing the pre-zero operational time τ^{inf} and associated non-zero rate w^{inf}. When the channel is re-activated we compute the next waiting time τ^{new} using the rescaling (essentially proposed in [7]),

$$\tau^{\text{new}} = t_{\text{current}} + \left(\tau^{\text{inf}} - t_{\text{current}}\right) w^{\text{inf}}/w^{\text{new}}. \tag{17}$$

For more information on implementation of solvers in URDME, consult [2].

3 Sample Applications

We shall now consider two sample applications of our URDME solver; one example in the 'forward' mode, i.e. propagating a definite perturbation, and one example in the 'backward' (or inverse) setting. Due to the computational complexity involved, the inverse problem we choose to consider is non-spatial. However, it is clearly possible to, at an increased computational cost, also target fully spatial formulations.

3.1 Spatial Stochastic Focusing

As a basic but informative example we consider the following enzymatic law,

$$C + E \xrightarrow{k\,c\cdot e} P + E, \tag{18}$$

in which E is an enzyme and C an intermediate complex which matures into a product P. The model is completed by adding the in- and outflow laws

$$\emptyset \underset{\beta_C c}{\overset{\alpha_C}{\rightleftharpoons}} C, \qquad \emptyset \underset{\beta_E e}{\overset{\alpha_E}{\rightleftharpoons}} E, \qquad P \xrightarrow{\beta_P p} \emptyset. \tag{19}$$

Stochastic focusing [11] is a non-linear stochastic effect under which an input signal is strongly amplified, and notably much more effectively so than for the corresponding mean field model. In the present case this effect can be observed in the response of the number of intermediate complexes C when the birth rate α_E is perturbed according to $\alpha_E \to \alpha_E(1 - \delta)$ (Fig. 1, left). A *spatial* version of (18) and (19) can be defined in the geometry $\Omega = [0, 1]$ with diffusion of the species. We generate an 'unperturbed' trajectory $C_1(t)$ for which $\alpha_E = c$ is constant and a 'perturbed case' $C_2(t)$ for which α_E is replaced by the space dependent function $\alpha_E(x) = c(1/2 + x)$. Note that this preserves the total production rate in the sense that

$$\int_\Omega \alpha_E(x)dV = c. \tag{20}$$

We combine the reactions with varying diffusion ε and observe a phenomenon which can be referred to as *Spatial stochastic focusing* (Fig. 1, center/right).

In the table below we determine at two different perturbations δ and for several values of TOL, the number of realizations $N = 10, 20, \ldots$ needed to bring the standard Monte Carlo error estimate std$/\sqrt{N}$ below TOL. This for the case of estimating $E[C_2(1) - C_1(1)]$ using either the Next Subvolume Method [3] or the solver proposed by us.

	NSM				AEM			
$\delta\backslash$ TOL	1/4	1/8	1/16	1/32	1/4	1/8	1/16	1/32
1/2	1,480	3,470	6,990	33,010	30	600	3,630	12,870
1/32	1,350	2,780	5,630	14,970	10	20	60	3,190

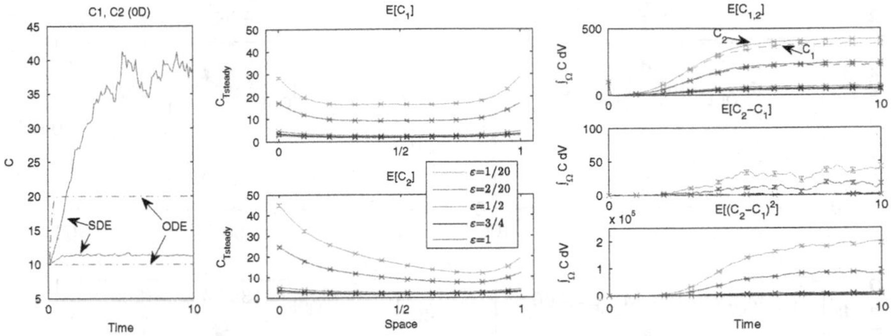

Fig. 1 *Left*: ODE and mean SDE solutions of the unperturbed (C_1, *red, lower*) and perturbed (C_2, *blue, upper*) model (18)–(19) in the well-stirred case ($\delta = 1/2$). *Center*: SDE solutions with spatial perturbation and varying diffusion. Each point represents the mean of C at steady-state. From top to bottom as in legend (Colors online). *Right*: traces of C_1 (*dashed*) and C_2 (*solid*) integrated over space and plotted over time. All SDE solutions are averages of $N = 10^4$ trajectories, error bars are std/$\sqrt{N}$

3.2 Enzymatic Control

Consider again the model (18)–(19) but with the enzyme E under *control*,

$$\emptyset \underset{\beta_E e}{\overset{s(t)}{\rightleftharpoons}} E, \tag{21}$$

with $s(t)$ a time-dependent signal. We define a *payoff function* $\varphi(P)$ by

$$\varphi(P) = (P - c_-)[c_- < P \leq C_+] + (C_+ - c_-)[C_+ < P] \tag{22}$$

with c_-/C_+ suitable cutoff values. Reasonable constraints are that $\|s(t)\|_\infty$ and $\|s(t)\|_1$ are bounded. After adding a regularization term the target functional becomes

$$\mathscr{M}[P] := \int_0^T \varphi(P_t)\,dt + \epsilon[s(t)]_{0 \leq t \leq T}, \tag{23}$$

with $[\cdot]$ the total variation. Thus the overall formulation is "*Find* $s(t)$ *such that in expectation,* $\mathscr{M}[P]$ *attains it maximum subject to the constraints*". Here $P = P(t)$ is the solution to (18)–(19) and (21) with $P(0) = E(0) = 0$. We solve the optimization problem both in the deterministic ODE setting and in the stochastic setting using URDME/AEM. The optimization algorithm applied was the Nelder-Mead simplex method [10] and results for two sets of cutoff-values are shown in Fig. 2.

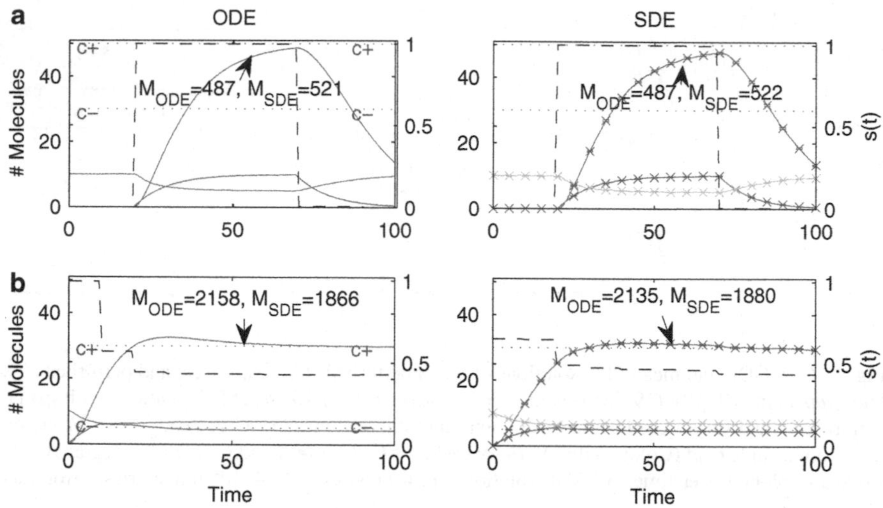

Fig. 2 (**a**) Optimal solution $s(t)$ for cutoff values $c_- = 30$ and $C_+ = 50$, for the case of a deterministic ODE (*left*), and an SDE (*right*). These cutoff values yield similar "all-or-nothing" optimal strategies in both cases. (**b**) Here $c_- = 5$ and $C_+ = 30$, and the optimal solutions are clearly different for the two cases. *Legend*: From top to bottom (colors online), P (*red*), C (*green*), E (*blue*), signal $s(t)$ (*black, dashed*). Values of the target functional for the optimal $s(t)$ are also indicated

Conclusions

We have presented a viable simulation algorithm for continuous-time Markov chains which relies upon a self-consistent use of Poisson processes. This computationally intensive technique enables perturbations in the input parameters to be propagated and opens up for several relevant applications. To the best of our knowledge none of the applications considered here have been addressed previously.

Through straightforward perturbation calculations in the 'forward' mode we have reported results for *spatial stochastic focusing*, where the strong focusing effect can be *uniquely* attributed to the spatial dimension. In a nutshell, the existence of a gradient implies an increase of outgoing products which cannot be explained through well-stirred and/or deterministic analysis.

As an example of an interesting inverse formulation we studied a simple chemical network and defined an arguably quite open criterion for optimality. By wrapping our simulator with a simple external optimizing routine we were able to find optimal control signals which realizes this optimality. In one case the signals found clearly differ from their deterministic versions.

(continued)

While developing computational algorithms simultaneously with challenging applications requires some care, it is our hope that this report shows the value of this approach.

References

1. D. Applebaum, *Lévy Processes and Stochastic Calculus*. Studies in Advanced Mathematics, vol. 93 (Cambridge University Press, Cambridge, 2004)
2. B. Drawert, S. Engblom, A. Hellander, URDME: a modular framework for stochastic simulation of reaction-transport processes in complex geometries. BMC Syst. Biol. **6**, 76 (2012)
3. J. Elf, M. Ehrenberg, Spontaneous separation of bi-stable biochemical systems into spatial domains of opposite phases. Syst. Biol. **1**, 2 (2004)
4. S. Engblom, On the stability of stochastic jump kinetics. Available at http://arxiv.org/abs/1202.3892 (2012)
5. S. Engblom, L. Ferm, A. Hellander, P. Lötstedt, Simulation of stochastic reaction–diffusion processes on unstructured meshes. SIAM J. Sci. Comput. **31**, 3 (2009)
6. S.N. Ethier, T.G. Kurtz, *Markov Processes: Characterization and Convergence*. Wiley Series in Probability and Mathematical Statistics (Wiley, New York, 1986)
7. M.A. Gibson, J. Bruck, Efficient exact stochastic simulation of chemical systems with many species and many channels. J. Phys. Chem. **104**(9), (2000)
8. D.T. Gillespie, Exact stochastic simulation of coupled chemical reactions. J. Phys. Chem. **81**, 25 (1977)
9. P. Glasserman, D.D. Yao, Some guidelines and guarantees for common random numbers. Manage. Sci. **38**, 6 (1992)
10. J.A. Nelder, R. Mead, A simplex method for function minimization. Comput. J. **7**, 4 (1965)
11. J. Paulsson, O.G. Berg, M. Ehrenberg, Stochastic focusing: fluctuation-enhanced sensitivity of intracellular regulation. Proc. Natl. Acad. Sci. U.S.A. **97**, 13 (2000)
12. P.S.M. Rathinam, M. Khammash, Efficient computation of parameter sensitivities of discrete stochastic chemical reaction networks. J. Chem. Phys. **132**, 3 (2010)

A Bayesian Approach to Physics-Based Reconstruction of Incompressible Flows

Iliass Azijli, Richard Dwight, and Hester Bijl

Abstract To reconstruct smooth velocity fields from measured incompressible flows, we introduce a statistical regression method that takes into account the mass continuity equation. It is based on a multivariate Gaussian process and formulated within the Bayesian framework, which is a natural framework for fusing experimental data with prior physical knowledge. The robustness of the method and its implementation to large data sets are addressed and compared to a method that does not include the incompressibility constraint. A two-dimensional synthetic test case is used to investigate the accuracy of the method and a real three-dimensional experiment of a circular jet in water is used to investigate the method's ability to fill up a gap containing a vortex ring.

1 Introduction

Measuring the velocity field of a flow has become quite common in many branches of science and engineering (e.g., oceanography, meteorology and wind tunnel testing) thanks to improvements in flow measurement techniques such as particle image velocimetry (PIV). To obtain a smooth field in the presence of measurement uncertainty and regions where information is absent (i.e., gaps), the experimental data has to be processed further. Additionally, it may be required to obtain unmeasured quantities such as vorticity and helicity.

To this end, the present work considers a statistical regression method based on Gaussian processes. Its statistical interpretation allows for a natural inclusion of measurement uncertainty. Past works have applied the method in the context of reconstructing measured flow fields [2, 5, 9]. The common approach in these applications is to handle the velocity components *independently* from each other in the reconstruction.

In reality however, the velocity components are related through the governing equations of fluid dynamics. In the present paper, it is assumed that the flow

I. Azijli (✉) • R. Dwight • H. Bijl
Delft University of Technology, Kluyverweg 2, Delft, The Netherlands
e-mail: i.azijli@tudelft.nl; r.p.dwight@tudelft.nl; h.bijl@tudelft.nl

A. Abdulle et al. (eds.), *Numerical Mathematics and Advanced Applications ENUMATH 2013*, Lecture Notes in Computational Science and Engineering 103,
DOI 10.1007/978-3-319-10705-9_52

of interest is incompressible and the standard method is modified such that the reconstructed velocity field satisfies the mass continuity equation

This paper is organized as follows. In Sect. 2, we present Gaussian process regression from a Bayesian perspective. The standard approach of handling the velocity components independently from each other is discussed and the method of incorporating the incompressibility constraint is introduced. In Sect. 3, we discuss two important issues related to the practical implementation of the methods, namely computational cost and robustness. In Sect. 4, we apply and compare the two methods to a two-dimensional synthetic test case and a real three-dimensional experiment. Finally, the conclusions are presented in section "Conclusions".

2 Gaussian Process Regression

Consider an unobservable Gaussian process $\mathscr{F} \sim GP\left(\mu\left(\mathbf{x}\right), c\left(\mathbf{x}, \mathbf{x}'; \theta\right)\right)$, where $\mathbf{x} \in \mathbb{R}^3$ and $\theta \in \mathbb{R}^d$ are the hyperparameters. If we discretize the space at n locations of interest, we get $\mathbf{F} \sim N\left(\boldsymbol{\mu}, P\right)$, where $\mu_i = \mu\left(\mathbf{x}^i\right)$ and $P^{ij} = c\left(\mathbf{x}^i, \mathbf{x}^j; \theta\right)$. Its realization is the state vector $\mathbf{f} \in \mathbb{R}^n$. The statistical model for the observations is defined as $\mathbf{y} = H\mathbf{f} + \boldsymbol{\epsilon}$, where $\boldsymbol{\epsilon} \sim N\left(0, R\right)$. The matrix R is as the observation error covariance matrix, representing the measurement uncertainty. Therefore, $\mathbf{Y}|\mathbf{f} \sim N\left(H\mathbf{f}, R\right)$, also known as the likelihood. We are interested in the distribution of the true state given the observed data $p\left(\mathbf{f}|\mathbf{y}\right)$, known as the posterior. According to Bayes' Rule, the posterior distribution is proportional to the prior $p\left(\mathbf{f}\right)$ times the likelihood $p\left(\mathbf{y}|\mathbf{f}\right)$. The mean and variance of the posterior are therefore normally distributed as well [12]. In this paper we are only interested in the posterior mean, given by (1). The term $R + HPH^T$ is called the gain matrix A.

$$E\left(\mathbf{F}|\mathbf{y}\right) = \boldsymbol{\mu} + PH^T\left(R + HPH^T\right)^{-1}\left(\mathbf{y} - H\boldsymbol{\mu}\right) \tag{1}$$

2.1 Standard Approach

The state is set to $\mathbf{f} = \left(\mathbf{u}_1\ \mathbf{u}_2\ \mathbf{u}_3\right)^T \in \mathbb{R}^{3n}$. The m observations are defined in $\mathbf{y} = \left(\mathbf{u}_1^o\ \mathbf{u}_2^o\ \mathbf{u}_3^o\right)^T \in \mathbb{R}^{3m}$. The covariance matrix $P \in \mathbb{R}^{3n\times 3n}$ is a 3×3 block diagonal matrix, with elements P_k, $k = 1, 2, 3$ on the diagonal. The off-diagonal blocks are zero because the method assumes that the velocity components are uncorrelated. The i, j entry of P_k is the covariance function $c\left(r^{ij}\right)$, $i, j = 1, \ldots, n$, where r^{ij} is the distance between the points $\mathbf{x}^i$ and $\mathbf{x}^j$, defined as $(r^{ij})^2 = \sum_{k=1}^{3}((x_k^i - x_k^j)/\theta_k)^2$. The parameter θ_k is the correlation length in the direction k.

2.2 Divergence-Free Approach

The divergence-free approach takes into account the mass continuity equation. For an incompressible flow, the density is constant in space and time, reducing the equation to $\nabla \cdot \mathbf{u} = 0$, i.e., a divergence-free velocity field. From vector calculus it is known that a divergence-free vector field can be obtained by taking the curl of a vector potential $\mathbf{a}$. The state vector is now defined as:

$$\mathbf{f} = \left(\mathbf{a}_1\ \partial_1\mathbf{a}_1\ \partial_2\mathbf{a}_1\ \partial_3\mathbf{a}_1\ \mathbf{a}_2\ \partial_1\mathbf{a}_2\ \partial_2\mathbf{a}_2\ \partial_3\mathbf{a}_2\ \mathbf{a}_3\ \partial_1\mathbf{a}_3\ \partial_2\mathbf{a}_3\ \partial_3\mathbf{a}_3\right)^T$$

where ∂_i represents the partial derivative with respect to x_i. Contrary to the standard approach, the state vector is now an unobservable; neither the vector potential nor the individual first partial derivatives are observed. Instead, we measure a particular *linear* combination of the first derivatives. To be more specific, the curl of the vector potential. The observation matrix should therefore be set up such that it will convert the state vector into the curl of the vector potential. The covariance matrix is again a block diagonal matrix. Equation (2) shows how the diagonal blocks are defined, with $k = 1, 2, 3$ and $\partial_{i,j}$ the second partial derivative with respect to x_i and x_j [8].

$$\dot{P}_k = \begin{bmatrix} P_k & \partial_1 P_k & \partial_2 P_k & \partial_3 P_k \\ -\partial_1 P_k & \partial_{1,1} P_k & \partial_{1,2} P_k & \partial_{1,3} P_k \\ -\partial_2 P_k & \partial_{1,2} P_k & \partial_{2,2} P_k & \partial_{2,3} P_k \\ -\partial_3 P_k & \partial_{1,3} P_k & \partial_{2,3} P_k & \partial_{3,3} P_k \end{bmatrix} \tag{2}$$

Again, the off-diagonals of the covariance matrix are zero. But the reason for this now is that the different components of the vector potential, not the velocity, are assumed to be uncorrelated. In the most general case there is no a-priori physical knowledge to assume that there *is* a relation between them. The gain matrix is given by (3).

$$A = R + HPH^T = R + \begin{bmatrix} \partial_{3,3} P_2 + \partial_{2,2} P_3 & -\partial_{1,2} P_3 & -\partial_{1,3} P_2 \\ -\partial_{1,2} P_3 & \partial_{3,3} P_1 + \partial_{1,1} P_3 & -\partial_{2,3} P_1 \\ -\partial_{1,3} P_2 & -\partial_{2,3} P_1 & \partial_{2,2} P_1 + \partial_{1,1} P_2 \end{bmatrix} \tag{3}$$

It can easily be verified that the columns and rows of the gain matrix are divergence-free. By taking the same covariance function for all directions of the vector potential ($P_1 = P_2 = P_3$) and assuming perfect measurements ($R = 0$), (3) reduces to:

$$A = \left(\Delta I - \nabla\nabla^T\right) P_1 \tag{4}$$

which turns out to be proportional to the operator constructed by Narcowich and Ward [6]. We have therefore generalized their method by allowing different covariance functions to be used for the different vector potential components and by including measurement uncertainty in the reconstruction. The latter was natural because our derivation was carried out from a Bayesian perspective.

3 Practical Implementation

The practical implementation of Gaussian process regression is discussed in this section, which is important considering the fact that data sets can be large. We compare the standard and divergence-free approach in terms of computational cost and conditioning.

3.1 Computational Cost

To evaluate the posterior mean given by (1), one first has to solve the linear system $A\mathbf{c} = \mathbf{y} - H\boldsymbol{\mu}$. The gain matrices for the standard and divergence-free cases are symmetric and additionally positive definite if the covariance function is positive definite [6]. The Cholesky factorization can therefore be used for a direct solver and the conjugate gradient method can be used as an iterative solver.

Reconstructing a velocity field with the standard approach can be split up into three separate problems since the gain matrix is a 3×3 block diagonal matrix. This is not possible for the divergence-free approach because the velocity components are related. The divergence-free approach is therefore nine times more expensive than the standard approach if the Cholesky factorization is used and three times more expensive if the conjugate gradient method is used, assuming similar convergence.

If the gain matrix has structure then the solution can be obtained cheaper. If the measurement grid is regular and all measurements have equal measurement uncertainty, the matrix has Toeplitz structure. The standard approach renders a 3-level Toeplitz matrix and the divergence-free approach renders a 3×3 block matrix with each block a 3-level Toeplitz matrix. Fast and superfast direct solvers of complexity $O\left(N^2\right)$ and $O\left(N \log^2 N\right)$, respectively, can then be used [1]. The conjugate gradient method can be accelerated by embedding a Toeplitz matrix into a circulant matrix [3]. By making use of the Fast Fourier Transform, the matrix-vector multiplication can be performed in $O\left(N \log N\right)$. It can be shown that the divergence-free approach will be $O\left(1 + \log_m 3\right)$ times more expensive than the standard approach, making the cost comparable, especially for larger system sizes. In case of unequal measurement uncertainty, there exist no fast *direct* solvers for the resulting Toeplitz-plus-diagonal systems [7]. However, the conjugate gradient method can still be used in combination with the FFT.

Table 1 Relative correlation length θ/L at which $\kappa(A) = 10^{15}$. The spatial dimension is 2 and the data points are defined on a regular 33 × 33 grid

	$\phi_{2,2}$	$\phi_{2,3}$	$\phi_{2,4}$	*gauss*
Standard	23.899	8.6501	2.0876	0.2848
Divergence free	919.28	24.637	3.8032	0.2117

3.2 *Conditioning*

The most important factors that influence the condition number of the gain matrix, $\kappa(A)$, are the covariance function, the separation distance of the data, the correlation length and the observation error [4]. The divergence-free gain matrix is influenced by the same parameters [6]. Decreasing the separation distance, increasing the correlation length and decreasing the observation error all increase the condition number. An ill-conditioned gain matrix causes inaccurate results, even if the algorithm is stable. In fact, the gain matrix can become so ill-conditioned that it stops being *numerically* positive definite, even though it will be *analytically* if its covariance function is positive definite. In that case, a Cholesky factorization cannot be carried out and the conjugate gradient method should in principle not be used.

Table 1 summarizes at which correlation range the various covariance functions reach a condition number of 10^{15}. The functions $\phi_{2,k}$, $k = 2, 3, 4$ are the Wendland functions with smoothness C^{2k} [11]. The Gaussian (*gauss*), an infinitely smooth function, is defined as $\exp\left(-\alpha^2 r^2\right)$. The constant α was set to 3.3 to make it resemble $\phi_{2,3}$ as closely as possible. Contrary to the Gaussian, the Wendland functions have compact support: $\phi_{2,k} = 0$ for $r \geq 1$. The Gaussian covariance function appears to be the most ill-conditioned, a property attributed to the fact that it is infinitely differentiable [4]. The smoother the Wendland function, the more it approaches the Gaussian. The idea that the level of differentiability is related to the conditioning of the gain matrix is supported by the observation that for the Wendland functions, the divergence-free gain matrix becomes numerically not-positive definite at a larger θ. The reverse happens for the Gaussian. For the divergence-free approach, the covariance function is differentiated twice, reducing the level of differentiability of the resulting function for the Wendland functions, explaining the better conditioning of the matrix.

4 Results

We apply the standard and divergence-free approach to a 2D synthetic test case and to a data set obtained from an actual 3D PIV experiment. The mean and standard deviation of the prior are estimated from the data. Furthermore, it is assumed that the correlation range in all directions is the same

4.1 2D Synthetic Test Case

We consider the 2D incompressible flow of a counter-rotating vortex pair. For the numerical experiments, the sample points are taken on uniform grids with spacing h and the RMSE is calculated using 5,000 validation points, distributed over the domain using Latin hypercube sampling. Table 2 shows how the minimum RMSE changes as a function of the sample density.

Three observations are made. First of all, as expected, the RMSE decreases with increasing sample density, both for the standard approach and the divergence-free approach. Secondly, the velocity field reconstructed with the divergence-free approach is more accurate than the standard approach. One exception is for $h = 2^{-5}$, when using $\phi_{2,2}$. However, it can be argued that since the Wendland functions have finite smoothness, the results for the covariance function of the standard approach $\phi_{2,k}$ should be compared with $\phi_{2,k+1}$ of the divergence-free approach since the smoothness decreases with two orders for the divergence-free approach. In that case, we observe that the divergence-free approach always produces a more accurate field. So it seems like introducing the physical knowledge of incompressibility indeed improves the reconstructed velocity field. Finally, a close inspection reveals that the spread in RMSE between the different covariance functions used is larger for the divergence-free approach, indicating that the reconstructed velocity field is more sensitive to the covariance function used.

It is important to note that the results in Table 2 show the *minimum* RMSE. The divergence-free approach does not reconstruct a more accurate velocity field for *every* θ. Figure 1 shows that for the small correlation length values, the standard approach is more accurate. This behavior is quite unfortunate since it decreases the sparseness of the gain matrix when using compactly supported covariance functions. Knowing that the divergence-free gain matrix is more expensive to solve, obtaining a sparse matrix is desired to decrease computational cost.

Table 2 RMSE as a function of sample density, covariance function and reconstruction approach used. (st) stands for standard and (df) stands for divergence-free

	$h = 2^{-1}$	$h = 2^{-2}$	$h = 2^{-3}$	$h = 2^{-4}$	$h = 2^{-5}$
gauss (st)	5.30e−1	2.76e−1	8.11e−2	9.43e−3	2.20e−4
gauss (df)	3.45e−1	1.95e−1	4.25e−2	6.14e−3	1.53e−4
$\phi_{2,2}$ (st)	5.31e−1	2.78e−1	8.16e−2	8.10e−3	9.77e−5
$\phi_{2,2}$ (df)	3.34e−1	2.30e−1	5.85e−2	6.81e−3	1.05e−4
$\phi_{2,3}$ (st)	5.31e−1	2.76e−1	8.12e−2	8.36e−3	1.04e−4
$\phi_{2,3}$ (df)	3.41e−1	2.00e−1	4.58e−2	5.50e−3	9.12e−5
$\phi_{2,4}$ (st)	5.31e−1	2.76e−1	8.11e−2	8.86e−3	1.10e−4
$\phi_{2,4}$ (df)	3.44e−1	1.94e−1	4.28e−2	4.36e−3	7.57e−5

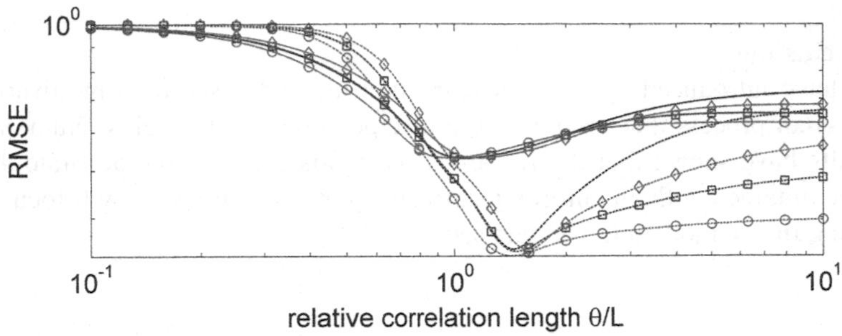

Fig. 1 RMSE of the velocity vs. θ/L for $h = 2^{-1}$. *Solid lines*: standard approach, *dotted lines*: divergence-free approach. No marker: Gaussian, ○ : $\phi_{2,2}$, □ : $\phi_{2,3}$, ◇ : $\phi_{2,4}$

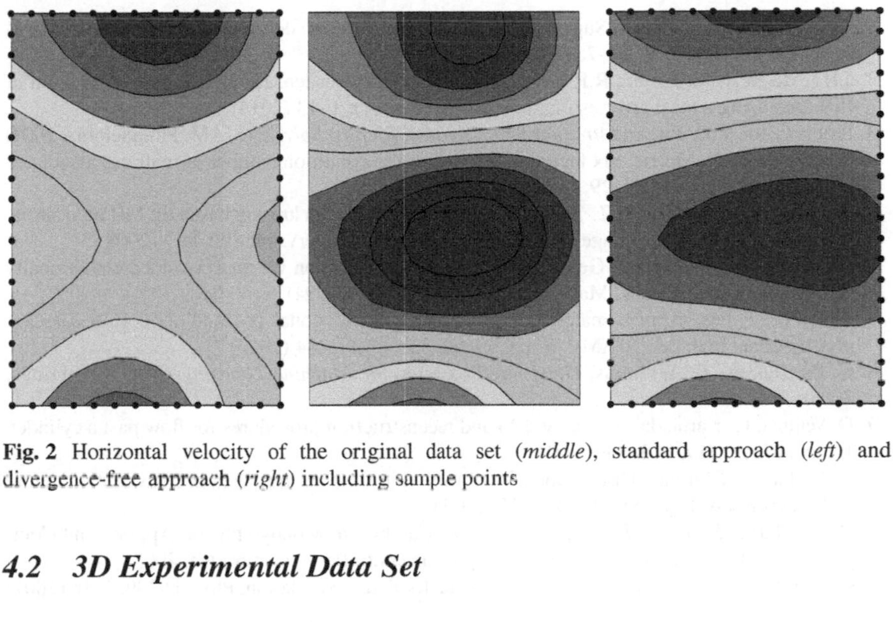

Fig. 2 Horizontal velocity of the original data set (*middle*), standard approach (*left*) and divergence-free approach (*right*) including sample points

4.2 3D Experimental Data Set

The velocity field considered in this section is that of a circular jet in water [10]. It was obtained using tomographic PIV, a measurement technique that extracts the three components of velocity in a volume. The jet velocity at the nozzle exit is 0.5 m/s, so assuming incompressible flow is an excellent approximation. We want to illustrate the ability of the divergence-free approach to reconstruct vortices. To this end, we took a data set of $37 \times 29 \times 5$ velocity vectors. From this set, we introduced a small gap in the shear layer of the jet, where vortex rings are formed, containing $20 \times 14 \times 5$ vectors. Figure 2 shows the horizontal velocity in the gap. The RMSE of the standard approach is 0.7634, while the RMSE of the divergence-free approach is 0.5881. The divergence-free approach is able to resolve the vortex, whereas the standard approach misses it. The divergence field for the standard approach, not plotted here, contains strong sources and sinks. The divergence field of the divergence-free approach is of course per definition zero.

Conclusions
We have introduced a statistical regression method based on a multivariate Gaussian process that enforces zero divergence on vector fields. Numerical results have shown that the reconstructed fields can be more accurate than those obtained without enforcing zero divergence. Future work will focus on testing the method on larger data sets.

References

1. G.S. Ammar, W.B. Gragg, Superfast solution of real positive definite Toeplitz systems. SIAM J. Matrix Anal. Appl. **9**, 61–76 (1988)
2. J.H.S. de Baar, M. Percin, R.P. Dwight, B.W. van Oudheusden, H. Bijl, Kriging regression of PIV data using a local error estimate. Exp. Fluids **55**(1), 1–13 (2014)
3. R.H.F. Chan, X.Q. Jin, *An Introduction to Iterative Toeplitz Solvers* (SIAM, Philadelphia, 2007)
4. G.J. Davis, M.D. Morris, Six factors which affect the condition number of matrices associated with Kriging. Math. Geol. **29**, 669–683 (1997)
5. S.L. Lee, A. Huntbatch, G.Z. Yang, Contractile analysis with kriging based on MR myocardial velocity imaging. Med. Image Comput. Comput. Assist. Interv. **11**, 892–899 (2008)
6. F.J. Narcowich, J.D. Ward, Generalized Hermite interpolation via matrix-valued conditionally positive definite functions. Math. Comput. **63**, 661–687 (1994)
7. M.K. Ng, J. Pan, Approximate inverse circulant-plus-diagonal preconditioners for toeplitz-plus-diagonal matrices. SIAM J. Sci. Comput. **32**, 1442–1464 (2010)
8. C. Rasmussen, C. Williams, *Gaussian Processes for Machine Learning* (MIT, Cambridge, 2006)
9. D. Venturi, G. Karniadakis, Gappy data and reconstruction procedures for flow past a cylinder. J. Fluid Mech. **519**, 315–336 (2004)
10. D. Violato, F. Scarano, Three-dimensional evolution of flow structures in transitional circular and chevron jets. Phys. Fluids **23**, 1–25 (2011)
11. H. Wendland, *Scattered Data Approximation*. Cambridge Monographs on Applied and Computational Mathematics, vol. 17 (Cambridge University Press, Cambridge, 2005)
12. C.K. Wikle, L.M. Berliner, A Bayesian tutorial for data assimilation. Phys. D **230**, 1–16 (2007)

Improved Stabilized Multilevel Monte Carlo Method for Stiff Stochastic Differential Equations

Assyr Abdulle and Adrian Blumenthal

Abstract An improved stabilized multilevel Monte Carlo (MLMC) method is introduced for stiff stochastic differential equations in the mean square sense. Using S-ROCK2 with weak order 2 on the finest time grid and S-ROCK1 (weak order 1) on the other levels reduces the bias while preserving all the stability features of the stabilized MLMC approach. Numerical experiments illustrate the theoretical findings.

1 Introduction

Estimating the expectation of a functional depending on a stochastic process is essential in many applications ranging from biology, chemistry, physics to economics [8, 9, 12, 13]. A popular approach for such problems is the use of classical and improved Monte Carlo (MC) techniques, in particular the multilevel Monte Carlo (MLMC) method using Euler-Maruyama (EM) [10] as numerical integrator [7]. An *explicit* stabilized multilevel Monte Carlo method has been proved to be useful and efficient for stiff problems in a mean square sense and specially attractive for problems of large dimensions (as it avoids solving nonlinear systems typically arising with implicit methods) [1]. Here we present an improved version of the stabilized MLMC method by using a higher weak order scheme on the finest time grid.

We consider the Itô stochastic differential equation (SDE)

$$\mathrm{d}X(t) = f(X(t))\mathrm{d}t + \sum_{r=1}^{m} g^r(X(t))\mathrm{d}W_r(t), \quad 0 \le t \le T, \qquad X(0) = X_0, \tag{1}$$

A. Abdulle (✉) • A. Blumenthal
ANMC, Section de Mathématiques, École Polytechnique Fédérale de Lausanne, Lausanne, Switzerland
e-mail: assyr.abdulle@epfl.ch; adrian.blumenthal@epfl.ch

A. Abdulle et al. (eds.), *Numerical Mathematics and Advanced Applications ENUMATH 2013*, Lecture Notes in Computational Science and Engineering 103,
DOI 10.1007/978-3-319-10705-9_53

where $X(t) \in \mathbb{R}^d$ is a random variable, $f : \mathbb{R}^d \to \mathbb{R}^d$ the drift function, $g^r : \mathbb{R}^d \to \mathbb{R}^d$ the diffusion functions and $W_r(t)$ independent one-dimensional Wiener processes (with $r = 1, 2, \ldots, m$). Further, we take into account a numerical approximation of the solution of (1) by using a discrete map $X_{n+1} = \Psi(X_n, h, \xi_n)$, where $\Psi(\cdot, h, \xi_n) : \mathbb{R}^d \to \mathbb{R}^d$, $X_n \in \mathbb{R}^d$ for $n \in \mathbb{N}$, h a time stepsize and ξ_n some random vector. Let us now consider any $\tau_n = nh \in [0, T]$ for h sufficiently small. The numerical approximation is said to be of strong order of convergence θ_s if $\max_{0\leq n\leq T/h} \mathbb{E}|X_n - X(\tau_n)| \leq Ch^{\theta_s}$ for a constant C (independent of h). It is said to be of weak order θ_w if for any function $\phi \in C_P^{2(\gamma+1)}(\mathbb{R}^d, \mathbb{R})$[1] there exists a constant C (independent of h) such that $|\mathbb{E}[\phi(X_n)] - \mathbb{E}[\phi(X(\tau_n))]| \leq Ch^{\theta_w}$.

The stability of a numerical method is another important issue for computations. A stochastic process $(X(t))_{t\geq 0}$ is said to be mean square stable if $\mathbb{E}\left[X(t)^2\right]$ tends to zero as t goes to infinity. The scalar linear SDE $\mathrm{d}X(t) = \lambda X(t)\mathrm{d}t + \mu X(t)\mathrm{d}W(t)$, $X(0) = 1$, with $\lambda \in \mathbb{C}$ and $\mu \in \mathbb{C}$ is commonly used to figure as test problem [9]. The stability domain of the exact solution is given by $\mathscr{S}_{exact} := \left\{(\lambda, \mu) \in \mathbb{C}^2 \mid \Re\{\lambda\} + \frac{1}{2}|\mu|^2 < 0\right\}$. Similarly a numerical method is mean square stable if $\mathbb{E}\left[X_n^2\right]$ tends to zero as n goes to infinity. The stability domain of the EM method (that has strong order 1/2 and weak order 1) is specified by $\mathscr{S}_{EM} = \left\{(p, q) \in \mathbb{C}^2 \mid |1 + p|^2 + q^2 < 1\right\}$ with $(p, q) = (h\lambda, \sqrt{h}|\mu|)$ (see [9]).

Stabilized stochastic methods were introduced in [2,3] with the aim of improving the stability behavior of the EM method, while staying explicit. These methods are s-stage *explicit* methods with fixed order based on

- a deterministic stabilization procedure;
- a finishing stochastic procedure to achieve the desired accuracy.

We will consider the s-stage S-ROCK1 method [3] defined by (for $s \geq 2$)

$$
\begin{aligned}
K_0 &= X_n, \quad K_1 = X_n + h\tfrac{\omega_1}{\omega_0} f(K_0), \\
K_i &= 2h\omega_1 \tfrac{T_{i-1}(\omega_0)}{T_i(\omega_0)} f(K_{i-1}) + 2\omega_0 \tfrac{T_{i-1}(\omega_0)}{T_i(\omega_0)} K_{i-1} - \tfrac{T_{i-2}(\omega_0)}{T_i(\omega_0)} K_{i-2}, \\
K_s &= 2h\omega_1 \tfrac{T_{s-1}(\omega_0)}{T_s(\omega_0)} f(K_{s-1}) + 2\omega_0 \tfrac{T_{s-1}(\omega_0)}{T_s(\omega_0)} K_{s-1} - \tfrac{T_{s-2}(\omega_0)}{T_s(\omega_0)} K_{s-2} \\
&\quad + \sum_{r=1}^{m} g^r(K_{s-1}) \Delta W_{n+1,r},
\end{aligned}
\tag{2}
$$

with $i = 2, 3, \ldots, s - 1$ and where $\omega_0 = 1 + \frac{\eta}{s^2}$ with η a damping parameter, $\omega_1 = \frac{T_s(\omega_0)}{T_s'(\omega_0)}$ with $(T_i(x))_{i\geq 0}$ the orthogonal Chebyshev polynomials, $\Delta W_{n+1,r} \sim \mathscr{N}(0, h)$ and $X_{n+1} = K_s$. Here the first $s - 1$ stages represent the stabilization procedure and the last stage a finishing procedure to achieve strong order 1/2 and weak order 1 [2, 3]. We will also consider the S-ROCK2 method introduced in [5].

[1]Here $C_P^{2(\gamma+1)}$ denotes the space of $2(\gamma + 1)$ times continuously differentiable functions with all partial derivatives bounded by a term of order $1 + |x|^{2u}$ with $u \in \mathbb{N}$ (polynomial growth).

Similar to S-ROCK1 this scheme uses a stabilization procedure (in this case ROCK2 [4]) on the first $s-2$ stages and then a finishing procedure on the last two stages to obtain a weak order of 2 and a strong order of $1/2$. While remaining explicit the two S-ROCK methods have an extended stability domain, which can be characterized as follows. Let $\mathcal{S}_{SDE,a} = \{(p,q) \in [-a,0] \times \mathbb{R} \mid |q| \leq \sqrt{-2p}\}$ be a "portion" of $\mathcal{S}_{exact}$ and $a^* = \sup\{a > 0 \mid \mathcal{S}_{SDE,a} \subset \mathcal{S}_{num}\}$ with $\mathcal{S}_{num}$ denoting the stability domain of the numerical method. It can be shown that for S-ROCK1 and S-ROCK2 $a_s^* = c_{SR1}(s)s^2$ and $a_s^* = c_{SR2}(s)(s+2)^2$, respectively. As s increases the constants $c_{SR1}(s)$ and $c_{SR2}(s)$ quickly reach a value independent of the stage number that can be estimated numerically as $c_{SR1} = 0.33$ (S-ROCK1) and $c_{SR2} = 0.42$ (S-ROCK2) [2,3,5].

2 Multilevel Monte Carlo Method for Stiff SDEs

We are interested in estimating $E := \mathbb{E}[\phi(X(T))]$, the expectation of some Lipschitz continuous functional $\phi : \mathbb{R}^d \to \mathbb{R}$ depending on the stochastic process $(X(t))_{t\in[0,T]}$ specified through (1). The classical Monte Carlo method uses Euler-Maruyama [10] as numerical integrator with a fixed time stepsize to approximate the stochastic process and sample averages to estimate the expectation. This approach is easy to implement, but computationally expensive. In fact defining the computational cost (or complexity) by the number of function evaluations, one can show that to achieve a mean square accuracy of $\mathcal{O}(\varepsilon^2)$ a computational cost of $\mathcal{O}(\varepsilon^{-3})$ is required (with $\varepsilon > 0$) (see e.g. [8]).

One way to improve the performance of standard Monte Carlo techniques is to use the multilevel Monte Carlo method, which is based on hierarchical sampling. In this approach Monte Carlo is applied to a sequence of nested time stepsizes. Simultaneously the number of samples is balanced according to the stepsize. Combining many samples of computationally cheap approximations (the ones using a large stepsize) with a few samples of computationally expensive approximations (the ones based on a fine time grid) reduces the cost significantly to $\mathcal{O}\left(\varepsilon^{-2}(\log(\varepsilon))^2\right)$ while maintaining the same mean square accuracy of $\mathcal{O}(\varepsilon^2)$ [7].

For stiff problems, using the standard MLMC approach with EM, there is some time stepsize restriction due to stability issues. Let $k \geq 2$ be some positive integer indicating the refinement factor. We consider the time stepsizes

$$h_\ell = \frac{T}{M_\ell}, \quad \ell = 0, 1, \ldots, L, \tag{3}$$

with L the total number of levels and $M_\ell = k^\ell$ denoting the number of time steps at level ℓ. Suppose there is some stability constraint given by $k^{-\ell_{EM}}\rho \leq 1$ with ℓ_{EM} corresponding to the largest possible stepsize such that the EM scheme is mean square stable and ρ a given stiffness parameter. Furthermore suppose that a root mean square accuracy of $\varepsilon = k^{-L}$ is desired. One has to distinguish between two cases:

1. $\ell_{EM} > L$.
   MLMC cannot be applied because none of the levels up to level L satisfies the stability constraint. Though one can use classical Monte Carlo with $h = T/k^{\ell_{EM}}$ which results in a computational cost of $\mathcal{O}\left(\varepsilon_{MC}^{-3}\right)$ and a precision $\mathcal{O}\left(\varepsilon_{MC}^{2}\right)$, where $\varepsilon_{MC} = k^{-\ell_{EM}}$.
2. $0 < \ell_{EM} \leq L$.
   MLMC can only be applied to the levels $\ell_{EM}, \ell_{EM}+1, \ldots, L$. The resulting computational complexity is $\mathcal{O}\left(\varepsilon^{-2}\left((\log(\varepsilon))^2 + \varepsilon^{-\ell_{EM}/L}\right)\right)$.

In [1] a stabilized multilevel Monte Carlo method, whose estimator we denote by $\hat{E}$, is introduced which uses as numerical integrator S-ROCK1 (2), an explicit Runge-Kutta method based on orthogonal Chebyshev polynomials. The stability constraint of this scheme is given by $\frac{k^{-\ell}\rho}{c_{SR1}s_\ell^2} \leq 1$, where s_ℓ is the number of stages at level ℓ and c_{SR1} a positive constant. Due to the extended stability domain of S-ROCK1 all levels are accessible in the stabilized MLMC approach. A computational cost of $\mathcal{O}\left(\varepsilon^{-2}(\log(\varepsilon))^2\left(1+\frac{\sqrt{\rho}}{|\log(\varepsilon)|}\right)\right)$ is necessary to attain a mean square precision of $\mathcal{O}\left(\varepsilon^2\right)$. For stiff problems as well as for nonstiff problems with no small noise, this means a significant improvement over the standard MLMC approach (which uses EM) as it is shown in [1].

3 Improved Stabilized Multilevel Monte Carlo Method for Stiff SDEs

In this section we describe how the stabilized multilevel Monte Carlo method can further be improved. As mentioned in the introduction, the EM method as well as the S-ROCK1 method are both of weak order 1 and strong order 1/2. The idea is to use a numerical integrator of higher weak order for the finest time grid (see [6]), in our case S-ROCK2 [5] with weak order 2, which leads to a reduction of the bias. In fact due to the telescopic sum representation of the multilevel estimator, only the estimator based on the smallest time stepsize (which uses S-ROCK2) appears in the bias. A smaller bias yields a reduction of the total number of levels, and thus a reduced computational cost, without decreasing the accuracy. Note that in the following we focus on problems that are either stiff or nonstiff but with significant noise. Problems with no stability issues can be treated in a similar way.

Recall the sequence of nested stepsizes (3). For $\ell = 0, 1, \ldots, L-1$ we denote by ϕ_ℓ the approximation of $\phi(X(T))$ using S-ROCK1 with time stepsize h_ℓ. The approximation of $\phi(X(T))$ using S-ROCK2 on the finest time grid which is based on h_L is indicated by ϕ_L. The improved stabilized multilevel Monte Carlo estimator is defined by

$$\tilde{E} := \sum_{\ell=0}^{L} \frac{1}{N_\ell} \sum_{i=1}^{N_\ell} \left(\phi_\ell^{(i)} - \phi_{\ell-1}^{(i)}\right) \quad \text{with } \phi_{-1} \equiv 0, \tag{4}$$

a sum of sample averages over N_ℓ independent and identically distributed samples. Note that $\phi_\ell^{(i)}$ and $\phi_{\ell-1}^{(i)}$ are based on the same Wiener path. The accuracy of the estimator $\tilde{E}$ can be measured, e.g., by the mean square error (see e.g. [8]), which can be split into bias and variance as follows:

$$\text{MSE}\left(\tilde{E}\right) = \mathbb{E}\left[\left(\tilde{E} - E\right)^2\right] = \text{Var}\left(\tilde{E}\right) + \left(\text{bias}\left(\tilde{E}\right)\right)^2.$$

Using the properties of the expectation we obtain

$$\mathbb{E}\left[\tilde{E}\right] = \sum_{\ell=0}^{L} \left(\mathbb{E}\left[\phi_\ell\right] - \mathbb{E}\left[\phi_{\ell-1}\right]\right) = \mathbb{E}\left[\sum_{\ell=0}^{L} \left(\phi_\ell - \phi_{\ell-1}\right)\right] = \mathbb{E}\left[\phi_L\right].$$

Hence the bias satisfies

$$\text{bias}\left(\tilde{E}\right) = \mathbb{E}\left[\tilde{E}\right] - E = \mathbb{E}\left[\phi_L\right] - E = \mathcal{O}\left(k^{-2L}\right) \tag{5}$$

since the S-ROCK2 method, on which ϕ_L is based, is of weak order 2. Furthermore, for the variance we use the Cauchy-Schwarz inequality to obtain

$$\text{Var}\left(\phi_\ell - \phi_{\ell-1}\right) \leq \left(\text{Var}\left(\phi_\ell - E\right)^{1/2} + \text{Var}\left(\phi_{\ell-1} - E\right)^{1/2}\right)^2.$$

Both numerical integrators, S-ROCK1 and S-ROCK2, are of strong order $1/2$ and ϕ is Lipschitz continuous by assumption. Thus

$$\text{Var}\left(\phi_\ell - E\right) \leq \mathbb{E}\left[\left(\phi_\ell - E\right)^2\right] \leq \mathbb{E}\left[\left(\phi\left(X_{M_\ell}\right) - \phi\left(X(T)\right)\right)^2\right] \leq Ck^{-\ell}$$

and therefore

$$\text{Var}\left(\tilde{E}\right) = \sum_{\ell=0}^{L} \frac{\text{Var}\left(\phi_\ell - \phi_{\ell-1}\right)}{N_\ell} \leq C\left(\sum_{\ell=0}^{L-1} \frac{k^{-\ell}}{N_\ell} + \frac{k^{-L}}{N_L}\right), \tag{6}$$

where C is a positive constant.

Assume now a mean square precision of $\mathrm{MSE}\left(\tilde{E}\right) = \mathcal{O}\left(\varepsilon^2\right)$ is desired for some $\varepsilon > 0$. Considering (5) we obtain a total number of levels $L = -\frac{1}{2}\frac{\log(\varepsilon)}{\log(k)}$ (or equivalently $\varepsilon = k^{-2L}$). Inspired by (6) the number of simulations per level ℓ is set to $N_\ell = k^{-\ell}k^{4L}(L-1)$ for $\ell = 0, 1, \ldots, L-1$ and $N_L = k^{-L}k^{4L}$, which yields $\mathrm{Var}\left(\tilde{E}\right) \leq Ck^{-4L}\left(2 + \frac{1}{L-1}\right) = \mathcal{O}\left(\varepsilon^2\right)$.

As mentioned above the stability constraint of S-ROCK1 is given by $\frac{k^{-\ell}\rho}{c_{SR1}s_\ell^2} \leq 1$. In a similar way one can define a stability criterion for S-ROCK2 $\frac{k^{-L}\rho}{c_{SR2}(s_L+2)^2} \leq 1$ with $s_L \geq 2$ and with c_{SR2} as defined above.

Theorem 1 *Let $\tilde{E}$ be the improved stabilized MLMC estimator introduced in* (4). *For a desired mean square accuracy of $\mathrm{MSE}\left(\tilde{E}\right) = \mathcal{O}\left(\varepsilon^2\right)$ the computational cost of $\tilde{E}$ is given by*

$$\mathit{Cost}\left(\tilde{E}\right) = \frac{1}{4}\varepsilon^{-2}\left(\frac{\log(\varepsilon)}{\log(k)}\right)^2\tilde{\alpha},$$

where $\tilde{\alpha} = \left(m\frac{L-1}{L} + \frac{1}{L}\left(\frac{\sqrt{k}}{\sqrt{k}-1}\right)\sqrt{\frac{\rho}{c_{SR1}}}\right) - \left(d_1\frac{\varepsilon^{1/4}\sqrt{\rho}}{L} + d_2\frac{(\sqrt{\rho}-d_3)}{L^2}\right)$ *with d_1, d_2, d_3 some positive constants.*

Proof For the computational cost of $\tilde{E}$ we obtain $\mathrm{Cost}\left(\tilde{E}\right)$

$$\begin{aligned}
&= \sum_{\ell=0}^{L-1} N_\ell M_\ell\left(s_\ell + m\right) + N_L M_L\left(s_L + 8 + 2m\right)\\
&= \sum_{\ell=0}^{L-1} k^{4L}(L-1)\left(\sqrt{\frac{\rho}{c_{SR1}}}k^{-\ell/2} + m\right) + k^{4L}\left(\sqrt{\frac{\rho}{c_{SR2}}}k^{-L/2} + 6 + 2m\right)\\
&= k^{4L}(L-1)\left(\sqrt{\frac{\rho}{c_{SR1}}}\frac{\sqrt{k}-k^{-L/2+1/2}}{\sqrt{k}-1} + mL\right) + k^{4L}\left(\sqrt{\frac{\rho}{c_{SR2}}}k^{-L/2} + 6 + 2m\right).
\end{aligned}$$

Using $\varepsilon = k^{-2L}$ and rearranging terms yields $\mathrm{Cost}\left(\tilde{E}\right) = \frac{1}{4}\varepsilon^{-2}\left(\frac{\log(\varepsilon)}{\log(k)}\right)^2\tilde{\alpha}$ with $\tilde{\alpha}$ as defined above. □

In comparison, for a same mean square accuracy, the cost of the stabilized MLMC estimator $\hat{E}$ of [1] is given by $\mathrm{Cost}\left(\hat{E}\right) = \varepsilon^{-2}\left(\frac{\log(\varepsilon)}{\log(k)}\right)^2\hat{\alpha}$, where $\hat{\alpha} = \left(m\frac{L+1/2}{L} + \frac{1}{2L}\left(\frac{\sqrt{k}}{\sqrt{k}-1}\right)\sqrt{\frac{\rho}{c_{SR1}}}\right) - d_4\frac{\varepsilon^{1/2}\sqrt{\rho}}{L}$ with d_4 a positive constant.

Asymptotically we observe that for both estimators

$$\mathrm{Cost}\left(\tilde{E}\right) = \mathrm{Cost}\left(\hat{E}\right) = \mathcal{O}\left(\varepsilon^{-2}\left(\log(\varepsilon)\right)^2\left(1 + \frac{\sqrt{\rho}}{|\log(\varepsilon)|}\right)\right),$$

however with a smaller constant prefactor for $\tilde{E}$ allowing for a cost reduction by a factor roughly between 0.25 (nonstiff problems but significant noise) and 0.5 (stiff problems). This can be seen by comparing $\tilde{\alpha}$ and $\hat{\alpha}$.

4 Numerical Experiments

In this section we investigate a two-dimensional nonlinear noncommutative SDE inspired by the one-dimensional population dynamics model (see [11])

$$\mathrm{d}\begin{pmatrix} X_1(t) \\ X_2(t) \end{pmatrix} = \begin{pmatrix} \alpha a_2(t) - \lambda_1 b_1(t) \\ -\lambda_2 b_2(t) \end{pmatrix} \mathrm{d}t + \begin{pmatrix} -\mu_1 b_1(t) & \mu_2 a_1(t) \\ -\mu_2 b_2(t) & 0 \end{pmatrix} \begin{pmatrix} \mathrm{d}W_1(t) \\ \mathrm{d}W_2(t) \end{pmatrix}$$

for $0 \leq t \leq 1$, where $a_i(t) = X_i(t) - 1$ and $b_i(t) = X_i(t)(1 - X_i(t))$ for $i \in \{1, 2\}$. The initial condition is given by $(X_1(0), X_2(0)) = (0.95, 0.95)$ and $(W_1(t))_{t\in[0,1]}$ and $(W_2(t))_{t\in[0,1]}$ are two independent Wiener processes. We consider two different scenarios. First a stiff problem with drift term $\lambda_1 \in \{-1, -100, -10{,}000\}$ and noise term $\mu_1 = \sqrt{|\lambda_1|}$. And then a nonstiff problem with no small noise by fixing $\lambda_1 = -1$ and varying $\mu_1 = \sqrt{-2\lambda_1 - \delta}$ with $\delta \in \{10^{-1}, 10^{-2}, 10^{-4}\}$. In addition we pick $\alpha = 2$, $\lambda_2 = -1$, $\mu_2 = 0.5$, $k = 2$. As root mean square accuracy we choose k^{-2L} with $L \in \{1, 2, \ldots, 5\}$. Stability is guaranteed by assessing the second moment at the time end point. In Fig. 1 we compare the number of function evaluations (by counting the drift and diffusion evaluations) of the improved stabilized (using S-ROCK1 and S-ROCK2), the stabilized (using S-ROCK1) and the standard (using EM) MLMC method. As expected the improved stabilized approach yields a cost reduction over the other two methods (see also Table 1).

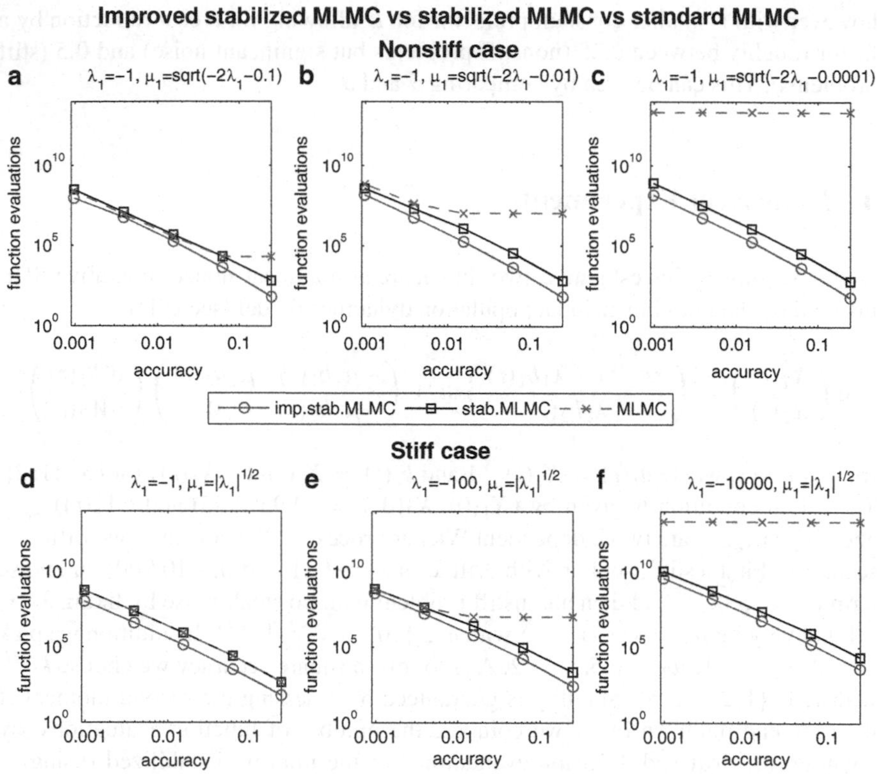

Fig. 1 Function evaluations against root mean square accuracy comparing the improved stabilized MLMC method using S-ROCK1 and S-ROCK2 with the stabilized (S-ROCK1) and the standard (EM) MLMC method

Table 1 Number of function evaluations of the improved stabilized MLMC (using S-ROCK1 and S-ROCK2), the stabilized MLMC (using S-ROCK1) and standard MLMC (using EM) for different values of the root mean square error. As parameters we take $\lambda_1 = -1, \mu_1 = \sqrt{-2\lambda_1 - 0.01}$ (b) and $\lambda_1 = -100, \mu_1 = \sqrt{|\lambda_1|}$ (e)

	Precision	2^{-2}	2^{-4}	2^{-6}	2^{-8}	2^{-10}
(b)	imp.stab.MLMC	64	4,352	184,320	5.70×10^6	14.99×10^7
	stab.MLMC	672	35,840	1,204,224	19.92×10^6	37.12×10^7
	MLMC	10.49×10^6	10.49×10^6	10.49×10^6	42.27×10^6	70.25×10^7
(e)	imp.stab.MLMC	256	16,896	614,400	16.91×10^6	39.53×10^7
	stab.MLMC	2,272	95,232	2,629,632	42.73×10^6	73.61×10^7
	MLMC	10.49×10^6	10.49×10^6	10.49×10^6	42.27×10^6	70.25×10^7

Acknowledgements This work is partially supported by the Swiss National Foundation Grant 200021_140692.

References

1. A. Abdulle, A. Blumenthal, Stabilized multilevel Monte Carlo method for stiff stochastic differential equations. J. Comput. Phys. **251**, 445–460 (2013)
2. A. Abdulle, S. Cirilli, Stabilized methods for stiff stochastic systems. C. R. Math. Acad. Sci. Paris **345**(10), 593–598 (2007)
3. A. Abdulle, T. Li, S-ROCK methods for stiff Itô SDEs. Commun. Math. Sci. **6**(4), 845–868 (2008)
4. A. Abdulle, A. Medovikov, Second order Chebyshev methods based on orthogonal polynomials. Numer. Math. **90**(1), 1–18 (2001)
5. A. Abdulle, G. Vilmart, K. Zygalakis, Weak second order explicit stabilized methods for stiff stochastic differential equations. SIAM J. Sci. Comput. **35**(4), A1792–A1814 (2013)
6. K. Debrabant, A. Rößler, On the acceleration of the multi–level Monte Carlo method. ArXiv preprint 1301.7650 (2013)
7. M. Giles, Multilevel Monte Carlo path simulation. Oper. Res. **56**(3), 607–617 (2008)
8. P. Glasserman, *Monte Carlo Methods in Financial Engineering*. Applications of Mathematics, vol. 53 (Springer, New York, 2004)
9. P. Kloeden, E. Platen, *Numerical Solution of Stochastic Differential Equations* (Springer, Berlin/New York, 1992)
10. G. Maruyama, Continuous Markov processes and stochastic equations. Rend. Circ. Mat. Palermo **4**, 48–90 (1955)
11. J. Murray, *Mathematical Biology I: An Introduction* (Springer, Seattle, 2002)
12. H.C. Öttinger, *Stochastic Processes in Polymeric Fluids* (Springer, Berlin, 1996). Tools and examples for developing simulation algorithms
13. P. Rué, J. Villà-Freixa, K. Burrage, Simulation methods with extended stability for stiff biochemical kinetics. BMC Syst. Biol. **4**(110), 1–13 (2010)

Adaptive Polynomial Approximation by Means of Random Discrete Least Squares

Giovanni Migliorati

Abstract We address adaptive multivariate polynomial approximation by means of the discrete least-squares method with random evaluations, to approximate in the L^2 probability sense a smooth function depending on a random variable distributed according to a given probability density. The polynomial least-squares approximation is computed using random noiseless pointwise evaluations of the target function. Here noiseless means that the pointwise evaluation of the function is not polluted by the presence of noise. Recent works Migliorati et al. (Found Comput Math 14:419–456, 2014), Cohen et al. (Found Comput Math 13:819–834, 2013), and Chkifa et al. (Discrete least squares polynomial approximation with random evaluations – application to parametric and stochastic elliptic PDEs, EPFL MATHICSE report 35/2013, submitted) have analyzed the univariate and multivariate cases, providing error estimates for (a priori) given sequences of polynomial spaces. In the present work, we apply the results developed in the aforementioned analyses to devise adaptive least-squares polynomial approximations. We build a sequence of quasi-optimal best n-term sets to approximate multivariate functions that feature strong anisotropy in moderately high dimensions. The adaptive approximation relies on a greedy selection of basis functions, which preserves the downward closedness property of the polynomial approximation space. Numerical results show that the adaptive approximation is able to catch effectively the anisotropy in the function.

1 Random Discrete Least Squares

The approximation of multivariate functions depending on several random variables is a challenging task. Different approaches have been proposed, such as Monte Carlo and quasi-Monte Carlo methods or stochastic collocation on sparse grids. Recent works have proven that the univariate polynomial approximation based on discrete least squares with noiseless pointwise evaluations in random uniformly distributed points is stable and optimally convergent in expectation [2] and in probability [5],

G. Migliorati (✉)
MATHICSE-CSQI, École Polytechnique Fédérale de Lausanne, Lausanne CH-1015, Switzerland
e-mail: giovanni.migliorati@epfl.ch

A. Abdulle et al. (eds.), *Numerical Mathematics and Advanced Applications ENUMATH 2013*, Lecture Notes in Computational Science and Engineering 103,
DOI 10.1007/978-3-319-10705-9_54

under the condition that the number of evaluations is proportional to the square of the dimension of the polynomial space. The analysis has been extended to the multivariate case in [1]: in any dimension, it has been proven that discrete least squares on polynomial spaces with random evaluations are stable and optimally convergent, again under the condition that the number of evaluations is proportional to the square of the dimension of the polynomial space, irrespectively of the "shape" of the polynomial space as long as it remains downward closed. Numerical observations have shown that a quasi-optimal convergence can still be achieved in case of a linear proportionality [1,5,6]. In this section we recall the construction of the random discrete least-squares approximation and the main results achieved in [1,2,5].

Let $d \in \mathbb{N}$ and $\Gamma := \prod_{k=1}^{d} \Gamma_k \subseteq \mathbb{R}^d$ denote the d-dimensional parameter set. For each $k = 1, \dots, d$ we define the probability density $\rho_k : \Gamma_k \to \mathbb{R}$ and the family of univariate polynomials $\{\varphi_i^k\}_{i \geq 0}$ orthonormal w.r.t. the corresponding density ρ_k, i.e. $\int_{\Gamma_k} \varphi_i^k(y)\varphi_j^k(y)\rho_k(y)dy = \delta_{ij}$. We will confine to the choice $\Gamma_k = [-1, 1]$ for all $k = 1, \dots, d$. Assume that the probability density $\rho : \Gamma \to \mathbb{R}$ has a product form $\rho = \prod_{i=1}^{d} \rho_i$. Given a finite multi-index set $\Lambda \subset \mathbb{N}_0^d$, for each $\boldsymbol{\nu} \in \Lambda$ we define the corresponding multivariate polynomial basis function

$$\psi_{\boldsymbol{\nu}}(\boldsymbol{y}) = \prod_{k=1}^{d} \varphi_{\nu_k}^k(y_k), \quad \boldsymbol{y} \in \Gamma. \tag{1}$$

The polynomial space $\mathbb{P}_\Lambda(\Gamma)$ associated with the multi-index set Λ is defined as $\mathbb{P}_\Lambda(\Gamma) := \text{span}\{\psi_{\boldsymbol{\nu}} : \boldsymbol{\nu} \in \Lambda\}$, and of course $\dim(\mathbb{P}_\Lambda) = \#(\Lambda)$. We denote by $\boldsymbol{Y}$ a d-dimensional random variable distributed according to the density ρ, and by $\phi = \phi(\boldsymbol{Y}) : \Gamma \to \mathbb{R}$ a smooth function (at least continuous) depending on $\boldsymbol{Y}$. Given m independent and identically distributed random variables $\boldsymbol{Y}_1, \dots, \boldsymbol{Y}_m \overset{\text{i.i.d.}}{\sim} \rho$ we introduce the following L^2_ρ inner product and its discrete counterpart

$$\langle u, v \rangle_{L^2_\rho} = \int_\Gamma u(\boldsymbol{y})v(\boldsymbol{y})\rho(\boldsymbol{y})d\boldsymbol{y}, \qquad \langle u, v \rangle_m = \frac{1}{m}\sum_{j=1}^{m} u(\boldsymbol{Y}_j)v(\boldsymbol{Y}_j), \tag{2}$$

which induce on $\mathbb{P}_\Lambda(\Gamma)$ the corresponding norm $\|\cdot\|_{L^2_\rho} := \langle \cdot, \cdot \rangle_{L^2_\rho}^{1/2}$ and the seminorm $\|\cdot\|_m := \langle \cdot, \cdot \rangle_m^{1/2}$. We focus on multi-index sets Λ featuring the following property.

Definition 1 The finite multi-index set $\Lambda \subset \mathbb{N}_0^d$ is downward closed (or it is a lower set) if whenever $\boldsymbol{\nu} \in \Lambda$ and $\nu'_j \leq \nu_j \ \forall j = 1, \dots, d$ then $\boldsymbol{\nu}' \in \Lambda$.

We define the discrete least-squares projection $\Pi_\Lambda^m \phi$ of the function ϕ over the space $\mathbb{P}_\Lambda(\Gamma)$ as

$$\Pi_\Lambda^m \phi := \underset{v \in \mathbb{P}_\Lambda(\Gamma)}{\text{argmin}} \|\phi - v\|_m^2. \tag{3}$$

The discrete projection (3) approximates the L^2 continuous projection

$$\Pi_\Lambda \phi := \underset{v\in\mathbb{P}_\Lambda(\Gamma)}{\operatorname{argmin}} \|\phi - v\|^2_{L^2_\rho}, \tag{4}$$

which in general cannot be exactly computed. We denote by $\{\beta^m_\nu\}_{\nu\in\Lambda}$ and $\{\beta_\nu\}_{\nu\in\Lambda}$, respectively, the coefficients in the expansion of the two projections over the polynomial space:

$$\Pi^m_\Lambda \phi = \sum_{\nu\in\Lambda} \beta^m_\nu \psi_\nu \quad \text{and} \quad \Pi_\Lambda \phi = \sum_{\nu\in\Lambda} \beta_\nu \psi_\nu. \tag{5}$$

Moreover, we introduce the following quantity $K(\Lambda) := \sup_{y\in\Gamma} \sum_{\nu\in\Lambda} |\psi_\nu(y)|^2$, which depends only on Λ and ρ, see [1]. For a given $\tau > 0$, we assume that the target function satisfies a uniform bound $|\phi(y)| \leq \tau$ for any $y \in \Gamma$. In addition, we introduce the truncation operator $T_\tau(t) := \operatorname{sign}(t)\min\{\tau, |t|\}$ and define the truncated discrete least-squares projector $\tilde{\Pi}^m_\Lambda := T_\tau \circ \Pi^m_\Lambda$. We recall from [1] the following result, specifically targeted to the case of polynomial approximation.

Theorem 1 (From [1]) *For any $\gamma > 0$, if m is such that $K(\Lambda)$ satisfies*

$$K(\Lambda) \leq \frac{0.15}{(1+\gamma)} \frac{m}{\ln m}, \tag{6}$$

then, for any $\phi \in L^\infty(\Gamma)$ with $\|\phi\|_{L^\infty} \leq \tau$, the following estimates hold

$$\mathbb{E}\left(\|\phi - \tilde{\Pi}^m_\Lambda \phi\|^2_{L^2_\rho} \right) \leq \left(1 + \frac{0.6}{(1+\gamma)\ln m} \right) \|\phi - \Pi_\Lambda \phi\|^2_{L^2_\rho} + 8\tau^2 m^{-\gamma},$$

$$Pr\left(\|\phi - \Pi^m_\Lambda \phi\|^2_{L^2_\rho} \leq (1+\sqrt{2}) \inf_{v\in\mathbb{P}_\Lambda} \|\phi - v\|_{L^\infty} \right) \geq 1 - 2m^{-\gamma}.$$

Given a finite multi-index set Λ, the quantity $K(\Lambda)$ can be directly computed, to precisely quantify the value of m which satisfies condition (6). Nonetheless, the following upper bounds have been derived.

Lemma 1 (From [1]) *For any lower set Λ, the quantity $K(\Lambda)$ satisfies*

$$K(\Lambda) \leq (\#(\Lambda))^2, \quad \textit{with tensorized Legendre polynomials}, \tag{7}$$

$$K(\Lambda) \leq (\#(\Lambda))^{\ln 3/\ln 2}, \quad \textit{with tensorized Chebyshev 1st kind polynomials}. \tag{8}$$

These bounds are sharp for the Tensor Product space, and their accuracy is discussed in more detail in [1]. More general bounds with tensorized Jacobi polynomials have been proven in [4]. The case of Hermite polynomials has been analyzed in [3, chap. 3].

2 Adaptive Random Discrete Least Squares

In general the knowledge of the best n-term sets, i.e. the sets of the n largest coefficients, is not available when approximating a given anisotropic function ϕ. The aim of the adaptive approximation approach consists in building a sequence $\{\Lambda_k\}_{k\geq 0}$ of lower multi-index sets, with $\Lambda_0 = \{\mathbf{0}\}$ and $\Lambda_k \subset \Lambda_{k+1}$ for any $k \geq 0$. The sequence is adaptively computed with random discrete least squares driven by a greedy selection. The following objects are useful to describe the adaptive algorithm.

Definition 2 The margin $\mathscr{M}(\Lambda)$ of a lower multi-index set Λ is

$$\mathscr{M}(\Lambda) := \{\boldsymbol{\nu} \in \mathbb{N}_0^d \ : \ \boldsymbol{\nu} \notin \Lambda \wedge \exists j > 0 \ : \ \boldsymbol{\nu} - \boldsymbol{e}_j \in \Lambda\}.$$

Definition 3 The reduced margin $\mathscr{R}(\Lambda)$ of a lower multi-index set Λ is

$$\mathscr{R}(\Lambda) := \{\boldsymbol{\nu} \in \mathbb{N}_0^d \ : \ \boldsymbol{\nu} \notin \Lambda \wedge \forall j = 1, \ldots, d \ : \ \boldsymbol{\nu}_j \neq 0 \Rightarrow \boldsymbol{\nu} - \boldsymbol{e}_j \in \Lambda\} \subseteq \mathscr{M}(\Lambda).$$

We devise an adaptive algorithm that computes at each iteration k the corresponding multi-index set Λ_k. At the kth iteration of the algorithm the set Λ_{k-1} is given, and we have to select the new multi-indices to add and form Λ_k. The new multi-indices are picked among the elements of $\mathscr{R}(\Lambda_{k-1})$, because this preserves the property of downward closedness of the set Λ_k at any iteration $k \geq 0$.

Since at every iteration of the adaptive algorithm we have to solve a least-squares problem, it can be beneficial from a computational standpoint to select a fraction of the multi-indices in the reduced margin rather than to pick them one at a time. To perform this multiple selection, we exploit the idea of the Dörfler marking, which has been originally proposed in the context of Adaptive Finite Elements. Given a multi-index set Λ, a subset $R \subseteq \mathscr{R}(\Lambda)$ of its reduced margin, a nonnegative function $e : R \to \mathbb{R}$ and a parameter $\theta \in (0, 1]$, we define the procedure Dörfler := Dörfler(R, e, θ) that computes a set $F \subseteq R \subseteq \mathscr{R}(\Lambda)$ of minimal positive cardinality such that

$$\sum_{\boldsymbol{\nu} \in F} e(\boldsymbol{\nu}) \geq \theta \sum_{\boldsymbol{\nu} \in R} e(\boldsymbol{\nu}). \tag{9}$$

For any $\boldsymbol{\nu} \in R$, the function $e(\boldsymbol{\nu})$ estimates the absolute value of the coefficient $\beta_{\boldsymbol{\nu}}$ appearing in the expansion (5) of $\Pi_\Lambda \phi$, i.e. $e(\boldsymbol{\nu}) \approx |\beta_{\boldsymbol{\nu}}|$. Therefore, the selection (9) corresponds to choose a fraction θ of the sum of absolute values associated with the (estimates of the) coefficients in the set R. Using an estimator $\tilde{e}$ of the square of the coefficients, i.e. $\tilde{e}(\boldsymbol{\nu}) \approx (\beta_{\boldsymbol{\nu}})^2$, the selection (9) with e replaced by $\tilde{e}$ would catch a fraction θ of the energy associated with the coefficients in R, but we will stick to the case $e(\boldsymbol{\nu}) \approx |\beta_{\boldsymbol{\nu}}|$. A natural choice of the estimator function e is the use of the estimates $\{\beta_{\boldsymbol{\nu}}^m\}_{\boldsymbol{\nu} \in \Lambda}$ of the coefficients coming from the discrete least-squares projection (3) at the kth iteration. Another further selection can be made, at each

Algorithm 2.1: Adaptive polynomial random discrete least-squares approximation

Set $r_0 = \phi(Y)$, $\Lambda_0 = \{\mathbf{0}\}$ and choose $k_{\mathrm{sg}}, k_{\max}$ such that $1 \leq k_{\mathrm{sg}} \leq k_{\max}$.
for $k = 1$ to $k_{\max}$ **do**
  $F_1 \leftarrow \text{Dörfler}(\mathscr{R}(\Lambda_{k-1}), |\langle r_{k-1}, \psi_{\boldsymbol{\nu}} \rangle_m|, \theta_1)$
  $\widetilde{\Lambda}_k \leftarrow \Lambda_{k-1} \cup F_1$
  $\Pi^m_{\widetilde{\Lambda}_k} \phi = \sum_{\boldsymbol{\nu} \in \widetilde{\Lambda}_k} \beta^m_{\boldsymbol{\nu}} \psi_{\boldsymbol{\nu}} \leftarrow \operatorname{argmin}_{u \in \mathbb{P}_{\widetilde{\Lambda}_k}} \|\phi - u\|_m$
  $F_2 \leftarrow \text{Dörfler}(F_1, |\beta^m_{\boldsymbol{\nu}}|, \theta_2)$
  $\Lambda_k \leftarrow \Lambda_{k-1} \cup F_2$
  if $k \bmod k_{\mathrm{sg}} = 0$ **then**
    $\Lambda_k \leftarrow \Lambda_k \cup \boldsymbol{\mu}$, with $\boldsymbol{\mu}$ being the most ancient multi-index in $\{\mathscr{R}(\Lambda_{k-1}) \setminus F_2\}$
  end if
  $r_k \leftarrow \phi - \Pi^m_{\widetilde{\Lambda}_k} \phi|_{\Lambda_k}$
end for

iteration k, by using the estimator related to the correlations between the residual r_{k-1} at the $(k-1)$th iteration and the polynomial basis functions associated with the multi-indices in $\mathscr{R}(\Lambda_{k-1})$:

$$e(\boldsymbol{\nu}) = |\tilde{\beta}^{m,k}_{\boldsymbol{\nu}}| := |\langle r_{k-1}, \psi_{\boldsymbol{\nu}} \rangle_m|, \quad \text{for any } \boldsymbol{\nu} \in \mathscr{R}(\Lambda_{k-1}). \tag{10}$$

At each iteration the correlations are computed using the same pointwise evaluations employed to compute the discrete least-squares projection, and their calculations are cheap compared with the computational cost required to calculate the least-squares approximation.

By combining the Dörfler marking with the two aforementioned estimators, we propose the adaptive algorithm described in Algorithm 2.1. The first Dörfler marking with parameter θ_1 uses the estimator (10) based on the correlations, and returns a set $F_1 \subseteq \mathscr{R}(\Lambda_{k-1})$ which is added to Λ_{k-1} to form $\widetilde{\Lambda}_k := \Lambda_{k-1} \cup F_1$. Then the discrete least-squares projection is computed over the polynomial space associated with $\widetilde{\Lambda}_k$. Afterwards, a second selection based on the Dörfler marking with parameter θ_2 is performed, using the more accurate discrete least-squares estimator, namely $e(\boldsymbol{\nu}) = |\beta^m_{\boldsymbol{\nu}}|$ for any $\boldsymbol{\nu} \in F_1$. This selection identifies a set $F_2 \subseteq F_1$ to allow the final update $\Lambda_k = \Lambda_{k-1} \cup F_2$.

The choice of the values of the parameters θ_1 and θ_2 should somehow reflect the reliability of the two Dörfler selections. The purpose of the first selection is to perform a rough screening of $\mathscr{R}(\Lambda_{k-1})$ and discard the less-promising multi-indices that would unnecessarily overload the computational cost of the least-squares projection. Therefore it is reasonable to choose permissive values of $\theta_1 > \theta_2$, since the correlation estimation of the coefficients $\{\beta_{\boldsymbol{\nu}}\}_{\boldsymbol{\nu} \in \mathscr{R}(\Lambda_{k-1})}$ is less accurate than their estimation by means of the least-squares projection. The value of the parameter θ_2 should be carefully chosen depending on d, $\#(\Lambda_k)$ and any available a priori information on the speed of decay of the coefficients of the target function ϕ. One could set the parameters $\theta_1 = \theta_1(k)$ and $\theta_2 = \theta_2(k)$ depending on the iteration counter k as well.

A safeguard mechanism prevents the advancement of the algorithm from getting stuck into null coefficients: once every k_{sg} iterations the most ancient element in the set $\{\mathscr{R}(\Lambda_{k-1}) \setminus F_2\}$ is added to the current set Λ_k. If the number of evaluations m in the discrete projection satisfies (6) at each iteration, then the approximation error of Algorithm 2.1 asymptotically converges to zero, i.e. $\lim_{k\to+\infty} \|\phi - \Pi^m_{\Lambda_k}\phi\|_{L^2_\rho} = 0$. In general the precise rate of convergence depends on the parameters θ_1, θ_2, k_{sg} and on how m is updated.

The key point is how to choose the number of evaluations of the function ϕ depending on $\#(\Lambda_k)$ at each step k of the algorithm. On the one hand, from Theorem 1 we obtain the stability and optimal convergence of the algorithm at every iteration k, if the number of evaluations m in the discrete least-squares projection satisfies condition (6). As observed in [1, 5] the quadratic proportionality can be relaxed to a linear proportionality without a loss in accuracy, when the function is smooth and the dimension d is not small. This choice has been tested in many examples involving PDEs with stochastic data in [6].

On the other hand, it is clear that if we are given a prescribed number of evaluations, or we know that we will reach a given size of the dimension of the polynomial space, then we can employ all the available evaluations starting from the first iteration of the adaptive algorithm. We will focus on this case: we consider a fixed number m of evaluations (our available resource) and use all of them at each iteration k of the adaptive algorithm until the set Λ_k becomes too large and the stability constraint (6) is violated.

3 Numerical Results

In this section we present some numerical tests in moderately high dimension $d = 16$ to check the capabilities of the adaptive approximation method outlined in Algorithm 2.1. We consider the following meromorphic function

$$\phi(\mathbf{y}) = \frac{1}{1 + \boldsymbol{\gamma} \cdot \mathbf{y}}, \quad \mathbf{y} \in \Gamma = [-1, 1]^d. \tag{11}$$

The point $\mathbf{y}$ is a realization of the random variable $\mathbf{Y}$ distributed according to the uniform or arcsine density over Γ, i.e. $\rho = \mathbb{U}(\Gamma)$ or $\rho = \text{Beta}(1/2, 1/2)$ respectively. The vector $\boldsymbol{\gamma}$ contains d positive weights which govern the anisotropy, and is defined as $\boldsymbol{\gamma} := \hat{\boldsymbol{\gamma}}(2\|\hat{\boldsymbol{\gamma}}\|_1)^{-1}$, with $\hat{\boldsymbol{\gamma}} = (1, 5 \times 10^{-1}, 10^{-1}, 5 \times 10^{-2}, \ldots, 5 \times 10^{-8})$.

The size of the best n-term set that we wish to approximate is kept fixed and equal to 500. Therefore the maximal value k_{max} of the iteration counter k in Algorithm 2.1 is not prescribed in advance, but k_{max} is the smallest $k \geq 1$ such that $\#(\Lambda_k) \geq 500$, so that $\#(\Lambda_{k_{max}}) \approx 500$. As an example, Fig. 1 shows an instance of the quasi-optimal set $\Lambda_{k_{max}}$. For each one of the two densities we choose either the number of points prescribed by the theoretical analysis to preserve stability in

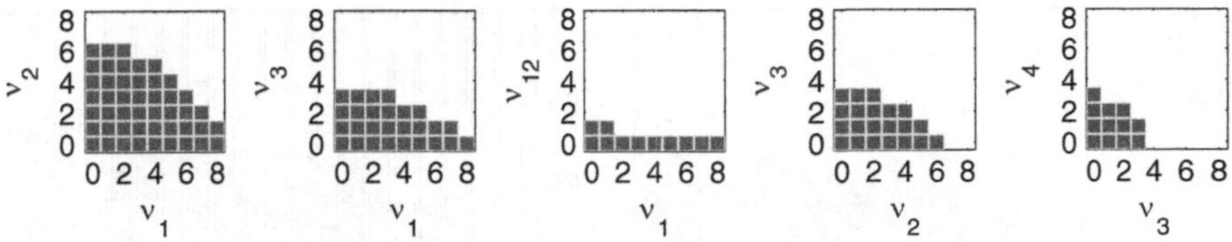

Fig. 1 Sections of the multi-index set Λ_k associated with the approximated best 500-term set of the function (11) with $\rho = \text{Beta}(1/2, 1/2)$. The set $\Lambda_{k_{\max}}$ has been computed by Algorithm 2.1, with $m = (\#(\Lambda))^{\ln 3/\ln 2}$, $\theta_1 = 0.5$ and $\theta_2 = 0.2$

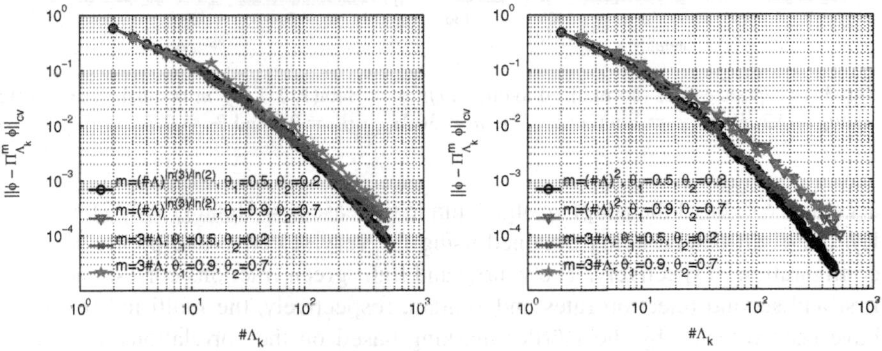

Fig. 2 Errors $\|\phi - \Pi^m_\Lambda \phi\|_{\text{cv}}$ with the function (11). *Left*: $\rho = \text{Beta}(1/2, 1/2)$ and Chebyshev polynomials of the 1st kind. *Right*: $\rho = \mathbb{U}(\Gamma)$ and Legendre polynomials

$\Lambda_{\max}$, namely $m = (\#(\Lambda_{k_{\max}}))^{\ln 3/\ln 2}$ with the Chebyshev polynomials of the 1st kind or $m = (\#(\Lambda_{k_{\max}}))^2$ with the Legendre polynomials, or a linear proportionality $m = 3\#(\Lambda_{k_{\max}})$. Moreover, we test two combinations of the values of the parameters θ_1 and θ_2, which make the algorithm very selective ($\theta_1 = 0.5$ and $\theta_2 = 0.2$) or poorly selective ($\theta_1 = 0.9$ and $\theta_2 = 0.7$). The parameter k_{sg} is set to a large value such that the safeguard mechanism is never activated, because it is not needed in this academic example with the function (11).

To estimate the approximation error we employ the cross-validation procedure described in [5], and denote by $\|\phi - \Pi^m_\Lambda \phi\|_{\text{cv}} := \|\phi - \Pi^m_\Lambda \phi\|_{L^\infty(\mathscr{C})}$ the cross-validated error estimated over a set $\mathscr{C}$ of 10^4 independent points distributed according to the underlying density ρ. Figure 2 shows the error obtained when approximating the function (11), in the two cases $\rho = \mathbb{U}(\Gamma)$ or $\rho = \text{Beta}(1/2, 1/2)$. Clearly, in the case $\rho = \text{Beta}(1/2, 1/2)$ the linear proportionality $m = 3\#(\Lambda_{k_{\max}})$ performs similarly to $m = (\#(\Lambda_{k_{\max}}))^{\ln 3/\ln 2}$, also for the poorly selective configuration of θ_1 and θ_2. In the case $\rho = \mathbb{U}(\Gamma)$ the difference between the linear proportionality $m = 3\#(\Lambda_{k_{\max}})$ and $m = (\#(\Lambda_{k_{\max}}))^2$ is more evident, but still provides a quasi-optimal convergence rate. In any case, the poorly selective configuration of the adaptive algorithm is able to catch the anisotropic decay of the coefficients, providing an approximation error which is very close to the error of the more selective (and more robust) configuration. Figure 3 shows the

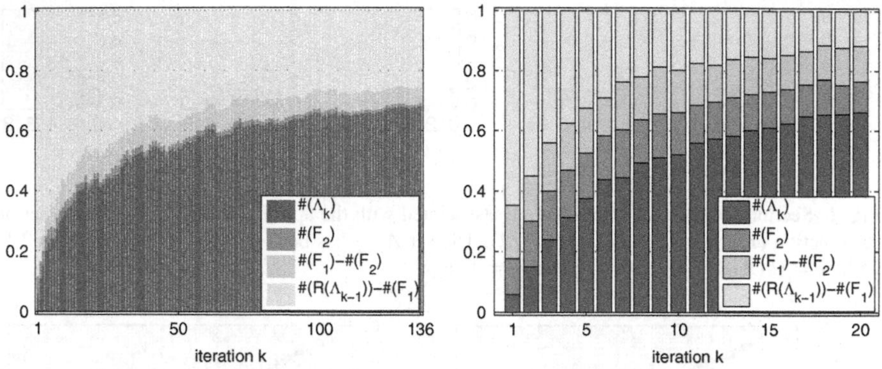

Fig. 3 Acceptance rates (normalized to one). *Left*: $\rho = \text{Beta}(1/2, 1/2)$, $m = 3\#\Lambda_{k_{\max}}$, $\theta_1 = 0.5$ and $\theta_2 = 0.2$. *Right*: $\rho = \text{Beta}(1/2, 1/2)$, $m = 3\#\Lambda_{k_{\max}}$, $\theta_1 = 0.9$ and $\theta_2 = 0.7$

acceptance rates of the adaptive algorithm in the case $\rho = \text{Beta}(1/2, 1/2)$, with the left and right results being obtained using the very selective and poorly selective configuration, respectively. The yellow and light green fractions correspond to the first and second rejection rates and contain, respectively, the multi-indices which have been rejected by the Dörfler marking based on the correlations and on the least-squares projection. The dark green fraction shows the growth of $\#(\Lambda_k)$ at each iteration k. The remaining green fraction is the acceptance rate of the algorithm, and corresponds to the multi-indices which have been selected at each iteration by the adaptive enrichment. Clearly the more selective the algorithm, the larger the total number of iterations.

References

1. A. Chkifa, A. Cohen, G. Migliorati, F. Nobile, R. Tempone, Discrete least squares polynomial approximation with random evaluations – application to parametric and stochastic elliptic PDEs, EPFL MATHICSE report 35/2013 (submitted)
2. A. Cohen, M. Davenport, D. Leviatan, On the stability and accuracy of least squares approximations. Found. Comput. Math. **13**, 819–834 (2013)
3. G.Migliorati, Polynomial approximation by means of the random discrete L^2 projection and application to inverse problems for PDEs with stochastic data. Ph.D. thesis, Dipartimento di Matematica "Francesco Brioschi", Politecnico di Milano, Milano, Italy, and Centre de Mathématiques Appliquées, École Polytechnique, Palaiseau, France, 2013
4. ———, Multivariate Markov-type and Nikolskii-type inequalities for polynomials associated with downward closed multi-index sets, to appear in J. Approx. Theory
5. G. Migliorati, F. Nobile, E. von Schwerin, R. Tempone, Analysis of the discrete L^2 projection on polynomial spaces with random evaluations. Found. Comput. Math. **14**, 419–456 (2014)
6. ——— Approximation of quantities of interest in stochastic PDEs by the random discrete L^2 projection on polynomial spaces. SIAM J. Sci. Comput. **35**, A1440–A1460 (2013)

Part IX
Solvers, High Performance Computing and Software Libraries

Schwarz Domain Decomposition Preconditioners for Plane Wave Discontinuous Galerkin Methods

Paola F. Antonietti, Ilaria Perugia, and Davide Zaliani

Abstract We construct Schwarz domain decomposition preconditioners for plane wave discontinuous Galerkin methods for Helmholtz boundary value problems. In particular, we consider additive and multiplicative non-overlapping Schwarz methods. Numerical tests show good performance of these preconditioners when solving the linear system of equations with GMRES.

1 Introduction

Over the last years, finite element methods based on non-polynomial shape functions for time harmonic wave propagation problems have become increasingly popular. The idea behind these methods is to incorporate information on the oscillatory behaviour of the solutions directly within the approximating spaces by using, instead of polynomial basis functions, Trefftz basis functions, namely, local solutions to the differential operator. For the Helmholtz equation these functions can be, for instance, plane waves or circular/shperical waves, with the same frequency as the original problem. Although these methods are not pollution-free, they can deliver more accurate results, for a given number of degrees of freedom, than standard polynomial finite element methods. The way of imposing continuity at interelement boundaries generates different Trefftz-type methods: ultra weak variational formulation/Trefftz-discontinuous Galerkin methods [8, 13, 14, 22, 23], partition of unity [6, 33], least squares methods [34] or Lagrange

P.F. Antonietti
Dipartimento di Matematica, Politecnico di Milano, Piazza Leonardo da Vinci 32, I-20133 Milano, Italy
e-mail: paola.antonietti@polimi.it

I. Perugia (✉)
Fakultät für Mathematik, Univeristät Wien, Oskar-Morgenstern-Platz 1, A-1090 Wien, Austria
e-mail: ilaria.perugia@univie.ac.at

D. Zaliani
Dipartimento di Matematica, Università di Pavia, Via Ferrata 1, I-27100 Pavia, Italy
e-mail: davidezal@yahoo.it

A. Abdulle et al. (eds.), *Numerical Mathematics and Advanced Applications ENUMATH 2013*, Lecture Notes in Computational Science and Engineering 103,
DOI 10.1007/978-3-319-10705-9_55

multiplier methods [1, 37]. These methods have been extended from acoustic to electromagnetic [12, 26, 27] and elastic [28, 32] time-harmonic wave propagation problems. Here, we consider plane wave discontinuous Galerkin (PWDG) methods, of which the ultra-weak variational formulation can be seen as a particular case, for the Helmholtz equation, and we attack the problem of preconditioning the arising algebraic linear systems by domain decomposition methods (for plane wave methods with Lagrange multipliers, domain decomposition preconditioners have been introduced in [20]).

The solution of algebraic linear systems arising from discretizations of the Helmholtz equation is a difficult problem [19] (see also the references therein for a bibliography on this topic). In case of Dirichlet or Neumann boundary conditions, classical discretization methods for the Helmholtz equation based on polynomial spaces result into real, symmetric and indefinite linear systems of equations that can be preconditioned, for example, by overlapping Schwarz, multigrid or substructuring type methods, see for example [9, 10, 18] and the references therein. A complete theory generalizing the classical Schwarz analysis for symmetric, positive definite problems to indefinite problems has been provided in [9, 10] where, exploiting GMRES converge bounds [16, 35], it is proven that GMRES converges uniformly (with respect to the meshsize and the number of subdomains) provided that (i) the subdomain and coarse partitions are sufficiently fine; (ii) the low-order term of the differential operator is a relatively compact perturbation of the second order term. For the Helmholtz problem, these methods are in general not scalable with respect to the wavenumber and become less and less effective as the wavenumber increases, unless a sufficient number of coarse points per wavelength is employed. Such a requirement may become unfeasible for practical applications. Nevertheless, they are currently employed for large scale computations, although a comprehensive and sharp theory is still missing (because it also relies on GMRES convergence bounds which are not sharp).

Using PWDG leads to a different situation: the resulting linear system of equations is complex (independently of the considered boundary conditions) and non-hermitian. Several strategies have already beed studied to cope with the severe ill-conditioning; see [7, 29, 31].

The aim of this contribution is to preliminarily explore the performance of a class of Schwarz methods to precondition the linear system of equations arising from PWDG approximation of the Helmholtz equation in a 2D cavity with impedance boundary condition. To keep as low as possible the computational effort without loosing effectiveness, we take advantage of the DG framework and consider the non-overlapping version of the classical Schwarz preconditioners. Indeed, according to [2–4, 21], non-overlapping preconditioners for DG methods converge as fast as overlapping solvers with minimal overlap for continuous discretizations. Our numerical experiments indicate that our preconditioners work well in reducing the computation effort in the solution of the resulting linear system of equations, and that the PWDG method seems to be particularly well suited for the development of solvers that are scalable with respect to the wavenumber.

2 The PWDG Method for the Helmholtz Problem

We consider the homogeneous Helmholtz problem in a bounded Lipschitz domain $\Omega \subset \mathbb{R}^2$, with impedance boundary condition along $\partial\Omega$. Given a wavenumber k (the corresponding wavelength is $\lambda = 2\pi/k$) such that $k \geq k_0 > 0$, the problem reads:

$$\begin{cases} -\Delta u - k^2 u = 0 & \text{in } \Omega, \\ \nabla u \cdot \mathbf{n} + iku = g & \text{on } \partial\Omega, \end{cases} \tag{1}$$

where i is the imaginary unit, $\mathbf{n}$ is the outer normal unit vector to $\partial\Omega$, and $g \in L^2(\partial\Omega)$ is given. The variational formulation of the problem reads as follows: find $u \in H^1(\Omega)$ such that, for all $v \in H^1(\Omega)$, it holds

$$\int_{\Omega} (\nabla u \cdot \nabla \overline{v} - k^2 u \overline{v})\, \mathrm{d}V + ik \int_{\partial\Omega} u\overline{v}\, dS = \int_{\partial\Omega} g\overline{v}\, \mathrm{d}V. \tag{2}$$

By Fredholm alternative, problem (2) is well posed, and stability estimates are given by [33, Proposition 8.1.4].

In order to derive the PWDG method, we consider a shape-regular, quasi-uniform family of finite element partitions $\{\mathcal{T}_h\}$ of Ω, possibly featuring hanging nodes. We assume, for simplicity, that the elements K of $\mathcal{T}_h$ are convex polygons. We write h for the mesh width of $\mathcal{T}_h$, i.e., $h = \max_{K\in\mathcal{T}_h} h_K$, with $h_K := \operatorname{diam}(K)$. We define the mesh skeleton $\mathcal{F}_h = \bigcup_{K\in\mathcal{T}_h} \partial K$, and set $\mathcal{F}_h^I = \mathcal{F}_h \setminus \partial\Omega$.

Given an element $K \in \mathcal{T}_h$, we denote by $PW_p(K)$ the plane wave space on K:

$$PW_p(K) = \{v \in L^2(K) : v(\mathbf{x}) = \sum_{j=1}^{p} \alpha_j \exp(ik\,\mathbf{d}_j \cdot (\mathbf{x} - \mathbf{x}_K)),\ \alpha_j \in \mathbb{C}\},$$

where $\mathbf{x}_K$ is the mass center of K, and $\mathbf{d}_j$, $|\mathbf{d}_j| = 1$, $1 \leq j \leq p$, are p different directions. We assume these directions to be uniformly spaced. We define the plane wave discontinuous finite element spaces on $\mathcal{T}_h$ as follows:

$$PW_p(\mathcal{T}_h) = \{v_{hp} \in L^2(\Omega) : v_{hp}|_K \in PW_{p_K}(K)\ \forall K \in \mathcal{T}_h\}.$$

The functions in $PW_p(\mathcal{T}_h)$ possess the local *Trefftz property*

$$-\Delta v_{hp} - k^2 v_{hp} = 0 \qquad \forall v_{hp} \in PW_{p_K}(K). \tag{3}$$

In this paper, we assume uniform local resolution, i.e., $p_K = p$ for all $K \in \mathcal{T}_h$, and we use the same directions $\mathbf{d}_j$, $1 \leq j \leq p$, in every element.

We briefly recall the derivation of the PWDG methods following [25]. We multiply the first equation of (1) by smooth test functions v and integrate by parts on each $K \in \mathcal{T}_h$ obtaining

$$\int_K (\nabla u \cdot \nabla \overline{v} - k^2 u\overline{v})\, \mathrm{d}V - \int_{\partial K} \nabla u \cdot \mathbf{n}_K \, \overline{v} \, \mathrm{d}S = 0.$$

Then, we integrate by parts a second time, and replace u and v by discrete functions $u_{hp}, v_{hp} \in PW_p(\mathcal{T}_h)$, and the traces of u and ∇u at ∂K by *numerical fluxes* to be defined ($u \to \hat{u}_{hp}$ $\nabla u \to ik\hat{\boldsymbol{\sigma}}_{hp}$). Taking into account the Trefftz property (3) of the test functions v_{hp}, we obtain the elemental formulation of the PWDG method:

$$\int_{\partial K} \hat{u}_{hp} \, \nabla \overline{v}_{hp} \cdot \mathbf{n}_K \, \mathrm{d}S - \int_{\partial K} ik\hat{\boldsymbol{\sigma}}_{hp} \cdot \mathbf{n}_K \, \overline{v}_{hp} \, \mathrm{d}S = 0 \, .$$

In order to complete the definition of the method, like in [23], we mimic the general form of the fluxes defined in [11]. Using the standard DG notation [5] for averages $\{\{\cdot\}\}$ and normal jumps $[\![\cdot]\!]_N$ across interelement boundaries, and denoting by ∇_h the elementwise application of ∇, we set

$$ik\hat{\boldsymbol{\sigma}}_{hp} = \begin{cases} \{\{\nabla_h u_{hp}\}\} - \alpha \, ik \, [\![u_{hp}]\!]_N & \text{on faces in } \mathcal{F}_h^I, \\ \nabla_h u_{hp} - (1-\delta)\left(\nabla_h u_{hp} + iku_{hp}\mathbf{n} - g_R \mathbf{n}\right) & \text{on faces on } \partial\Omega, \end{cases}$$

$$\hat{u}_{hp} = \begin{cases} \{\{u_{hp}\}\} - \beta \, (ik)^{-1} [\![\nabla_h u_{hp}]\!]_N & \text{on faces in } \mathcal{F}_h^I, \\ u_{hp} - \delta\left((ik)^{-1} \nabla_h u_{hp} \cdot \mathbf{n} + u_{hp} - (ik)^{-1} g_R\right) & \text{on faces on } \partial\Omega, \end{cases}$$

where the so-called flux parameters $\alpha, \beta, \delta > 0$ here are assumed to be *constant*, with $\delta \leq 1/2$ (taking $\alpha = \beta = \delta = 1/2$ gives the ultra weak variational formulation [13]).

Remark 1 More general choices of flux parameters can be useful, for instance, for improving some convergence properties in the h-version of the method [22], or when non quasi-uniform meshes are used [24, 25].

Adding over all elements (and multiplying by $-i$), we obtain the following formulation of the PWDG method: find $u_{hp} \in PW_p(\mathcal{T}_h)$ such that, for all $v_{hp} \in PW_p(\mathcal{T}_h)$,

$$\mathcal{A}_h(u_{hp}, v_{hp}) := \mathcal{B}_h(u_{hp}, v_{hp}) + \mathcal{S}_h(u_{hp}, v_{hp}) = \ell_h(v_{hp}) \, , \tag{4}$$

where

$$\mathscr{B}_h(u,v) = i\left[-\int_{\mathscr{F}_h^I}\{\{u\}\}[\![\nabla_h\overline{v}]\!]_N\,\mathrm{d}S + \int_{\mathscr{F}_h^I}\{\{\nabla_h u\}\}\cdot[\![\overline{v}]\!]_N\,\mathrm{d}S\right.$$

$$\left.-\int_{\partial\Omega}(1-\delta)\,u\,\nabla_h\overline{v}\cdot\boldsymbol{n}\,\mathrm{d}S + \int_{\partial\Omega}\delta\,\nabla_h u\cdot\boldsymbol{n}\,\overline{v}\,\mathrm{d}S\right],$$

$$\mathscr{S}_h(u,v) = \int_{\mathscr{F}_h^I}\beta\,k^{-1}[\![\nabla_h u]\!]_N[\![\nabla_h\overline{v}]\!]_N\,\mathrm{d}S + \int_{\mathscr{F}_h^I}\alpha\,k\,[\![u]\!]_N\cdot[\![\overline{v}]\!]_N\,\mathrm{d}S$$

$$+\int_{\partial\Omega}\delta\,k^{-1}(\nabla_h u\cdot\boldsymbol{n})\,(\nabla_h\overline{v}\cdot\boldsymbol{n})\,\mathrm{d}S + \int_{\partial\Omega}(1-\delta)\,k\,u\,\overline{v}\,\mathrm{d}S,$$

and

$$\ell_h(v) = \int_{\partial\Omega}\delta\,k^{-1}g\,\nabla_h\overline{v}\cdot\boldsymbol{n}\,\mathrm{d}S - i\int_{\partial\Omega}(1-\delta)\,g\,\overline{v}\,\mathrm{d}S\ .$$

The PWDG method (4) is *unconditionally* well-posed and stable; see, e.g., [8, 13, 23, 25], where error estimates were also derived. We only recall here coercivity and continuity properties of the sesquilinear form $\mathscr{A}_h(\cdot,\cdot)$ from [23, 25]. To this aim, on the mesh $\mathscr{T}_h$, we define the *Trefftz space*

$$T(\mathscr{T}_h) := \left\{v\in L^2(\Omega):\ \exists s>0\text{ s.t. } v\in H^{\frac{3}{2}+s}(\mathscr{T}_h)\text{ and } \Delta v + k^2 v = 0\text{ in each } K\in\mathscr{T}_h\right\},$$

where $H^r(\mathscr{T}_h)$ is a shorthand notation for elementwise H^r-spaces on $\mathscr{T}_h$; the solution u of problem (1) actually belongs to $T(\mathscr{T}_h)$.

The mesh-dependent quantity

$$|||v|||^2_{\mathscr{A}_h} := \mathrm{Re}[\mathscr{A}_h(v,v)] - \mathscr{S}_h(v,v) \tag{5}$$

defines a norm in $T(\mathscr{T}_h)$ (coercivity). Moreover, setting

$$|||v|||^2_{\mathscr{A}_h^+} = |||v|||^2_{\mathscr{A}_h} + k\left\|\beta^{-\frac{1}{2}}\{\{v\}\}\right\|^2_{0,\mathscr{F}_h^I} + k^{-1}\left\|\alpha^{-\frac{1}{2}}\{\{\nabla_h v\}\}\right\|^2_{0,\mathscr{F}_h^I} + k\left\|\delta^{-\frac{1}{2}}v\right\|^2_{0,\partial\Omega},$$

for all $v, w\in T(\mathscr{T}_h)$, we have (continuity)

$$|\mathscr{A}_h(v,w)| \le 2\,\|v\|_{\mathscr{A}_h^+}\,\|w\|_{\mathscr{A}_h}\,.$$

In order to give an idea of the error behavior, we report in Fig. 1 the diagram of the L^2-error for increasing local number p of plane waves, for two different values of the wavenumber k, for a test case with smooth analytical solution. After a preasymptotic region of amplitude proportional to k, the convergence is exponential,

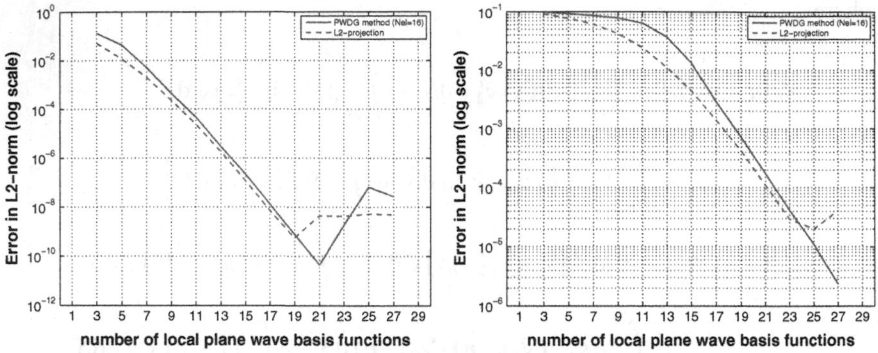

Fig. 1 L^2-error of PWDG method versus the local number p of plane waves (loglog scale). Test problem: $\Omega = (0,1)^2$, g such that the analytical solution is given, in polar coordinates $\mathbf{x} = (r\cos\theta, r\sin\theta)$, by $u(\mathbf{x}) = J_1(kr)\cos\theta$, with wavenumber $k = 10$ (*left*) and $k = 40$ (*right*), on a uniform mesh of 16 squares

until onset of numerical instability, which due to the fact that, for high p, the local basis functions are close to be linearly dependent (this region is delayed for high k). Setting $N_h := \dim(PW_p(\mathcal{T}_h))$, the algebraic linear system associated with the PWDG method (4) on the mesh $\mathcal{T}_h$ is

$$A\mathbf{u} = \mathbf{b}, \tag{6}$$

where $A \in \mathbb{C}^{N_h \times N_h}$ is the matrix associated with the sesquilinear form $\mathcal{A}_h(\cdot,\cdot)$, and $\mathbf{b} \in \mathbb{C}^{N_h}$ is the vector associated with the functional $\ell_h(\cdot)$. The GMRES iteration counts to a given tolerance, for a fixed mesh and local number of plane wave directions p, increases with k, provided that the mesh is fine enough.

We consider the same test case as in the caption of Fig. 1, fixing the mesh and varying k. In each test, p is selected as the smallest value for which the L^2-error of the PWDG method (numerical solution computed by a direct solver) is $<10^{-3}$. We have accelerated the GMRES choosing as preconditioner the incomplete LU factorization of A with no fill-in and no pivoting (PGMRES). We report in Table 1 and in Table 2 the number of GMRES and PGMRES iterations needed to achieve convergence up to a (relative) tolerance of 10^{-8}. The number of GMRES iterations remains of the same order of magnitude and that of PGMRES is constant, and much lower than that of GMRES.

While we leave the preconditioning of the p-version of PWDG to future investigation, we develop in the following sections Schwarz domain decomposition preconditioners for the h-version, addressing the issue of their scalability. Specific features of PWDG spaces (or of more general Trefftz spaces) could also be considered in order to improve the condition number of the linear systems. We refer to [29, 31] for results in this direction, but we do not elaborate on that here.

Table 1 GMRES and PGMRES iteration counts: uniform mesh of 16 squares

	$k = 10$ ($p = 9$)	$k = 20$ ($p = 13$)	$k = 30$ ($p = 17$)	$k = 40$ ($p = 19$)	$k = 50$ ($p = 23$)
Syst size	144	208	272	304	368
nnz(A)	5,119	10,747	18,379	23,014	33,677
GMRES	114	172	233	186	233
PGMRES	17	16	15	15	15

Table 2 GMRES and PGMRES iteration counts: uniform mesh of 64 squares

	$k = 10$ ($p = 7$)	$k = 20$ ($p = 9$)	$k = 30$ ($p = 11$)	$k = 40$ ($p = 13$)	$k = 50$ ($p = 15$)
Syst size	448	576	704	832	960
nnz(A)	13,861	23,052	34,461	48,347	64,342
GMRES	287	384	444	554	606
PGMRES	28	28	28	27	28
	$k = 60$ ($p = 17$)	$k = 70$ ($p = 17$)	$k = 80$ ($p = 19$)	$k = 90$ ($p = 21$)	$k = 100$ ($p = 23$)
Syst size	1,088	1,088	1,216	1,344	1,472
nnz(A)	82,643	82,802	103,561	126,460	151,561
GMRES	716	359	371	456	477
PGMRES	27	27	27	27	27

3 Domain Decomposition Preconditioners

To solve efficiently (6), we consider two-level Schwarz domain decomposition preconditioners. Let $\mathscr{T}_S$ be a partition of Ω into N_S non-overlapping subdomains: $\overline{\Omega} = \cup_{j=1}^{N_S} \overline{\Omega}_j$, and let $\{\mathscr{T}_H\}_{H>0}$ and $\{\mathscr{T}_h\}_{h>0}$ be two families of coarse and fine partitions, respectively. We assume all the partitions to be shape-regular and quasi-uniform, and such that $\mathscr{T}_S \subseteq \mathscr{T}_H \subseteq \mathscr{T}_h$, i.e., each $\Omega_j \in \mathscr{T}_S$ is union of elements $D \in \mathscr{T}_H$, and in turn each element $D \in \mathscr{T}_H$ is union of elements $K \in \mathscr{T}_h$. From here on, we omit the index p and set, for brevity, $PW_h = PW_p(\mathscr{T}_h)$; we recall that $N_h = \dim(PW_h)$.

3.1 Local and Coarse Spaces, Prolongation and Restriction Operators

We define, for each subdomain $\Omega_j \in \mathscr{T}_S$, the local PWDG space PW_h^j defined as

$$PW_h^j = \{v \in L^2(\Omega_j) : v|_K \in PW(K)\ \forall K \in \mathscr{T}_h\}, \qquad N_j := \dim(PW_h^j).$$

We denote by $\mathscr{R}_j^T : PW_h^j \hookrightarrow PW_h$ the inclusion operator (*prolongation operator*) and by $R_j^T \in \mathbb{R}^{N_h \times N_j}$ its matrix representation. The *restriction operator* $\mathscr{R}_j : PW_h \to PW_h^j$ is defined as the operator whose matrix representation is $\overline{R}_j \in \mathbb{R}^{N_j \times N_h}$, the conjugate transpose of R_j^T. Clearly, $R_j^T(r,s) = 1$ whenever the r-th basis function of PW_h coincides with the s-th basis function of PW_h^j, and $R_j^T(r,s) = 0$ otherwise (and consequently $\overline{R}_j = R_j$).

We define the local sesquilinear forms $\mathscr{A}_h^j(\cdot,\cdot) : PW_h^j \times PW_h^j \to \mathbb{C}$ by

$$\mathscr{A}_h^j(u_j, v_j) = \mathscr{A}_h(\mathscr{R}_j^T u_j, \mathscr{R}_j^T v_j) \quad \forall\, u_j, v_j \in PW_h^j; \tag{7}$$

their associated matrices are

$$A_j = R_j A R_j^T \in \mathbb{C}^{N_j \times N_j},$$

We observe that the restriction of the formulation to each subdomain coincides with the PWDG formulation of the Helmholtz problem with impedance boundary condition on the subdomain boundary. Therefore, in the present situation, *exact* local solvers [4, 21] and *inexact* local solvers [2, 21] coincide.

Now the coarse mesh $\mathscr{T}_H$ comes into play. We define the coarse PWDG space as

$$PW_h^0 = PW(\mathscr{T}_H) = \{v \in L^2(\Omega) : \; v|_D \in PW(D) \;\forall D \in \mathscr{T}_D\}, \quad N_0 := \dim(PW_h^0).$$

We also introduce the coarse space prolongation and restriction operators: $\mathscr{R}_0^T : PW_h^0 \hookrightarrow PW_h$, with associated matrix $R_0^T \in \mathbb{C}^{N_h \times N_0}$, and $\mathscr{R}_0 : PW_h \to PW_0$, which is the operator whose associated matrix is $\overline{R}_0 \in \mathbb{C}^{N_0 \times N_h}$. We also define

$$A_0 = \overline{R}_0 A R_0^T \in \mathbb{C}^{N_0 \times N_0},$$

associated with the sesquilinear form

$$\mathscr{A}_h^0(u_0, v_0) = \mathscr{A}_h(\mathscr{R}_0^T u_0, \mathscr{R}_0^T v_0) \quad \forall\, u_0, v_0 \in PW_h^0. \tag{8}$$

We show how the operator R_0^T is constructed. Let $D \in \mathscr{T}_H$ be an element of the coarse mesh, and let $\{K_r\}_{r=1}^{N_D}$ be the elements of the fine mesh $\mathscr{T}_h$ contained in D. Consider one of the p basis function of PW_h^0 supported within D: $\Phi_\ell(\mathbf{x}) = exp(ik\,\mathbf{d}_\ell \cdot (\mathbf{x} - \mathbf{x}_D))$. Now, we express $\Phi_\ell(\mathbf{x})$ as linear combination of the basis functions of PW_h supported within the elements $\{K_r\}_{r=1}^{N_D}$, i.e., we compute the coefficients $\alpha_j^r \in \mathbb{C}$ such that

$$\Phi_\ell(\mathbf{x}) = \sum_{r=1}^{N_D} \sum_{j=1}^{p} \alpha_j^r \phi_h^{j,r}(\mathbf{x}), \tag{9}$$

where $\phi_h^{j,r}(\mathbf{x}) = exp(ik\,\mathbf{d}_j \cdot (\mathbf{x} - \mathbf{x}_{K_r}))$ for $\mathbf{x} \in K_r$, and $\phi_h^{j,r}(\mathbf{x}) = 0$ outside K_r. Clearly, $\alpha_j^r = 0$ for every $j \neq \ell$, i.e., the basis functions of PW_h with direction different from $\mathbf{d}_\ell$ do not enter the expression (9); therefore

$$\Phi_\ell(\mathbf{x}) = \sum_{r=1}^{N_D} \alpha_\ell^r \phi_h^{\ell,r}(\mathbf{x}),$$

with

$$\alpha_\ell^r = \exp(ik\,\mathbf{d}_\ell \cdot (\mathbf{x}_{K_r} - \mathbf{x}_D)).$$

Requiring that $\overline{R}_{0,D} R_{0,D}^T = I$, where I is the $N_0 \times N_0$ identity matrix, assuming that the unknowns are ordered by an external loop over all elements, and an internal loop over the p directions, the elemental contribution $R_{0,D}^T \in \mathbb{C}^{(r*p)\times p}$ to the matrix R_0^T is given by

$$R_{0,D}^T((r-1)*p+\ell,\ell) = \frac{1}{\sqrt{N_D}}\alpha_\ell^r, \qquad 1 \leq \ell \leq p.$$

3.2 Schwarz Operators

For $j = 0,\ldots,N_S$, we define $\tilde{\mathscr{P}}_j : PW_h \to PW_h^j$ as the (unique) solution of the following problem

$$\mathscr{A}_h^j(\tilde{\mathscr{P}}_j u, v_j) = \mathscr{A}_h(u, \mathscr{R}_j^T v_j) \quad \forall v_j \in PW_h^j.$$

Well-posedness of the local sesquilinear forms $\mathscr{A}_h^j(\cdot,\cdot)$, $0 \leq j \leq N_S$, defined in (7) and (8) follows from the fact that $\mathrm{Re}[\mathscr{A}_h^j(v,v)]$ are norms in the spaces PW_h^j. Therefore the projection operators $\tilde{\mathscr{P}}_j$ are well defined. We define the operators $\mathscr{P}_j = \mathscr{R}_j^T \tilde{\mathscr{P}}_j : PW_h \to PW_h$, $j = 0,\ldots,N_S$, denote by P_j their matrix representations, and observe that

$$P_j = R_j^T A_j^{-1} \overline{R}_j A \quad 0 \leq j \leq N_S.$$

Since $A_j = \overline{R}_j A R_j^T$, $0 \leq j \leq N_S$, $P_j^2 = P_j$, i.e., the P_j's are projectors.

We define the *additive* and *multiplicative* Schwarz operators as

$$\mathscr{Q}_{ad} := \sum_{j=0}^{N_S} \mathscr{P}_j, \quad \mathscr{Q}_{mu} := I - \mathscr{E}_{mu},$$

where the error propagation operator $\mathscr{E}_{mu}$ is defined as

$$\mathscr{E}_{mu} = (I - \mathscr{P}_{N_S}) \cdots (I - \mathscr{P}_1)(I - \mathscr{P}_0).$$

From the algebraic point of view, the Schwarz operators can be seen as preconditioned operators for the original operator A, and can be written as the product of a suitable preconditioner and A. For example, for the additive operator Q_{ad} we have

$$Q_{ad} = P_{ad}^{-1} A, \qquad P_{ad}^{-1} = \sum_{j=0}^{N_S} R_j^T A_j^{-1} \overline{R}_j;$$

analogously we can write Q_{mu} as $Q_{mu} = P_{mu}^{-1} A$.

Then, the preconditioned linear system of equations we are interested in solving is

$$Q\mathbf{u} = \mathbf{g}, \tag{10}$$

where $Q = P^{-1}A$ and $\mathbf{g} = P^{-1}\mathbf{b}$, with either $P = P_{ad}$ or $P = P_{mu}$.

4 Numerical Results

We investigate the performance of our preconditioners when varying the fine and coarse grids, the number of subdomains N_S, as well as the wavenumber k and the number of directions p. We use a uniform subdomain partition of $\Omega = (0, 1)^2$ consisting of $N_S = 4, 16, 64$ square subdomains. We have solved our problem on a sequence of Cartesian grids with $1/h = 4, 8, 16, 32$; for the coarse grids we have also considered Cartesian partitions with $1/H = 2, 4, 8, 16, 32$.

We choose the impedance datum g so that the analytical solution of problem (1) is given, in polar coordinates $\mathbf{x} = (r\cos\theta, r\sin\theta)$, by $u(\mathbf{x}) = J_1(kr)\cos(\theta)$, where $J_1(\cdot)$ is the Bessel function of the first type. The flux parameters have been chosen as $\alpha = \beta = \delta = \frac{1}{2}$, cf. Sect. 2.

Throughout this section, the linear systems of equations have been solved by GMRES with a (relative) tolerance set equal to 10^{-6}. We report the results obtained by taking as initial guess the null vector. We point out that completely analogous results have been obtained with normalized random initial guess vectors.

We first investigate whether the additive and multiplicative Schwarz preconditioners are scalable, i.e., the iteration counts needed to reduce the residual up to a (user defined) tolerance are independent of the number of subdomains. In Tables 3 and 4 we report the iteration counts for $k = 30$, $p = 15$ and $N_S = 4, 16, 64$ computed with the additive and multiplicative preconditioners, respectively. The

Table 3 Additive preconditioner: GMRES iteration counts. Wavenumber $k = 30$, number of directions $p = 15$

	$N_S = 4$				$N_S = 16$				$N_S = 64$			
H^{-1} \ h^{-1}	4	8	16	32	4	8	16	32	4	8	16	32
2	22	35	48	64	–	–	–	–	–	–	–	–
4	13	29	41	56	15	34	46	63	–	–	–	–
8	–	16	33	46	–	17	37	53	–	19	41	56
16	–	–	16	39	–	–	17	43	–	–	17	46
iter(A)	134	670	1,510	3,552	134	670	1,510	3,552	134	670	1,510	3,552

Table 4 Multiplicative preconditioner: GMRES iteration counts. Wavenumber $k = 30$, number of directions $p = 15$

	$N_S = 4$				$N_S = 16$				$N_S = 64$			
H^{-1} \ h^{-1}	4	8	16	32	4	8	16	32	4	8	16	32
2	12	19	25	34	–	–	–	–	–	–	–	–
4	2	14	21	30	2	17	25	34	–	–	–	–
8	–	2	16	24	–	2	17	26	–	2	21	29
16	–	–	2	18	–	–	2	20	–	–	3	21
iter(A)	134	670	1,510	3,552	134	670	1,510	3,552	134	670	1,510	3,552

proposed preconditioners seem to be asymptotically scalable, indeed for $N_S \geq 16$ the number of iterations seems to be quite independent of the number of subdomains. Moreover, in all the cases the preconditioners are effective, as confirmed by a comparison with the iterations counts needed to solve the unpreconditioned system, cf. last line of Table 3, and, as expected, the multiplicative preconditioner performs much better than the additive one. In Tables 3–6, the symbol "–" refers to the case where the hypothesis $\mathcal{T}_S \subseteq \mathcal{T}_H \subseteq \mathcal{T}_h$ is not satisfied, and therefore the construction of the preconditioner is meaningless. The lower diagonals reported in Table 3 (and also in the tables below) correspond to the limit case $h = H$. In this case, the coarse component of the preconditioner is an exact solver for the original linear system (6), indeed, if $A_0 = A$, then $R_0 = I$ and the operator P_0 becomes $P_0 = R_0^T A^{-1} \overline{R}_0 A = I$. Thus, in principle, we should obtain convergence in one iteration, and the fact that this does not happen indicates that the local solutions "spoil" the result (see also [2]). The results for $h = H$ are reported in order to give an idea of the performance of the preconditioners also in such limit case.

We have repeated the same set of experiments taking the wavenumber $k = 50$, cf. Tables 5 and 6 for the additive and multiplicative preconditioners, respectively. We observe that the iteration counts of the preconditioned system do not seem to vary significantly as the wavenumber increases, whereas the iteration counts of the unpreconditioned one increase, at least when $1/h$ becomes sufficiently large. Also in this case the multiplicative preconditioner outperforms the additive one.

Table 5 Additive preconditioner: GMRES iteration counts. Wavenumber $k = 50$, number of directions $p = 15$

	$N_S = 4$				$N_S = 16$				$N_S = 64$			
H^{-1} \ h^{-1}	4	8	16	32	4	8	16	32	4	8	16	32
2	16	29	42	55	–	–	–	–	–	–	–	–
4	10	27	42	57	15	37	52	70	–	–	–	–
8	–	14	31	43	–	16	32	45	–	19	37	50
16	–	–	17	37	–	–	18	37	–	–	19	40
iter(A)	33	508	1,977	4,461	33	508	1,977	4,461	33	508	1,977	4,461

Table 6 Multiplicative preconditioner: GMRES iteration counts. Wavenumber $k = 50$, number of directions $p = 15$

	$N_S = 4$				$N_S = 16$				$N_S = 64$			
H^{-1} \ h^{-1}	4	8	16	32	4	8	16	32	4	8	16	32
2	8	15	23	30	–	–	–	–	–	–	–	–
4	2	14	22	30	2	18	27	36	–	–	–	–
8	–	2	15	22	–	2	16	24	–	2	20	37
16	–	–	1	16	–	–	2	17	–	–	2	19
iter(A)	33	508	1,977	4,461	33	508	1,977	4,461	33	508	1,977	4,461

Table 7 Additive preconditioner: GMRES iteration counts. Wavenumber $k = 30, 40, 50$, number of directions $p = 13, 15, 17, 19$ ($N_S = 16$ and $1/h = 16$)

	$k = 30$				$k = 40$				$k = 50$			
H^{-1} \ p	13	15	17	19	13	15	17	19	13	15	17	19
4	47	46	47	49	50	48	48	47	49	52	52	50
8	35	37	39	41	33	35	36	37	32	32	34	34
iter(A)	1,510	1,510	1,511	1,511	1,826	1,825	1,824	1,825	1,883	1,977	1,973	1,973

We next address the performance of our preconditioners when varying the number of directions p, fixing, for the sake of simplicity, $N_S = 16$ and $1/h = 16$. In Tables 7 and 8 we report the GMRES iteration counts for $p = 13, 15, 17, 19$ and $k = 30, 40, 50$ computed with the additive and multiplicative preconditioners, respectively. The last row of Tables 7 and 8 shows the corresponding iteration counts needed to solve the unpreconditioned systems. From the numerical results, we can observe that the iteration counts needed to solve the preconditioned systems, both with the additive and the multiplicative preconditioners, seem to be fairly independent of k, and that these preconditioners seem to be very effective in accelerating GMRES convergence.

Table 8 Multiplicative preconditioner: GMRES iteration counts. Wavenumber $k = 30, 40, 50$, number of directions $p = 13, 15, 17, 19$ ($N_S = 16$ and $1/h = 16$)

	$k = 30$				$k = 40$				$k = 50$			
H^{-1} \ p	13	15	17	19	13	15	17	19	13	15	17	19
4	25	25	26	26	25	25	26	26	27	27	27	27
8	17	17	19	20	16	17	18	19	16	16	17	18
iter(A)	1,510	1,510	1,511	1,511	1,826	1,825	1,824	1,825	1,883	1,977	1,973	1,973

Table 9 Additive, multiplicative and hybrid preconditioners: GMRES iteration counts. Wavenumber $k = 30$, number of directions $p = 15$ ($N_S = 4$)

	Additive				Multiplicative				Hybrid			
H^{-1} \ h^{-1}	4	8	16	32	4	8	16	32	4	8	16	32
2	22	35	46	64	12	19	25	34	17	32	43	60
4	13	29	41	56	2	14	21	30	2	24	35	49
8	–	16	33	46	–	2	16	24	–	2	25	37
16	–	–	16	39	–	–	2	18	–	–	3	30

Finally, we have tested a *hybrid* additive-multiplicative preconditioner (multiplicative coarse problem and additive local problems).[1] As reported in Table 9, the performance of this hybrid preconditioner is intermediate with respect to the pure additive and multiplicative ones.

5 The Issue of GMRES Convergence

In the following, we recall the GMRES convergence theory developed in [15,17,36] which provides sufficient conditions for non-stagnation of GMRES, i.e., the iterative method makes some progress in reducing the residual at each iteration step, and establishes upper bounds on the residual norm.

Given the matrix $Q \in \mathbb{C}^{N_h \times N_h}$, we define its associated field of values as

$$F(Q) = \left\{ \frac{\overline{\mathbf{x}}^T Q \mathbf{x}}{\overline{\mathbf{x}}^T \mathbf{x}}, \quad \mathbf{0} \neq \mathbf{x} \in \mathbb{C}^{N_h} \right\},$$

[1]We express our gratitude to the anonymous reviewer for suggesting us these tests.

and denote by $\nu(F(Q))$ the distance of $F(Q)$ from the origin. The theory proposed by [15, 36] states that the κ-th residual $\mathbf{r}_\kappa$ of GMRES satisfies

$$\frac{\|\mathbf{r}_\kappa\|}{\|\mathbf{r}_0\|} \leq \left(1 - \nu(F(Q))\,\nu(F(Q^{-1}))\right)^{\kappa/2}. \tag{11}$$

Denote by $H(Q)$ the hermitian part of Q, i.e., $H(Q) = \frac{Q+\overline{Q}^T}{2}$. If $H(Q)$ is positive definite, then $\nu(F(Q)) \geq \lambda_{\min}(H(Q))$, the minimum eigenvalue of $H(Q)$. In our case, if A is the coefficient matrix of the unpreconditioned system, $H(A)$ is positive definite, since it is associated with the bilinear form $\mathscr{S}_h(\cdot,\cdot)$ (see (4)), which is a scalar product in $PW_p(\mathscr{T}_h)$. If we denote by Q the left/right preconditioned matrix, i.e., $Q = (\overline{P}^T)^{-1/2}AP^{-1/2}$, with either $P = P_{ad}$ or $P = P_{mu}$, then $H(Q)$ is also positive definite. Moreover, one can bound $\nu(F(Q^{-1}))$ from below by

$$\nu(F(Q^{-1})) \geq \frac{\lambda_{\min}(H(Q))}{\|Q\|^2},$$

where $\|\cdot\|$ is the natural (complex) Euclidean matrix norm (see [30]), and write a weaker but more practical version of the bound (11):

$$\frac{\|\mathbf{r}_\kappa\|}{\|\mathbf{r}_0\|} \leq \left(1 - \frac{\lambda^2_{\min}(H(Q))}{\|Q\|^2}\right)^{\kappa/2},$$

which was firstly derived in [16, 17].

We have performed some experiments on the test problem described in Sect. 4 with $k = 30$ and number of plane wave directions per element $p = 11, 13, 15$, on the grid with $1/h = 16$. We have compared the values of $\lambda_{\min}(H(R))$ and $\lambda_{\min}(H(R^{-1}))$ for the original matrix ($R = A$) and for the left/right preconditioned matrix $Q = (\overline{P}^T)^{-1/2}AP^{-1/2}$ ($R = Q$), with $P = P_{ad}$ constructed with $N_S = 4$ subdomains and a coarse grid with $1/H = 2$. The results reported in Table 10 show that, while $\lambda_{\min}(H(R^{-1}))$ (which is a lower bound for $\nu(F(R^{-1}))$) is essentially the same for the unpreconditioned and the preconditioned matrices, and for all the considered values of p, $\lambda_{\min}(H(R))$ (which is a lower bound for $\nu(F(R))$) is definitely larger for the preconditioned matrix (and it is almost uniform in p).

Table 10 Values of $\lambda_{\min}(H(R))$ and $\lambda_{\min}(H(R^{-1}))$ (which are lower bound for $\nu(F(R))$ and $\nu(F(R^{-1}))$, respectively) for $R = A$ (unpreconditioned matrix) and $R = Q = (\overline{P}^T)^{-1/2}AP^{-1/2}$ (preconditioned matrix). Wavenumber $k = 30$, Cartesian mesh with $1/h = 16$; $P = P_{ad}$ constructed with $N_S = 4$ subdomains and a coarse grid with $1/H = 2$

	$p = 11$	$p = 13$	$p = 15$
$\lambda_{\min}(H(A))$	$1.3338 \cdot 10^{-5}$	$1.0956 \cdot 10^{-7}$	$1.0079 \cdot 10^{-9}$
$\lambda_{\min}(H(A^{-1}))$	$1.7385 \cdot 10^{-2}$	$1.4711 \cdot 10^{-2}$	$1.2749 \cdot 10^{-2}$
$\lambda_{\min}(H(Q))$	$5.8730 \cdot 10^{-2}$	$4.4976 \cdot 10^{-2}$	$3.5210 \cdot 10^{-2}$
$\lambda_{\min}(H(Q^{-1}))$	$3.1358 \cdot 10^{-2}$	$3.1919 \cdot 10^{-2}$	$3.1977 \cdot 10^{-2}$

Table 11 Values of $\lambda_{\min}(H(R))$ and $\lambda_{\min}(H(R^{-1}))$ (which are lower bound for $\nu(F(R))$ and $\nu(F(R^{-1}))$, respectively) for $R = A$ (unpreconditioned matrix) and $R = Q = (\overline{P}^T)^{-1/2}AP^{-1/2}$ (preconditioned matrix). Cartesian mesh with $1/h = 16$, number of plane wave directions per element $p = 15$; $P = P_{ad}$ constructed with $N_S = 4$ subdomains and a coarse grid with $1/H = 2$

	$k = 30$	$k = 40$	$k = 50$
$\lambda_{\min}(H(A))$	$1.0079 \cdot 10^{-9}$	$7.0787 \cdot 10^{-8}$	$1.8717 \cdot 10^{-6}$
$\lambda_{\min}(H(A^{-1}))$	$1.2749 \cdot 10^{-2}$	$1.2495 \cdot 10^{-2}$	$1.3699 \cdot 10^{-2}$
$\lambda_{\min}(H(Q))$	$3.5210 \cdot 10^{-2}$	$6.1622 \cdot 10^{-2}$	$3.6636 \cdot 10^{-2}$
$\lambda_{\min}(H(Q^{-1}))$	$3.1977 \cdot 10^{-2}$	$3.0694 \cdot 10^{-2}$	$2.9770 \cdot 10^{-2}$

A similar situation is observed when varying k ($k = 30, 40, 50$); the results for $p = 15$, $1/h = 16$, $N_S = 4$ subdomains and $1/H = 2$ are reported in Table 11. Theoretical estimates of these quantities are under investigation.

References

1. M. Amara, R. Djellouli, C. Farhat, Convergence analysis of a discontinuous Galerkin method with plane waves and lagrange multipliers for the solution of Helmholtz problems. SIAM J. Numer. Anal. **47**(2), 1038–1066 (2009)
2. P.F. Antonietti, B. Ayuso, Schwarz domain decomposition preconditioners for discontinuous Galerkin approximations of elliptic problems: non-overlapping case. M2AN Math. Model. Numer. Anal. **41**(1), 21–54 (2007)
3. P.F. Antonietti, S. Giani, P. Houston, Domain decomposition preconditioners for discontinuous Galerkin methods for elliptic problems on complicated domains. J. Sci. Comput. **60**(1), 203–227 (2014)
4. P.F. Antonietti, P. Houston, A class of domain decomposition preconditioners for *hp*-discontinuous Galerkin finite element methods. J. Sci. Comput. **46**(1), 124–149 (2011)
5. D. Arnold, F. Brezzi, B. Cockburn, L. Marini, Unified analysis of discontinuous Galerkin methods for elliptic problems. SIAM J. Numer. Anal. **39**(5), 1749–1779 (2002)
6. I. Babuška, J.M. Melenk, The partition of unity method. Int. J. Numer. Methods Eng. **40**(4), 727–758 (1997)
7. T. Betcke, M.J. Gander, J. Phillips, Block jacobi relaxation for plane wave discontinuous galerkin methods, In: Erhel, J., Gander, M.J., Halpern, L., Pichot, G., Sassi, T., Widlund, O.B. (eds.) *Domain Decomposition Methods in Science and Engineering XXI*. Lecture Notes in Computational Science and Engineering, vol. 98 (Springer, Berlin/New York, 2014)
8. A. Buffa, P. Monk, Error estimates for the ultra weak variational formulation of the Helmholtz equation. Math. Mod. Numer. Anal. **42** 925–940 (2008). Published online August 12, 2008. doi: 10.1051/m2an:2008033
9. X.-C. Cai, O.B. Widlund, Domain decomposition algorithms for indefinite elliptic problems. SIAM J. Sci. Stat. Comput. **13**(1), 243–258 (1992)
10. ———, Multiplicative Schwarz algorithms for some nonsymmetric and indefinite problems. SIAM J. Numer. Anal. **30**(4), 936–952 (1993)
11. P. Castillo, B. Cockburn, I. Perugia, D. Schötzau, An a priori error analysis of the local discontinuous Galerkin method for elliptic problems. SIAM J. Numer. Anal. **38**(5), 1676–1706 (2000)
12. O. Cessenat, Application d'une nouvelle formulation variationnelle aux équations d'ondes harmoniques, Problèmes de Helmholtz 2D et de Maxwell 3D, Ph.D. thesis, Université Paris IX Dauphine, 1996

13. O. Cessenat, B. Després, Application of an ultra weak variational formulation of elliptic PDEs to the two-dimensional Helmholtz equation. SIAM J. Numer. Anal. **35**(1), 255–299 (1998)
14. ———, Using plane waves as base functions for solving time harmonic equations with the ultra weak variational formulation. J. Comput. Acoust. **11**, 227–238 (2003)
15. M. Eiermann, O.G. Ernst, Geometric aspects of the theory of Krylov subspace methods. Acta Numer. **10**, 251–312 (2001)
16. S.C. Eisenstat, H.C. Elman, M.H. Schultz, Variational iterative methods for nonsymmetric systems of linear equations. SIAM J. Numer. Anal. **20**(2), 345–357 (1983)
17. H. Elman, Iterative methods for large, sparse, nonsymmetric systems of linear equations, Ph.D. thesis, Yale University, 1982
18. H.C. Elman, O.G. Ernst, D.P. O'Leary, A multigrid method enhanced by Krylov subspace iteration for discrete Helmhotz equations. SIAM J. Sci. Comput. **23**(4), 1291–1315 (2001). (electronic)
19. O.G. Ernst, M.J. Gander, Why it is difficult to solve Helmholtz problems with classical iterative methods, in *Numerical Analysis of Multiscale Problems*. Lecture Notes in Computational Science and Engineering, vol. 83 (Springer, Heidelberg, 2012), pp. 325–363
20. C. Farhat, R. Tezaur, J. Toivanen, A domain decomposition method for discontinuous Galerkin discretizations of Helmholtz problems with plane waves and Lagrange multipliers. Int. J. Numer. Methods Eng. **78**(13), 1513–1531 (2009)
21. X. Feng, O.A. Karakashian, Two-level additive Schwarz methods for a discontinuous Galerkin approximation of second order elliptic problems. SIAM J. Numer. Anal. **39**(4), 1343–1365 (2001). (electronic)
22. C.J. Gittelson, R. Hiptmair, I. Perugia, Plane wave discontinuous Galerkin methods: analysis of the h–version, M2AN Math. Model. Numer. Anal. **43**(2), 297–332 (2009)
23. R. Hiptmair, A. Moiola, I. Perugia, Plane wave discontinuous Galerkin methods for the 2D Helmholtz equation: analysis of the p-version. SIAM J. Numer. Anal. **49**, 264–284 (2011)
24. ———, Plane wave discontinuous Galerkin method: exponential convergence of the *hp*-version, Technical Report 2013-31, SAM-ETHZ, Zürich, 2013
25. ———, Trefftz discontinuous Galerkin methods for acoustic scattering on locally refined meshes. Appl. Numer. Math. (2013). doi: 10.1016/j.apnum.2012.12.004
26. R. Hiptmair, A. Moiola, I. Perugia, Error analysis of Trefftz-discontinuous Galerkin methods for the time-harmonic Maxwell equations. Math. Comput. **82**(281), 247–268 (2013)
27. T. Huttunen, M. Malinen, P. Monk, Solving Maxwell's equations using the ultra weak variational formulation. J. Comput. Phys. **223**, 731–758 (2007)
28. T. Huttunen, P. Monk, F. Collino, J.P. Kaipio, The ultra-weak variational formulation for elastic wave problems. SIAM J. Sci. Comput. **25**(5), 1717–1742 (2004)
29. T. Huttunen, P. Monk, J.P. Kaipio, Computational aspects of the ultra-weak variational formulation. J. Comput. Phys. **182**(1), 27–46 (2002)
30. J. Liesen, P. Tichý, The field of values bound on ideal GMRES. (2013) arXiv:1211.5969v2
31. T. Luostari, T. Huttunen, P. Monk, Improvements for the ultra weak variational formulation. Int. J. Numer. Methods Eng. **94**(6), 598–624 (2013)
32. T. Luostari, T. Huttunen, P. Monk, Error estimates for the ultra weak variational formulation in linear elasticity. ESAIM Math. Model. Numer. Anal. **47**(1), 183–211 (2013)
33. J.M. Melenk, On generalized finite element methods, Ph.D. thesis, University of Maryland, 1995
34. P. Monk, D. Wang, A least squares method for the Helmholtz equation. Comput. Methods Appl. Mech. Eng. **175**(1/2), 121–136 (1999)
35. Y. Saad, M.H. Schultz, GMRES: a generalized minimal residual algorithm for solving nonsymmetric linear systems. SIAM J. Sci. Stat. Comput. **7**(3), 856–869 (1986)
36. G. Starke, Field-of-values analysis of preconditioned iterative methods for nonsymmetric elliptic problems. Numer. Math. **78**(1), 103–117 (1997)
37. R. Tezaur, C. Farhat, Three-dimensional discontinuous Galerkin elements with plane waves and Lagrange multipliers for the solution of mid-frequency Helmholtz problems. Int. J. Numer. Meth. Eng. **66**(5), 796–815 (2006)

A Deflation Based Coarse Space in Dual-Primal FETI Methods for Almost Incompressible Elasticity

Sabrina Gippert, Axel Klawonn, and Oliver Rheinbach

Abstract A new coarse space for FETI-DP domain decomposition methods for mixed finite element discretizations of almost incompressible linear elasticity problems in 3D is presented. The mixed finite element discretization uses continuous piecewise triquadratic displacements and discontinuous piecewise constant pressures. The piecewise constant pressure variables are statically condensated on the element level. The new coarse space is significantly smaller than earlier known coarse spaces for FETI-DP or BDDC methods for the equations of almost incompressible elasticity or Stokes' equations. For discretizations with discontinuous pressure elements it is well-known that a zero net flux condition on each subdomain is needed to ensure a good condition number. Usually, this constraint is enforced for each vertex, edge, and face of each subdomain separately. Here, a coarse space is discussed where all vertex and edge constraints are treated as usual but where all faces of each subdomain contribute only a single constraint. This approach is presented within a deflation based framework for the implementation of coarse spaces into FETI-DP methods.

1 Introduction

It is well known that in order to obtain a good condition number bound for FETI-DP or BDDC methods [1, 3, 10, 13] a zero net flux condition for each subdomain has to be enforced; cf. [2, 8, 9, 11, 12]. This is usually done using one constraint for each vertex, edge, and face of each subdomain. In our approach, all vertices are chosen as primal constraints and the edge constraints are enforced using a transformation of basis approach with partial assembly; see, e.g., [7, 10]. Here, we

S. Gippert • A. Klawonn (✉)
Mathematisches Institut, Universität zu Köln, Weyertal 86-90, 50931 Köln, Germany
e-mail: sgippert@math.uni-koeln.de; klawonn@math.uni-koeln.de

O. Rheinbach
Institut für Numerische Mathematik und Optimierung, Technische Universität Bergakademie Freiberg, Akademiestraße 6, 09596 Freiberg, Germany
e-mail: oliver.rheinbach@math.tu-freiberg.de

A. Abdulle et al. (eds.), *Numerical Mathematics and Advanced Applications ENUMATH 2013*, Lecture Notes in Computational Science and Engineering 103,
DOI 10.1007/978-3-319-10705-9_56

focus on establishing the zero net flux condition for faces using a deflation method; see, e.g., [8]. But while the traditional way of implementing this condition for face terms is to use one constraint for each face, in our approach it is sufficient to establish one single face constraint for each subdomain. Therefore, this new coarse space is significantly smaller than earlier known coarse spaces for FETI-DP or BDDC methods for almost incompressible elasticity or Stokes' equations. We give a brief summary of the deflation method in Sect. 2 before we describe the construction of the new coarse space in Sect. 3. Finally, we present some numerical results, see Sect. 4, which confirm our theoretical considerations.

2 The Deflation Method

The standard coarse space of FETI-DP and BDDC domain decomposition methods can be complemented by the use of projections; see, e.g., [8]. Using this approach, a second, independent coarse space can thus be implemented for FETI-DP and BDDC methods. The deflation approach can also be used to obtain robustness with respect to heterogeneities inside subdomains; see, e.g., [6]. Such projection approaches are known as projector preconditioning or as the deflation method, closely related is the balancing preconditioner. Due to space limitations, we refer to [8] for a more complete list of references for deflation, projector preconditioning, and balancing methods. We briefly recall the projection framework for FETI-DP and BDDC methods as presented in [8]. The solution of a symmetric positive definite (or semidefinite) system

$$F\lambda = d \tag{1}$$

using the deflation method consists of solving

$$M^{-1}(I-P)^T F\lambda = M^{-1}(I-P)^T d$$

with respect to λ using the conjugate gradient method and a projection

$$P = U(U^T F U)^{-1} U^T F.$$

Here, M^{-1} is a symmetric positive definite preconditioner. This is equivalent to solving (1) by conjugate gradients using the symmetric preconditioner

$$M_{PP}^{-1} = (I-P)M^{-1}(I-P)^T.$$

We set $\overline{\lambda} := PF^{-1}d$. The solution λ^* of the original problem (1) is then

$$\lambda^* = \overline{\lambda} + \lambda.$$

If we include the computation of $\overline{\lambda} = PF^{-1}d = U(U^TFU)^{-1}U^Td$ into every step of the iteration, we obtain the balancing preconditioner

$$M_{BP}^{-1} = (I-P)M^{-1}(I-P)^T + U(U^TFU)^{-1}U^T.$$

We then obtain the solution of (1) directly without an additional correction.

For details on the deflation method or the balancing preconditioner applied to the FETI-DP or BDDC method, see [8]. In [8] it was shown that for every FETI-DP method using a standard coarse problem based on partial assembly and a transformation of basis there exists a corresponding FETI-DP method using deflation that has essentially identical eigenvalues. Note that the reverse is not true. Indeed the FETI-DP method presented in this paper is an example of a FETI-DP method using deflation where no corresponding standard FETI-DP method exists.

3 A New Coarse Space for Almost Incompressible Linear Elasticity

For almost incompressible linear elasticity in 3D, we use the variational form of the mixed problem: Find $(u,p) \in H_0^1(\Omega, \partial\Omega_D) \times L_2(\Omega)$, such that

$$\int_\Omega G\varepsilon(u) : \varepsilon(v)\, dx + \int_\Omega \mathrm{div}(v)\, p\, dx = < F, v > \quad \forall v \in H_0^1(\Omega, \partial\Omega_D)$$

$$\int_\Omega \mathrm{div}(u)\, q\, dx - \int_\Omega (G\,\beta)^{-1}\, p\, q\, dx = 0 \quad \forall q \in L_2(\Omega),$$

where the pressure $p := G\,\beta\,\mathrm{div}(u) \in L_2(\Omega)$ is introduced as an additional variable. We use the expressions $G = \frac{E}{1+\nu}$ and $\beta = \frac{\nu}{1-2\nu}$, using Young's modulus E and Poisson's ratio ν. This mixed formulation is discretized by $Q_2 - P_0$ mixed finite elements. Static condensation of the piecewise discontinuous pressure leads to a symmetric positive definite problem. We use a FETI-DP algorithm with primal vertex constraints and primal edge averages to solve this reduced linear system. The primal edge constraints are enforced using a transformation of basis with partial assembly; see, e.g., [7]. It is necessary to establish a zero net flux condition on each subdomain to ensure a good condition number for FETI-DP or BDDC methods; see also [2, 8, 9, 11, 12]. We note that almost incompressible elasticity was considered in Neumann-Neumann preconditioners already in [5]. We will use primal edge averages in the normal directions to enforce the zero net flux condition for the edge terms. Traditionally, one normal constraint for each face is used to enforce the zero net flux. We will show that, instead of one constraint for each face, a single constraint for each subdomain is sufficient, obtained by summing up the face contributions. This constraint is implemented using deflation or balancing.It cannot be implemented by a transformation of basis and partial assembly. Configurations with almost incompressible components, i.e., almost incompressible inclusions inside subdomains, which can be enclosed in a compressible hull, have been analyzed in [4]. There, an enhanced coarse space is not needed. We need to enforce a zero net flux condition on each subdomain, cf. [12]. Due to space limitations, for the definitions of the FETI-DP jump operator B and the scaled jump operator B_D, we refer to [7, p. 1892, bottom (def. of B); p. 1893, bottom, (def. of B_D)]. We also need a local assembly operator $R^{(i)T}$, cf., e.g., [4, p. 2221, bottom] or [10, Sect. 4.1, p. 1533], which assembles in the primal variables. A central role in the analysis of FETI-DP methods is played by the operator $P_D := B_D^T B$; cf., e.g., [13, Sect. 6.4] or [10, Sect. 8.1]. The zero net flux condition can be written as

$$\int_{\partial\Omega_i} \underbrace{(R^{(i)} P_D w)}_{=:v^{(i)}} \cdot n \, ds = 0.$$

Let us now denote by $\mathscr{F}^{ij}$ the face shared by the subdomains Ω_i and Ω_j. Correspondingly, we denote by $\mathscr{E}^{ik}$ the edge shared by the subdomains Ω_i, Ω_j, Ω_k, and Ω_l. We assume that all subdomain vertices are primal. By $\theta_{\mathscr{F}^{ij}}$ and $\theta_{\mathscr{E}^{ij}}$, we denote finite element partition of unity functions which are one on the face and edge, respectively, and zero elsewhere on the interface. Then, we have the following representation of $v^{(i)}$, cf., [4, p. 2224] or [10, p. 1555, (8.7)],

$$\begin{aligned} v^{(i)} &= \sum_{\mathscr{F}^{ij} \subset \partial\Omega_i} I^h\left(\theta_{\mathscr{F}^{ij}} v^{(i)}\right) + \sum_{\mathscr{E}^{ik} \subset \partial\Omega_i} I^h\left(\theta_{\mathscr{E}^{ik}} v^{(i)}\right) \qquad (2) \\ &= \sum_{\mathscr{F}^{ij} \subset \partial\Omega_i} I^h\left(\theta_{\mathscr{F}^{ij}} \delta_j^\dagger (w^{(i)} - w^{(j)})\right) + \sum_{\mathscr{E}^{ik} \subset \partial\Omega_i} \left\{ I^h\left(\theta_{\mathscr{E}^{ik}} \delta_j^\dagger (w^{(i)} - w^{(j)})\right) \right. \\ &\qquad \left. + I^h\left(\theta_{\mathscr{E}^{ik}} \delta_k^\dagger (w^{(i)} - w^{(k)})\right) + I^h\left(\theta_{\mathscr{E}^{ik}} \delta_l^\dagger (w^{(i)} - w^{(l)})\right) \right\}. \end{aligned}$$

Thus, we can write $\int_{\partial\Omega_i} (R^{(i)} P_D w) \cdot n \, ds = 0$ as

$$\begin{aligned} &\sum_{\mathscr{F}^{ij} \subset \partial\Omega_i} \int_{\partial\Omega_i} [I^h\left(\theta_{\mathscr{F}^{ij}} \delta_j^\dagger (w^{(i)} - w^{(j)})\right)] \cdot n \, ds \\ &+ \sum_{\mathscr{E}^{ik} \subset \partial\Omega_i} \left\{ \int_{\partial\Omega_i} [I^h\left(\theta_{\mathscr{E}^{ik}} \delta_j^\dagger (w^{(i)} - w^{(j)})\right)] \cdot n \, ds \right. \\ &+ \left. \int_{\partial\Omega_i} [I^h\left(\theta_{\mathscr{E}^{ik}} \delta_k^\dagger (w^{(i)} - w^{(k)})\right)] \cdot n \, ds + \int_{\partial\Omega_i} [I^h\left(\theta_{\mathscr{E}^{ik}} \delta_l^\dagger (w^{(i)} - w^{(l)})\right)] \cdot n \, ds \right\} = 0. \end{aligned}$$

To satisfy this condition, we consider the face terms and the construction of the matrix U^T. The edge constraints are established using a transformation of basis with partial assembly. We have

$$\sum_{\mathcal{F}^{ij}\subset\partial\Omega_i} \int_{\partial\Omega_i} [I^h\left(\theta_{\mathcal{F}^{ij}}\delta_j^\dagger(w^{(i)} - w^{(j)})\right)] \cdot n \, ds$$

$$= \sum_{\mathcal{F}^{ij}\subset\partial\Omega_i} \int_{\partial\Omega_i} \sum_{x\in\mathcal{F}_h^{ij}} \delta_j^\dagger \left(w^{(i)} - w^{(j)}\right)(x)\varphi_x \cdot n \, ds$$

$$= \sum_{\mathcal{F}^{ij}\subset\partial\Omega_i} \sum_{x\in\mathcal{F}_h^{ij}} \left\{ \int_{\partial\Omega_i} \delta_j^\dagger \, n_1 \left(w_1^{(i)} - w_1^{(j)}\right)(x)\, \varphi_x^{(1)} \, ds \right.$$

$$\left. + \int_{\partial\Omega_i} \delta_j^\dagger \, n_2 \left(w_2^{(i)} - w_2^{(j)}\right)(x)\, \varphi_x^{(2)} \, ds + \int_{\partial\Omega_i} \delta_j^\dagger \, n_3 \left(w_3^{(i)} - w_3^{(j)}\right)(x)\, \varphi_x^{(3)} \, ds \right\}.$$

Here, we have used the notation $w^{(i)} = (w_1^{(i)}, w_2^{(i)}, w_3^{(i)})^T$ and $\varphi_x = (\varphi_x^{(1)}, \varphi_x^{(2)}, \varphi_x^{(3)})^T$. For all of those integrals, we just integrate for all $x \in \mathcal{F}_h^{ij}$ over $\partial\Omega_i \cap \text{supp}(\varphi_x) \subset \mathcal{F}^{ij}$; see Fig. 1. Thus, we obtain

$$0 = U^T B_B u_B := \sum_{\mathcal{F}^{ij}\subset\partial\Omega_i} \sum_{x\in\mathcal{F}_h^{ij}} \left\{ \int_{\mathcal{F}^{ij}} \delta_j^\dagger \, n_1^{(j)} \left(w_1^{(i)} - w_1^{(j)}\right)(x)\, \varphi_x^{(1)} \, ds \right.$$

$$\left. + \int_{\mathcal{F}^{ij}} \delta_j^\dagger \, n_2^{(j)} \left(w_2^{(i)} - w_2^{(j)}\right)(x)\, \varphi_x^{(2)} \, ds + \int_{\mathcal{F}^{ij}} \delta_j^\dagger \, n_3^{(j)} \left(w_3^{(i)} - w_3^{(j)}\right)(x)\, \varphi_x^{(3)} \, ds \right\},$$

where $n^{(j)}$ specifies the outer normal in the direction of subdomain Ω_j and

$$\sum_{\mathcal{F}^{ij}\subset\partial\Omega_i} \sum_{x\in\mathcal{F}_h^{ij}} \left(\int_{\mathcal{F}^{ij}} \delta_j^\dagger \, n_1^{(j)} \, \varphi_x^{(1)} \, ds, \int_{\mathcal{F}^{ij}} \delta_j^\dagger \, n_2^{(j)} \, \varphi_x^{(2)} \, ds, \int_{\mathcal{F}^{ij}} \delta_j^\dagger \, n_3^{(j)} \, \varphi_x^{(3)} \, ds \right)$$

defines one row in U^T. Therefore, the face condition is enforced by one constraint for each subdomain. It is also possible to enforce a stronger condition, i.e., instead

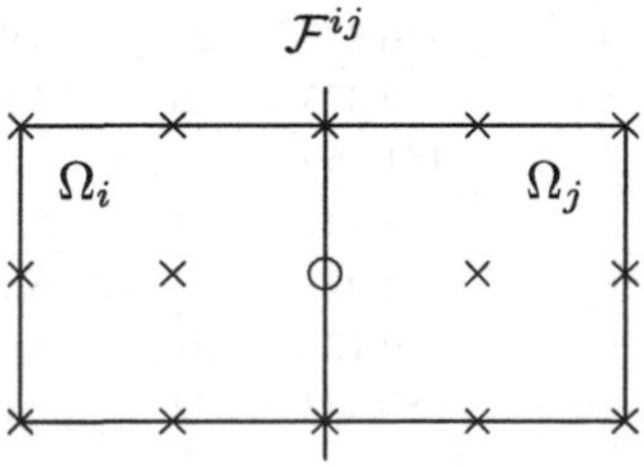

Fig. 1 A cross section of a face shared by Ω_i and Ω_j. The support of a shape function corresponding to a node on a face is a subset of the face

of ensuring the sum over all faces belonging to one subdomain to be zero, we make each additive term zero; see, e.g., [2,9,11,12]. We will refer to this approach as the standard coarse space. We then obtain one constraint for each face.

4 Numerical Results

We consider the mixed formulation of almost incompressible linear elasticity on the unit cube $\Omega = [0,1]^3$ with a Young modulus $E = 210$; see Sect. 3. The problem is discretized with $Q_2 - P_0$ mixed finite elements and the pressure is statically condensated on the element level. We apply a constant volume force, use homogeneous Dirichlet boundary conditions on $\partial\Omega$, and choose all vertices as primal constraints. The zero net flux condition for edges is established using a transformation of basis, see, e.g. [7], and for the face constraints we use projector preconditioning/deflation as described before. We run all tests for the standard coarse space and also using one constraint for each subdomain, i.e., we sum up all face constraints belonging to one subdomain and refer to this strategy as the new coarse space. All experiments were carried out using Matlab R2011b. In the first set of tests, we use a fixed Poisson ratio of $\nu = 0.499999$ and increase the number of elements, while having a constant number of subdomains, i.e., $N = 27$; see Table 1. We only observe a slight difference in the condition number, but we save 50 % of the constraints when using our new coarse space. Here, our stopping criterion for the cg method is the relative reduction of the preconditioned residual to 10^{-14}. We have chosen a very small tolerance in order to obtain an accurate eigenvalue estimate. A least square fit of the condition number estimates with a quadratic polynomial in $\log(H/h)$ confirms our theoretical condition number

Table 1 For $\nu = 0.499999$, $E = 210$, and a constant number of subdomains with $1/H = 3$, the subdomain size is increased, resulting in an increased overall number of degrees of freedom (dof)

		New coarse space (one constraint for each subdomain) # constraints: 27		**Standard coarse space** (one constraint for each face) # constraints: 54	
H/h	dof	# its	Cond	# its	Cond
2	6,591	16	2.2118	15	1.9679
3	20,577	20	3.2485	19	3.0076
4	46,875	23	3.9686	22	3.6786
5	89,373	26	4.6184	24	4.2866
6	151,959	27	5.2073	26	4.8374
7	238,521	28	5.7442	27	5.3401
8	352,947	29	6.2369	29	5.8019
9	499,125	31	6.6920	30	6.2290
10	680,943	32	7.1150	31	6.6262

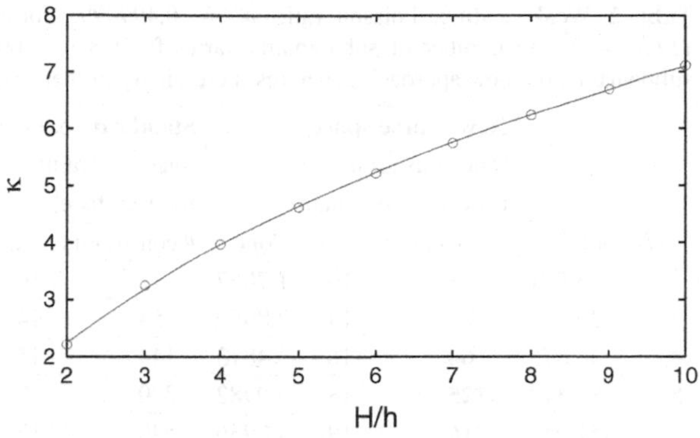

Fig. 2 Least square fit of a quadratic polynomial in $\log\left(\frac{H}{h}\right)$ to the data from Table 1 (left, new coarse space)

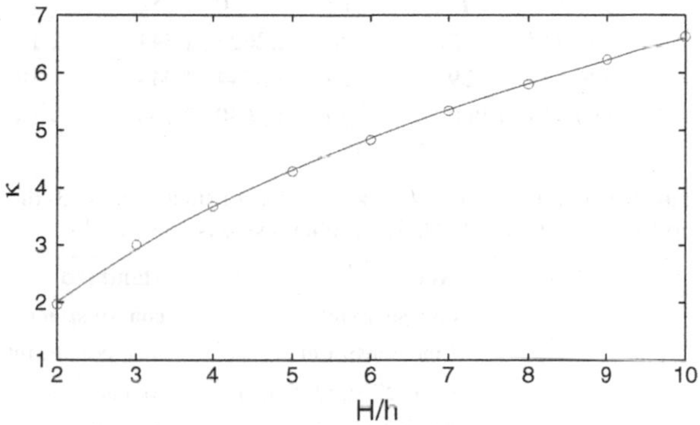

Fig. 3 Least square fit of a quadratic polynomial in $\log\left(\frac{H}{h}\right)$ to the data from Table 1 (right, standard coarse space)

estimate of $C\,(1 + \log(H/h))^2$; see Figs. 2 and 3. In Table 2, also for a fixed Poisson ratio of $\nu = 0.499999$, we present the results for weak scaling, i.e., we increase the number of subdomains from 8 to 1,000, but keep the subdomain size fixed with $H/h = 3$. The condition number does not vary significantly, but for an increasing number of subdomains the new approach is increasingly advantageous. In a last set of experiments, we consider a variable incompressibility on the whole domain, varying the Poisson ratio from $\nu = 0.4$ up to $\nu = 0.4999999999$. We fix the number of subdomains to $N = 27$ and use $H/h = 8$ elements in each direction of each subdomain; see Table 3. The condition number is bounded independently of the almost incompressibility of the material, as expected from our theory. In Table 3 (right) we also present numerical results using the coarse space for compressible linear elasticity, i.e., primal vertices and edge averages. We see that the system becomes ill-conditioned while increasing the incompressibility. In Tables 2 and 3, our stopping criterion for the cg method is the relative reduction of the preconditioned residual by 10^{-10}.

Table 2 Weak scaling. Poisson ratio $\nu = 0.499999$. For a constant subdomain size with $H/h = 3$, the number of subdomains varies from 8 to 1,000. For an increasing number of subdomains, the new approach generates increasingly higher savings

		New coarse space (one constraint for each subdomain)			**Standard coarse space** (one constraint for each face)			
$1/H$	dof	# constraints	# its	Cond	# constraints	# its	Cond	Constraints saved (%)
2	6,591	**8**	10	1.7057	12	10	1.7057	33.3
3	20,577	**27**	14	2.8989	54	14	2.5185	50.0
4	46,875	**64**	16	3.4813	144	15	2.9639	55.6
5	89,373	**125**	18	3.9982	300	17	3.3369	58.3
6	151,959	**216**	19	4.0936	540	17	3.5887	60.0
7	238,521	**343**	19	4.2320	882	18	3.7705	61.1
8	352,947	**512**	20	4.2627	1,344	18	3.8973	61.9
9	499,125	**729**	20	4.3544	1,944	18	3.9884	62.5
10	680,943	**1,000**	20	4.3630	2,700	18	4.0554	63.0

Table 3 For a constant $H/h = 8$ and a constant $1/H = 3$, the Poisson ratio varies from $\nu = 0.4$ up to $\nu = 0.4999999999$. The problem size is 352,947 dof

	New coarse space (one constraint for each subdomain) # constraints: 27		**Standard coarse space** (one constraint for each face) # constraints: 54		**Vertices + edge averages**	
ν	# its	Cond	# its	Cond	# its	Cond
0.4	19	4.3946	19	4.2827	19	5.1127
0.49	21	5.7142	20	4.2219	26	18.1360
0.4999	21	6.2180	21	5.7185	48	1.5680e+03
0.499999	22	6.2367	21	5.7954	65	1.5657e+05
0.49999999	22	6.2369	21	5.8018	76	1.5657e+07
0.4999999999	22	6.2369	22	5.8019	87	1.5657e+09

References

1. C. Dohrmann, A preconditioner for substructuring based on constrained energy minimization. SIAM J. Sci. Comput. **25**, 246–258 (2003)
2. C. Dohrmann, A substructuring preconditioner for nearly incompressible elasticity problems, Technical Report Sandia National Laboratories (2004)
3. C. Farhat, M. Lesoinne, P. LeTallec, K. Pierson, D. Rixen, FETI-DP: a dual-primal unified FETI method – part i: a faster alternative to the two-level FETI method. Int. J. Numer. Methods Eng. **50**, 1523–1544 (2001)
4. S. Gippert, A. Klawonn O. Rheinbach, Analysis of FETI-DP and BDDC for linear elasticity in 3D with almost incompressible components and varying coefficients inside subdomains. SIAM J. Numer. Anal. **50**, 2208–2236 (2012)

5. P. Goldfeld, Balancing Neumann-Neumann preconditioners for the mixed formulation of almost-incompressible linear elasticity, Ph.D. thesis, New York University (2003)
6. A. Klawonn, M. Lanser, P. Radtke, O. Rheinbach, On an Adaptive Coarse Space and on Nonlinear Domain Decomposition, in *Proceedings of the 21st International Conference on Domain Decomposition Methods*, Rennes, 25–29 June 2012. Lecture Notes in Computational Science and Engineering, vol. 98 (Springer, 2014, To appear), 12p.
7. A. Klawonn, O. Rheinbach, A parallel implementation of dual-primal FETI methods for three-dimensional linear elasticity using a transformation of basis. SIAM J. Sci. Comput. **28**, 1886–1906 (2006)
8. A. Klawonn, O. Rheinbach, Deflation, projector preconditioning, and balancing in iterative substructuring methods: connections and new results. SIAM J. Sci. Comput. **45**, A459–A484 (2012)
9. A. Klawonn, O. Rheinbach, B. Wohlmuth, Dual-primal iterative substructuring for almost incompressible elasticity, in *Domain Decomposition Methods in Science and Engineering XVI*, Lecture Notes in Computational Science and Engineering, vol. 55 (Springer, Berlin/Heidelberg, 2007), pp. 397–404
10. A. Klawonn, O.B. Widlund, Dual-primal FETI methods for linear elasticity. Commun. Pure Appl. Math. **59**, 1523–1572 (2006)
11. J. Li, A dual-primal FETI method for incompressible Stokes equations. Numer. Math. **102**, 257–275 (2005)
12. J. Li, O. Widlund, BDDC algorithms for incompressible Stokes equations. SIAM J. Numer. Anal. **44**, 2432–2455 (2006)
13. A. Toselli, O. Widlund, *Domain Decomposition Methods-Algorithms and Theory*, vol. 34 (Springer, New York, 2004)

Scalable Hybrid Parallelization Strategies for the DUNE Grid Interface

Christian Engwer and Jorrit Fahlke

Abstract The DUNE framework provides a PDE toolbox which is both flexible and efficient. Integration of hardware oriented techniques into DUNE will be necessary to maintain performance on modern and future architectures. We present the current effort to add hybrid parallelization to the DUNE grid interface, which up to now only supports MPI parallelization. In current hardware trends, we see a transition from multi-core to many-core architectures, like the Intel PHI. Techniques which worked well on traditional multi-core CPUs don't scale anymore on many-core systems. We compare different strategies to add a thread parallel layer to DUNE and discuss their scalability and performance.

1 Introduction

Numerical software currently undergoes a dramatic change. We discuss this change in the context of simulations of partial differential equations (PDEs). Since mathematical models are growing in complexity, we seek coupled multi-physics applications, and advanced numerical methods. This calls for flexible general purpose frameworks, with a large body of functionality.

At the same time the underlying hardware is posing orthogonal challenges. The memory and power wall problems are becoming hard limitations, and further performance improvements are only achieved by using all levels of parallelism and heterogeneity. Many people believe this can only be achieved by hardware-software co-design, which contradicts the flexibility goal [5].

Simulation Software becomes more specialised and at the same time generalised. On one hand, applications need to specialise toward advanced models in order to answer more detailed questions. On the other hand, it is infeasible to write such an application from scratch, so a general basis is needed to build upon. Here *Software frameworks* play an important role to provide the required flexibility. As applications

C. Engwer (✉) • J. Fahlke
Institute for Computational und Applied Mathematics, WWU Münster, Münster, Germany
e-mail: christian.engwer@uni-muenster.de; jorrit.fahlke@uni-muenster.de

A. Abdulle et al. (eds.), *Numerical Mathematics and Advanced Applications ENUMATH 2013*, Lecture Notes in Computational Science and Engineering 103,
DOI 10.1007/978-3-319-10705-9_57

continue to grow in complexity, the need for sustainable development of software for PDEs is increasing rapidly: Modern numerical ingredients such as unstructured grids, adaptivity, high-order discretizations and fast and robust multilevel solvers are required to achieve high numerical efficiency, and several physical models must be combined in challenging applications. This is beyond the scope of an individual simulation application, but requires additional support. Frameworks like Deal.II [1], Fenics [8], or DUNE [2, 3] (the one we are focusing on) support developers by providing a rich set of numerical algorithms and mathematical models. Such frameworks are designed from the beginning for flexibility and generality. Thus users can easily extend the generic framework code with their own algorithms and models. Using modern C++ techniques DUNE supports this fusion of user and framework code at compile time, which enables many compiler optimisations and thus grants flexibility and efficiency.

Hardware is undergoing a dramatic change. Current peta-scale systems in general still follow the old paradigms of high performance compute nodes, linked by fast interconnects. Both on the low-power end and at the high performance end, future systems will differ significantly: Typical workstations and cluster nodes now comprise at least two multicore CPUs and potentially several manycore accelerators such as GPUs, and energy-efficient designs such as ARM+GPU or BlueGene-Q are gaining ground. Future systems will offer much less memory per node, and show a massive increase of parallelism inside a single node, either with a 'many conventional core' approach or by combining fewer cores with specialised accelerator designs like GPUs [6]. This is a 100 to 1,000-fold increase of parallelism within each node, combined with an ever increasing impact of the memory wall problem. While message passing will still be the communication of choice between NUMA-nodes, dedicated hierarchic layers of hybrid parallelism will be necessary to exploit instruction level parallelism (ILP) and short-vector units (SIMD).

The Challenge posed for frameworks is the adoption of these new hardware paradigms. As frameworks allow for thorough user extensions at a very fine grained level, it is much harder to support modern hardware than it is for classic coarse grained libraries like BLAS. A complete rewrite of the framework for every change in hardware is not feasible and contradicts the concept of fine grained user interfaces. Thus all changes in the framework should be hidden from the user code, or at least require only moderate changes. Keeping the generality and flexibility of software frameworks while adapting them to the hardware revolution to make use of the advertised performance improvements in a transparent way is the main challenge today.

Our aim is to combine the flexibility, generality and application base of DUNE [2,3] with the concepts of 'hardware-oriented numerics' as developed in the FEAST project [9]. The hypothesis is that advanced numerical methods are the key to enable efficient use of the underlying hardware and to maintain generality alike. The work presented in this paper is part of the EXA-DUNE[1] project.

[1] http://www.sppexa.de/general-information/projects.html#EXADUNE

2 Concepts

For PDE-based simulations the computation time is dominated by two phases, the assembly and the solving of sparse linear problems. We define a partition $\mathscr{T}(\Omega)$ of the computational domain, which we refer to as the mesh or grid. This mesh induces our FEM function space and our degrees of freedom. On each cell of the mesh local contributions to the global system matrix are computed and collected in a local matrix, then these local matrices are used to update the global matrix.

To maintain performance assembling of the linear system and the linear solver need to be accelerated homogeneously. In the following we discuss the necessary changes to the DUNE grid interface and to the assembler in DUNE-PDELab. Changes to the linear algebra are not discussed in this paper and will be incorporated later.

3 Design and Implementation

The mesh is one of the key components of DUNE. The grid interface [2] follows a generic definition [3], which can be implemented in many different ways and also allows to use existing external mesh libraries through this interface. Based on this grid interface and on the linear algebra library (DUNE-ISTL) the discretization module DUNE-PDELab provides many choices of function spaces and many different discretization schemes, which the user can easily extend or combine into complete discretizations.

Up to now DUNE only considered MPI parallelization, as suitable data decomposition is directly supported by the DUNE grid interface. As DUNE supports external grid managers and many of these were only designed for MPI and don't support hybrid parallelization, we are seeking a hybrid approach which can be implemented on top of the existing grid interface. Such an interface can be implemented in a generic fashion, but can be specialised if a specific grid implementation provides additional information.

Levels of Parallelism We plan for three levels of parallelism. For an efficient assembly of the stiffness matrix and the right hand side vector, the key is concurrent access to grid information and to associated data.

Globally the grid is partitioned using the existing MPI layer. This gives coarse grained parallelism on the level of UMA nodes, where all cores within one MPI node have uniform memory access.

Within each UMA node system threads are used to share the workload among all cores. For a user-defined number of concurrent threads the grid will be locally partitioned such that each thread handles the same amount of work.

On the finest layer future extensions will make use of vectorisation (SIMD, ILP) by adapting the internal data structures used in DUNE and especially in the assembler.

Shared Memory parallelization using system threads is the main focus of our following experiments. The coarse grained message-passing level is used as it is and finer grained vectorisation level will be investigated in future work.

We introduce the concept of `EntitySets` to define iterator ranges, which describe different mesh partitions. An `EntitySet` describes a set of grid objects, e.g. a set of grid cells, which can be iterated over. For each cell and the associated sub-entities we compute indices to store data consecutively; using these indices we can directly access linear algebra vectors or matrices. As the `EntitySet` lives outside the original mesh, it can take locally varying computational costs into account.

For the local partitioning of a mesh $\mathcal{T}(\Omega)$ we consider three different strategies, where the first two are directly based on the induced linear ordering of all mesh cells $e \in \mathcal{T}(\Omega)$.

Strided: For P threads each thread p iterates over the whole set of cells $e \in \mathcal{T}(\Omega)$, but stops only at cells where $e \bmod P = p$ holds. As all P threads have to iterate over the whole grid simultaneously, they might start competing for bandwidth.

Ranged: We define consecutive iterator ranges of the size $|\mathcal{T}|/P$. This is efficiently implemented using entry points in the form of begin and end iterators. The memory requirement is $O(P)$ and thus will not strain the bandwidth.

General: Technically all other partitioning strategies will be handled in the same way. On structured meshes we can directly define geometric partitions, e.g. equidistant partitions along one or all coordinate axes (later called *sliced* or *tensor*, repectively). For unstructured meshes graph partitioning libraries like METIS or SCOTCH offer different strategies. We support all these by storing copies of all cells in an `EntitySet`. While this approach is the most flexible one, it is memory intensive, which might lead to cache trashing. The additional memory requirement is $O(|\mathcal{T}|)$, but the constant can be big, depending on the actual grid implementation.

Data Access is the other critical component. During assembly data races can occur, as different local vectors and local matrices contribute to the same global entries. Two approaches are possible to avoid race conditions: locking and colouring. As global locking is known to diminish performance as all threads are competing for this single lock, we discard this option right away and consider three different strategies:

Elock: *entity-wise locks* are expected to give very good performance, as they correspond to the granularity of the critical sections. The downside is the additional memory requirement of $O(|\mathcal{T}|)$.

Batched: *batched write* operations are a compromise between global and entity-wise locking. Threads still compete for a global lock, but the frequency of locking attempts is reduced by collecting updates in a temporary buffer. A lock is acquired when the buffer is full and all buffered updates are performed at once. The additional memory is $O(P)$ with a large constant.

Colouring: *colouring* avoids competing data access by assigning each partition to a colour such that there is no overlap between partitions of the same color. Different colors must be handled strictly in sequence, but partitions of the same color may be handled concurrently. Colouring is not meaningful for some partitioning strategies, e.g. strided and ranged. In general colouring may add to the set-up time, but requires very little memory, $O(P)$ with a small constant.

4 Performance Evaluation

Our goal is to evaluate the different strategies for system-level thread parallelization. We validate the cross-architecture scalability and formulate best practice suggestions. We restrict ourselves to the test problem of a stationary advection-diffusion equation in two dimensions

$$\nabla \cdot \{-A(x)\nabla u + b(x)u\} = f \qquad \text{in } \Omega \tag{1}$$

with Dirichlet and outflow boundary conditions and compare the performance for assembling the stiffness matrix and the residual. These are the most expensive mesh-related operations in the FE method. Equation (1) is discretized using the weighted SIPG discontinuous Galerkin method [4]. Ansatz and test space are discretized using an orthonormal P^k basis of degree k. While this scheme can actually be implemented in a completely race-free manner, doing so means that fluxes have to be computed twice.

To get a worst case estimate on the impact of different partitioning and data access strategies, we use a lightweight MPI-parallel structured mesh (`Dune::YaspGrid`) for our performance evaluations. For unstructured meshes the relative overhead will be considerably smaller and thus we expect better parallel efficiency.

We discuss timing results and scalability for the different hybridisation strategies and compare the results for a multi-core system with many-core architecture. Our experiments for the multi-core system are performed on a 4-socket (i.e. 4-UMA-node) Intel Xeon E7-4850 system at 2.00 GHz, with 10 cores per socket, 2 hyperthreads per core, 198 GB DDR3 total memory and 4 × 25.6 GB/s transfer rate. The many-core system is an Intel Xeon-PHI 5110P with 60 compute cores at 1.05 GHz, 4 hyperthreads per core, 8 GB total memory and 320 GB/s bandwidth. Performance is measured for the assembly of the residual and of the Jacobian, i.e. the right-hand side and the stiffness matrix. Additionally we compare results for different polynomial degrees.

On the CPU we observe good scalability for all partitioning strategies, see Fig. 1. This is in correspondence to the experimental hybrid parallelization discussed in [7]. The parallel efficiency of the residual drops down to ~50 %, whereas the Jacobian keeps an efficiency ≥60 % up to the 10 physical cores. Hyper-threading improves the run times further, even though the efficiency drops significantly. The better

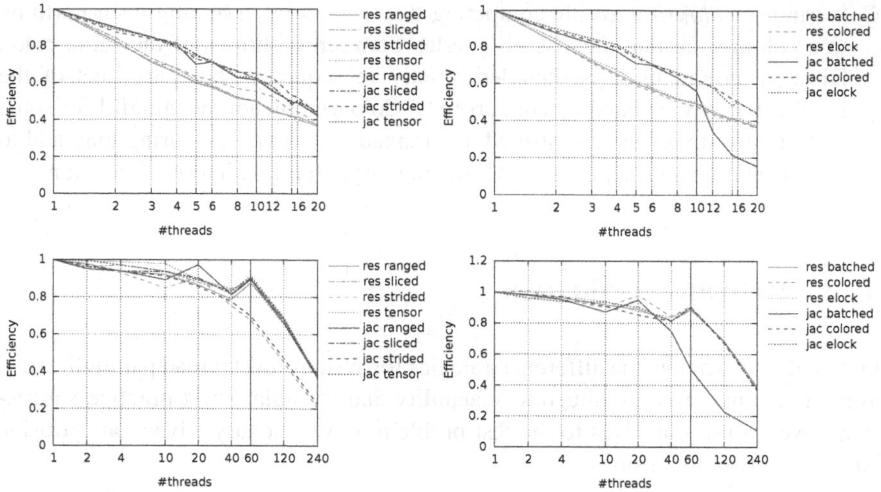

Fig. 1 Parallel efficiency of the assembly of residual (*res*) and Jacobian (*jac*) for different partitioning (*left*, entity-wise locking) and locking (*right*, sliced partitioning) strategies on CPU and PHI, polynomial degree $k = 1$

efficiency of the Jacobian is due to the increased algorithmic intensity for higher polynomial degrees.

For data access we compare batched writes, entity-wise locks and a lock-free strategy, via colouring. In our experiments the performance of entity-wise locking and colouring is comparable. Batched writes pose difficulties when assembling the Jacobian: the performance of batched writes depends severely on the size of the temporary buffer and the sparsity pattern, which means that good performance can only be achieved by tuning buffer sizes.

By comparing different partitioning strategies possible performance issues become more visible. While strided partitioning is attractive in multi-core CPU systems, it does not scale to the larger numbers of cores on the PHI. This is to be expected: for this kind of partitioning, all threads will usually operate on nearby mesh cells at any given time. On the CPU they benefit from the level 3 cache shared by all 10 cores: one thread is likely to access data that a different thread has just loaded. On the PHI each of the 60 cores has its own cache: the cores compete for the memory bandwidth to transfer cache lines, but only a small part of each cache line is actually used for computation.

Apart from strided partitioning, the choice of partitioning strategy has very little effect. This means that the memory bandwidth has not been reached for these schemes; we expect this effect to become important when element local computations are vectorized. In this case it will be necessary to further increase the algorithmic intensity—either by the use of significantly higher polynomial orders, or by using locally structured low-order computations.

Table 1 Comparison of different polynomial degrees k, number of threads P, and hardware X. Time per DOF t_P^X [μs] and efficiency E_P^X of the Jacobian assembly using sliced partitioning and entity-wise locking. We see a clear benefit from higher order discretizations, due to the increased algorithmic intensity

k	t_1^{CPU}	t_{10}^{CPU}	t_{20}^{CPU}	E_{10}^{CPU}	E_{20}^{CPU}	t_1^{PHI}	t_{60}^{PHI}	t_{120}^{PHI}	t_{240}^{PHI}	E_{60}^{PHI}	E_{120}^{PHI}	E_{240}^{PHI}
0	4.59	0.74	0.54	62%	42%	59.57	1.33	1.17	1.20	75%	43%	21%
1	1.38	0.22	0.17	62%	42%	18.92	0.37	0.27	0.26	84%	57%	30%
2	1.10	0.15	0.12	72%	46%	17.12	0.32	0.21	0.19	90%	69%	38%
3	1.29	0.16	0.13	79%	50%	19.84	0.36	0.23	0.20	92%	72%	41%
4	1.52	0.18	0.15	87%	49%							
5	1.81	0.21	0.18	88%	51%							

Table 1 shows the computation time per DOF and the obtained efficiency for different polynomial degrees in the DG discretization. Due to increased algorithmic intensity the efficiency increases for higher order discretizations. Hyperthreading does diminish the efficiency, but still gives a slight improvement in computation time. In computation time the Xeon PHI does not pay of, as the current implementation does not use vectorisation, which necessary to unlock the potential of the PHI.

Summary and Conclusions

We have shown that many-core architectures require additional care in designing the thread parallelism for hybrid simulations. As many mesh libraries were only designed for distributed memory, using MPI parallelization, we designed the extensions such that thread parallelism can be implemented ontop of an existing DUNE grid. Support for this additional layer was added to the DUNE-PDELab module. We emphasise that from the user point of view the changes are totally transparent and hidden underneath the discretization interface.

We demonstrated performance tests on an Intel Xeon PHI and compared with results for a 4-socket Intel Xeon E7-4850 system. With a ranged partitioning and entity-wise locking, or with colouring and the according partitioning, it is possible to provide a low overhead thread parallelization layer, which shows good performance on classic multi-core CPUs and on modern many-core systems alike. The performance gain from coloring is negligible, but increases code complexity, so that this approach is less favourable. We increased the efficiency further to ~90 % by the use of higher order methods.

Supporting modern hardware paradigms is possible within a general purpose interface, without sacrificing flexibility and still obtain good performance. From the user's perspective all changes are completely transparent.

(continued)

Future work will investigate how to add SIMD and wide-SIMD support during the assembly of stiffness matrices and residuals and incorporate SIMD support in the linear algebra. Adding SIMD support is still an open issue, because in order to keep flexibility, it is no option to directly use intrinsics in the user code.

Acknowledgements This work was supported by the German Research Foundation (DFG) through the Priority Programme 1648 'Software for Exascale Computing' (SPPEXA).

References

1. W. Bangerth, C. Burstedde, T. Heister, M. Kronbichler, Algorithms and data structures for massively parallel generic adaptive finite element codes. ACM Trans. Math. Softw. **38**(2), 14:1–14:28 (2012)
2. P. Bastian, M. Blatt, A. Dedner, C. Engwer, R. Klöfkorn, R. Kornhuber, M. Ohlberger, O. Sander, A generic grid interface for parallel and adaptive scientific computing. part II: implementation and tests in DUNE. Computing **82**(2–3), 121–138 (2008)
3. P. Bastian, M. Blatt, A. Dedner, C. Engwer, R. Klöfkorn, M. Ohlberger, O. Sander, A generic grid interface for parallel and adaptive scientific computing. part I: abstract framework. Computing **82**(2–3), 103–119 (2008)
4. A. Ern, A.F. Stephansen, P. Zunino, A discontinuous galerkin method with weighted averages for advection–diffusion equations with locally small and anisotropic diffusivity. IMA J. Numer. Anal. **29**(2), 235–256 (2009)
5. X.S. Hu, R.C. Murphy, S. Dosanjh, K. Olukotun, S. Poole, Hardware/software co-design for high performance computing: challenges and opportunities, in *2010 IEEE/ACM/IFIP International Conference on Hardware/Software Codesign and System Synthesis (CODES+ISSS)*, Scottsdale (IEEE, 2010), pp. 63–64
6. D.E. Keyes, Exaflop/s: the why and the how. Comptes Rendus Mécanique **339**(2–3), 70–77 (2011)
7. R. Klöfkorn, Efficient matrix-free implementation of discontinuous galerkin methods for compressible flow problems, in *ALGORITMY 2012, Proceedings of Contributed Papers and Posters*, Podbanske, ed. by A. Handlovičová, Z. Minarechová, D. Ševčovič (Publishing House of STU, 2012), pp. 11–21. http://www.iam.fmph.uniba.sk/algoritmy2012/
8. A. Logg, K.-A. Mardal, G. Wells, *Automated Solution of Differential Equations by the Finite Element Method* (Springer, Berlin/New York, 2012)
9. S. Turek, D. Göddeke, C. Becker, S. Buijssen, S. Wobker, FEAST – Realisation of hardware-oriented numerics for HPC simulations with finite elements. Concurr. Comput.: Pract. Experience **22**(6), 2247–2265 (2010)

Unveiling WARIS Code, a Parallel and Multi-purpose FDM Framework

Raúl de la Cruz, Mauricio Hanzich, Arnau Folch, Guillaume Houzeaux, and José María Cela

Abstract WARIS is an in-house multi-purpose framework focused on solving scientific problems using Finite Difference Methods as numerical scheme. Its framework was designed from scratch to solve in a parallel and efficient way Earth Science and Computational Fluid Dynamic problems on a wide variety of architectures. WARIS uses structured meshes to discretize the problem domains, as these are better suited for optimization in accelerator-based architectures. To succeed in such challenge, WARIS framework was initially designed to be modular in order to ease development cycles, portability, reusability and future extensions of the framework. In order to assess its performance, a code that solves the vectorial Advection-Diffusion-Sedimentation equation has been ported to the WARIS framework. This problem appears in many geophysical applications, including atmospheric transport of passive substances. As an application example, we focus on atmospheric dispersion of volcanic ash, a case in which operational code performance is critical given the threat posed by this substance on aircraft engines. Preliminary results are very promising, performance has been improved by 8.2× with respect to the baseline code using a realistic case. This opens new perspectives for operational setups, including efficient ensemble forecast.

1 Introduction

Many relevant problems arising in geoscience and Computational Fluid Dynamics (CFD) can be solved numerically using Finite Difference Methods (FDM) on structured computational meshes. Examples include, seismic wave propagation, numerical weather prediction or atmospheric transport. FDM numerical schemes on structured meshes allow peak performances of $\approx 20 - 30\,\%$, about 3 times larger than analogous FE (Finite Element) methods.

R. de la Cruz (✉) • M. Hanzich • A. Folch • G. Houzeaux • J.M. Cela
CASE Department, Barcelona Supercomputing Center, Barcelona, Spain
e-mail: delacruz@bsc.es; mauricio.hanzich@bsc.es; arnau.folch@bsc.es; guillaume.houzeaux@bsc.es; josem.cela@bsc.es

A. Abdulle et al. (eds.), *Numerical Mathematics and Advanced Applications ENUMATH 2013*, Lecture Notes in Computational Science and Engineering 103,
DOI 10.1007/978-3-319-10705-9_58

WARIS is a brand-new multi-purpose framework aimed at solving efficiently in a parallel way this kind of scientific computing problems. The design requirements were to obtain a portable framework (i.e. able to run on any hardware platform) suited for accelerated-based architectures, with reusable software components, easily extendible, and able to solve the physical problems on structured meshes explicitly, implicitly or semi-implicitly.

The next sections are organized as follows: Sect. 2 describes the procedure followed during the development stages of WARIS, and introduces the current state-of-the-art regarding FDM optimizations as well. Section 3 elaborates on the final design chosen for the WARIS framework, detailing also its internals. Finally, Sect. 4 and section "Conclusions" expose a case of study and the conclusions respectively.

2 Software Engineering

Software engineering plays an important role when achieving a good software design. The desirable aspects of a software package are reliability, efficiency, and robustness; bestowing accurate results, high performance and solid components respectively. Meeting all those requirements lead to a successful project. In order to succeed in such a challenge, an application development life-cycle methodology must be applied [11]. Besides, life-cycle methodology is particularly important when HPC software comes into play. This importance is due to several inherent needs of the HPC software.

Firstly, given that the numerical software is clearly expensive to develop due to the involved research process, it is intended to reuse code as much as possible. A modular and flexible design of the software framework may help in this approach owing to many important reasons. Software must be flexible and modular whether same code is wanted to be used for different scientific problems and likewise, be adapted to a novel hardware architecture or a different programming model. In this regard, the HPC is far more dynamic in terms of adaptability requirements than the rest of the computational areas, and therefore these reasons have a greater influence. Secondly, the HPC involves large numerical simulations which may require hundreds or thousands of computational nodes during days or weeks. Unsuccessful executions (abnormal end) might arise from time to time, leading to unfruitful runs and a waste of time and resources. Hence, the cost of these executions is utterly high in terms of power consumption and resource usage.

Finally, a modular and robust infrastructure is crucial, but performing the numerical simulations in an efficient way is also a key point. Additionally, considering that each simulation run can last several days or weeks, even a mild optimization in performance of only 5–10 % may lead to days of savings in core-hours of computational resources. In order to do so, unlike ordinary development life-cycle, an additional stage for the optimization process of the modules is also included. This stage is repeatedly performed in order to successively optimize the performance of each module.

2.1 *Boosting Numerical Codes*

Irregular codes (stencil computations and sparse algebra) are usually limited by memory access (memory bound). Therefore, the ratio of floating-point operations to memory access is low compared with regular codes (FFT and dense linear algebra), which are mainly compute bound. In explicit FDM schemes, the basic structure of stencil computation is that the central point accumulates the contribution of neighbor points in every axis of the Cartesian system. The number of neighbor points in every axis relates to the accuracy level of the stencil, where more neighbor points lead to higher accuracy. The stencil computation is then repeated for every point in the computational domain, thus solving the spatial differential operator.

Two inherent problems can be identified from the structure of the stencil computation. Firstly, the non-contiguous memory access pattern. In order to compute the central point of the stencil, a set of neighbors has to be accessed. Some of these neighbor points are distant in the memory hierarchy, requiring many cycles in latencies to be accessed [6]. Secondly, the low computational-intensity and reuse ratios. After gathering the set of data points, just one central point is computed and only the accessed data points in the sweep direction might be reused for the computation of the next central point [5].

The state-of-the-art in performance optimizations for stencil computation is very prolific. The contributions can be divided into three dissimilar groups: space blocking, time blocking and pipeline optimizations. Space blocking algorithms promote data reuse by traversing data in a specific order. Space blocking is especially useful when the dataset structure does not fit into the memory hierarchy [6, 13]. Time blocking algorithms [9] perform loop unrolling over time-step sweeps to exploit the grid points as much as possible, and thus increase data reuse. Finally, low level optimizations at the CPU pipeline include several well-known techniques. These techniques may be categorized into loop transformations [10], data access [2] and streaming optimizations (SMP, SIMD and MIMD).

3 System Architecture

As the number of physical problems that should be supported could be in the order of tenths, the primary system (named Physical Simulator Kernel, PSK) should be flexible enough to accommodate new problems reusing as much code as possible.The PSK framework is divided in two main sets of components. On one hand, there is a *framework* responsible for those tasks that are common to any physical simulation being solved, such as domain decomposition, communications and I/O operations. On the other hand there is a set of *specializations* that are used to configure the framework in order to have a complete solution for a given physical problem. Those specializations depend on aspects such as: the physical problem, the hardware platform (e.g. general purpose, GPU, FPGA, Xeon Phi) and the programming model being used for development. As the PSK main goal is to generalize the way a physical simulation is built, some aspects have to be fixed and restricted, in order to bound the framework functionality and limits.

3.1 Hardware Architecture Model

The first aspect to be considered is the computational architecture model that will be supported by the PSK. Figure 1 shows the concepts in such model and their relation. The main building block of the architecture model is the *Computational Node* (CN), which is built using both the *host* and *device* elements that communicate through a *Common Address Space* (CAS) memory. Examples of such devices include GPUs [15] and brand-new Intel's Xeon Phi [12]. Following this model the PSK framework is executed in the *host* while the *device* processes the specifics of the physical problem being simulated.

In order to run a physical simulation the PSK will construct a defined structure of MPI processes and threads across the CNs to be used for such simulation. Figure 2. Left shows the structure for a simulator using two CNs: CN0 and CN1. As the memory address space of CN0 and CN1 are disjoint, the PSK system must provide *Domain Decomposition* (DD) between both CNs (known as extra-node DD), in order to coordinate the simulation process. Moreover, there are some cases (e.g. multi- and many-core architectures), for which the PSK have to provide DD also inside a CN (known as intra-node DD). Notice, that each small node into Fig. 2. Left represents threads and I/O devices, whereas the small and large boxes are the MPI processes running on each CN and the intra-node domains in a CN respectively.

3.2 Software Architecture Model

Once the hardware architecture model is defined, this subsection will depict the software architecture model. That is, which is the PSK framework architecture

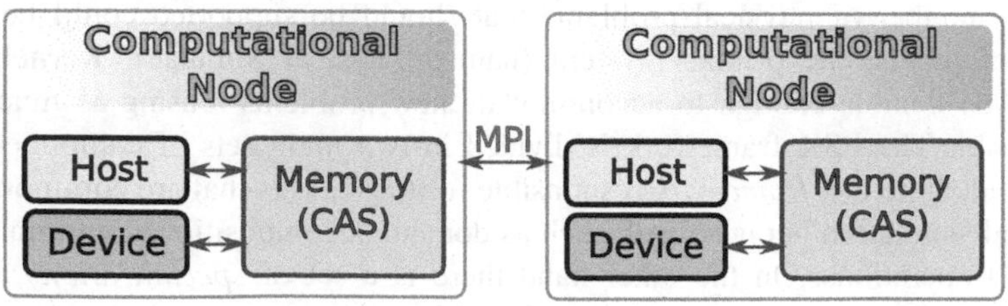

Fig. 1 Hardware architecture model supported by the PSK

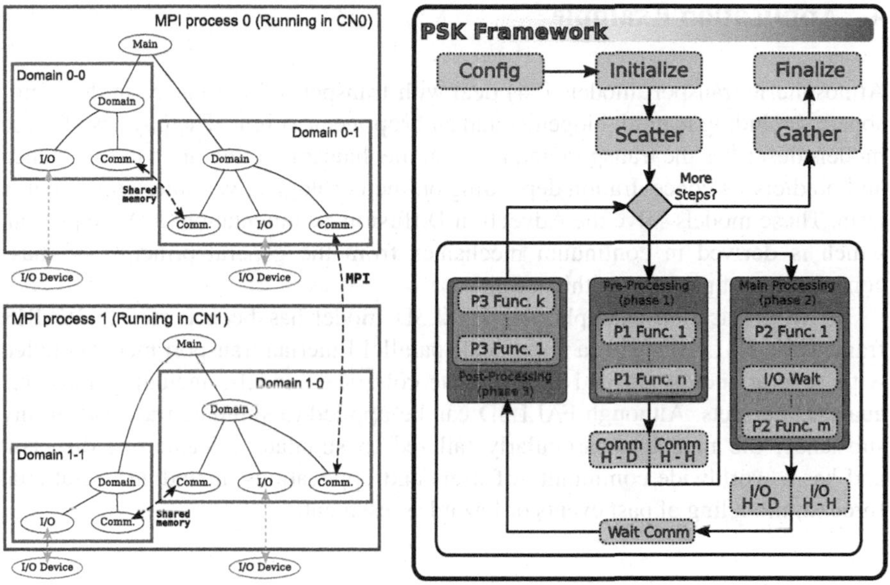

Fig. 2 *Left*: domain decomposition model. *Right*: PSK software model

and what is to be done in order to use it and extend or specialize it for a specific physical problem. The specialization process is done by implementing an interface defined by the PSK. Each of the functions to be implemented are known as a *specialized functions*. Among this functions there are: initialization and finalization routines for managing data structures that belongs to the physical problem at hands, proper functions for the processing at each iteration of the simulation process, or some functions for scattering and gathering data among different domains if they are needed. Figure 2. Right shows the PSK framework structure. The dashed boxes represent the functions to be provided by the user in order to specialize the framework for a specific physical problem.

Regarding the PSK main structure (i.e. inside the main loop), it is divided in three different phases (known as Pre- Main- and Post-Processing), which are separated by some stages such as communications or I/O. The aim of such structure is to provide an environment capable of overlapping computation with communication and I/O. In order to do so, the functions provided by the user for *phase 1* (P1) must process all the physical problem involved in the exchange stage for the current iteration step. Then, an asynchronous communication of these areas is started while the computation in *phase 2* (P2) processes the remaining domain for the current iteration. Likewise, an I/O operation may be started asynchronously at the end of *phase 2*, enabling as well overlapping with main computation. Finally, an optional *phase 3* (P3) is also considered for such cases where some processing is needed after the communication, but prior to the next iteration.

4 Application Example

Atmospheric transport models [14] deal with transport of substances in the atmosphere, including natural, biogenic, and anthropogenic origin. The physics of these models describes the transport and removal mechanisms acting upon the substance and predicts its concentration depending on meteorological variables and a source term. These models solve the Advection-Diffusion-Sedimentation (ADS) equation, which is derived in continuum mechanics from the general principle of mass conservation of particles within a fluid.

As an application example, the FALL3D model has been ported to WARIS framework. FALL3D [4] is a multi-scale parallel Eulerian transport model coupled with several mesoscale and global meteorological models, including most reanalyses datasets. Although FALL3D can be applied to simulate transport of any substance, the model is particularly tailored to simulate volcanic ash dispersal and has a worldwide community of users and applications, including operational forecast, modeling of past events or hazard assessment.

4.1 Volcanic Ash Dispersal

Volcanic ash generated during explosive volcanic eruptions can be transported by the prevailing winds thousands of kilometers downwind posing a serious threat to civil aviation. FALL3D models run worldwide operationally to provide advice to the civil aviation authorities, which need to react promptly in order to prevent in-flight aircraft encounters with clouds. Here, we investigate to which extent WARIS-Transport can accelerate model forecasts and could be used for ensemble forecasting [1]. We focus on a paradigmatic case occurred during April-May 2010, when ash clouds from the Eyjafjallajökull volcano in Iceland disrupted the European airspace for almost 1 week, resulting in thousands of flight cancellations and millionaire economic loss [7].

The following results include several performance techniques carried out in the WARIS-Transport module. These techniques are: *SIMDization*, blocking and pipeline optimizations of the explicit kernels. Furthermore, the parallel I/O operations have been dramatically improved by implementing an active buffering strategy [8] with two-phase collective I/O calls [3]. As an example, Fig. 3 shows how a wise choice of the blocking parameter is crucial to reduce the execution time of the explicit kernel. In this particular case ($256 \times 2048 \times 64$ domain size), the kernel execution time has been reduced by 24.1 %.

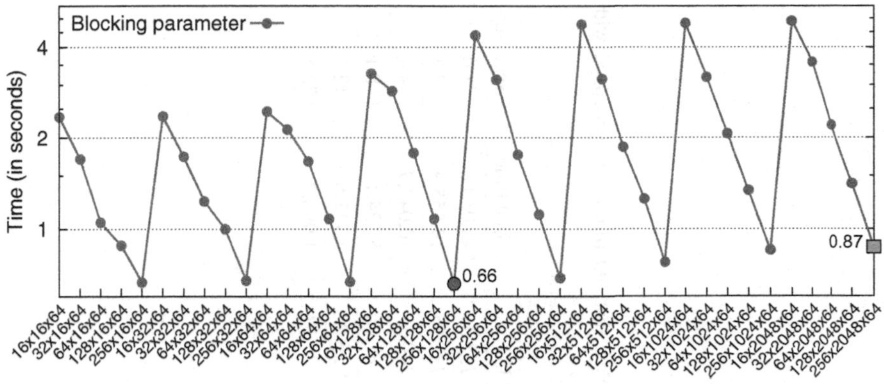

Fig. 3 Blocking impact on the kernel execution time. The small rectangle at the right shows the performance of the naive implementation, whereas the circle at the middle depicts the best blocking parameter. WARIS automatically selects the best parameter

All the tests have been conducted in MareNostrum supercomputer (Intel SandyBridge-EP E5-2670) with different number of processors. The Eyjafjallajökull case involves an input dataset of 9 GB of meteorological data, corresponding to 8 days of eruption, and 370 MB of concentration and dispersal simulated output data for the coarse-grain mesh (41 × 241 × 141). The FALL3D code taken as a reference requires 1 h and 58 min to complete this simulation on 16 processors with MPI. On the other hand, the WARIS-Transport module only took 14.4 min to process it. These times make our implementation 8.2× faster than Fall3D code. Additionally, strong scaling results were obtained for WARIS-Transport using a fine-grain mesh (41 × 480 × 280) of the Eyjafjallajökull case. Scalability results obtained for different number of processors (up to 32) are broken down in Table 1. Results are categorized in four groups, explicit kernel (P1 and P2 stages), other computations (P3 stage), meteorological input and ash dispersal output. P1 (boundary elements), P2 (internal elements) and P3 stages refer to the kernel functions in the PSK framework structure. Finally, postprocess and preprocess columns consider the data arrangement computation required after reading meteorological data and before writing ash dispersal results respectively.

Table 1 WARIS-Transport module times (in seconds) with fine-grain mesh of Eyjafjallajökull volcano case. Note that I/O times are overlapped with computation thanks to a double-buffering system and a dedicated I/O thread

	Explicit kernel		Others	Meteo data		Output			
Num. proc.	P1 stage	P2 stage	P3 stage	I/O	Postprocess	Preprocess	I/O	Total time	Speed-up
1	0.0	28,932.83	4,718.99	*302.75*	11,108.91	250.06	*133.30*	45,010.8	1.0
2	202.21	12,004.14	1,854.76	*142.79*	5,408.69	86.14	*131.25*	19,555.9	2.3
4	208.52	6,076.23	942.60	*106.75*	2,781.53	39.11	*140.52*	10,047.9	4.4
8	233.89	3,170.07	528.48	*71.08*	1,573.25	18.44	*81.86*	5,524.1	8.1
16	291.52	2,099.81	391.37	*70.15*	917.97	10.93	*29.09*	3,711.6	12.1
32	287.49	967.23	192.83	*84.29*	546.03	4.66	*11.81*	1,998.2	22.5

Conclusions
WARIS framework has shown appealing capabilities by providing successful support for scientific problems using FDM. In the foreseeable future, as the amount of computational resources will increase, more sophisticated physics may be simulated. Furthermore, it provides support for a wide-range of hardware platforms. Therefore, as the computational race keeps the hardware changing every day, support for specific platforms that will give the best performance results will be supplied for the different simulated physics.

References

1. C. Bonadonna, A. Folch, S. Loughlin, H. Puempel, Future developments in modelling and monitoring of volcanic ash clouds: outcomes from the first IAVCE-WMO workshop on ash dispersal forecast and civil aviation. Bull. Volcanol. **74**, 1–10 (2012)
2. D. Callahan, K. Kennedy, A. Porterfield, Software prefetching, in *ASPLOS-IV: Proceedings of the Fourth International Conference on Architectural Support for Programming Languages and Operating Systems*, Santa Clara (ACM, New York, 1991), pp. 40–52
3. J.M. del Rosario, R. Bordawekar, A. Choudhary, Improved parallel I/O via a two-phase run-time access strategy. SIGARCH Comput. Archit. News **21**(5), 31–38 (1993)
4. A. Folch, A. Costa, G. Macedonio, Fall3d: a computational model for transport and deposition of volcanic ash. Comput. Geosci. **35**(6), 1334–1342 (2009)
5. M. Frigo, V. Strumpen, Cache oblivious stencil computations, in *19th ACM International Conference on Supercomputing*, Cambridge, June 2005, pp. 361–366
6. S. Kamil, P. Husbands, L. Oliker, J. Shalf, K. Yelick, Impact of modern memory subsystems on cache optimizations for stencil computations, in *MSP '05: Proceedings of the 2005 workshop on Memory System Performance*, Chicago (ACM, New York, 2005), pp. 36–43
7. B. Langmann, A. Folch, M. Hensch, V. Matthias, Volcanic ash over europe during the eruption of Eyjafjallajökull on iceland, April–May 2010. Atmos. Environ. **48**, 1–8 (2012)
8. X. Ma, M. Winslett, J. Lee, S. Yu, *Improving MPI-IO Output Performance with Active Buffering Plus Threads*, IPDPS '03, Nice (IEEE Computer Society, Washington, DC, 2003), pp. 68.2–
9. J. McCalpin, D. Wonnacott, Time skewing: a value-based approach to optimizing for memory locality, Technical Report DCS-TR-379, Department of Computer Science, Rutgers University, 1999
10. K.S. McKinley, S. Carr, C.-W. Tseng, Improving data locality with loop transformations. ACM Trans. Program. Lang. Syst. **18**, 424–453 (1996)
11. W.L. Oberkampf, C.J. Roy, *Verification and Validation in Scientific Computing*, 1st edn. (Cambridge University Press, New York, 2010)
12. J. Reinders, *An Overview of Programming for Intel Xeon processors and Intel Xeon Phi coprocessors*, (Intel, 2012)
13. G. Rivera, C.W. Tseng, Tiling optimizations for 3D scientific computations, in *Proceedings of ACM/IEEE Supercomputing Conference (SC 2000)*, Dallas (IEEE Computer Society, Washington, DC, Nov 2000), p. 32
14. A. Russell, R. Dennis, Narsto critical review of photochemical models and modelling. Atmos. Environ. **34**, 2261–2282 (2000)
15. J. Sanders, E. Kandrot, *CUDA by Example: An Introduction to General-Purpose gpu Programming* (Addison-Wesley Professional, Upper Saddle River, 2010)

Integrating Multi-threading and Accelerators into DUNE-ISTL

Steffen Müthing, Dirk Ribbrock, and Dominik Göddeke

Abstract A major challenge in PDE software is the balance between user-level flexibility and performance on heterogeneous hardware. We discuss our ideas on how this challenge can be tackled, exemplarily for the DUNE framework and in particular its linear algebra and solver components. We demonstrate how the former MPI-only implementation is modified to support MPI+[CPU/GPU] threading and vectorisation. To this end, we devise a novel block extension of the recently proposed SELL-C-σ format. The efficiency of our approach is underlined by benchmark computations that exhibit reasonable speedups over the CPU-MPI-only case.

1 Introduction

Software development, in the scope of our work for the numerical solution of a wide range of PDE (partial differential equations) problems, faces contradictory challenges. On the one hand, users and developers prefer flexibility and generality, on the other hand, the changing hardware landscape requires algorithmic adaptation and specialisation to be able to exploit a large fraction of peak performance.

1.1 Software Frameworks

A framework approach for entire application domains rather than distinct problem instances targets the first challenge. We are particularly interested in frameworks

S. Müthing
Interdisciplinary Center for Scientific Computing, University of Heidelberg, Heidelberg, Germany
e-mail: steffen.muething@iwr.uni-heidelberg.de

D. Ribbrock • D. Göddeke (✉)
Institute of Applied Mathematics (LS3), TU Dortmund, Dortmund, Germany
e-mail: dirk.ribbrock@math.tu-dortmund.de; dominik.goeddeke@math.tu-dortmund.de

A. Abdulle et al. (eds.), *Numerical Mathematics and Advanced Applications ENUMATH 2013*, Lecture Notes in Computational Science and Engineering 103,
DOI 10.1007/978-3-319-10705-9_59

for the solution of PDE problems with grid-based discretisation techniques. In contrast to the more conventional approach of developing in a 'bottom-up' fashion starting with only a limited set of problems (likely, a single problem) and solution methods in mind, frameworks are designed from the beginning with flexibility and general applicability in mind so that new physics and new mathematical methods can be incorporated more easily. In a software framework the generic code of the framework is extended by the user to provide application specific code instead of just calling functions from a library. Template meta-programming in C++ supports this extension step in a very efficient way, performing the fusion of framework and user code at compile time which reduces granularity effects and enables a much wider range of optimisations by the compiler.

1.2 Target Applications and Numerical Approach

Our work within the EXA-DUNE project ultimately targets applications in the field of porous media simulations. These problems are characterised by strongly varying coefficients and extremely anisotropic meshes, which mandate powerful and robust solvers and thus do not lend themselves to the current trend in HPC towards matrix-free methods with their beneficial properties in terms of memory bandwidth and / or FLOPs/DOF ratio; typical matrix-free techniques like Cholesky preconditioning and stencil-based geometric multigrid are not suited to those types of problems. For that reason we aim at algebraic multigrid (AMG) preconditioners known to work well in this context, and work towards further improving their scalability and (hardware) performance.

1.3 Hardware Development

Future exascale systems are characterised by a massive increase in node-level parallelism and heterogeneity. Current examples include nodes with multiple conventional CPU cores arranged in different sockets. GPUs require much more fine-grained parallelism, and Intel's Xeon Phi design shares similarities with both these extremes. One important common feature of all these architectures is that reasonable performance can only be achieved by explicitly using their (wide-) SIMD capabilities. The situation becomes more complicated as different programming models, APIs and language extensions are needed, which lack performance portability. Instead, different data structures and memory layouts are required for different architectures. In addition, it is no longer possible to view the available off-chip DRAM memory within one node as globally shared in terms of performance. Firstly, accelerators are typically equipped with dedicated memory, which improves accelerator-local latency and bandwidth substantially, but at the same time suffers from a (relatively) slow connection to the host. Due to NUMA (non-uniform memory access) effects, a similar (albeit less dramatic in absolute numbers) imbalance can already be observed on multi-socket multi-core CPU systems. There is common agreement in the community that the existing MPI-only programming model has reached its limitations. The most prominent successor will likely be 'MPI+X', so that MPI can still be used for coarse-grained communication, while some kind of shared memory abstraction is used within MPI processes.

1.4 Consequences and Challenges

Obviously, the necessary adaptations to the changing hardware landscape should be hidden as much as possible from the user. The conventional library-based approach to software development only addresses this challenge at the component level, forcing users to manually integrate those changes into their applications. Software frameworks aid the user with higher abstraction and integration levels, which isolate applications from hardware-specific implementation details. Nonetheless, frameworks still face the conflict between generality, flexibility and API stability on the user side and the need to adapt to new hardware and its potentially disruptive programming models 'under the hood' for optimal performance. Finding the right balance between those extremes defines the challenge of effective framework development in HPC.

1.5 Paper Contribution

In the EXA-DUNE project, we pursue different avenues to preparing the DUNE framework [3, 4] for the exascale era.[1] The goal is to combine the flexibility, generality and application base of DUNE with the concepts of hardware-oriented numerics as developed in the FEAST project [12]. In this paper, we report on our first results and design decisions. We focus on extending DUNE's linear solver module ISTL with architecture-aware backends for low-level constructs like vectors, matrices and preconditioners.

2 Design and Implementation

Our approach to enable hybrid parallelism and memory heterogeneity within the DUNE package can be categorised as follows: (1) We redesign the linear algebra part of ISTL (DUNE's linear solver library) around a novel block extension of

[1] http://www.sppexa.de/general-information/projects.html#EXADUNE

the SELL-C-σ format [8] so that existing solvers need not be modified to benefit from CPU/GPU threading. (2) We equip PDElab (DUNE's 'user interface') with support for this format so that existing user-level code and other DUNE components can, e.g., directly assemble into this format without the need for costly format conversions. In the following, we describe our changes in a bottom-up fashion.

2.1 *Operations: Linear Algebra Kernels*

Independent of the architecture, performance improvements not related to mathematically superior algorithms stem from threading and vectorisation (SIMD units, UMA domains). This distinction is explicit on modern multicore CPUs and the Xeon Phi, and implicit on GPUs. We tackle the vectorisation level by creating two collections of linear algebra kernels, one based on CUDA for NVIDIA GPUs and a shared one based on Intel's Threading Building Blocks (TBB) for multi-core CPUs and Xeon Phi. Our design can easily encorporate other specialisations. Following the DUNE philosophy, all new kernels are integrated into ISTL via (new) C++ template interfaces, enabling standard architecture-dependent optimisations for data types, SIMD block sizes and data alignment independent of the interfaces. We choose not to manually program CPU vector units using compiler intrinsics because of the large maintenance overhead every time a new SIMD instruction set is released. Instead, we rely on the auto-vectorisers of modern compilers, which we feed with explicit aliasing and alignment hints for the data arrays: Both GCC 4.8 and ICC 14.0 are able to generate vectorised code of sufficient quality, as verified by inspecting the generated assembly code. The actual kernel implementation follows standard approaches as reported elsewhere in abundance.

2.2 *Containers: Matrices and Vectors*

One central design choice for implementing numerical linear algebra on heterogeneous architectures is the issue of matrix (and associated vector) formats. On CPUs, (block) CRS is the general format of choice so far [1, 7], while the GPGPU community prefers ELL-like formats that enable a more efficient use of wide-SIMD [8]. Format conversions between architecture-optimised formats pose a severe bottleneck and should thus be avoided. Recent work by Kreutzer et al. [8] indicates a feasible solution: Their SELL-C-σ format potentially constitutes a 'best compromise' across architectures. Note that we do not implement a matrix reordering step in our adoption of the format, as our matrices have a very uniform row length distribution and thus, padding is not excessive and need not be avoided.

Other high-level PDE solver frameworks like FEniCS [9] and deal.II [2] support hybrid parallelism via MPI + OpenMP / TBB, but rely on existing libraries at the linear algebra level and do not fully support other accelerators. The linear algebra packages PETSc [1], Trilinos [7] and MTL [10] all support (at least in part) threading and/or GPUs on top of MPI, but exploiting these features from higher-level projects can prove difficult.

2.3 Block Matrices

As part of EXA-DUNE, we are investigating Discontinuous Galerkin (DG) discretisations for porous media simulations. Vectors and matrices associated with DG discretisations exhibit a natural block structure at the mesh cell level, which can be exploited by storing only block sparsity patters. As almost all of the memory required by a CRS or ELL matrix is taken up by two arrays storing the non-zero entries and their associated column indices, storing only block column indices implies a factor of $\approx \frac{1+b^2}{2}$ for a block size b in storage and bandwidth (cf. Fig. 1).

Block CRS matrices work by storing the individual blocks as small, dense matrices, which is a well-known and widely-used optimisation technique that can be implemented fairly easily [1, 7]. The efficient implementation proves more difficult in our setting: ELL-like formats and SIMD-awareness require coalescing a number of matrix entries that corresponds to the SIMD width. Most implementations thus introduce blocks that match the SIMD width [11], which works fine in the typical context of (GPU) performance studies, where the block structure tends to be artificially introduced as an optimization parameter, accompanied by standard padding. In our setting the block size is a property of the DG basis that we cannot influence. Thus, our implementation needs to work with arbitrary block sizes. We follow an alternative approach [5], which performs SIMD coalescing at the level

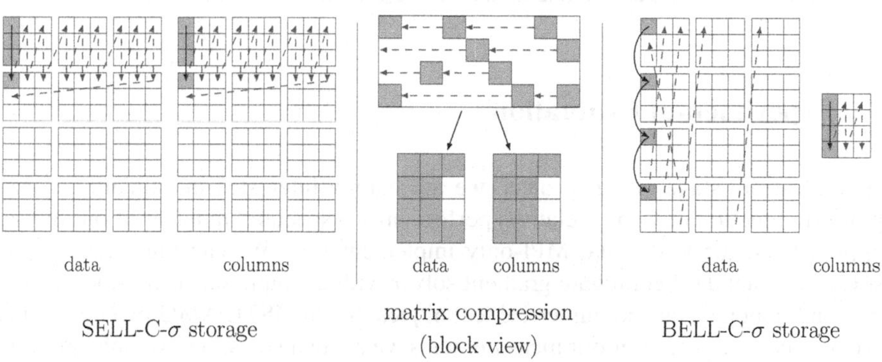

Fig. 1 Data layout of SELL-C-σ and the blocked version BELL-C-σ for a single chunk of a matrix with block size 3 and SIMD width 4. The columns are compressed and padded up to a uniform width (*center*). For each scheme, the figure shows the in-memory data layout of those compressed arrays as a path along the *arrows*. Note that SIMD chunks are not block-aligned for SELL-C-σ, while BELL-C-σ coalesces storage from four blocks to allow vectorisation of operations across those blocks

of entire matrix blocks as illustrated in Fig. 1. Our kernels then operate on several blocks at a time, only requiring vertical padding with empty blocks at the end of the matrix.

In order to further exploit the mathematically motivated blocking structure in our setting, we also implement a block Jacobi preconditioner on top of the blocked matrix, which performs an exact inversion of the diagonal blocks. The diagonal blocks within a SIMD group lack alignment, so the preconditioner extracts the diagonal block band in a preprocessing step and operates on this auxiliary data.

2.4 *Solvers and PDELab*

Due to the clear separation of algorithms and data structures in ISTL, we transparently reuse existing solver implementations on top of our new container formats without any code changes in the framework. All modifications are restricted to components that directly interact with the matrix structure, i.e., the containers themselves and the preconditioners. In keeping with the DUNE framework approach, we fully integrate our new containers into the high-level PDE toolbox PDELab by implementing a new backend interface that encapsulates the translation of user-space (i, j) indexing to the underlying data layout. As all high-level access to the containers happens via this backend interface, DUNE's existing grid and system assembly infrastructure can directly operate on the new containers, and we avoid using an intermediate matrix format that would then have to be explicitly converted to (B/S)ELL-C-σ. As a direct consequence of this tight integration with the existing solver library and high-level infrastructure, porting PDELab programs is a very straightforward process that only requires modifying the two or three lines of source code which define the active backends for vectors, matrices and preconditioners.

3 Experimental Evaluation

At the current stage of our project, we are mainly interested in validating cross-platform functionality and (relative) performance, especially in terms of our hybrid approach vs. the traditional MPI-only implementation. We can thus restrict ourselves to a standard conjugate gradient solver with a simple scalar or block Jacobi preconditioner, advanced numerical techniques like the ISTL AMG preconditioner are not necessary yet. For our measurements we adapt an existing example program from PDELab that solves a stationary diffusion problem:

$$\nabla \cdot (K \nabla u) = f \text{ in } \Omega \subset \mathbb{R}^3$$
$$u = g \text{ on } \Gamma = \partial \overline{\Omega}$$

with $f = (6 - 4|x|^2)\exp(-|x|^2)$ and $g = \exp(-|x|^2)$. A Discontinuous Galerkin discretisation is used with a weighted SIPG scheme [6]. We restrict our experiments to the unit cube $\Omega = (0, 1)^3$ and unit permeability $K = 1$.

Most of our experiments are executed on a single-socket Intel Sandy Bridge machine (8 GB DDR3-1333 RAM, 2 GHz 4-core Intel Core i7-2635QM, no Hyper-Threading) which supports 256-bit wide SIMD using AVX instructions. As this baseline machine only has a single UMA memory domain, we also run larger benchmarks on a 4-socket server with AMD Opteron 6172 12-core processors and 128 GB RAM. These CPUs internally comprise two dies with separate cache hierarchies and memory controllers, creating a fairly complex memory layout with 8 6-core UMA domains. This larger platform allows us to test the feasibility of our fundamental approach to parallelisation, drawing the line between classical message passing parallelism (MPI) and shared memory approaches at the level of a single UMA domain, but is limited to two-way SIMD. For the GPU measurements, we use an NVIDIA Tesla C2070 which comes from the same hardware generation as the AMD server. While the host platform differs from our CPU testbeds, this does not influence the results shown below as we only benchmark the CG solver, whose compute- and bandwidth-intensive components run entirely on the GPU.

We use a structured grid with a uniform mesh size h and DG spaces of order $p = 1, 2, 3$, which in 3D translates into block sizes of 8, 27 and 64, and to matrix densities with ≈ 56, 189 and 448 non-zero entries per row, respectively. For each space, we choose problem sizes that stretch up to the limit of available memory on our test systems, which is about 6 GB for both the multicore CPU system and the GPU card. In order to enable a fair comparison with a non-multithreaded pure MPI version of the code, we take care to choose problem sizes that can be decomposed into evenly sized subdomains without excessively large surfaces to avoid load balancing issues in the MPI version. Due to the large number of DOFs per cell for the higher-order spaces, this restriction limits us to a smaller number of samples for larger values of p. All computations are carried out in double precision floating point arithmetic.

The blocked format is currently only available on the CPU, here we also investigate the impact of switching from a scalar preconditioner to a block version that performs an exact inversion of the diagonal blocks. All of the compared implementations use common data structures and mostly perform identical basic operations (with the exception of the block matrix version), so we can expect all of them to require the same number of CG iterations to achieve a given error reduction. Moreover, due to the constant number of matrix entries per DOF for a fixed value of p (apart from boundary effects), we can actually expect a constant time per iteration and DOF for large problem sizes (after saturating either the compute or the memory bandwidth of the system). Figure 2 illustrates that this assumption holds very well, with all shared-memory implementations quickly reaching a saturation plateau. The MPI version appears to actually become faster with growing problem sizes, this is because it is based on an overlapping domain decomposition and thus benefits from the reduction of the relative amount of overlap in the bigger problems. We can see that the dominant computation kernels of the CG solver (SpMV and the

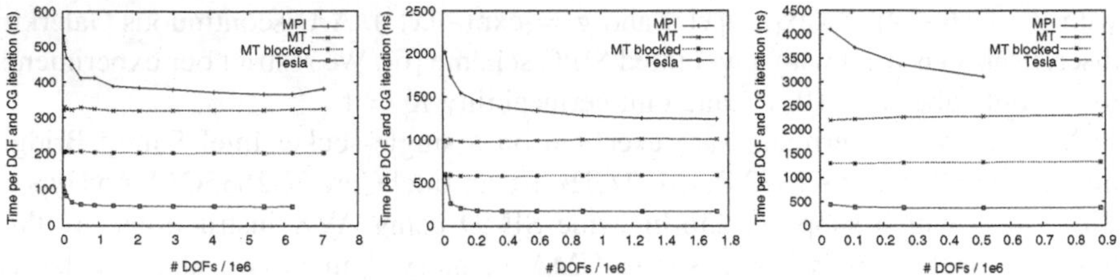

Fig. 2 Normalised execution time of the (Block-) Jacobi preconditioned CG solver for polynomial degrees $p = 1, 2, 3$ (left to right) of the DG discretisation. The multithreaded (MT) versions use a SIMD block size of 8. Missing data points indicate insufficient memory

Table 1 Comparison of different MPI / shared memory partition models for varying degree p of the DG discretisation and mesh width h. For each configuration, we list timings for 100 iterations of the CG solver ($t_{M/T}$, where M is the number of MPI ranks and T the number of threads per process), and the speedups compared to the MPI-only (48/1) case

p	h^{-1}	$t_{48/1}[s]$	$t_{8/6}[s]$	$\frac{t_{48/1}}{t_{8/6}}$	$t_{4/12}[s]$	$\frac{t_{48/1}}{t_{4/12}}$	$t_{1/48}[s]$	$\frac{t_{48/1}}{t_{1/48}}$
1	192	262.8	259.5	1.01	622.6	0.42	1,695.0	0.16
1	256	645.1	600.2	1.07	1,483.3	0.43	2,491.7	0.26
2	96	345.8	318.3	1.09	814.8	0.42	1,639.5	0.21
2	128	999.5	785.7	1.27	1,320.7	0.76	2,619.0	0.38
3	32	120.7	70.1	1.72	183.0	0.66	622.8	0.19
3	64	709.6	502.9	1.41	1,237.2	0.57	1,958.2	0.36

(block-) Jacobi preconditioner) are entirely limited by memory bandwidth across all examined architectures. The threaded implementation is approximately 25 % faster than the MPI baseline, in line with common expectations. Switching from scalar to blocked containers yields a speedup of ≈1.6–2.0 depending on p, which is in line with the bandwidth savings of moving to a block-level column index array (the additional work by the block inversion in the CPU preconditioner has negligible impact). Finally, the GPU gives a speedup of 3–4 over the best CPU implementation.

In order to validate our assumption that shared memory parallelism should be limited to individual UMA domains, we measure the runtime of several large benchmark problems on the AMD server with four different parallel setups, an MPI-only version with 48 single-thread processes, an optimal configuration with 8 MPI processes that each span a complete UMA domain (6 cores), and 2 suboptimal configurations employing 4 processes with 12 cores each (one process per socket) and one process with 48 threads (using only shared memory). For all measurements, we enforce process pinning. Ignoring the NUMA issue, we expect the timings to improve for smaller numbers of MPI processes because there is less domain overlap, reducing the effective problem size. The results in Table 1 clearly show a noteworthy improvement from the MPI-only setting to our UMA-domain approach (columns $t_{48/1}$ vs. $t_{8/6}$), with better results for higher polynomial degrees due to slightly worse surface-to-volume ratios in the MPI case. The remaining columns show that extending shared-memory parallelism across the UMA domain boundary causes a major performance breakdown by a factor of 2 in the intermediate setting and up to 6 for the worst case.

We additionally note that we do not observe relevant differences when switching off compiler vectorisation for the CPU in our experiments: As the benchmarks are entirely bandwidth bound and as our blocking scheme ensures cache line reuse, performing the actual computations in SIMD instructions has negligible impact.

Acknowledgements This work was supported (in part) by the German Research Foundation (DFG) through the Priority Programme 1648 'Software for Exascale Computing' (SPPEXA), grants GO 1758/2-1 and BA 1498/10-1.

References

1. S. Balay, J. Brown, K. Buschelman, W. Gropp, D. Kaushik, M. Knepley, L. McInnes, B. Smith, H. Zhang, *PETSc Web page*, 2011. http://www.mcs.anl.gov/petsc
2. W. Bangerth, C. Burstedde, T. Heister, M. Kronbichler, Algorithms and data structures for massively parallel generic adaptive finite element codes. ACM Trans. Math. Softw. **38**(2), 14:1–14:28 (2012)
3. P. Bastian, M. Blatt, A. Dedner, C. Engwer, R. Klöfkorn, R. Kornhuber, M. Ohlberger, O. Sander, A generic grid interface for parallel and adaptive scientific computing. Part II: implementation and tests in DUNE. Computing **82**(2–3), 121–138 (2008)
4. P. Bastian, M. Blatt, A. Dedner, C. Engwer, R. Klöfkorn, M. Ohlberger, O. Sander, A generic grid interface for parallel and adaptive scientific computing. Part I: abstract framework. Computing **82**(2–3), 103–119 (2008)
5. J. Choi, A. Singh, R. Vuduc, Model-driven autotuning of sparse matrix-vector multiply on GPUs, in *Principles and Practice of Parallel Programming*, (Bangalore, India, 2010), pp. 115–126
6. A. Ern, A. Stephansen, P. Zunino, A discontinuous Galerkin method with weighted averages for advection-diffusion equations with locally small and anisotropic diffusivity. IMA J. Numer. Anal. **29**(2), 235–256 (2009)
7. M. Heroux, R. Bartlett, V. Howle, R. Hoekstra, J. Hu, T. Kolda, R. Lehoucq, K. Long, R. Pawlowski, E. Phipps, A. Salinger, H. Thornquist, R. Tuminaro, J. Willenbring, A. Williams, K. Stanley, An overview of the Trilinos project. ACM Trans. Math. Softw. **31**(3), 397–423 (2005)
8. M. Kreutzer, G. Hager, G. Wellein, H. Fehske, A. Bishop, *A unified sparse matrix data format for modern processors with wide SIMD units*. SIAM J. Sci. Comput. **36**(5), C401–C423 (2014)
9. A. Logg, K.-A. Mardal, G. Wells, *Automated Solution of Differential Equations by the Finite Element Method* (Springer, Berlin/New York, 2012)
10. J. Sick and A. Lumsdale, A modern framework for portable high-performance numerical linear algebra, in *Advances in Software Tools for Scientific Computing*. Lecture Notes in Computational Science and Engineering, vol. 10 (Springer, Berlin/Heidelberg, 2000), pp. 1–56
11. B. Su and K. Keutzer, clSpMV: a cross-platform OpenCL SpMV framework on GPUs, in *ISC'12*, (Venice, Italy, 2012), pp. 353–364
12. S. Turek, D. Göddeke, C. Becker, S. Buijssen, and S. Wobker, FEAST – realisation of hardware-oriented numerics for HPC simulations with finite elements. Concurr. Comput.: Pract. Experience **22**(6), 2247–2265 (2010)

FMMTL: FMM Template Library
A Generalized Framework for Kernel Matrices

Cris Cecka and Simon Layton

Abstract In response to two decades of development in structured dense matrix algorithms and a vast number of research codes, we present designs and progress towards a codebase that is abstracted over the primary domains of research. In the domain of mathematics, this includes the development of interaction kernels and their low-rank expansions. In the domain of high performance computing, this includes the optimized construction, traversal, and scheduling algorithms for the appropriate operations. We present a versatile system that can encompass the design decisions made over a decade of research while providing an abstracted, intuitive, and usable front-end that can integrated into existing linear algebra libraries.

1 Introduction

Structured dense matrices arise in a broad range of engineering applications including the discretization of Fredholm integral equations, boundary elements methods, N-body problems, signal processing, statistics, and machine learning. This research concerns kernel matrix equations of the form

$$r_i = \sum\nolimits_j K(t_i, s_j)\, c_j \tag{1}$$

where we refer to K as the *kernel* generating elements of the dense matrix, the s_j as *sources* and c_j as a source's associated *charge*, and the t_i as *targets* and r_i as a target's associated *result*. Of course, a direct computation of a kernel matrix-vector product requires $\mathcal{O}(N^2)$ computations, where N is the cardinality of the source set and target sets. Fast multipole methods (FMMs) and tree-codes allow

C. Cecka (✉)
Harvard University, Cambridge, MA, USA
e-mail: ccecka@seas.harvard.edu; ccecka@stanford.edu

S. Layton
Boston University, Boston, MA, USA
e-mail: slayton@bu.edu

A. Abdulle et al. (eds.), *Numerical Mathematics and Advanced Applications ENUMATH 2013*, Lecture Notes in Computational Science and Engineering 103,
DOI 10.1007/978-3-319-10705-9_60

Table 1 A table of common kernels emphasizing the varied domain and ranges of the operators

Name/equation	$K(x, y)$	Domain	Range
Laplace, Poisson	$1/\|\mathbf{x}-\mathbf{y}\|$	$\mathbb{R}^3 \times \mathbb{R}^3$	$\mathbb{R}$
Yukawa, Helmholtz	$e^{k\|\mathbf{x}-\mathbf{y}\|}/\|\mathbf{x}-\mathbf{y}\|$	$\mathbb{R}^3 \times \mathbb{R}^3$	$\mathbb{C}$
Stokes	$\frac{1}{\|\mathbf{x}-\mathbf{y}\|}\left(\mathbf{I}+\frac{(\mathbf{x}-\mathbf{y})(\mathbf{x}-\mathbf{y})^T}{\|\mathbf{x}-\mathbf{y}\|^2}\right)$	$\mathbb{R}^3 \times \mathbb{R}^3$	$\mathbb{R}^{3,3}$
Gaussian	$e^{-\varepsilon\|\mathbf{x}-\mathbf{y}\|^2}$	$\mathbb{R}^n \times \mathbb{R}^n$	$\mathbb{R}$
Multiquadric	$(1+\|\mathbf{x}-\mathbf{y}\|^2)^{\pm 1/2}$	$\mathbb{R} \times \mathbb{R}$	$\mathbb{R}$

for an approximate evaluation of this matrix-vector product with only $\mathcal{O}(N \log^{\alpha} N)$ complexity, where $\alpha = 0, 1$, or 2 depending on specifics regarding the kernel, the expansions, the traversal algorithms, and the distribution of sources and targets. Common kernels in physics and statistics and their domains and ranges are listed in Table 1.

FMMs require multiple, carefully optimized steps and numerical analysis in order to achieve the improved asymptotic performance and required accuracy. These research areas span tree construction, tree traversal, numerical and functional analysis, and complex heterogeneous parallel computing strategies for each stage. Unfortunately, many FMM codes are written with a particular application (an interaction kernel and/or compute environment) in mind [2,4,10]. It is often difficult to extract out advances from one research area and apply them to another code or application. Indeed, Yokota et al. [13] discuss recent developments and comparisons to note the disappointing lack of fair benchmarking comparisons between kernel expansions, data structures, traversal algorithms, and parallelization strategies.

In this paper, we review recent development of a parallel, generalized framework and repository for kernel matrices of the form (1). This overarching goal of this library, called FMMTL, is to separate academic concerns in research and development of fast structured dense matrix algorithms. Using advanced C++ techniques and design, we are able to develop the code at a high level, isolate development hurdles and choices, and collect a repository of kernels and their associated expansions for rapid application deployment in any of the above domain areas. This is accomplished by defining generalized interfaces for kernels independent of algorithmic concerns with tree construction and traversal and presenting a coherent front-end for working with kernel matrices as abstract data types. Problems of the form (1) can be constructed and manipulated in an intuitive way and should be able to take advantage of existing solvers or provide their own. The library has already seen use in simple Poisson problems, more advanced boundary element solvers, and the use of FMM as a preconditioner.

2 Background

Fast methods for kernel matrices define or compute a low-rank approximation to the kernel valid for some set of the sources, S, and targets, T:

$$K(T, S) \approx U(T)\, \tilde{K}\, V^T(S).$$

These approximations can be computed analytically with series expansion or interpolations of the known kernel function [4, 5, 7] or algebraically by rank-reducing operations on samples of the kernel function [6, 11, 12]. This allows off-diagonal blocks of the kernel matrix to be approximated and computed quickly and defines the class of Hierarchically Off-Diagonally Low-Rank (HODLR) matrices. The fundamental operations used in working with HODLR matrices are:

$$\text{S2M: } M = V^T(S) * C \qquad \text{M2L: } L = \tilde{K} * M \qquad \text{L2T: } R \approx U(T) * \tilde{K}$$

where R are the results associated with the targets T, C are the charges associated with sources S, and we call M a *multipole expansion* and L a *local expansion*.

Hierarchically SemiSeparable matrices (HSS) allow the multipoles of sets of sources to be computed from the multipoles of subsets to form a hierarchy of low-rank approximations. The operations involved are extended to include:

$$\text{M2M: } M' = \tilde{V}^T * M \qquad \text{L2L: } L' = \tilde{U} * L$$

Convenient operators to add to this pool are found in most often in tree-codes and can be written as:

$$\text{S2L: } L = \tilde{K} * V^T(S) * c \qquad \text{M2T: } R \approx U * \tilde{K} * M$$

Finally, the sets of sources and targets whose block in the kernel matrix is approximated in this way are chosen with a rule called the multipole acceptance criteria (MAC). Whether this rule accepts "nearby clusters" of sources and targets differentiates $\mathscr{H}$ matrices from HOLDR matrices and $\mathscr{H}^2$ matrices from HSS matrices.

In practice, these operations are often built into research implementations and can be difficult to extract, understand, and modify. It is these operators that we wish to classify and fully abstract in order to develop a library that can be used, without modification, for any kernel matrix and any definition of the above operators. Additional goals of FMMTL are to isolate algorithmic features such as tree construction and traversal that are also too often entangled with problem-specific data or algorithms.

3 Design Considerations

The design of FMMTL attempts to make kernels and their expansions independent citizens in that their implementation should not depend on or know about trees, clusters of sources or targets, traversals, or parallelism. Furthermore, it should be portable and easy to use and install. For this reason, the only dependency is a modern C++ compiler with C++11 support and the renowned C++ Boost library (headers only). Additionally, if CUDA is installed and available, the library will attempt to use GPU acceleration.

In this section, we offer a brief overview of the features and design considerations in FMMTL.

3.1 Kernels

Kernels are simply function objects used to generate elements of the matrix, but should also define the domain of the problem. The fundamental types required are the domain of the kernel (the `source_type` and the `target_type`) and the range of the kernel (the `kernel_value_type`).

In Listing 1, `Vec` is a statically sized abstract vector type designed to work on multiple architectures. In addition, note the transpose method labeled optional. In many cases, the kernel satisfies a symmetry property

$$K(s,t) = \mathscr{T} \circ K(t,s)$$

that can be computed much more efficiently than evaluation of the kernel and may be used to accelerate the computation in the case that the source and target sets are the same. The library uses advanced SFINAE – Substitution Failure Is Not An Error – compiler techniques to statically detect whether this optional method is defined at compile time and will use it if appropriate.

```
struct MyKernel : public fmmtl::Kernel<MyKernel> {
  typedef Vec<3,double> source_type;
  typedef Vec<3,double> target_type;
  typedef double kernel_value_type;

  FMMTL_INLINE kernel_value_type
  operator()(const target_type& t, const source_type& s) const {
    return norm(s-t);
  }
  /** Optional transpose operation for optimization **/
  FMMTL_INLINE kernel_value_type
  transpose(const kernel_value_type& kts) const {
    return kts;
  }
};
```

Listing 1 Example kernel typedefs and members

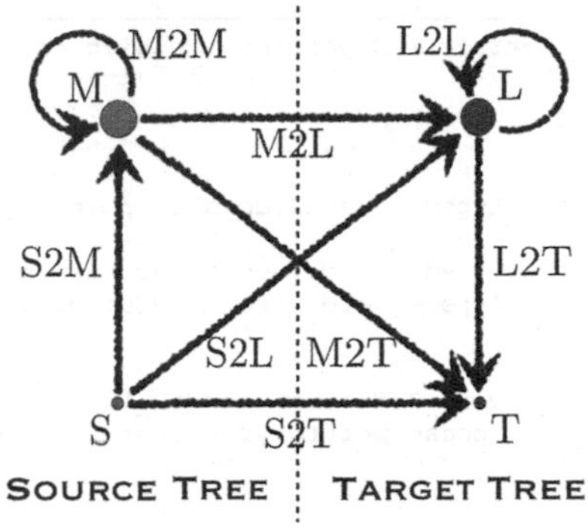

Fig. 1 The "algebraic system" of hierarchical methods is composed of sources S, multipole expansions M, local expansions L, and targets T. The S2T operation is defined by the kernel and all other operators may be defined by the expansion

3.2 Expansions

An expansion is the low-rank approximation of a kernel that can be used to accelerate the kernel matrix operations. The primary role of expansions is to declare the type of the multipole and local objects and provide methods for transforming between the `source_type`, `target_type`, `multipole_type`, and `local_type`. All possible conversions are shown in Fig. 1.

Nearly all of the functions in an expansion are optional and their availability is statically detected. This information can be used to determine computational pathways and potentially choose the most efficient. Additionally, this feature is attractive when expansions cannot define a certain operator or some traversal algorithms do not consider some operators. For example, the two primary types of tree-codes both use subsets of the operations shown in Fig. 1. The particle-cluster tree-codes [3] use the S2M, M2M, and M2T operations while the cluster-particle tree-codes use the S2L, L2L, and L2T operations. Similarly, many fast multipole methods neglect the S2L and M2T operators while others are beginning to consider a larger set of operations to dynamically determine the cheapest computational pathway [1, 8, 14].

The expansion also defines a `point_type` which is used as the spacial embedding of the sources and targets for clustering and hierarchical constructions. Because `source_type` and `target_type` need only be convertible to this `point_type`, they are free to be much more complicated objects. For example, in boundary element methods, the source and target types are naturally triangles, patches, or basis functions. These may have a spacial center that can be used for the construction of the tree, but should remain independent entities for simplifying the definition of the kernel function.

```
struct MyExpansion : public fmmtl::Expansion<MyKernel, MyExpansion> {
  /** Spacial type to provide an interpretation for clustering.
  * @note source_type and target_type must be either
  * (1) convertible to point_type, or
  * (2) S2P and/or T2P shall be defined to provide a conversion. */
  typedef Vec<3,double> point_type;

  typedef std::vector<double> multipole_type;
  typedef std::vector<double> local_type;

  /** Optional S2M Operator **/
  void S2M(const source_type& s, const charge_type& c,
   const point_type& center, multipole_type& M) const {
  // Compute M += V^T(s) * c
  }
  ...
};
```

Listing 2 Example expansion typedefs and members

3.3 Tree and Traversals

The lightweight tree data structure is constructed on any `point_type`. Depending on the dimension D of the `point_type`, another compile-time constant, a D-dimensional binary tree ($D = 2$ is a quadtree, $D = 3$ is an octree, etc) is constructed. This is accomplished via partially sorting the points on a space-filling curve. However, this implementation detail is hidden behind an interface that any reasonable tree structure should provide, so alternate tree types and representations, such as a k-d tree, can be swapped in at will. A brief interface for a tree is given in Listing 3.

```
struct Tree {
  struct Box {
  unsigned index() const;
  body_iterator body_begin() const;
  body_iterator body_end() const;
  box_iterator child_begin() const;
  box_iterator child_end() const;
  Box parent() const;
  };
  struct Body {
  unsigned number() const; // Original index this body was added
  unsigned index() const;  // Index within the tree
  };

  body_iterator body_begin() const;
  body_iterator body_end() const;
  box_iterator box_begin(int level) const;
  box_iterator box_end(int level) const;
};
```

Listing 3 A truncated interface for a general tree data structure

The tree need not store the points or the expansions as other implementations of hierarchical algorithms appear to. Instead, each box and point in the tree have an immutable identification index that can be used to manage arbitrary data outside of the data structure. This allows the tree structure to be lightweight and allows a more context-aware data structure to manage source, target, and expansion data independent of the tree.

The traversals are implemented as a dual tree traversal and are templated on the `Box` type, requiring a `Box` to implement a small number of reasonable methods such as those in Listing 3. The dual tree traversal is often used in tree-codes, but not fast multipole methods. However, Yokota et al. [13] take advantage of its versatility for hierarchical problems to generalize their codes. We would like to note that with a sufficiently generalized MAC, the dual tree traversal can produce the same interaction lists as the classic FMM, the adaptive UVWX schemes [12], and modern tree-codes. We find that there are two types of MAC: static MACs which depend only the position and size of the boxes, and dynamic MACs which depend on the expansions or are otherwise dependent on the source and target distributions [9, 13].

3.4 *Optimizations*

When an operator is determined to be required, the operation may be dispatched immediately or scheduled for later use and reuse. In general, making these choices has been an algorithmic option in the development of hierarchical methods. Recent studies have shown that event-driven parallel runtime systems provide success in dynamically resolving the dependencies within fast multipole methods and tree-codes. Ltaief and Yokota [8] use QUARK to schedule threads dynamically to accommodate the data flow of exaFMM's dual tree traversal. Agullo et al. [1] apply a similar approach with StarPU to their black-box FMM using parallelism on both the CPU and the GPU. These approaches appear promising and could result in an efficient and easier to apply parallelism strategy than a statically implemented distributed algorithm.

FMMTL provides a generic way to implement kernels so that if a CUDA compiler is installed and a GPU is available, the library will use the GPU to accelerate the costly S2T computation. This requires an accumulation of interacting source and target sets and a compression of the interaction list to a form that is suitable for the GPU. Dynamic parallelization studies have found that the methods benefited the most from assigning the S2T operations exclusively to the GPU [1] due to the structured nature and high flop-to-byte ratio of S2T operations. The first major FMMTL parallelization step of executing a generalized S2T on the GPU is motivated by these results.

```
MyExpansion K(...);     // Expansion order, error eps, etc
..
std::vector<source_type> s = ...  // Define sources
std::vector<charge_type> c = ...  // Define charges
std::vector<target_type> t = ...  // Define targets

fmmtl::kernel_matrix<MyExpansion> M = K(t,s);  // Construct
fmmtl::set_options(M, opts);      // Set options
...
std::vector<result_type> r_apprx = M * c;  // FMM/Treecode
std::vector<result_type> r_exact = fmmtl::direct(M * c); // O(N^2)
```

Listing 4 A use case of a kernel matrix abstract data type

4 Usage

Providing a kernel matrix data type allows the abstraction level of our code to remain high while retaining generality and efficiency. Integration into existing efficient linear algebra libraries such as MKL, Eigen, and/or ViennaCL for use of solvers and preconditioners is an attractive option. Additionally, higher level linear algebra and computer science concepts such as submatrix blocking, lazy evaluation, and template expressions become a possibility.

4.1 Preliminary Benchmark

While FMMTL remains relatively new and the primary focus has been on design rather than performance, we provide preliminary results in this section to show that the abstractions and design decisions do not significantly impact raw performance of the algorithm.

In Fig. 2, the performance of FMMTL for the benchmark case of the Laplace kernel (potential+force) with expansion order $p = 8$ and varying number of

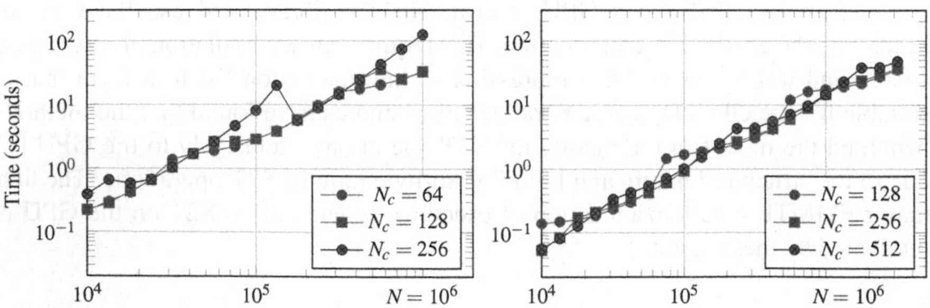

Fig. 2 Timings for the Laplace kernel (potential+force) showing the performance of an FMM with expansion order $p = 8$ and N particles uniformly distributed in a cube. These show (*left*) CPU only and (*right*) with GPU acceleration in S2T

sources. The hardware used was an Intel Xeon W3670 3.2 GHz CPU and an Nvidia GTX580 GPU. This closely follows the benchmark presented by Yokota et al. in [14].

These results show that the performance of FMMTL for this benchmark is on par with that of the more well-tuned ExaFMM [14]. To alleviate the dependence of the performance on the value of N_c, the maximum number of bodies per leaf box, similar auto-tuning procedures may be performed as suggested in [1, 13, 14], which is made easier and more general due to the encapsulation of operators in FMMTL. In the current state of the library, much more static information about the kernel and expansion may be taken advantage of in the tree traversal and operator evaluation which will be discussed in a forthcoming optimization, performance analysis, and application study publication.

Conclusion

Reviewing the literature and codes produced for hierarchical matrix algorithms reveals a large number of difficult to use and modify research codes. In the FMMTL library, we attempt to separate out the needs of a kernel from the needs of an expansion and isolate the tree construction and tree traversal. By doing so, continuing research into optimal kernel expansions, tree data structures, tree traversals, cluster interaction, and parallel computing strategies can continue independently of one another. At the very least, by growing a repository of kernels and expansions in a uniform format allows research to conduct fair comparisons – a requirement that is needed in the short-term to determine where and why to allocate research resources.

For the time being, careful design has been critical to the development of FMMTL to ensure that dependencies between components is low. Despite this, the serial performance is on par with hand-tuned lower-level research codes and higher performance already possible with OpenMP and GPU acceleration.

References

1. E. Agullo, B. Bramas, O. Coulaud, E. Darve, M. Messner, T. Takahashi, Pipelining the Fast Multipole Method over a Runtime System. Research Report RR-7981, INRIA, May 2012
2. J. Bédorf, E. Gaburov, S. Portegies Zwart, A sparse octree gravitational n-body code that runs entirely on the GPU processor. J. Comput. Phys. **231**(7), 2825–2839 (2012)
3. H.A. Boateng, R. Krasny, Comparison of treecodes for computing electrostatic potentials in charged particle systems with disjoint targets and sources. J. Comput. Chem. **34**(25), 2159–2167 (2013)
4. C. Cecka, E. Darve, Fourier-based fast multipole method for the helmholtz equation. SIAM J. Sci. Comput. **35**(1), A79–A103 (2013)

5. H. Cheng, L. Greengard, V. Rokhlin, A fast adaptive multipole algorithm in three dimensions. J. Comput. Phys. **155**(2), 468–498 (1999)
6. W. Fong, E. Darve, The black-box fast multipole method. J. Comput. Phys. **228**(23), 8712–8725 (2009)
7. J. Kurzak, B.M. Pettitt, Fast multipole methods for particle dynamics. Mol. Simul. **32**(10–11), 775–790 (2006)
8. H. Ltaief, R. Yokota, Data-driven execution of fast multipole methods. Concurr. Comput.: Pract. Experience, pp. n/a–n/a (2013)
9. J.K. Salmon, M.S. Warren, Skeletons from the treecode closet. J. Comput. Phys. **111**(1), 136–155 (1994)
10. T. Takahashi, C. Cecka, W. Fong, E. Darve, Optimizing the multipole-to-local operator in the fast multipole method for graphical processing units. Int. J. Numer. Methods Eng. **89**(1), 105–133 (2012)
11. L. Ying, A kernel-independent fast multpole algorithm for radial basis functions. J. Comput. Phys. **213**, 451–457 (2006)
12. L. Ying, G. Biros, D. Zorin, A kernel-independent adaptive fast multipole algorithm in two and three dimensions. J. Comput. Phys. **196**(2), 591–626 (2004)
13. R. Yokota, An fmm based on dual tree traversal for many-core architectures. J. Algorithms Comput. Technol. **7**, 301–324 (2013)
14. R. Yokota, L.A. Barba, Hierarchical n-body simulations with autotuning for heterogeneous systems. Comput. Sci. Eng. **14**(3), 30–39 (2012)

Part X
Computational Fluid and Structural Mechanics

Modified Pressure-Correction Projection Methods: Open Boundary and Variable Time Stepping

Andrea Bonito, Jean-Luc Guermond, and Sanghyun Lee

Abstract In this paper, we design and study two modifications of the first order standard pressure increment projection scheme for the Stokes system. The first scheme improves the existing schemes in the case of open boundary condition by modifying the pressure increment boundary condition, thereby minimizing the pressure boundary layer and recovering the optimal first order decay. The second scheme allows for variable time stepping. It turns out that the straightforward modification to variable time stepping leads to unstable schemes. The proposed scheme is not only stable but also exhibits the optimal first order decay. Numerical computations illustrating the theoretical estimates are provided for both new schemes.

1 Introduction

We consider the time-dependent Stokes system on a bounded domain $\Omega \subset \mathbb{R}^d$, $d = 2, 3$, with Lipschitz boundary $\partial\Omega$ and over a finite time interval $[0, T]$. For a given force $\mathbf{f} : \Omega \times [0, T] \to \mathbb{R}^d$, the velocity $\mathbf{u} : \Omega \times [0, T] \to \mathbb{R}^d$ and the pressure $p : \Omega \times [0, T] \to \mathbb{R}$ are related via the following system

$$\rho\partial_t \mathbf{u} - 2\mathrm{div}\left(\mu\nabla^S \mathbf{u}\right) + \nabla p = f \quad \text{and} \quad \mathrm{div}(\mathbf{u}) = 0 \qquad \text{in } \Omega \times [0, T], \tag{1}$$

where ρ and μ are the fluid density and viscosity of the fluid assumed to be constant (and positive) and $\nabla^S := \frac{1}{2}\left(\nabla + \nabla^T\right)$ denotes the symmetric part of the gradient. Relations (1) is supplemented by a boundary condition either prescribing the velocity or the force at the boundary. In order to simplify the presentation, we consider homogeneous cases

$$\mathbf{u} = 0 \qquad \text{on } \partial\Omega \times [0, T] \tag{2}$$

A. Bonito (✉) • J.-L. Guermond • S. Lee
Texas A&M University, College Station, TX 77843, USA
e-mail: bonito@math.tamu.edu; guermond@math.tamu.edu; shlee@math.tamu.edu

A. Abdulle et al. (eds.), *Numerical Mathematics and Advanced Applications ENUMATH 2013*, Lecture Notes in Computational Science and Engineering 103,
DOI 10.1007/978-3-319-10705-9_61

or

$$(2\mu\nabla^S\mathbf{u} - p)\boldsymbol{\nu} = 0 \qquad \text{on } \partial\Omega \times [0, T], \tag{3}$$

where $\boldsymbol{\nu}$ is the unit, outward pointing normal of $\partial\Omega$. In addition, the initial velocity $\mathbf{u}_0 : \Omega \to \mathbb{R}^d$ is prescribed, i.e. $\mathbf{u}(0, .) := \mathbf{u}_0$. At this point, we note that the extension to the Navier-Stokes system is treated similarly with the additional, but well known, techniques used to cope with the additional nonlinearity.

Most projection methods are based on the original ideas of Chorin [1] and Temam [9], see also Goda [2]. We refer to [4] for an overview of projection methods.

In this work, we obtain two different results regarding the so-called incremental pressure correction schemes studied for instance in [3, 5–7]:

- The scheme proposed in [3] when the system is subject to open boundary conditions, see (3), is suboptimal with respect to the time discretization parameter. We propose and study a new scheme able to recover the optimal convergence rate, see Fig. 1.
- We analyze a new scheme allowing for variable time stepping. It turns out that the straightforward generalization of constant time stepping to variable time stepping is unstable, see Fig. 2. To the best of our knowledge, projection schemes with variable time stepping have not been studied in the literature. Notice however, that no additional difficulty arises from having variable time stepping in the non-incremental scheme setting.

Given a positive integer N, let $0 = t^0 < t^1 < t^2 < \cdots < t^N = T$ be a subdivision of the time interval $[0, T]$ and set $\delta t^n := t^n - t^{n-1}$. The norm in $L_2(\Omega)$ is denoted by $\|.\|_0$ and we equip $H^1(\Omega)$ with the norm $\|.\|_1 := \left(\|.\|_0^2 + \|\nabla.\|_0^2\right)^{1/2}$.

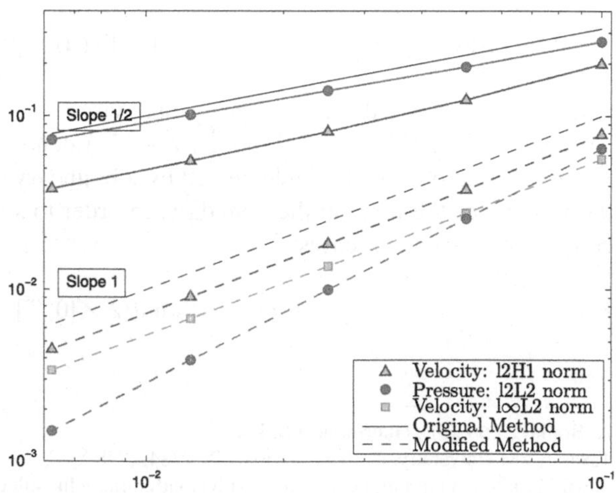

Fig. 1 Decay of different error norms versus δt for the original and modified standard pressure correction projection method. Suboptimal order of convergence $\mathcal{O}(\delta t^{1/2})$ is observed for the standard method while the optimal order of convergence $\mathcal{O}(\delta t)$ is recovered using the proposed scheme

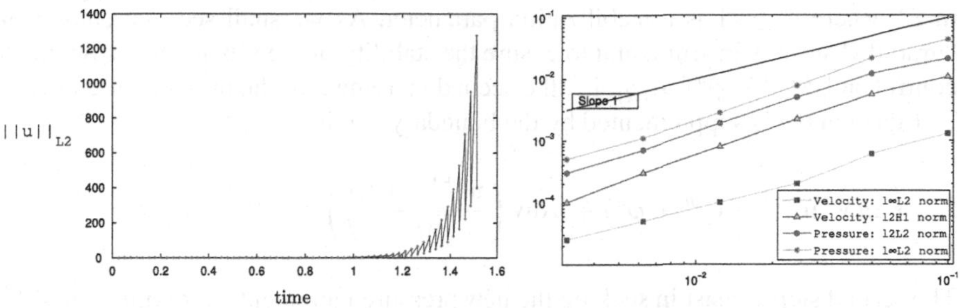

Fig. 2 (*Left*) Evolution of $\|\mathbf{u}(t^n,.)\|_{L_2(\Omega)}$ when using the standard scheme with $\delta t^1 = 0.025$ and δt^n given by (16). (*Right*) Decay of the velocity and pressure errors versus $\overline{\delta t}$ and with the time steps δt^n given by (16) when using the proposed scheme. The optimal order of convergence $\mathcal{O}(\overline{\delta t})$ is observed

In addition, given a sequence of function $\varphi_{\delta t} := \{\varphi^n\}_{n=0}^N$, we define the following discrete (in time) norms:

$$\|\varphi_{\delta t}\|_{l^2(E)} := \left(\sum_{n=0}^{N} \delta t^n \|\varphi^n\|_E^2\right)^{1/2}, \qquad \|\varphi_{\delta t}\|_{l^\infty(E)} := \max_{0\le n\le N}(\|\varphi^n\|_E). \tag{4}$$

for $E := L_2(\Omega)$ or $H^1(\Omega)$.

2 Optimal Incremental Projection Scheme for Open Boundary Problem

We consider the system (1) supplemented with the force condition at the boundary (3) and focus on the case of uniform (constant) time steps, i.e. $\delta t := \frac{T}{N} = \delta t^n$, $n = 0, \cdots, N$. The case of variable time steps is discussed in Sect. 3. The approximations of $\mathbf{u}(t^n,.)$ and $p(t^n,.)$, $n = 0, \ldots, N$, are denoted $\mathbf{u}^n$ and p^n respectively. For clarity, we also denote by ϕ^n the pressure increment approximation, i.e.

$$p^n = p^{n-1} + \phi^n. \tag{5}$$

Together with the initial condition on the velocity $\mathbf{u}^0 = \mathbf{u}_0$, the algorithm requires initial pressure $p(0)$ and we set $p^{-1} := p^0 := p(0)$, and so $\phi^0 := 0$. We seek recursively the velocity $\mathbf{u}^{n+1}$ and the pressure p^{n+1} in three steps. First, given $\mathbf{u}^n$, ϕ^n and p^n, the velocity approximation at t^{n+1} is given by

$$\rho\frac{\mathbf{u}^{n+1}-\mathbf{u}^n}{\delta t} - 2\mathrm{div}(\mu\nabla^S\mathbf{u}^{n+1}) + \nabla(p^n+\phi^n) - \alpha\nabla\mathrm{div}\left(\frac{\mathbf{u}^{n+1}-\mathbf{u}^n}{\delta t}\right) = f(t^{n+1},.), \tag{6}$$

in Ω, where $\alpha \geq 1$ is a stabilization parameter. As we shall see, the consistent "grad-div" term is instrumental to ensure the stability of the scheme by providing a control on $\|\phi^{n+1} - \phi^n\|_{H^1(\Omega)}$, i.e. the second increment of the pressure; see (13).

Equation (6) is supplemented by the boundary condition

$$\left(2\mu\nabla^S \mathbf{u}^{n+1} - (p^n + \phi^n) + \alpha \mathrm{div}\left(\frac{\mathbf{u}^{n+1} - \mathbf{u}^n}{\delta t}\right)\right)\nu = 0 \qquad \text{on } \partial\Omega. \tag{7}$$

The second step consist in seeking the new pressure increment approximation ϕ^{n+1} as the solution to

$$-\delta t \Delta\phi^{n+1} + \delta t \phi^{n+1} = -\mathrm{div}(\mathbf{u}^{n+1}) \quad \text{in } \Omega \tag{8}$$

together with the boundary condition

$$\frac{\partial}{\partial \nu}\phi^{n+1} = 0 \quad \text{on } \partial\Omega. \tag{9}$$

Finally, the new pressure approximation is then given by (5).

The novelty of this projection scheme is to impose a Neuman boundary condition on the pressure increment (and therefore on the pressure). Its aim is to reduce the boundary layer on the pressure and improve the convergence rate. Compare with [3] where a Dirichlet condition $p^{n+1} = p^n$ is proposed on the pressure. This is at the expense of adding (i) an harmless zero order term $\delta t \phi^{n+1}$ in (8) to be able to recover the full $l^2(H^1(\Omega))$ norm for the pressure and (ii) the more serious "grad-div" stabilization term in (6), which complicates the linear algebra. Notice that the boundary condition (9) proposed here corresponds to the standard boundary condition when the velocity is imposed at the boundary; refer to [3].

We now briefly discuss the stability and error estimates for the scheme (6)–(9).

Theorem 1 (Velocity Stability) *Set* $\mathbf{f} \equiv 0$ *and assume* $\alpha \geq 1$, *then there holds*

$$\rho\|\mathbf{u}_{\delta t}\|^2_{l^\infty(L_2(\Omega))} + 4\mu\|\nabla^S \mathbf{u}_{\delta t}\|^2_{l^2(L_2(\Omega))} + \alpha\|\mathrm{div}(\mathbf{u}_{\delta t})\|^2_{l^\infty(L_2(\Omega))} + (\delta t)^2\|p_{\delta t}\|^2_{l^\infty(H^1(\Omega))}$$
$$\leq \rho\|\mathbf{u}_0\|_0^2 + \alpha\|\mathrm{div}(\mathbf{u}_0)\|_0^2 + (\delta t)^2\|p_0\|_1^2$$

provided $\mathbf{u}_0 \in L_2(\Omega)^d$, $\mathrm{div}(\mathbf{u}_0) \in L_2(\Omega)$ *and* $p_0 \in H^1(\Omega)$.

Proof Multiplying (6) by $2\delta t \mathbf{u}^{n+1}$ and integrating over Ω one gets after integrating by parts and using the boundary condition (7)

$$\begin{aligned}
&\rho\left(\|\mathbf{u}^{n+1}\|_0^2 + \|\mathbf{u}^{n+1} - \mathbf{u}^n\|_0^2 - \|\mathbf{u}^n\|_0^2\right) + 4\delta t\mu\|\nabla^S \mathbf{u}^{n+1}\|_0^2\\
&+ \alpha\left(\|\mathrm{div}(\mathbf{u}^{n+1})\|_0^2 + \|\mathrm{div}(\mathbf{u}^{n+1} - \mathbf{u}^n)\|_0^2 - \|\mathrm{div}(\mathbf{u}^n)\|_0^2\right)\\
&- 2\delta t \int_\Omega (p^n + \phi^n)\mathrm{div}(\mathbf{u}^{n+1}) d\mathbf{x} = 0.
\end{aligned} \tag{10}$$

The last term in the left hand side of the above relation is estimated upon multiplying (8) by $2\delta t(p^n + \phi^n)$, integrating over Ω and using the boundary condition (9)

$$-2\delta t \int_\Omega (p^n + \phi^n)\mathrm{div}(\mathbf{u}^{n+1})d\mathbf{x} = 2(\delta t)^2 \int_\Omega \nabla\phi^{n+1} \cdot \nabla(p^n + \phi^n)d\mathbf{x} + 2(\delta t)^2 \int_\Omega \phi^{n+1}(p^n + \phi^n)d\mathbf{x}.$$

In view of (5), we write $p^n + \phi^n = \phi^n - \phi^{n+1} + p^{n+1}$ and realize that

$$-2\delta t \int_\Omega (p^n + \phi^n)\mathrm{div}(\mathbf{u}^{n+1})d\mathbf{x} = (\delta t)^2\|\phi^n\|_1^2 - (\delta t)^2\|\phi^{n+1} - \phi^n\|_1^2 + (\delta t)^2\|p^{n+1}\|_1^2 - (\delta t)^2\|p^n\|_1^2. \tag{11}$$

It remains to derive a bound for $\|\phi^{n+1} - \phi^n\|_1$. Multiplying by $\phi^{n+1} - \phi^n$ the difference of two successive relations (8) and integrating over Ω yield

$$\delta t\|\phi^{n+1} - \phi^n\|_1^2 = -\int_\Omega \mathrm{div}(\mathbf{u}^{n+1} - \mathbf{u}^n)(\phi^{n+1} - \phi^n)d\mathbf{x}, \tag{12}$$

after an integration by parts and taking advantage of the boundary condition (9). Hence, we deduce that

$$\delta t\|\phi^{n+1} - \phi^n\|_1 \le \|\mathrm{div}(\mathbf{u}^{n+1} - \mathbf{u}^n)\|_0. \tag{13}$$

Gathering the estimate (13), (11) and (10), we obtain

$$\begin{aligned}&\rho\left(\|\mathbf{u}^{n+1}\|_0^2 + \|\mathbf{u}^{n+1} - \mathbf{u}^n\|_0^2 - \|\mathbf{u}^n\|_0^2\right) + 4\delta t\mu\|\nabla^S\mathbf{u}^{n+1}\|_0^2\\ &\quad + \alpha\left(\|\mathrm{div}(\mathbf{u}^{n+1})\|_0^2 - \|\mathrm{div}(\mathbf{u}^n)\|_0^2\right) + (\alpha - 1)\|\mathrm{div}(\mathbf{u}^{n+1} - \mathbf{u}^n)\|_0^2\\ &\quad + (\delta t)^2\left(\|p^{n+1}\|_1^2 - \|p^n\|_1^2 + \|\phi^n\|_1^2\right) \le 0.\end{aligned}$$

The desired bound follows after summing for $n = 0$ to $N - 1$. □

We emphasize that the above proof is closely related to the case where Dirichlet boundary conditions are imposed on the velocity; refer for instance to [4, 8]. The difference resides on the fact that (13) can be circumvented using an integration by parts in (12). Hence following the techniques developed for the Dirichlet case together with the argumentation leading to (13) yields the optimal convergence rates

$$\max_{n=1,\dots,N} \|\mathbf{u}(t^n,.) - \mathbf{u}^n\|_{L^2(\Omega)} + \left(\sum_{n=1}^{N} \delta t \|\nabla^S(\mathbf{u}(t^n,.) - \mathbf{u}^n)\|^2_{L_2(\Omega)}\right)^{1/2}$$

$$+ \alpha \max_{n=1,..,N} \|\mathrm{div}(\mathbf{u}(t^n,.) - \mathbf{u}^n)\|_{L_2(\Omega)} + \left(\sum_{n=1}^{N} \delta t \|p(t^n) - p^n\|^2_{L_2(\Omega)}\right)^{1/2} \leq C\delta t,$$

with a constant C independent of N and provided the exact velocity $\mathbf{u}$ and pressure p satisfy the appropriate regularity conditions.

To illustrate the optimality of the proposed algorithm, we consider the exact solution

$$\mathbf{u}(t,x,y) := \begin{pmatrix} \sin(t+x)\sin(t+y) \\ \cos(t+x)\cos(t+y) \end{pmatrix}, \quad p(t,x,y) = \sin(t+x-y)$$

defined $\Omega := (0,1)^2$. The behavior of the errors in velocity and pressure approximations versus the time step δt used are depicted in Fig. 1. Suboptimal order of convergence $\mathcal{O}(\delta t^{1/2})$ is observed for the standard method while the optimal order of convergence $\mathcal{O}(\delta t)$ is recovered using the proposed scheme. The space discretization is chosen fine enough not to interfere with the time discretization error.

3 Variable Time Stepping

We now consider variable time steps δt^n satisfying

$$\delta t^n \leq \overline{\delta t}, \qquad 1 \leq n \leq N,$$

for a positive constant $\overline{\delta t}$ independent of n. The incremental projection scheme with variable time stepping reads as follow. Given $\mathbf{u}^n$, ϕ^n and p^n, the velocity approximation at t^{n+1} is defined by the relation

$$\rho\frac{\mathbf{u}^{n+1} - \mathbf{u}^n}{\delta t^{n+1}} - 2\mathrm{div}(\mu\nabla^S\mathbf{u}^{n+1}) + \nabla(p^n + \frac{(\overline{\delta t})^2}{\delta t^n \delta t^{n+1}}\phi^n) = f(t^{n+1},.). \quad \text{in } \Omega \tag{14}$$

For simplicity, we consider the boundary condition $\mathbf{u} = 0$ on $\partial\Omega$ but the techniques presented in Sect. 2 for the open boundary condition case apply in this context as well. The pressure increment ϕ^{n+1} solves

$$-\frac{(\overline{\delta t})^2}{\delta t^{n+1}}\Delta\phi^{n+1} = -\rho\text{div}(\mathbf{u}^{n+1}) \quad \text{in } \Omega \qquad \text{and} \qquad \frac{\partial}{\partial\nu}\phi^{n+1} = 0 \quad \text{on } \partial\Omega. \tag{15}$$

Finally, the pressure is updated according to relation (5).

The standard pressure correction schemes are derived from the original velocity prediction – projection scheme, see for instance [4]. When the same time step value is used for the velocity prediction and correction, the factors multiplying the increment ϕ^n in (14) and (15) becomes $\frac{\delta t^n}{\delta t^{n+1}}$ and δt^{n+1} instead of $\frac{\overline{\delta t}^2}{\delta t^n \delta t^{n+1}}$ and $\frac{\overline{\delta t}^2}{\delta t^{n+1}}$ as in the proposed scheme (14)–(15). This alternative is referred as the standard scheme but we emphasize that there is no reason for the projection step to use the velocity prediction time step as projection parameter. In fact, this choice turns out to be numerically unstable as illustrated now. We consider the same setting as in Sect. 2 but with variable time steps given by

$$\delta t^n = \delta t^1 \times \begin{cases} 1 & \text{when } n \text{ is odd,} \\ 10^{-2} & \text{when } n \text{ is even,} \end{cases} \tag{16}$$

for different values of δt^1. In this case, we set $\overline{\delta t} := \delta t^1$. Figure 2 (left) illustrates the unstable behavior of $\|\mathbf{u}^n\|_{L_2(\Omega)}$ for $n = 0, \ldots, N$ when using the standard scheme with $\delta t^1 = 0.025$. However, the $l^2(H^1(\Omega))$ and $l^\infty(L_2(\Omega))$ errors on the velocity decay like $\overline{\delta t}$ when the proposed scheme (14)–(15) is used, see Fig. 2 (right).

We now briefly discuss the stability and error estimates for the scheme (14)–(15).

Theorem 2 (Velocity Stability) *Set* $\mathbf{f} \equiv 0$, *and assume* $\delta t^n \le \overline{\delta t}$, $n = 1, .., N$, *then there holds*

$$\rho\|\mathbf{u}_{\delta t}\|^2_{l^\infty(L_2(\Omega))} + 4\mu\|\nabla^S\mathbf{u}_{\delta t}\|^2_{l^2(L_2(\Omega))} + \frac{1}{\rho}(\overline{\delta t})^2\|p_{\delta t}\|^2_{l^\infty(H^1(\Omega))}$$
$$\le \rho\|\mathbf{u}_0\|^2_0 + (\overline{\delta t})^2\|p_0\|^2_1$$

provided $\mathbf{u}_0 \in L_2(\Omega)^d$ *and* $p_0 \in H^1(\Omega)$.

Proof Multiplying (14) by $2\delta t^{n+1}\mathbf{u}^{n+1}$ and integrating over Ω one gets after integrating by parts and using the boundary condition $\mathbf{u} = 0$,

$$\begin{aligned}\rho\left(\|\mathbf{u}^{n+1}\|^2_0 + \|\mathbf{u}^{n+1} - \mathbf{u}^n\|^2_0 - \|\mathbf{u}^n\|^2_0\right) + 4\delta t^{n+1}\mu\|\nabla^S\mathbf{u}^{n+1}\|^2_0 \\ - 2\int_\Omega\left(\delta t^{n+1}p^n + \frac{(\overline{\delta t})^2}{\delta t^n}\phi^n\right)\text{div}(\mathbf{u}^{n+1})d\mathbf{x} = 0.\end{aligned} \tag{17}$$

The pressure increment relation (15) is invoked to derive a bound for the last term in the left hand side of the above relation. More precisely, multiplying (15) by

$2(\delta t^{n+1} p^n + \frac{(\overline{\delta t})^2}{\delta t^n}\phi^n)$, integrating over Ω and using the boundary condition (9) we realize that

$$-2\rho \int_{\Omega} (\delta t^{n+1} p^n + \frac{(\overline{\delta t})^2}{\delta t^n}\phi^n)\text{div}(\mathbf{u}^{n+1})d\mathbf{x}$$
$$= 2(\overline{\delta t})^2 \int_{\Omega} \nabla\phi^{n+1} \cdot \nabla p^n d\mathbf{x} + 2\int_{\Omega} \nabla\left(\frac{(\overline{\delta t})^2}{\delta t^{n+1}}\phi^{n+1}\right) \cdot \nabla\left(\frac{(\overline{\delta t})^2}{\delta t^n}\phi^n\right) d\mathbf{x}.$$

Relation (5) allows us to rewrite the right hand side of the above expression as

$$(\overline{\delta t})^2 \left(\|\nabla p^{n+1}\|_0^2 - \|\nabla p^n\|_0^2 - \|\nabla\phi^{n+1}\|_0^2\right)$$
$$+ \frac{(\overline{\delta t})^4}{(\delta t^{n+1})^2}\|\nabla\phi^{n+1}\|_0^2 + \frac{(\overline{\delta t})^4}{(\delta t^n)^2}\|\nabla\phi^n\|_0^2 - \left\|\nabla\left(\frac{(\overline{\delta t})^2}{\delta t^{n+1}}\phi^{n+1} - \frac{(\overline{\delta t})^2}{\delta t^n}\phi^n\right)\right\|_0^2.$$

Going back to (17), we get

$$\rho\left(\|\mathbf{u}^{n+1}\|_0^2 + \|\mathbf{u}^{n+1} - \mathbf{u}^n\|_0^2 - \|\mathbf{u}^n\|_0^2\right) + 4\delta t^{n+1}\mu\|\nabla^S\mathbf{u}^{n+1}\|_0^2$$
$$+ \frac{1}{\rho}(\overline{\delta t})^2\left(\|\nabla p^{n+1}\|_0^2 - \|\nabla p^n\|_0^2\right) + \frac{1}{\rho}(\overline{\delta t})^2\left(\frac{(\overline{\delta t})^2}{(\delta t^{n+1})^2} - 1\right)\|\nabla\phi^{n+1}\|_0^2$$
$$+ \frac{1}{\rho}\frac{(\overline{\delta t})^4}{(\delta t^n)^2}\|\nabla\phi^n\|_0^2 = \frac{1}{\rho}\left\|\nabla\left(\frac{(\overline{\delta t})^2}{\delta t^{n+1}}\phi^{n+1} - \frac{(\overline{\delta t})^2}{\delta t^n}\phi^n\right)\right\|_0^2.$$

The difference of two successive relations (15) together with the boundary condition $\mathbf{u}^n = \mathbf{u}^{n+1} = 0$ on $\partial\Omega$ guarantee that

$$\left\|\nabla\left(\frac{(\overline{\delta t})^2}{\delta t^{n+1}}\phi^{n+1} - \frac{(\overline{\delta t})^2}{\delta t^n}\phi^n\right)\right\|_0 \leq \rho\|\mathbf{u}^{n+1} - \mathbf{u}^n\|_0.$$

Hence, using the assumption $\delta t^{n+1} \leq \overline{\delta t}$,

$$\rho\left(\|\mathbf{u}^{n+1}\|_0^2 - \|\mathbf{u}^n\|_0^2\right) + 4\delta t^{n+1}\mu\|\nabla^S\mathbf{u}^{n+1}\|_0^2 + \frac{1}{\rho}(\overline{\delta t})^2\left(\|\nabla p^{n+1}\|_0^2 - \|\nabla p^n\|_0^2\right)$$
$$+ \frac{1}{\rho}\frac{(\overline{\delta t})^4}{(\delta t^n)^2}\|\nabla\phi^n\|_0^2 \leq 0,$$

and the desired bound follows after summing for $n = 0$ to $N - 1$. □

Regarding the error decay we have that under the assumption $\delta t^n \leq \overline{\delta t}$, $n = 1, \ldots, N$, there exists a constant C independent of n and $\overline{\delta t}$ such that

$$\max_{n=1,\ldots,N} \|\mathbf{u}(t^n,.) - \mathbf{u}^n\|_{L^2(\Omega)} + \left(\sum_{n=1}^{N} \delta t^n \|\mathbf{u}(t^n,.) - \mathbf{u}^n\|^2_{H^1(\Omega)}\right)^{1/2} \leq C\overline{\delta t},$$

provided $\mathbf{u}$ and p are smooth enough and $\overline{\delta t}$ is sufficiently small. The proof of the above claim is omitted but relies on the argumentations provided in the proof of Theorem 2. In addition, we emphasize that scheme (14)–(15) does not optimize the choice of δt^n in order to equi-distribute the time discretization errors and explain that the decay rate is dictated by $\overline{\delta t}$ (and not δt^n, $n = 1, \ldots, N$). Including such mechanism is out of the scope of this work. Moreover, the decay rate for the $l^2(L_2(\Omega))$ error on the pressure is still an open problem but the numerical results provided in Fig. 2 indicate an optimal rate.

Acknowledgements Partially supported by NSF grant DMS-1254618 and award No. KUS-C1-016-04 made by King Abdullah University of Science and Technology (KAUST).

References

1. A.J. Chorin, On the convergence of discrete approximations to the Navier-Stokes equations. Math. Comput. **23**, 341–353 (1969)
2. K. Goda, A multistep technique with implicit difference schemes for calculating two- or three-dimensional cavity flows. J. Comput. Phys. **30**, 76–95 (1979)
3. J.L. Guermond, P. Minev, J. Shen, Error analysis of pressure-correction schemes for the time-dependent stokes equations with open boundary conditions. SIAM J. Numer. Anal. **43**(1), 239–258 (2005). (electronic)
4. J.L. Guermond, P. Minev, J. Shen, An overview of projection methods for incompressible flows. Comput. Methods Appl. Mech. Eng. **195**(44–47), 6011–6045 (2006)
5. J.-L. Guermond, L. Quartapelle, On the approximation of the unsteady Navier-Stokes equations by finite element projection methods. Numer. Math. **80**(2), 207–238 (1998)
6. J.L. Guermond, J. Shen, Velocity-correction projection methods for incompressible flows. SIAM J. Numer. Anal. **41**(1), 112–134 (2003). (electronic)
7. J.L. Guermond, J. Shen, On the error estimates of rotational pressure-correction projection methods. Math. Comput. **73**, 1719–1737 (2004)
8. J.-L. Guermond, A.J. Salgado, A splitting method for incompressible flows with variable density based on a pressure poisson equation. J. Comput. Phys. **228**(8), 2834–2846 (2009)
9. R. Témam, Sur l'approximation de la solution des équations de Navier-Stokes par la méthode des pas fractionnaires. II. Arch. Rational Mech. Anal. **33**, 377–385 (1969)

An ALE-Based Method for Reaction-Induced Boundary Movement Towards Clogging

Kundan Kumar, Tycho L. van Noorden, Mary F. Wheeler, and Thomas Wick

Abstract In this study, reaction-induced boundary movements in a thin channel are investigated. Here, precipitation-dissolution reactions taking place at the boundaries of the channel resulting in boundary movements act as a precursor to the clogging process. The resulting problem is a coupled flow-reactive transport process in a time-dependent geometry. We propose an ALE-based method (ALE – arbitrary Lagrangian-Eulerian) to perform full 2D computations. We derive a 1D model that approximates the 2D solution by integrating over the thickness of the channel. The boundary movements lead in the limit to clogging when the flow gets choked for a given pressure gradient applied across the channel. Numerical tests of the full 2D model are consulted to confirm the theory.

1 Motivation

Reactive flows are of great importance in a variety of fields including but not limited to porous media, biomedical applications, and biofilm growth [2, 3, 6–8]. Reactive processes such as precipitation-dissolution lead to geometry changes leading to changes in the flow which in turn affects the transport. Hence, the resulting model

K. Kumar • M. F. Wheeler
The Institute for Computational Engineering and Sciences, Center for Subsurface Modeling, The University of Texas at Austin, Austin, TX 78712, USA
e-mail: kkumar@ices.utexas.edu; mfw@ices.utexas.edu

T. L. van Noorden
Comsol B.V., Zoetermeer, The Netherlands
e-mail: t.l.v.noorden@gmail.com

T. Wick (✉)
The Institute for Computational Engineering and Sciences, The University of Texas at Austin, Austin, TX 78712, USA
e-mail: twick@ices.utexas.edu

A. Abdulle et al. (eds.), *Numerical Mathematics and Advanced Applications ENUMATH 2013*, Lecture Notes in Computational Science and Engineering 103,
DOI 10.1007/978-3-319-10705-9_62

must consider geometry changes, reactive processes, transport and flow problems in a coupled manner. In this work, we consider flow in a thin channel where ions are transported by flow and undergo molecular diffusion. The ions react to each other at the boundaries of the channel leading to the deposition of the crystalline material. We consider both the precipitation and dissolution processes as prototypes of reactions. The particularity is in the reactions, namely, precipitation and dissolution processes taking place at the boundaries of the channel leading to the deposition of the crystal material.

We employ an ALE-based method (ALE – arbitrary Lagrangian-Eulerian) to study coupled flow-transport phenomena in a time-dependent geometry. The initial geometry is quite simple and taken to be a thin channel which is a representative pore scale geometry. The changes in the geometry, as already stated, result from the reactions which are themselves functions of concentration and geometry. Hence, the time-dependent configuration remains an unknown and hence part of the solution variable. Since full 2D computations for the channel are expensive, we consider a 1D upscaled model derived in [5]. Both the 2D model and its approximate 1D model predict decreasing of strength of the flow as the channel progressively gets narrower. We term the limit of the narrowing of the channel as clogging, which is consistent with the intuitive notion.

The ALE method presented in this paper has been discussed in detail in [4] whereas the upscaled 1D model has been derived in [5] for a different set of boundary conditions for flow. We consider pressure boundary conditions for the flow which allows us to investigate the choking of the flow when the channel becomes constricted. These earlier studies did not consider the clogging process due to their choice for the boundary conditions for the flow. Our motivation for the present investigation stems from studying processes preceding the clogging and its effect on the flow and transport and further being able to define both 2D and upscaled 1D equations describing the behavior of *post-clogging*. Consequently, this work is a beginning in this direction.

The outline of the article is as follows: In Sect. 2, we recapitulate the underlying partial differential equations for the thin strip. Then, in Sect. 3, we state a 1D upscaled model. Next, the ALE method and discretization schemes for solving the free and moving boundary problem in the thin strip are described in Sect. 4. It is followed by Sect. 5 where clogging is discussed. Finally, in Sect. 6, the numerical experiments are conducted for full 2D model and we conclude by commenting on the consistence of 1D model with full 2D computations.

2 Equations

Let Ω_0 be a bounded domain in $\mathbb{R}^2$ representing a thin strip. The region occupied by the flow is $\Omega(t) \subset \Omega_0$, the precipitate layer is described by $\Gamma(t)$, with the inlet and outlet denoted by $\Gamma_{\rm i}$ and $\Gamma_{\rm o}$. The geometry description in which the flow and

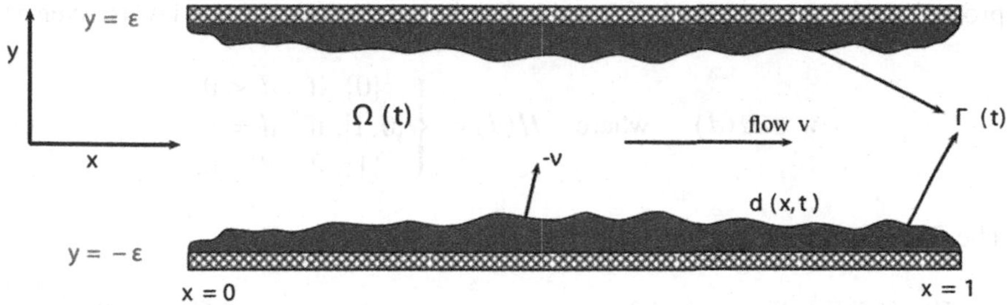

Fig. 1 Schematic of a thin channel showing the geometry changes due to precipitate being formed at the boundaries. The flow and transport takes place in $\Omega(t)$ and the reactions take place at lateral boundaries $\Gamma(t)$

transport processes take place is given by:

$$
\begin{aligned}
\Omega(t) &:= \{(x, y) \in \mathbb{R}^2 \mid 0 \le x \le 1,\ -(\varepsilon - d(x,t)) \le y \le (\varepsilon - d(x,t))\},\\
\Gamma(t) &:= \{(x, y) \in \mathbb{R}^2 \mid 0 \le x \le 1,\ y \in \{-(\varepsilon - d(x,t)), (\varepsilon - d(x,t))\}\},\\
\Gamma_{\rm i}(t) &:= \{(x, y) \in \mathbb{R}^2 \mid x = 0,\ -(\varepsilon - d(0,t)) \le y \le (\varepsilon - d(0,t))\},\\
\Gamma_{\rm o}(t) &:= \{(x, y) \in \mathbb{R}^2 \mid x = 1,\ -(\varepsilon - d(1,t)) \le y \le (\varepsilon - d(1,t))\}.
\end{aligned}
$$

Due to the reactions at the boundaries, Ω and the boundaries Γ's are time-dependent. The schematic illustration for the thin strip is displayed in Fig. 1.

The flow and transport of the solutes (the ions) are described by the following system of equations. The transport equation reads:

$$
\begin{aligned}
\partial_t c &= \nabla \cdot (D\nabla c - vc), && \text{in } \Omega(t) \times (0, T),\\
\rho_s \partial_t d &= f(c, \rho_s d)\sqrt{1 + (\partial_x d)^2}, && \text{on } \Gamma(t) \times (0, T),\\
f(c, \rho_s d) &= r(c) - w, && \text{on } \Gamma(t) \times (0, T).
\end{aligned} \tag{1}
$$

Here, the unknowns are: $c(x, y, t)$, concentration of the charged ions, $d(x,t)$ free and moving boundary resulting due to reactions, and $v(x, y, t)$ the flow field. The known physical parameters are: $D > 0$, diffusion constant, ρ_s, the density of ions in the precipitate. Equation $(1)_1$ describes the transport of solutes due to convection and molecular diffusion processes, whereas $(1)_2$ describes the movement of the boundary due to reaction term f. According to $(1)_3$, the reaction rate f is imposed by the following structure:

$$
f(c, \rho_s d) = r(c) - w, \tag{2}
$$

where $r(\cdot)$ describes the precipitation part whereas w models the dissolution process. Additionally, we assume that $r(\cdot) : \mathbb{R} \to [0, \infty)$, is monotone and locally Lipschitz continuous in $\mathbb{R}$. The usual mass-action kinetics laws governing the precipitation process satisfy this assumption. For the dissolution process, the rate law is given as

$$w \in H(d), \quad \text{where} \quad H(d) = \begin{cases} \{0\}, \text{if} & d < 0, \\ [0,1], \text{if} & d = 0, \\ \{1\}, \text{if} & d > 0. \end{cases} \tag{3}$$

The flow equations read:

$$\begin{array}{lll} \text{Continuity:} & \nabla \cdot v \quad = 0, & \text{in } \Omega(t) \times (0,T), \\ \text{Momentum:} & \rho_f v \cdot \nabla v = \nabla \cdot \left(\mu_f (\nabla v + (\nabla v)^T)\right) - \nabla p, & \text{in } \Omega(t) \times (0,T), \end{array} \tag{4}$$

where p is the pressure field and $\mu_f = \rho_f \nu_f$ is the dynamic viscosity. The flow and transport equations are complemented by the initial and boundary conditions. The initial conditions read:

$$c(x,y,0) = c_o, \qquad d(x,0) = d_o. \tag{5}$$

The boundary conditions read:

$$\begin{array}{llll} c & = c_b, & p(0,y,t) \;\; = 1, & \text{on } \Gamma_i(t) \times (0,T), \\ \partial_x c & = 0, & p(L,y,t) \;\; = 0, & \text{on } \Gamma_o(t) \times (0,T), \\ v & = 0, & v \cdot (-D\nabla c)\sqrt{1 + (\partial_x d)^2} = \partial_t d(\rho_s - c) & \text{on } \Gamma(t) \times (0,T). \end{array} \tag{6}$$

As stated above, at the inlet and outlet, we prescribe the pressures and further impose that the flow takes place normal to the boundaries.

3 A 1D Averaged Model

An upscaled model is obtained by integrating the equations in the y-direction. We consider a sequence of problems depending upon the thickness of strip ε and using formal asymptotic expansions, the unknowns are assumed to be of the form

$$z^\varepsilon = z_0 + \varepsilon z_1 + O(\varepsilon^2),$$

with z^ε denoting any of $c^\varepsilon, d^\varepsilon, v^\varepsilon$. Following the procedure in [5], the following upscaled equations are derived

$$\begin{array}{ll} \partial_x v_0 = 0, \qquad v_0 - \frac{(1-d_0)^3}{3\mu} \partial_x P_0 = 0, & \\ \partial_t \left((1-d_0)c_0\right) + \partial_t(\rho_s d_0) & = \partial_x \left(D(1-d_0)\partial_x c_0\right) - \partial_x(v_0 c_0), \\ \partial_t d_0 - f(u_0, \rho_s d_0) & = 0. \end{array} \tag{7}$$

As our interest is in the case of closing of the channel, the above system of equations degenerates as $d_0 \to 1$. In this work, we consider only the flow equations, that is, $(7)_1$. The limit case for the reactive transport will be treated in future studies.

4 ALE Based Method for Full Thin-Strip Computations and Discretization

The moving boundary problem is computed with the help of the arbitrary Lagrangian-Eulerian (ALE) approach that is mostly well-known from fluid-structure interaction computations. Here, rather than computing the equations on the physical mesh (bottom figure in Fig. 2), the equations are solved on a reference mesh (top figure in Fig. 2) by transforming them with the ALE-mapping. The discretization is based on Rothe's method: first in time and than in space. A one-step-θ scheme is employed for temporal discretization and a Galerkin finite element method for spatial discretization including local mesh refinement with hanging nodes. Since we are solving the incompressible Navier-Stokes equations and due to the ALE-mapping, we deal with a nonlinear system of equations, which is solved in a monolithic fashion. The linear equations are treated with a direct solver. Rather than providing all necessary information and all important references of this section, we would like to refer the friendly reader to [4], where all details are given. The discretization is realized with the multiphysics template [10] based on the finite element software deal.II [1].

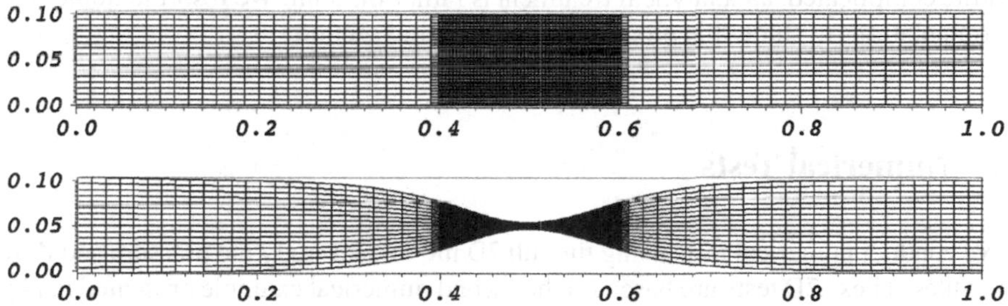

Fig. 2 Initial (and also the reference) mesh and the deformed mesh at end time step $T = 14$. Local mesh refinement with hanging nodes is used in the middle of the channel

5 Clogging of the Channel

When the channel starts getting narrower, the flow profile alters because of changing geometry. However, as the channel starts getting clogged, the flow is expected to decrease and eventually, the channel should be closed. For the upscaled model, following calculations show that the flow becomes zero as the channel closes. Using $(7)_1$

$$\partial_x\left(\frac{(1-d_0)^3}{3\mu}\partial_x P_0\right) = 0, \text{ leading to } \frac{(1-d_0)^3}{3\mu}\partial_x P_0 = C,$$

and hence,

$$P_0(x,t) = \int_x^1 \frac{C}{(1-d_0(\xi,t))^3} d\xi,$$

where C is obtained by using the boundary conditions for P_0,

$$v_0(t) = C(t) = \left\{ \int_0^1 \frac{1}{(1-d_0(\xi,t))^3} d\xi \right\}^{-1}. \tag{8}$$

Now considering (8), formally, one sees that the integral is dominated by the regions where $1 - d_0$ is small and the flow $v_0(t)$ decreases as $(1-d_0)^3$. Hence, wherever locally $d_0 \to 1$, we get that the flow in the channel tends to zero, allowing us to conclude that in the limit (clogging), the flow becomes zero. Since the 2D model is quite complicated, an analytical treatment is rather difficult. We resort to numerical computations to study this process in the following section.

6 Numerical Tests

We conduct numerical tests using the full 2D model and study the pressure and flow profiles. These 2D tests are based on the second numerical example presented in [4]. Specifically, we have a right-hand side force function (representing an analytical expression for a point source)

$$f(x,y) = a\exp(-b(x-x_m)^2 - c(y-y_m)^2),$$

where $a = 1{,}000, b = c = 100$ and $x_m = 0.5, y_m = 0.05$, representing a source with maximum strength at (x_m, y_m) and having an exponential decay and causing the precipitation in the middle of the thin channel $\Omega := [0,1] \times [0,0.1]$. All material parameters and geometry information are described in the previously mentioned article [4]. In contrast, the flow is now driven by pressure difference such that we have $p = 1$ on the inflow (left boundary) and $p = 0$ at the right (outflow) boundary. The initial concentration is $c = 1$ for all $x \in \Omega$. In addition, we prescribe $c = 0$ at the left boundary. The goal of our present study is now different from [4]. We are specifically interested in the pressure behavior along the x-axis and the validity of approximating the behavior through the lower-dimensional lubrication equation $(7)_1$.

Figure 3 shows the pressure and the pressure gradient at $T = 14$ when the channel has closed by ≈92 %. Furthermore, the two bottom figures show $w^2 \partial_x p$ (w is the width of flow domain) and the v_x velocity with respect to time. The choice of this scaling w^2 is motivated by considering $(7)_1$; since the total flow follows the cubic law, the average flow obeys a square law. For the 2D model, achieving the limit is not possible since the mesh will degenerate as the channel is closed. (This drawback in the numerics is investigated in terms of a nonstandard fluid-structure interaction framework in [9]). However, the amount of channel constriction is pretty close to the process of clogging. The profile shows that the pressure gradients are blowing up as the channel gets smaller. However, when this is weighted with $2(\varepsilon - d)^2$, that is with square of the opening width of the channel, the resulting quantity goes to zero. This quantity is proportional to the flow and showing similar behavior as displayed in Fig. 3. This suggests that the flow strength vanishes as the channel progressively gets clogged. This is consistent with the case of upscaled model. Figure 4 displays the plot of concentration for two different times.

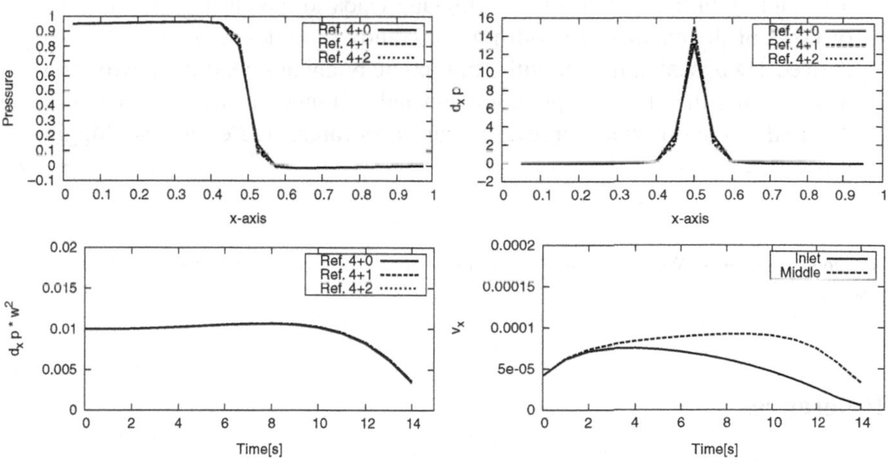

Fig. 3 Results of the 2D numerical simulation: Profiles at final time $T = 14$ for the pressure, its gradient, and the key observation quantity $\partial_x pw^2$ shown in the first three figures. Each of the quantities of interest is computed on a sequence of three locally refined meshes to have numerical evidence of convergence. In the final figure, the velocity component in normal flow direction integrated over the cross section is shown on the finest mesh for the inlet and the middle (narrow part of the channel)

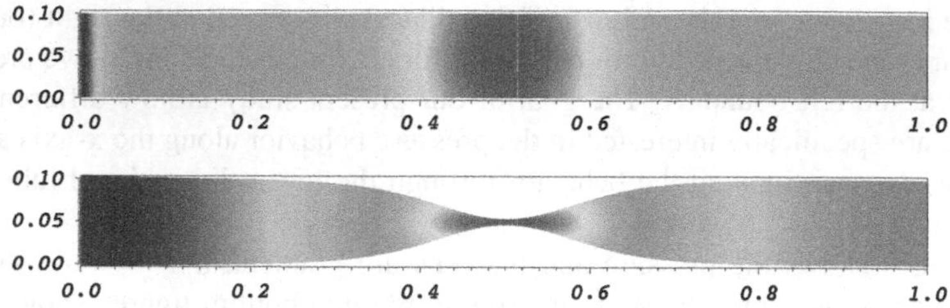

Fig. 4 Results of the 2D numerical simulation: Concentration at the first time step and the end time step $T = 14$. At the initial time, the concentration is $c = 1$ in the whole channel. Starting the simulation, $c = 0$ is applied at the inlet boundary and the source term f increases the concentration in the middle

Conclusion

In this work, we investigated a coupled flow-reactive transport model in a time-dependent thin channel where the geometry changes are induced by reactive boundary conditions. The 2D model is solved using an ALE-based method. A pressure gradient is applied across the channel and as the channel gets constricted, the flow strength diminishes so that in the limit we get no-flow across the channel. The approximating 1D model can be analytically studied and formal arguments are employed to obtain the same observations. The study highlights that a local clogging leads to the closing (in the sense of flow) of the channel. In addition, this study provides some hints that the derived 1D upscaled model with appropriate boundary conditions will allow us to capture the clogging phenomenon and continue the solution thereafter. The findings of this work serve as precursor for future studies of post-clogging processes.

Acknowledgements We would like to thank Prof. Sorin Pop (Univ. Eindhoven) for helpful discussions.

References

1. W. Bangerth, T. Heister, G. Kanschat, Differential equations analysis library deal. II (2014)
2. S. Čanić, A. Mikelić, Effective equations modeling the flow of a viscous incompressible fluid through a long elastic tube arising in the study of blood flow through small arteries. SIAM J. Appl. Dyn. Syst. **2**(3), 431–463 (2003). (electronic)
3. C. D'Angelo, P. Zunino, Robust numerical approximation of coupled Stokes' and Darcy's flows applied to vascular hemodynamics and biochemical transport. M2AN **45**, 447–476 (2011)
4. K. Kumar, M. Wheeler, T. Wick, Reactive flow and reaction-induced boundary movement in a thin channel. SIAM J. Sci. Comput. **35**(6), B1235–1266 (2013)

5. T.L. van Noorden, Crystal precipitation and dissolution in a thin strip. Eur. J. Appl. Math. **20**(1), 69–91 (2009)
6. T.L. van Noorden, I.S. Pop, A. Ebigbo, R. Helmig, An upscaled model for biofilm growth in a thin strip. Water Resour. Res. **46**, W06505 (2010)
7. A. Quarteroni, A. Veneziani, P. Zunino, A domain decomposition method for advection-diffusion processes with application to blood solutes. SISC **23**(6), 1959–1980 (2002)
8. ———, Mathematical and numerical modeling of solute dynamics in blood flow and arterial walls. SINUM **39**(5), 1488–1511 (2002)
9. S. Frei, T. Richter, T. Wick, Solid growth and clogging in fluid-structure interaction using ALE and fully Eulerian coordinates (2014, Preprint, in review). usevs.ices.utexas.edu/~twick/
10. T. Wick, Solving monolithic fluid-structure interaction problems in arbitrary Lagrangian Eulerian coordinates with the deal.ii library. Arch. Numer. Softw. **1**, 1–19 (2013)

A Unified Approach for Computing Tsunami, Waves, Floods, and Landslides

Alexander Danilov, Kirill Nikitin, Maxim Olshanskii, Kirill Terekhov, and Yuri Vassilevski

Abstract The prediction of large-scale hydrodynamic events such as tsunami spread and run-up, dam break, flood, or landslide run-out is a challenging and important problem of applied mathematics and scientific computing. The paper presents a computational approach based on free surface flow models for fluids of complex rheology to simulate such events and phenomena with detail and prediction confidence typically not achievable by simplified models. Using nonlinear defining relations for stress and rate of strain tensors allows a unified approach to simulate events described by both the Newtonian model (tsunami, dam break) and non-Newtonian models (landslide, snow avalanches, lava flood, mud flow). The computational efficiency of the numerical approach owes to the level-set method for free surface capturing and to an accurate and stable FV/FD method on dynamically adapted octree meshes for discretization of flow and level set equations. In this paper we briefly describe the numerical method and present results of several simulations of hydrodynamic events: a dam break, a landslide and tsunami spread and run-up.

1 Introduction

Numerical simulations became a standard tool for the study and prediction of disasters and events involving water and mud flows, landslides, avalanches and other phenomena described by equations of continuous medium. Since the computational complexity of full-scale simulations of such events is highly demanding for computer resources, it is common to describe phenomena by reduced-order approaches, for example, based on 2D and even 1D equations and simplified rheological models. A few examples are the use of the shallow water equations [14, 19] to simulate the

---

A. Danilov • K. Nikitin (✉) • K. Terekhov • Y. Vassilevski
Institute of Numerical Mathematics RAS, Gubkin str. 8, 119333, Moscow, Russia
e-mail: a.a.danilov@gmail.com; nikitin.kira@gmail.com; kirill.terehov@gmail.com; yuri.vassilevski@gmail.com

M. Olshanskii
Department of Mathematics, University of Houston, Houston, TX 77204-3008, USA
e-mail: molshan@math.uh.edu

A. Abdulle et al. (eds.), *Numerical Mathematics and Advanced Applications ENUMATH 2013*, Lecture Notes in Computational Science and Engineering 103, DOI 10.1007/978-3-319-10705-9_63

spread of tsunami or modified hydrodynamics equations to compute landslide and debris flows [5].

In this paper, we describe a computational approach that allows one to simulate many of complex hydrodynamic events using fewer simplifying assumptions to describe the physics of flows and accounting for their three-dimensional nature and real environment. Instead of models based on shallow water theory or other approaches using simplified physics, we use the full system of incompressible Navier-Stokes equations for flows with free surface. The necessary model order reduction is performed on the numerical method stage of simulations and is essentially due to the use of accurate discretization schemes on dynamically adapted octree meshes. The phenomenological diversity of the processes we are interested in is accounted by the choice of proper nonlinear relations between the stress and the strain rate tensors. We consider both Newtonian model to describe dam break and ocean tsunami flows and viscoplastic Hershel-Bulkley model for landslides. However, the developed numerical technology makes it easy for a researcher to incorporate different, even more complicated, defining relations.

In all flows considered here, finding free surface dynamics is critical. To capture the free surface evolution, we apply the level set method [17]. We discretize flow and level set equations using octree meshes. Discretizations on octree meshes benefit from their regular orthogonal structure on one hand and the embedded hierarchy on the other hand, which make the reconstruction and adaptation process as well as data access fast and easy. Nowadays such grids are widely used in simulations and visualization, see, e.g., [6, 8, 9, 11, 15, 18]. However, building accurate and stable discretizations on such grids is a challenging task. For the fluid equations we use the second order accurate finite difference/finite volume method with compact nodal stencils developed in [12]. This method is a stable extension of the classical MAC scheme on octree meshes. Semi-Lagrangian particle level set method [13] is used to solve the transport equation for the level set function. To integrate in time, we apply a second order splitting scheme of Chorin type. This scheme decouples one time step on convection, diffusion and plasticity, pressure correction, and the level set function advection substeps. Numerical stability and accuracy of the whole method were verified for the case of Newtonian flows in [10, 12] by computing analytical solutions and benchmark flows in a cubic cavity and over a 3D cylinder. Computed results for a collapsing water column perfectly match available experimental data. The efficiency of the approach for non-Newtonian flows was accessed in [11, 18], where the flow of Hershel-Bulkley fluid from a reservoir over inclined planes was computed and compared to documented experimental results with Carbopol Ultrez 10 gel. These results let us believe that the entire approach is computationally efficient, highly predictive and thus it is a reliable tool for simulation of large-scale hydrodynamic events.

The remainder of the paper is organized as follows. The mathematical model is given in Sect. 2. In Sect. 3, we sketch the basics of the numerical approach. In Sect. 4, we discuss our approach to the topography-specific design of the computational domain. Section 5 presents results of numerical simulations of three hypothetical scenarios: the break of the Sayano-Shushenskaya dam, landslide in the Sayan mountains and tsunami run-up in the Bay of Bengal. The computations were done using real topography and bathymetry. It is remarkable that such different natural disaster scenarios can be simulated in a unified framework of computational non-Newtonian fluid dynamics.

2 Mathematical Model

Conservation of mass and momentum for an incompressible viscous fluid leads to the Navier-Stokes equations for the unknown velocity field $\mathbf{u}$ and stress tensor $\boldsymbol{\tau}$:

$$\begin{cases} \rho\left(\dfrac{\partial \mathbf{u}}{\partial t} + (\mathbf{u}\cdot\nabla)\mathbf{u}\right) - \mathbf{div}\ \boldsymbol{\tau} = \mathbf{f} \\ \nabla\cdot\mathbf{u} = 0 \end{cases} \quad \text{in } \Omega(t), \tag{1}$$

where $\mathbf{f}$ are given mass forces, $\Omega(t) \in \mathbb{R}^3$ is a spatial domain occupied by fluid and dependent on time, ρ is the density. For the strain rate tensor $\mathbf{Du} = \frac{1}{2}[\nabla\mathbf{u} + (\nabla\mathbf{u})^T]$ and the stress tensor we consider the Hershel-Bulkley defining relations [3]:

$$\begin{aligned} \boldsymbol{\tau} = -p\,\mathbf{I} + \left(K\,|\mathbf{Du}|^{n-1} + \tau_s|\mathbf{Du}|^{-1}\right)\mathbf{Du} \;\Leftrightarrow\; |\boldsymbol{\tau}| > \tau_s, \\ \mathbf{Du} = \mathbf{0} \;\Leftrightarrow\; |\boldsymbol{\tau}| \le \tau_s, \end{aligned} \tag{2}$$

where K is the consistency parameter, τ_s is the yield stress parameter, n is the fluid index, p is the pressure. Newtonian flows correspond to the choice $\tau_s = 0, n = 1$. In computations we use a regularized model [11, 18].

A volume occupied by fluid and the velocity field at $t = 0$ are assumed to be given:

$$\Omega(0) = \Omega_0, \quad \mathbf{u}|_{t=0} = \mathbf{u}_0. \tag{3}$$

Finding $\Omega(t)$ for $t > 0$ is a part of the problem which is solved together with Eqs. (1). To formulate this more precisely, let us divide the boundary of the whole volume into the static boundary Γ_D (for instance, the rigid walls or the bottom of a bassin) and the free boundary $\Gamma(t)$ (in practice it usually models an interface between fluid and air), i.e. $\overline{\partial\Omega(t)} = \overline{\Gamma_D} \cup \overline{\Gamma(t)}$. Generally speaking, the boundary $\overline{\partial\Omega(t)}$ depends on time.

We assume the non-penetration condition on Γ_D and, depending on a physical setup, the no-slip or slip with friction conditions. The free boundary evolves with the normal velocity of fluid, which can be written as the kinematic condition

$$v_\Gamma = \mathbf{u}\cdot\mathbf{n}_\Gamma,$$

where $\mathbf{n}_\Gamma$ is the outer unit normal to the surface $\Gamma(t)$, v_Γ is the normal velocity of the surface $\Gamma(t)$. Normal stresses on the free surface are balanced by the surface tension forces. This leads to the boundary condition

$$\boldsymbol{\tau}\mathbf{n}_\Gamma = \varsigma\kappa\mathbf{n}_\Gamma - p_{\text{ext}}\mathbf{n}_\Gamma \quad \text{on } \Gamma(t), \tag{4}$$

where κ is the sum of the principal curvatures of the surface, ς is the surface tension coefficient, p_{ext} is the external pressure. If the surface tension forces are not taken into account, we may assume $\varsigma = 0$.

In order to find the position of the free boundary at each time moment we use the implicit definition of $\Gamma(t)$ as the zero level set of the globally defined indicator function $\phi(t, \mathbf{x})$ instead of the kinematic condition

$$\phi(t,\mathbf{x}) = \begin{cases} < 0 & \text{if } \mathbf{x} \in \Omega(t) \\ > 0 & \text{if } \mathbf{x} \in \mathbb{R}^3 \setminus \overline{\Omega(t)} \\ = 0 & \text{if } \mathbf{x} \in \Gamma(t) \end{cases} \qquad \text{for all } t \in [0, T].$$

The function ϕ is called the level set function. The level set function satisfies the following transport equation [13]:

$$\frac{\partial \phi}{\partial t} + \tilde{\mathbf{u}} \cdot \nabla \phi = 0 \quad \text{in } \mathbb{R}^3 \times (0, T], \tag{5}$$

where $\tilde{\mathbf{u}}$ is the fluid velocity field extended outside $\Omega(t)$. Initial condition (3) is used to define $\phi(0, \mathbf{x})$. One can impose the additional restriction

$$|\nabla \phi| = 1 \tag{6}$$

onto the level set function to ensure numerical stability, i.e. ϕ is the signed distance function. Given ϕ, the outer normal and the curvature of the free boundary can be calculated from $\mathbf{n}_\Gamma = \nabla\phi/|\nabla\phi|$, $\kappa = \nabla \cdot \mathbf{n}_\Gamma$.

The mathematical model used in our calculations consists of Eqs. (1)–(6) and appropriate boundary conditions on static boundaries.

3 Numerical Method

For the numerical time integration of (1)–(5) we use a second order splitting scheme of Chorin type with the BDF2 approximation of time derivative. For some given $\mathbf{u}(t)$, $p(t)$, $\phi(t)$ at time t, one time step consists in finding $\mathbf{u}(t + \Delta t)$, $p(t + \Delta t)$, $\phi(t + \Delta t)$, where Δt is a time increment. This is done in few substeps. First, using the second order semi-Lagrangian method [16], we integrate back in time the level set transport equation (5) along characteristics interpolated from previous time steps.

At this step we find the new volume occupied by a fluid, $\Omega(t) \rightarrow \Omega(t + \Delta t)$. Approximation errors may lead to a non-physical loss or gain of fluid volume, i.e. $|\Omega(t)| \neq |\Omega(t + \Delta t)|$ even in the absence of sources or sinks. To reduce this loss/gain of volume, we use the grid adaptation towards the free surface and the particle method [13]. This enhances the conservation properties of the semi-Lagrangian method significantly. If a further correction is still needed, one solves for δ the equation

$$\text{meas}\{\mathbf{x} : \phi(\mathbf{x}) \leq \delta\} = Vol^{\text{reference}}$$

and correct $\phi^{new} = \phi - \delta$. The above equation is solved for δ by the secant method for root finding and Monte-Carlo method to approximate meas$\{\mathbf{x} : \phi(\mathbf{x}) < \delta\}$. For stability reasons, the substep is accomplished be the re-initialization of ϕ^{new} such that (6) holds, see [10, 11]. After $\Omega(t + \Delta t)$ is found, we rebuild the octree mesh and re-interpolate all unknowns to the new mesh. If an extension of velocity (or pressure) to the exterior of $\Omega(t)$ is needed during the stages of re-interpolation or numerical integration of (5), then we perform the extension of unknowns along normals to free surface, i.e. the extension of velocity field satisfies $(\nabla\phi) \cdot \nabla\mathbf{u}(t) = 0$ in the exterior of $\Omega(t)$.

Further, one finds new values of hydrodynamic variables in $\Omega(t + \Delta t)$: $\{\mathbf{u}(t), p(t)\} \rightarrow \{\mathbf{u}(t + \Delta t), p(t + \Delta t)\}$. This is done in the following substeps: (i) The convective terms are approximated using the second order accurate compact upwind method on octree grids from [12] and the momentum equation is solved for an intermediate velocity, while the plasticity terms are treated explicitly [11]; (ii) The intermediate velocity field is projected onto the subspace of discrete divergence free functions and the pressure is updated by the solution of the discrete Poisson equation.

The time step is variable and controlled by a CFL type condition.

4 Design of Computational Domain

We use the Google SketchUp tool to generate a polygonal approximation of the earth surface and facilities on it in the area of interest. Depending on the problem the topographic data can be obtained from the Shuttle Radar Topography Mission (NASA), or the Google Maps, or the Google Earth projects. Ocean bathymetry can be retrieved from ETOPO data. As an example, the left picture of Fig. 1 shows the modeled 'broken' dam in a real life topography of Sayan mountains and Yenisei river. The scene of a rocky bay in the right picture is purely hand-made.

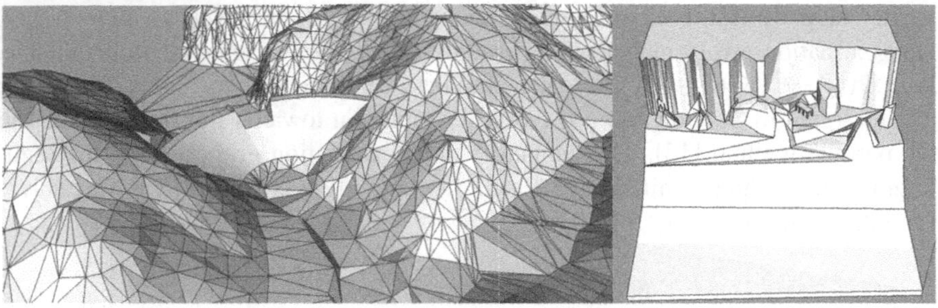

Fig. 1 Design of the computational domain using Google SketchUp software: environment of a dam (*left*), a rocky bay (*right*)

5 Simulation of Hydrodynamic Events

In this section all variables are dimensional and given in the SI system.

First we present the results of simulations of the break of the Sayano-Shushenskaya dam. The calculations presented do not necessarily show any actual or feasible scenario for this hydro power plant. Rather, they show that with the present approach a prediction is practically possible, if more detailed geophysical data for the riverside area and the dam conditions are supplied. The computational domain and the part of the dam simulated as 'broken' are shown in green in Fig. 2 (left). The maximal number of active cells in the computational octree meshes was about 520,000. The water rise levels at given points are shown as graphs in Fig. 2 (right). The steplike graphs of the water level reflect the appearance of waves which are clearly seen on animation available from [4].

Next, we simulate a rock landslide on the right-bank slope near the same dam. Due to the absence of the rheological data for the considered slope, we adopted the coefficients K, τ_s, n of the Hershel-Bulkley model from [1] measured for rock landslides in the south of Italy. Figure 3 (left) shows the top view of the landslide at intermediate time moment and the velocities of the fluid. The graph of the maximal pressure acting on the dam structures at the place of landslide is given in Fig. 3 (right). The maximal number of cells in the computational meshes approached 560,000.

Finally, we simulate a tsunami run-up in two stages. At the first stage we solve the shallow water equations [19] for horizontal velocity components and water level defined in the World Ocean. The solution is based on the mixed finite element method [2] on unstructured triangulation of the World Ocean and the fractional time stepping scheme [7]. At the second stage, when the tsunami wave approaches the shore, we adopt the elevation and horizontal velocity in the boundary conditions of the 3D free surface flow model presented in Sect. 2.

Figure 4 (left) demonstrates the tsunami wave elevation in 100 min after a hypothetical earthquake with epicenter at 9°40′ north latitude and 92°30′ east longitude. The epicenter is located in seismically active region between Andaman

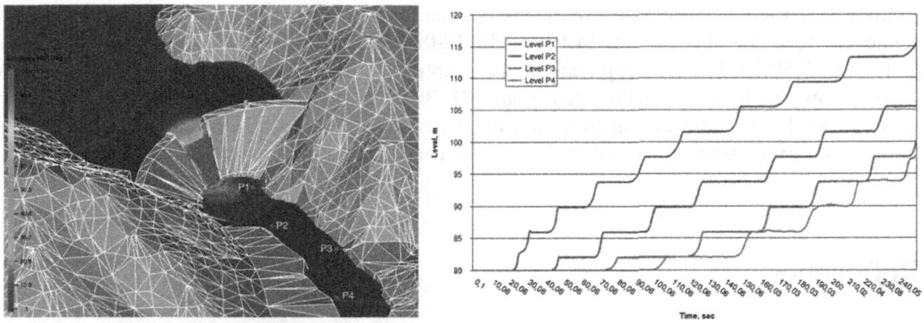

Fig. 2 *Left*: Mesh representation of the dam and the surrounding area. The monitoring of the water level variation is performed in the marked points. *Right*: Dependence of water level on time at points P1–P4

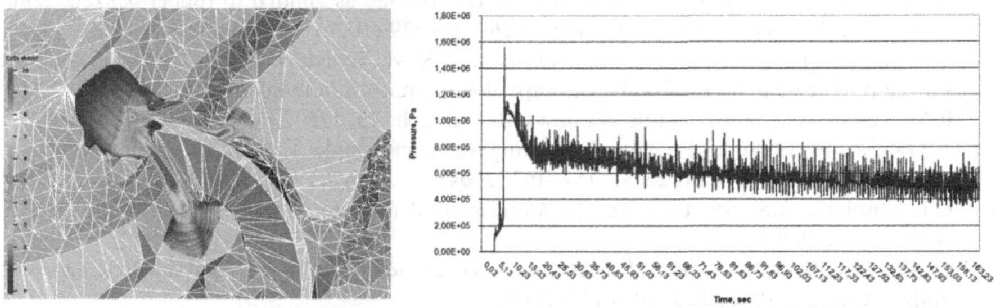

Fig. 3 *Left*: Migration of the landslide at time moment $t = 100$ s. *Right*: The pressure experienced by the dam at the place of the landslide

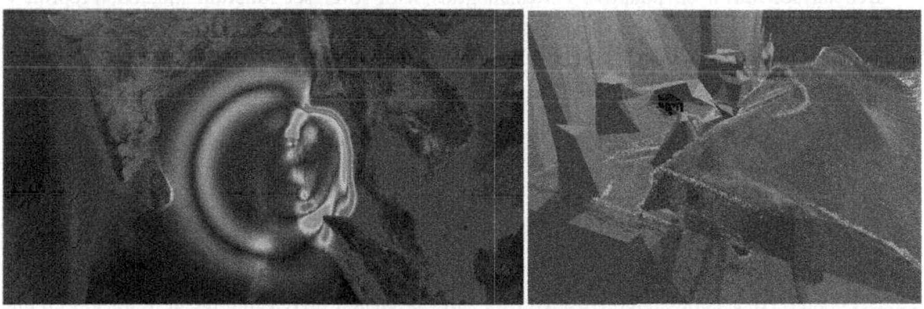

Fig. 4 *Left*: Tsunami elevation in 100 min after an earthquake in the Bay of Bengal. *Right*: Tsunami run-up in a bay with rocky relief

and Nicobar islands in the Bay of Bengal. The colored field of ocean elevation is superposed on the Google Earth images of the Bay of Bengal. The features of tsunami run-up depend on the ocean bed and the terrain. The terrain shown in Fig. 1 (right) is capable to produce wave breaking, which is dangerous for boats and facilities, see Fig. 4 (right). Development and breaking of waves in the run-up can be inspected in [4].

Acknowledgement Development of the viscous and viscoplastic free surface flow model has been supported in part by RFBR grants 14-01-00830, 12-01-33084, 12-01-00283, the Russian President grant MK 7159.2013.1. Development of the adaptive computational grids generator has been supported in part by Russian President grant MK 3675.2013.1. Interaction between the shallow water model for tsunami propagation and the 3D model of water flow with free boundary for tsunami run-up has been supported by Russian Science Foundation grant 14-11-00434.

References

1. T. Bisantino, P. Fischer, F. Gentile, Rheological characteristics of debris-flow material in South-Gargano watersheds. Nat. Hazards **54**, 209–223 (2010)
2. C. Cotter, D. Ham, C. Pain, A mixed discontinuous/continuous finite element pair for shallow-water ocean modeling. Ocean Model. **26**, 86–90 (2009)
3. W.H. Herschel, R. Bulkley, Measurement of consistency as applied to rubber-benzene solutions. Proc. Am. Assoc. Test Mater. Part II **26**, 621–629 (1926)
4. Y.V. Vassilevski, K.D. Nikitin, M.A. Olshanskii, K.M. Terekhov, A.Y. Chernyshenko, Free surface flows research group web site. http://dodo.inm.ras.ru/research/freesurface
5. R.M. Iverson, The physics of debris flows. Rev. Geoph. **35**, 245–296 (1997)
6. F. Losasso, F. Gibou, R. Fedkiw, Simulating water and smoke with an octree data structure. ACM Trans. Graph. (TOG) **23**(3), 457–462 (2004)
7. G.I. Marchuk, A.S. Sarkisyan (eds.), *Mathematical Models of Ocean Circulation* (Nauka, Novosibirsk, 1980)
8. C. Min, F. Gibou, A second order accurate level set method on non-graded adaptive cartesian grids. J. Comput. Phys. **225**, 300–321 (2007)
9. K.D. Nikitin, Y.V. Vassilevski, Free surface flow modelling on dynamically refined hexahedral meshes. Russ. J. Numer. Anal. Math. Model. **23**, 469–485 (2008)
10. K.D. Nikitin, M.A. Olshanskii , K.M. Terekhov, Y.V. Vassilevski, Numerical simulations of free surface flows on adaptive cartesian grids with level set function method, Preprint is available online, (2010)
11. K.D. Nikitin, M.A. Olshanskii, K.M. Terekhov, Y.V. Vassilevski, A numerical method for the simulation of free surface flows of viscoplastic fluid in 3D. J. Comput. Math. **29**, 605–622 (2011)
12. M.A. Olshanskii, K.M. Terekhov, Y.V. Vassilevski, An octree-based solver for the incompressible Navier-Stokes equations with enhanced stability and low dissipation. Comput. Fluids **84**, 231–246 (2013)
13. S. Osher, R. Fedkiw, *Level Set Methods and Dynamic Implicit Surfaces* (Springer, New York, 2002)
14. E.N. Pelinovskii, *Hydrodynamics of Tsunami Waves* (IAP RAS, Nizhny Novgorod, 1996)
15. S. Popinet, An accurate adaptive solver for surface-tension-driven interfacial flows. J. Comput. Phys. **228**, 5838–5866 (2009)
16. J. Strain, Semi-Lagrangian methods for level set equations. J. Comput. Phys. **151**, 498–533 (1999)
17. M. Sussman, P. Smereka, S. Osher, A level set approach for computing solutions to incompressible two-phase flow. J. Comput. Phys. **114**, 146–159 (1994)
18. Y.V. Vassilevski, K.D. Nikitin, M.A. Olshanskii , K.M. Terekhov, CFD technology for 3D simulation of large-scale hydrodynamic events and disasters. Russ. J. Numer. Anal. Math. Model. **27**, 399–412 (2012)
19. C.B. Vreugdenhil, *Numerical Methods for Shallow-Water Flow* (Kluwer, Dordrecht/Boston, 1994)

Newton-Multigrid Least-Squares FEM for S-V-P Formulation of the Navier-Stokes Equations

Abderrahim Ouazzi, Masoud Nickaeen, Stefan Turek, and Muhammad Waseem

Abstract Least-squares finite element methods are motivated, beside others, by the fact that in contrast to standard mixed finite element methods, the choice of the finite element spaces is not subject to the LBB stability condition and the corresponding discrete linear system is symmetric and positive definite. We intend to benefit from these two positive attractive features, on one hand, to use different types of elements representing the physics as for instance the jump in the pressure for multiphase flow and mass conservation and, on the other hand, to show the flexibility of the geometric multigrid methods to handle efficiently the resulting linear systems. With the aim to develop a solver for non-Newtonian problems, we introduce the stress as a new variable to recast the Navier-Stokes equations into first order systems of equations. We numerically solve S-V-P, Stress-Velocity-Pressure, formulation of the incompressible Navier-Stokes equations based on the least-squares principles using different types of finite elements of low as well as higher order. For the discrete systems, we use a conjugate gradient (CG) solver accelerated with a geometric multigrid preconditioner. In addition, we employ a Krylov space smoother which allows a parameter-free smoothing. Combining this linear solver with the Newton linearization results in a robust and efficient solver. We analyze the application of this general approach, of using different types of finite elements, and the efficiency of the solver, geometric multigrid, throughout the solution of the prototypical benchmark configuration 'flow around cylinder'.

A. Ouazzi (✉) • M. Nickaeen • S. Turek • M. Waseem
Institut für Angewandte Mathematik, LSIII, TU Dortmund, Vogelpothsweg 87, D-44227 Dortmund, Germany
e-mail: Abderrahim.Ouazzi@math.tu-dortmund.de; masoud.nickaeen@rutgers.edu; ture@featflow.de; Muhammad.Waseem@math.tu-dortmund.de

A. Abdulle et al. (eds.), *Numerical Mathematics and Advanced Applications ENUMATH 2013*, Lecture Notes in Computational Science and Engineering 103,
DOI 10.1007/978-3-319-10705-9_64

1 Introduction

Least-Squares FEM (LSFEM) is generally motivated by the desire to recover the advantageous features of Rayleigh-Ritz methods, as for instance, the choice of the approximation spaces is free from discrete compatibility conditions and the corresponding discrete system is symmetric and positive definite [1].

In this paper, we solve the incompressible Navier-Stokes (NS) equations with LSFEM. Direct application of the LSFEM to the second-order NS equations requires the use of quite impractical C^1 finite elements [1]. Therefore, we introduce the stress as a new variable, on one hand to recast the Navier-Stokes equations to a first-order system of equations, and on the other hand to develop the basic solver for non-newtonian problems, i.e. the stress-velocity-pressure (S-V-P) formulation.

The resulting LSFEM system is symmetric and positive definite [1]. This permits the use of the conjugate gradient (CG) method and efficient multigrid solvers for the solution of the discrete system. In order to improve the efficiency of the solution method, the multigrid and the Krylov subspace method, here CG, can be combined with two different strategies. The first strategy is to use the multigrid as a preconditioner for the Krylov method [2]. The advantage of this scheme is that the Krylov method reduces the error in eigenmodes that are not being effectively reduced by multigrid. The second strategy is to employ Krylov preconditions methods as multigrid smoother. The Krylov methods appropriately determine the size of the solution updates at each smoothing step. This leads to smoothing sweeps which, in contrast to the standard SOR or Jacobi smoothing, are free from predefined damping parameters.

We develop a geometric multigrid solver as a preconditioner for the CG (MPCG) iterations to solve the S-V-P system with LSFEM. The MPCG solver has been first introduced and successfully used by the authors for the solution of the vorticity-based Navier-Stokes equations [4]. We use a CG pre/post-smoother to obtain efficient and parameter-free smoothing sweeps. We demonstrate a robust and grid independent behavior for the solution of different flow problems with both bilinear and biquadratic finite elements. Moreover, we show through the 'flow around cylinder' benchmark that accurate results can be obtained with LSFEM provided that higher order finite elements are used.

Therefore, the paper is organized as follows: in the next section we introduce the incompressible NS equations, the Newton linearization, the continuous and the discrete least-square principles with their properties and the designed LSFEM solver. In the third section, we present the general MPCG solver settings and the detailed results of the flow parameters in the 'flow around cylinder' problem. Finally, we give a conclusion and an outlook in the last section.

2 LSFEM for the Navier-Stokes Equations

The incompressible NS equations for a stationary flow are given by

$$\begin{cases} \boldsymbol{u}\cdot\nabla\boldsymbol{u}+\nabla p-\nu\Delta\boldsymbol{u}=f & \text{in } \Omega \\ \nabla\cdot\boldsymbol{u}=0 & \text{in } \Omega \\ \boldsymbol{u}=\boldsymbol{g}_D & \text{on } \Gamma_D \\ \boldsymbol{n}\cdot\boldsymbol{\sigma}=\boldsymbol{g}_N & \text{on } \Gamma_N \end{cases} \tag{1}$$

where $\Omega \subset \mathbb{R}^2$ is a bounded domain, p is the normalized pressure $p = P/\rho$, $\nu = \mu/\rho$ is the kinematic viscosity, f is the source term, $\boldsymbol{g}_D$ is the value of the Dirichlet boundary conditions on the Dirichlet boundary Γ_D, $\boldsymbol{g}_N$ is the prescribed traction on the Neumann boundary $\boldsymbol{\Gamma}_N$, $\boldsymbol{n}$ is the outward unit normal on the boundary, $\boldsymbol{\sigma}$ is the stress tensor and $\Gamma = \Gamma_D \cup \Gamma_N$ and $\Gamma_D \cap \Gamma_N = \emptyset$. The kinematic viscosity and the density of the fluid are assumed to be constant. The first equation in (1) is the momentum equation where velocities $\boldsymbol{u} = [u\,,\,v]^T$ and pressure p are the unknowns and the second equation represents the continuity equation.

2.1 *First-Order Stress-Velocity-Pressure System*

The straightforward application of the LSFEM to the second-order NS equations requires C^1 finite elements [1]. To avoid the practical difficulties in the implementation of such FEM, we first recast the second-order equation to a system of first-order equations. Another important reason for not using the straightforward LSFEM is that the resulting system matrix will be ill-conditioned.

To derive the S-V-P formulation, the Cauchy stress, $\boldsymbol{\sigma}$, is introduced as a new variable

$$\sigma = 2\nu\mathbf{D}(\boldsymbol{u}) - p\mathbf{I} \tag{2}$$

where $2\mathbf{D} := \nabla + \nabla^T$. Using the NS equations and the stress equation (2) we obtain the first-order Stress-Velocity-Pressure (S-V-P) system of equations

$$\begin{cases} \boldsymbol{u}\cdot\nabla\boldsymbol{u}-\nabla\cdot\boldsymbol{\sigma}=f & \text{in } \Omega \\ \nabla\cdot\boldsymbol{u}=0 & \text{in } \Omega \\ \sigma+p\mathbf{I}-2\nu\mathbf{D}(\boldsymbol{u})=0 & \text{in } \Omega \\ \boldsymbol{u}=\boldsymbol{g}_D & \text{on } \Gamma_D \\ \boldsymbol{n}\cdot\boldsymbol{\sigma}=\boldsymbol{g}_N & \text{on } \Gamma_N \end{cases} \tag{3}$$

2.2 Continuous Least-Squares Principle

We introduce the spaces of admissible functions based on the residuals of the first-order system (3)

$$V := \mathbf{H}(\mathrm{div}, \Omega) \cap \mathbf{H}^s(\Omega) \times \mathbf{H}^1_{g_D,D}(\Omega) \times L^2_0(\Omega) \tag{4}$$

and we define the S-V-P least-squares energy functional in the L^2-norm

$$\begin{aligned}\mathcal{J}(\sigma, \boldsymbol{u}, p; f) =& \frac{1}{2}\bigg(\int_\Omega |\sigma + p\mathbf{I} - 2\nu\mathbf{D}(\boldsymbol{u})|^2 \, d\Omega \\ &+ \int_\Omega |\nabla \cdot \boldsymbol{u}|^2 \, d\Omega + \int_\Omega |\boldsymbol{u} \cdot \nabla \boldsymbol{u} + \nabla \cdot \sigma - f|^2 \, d\Omega \\ &+ \int_{\Gamma_N} |\boldsymbol{n} \cdot \sigma - \boldsymbol{g}_N|^2 \, ds\bigg) \qquad \forall (\sigma, \boldsymbol{u}, p) \in V\end{aligned} \tag{5}$$

where we have assumed extra regularity for the stress to define the functional on the boundary Γ_N in $L^2(\Gamma_N)$. The minimization problem associated with the least-squares functional in (5) is to find $\tilde{\boldsymbol{u}} \in V, \tilde{\boldsymbol{u}} := (\sigma, \boldsymbol{u}, p)$, such that

$$\tilde{\boldsymbol{u}} = \operatorname*{argmin}_{\tilde{v} \in V} \mathcal{J}(\tilde{v}; f) \tag{6}$$

2.3 Newton Linearization

The S-V-P system (3) is nonlinear, due to the presence of the convective term, $\boldsymbol{u} \cdot \nabla \boldsymbol{u}$, in the momentum equation. Let $\mathcal{R}$ denote the residuals for the S-V-P system (3). We use the Newton method to approximate the nonlinear residuals. The nonlinear iteration is updated with the correction $\delta\tilde{\boldsymbol{u}}$, $\tilde{\boldsymbol{u}}^{k+1} = \tilde{\boldsymbol{u}}^k + \delta\tilde{\boldsymbol{u}}$. Then, the Newton linearization gives the following approximation for the residuals:

$$\begin{aligned}\mathcal{R}(\tilde{\boldsymbol{u}}^{k+1}) =& \mathcal{R}(\tilde{\boldsymbol{u}}^k + \delta\tilde{\boldsymbol{u}}) \\ \simeq& \mathcal{R}(\tilde{\boldsymbol{u}}^k) + \left[\frac{\partial \mathcal{R}(\tilde{\boldsymbol{u}}^k)}{\partial x}\right] \delta\tilde{\boldsymbol{u}}\end{aligned} \tag{7}$$

Using the least-squares principle, the resulting quadratic linearized functional, $\mathcal{L}$, is given in terms of L^2-norms as:

$$\mathcal{L}(\boldsymbol{u}^k; \delta\tilde{\boldsymbol{u}}) = \frac{1}{2}\int_\Omega \left|\mathcal{R}(\tilde{\boldsymbol{u}}^k) + \left[\frac{\partial \mathcal{R}(\tilde{\boldsymbol{u}}^k)}{\partial x}\right] \delta\tilde{\boldsymbol{u}}\right|^2 d\Omega \tag{8}$$

where we omitted the residual on the Neumann boundary for alluding briefly the main points. Minimizing the quadratic linearized functional (8) is equivalent to find $\delta\tilde{\boldsymbol{u}}$ such that:

$$\int_{\Omega} \left(\mathscr{R}(\tilde{\boldsymbol{u}}^k) + \left[\frac{\partial \mathscr{R}(\tilde{\boldsymbol{u}}^k)}{\partial x} \right] \delta\tilde{\boldsymbol{u}} \right) \cdot \left(\left[\frac{\partial \mathscr{R}(\tilde{\mathbf{u}^k})}{\partial x} \right] \tilde{\boldsymbol{v}} \right) d\Omega = 0 \quad \forall \tilde{\boldsymbol{v}} \tag{9}$$

In the operator form, let $\mathscr{A}$ and $\mathscr{F}$ defined as follows:

$$\mathscr{A}(\tilde{\boldsymbol{u}}^k) := \left[\frac{\partial \mathscr{R}(\tilde{\boldsymbol{u}}^k)}{\partial x} \right]^* \left[\frac{\partial \mathscr{R}(\tilde{\boldsymbol{u}}^k)}{\partial x} \right], \quad \mathscr{F}(\tilde{\boldsymbol{u}}^k) := - \left[\frac{\partial \mathscr{R}(\tilde{\boldsymbol{u}}^k)}{\partial x} \right]^* \mathscr{R}(\tilde{\boldsymbol{u}}^k). \tag{10}$$

Then, the linear system to solve at each nonlinear iteration is:

$$\mathscr{A}(\tilde{\boldsymbol{u}}^k)\delta\tilde{\boldsymbol{u}} = \mathscr{F}(\tilde{\boldsymbol{u}}^k) \tag{11}$$

The resulting Newton iteration for the least-squares formulation is given as follows:

$$\tilde{\boldsymbol{u}}^{k+1} = \tilde{\boldsymbol{u}}^k - \left(\left[\frac{\partial \mathscr{R}(\tilde{\boldsymbol{u}}^k)}{\partial x} \right]^* \left[\frac{\partial \mathscr{R}(\tilde{\boldsymbol{u}}^k)}{\partial x} \right] \right)^{-1} \left[\frac{\partial \mathscr{R}(\tilde{\boldsymbol{u}}^k)}{\partial x} \right]^* \mathscr{R}(\tilde{\boldsymbol{u}}^k) \tag{12}$$

2.4 Variational Formulation

The variational formulation problem based on the optimality condition of the minimization problem (6), considering the Newton Linearization in Sect. 2.3, reads

$$\begin{cases} \quad \textit{Find}\, (\boldsymbol{\sigma}, \boldsymbol{u}, p) \in V \quad \textit{s.t.} \\ \langle \mathscr{A}(\boldsymbol{\sigma}, \boldsymbol{u}, p), (\boldsymbol{\tau}, \boldsymbol{v}, q) \rangle = \mathscr{F}(\boldsymbol{\tau}, \boldsymbol{v}, q) \end{cases} \tag{13}$$

where $\mathscr{A}$ is the bilinear form defined on $V \times V \to \mathbb{R}$ as follows

$$\begin{aligned} \langle \mathscr{A}(\boldsymbol{\sigma}^k, \boldsymbol{u}^k, p^k)(\boldsymbol{\sigma}, \boldsymbol{u}, p), (\boldsymbol{\tau}, \boldsymbol{v}, q) \rangle &= \int_{\Omega} (\sigma + p\mathbf{I} - 2\nu \mathbf{D}(\boldsymbol{u})) : (\tau + q\mathbf{I} - 2\nu \mathbf{D}(\boldsymbol{v}))\, d\Omega \\ &+ \int_{\Gamma_N} (\boldsymbol{n} \cdot \boldsymbol{\sigma}) \cdot (\boldsymbol{n} \cdot \boldsymbol{\tau})\, ds + \int_{\Omega} (\nabla \cdot \boldsymbol{u})\, (\nabla \cdot \boldsymbol{v})\, d\Omega \\ + \int_{\Omega} (\boldsymbol{u} \cdot \nabla \boldsymbol{u}^k &+ \boldsymbol{u}^k \cdot \nabla \boldsymbol{u} + \nabla \cdot \boldsymbol{\sigma}) \cdot (\boldsymbol{v} \cdot \nabla \boldsymbol{u}^k + \boldsymbol{u}^k \cdot \nabla \boldsymbol{v} + \nabla \cdot \boldsymbol{\tau})\, d\Omega \end{aligned} \tag{14}$$

and the bilinear form $\mathscr{F}$ is defined on $V \to \mathbb{R}$ as follows:

$$\mathscr{F}(\boldsymbol{\sigma}^k, \boldsymbol{u}^k, p^k)(\boldsymbol{\tau}, \boldsymbol{v}, q) = \int_\Omega \left(\frac{\partial \mathscr{R}(\boldsymbol{\sigma}^k, \boldsymbol{u}^k, p^k)}{\partial x}(\boldsymbol{\tau}, \boldsymbol{v}, q) \right) \cdot \left(\mathscr{R}(\boldsymbol{\sigma}^k, \boldsymbol{u}^k, p^k) \right) d\Omega \tag{15}$$

2.5 Operator Form of the Problem

To analyze the properties of the least-squares problem, let us write the bilinear form (14) as in (10). Then, the S-V-P operator reads:

$$\begin{aligned}\mathscr{A}(\boldsymbol{\sigma}, \boldsymbol{u}, p) &= \left[\frac{\partial \mathscr{R}(\boldsymbol{\sigma}, \boldsymbol{u}, p)}{\partial x} \right]^* \left[\frac{\partial \mathscr{R}(\boldsymbol{\sigma}, \boldsymbol{u}, p)}{\partial x} \right] \\ &= \begin{pmatrix} \mathbf{I} - \nabla\nabla \cdot + \boldsymbol{n}_{\Gamma_N} \boldsymbol{n}_{\Gamma_N} \cdot & -2\nu \mathbf{D} - \nabla C(\boldsymbol{u}) & I \\ 2\nu \nabla \cdot + C^*(\boldsymbol{u}) \nabla \cdot & -4\nu^2 \nabla \cdot \mathbf{D} - \nabla\nabla \cdot + C^*(\boldsymbol{u}) C(\boldsymbol{u}) & 2\nu \nabla \cdot \\ \mathbf{I} & -2\nu \mathbf{D} & I \end{pmatrix}\end{aligned} \tag{16}$$

Here, the term $C(\boldsymbol{u})$ is defined as follows:

$$C(\boldsymbol{u})\boldsymbol{v} = \boldsymbol{v} \cdot \nabla \boldsymbol{u} + \boldsymbol{u} \cdot \nabla \boldsymbol{v} \tag{17}$$

The resulting matrix, from Eqs. (16), is symmetric and positive definite. So, after discretization, we are able to use the CG method to efficiently solve the system of equations. Our aim is to design an efficient solver which exploits the properties of the least-squares system with respect to both the CG and the multigrid methods. Therefore, we use CG as the main solver and accelerate it with the multigrid preconditioning.

2.6 Discrete Least-Squares Principle

Let the bounded domain $\Omega \subset \mathbb{R}^d$ be partitioned by a *grid* $\mathscr{T}_h$ consisting of elements $K \in \mathscr{T}_h$ which are assumed to be open quadrilaterals or hexahedrons such that $\Omega = \mathrm{int}\big(\bigcup_{K \in \mathscr{T}_h} \overline{K} \big)$. Furthermore, let $H^{1,h}(\Omega)$, $H^{s,h}(\Omega)$, and $H^{\mathrm{div},h}(\Omega)$ denote the spaces of elementwise H^1, H^s, and $H(\mathrm{div})$ functions with respect to $\mathscr{T}_h$ [3].

Now, we turn to the approximation of the problem (13) with the finite element method. So, we introduce the approximation spaces V^h such that

$$V^h \subset \mathbf{H}^{\mathrm{div},h}(\Omega) \cap \mathbf{H}^{s,h}(\Omega) \times \mathbf{H}^{1,h}_{g_D,D}(\Omega) \times L^2_0(\Omega) \tag{18}$$

and we consider the approximated problems

$$\begin{cases} Find\,(\boldsymbol{\sigma}_h, \boldsymbol{u}_h, p_h) \in V^h \quad s.t. \\ \langle \mathscr{A}^h(\boldsymbol{\sigma}, \boldsymbol{u}, p), (\boldsymbol{\tau}_h, \boldsymbol{v}_h, q_h) \rangle = \mathscr{F}^h(\boldsymbol{\tau}_h, \boldsymbol{v}_h, q_h) \end{cases} \tag{19}$$

where $\mathscr{A}^h$ is an approximate bilinear form of (14) defined on $V^h \times V^h \to \mathbb{R}$.

The least-squares formulation allows a free choice of FE spaces [1]. So, we are able to use different combinations of FE approximations, as for instance, discontinuous P_0^{dc}, P_1^{dc}, H^1-nonconforming $\tilde{Q}_1$ and $\tilde{Q}_2$, H^1-conforming Q_1 and Q_2, or from $H(\mathrm{div})$. Here we use different combinations of finite element spaces allowing better comparison with the standard mixed finite element for velocity and pressure. Therefore, we set $V^h \subset V$, and $\mathscr{A}^h = \mathscr{A}$.

2.7 MPCG Solver

The discrete linear system of equations resulting from the least-squares finite element method (16) has a symmetric and positive definite (SPD) coefficient matrix i.e.

$$\mathscr{A} = \begin{pmatrix} A_{\sigma\sigma} & A_{\sigma u} & A_{\sigma p} \\ A_{\sigma u} & A_{uu} & A_{up} \\ A_{\sigma p} & A_{up} & A_{pp} \end{pmatrix} \tag{20}$$

Therefore, it is appropriate to take full advantage of the symmetric positive definiteness by using solvers specially designed for such systems. In addition, the resulting system matrix is sparse due to the properties of the interpolation functions used in the finite element discretization. Our main focus is on the iterative solvers. We specifically employ the conjugate gradient method as a Krylov subspace solver suitable for the SPD systems. In addition, we use multigrid method as a highly efficient defect correction scheme for sparse linear systems arising in the discretization of (elliptic) partial differential equations [4].

3 Numerical Results and Discussions

We investigate the performance of the MPCG solver for the system (20) for a wide range of parameters, using the benchmarks quantities drag, lift, pressure drop, and the Global Mass Conservation (GMC) (see [4]). Moreover, we analyze the performance of the MPCG solver for the solution of the S-V-P. Figure 1 shows the computational mesh of the coarsest level.

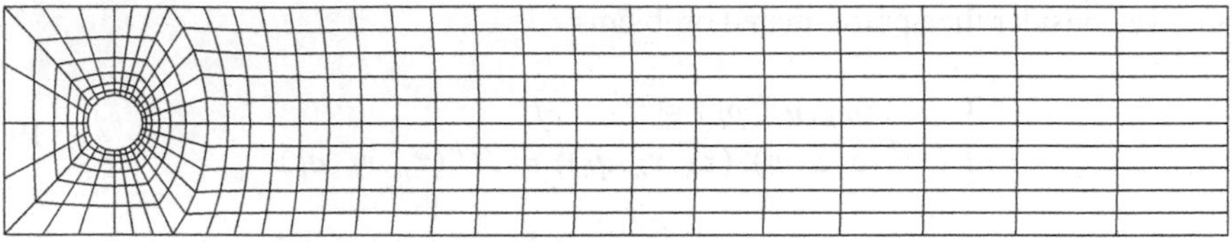

Fig. 1 Flow around cylinder: computational grid on level 1

Table 1 S-V-P Formulation: Benchmark quantities for flow around cylinder at Re = 20

	Level	d.o.f.	C_D	C_L	Δp	$GMC_{\vert x=2.2}$	NL/MG
$Q_1/Q_1/Q_1$	4	135,024	5.1716353	0.0210522	0.0103135	1.114501	7/19
	5	535,776	5.4440131	0.0142939	0.1117922	0.299773	7/17
	6	2,134,464	5.5415463	0.0117584	0.1152451	0.077866	7/17
$Q_2/Q_2/Q_2$	3	135,024	5.5588883	0.0101360	0.1165546	0.022791	6/12
	4	535,776	5.5769755	0.0105355	0.1173265	0.003022	6/12
	5	2,134,464	5.5792424	0.0106064	0.1174766	0.000556	6/12
$Q_2/Q_2/P_1^{dc}$	3	129,128	5.5586141	0.0101405	0.1168068	0.0320698	6/13
	4	512,912	5.5769573	0.0105351	0.1173867	0.0039341	6/13
	5	2,044,448	5.5792414	0.1060618	0.1174911	0.0004692	6/13
Ref.:		$C_D = 5.57953523384, C_L = 0.010618948146, \Delta p = 0.11752016697$					

We present the drag and the lift coefficients, the pressure drop across the cylinder, and the $GMC_{|x=2.2}$ values at the outflow ($x = 2.2$) at Reynolds number Re = 20 for the S-V-P formulation in Table 1 which also shows the number of nonlinear iterations and the corresponding averaged linear solver (MPCG solver) iterations for different levels.

Using higher order finite elements, the method shows excellent convergence towards the reference solution. We observe a grid-independent convergence behavior.

4 Summary

We presented a numerical study regarding the accuracy and the efficiency of least-squares finite element formulation of the incompressible Navier-Stokes equations. The first-order system is introduced using the stress, velocity, and pressure, known as the S-V-P formulation. We investigated different finite element spaces of higher and low order. Using the Newton scheme, the linearization is performed on the continuous operators. Then, the least-squares minimization is applied. The resulting linear system is solved using an extended multigrid-preconditioned conjugate gradient solver. The flow accuracy and the mass conservation of the LSFEM formulations are investigated using the incompressible steady-state laminar 'flow around cylinder' problem.

On the accuracy aspect, we have shown that highly accurate results can be obtained with higher order finite elements. More importantly, we have obtained more accurate results with the higher-order finite elements with less number of degrees of freedom as compared to the lower-order elements. This obviously amounts to less computational costs. On the efficiency aspect, we have shown that the MPCG solver performs efficiently for LSFEM formulation.

Having the basic S-V-P LSFEM solver, our main objective is the investigation of generalized Newtonian fluids with the nonlinearity due to the stress $\boldsymbol{\sigma} = 2\nu(\dot{\gamma})\mathbf{D}(\boldsymbol{u}) - p\mathbf{I}$, and multiphase flow problems with the jump in the stress and discontinuous pressure.

Acknowledegement This work was supported by the Mercator Research Center Ruhr (MERCUR) grant Pr-2011-0017, the DFG and STW through the grants DFG/STW: TU 102/44-1, and the HEC-DAAD program.

References

1. P. Bochev, M. Gunzburger, Least-squares finite element methods (Springer, New York, 2009)
2. J.J. Heys, T.A. Manteuffel, S.F. McCormick, L.N. Olson, Algebraic multigrid for higher-order finite elements. J. Comput. Phys. **204**(2), 520–532 (2005)
3. M. Köster, A. Ouazzi, F. Schieweck, S. Turek, P. Zajac, New robust nonconforming finite elements of higher order. Appl. Numer. Math. **62**, 166–184 (2012)
4. M. Nickaeen, A. Ouazzi, S. Turek, Newton multigrid least-squares FEM for the V-V-P formulation of the Navier-Stokes equations. J. Comput. Phys. **256**, 416–427 (2014)

Compressible Flows of Viscous Fluid in 3D Channel

Petra Pořízková, Karel Kozel, and Jaromír Horáček

Abstract This study deals with the numerical solution of a 3D compressible flow of a viscous fluid in a channel for low inlet airflow velocity. The channel is a simplified model of the glottal space in the human vocal tract. The system of Navier-Stokes equations has been used as mathematical model of laminar flow of the compressible viscous fluid in a domain. The numerical solution is implemented using the finite volume method (FVM) and the predictor-corrector MacCormack scheme with artificial viscosity using a grid of hexahedral cells. The numerical simulations of flow fields in the channel, acquired from a developed program, are presented for inlet velocity $\hat{u}_\infty = 4.12\,\mathrm{ms}^{-1}$ and Reynolds number $\mathrm{Re}_\infty = 4{,}481$.

1 Introduction

A current challenging question is a mathematical and physical description of the mechanism for transforming the airflow energy in the glottis into the acoustic energy representing the voice source in humans. The voice source signal travels from the glottis to the mouth, exciting the acoustic supraglottal spaces, and becomes modified by acoustic resonance properties of the vocal tract [1].

Acoustic wave propagation in the vocal tract is usually modeled from incompressible flow models separately using linear acoustic perturbation theory, the wave equation for the potential flow [2] or the Light-hill approach on sound generated aerodynamically [3]. In reality, the airflow coming from the lungs causes self-oscillations of the vocal folds, and the glottis completely closes in normal phonation regimes, generating acoustic pressure fluctuations. The goal is numerical simulation of flow in the channel which involves attributes of real flow causing acoustic perturbations.

P. Pořízková (✉)
Czech Technical University in Prague, Karlovo náměstí 13, 121 35, Prague 2, Czech Republic
e-mail: puncocha@marian.fsik.cvut.cz

K. Kozel • J. Horáček
Institute of Thermomechanics Academy of Sciences, Dolejškova 5, Prague 8, Czech Republic
e-mail: Karel.Kozel@fs.cvut.cz; jaromirh@it.cas.cz

A. Abdulle et al. (eds.), *Numerical Mathematics and Advanced Applications ENUMATH 2013*, Lecture Notes in Computational Science and Engineering 103,
DOI 10.1007/978-3-319-10705-9_65

2 Mathematical Model

The system of Navier-Stokes equations has been used as mathematical model to describe the unsteady laminar flow of the compressible viscous fluid in a domain. The system is expressed in non-dimensional conservative form [4]:

$$\frac{\partial \mathbf{W}}{\partial t}+\frac{\partial \mathbf{F}}{\partial x}+\frac{\partial \mathbf{G}}{\partial y}+\frac{\partial \mathbf{H}}{\partial z}=\frac{1}{\mathrm{Re}}\left(\frac{\partial \mathbf{R}}{\partial x}+\frac{\partial \mathbf{S}}{\partial y}+\frac{\partial \mathbf{T}}{\partial z}\right). \tag{1}$$

$\mathbf{W} = [\rho, \rho u, \rho v, \rho w, e]^T$ is the vector of conservative variables where ρ denotes density, (u, v, w) is velocity vector and e is the total energy per unit volume. **F**, **G**, **H** are the vectors of inviscid fluxes and **R**, **S**, **T** are the vectors of viscous fluxes. The static pressure p in inviscid fluxes is expressed by the state equation in the form

$$p=(\kappa-1)\left[e-\frac{1}{2}\rho\left(u^2+v^2+w^2\right)\right], \tag{2}$$

where $\kappa = 1.4$ is the ratio of specific heats.

The reference variables for transformation are inflow variables (marked with the infinity subscript): the speed of sound $\hat{c}_\infty$=343 ms^{-1}, density $\hat{\rho}_\infty$=1.225 kg m^{-3}, dynamic viscosity $\hat{\eta}_\infty=18 \cdot 10^{-6}$ Pa · s (for temperature $\hat{T}_\infty$=293.15 K) and a reference length $\hat{L}_r$=0.02 m. General Reynolds number in (1) is computed from reference variables $\mathrm{Re}=\hat{\rho}_\infty\hat{c}_\infty\hat{L}_r/\hat{\eta}_\infty$. The non-dimensional dynamic viscosity in the dissipative terms is a function of temperature in the form $\eta = (T/T_\infty)^{3/4}$.

3 Computational Domain and Boundary Conditions

The bounded computational domain D used for the numerical solution of flow field in the channel is shown in Fig. 1. The domain is symmetric channel in y and z directions, the shape of which is inspired by the shape of the trachea (inlet part), vocal folds, false vocal folds and supraglottal spaces (outlet part) in human vocal tract. The gap width is the narrowest part of the channel in y direction and is set in middle position $\hat{g}_{mid} = 1.6$ mm.

The boundary conditions are considered in the following formulation:

1. Upstream conditions: $u_\infty = \frac{\hat{u}_\infty}{\hat{c}_\infty}$; flow rate at the inlet is constant $H^2 \cdot u_\infty$; $\rho_\infty = 1$; p_∞ is extrapolated from domain.
2. Downstream conditions: $p_2 = 1/\kappa$; $(\rho, \rho u, \rho v, \rho w)$ are extrapolated from domain.
3. Flow on the wall: $(u, v, w) = (u_{wall}, v_{wall}, w_{wall})$ – velocity of the channel walls and for temperature $T = \kappa p/\rho$ is $\frac{\partial T}{\partial \mathbf{n}} = 0$.

The general Reynolds number in (1) is multiply with non-dimensional value $u_\infty H$ which represents kinematic viscosity scale at inlet. For computation of real problem inlet Reynolds number $\mathrm{Re}_\infty = \mathrm{Re} \cdot \frac{\hat{u}_\infty}{\hat{c}_\infty}\frac{\hat{H}}{\hat{L}_r} = \mathrm{Re} \cdot u_\infty H$ is used in (3).

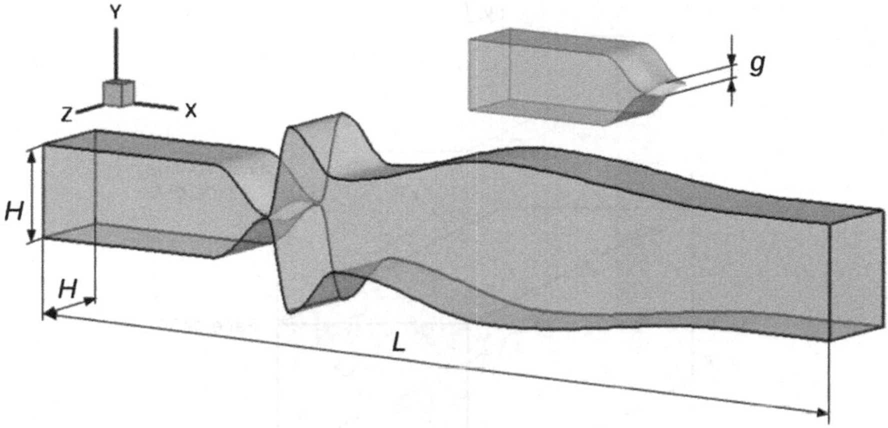

Fig. 1 The computational domain D. $L = 8\,(160\,\text{mm})$, $H = 0.8\,(16\,\text{mm})$, $g_{mid} = 0.08\,(1.6\,\text{mm})$ – middle position

4 Numerical Solution

The numerical solution uses finite volume method (FVM) in cell centered form on the grid of hexahedral cells. The bounded domain is divided into mutually disjoint sub-domains $D_{i,j,k}$ (i.e. hexahedral cells). The system of Eqs. (1) is integrated over the sub-domains $D_{i,j,k}$ using the Green formula and the Mean value theorem. The explicit predictor-corrector MacCormack (MC) scheme in the domain is used. The scheme is second order accurate in time and space (on orthogonal grid):

$$
\begin{aligned}
\mathbf{W}^{n+1/2}_{i,j,k} &= \frac{\mu^n_{i,j,k}}{\mu^{n+1}_{i,j,k}} \mathbf{W}^n_{i,j,k} - \frac{\Delta t}{\mu^{n+1}_{i,j,k}} \sum_{q=1}^{6} A_q \Bigg[\left(\tilde{\mathbf{F}}^n_q - s_{1_q} \mathbf{W}^n_q - \frac{1}{\mathrm{Re}_\infty} \tilde{\mathbf{R}}^n_q \right) n_{1_q} \\
&+ \left(\tilde{\mathbf{G}}^n_q - s_{2_q} \mathbf{W}^n_q - \frac{1}{\mathrm{Re}_\infty} \tilde{\mathbf{S}}^n_q \right) n_{2_q} + \left(\tilde{\mathbf{H}}^n_q - s_{3_q} \mathbf{W}^n_q - \frac{1}{\mathrm{Re}_\infty} \tilde{\mathbf{T}}^n_q \right) n_{3_q} \Bigg], \\
\overline{\mathbf{W}}^{n+1}_{i,j,k} &= \frac{\mu^n_{i,j,k}}{\mu^{n+1}_{i,j,k}} \frac{1}{2} \left(\mathbf{W}^n_{i,j,k} + \mathbf{W}^{n+1/2}_{i,j,k} \right) - \frac{\Delta t}{2\mu^{n+1}_{i,j,k}} \sum_{q=1}^{6} A_q \Bigg[\Bigg(\tilde{\mathbf{F}}^{n+1/2}_q - s_{1_q} \mathbf{W}^{n+1/2}_q \\
&- \frac{1}{\mathrm{Re}_\infty} \tilde{\mathbf{R}}^{n+1/2}_q \Bigg) n_{1_q} + \left(\tilde{\mathbf{G}}^{n+1/2}_q - s_{2_q} \mathbf{W}^{n+1/2}_q - \frac{1}{\mathrm{Re}_\infty} \tilde{\mathbf{S}}^{n+1/2}_q \right) n_{2_q} \\
&+ \left(\tilde{\mathbf{H}}^{n+1/2}_q - s_{3_q} \mathbf{W}^{n+1/2}_q - \frac{1}{\mathrm{Re}_\infty} \tilde{\mathbf{T}}^{n+1/2}_q \right) n_{3_q} \Bigg],
\end{aligned}
\tag{3}
$$

where $\Delta t = t^{n+1} - t^n$ is the time step, $\mu_{i,j,k} = \int\int_{D_{i,j,k}} dx\,dy\,dz$ is the volume of cell $D_{i,j,k}$, $\mathbf{n}_q = (n_1, n_2, n_3)_q$ is outlet normal vector on face q (see Fig. 2), A_q is area of the face and vector $\mathbf{s}_q = (s_1, s_2, s_3)_q$ represents the speed of the face. The physical

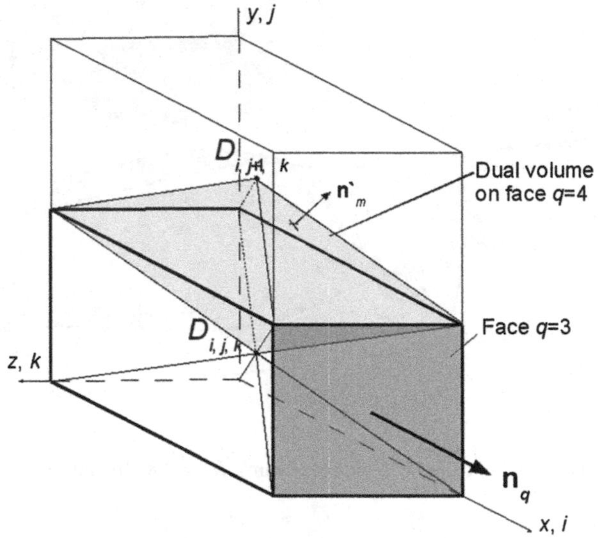

Fig. 2 Finite volume of cell $D_{i,j,k}$ and dual volume V_q' on face q

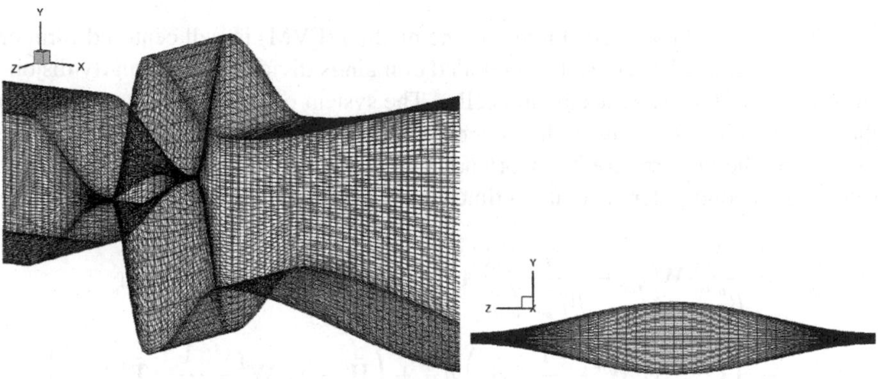

Fig. 3 Mesh of the hexahedral cells $D_{i,j,k}$ in part of the domain D and detail of mesh in the gap

fluxes $\mathbf{F}$, $\mathbf{G}$, $\mathbf{H}$, $\mathbf{R}$, $\mathbf{S}$, $\mathbf{T}$ on the face q of the cell $D_{i,j,k}$ are replaced by numerical fluxes (marked with tilde) $\tilde{\mathbf{F}}$, $\tilde{\mathbf{G}}$, $\tilde{\mathbf{H}}$, $\tilde{\mathbf{R}}$, $\tilde{\mathbf{S}}$, $\tilde{\mathbf{T}}$ as approximations of the physical fluxes. The higher partial derivatives of velocity and temperature in $\tilde{\mathbf{R}}_q$, $\tilde{\mathbf{S}}_q$, $\tilde{\mathbf{T}}_q$ are approximated using dual volumes V_q' as shown in Fig. 2.

The last term used in the MC scheme is the Jameson artificial dissipation $AD(W_{i,j,k})^n$ [5], then the vector of conservative variables $\mathbf{W}$ can be computed at a new time level $\mathbf{W}_{i,j,k}^{n+1} = \overline{\mathbf{W}}_{i,j,k}^{n+1} + AD(W_{i,j,k})^n$.

The grid of the channel have successive refinement cells near the wall, the minimum cell size in y and z directions is $\Delta y_{min}, \Delta z_{min} \approx 1/\sqrt{\mathrm{Re}_\infty}$ to resolve capture boundary layer effects (see Fig. 3).

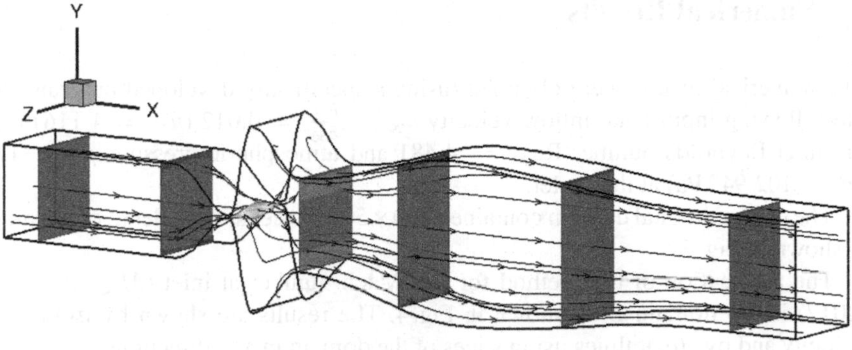

Slices in x direction

Detail of flow in the gap

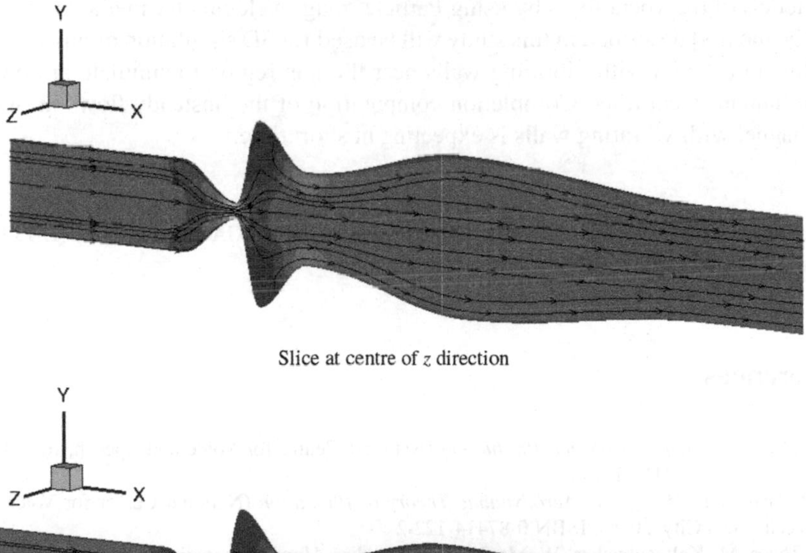

Slice at centre of z direction

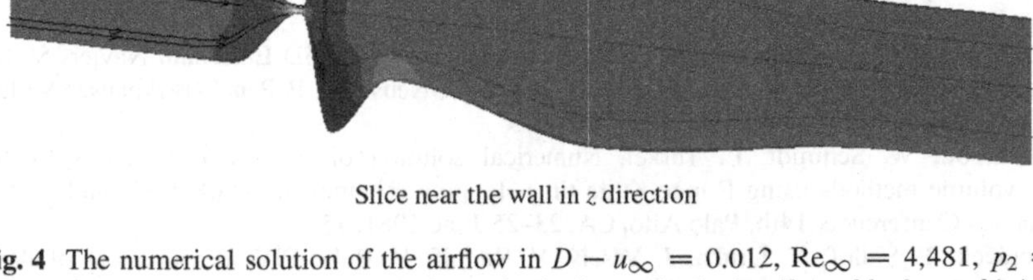

Slice near the wall in z direction

Fig. 4 The numerical solution of the airflow in $D - u_\infty = 0.012$, $\mathrm{Re}_\infty = 4{,}481$, $p_2 = 1/\kappa$, $200 \times 50 \times 50$ cells. The results are shown by isolines of velocity (from *black* to *white*) and by streamlines

5 Numerical Results

The numerical results were obtained (using a specifically developed program) for the following input data: inflow velocity $u_\infty = \frac{\hat{u}_\infty}{\hat{c}_\infty} = 0.012$ ($\hat{u}_\infty = 4.116\,\mathrm{m/s}$), the inlet Reynolds number $\mathrm{Re}_\infty = 4{,}481$ and atmospheric pressure $p_2 = 1/\kappa$ ($\hat{p}_2 = 102{,}942\,\mathrm{Pa}$) at the outlet.

The computational domain contained $200 \times 50 \times 50$ cells in D, detail of the mesh is shown in Fig. 3.

The application of the method for low Mach number at inlet ($M_\infty = u_\infty = 0.012$) in 3D domain D is shown in Fig. 4. The results are shown by isolines of velocity and by streamlines using slices of the domain in x, z directions.

Discussion and Conclusions

The governing system (1) for flow of viscous compressible fluid based on Navier-Stokes equations for laminar flow is tested in 3D domain. A similar generation of large-scale vortices, vortex convection and diffusion, jet flapping, and general flow patterns were experimentally obtained in physical models of the vocal folds by using Particle Image Velocimetry method in [6]. The method described in this study will be used for 3D simulation of unsteady flow in domain with vibrating walls near the gap region to simulate airflow in human vocal tract. Completion computation of the unsteady flows in 3D channel with vibrating walls is expecting in short time.

Acknowledgement This contribution was partially supported by GAČR P101/11/0207, 13-005-22S and P101/10/1329.

References

1. I.R. Titze, *Principles of Voice Production* (National Centre for Voice and Speech, Iowa City, 2000). ISBN 0-87414-122-2
2. I.R. Titze, *The Myoelastic Aerodynamic Theory of Phonation* (National Center for Voice and Speech, Iowa City, 2006). ISBN 0-87414-122-2
3. S. Zöner, M. Kalteenbacher, W. Mattheus, C. Brücker, Human phonation analysis by 3d aero-acoustic computation, in *Proceedings of the International Conference on Acoustic NAG/DAGA*, Rotterdam (2009), pp. 1730–1732
4. J. Fürst, M. Janda, K. Kozel, Finite volume solution of 2D and 3D Euler and Navier- Stokes equations, in *Mathematical Fluid Mechanics*, ed. by J. Neustupa, P. Penel (Birkhäuser Verlag, Berlin, 2001), pp. 173–194. ISBN 3-7643-6593-5
5. A. Jameson, W. Schmidt, E. Turkel, Numerical solution of the Euler equations by the finite volume methods using Runge-Kutta time-stepping schemes, in *AIAA*. Fluid and Plasma Dynamics Conference, 14th, Palo Alto, CA, 23–25 June 1981, 15 p.
6. J. Horáček, P. Šidlof, V. Uruba, J. Veselý, V. Radolf, V. Bula, PIV measurement of flow-patterns in human vocal tract model, in *Proceedings of the International Conference on Acoustic NAG/DAGA 2009*, Rotterdam (2009), pp. 1737–1740

Numerical Investigation of Network Models for Isothermal Junction Flow

Gunhild Allard Reigstad and Tore Flåtten

Abstract This paper deals with the issue of how to properly model fluid flow in pipe junctions. In particular we investigate the numerical results from three alternative network models, all three based on the isothermal Euler equations. Using two different test cases, we focus on the physical validity of simulation results from each of the models. Unphysical solutions are characterised by the presence of energy production in junctions. Our results are in accordance with previous conclusions; that only one of the network models yields physical solutions for all subsonic initial conditions. The last test case shows in addition how the three models may predict fundamentally different waves for a given set of initial data.

1 Introduction

A network model describes the global weak solution of hyperbolic conservation laws defined on N segments of the real line that are connected at a common point. In addition to fluid flow in pipeline junctions, such models are used to describe for example traffic flow, data networks, and supply chains [4].

For fluid flow, the model describes a junction that connects N pipe sections. Each section is modelled along a local axis ($x \in \mathbb{R}^+$) and $x = 0$ at the pipe-junction interface. The problem is investigated by defining a generalized Riemann problem at the junction, and thus the condition of constant initial conditions in each pipe section is presupposed.

G.A. Reigstad (✉)
Department of Energy and Process Engineering, Norwegian University of Science and Technology (NTNU), NO-7491 Trondheim, Norway
e-mail: Gunhild.Reigstad@ntnu.no

T. Flåtten
SINTEF Materials and Chemistry, P.O. Box 4760 Sluppen, NO-7465 Trondheim, Norway
e-mail: Tore.Flatten@sintef.no

A. Abdulle et al. (eds.), *Numerical Mathematics and Advanced Applications ENUMATH 2013*, Lecture Notes in Computational Science and Engineering 103,
DOI 10.1007/978-3-319-10705-9_66

The flow condition in each pipe section is found as the solution to the half-Riemann problem

$$\frac{\partial \boldsymbol{U}_k}{\partial t} + \frac{\partial}{\partial x}\boldsymbol{F}(\boldsymbol{U}_k) = 0,$$
$$\boldsymbol{U}_k(x,0) = \begin{cases} \bar{\boldsymbol{U}}_k & \text{if } x > 0 \\ \boldsymbol{U}_k^* & \text{if } x < 0, \end{cases} \tag{1}$$

restricted to $x \in \mathbb{R}^+$. $\boldsymbol{U}_k^*$ is a constructed state, defined as

$$\boldsymbol{U}_k^*\left(\bar{\boldsymbol{U}}_1,\ldots,\bar{\boldsymbol{U}}_N\right) = \lim_{x\to 0+} \boldsymbol{U}_k(x,t). \tag{2}$$

$\boldsymbol{U}_k^*$ is per definition connected to the initial condition, $\bar{\boldsymbol{U}}_k$, by waves of non-negative speed only. This ensures that the constructed state propagates into the pipe section.

In the present paper we consider the isothermal Euler equations, which are described by the isentropic conservation law

$$\frac{\partial}{\partial t}\begin{bmatrix}\rho \\ \rho v\end{bmatrix} + \frac{\partial}{\partial x}\begin{bmatrix}\rho v \\ \rho v^2 + p(\rho)\end{bmatrix} = \begin{bmatrix}0 \\ 0\end{bmatrix}, \tag{3}$$

together with the pressure law

$$p(\rho) = a^2\rho. \tag{4}$$

Here ρ and v are the fluid density and velocity, respectively, $p(\rho)$ is the pressure and a is the constant fluid speed of sound. Initial conditions of standard Riemann problems are, for this set of equations, connected by two waves. Only waves of the second family have non-negative speed at subsonic conditions. Therefore $\boldsymbol{U}_k^*$ and $\bar{\boldsymbol{U}}_k$ are connected by either a rarefaction or a shock wave of this family [7].

In addition to the wave-equation describing the relation between $\boldsymbol{U}_k^*$ and $\bar{\boldsymbol{U}}_k$, a set of equations is needed for $\boldsymbol{U}_k^*$ to be uniquely defined. The equations are denoted *coupling conditions*, and for the isothermal Euler equations, they are related to mass and momentum:

CC1: Mass is conserved at the junction

$$\sum_{k=1}^{N} \rho_k^* v_k^* = 0. \tag{5}$$

CC2: There is a unique, scalar momentum related coupling constant at the junction

$$\mathcal{H}_k^*(\rho_k^*, v_k^*) = \tilde{\mathcal{H}} \qquad \forall k \in \{1,\ldots,N\}. \tag{6}$$

Three different expressions for the momentum related coupling constant are considered in this paper. Pressure

$$\mathscr{H}_k^*(\rho_k^*, v_k^*) = \rho_k^*, \tag{7}$$

and momentum flux

$$\mathscr{H}_k^*(\rho_k^*, v_k^*) = \rho_k^* \left(1 + \left(\frac{v_k^*}{a}\right)^2\right), \tag{8}$$

have been frequently used in the literature [1–3, 5]. The Bernoulli invariant

$$\mathscr{H}_k^*(\rho_k^*, v_k^*) = \ln\left(\rho_k^*\right) + \frac{1}{2}\left(\frac{v_k^*}{a}\right)^2 \tag{9}$$

was recently proposed [7]. The constant may also be stated as $\mathscr{H}(\rho_k^*, M_k^*)$ where M_k^* is the Mach number, $M_k^* = v_k^*/a$.

The suitability of a suggested momentum related coupling constant is evaluated according to two criteria. First, a standard Riemann problem in a pipe section of uniform cross sectional area may be modelled as two pipe sections connected at a junction. The resulting network model must then have a solution equal to the solution of the standard Riemann problem. This imposes a symmetry- and a monotonicity constraint on the momentum related coupling constant [6]. Second, the solutions of the network model must be physically reasonable. This is determined by the entropy condition (10), which states that energy production does not occur in a junction if the solution is physical.

$$E_{\text{crit}} = \sum_{k=1}^{N} \rho_k^* v_k^* \left(\frac{1}{2}(v_k^*)^2 + a^2 \ln \frac{\rho_k^*}{\rho_0}\right) \leq 0, \tag{10}$$

where ρ_0 is some reference density.

The entropy condition was first used by Colombo and Garavello [3] and is based on the mechanical energy flux function. The presented condition (10) is derived for the isothermal Euler equations.

An analytical investigation on the relation between the entropy condition and the momentum related coupling constant was previously performed for the special case of three pipe sections connected at a junction [7]. The analysis showed that for certain flow rates within the subsonic domain, both pressure (7) and momentum flux (8) as coupling constant yield unphysical solutions. Physical solutions for all subsonic flow rates were only guaranteed when the Bernoulli invariant (9) was used as coupling constant. In the present paper, two numerical test cases will be used to verify this analysis and to explore the behaviour of the different models.

The first test case consists of five pipe sections connected at a junction. The case illustrates how the network model easily may be applied to a junction connecting a large number of pipe sections. We will as well evaluate the results in terms of physical soundness using the entropy condition (10).

The second case consists of three pipe sections connected by two junctions such that a closed system is constructed. We will show how the different models produce fundamentally different results in terms of rarefaction and shock waves. The total energy of the system as a function of time will as well be presented in order to display the effect of having unphysical solutions.

2 Numerical Results

The fluid flow in each pipe section is solved by a classical approximate Riemann solver of Roe as described by Reigstad et al. [6]. In the two cases, the fluid speed of sound is set to $a = 300$ m/s and the Courant-Friedrichs-Lewy condition is set to $C = 0.5$.

2.1 *Case 1: Five Pipe Sections Connected at a Junction*

Five pipe sections, each of length $L = 50$ m are connected at a single junction. The initial conditions of each pipe section are given in Table 1. Interaction between the fluids in the pipe sections first occur at $T = 0.0$ s and immediately afterwards one wave enters each section.

For a given set of initial conditions we may calculate the constructed states, U_k^*, and the analytical velocity and pressure profiles at a given time as function of distance through the pipe section. The analytical profiles are compared to numerical results derived at different grid resolutions in order to identify the suitable grid cell size [6]. For the present case, a grid resolution of $\Delta x = 5.0 \times 10^{-2}$ m was chosen.

Table 1 presents the constructed state densities and velocities for the three different models. As seen, the values are different for the three models. However,

Table 1 Initial conditions and constructed states of the network models

	$T = 0.0$ s		$\mathcal{H} = \rho$		$\mathcal{H} = \rho\left(1 + M^2\right)$		$\mathcal{H} = \ln(\rho) + \frac{1}{2}M^2$	
Pipe section	$\bar{p}$ (bar)	$\bar{v}$ (m/s)	p^* (bar)	v^* (m/s)	p^* (bar)	v^* (m/s)	p^* (bar)	v^* (m/s)
1	1.00	0.00	1.30	79.4	1.27	71.0	1.28	74.3
2	1.20	0.00	1.30	24.5	1.32	29.6	1.31	27.4
3	1.30	0.00	1.30	0.453	1.34	8.28	1.32	4.58
4	1.50	0.00	1.30	−42.5	1.31	−39.6	1.31	−41.1
5	1.60	0.00	1.30	−61.8	1.27	−69.5	1.29	−64.6

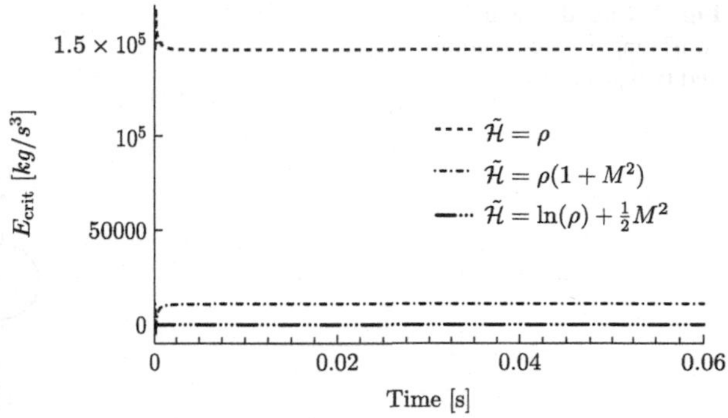

Fig. 1 Entropy function values for the three different network models

the predictions of rarefaction- and shock waves are consistent, with shock waves in pipe section 1–3 and rarefaction waves in pipe section 4 and 5.[1]

Our main focus is to evaluate the simulation results with the aid of the entropy condition (10). The results are shown in Fig. 1. As expected, Bernoulli invariant as coupling constant yields energy conservation at the junction. The two other options lead to energy production at the junction for the given set of initial data. That is, the solutions are unphysical.

Analytically, the E_{crit} profile for a given set of initial conditions is a constant value. The deviation seen in Fig. 1 is due to the numerical implementation, where the constructed state, $\boldsymbol{U}_k^*$, at a new time-step is calculated based on the calculated conditions in the inner grid cell closest to the interface, at the previous time-step. As the waves propagate into the pipe sections, the numerical $\boldsymbol{U}_k^*$ values will deviate from the analytical ones. However, the impact is temporary and the entropy function soon regains its initial value.

2.2 *Case 2: A Closed System of Three Pipe Sections and Two Junctions*

An outline of the closed system is shown in Fig. 2. Three pipe sections, labelled S1 to S3, each of length $L = 50$ m, are connected by two junctions. At the bottom of the figure, the local axis along each pipe as assumed by the network theory is shown for the junctions. The global axis direction is from L to H.

Initially, the pipe sections are filled with stagnant fluid of uniform pressure. At $T = 0.0$ s two waves enter each pipe section as the interaction between the fluids is initiated. The initial conditions are summarised in Table 2.

The grid cell size was determined based on a comparison between analytical and numerical results obtained at different grid cell sizes. The results were compared for

[1]For examples of rarefaction- and shock waves, see Fig. 3

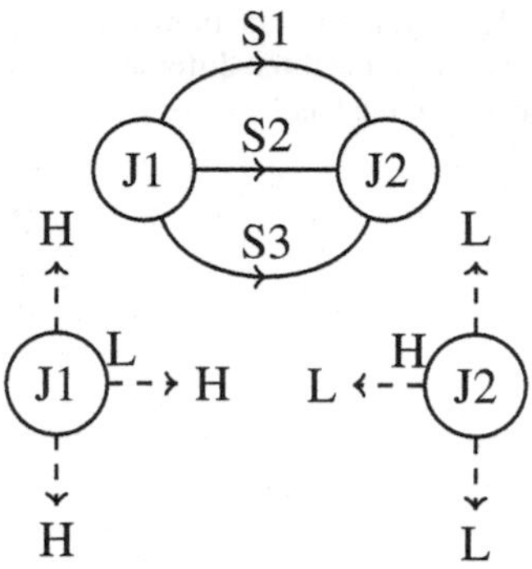

Fig. 2 Closed system consisting of three sections and two junctions

Table 2 Initial conditions

Section	p (bar)	v (m/s)
1	1.0	0.0
2	1.5	0.0
3	1.9	0.0

simulations run until $T = 0.06$ s. At this time there has been no interaction between the two waves that entered the pipe sections initially. Therefore it is possible to derive analytical profiles for total energy as a function of time. Pressure and velocity profiles were also compared. The comparison showed that a resolution of $\Delta x = 5.0 \times 10^{-3}$ m was needed to obtain accurate numerical energy profiles. A resolution of $\Delta x = 5.0 \times 10^{-2}$ m was sufficient if only pressure and velocity were considered. Thus the grid cell size was set to $\Delta x = 5.0 \times 10^{-3}$ m. A similar comparison was performed by Reigstad et al. [6].

Pressure- and velocity profiles for each of the three pipe sections and each of the momentum related coupling constants are presented in Fig. 3. In the first pipe section, $S1$, the three coupling constants all predict that two shock waves will enter. Similarly, two rarefaction waves are predicted to propagate into the third pipe section. In the second pipe section, the three models yield different kind of waves. The models using pressure and Bernoulli invariant as momentum related coupling constant predict two rarefaction waves to enter, while the model using momentum flux predicts shock waves. This is due to the predicted pressure at the pipe-junction boundary, p_2^*. Momentum flux as coupling constant results in a pressure which is larger than the pressure within the pipe, $p_2^* > \bar{p}_2$. The two other models predict pressures that are lower. Correspondingly, the Lax-criterion for shock- and rarefaction waves results in the difference in predicted wave type [3].

Total energy as function of time is shown in Fig. 4. In Fig. 4a numerical results are compared to analytical profiles derived under the constraint of energy conservation at the junctions. Figure 4b presents long term numerical results, for which no analytical profiles are available.

The physical soundness of the numerical solutions showed in Fig. 4a is determined by a comparison with the profiles denoted "Analytic 2". If the numerical

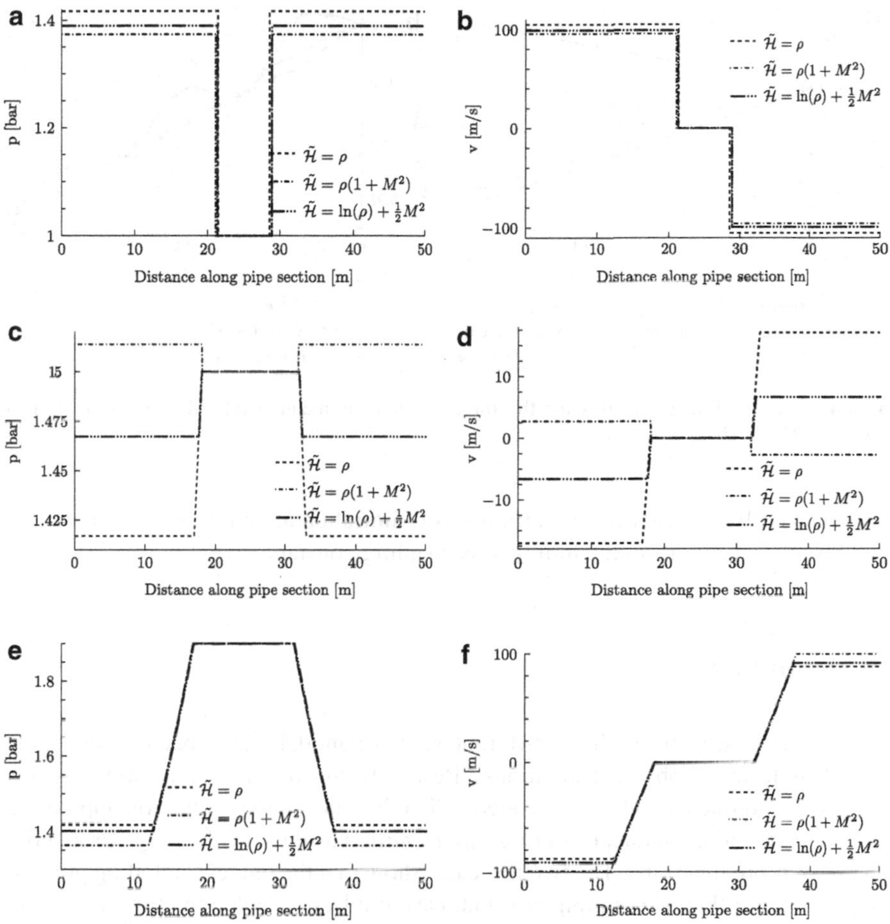

Fig. 3 Pressure and velocity profiles at $T = 0.06$ s for the three different momentum related coupling constants. (**a**) Pressure – pipe section 1. (**b**) Velocity – pipe section 1. (**c**) Pressure – pipe section 2. (**d**) Velocity – pipe section 2. (**e**) Pressure – pipe section 3. (**f**) Velocity – pipe section 3

profiles show a larger total energy than the corresponding analytic curve, energy production is present in the numerical results, and thus the solutions are unphysical [6].

As earlier predicted, models with momentum flux or pressure as momentum related coupling constant yield unphysical solutions for the selected set of initial data [7]. Using the Bernoulli invariant as coupling constant results in energy conservation at the junctions.

In Fig. 4b, a net reduction in total energy is observed for all three network models. The influence of the energy production in the junctions is clearly seen for pressure as momentum related coupling constant, as the profile does not decrease

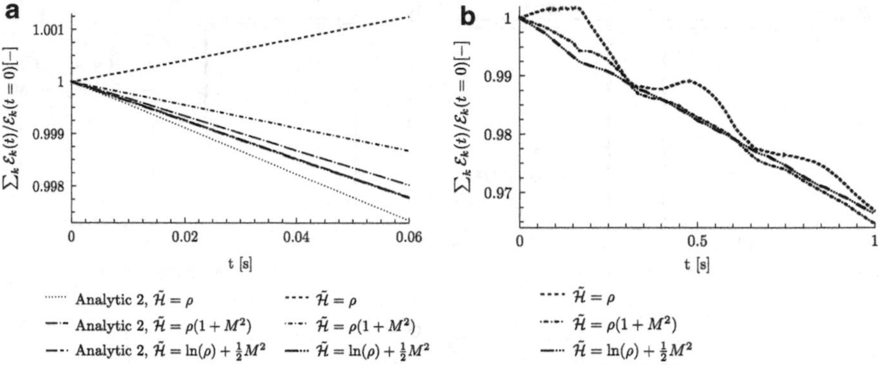

Fig. 4 Case 2 – Energy-profiles for the three different momentum related coupling constants. (**a**) $T = 0.06$ s. (**b**) $T = 1.0$ s

monotonically. In general, for certain sets of initial data, non-monotonicity will as well be observed for momentum flux as coupling constant.

3 Summary

Numerical results from three different network models have been investigated, mainly in terms of physical soundness. Results from two different network layouts, one open and one closed, are considered. Two layout-related evaluation approaches are applied, and unphysical solutions are identified as those with energy production in one or more junctions. The two test cases show that the models including pressure or momentum flux as coupling constant have unphysical solutions for the selected initial data. The network model which uses Bernoulli invariant as coupling constant has physical solutions, as energy is conserved at the junctions.

This is in accordance with analytical results; only Bernoulli invariant yields physical solutions for all subsonic initial conditions [7].

Acknowledgement The work of the first author was financed through the research project *Enabling low emission LNG systems*. The author acknowledges the project partners; Statoil and GDF SUEZ, and the Research Council of Norway (193062/S60) for support through the *Petromaks* programme.

References

1. M.K. Banda, M. Herty, A. Klar, Coupling conditions for gas networks governed by the isothermal Euler equations. Netw. Heterog. Media **1**(2), 295–314 (2006)
2. ———, Gas flow in pipeline networks. Netw. Heterog. Media **1**(1), 41–56 (2006)

3. R.M. Colombo, M. Garavello, A well posed Riemann problem for the p-system at a junction. Netw. Heterog. Media **1**(3), 495–511 (2006)
4. M. Garavello, A review of conservation laws on networks. Netw. Heterog. Media **5**(3), 565–581 (2010)
5. M. Herty, M. Seaïd, Simulation of transient gas flow at pipe-to-pipe intersections. Int. J. Numer. Methods Fluids **56**, 485–506 (2008)
6. G.A. Reigstad, T. Flåtten, Numerical network models and entropy principles for isothermal junction flow. Netw. Heterog. Media **9**(1), 65–95 (2014)
7. G.A. Reigstad, T. Flåtten, N.E. Haugen, T. Ytrehus, Coupling constants and the generalized Riemann problem for isothermal junction flow (2013, submitted)

Numerical Simulation of Compressible Turbulent Flows Using Modified EARSM Model

Jiří Holman and Jiří Fürst

Abstract This work describes the numerical solution of compressible turbulent flows. Turbulent flows are modeled by the system of averaged Navier-Stokes equations closed by the Explicit Algebraic Reynolds Stress Model (EARSM) of turbulence. EARSM model used in this work is based on the Kok's TNT model equations. New set of model constants which is more suitable for conjunction with EARSM model has been derived. Recalibrated model of turbulence together with the system of averaged Navier-Stokes equations is then discretized by the finite volume method and used for the solution of some realistic problems in external and internal aerodynamics.

1 Governing Equations

Compressible turbulent flows are modeled by the system of averaged Navier-Stokes equations in a vector form

$$\frac{\partial W}{\partial t} + \frac{\partial F_j(W)}{\partial x_j} = \frac{\partial R_j(W, \nabla W)}{\partial x_j}, \tag{1}$$

where $W = (\rho, \rho\mathbf{u}, \rho E)^T$ is vector of unknown conservative variables,[1] F_j are inviscid fluxes and R_j are viscous fluxes [6].

System (1) is also equipped with the equation of perfect gas in form

$$p = (\kappa - 1)\left[\rho E - \frac{1}{2}\rho u_j u_j - \rho k\right]. \tag{2}$$

[1]Where ρ is density, $\mathbf{u}$ is velocity vector, E is specific total energy and p is pressure.

J. Holman (✉) • J. Fürst
Faculty of Mechanical Engineering, Department of Technical Mathematics, CTU in Prague, Karlovo náměstí 13, Prague 2, Czech Republic
e-mail: jirka.holman@seznam.cz; Jiri.Holman@fs.cvut.cz; Jiri.Furst@fs.cvut.cz

A. Abdulle et al. (eds.), *Numerical Mathematics and Advanced Applications ENUMATH 2013*, Lecture Notes in Computational Science and Engineering 103,
DOI 10.1007/978-3-319-10705-9_67

1.1 EARSM Model of Turbulence

Averaged Navier-Stokes equations are closed by the EARSM model of turbulence developed by Wallin and Johanson [5] (designated here as EARSM-TNT). This model is derived as a simplified solution of full differential Reynolds stress transport model where both advection and diffusion are neglected.

Tensor of Reynolds stresses is approximated as

$$\tau^t_{ij} = 2\mu_T S_{ij} - \frac{2}{3}\delta_{ij}\rho k - \rho k a_{ij}, \tag{3}$$

where S_{ij} is strain-rate tensor defined as

$$S_{ij} = \frac{1}{2}\left(\frac{\partial u_i}{\partial x_j} + \frac{\partial u_j}{\partial x_i}\right) - \frac{1}{3}\delta_{ij}\frac{\partial u_k}{\partial x_k}, \tag{4}$$

μ_T is turbulent viscosity,

$$\mu_T = -\frac{1}{2}(\beta_1 + II_\Omega \beta_6)\rho k \tau \tag{5}$$

and a_{ij} is extra anisotropy given by

$$\begin{aligned} a_{ij} &= \beta_3\left(\Omega^*_{ik}\Omega^*_{kj} - \frac{1}{3}II_\Omega \delta_{ij}\right) \\ &+ \beta_4(S^*_{ik}\Omega^*_{kj} - \Omega^*_{ik}S^*_{kj}) \\ &+ \beta_6\left(S^*_{ik}\Omega^*_{kl}\Omega^*_{lj} + \Omega^*_{ik}\Omega^*_{kl}S^*_{lj} - II_\Omega S^*_{ij} - \frac{2}{3}IV\delta_{ij}\right) \\ &+ \beta_9(\Omega^*_{ik}S^*_{kl}\Omega^*_{lm}\Omega^*_{mj} - \Omega^*_{ik}\Omega^*_{kl}S^*_{lm}\Omega^*_{mj}). \end{aligned} \tag{6}$$

Both turbulent viscosity μ_T and extra anisotropy a_{ij} are strongly nonlinear terms which depends on normalized strain-rate tensor $S^*_{ij} = \tau S_{ij}$, normalized tensor of rotation $\Omega^*_{ij} = \tau \Omega_{ij}$ and related invariants $II_S = S^*_{kl}S^*_{lk}$, $II_\Omega = \Omega^*_{kl}\Omega^*_{lk}$ and $IV = S^*_{kl}\Omega^*_{lm}\Omega^*_{mk}$. Beta coefficients $\beta_1 - \beta_9$ are also nonlinear terms which depends on invariants II_S, II_Ω, IV and also on $V = S^*_{kl}S^*_{lm}\Omega^*_{mn}\Omega^*_{nk}$. For more details see [5]. Quantity τ is turbulent time scale defined as

$$\tau = \max\left(\frac{1}{\beta^*\omega},\ C_\tau\sqrt{\frac{\mu}{\beta^*\rho k\omega}}\right), \tag{7}$$

with constant $C_\tau = 6$.

In case of two-dimensional mean flows there are only two nonzero beta coefficients β_1 and β_4 and relations (5) and (6) are simplified to

$$\mu_T = -\frac{1}{2}\beta_1 \rho k \tau, \qquad a_{ij} = \beta_4 (S^*_{ik}\Omega^*_{kj} - \Omega^*_{ik}S^*_{kj}). \tag{8}$$

This version of the EARSM model is based on the Kok's TNT $k - \omega$ model of turbulence [4]. TNT model includes transport equations for the turbulent kinetic energy k and the specific dissipation rate ω in form:

$$\frac{\partial(\rho k)}{\partial t} + \frac{\partial(\rho k u_j)}{\partial x_j} = P - \beta^* \rho k \omega + \frac{\partial}{\partial x_j}\left[\left(\mu + \sigma^* \mu_T\right)\frac{\partial k}{\partial x_j}\right] \tag{9}$$

$$\frac{\partial(\rho \omega)}{\partial t} + \frac{\partial(\rho \omega u_j)}{\partial x_j} = \alpha \frac{\omega}{k} P - \beta \rho \omega^2 + \sigma_d \frac{\rho}{\omega} \max\left(\frac{\partial k}{\partial x_j}\frac{\partial \omega}{\partial x_j}, 0\right) + \frac{\partial}{\partial x_j}\left[\left(\mu + \sigma \mu_T\right)\frac{\partial \omega}{\partial x_j}\right], \tag{10}$$

where P is production term, μ is molecular viscosity and $\alpha, \beta, \beta^*, \sigma, \sigma^*$ and σ_d are model constants.[2]

2 Calibration of Model Constants

Model constants of the original TNT model were derived under the assumption of linear relation between tensor of Reynolds stresses and strain-rate tensor (so called Boussinesq hypothesis). On the other hand, EARSM model is nonlinear and Hellsten shows that not every two-equation model of turbulence is suitable for conjunction with EARSM constitutive relations [2]. Therefore our goal in this section is revision of the TNT model constants.

The first relation for the model constants can be obtained from the decaying homogeneous isotropic turbulence. Transport equations for the turbulent kinetic energy k and for the specific dissipation rate ω are reduced to the set of ordinary differential equations

$$\frac{dk}{dt} = -\beta^* k \omega, \qquad \frac{d\omega}{dt} = -\beta^* \omega^2 \tag{11}$$

for this case. Solution of system (11) reads

$$k \sim t^{-\beta^*/\beta}. \tag{12}$$

[2]Original values of TNT model are $\alpha = 0.553, \beta = 0.075, \beta^* = 0.09, \sigma = 0.5, \sigma^* = 0.667$ and $\sigma_d = 0.5$.

Experimental observations indicate $k \sim t^{-n}$ with $n = 1.25 \pm 0.06$ and after comparison with (12) we can obtain relation for the ratio of β^* and β

$$\frac{\beta^*}{\beta} = \frac{6}{5}. \tag{13}$$

Next relations are derived from the flat plate zero pressure gradient boundary layer, more specifically from the logarithmic part of boundary layer which is far enough from wall so that molecular viscosity is negligible but close enough so that convection can be neglected too. Transport equations of the TNT model together with the averaged Navier-Stokes equations are reduced to system:

$$\begin{aligned}
0 &= \frac{\partial}{\partial y}\left[\nu_T \frac{\partial u}{\partial y}\right] \\
0 &= \nu_T \left(\frac{\partial u}{\partial y}\right)^2 - \beta^* k\omega + \sigma^* \frac{\partial}{\partial y}\left[\nu_T \frac{\partial k}{\partial y}\right] \\
0 &= \alpha \left(\frac{\partial u}{\partial y}\right)^2 - \beta\omega^2 + \sigma \frac{\partial}{\partial y}\left[\nu_T \frac{\partial \omega}{\partial y}\right].
\end{aligned} \tag{14}$$

Solution of system (14) is

$$u = \frac{u_\tau}{\kappa} \ln y + const., \quad k = \frac{u_\tau^2}{\sqrt{\beta^*}}, \quad \omega = \frac{u_\tau}{\sqrt{\beta^*}\kappa y}, \tag{15}$$

where $u_\tau = \sqrt{\tau_{wall}/\rho}$ is friction velocity and $\kappa = 0.41$ is Von Kármán constant. After substituting solution (15) back to system (14) we obtain relations for the model constant α,

$$\alpha = \frac{\beta}{\beta^*} - \frac{\sigma\kappa^2}{\sqrt{\beta^*}} \tag{16}$$

and for the component of Reynolds stress τ_{xy}^t,

$$\tau_{xy}^t = u_\tau^2. \tag{17}$$

Measurements indicate the ratio of τ_{xy}^t to the turbulent kinetic energy k is about 3/10 in the logarithmic layer and therefore value of the model constant β^* is 0.09 (the value is obtained using relation (17) and second relation in (15)). Value of the model constant $\beta = 0.075$ is consequently obtained from relation (13).

Analysis of defect layer and sublayer indicate optimum[3] choice of the model constant $\sigma = 0.5$ [6]. Finally, model constant $\alpha = 0.553$ is obtained from relation (16) using Von Kármán constant $\kappa = 0.41$.

[3]Optimum in a sense that turbulence model equations can be integrated to the wall without any damping functions and without distance to the nearest wall information.

Last two model constants have small effect in the inner layer and their values are determined form the behavior near the edges of shear layers. Transport equations of the TNT model together with the averaged Navier-Stokes equations are simplified to the system:

$$\begin{aligned} \nu_C \frac{\partial u}{\partial y} &= \frac{\partial}{\partial y}\left[\nu_T \frac{\partial u}{\partial y}\right] \\ \nu_C \frac{\partial k}{\partial y} &= \sigma^* \frac{\partial}{\partial y}\left[\nu_T \frac{\partial k}{\partial y}\right] \\ \nu_C \frac{\partial \omega}{\partial y} &= \sigma \frac{\partial}{\partial y}\left[\nu_T \frac{\partial \omega}{\partial y}\right] + \sigma_d \frac{\nu_T}{k} \frac{\partial k}{\partial y} \frac{\partial \omega}{\partial y}, \end{aligned} \tag{18}$$

where convection speed ν_C is assumed to be constant. Kok found that the following power functions form at least a weak solution to the system (18) if σ^*, σ, and σ_d are suitably selected [4]:

$$\begin{aligned} u(y) &= u_0 f^{\sigma^* \sigma/(\sigma - \sigma^* + \sigma_d)} \\ k(y) &= k_0 f^{\sigma/(\sigma - \sigma^* + \sigma_d)} \\ \omega(y) &= \omega_0 f^{(\sigma^* - \sigma_d)/(\sigma - \sigma^* + \sigma_d)}, \end{aligned} \tag{19}$$

where function $f(y)$ reads

$$f(y) = \max\left(\frac{\delta_0 - y}{\delta_0}, 0\right) \tag{20}$$

with index 0 indicating characteristic scales of the problem. Solution (19) must be non-singular to form at least a weak solution to the problem and that the slope of the velocity remains bounded at the edge. The resulting conditions are:

$$\begin{aligned} \sigma - \sigma^* + \sigma_d &> 0 \\ \sigma^* - \sigma_d &> 0 \\ \sigma - \sigma^* + \sigma_d &\leq \sigma^* \sigma \end{aligned} \tag{21}$$

with $\sigma^* > 0.5$ and $\sigma > 0$. Requirements (21) hold for the models of turbulence based on the Boussinesq eddy-viscosity hypothesis. Hellsten modified this requirements for the nonlinear constitutive relations [2]:

$$\begin{aligned} \sigma - \sigma^* + \sigma_d &> 0 \\ \sigma^* - \sigma_d &> 0 \\ \sigma^* &> 1 \end{aligned} \tag{22}$$

We have chosen $\sigma^* = 1.01$ according to the last condition in (22). Substituting σ^* to second inequation in (22) we obtain condition for the last constant $\sigma_d > 0.51$. Our choice is lower limit $\sigma_d = 0.52$. Note that σ^* and σ_d are the only recalibrated model constants and modified set of model constants also satisfies original requirements derived by Kok.

3 Numerical Methods

We have developed in-house solver for two-dimensional averaged Navier-Stokes equations (together with modified EARSM model) based on the finite volume method [1]. Inviscid numerical fluxes are approximated by the HLLC scheme with the piecewise linear MUSCL or WENO reconstruction of second order accuracy [3]. Viscous numerical fluxes are approximated by the central differencing with aid of dual mesh [3]. The resulting system of ordinary differential equations is then solved by the explicit two-stage TVD Runge-Kutta method with local time-step and point implicit treatment of source terms [3].

We also used freely available OpenFOAM software based on the in-house modified rhoSimpleFOAM solver with segregated approach (SIMPLE loop) and second order interpolations. This software was also equipped with EARSM model of turbulence based on the recalibrated TNT model equations. This software was used for three-dimensional cases.

4 Numerical Solution of Compressible Turbulent Flows

This chapter presents some numerical solutions of the compressible turbulent flows. For validation purpose, the subsonic flow around the flat plate was solved first. Next test cases represents realistic problems from both external and internal aerodynamics.

4.1 Subsonic Flow Around the Flat Plate

The first solved case was subsonic flow around the flat plate. This case is characterized by inlet Mach number $M_\infty = 0.2$, zero angle of attack and Reynolds number $Re = 8 \cdot 10^5$. We used rectangular computational domain $[-1, 16.67] \times [0, 3]$ where flat plate starts at point $x = 0$. Computational mesh was structured H-type grid with 110×80 cells with 95 cells at flat plate itself.

From Fig. 1 one can see that EARSM model with original Kok's model constants has qualitatively wrong shape of velocity profile. On the other hand, recalibrated model is in good agreement with Hellsten model which was designed

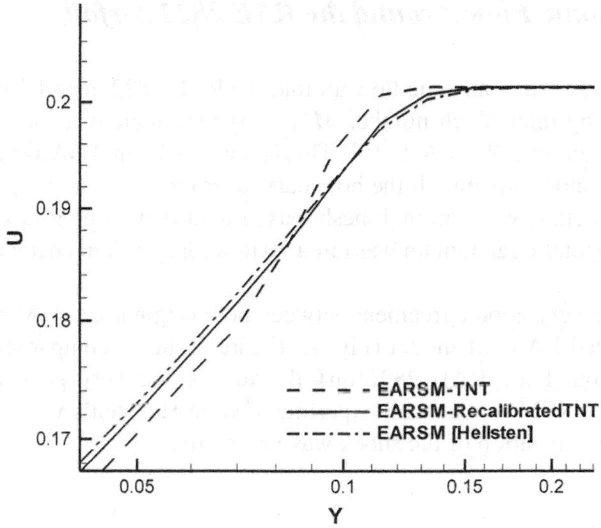

Fig. 1 Detail of profiles of velocity in point x = 12.68

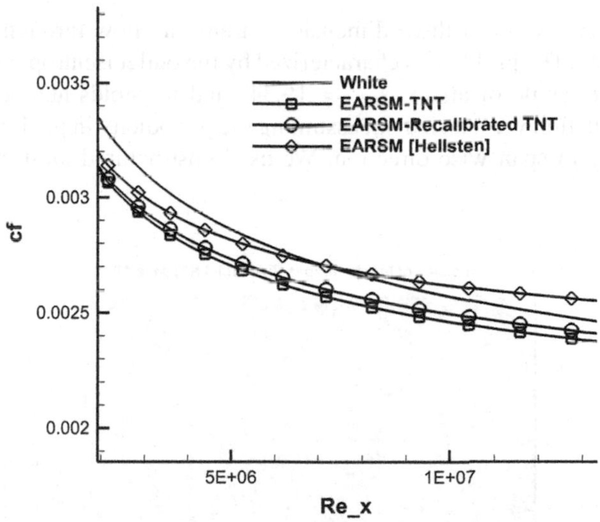

Fig. 2 Comparison of distributions of friction coefficient

especially for the conjunction with EARSM constitutive relations. Figure 2 shows a good agreement of all models with experiment but we can see that recalibrated EARSM model archived slightly better results then original EARSM-TNT model.

4.2 Transonic Flow Around the RAE 2822 Airfoil

Next solved case was transonic flow around the RAE 2822 airfoil. This problem is characterized by inlet Mach number $M_\infty = 0.754$, angle of attack $\alpha_\infty = 2.57°$ and Reynolds number $Re = 6.2 \cdot 10^6$. This is a well known AGARD case 10 where interaction of shock wave with the boundary layer create a small separation region behind the shock. Computational mesh was structured C-type grid with 300×70 cells and computational domain was circa 30 times larger then characteristic size of the airfoil.

We can see very good agreement between both original EARSM model (Fig. 3) and recalibrated EARSM model (Fig. 4). Figure 5 shows comparison of pressure coefficient around the RAE 2822 airfoil. We can see very good agreement of recalibrated EARSM model with experiment while Hellsten's version of EARSM failed to capture position of the shock wave correctly.

4.3 Transonic Flow Through the SE 1050 Turbine Cascade

Last solved case was full three dimensional transonic flow through the SE 1050 turbine cascade. The problem is characterized by the outlet isentropic Mach number $M_{2is} = 1.198$, angle of attack $\alpha_\infty = 19.34°$ and Reynolds number $Re = 1.5 \cdot 10^6$. The simulation was carried out assuming the periodicity in pitch-wise direction and symmetry in span-wise direction. We used unstructured mesh with $3.3 \cdot 10^6$ cells.

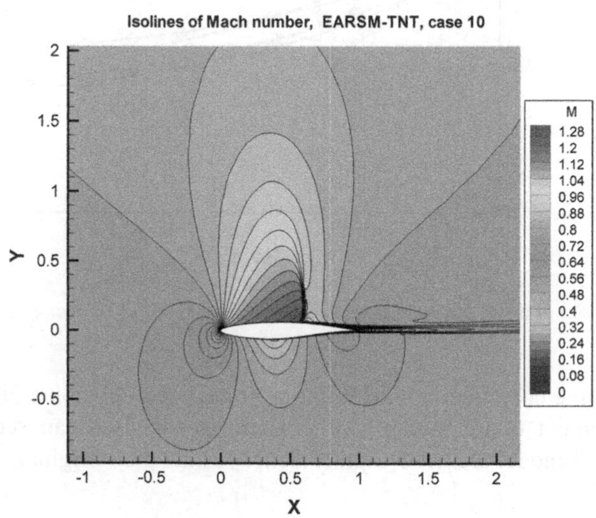

Fig. 3 Flow around the RAE 2822

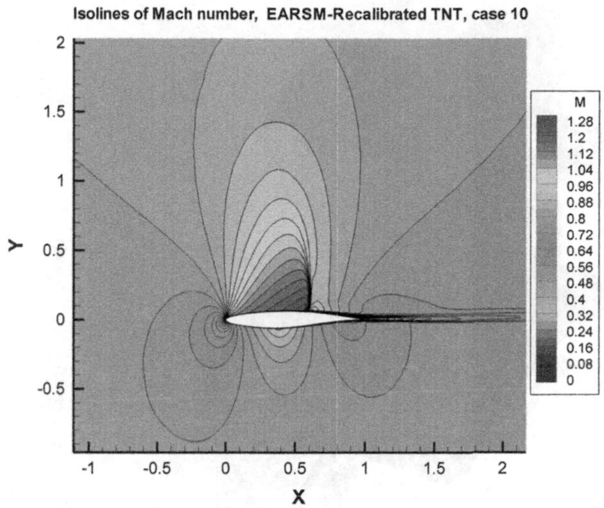

Fig. 4 Flow around the RAE 2822

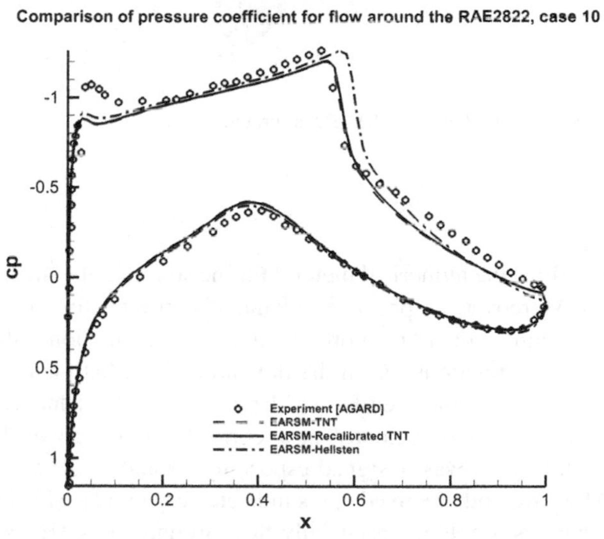

Fig. 5 Transonic flow around the RAE 2822

Our simulations have archived very good match between original and recalibrated (Fig. 6) EARSM models and further with well known SST $k - \omega$ model.

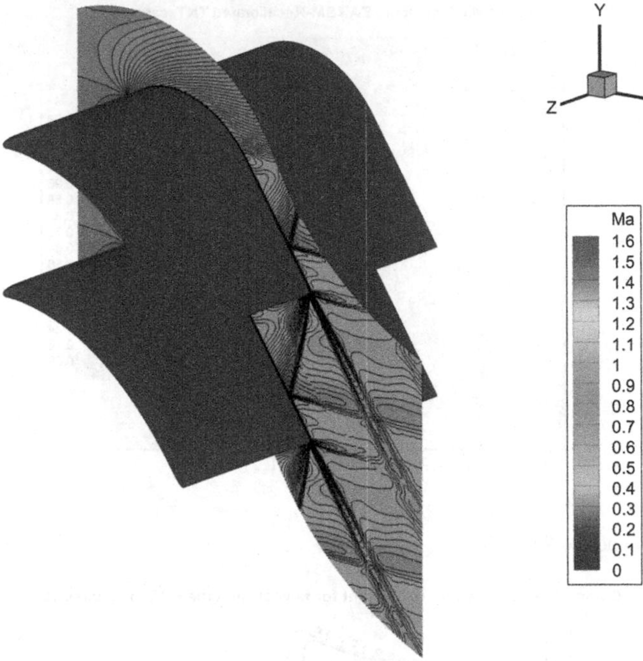

Fig. 6 Flow through the SE 1050, EARSM-Recalibrated

Conclusion

We have described the numerical method for the solution of compressible turbulent flows. Moreover, we propose new model constants for the TNT model transport equations which are more suitable for conjunction with EARSM model constitutive relations. From the flow around the flat plate one can see an improvement of recalibrated EARSM model over the standard EARSM-TNT version. Profiles of velocity now corresponding very well with the Hellsten model which was designed especially for the EARSM constitutive relations. Moreover, other solved cases indicate very good performance in the real aerodynamics problems (especially flow around the RAE 2822 airfoil). Finally, recalibrated EARSM model remains as simple as original EARSM-TNT model and does not require any damping functions or wall distance information like Hellsten model.

Acknowledgment The work was supported by the Grant no. GAP101/12/1271 and Grant no. P101/10/1329 of the Grant Agency of Czech Republic.

References

1. J.H. Ferziger, M. Peric, *Computational Methods for Fluid Dynamics* (Springer, Berlin/ New York, 1999)
2. A. Hellsten, New two-equation turbulence model for aerodynamics applications, Report A-21, Helsinki University of Technology, 2004
3. J. Holman, Numerical solution of compressible turbulent flows in external and internal aerodynamics, Diploma thesis CTU in Prague, 2007 (in czech)
4. J.C. Kok, Resolving the dependence on freestream values for $k-\omega$ turbulence model. AIAA J. **38**(7), 1292–1295 (2000)
5. S. Wallin, Engineering turbulence modeling for CFD with focus on explicit algebraic Reynoldes stress models, Ph.D. thesis, Royal Institute of Technology, 2000
6. D.C. Wilcox, *Turbulence Modeling for CFD*, 2nd edn. (DWC Industries, La Cãnada, 1994)

Steady Mixed Convection in a Heated Lid-Driven Square Cavity Filled with a Fluid-Saturated Porous Medium

Bengisen Pekmen and Munevver Tezer-Sezgin

Abstract Steady mixed convection flow in a porous square cavity with moving side walls is studied numerically using the dual reciprocity boundary element method (DRBEM). The equations governing the two-dimensional, steady, laminar mixed convection flow of an incompressible fluid are solved for various values of parameters as Darcy (Da), Grashof (Gr), and Prandtl (Pr) numbers. The results are given in terms of vorticity contours, streamlines and isotherms. Further, average Nusselt number variations with respect to the problem parameters are also presented. The fluid flows slowly as Da decreases since the permeability of the medium decreases, and the increase in Grashof number causes the flow to pass to the natural convective behavior. DRBEM has the advantage of using considerably small number of grid points due to the boundary only nature of the method. This provides the numerical procedure computationally cheap and efficient.

1 Introduction

In many fundamental heat transfer analyses, convective flows in porous media have received much attention and played the central role due to the important applications as in packed sphere beds, insulation for buildings, grain storage, chemical catalytic reactors, and geophysical problems. The underground spread of pollutants, solar power collectors, and geothermal energy systems include porous media.

The theoretical and analytical details of heat transfer in porous medium may be found in the books [5, 6]. Also, a lot of numerical studies concerning heat

B. Pekmen (✉)
Department of Mathematics, Atılım University, 06836, Ankara, Turkey

Institute of Applied Mathematics, Middle East Technical University, 06800, Ankara, Turkey
e-mail: e146303@metu.edu.tr

M. Tezer-Sezgin
Department of Mathematics, Institute of Applied Mathematics, Middle East Technical University, 06800, Ankara, Turkey
e-mail: munt@metu.edu.tr

A. Abdulle et al. (eds.), *Numerical Mathematics and Advanced Applications ENUMATH 2013*, Lecture Notes in Computational Science and Engineering 103,
DOI 10.1007/978-3-319-10705-9_68

transfer in a porous medium are reported in the last decade. Among these, numerical solutions obtained by DRBEM [9], finite element method (FEM) [2], penalty FEM with biquadratic elements [10], finite volume method (FVM) [1, 11], and the finite difference method (FDM) [4, 7] may be mentioned.

In this study, steady mixed convection flow in a porous square cavity with differentially heated and moving side walls is studied numerically using the DRBEM. An isotropic, homogeneous porous medium saturated with an incompressible, viscous fluid is considered. The thermal and physical properties of the fluid are assumed to be constant, but the fluid density varies according to Boussinesq approximation. The fluid and the solid particles are also assumed to be in local thermal equilibrium. Viscous dissipation, and Forchheimer terms (quadratic drag terms) in the momentum equations are neglected.

The two-dimensional, steady, laminar mixed convection flow of an incompressible fluid is taken into account. The non-dimensional governing equations in terms of stream function ψ-temperature T-vorticity w are [2]

$$\nabla^2\psi = -w \tag{1a}$$

$$\frac{1}{\epsilon_p Re}\nabla^2 w = \frac{1}{\epsilon_p^2}\left(u\frac{\partial w}{\partial x} + v\frac{\partial w}{\partial y}\right) - \frac{Gr}{Re^2}\frac{\partial T}{\partial x} + \frac{1}{Da\,Re}w \tag{1b}$$

$$\frac{1}{Pr\,Re}\nabla^2 T = u\frac{\partial T}{\partial x} + v\frac{\partial T}{\partial y} \tag{1c}$$

where ϵ_p is the porosity of the porous medium, $u = \partial\psi/\partial y$, $v = -\partial\psi/\partial x$, $w = \partial v/\partial x - \partial u/\partial y$. Non-dimensional physical parameters are Reynolds, Grashof, Darcy and Prandtl numbers, respectively, given as

$$Re = \frac{U_0 L}{\nu_e}, \quad Gr = \frac{g\beta\Delta T L^3}{\nu_e^2}, \quad Da = \frac{\kappa}{L^2}, \quad Pr = \frac{\nu_e}{\alpha_e}, \tag{2}$$

with characteristic velocity U_0, characteristic length L, gravitational acceleration g, effective kinematic viscosity ν_e, permeability of the porous medium κ, thermal expansion coefficient β, temperature difference $\Delta T = T_h - T_c$, effective thermal diffusivity α_e of the porous medium.

We consider the problem geometry consisting of the cross-section of a unit square cavity which has the moving lids on the left and right walls (Fig. 1). The boundary conditions are as follows. The velocity $v = 1$ on the vertical walls with $u = \psi = 0$; and $u = v = \psi = 0$ on the horizontal walls. The right wall is the hot ($T_h = 1$), and the left wall is the cold ($T_c = 0$) wall while the top and bottom walls are adiabatic ($\partial T/\partial n = 0$). Vorticity boundary conditions are unknown, and are going to be derived with the help of DRBEM coordinate matrix during the iterative solution procedure.

Fig. 1 Problem configuration

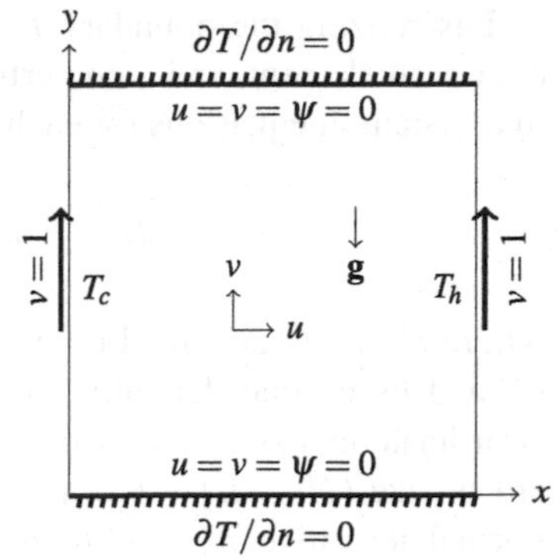

2 DRBEM Application

DRBEM treats all the right hand side terms of Eqs. 1 as inhomogeneity, and an approximation for this inhomogeneous term is proposed [8] as

$$b \approx \sum_{j=1}^{N+L} \alpha_j f_j = \sum_{j=1}^{N+L} \alpha_j \nabla^2 \hat{u}_j \tag{3}$$

where N is the number of boundary nodes, L is the number of internal collocation points, α_j's are sets of initially unknown coefficients, and the f_j's are approximating functions which are related to particular solutions $\hat{u}_j$ with $\nabla^2 \hat{u}_j = f_j$. The radial basis functions f_j's are usually chosen as polynomials of radial distance $r_{ij} = \sqrt{(x_i - x_j)^2 + (y_i - y_j)^2}$ as $f_{ij} = 1 + r_{ij} + r_{ij}^2 + \ldots + r_{ij}^n$ where i and j correspond to the source(fixed) (x_i, y_i) and the field(variable) (x_j, y_j) points, respectively.

DRBEM transforms differential equations defined in a domain Ω to integral equations on the boundary Γ. For this, differential equation is multiplied by the fundamental solution $u^* = -ln(r)/(2\pi)$ of Laplace equation and integrated over the domain. In Eqs. 1, the right hand sides are approximated using Eq. 3 giving Laplacian terms on both sides. Using Divergence theorem for the Laplacian terms on both sides of the equation, domain integrals are transformed to boundary integrals as follows

$$c_i u_i + \int_\Gamma u q^* d\Gamma - \int_\Gamma q u^* d\Gamma = \sum_{j=1}^{N+L} \alpha_j \left(c_i \hat{u}_{ij} + \int_\Gamma q^* \hat{u}_{ij} d\Gamma - \int_\Gamma u^* \hat{q}_{ij} d\Gamma \right), \tag{4}$$

where $c_i = 0.5$ if the boundary Γ is a straight line and $i \in \Gamma$, and $c_i = 1$ when node i is inside, $\hat{q}_{ij} = \partial \hat{u}_{ij}/\partial n$ with the outward unit normal $\mathbf{n}$ to Γ.

Discretizing the boundary Γ by using N linear elements, evaluating integrals over each element, and then performing assembly procedure for all elements result in a system of equations for each of the Eqs. 1 as

$$Hu - Gu_q = \left(H\hat{U} - G\hat{Q}\right) F^{-1}b, \tag{5}$$

where H and G are BEM matrices contain integral values of fundamental solution u^* and its normal derivative over the boundary elements, respectively. F is the coordinate matrix formed from the radial basis functions f_j's. $\hat{U}$ and $\hat{Q}$ matrices are of size $(N + L) \times (N + L)$, and are built from particular solution $\hat{u}$ and its normal derivative $\hat{q} = \partial\hat{u}/\partial n$ at the $(N + L)$ source and field points. The vector b is formed from the right hand sides of Eqs. 1.

Matrix-vector form for Eqs. 1 are written as

$$H\psi^{m+1} - G\psi_q^{m+1} = -Sw^m \tag{6a}$$

$$(H - PrReSM)\,T^{m+1} - GT_q^{m+1} = 0 \tag{6b}$$

$$\left(H - \frac{Re}{\epsilon_p}SM - \frac{\epsilon_p}{Da}S\right) w^{m+1} - Gw_q^{m+1} = -\epsilon_p\frac{Gr}{Re}S\frac{\partial F}{\partial x}F^{-1}T^{m+1} \tag{6c}$$

where $S = (H\hat{U} - G\hat{Q})F^{-1}$, $u^{m+1} = (\partial F/\partial y)F^{-1}\psi^{m+1}, v^{m+1} = -(\partial F/\partial x)F^{-1}\psi^{m+1}$,
$M = \left([u]_d^{m+1}\frac{\partial F}{\partial x}F^{-1} + [v]_d^{m+1}\frac{\partial F}{\partial y}F^{-1}\right)$, the subscript d shows the diagonal matrix, and m is the iteration level.

Unknown vorticity boundary conditions are obtained from the definition of w as

$$w = \frac{\partial v}{\partial x} - \frac{\partial u}{\partial y} = \frac{\partial F}{\partial x}F^{-1}v - \frac{\partial F}{\partial y}F^{-1}u, \tag{7}$$

with the help of coordinate matrix F. Also, all the space derivatives in b are computed by using DRBEM coordinate matrix F, i.e.

$$\frac{\partial T}{\partial x} = \frac{\partial F}{\partial x}F^{-1}T, \quad \frac{\partial w}{\partial y} = \frac{\partial F}{\partial y}F^{-1}w. \tag{8}$$

Systems of Eqs. 6a–6c are solved iteratively for the unknowns ψ, T, w, and normal derivatives ψ_q, T_q, w_q. Initially, ψ, T and w are taken as zero except on the boundary. First, Eq. 6a is solved for stream function. Then, stream function is used to compute velocity components u and v inserting their boundary conditions.

The energy and vorticity transport equations are then solved by using u and v, respectively. The iterations continue until the criterion [4]

$$\frac{\|\psi^{m+1}-\psi^{m}\|_{\infty}}{\|\psi^{m+1}\|_{\infty}}+\frac{\|T^{m+1}-T^{m}\|_{\infty}}{\|T^{m+1}\|_{\infty}}+\frac{\|w^{m+1}-w^{m}\|_{\infty}}{\|w^{m+1}\|_{\infty}}<\epsilon \tag{9}$$

is satisfied where $\epsilon = 10^{-5}$ is the tolerance to stop the iterations.

In order to accelerate the convergence for large values of problem parameters a relaxation parameter $0 < \gamma \leq 1$ is used for the vorticity as $w^{m+1} \leftarrow \gamma w^{m+1} + (1-\gamma)w^{m}$. Further, average Nusselt number through the heated wall is computed by $\overline{Nu} = \int_0^1 (\partial T/\partial x)dy$.

3 Numerical Results

As a validation case, a non-porous unit square cavity with heated bottom, cold top wall, adiabatic left and right walls and moving top lid is considered. As is seen in Table 1, present results using considerably small number of grid points are in good agreement with the results in [10] where 57×57 grid points are used.

In the numerical computations of stream function, vorticity and temperature in a square cavity with heated and upwards moving vertical walls, radial basis function $f = 1 + r$, and 8-point Gaussian quadrature are used for the construction of F, H and G BEM matrices. $N = 96$, $L = 625$ are taken, and $Re = 100$ is fixed. Cavity contains a fluid saturated porous medium with $\epsilon_p \leq 1$. Mixed convection flow behavior in this porous medium is depicted in terms of streamlines, isotherms, and vorticity contours for various values of Da, Gr and Pr.

As Da decreases (Fig. 2), permeability decreases and causes a force opposite to the flow direction which tends to resist the flow. This means that the fluid flows slowly. While the center of streamlines is in the direction of moving lids, they cluster along the left and right boundaries forming boundary layers, and the effects of moving walls almost disappear. Isotherms become almost perpendicular to the top and bottom walls pointing to the increase in conduction dominated effect. Circulation in the vorticity through the upper corners due to the effect of moving

Table 1 $Re = 500$, $\gamma = 0.1$, $\overline{Nu}$ comparison with various Pr numbers

		[10]	Present		
Pr	Gr	$\overline{Nu}$	$\overline{Nu}$	N,L	CPU(sec.)
0.01	10^4	1.0431	1.0372	136,529	139.9
0.01	10^5	1.0721	1.0733	136,529	129.3
0.1	10^4	2.3815	2.3711	96,529	110.7
0.1	10^5	2.8704	2.8731	96,576	143.9
1	10^4	5.5695	5.5661	96,729	256.2
1	10^5	6.3313	6.3242	96,900	591.7

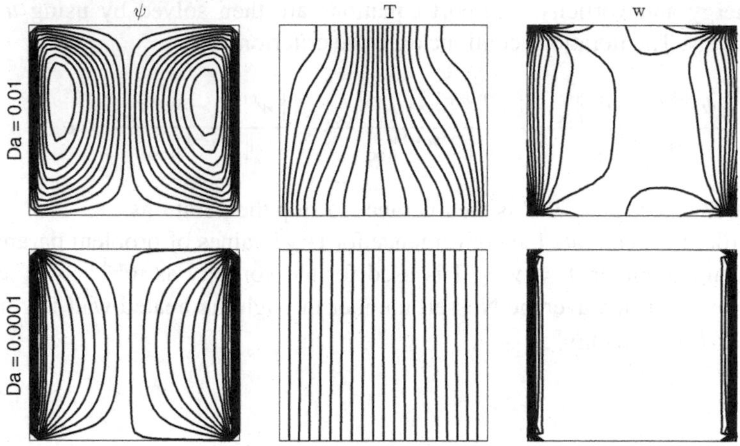

Fig. 2 $Pr = 0.71,\ Gr = 10^3,\ \epsilon_p = 1$

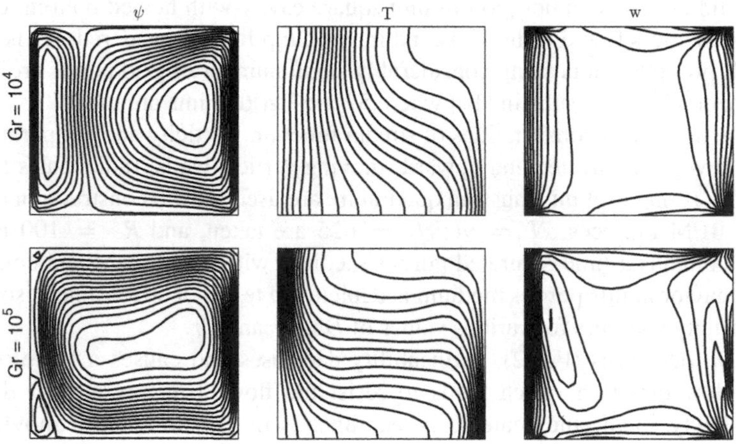

Fig. 3 $Pr = 0.71,\ Da = 0.01,\ \epsilon_p = 1$

lids diminishes, and strong boundary layers are formed through the right and left walls leaving a stagnant region at the center.

As Gr increases, the left counter-clockwise secondary cell starts to be squeezed through the left wall, and the clockwise primary cell is centered. Buoyancy effect is pronounced due to the increase in $Ri = Gr/Re^2$. That is, natural convection is high. Actually, this can be seen in isotherms at $Gr = 10^5$. While the isotherms pronounce the forced convection with $Gr = 10^3,\ Da = 0.01(Ri = 0.1)$ in Fig. 2, they cluster through the left and right walls forming strong temperature gradients for $Gr = 10^5$ (Fig. 3). Even though there is a Darcy effect with strength $Da = 0.01$, one is able to observe the characteristics of mixed convection flow in a non-porous medium in

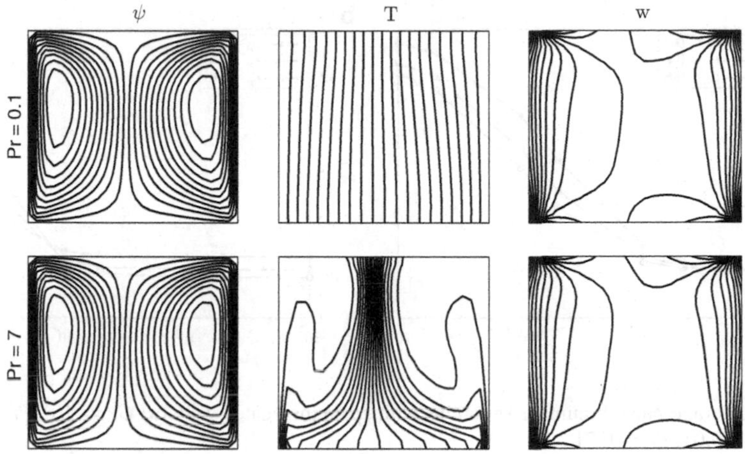

Fig. 4 $Da = 0.01,\ Gr = 10^3,\ \epsilon_p = 1$

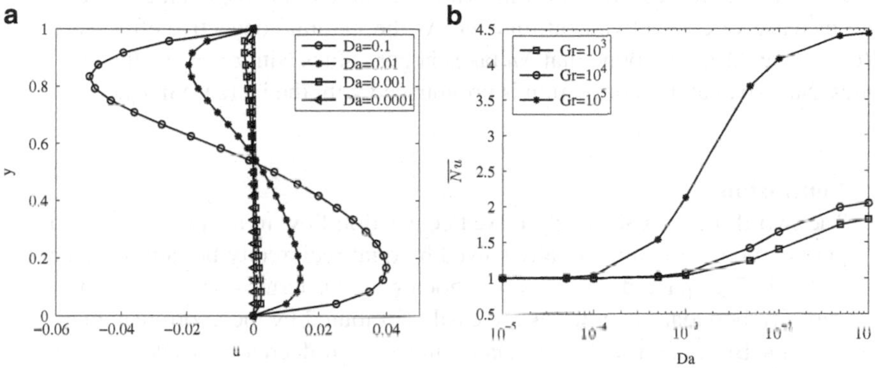

Fig. 5 Mid-u-velocity profile and average Nusselt Number on the heated wall. (**a**) $Gr = 10^3,\ Pr = 0.71,\ \epsilon_p = 1$. (**b**) $Pr = 0.71,\ \epsilon_p = 1$

the cavity [3]. Vorticity almost covers the cavity with new cells through the left and right walls, and spreads also along the top and bottom walls.

The increase in Pr only affects the isotherms (as is seen in Fig. 4) due to the dominance of convection terms in the temperature equation.

The decrease in the velocity of the fluid with the decrease in Da number is shown in Fig. 5a with the u-velocity profile through $x = 0.5$. The dominance of natural convection with high Gr is depicted in Fig. 5b. When Gr is increased, $\overline{Nu}$ values also increase. Average Nusselt number is almost the same for all values of Grashof number with $Da \leq 10^{-4}$ due to the dominance of conduction. However, $\overline{Nu}$ increases as Da increases showing the increase in the heat transfer.

Finally, we show how the heat transfer is affected by different values of porosity. As is seen in Fig. 6a ($Ri < 1$, forced convection is dominant), $\overline{Nu}$ increases at all ϵ_p

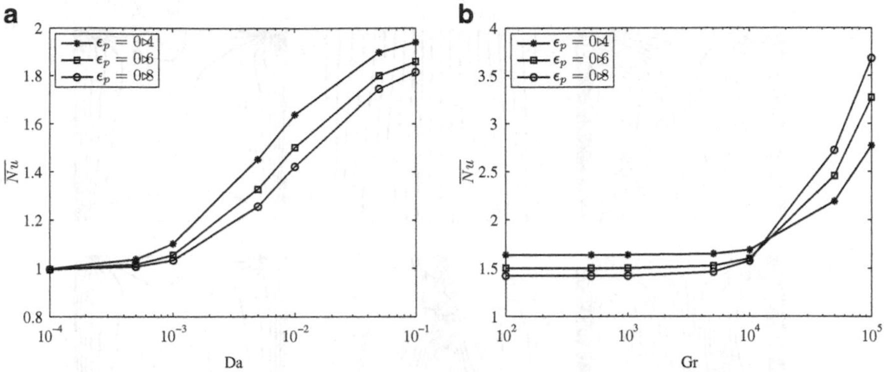

Fig. 6 Average Nusselt number variations with ϵ_p on the heated wall. (**a**) $Gr = 10^3, Pr = 0.71$. (**b**) $Da = 0.01, Pr = 0.71$

values as Da increases. High $\overline{Nu}$ values are obtained by small ϵ_p values which yields the increase in convective heat transfer. As the natural convective effect increases $Ri > 1$ (Fig. 6b), it is found that $\overline{Nu}$ takes larger values with $\epsilon_p = 0.8$ than the other ones. Namely, natural convection is pronounced with the increase in ϵ_p.

Conclusion

The two-dimensional, steady mixed convection flow in a square cavity with porous medium is numerically solved by dual reciprocity boundary element method. The space derivatives in inhomogeneous terms as well as unknown vorticity boundary conditions are easily computed by the coordinate matrix. For this Brinkmann-extended Darcy model, the decrease in Darcy number causes the fluid to flow slowly, and the heat to transfer in conductive mode. Natural convection is pronounced with the increase in Grashof number. In natural convection mode ($Ri > 1$), convective heat transfer increases in a high porosity of the medium.

References

1. M. Bhuvaneswari, S. Sivasankaran, Y.J. Kim, Effect of aspect ratio on convection in porous enclosure with partially active thermal walls. Comput. Math. Appl. **62**, 3844–3856 (2011)
2. S. Das, R.K. Sahoo, Effect of Darcy, fluid Rayleigh and heat generation parameters on natural convection in a porous square enclosure: a Brinkman-extended Darcy model. Int. Commun. Heat Mass Transf. **26**, 569–578 (1999)
3. R. Iwatsu, J.M. Hyun, K. Kuwahara, Mixed convection in a driven cavity with a stable vertical temperature gradient. Int. J. Heat Mass Transf. **36**, 1601–1608 (1993)

4. K.M. Khanafer, A.J. Chamkha, Mixed convection flow in a lid-driven enclosure filled with a fluid-saturated porous medium. Int. J. Heat Mass Transf. **42**, 2465–2481 (1999)
5. A. Narasimhan, *Essentials of Heat and Fluid Flow in Porous Media* (CRC/Taylor & Francis Group, Boca Raton, 2013)
6. D.A. Nield, A. Bejan, *Convection in Porous Media* (Springer, New York, 2006)
7. P. Nithiarasu, K.N. Seetharamu, T. Sundararajan, Natural convective heat transfer in a fluid saturated variable porosity medium. Int. J. Heat Mass Transf. **40**, 3955–3967 (1997)
8. P.W. Partridge, C.A. Brebbia, L.C. Wrobel, *The Dual Reciprocity Boundary Element Method* (Computational Mechanics Publications/Elsevier Science, Southampton, Boston/London, New York, 1992)
9. B. Pekmen, M. Tezer-Sezgin, DRBEM solution of free convection in porous enclosures under the effect of a magnetic field. Int. J. Heat Mass Transf. **56**, 454–468 (2013)
10. D. Ramakrishna, T. Basak, S. Roy, I. Pop, Numerical study of mixed convection within porous square cavities using Bejan's heatlines: effects of thermal aspect ratio and thermal boundary conditions. Int. J. Heat Mass Transf. **55**, 5416–5448 (2012)
11. E. Vishnuvardhanarao, M.K. Das, Laminar mixed convection in a parallel two-sided lid-driven differentially heated square cavity filled with a fluid-saturated porous medium. Numer. Heat Transf. A **53**, 88–110 (2007)

The Influence of Boundary Conditions on the 3D Extrusion of a Viscoelastic Fluid

Marco Picasso

Abstract The influence of slip/no-slip boundary conditions on the shape of a jet flowing out of an axisymmetric capillary die is investigated. Numerical results are obtained using the numerical model presented in [1]. It is shown that with no-slip boundary conditions along the tube, a large swelling is obtained past the die, even with a small Weissenberg number. On the other side, no swelling occurs when perfect slip boundary conditions apply along the die, but the stress induced by the contraction flow prior to the die has strong effects on the shape of the viscoleastic jet. Finally, these strong memory effects vanish when the Phan-Thien Tanner model is considered instead of Oldroyd-B.

1 Introduction

Numerical simulation of extrusion for viscoelastic flows is of great importance for industrial processes involving pasta dough. A free surface flow is involved, since the location of the interface between the surrounding air and the dough is unknown. Few 3D models have been proposed for such simulations [1,2,5]. Moreover, the question of rheology and boundary conditions is central in order to reproduce experiments with accuracy. The goal of this paper is to discuss the role of boundary conditions. This point is important in practice since the die is designed either to give a smooth finish (Teflon dies) or a matt surface (bronze die).

2 The Model

Let Λ be a cavity of $\mathbb{R}^3$ partially filled with liquid. Let T be the final time of the simulation. Let φ be volume fraction of liquid, $\varphi : \Lambda \times (0,T) \rightarrow \mathbb{R}$, $\varphi = 1$ in the liquid, $\varphi = 0$ in the surrounding air, see Fig. 1. Let Q_T be the liquid space-time

M. Picasso (✉)
MATHICSE, Station 8, Ecole Polytechnique Fédérale de Lausanne, 1015 Lausanne, Switzerland
e-mail: marco.picasso@epfl.ch

A. Abdulle et al. (eds.), *Numerical Mathematics and Advanced Applications ENUMATH 2013*, Lecture Notes in Computational Science and Engineering 103,
DOI 10.1007/978-3-319-10705-9_69

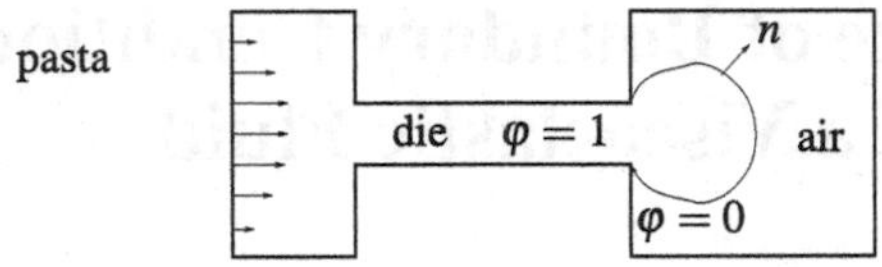

Fig. 1 The extrusion process. A pasta dough enters the cavity Λ. The volume fraction of liquid φ indicates the presence or absence of liquid

domain

$$Q_T = \{(x,t) \in \Lambda \times (0,T) \text{ such that } \varphi(x,t) = 1, 0 \le t \le T\}.$$

The unknowns of the model are the volume fraction of liquid $\varphi : \Lambda \times (0,T) \to \mathbb{R}$, the velocity $v : Q_T \to \mathbb{R}^3$, the pressure $p : Q_T \to \mathbb{R}$ and the extra-stress $\sigma : Q_T \to \mathbb{R}^{3\times 3}$ such that

$$\frac{\partial \varphi}{\partial t} + v \cdot \nabla \varphi = 0 \qquad \text{in } \Lambda \times (0,T), \tag{1}$$

$$\rho \frac{\partial v}{\partial t} + \rho (v \cdot \nabla) v - 2\eta_s \operatorname{div} \epsilon(v) + \nabla p - \operatorname{div} \sigma = \rho g \qquad \text{in } Q_T, \tag{2}$$

$$\operatorname{div} v = 0 \qquad \text{in } Q_T, \tag{3}$$

$$\left(1 + \alpha \frac{\lambda}{\eta_p} tr(\sigma)\right) \sigma + \lambda \Big(\frac{\partial \sigma}{\partial t} + (v \cdot \nabla)\sigma - \nabla v \sigma - \sigma \nabla v^T \Big) - 2\eta_p \epsilon(v) = 0 \text{ in } Q_T. \tag{4}$$

Here, ρ is the fluid density, η_s and η_p are the solvent and polymer viscosities, respectively, $\epsilon(v) = 1/2(\nabla v + \nabla v^T)$ is the strain rate tensor, g the gravity, λ the relaxation time, α the extensibility parameter, $tr(\cdot)$ the trace operator. Note that the velocity, pressure and extra-stress are only defined in the liquid domain whereas the volume fraction of liquid is defined in the whole cavity. When $\alpha = 0$ the above model corresponds to the so-called Oldroyd-B model.

Initial conditions for the volume fraction of liquid $\varphi(\cdot, 0)$, the velocity $v(\cdot, 0)$ and the extra-stress $\sigma(\cdot, 0)$ have to be prescribed. Concerning the boundary conditions, the volume fraction of liquid φ, the velocity v and the extra-stress σ are prescribed at the inflow region of the cavity Λ. A zero force condition applies on the liquid-air free surface

$$2\eta_s \epsilon(v) n + \sigma n - p n = 0,$$

whereas either slip or no-slip conditions apply along the die

$$(v \cdot n = 0 \text{ and } (2\eta_s \epsilon(v) - pI + \sigma)n.t_i = 0,\ i = 1, 2) \quad \text{or} \quad v = 0,$$

where n, t_1, t_2 are the unit outer normal and two tangent vectors at the boundary of the die, respectively.

3 Numerical Method

In [1] an implicit order one splitting scheme was advocated for the time discretization. Let τ be the time step, $t^n = n\tau$, $n = 0, 1, 2, \ldots$ and let $\varphi^{n-1} : \Lambda \to \mathbb{R}$ be an approximation of φ at time t^{n-1}. Let $\Omega^{n-1} = \{x \in \Lambda; \varphi^{n-1}(x) = 1\}$ be the corresponding liquid domain and let $v^{n-1} : \Omega^{n-1} \to \mathbb{R}^3$, $\sigma^{n-1} : \Omega^{n-1} \to \mathbb{R}^{3\times3}$ be approximations of the velocity and extra-stress at time t^{n-1}, respectively. Then, the new approximations φ^n, v^n, σ^n are computed as follows, see Fig. 2.

The prediction step consists in solving three convections problems between t^{n-1} and t^n, starting from $\varphi^{n-1}, v^{n-1}, \sigma^{n-1}$:

$$\frac{\partial \varphi}{\partial t} + v \cdot \nabla \varphi = 0, \quad \frac{\partial v}{\partial t} + (v \cdot \nabla)v = 0, \quad \frac{\partial \sigma}{\partial t} + (v \cdot \nabla)\sigma = 0. \tag{5}$$

We denote φ^n the obtained volume fraction of liquid at time t^n and $\Omega^n = \{x \in \Lambda; \varphi^n(x) = 1\}$ the new liquid domain. The obtained velocity and extra-stress at time t^n are denoted as $v^{n-1/2} : \Omega^n \to \mathbb{R}^3$ and $\sigma^{n-1/2} : \Omega^n \to \mathbb{R}^{3\times3}$, respectively; they are predictions of the velocity and extra-stress.

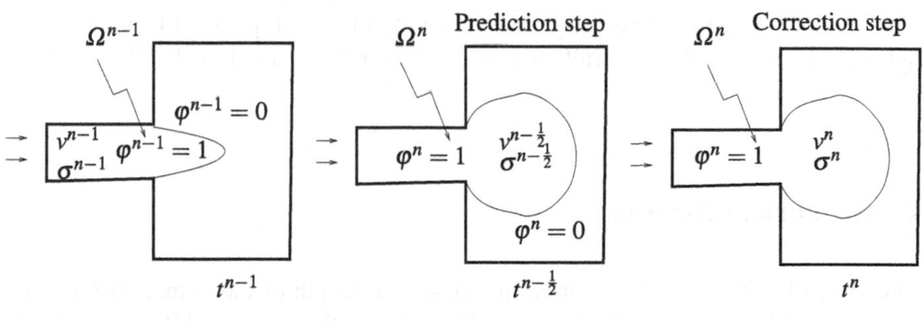

Fig. 2 Time discretization: the splitting algorithm

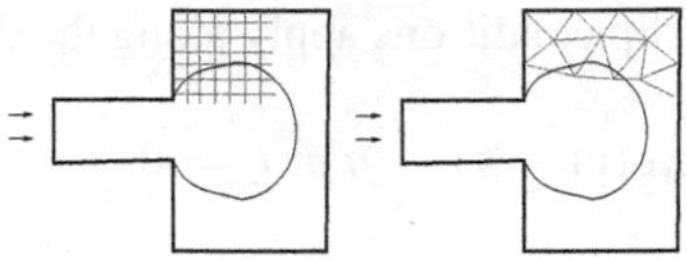

Fig. 3 The two grids used for solving problem (5) and (6). In the *left* figure, the small cubic cells of size h, in the *right* figure, the coarse tetrahedral finite elements with size H. In practice, $H \simeq 4h$

In the correction step, a viscoelastic problem without convection is solved in $\Omega^n \times (t^{n-1}, t^n)$:

$$\begin{aligned}
&\rho\frac{\partial v}{\partial t} - 2\eta_s \operatorname{div} \epsilon(v) + \nabla p - \operatorname{div} \sigma = \rho g, \\
&\operatorname{div} v = 0, \\
&\left(1 + \alpha\frac{\lambda}{\eta_p} tr(\sigma)\right)\sigma + \lambda\left(\frac{\partial \sigma}{\partial t} - \nabla v \sigma - \sigma \nabla v^T\right) - 2\eta_p \epsilon(v) = 0,
\end{aligned} \qquad (6)$$

starting from the predictions $v^{n-1/2}$ and $\sigma^{n-1/2}$. We then set v^n and σ^n to the obtained velocity and extra-stress at time t^n, respectively.

Two fixed grids are used to solve problems (5) and (6), see Fig. 3. A structured grid of small cubic cells (size h) is used to solve the convection problems (5), with goal to reduce numerical diffusion of the volume fraction of liquid φ as much as possible. An unstructured finite element grid with coarse tetrahedrons (size H) is used to solve problem (6). A trade-off between accuracy (h as small as possible) and computational complexity (H as large as possible) is to use $H \simeq 4h$. Since all the methods used are implicit, no stability condition occurs between the time step τ and the mesh spacing h. However, the precision depends on the CFL number (maximum velocity times τ divided by h), typical CFL numbers are between 1 and 10. The overall method is then $O(h + \tau)$. For details, we refer to [1].

4 Numerical Results

The finite element mesh is reported in Fig. 4. The length of the domain is 0.035 m, the biggest diameter is 0.012 m, the diameter of the die is 0.001 m, the finite element mesh size in the die being $H = 0.0001$ m, the size of the cells being $h = 0.000025$ m.

The physical data are $\rho = 1{,}300\,\text{kg/m}^3$, $\eta_s = 0$, $\eta_p = 1{,}500$ kg/(ms), we start with an Oldroyd-B fluid, thus $\alpha = 0$ in (4). The relaxation time λ will range from 0.001 to 1 s depending on the boundary conditions along the die, either slip or no-slip. The velocity at the inlet is such that the maximum velocity in the die is 0.05 m/s,

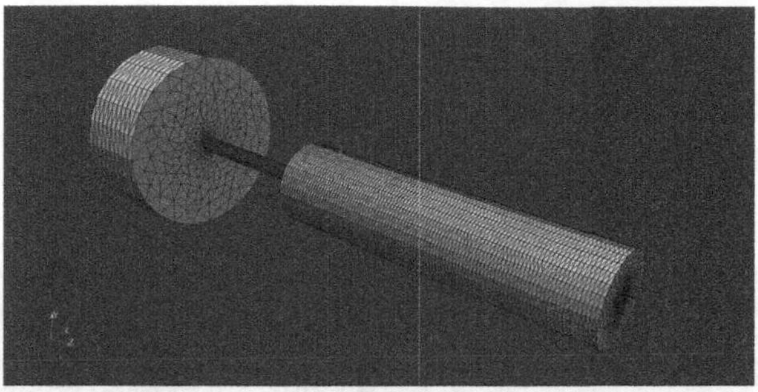

Fig. 4 The finite element mesh

thus the Weissenberg number (relaxation time times maximum velocity divided by the die diameter) ranges from 0.05 to 50. Note that the wall shear rate is not used in the definition of the Weissenberg number since it is zero when slip boundary conditions apply. At time $T = 0.8$ s the extruded jet is about to reach the end of the computational domain and the simulation is stopped, the time step is $\tau = 0.005$ s, so that the maximum CFL number is $0.05\tau/h = 10$.

We first present some numerical results when no-slip conditions apply along the die wall. In Fig. 5 the shape of the jet, the norm of the velocity and the σ_{33} component of the extra-stress are presented for $\lambda = 0.02$ s. The jet swelling can clearly be observed. The shape of the extruded jet at final time $T = 0.8$ is shown in Fig. 6 with $\lambda = 0.002$, 0.005 and 0.01 s. When $\lambda = 0.01$ s the jet buckles, which does not correspond to experiments. Indeed, experiments have shown that the jet does not buckle [3] although the measured relaxation time of pasta ranges from 0.001 to 10 s [4]; thus we conclude that the numerical results of Fig. 6 do not correspond to experiments; we believe this discrepancy can be explained by the fact that no-slip boundary conditions may not be physically relevant.

We now present numerical results with slip boundary conditions along the die, the relaxation time ranging from $\lambda = 0.1$ to 1 s. It should be stressed that when slip boundary conditions apply along the die wall, the shear stress is small since the velocity remains constant along the diameter. Also note that there is no more singularity at the die exit. The results corresponding to $\lambda = 0.1$ s are reported in Fig. 7. Clearly, looking at the extra-stress σ_{33}, there is a competition between the elastic effects due to the contraction at the entrance of the die and the extrusion at the exit of the die. However, there is no jet swelling. When $\lambda = 1$ s, the jet buckles, see Fig. 8, which again does not correspond to experiments. Finally, we have considered a Phan-Thien Tanner model instead of Oldroyd-B with $\alpha = 15$ in (4). The results are reported in Fig. 9 and show that the jet does not swell when $\lambda = 1$ s, which better fits the experiments.

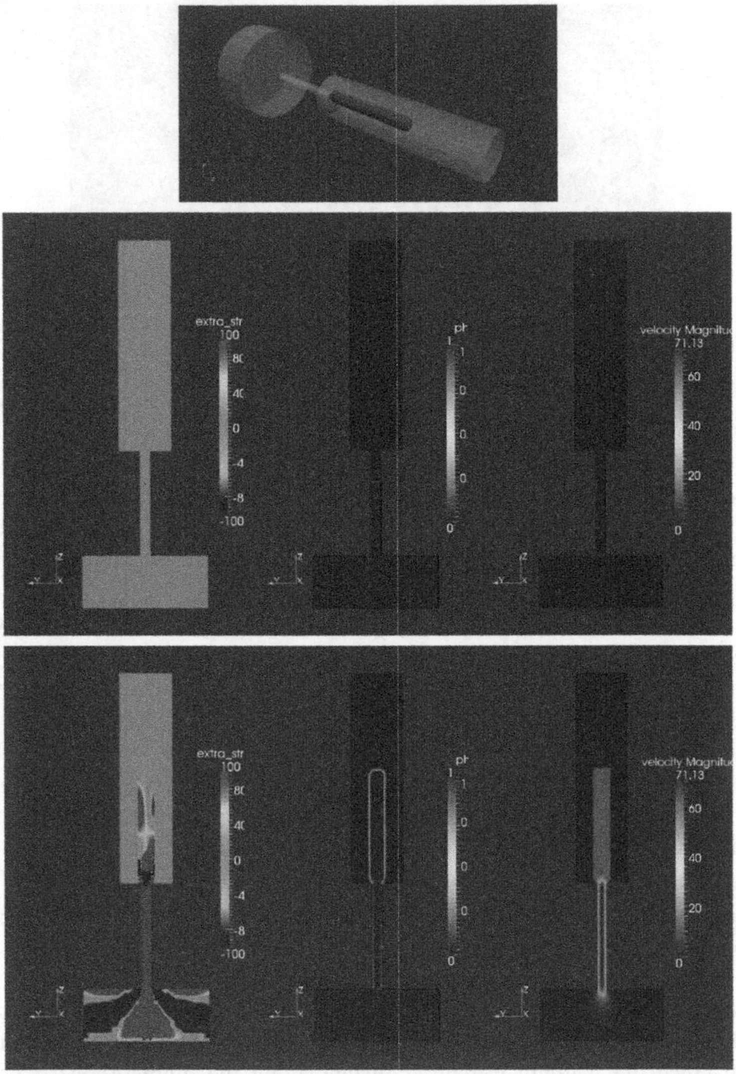

Fig. 5 Initial and final shape of the extruded jet with no-slip conditions and $\lambda = 0.002$ s

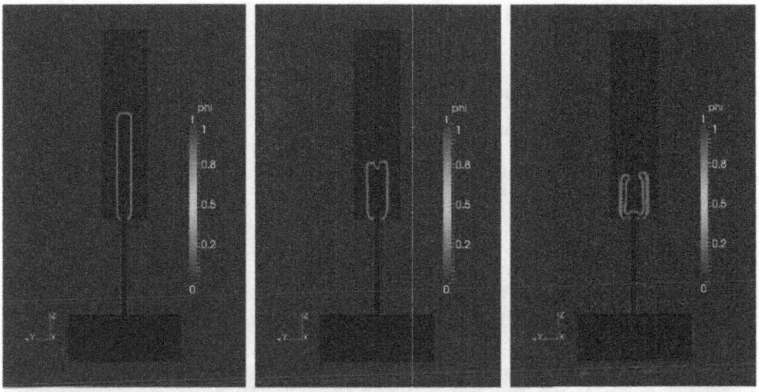

Fig. 6 Final shape of the extruded jet with no-slip conditions and $\lambda = 0.002$, 0.005 and 0.01 s

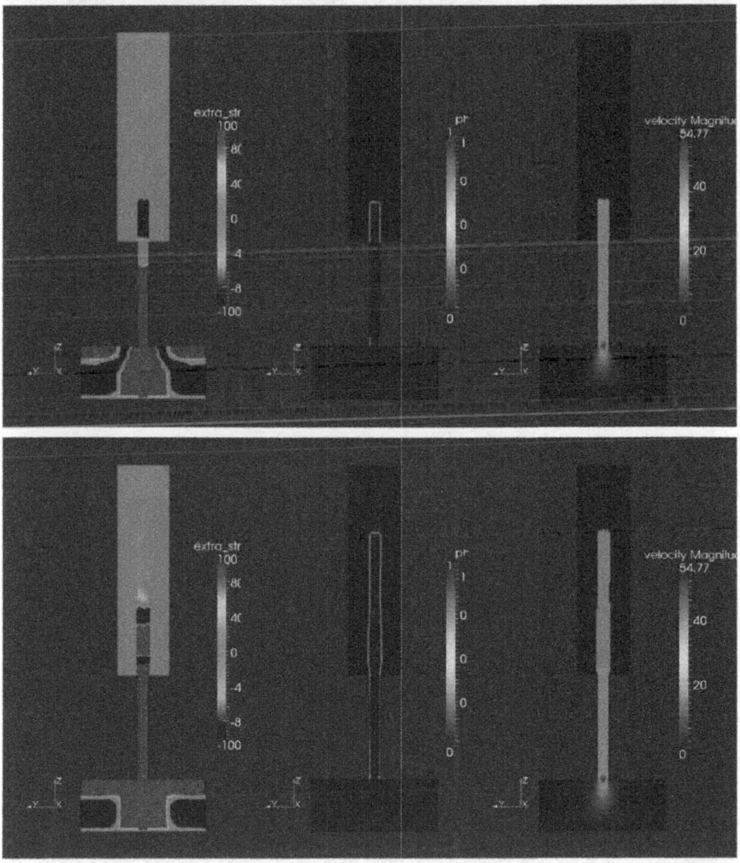

Fig. 7 Shape of the extruded jet with slip conditions and $\lambda = 0.1$ s at time $t = 0.2$ s and $t = 0.6$ s

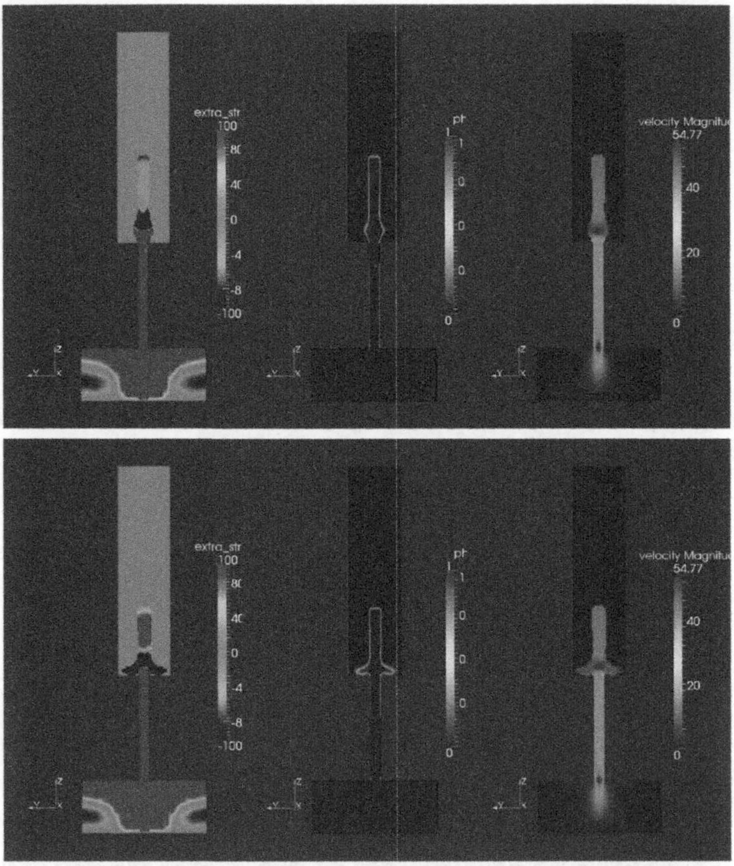

Fig. 8 Shape of the extruded jet with slip conditions and $\lambda = 1$ s at time $t = 0.5$ s and $t = 0.6$ s

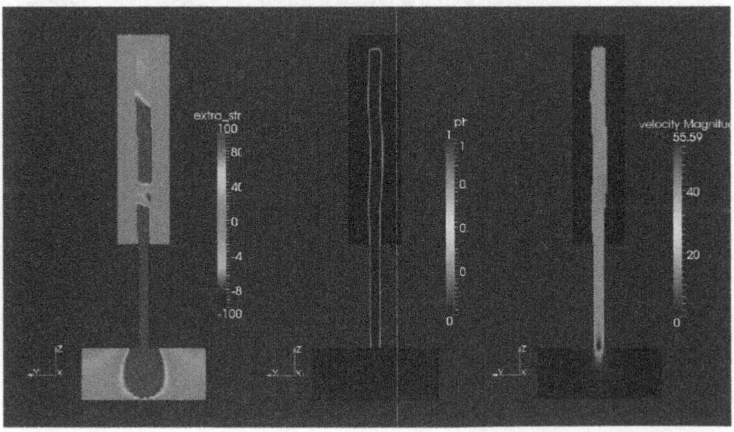

Fig. 9 Shape of the extruded jet with slip conditions, $\lambda = 1$ s and the Phan-Thien Tanner model. at time $t = 0.7$ s

Conclusions and Perspectives
We have reported numerical results corresponding to the 3D extrusion of a viscoelastic fluids with slip or no-slip boundary conditions. Numerical results indicate that the Phan-Thien Tanner model with slip boundary conditions seems to produce realistic results. We propose to investigate partial slip boundary conditions, the sliding parameter being tuned with experiments.

Acknowledgements All the computations have been performed with the cfsFlow software developed by EPFL and Ycoor Systems SA (`ycoorsystems.com`). The author would like to thank Alexandre Masserey and Gilles Steiner for support.

References

1. A. Bonito, M. Picasso, M. Laso, Numerical simulation of 3D viscoelastic flows with free surfaces. J. Comput. Phys. **215**(2), 691–716 (2006)
2. T. Coupez, Metric construction by length distribution tensor and edge based error for anisotropic adaptive meshing. J. Comput. Phys. **230**(7), 2391–405 (2011)
3. A. Kratzer, Hydration, dough formation and structure development in durum wheat pasta processing, Ph.D. thesis, ETHZ, 2011
4. A. Kratzer, S. Handschin, V. Lehmann, D. Gross, F. Escher, B. Conde-Petit, Hydration dynamics of durum wheat endosperm as studied by magnetic resonance imaging and soaking experiments. Cereal Chem. **85**(5), 660–666 (2008)
5. M. Tome, A. Castelo, V. Ferreira, S. McKee, A finite difference technique for solving the Oldroyd-B model for 3d-unsteady free surface flows. J. Non-Newtonian Fluid Mech. **154**(2–3), 179–206 (2008)

Numerical Simulation of Polymer Film Stretching

Hogenrich Damanik, Abderrahim Ouazzi, and Stefan Turek

Abstract We present numerical simulations of a film stretching process between two rolls of different temperature and rotational velocity. Film stretching is part of the industrial production of sheets of plastics which takes place after the extrusion process. The goal of the stretching of the sheet material is to rearrange the orientation of the polymer chains. Thus, the final products have more smooth surfaces and homogeneous properties. In numerical simulation, the plastic sheet is modelled geometrically as a membrane and rheologically as a polymer melt. The thickness of the membrane is not assumed to be constant but rather depends on the rheology of the polymer and the heat transfer. The rheology of the sheet material is governed by a viscoelastic fluid and is coupled to the flow model. An Λ-stable time integrator is applied to the systems in which the continuous spatial system is discretized within the FEM framework at each time step. The resulting discrete systems are solved via Newton-multigrid techniques. Moreover, a level set method is used to capture the free surface. We obtain similar results for test configurations with available results from literature and present "neck-in" as well as "dog-bone" effects.

1 Introduction

Film casting processes are widely practiced in industry. The purpose is to produce thin sheets of polymer. They are mainly used for food packaging, drugs, coating, etc. Having extruded from the die below the melting temperature, the film sheet needs to be further oriented on the molecular level to obtain more smooth material properties at desired thickness. This is done by several rolls stretching the sheet material, see Fig. 1. The first several rolls warm up the temperature of the sheet material with constant heat source. Then, the two middle rolls stretch the sheet with different velocity and temperature. The last rolls cool down the temperature of the sheet material. Industrial objectives are to improve the properties of end products

H. Damanik (✉) • A. Ouazzi • S. Turek
Institut für Angewandte Mathematik, TU Dortmund, Vogelpothsweg 87, Dortmund, Germany
e-mail: hdamanik@math.tu-dortmund.de; ouazzi@math.tu-dortmund.de; ture@featflow.de

A. Abdulle et al. (eds.), *Numerical Mathematics and Advanced Applications ENUMATH 2013*, Lecture Notes in Computational Science and Engineering 103,
DOI 10.1007/978-3-319-10705-9_70

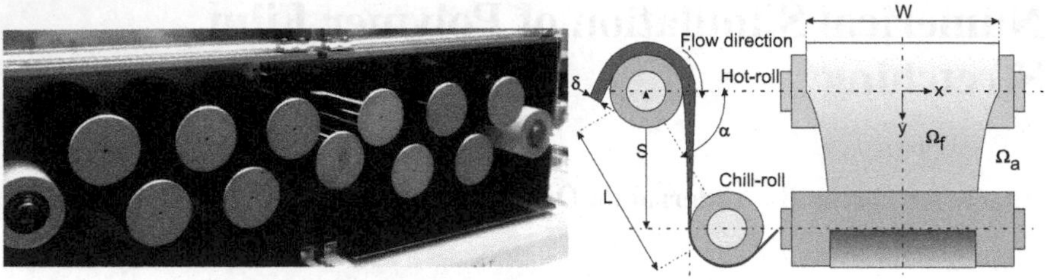

Fig. 1 *Left*: Laboratory tools (Courtesy of Schöppner, Wibbeke). *Right*: The numerical setup

at higher production rates and to reduce production cost. In reality, the higher the rate of production process is the poorer the quality of the end product becomes. The reduced qualities are well-known as "neck-in" and "dog-bone/edge-bead" effects as studied in [8] for the extrusion process.

Numerical treatment of such process has been studied in the work of [7] where a 2D membrane model is introduced together with isothermal Newtonian flow. Furthermore, a viscoelastic model is clearly of importance for the corresponding process, as shown in [13, 14].

2 Membrane Model, Viscoelasticity and Temperature Coupling

The thickness of the sheet material film, which is roughly 0.1 % of the width, makes it possible to use a 2D membrane model as described in the work of [7]. This is numerically more advantageous, but we are aware that this model may not be able to predict all viscoelastic physical phenomena such as rod-climbing, delayed die-swell, Kaye effect, siphon effect and melt-fracture, to name a few, see [10]. Following [7], the 2D membrane model starts with the assumption that inertia can be neglected and that incompressibility holds, such as

$$\begin{cases} \dfrac{\partial \mathbf{u}}{\partial t} + \nabla \cdot e\mathbf{T} + \rho \mathbf{g} e = 0 \\ \dfrac{\partial \mathbf{u}}{\partial t} + \nabla \cdot e\mathbf{u} = 0 \end{cases} \tag{1}$$

where $\mathbf{T} = 2\eta_s \operatorname{tr}(\mathbf{D})\mathbf{I} + 2\eta_s \mathbf{D} + \frac{\eta_p}{\Lambda}(\boldsymbol{\tau} - \mathbf{I})$. Here, the material parameters are density ρ, solvent viscosity η_s, polymer viscosity η_p and polymer relaxation time Λ. The later mentioned introduces the Weissenberg number $\mathsf{We} = \Lambda \frac{u_c}{l_c}$ with characteristic velocity u_c and length l_c. The thickness e appears as new unknown and the hydrostatic pressure is replaced by the trace of the 2D velocity gradient, which is a consequence of the 2D simplification of the third direction in 3D. In the presence of viscoelasticity, the conformation stress tensor $\boldsymbol{\tau}$ is governed by the following viscoelastic constitutive law,

$$\frac{\partial \boldsymbol{\tau}}{\partial t} + (\mathbf{u} \cdot \nabla)\boldsymbol{\tau} - \nabla \mathbf{u} \cdot \boldsymbol{\tau} - \boldsymbol{\tau} \cdot \nabla \mathbf{u}^T = \frac{1}{\Lambda} f(\boldsymbol{\tau}). \tag{2}$$

Depending on how one sets $f(\boldsymbol{\tau})$, many viscoelastic models as described in [9] can be included. In this study, we use the Oldroyd-B model with 59 % solvent contribution. Furthermore, the non-isothermal condition is treated via a transport-diffusion equation for the temperature

$$\frac{\partial \theta}{\partial t} + (\mathbf{u} \cdot \nabla)\theta = k_1 \nabla^2 \theta, \tag{3}$$

which influences the viscosity and the relaxation time of the fluid, for example by the well-known Arrhenius dependence

$$\eta_{s/p} = \eta_{s0/p0} \exp \frac{E}{R}\left(\frac{1}{\theta} - \frac{1}{\theta_0}\right) \text{ and } \Lambda = \Lambda_0 \exp \frac{E}{R}\left(\frac{1}{\theta} - \frac{1}{\theta_0}\right). \tag{4}$$

Here, the parameters k_1, E, R are the heat diffusion coefficient, activation of energy and ideal gas constant. The subscript zero denotes the constant value of the corresponding parameter.

The system of Eqs. (1)–(3) is solved with the following boundary conditions: Dirichlet data for the velocity and temperature at both inflow and outflow (u_{in},u_{out},θ_{in},θ_{out}), while a Neumann condition is set for the viscoelastic stress. The thickness Dirichlet data is set for the inflow only (δ). We assume that there is no force on the interface, $\mathbf{T} \cdot \mathbf{n} = 0$, where $\mathbf{n}$ is the unit normal at the interface pointing into Ω_a.

3 Multiphase Treatment

In the presence of the free surface for the above configuration, a surface tracking method is possible. However, it is more convenient to use a single mesh without having to update the mesh at every time step. Furthermore, ALE formulations make also sense since the deformation of the mesh is small. On the other hand, it is computationally cheaper, in view of non-isothermal situations, to avoid additional numerical variables from ALE formulations. Thus, a level set approach [12] is a good candidate,

$$\frac{\partial \varphi}{\partial t} + (\mathbf{u} \cdot \nabla)\varphi = 0, \tag{5}$$

to capture the free surface as implemented in [6] for multiphase viscoelastic flow. One needs to take care that the function should approximate the distance property of $\|\nabla\varphi\| = 1$ at each time step which requires an additional reinitialization procedure, also implemented in [6].

4 Numerical Treatment

The numerical strategy to deal with the multiphase character, that means where and what to solve, is based on the sign of the level set function. In the following we describe the numerical treatment via the backward Euler scheme, for simplicity. We proceed, also implemented in [3, 4] and by neglecting the gravity (also for simplicity), as follows: Given initial solutions $(\mathbf{u}^n, e^n, \boldsymbol{\tau}^n, \theta^n)$ and interface φ^n in each time step, the coupled weak formulation of the above system of equations is to find $(\mathbf{u} = \mathbf{u}^{n+1}, e = e^{n+1}, \boldsymbol{\tau} = \boldsymbol{\tau}^{n+1}, \theta = \theta^{n+1})$ for the next time step with $\Delta t = t^{n+1} - t^n$ so that

$$\frac{1}{\Delta t}\langle \mathbf{u}, \phi\rangle_\Omega - \langle e\,\mathbf{T}, \nabla\phi\rangle_\Omega + \langle (e\mathbf{T})\cdot\mathbf{n}, \phi\rangle_{\partial\Omega} = \frac{1}{\Delta t}\langle \mathbf{u}^n, \phi\rangle_\Omega \tag{6}$$

$$\frac{1}{\Delta t}\langle e, \phi\rangle_\Omega + \langle (e\nabla\cdot\mathbf{u} + (\mathbf{u}\cdot\nabla)e), \phi\rangle_\Omega = \frac{1}{\Delta t}\langle e^n, \phi\rangle_\Omega \tag{7}$$

$$\frac{1}{\Delta t}\langle \boldsymbol{\tau}, \phi\rangle_\Omega + \langle (\mathbf{u}\cdot\nabla)\boldsymbol{\tau} - \nabla\mathbf{u}\cdot\boldsymbol{\tau} - \boldsymbol{\tau}\cdot\nabla\mathbf{u}^T - \frac{1}{\Lambda}f(\boldsymbol{\tau}), \phi\rangle_\Omega = \frac{1}{\Delta t}\langle \boldsymbol{\tau}^n, \phi\rangle_\Omega \tag{8}$$

$$\frac{1}{\Delta t}\langle \theta, \phi\rangle_\Omega + \langle (\mathbf{u}\cdot\nabla)\theta, \phi\rangle_\Omega + k_1\langle \nabla\theta, \nabla\phi\rangle_\Omega = \frac{1}{\Delta t}\langle \theta^n, \phi\rangle_\Omega \tag{9}$$

with an admissible inner product $\langle\cdot,\cdot\rangle$, and with test functions $\phi \in Q_2$ as higher order finite element functions. Then, given a current solution $\mathbf{u}$, one seeks a solution for φ of the next time step via

$$\frac{1}{\Delta t}\langle \varphi, \phi\rangle_\Omega + \langle (\mathbf{u}\cdot\nabla)\varphi, \phi\rangle_\Omega = \frac{1}{\Delta t}\langle \varphi^n, \phi\rangle_\Omega \tag{10}$$

also here with $\phi \in Q_2$ as higher order finite element approximation. The material parameters are set to be level set dependent, denoting Ω_f and Ω_a, where $\Omega = \Omega_f \cup \Omega_a$ with $\Omega_f \cap \Omega_a = 0$. Next, the same redistancing procedure, as in [6], is applied to maintain the distanced property of the level set function in each time step. The process is then repeated for the next time steps.

One expects that the solution of Eq. (2) may not be smooth even at lower Weissenberg numbers [11]. This fact introduces problems with the Galerkin formulation. Unlike in the Stokes problem where a pair of Q_2P_1 FEM satisfies the so-called LBB condition for velocity-pressure [1], in the presence of viscoelasticity, the same space approximation of velocity-stress violates the inf-sup stability condition. A remedy can be obtained by adding a consistent stabilization term which penalizes the jump of the solution gradient over element edges E (with h_E denoting the length of the edge, see also [2]). This jump term "smooths" the spurious velocity components, as presented in [15], as well as the stress components, as in [4]. Thus, it avoids unnecessary numerical artifacts. This term can be written in the following form (see [15] for more details and also [4] in the case of viscoelasticity):

$$J_\sigma = \sum_{\text{edge E}} \gamma h_E^2 \int_E [\nabla \boldsymbol{\tau}] : [\nabla \phi] ds. \tag{11}$$

Regarding the numerical solvers, the obtained discrete system of Eqs. (6)–(11) is nonlinear and fully coupled. Therefore, a damped Newton interation is applied to the solution vector

$$\mathbf{x}^{n+1} = \mathbf{x}^n + \omega^n \left[\frac{\partial \mathscr{R}(\mathbf{x}^n)}{\partial \mathbf{x}} \right]^{-1} \mathscr{R}(\mathbf{x}^n), \tag{12}$$

where $\mathbf{x}$ represents the vector of coefficients corresponding to the above physical unknowns, with a damping parameter ω^n. The resulting linear system is solved via a monolithic multigrid solver, see [5].

5 Numerical Results

There exist several numerical attempts for similar problems of film casting, as for example in [7, 13, 14]. Unfortunately there is no common benchmark on this issue. Numerical parameters of the corresponding fluids model are not easily available, thus new numerical techniques are hard to validate. So, here we try the dimensionless numbers of numerical attempts from the following Table 1, which use the same geometry as the one in the work of [14] with similar fluid parameters. Qualitative comparison with reference is still possible, as shown in Fig. 2. We simulate several mesh levels (12 × 8, 24 × 16, 48 × 32, denoted as L2, L3, L4 accordingly) to be sure that the solutions are converging. The following Fig. 2 shows that our numerical results lead to converged solutions with mesh refinement. The end width of the free surface is wider than that of [14]. As a consequence, the thickness along the symmetry line is thinner than that of [14] accordingly. In general, the

Table 1 Film stretching condition

Case	Dist.(S)	(1/2 W)	Thick. (δ)	u_{in}	u_{out}	θ_{in}	θ_{out}	η_0	We
Newtonian	5	12.5	0.07	0.1	1.5	–	–	1	0
Viscoelastic	5	12.5	0.07	0.1	1.5	–	–	1	0.03

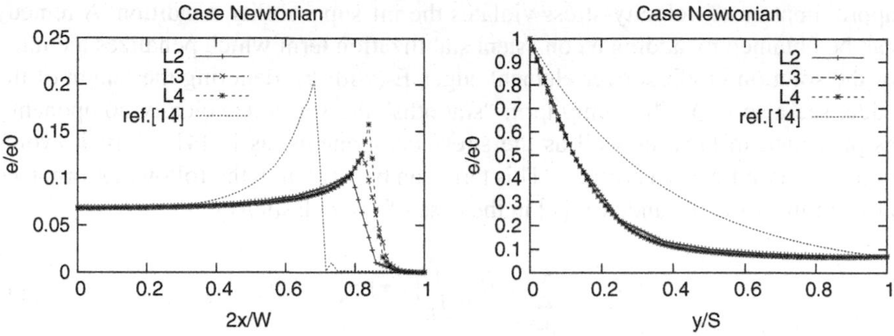

Fig. 2 Thickness profile at the chill roll and the symmetry line

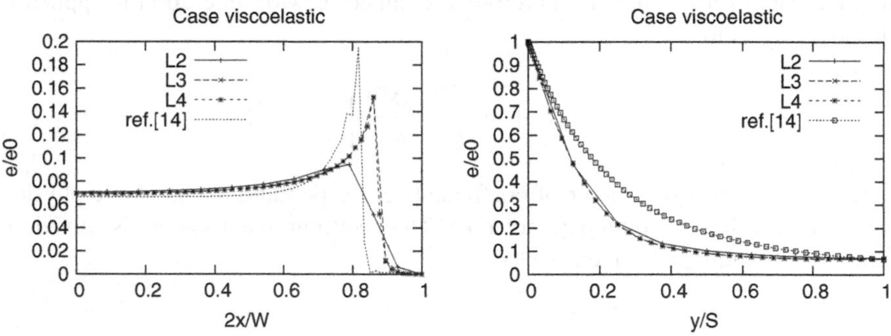

Fig. 3 Thickness profile at the chill roll and the symmetry line

Table 2 Film stretching condition of Fig. 1

Dist.(S)	(1/2 W)	Thick. (δ)	u_{in}	u_{out}	θ_{in}	θ_{out}	η_0	We	E	R	k_1
3.487	7	0.02	5/3	5	433	413	1	0.04779	45	8.31	0

results show a similar behaviour when the fluid is stretched. Here, the "dog-bone" effect is clearly visible from the left of Fig. 2. In the presence of viscoelasticity, see Fig. 3, the shear thinning effect in the direction of the elongational flow makes the end width of the film to be wider than that in the case of Newtonian which is also qualitatively shown in [14] with the Upper Convective Maxwell model. As in the Newtonian case, one sees clearly "neck-in" effects for the corresponding numerical setup.

Having qualitatively compared the numerical results, we do simulations for the setup in Fig. 1 with the conditions as in Table 2. The geometry and flow conditions are slightly adjusted from the one in Table 1. Here, the temperature of the two rolls is taken into account. The rest of the data is meant for numerical tests only. For this setup, we are quite flexible to choose the time step size ($\Delta t = 0.01$ is also used in Table 1) due to the monolithic treatment of velocity, thickness and stress. Care has to be taken that the time step size is not too big, not to deteriorate the

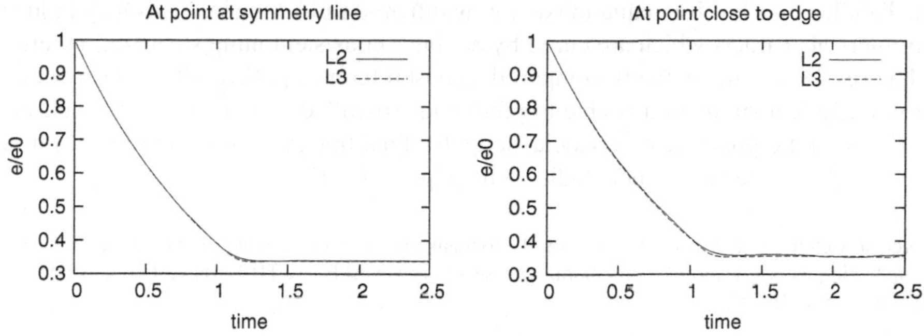

Fig. 4 Thickness evolution at two points at the chill roll (exit/take up)

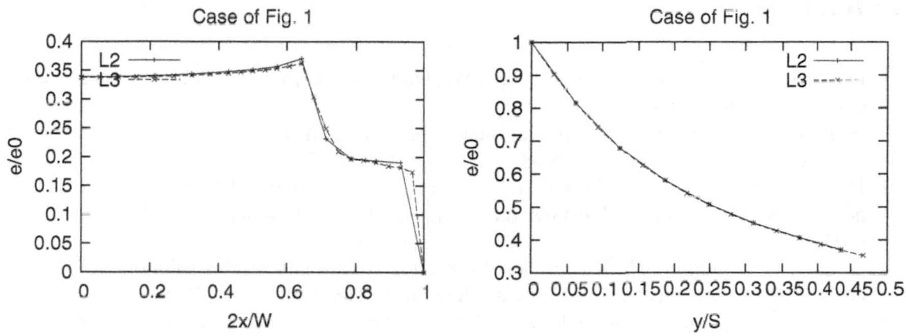

Fig. 5 Thickness profile at the chill roll (exit/take up)

decoupling of the level set function. We found that the above time step size is quite optimal in this case. One can see in Fig. 4 that the flow of the system gets steady. This is shown by the thickness evolution at two points: one in the middle of the exit, and one is close to the edge of the free surface. For two levels of computation, the solutions seem to reach mesh convergence. In the region close to the free surface, care has to be taken that the numerical parameters do not lead to additional "jumps" across the element. However this is not the case in the region (Ω_f) far from the free surface which shows clearly mesh converged solutions, see Fig. 5. Furthermore, the simulation of the complete system shows the same behaviour of "dog-bone" effect. Here, the thickness in the middle is relatively thinner than that near the free surface.

6 Summary

A membrane model for simulating the stretching of viscoelastic flow is presented. The total governing system includes the Stokes equations as well as non-isothermal viscoelastic constitutive laws and they are fully coupled. The multiphase behaviour is handled by a level set equation denoting different fluid domains. This may induce numerical artifacts which are cured by adding a consistent jump stabilization term. The results are qualitatively compared against reference [14] results and show that the numerical simulation is able to predict the so-called "dog-bone" and "neck-in" effects. In the presence of viscoelasticity, the final thickness at the chill roll shows less of these effects. Further studies will be performed.

Acknowledgement The authors would like to thank the German Reasearch Foundation (DFG) for supporting the work through collaborative research center SFB/TR TRR 30 (TP C3) and BMBF-Project 05M13RDC.

References

1. D.N. Arnold, D. Boffi, R.S. Falk, Approximation by quadrilateral finite elements. Math. Comput. **71**(239), 909–922 (2002)
2. A. Bonito, E. Burman, A continuous interior penalty method for viscoelastic flows. SIAM J. Sci. Comput. **30**, 1156–1177 (2008)
3. H. Damanik, J. Hron, A. Ouazzi, and S. Turek, A monolithic FEM–multigrid solver for non-isothermal incompressible flow on general meshes. J. Comput. Phys. **228**, 3869–3881 (2009)
4. ______, A monolithic FEM approach for the log-conformation reformulation (lcr) of viscoelastic flow problems. J. Non-Newton. Fluid Mech. **165**, 1105–1113 (2010)
5. ______, Monolithic newton-multigrid solution techniques for incompressible nonlinear flow models. Intern. J. Numer. Methods Fluids **71**, 208–222 (2012)
6. H. Damanik, A. Ouazzi, and S. Turek, *Numerical Simulation of a Rising Bubble in Viscoelastic Fluids*. Numerical Mathematics and Advanced Applications Enumath 2011 (Springer, Berlin, 2011), pp. 489–497. ISBN 978-3-642-33133-6
7. S. d'Halewyn, Y. Demay, J.F. Agassant, Numerical simulation of the cast film process. Polym. Eng. Sci. **30**, 335–340 (1990)
8. T. Dobroth, L. Erwin, Causes of edge beads in cast films. Polym. Eng. Sci. **26**, 462–467 (1986)
9. D.D. Joseph, *Fluid Dynamics of Viscoelastic Liquids*. Applied Mathematical Sciences, vol. 84 (Springer, New York, 1990)
10. R. Keunings, An algorithm for the simulation of transient viscoelastic flows with free surfaces. J. Comput. Phys. **62**, 199–220 (1986)
11. M. Renardy, A comment on smoothness of viscoelastic stresses. J. Non-Newton. Fluid Mech. **138**, 204–205 (2006)
12. J. Sethian, *Level Set Methods and Fast Marching Methods*, 2nd edn. (Cambridge University Press, Cambridge, 1999)
13. D. Silagy, Y. Demay, J.F. Agassant, Stationary and stability analysis of the film casting process. J. Non-Newton. Fluid Mech. **79**, 563–583 (1998)
14. C. Sollogoub, Y. Demay, J.F. Agassant, Non-isothermal viscoelastic numerical model of the cast-film process. J. Non-Newton. Fluid Mech. **138**, 76–86 (2006)
15. S. Turek, A. Ouazzi, Unified edge–oriented stabilization of nonconforming FEM for incompressible flow problems: numerical investigations. J. Numer. Math. **15**, 299–322 (2007)

Numerical Modelling of Viscoelastic Fluid-Structure Interaction and Its Application for a Valveless Micropump

Xingyuan Chen, Michael Schäfer, and Dieter Bothe

Abstract An implicit partitioned coupling algorithm is used to simulate and investigate the interaction between a viscoelastic fluid and an elastic structure. As a test case, a lid-driven cavity with flexible bottom is studied. It is found that the amplitude, the frequency, and the equilibrium position of the structural oscillation are different from the Newtonian case. As a potential application, a two-dimensional valveless micropump pumping a viscoelastic fluid is studied. The simulation shows that the pumping average flow rate is different when the pumping medium is a viscoelastic fluid instead of a Newtonian fluid.

1 Introduction

Micropumps have become indispensable in many biomedical applications, such as sampling and drug delivery. In these applications the transported fluids are usually non-Newtonian, in particular viscoelastic, and are flowing in deformable domains along with the interaction with elastic solids. Thus one is faced with a viscoelastic fluid-structure interaction (FSI) problem.

Among different types of micropumps, the valveless one is popular. A typical valveless micropump consists of a large chamber with an oscillating diaphragm, where FSI occurs, and two diffuser/nozzle elements at inlet and outlet, which control

X. Chen
Center of Smart Interfaces and Graduate School of Computational Engineering, Technische Universität Darmstadt, Darmstadt, Germany
e-mail: chen@csi.tu-darmstadt.de

M. Schäfer
Department of Numerical Methods in Mechanical Engineering, Technische Universität Darmstadt, Darmstadt, Germany
e-mail: schaefer@fnb.tu-darmstadt.de

D. Bothe (✉)
Department of Mathematics and Center of Smart Interfaces, Technische Universität Darmstadt, Darmstadt, Germany
e-mail: bothe@csi.tu-darmstadt.de

A. Abdulle et al. (eds.), *Numerical Mathematics and Advanced Applications ENUMATH 2013*, Lecture Notes in Computational Science and Engineering 103,
DOI 10.1007/978-3-319-10705-9_71

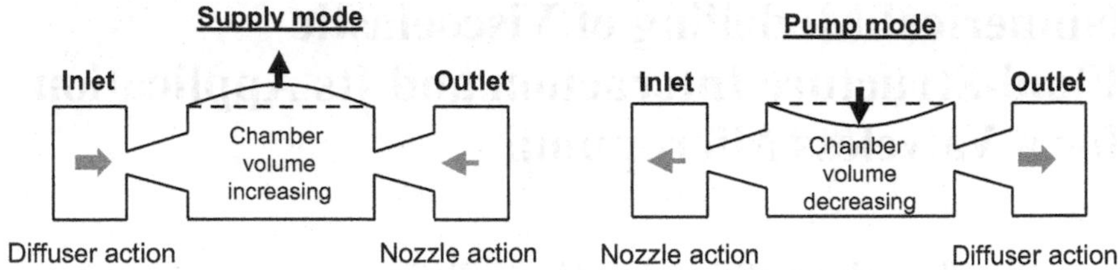

Fig. 1 Pumping principle of a valveless micropump

the flow direction. The working principle is illustrated in Fig. 1. In supply mode, the chamber volume increases, and the fluid flows into both inlet and outlet. The inlet element acts as a diffuser, which has lower pressure loss than the outlet element acting as a nozzle. Thus more volume flux comes from inlet than outlet. In pump mode, the pump works analogously.

In the past decades the simulation techniques for FSI and viscoelastic fluid flow have been developed quickly. However, not much work has been focused on the combination of them. Although there are still many challenges, e.g. achieving quantitative accurate results, in both of the problems, the developed techniques provide the possibility to numerically investigate the viscoelastic FSI problem qualitatively.

In the present work we consider an incompressible Oldroyd-B fluid interacting with an elastic structure, using an implicit partitioned coupling algorithm. As a prototype test case, a lid-driven cavity with flexible bottom is employed for a preliminary study. Then a two-dimensional valveless micropump is simulated. We show the differences in the behaviours between the Newtonian and viscoelastic FSI in the prototype test case and the different pumping performances of the micropump.

2 Governing Equations

We consider the FSI problem consisting of a fluid domain Ω_f and a structural domain Ω_s. These two domains share a common moving interface Γ. To account for the moving fluid domain, the Arbitrary-Lagrangian-Eulerian (ALE) formulation is applied to describe the fluid motion. Omitting the gravitational force, the governing equations for an incompressible fluid are

$$\nabla \cdot \mathbf{u} = 0 \quad \text{in} \quad \Omega_f \times (0, T)\,, \tag{1}$$

$$\rho^f \left.\frac{\partial \mathbf{u}}{\partial t}\right|_{\chi} + \rho^f (\mathbf{u} - \mathbf{u}^g) \cdot \nabla \mathbf{u} = -\nabla p + \nabla \cdot \boldsymbol{\tau} \quad \text{in} \quad \Omega_f \times (0, T)\,, \tag{2}$$

where ρ^f is the density of the fluid, $\mathbf{u}$ is the velocity of the fluid, $\mathbf{u}^g$ is the velocity field of the mesh, p is the pressure, and $\boldsymbol{\tau}$ is the extra stress. The term $\frac{\partial \mathbf{u}}{\partial t}\big|_\chi$ represents the time derivative of the velocity in the referential configuration (here the mesh configuration). To model the viscoelastic fluid, the Oldroyd-B model is employed. The extra stress $\boldsymbol{\tau}$ is separated into the solvent contribution $\boldsymbol{\tau}_1$ and the polymer contribution $\boldsymbol{\tau}_2$ according to $\boldsymbol{\tau} = \boldsymbol{\tau}_1 + \boldsymbol{\tau}_2$. The solvent contribution is modeled by $\boldsymbol{\tau}_1 = 2\eta_1\mathbf{D}$, where η_1 is the solvent viscosity and $\mathbf{D} = \frac{1}{2}\left(\nabla\mathbf{u} + (\nabla\mathbf{u})^\mathsf{T}\right)$. The polymer contribution fulfills the constitutive equation in a moving domain according to

$$\left.\frac{\partial \boldsymbol{\tau}_2}{\partial t}\right|_\chi + (\mathbf{u}-\mathbf{u}^g)\cdot\nabla\boldsymbol{\tau}_2 = \boldsymbol{\tau}_2\nabla\mathbf{u} + (\nabla\mathbf{u})^\mathsf{T}\boldsymbol{\tau}_2 + \frac{1}{\lambda}\left(2\eta_2\mathbf{D} - \boldsymbol{\tau}_2\right) \quad \text{in} \quad \Omega_f \times (0,T), \tag{3}$$

where η_2 is the polymer viscosity and λ is the relaxation time of the fluid. The constitutive equation can be formulated using the conformation tensor $\mathbf{C} = \lambda\boldsymbol{\tau}_2/\eta_2 + \mathbf{I}$. The evolution equation for $\mathbf{C}$ in the ALE description reads

$$\left.\frac{\partial \mathbf{C}}{\partial t}\right|_\chi + (\mathbf{u}-\mathbf{u}^g)\cdot\nabla\mathbf{C} = \mathbf{C}\nabla\mathbf{u} + (\nabla\mathbf{u})^\mathsf{T}\mathbf{C} + \frac{1}{\lambda}(\mathbf{I}-\mathbf{C})\,. \tag{4}$$

To cope with the High Weissenberg Number Problem (HWNP) in the simulation of viscoelastic fluids, there exist several stabilization approaches in the finite volume framework, as we described and compared them in [1]. In the present work, the approach of symmetry factorization of the conformation tensor [2] is applied. The constitutive equation (4) is formulated in the form of its square root $\mathbf{s} = \mathbf{C}^{1/2}$:

$$\left.\frac{\partial \mathbf{s}}{\partial t}\right|_\chi + (\mathbf{u}-\mathbf{u}^g)\cdot\nabla\mathbf{s} = \mathbf{s}\,\nabla\mathbf{u} + \mathbf{M}\mathbf{s} + \frac{1}{2\lambda}(\mathbf{s}^{-\mathsf{T}} - \mathbf{s})\,, \tag{5}$$

where $\mathbf{M}$ is an anti-symmetric tensor. The computation of the components of the matrix $\mathbf{M}$ is detailed in [2] and [1].

For the problem under consideration, there are two dimensionless numbers in the description of the fluid domain: the Reynolds number $\mathrm{Re} = \rho UL/\eta$ and the Weissenberg number $\mathrm{Wi} = \lambda U/L$, where $\eta = \eta_1 + \eta_2$ is the total viscosity, U is the characteristic velocity and L is the characteristic length.

In the structural domain, the displacement of a material point of the structure is evaluated by $\mathbf{d} = \mathbf{x} - \mathbf{X}$, where $\mathbf{x}$ represents the current position of the material point originally at $\mathbf{X}$. The balance of momentum for the solid domain Ω_s reads

$$\rho^s\frac{\partial^2\mathbf{d}}{\partial t^2} = \nabla\cdot(\mathbf{F}\,\mathbf{S}^\mathsf{T}) + \rho^s\mathbf{f}^s \quad \text{in} \quad \Omega_s \times (0,T)\,, \tag{6}$$

where $\mathbf{F} = \partial\mathbf{x}/\partial\mathbf{X}$ denotes the deformation gradient, $\mathbf{S}$ is the second Piola-Kirchhoff stress tensor, ρ^s is the density of the solid, and $\mathbf{f}^s$ are external volume forces acting on the solid. In the present investigation we consider the Saint Venant-Kirchhoff material law $\mathbf{S} = \lambda^s(\text{tr}\mathbf{E})\mathbf{I} + 2\mu^s\mathbf{E}$, with the Green-Lagrangian strain tensor $\mathbf{E} = \frac{1}{2}(\mathbf{F}^\mathsf{T}\mathbf{F} - \mathbf{I})$, as kinematic property. The parameters λ^s and μ^s are the two Lamé constants which can also be expressed with Young's modulus E and Poisson ratio ν^s by

$$E = \frac{\mu^s(3\lambda^s + 2\mu^s)}{\lambda^s + \mu^s} \quad \text{and} \quad \nu^s = \frac{\lambda^s}{2(\lambda^s + \mu^s)}. \tag{7}$$

The problem formulation is closed by prescribing suitable boundary and interface conditions. On solid and fluid boundaries Γ^s and Γ^f, standard conditions as for individual solid and fluid problems can be prescribed. For the velocities and the stresses on a fluid-solid interfaces Γ, we have

$$\mathbf{u} = \frac{\partial \mathbf{d}}{\partial t} = \mathbf{u}^b \quad \text{and} \quad (-p\mathbf{I} + \boldsymbol{\tau})\mathbf{n} = \mathbf{T}^s\mathbf{n}, \tag{8}$$

where $\mathbf{u}^b$ is the velocity of the interfaces and $\mathbf{T}^s = \mathbf{F}\,\mathbf{S}\,\mathbf{F}^\mathsf{T}/\det(\mathbf{F})$.

3 Numerical Methods

A schematic view of the implicit partitioned algorithm for the viscoelastic fluid-structure coupling is illustrated in Fig. 2. The structural part is treated by the finite-element solver FEAP (see [3]), whereas the fluid part is treated by our in-house collocated finite-volume solver FASTEST. The detailed discretization of the governing equations for viscoelastic fluid flow can be read in [1]. In the fluid solver the transfinite interpolation method is used for grid movement.

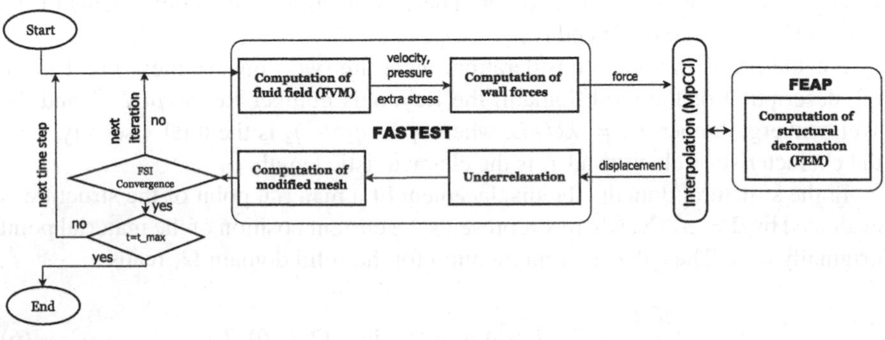

Fig. 2 Solution flowchart of the viscoelastic fluid-structure interaction

For the ALE description, a discrete form of the space conservation law

$$\frac{d}{dt}\int_{V_f} dV = \int_{S_f} \mathbf{u}^g \cdot \mathbf{n} dS \tag{9}$$

is taken into account to compute the additional convective fluxes in equations for the blocks which are moving. This is done via the swept volumes δV_c of the control volume faces for which one has the relation (see [4])

$$\sum_c \frac{\delta V_c^n}{\Delta t_n} = \frac{V_f^n - V_f^{n-1}}{\Delta t_n} = \sum_c (\mathbf{u}^g \cdot \mathbf{n} S_f)_c^n, \tag{10}$$

where the summation index c runs over the faces of the control volume, the index n denotes the time level t_n and Δt_n is the time step size. The fluid-structure interaction loop is repeated until the convergence criterion is reached. The latter is defined via the change of the mean displacements and reads as

$$\frac{1}{N}\sum_{k=1}^{N} \frac{\|\mathbf{d}^{k,m-1} - \mathbf{d}^{k,m}\|_\infty}{\|\mathbf{d}^{k,m}\|_\infty} < \varepsilon_{FSI}, \tag{11}$$

where m denotes the FSI iteration counter, N is the number of interface nodes, and $\| \cdot \|_\infty$ denotes the maximum norm. The data transfer between the flow and solid solvers within the partitioned algorithm is performed via an interface realized by the coupling library MpCCI (see [5]) which controls the data communication and carries out the interpolations of the data from the fluid and solid grids. To stabilize the coupled solution algorithm, an under-relaxation is employed, i.e., the actually computed displacements $\mathbf{d}^{act}$ are weighted with the values $\mathbf{d}^{old}$ from the preceding iteration to give the new displacement

$$\mathbf{d}^{new} = \alpha_{FSI}\mathbf{d}^{act} + (1 - \alpha_{FSI})\mathbf{d}^{old} \quad \text{with} \quad 0 < \alpha_{FSI} \leq 1. \tag{12}$$

4 Test Cases and Results

4.1 Lid-Driven Cavity with Flexible Bottom

This test case with Newtonian fluid has been investigated applying the same algorithm as in [6]. The geometry and boundary conditions of the problem are shown in Fig. 3. We use the same physical parameters as in [6]. The fluid parameters are $\rho^f = 1\,\text{kg/m}^3$ and $\eta^f = 0.01$ Pa · s for the Newtonian case. When the fluid is viscoelastic the solvent and polymer viscosities are $\eta_1 = \eta_2 = 0.005$ Pa · s. The fluid relaxation times are $\lambda = 0.5, 1, 2$ and 3 s, respectively. The corresponding Reynolds number is Re = 100 and Weissenberg numbers are Wi = 0.5, 1, 2 and 3.

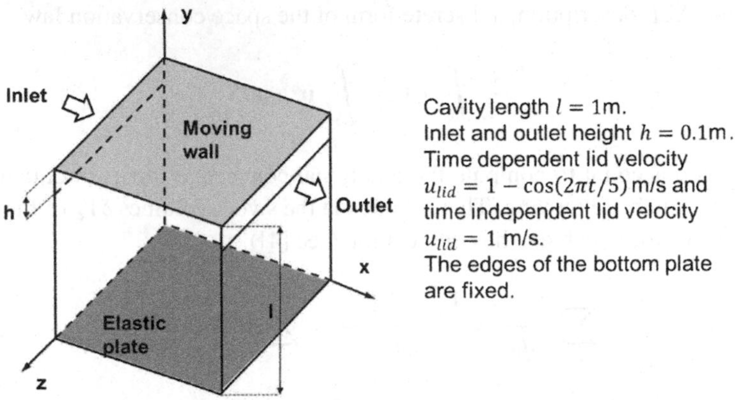

y
Inlet
Moving wall
h
Outlet
x
Elastic plate
l
z
Cavity length $l = 1$m.
Inlet and outlet height $h = 0.1$m.
Time dependent lid velocity
$u_{lid} = 1 - \cos(2\pi t/5)$ m/s and
time independent lid velocity
$u_{lid} = 1$ m/s.
The edges of the bottom plate
are fixed.

Fig. 3 Geometry and boundary conditions for lid-driven cavity with flexible bottom

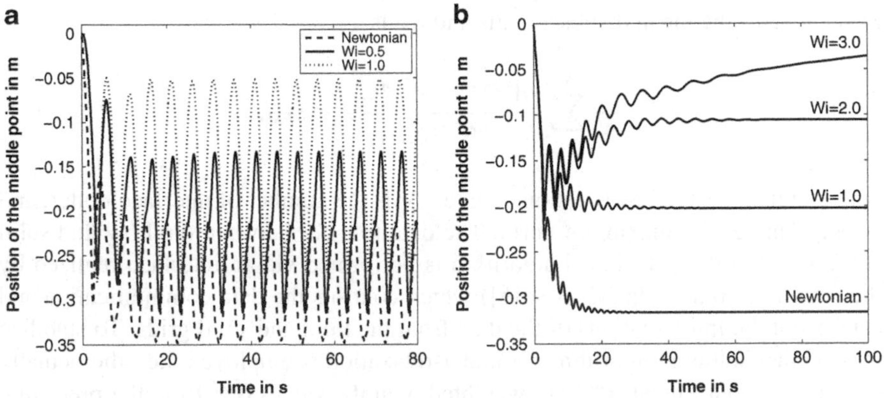

Fig. 4 Position of the plate middle point in lid-driven cavity with flexible bottom with (**a**) time dependent lid velocity (**b**) time independent velocity

The plate thickness is $d = 0.002$ m with Young's modulus $E = 250\,\text{N/m}^2$, Poisson ratio $\nu = 0$ and density $\rho_s = 500\,\text{kg/m}^3$. For the spatial discretization of the flow domain 32 control volumes in each spatial direction are employed. The plate is discretized with $20 \times 20 \times 4$ trilinear 8-node solid hexahedral solid elements. The time step size is 0.1 s. The FSI under-relaxation factor is fixed at 0.5.

The position histories of the plate middle point are shown in Fig. 4a for time dependent lid velocity and in Fig. 4b for time independent lid velocity. In both cases, the equilibrium position moves upwards and the oscillation amplitude increases, when the Weissenberg number increases. The oscillation frequency is the same as the lid velocity oscillation frequency when the lid velocity is time dependent. The frequency increases slightly when the Weissenberg number increases in case the lid velocity is time independent.

4.2 Valveless Micropump

The pumping net flow rate in a valveless micropump greatly depends on the diffuser efficiency ratio $\eta = \xi_n/\xi_d$, where ξ_n and ξ_d are the pressure loss coefficients of the nozzle and diffuser, respectively. Before simulating the whole pump, we investigate the influence of fluid relaxation time on the diffuser efficiency ratio by a pure two-dimensional CFD simulation. The geometry of the element is illustrated in Fig. 5. A constant volume flux is given at the narrow side for the diffuser direction or at the wide side for the nozzle direction. The Weissenberg number is calculated with the velocity and the height at the narrow side.

From Fig. 6a we see that for a fixed opening angle, the diffuser efficiency ratio decreases as the fluid relaxation time (Weissenberg number) increases. From Fig. 6b we see that for a fixed relaxation time (Weissenberg number) there is an optimal angle.

Considering the influence of viscoelasticity of the fluid on FSI and on the diffuser coefficient ratio of the diffuser/nozzle element, we simulate the whole two-dimensional valveless micropump. Its geometry and material properties are shown in Fig. 7a. The external pressure on the plate has amplitude $p = 2$ kPa and frequency $f = 10$ Hz. The pumping net flow rates for different fluid relaxation times are

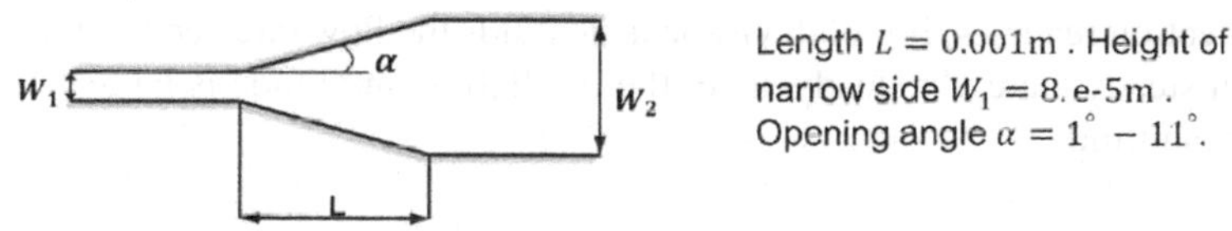

Fig. 5 Geometry of diffuser/nozzle element

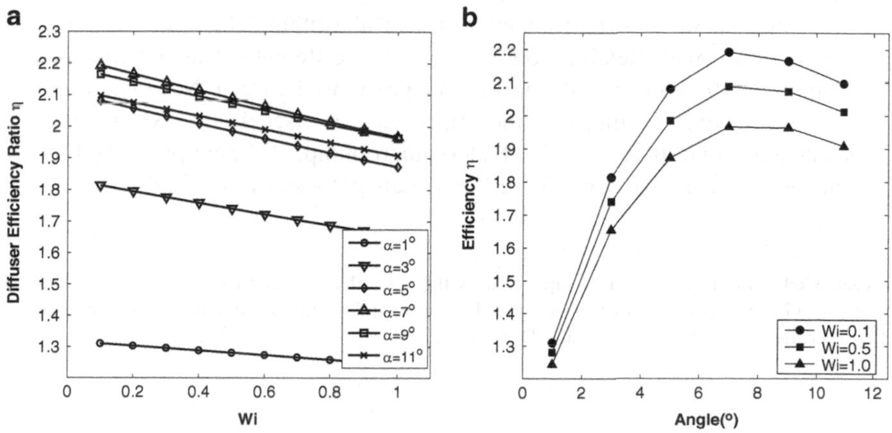

Fig. 6 Diffuser efficiency ratio of diffuser/nozzle element for different (**a**) Weissenberg numbers and (**b**) opening angles

a

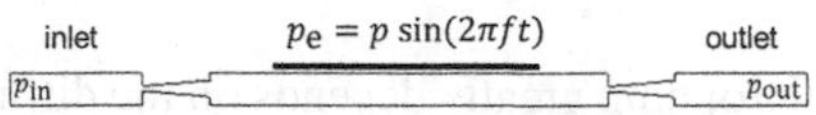

Pump geometry	
Length of inlet /outlet	3.95e-3 m
Length of chamber	7.0e-3 m
Height of chamber	1.0e-3 m

Fluid property	
Density	1000 kg/m^3
Viscosity	1.e-3 pa.s
Relaxation time	0, 0.01s and 1.0 s

Diffuser/nozzle geometry	
Length	1.e-3 m
Open angle	12.95°

Plate property	
Length	5.0e-3 m
Height	1.0e-4 m
Density	8000 kg/m^3
Elastic modulus	1.e11 pa

b

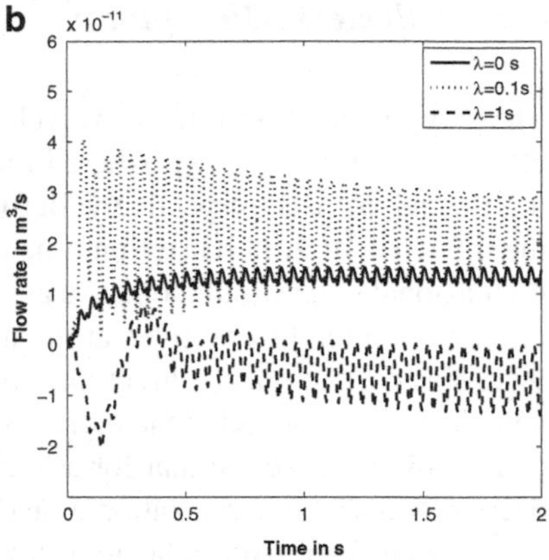

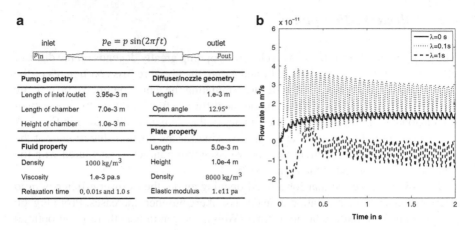

Fig. 7 (**a**) Geometry and material property and (**b**) Net flow rate for a two-dimensional valveless micropump

shown in Fig. 7b. When the fluid relaxation time changes, the average flow rate varies as expected. For viscoelastic fluid with $\lambda = 0.1$ s, the average flow rate is larger than for a Newtonian fluid. For viscoelastic fluid with $\lambda = 1$ s, the average flow rate is smaller than for Newtonian fluid and, more importantly, the average flow direction reverses. For both viscoelastic fluids the flow rate needs a longer time to reach steady state. At steady state, the oscillation amplitude is larger than for a Newtonian fluid.

Conclusion

In this work we have shown the possibility of investigating viscoelastic FSI using an implicit partitioned coupling algorithm. The study of a lid-driven cavity with flexible bottom shows the different dynamic properties (amplitude, frequency and equilibrium position) between Newtonian and viscoelastic FSI. To illustrate the importance of viscoelastic FSI, we have simulated a two-dimensional valveless micropump. Different pumping flow rates are found between pumping Newtonian and viscoelastic fluids.

Acknowledgement This work is supported by the 'Excellence Initiative' of the German Federal and State Governments, the Center of Smart Interfaces and the Graduate School of Computational Engineering at Technische Universität Darmstadt.

References

1. X. Chen, H. Marschall, H. Schäfer, D. Bothe, A comparison of stabilisation approaches for finite-volume simulation of viscoelastic fluid flow. Int. J. Comput. Fluid Dyn. **27**(6–7), 229–250 (2013)
2. N. Balci, B. Thomases, M. Renardy, C.R. Doering, Symmetric factorization of the conformation tensor in viscoelastic fluid models. J. Non-Newton. Fluid **166**(11), 546–553 (2011)
3. R.L. Taylor, FEAP - a finite element analysis program. Version 7.5 user manual. University of California, Berkeley (2004)
4. I. Demirdžić, M. Perić, Space conservation law in finite volume calculations of fluid flow. Int. J. Numer. Methods Fluids **8**(9), 1037–1050 (1988)
5. MpCCI Mesh-based parallel code coupling interface, Version 3.0.5 user guide, Fraunhofer institute for algorithms and scientific computing SCAI, Sankt Augustin (2006)
6. M. Schäfer, H. Lange, M. Heck, Implicit partitioned fluid-structure interaction coupling, in *Proceedings of 1er Colloque du GDR Interactions Fluide-Structure*, Sophia Antipolis, 2005, pp. 31–38

Numerical Investigation of Convergence Rates for the FEM Approximation of 3D-1D Coupled Problems

Laura Cattaneo and Paolo Zunino

Abstract We consider the numerical approximation of second order elliptic equations with singular forcing terms. In particular we investigate the case where a Dirac measure on a one-dimensional (1D) manifold is the forcing term for a three-dimensional (3D) problem. A partial differential equation is also defined on the manifold. The two problems are coupled by means of the intensity of the Dirac measure, which depends on both solutions. Such a problem is used to model the interaction of microcirculation and interstitial flow at the microscale, where the complicated geometrical configuration of the capillary network is taken into account. In order to facilitate the numerical discretization, the capillary bed is modeled as a collection of connected one-dimensional manifolds able to carry blood flow. We apply the finite element method (FEM) to discretize the equations in the interstitial volume and the capillary network. Because of the singular forcing terms, the solution of the coupled problem is not regular enough to apply the standard error analysis. A novel theoretical framework has been recently proposed to analyze elliptic problems with Dirac right hand sides. Using numerical experiments, in this work we investigate the validity of the available error estimates in the more general case of 3D-1D coupled problems, where the 1D problem acts as a concentrated source embedded in the surrounding volume.

1 Introduction

The microcirculation is a fundamental part of the cardiovascular system [8], because it is responsible for mass transfer from blood to organs. Theoretical models, with a variety of different approaches, help to understand and quantify the main

L. Cattaneo (✉)
MOX – Department of Mathematics – Politecnico di Milano, via Bonardi 9, 20133 Milano, Italy
e-mail: laura1.cattaneo@polimi.it; attac85@gmail.com

P. Zunino
Department of Mechanical Engineering and Materials Science, University of Pittsburgh, 3700 O'Hara Street, Pittsburgh, PA 15261, USA
e-mail: paz13@pitt.edu

A. Abdulle et al. (eds.), *Numerical Mathematics and Advanced Applications ENUMATH 2013*, Lecture Notes in Computational Science and Engineering 103,
DOI 10.1007/978-3-319-10705-9_72

mechanisms at the basis of these phenomena. Our work stems from the idea of representing vessels as one-dimensional sources embedded into the surrounding tissue, originally introduced in [6, 10–12]. We aim to study the interplay of microcirculation and interstitial flow on a space scale that is sufficiently small to clearly separate the capillary bed from the interstitial tissue. To solve the governing equations of the flow in the interstitial and capillary domains, we adopt the finite element method. More precisely, we apply the immersed finite element method [9, 13] to reduce the computational cost of the model. The capillary bed is modeled as a network of one-dimensional channels. Due to the natural leakage of capillaries, it acts as a concentrated source of flow immersed into the interstitial volume. This reduced modelling approach significantly simplifies the issues related to the simulation of the flow in the microvessels, [3–5]. However, this approach ends up with a coupled system of equations where the microvasculature is represented as a Dirac measure source term coupled with the solution in the surrounding volume. The approximation of such problem using finite elements poses significant questions with respect to the accuracy of the method. When only an L^2 control on the error is required, the a priori error estimates recently published in [7] represent a very interesting theoretical framework to analyze our problem. However, these theoretical results only apply to a two-dimensional (2D) elliptic problem with Dirac source terms, while our model turns out to be a 3D-1D coupled system of elliptic equations. After summarizing the problem formulation, in this work we aim at investigating whether the convergence properties proved in [7] are valid in a more general case, where the Dirac measure term is in turn coupled to the solution in the surrounding volume through a partial differential equation.

2 Model Set Up

Following the approach presented in [1, 2], we consider a domain Ω that is decomposed into two parts, Ω_v and Ω_t, the capillary bed and the tissue interstitium, respectively. Assuming that the capillaries can be described as cylindrical vessels, we denote with Γ the outer surface of Ω_v, with R its radius and with Λ the centerline of the capillary network, as reported in Fig. 1. We consider R constant and any physical quantity of interest, such as the blood pressure p and the blood velocity $\mathbf{u}$, is a function of space, being $\mathbf{x} \in \Omega$ the spatial coordinates, and time t. These quantities obey to different balance laws, depending on the portion of the domain of interest and, in general, they are not continuous at the interface between subdomains. We consider the tissue interstitium Ω_t as an isotropic porous medium, such as the Darcy's law applies. To set up the microcirculation model we rely on the following assumptions: (i) the displacement of the capillary walls can be neglected, because the pressure pulsation at the level of capillaries is small; (ii) the convective effects can be neglected, because the flow in each capillary is slow; (iii) the flow almost instantaneously adapts to the changes in pressure at the network boundaries, because the resistance of the network is large with respect to its inductance. This means that

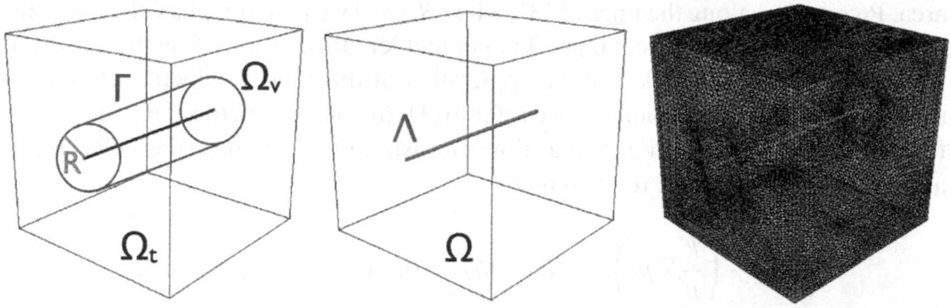

Fig. 1 Starting from the left: interstitial tissue with one embedded capillary; reduction from 3D to 1D description of the capillary vessel; computational mesh used to define the reference solution p_t

the quasi-static approximation is acceptable. As a result of that, the blood flow along each branch of the capillary network can be described by means of Poiseuille's law for laminar stationary flow of incompressible viscous fluid through a cylindrical tube with radius R. As a consequence of all these modelling assumptions, the fluid problem in the entire domain Ω reads as follows:

$$\begin{cases} \nabla \cdot \mathbf{u}_t = 0, & \mathbf{u}_t = -\dfrac{k}{\mu}\nabla p_t \qquad \text{in } \Omega_t \\ -\dfrac{\pi R^4}{8\mu}\Delta p_v = 0, & \mathbf{u}_v = -\dfrac{R^2}{8\mu}\nabla p_v \quad \text{in } \Omega_v \end{cases} \tag{1}$$

where μ and k denote the dynamic blood viscosity and the constant tissue permeability, respectively. At the interface $\Gamma = \partial\Omega_v \cap \partial\Omega_t$ we impose continuity of the flow:

$$\mathbf{u}_t \cdot \mathbf{n} = L_p(p_v - p_t) \quad \mathbf{u}_t \cdot \boldsymbol{\tau} = 0, \qquad \text{on } \Gamma \tag{2}$$

where $\mathbf{n}$ is the outward unit vector normal to the capillary surface. The fluid flux across the capillary wall can be obtained on the basis of the Starling law and L_p represents the hydraulic conductivity of the vessel wall. Problem (1) represents a simpler setting than the one studied in [1], where the role of the lymphatic system was also taken into account. However, this simplification will not compromise the generality of the results addressed later on. Finally, to be uniquely solvable, problem (1) must be complemented by boundary conditions on $\partial\Omega_t$ and $\partial\Omega_v$. We will fix it for our numerical example in Sect. 3. To avoid solving the complex three-dimensional (3D) geometry of the capillary network, we exploit the immersed boundary method combined with the assumption of large aspect ratio between vessel radius and capillary axial length. More precisely, we apply a suitable rescaling of the equations and let the capillary radius, R, go to zero. By this way, we replace the immersed interface and the related interface conditions with an equivalent mass source. We denote with f the flux released by the surface Γ, which is a flux per unit area. Proceeding along the lines of [3], when $R \to 0$ we aim to replace the mass flux per unit area by an equivalent mass flux per unit length, distributed on the centerline Λ of the capillary network. For the application of this model reduction strategy to the particular case of problem (1) we refer to [1]. In conclusion, the coupled problem for microcirculation and interstitial flow consists to find the pressure fields p_t, p_v and the velocity fields $\mathbf{u}_t$, $\mathbf{u}_v$ such that

$$\begin{cases} -\nabla \cdot \left(\frac{k}{\mu} \nabla p_t \right) - f(p_{t/v})\delta_\Lambda = 0, \ \mathbf{u}_t = -\frac{k}{\mu}\nabla p_t & \text{in } \Omega \\ -\frac{\pi R^4}{8\mu}\frac{\partial^2 p_v}{\partial s^2} + f(p_{t/v}) = 0, \ \mathbf{u}_v = -\frac{R^2}{8\mu}\frac{\partial p_v}{\partial s}\boldsymbol{\lambda} & s \in \Lambda \end{cases} \tag{3}$$

where the term $f(p_{t/v})$ accounts for the blood flow leakage from vessels to tissue and it has to be understood as the Dirac measure concentrated on Λ and having line density f. The expression of $f(p_{t/v})$ is provided using the Starling equation (2), suitably rescaled as discussed above,

$$f(p_{t/v}) = 2\pi R L_p (p_v - \overline{p}_t) \quad \text{with} \quad \overline{p}_t(s) = \frac{1}{2\pi R}\int_0^{2\pi} p_t(s,\theta) R d\theta. \tag{4}$$

3 Numerical Approximation

For complex geometrical configurations explicit solutions of problem (3) are not available. Numerical simulations are the only way of applying the model to real cases. Moreover, the solution of problem (3) does not satisfy standard regularity estimates, because the forcing term of Eq. (3)(*a*) is a Dirac measure. To characterize the regularity of the trial and test spaces we do not resort to weighted Sobolev spaces, as proposed in [4]. Indeed, that approach to analyze and discretize the problem naturally ends up with error estimates requiring finite element approximation on graded meshes, a necessary condition to capture the solution gradients in the neighborhood of the singularity. Here, we investigate the validity of weaker error estimates, which provide control on the approximation error under less restrictive requirements on the scheme. This objective is achieved by following [7], where the error analysis is based on the fact that p_t does not belong to H^1 but $p_t \in W_0^{1,p}$, $p \in [1, 3 - d/2)$ holds instead. Then, we consider the solution p_t of (3) as an element of $W_0^{1,p}$, $p \in [1, 3 - d/2)$ and we define our test space $W_0^{1,q}$ as:

$$W_0^{1,q} = \{ v \in W^{1,q} : v = 0 \text{ on } \partial\Omega \}, \qquad \frac{1}{p} + \frac{1}{q} = 1$$

The discretization of problem (3) is achieved by means of the finite element method that arises from the variational formulation of the problem, obtained by multiplying the first equation by a test function $q_t \in W_0^{1,q}$ and integrating over Ω. We choose test functions for the pressure field on the capillary bed that are continuous on the entire network, namely $q_v^h \in V_{v,0} \subset C^0(\Lambda)$ on 1D manifolds and we obtain

$$\kappa_v\big(\partial_s p_v, \partial_s q_v\big)_\Lambda + 2\pi R L_p\big(p_v - \overline{p}_t, q_v\big)_\Lambda = \big(p_{v,0}, q_v\big)_\Lambda, \ \forall q_v \in V_{v,0}.$$

Then, the weak formulation of (3) consists to find $p_t \in W_0^{1,p}$, and $p_v \in V_{v,0}$ such that,

$$\begin{cases} a_t(p_t, q_t) + b_\Lambda(\overline{p}_t, q_t) = F_t(q_t) + b_\Lambda(p_v, q_t), \ \forall q_t \in W_0^{1,q}, , \\ a_v(p_v, q_v) + b_\Lambda(p_v, q_v) = F_v(q_v) + b_\Lambda(\overline{p}_t, q_v), \ \forall q_v \in V_{v,0}, \end{cases} \quad (5)$$

with the following bilinear forms and right hand sides,

$$a_t(p_t, q_t) := \kappa_t\big(\nabla p_t, \nabla q_t\big)_\Omega, \ a_v(p_v, q_v) := \kappa_v\big(\partial_s p_v, \partial_s q_v\big)_\Lambda,$$
$$b_\Lambda(p_v, q_v) := 2\pi R L_p\big(p_v, q_v\big)_\Lambda,$$
$$F_t(q_t) := 0, \ F_v(q_v) := \big(p_{v,0}, q_v\big)_\Lambda.$$

where $\kappa_t = k/\mu$ and $\kappa_v = \pi R^4/(8\mu)$.

The main advantage of the reduced model formulation (3) is that at the discrete level the partition of the domains Ω and Λ into elements are completely independent. We denote with $\mathcal{T}_t^h$ an admissible family of partitions of Ω into tetrahedrons $K \in \mathcal{T}_t^h$, where the apex h denotes the mesh characteristic size. Let $V_t^h := \{v \in C^0(\Omega) : v|_K \in \mathbb{P}^1(K), \ \forall K \in \mathcal{T}_t^h\}$ be the space of piecewise linear continuous finite elements on $\mathcal{T}_t^h$. For the discretization of the capillary bed, we partition each branch Λ_i of Λ into a sufficiently large number of linear segments E, whose collection is Λ_i^h, which represents a finite element mesh on a one-dimensional manifold. Then, we will solve our equations on $\Lambda^h := \cup_{i=1}^N \Lambda_i^h$ that is a discrete model of the true capillary bed. Let $V_{v,i}^h := \{v \in C^0(\Lambda_i) : v|_E \in \mathbb{P}^1(E), \ \forall E \in \Lambda_i^h\}$ be the piecewise linear and continuous finite element space on Λ_i. The numerical approximation of the equation posed on the capillary bed is then achieved using the space $V_v^h := \big(\cup_{i=1}^N V_{v,i}^h\big) \cap C^0(\Lambda)$. We observe that the continuity of the discrete pressure at the junctions of the network is enforced by construction, by means of the approximation space. More precisely, we will use $V_{v,0}^h$, that is the restriction of V_v^h to functions that vanish on the boundary of Λ, to enforce essential boundary conditions on the pressure, at the inflow and outflow sections of the capillary bed. The mesh characteristic size is denoted with a single parameter h, because we will proportionally refine both finite element spaces V_t^h, V_v^h. The discrete problem arising from (5) requires to find $p_t^h \in V_t^h$ and $p_v^h \in V_{v,0}^h$ such that

$$\begin{cases} a_t(p_t^h, q_t^h) + b_{\Lambda^h}(\overline{p}_t^h, q_t^h) = F_t(q_t^h) + b_{\Lambda^h}(p_v^h, q_t^h), \ \forall q_t^h \in V_t^h, \\ a_v(p_v^h, q_v^h) + b_{\Lambda^h}(p_v^h, q_v^h) = F_v(q_v^h) + b_{\Lambda^h}(\overline{p}_t^h, q_v^h), \ \forall q_v^h \in V_{v,0}^h, \end{cases} \quad (6)$$

where the bilinear forms $a_t(\cdot,\cdot)$, $a_v(\cdot,\cdot)$, $b_\Lambda(\cdot,\cdot)$ are the same as before, with the only difference that $b_{\Lambda^h}(\cdot,\cdot)$ is now defined over the discrete representation of the network Λ_h. In particular, the evaluation of the bilinear forms $b_{\Lambda^h}(\overline{p}_t^h, q_t^h)$, $b_{\Lambda^h}(\overline{p}_t^h, q_v^h)$ involves interpolation and average operators. For every node $s_k \in \Lambda^h$ we define $\mathscr{T}_\gamma^h(s_k)$ as the discretization of the perimeter of the vessel, denoted by $\gamma(s_k)$. For simplicity, we assume that $\gamma(s_k)$ is a circle of radius R defined on the orthogonal plane to Λ^h at point s_k. After restricting p_t^h onto the nodes of $\mathscr{T}_\gamma^h(s_k)$, we calculate $\overline{p}_t^h$ using the composite trapezoidal formula. The related quadrature error will be neglected in the forthcoming convergence analysis, because the partition $\mathscr{T}_\gamma^h(s_k)$ is significantly finer than $\mathscr{T}_t^h$ and Λ_h. Prescribed boundary conditions are a pressure drop along the capillary and an imposed pressure value on the outer tissue domain, that is, $p_v(s = in) = 1$, $p_v(s = out) = 0.5$ and $p_t = p_0 = 0$ on $\partial\Omega$. The tissue domain is a cube of side $L = 50\,\mu\text{m}$ and the fixed radius of the capillary is $R = 7.5\,\mu\text{m}$. We solve the non-dimensional form of (6), fixing the characteristic length $d = 50\,\mu\text{m}$, so that $L = 1$ and $R = 0.15$. For the values of all the other equation parameters we refer to [1].

4 Error Analysis, Numerical Results and Discussion

The solution of the problem (5) is characterized by a low regularity. For this reason, studying the convergence properties of (5) to (6) is a challenging task. A novel approach for the a priori error analysis of an elliptic problem with a Dirac measure source term has recently been developed in [7], where the authors derive a quasi-optimal a priori estimate for first order finite elements approximation and optimal error bounds for higher order approximations, on a family of quasi-uniform meshes in a L^2-seminorm. Graded meshes are no longer needed to achieve optimality in this new theoretical context. The aim of this work is to perform numerical experiments to investigate whether the same a priori estimates are still valid for coupled problems such as (3). For the sake of clarity, we report below the main results of [7], adapted to this particular case.

Let $p_t \in W_0^{1,p}$ the weak solution of (5) *and let $p_t^h \in V_t^h$ be the finite element approximation given by* (6). *Then, the following upper bound for the L^2-error holds:*

$$\|p_t - p_t^h\|_{L^2(\Omega\setminus\Omega_v)} \lesssim h^2|\log h|. \tag{7}$$

We investigate with numerical experiments the validity of (7) in the model introduced above. Since we do not know an analytic expression for the exact solution p_t of the considered problem, in order to evaluate the norm of the error (7), we construct a reference solution on a very fine mesh, reported in Fig. 1. We fix in this case $h = 1/40$. From now on, we will identify the exact solution p_t with the numerical solution p_t^h obtained on this fine mesh. Then we compute the numerical solution p_t^h on coarser meshes, fixing $h = 1/20$, $h = 1/10$ and $h = 1/5$

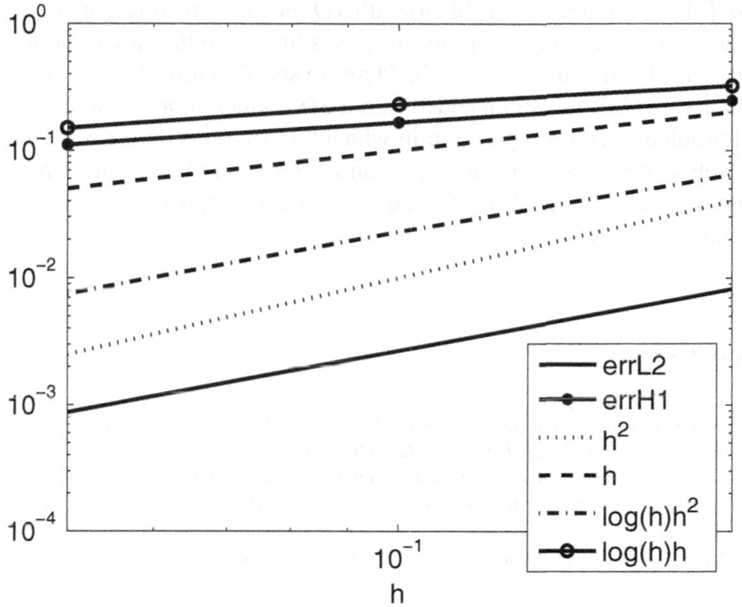

Fig. 2 L^2 and H^1 seminorm error decay

respectively. We compute the L^2 norm of the error, $\|p_t - p_t^h\|_{L^2(\Omega\setminus\Omega_v)}$, fixing Ω_v as the cylinder of radius R surrounding the vessel and using the three-dimensional composite trapezoidal formula based on the vertices of the fine mesh ($h = 1/40$). This is a second-order formula which does not pollute the convergence rate of the method. For the sake of completeness we also assess the error using the H^1-seminorm. The numerical results are reported in Fig. 2. These tests confirm that error bound (7) is still valid for (5). Furthermore, the following behavior of the H^1-seminorm error is observed:

$$\|p_t - p_t^h\|_{H^1(\Omega\setminus\Omega_v)} \lesssim h|\log h|. \tag{8}$$

Although still preliminary, these findings are encouraging at multiple levels. From the standpoint of applications, they confirm that graded meshes are not required to accurately approximate the pressure field in the volume outside Ω_v, where it has a precise physical meaning according to problem (1). A standard finite element formulation, as in (6), combined with suitably refined quasi-uniform partition, $\mathscr{T}_t^h$, will capture the main features of the pressure field, as prescribed by (7). As a result, the 3D to 1D model reduction technique discussed here turns out to be a very effective approach, which brings significant simplifications to handle the network of capillaries at the computational level, without compromising the accuracy of the discretization method. From the theoretical standpoint, we believe that these results shed light on possible directions to extend the analysis of [7]. The results of [7] apply to elliptic problems with a Dirac measure as forcing term and the numerical evidence in support of the analysis is limited to the case of point sources distributed in a bi-dimensional domain. Here we study a more challenging problem, in which a 1D manifold is embedded into a 3D domain. Furthermore, we solve a coupled problem, namely a problem in which the solutions on the 3D domain and 1D manifold depend on each other. Also, an H^1 error estimate seems to be satisfied too. Ongoing work is therefore oriented to extend the analysis of [7] to a problem setting equivalent to (5).

References

1. L. Cattaneo, P. Zunino, Computational models for fluid exchange between microcirculation and tissue interstitium, Netw. Heterog. Media **9**(1), 135–159 (2014)
2. L. Cattaneo, P. Zunino, A computational model of drug delivery through microcirculation to compare different tumor treatments, Int. J. Numer. Meth. Biomed. Engng. (2014) DOI: 10.1002/cnm.2661
3. C. D'Angelo, Multiscale modeling of metabolism and transport phenomena in living tissues, Phd thesis, 2007
4. C. D'Angelo, Finite element approximation of elliptic problems with dirac measure terms in weighted spaces: applications to one- and three-dimensional coupled problems. SIAM J. Numer. Anal. **50**(1), 194–215 (2012)
5. C. D'Angelo, A. Quarteroni, On the coupling of 1D and 3D diffusion-reaction equations. Application to tissue perfusion problems. Math. Models Methods Appl. Sci. **18**(8), 1481–1504 (2008)
6. K.O. Hicks, F.B. Pruijn, T.W. Secomb, M.P. Hay, R. Hsu, J.M. Brown, W.A. Denny, M.W. Dewhirst, W.R. Wilson, Use of three-dimensional tissue cultures to model extravascular transport and predict in vivo activity of hypoxia-targeted anticancer drugs. J. Natl. Cancer Inst. **98**(16), 1118–1128 (2006)
7. T. Koeppl, B. Wohlmuth, Optimal a priori error estimates for an elliptic problem with Dirac right-hand side. SIAM J. Numer. Anal. **52**(4), 1753–1769 (2014)
8. J. Lee, T.C. Skalak, *Microvascular Mechanics: Hemodynamics of Systemic and Pulmonary Microcirculation* (Springer, New York, 1989)
9. W.K. Liu, Y. Liu et al., Immersed finite element method and its applications to biological systems. Comput. Methods Appl. Mech. Eng. **195**(13–16), 1722–1749 (2006)
10. T.W. Secomb, R. Hsu, R.D. Braun, J.R. Ross, J.F. Gross, M.W. Dewhirst, Theoretical simulation of oxygen transport to tumors by three-dimensional networks of microvessels. Adv. Exp. Med. Biol. **454**, 629–634 (1998)
11. T.W. Secomb, R. Hsu, E.Y.H. Park, M.W. Dewhirst, Green's function methods for analysis of oxygen delivery to tissue by microvascular networks. Ann. Biomed. Eng. **32**(11), 1519–1529 (2004)
12. Q. Sun, G.X. Wu, Coupled finite difference and boundary element methods for fluid flow through a vessel with multibranches in tumours. Int. J. Numer. Methods Biomed. Eng. **29**(3), 309–331 (2013)
13. L. Zhang, A. Gerstenberger, X. Wang, W.K. Liu, Immersed finite element method. Comput. Methods Appl. Mech. Eng. **193**(21–22), 2051–2067 (2004)

The Interaction of Compressible Flow and an Elastic Structure Using Discontinuous Galerkin Method

Adam Kosík, Miloslav Feistauer, Martin Hadrava, and Jaromír Horáček

Abstract In this paper we are concerned with the numerical simulation of the interaction of fluid flow and an elastic structure in a 2D domain. For each individual problem we employ the discretization by the discontinuous Galerkin finite element method (DGM). We describe the application of the DGM to the problem of compressible fluid flow in a time-dependent domain and also to the dynamic problem of the deformation of an elastic body. Finally, we present our approach to the coupling of these two independent problems: both are solved separately at a given time instant, but we require the approximate solutions to satisfy certain transient conditions. These transient conditions are met through several inner iterations. In each iteration a calculation of both the elastic body deformation problem and the problem of the compressible fluid flow is performed. The presented method can be applied to solve a selection of problems of biomechanics and aviation. Our numerical experiments are inspired by the simulation of airflow in human vocal folds, which implies the choice of the properties of the flowing fluid and the material properties of the elastic body. The results are post-processed in order to get a visualization of the approximate solution. We are especially interested in the visualization of the elastic body deformation and the visualization of some chosen physical quantities of the flow.

A. Kosík (✉) • M. Feistauer • M. Hadrava
Charles University in Prague, Faculty of Mathematics and Physics, Sokolovská 83, 186 75 Praha 8, Czech Republic
e-mail: adam.kosik@atlas.cz; feist@karlin.mff.cuni.cz; martin@hadrava.eu

J. Horáček
Institute of Thermomechanics, The Academy of Sciences of the Czech Republic, v. v. i., Dolejškova 1402/5, 182 00 Praha 8, Czech Republic
e-mail: jaromirh@it.cas.cz

A. Abdulle et al. (eds.), *Numerical Mathematics and Advanced Applications ENUMATH 2013*, Lecture Notes in Computational Science and Engineering 103,
DOI 10.1007/978-3-319-10705-9_73

1 Introduction

The area of the fluid-structure interaction includes many specific problems, which (after a reasonable simplification) can be described by a mathematical model of partial differential equations and subsequently solved by appropriate numerical methods. The problem of our interest consists of the fluid-structure interaction where an elastic structure changes the domain of the fluid flow due to its deformation. We especially focus on airflow problems, which are described by a compressible viscous flow model.

The applicability of the discontinuous Galerkin method (DGM) to simulation of compressible viscous flow is now well-known, but the method also seems to be a good choice when solving the problem of dynamic elasticity, see, e.g., [4]. The coupled problem is solved by the so-called staggered approach, which means that both problems are solved at a given time instant separately. The approximate solutions are required to fulfill certain transient conditions, which are met through several inner iterations.

In this paper the application of the DGM to both problems is described. It is applied to the spatial discretization of both problems. The time discretization is based either on finite-difference methods or on the space-time discontinuous Galerkin method (STDGM). The STDGM applies the main concept of the DGM (piecewise polynomial but in general discontinuous approximation) both to the time and to the space semi-discretizations.

Flow of viscous compressible fluid is described by the Navier-Stokes equations. The time-dependent computational domain and a moving grid are taken into account employing the arbitrary Lagrangian-Eulerian (ALE) formulation of the Navier-Stokes equations.

2 Mathematical Model

Let us first define the problem of compressible flow in a time-dependent bounded domain $\Omega_t \subset \mathbb{R}^2$ with $t \in [0, T]$. The boundary of Ω_t is formed by three disjoint parts: $\partial\Omega_t = \Gamma_I \cup \Gamma_O \cup \Gamma_{W_t}$, where Γ_I is the inlet, Γ_O is the outlet and Γ_{W_t} represents impermeable fixed or elastic walls.

The time dependence of the domain Ω_t is taken into account with the aid of the *Arbitrary Lagrangian-Eulerian* (ALE) method, see, e.g., [5]. The method is based on a regular one-to-one ALE mapping $\mathcal{A}_t$ of the reference configuration Ω_0 onto the current configuration Ω_t, i.e. $\mathcal{A}_t : \overline{\Omega}_0 \longrightarrow \overline{\Omega}_t$, where $\boldsymbol{X} \in \overline{\Omega}_0 \longmapsto \boldsymbol{x} = \boldsymbol{x}(\boldsymbol{X}, t) = \mathcal{A}_t(\boldsymbol{X}) \in \overline{\Omega}_t$. We define the domain velocity both in the reference and the current configuration:

$$\tilde{z}(\boldsymbol{X}, t) = \frac{\partial}{\partial t}\mathcal{A}_t, \quad t \in [0, T], \ \boldsymbol{X} \in \Omega_0, \tag{1}$$

$$z(\boldsymbol{x}, t) = \tilde{z}(\mathcal{A}_t^{-1}(\boldsymbol{x}), t), \ t \in [0, T], \ \boldsymbol{x} \in \Omega_t$$

and the so-called ALE derivative of the state vector function $w = w(x,t)$ defined for $x \in \Omega_t$ and $t \in [0,T]$:

$$\frac{D^{\mathcal{A}}}{Dt} w(x,t) = \frac{\partial \tilde{w}}{\partial t}(X,t), \tag{2}$$

where $\tilde{w}(X,t) = w(\mathcal{A}_t(X),t)$, $X \in \Omega_0$, $X \in \Omega_0$ and $x = \mathcal{A}_t(X)$. Using the chain rule we are able to express the ALE derivative in the form

$$\frac{D^{\mathcal{A}} w_i}{Dt} = \frac{\partial w_i}{\partial t} + \operatorname{div}(z w_i) - w_i \operatorname{div} z, \quad i = 1, \ldots, 4. \tag{3}$$

Application of (3) to the continuity equation, the Navier-Stokes equations and the energy equation leads to the governing system in the ALE form

$$\frac{D^{\mathcal{A}} w}{Dt} + \sum_{s=1}^{2} \frac{\partial g_s(w)}{\partial x_s} + w \operatorname{div} z = \sum_{s=1}^{2} \frac{\partial R_s(w, \nabla w)}{\partial x_s}, \tag{4}$$

where $w = (\rho, \rho v_1, \rho v_2, E)^T \in \mathbb{R}^4$, $g_s(w) = f_s - z_s w$, $s = 1,2$, $f_s = (\rho v_s, \rho v_1 v_s + \delta_{1s} p, \rho v_2 v_s + \delta_{2s} p, (E+p) v_s)^T$, $R_s(w, \nabla w) = (0, \tau_{s1}^V, \tau_{s2}^V, \tau_{s1}^V v_1 + \tau_{s2}^V v_2 + k \frac{\partial \theta}{\partial x_s})^T$, $s = 1,2$, $\tau_{ij}^V = \lambda \delta_{ij} \operatorname{div} v + 2\mu d_{ij}(v)$, $d_{ij}(v) = \frac{1}{2}\left(\frac{\partial v_i}{\partial x_j} + \frac{\partial v_j}{\partial x_i}\right)$, $i, j = 1, 2$. For a detailed description, see, for example, [3]. The following notation is used:

- ρ – fluid density
- p – pressure
- E – total energy
- $v = (v_1, v_2)$ – velocity vector
- θ – absolute temperature
- $c_v > 0$ – specific heat at constant volume
- $\gamma > 1$ – Poisson adiabatic constant
- $\mu > 0, \lambda = -2\mu/3$ – viscosity coefficients
- $k > 0$ – heat conduction coefficient
- τ_{ij}^V – components of the viscous part of the stress tensor
- $\tau_{ij} = -p\delta_{ij} + \tau_{ij}^V, i, j = 1, 2$ – components of the stress tensor τ.

The vector-valued function w is called the state vector, f_s are inviscid fluxes and R_s represent viscous terms. The system (4) is completed by the thermodynamical relations $p = (\gamma - 1)\left(E - \rho \frac{|v|^2}{2}\right)$, $\theta = \frac{1}{c_v}\left(\frac{E}{\rho} - \frac{1}{2}|v|^2\right)$ and equipped with the initial condition $w(x,0) = w^0(x)$, $x \in \Omega_0$ and the boundary conditions:

Γ_I: $\quad \rho = \rho_D, \quad v = v_D, \quad \sum_{j=1}^{2}\left(\sum_{i=1}^{2} \tau_{ij}^V n_i\right) v_j + k \frac{\partial \theta}{\partial n} = 0$, on inlet

Γ_{W_t}: $\quad v = z_D(t) =$ velocity of a moving wall, $\quad \frac{\partial \theta}{\partial n} = 0$, on moving wall

Γ_O: $\quad \sum_{j=1}^{2} \tau_{ij}^V n_j = 0, \quad \frac{\partial \theta}{\partial n} = 0, \quad i = 1, 2$, on outlet, with prescribed data ρ_D, $\boldsymbol{v}_D$ and z_D. By $\boldsymbol{n}$ we denote the unit outer normal to Ω_t.

The elastic structure deformation is described by equations of dynamic linear elasticity. We consider an elastic body $\Omega^b \subset \mathbb{R}^2$, which has a common boundary with the reference domain Ω_0 occupied by the fluid at the initial time. The boundary of Ω^b is formed by two disjoint parts $\partial\Omega^b = \Gamma_N \cup \Gamma_D$, where $\Gamma_N \cap \Gamma_D = \emptyset$, $\Gamma_N \subset \Gamma_{W_0}$ and Γ_D is a fixed part of the boundary. We denote the displacement of the body by $\boldsymbol{u} = \boldsymbol{u}(\boldsymbol{X}, t)$, $\boldsymbol{X} \in \Omega^b$, $t \in (0, T)$. The dynamic elasticity problem is given as follows: find $\boldsymbol{u} : \Omega^b \to \mathbb{R}^2$ such that

$$\rho^b \frac{\partial^2 \boldsymbol{u}}{\partial t^2} + c_M \rho^b \frac{\partial \boldsymbol{u}}{\partial t} - \operatorname{div} \boldsymbol{\sigma}(\boldsymbol{u}) - c_K \frac{\partial}{\partial t} \operatorname{div} \boldsymbol{\sigma}(\boldsymbol{u}) = \boldsymbol{0} \quad \text{in } \Omega^b \times (0, T), \tag{5}$$

$$\boldsymbol{u} = \boldsymbol{u}_D \quad \text{in } \Gamma_D \times (0, T), \quad -\boldsymbol{\sigma}(\boldsymbol{u}) \cdot \boldsymbol{n} = \boldsymbol{g}_N \quad \text{in } \Gamma_N \times (0, T), \tag{6}$$

$$\boldsymbol{u}(x, 0) = \boldsymbol{u}_0(x), \quad x \in \Omega^b, \quad \frac{\partial \boldsymbol{u}}{\partial t}(x, 0) = z_0(x), \quad x \in \Omega^b. \tag{7}$$

Here $\boldsymbol{u}_D : \Gamma_D \times (0, T) \to \mathbb{R}^2$ – boundary displacement, $\boldsymbol{g}_N : \Gamma_N \times (0, T) \to \mathbb{R}^2$ – acting surface force, $\boldsymbol{u}_0 : \Omega^b \to \mathbb{R}^2$ – initial displacement, $z_0 : \Omega^b \to \mathbb{R}^2$ – initial displacement velocity are given functions and $\rho^b > 0$ is a given constant material density. We assume a linear dependence between the stress tensor $\boldsymbol{\sigma}(\boldsymbol{u})$ and the strain tensor $\boldsymbol{e}(\boldsymbol{u})$ and that the material is isotropic and homogeneous (the details can be found in [4]). The expressions $c_M \rho \frac{\partial \boldsymbol{u}}{\partial t}$ and $c_K \frac{\partial}{\partial t} \operatorname{div} \boldsymbol{\sigma}(\boldsymbol{u})$ represent the damping terms, with $c_M, c_K \geq 0$.

3 Discretization

In this section we shall briefly describe the discretization. A more thorough description of the discretization of the dynamical elasticity system can be found in [4], while the discretization of the system for the fluid flow problem can be found in [1].

3.1 Space Discretization

We begin with the main idea of the space discretization technique. In both problems we consider a polygonal computational domain and a triangulation of the domain consisting of triangular elements, which have the standard properties known from the finite element method. We define a finite-dimensional space of piecewise polynomial functions, which are in general discontinuous at the edges of the triangulation.

The equations are multiplied by test functions of this finite-dimensional space and integrated over each element of the triangulation. Subsequently Green's theorem is employed, a suitable "stabilization" and interior and boundary penalty terms (vanishing for the exact regular solution) are added and finally the resulting equations are summed over all elements of the triangulation.

Moreover, the resulting discrete problem for the fluid flow is partially linearized, which leads to a semi-implicit scheme. We also apply the concept of an artificial viscosity in the vicinity of internal and boundary layers.

Both problems are discretized in time by two different approaches: by a finite-difference method and by the space-time discontinuous Galerkin method (STDGM).

The finite-difference method is based on a second order backward-difference formula for the approximation of the time derivative:

$$\frac{\partial \boldsymbol{u}}{\partial t}(t) \approx \frac{3\boldsymbol{u}(t) - 4\boldsymbol{u}(t-\tau) + \boldsymbol{u}(t-2\tau)}{2\tau}, \tag{8}$$

where $\tau > 0$ is a chosen time step.

The STDGM is a fundamentally different approach to the time discretization compared to methods based on finite-difference approximations. The sought solution is approximated by a discontinuous piecewise polynomial function both in space and in time. The method allows an arbitrary choice of the polynomial degree both in space and in time and yields a robust and a very accurate scheme.

4 Implementation Remarks

In this section we shall mention some important topics of the definition and implementation of the discrete solver for the coupled problem of the fluid-structure interaction. We shall describe the coupling procedure and explain how to get an appropriate ALE mapping.

The construction of the ALE mapping and the domain velocity is following. For the construction of the ALE mapping and the domain velocity we employ a static linear elasticity model. We consider an elastic body represented by a bounded domain $\Omega_0 \subset \mathbb{R}^2$, which is the reference domain occupied by the fluid. The ALE mapping $\mathscr{A}_t$ is obtained as the solution of the following problem:

$$-\operatorname{div} \boldsymbol{\sigma}(\mathscr{A}_t) = \boldsymbol{0} \text{ in } \Omega_0, \quad \mathscr{A}_t = \boldsymbol{u}(t) \text{ on } \Gamma_{W_t}, \quad \mathscr{A}_t = \boldsymbol{0} \text{ on } \partial\Omega_0 \setminus \Gamma_{W_t}. \tag{9}$$

The two Dirichlet boundary conditions mean that on the common boundary the displacement of the flow domain Ω_0 is equal to the displacement of the elastic bodies represented by the domain Ω^b, and that the displacement of the elastic bodies is zero otherwise. We again assume a linear dependence between the *stress tensor* $\boldsymbol{\sigma}(\mathscr{A}_t)$ and the *infinitesimal strain tensor* $\boldsymbol{e}(\mathscr{A}_t)$ and that the material is isotropic and piecewise homogeneous. The applied model implies that the flow domain is represented by some elastic material. Satisfactory results are obtained with the aid of non-physical material parameters, see [6]. In our experiments we prescribed piecewise constant Lamè parameters by setting $\lambda = -c \cdot \text{avgdiam}\, \mathscr{T}_h/\text{diam}\, K$, $\mu = -\lambda$, where $c > 0$ is a suitable constant, diam K is the diameter of an element K and avgdiam $\mathscr{T}_h$ is the average element diameter of the triangulation $\mathscr{T}_h$. In contrast to a material with physical Lamè parameters, where $\lambda, \mu > 0$, we set $\lambda + \mu = 0$. Moreover, on each element K we divide the Lamè parameters by the relative element size. Finally, we choose an appropriate constant $c > 0$. Our numerical experiments suggest that smaller values of c yield more suitable mesh deformations.

Necessary to apply a suitable coupling procedure. Let us introduce the following algorithm:

1. Assume that the approximate solution of the flow problem and the deformation of the structure $\boldsymbol{u}$ on the time level t_k are known.
2. Set the deformation on the time level t_k as deformation $\boldsymbol{u}^0$ on the time level $t_{k+1}, l := 1$ and apply the following iterative process. We use the upper index to distinct the approximate solution obtained during the iterative process:
   (a) Compute the aerodynamic stress tensor τ^l and the acting aerodynamical force transformed to the interface Γ_N by the transmission conditions $\boldsymbol{g}_N = -\tau^l \cdot \boldsymbol{n}$
   (b) Solve the elasticity problem (5)–(7), compute the deformation $\boldsymbol{u}^l$ on the time level t_{k+1} and the approximation $\Omega^l_{t_{k+1}}$ of the domain occupied by the fluid at time t_{k+1}.
   (c) Determine the approximation of the ALE mapping $\mathscr{A}^l_{t_{k+1}}$ at time t_{k+1} and approximate the domain velocity z^l_{k+1}.
   (d) Solve the flow problem (4) on the approximation $\Omega^l_{t_{k+1}}$.
   (e) If the variation $|\boldsymbol{u}^l_{k+1} - \boldsymbol{u}^{l-1}_{k+1}|$ is larger than prescribed, go to (a) and $l := l + 1$. Else $k := k + 1$ and go to (2).

A simplified so-called "weakly coupled" scheme is defined by the same algorithm with the exception that we only perform a single inner iteration, i.e. we set $k = k + 1$ and go to (2) already in the case when $l = 1$.

For the numerical solution of the dynamic 2D linear elasticity problem with mixed boundary conditions we developed a .NET library written in C#. The flow problem is solved by a library developed by J. Česenek, which is written in the C language. (For details, see [1].) Both libraries support several time discretization techniques, built on top of the DG discretization in space with an arbitrary choice of the degree of the polynomial approximation. The time discretizations are based on the backward Euler formula, the second-order backward difference formula and the STDGM with an arbitrary choice of the degree of the polynomial approximation in time. The resulting linear systems are solved using the direct solver UMFPACK [2] or by the iterative solver GMRES with block diagonal preconditioning.

The presented method can be applied to solve a selection of problems of biomechanics and aviation. Specifically, in this paper we are focused to model a simplified 2D simulation of vibrations of vocal folds, which are caused by the airflow originating in human lungs.

5 Numerical Results

We consider the model of flow through a channel with two elastic bodies (see Fig. 1). The numerical experiments were carried out for the following data:

- Magnitude of the inlet velocity $v_{in} = 4\,\mathrm{m.s^{-1}}$
- The viscosity $\mu = 15 \cdot 10^{-6}\,\mathrm{kg.m^{-1}.s^{-1}}$
- The inlet density $\rho_{in} = 1.225\,\mathrm{kg.m^{-3}}$
- The outlet pressure $p_{out} = 97{,}611\,\mathrm{Pa}$
- The Reynolds number $Re = \rho_{in} v_{in} h / \mu = 5{,}227$
- Heat conduction coefficient $k = 2.428 \cdot 10^2\,\mathrm{kg.m.s^{-2}.K^{-1}}$
- The specific heat $c_v = 721.428\,\mathrm{m^2.s^{-2}.K^{-1}}$
- The Poisson adiabatic constant $\gamma = 1.4$

The relative tolerance for the GMRES solver was set to 10^5. The Young modulus and the Poisson ratio have values $E^b = 25{,}000\,\mathrm{Pa}$ and $\nu^b = 0.4$, respectively, the structural damping coefficient is equal to the constant $c_M = 100\,\mathrm{s^{-1}}$ and the material density $\rho^b = 1{,}040\,\mathrm{kg.m^{-3}}$. We used the time step $\tau = 2 \cdot 10^{-5}\,\mathrm{s}$. Example of the computed results is shown in Fig. 2.

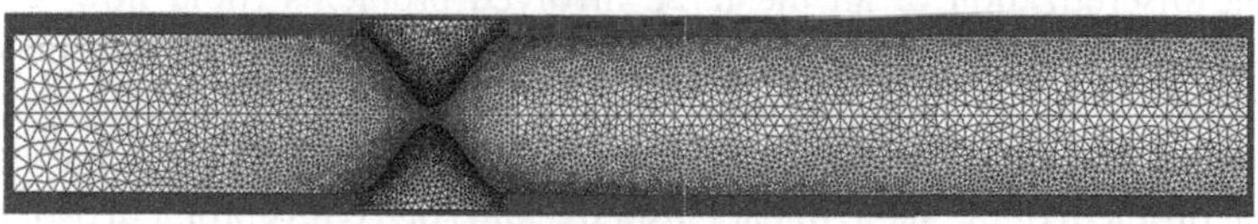

Fig. 1 The computational mesh of the fluid domain and the structure domain

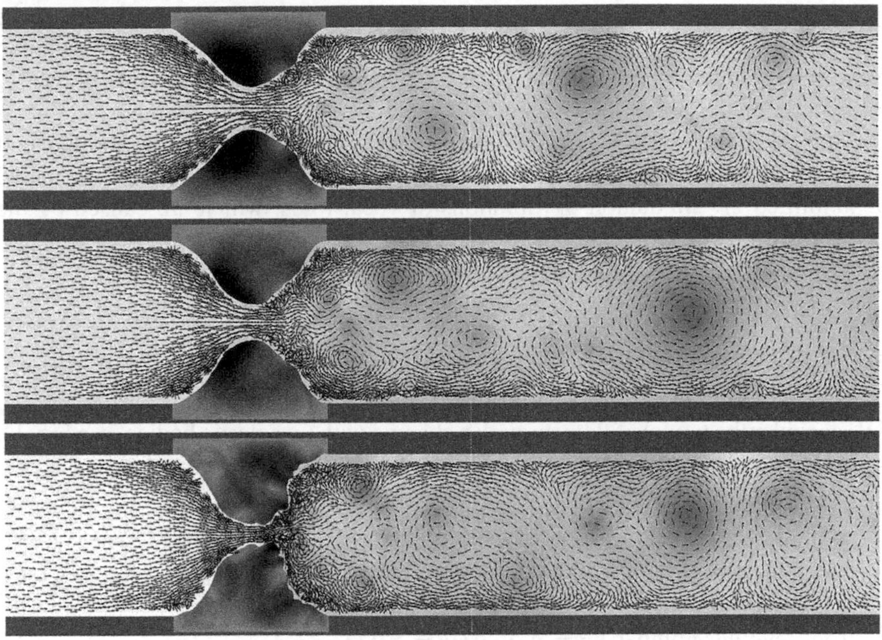

Fig. 2 The velocity vectors and the pressure field (the values of pressure dark to light) at several time instants

Conclusion

In this paper we presented a method for the numerical solution of the interaction of the viscous compressible fluid flow and an elastic body in 2D. The work on this topic is motivated by the problems of biomechanics of vocal tract. The main goal was to show the applicability of the DGM to the discretization of all the three involved problems (fluid flow, elastic material deformation and mesh movement) both in space and in time. Selected elements of the coupling procedure and the implementation techniques were presented. We also described a construction of the ALE mapping, which leads to the problem of the static linear elasticity, and solved this problem by means of the DGM. Finally we presented the results of our numerical simulations.

Acknowledgment This work was supported by the grants P101/11/0207 (J. Horáček) and 13-00522S (M. Feistauer) of the Czech Science Foundation. The work of A. Kosík was supported by the grant GAChU 549912, the work of M. Hadrava was supported by the grant SVV-2014-260106, both financed by the Charles University in Prague.

References

1. J. Česenek, M. Feistauer, A. Kosík, DGFEM for the analysis of airfoil vibrations induced by compressible flow. Z. Angew. Math. Mech. **93**(6–7), 387–402 (2013)
2. T.A. Davis, I.S. Duff, An unsymmetric-pattern multifrontal method for sparse LU factorization. SIAM J. Matrix Anal. Appl. **18**(1), 140–158 (1997)
3. M. Feistauer, J. Horáček, V. Kučera, J. Prokopová, On numerical solution of compressible flow in time-dependent domains. Math. Bohemica **137**, 1–16 (2011)
4. M. Hadrava, M. Feistauer, J. Horáček, A. Kosík, Space-time discontinuous Galerkin method for the problem of linear elasticity, in *Numerical Mathematics and Advanced Applications – ENUMATH 2013: Proceedings of ENUMATH 2013, the 10th European Conference on Numerical Mathematics and Advanced Applications*, Lausanne, Aug 2013 (Springer, Cham, 2014)
5. T. Nomura, T.J.R. Hughes, An arbitrary Lagrangian-Eulerian finite element method for interaction of flow and a rigid body. Comput. Methods Appl. Mech. Eng. **95**, 115–138 (1992)
6. K. Stein, T. Tezduyar, R. Benney, Mesh moving techniques for fluid-structure interactions with large displacements. J. Appl. Mech. **70**(1), 58–63 (2003)

Eulerian Techniques for Fluid-Structure Interactions: Part I – Modeling and Simulation

Stefan Frei, Thomas Richter, and Thomas Wick

Abstract This contribution is the first part of two papers on the Fully Eulerian formulation for fluid-structure interactions. We derive a monolithic variational formulation for the coupled problem in Eulerian coordinates. Further, we present the Initial Point Set method for capturing the moving interface. For the discretization of this interface problem, we introduce a modified finite element scheme that is locally fitted to the moving interface while conserving structure and connectivity of the system matrix when the interface moves. Finally, we focus on the time-discretization for this moving interface problem.

1 Introduction

The underlying difficulty of fluid-structure interactions (fsi) is the free boundary character of the coupled system: as the deformation or motion of the solid determines the interface to the fluid problem, the domains (fluid as well as solid) are subject to change. In problems of solid mechanics, the displacements are usually represented in Lagrangian coordinates, such that the computational domain is always fixed. The shape of the current configuration is expressed by the displacement field. This concept does not directly transfer to coupled fsi problems, as fluid flows are usually considered in Eulerian coordinates. A direct coupling between the fixed Lagrangian and the moving Eulerian domain is not possible.

For stiffly coupled problems, monolithic formulations of the coupled system are required for robust implicit discretization and solution techniques. A simple approach is to reformulate the flow problem on a fixed coordinate system, that

S. Frei • T. Richter (✉)
Institut für Angewandte Mathematik, Universität Heidelberg, INF 294, 69120 Heidelberg, Germany
e-mail: thomas.richter@iwr.uni-heidelberg.de; stefan.frei@iwr.uni-heidelberg.de

T. Wick
The Institute for Computational Engineering and Sciences, The University of Texas at Austin, Austin, TX 78712, USA
e-mail: twick@ices.utexas.edu

A. Abdulle et al. (eds.), *Numerical Mathematics and Advanced Applications ENUMATH 2013*, Lecture Notes in Computational Science and Engineering 103,
DOI 10.1007/978-3-319-10705-9_74

matches the fluid-problem. By introducing a reference domain and a mapping between this reference domain and the current configuration, the fluid problem can be expressed on a fixed domain. All motion is hidden in the transformation, which is now an unknown part of the system. This Arbitrary Lagrangian Eulerian (ALE) formulation is one possibility out of two and is often used and highly successful (see, e.g. the survey [2]), mostly due to the simple structure and the very good accuracy, that can be achieved. We notice that the reference system for the fluid problem is artificial. Problems appear, if the fluid domain undergoes a very large deformation. The mapping between artificial reference domain and current configuration must be invertible and differentiable. If the deformation gets too large, e.g. if the topology of the domain is changed (by contact), the ALE approach will fail. By remeshing and definition of a new reference domain, one can overcome this limitation, however at the cost of loosing a strictly monolithic formulation.

Here, we present an Eulerian formulation for the coupled problem, which is similar to the ALE approach, as coupling will be realized in a monolithic variational formulation. The fluid problem is given in its natural Eulerian framework, and the solid problem will also be mapped to Eulerian coordinates, such that both sub-problems are formulated in the moving current configuration. This approach has first been introduced by Dunne [3] and then been further analyzed and developed into a computational method [4, 13, 14, 16]. Two major differences between the Eulerian and the ALE approach are of importance: First, we do not have to use artificial reference domains. The mapping between Lagrangian and Eulerian systems is natural and will never be the cause for a breakdown of the approach. Large motion, deformation and contact are possible. Second, as the problems are given in the moving current configuration on a fixed spatial coordinate system, the formulation is of front-capturing type. The position of the interface must be carefully followed and achieving good interface accuracy will be challenging.

The Fully Eulerian approach must be distinguished from other techniques like Euler-Lagrange schemes based on Level-Sets [9], the XFEM dual mortar approach [10], or Peskin's immersed boundary method [12] where two different meshes are used and the information is provided by smoothed delta-functions. The key difference of these methods to the Fully Eulerian approach is that we neither need Lagrange-multipliers, and that we work on one common fixed background mesh, that allows us to realize the coupling by variational techniques.

The following second section is devoted to an introduction of the Fully Eulerian formulation for fluid-structure interactions. Then, in Sect. 3 we describe a spatial finite element discretization that is able to locally resolve the interface. Section 4 discusses the temporal discretization of the coupled system. Numerical test-cases and different applications of the Fully Eulerian formulation are presented in the second part of this series [6].

2 Fluid-Structure Interactions in Eulerian Coordinates

Let $\Omega \subset \mathbb{R}^d$ be a two- or three-dimensional domain, that is split into a fluid-domain $\mathscr{F}$ and a solid-domain $\mathscr{S}$ and a common interface $\mathscr{I}$ by $\Omega = \mathscr{F} \cup \mathscr{I} \cup \mathscr{S}$. By $\Omega = \Omega(0)$, $\mathscr{F} = \mathscr{F}(0)$ and $\mathscr{I} = \mathscr{I}(0)$ we denote the stress-free reference configuration. On the sub-domain $\mathscr{F}$ we prescribe the incompressible Navier-Stokes equations, while $\mathscr{S}$ is governed by an elastic structure. The two problems are coupled on the common interface by prescribing continuity of velocities $\mathbf{v}_f = \mathbf{v}_s$ as well as continuity of normal stresses $\boldsymbol{\sigma}_f \mathbf{n} = \boldsymbol{\sigma}_s \mathbf{n}$, where by $\boldsymbol{\sigma}_f$ and $\boldsymbol{\sigma}_s$ we denote the Cauchy stresses of fluid and solid and $\mathbf{n}$ denotes the normal vector. By the dynamics of the coupled problem, the solid domain will undergo a motion or deformation $\mathscr{S} \to \mathscr{S}(t)$ and the fluid-domain will move along, such that the joint domain $\Omega(t) = \mathscr{F}(t) \cup \mathscr{I}(t) \cup \mathscr{S}(t)$ will neither overlap nor produce gaps. The main task for a monolithic variational formulation of the coupled problem is to state the solid equations on this moving Eulerian domain $\mathscr{S}(t)$. Details on the derivation of the equations as well as differences to the traditional ALE formulation are presented in detail in the literature, see e.g. [4].

Here, by $\mathbf{v}_s$ and $\mathbf{u}_s$ we denote solid velocity and displacement in the Eulerian framework. By the relation $\hat{x} := x - \mathbf{u}(x,t)$ we define the mapping of a Eulerian coordinate $x \in \mathscr{S}(t)$ back the reference coordinate $x \in \mathscr{S} = \mathscr{S}(0)$ of the particle. By $\mathbf{F} := I - \nabla \mathbf{u}$ we denote the Eulerian displacement gradient with determinant $J := \det \mathbf{F}$. It holds $\mathbf{F} = \hat{\mathbf{F}}^{-1}$, where $\hat{\mathbf{F}}$ is the usual Lagrangian displacement gradient [4]. Finally, the Green Lagrange strain tensor has the Eulerian representation $\mathbf{E} := \frac{1}{2}(\mathbf{F}^{-T}\mathbf{F}^{-1} - I)$. This notation allows to state various constitutive laws of elastic materials in Eulerian coordinates. For simplicity, we restrict all considerations to the St. Venant-Kirchhoff material, where the Cauchy stresses are given by

$$\boldsymbol{\sigma}_s := J\mathbf{F}^{-1}\left(2\mu_s \mathbf{E} + \lambda_s \operatorname{tr}(\mathbf{E}) I\right)\mathbf{F}^{-T},$$

with Lamé coefficients μ_s and λ_s.

2.1 Variational Formulation in Eulerian Coordinates

We start by defining the correct functional spaces for the solution of the coupled problem. As velocities of fluid and solid are continuous on the complete domain $\Omega(t) = \mathscr{F}(t) \cup \mathscr{I}(t) \cup \mathscr{S}(t)$, we define a global function space that directly incorporates the kinematic coupling condition

$$\mathbf{v} \in \mathbf{v}^D + \mathscr{V}, \quad \mathscr{V} := H_0^1(\Omega(t); \Gamma^D(t))^d,$$

where $\Gamma^D(t)$ is that part of the domain's boundary, where Dirichlet conditions are prescribed and $\mathbf{v}^D \in H^1(\Omega(t))^d$ is an extension of the Dirichlet data into the domain. Fluid and solid velocities are given by restriction of $\mathbf{v}$ to the subdomains $\mathbf{v}_f := \mathbf{v}|_{\mathcal{F}(t)}$ and $\mathbf{v}_s := \mathbf{v}|_{\mathcal{S}(t)}$, respectively. Considering compressible elastic structures, the pressure is only given in the fluid domain

$$p_f \in \mathcal{L}_f := L^2(\mathcal{F}(t)).$$

As the Eulerian formulation does not involve transformation of the fluid-domain, no additional displacement variable (like in the ALE approach) is required. We find the solid displacement in the form

$$\mathbf{u}_s \in \mathbf{u}_s^D + \mathcal{W}_s, \quad \mathcal{W}_s := H_0^1(\mathcal{S}(t); \Gamma_s^D(t))^d,$$

where by $\Gamma_s^D(t)$ we denote the Dirichlet part of the solid boundary and by $\mathbf{u}_s^D \in H^1(\mathcal{S}(t))^d$ an extension of the Dirichlet values into the solid domain. Finally, velocities $\mathbf{v} \in \mathbf{v}^D + \mathcal{V}$, displacement $\mathbf{u}_s \in \mathbf{u}_s^D + \mathcal{W}_s$ and pressure $p_f \in \mathcal{L}_f$ are defined by the system:

$$\begin{aligned}
&(\rho_f(\partial_t \mathbf{v}_f + \mathbf{v}_f \cdot \nabla \mathbf{v}_f), \boldsymbol{\phi}_f)_{\mathcal{F}(t)} + (J_s \rho_s^0(\partial_t \mathbf{v}_s + \mathbf{v}_s \cdot \nabla \mathbf{v}_s), \boldsymbol{\phi}_s)_{\mathcal{S}(t)} \\
&+(\boldsymbol{\sigma}_f, \nabla \boldsymbol{\phi}_f)_{\mathcal{F}(t)} + (\boldsymbol{\sigma}_s, \nabla \boldsymbol{\phi}_s)_{\mathcal{S}(t)} = (\rho_f \mathbf{f}_f, \boldsymbol{\phi}_f)_{\mathcal{F}(t)} + (J\rho_s^0 \mathbf{f}_s, \boldsymbol{\phi}_s)_{\mathcal{S}(t)} \quad \forall \boldsymbol{\phi} \in \mathcal{V} \\
&\qquad (\partial_t \mathbf{u}_s + \mathbf{v}_s \cdot \nabla \mathbf{u}_s, \boldsymbol{\psi}_s)_{\mathcal{S}(t)} = (\mathbf{v}_s, \boldsymbol{\psi}_s)_{\mathcal{S}(t)} \quad \forall \boldsymbol{\psi}_s \in \mathcal{W}_s \\
&\qquad (\operatorname{div} \mathbf{v}_f, \xi_f)_{\mathcal{F}(t)} = 0 \quad \forall \xi_f \in \mathcal{L}_f,
\end{aligned} \tag{1}$$

where by ρ_f and ρ_s^0 we denote the densities of fluid and solid in reference state, by $\boldsymbol{\sigma}_f := -p_f I + \rho_f \nu_f (\nabla \mathbf{v}_f + \nabla \mathbf{v}_f^T)$ the fluid stresses with kinematic viscosity ν_f. The global definition of the test-function $\boldsymbol{\phi} \in \mathcal{V}$ ensures the dynamic coupling condition of the normal stresses. As for the velocities, we use the notation $\boldsymbol{\phi}_f := \boldsymbol{\phi}|_{\mathcal{F}(t)}$ and $\boldsymbol{\phi}_s := \boldsymbol{\phi}|_{\mathcal{S}(t)}$.

This system of equations in not closed, as the motion of the domains is determined in an implicit sense only. Without knowledge of the solution, the affiliation of a coordinate $x \in \Omega(t)$ to either solid- or fluid-domain is not immediately possible. The next section will focus on this issue.

2.2 *The Initial Point Set Method*

One common possibility to capture the interface in fixed mesh methods is to use Level-Set functions [15] that transport the interface as zero contour of a signed distance function with the fluid and solid velocity. Eulerian Level-Set methods for fsi problems are discussed in the literature [7, 8]. Here, we refrain from using

Level-Sets due to two reasons: first, Level-Sets have difficulties capturing sharp edges. And second, an additional equation has to be solved and the problem complexity increases. Instead, we base the interface capturing on a transportation of the complete reference domain instead of the interface:

$$\partial_t \Omega(t) + \mathbf{v} \cdot \nabla \Omega(t) = 0.$$

Within the solid domain, the displacement $\mathbf{u}_s$ exactly takes this role. For $x \in \mathcal{S}(t)$, the displacement vector points back to the reference domain $x - \mathbf{u}_s(x,t) \in \mathcal{S} = \mathcal{S}(0)$. Hence, if x and $\mathbf{u}$ are available, we can decide, whether $x - \mathbf{u}$ is part of the reference solid or not. To apply this concept, we must define a displacement field $\mathbf{u}$ on the complete domain $\Omega(t)$. Then, the Initial Point Set [3, 13] is given as

$$\Phi_{\text{IPS}}(x,t) := \begin{cases} x - \mathbf{u}_s(x,t) & x \in \mathcal{S}(t), \\ x - \text{ext}(\mathbf{u}_s)(x,t) & x \in \mathcal{F}(t). \end{cases}$$

The extension of the solid displacement is only required in a close neighborhood of the interface [13]. Given the initial point set, the domain affiliation of $x \in \Omega(t)$ is determined by $\Phi_{\text{IPS}}(x,t) \in \mathcal{S}(0)$ for the solid domain and $\Phi_{\text{IPS}}(x,t) \notin \mathcal{S}(0)$ for coordinates in the fluid domain $\mathcal{F}(t)$. Here, we stress one detail in the realization: a coordinate $x \in \Omega(t)$ belongs to the fluid part, if the Initial Point Set Φ_{IPS} maps out of the reference solid domain. No mapping between $\mathcal{F}(0)$ and $\mathcal{F}(t)$ is required, see [13] for a discussion. The extension can be embedded into the variational system and the coupling condition $\mathbf{u}_f = \mathbf{u}_s$ is realized by finding a global displacement field on the whole domain $\mathbf{u} \in \mathbf{u}^D + \mathcal{W}$, where $\mathcal{W} := H_0^1(\Omega(t); \Gamma_s^D)^d$.

3 Finite Element Discretization

Typically, in fluid-structure interaction problems the overall dynamics of the system strongly depend on the dynamics in the interface region. Hence, one key ingredient for both stability and accuracy reasons is to capture the interface accurately. The combined velocity consisting of solid and fluid part typically shows a kink at the interface. It is important to resolve this kink accurately in our discretization scheme. One standard approach to include jumps or kinks into the discrete space is the Extended Finite Element Method [11]. A drawback of the XFEM method is the addition and elimination of degrees of freedom which leads to a local distortion of the connectivity and structure of the system matrix. Furthermore, one may have to deal with so called "blending" cells lying next to the interface cells that might distort the method's accuracy. Finally, the condition number of the system matrix does not necessarily remain bounded. Here, we present a method [5], that avoids these issues. The idea is to use a fixed background mesh consisting of patches that remains unchanged for all time steps. Inside the patches we adjust degrees of freedom locally by choosing a special parametric finite element space.

Locally Modified Parametric Finite Element Scheme Let Ω_h be a form and shape-regular decomposition of the domain $\Omega \subset \mathbb{R}^2$ into open quadrangles. The mesh Ω_h does not necessarily resolve the partitioning $\Omega(t) = \mathscr{F}(t) \cup \mathscr{I}(t) \cup \mathscr{S}(t)$ and the interface $\mathscr{I}(t)$ can cut the elements $K \in \Omega_h$. We further assume, that the mesh Ω_h has a patch-hierarchy in such a way, that each four adjacent quads arise from uniform refinement of one common father-element, see Fig. 1. The interface $\mathscr{I}$ may cut the patches in the following way: Each (open) patch $P \in \Omega_h$ is either not cut $P \cap \mathscr{I} = \emptyset$ or cut in exactly two points on its boundary: $P \cap \mathscr{I} \neq \emptyset$ and $\partial P \cap \mathscr{I} = \{x_1^P, x_2^P\}$.

We define the finite element trial space $V_h \subset H_0^1(\Omega)$ as iso-parametric space on the triangulation Ω_h. If a patch is not cut by the interface, we use the standard space of bilinear functions $\hat{Q}$ (bilinear on each of the four sub-quads) for both reference element transformation and finite element basis. If a patch $P \in \Omega_h$ however is cut, we use the space $\hat{Q}_{\text{mod}}$ of piecewise linear functions (linear on each of the eight triangles) for transformation and basis. Depending on the position of the interface $\mathscr{I}$ in the patch P, three different reference configurations are considered, see the right sketch in Fig. 1. Note that the functions in $\hat{Q}$ and $\hat{Q}_{\text{mod}}$ are all piecewise linear on the edges ∂P, such that mixing different element types does not affect the continuity of the global finite element space.

Next, we present the subdivision of interface patches P into eight triangles each. We distinguish four different types of interface cuts, see Fig. 2: Configurations A and B are based on the reference patches $\hat{P}_2$ and $\hat{P}_3$, configurations C and D use the

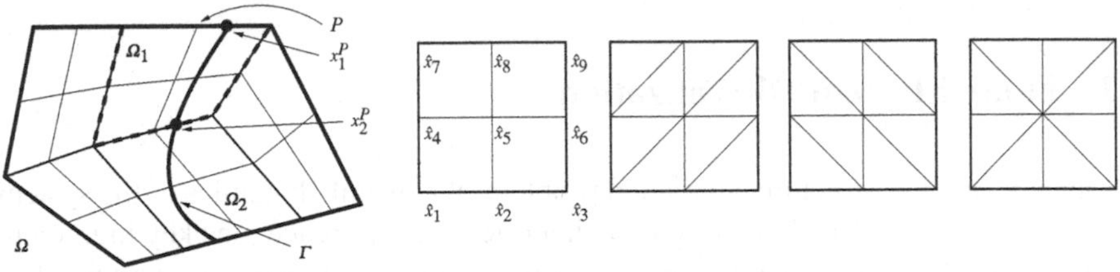

Fig. 1 *Left*: triangulation Ω_h with interface $\mathscr{I}$. Patch P is cut by $\mathscr{I}$ at x_1^P and x_2^P. *Right*: subdivision of reference patches $\hat{P}_1, \ldots, \hat{P}_4$ into eight triangles each

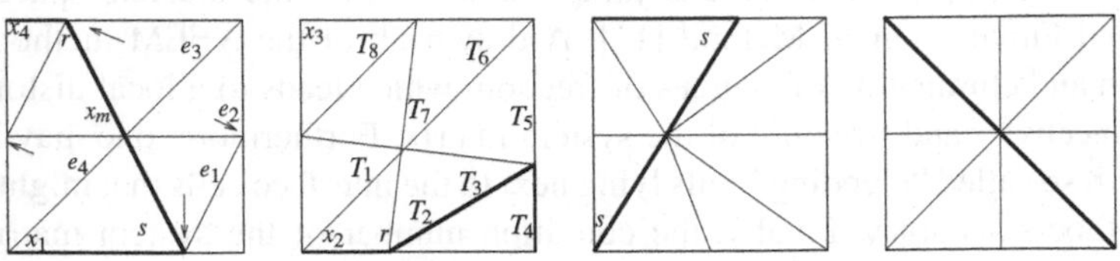

Fig. 2 Different types of cut patches. From left to right: A, B, C and D. The subdivision can be anisotropic with $r, s \in (0, 1)$ arbitrary

reference patch $\hat{P}_4$, see Fig. 1. If an edge is intersected by the interface we move the corresponding point e_i on this edge to the point of intersection. The position of the midpoint x_m depends on the specific configuration. As the cut of the elements can be arbitrary with $r, s \to 0$, the triangle's aspect ratio can be very large, considering $h \to 0$ it is not necessarily bounded. We can however guarantee, that the maximum angles in all triangles will be bounded away from 180°. This result allows us to define stable interpolation operators and to derive error estimates [5].

To cope with the condition number of the system matrix, that can be unbound for some configurations $r, s \to 0$, we modify the parametric basis in a hierarchical way. By splitting of the finite element space $V_h = V_{2h} + V_b$, where V_{2h} is the standard space of linear functions on the patches $P \subset \Omega_h$ and V_b is the space with only local contributions, the effect of the interface motion is kept locally. This modification allows us to show an interface-independent condition number for the system matrix of elliptic problems [5].

4 Outlook: Accurate Temporal Discretization

As time-stepping scheme we use the implicit Euler method. The implicit Euler method has excellent stability properties, may suffer from strong dissipation, however. Due to the hyperbolic character of the structure equation, it is desirable to use a scheme with better dissipation properties. Furthermore, for stability and accuracy reasons, it is important to capture the interface movement accurately. The combined functions v and u both typically show kinks, their gradients are typically discontinuous across the interface. A standard time-stepping scheme for the first equation in (1) reads

$$k^{-1}(\rho_f(\mathbf{v}_f^m - \mathbf{v}_f^{m-1}), \boldsymbol{\phi}_f)_{\mathscr{F}(t_m)} + (\theta \mathbf{v}_f^m \cdot \nabla \mathbf{v}_f^m + (1-\theta)\mathbf{v}_f^{m-1} \cdot \nabla \mathbf{v}_f^{m-1}, \boldsymbol{\phi}_f)_{\mathscr{F}(t_m)} + \dots$$

Implementation of this scheme is not straightforward, however, as the domains $\mathscr{F}$ and $\mathscr{S}$ change with time. Points belonging to $\mathscr{S}$ at time t_{m-1} might lie in $\mathscr{F}$ at time t_m. In this case the fluid velocity v_f^{m-1} is not defined in some parts of $\mathscr{F}(t_m)$.

In order to capture the velocity kinks accurately and not depend on artificial extensions, we propose the use of a moving mesh technique at each time step in the interface region. Similar to the ALE Method, we define a transformation T_m from a fixed reference domain (e.g. $\Omega(t_m)$) back in time to the time slab $Q(t) = \{(x,t) \mid t \in (t_{m-1}, t_m), x \in \Omega(t)\}$ that maps $\mathscr{F}(t_m)$ to $\mathscr{F}(t)$, $\mathscr{S}(t_m)$ to $\mathscr{S}(t)$ and $\mathscr{I}(t_m)$ to $\mathscr{I}(t)$. We use this transformation in a neighborhood of the interface $\mathscr{I}(t_m)$ only, outside we set $T_m = \text{id}$ the identity (cf. Fig. 3). The reference domain (e.g. $\Omega(t_m)$) changes in every time step. A similar method has been proposed by Baiges and Codina [1]. In order to avoid the need for remeshing around the interface, we use the same mesh in all time steps, with the only difference that – as explained in Sect. 3 – patches cut by the interface are arranged in such a way that the interface is

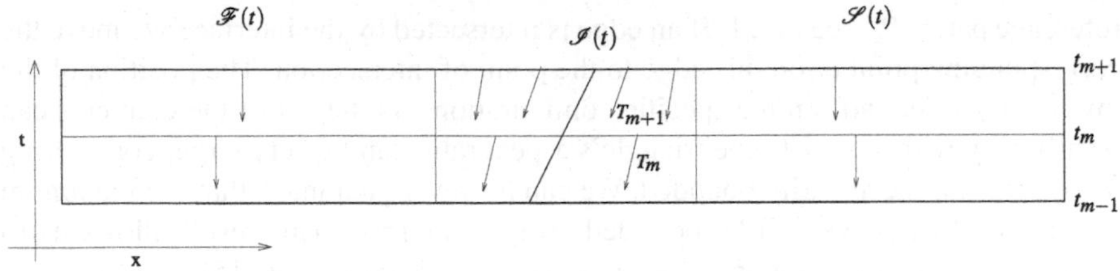

Fig. 3 Extract of the space-time domain with moving interface $\mathscr{I}(t)$. We use an ALE time stepping scheme near the interface to track the interface movement accurately. The transformations T_m and T_{m+1} are indicated by *arrows*. Outside of the interface region, we use a standard θ-scheme

captured. Note that with this technique the interface motion is tracked accurately by a moving mesh line that moves with the interface.

References

1. J. Baiges, R. Codina, The fixed-mesh ALE approach applied to solid mechanics and fluid-structure interaction problems. Int. J. Numer. Methods Eng. **81**, 1529–1557 (2010)
2. H.-J. Bungartz, M. Schäfer (eds.), *Fluid-Structure Interaction II: Modelling, Simulation, Optimisation*. Lecture Notes in Computational Science and Engineering, vol. 73 (Springer, New York, 2010)
3. T. Dunne, An Eulerian approach to fluid-structure interaction and goal-oriented mesh refinement. Int. J. Numer. Math. Fluids. **51**, 1017–1039 (2006)
4. T. Dunne, R. Rannacher, T. Richter, Numerical simulation of fluid-structure interaction based on monolithic variational formulations, in *Comtemporary Challenges in Mathematical Fluid Mechanics*, ed. by G. Galdi, R. Rannacher (World Scientific, Singapore, 2010)
5. S. Frei, T. Richter, A locally modified parametric finite element method for interface problems. SIAM J. Numer. Anal. **52**(5), 2315–2334 (2013)
6. S. Frei, T. Richter, T. Wick, Eulerian techniques for fluid-structure interactions – part II: applications, in *ENUMATH 2013 Proceedings*, Lausanne. Lecture Notes in Computational Science and Engineering (Springer, 2014)
7. P. He, R. Qiao, A full-Eulerian solid level set method for simulation of fluid – structure interactions. Microfluid. Nanofluidics **11**, 557–567 (2011)
8. A. Laadhari, R. Ruiz-Baier, A. Quarteroni, Fully Eulerian finite element approximation of a fluid-structure interaction problem in cardiac cells. Int. J. Numer. Methods Eng. **96**, 712–738 (2013)
9. A. Legay, J. Chessa, T. Belytschko, An Eulerian-Lagrangian method for fluid-structure interaction based on level sets. Comput. Methods Appl. Mech. Eng. **195**, 2070–2087 (2006)
10. U. Mayer, A. Popp, A. Gerstenberger, W. Wall, 3d fluid – structure – contact interaction based on a combined XFEM FSI and dual mortar contact approach. Comput. Mech. **46**(1), 53–67 (2010)
11. N. Moes, J. Dolbow, T. Belytschko, A finite element method for crack growth without remeshing. Int. J. Numer. Methods Eng. **46**, 131–150 (1999)
12. C. Peskin, The immersed boundary method, in *Acta Numerica 2002*, vol. 1–39 (Cambridge University Press, Cambridge, 2002), pp. 1–39
13. T. Richter, A fully Eulerian formulation for fluid-structure interaction problems. J. Comput. Phys. **233**, 227–240 (2013)

14. T. Richter, T. Wick, Finite elements for fluid-structure interaction in ALE and fully Eulerian coordinates. Comput. Methods Appl. Mech. Eng. **199**, 2633–2642 (2010)
15. J. Sethian, *Level Set Methods and Fast Marching Methods Evolving Interfaces in Computational Geometry* (Cambridge University Press, Cambridge, 1999)
16. T. Wick, Fully Eulerian fluid-structure interaction for time-dependent problems. Comput. Methods Appl. Mech. Eng. **255**, 14–26 (2013)

Eulerian Techniques for Fluid-Structure Interactions: Part II – Applications

Stefan Frei, Thomas Richter, and Thomas Wick

Abstract This contribution is the second part of two papers on the Fully Eulerian formulation for fluid-structure interactions (fsi). We present different fsi applications using the Fully Eulerian scheme, where traditional interface-tracking approaches like the Arbitrary Lagrangian-Eulerian (ALE) framework show difficulties. Furthermore, we present examples where parts of the geometry undergo a large motion or deformation that might lead to contact and/or topology changes. Finally, we present an application of the scheme for growing structures. The verification of the framework is performed with mesh convergence studies and comparisons to ALE techniques.

1 Introduction

In this second part of the series on the Fully Eulerian formulation for fluid-structure interactions (fsi), we present different test-cases and applications to highlight the potential of this novel formulation. We focus on specific difficulties like large deformations, motion and contact, where interface-tracking approaches such as the Arbitrary Lagrangian Eulerian (ALE) formulation tend to fail without remeshing. Details on the derivation of the model, as well as the finite element discretization of the resulting equations are given in the first part [7].

The paper is organized as follows: In Sect. 2 we verify the Eulerian formulation by means of benchmark problems and comparison with ALE computations. Further, by modification of the benchmark descriptions, we go beyond the limit that can be reached by ALE techniques. Next, Sect. 3 presents test-cases, where the solid

S. Frei • T. Richter
Institut für Angewandte Mathematik, Universität Heidelberg, INF 294, 69120 Heidelberg, Germany
e-mail: stefan.frei@iwr.uni-heidelberg.de; thomas.richter@iwr.uni-heidelberg.de

T. Wick (✉)
The Institute for Computational Engineering and Sciences, The University of Texas at Austin, Austin, TX 78712, USA
e-mail: twick@ices.utexas.edu

A. Abdulle et al. (eds.), *Numerical Mathematics and Advanced Applications ENUMATH 2013*, Lecture Notes in Computational Science and Engineering 103,
DOI 10.1007/978-3-319-10705-9_75

undergoes a very large motion. In Sect. 4 we focus on problems with contact and break of contact. In Sect. 5, we apply the locally modified finite element technique described in the first part to a simple fsi problem. Finally, in Sect. 6 we discuss applications with growing structures, as they appear in the growth and rupture of plaque in blood vessels. We conclude in section "Conclusion", where we also discuss some open topics and shortcomings of the Fully Eulerian approach for fluid-structure interactions. All tests are computed either with the finite software library Gascoigne [2] or with the fsi-code [12] based on deal.II [1].

2 Numerical Validation: Benchmark Problems

In this section, we present two test cases. The first test is based on a Computational Structure Mechanics (CSM) benchmark in which a gravitational force acts on an elastic beam deflecting it towards the bottom of the configuration (see the results in Fig. 1). The second example is an extension to fluid-structure interaction. Although both tests reach a stationary limit, important issues such as interface cuts are already present. The CSM test case is split into two sub-cases. The first case is a widely used benchmark [3], in the second one we increase the force acting on the beam such that it touches the lower wall. This test case is motivated by studies of Dunne [4] and it shows the potential of the fully Eulerian formulation. The results of the first case are summarized in Table 1 (left) and compared to results obtained with an ALE code. In the second test [11] we are able to simulate the situation where the beam touches the lower wall (up to one mesh cell because otherwise the fluid continuum equations are no longer valid), see Fig. 1. For the y-displacement, the results are very similar to results obtained by Richter [9] (see Table 1 at right). For the x-displacements, we observe a slight difference, however, our findings are in reasonable agreement.

In the second test, the FSI 1 benchmark [3] is considered in which a parabolic inflow is prescribed. The elastic beam deforms caused by a pressure difference because of the non-symmetric location of the cylinder. First results using a stationary code were presented in [10]. Now, we use our recent advances [9, 11] to recompute

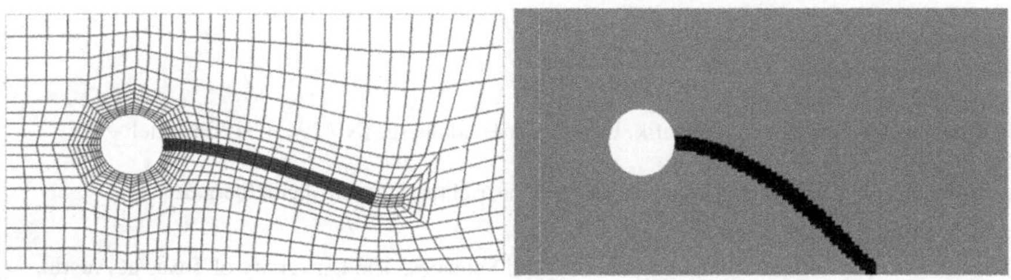

Fig. 1 CSM-1 benchmark test (*left*) and CSM touching the lower wall for $g_s = 8\,\mathrm{ms}^{-2}$ (*right*). The *left figure* displays the ALE computation with the moving mesh whereas the *right* shows the Eulerian result

Table 1 Results for CSM 1 test in Eulerian coordinates. Force $g_s = 2\,\mathrm{ms}^{-2}$ (left) and $g_s = 8\,\mathrm{ms}^{-2}$ (right)

DoF	$u_x(A)[\times 10^{-3}]$	$u_y(A)[\times 10^{-3}]$
4,004	−4.6371	−65.7587
12,868	−8.7366	−67.5678
48,768	−7.8551	−66.5940
179,936	−7.0841	−66.2150
ALE	−7.1455	−65.8808

DoF	$u_x(A)[\times 10^{-3}]$	$u_y(A)[\times 10^{-3}]$
20,988	−25.26	−192.61
54,744	−54.55	−195.26
184,629	−53.33	−196.11
691,233	−55.04	−196.89
(Ref. [9])	−66.857	−192.35

Table 2 Results for the FSI 1 benchmark

Level	DoF	$u_x(A)[\times 10^{-3}]$	$u_y(A)[\times 10^{-3}]$	Drag	Lift
3	131,976	0.0236	0.8146	18.831	0.7784
	(Ref. [8])	0.0227	0.8209	14.295	0.7638

this example with a nonstationary code version. Our findings are summarized in Table 2. In order to keep the computational cost reasonable, local mesh refinement around the elastic beam is applied.

3 Large Motion: 360° Rotation

Usually, the reason for the break-down of the ALE approach is loss of regularity in the ALE transformation map and not the large deformation or motion of the structure itself.

In Fig. 2, we show a prototypical configuration: an (elastic or rigid) body $\mathcal{S} = \mathcal{S}(0)$ is centered in a flow container $\mathcal{S} \subset \Omega$. By a rotational flow in the fluid-domain $\mathcal{F} = \mathcal{F}(0) = \Omega \setminus \mathcal{S}(0)$ the object starts to rotate. Here, it is not a deformation of the solid but a rigid body rotation that causes severe problems in ALE computations (top line of Fig. 2).

In the Fully Eulerian formulation, the large motion and rotation does not cause any problems, as the fluid problem is given in the Eulerian framework, see the bottom line of Fig. 2. In this framework, the interface moves through the domain and must be captured by the Initial Point Set. As these computations have been done without a specially fitted interface finite element method, we observe a strong loss of accuracy. For the Eulerian framework, meshes with a finer resolution are required to reach the same accuracy as with an ALE approach. In order to obtain approximately the same accuracy, we used about 4,000 locally refined elements for the Eulerian approach vs. 400 elements in the ALE case (cf. Fig. 2).

As an interface-capturing technique, the Fully Eulerian approach is not strictly mass-conserving, the solid mass

$$m_s(u) = \int_{\mathcal{S}(t)} J(u)\rho_s^0\,\mathrm{d}x$$

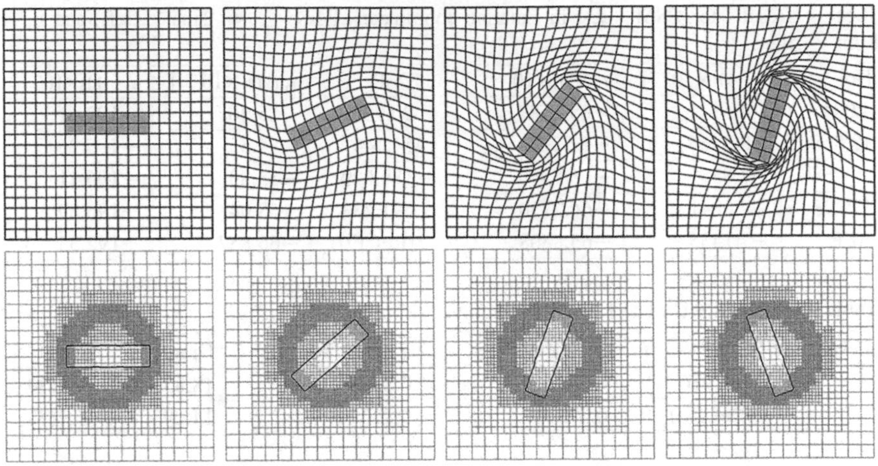

Fig. 2 Rotational flow around an unmounted obstacle at different time steps. *Top row*: ALE computation. *Bottom row*: Fully Eulerian

depends on the accuracy of the captured interface. For this test-case, we observe, that linear finite elements (even without fitted interface modifications) show second order in capturing the solid mass

$$|m_s(u) - m_s(u_h)| = O(h^2).$$

4 Touching the Boundary

We consider a test problem, that has been introduced in [9]. An elastic ball "falls" due to gravity in a viscous fluid until it touches the bottom of the rigid fluid domain. Due to elasticity, the ball bounces off and is elevated, then falls down again for several times until the motion is finally damped by viscous effects.

The fixed computational domain is set to $\Omega = (-1, 1)^2$ m, and at reference time $t = 0$ the system is at rest, with the ball being centered in the origin $\mathcal{S}(0) = \{x \in \Omega : |x| < 0.4\,\text{m}\}$. The fluid-domain $\mathcal{F}(0) = \Omega \setminus \mathcal{S}(0)$ is governed by a viscous incompressible Navier-Stokes fluid with density $\rho_f = 10^3\,\text{kg}\,\text{m}^{-3}$ and kinematic viscosity $\nu_f = 10^{-2}\,\text{m}^2\,\text{s}^{-1}$. The elastic ball has density $\rho_s^0 = 10^3\,\text{kg}\,\text{m}^{-3}$ and the Lamé coefficients $\mu_s = 10^4\,\text{kg}\,\text{m}^{-1}\,\text{s}^{-2}$ and $\lambda_s = 4 \cdot 10^4\,\text{kg}\,\text{m}^{-1}\,\text{s}^{-2}$. The problem is driven by a right hand side, that acts on the solid-domain only, $\mathbf{f}_s = -1\,\text{m}\,\text{s}^{-2}$ and $\mathbf{f}_f = 0$. On the bottom part of the boundary $\Gamma_{\text{bot}} = \{(-1, 1) \times \{-1\}\}$ we prescribe homogenous Dirichlet conditions for the velocities $\mathbf{v} = 0$. On all other parts of the boundary $\partial\Omega \setminus \Gamma_{\text{bot}}$ we prescribe the "do-nothing" outflow condition for the fluid $-p_f\mathbf{n} + \rho_f \nu_f \partial_n \mathbf{v}_f = 0$. Finally, we prescribe homogenous Neumann conditions for the displacement $\partial_n \mathbf{u} = 0$ on the complete boundary $\partial\Omega$.

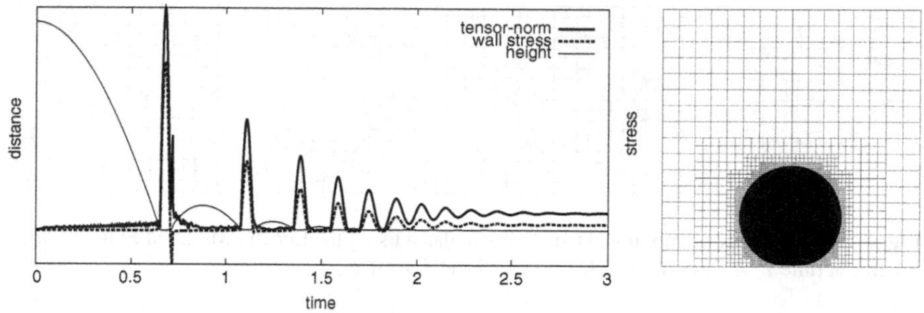

Fig. 3 *Left*: height of the falling ball over the ground, wall-stress and solid Cauchy stress. *Right*: configuration at the time of closest contact $t = 0.675$. Small undershoots in the wall-stress are due to the forces of the incompressible fluid on the boundary at separation time

The main interest of this test case is the contact of the structure with the boundary of the domain. In Fig. 3, we plot the distance of the ball to the ground. First we observe, that the ball touches the ground for a short time-interval and then bounces off. The maximum elevation is reduced after each contact with the domain's boundary. Further, we show the normal wall stress on the lower boundary and the norm of the Cauchy stresses within the solid:

$$J_{\text{wall-stress}} = \int_{\Gamma_{\text{bot}}} \mathbf{n} \cdot \boldsymbol{\sigma}_f \mathbf{n} \, \mathrm{d}o, \quad J_{\text{solid-stress}} = \int_{\mathcal{S}(t)} \boldsymbol{\sigma}_s : \boldsymbol{\sigma}_s \, \mathrm{d}x.$$

There will always be a thin layer of fluid around the structure, such that there is no real "contact" between both phases. The right part in Fig. 3 shows a plot of the elastic ball on the fixed background mesh at the time where ball and boundary are closest. The forces are transferred via the remaining small layer of fluid. These results are stable under refinement of the temporal and spatial discretization. It however still remains to show, that the realization of contact, which is modeled by the discretization only, gives realistic results. Here, comparisons to experiments and numerical benchmarking with alternative formulations are necessary steps in future work.

5 Locally Modified Finite Element Scheme

In this section, we present first nonstationary results using the Locally Modified Finite Element scheme described in the first part [5, 7] applied to a simplified fluid-structure interaction problem. The problem under consideration consists of an elastic ball in the middle of a fluid governed by the linear Stokes equation. The elastic solid is governed by a fully linearized elasticity model. The flow field is driven by a prescribed parabolic inflow on the left-hand side of the domain. This

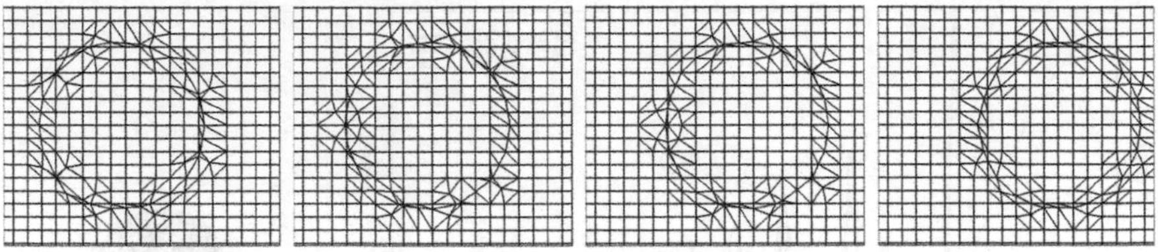

Fig. 4 Screen-shots of a moving elastic ball simulated using the Locally Modified Finite Element scheme at time $T = 0.05$, $T = 0.1$, $T = 0.15$ and $T = 0.2$

causes the elastic ball to move towards the right side where we prescribe the do-nothing outflow condition. In Fig. 4, we show screen-shots of simulation results at four different time steps. The time step was chosen $k = 10^{-3}$.

As described in [5, 7], we use a fixed patch mesh for all time steps. Outside the interface region, we split each patch $P \in \Omega_h$ into four quadrilaterals (type 1). Patches cut by the interface are split into 8 triangles that resolve the interface with a linear approximation (types 2–4). The type assigned to a patch may vary in every time step depending on the position of the interface. Although the aspect ratio of the triangles can get arbitrarily bad, we can make sure, that all triangles have interior angles bound away from π. This guarantees robust interpolation estimates. To cope with the bad conditioning of the system matrix, we use a hierarchical basis on those patches, that are cut by the interface.

This approach is equivalent to a fitted finite element method using a mixed triangular-quadrilateral mesh, which is well known to give optimal approximation properties. However, instead of modifying the mesh, we locally modify the finite element basis. The number of unknowns and the connectivity of the system matrix does not depend on the interface location.

6 Growing Structures and Clogging Phenomena

In this final example, we present results showing our recent efforts in modeling and simulating growing solids and clogging phenomena. The key idea relies on a multiplicative decomposition of the displacement gradient into an elastic and a growth part. The fundamental relation (see Part I [7]) is given by

$$F = \hat{F}^{-1}. \tag{1}$$

Now, the Eulerian displacement gradient is split into a growth part and an elastic part using relation (1):

$$F = \hat{F}^{-1} = \hat{F}_g^{-1} \hat{F}_e^{-1} =: \hat{F}_g^{-1} F_e, \quad F_e := \hat{F}_e^{-1},$$

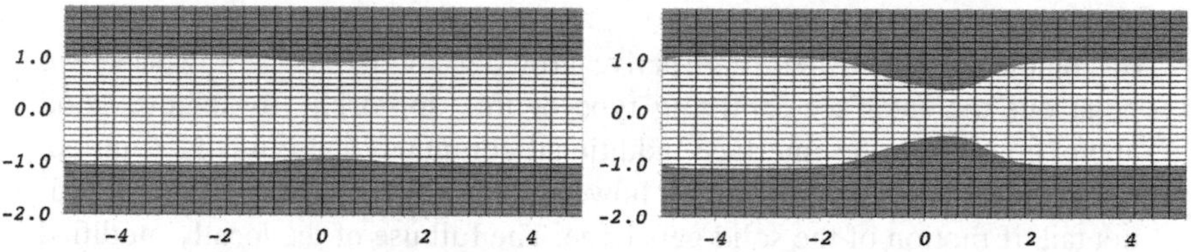

Fig. 5 Configuration 2: Deformation at times $T = 10, 40$ in the time interval [0, 80]

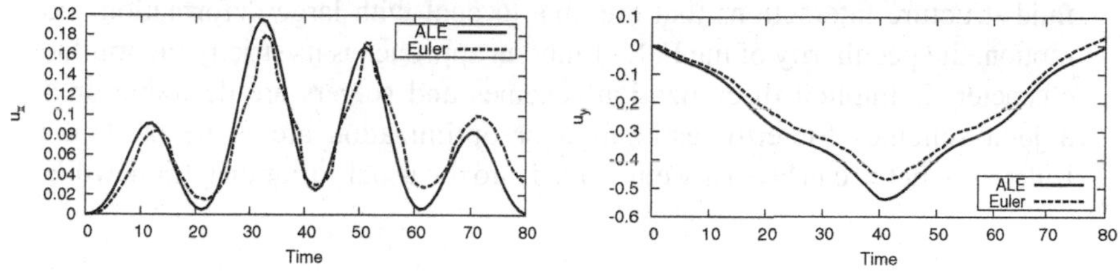

Fig. 6 Deformation of the top solid at channel's mid-point at times $T = 10, 40$ in the time interval [0, 80]

and

$$J := \hat{J}^{-1} = \hat{J}_g^{-1}\hat{J}_e^{-1} =: \hat{J}_g^{-1} J_e, \quad J_e := \det F_e = \hat{J}_e^{-1}.$$

The growth tensor is defined as $\hat{F}_g^{-1} := \hat{g}(\hat{x}, t)I$.

The problem is driven by a pressure difference described as cosine function and simultaneously by growth of the structure. Consequently, we consider two effects, namely, fsi-interaction and solid-growth. The configuration and the material parameters are taken from [6], Example 2. In addition to these tests, we now consider growth and back-growth of the solid (for instance when a plaque-disease occurs and vanishes after treatment). By observing Figs. 5 and 6, we see that the findings in both frameworks show similar qualitative behavior. The results of the Eulerian approach are less accurate and stable, as this test-case is computed using the standard non-fitted finite element approach without the modifications described in Sect. 5.

Conclusion

As shown in our studies, the Fully Eulerian formulation for fluid-structure interactions offers an alternative modeling approach, that can be preferable for certain classes of problems. As a monolithic model, one can use strong implicit discretization schemes with large time-steps, independent of the problem's stiffness. One benefit of the Eulerian scheme is the simple incorporation

(continued)

of complex models, like, e.g. active growing solids. As a drawback, we point out the interface-capturing type of this approach. To achieve good approximation at the interface, additional computational effort is required. Often, finer meshes are required, however, the Fully Eulerian approach will not fail, if motion of the solid gets large. The full use of the locally modified finite element scheme [5, 7] for nonstationary problems will essentially remove this drawbacks. Among the large variety of different schemes for fluid-structure interactions that are able to deal with large deformation and motion, the peculiarity of the Fully Eulerian approach is its strictly monolithic character. If implicit discretization schemes and solvers are desirable or if adjoint schemes for error estimation or optimization are to be used, the Eulerian scheme can be easily embedded into the usual variational framework.

References

1. W. Bangerth, R. Hartmann, G. Kanschat, deal.II – a general purpose object oriented finite element library. ACM Trans. Math. Softw. **33**(4), 24/1–24/27 (2007)
2. R. Becker, M. Braack, D. Meidner, T. Richter, B. Vexler, The finite element toolkit GASCOIGNE. (2014). http://www.gascoigne.uni-hd.de
3. H.-J. Bungartz, M. Schäfer, *Fluid-Structure Interaction: Modelling, Simulation, Optimization*. Lecture Notes in Computational Science and Engineering, vol. 53 (Springer, Berlin/Heidelberg, 2006)
4. T. Dunne, An Eulerian approach to fluid-structure interaction and goal-oriented mesh refinement. Int. J. Numer. Math. Fluids. **51**, 1017–1039 (2006)
5. S. Frei, T. Richter, A locally modified parametric finite element method for interface problems. SIAM J. Numer. Anal. **52**(5), 2315–2334 (2013)
6. S. Frei, T. Richter, T. Wick, Solid growth and clogging in fluid-structure interaction using ALE and fully Eulerian coordinates (2014, Preprint, in review)
7. ———, Eulerian techniques for fluid-structure interactions – part I: modeling and simulation, in *ENUMATH 2013 Proceedings*, Lausanne. Lecture Notes in Computational Science and Engineering (Springer, 2014)
8. J. Hron, S. Turek, *Proposal for Numerical Benchmarking of Fluid-Structure Interaction Between an Elastic Object and Laminar Incompressible Flow*. In Fluid-Structure Interaction: Modeling, Simulation, Optimization, eds. by H.-J. Bungartz, M. Schäfer, vol. 53 (Springer, Berlin/Heidelberg, 2006), pp. 146–170
9. T. Richter, A fully Eulerian formulation for fluid-structure-interaction problems. J. Comput. Phys. **233**, 227–240 (2013)
10. T. Richter, T. Wick, Finite elements for fluid-structure interaction in ale and fully Eulerian coordinates. Comput. Methods Appl. Mech. Eng. **199**, 2633–2642 (2010)
11. T. Wick, Fully Eulerian fluid-structure interaction for time-dependent problems. Comput. Methods Appl. Mech. Eng. **255**, 14–26 (2013)
12. ———, Solving monolithic fluid-structure interaction problems in arbitrary Lagrangian Eulerian coordinates with the deal.ii library. Arch. Numer. Softw. **1**, 1–19 (2013)

Part XI
Computational Electromagnetics

A Local Projection Stabilization FEM for the Linearized Stationary MHD Problem

Benjamin Wacker and Gert Lube

Abstract We present a local projection stabilization (LPS) type finite element (FE) method for the linearized stationary magnetohydrodynamics (MHD) problem. In contrast to the residual-based stabilization in Badia et al. (J Comput Phys 234:399–416, 2013; Analysis of an unconditionally convergent stabilized finite element formulation for incompressible magnetohydrodynamics, submitted), we investigate a symmetric LPS method comparable to the term-by-term stabilization in Badia et al. (Int J Numer Methods Eng 93:302–328, 2013).

1 Introduction

Following the time discretization and linearization approach in [1–3], we consider the stationary MHD model

$$-\nu\Delta\mathbf{u} + (\mathbf{a}\cdot\nabla)\mathbf{u} + \nabla p - (\nabla\times\mathbf{b})\times\mathbf{d} = \mathbf{f_u} \;,\; \nabla\cdot\mathbf{u} = 0, \tag{1}$$

$$\lambda\nabla\times(\nabla\times\mathbf{b}) + \nabla r - \nabla\times(\mathbf{u}\times\mathbf{d}) = \mathbf{f_b} \;,\; \nabla\cdot\mathbf{b} = 0, \tag{2}$$

in a bounded Lipschitz domain $\Omega \subset \mathbb{R}^d$, $d \in \{2,3\}$ with $\nabla\cdot\mathbf{a} = 0$. $\mathbf{a}$ and $\mathbf{d}$ are the vector-fields for the velocity and magnetic field at linearization. For the unknown velocity field $\mathbf{u}$, magnetic field $\mathbf{b}$, pressure p and magnetic pseudo-pressure r (vanishing in the continuous case), we introduce the function spaces

$$V = \left\{\mathbf{v}\in\left[H^1(\Omega)\right]^d \;:\; \mathbf{v} = 0 \text{ on } \partial\Omega\right\} \quad,\quad Q = L_0^2(\Omega),$$

$$C = \{\mathbf{c}\in H(curl\,;\Omega) \;:\; \mathbf{n}\times\mathbf{c} = 0 \text{ on } \partial\Omega\} \quad,\quad S = H_0^1(\Omega)$$

B. Wacker • G. Lube (✉)
Institute for Numerical and Applied Mathematics, University of Göttingen, Lotzestr. 16-18, 37073 Göttingen, Germany
e-mail: b.wacker@math.uni-goettingen.de; lube@math.uni-goettingen.de

A. Abdulle et al. (eds.), *Numerical Mathematics and Advanced Applications ENUMATH 2013*, Lecture Notes in Computational Science and Engineering 103,
DOI 10.1007/978-3-319-10705-9_76

where $(\cdot,\cdot)$ and $\langle\cdot,\cdot\rangle$ are appropriate inner and dual products. The variational problem reads: Find $\mathbf{U} := (\mathbf{u}, \mathbf{b}, p, r) \in V \times C \times Q \times S$ such that

$$\mathscr{A}_G(\mathbf{U}, \mathbf{V}) = \mathscr{F}_G(\mathbf{V}), \quad \forall \mathbf{V} := (\mathbf{v}, \mathbf{c}, q, s) \in V \times C \times Q \times S \tag{3}$$

with

$$\begin{aligned}\mathscr{A}_G(\mathbf{U}, \mathbf{V}) = {} & \nu(\nabla\mathbf{u}, \nabla\mathbf{v}) + \langle \mathbf{a}\cdot\nabla\mathbf{u}, \mathbf{v}\rangle - (p, \nabla\cdot\mathbf{v}) - \langle(\nabla\times\mathbf{b})\times\mathbf{d}, \mathbf{v}\rangle \\ & + (\nabla\cdot\mathbf{u}, q) - (\mathbf{b}, \nabla s) \qquad (4) \\ & + \lambda(\nabla\times\mathbf{b}, \nabla\times\mathbf{c}) + (\nabla r, \mathbf{c}) - \langle\nabla\times(\mathbf{u}\times\mathbf{d}), \mathbf{c}\rangle, \\ \mathscr{F}_G(\mathbf{V}) = {} & \langle \mathbf{f_u}, \mathbf{v}\rangle + \langle \mathbf{f_b}, \mathbf{c}\rangle. \qquad (5)\end{aligned}$$

Let $\mathscr{T}_h$ be the primal grid with FE spaces of Taylor-Hood type

$$V_h \times Q_h / C_h \times S_h = \mathbb{P}^k_{\mathscr{T}_h} \times \mathbb{P}^{k-1}_{\mathscr{T}_h} \text{ or } \mathbb{Q}^k_{\mathscr{T}_h} \times \mathbb{Q}^{k-1}_{\mathscr{T}_h}, \quad k \in \mathbb{N} \setminus \{1\}. \tag{6}$$

The pair $V_h \times Q_h$ is discretely-divergence-free, thus

$$V_h^{div} := \{\mathbf{v_h} \in V_h : (\nabla\cdot\mathbf{v}_h, q_h) = 0 \ \forall q_h \in Q_h\} \neq \{0\}.$$

Let $\mathscr{M}_h = \mathscr{T}_h$ or $\mathscr{M}_h = \mathscr{T}_{2h}$ be the macro grid with discontinuous FE spaces $D_h^{u/b} \subset [L^2(\Omega)]^d$. The local orthogonal L^2-projectors are denoted as $\pi_M^{u/b} : [L^2(M)]^d \to D_h^{u/b}|_M$. The global projections $\pi_h^{u/b} : [L^2(\Omega)]^d \to D_h^{u/b}$ are given as $(\pi_h^{u/b}\mathbf{w})|_M := \pi_M^{u/b}(\mathbf{w}|_M)$. The fluctuation operator $\kappa_h^{u/b} : [L^2(\Omega)]^d \to [L^2(\Omega)]^d$ with $\kappa_h^{u/b}\mathbf{w} := \left((id - \pi_h^{u/b})w_i\right)_{i=1}^d$ is assumed to have the approximation property

$$\|\kappa_h^{u/b}\mathbf{v}\|_{0,M} \le Ch_M^l \|\mathbf{v}\|_{l,M} \quad \forall \mathbf{v} \in [W^{l,2}(M)]^d, l = 0, \dots, s, \ s \in \{0, \dots, k\}. \tag{7}$$

Let $\mathbf{U_h} = (\mathbf{u_h}, \mathbf{b_h}, p_h, r_h), \mathbf{V_h} = (\mathbf{v_h}, \mathbf{c_h}, q_h, s_h) \in V_h \times C_h \times Q_h \times S_h \subset V \times C \times Q \times S$. Then the LPS terms read

$$\begin{aligned}\mathscr{S}_{lps}(\mathbf{U_h}, \mathbf{V_h}) = \sum_M \Big\{ & \tau_1 \left(\kappa_h^u((\mathbf{a_M}\cdot\nabla)\mathbf{u_h}), \kappa_h^u((\mathbf{a_M}\cdot\nabla)\mathbf{v_h})\right)_M + \tau_2(\nabla\cdot\mathbf{u_h}, \nabla\cdot\mathbf{v_h})_M \\ & + \tau_3 \left(\kappa_h^b((\nabla\times\mathbf{b_h})\times\mathbf{d_M}), \kappa_h^b((\nabla\times\mathbf{c_h})\times\mathbf{d_M})\right)_M \\ & + \tau_4 \left(\kappa_h^b((\nabla\times(\mathbf{u_h}\times\mathbf{d_M})), \kappa_h^b(\nabla\times(\mathbf{v_h}\times\mathbf{d_M}))\right)_M \\ & + \tau_5(\nabla r_h, \nabla s_h)_M + \tau_6(\nabla\cdot\mathbf{b_h}, \nabla\cdot\mathbf{c_h})_M \Big\}\end{aligned}$$

where $(\cdot,\cdot)_M$ is the L^2 scalar product on cell M. Here, $\mathbf{a_M}$ and $\mathbf{d_M}$ are elementwise constant approximations of $\mathbf{a}|_\mathbf{M}$ and $\mathbf{d}|_\mathbf{M}$. The LPS problem consists of finding $\mathbf{U_h} \in V_h \times C_h \times Q_h \times S_h$ such that for all $\mathbf{V_h} \in V_h \times C_h \times Q_h \times S_h$:

$$\mathscr{A}_{stab}(\mathbf{U_h},\mathbf{V_h}) = \mathscr{A}_G(\mathbf{U_h},\mathbf{V_h}) + \mathscr{S}_{lps}(\mathbf{U_h},\mathbf{V_h}) = \mathscr{F}_G(\mathbf{V_h}). \tag{8}$$

2 Stability of the Proposed Method

For $k \in \mathbb{N}_0$ and $D \subseteq \Omega$, we use the notation $|\cdot|_{k,D} := |\cdot|_{H^k(D)}$ and $\|\cdot\|_{p,D} := \|\cdot\|_{L^p(D)}$ with $1 \le p \le \infty$. In case of $D = \Omega$, we omit index D.

Lemma 1 *For* $\mathbf{U},\mathbf{V} \in V \times C \times Q \times S$, *it holds for the symmetric LPS terms*

$$(i)\ \mathscr{S}_{lps}(\mathbf{U},\mathbf{U}) \ge 0,\quad (ii)\ \left|\mathscr{S}_{lps}(\mathbf{U},\mathbf{V})\right| \le \left(\mathscr{S}_{lps}(\mathbf{U},\mathbf{U})\right)^{\frac{1}{2}} \left(\mathscr{S}_{lps}(\mathbf{V},\mathbf{V})\right)^{\frac{1}{2}}.$$

Let $\mathbf{V} = (\mathbf{v},\mathbf{c},q,s) \in V \times C \times Q \times S$. Integration by parts yields $\langle \mathbf{a}\cdot\nabla\mathbf{v},\mathbf{v}\rangle = 0$ and $\langle(\nabla\times\mathbf{c})\times\mathbf{d},\mathbf{v}\rangle = -\langle\nabla\times(\mathbf{v}\times\mathbf{d}),\mathbf{c}\rangle$, hence

$$\mathscr{A}_G(\mathbf{V},\mathbf{V}) = \nu\|\nabla\mathbf{v}\|_0^2 + \lambda\|\nabla\times\mathbf{c}\|_0^2.$$

We define the following expressions

$$\|\mathbf{V}\|_{\mathrm{G}}^2 = \nu\|\nabla\mathbf{v}\|_0^2 + \lambda\|\nabla\times\mathbf{c}\|_0^2, \qquad \|\mathbf{V}\|_{\mathrm{lps}}^2 = \mathscr{S}_{lps}(\mathbf{V},\mathbf{V}). \tag{9}$$

Symmetric testing $\mathbf{V_h} = \mathbf{U_h}$ yields

$$\left(\mathscr{A}_G + \mathscr{S}_{\mathrm{lps}}\right)(\mathbf{v_h},\mathbf{v_h}) = \|\mathbf{v_h}\|_{\mathrm{G}}^2 + \|\mathbf{v_h}\|_{\mathrm{lps}}^2 = \langle\mathbf{f_u},\mathbf{u_h} > + < \mathbf{f_b},\mathbf{b_h}\rangle. \tag{10}$$

By the discrete Babuška-Brezzi-condition, we have for all $p_h \in Q_h$ the unique existence of $\mathbf{v_h} \in V_h$ with

$$\nabla\cdot\mathbf{v_h} = -p_h\ ,\ |\mathbf{v_h}|_1 \le \beta_u^{-1}\|p_h\|_0. \tag{11}$$

Examining the term $\left(\mathscr{A}_G + \mathscr{S}_{lps}\right)(\mathbf{U_h},(\mathbf{v_h},\mathbf{0},0,0))$, we end up with

$$\begin{aligned}\|p_h\|_0^2 \le \|\mathbf{f_u}\|_{-1}|\mathbf{v_h}|_1 + \nu\|\nabla\mathbf{u_h}\|_0|\mathbf{v_h}|_1 + \left|\mathscr{S}_{lps}((\mathbf{u_h},\mathbf{0},0,0),(\mathbf{v_h},\mathbf{0},0,0))\right| \\ -(\mathbf{a}\cdot\nabla\mathbf{u_h},\mathbf{v_h}) + ((\nabla\times\mathbf{b_h})\times\mathbf{d},\mathbf{v_h})\end{aligned}$$

by using that $\mathbf{u_h} \in V_h^{div}$. Based on the inequalities

$$-(\mathbf{a}\cdot\nabla\mathbf{u_h},\mathbf{v_h}) = (\mathbf{a}\cdot\nabla\mathbf{v_h},\mathbf{u_h}) \le C_p\|\mathbf{a}\|_\infty|\mathbf{u_h}|_1|\mathbf{v_h}|_1, \tag{12}$$

$$((\nabla\times\mathbf{b_h})\times\mathbf{d},\mathbf{v_h}) \le C_p\|\mathbf{d}\|_\infty\|\nabla\times\mathbf{b_h}\|_0|\mathbf{v_h}|_1, \tag{13}$$

$$\|(\mathbf{v_h}, \mathbf{0}, 0, 0)\|_{\text{lps}}^2 \le \max_M \left(\tau_1 |\mathbf{a_M}|^2 + \tau_2 d + \tau_4 |\mathbf{d}_M|^2\right) |\mathbf{v_h}|_1^2 \tag{14}$$

together with (11), we obtain after some calculation an estimate of the fluid pressure

$$\beta_u \|p_h\|_0 \le \|\mathbf{f_u}\|_{-1} + \left(\sqrt{\nu} + \frac{C_p \|\mathbf{a}\|_\infty}{\sqrt{\nu}} + \frac{C_p \|\mathbf{d}\|_\infty}{\sqrt{\lambda}}\right) \|\mathbf{U_h}\|_G$$
$$+ \left(\max_M \left(\sqrt{\tau_1} |\mathbf{a_M}|\right) + \max_M \sqrt{\tau_2 d} + \sqrt{\tau_4} |\mathbf{d}_M|\right) \|\mathbf{U_h}\|_{\text{lps}}. \tag{15}$$

We define the norms

$$\|\mathbf{c}\|_C = \sqrt{\lambda}\left(L_0^{-1}\|\mathbf{c}\|_0 + \|\nabla \times \mathbf{c}\|_0\right), \quad \|s\|_S = \left(\|s\|_0 + L_0 \|\nabla s\|_0\right)/\sqrt{\lambda} \tag{16}$$

on the spaces C and S with a length-scale $L_0 \sim \text{diam}(\Omega)$. Using integration by parts, we define the bilinear form of the Maxwell problem

$$\mathscr{C}_{Max}\left((\mathbf{b}, r), (\mathbf{c}, s)\right) = \mathscr{A}_G\left((\mathbf{0}, \mathbf{b}, 0, r), (\mathbf{0}, \mathbf{c}, 0, s)\right).$$

For this problem, the continuous Babuška-Brezzi-condition

$$\inf_{(\mathbf{b},r)\in C\times S} \sup_{(\mathbf{c},s)\in C\times S} \frac{\mathscr{C}_{Max}\left((\mathbf{b}, r), (\mathbf{c}, s)\right)}{\left(\|\mathbf{b}\|_C + \|r\|_S\right)\left(\|\mathbf{c}\|_C + \|s\|_S\right)} \ge \beta_m \tag{17}$$

holds. Let $(\mathbf{b_h}, 0) \in C_h \times S_h \subset C \times S$. By (17) there exists a unique $(\overline{\mathbf{c}}, \overline{s}) \in C \times S$ with $\|\overline{\mathbf{c}}\|_C + \|\overline{s}\|_S = 1$ such that

$$\beta_m \|\mathbf{b_h}\|_C \le \mathscr{C}_{Max}\left((\mathbf{b_h}, 0), (\overline{\mathbf{c}}, \overline{s})\right) \tag{18}$$
$$= \mathscr{C}_{Max}\left((\mathbf{b_h}, 0), (\overline{\mathbf{c}}, 0)\right) + \mathscr{C}_{Max}\left((\mathbf{b_h}, 0), (0, \overline{s})\right) =: I + II$$

holds. We get with $\|\overline{s}\|_S \le 1$ and $\|\overline{\mathbf{c}}\|_C \le 1$ the estimates

$$I = \lambda\left(\nabla \times \mathbf{b_h}, \nabla \times \overline{\mathbf{c}}\right) \le \sqrt{\lambda}\|\nabla \times \mathbf{b_h}\|_0 \sqrt{\lambda}\|\nabla \times \overline{\mathbf{c}}\|_0 \le \|\mathbf{U_h}\|_G, \tag{19}$$

$$II = |(\nabla \cdot \mathbf{b_h}, \overline{s})| \le \left(\sum_M \tau_6 \|\nabla \cdot \mathbf{b_h}\|_{0,M}^2\right)^{\frac{1}{2}} \left(\sum_M \frac{1}{\tau_6} \|\overline{s}\|_{0,M}^2\right)^{\frac{1}{2}}$$

$$\le \|\mathbf{U_h}\|_{lps} \cdot \left(\max_M \frac{1}{\sqrt{\tau_6}}\right) \cdot \sqrt{\lambda} \frac{\|\overline{s}\|_0}{\sqrt{\lambda}} = \max_M \sqrt{\frac{\lambda}{\tau_6}} \cdot \|\mathbf{U_h}\|_{\text{lps}} \tag{20}$$

by the Cauchy-Schwarz inequality. Putting (19) and (20) into (18) and requiring $\max_M \lambda/\tau_6 \le 1$ leads to

$$\beta_M \|\mathbf{b_h}\|_C \leq \left(\|\mathbf{U_h}\|_G + \max_M \sqrt{\frac{\lambda}{\tau_6}} \|\mathbf{U_h}\|_{\text{lps}}\right) \leq \|\mathbf{U_h}\|_G + \|\mathrm{U}\|_{\text{lps}} \tag{21}$$

Adding (10) and squared inequality (20) and applying Young's inequality leads after some calculation to

$$\beta_M^2 \|\mathbf{b_h}\|_C^2 + \nu \|p_{\mathbf{u_h}}\|_0^2 \leq \frac{g}{\nu} \|\mathbf{f_u}\|_{-1}^2 + \frac{g}{\beta_M^2} \|\mathbf{b_h}\|_{C^*}^2$$

where C^* in the dual of the space C. This gives the uniqueness and existence of the discrete velocity and magnetic fields $\mathbf{u_h}$ and $\mathbf{b_h}$ and inequality. (15) implies the unique existence of the discreate kinematic pressure.

Finally, from the equation $\left(\mathscr{A}_G + \mathscr{S}_{lps}\right)(\mathbf{U_h}, (\mathbf{0}, \mathbf{0}, 0, r_h)) = -(\mathbf{b_h}, \nabla r_h) + \sum_M \tau_5 \|\nabla r_h\|_{0,M}^2 = 0$, we conclude by using (16) that

$$\|\nabla r_h\|_0 \leq \left(\min_M \sqrt{\tau_5}\right)^{-1} \|\mathbf{b_h}\|_0 \leq L_0 \left(\sqrt{\lambda} \min_M \sqrt{\tau_5}\right)^{-1} \|\mathbf{b_h}\|_C. \tag{22}$$

This implies existence of the unique discrete magnetic pseudo-pressure. As full control of ∇r_h is essential to enforce condition $\nabla \cdot \mathbf{b}_h = 0$, (22) and (21) suggest with $C \geq 1$

$$\tau_5 \sim L_0^2/\lambda, \qquad \tau_6 \geq C\lambda. \tag{23}$$

3 Error Analysis for Smooth Solutions

Subtracting (3) and (8) gives the approximate Galerkin orthogonality.

Lemma 2 *Let* $\mathbf{U}$ *and* $\mathbf{U_h}$ *be the solutions of (3) and (8). Then*

$$\left(\mathscr{A}_G + \mathscr{S}_{lps}\right)(\mathbf{U} - \mathbf{U_h}, \mathbf{V_h}) = \mathscr{S}_{lps}(\mathbf{U}, \mathbf{V_h}), \quad \forall \mathbf{V_h} \in V_h \times C_h \times Q_h \times S_h. \tag{24}$$

Let $\mathbf{J} = \left(\mathbf{j^u}, \mathbf{j^b}, j^p, j^r\right)$ be appropriate interpolation operators. In particular, we assume that $\mathbf{j^u u} \in V_h^{div}$. We therefore decompose the error as

$$\mathbf{U} - \mathbf{U_h} = (\mathbf{U} - \mathbf{JU}) + (\mathbf{JU} - \mathbf{U_h}) = \boldsymbol{\varepsilon} + \mathbf{E_h} \equiv \left(\boldsymbol{\varepsilon}_\mathbf{u}, \boldsymbol{\varepsilon}_\mathbf{b}, \varepsilon_p, \varepsilon_r\right) + \left(\mathbf{e_u}, \mathbf{e_b}, e_p, e_r\right).$$

Set now $\mathbf{V_h} = \mathbf{E_h}$ in (24), thus

$$\|\mathbf{E_h}\|_G^2 + \|\mathbf{E_h}\|_{lps}^2 = \underbrace{\mathscr{S}_{lps}\left(\mathbf{U}, \mathbf{E_h}\right)}_{=I} \underbrace{-\mathscr{A}_G\left(\boldsymbol{\varepsilon}, \mathbf{E_h}\right)}_{=II} \underbrace{-\mathscr{S}_{lps}\left(\boldsymbol{\varepsilon}, \mathbf{E_h}\right)}_{=-III}. \tag{25}$$

We obtain

$$I \leq \left(\mathscr{S}_{lps}(\mathbf{U},\mathbf{U})\right)^{\frac{1}{2}}\left(\mathscr{S}_{lps}(\mathbf{E_h},\mathbf{E_h})\right)^{\frac{1}{2}} = \|\mathbf{U}\|_{lps}\|\mathbf{E_h}\|_{lps}, \tag{26}$$

$$|III| = \mathscr{S}_{lps}(\boldsymbol{\varepsilon},\mathbf{E_h}) \leq \|\boldsymbol{\varepsilon}\|_{lps}\|\mathbf{E_h}\|_{lps}, \tag{27}$$

$$-II \leq \|\boldsymbol{\varepsilon}\|_G\|\mathbf{E_h}\|_G + IV \tag{28}$$

$$\begin{aligned} IV &= (\mathbf{a}\cdot\nabla\boldsymbol{\varepsilon}_{\mathbf{u}},\mathbf{e_u}) - ((\nabla\times\boldsymbol{\varepsilon}_{\mathbf{b}})\times\mathbf{d},\mathbf{e_u}) - (\nabla\times(\boldsymbol{\varepsilon}_{\mathbf{u}}\times\mathbf{d}),\mathbf{e_b}) \\ &\quad -\left(\boldsymbol{\varepsilon}_p,\nabla\cdot\mathbf{e_u}\right) + \left(\nabla\cdot\boldsymbol{\varepsilon}_{\mathbf{u}},\mathbf{e}_p\right) + \left(\nabla\boldsymbol{\varepsilon}_r,\mathbf{e_p}\right) - (\boldsymbol{\varepsilon}_{\mathbf{b}},\nabla\mathbf{e}_r). \end{aligned} \tag{29}$$

Then we can summarize estimates (25)–(29) as

$$\|\mathbf{E_h}\|_G^2 + \|\mathbf{E_h}\|_{lps}^2 \leq \left(\|\boldsymbol{\varepsilon}\|_{lps} + \|\mathbf{U}\|_{lps}\right)\|\mathbf{E_h}\|_{lps} + \|\boldsymbol{\varepsilon}\|_G\|\mathbf{E_h}\|_G + |IV|. \tag{30}$$

Integration by parts and Cauchy-Schwarz inequality give for the terms in IV:

$$(\mathbf{a}\cdot\nabla\boldsymbol{\varepsilon}_{\mathbf{u}},\mathbf{e_u}) = -\left(\mathbf{a}\cdot\nabla\mathbf{e_u},\boldsymbol{\varepsilon}_{\mathbf{u}}\right) \leq \left(\sum_M \frac{\|\mathbf{a}\|_{\infty,M}^2}{\nu}\|\boldsymbol{\varepsilon}_{\mathbf{u}}\|_{0,M}^2\right)^{\frac{1}{2}}\|\mathbf{E_h}\|_G,$$

$$-\left(\boldsymbol{\varepsilon}_p,\nabla\cdot\mathbf{e_u}\right) \leq \left(\sum_M \min\left(\frac{d}{\nu};\frac{1}{\tau_2}\right)\|\boldsymbol{\varepsilon}_p\|_{0,M}^2\right)^{\frac{1}{2}}\|\mathbf{E_h}\|_{lps},$$

$$-(\boldsymbol{\varepsilon}_{\mathbf{b}},\nabla e_r) \leq \left(\sum_M \frac{1}{\tau_5}\|\boldsymbol{\varepsilon}_{\mathbf{b}}\|_{0,M}^2\right)^{\frac{1}{2}}\|\mathbf{E_h}\|_{lps}, \tag{31}$$

$$-(\nabla\times(\boldsymbol{\varepsilon}_{\mathbf{u}}\times\mathbf{d}),\mathbf{e_b}) = (\varepsilon_{\mathbf{u}},(\nabla\times\mathbf{e_b})\times\mathbf{d}) \leq \left(\sum_M \frac{\|\mathbf{d}\|_{\infty,M}^2}{\lambda}\|\boldsymbol{\varepsilon}_{\mathbf{u}}\|_{0,M}^2\right)^{\frac{1}{2}}\|\mathbf{E_h}\|_G.$$

The term $(\nabla\varepsilon_r,\mathbf{e_b})$ vanishes since $r = j^r r \equiv 0$. Moreover, term $-(e_p,\nabla\cdot\varepsilon_{\mathbf{u}})$ vanishes via $\nabla\cdot\mathbf{u} = 0$ and since $\mathbf{j^u u} \in V_h^{div}$. Let $\mathbf{d} \in \left[\mathrm{W}^{1,\infty}(\Omega)\right]^d$. By formula $\nabla\times(\mathbf{e}\times\mathbf{f}) = \mathbf{f}\cdot\nabla\mathbf{e} - \mathbf{f}(\nabla\cdot\mathbf{e}) - \mathbf{e}\cdot\nabla\mathbf{f} + \mathbf{e}(\nabla\cdot\mathbf{f})$, the inequalities of Cauchy, Schwarz and Poincare, it follows

$$-((\nabla\times\boldsymbol{\varepsilon}_{\mathbf{b}})\times\mathbf{d},\mathbf{e_u}) = \sum_M (\boldsymbol{\varepsilon}_{\mathbf{b}},\nabla\times(\mathbf{e_u}\times\mathbf{d}))_M$$

$$\leq \left(\sum_M \nu^{-1}(1+\sqrt{d})^2\left(\|\mathbf{d}\|_{\infty,M} + \|\nabla\mathbf{d}\|_{\infty,M}\right)^2\|\boldsymbol{\varepsilon}_{\mathbf{b}}\|_{0,M}^2\right)^{\frac{1}{2}}\|\mathbf{E_h}\|_G. \tag{32}$$

We then summarize Eqs. (30)–(32). Using Young's inequality, we obtain

$$\|\mathbf{E_h}\|_G^2 + \|\mathbf{E_h}\|_{lps}^2 \leq S_1^2 + S_2^2,$$

with

$$S_1 := \|\varepsilon\|_G + \Big(\sum_M \frac{1}{\nu}\|\mathbf{a}\|^2_{\infty,M}\|\boldsymbol{\varepsilon}_{\mathbf{u}}\|^2_{0,M}\Big)^{\frac{1}{2}} + \Big(\sum_M \frac{1}{\lambda}\|\mathbf{d}\|^2_{\infty,M}\|\boldsymbol{\varepsilon}_{\mathbf{u}}\|^2_{0,M}\Big)^{\frac{1}{2}}$$

$$+\Big(\sum_M \nu^{-1}(1+\sqrt{d})^2\big(\|\mathbf{d}\|_{\infty,M} + \|\nabla\mathbf{d}\|_{\infty,M}\big)^2\|\boldsymbol{\varepsilon}_{\mathbf{b}}\|^2_{0,M}\Big)^{\frac{1}{2}},$$

$$S_2 := \|\boldsymbol{\varepsilon}\|_{lps} + \|\mathbf{U}\|_{lps} + \Big(\sum_M \min\Big(\frac{d}{\nu};\frac{1}{\tau_2}\Big)\|\boldsymbol{\varepsilon}_{\mathbf{p}}\|^2_{0,M}\Big)^{\frac{1}{2}} + \Big(\sum_M \frac{1}{\tau_5}\|\boldsymbol{\varepsilon}_{\mathbf{b}}\|^2_{0,M}\Big)^{\frac{1}{2}}.$$

The approximation properties of the FE spaces, see [5], and the local L^2-projector yield for $\mathbf{U} \in [H^{k+1}(\Omega)]^d \times [H^{k+1}(\Omega)]^d \times H^k(\Omega) \times H^k(\Omega)$ that

$$S_1^2 \le C\sum_M h_M^{2k}\Big[\Big(\nu\big(1+\frac{\|\mathbf{a}\|^2_{\infty,M}h_M^2}{\nu^2}\big) + \lambda\frac{\|\mathbf{d}\|^2_{\infty,M}h_M^2}{\lambda^2}\big)\Big)|\mathbf{u}|^2_{k+1,\omega_M}$$

$$+\Big(\lambda + \frac{h_M^2}{\nu}\big(\|\mathbf{d}\|_{\infty,M} + \|\nabla\mathbf{d}\|_{\infty,M}\big)^2\Big)|\mathbf{b}|^2_{k+1,\omega_M}\Big], \tag{33}$$

$$S_2^2 \le C\sum_M h_M^{2s}\Big[\big(\tau_2 d^2 + \tau_1|\mathbf{a_M}|^2 + \tau_4|\mathbf{d_M}|^2\big)|\mathbf{u}|^2_{s+1,\omega_M}$$

$$+\min\Big(\frac{d}{\nu};\frac{1}{\tau_2}\Big)|p|^2_{s,\omega_M} + \big(\tau_3|\mathbf{d_M}|^2 + \tau_6 d^2 + \frac{h_M^2}{\tau_5}\big)|\mathbf{b}|^2_{s+1,\omega_M}\Big] \tag{34}$$

where ω_M denotes an appropriate patch around cell M.

Denote the local fluid and magnetic Reynolds numbers by

$$Re_{f,M} := \|\mathbf{a}\|_{\infty,M}h_M/\nu, \qquad Re_{m,M} := \|\mathbf{d}\|_{\infty,M}h_M/\lambda.$$

respectively. We will call an error estimate to be of order k if the coefficients multiplying corresponding Sobolev norms of the solutions are of order h^k uniformly w.r.t. the problem data. In this case, sufficient conditions can be found by the following (mild) restrictions on the local mesh width h_M

$$\sqrt{\nu}Re_{f,M} \le C,\ \ \sqrt{\lambda}Re_{m,M} \le C,\ \ h_M(\|\mathbf{d}\|_{\infty,M} + \|\nabla\mathbf{d}\|_{\infty,M}) \le C\sqrt{\nu} \tag{35}$$

and on the stabilization parameters (by using (7))

$$0 \le \tau_1 \le Ch_M^{2(k-s)}/|\mathbf{a_M}|^2,\ \ 0 \le \tau_3,\tau_4 \le Ch_M^{2(k-s)}/|\mathbf{d_M}|^2,\ \ Ch_M^2 \le \tau_5. \tag{36}$$

Condition (23) implies the latter condition on τ_5. Moreover, (34) suggests the balance $\tau_5\tau_6 \sim h_M^2$, thus (see also [1, 2])

$$\tau_5 \sim L_0^2/\lambda, \qquad \tau_6 \sim h_M^2 \lambda / L_0^2. \tag{37}$$

A balance of the terms with the div-div parameter τ_2 leads to the practically unfeasible formula $\tau_2 \sim \max\big(0; |p|_{k,M}|/|\mathbf{u}|_{k+1,M} - \nu\big)$. A reasonable compromise is to set

$$\tau_2 \sim 1. \tag{38}$$

Theorem 1 *Assume that the solution* $(\mathbf{u}, \mathbf{b}, p)$ *of (3) belongs to* $[H^{k+1}(\Omega)]^d \times [H^{k+1}(\Omega)]^d \times H^k(\Omega)$ *and that* $\mathbf{j}^{\mathbf{u}}\mathbf{u} \in V_h^{div}$. *Further, let the LPS parameters be chosen according to condition (36)–(37) and that the local mesh width* h_M *is chosen such that (35) is valid. Then we obtain (using* $r \equiv 0$*)*

$$\|\mathbf{U_h} - \mathbf{JU}\|_{\mathrm{G}}^2 + \|\mathbf{U_h} - \mathbf{JU}\|_{\mathrm{lps}}^2 \le C \sum_M h_M^{2k} \Big(|\mathbf{u}|_{k+1,\omega_M}^2 + |\mathbf{b}|_{k+1,\omega_M}^2 + |p|_{k,\omega_M}^2 \Big).$$

Numerical results for the magnetic part, i.e. $\mathbf{u} \equiv 0$, $p \equiv 0$, show the relevance of the parameter design (37) for Taylor-Hood type pairs $C_h \times S_h$. In particular, this is valid if the magnetic field $\mathbf{b}$ does not belong to $[H^1(\Omega)]^d$. Such singular solutions can be well approximated on meshes with suitable macro-element structure, like cross-box elements, see [1]. Our results confirm this for Taylor-Hood type pairs $C_h \times S_h$ as well.

Numerical experiments for the fluid part, i.e. $\mathbf{b} = \mathbf{0}, r = 0$, see [4], show: The mesh conditions (35) are much less restrictive than the typical ones on the local Peclet number $Pe_M := h_M \|\mathbf{a}\|_{\infty,M}/\nu \le 1$ in the Galerkin method for advection-diffusion problems. The div-div stabilization term is very important for robust estimates in case of Taylor-Hood elements. Compared to the Galerkin method, much better local mass conservation clearly improves the H^1- and L^2-error rates for velocity $\mathbf{u}_h$. Increasing values of $Re_f := \|\mathbf{a}\|_\infty C_P/\nu$ can lead to order reduction. Nevertheless, the choice of the div-div parameters τ_2 is still a question of ongoing discussion. It turns out that the SUPG-stabilization is much less important than div-div stabilization, thus showing the surprising robustness of the Galerkin-FEM with div-div stabilization in case of inf-sup stable pairs $V_h \times Q_h$.

4 Improved Error Estimates

The restrictions (35) on the mesh width are not convincing. Let us assume the following orthogonality conditions

$$(\mathbf{v} - \mathbf{j}^{\mathbf{u}}\mathbf{v}, \zeta_{\mathbf{h}}) = 0 \quad \forall \mathbf{v} \in V \text{ and } \forall \zeta_{\mathbf{h}} \in \big[D_h^{\mathbf{u}}(M)\big], \tag{39}$$

$$\big(\mathbf{c} - \mathbf{j}^{\mathbf{b}}\mathbf{c}, \eta_{\mathbf{h}}\big) = 0 \quad \forall \mathbf{c} \in C \text{ and } \forall \eta_{\mathbf{h}} \in \big[D_h^{\mathbf{b}}(M)\big]. \tag{40}$$

Sufficient conditions on $\mathscr{T}_h, \mathscr{M}_h$, the FE and projection spaces for (39)–(40) can be found in [6] or [4]. In particular, for the one-level approach with $\mathscr{T}_h = \mathscr{M}_h$, one has to enrich the velocity space by local bubble functions [6]. Another implication is that $\mathbf{j^u u} \notin V_h^{div}$, hence the mixed term $(e_p, \nabla \cdot \varepsilon_{\mathbf{u}})$ has to be considered. Moreover, a careful selection of the pressure spaces Q_h is required. The critical mixed term vanishes for continuous pressure space $Q_h = \mathbb{P}_{k-1}$. In case of discontinuous space $Q_h = \mathbb{P}_{-(k-1)}$, one can introduce additional pressure jump terms across interior edges to handle it, see [4].

Equations (39)–(40) allow modified estimates of the skew-symmetric terms

$$(\mathbf{a} \cdot \nabla \boldsymbol{\varepsilon}_{\mathbf{u}}, \mathbf{e_u}) = -\left(\kappa_h^u (\mathbf{a} \cdot \nabla \mathbf{e_u}), \boldsymbol{\varepsilon}_{\mathbf{u}}\right) \leq \Big(\sum_M \frac{1}{\tau_1} \|\boldsymbol{\varepsilon}_{\mathbf{u}}\|_{0,M}^2\Big)^{\frac{1}{2}} \|\mathbf{E_h}\|_{lps},$$

$$-\left((\nabla \times \boldsymbol{\varepsilon}_{\mathbf{b}}) \times \mathbf{d}, \mathbf{e_u}\right) = \left(\boldsymbol{\varepsilon}_{\mathbf{b}}, \kappa_h^b (\nabla \times (\mathbf{e_u} \times \mathbf{d}))\right) \leq \Big(\sum_M \frac{1}{\tau_4} \|\boldsymbol{\varepsilon}_{\mathbf{b}}\|_{0,M}^2\Big)^{\frac{1}{2}} \|\mathbf{E_h}\|_{lps},$$

$$-\left(\nabla \times (\boldsymbol{\varepsilon}_{\mathbf{u}} \times \mathbf{d}), \mathbf{e_b}\right) = \left(\boldsymbol{\varepsilon}_{\mathbf{u}}, \kappa_h^b ((\nabla \times \mathbf{e_b}) \times \mathbf{d})\right) \leq \Big(\sum_M \frac{1}{\tau_3} \|\boldsymbol{\varepsilon}_{\mathbf{u}}\|_{0,M}^2\Big)^{\frac{1}{2}} \|\mathbf{E_h}\|_{lps}.$$

Then, a modification of (33) leads to

$$S_1^2 \leq C \sum_M h_M^{2k} \Big[\Big(\nu + \frac{h_M^2}{\tau_1} + \frac{h_M^2}{\tau_3}\Big) |\mathbf{u}|_{k+1,M}^2 + \Big(\lambda + \frac{h_M^2}{\tau_4}\Big) |\mathbf{b}|_{k+1,M}^2 \Big]. \qquad (41)$$

Preserving the choice of div-div parameters according to (38) and of (37), a calibration of the parameters in (41) and (34) gives

$$C h_M^2 \leq \tau_1 \leq C/|\mathbf{a}_M|^2, \qquad C h_M^2 \leq \tau_3, \tau_4 \leq C/|\mathbf{d}_M|^2, \qquad (42)$$

and allows to omit the restrictions (35). A careful estimation has to consider the approximation of $\mathbf{a_M} \sim \mathbf{a}$ and $\mathbf{d_M} \sim \mathbf{d}$. For simplicity, we assume here elementwise constant fields $\mathbf{a}|_M = \mathbf{a_M}$ and $\mathbf{d}|_M = \mathbf{b_M}$.

Theorem 2 *Let the orthogonality conditions (39)–(40) be valid. Assume that the solution $(\mathbf{u}, \mathbf{b}, p)$ of (3) belongs to $[H^{k+1}(\Omega)]^d \times [H^{k+1}(\Omega)]^d \times H^k(\Omega)$. Further, let the LPS parameters be chosen according to conditions (38), (37) and (42). Then we obtain the quasi-optimal error estimate in Theorem 1 without the mesh-width restrictions (35).*

Numerical experiments for the fluid part, i.e. $\mathbf{b} = \mathbf{0}, r = 0$, show: One can omit restriction (35) if conditions (39)–(40) are valid, see [4]. The experiments indicate that optimal error estimates for the H^1- and L^2-error rates for the velocity $\mathbf{u}_h$ are obtained which are robust with respect to Re_f.

Corresponding numerical experiments for the magnetic part and the full MHD problem are in preparation and will be reported elsewhere.

References

1. S. Badia, R. Codina, R. Planas, On an unconditionally convergent stabilized finite element approximation of resistive magnetohydrodynamics. J. Comput. Phys. **234**, 399–416 (2013)
2. S. Badia, R. Codina, R. Planas, Analysis of an unconditionally convergent stabilized finite element formulation for incompressible magnetohydrodynamics. Ach. Comp. Meth. Eng. (2014)
3. S. Badia, R. Planas, J.V. Gutierrez-Santacreu, Unconditionally stable operator splitting algorithms for the incompressible magnetohydrodynamics (MHD) system discretized by a stabilized finite element formulation based on projections. Int. J. Numer. Methods Eng. **93**, 302–328 (2013)
4. H. Dallmann, D. Arndt, G. Lube, Some remarks on local projection stabilization for the Oseen problem. NAM-Preprint, University of Göttingen. (2014). http://num.math.un-goettingen.de/lube/DAL-Oseen-final.pdf
5. V. Girault, R. Scott, A quasi-local interpolation operator preserving the discrete divergence. Calcolo **40**, 1–19 (2003)
6. G. Matthies, P. Skrzypacz, L. Tobiska, A unified convergence analysis for local projection stabilization applied to the Oseen problem. Math. Model. Numer. Anal. **41**(4), 713–742 (2007)

Domain Decomposition for Computing Ferromagnetic Effects

Michel Flueck, Ales Janka, and Jacques Rappaz

Abstract We present here a model for simulating the ferromagnetic screening effect in thin steel plates. We exhibit a domain decomposition method to solve this problem by using only Laplace equations. We then apply this on an academic situation of a steel plate placed in front of a linear conductor and on an industrial application in aluminum production. More details and proofs can be found in Flück et al. (Numerical methods for ferromagnetic plates. Scientific report, SB/MATH-ICSE, EPFL).

1 The Ferromagnetic Screening Effect and Its Industrial Application

We briefly present an academic situation illustrating ferromagnetic screening and an industrial application as well.

1.1 *The Ferromagnetic Screening Effect in General*

Let us consider an infinite linear electric conductor with a given steady current. The well known Biot-Savart formula gives the induction produced by this current in the whole space. Figure 1(left) shows this induction field due to a wire orthogonal to the plane of the Figure.

Let us now place a rectangular plate in front of the conductor. Figure 1(right) shows the induction field as modified due to the ferromagnetic effects. As we can see

M. Flueck (✉) • J. Rappaz
Ecole Polytechnique Fédérale de Lausanne, Lausanne, Switzerland
e-mail: michel.flueck@epfl.ch; jacques.rappaz@epfl.ch

A. Janka
Ecole d'Ingénieurs et d'Architectes de Fribourg, Fribourg, Switzerland
e-mail: ales.janka@hefr.ch

A. Abdulle et al. (eds.), *Numerical Mathematics and Advanced Applications ENUMATH 2013*, Lecture Notes in Computational Science and Engineering 103,
DOI 10.1007/978-3-319-10705-9_77

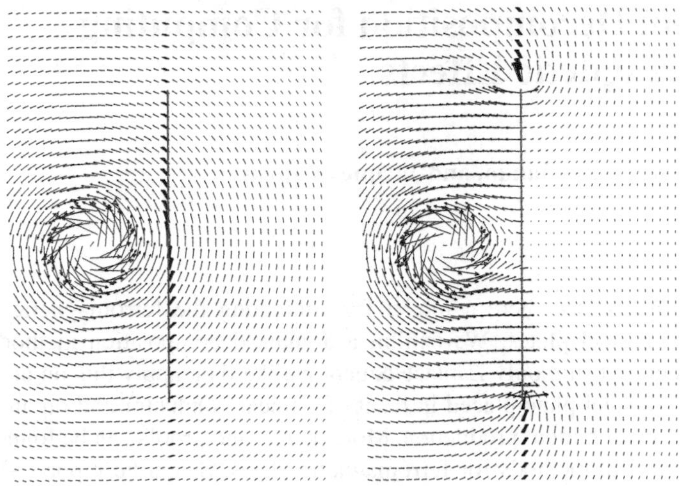

Fig. 1 *Left*: the induction field generated by an infinite linear conductor. *Right*: the induction field modified by ferromagnetic effects. The view is on a plane perpendicular to the conductor using same scale for both plots

induction behind the plate is much smaller in intensity. This is called ferromagnetic screening effect. It is precisely this effect we want to model and compute here.

1.2 The Industrial Application: An Aluminum Electrolysis Cell

An aluminum electrolysis cell is a 15 m long, 4 m wide and 1 m high cell that contains smelted aluminum (about 20 cm high) at the bottom and liquid electrolyte (about 5 cm) placed above the aluminum, see Fig. 2.

A strong direct current is forced through these liquids using carbon anodes in the electrolyte and a carbon cathode under the aluminum. The temperature is about 950 °C. The whole system is built in a big steel container 3 cm thick (not shown in Fig. 2 but shown in Fig. 3).

To feed the cell with current (it can reach 500,000A) big conductors come into the cell from outside the steel container. This current crosses vertically the cell allowing the electrolysis process and production of aluminium. It is then taken off the cell by other conductors outside the steel container and send to the next cell.

The two fluids, aluminium and electrolyte undergo Lorentz forces. This produces movement of these fluids and a deformation of their common interface. In order to precisely model these movements, we need to know the induction field in the fluids. This induction is generated partly by the conductors outside the container and it is attenuated by the screening effect of the steel container, see Fig. 3. It is then of big importance to have a good idea of the screening effect if we want to precisely model the movements. This is the industrial application of our model.

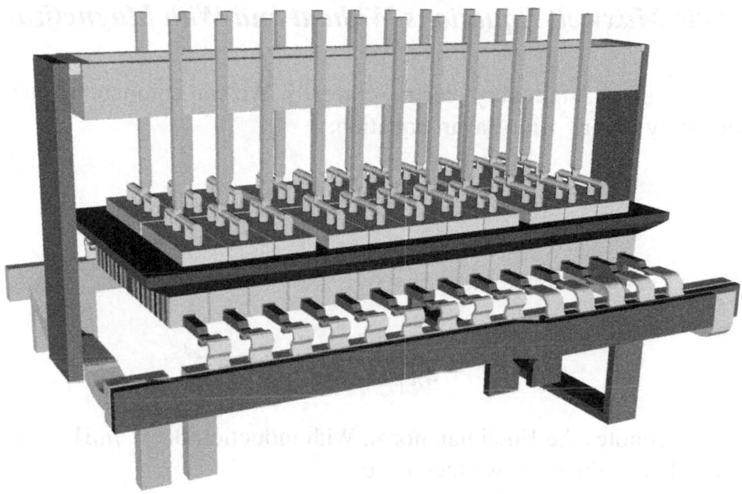

Fig. 2 3D view of an aluminum electrolysis cell. The fluids are in dark. Container of the fluids is not visible here. The anodes are immersed in the fluids. The cathode lies under the fluids. All the rest are conductors

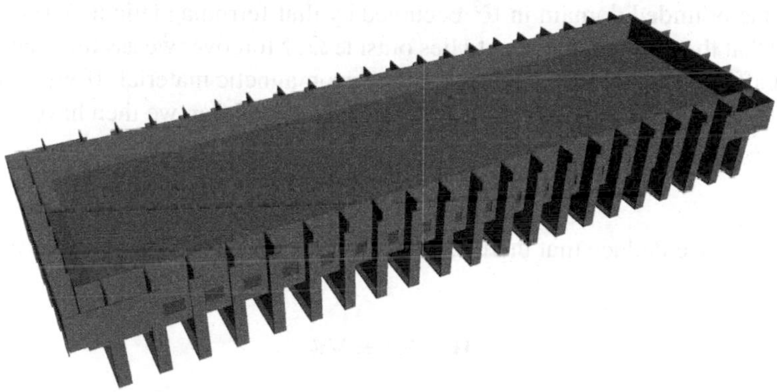

Fig. 3 The steel container with its vertical supports. The fluids are represented in dark

2 Modelling of Ferromagnetism

We first recall the Maxwell equation for a static situation. Then we present our model for ferromagnetism already used in [1]. Finally we give an existence and unicity theorem to our problem. See also [4].

2.1 Static Maxwell Equations Without and With Magnetization

We consider a given, stationary electric current $\mathbf{j}$. We can compute magnetic field $\mathbf{H}_0$ produced by $\mathbf{j}$ using Biot-Savart equation:

$$\mathbf{H}_0(\mathbf{x}) = \int_{\mathbb{R}^3} \nabla_{\mathbf{x}} G(\mathbf{x}, \mathbf{y}) \wedge \mathbf{j}(\mathbf{y}) d\mathbf{y}$$

where $G(\mathbf{x}, \mathbf{y})$ is the Green kernel given by:

$$G(\mathbf{x}, \mathbf{y}) = \frac{1}{4\pi \|\mathbf{x} - \mathbf{y}\|}, \quad \mathbf{x}, \mathbf{y} \in \mathbb{R}^3$$

where $\| \dots \|$ denotes the Euclidian norm. With induction $\mathbf{B}_0 = \mu_0 \mathbf{H}_0$ where μ_0 is the permeability of the void, we then have:

$$\mathrm{div}(\mathbf{B}_0) = 0, \qquad \mathrm{curl}(\mathbf{H}_0) = \mathbf{j}.$$

Let us now introduce a ferromagnetic material in the situation of Fig. 3. We call Ω the bounded domain in $\mathbb{R}^3$ occupied by that ferromagnetic material and we assume that the support of current $\mathbf{j}$ lies outside Ω. Moreover we assume that current $\mathbf{j}$ is not affected by the introduction of the ferromagnetic material. If we call $\mathbf{H}$ the magnetic field and $\mathbf{B}$ the induction field in the new situation we then have:

$$\mathrm{div}(\mathbf{B}) = 0, \qquad \mathrm{curl}(\mathbf{H}) = \mathbf{j}.$$

From which we deduce that there exists a scalar function ψ defined in $\mathbb{R}^3$ and such that:

$$\mathbf{H} - \mathbf{H}_0 = \nabla \psi.$$

We then have the constitutive relation taking into account magnetization in the material

$$\mathbf{B} = \mu \mathbf{H}$$

where

$$\mu = \mu_0 \text{ outside } \Omega \text{ and } \mu = \mu_0 \mu_r(\|\mathbf{H}\|) \text{ inside } \Omega$$

which makes the problem non-linear. The function μ_r is a property of the material. Figure 4 gives a typical example for the graph of μ_r. In particular, we can see that with very strong currents, ferromagnetic effects vanish due to saturation.

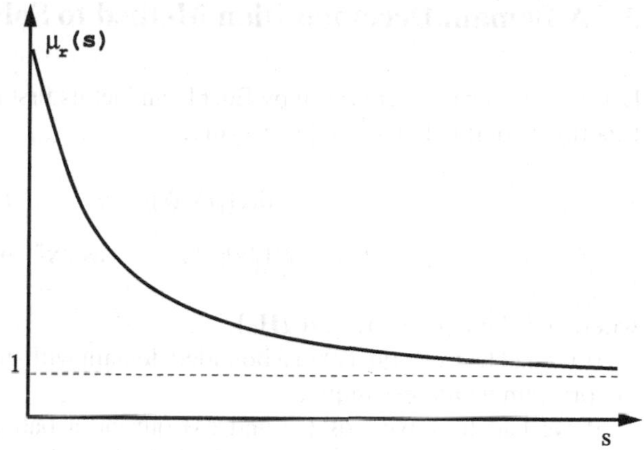

Fig. 4 The graph of μ_r function in a ferromagnetic material, $s = \|\mathbf{H}\|$

With an additional condition for finite energy, $\psi(\mathbf{x}) = \mathcal{O}(\|\mathbf{x}\|^{-1})$ at infinity, we finally get for ψ the following equation in $\mathbb{R}^3$:

$$\operatorname{div}\bigl(\mu(\mathbf{H}_0 - \nabla\psi)\bigr) = 0. \tag{1}$$

This is a non-linear problem for function ψ which is defined in the whole space $\mathbb{R}^3$. Let us show now how we solve this problem.

2.2 A Theorem of Existence and Uniticity

Introducing the Beppo-Levi [3] space

$$W^1(\mathbb{R}^3) = \left\{\varphi : \mathbb{R}^3 \mapsto \mathbb{R} : \frac{\varphi(\mathbf{x})}{1 + \|\mathbf{x}\|} \in L^2(\mathbb{R}^3), \nabla\varphi \in L^2(\mathbb{R}^3)\right\}$$

and using a variational formulation, problem for ψ writes: find $\psi \in W^1(\mathbb{R}^3)$ such that forall $\varphi \in W^1(\mathbb{R}^3)$ we have

$$\int_{\mathbb{R}^3} \mu(\mathbf{H}_0 + \nabla\psi) \cdot \nabla\varphi \, d\mathbf{x} = 0$$

μ being the function defined above. Recall that this makes the problem non-linear. It can be shown [2] that, under certain hypothesis on function μ_r, this problem has a unique solution.

3 A Domain Decomposition Method to Solve Problem

Let us consider problem given by Eq. (1) and let us first rewrite it. We seek a scalar function ψ defined in $\mathbb{R}^3$ and satisfying:

$$-\operatorname{div}(\mu\nabla\psi) = f, \qquad \text{in } \mathbb{R}^3 \tag{2}$$

$$\psi(\mathbf{x}) = \mathcal{O}(\|\mathbf{x}\|^{-1}), \qquad \text{as } \|\mathbf{x}\| \to \infty \tag{3}$$

where f is here given by $\operatorname{div}(\mu\mathbf{H}_0)$.

If we had to solve Eq. (2) in a bounded domain with given values on its boundary, the problem would be simple.

If we had to solve Eqs. (2) and (3) outside a ball centered at the origin and containing Ω, with given values on its boundary, the problem would be solved by using Poisson's formula since $\mu = \mu_0$ outside the ball.

So we present an iterative process which alternates between both problems.

3.1 The Domain Decomposition Algorithm

We fix two balls B_r and B_R centered at the origin, of radius r and R resp. with $0 < r < R$ choosen such that $\overline{\Omega} \subset B_r$.

Figure 5 gives a simple view of the situation. Then, given $\psi_0 = 0$ on ∂B_R, the algorithm is defined as follows:

- If ψ_k is given on ∂B_R (say $\psi_k \in H^{1/2}(\partial B_R)$)

  1. We compute $\psi_{k+\frac{1}{2}} \in H^1(B_R)$ satisfying $\psi_{k+\frac{1}{2}} = \psi_k$ on ∂B_R and

$$\int_{B_R} \mu\nabla\psi_{k+\frac{1}{2}}.\nabla v\, dx = \int_{B_R} f v\, dx, \quad \forall v \in H_0^1(B_R).$$

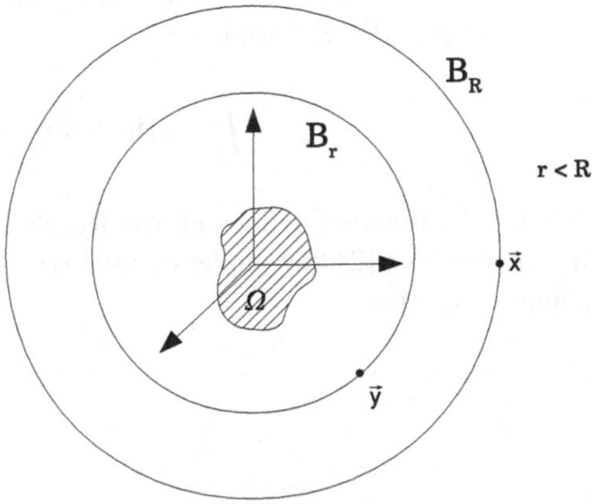

Fig. 5 Schematic representation of domain Ω and the two balls B_r and B_R

- $\psi_{k+\frac{1}{2}}$ is computed by a $\mathbb{P}_1$-finite element method on a tetrahedral mesh on B_R, the linear problem is solved by a CG(diag) method.
- Since $f = \mathrm{div}(\mu \mathbf{H}_0) = \mathrm{div}(\mu - \mu_0)\mathbf{H}_0$ and $\mu = \mu_0$ outside Ω we have

$$\int_{B_R} fv dx = -\int_{\Omega} (\mu - \mu_0)\mathbf{H}_0 \,.\nabla v \, dx, \quad \forall v \in H_0^1(B_R).$$

2. We compute $\psi_{k+1} \in H^{1/2}(\partial B_R)$ given by Poisson's formula

$$\psi_{k+1}(\mathbf{x}) = \frac{|\mathbf{x}|^2 - r^2}{4\pi r} \int_{\partial B_r} \frac{\psi_{k+\frac{1}{2}}(\mathbf{y})}{|\mathbf{x} - \mathbf{y}|^3} \, ds(\mathbf{y}), \quad \forall \mathbf{x} \in \partial B_R.$$

- $\psi_{k+1}(\mathbf{x})$ is computed by Gauss integration on the triangles which discretize ∂B_r.
- Notice that this integral implies no singularity.

It can be shown that this algorithm converges to the solution ψ of our problem.

4 Numerical Results

We give here some numerical results for the academic problem and for the industrial one as well, using the algorithm just described above.

4.1 *The Academic Case*

Figure 6 shows the linear conductor and the ferromagnetic plate introduced at the beginning. Note that it is not necessary to center the big ball at the origin. The center is choosen in order to include measurement line. We plot in Fig. 7 the vertical component of induction along line 1 which is perpendicular to the plate and on the opposite side to the conductor: on the left, the induction without plate; on the right the effect of the plate can be seen.

The screening effect is clear. Note that, if we take a much bigger current in the conductor, this effect completely disappears, due to saturation of the magnetic field in the plate.

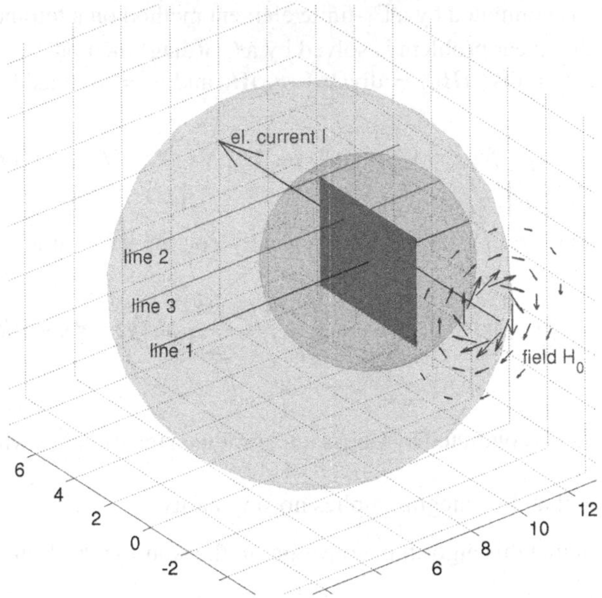

Fig. 6 Academic test of a ferromagnetic plate near a linear conductor

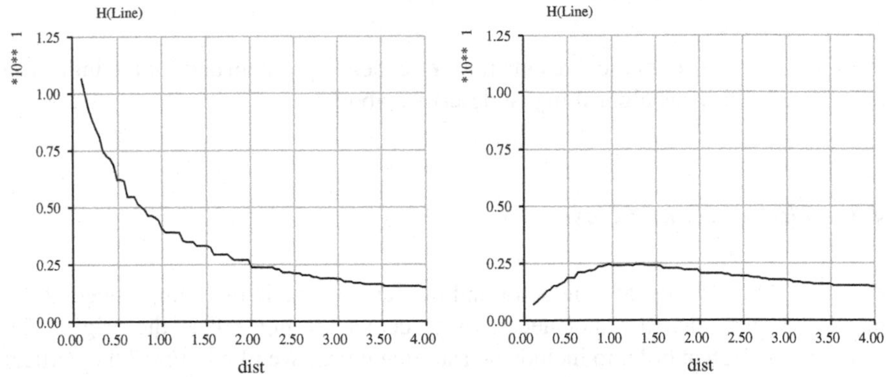

Fig. 7 *Left*: Vertical component of induction on line 1, without ferromagnetic effects *Right*: Vertical component of induction on line 1, with ferromagnetic effects

4.2 The Industrial Case

Let us come back now to the industrial situation described in Sect. 1.2. To exhibit the effect of ferromagnetic steel container we plot here the stationary velocity field computed on an horizontal plane placed in the aluminum. Figure 8 gives the velocity field without screening effect while Fig. 9 gives the velocity field with screening effect both figures being at the same scale.

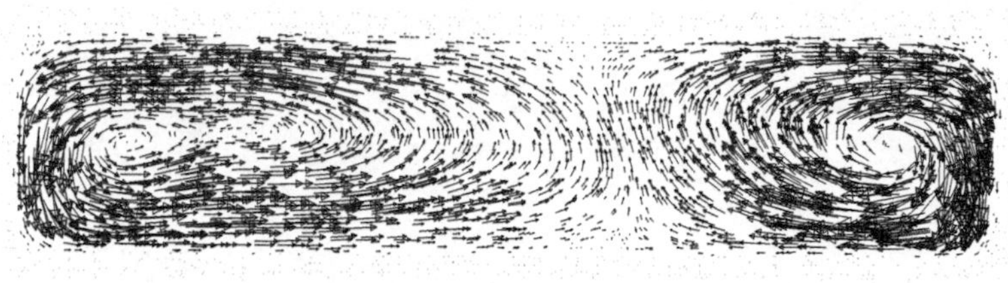

Fig. 8 Velocity field at the interface without ferromagnetism

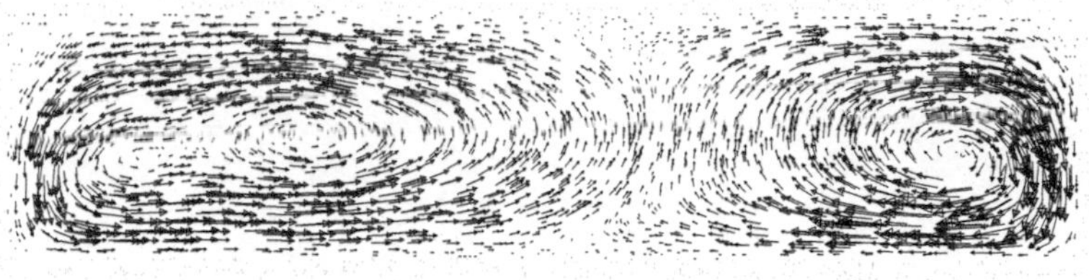

Fig. 9 Velocity field at the interface with ferromagnetism

It is clear that velocity at the interface is reduced when we take into account ferromagnetic effects in the induction field.

In order to conclude, we showed that our domain decomposition method for computing ferromagnetic screen effect is very efficient and can easily be used in industrial applications.

References

1. J. Descloux, M. Flück, M.V. Romerio, A problem of magnetostatics related to thin plates. RAIAO Modél. Math. Anal. Numér. **32**(7), 859–876 (1998)
2. M. Flück, A. Janka, J. Rappaz, Numerical methods for ferromagnetic plates. Scientific report, SB/MATHICSE, EPFL
3. J-C. Nédélec, *Acoustic and Electromagnetic Equations, Integral Representations for Harmonic Problems*. Applied Mathematical Sciences, vol. 144 (Springer, New York, 2001)
4. R. Touzani, J. Rappaz, *Mathematical Models for Eddy Currents and Magnetostatics*, chap 11. Scientific Computation (Springer, Dordrecht, 2013)

Mortar FEs on Overlapping Subdomains for Eddy Current Non Destructive Testing

Alexandra Christophe, Francesca Rapetti, Laurent Santandrea, Guillaume Krebs, and Yann Le Bihan

Abstract We focus on the analysis of a magnetostatic problem with discontinuous coefficients and on its non-conforming mortar finite element discretization on overlapping grids. The global air-filled domain and the nested ferromagnetic subdomain are indeed discretized by two independent triangulations, as the latter subdomain can change position inside the global domain. The two directional exchanges of information between the grids are established by using stable mortar-like projection operators. Numerical results are presented and commented both from a mathematical and physical point of view.

1 Introduction

In the non-destructive testing (NDT) of a conductor by eddy currents, an inducting coil is fed with an alternative electrical current and moves over the underlying conductor to induce eddy currents on it. In proximity of a defect, the circulation of the eddy currents in the conductor is modified with respect to the one associated with an unflawed conductor [9]. The modification of the eddy current distribution induces a modification of the total magnetic field in the coil (combination of the inducting field created by the coil current and the induced one created by the eddy currents) which yields a variation of the coil impedance. The coil impedance brings then information on the structural health of the conductor. The coil is generally filled with a ferromagnetic material with the purpose of concentrating and increasing the intensity of the inducting magnetic field on the conductor (see Fig. 1) in order to

A. Christophe (✉) • L. Santandrea • G. Krebs • Y. Le Bihan
Lab. de Génie Electrique de Paris, UMR 8507 CNRS, SUPELEC, Paris-Sud and Paris VI Universities, 11 Rue Joliot, Plateau de Moulon, 91192 Gif sur Yvette, France
e-mail: alexandra.christophe@lgep.supelec.fr; santandrea@lgep.supelec.fr; Guillaume.Krebs@supelec.fr; yann.lebihan@supelec.fr

F. Rapetti
Dept. de Mathématiques "J.-A. Dieudonné", UMR 7351 CNRS, Univ. de Nice Sophia-Antipolis, Parc Valrose, 06108 Nice 02, France
e-mail: frapetti@unice.fr; Francesca.RAPETTI@unice.fr

A. Abdulle et al. (eds.), *Numerical Mathematics and Advanced Applications ENUMATH 2013*, Lecture Notes in Computational Science and Engineering 103,
DOI 10.1007/978-3-319-10705-9_78

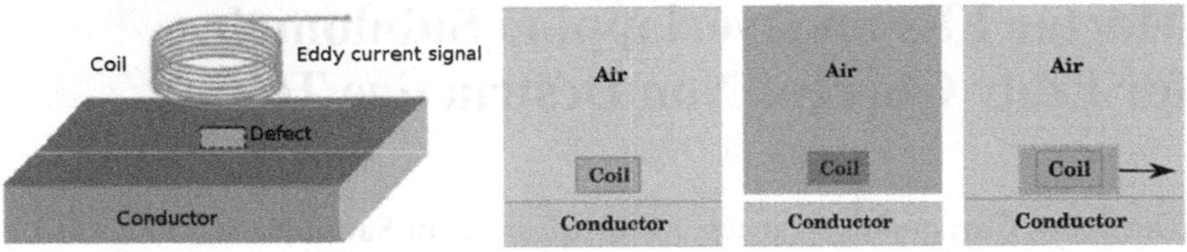

Fig. 1 From left to right: Eddy current non-destructive testing configuration; A unique subdomain for the whole configuration (a computational expensive global re-meshing is thus necessary at each new position of the coil); Two non-overlapping subdomains (the global re-meshing is avoided but the interface between the subdomains must be invariant with respect to the coil motion); Two overlapping subdomains, the one containing the coil and a thin layer of air around the coil, the other the whole configuration with some air at the place of the coil (no constraint neither on the meshing nor on the coil motion)

have more intense eddy currents thus more visible perturbation effects associated to the presence of the defect [7]. It has to be remarked that, even if the coil moves rather quickly over the conductor, there is no quasi-static effect associated with this movement. Therefore, there is no relation between the solutions of the NDT problem for two different positions of the testing coil. At the continuous level, we consider two non-overlapping subdomains, each subdomain being associated with a fixed value of the magnetic permeability and one subdomain (the one containing the field source) changing position inside the other. At the discrete level, we deal instead with two overlapping non-nested (thus non-matching) meshes, and a new method generalizing the pioneering approach presented in [8] and its successive applications to linear elasticity in [5] and electrodynamic levitation in [10]. With the proposed discrete method, the meshes in the global (say, background) domain and in the ferromagnetic (say, foreground) domain containing the coil, are created independently of each other and only once, even if the coil occupies different positions in the global domain. Moreover, the coupling between the meshes is bilateral in order to simulate the reaction of the materials to the source field. The bilateral exchange is required by the very basis of eddy current non-destructive testing which involves a signal transfer from the coil to the conductor (eddy currents induced in the conductor by the coil source) and a signal transfer from the conductor to the coil (modification of the magnetic field in the coil due to the eddy currents).

In this paper, we wish to consider the magnetostatic problem only, as its analysis and discretization already present some difficulties. However, the discrete method proposed here for a magnetostatic problem is the same that would be used in a complete NDT situation, as the conductor is included in the background domain and it can be managed that the mesh elements in the conductor are never overlapped by the foreground mesh, since the coil is never in contact with the conductor but stays at a certain height, called lift-off. Finally, it must be noticed that the magnetic materials used in eddy current testing as coil cores are in ferrite which exhibit a very low and completely negligible electrical conductivity.

The paper is organized as follows. In Sect. 2 we describe the model problem of magnetostatics with non-constant coefficients. In Sect. 3 we introduce the discrete formulation obtained by using the generalized overlapping mortar element method and the corresponding algebraic formulation. Some numerical results are presented and discussed in Sect. 4. Conclusions are drawn in section “Conclusions and Acknowledgements”.

2 Continuous Problem Formulation

Let Ω be an open bounded domain in $\mathbb{R}^2$ with polygonal boundary $\partial\Omega = \Gamma_D \cup \Gamma_N$ and $\mathrm{mes}(\Gamma_D) > 0$. We may consider Ω as the transverse (x, y)-section of an infinitely long z-axed cylinder $\mathscr{C} \subset \mathbb{R}^3$. In this work, we restrict the analysis to magnetostatics: the problem equations in $\mathscr{C}$ are

$$\nabla \times H = J_s, \quad B = \mu H, \quad \nabla \cdot B = 0, \tag{1}$$

where H (resp. B) is the magnetic field (resp. induction) and J_s the source current density. The problem is well posed by adding boundary conditions on $\partial\mathscr{C}$ and suitable interface conditions at any surface where the magnetic permeability μ is discontinuous (in particular, $[H]\times n = 0$ and $[B]\cdot n = 0$, where n is the normal to the interface and $[v]$ stands for the jump of v at the interface). The magnetic permeability μ is supposed to be a bounded function which respects the symmetry of the domain, that is $\mu(x, y, z) = \mu(x, y) \geq \mu_0$, with $\mu_0 = 4\pi 10^{-7}$ H/m the permeability of the air. The source current density J_s has the support strictly contained in ω, a polygonal subdomain of Ω, with complement in Ω denoted by $\omega^c = \Omega \setminus \bar{\omega}$, and boundary Γ. We consider $J_s = f(x, y)\mathbf{e}_z$ so that the problem can be reduced to a scalar PDE (with discontinuous coefficients). To this purpose, we introduce a magnetic vector potential $A = u(x, y)\mathbf{e}_z$ related to B by the well-known condition $B = \nabla \times A$. Note that by construction, $\nabla \cdot A = 0$, and this yields uniqueness of A. In the following, the Dirichlet condition on $\partial\Omega$ is represented by a function $g_D \in H^{1/2}(\Gamma_D)$ whose $H^1(\Omega)$ lifting is denoted by $\overline{g_D}$, the source $f \in L^2(\omega)$ extended by zero in ω^c (see [2] for the definition of the involved functional spaces). The magnetic permeability is a piecewise constant function $\mu = \mu_\omega \mathscr{I}_\omega + \mu_{\omega^c} \mathscr{I}_{\omega^c}$, with $\mathscr{I}_E$ the characteristic function of the domain E where $E = \omega$ or $E = \omega^c$. In the case of the NDT application, we generally have $\mu_\omega > \mu_{\omega^c}$ and the position of ω inside Ω changes (indeed, ω contains the testing coil and Ω the tested object). We wish to compute the solution of problem (1) for different positions of ω in Ω by a finite element method with independent meshes in Ω and ω. We thus have to find $u := (u_\omega, u_{\omega^c})$ such that

$$-\nabla \cdot \left(\frac{1}{\mu_{\omega^c}} \nabla u_{\omega^c}\right) = 0, \quad \text{in } \omega^c, \tag{2}$$

$$\nabla \cdot \left(\frac{1}{\mu_\omega} \nabla u_\omega\right) = f, \quad \text{in } \omega, \tag{3}$$

$$u_{\omega^c} = u_\omega, \quad \text{on } \Gamma, \tag{4}$$

$$\frac{1}{\mu_{\omega^c}}\partial_n u_{\omega^c} = \frac{1}{\mu_\omega}\partial_n u_\omega, \quad \text{on } \Gamma, \tag{5}$$

$$u_{\omega^c} = g_D, \quad \text{on } \Gamma_D, \tag{6}$$

$$\partial_n u_{\omega^c} = 0, \quad \text{on } \Gamma_N. \tag{7}$$

The weak problem consists in finding $(u_\omega, u_{\omega^c}) \in \mathcal{U}_{D,\Gamma_D}$ such that

$$a_g((u_\omega, u_{\omega^c}), (w, v)) = ((f, 0), (w, v))_g, \quad \forall (w, v) \in \mathcal{U}_{0,\Gamma_D} \tag{8}$$

with $(u_\omega, u_{\omega^c} - \overline{g_D}) \in \mathcal{U}_{0,\Gamma_D}$. Here, $\mathcal{U}_{D,\Gamma_D}$ is the functional space $\{u := (u_\omega, u_{\omega^c}) \in H^1(\omega) \times H^1(\omega^c),\ u_{\omega^c|\Gamma_D} = g_D,\ u_{\omega|\Gamma} = u_{\omega^c|\Gamma}\}$ and $\mathcal{U}_{0,\Gamma_D}$ is $\mathcal{U}_{D,\Gamma_D}$ with g_D replaced by 0. Note that $\mathcal{U}_{D,\Gamma_D}$ (resp. $\mathcal{U}_{0,\Gamma_D}$) is isomorphic to $H^1_{D,\Gamma_D}(\Omega) = \{u \in H^1(\Omega),\ u_{|\Gamma_D} = g_D\}$ (resp. $H^1_{0,\Gamma_D}(\Omega)$), as a consequence, it is a Hilbert space endowed with the broken norm. In (8) we have set $((f, 0), (w, v))_g := (f, w)_\omega = \int_\omega f\, w$ and $a_g((u_\omega, u_{\omega^c}), (w, v)) := a_\omega(u_\omega, w) + a_{\omega^c}(u_{\omega^c}, v)$, with $a_E(u_E, z) := \int_E \frac{1}{\mu_E} \nabla u_E \cdot \nabla z$ the standard bilinear form over a domain E. Problem (8) has a unique solution by the Lax-Milgram lemma, since the linear form $(w, v) \mapsto ((f, -\overline{g_D})_g, (w, v))_g$ is continuous on $\mathcal{U}_{0,\Gamma_D}$ and the bilinear form $a_g(.,.)$ is continuous and elliptic on $\mathcal{U}_{0,\Gamma_D}$. Indeed, the continuity of $a_g(.,.)$ with constant $\overline{\nu} = \max\{\frac{1}{\mu_\omega}, \frac{1}{\mu_{\omega^c}}\}$ follows from the continuity of $a_\omega(.,.)$ and $a_{\omega^c}(.,.)$. Similar reasoning for the ellipticity of $a_g(.,.)$ where the constant is now $\underline{\nu} = \min\{\frac{1}{\mu_\omega}, \frac{1}{\mu_{\omega^c}}\}$.

3 Discrete Problem Formulation

For the discretization of problem (8) we adopt piecewise linear finite elements, separately, all over $\bar{\Omega}$ and $\bar{\omega}$. Indeed, as the position of ω can change in Ω and we do not wish to re-create a computational mesh over $\bar{\omega^c}$ for each new position of ω, we use two different shape regular triangulations τ_H on $\bar{\Omega}$ and τ_h on $\bar{\omega}$, with H and h indicating the maximum element diameters, respectively. Moreover, in presence of discontinuous coefficients, we should choose $H > h$ if $\mu_{\omega^c} < \mu_\omega$ and $H < h$ if $\mu_{\omega^c} < \mu_\omega$, for optimal convergence rates w.r.t. H and h of the approximation error. The latter condition is easy to fulfill with overlapping triangulations τ_H and τ_h, completely independent of each other. The flexibility has however a price: indeed, the transmission condition (4) on Γ which is contained in the definition of the space $\mathcal{U}_{D,\Gamma_D}$ cannot be imposed punctually (τ_H and τ_h do not match on Γ) but weakly, by involving a suitable projection operator which is well known in the framework of mortar finite elements.

We denote by $S_H(\Omega)$ the piecewise linear and H^1-conforming finite element space in Ω and let be $S_{0,D;H}(\Omega) = S_H(\Omega) \cap H^1_{0,\Gamma_D}(\Omega)$. Functions in $S_{0,D;H}(\Omega)$ are linear combinations of the classical hat functions, with degrees of freedom associated to the vertices of τ_H which are in $\Omega \cup \Gamma_N$. The space of standard conforming elements of first order associated with τ_h on ω is denoted by $S_h(\omega)$. We note that no boundary conditions are imposed on $S_h(\omega)$. Moreover, we assume that Γ can be written as union of edges in τ_h. The trace space of $S_h(\omega)$ on Γ is called $W_h(\Gamma)$.

We need to introduce a projection operator $\pi_\Gamma : S_H(\Omega) \to W_h(\Gamma)$, that passes the information from Ω to the boundary Γ of ω at the discrete level. Indeed, we have that $v_{H|\Gamma} \notin W_h(\Gamma)$, $\forall\, v_H \in S_H(\Omega)$. To get an optimal a priori estimate, the operator π_Γ has to be stable in the $H^{1/2}(\Gamma)$-norm (see [11]). We then introduce the weak form of (4) on Γ as follows:

$$b_\Gamma(u_h, z) = 0, \qquad b_\Gamma(w, z) := \int_\Gamma (w - \pi_\Gamma u_H) z, \qquad \forall\, z \in W_h(\Gamma). \tag{9}$$

To simulate the "reaction" of ω, we introduce a projection operator π_γ : $S_h(\omega) \to W_H(\gamma)$, stable in the $H^{1/2}(\gamma)$-norm, where γ is the polygonal line formed by the edges of τ_H lying under ω and bounding, on the one side, the strep of elements of τ_H containing Γ (see Fig. 2). Note that Γ does not change when ω moves whereas γ does. We introduce a second weak form of (4) on γ that reads:

$$b_\gamma(u_H, t) = 0, \qquad b_\gamma(v, t) := \int_\gamma (\pi_\gamma u_h - v) t, \qquad \forall\, t \in W_H(\gamma). \tag{10}$$

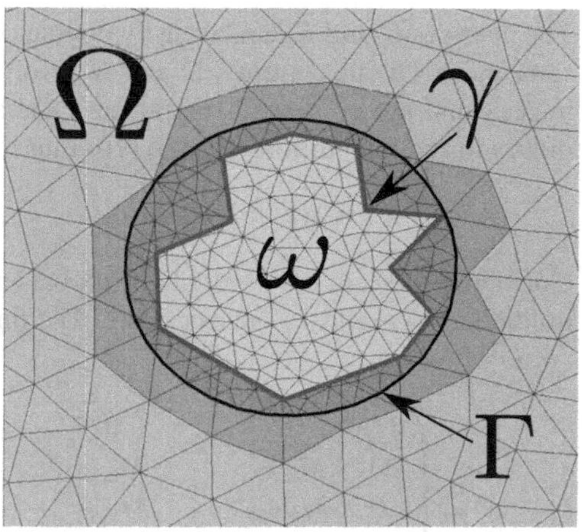

Fig. 2 The exchange form Ω to ω is realized on the boundary Γ. The interface of transmission γ from ω to Ω is chosen as the inner boundary of the elements of Ω overlapped by Γ

The discrete problem reads: find $(u_h, u_H) \in S_h(\omega) \times S_{g_H,D;H}(\Omega)$ such that

$$\begin{aligned} a_g((u_h,u_H),(w,v)) + b_g((w,v),(\lambda_h,\lambda_H)) &= ((f_h,0),(w,v))_g, \\ &\qquad \forall (w,v) \in S_h(\omega) \times S_{0,D;H}(\Omega), \\ b_g((u_h,u_H),(\lambda'_h,\lambda'_H)) = 0, &\qquad \forall (\lambda'_h,\lambda'_H) \in W_h(\Gamma) \times W_H(\gamma). \end{aligned} \tag{11}$$

where $b_g((u_h,u_H),(r,s)) := b_\Gamma(u_h,r) + b_\gamma(u_H,s)$. We have denoted by $f_h = I_h f$ (resp. $g_H = I_{H|\Gamma} g_D$) where I_h (resp. $I_{H|\Gamma}$) is the interpolation operator over τ_h (resp. τ_H, restricted to Γ). In this system, λ_h, λ_H play the role of the Lagrange multipliers associated with the constraints (9) and (10). By adapting the results developed in [1] for the non-overlapping case, we have that problem (11) has a unique solution.

In order to write the matrix form of problem (11), we adopt bold letters to indicate vectors of degrees of freedom (dof), namely, $\mathbf{u} = (\mathbf{u}_I, \mathbf{u}_B)$, with $\mathbf{u}_I$ for dofs associated to "interior" nodes and $\mathbf{u}_B$ for dofs associated to "boundary" nodes. For the latter, we have $\mathbf{u}_B = \mathbf{u}_\Gamma$ in ω and $\mathbf{u}_B = \mathbf{u}_\gamma$ in Ω. Let ϕ_i and ψ_i be the basis functions associated to nodes in τ_H and τ_h, respectively, and C, D, E, H be the matrices with entries detailed, on each edge of the concerned interfaces, here below:

$$\begin{aligned} C(i,j) &= \int_{e\in\Gamma} \psi_j \psi_i, && i,j \text{ nodes in } \tau_h \cap \Gamma \\ D(i,k) &= \int_{e\in\Gamma} \phi_k \psi_i, && k \text{ node in } \tau_H,\ i \text{ node in } \tau_h \cap \Gamma \\ E(i,j) &= \int_{e\in\gamma} \phi_j \phi_i, && i,j \text{ nodes in } \tau_H \cap \gamma \\ H(i,k) &= \int_{e\in\gamma} \psi_k \phi_i, && k \text{ node in } \tau_h,\ i \text{ node in } \tau_H \cap \gamma. \end{aligned} \tag{12}$$

The matrices C and E can be easily computed since both basis functions live on the same mesh. On the contrary, D and H involve discrete functions living on different meshes and are computed relying on quadrature formula, as detailed in [4]. The matrix form of conditions (9) and (10) thus read, respectively:

$$C\mathbf{u}_\Gamma - D\mathbf{u}_H = 0, \qquad E\mathbf{u}_\gamma - H\mathbf{u}_h = 0. \tag{13}$$

Conditions (13) are imposed, on the global linear system, with the help of Lagrange multipliers, λ_Γ and λ_γ, respectively. The final algebraic linear system to solve is $\mathscr{M}\mathbf{u} = \mathbf{f}$ where

$$\mathscr{M} = \begin{pmatrix} M_H & 0 & 0 & -D^{\mathrm{T}} & E^{\mathrm{T}} \\ 0 & M_{h,II} & M_{h,IB} & 0 & 0 \\ 0 & M_{h,BI} & M_{h,BB} & C^{\mathrm{T}} & (H_{\mathrm{I}} M_{h,II}^{-1} M_{h,IB} - H_{\mathrm{B}})^{\mathrm{T}} \\ -D & 0 & C & 0 & 0 \\ E & 0 & H_{\mathrm{II}} M_{h,II}^{-1} M_{h,IB} - H_{\mathrm{IB}} & 0 & 0 \end{pmatrix} \tag{14}$$

and M_h (resp. M_H) is the stiffness matrix associated to the discrete operator defined in ω (resp. in ω^c but extended to Ω). The global vector of unknowns is $\mathbf{u} = (\mathbf{u}_H, \mathbf{u}_{h,I}, \mathbf{u}_{h,B}, \lambda_\Gamma, \lambda_\gamma)^{\mathrm{T}}$ and the right-hand side $\mathbf{f} = (\mathbf{0}, \mathbf{f}_{h,I}, \mathbf{f}_{h,B}, \mathbf{0}, H_{\mathrm{II}} M_{h,II}^{-1} \mathbf{f}_{h,I})^{\mathrm{T}}$. The resultant system $\mathscr{M}\mathbf{u} = \mathbf{f}$ is symmetric and sparse, solved by iterative methods.

Remark 1 In condition (13), $\mathbf{u}_\gamma$ depends on $\mathbf{u}_{h,I}$ and $\mathbf{u}_{h,B}$. Hence, the method of Lagrange multipliers imposes a penalty on the vector of interior dofs $\mathbf{u}_{h,I}$. In order to remove this penalty, we choose to apply a Guyan condensation [3, 6], where $\mathbf{u}_\gamma$ is expressed only in terms of $\mathbf{u}_\Gamma$. Note that, in the extreme case where γ and Γ are two piecewise linear curves with nodes on $\partial\omega$, system (14) with condensation is equivalent to the one on independent non-overlapping meshes in $\bar{\omega}$ and $\bar{\omega}^c$.

4 Numerical Results and Analogy with Linear Elasticity

To illustrate an application of the presented method, we consider $\Omega = [-1, 1]^2$ and $\omega = [-0.5, 0.5] \times [-0.3, 0.3]$, with $\mu_\omega \neq \mu_{\omega^c}$. We impose homogeneous Neumann boundary conditions $\partial_n u = 0$ on the horizontal sides of Ω and non-homogeneous Dirichlet conditions on the vertical sides of Ω, and we take $f = 0$. If $\mu_\omega = \mu_{\omega^c}$, the magnetic equipotential lines would be all parallel to the horizontal sides of Ω. The presence of an intrusion (ω) with different magnetic permeability with respect to the surrounding material (ω^c) acts as either an attractor (when $\mu_\omega > \mu_{\omega^c}$) or a repeller (when $\mu_\omega < \mu_{\omega^c}$) for the field lines (see the left-side and left-center pictures of Fig. 3). Indeed, the magnetic field lines follow the way with minimal magnetic reluctivity $\nu = 1/\mu$. When $\mu_\omega > \mu_{\omega^c}$, the physical situation is analogous to having a paramagnetic material in ω with air whole around. When $\mu_\omega < \mu_{\omega^c}$, the physical situation is similar to having ω filled by a diamagnetic material.

The results shown in the left-side and left-center pictures of Fig. 3 can be interpreted also from an "elastic" point of view, relying on a so-called physical

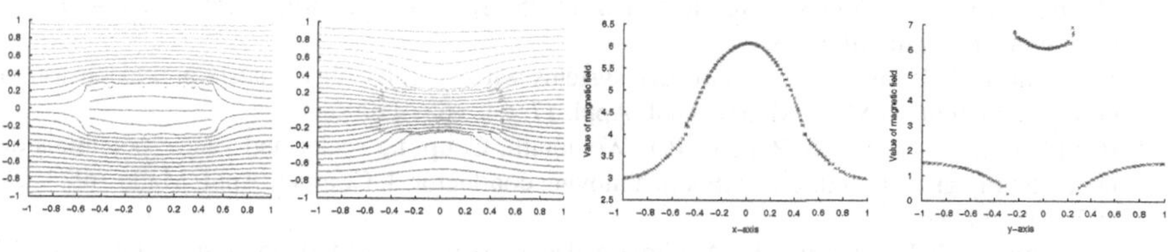

Fig. 3 Magnetic equipotential lines for $\mu_\omega < \mu_{\omega^c}$ (*left-side*) and $\mu_\omega > \mu_{\omega^c}$ (*left-center*). Case $\mu_\omega = 100\mu_{\omega^c}$. Value of the x-component of B along the line $y = 0$ (*right-center*) and $x = 0$ (*right-side*). We may remark the continuity of B_x across the vertical sides of Γ despite the weak constraint, as in this case $B_x = B \cdot n$. The jump of B_x is visible across the horizontal sides of Γ, as in this case we have $B_x \mathbf{e}_x = n \times (B \times n)$

analogy. More precisely, in electromagnetics we may define an impedance $\mathscr{Z}$ of the material as $\mathscr{Z}_{em} = \mu_0 c$ where c is the light speed in the material. The electromagnetic impedance is analogous to the one known in fluid dynamics such as $\mathscr{Z}_{fd} = \rho c$ where ρ is the fluid density and c denotes the sound speed in the fluid. Comparing the two expression, we remark that ρ is equivalent to μ_0. In the empty space, $c = 1/\sqrt{\mu_0 \epsilon_0}$ so that $\mathscr{Z}_{em} = \sqrt{\mu_0/\epsilon_0}$. Recalling that the velocity of a transverse elastic wave is equal to $\sqrt{E/\rho}$, with E the elastic modulus the inverse of which is a sort of electric permittivity ϵ_0, we come up with $\rho = \mu = \mu_0 \mu_r$ (with μ_r the relative magnetic permeability of the material). So, a diamagnetic material is equivalent from an "elastic" point of view to a material with higher density (e.g., a stone) whereas a paramagnetic material looks like an object with lower density (e.g., a bath sponge). The higher ρ, the smaller the deformation of the material in presence of an external load. When $\mu_\omega > \mu_{\omega^c}$ (resp. $\mu_\omega < \mu_{\omega^c}$), the situation looks like a stone (resp. a sponge) surrounded by a sponge (resp. a stone): under an external load, only the sponge deformates (left-center picture of Fig. 3).

Conclusions and Acknowledgements
The variant of the overlapping mortar element method developed in this paper has been motivated by the need of providing efficient modeling tools for eddy current NDT. Differences with previous existing works are the facts that the sub-domains have different material properties and the field source can be in the moving part. This work has the financial support of CEA-LIST.

References

1. F. Ben Belgacem, The mortar finite element method with Lagrange multipliers. Numer. Math. **84**(2), 173–197 (1999)
2. C. Bernardi, Y. Maday, F. Rapetti, *Discretisations variationnelles de problemes aux limites elliptiques*. Mathematiques & Applications, vol. 45 (Springer, Berlin/Heidelberg, 2004). http://math.unice.fr/~frapetti/publivre.html
3. S.P. Bhat, D.S. Bernstein, Second-order systems with singular mass matrix and extension of Guyan reduction. SIAM J. Matrix Anal. Appl. **17**, 649–657 (1996)
4. B. Flemisch, Y. Maday, F. Rapetti, B.I. Wohlmuth, Coupling scalar and vector potentials on nonmatching grids for eddy currents in a moving conductor. J. Comput. Appl. Math. **168**(1–2), 192–205 (2004)
5. B. Flemisch, B.I. Wohlmuth, A domain decomposition method on nested domains and nonmatching grids. Numer. Methods Partial Differ. Equ. **20**, 374–387 (2004)
6. R.J. Guyan, Reduction of stiffness and mass matrices. AIAA J. **3**, 3 (1964)
7. Y. Le Bihan, P.-Y. Joubert, D. Placko, Wall thickness evaluation of single-crystal hollow blades by eddy current sensor. NDT E Int. **34**(5), 363–368 (2001)
8. Y. Maday, F. Rapetti, B.I. Wohlmuth, Mortar element coupling between global scalar and local vector potentials to solve eddy current problems, in *Numerical Mathematics and Advanced Applications* ed. by F. Brezzi, A. Buffa, S. Corsaro, A. Murli (Springer, Milan, 2003), pp. 847–865

9. R.C. McMaster, F. Forster, H.L. Libby, *Nondestructive Testing Handbook Vol. 4: Electromagnetic Testing*, 2nd edn. (ASNT, Columbus, 1986)
10. F. Rapetti, An overlapping mortar element approach to coupled magneto-mechanical problems. Math. Comput. Simul. **80**(8), 1647–1656 (2010)
11. L.R. Scott, S. Zhang, Finite element interpolation of nonsmooth functions satisfying boundary conditions. Math. Comput. **54**, 483–493 (1990)

Solving a Maxwell Interface Problem by a Local L^2 Projected C^0 Finite Element Method

Huoyuan Duan, Ping Lin, and Roger C.E. Tan

Abstract In general, the solution of a Maxwell interface problem would not belong to H^1 space and the standard C^0 finite element method fails. With the help of local L^2 projections applied to both the curl and div operators, we propose a new C^0 finite element method which can correctly converge to the non H^1 space solution. Stability and error estimates are given.

1 Introduction

In this note we shall develop a C^0 finite element method for the Maxwell interface problem. Let $\Omega \subset \mathbb{R}^3$ denote a Lipschitz polyhedron, with boundary Γ. To model different anisotropic inhomogeneous materials filling Ω, we introduce two symmetric tensor fields $\mu, \varepsilon \in \mathbb{R}^{3\times 3}$ satisfying the boundedness and uniform ellipticity conditions as follows: let $\omega = (\omega_{ij}) \in \mathbb{R}^{3\times 3}$ ($\omega_{ij} = \omega_{ji}$) stand for either μ or ε, $\omega_{ij} \in L^\infty(\Omega), 1 \leq i, j \leq 3, \quad \xi \cdot \omega \cdot \xi \geq C\,|\xi|^2$ a.e. in $\Omega, \forall \xi \in \mathbb{R}^3$. Let ε and μ be given by piecewise Lipschitz continuous functions. This defines a partition of Ω into a finite number of subdomains Ω_j, $1 \leq j \leq J$, in each of which μ and ε are Lipschitz continuous. Denote by $\mathscr{E}_{\text{inter}}$ the set of all the material interfaces $\Gamma_{ij} = \Omega_i \cap \Omega_j$. We assume that each subdomain Ω_j is a Lipschitz polyhedron. Let $\mathbf{n}$ denote the outward unit normal vector to Γ or the unit normal vector to Γ_{ij} oriented from Ω_i to Ω_j, and let jumps $[\mathbf{v} \cdot \mathbf{n}]|_{\Gamma_{ij}} = \mathbf{v} \cdot \mathbf{n}|_{\Omega_i} - \mathbf{v} \cdot \mathbf{n}|_{\Omega_j}$ and $[\mathbf{v} \times \mathbf{n}]|_{\Gamma_{ij}} = \mathbf{v} \times \mathbf{n}|_{\Omega_i} - \mathbf{v} \times \mathbf{n}|_{\Omega_j}$. Let $\mathbf{u}$ denote an unknown field and $\mathbf{f}$, g and χ

H. Duan
School of Mathematics and Statistics, Collaborative Innovation Centre of Mathematics, Wuhan University, 430072 Wuhan, China
e-mail: hyduan.math@whu.edu.cn

P. Lin (✉)
Division of Mathematics, University of Dundee, DD1 4HN Dundee, UK
e-mail: plin@maths.dundee.ac.uk

R.C.E. Tan
Department of Mathematics, National University of Singapore, 117543 Singapore, Singapore
e-mail: scitance@nus.edu.sg

A. Abdulle et al. (eds.), *Numerical Mathematics and Advanced Applications ENUMATH 2013*, Lecture Notes in Computational Science and Engineering 103,
DOI 10.1007/978-3-319-10705-9_79

be given functions and $\kappa \geq 0$. The Maxwell interface problem [7, 14, 18] we shall consider is to find $\mathbf{u}$ such that

$$\mathbf{curl}\,(\mu^{-1}\,\mathbf{curl}\,\mathbf{u})+\kappa\,\varepsilon\,\mathbf{u}=\mathbf{f},\quad \mathrm{div}(\varepsilon\,\mathbf{u})=g \quad \text{in } \Omega,\quad \mathbf{u}\times\mathbf{n}=\chi \quad \text{on } \Gamma. \tag{1}$$

In general, it would often happen that the solution of (1) belong to fractional Hilbert space $\prod_{j=1}^{J}(H^r(\Omega_j))^3$ for some $r \leq 1$, whenever μ and ε are discontinuous, anisotropic and inhomogeneous and Ω possesses reentrant corners and edges. A most useful variational problem for studying problem (1) is the well-known plain regularization (PR) variational problem, which is an elliptic problem with curl-curl and div-div terms, suitable for theoretical and numerical purposes, see [2, 4, 5, 7, 12, 14, 19, 23, 24]. From the viewpoint of finite element discretizations, therefore, it is natural to use the very common C^0 elements to approximate the solution. Unfortunately, a direct discretization using C^0 elements cannot yield a correctly convergent finite element solution, see [4, 5, 19, 20].

In [15, 16] we proposed an element-local L^2 projected C^0 finite element method for solving problem (1) in homogeneous media (i.e., $\mu = \varepsilon = 1$). This method consists of element-local L^2 projectors which are applied to both div and curl operators in the PR formulation and that the C^0 linear element (enriched with suitable face-and element-bubbles) is used. Theoretical and numerical results for source and eigenvalue problems showed that this method yields correct and good C^0 finite element solutions. In this note we shall consider this type of method for the Maxwell interface problem (1). Precisely, the method and theory in [15] to problem (1) are generalized, where μ and ε are allowed to be discontinuous, anisotropic and inhomogeneous and the solution $\mathbf{u}$ is allowed to have piecewisely low regularity, i.e., $\mathbf{u} \in \prod_{j=1}^{J}(H^r(\Omega_j))^3$. For this purpose, we must construct a suitable C^0 finite element method in order to allow ε to be discontinuous. This has been done by incorporating ε into the finite element space. Thus, to some extent, the proposed finite element method is *heterogeneous*. Across the material interfaces where ε is discontinuous, the finite element space is also discontinuous, but elsewhere continuous. We have previously considered problem (1) in [17], but therein the *global* L^2 projection for the `div` operator is adopted. Here we instead use the element-local L^2 projection for the div operator. This will greatly simplify the implementation of the finite element method. In addition, to accommodate the case where ε discontinuous, we additionally introduce face-local L^2 projection on $\mathscr{E}_{\mathrm{inter}}$. Thus, the C^0 finite element method for the Maxwell interface problem in this note features element-local L^2 projections for both curl and div operators, a heterogeneous C^0 finite element space, and face-local L^2 projections associated with the interface discontinuity. Theoretical results of stability and error estimates can be obtained by adapting the argument from [15], and the details of proving these results will be omitted.

2 The Finite Element Method

Let $\mathscr{C}_h$ denote a regular triangulation [10] of $\bar{\Omega}$ into tetrahedra, with diameters h_K for all $K \in \mathscr{C}_h$ bounded by h. We assume that the closure of each material interface is the union of the closure of element faces in $\mathscr{C}_h$. For simplicity, *we assume that ε is piecewise constants with respect to* $\mathscr{C}_h$. Let $\mathscr{E}_h = \mathscr{E}_h^0 \bigcup \mathscr{E}_h^\partial$ denote the set of all element faces in $\mathscr{C}_h$, where $\mathscr{E}_h^0$ is the set of all the interior element faces in $\mathscr{C}_h$ and $\mathscr{E}_h^\partial$ the set of all the element faces on Γ. We now introduce element-and face-bubbles. Let K be a tetrahedron with vertices a_i, $1 \leq i \leq 4$, we denote by λ_i the barycentric-coordinate of a_i: λ_i is a linear polynomial on K and takes values '1' at a_i and '0' at other vertices, cf. [10]. Denote by F_i the face opposite a_i. Introduce the following face bubbles: $b_{F_1} = \lambda_2\lambda_3\lambda_4,\quad b_{F_2} = \lambda_3\lambda_4\lambda_1,\quad b_{F_3} = \lambda_4\lambda_1\lambda_2,\quad b_{F_4} = \lambda_1\lambda_2\lambda_3$ and the element-bubble: $b_K = \lambda_1\lambda_2\lambda_3\lambda_4 \in H_0^1(K)$. We have $b_{F_i} \in H_0^1(F_i),\quad b_{F_i}|_{F_j} = 0 \quad \forall j \neq i$. As usual, let $\mathscr{P}_n$ be the space of polynomials of degree not greater than $n \geq 0$. Given $K \in \mathscr{C}_h$ with boundary ∂K. On K, we introduce

$$\mathbf{P}(K) := \operatorname{span}\{\mathbf{p}_{F,l}\, b_F, 1 \leq l \leq m_F, \forall F \subset \partial K\}$$

where $\mathbf{p}_{F,l}$, $1 \leq l \leq m_F$, are chosen so that $(\mathscr{P}_1(F))^3 = \operatorname{span}\{\mathbf{p}_{F,l}|_F, 1 \leq l \leq m_F\}$. Define face-bubble space and element-bubble space:

$$\Phi_h := \left\{ \mathbf{v} \in \prod_{j=1}^{J} (H^1(\Omega_j))^3; \mathbf{v}|_K \in \mathbf{P}(K), \forall K \in \mathscr{C}_h \right\},$$

$$\Theta_h := \{\mathbf{v} \in (H_0^1(\Omega))^3; \mathbf{v}|_K \in (\mathscr{P}_0(K))^3\, b_K, \forall K \in \mathscr{C}_h\}.$$

Note that Φ_h is not defined as globally continuous in order to deal with the discontinuity of ε across the material interfaces. Let

$$\mathfrak{U}_h := \left\{\mathbf{v} \in (H^1(\Omega))^3; \mathbf{v}|_K \in (\mathscr{P}_1(K))^3, \forall K \in \mathscr{C}_h\right\}.$$

We define the finite element space for approximating the solution as follows:

$$U_h := \mathfrak{U}_h + \Phi_h + \Theta_h. \tag{2}$$

Note that U_h is continuous, only except for those discontinuous interfaces of ε. Let

$$V = \Big\{ \mathbf{v} \in (L^2(\Omega))^3; \mathbf{curl}\,\mathbf{v} \in (L^2(\Omega))^3, \operatorname{div}(\varepsilon\,\mathbf{v}) \in \prod_{j=1}^{J} L^2(\Omega_j),$$
$$[(\varepsilon\,\mathbf{v})\cdot\mathbf{n}]|_F \in L^2(F), \forall F \in \mathscr{E}_{\text{inter}}, \mathbf{v}\times\mathbf{n}|_F \in (L^2(F))^3, \forall F \in \mathscr{E}_h^\partial \Big\}.$$

Let

$$P_h^n := \{q \in L^2(\Omega); q|_K \in \mathscr{P}_n(K), \forall K \in \mathscr{C}_h\} \quad (n = 0, 1).$$

For any $\mathbf{v} \in V$, we define $\breve{R}_h(\mathrm{div}(\varepsilon\,\mathbf{v})) \in P_h^0$ and $R_h(\mu^{-1}\,\mathbf{curl v}) \in (P_h^1)^3$ as follows:

$$(\breve{R}_h(\mathrm{div}(\varepsilon\,\mathbf{v})), q) := \sum_{K\in\mathscr{C}_h} (\mathrm{div}(\varepsilon\,\mathbf{v}), q)_{0,K} \quad \forall q \in P_h^0,$$

$$(R_h(\mu^{-1}\,\mathbf{curl\,v}), \mathbf{q})_\mu := \sum_{K\in\mathscr{C}_h} (\mathbf{curl\,v}, \mathbf{q})_{0,K} - \sum_{F\in\mathscr{E}_h^\partial} \int_F (\mathbf{v}\times\mathbf{n})\cdot\mathbf{q} \quad \forall \mathbf{q} \in (P_h^1)^3,$$

where $||\mathbf{v}||_\mu^2 := ||\mu^{\frac{1}{2}}\mathbf{v}||_0^2$. We additionally define an L^2 projection: $R_h^{\mathrm{inter},\Gamma}([(\varepsilon\,\mathbf{v})\cdot\mathbf{n}])|_F \in \mathscr{P}_0(F)$ for any $F \in \mathscr{E}_{\mathrm{inter}}$, by

$$R_h^{\mathrm{inter}}([(\varepsilon\,\mathbf{v})\cdot\mathbf{n}])|_F := \frac{1}{|F|}\int_F [(\varepsilon\,\mathbf{v})\cdot\mathbf{n}] \quad \forall F \in \mathscr{E}_{\mathrm{inter}}.$$

In what follows we define some mesh-dependent bilinear and linear forms:

$$S_{h,\mathrm{div}}(\mathbf{u},\mathbf{v}) := \sum_{K\in\mathscr{C}_h} h_K^2(\mathrm{div}\,\mathbf{u}, \mathrm{div}\,\mathbf{v})_{0,K}, \; Z_{h,\mathrm{div}}(g;\mathbf{v}) := \sum_{K\in\mathscr{C}_h} h_K^2(g, \mathrm{div}\,\mathbf{v})_{0,K},$$

$$S_{h,\mathrm{inter}}(\mathbf{u},\mathbf{v}) := \sum_{F\in\mathscr{E}_{\mathrm{inter}}} h_F \int_F [(\varepsilon\,\mathbf{u})\cdot\mathbf{n}]\,[(\varepsilon\,\mathbf{v})\cdot\mathbf{n}],$$

$$S_{h,\mathbf{curl}}(\mathbf{u};\mathbf{v}) := \sum_{K\in\mathscr{C}_h} h_K^2(\mathbf{curl\,u}, \mathbf{curl\,v})_{0,K},$$

$$S_{h,\Gamma}(\mathbf{u},\mathbf{v}) := \sum_{F\in\mathscr{E}_h^\partial} h_F \int_F (\mathbf{u}\times\mathbf{n})\cdot(\mathbf{v}\times\mathbf{n}), \; \mathscr{Z}_h(\mathbf{v}) := Z_{h,\mathrm{div}}(g;\mathbf{v}).$$

We thus define the stabilization term and its right-hand side as follows:

$$\mathscr{S}_h(\mathbf{u},\mathbf{v}) := \sum_{F\in\mathscr{E}_{\mathrm{inter}}} (R_h^{\mathrm{inter}}([(\varepsilon\,\mathbf{u})\cdot\mathbf{n}]), R_h^{\mathrm{inter}}([(\varepsilon\,\mathbf{v})\cdot\mathbf{n}]))_{0,F}$$
$$+ S_{h,\mathrm{div}}(\mathbf{u},\mathbf{v}) + S_{h,\mathbf{curl}}(\mathbf{u},\mathbf{v}) + S_{h,\mathrm{inter}}(\mathbf{u},\mathbf{v}) + S_{h,\Gamma}(\mathbf{u},\mathbf{v}).$$

The finite element method we propose is to find $\mathbf{u}_h \in U_h$ such that

$$\mathscr{L}_h(\mathbf{u}_h,\mathbf{v}) = \mathscr{F}_h(\mathbf{v}) \quad \forall \mathbf{v} \in U_h,$$

where

$$\mathscr{L}_h(\mathbf{u},\mathbf{v}) := (R_h(\mu^{-1}\,\mathbf{curl}\,\mathbf{u}), R_h(\mu^{-1}\,\mathbf{curl}\,\mathbf{v}))_\mu$$
$$+(\breve{R}_h(\mathrm{div}(\varepsilon\mathbf{u})), \breve{R}_h(\mathrm{div}(\varepsilon\mathbf{v}))) + \mathscr{S}_h(\mathbf{u},\mathbf{v}), \qquad (3)$$

$$\mathscr{F}_h(\mathbf{v}) := (\mathbf{f},\mathbf{v}) + (g, \breve{R}_h(\mathrm{div}(\varepsilon\,\mathbf{v}))) + \mathscr{Z}_h(\mathbf{v}).$$

3 Coercivity and Condition Number

This subsection is devoted to the coercivity property of $\mathscr{L}_h$ and the condition number.

Theorem 1 *We have*

$$\mathscr{L}_h(\mathbf{v},\mathbf{v}) \geq C\,||\mathbf{v}||_0^2 \quad \forall \mathbf{v} \in U_h.$$

Theorem 2 *Assuming quasiuniform meshes, we have the condition number $\mathscr{O}(h^{-2})$ of the resulting linear system of problem (3).*

4 Error Estimates

In this section we establish the error bound in an energy norm between the exact solution and the finite element solution. This consists mainly of how to estimate the inconsistent errors caused by the L^2 projector R_h of the curl operator and how to construct an appropriate interpolant of the exact solution to eliminate the effects of the first order derivatives from both div and curl operators on the solution (not in H^1).

Hypothesis H1 Let $\mathbf{u}$ be the solution to problem (1), with $\mathbf{z} := \mu^{-1}\,\mathbf{curl}\,\mathbf{u} \in H(\mathbf{curl};\Omega) \cap H_0(\mathrm{div}^0;\mu;\Omega)$, where $H_0(\mathrm{div}^0;\mu;\Omega) = \{\mathbf{v} \in (L^2(\Omega))^3 : \mathrm{div}(\mu\mathbf{v}) = 0, (\mu\mathbf{v})\cdot\mathbf{n}|_\Gamma = 0\}$. We require that $\mathbf{z} \in \prod_{j=1}^J (H^r(\Omega_j))^3$ with some $r \leq 1$ and that $\mathbf{z}$ allows the regular-singular decomposition:

$$\mathbf{z} := \mathbf{z}_H + \nabla\varphi \quad \text{in } \Omega_j, 1 \leq j \leq J,$$

where $\mathbf{z}_H \in H(\mathbf{curl};\Omega)\cap\prod_{j=1}^J (H^{1+r}(\Omega_j))^3$ and $\varphi \in H^1(\Omega)\cap\prod_{j=1}^J H^{1+r}(\Omega_j)$ satisfy

$$\sum_{j=1}^J (||\mathbf{z}_H||_{1+r,\Omega_j} + ||\varphi||_{1+r,\Omega_j}) \leq C_*,$$

where C_* depends on the right-hand sides of problem (1).

Readers may refer to [12, 14, 25] for details of the regular-singular decompositions relating to Hypothesis H1. Under Hypothesis H1, the inconsistent errors caused by the L^2 projector R_h of the curl operator is given as follows.

Lemma 1 *Let* $\mathbf{u}$ *be the solution to problem (1). Assuming Hypothesis H1, we have*

$$(\mu^{-1}\,\mathbf{curl}\,\mathbf{u}, R_h(\mu^{-1}\,\mathbf{curl}\,\mathbf{v}_h))_\mu - \sum_{K\in\mathscr{C}_h}(\mu^{-1}\,\mathbf{curl}\,\mathbf{u}, \mathbf{curl}\,\mathbf{v}_h)_{0,K}$$
$$+ \sum_{F\in\mathscr{E}_h^\partial}\int_F(\mathbf{v}_h\times\mathbf{n})\cdot(\mathbf{n}\times(\mu^{-1}\,\mathbf{curl}\,\mathbf{u}\times\mathbf{n})) \le C\,h^r\,||R_h(\mu^{-1}\,\mathbf{curl}\,\mathbf{v}_h)||_\mu$$
$$+ C\,h^r\left(\sum_{K\in\mathscr{C}_h}h_K^2\,||\mathbf{curl}\,\mathbf{v}_h||_{0,K}^2 + \sum_{F\in\mathscr{E}_h^\partial}h_F\int_F|\mathbf{v}_h\times\mathbf{n}|^2\right)^{\frac{1}{2}}.$$

Theorem 3 *Let* $\mathbf{u}$ *and* $\mathbf{u}_h$ *be the exact solution to problem (1) and the finite element solution to problem (3). We have the following inconsistent error estimates*

$$|\mathscr{L}_h(\mathbf{u}_h - \mathbf{u}, \mathbf{v}_h)| \le C\,h^r\,|||\mathbf{v}_h|||_{\mathscr{L}_h} \quad \forall \mathbf{v}_h \in U_h,$$

where $|||\mathbf{v}_h|||_{\mathscr{L}_h}^2 := \mathscr{L}_h(\mathbf{v}_h, \mathbf{v}_h)$.

In what follows we shall construct interpolants $\tilde{\mathbf{u}}$ of the solution $\mathbf{u}$.

Lemma 2 *Let* $\mathbf{u}$ *be the exact solution. Assuming* $\mathbf{u} \in \prod_{j=1}^J (H^r(\Omega_j))^3$ *with* $r > \frac{1}{2}$. *Let* U_h *be defined as in (2). We have a* $\tilde{\mathbf{u}} \in U_h$ *such that*

$$||\breve{R}_h(\mathrm{div}\,(\varepsilon\,(\mathbf{u}-\tilde{\mathbf{u}})))||_0^2 = ||R_h(\mu^{-1}\,\mathbf{curl}\,(\mathbf{u}-\tilde{\mathbf{u}}))||_\mu^2$$
$$= \sum_{F\in\mathscr{E}_{\mathrm{inter}}}||R_h^{\mathrm{inter}}([(\varepsilon(\mathbf{u}-\tilde{\mathbf{u}})\cdot\mathbf{n}])||_{0,F}^2 = 0,$$

$$||\mathbf{u}-\tilde{\mathbf{u}}||_0 \le C\,h^r\sum_{j=1}^J||\mathbf{u}||_{r,\Omega_j}.$$

We introduce an energy norm as follows:

$$|||\mathbf{v}|||_{h,\mathbf{curl},\mathrm{div}}^2 := ||\mathbf{v}||_0^2 + \mathscr{L}_h(\mathbf{v},\mathbf{v})$$
$$= ||\mathbf{v}||_0^2 + ||\breve{R}_h(\mathrm{div}(\varepsilon\,\mathbf{v}))||_0^2 + ||R_h(\mu^{-1}\,\mathbf{curl}\,\mathbf{v})||_\mu^2 + \mathscr{S}_h(\mathbf{v},\mathbf{v}),$$

which corresponds to the usual energy norm $||\mathbf{v}||_{\mathbf{curl},\mathrm{div}}^2 = ||\mathbf{v}||_0^2 + ||\mathrm{div}(\varepsilon\,\mathbf{v})||_0^2 + ||\mathbf{curl}\,\mathbf{v}||_\mu^2$. Combining Lemma 2 and Theorem 3, we can obtain the error bound.

Theorem 4 *Let* $\mathbf{u}$ *and* $\mathbf{u}_h$ *be the solution to problem (1) and the finite element solution to (3). Assume that Hypotheses H1) holds and that* $\mathbf{u}, \mu^{-1}\,\mathbf{curl}\,\mathbf{u} \in \prod_{j=1}^J (H^r(\Omega_j))^3$ *with* $r > 1/2$. *Then*

$$||\mathbf{u}-\mathbf{u}_h||_{h,\mathbf{curl},\mathrm{div}} \le C\,h^r.$$

We require that the *piecewise* regularity $r > 1/2$, i.e, $\mathbf{u}, \mu^{-1}\,\mathbf{curl}\,\mathbf{u} \in \prod_{j=1}^{J}(H^r(\Omega_j))^3$ for some $r > 1/2$. This is because in proving Lemma 2 the element-face integrals of **u** are used. Such an assumption is also used elsewhere, see, e.g., [8]. In [21] it requires even more regularity, i.e., $r = 1$, and in [9] a piecewise regularity $W^{1,p}(\Omega_j)$ for some $p > 2$ is assumed. Meanwhile, the piecewise regularity $r > 1/2$ is shown in [6, 22].

Remark 1 In the use of C^0 elements for solving Maxwell equations (1), several other methods may be employed, i.e., weighted methods [11, 13], $H^{-\alpha}$-norm method [1, 3]. The weighted method lies in the introduction of a weight function which depends on the geometrical singularities of the solution itself to the divergence operator, while the $H^{-\alpha}$-norm method is to measure the divergence operator in the norm of $H^{-\alpha}(\Omega)$, the dual of $H_0^{\alpha}(\Omega)$, where $1/2 < \alpha \leq 1$. Both methods would lead to a rather low convergence rate far from r, if the C^0 element space of the solution does not include the gradient of a C^1 element. In sharp contrast with these methods, a priori geometrical singularities of the solution and the inclusion of the gradient of C^1 elements into the C^0 element space of the solution are not required in our method. The convergence rate in our method is optimal r.

Acknowledgments HD is partially supported by the National Natural Science Foundation of China under grants 11071132 and 11171168 and the Research Fund for the Doctoral Program of Higher Education of China under grants 20100031110002 and 20120031110026 and the Scientific Research Foundation for Returned Scholars, Ministry of Education of China, and the Wuhan University start-up fund from the Fundamental Research Funds for the Central Universities 2042014kf0218. PL is partially supported by the Fundamental Research Funds for the Central Universities under grants 06108038 and 06108137.

References

1. S. Badia, R. Codina, A nodal-based finite element approximation of the Maxwell problem suitable for singular solutions. SIAM J. Numer. Anal. **50**, 398–417 (2012)
2. M. Birman, M. Solomyak, L^2-theory of the Maxwell operator in arbitrary domains. Russ. Math. Surv. **42**, 75–96 (1987)
3. A. Bonito, J.-L. Guermond, Approximation of the eigenvalue problem for the time-harmonic Maxwell system by continuous Lagrange finite elements. Math. Comput. **80**, 1887–1910 (2011)
4. A.S. Bonnet-Ben Dhia, C. Hazard, S. Lohrengel, A singular field method for the solution of Maxwell's equations in polyhedral domains. SIAM J. Appl. Math. **59**, 2028–2044 (2000)
5. A. Bossavit, *Computational Electromagnetism: Variational Formulations, Complementarity, Edge Elements* (Academic, New York, 1998)
6. G. Caloz, M. Dauge, V. Péron, Uniform estimates for transmission problems with high contrast in heat conduction and electromagnetism. J. Math. Anal. Appl. **370**, 555–572 (2010)
7. M. Cessenat, *Mathematical Methods in Electrmagnetism: Linear Theory and Applications* (World Scientific, River Edge, 1996)

8. Z.M. Chen, Q. Du, J. Zou, Finite element methods with matching and nonmatching meshes for Maxwell equations with discontinuous coefficients. SIAM J. Numer. Anal. **37**, 1542–1570 (2000)
9. T. Chung, B. Engquist, Convergence analysis of fully discrete finite volume methods for Maxwell's equations in nonhomogeneous media. SIAM J. Numer. Anal. **43**, 303–317 (2005)
10. P.G. Ciarlet, Basic error estimates for elliptic problems, in *Handbook of Numerical Analysis, Vol. II. Finite Element Methods (Part 1)*, ed. by P.G. Ciarlet, J.L. Lions (North-Holland, Amsterdam, 1991)
11. P. Ciarlet Jr., F. Lefèvre, S. Lohrengel, S. Nicaise, Weighted regularization for composite materials in electromagnetism. M2AN Math. Model. Numer. Anal. **44**, 75–108 (2010)
12. M. Costabel, M. Dauge, Singularities of electromagnetic fields in polyhedral domains. Arch. Ration. Mech. Anal. **151**, 221–276 (2000)
13. M. Costabel, M. Dauge, Weighted regularization of Maxwell equations in polyhedral domains. Numer. Math. **93**, 239–277 (2002)
14. M. Costabel, M. Dauge, S. Nicaise, Singularities of Maxwell interface problem. M2AN Math. Model. Numer. Anal. **33**, 627–649 (1999)
15. H.Y. Duan, F. Jia, P. Lin, R.C.E. Tan, The local L^2-projected C^0 finite element method for the Maxwell problem. SIAM J. Numer. Anal. **47**, 1274–1303 (2009)
16. H.Y. Duan, P. Lin, R.C.E. Tan, Error estimates for a vectorial second-order elliptic eigenproblem by the local L^2 projected C^0 finite element method. SIAM J. Numer. Anal. **50**, 3208–3230 (2013)
17. ———, Analysis of a continuous finite element method for $H(\mathbf{curl}, \mathrm{div})$-elliptic interface problem. Numer. Math. **123**, 671–707 (2013)
18. P. Fernandes, G. Gilardi, Magnetostatic and electrostatic problems in inhomogeneous anisotropic media with irregular boundary and mixed boundary conditions. Math. Models Methods Appl. Sci. **7**, 957–991 (1997)
19. C. Hazard, M. Lenoir, On the solution of time-harmonic scattering problems for Maxwell's equations. SIAM J. Math. Anal. **27**, 1597–1630 (1996)
20. R. Hiptmair, Finite elements in computational electromagnetism. Acta Numer. **11**(1), 237–339 (2002)
21. R. Hiptmair, J. Li, J. Zou, Convergence analysis of finite element methods for H(div)-elliptic interface problems. J. Numer. Math. **18**, 187–218 (2010)
22. J.G. Huang, J. Zou, Uniform a pripori estimates for elliptic and static Maxwell interface problems. Disret. Contin. Dyn. Syst. Ser. B(DCDS-B) **7**, 145–170 (2007)
23. J.M. Jin, *The Finite Element Method in Electromagnetics*, 2nd edn. (Wiley, New York, 2002)
24. P. Monk, *Finite Element Methods for Maxwell Equations* (Clarendon Press, Oxford, 2003)
25. S. Nicaise, Edge elements on anisotropic meshes and approximation of the Maxwell equations. SAIM J. Numer. Anal. **39**, 784–816 (2001)

Editorial Policy

1. Volumes in the following three categories will be published in LNCSE:

i) Research monographs
ii) Tutorials
iii) Conference proceedings

Those considering a book which might be suitable for the series are strongly advised to contact the publisher or the series editors at an early stage.

2. Categories i) and ii). Tutorials are lecture notes typically arising via summer schools or similar events, which are used to teach graduate students. These categories will be emphasized by Lecture Notes in Computational Science and Engineering. **Submissions by interdisciplinary teams of authors are encouraged.** The goal is to report new developments – quickly, informally, and in a way that will make them accessible to non-specialists. In the evaluation of submissions timeliness of the work is an important criterion. Texts should be well-rounded, well-written and reasonably self-contained. In most cases the work will contain results of others as well as those of the author(s). In each case the author(s) should provide sufficient motivation, examples, and applications. In this respect, Ph.D. theses will usually be deemed unsuitable for the Lecture Notes series. Proposals for volumes in these categories should be submitted either to one of the series editors or to Springer-Verlag, Heidelberg, and will be refereed. A provisional judgement on the acceptability of a project can be based on partial information about the work: a detailed outline describing the contents of each chapter, the estimated length, a bibliography, and one or two sample chapters – or a first draft. A final decision whether to accept will rest on an evaluation of the completed work which should include

- at least 100 pages of text;
- a table of contents;
- an informative introduction perhaps with some historical remarks which should be accessible to readers unfamiliar with the topic treated;
- a subject index.

3. Category iii). Conference proceedings will be considered for publication provided that they are both of exceptional interest and devoted to a single topic. One (or more) expert participants will act as the scientific editor(s) of the volume. They select the papers which are suitable for inclusion and have them individually refereed as for a journal. Papers not closely related to the central topic are to be excluded. Organizers should contact the Editor for CSE at Springer at the planning stage, see *Addresses* below.

In exceptional cases some other multi-author-volumes may be considered in this category.

4. Only works in English will be considered. For evaluation purposes, manuscripts may be submitted in print or electronic form, in the latter case, preferably as pdf- or zipped ps-files. Authors are requested to use the LaTeX style files available from Springer at http://www.springer.com/gp/authors-editors/book-authors-editors/manuscript-preparation/5636 (Click on LaTeX Template → monographs or contributed books).

For categories ii) and iii) we strongly recommend that all contributions in a volume be written in the same LaTeX version, preferably LaTeX2e. Electronic material can be included if appropriate. Please contact the publisher.

Careful preparation of the manuscripts will help keep production time short besides ensuring satisfactory appearance of the finished book in print and online.

5. The following terms and conditions hold. Categories i), ii) and iii):

Authors receive 50 free copies of their book. No royalty is paid.
Volume editors receive a total of 50 free copies of their volume to be shared with authors, but no royalties.

Authors and volume editors are entitled to a discount of 33.3 % on the price of Springer books purchased for their personal use, if ordering directly from Springer.

6. Springer secures the copyright for each volume.

Addresses:

Timothy J. Barth
NASA Ames Research Center
NAS Division
Moffett Field, CA 94035, USA
barth@nas.nasa.gov

Michael Griebel
Institut für Numerische Simulation
der Universität Bonn
Wegelerstr. 6
53115 Bonn, Germany
griebel@ins.uni-bonn.de

David E. Keyes
Mathematical and Computer Sciences
and Engineering
King Abdullah University of Science
and Technology
P.O. Box 55455
Jeddah 21534, Saudi Arabia
david.keyes@kaust.edu.sa

and

Department of Applied Physics
and Applied Mathematics
Columbia University
500 W. 120 th Street
New York, NY 10027, USA
kd2112@columbia.edu

Risto M. Nieminen
Department of Applied Physics
Aalto University School of Science
and Technology
00076 Aalto, Finland
risto.nieminen@aalto.fi

Dirk Roose
Department of Computer Science
Katholieke Universiteit Leuven
Celestijnenlaan 200A
3001 Leuven-Heverlee, Belgium
dirk.roose@cs.kuleuven.be

Tamar Schlick
Department of Chemistry
and Courant Institute
of Mathematical Sciences
New York University
251 Mercer Street
New York, NY 10012, USA
schlick@nyu.edu

Editor for Computational Science
and Engineering at Springer:
Martin Peters
Springer-Verlag
Mathematics Editorial IV
Tiergartenstrasse 17
69121 Heidelberg, Germany
martin.peters@springer.com

Lecture Notes in Computational Science and Engineering

1. D. Funaro, *Spectral Elements for Transport-Dominated Equations.*
2. H.P. Langtangen, *Computational Partial Differential Equations.* Numerical Methods and Diffpack Programming.
3. W. Hackbusch, G. Wittum (eds.), *Multigrid Methods V.*
4. P. Deuflhard, J. Hermans, B. Leimkuhler, A.E. Mark, S. Reich, R.D. Skeel (eds.), *Computational Molecular Dynamics: Challenges, Methods, Ideas.*
5. D. Kröner, M. Ohlberger, C. Rohde (eds.), *An Introduction to Recent Developments in Theory and Numerics for Conservation Laws.*
6. S. Turek, *Efficient Solvers for Incompressible Flow Problems.* An Algorithmic and Computational Approach.
7. R. von Schwerin, *Multi Body System SIMulation.* Numerical Methods, Algorithms, and Software.
8. H.-J. Bungartz, F. Durst, C. Zenger (eds.), *High Performance Scientific and Engineering Computing.*
9. T.J. Barth, H. Deconinck (eds.), *High-Order Methods for Computational Physics.*
10. H.P. Langtangen, A.M. Bruaset, E. Quak (eds.), *Advances in Software Tools for Scientific Computing.*
11. B. Cockburn, G.E. Karniadakis, C.-W. Shu (eds.), *Discontinuous Galerkin Methods.* Theory, Computation and Applications.
12. U. van Rienen, *Numerical Methods in Computational Electrodynamics.* Linear Systems in Practical Applications.
13. B. Engquist, L. Johnsson, M. Hammill, F. Short (eds.), *Simulation and Visualization on the Grid.*
14. E. Dick, K. Riemslagh, J. Vierendeels (eds.), *Multigrid Methods VI.*
15. A. Frommer, T. Lippert, B. Medeke, K. Schilling (eds.), *Numerical Challenges in Lattice Quantum Chromodynamics.*
16. J. Lang, *Adaptive Multilevel Solution of Nonlinear Parabolic PDE Systems.* Theory, Algorithm, and Applications.
17. B.I. Wohlmuth, *Discretization Methods and Iterative Solvers Based on Domain Decomposition.*
18. U. van Rienen, M. Günther, D. Hecht (eds.), *Scientific Computing in Electrical Engineering.*
19. I. Babuška, P.G. Ciarlet, T. Miyoshi (eds.), *Mathematical Modeling and Numerical Simulation in Continuum Mechanics.*
20. T.J. Barth, T. Chan, R. Haimes (eds.), *Multiscale and Multiresolution Methods.* Theory and Applications.
21. M. Breuer, F. Durst, C. Zenger (eds.), *High Performance Scientific and Engineering Computing.*
22. K. Urban, *Wavelets in Numerical Simulation.* Problem Adapted Construction and Applications.
23. L.F. Pavarino, A. Toselli (eds.), *Recent Developments in Domain Decomposition Methods.*

24. T. Schlick, H.H. Gan (eds.), *Computational Methods for Macromolecules: Challenges and Applications.*

25. T.J. Barth, H. Deconinck (eds.), *Error Estimation and Adaptive Discretization Methods in Computational Fluid Dynamics.*

26. M. Griebel, M.A. Schweitzer (eds.), *Meshfree Methods for Partial Differential Equations.*

27. S. Müller, *Adaptive Multiscale Schemes for Conservation Laws.*

28. C. Carstensen, S. Funken, W. Hackbusch, R.H.W. Hoppe, P. Monk (eds.), *Computational Electromagnetics.*

29. M.A. Schweitzer, *A Parallel Multilevel Partition of Unity Method for Elliptic Partial Differential Equations.*

30. T. Biegler, O. Ghattas, M. Heinkenschloss, B. van Bloemen Waanders (eds.), *Large-Scale PDE-Constrained Optimization.*

31. M. Ainsworth, P. Davies, D. Duncan, P. Martin, B. Rynne (eds.), *Topics in Computational Wave Propagation.* Direct and Inverse Problems.

32. H. Emmerich, B. Nestler, M. Schreckenberg (eds.), *Interface and Transport Dynamics.* Computational Modelling.

33. H.P. Langtangen, A. Tveito (eds.), *Advanced Topics in Computational Partial Differential Equations.* Numerical Methods and Diffpack Programming.

34. V. John, *Large Eddy Simulation of Turbulent Incompressible Flows.* Analytical and Numerical Results for a Class of LES Models.

35. E. Bänsch (ed.), *Challenges in Scientific Computing - CISC 2002.*

36. B.N. Khoromskij, G. Wittum, *Numerical Solution of Elliptic Differential Equations by Reduction to the Interface.*

37. A. Iske, *Multiresolution Methods in Scattered Data Modelling.*

38. S.-I. Niculescu, K. Gu (eds.), *Advances in Time-Delay Systems.*

39. S. Attinger, P. Koumoutsakos (eds.), *Multiscale Modelling and Simulation.*

40. R. Kornhuber, R. Hoppe, J. Périaux, O. Pironneau, O. Wildlund, J. Xu (eds.), *Domain Decomposition Methods in Science and Engineering.*

41. T. Plewa, T. Linde, V.G. Weirs (eds.), *Adaptive Mesh Refinement – Theory and Applications.*

42. A. Schmidt, K.G. Siebert, *Design of Adaptive Finite Element Software.* The Finite Element Toolbox ALBERTA.

43. M. Griebel, M.A. Schweitzer (eds.), *Meshfree Methods for Partial Differential Equations II.*

44. B. Engquist, P. Lötstedt, O. Runborg (eds.), *Multiscale Methods in Science and Engineering.*

45. P. Benner, V. Mehrmann, D.C. Sorensen (eds.), *Dimension Reduction of Large-Scale Systems.*

46. D. Kressner, *Numerical Methods for General and Structured Eigenvalue Problems.*

47. A. Boriçi, A. Frommer, B. Joó, A. Kennedy, B. Pendleton (eds.), *QCD and Numerical Analysis III.*

48. F. Graziani (ed.), *Computational Methods in Transport.*

49. B. Leimkuhler, C. Chipot, R. Elber, A. Laaksonen, A. Mark, T. Schlick, C. Schütte, R. Skeel (eds.), *New Algorithms for Macromolecular Simulation.*

50. M. Bücker, G. Corliss, P. Hovland, U. Naumann, B. Norris (eds.), *Automatic Differentiation: Applications, Theory, and Implementations.*

51. A.M. Bruaset, A. Tveito (eds.), *Numerical Solution of Partial Differential Equations on Parallel Computers.*

52. K.H. Hoffmann, A. Meyer (eds.), *Parallel Algorithms and Cluster Computing.*

53. H.-J. Bungartz, M. Schäfer (eds.), *Fluid-Structure Interaction.*

54. J. Behrens, *Adaptive Atmospheric Modeling.*

55. O. Widlund, D. Keyes (eds.), *Domain Decomposition Methods in Science and Engineering XVI.*

56. S. Kassinos, C. Langer, G. Iaccarino, P. Moin (eds.), *Complex Effects in Large Eddy Simulations.*

57. M. Griebel, M.A Schweitzer (eds.), *Meshfree Methods for Partial Differential Equations III.*

58. A.N. Gorban, B. Kégl, D.C. Wunsch, A. Zinovyev (eds.), *Principal Manifolds for Data Visualization and Dimension Reduction.*

59. H. Ammari (ed.), *Modeling and Computations in Electromagnetics: A Volume Dedicated to Jean-Claude Nédélec.*

60. U. Langer, M. Discacciati, D. Keyes, O. Widlund, W. Zulehner (eds.), *Domain Decomposition Methods in Science and Engineering XVII.*

61. T. Mathew, *Domain Decomposition Methods for the Numerical Solution of Partial Differential Equations.*

62. F. Graziani (ed.), *Computational Methods in Transport: Verification and Validation.*

63. M. Bebendorf, *Hierarchical Matrices.* A Means to Efficiently Solve Elliptic Boundary Value Problems.

64. C.H. Bischof, H.M. Bücker, P. Hovland, U. Naumann, J. Utke (eds.), *Advances in Automatic Differentiation.*

65. M. Griebel, M.A. Schweitzer (eds.), *Meshfree Methods for Partial Differential Equations IV.*

66. B. Engquist, P. Lötstedt, O. Runborg (eds.), *Multiscale Modeling and Simulation in Science.*

67. I.H. Tuncer, Ü. Gülcat, D.R. Emerson, K. Matsuno (eds.), *Parallel Computational Fluid Dynamics 2007.*

68. S. Yip, T. Diaz de la Rubia (eds.), *Scientific Modeling and Simulations.*

69. A. Hegarty, N. Kopteva, E. O'Riordan, M. Stynes (eds.), *BAIL* 2008 – *Boundary and Interior Layers.*

70. M. Bercovier, M.J. Gander, R. Kornhuber, O. Widlund (eds.), *Domain Decomposition Methods in Science and Engineering XVIII.*

71. B. Koren, C. Vuik (eds.), *Advanced Computational Methods in Science and Engineering.*

72. M. Peters (ed.), *Computational Fluid Dynamics for Sport Simulation.*

73. H.-J. Bungartz, M. Mehl, M. Schäfer (eds.), *Fluid Structure Interaction II - Modelling, Simulation, Optimization.*

74. D. Tromeur-Dervout, G. Brenner, D.R. Emerson, J. Erhel (eds.), *Parallel Computational Fluid Dynamics 2008.*

75. A.N. Gorban, D. Roose (eds.), *Coping with Complexity: Model Reduction and Data Analysis.*

76. J.S. Hesthaven, E.M. Rønquist (eds.), *Spectral and High Order Methods for Partial Differential Equations.*

77. M. Holtz, *Sparse Grid Quadrature in High Dimensions with Applications in Finance and Insurance.*

78. Y. Huang, R. Kornhuber, O.Widlund, J. Xu (eds.), *Domain Decomposition Methods in Science and Engineering XIX.*

79. M. Griebel, M.A. Schweitzer (eds.), *Meshfree Methods for Partial Differential Equations V.*

80. P.H. Lauritzen, C. Jablonowski, M.A. Taylor, R.D. Nair (eds.), *Numerical Techniques for Global Atmospheric Models.*

81. C. Clavero, J.L. Gracia, F.J. Lisbona (eds.), *BAIL 2010 – Boundary and Interior Layers, Computational and Asymptotic Methods.*

82. B. Engquist, O. Runborg, Y.R. Tsai (eds.), *Numerical Analysis and Multiscale Computations.*

83. I.G. Graham, T.Y. Hou, O. Lakkis, R. Scheichl (eds.), *Numerical Analysis of Multiscale Problems.*

84. A. Logg, K.-A. Mardal, G. Wells (eds.), *Automated Solution of Differential Equations by the Finite Element Method.*

85. J. Blowey, M. Jensen (eds.), *Frontiers in Numerical Analysis - Durham 2010.*

86. O. Kolditz, U.-J. Gorke, H. Shao, W. Wang (eds.), *Thermo-Hydro-Mechanical-Chemical Processes in Fractured Porous Media - Benchmarks and Examples.*

87. S. Forth, P. Hovland, E. Phipps, J. Utke, A. Walther (eds.), *Recent Advances in Algorithmic Differentiation.*

88. J. Garcke, M. Griebel (eds.), *Sparse Grids and Applications.*

89. M. Griebel, M.A. Schweitzer (eds.), *Meshfree Methods for Partial Differential Equations VI.*

90. C. Pechstein, *Finite and Boundary Element Tearing and Interconnecting Solvers for Multiscale Problems.*

91. R. Bank, M. Holst, O. Widlund, J. Xu (eds.), *Domain Decomposition Methods in Science and Engineering XX.*

92. H. Bijl, D. Lucor, S. Mishra, C. Schwab (eds.), *Uncertainty Quantification in Computational Fluid Dynamics.*

93. M. Bader, H.-J. Bungartz, T. Weinzierl (eds.), *Advanced Computing.*

94. M. Ehrhardt, T. Koprucki (eds.), *Advanced Mathematical Models and Numerical Techniques for Multi-Band Effective Mass Approximations.*

95. M. Azaïez, H. El Fekih, J.S. Hesthaven (eds.), *Spectral and High Order Methods for Partial Differential Equations ICOSAHOM 2012.*

96. F. Graziani, M.P. Desjarlais, R. Redmer, S.B. Trickey (eds.), *Frontiers and Challenges in Warm Dense Matter.*

97. J. Garcke, D. Pflüger (eds.), *Sparse Grids and Applications – Munich 2012.*

98. J. Erhel, M. Gander, L. Halpern, G. Pichot, T. Sassi, O. Widlund (eds.), *Domain Decomposition Methods in Science and Engineering XXI.*

99. R. Abgrall, H. Beaugendre, P.M. Congedo, C. Dobrzynski, V. Perrier, M. Ricchiuto (eds.), *High Order Nonlinear Numerical Methods for Evolutionary PDEs - HONOM 2013.*

100. M. Griebel, M.A. Schweitzer (eds.), *Meshfree Methods for Partial Differential Equations VII.*

101. R. Hoppe (ed.), *Optimization with PDE Constraints - OPTPDE 2014.*

102. S. Dahlke, W. Dahmen, M. Griebel, W. Hackbusch, K. Ritter, R. Schneider, C. Schwab, H. Yserentant (eds.), *Extraction of Quantifiable Information from Complex Systems.*

103. A. Abdulle, S. Deparis, D. Kressner, F. Nobile, M. Picasso (eds.), *Numerical Mathematics and Advanced Applications - ENUMATH 2013.*

For further information on these books please have a look at our mathematics catalogue at the following URL: `www.springer.com/series/3527`

Monographs in Computational Science and Engineering

1. J. Sundnes, G.T. Lines, X. Cai, B.F. Nielsen, K.-A. Mardal, A. Tveito, *Computing the Electrical Activity in the Heart.*

For further information on this book, please have a look at our mathematics catalogue at the following URL: www.springer.com/series/7417

Texts in Computational Science and Engineering

1. H. P. Langtangen, *Computational Partial Differential Equations.* Numerical Methods and Diffpack Programming. 2nd Edition
2. A. Quarteroni, F. Saleri, P. Gervasio, *Scientific Computing with MATLAB and Octave.* 4th Edition
3. H. P. Langtangen, *Python Scripting for Computational Science.* 3rd Edition
4. H. Gardner, G. Manduchi, *Design Patterns for e-Science.*
5. M. Griebel, S. Knapek, G. Zumbusch, *Numerical Simulation in Molecular Dynamics.*
6. H. P. Langtangen, *A Primer on Scientific Programming with Python.* 4th Edition
7. A. Tveito, H. P. Langtangen, B. F. Nielsen, X. Cai, *Elements of Scientific Computing.*
8. B. Gustafsson, *Fundamentals of Scientific Computing.*
9. M. Bader, *Space-Filling Curves.*
10. M. Larson, F. Bengzon, *The Finite Element Method: Theory, Implementation and Applications.*
11. W. Gander, M. Gander, F. Kwok, *Scientific Computing: An Introduction using Maple and MATLAB.*

For further information on these books please have a look at our mathematics catalogue at the following URL: www.springer.com/series/5151

Printed in the United States
By Bookmasters